Frontier Studies in Soil Science

Series Editor

Avelino Núñez-Delgado, Lugo, Spain

This series aims to illuminate the latest research, insights, and advancements in the ever-evolving world of soil science. This series is dedicated to fostering a comprehensive understanding of soils, from their micro to macro characteristics, their role in ecosystems, to their significance in global challenges. Our commitment is to present groundbreaking studies, carried out by esteemed experts and emerging scholars, ensuring that readers are consistently provided with the most up-to-date and relevant information in the realm of soil science.

Muhammad Shaaban
Editor

Soils and Sustainable Agriculture

Interplay of Soil, Plant, Water and Environmental Systems for Sustainable Agriculture

Volume 1

Springer

Editor
Muhammad Shaaban
College of Agriculture
Henan University of Science
and Technology
Luoyang, China

ISSN 3004-9865 ISSN 3004-9873 (electronic)
Frontier Studies in Soil Science
ISBN 978-3-031-91113-2 ISBN 978-3-031-91114-9 (eBook)
https://doi.org/10.1007/978-3-031-91114-9

This Springer imprint is published by the registered company Springer Nature Switzerland AG
The registered company address is: Gewerbestrasse 11, 6330 Cham, Switzerland

Preface

In an era marked by rapid environmental changes and growing concerns about food security, the relationship between soil health and sustainable agriculture becomes increasingly crucial. *Soils and Sustainable Agriculture: Interplay of Soil, Plant, Water, and Environmental Systems for Sustainable Agriculture* aims to elucidate this intricate relationship and offer insights into sustainable agricultural practices that honor and enhance the natural systems.

Soil is the foundation of terrestrial life, serving as a dynamic medium in which plants grow and as a vital component of our ecosystems. It plays a pivotal role in nutrient cycling, water retention, and biological diversity, all of which are essential for robust agricultural systems. However, the pressures of modern agriculture, such as soil degradation, pollution, and climate change, have put immense strain on this precious resource, highlighting the need for a new paradigm in agricultural practices.

This book is the result of collaborative efforts from researchers, academics, and practitioners in the fields of soil science, agronomy, and environmental studies. Each chapter addresses various aspects of the interplay between soil, plants, water, and environmental systems, providing a comprehensive overview that integrates theory with practical applications. Readers will find discussions on soil management practices, the role of organic matter in soil health, water conservation techniques, and the impact of climate change on agricultural productivity.

Our goal with this publication is to equip practitioners, educators, and students with the knowledge and tools necessary to foster sustainable agricultural systems. By embracing an interdisciplinary approach, we highlight the importance of holistic management practices that not only enhance soil fertility and crop yields but also promote ecological balance and resilience.

We hope that this book inspires readers to recognize the indispensable role of soil in sustainable agriculture and motivates them to adopt practices that preserve and restore it for future generations. In our quest for food security and environmental health, let us remember that sustainable agriculture begins with a commitment to nurturing the very foundation of life, our soil.

Thank you for joining us on this journey towards a more sustainable future.

Luoyang, China
December 2024

Prof. Dr. Muhammad Shaaban

Contents

Why Soils Are Vital for Sustainable Agriculture?

Muhammad Shaaban

Abstract This chapter explores the critical role of soil in sustainable agriculture, emphasizing its multifaceted functions in nutrient supply, water retention, and carbon sequestration. It examines the relationship between soil and agricultural productivity, highlighting the importance of maintaining healthy soils for food security and economic stability. The chapter further investigates the impact of climate change on soil ecosystems and discusses sustainable practices, such as crop rotation, cover cropping, and reduced tillage, that enhance soil health. It also addresses the socioeconomic implications of soil management, outlining the economic benefits of sustainable practices and the necessity for comprehensive policy frameworks. Ultimately, the chapter underscores the importance of prioritizing soil conservation as a vital component of a resilient and sustainable agricultural system, ensuring long-term productivity and ecological balance.

1 Introduction

Sustainable agriculture is defined as a farming approach that meets current agricultural needs without compromising the ability of future generations to meet their own needs. It emphasizes practices that foster environmental health, economic profitability, and social equity. Central to sustainable agriculture is the concept of working within the ecological limits of each farming system, promoting the responsible use of resources while enhancing the resilience and productivity of agricultural landscapes.

Understanding the role of soil in agricultural systems is crucial to appreciating why it is so vital for sustainable agriculture. Soils are not merely inert mediums for plant growth; they are dynamic ecosystems that host a diverse array of organisms, support critical nutrient and water cycling processes, and provide structural support to plants. The health and quality of the soil directly influence crop yields, resilience to pests and diseases, and the sustainable management of water resources. Healthy soils

M. Shaaban (✉)
College of Agriculture, Henan University of Science and Technology, Luoyang, China
e-mail: shabanbzu@hotmail.com

M. Shaaban (ed.), *Soils and Sustainable Agriculture*, Frontier Studies in Soil Science,
https://doi.org/10.1007/978-3-031-91114-9_1

contribute to a robust agricultural system that can adapt to changing environmental conditions and demands (Shaaban & Nunez-Delgado, 2024).

This chapter focuses on the essential relationship between soil health and sustainable agriculture. By examining the various functions of soil, the impact of soil health on agricultural productivity, and the socioeconomic implications of soil management practices, we can gain a nuanced understanding of why maintaining healthy soils is crucial for achieving sustainable agricultural outcomes. This exploration not only highlights the significance of soil in farming practices but also calls attention to the importance of effective soil management in the face of contemporary challenges such as climate change, food security, and biodiversity loss. Ultimately, the focus of this chapter serves to underscore the need for ongoing commitment to soil conservation and sustainable management as foundational elements of responsible agriculture.

2 The Functions of Soil

Soil serves a multitude of essential functions that are critical to the success and sustainability of agricultural practices. One of the primary functions of soil is its role in nutrient supply and cycling. Soils provide essential nutrients that plants require for growth, including macro- and micronutrients. These nutrients are vital for various plant processes, including photosynthesis, energy transfer, and overall growth and development. The availability of these nutrients is influenced by several factors, including soil pH, organic matter content, and mineral composition, which can vary significantly across different soils (Liu et al., 2023; Wu et al., 2020). Soil microorganisms play a crucial role in nutrient cycling, enhancing the availability of these essential elements (Shaaban, 2024b). Bacteria, fungi, and other microorganisms decompose organic matter, releasing nutrients back into the soil in forms that plants can readily absorb (Shaaban & Nunez-Delgado, 2024). This biological activity not only contributes to nutrient availability but also supports the formation of soil aggregates, improving soil structure and resilience. Moreover, specific groups of microorganisms, such as mycorrhizal fungi, form symbiotic relationships with plant roots, enhancing nutrient uptake, particularly phosphorus, and improving plants' overall health and growth. Another vital function of soil is its ability to retain and drain water, a critical aspect for crop production. The structure of the soil plays a significant role in determining its water retention capacity and drainage characteristics. Well-structured soils, with a balance of sand, silt, and clay particles, possess the ability to hold sufficient moisture while allowing excess water to drain away. This balance is essential in preventing waterlogging, which can damage plant roots, as well as drought stress, which can limit growth. Effective soil management practices, such as maintaining organic matter levels and optimizing soil structure, help improve water retention and drainage, directly influencing irrigation efficiency and the resilience of crops to drought conditions (Weil & Brady, 2017). Furthermore, soil provides essential support for plant growth, serving as a foundation for root systems.

The composition of the soil, including its texture and structure, affects root development and overall plant stability. A well-balanced soil provides the physical space needed for roots to expand and access nutrients and water. Additionally, soil acts as a medium for plant anchorage; strong root systems are necessary for maintaining plant health and stability, especially in adverse weather conditions such as high winds or heavy rainfall. Healthy soils with good structure and adequate organic content will promote robust root development, enabling plants to thrive and efficiently utilize available resources (Weil & Brady, 2017).

3 Soil Health and Biodiversity

Soil health is defined as the capacity of soil to function effectively as a living ecosystem and to sustain plant and animal productivity, maintain environmental quality, and promote plant and animal health. Healthy soils are characterized by a rich diversity of organisms, a well-balanced nutrient profile, and a resilient structure that can withstand environmental stressors. The concept of soil health extends beyond just the chemical composition of the soil; it encompasses biological, physical, and ecological processes that contribute to the soil's ability to support sustainable agricultural systems. Maintaining and improving soil health is essential for enhancing productivity, ensuring ecological balance, and supporting broader agronomic goals (Weil & Brady, 2017). Healthy soils have a significant impact on agricultural productivity. They provide the necessary nutrients, water, and support systems that plants need to grow and thrive. When soils are healthy, they can effectively retain moisture, structure, and provide a balanced supply of essential nutrients, resulting in higher crop yields and better-quality produce. Moreover, healthy soils are more resilient to disturbances such as pests, diseases, and extreme weather conditions. They facilitate root penetration, enhance water infiltration, and improve nutrient availability, all of which lead to a more stable and productive agricultural system. In contrast, degraded soils often lead to lower productivity, increased susceptibility to erosion, and reduced ability to respond to environmental challenges, making soil health a critical component of sustainable agriculture.

The role of biodiversity in soil ecosystems is fundamental to soil health and, by extension, agricultural productivity. Soil biodiversity encompasses the variety of organisms living within the soil, including a vast array of microorganisms, fungi, and macrofauna such as earthworms and insects. Microbial diversity plays a crucial role in nutrient cycling, decomposing organic matter, and enhancing soil structure. Different groups of microbes perform various functions that contribute to soil health, such as nitrogen fixation, organic matter decomposition, and bioremediation (Shaaban, 2024b; Shaaban & Nunez-Delgado, 2024; Weil & Brady, 2017). A diverse microbial community can better withstand environmental pressures and maintain soil function, leading to enhanced nutrient availability and improved plant health.

Soil fauna, which includes organisms such as earthworms, nematodes, and arthropods, also contribute significantly to soil health (Shaaban et al., 2018; Wu et al., 2015,

2023). These organisms help decompose organic matter, creating humus, which enriches the soil and enhances its nutrient-holding capacity. Earthworms, in particular, are considered indicators of soil health due to their ability to aerate the soil, improve drainage, and stimulate microbial activity through their feeding processes (Na et al., 2022). Additionally, soil fauna plays a vital role in the formation of soil aggregates, enhancing soil structure and stability. This aggregation process is crucial for maintaining proper aeration and water infiltration rates, further supporting plant growth and ecosystem health.

4 Soil Conservation and Sustainable Practices

Soil conservation is critical for sustaining agricultural productivity and ensuring environmental health. The importance of soil conservation cannot be overstated, as it directly impacts various aspects of farming and land management. One of the primary concerns in agriculture is soil erosion, which leads to the loss of the topsoil layer that is rich in nutrients and organic matter. Erosion can occur due to wind, water, and human activities, and it significantly diminishes soil quality and agricultural potential. Implementing effective soil conservation strategies is essential to prevent erosion, maintain soil integrity, and sustain agricultural productivity (Weil & Brady, 2017). In addition to preventing erosion, soil conservation plays a key role in maintaining soil fertility. Fertile soils are essential for healthy crop production, as they provide plants with the nutrients necessary for growth (Shaaban, 2024a). Conservation practices help preserve the natural structure and composition of the soil, ensuring that it retains essential nutrients and moisture. This preservation allows for more efficient use of fertilizers and organic inputs, reducing the need for chemical amendments and supporting sustainable farming practices that minimize environmental impact. Several sustainable agricultural practices enhance soil health and contribute to effective soil conservation. One widely recognized practice is crop rotation, which involves alternating the types of crops grown on a specific piece of land in successive seasons (Wu et al., 2017). Crop rotation helps to break pest and disease cycles, improve soil structure, and enhance soil fertility through the incorporation of a diverse range of root systems and organic matter. For example, leguminous crops such as soybeans or clover can naturally fix atmospheric nitrogen, enriching the soil for subsequent crops. Another effective practice is cover cropping, where farmers plant specific crops during the off-season to protect and enrich the soil. Cover crops, such as rye, vetch, or clover, cover the soil surface, preventing erosion and reducing weed pressure. These crops also contribute organic matter to the soil as they decompose, enhancing nutrient cycling and improving soil structure. Additionally, cover crops can increase soil moisture retention and surface stability, making the soil more resilient to extreme weather events (Kaspar & Singer, 2011).

Reduced tillage is another important sustainable practice that significantly benefits soil health. Conventional tillage can disrupt soil structure, decrease organic matter, and increase erosion risks. By minimizing or eliminating tillage, farmers can preserve soil structure and protect the habitat of beneficial microorganisms and soil fauna. No-till or reduced tillage methods can improve water infiltration, enhance soil structure, and promote greater biodiversity, leading to healthier soils and more productive crop yields (Krauss et al., 2022). Organic amendments, such as compost, manure, or green manures, are valuable tools for improving soil health and fertility (Shaaban et al., 2013a, 2013b). These organic materials increase the nutrient content of the soil, enhance microbial activity, and improve soil aggregation and moisture retention (Wu et al., 2017). Organic amendments also contribute to the development of soil biodiversity by providing a food source for a variety of soil organisms. As these amendments decompose, they release nutrients slowly, ensuring a steady supply for plants over time and reducing the risk of nutrient leaching.

5 The Relationship Between Soil and Climate Change

The relationship between soil and climate change is complex and increasingly critical as global temperatures rise and weather patterns shift. One of the essential roles of soil in the context of climate change is its function as a carbon sink (Shaaban & Nunez-Delgado, 2024). Soils have the capacity to sequester carbon dioxide (CO_2) from the atmosphere through the process of photosynthesis and the subsequent decomposition of organic matter. When plants photosynthesize, they absorb CO_2 from the air, and when these plants die and decompose, a portion of that carbon is incorporated into the soil organic matter. This organic matter not only enhances soil structure and fertility but also acts as a reservoir for carbon, effectively removing it from the atmosphere and storing it in the soil. Practices that enhance soil carbon sequestration, such as improving soil health through organic amendments and adopting conservation tillage methods, can play a critical role in mitigating climate change. However, climate change also threatens soil health through various mechanisms. Increasing temperatures can alter soil microbial communities and their activities, potentially disrupting nutrient cycling and other essential processes. Changes in precipitation patterns, including more frequent and severe droughts or heavy rainfall, can lead to soil erosion, nutrient leaching, and degradation of soil structure. As soils become increasingly stressed, their ability to support plant growth diminishes, resulting in reduced crop yields and compromised food security. Additionally, the warming climate may exacerbate the prevalence of pests and diseases that affect soil health, further complicating agricultural resilience (Weil & Brady, 2017). To address the challenges posed by climate change and to harness the potential of soils as a solution, several strategies for mitigating climate change through effective soil management must be implemented. One key approach is adopting practices that enhance soil organic matter content. Increasing organic matter not only improves soil structure and fertility but also boosts the soil's capacity to sequester carbon. Practices such as adding compost, utilizing cover crops, and engaging in agroforestry can significantly enhance organic matter levels, thus benefiting both soil health and climate stability.

Another important strategy is implementing conservation tillage and reduced tillage practices. By minimizing soil disturbance, farmers can help maintain soil structure and enhance its ability to sequester carbon. This practice also reduces erosion and protects existing soil aggregates, which are essential for maintaining healthy soil ecosystems (Krauss et al., 2022). Implementing crop rotations and diversifying plantings can further enhance resilience against climate variability, improve nutrient cycling, and increase carbon storage within the soil (Bozal-Leorri et al., 2023). Moreover, scientists and practitioners are increasingly exploring the potential of regenerative agricultural practices, which emphasize the restoration and enhancement of soil health. This encompasses a suite of techniques aimed at rehabilitating degraded soils, increasing biodiversity, and promoting ecosystem services. Practices such as agroecology and integrated farming systems emphasize the interaction between soil health, biodiversity, and climate resilience, helping to create a holistic approach to agricultural production. In addition, employing precision agriculture technologies can enhance soil management. By utilizing data-driven approaches, farmers can make informed decisions regarding fertilizer application and water management, optimizing resource use and reducing environmental impacts. This not only helps in maintaining soil health but also supports efforts to mitigate climate change through targeted interventions.

6 Socioeconomic Implications

The socioeconomic implications of soil health and management are far-reaching, particularly regarding food security, economic stability, and environmental policies. One of the most critical roles of soil is its contribution to food security. Healthy soils are the foundation of successful agriculture, providing the necessary nutrients and supportive structures that enable crops to grow. As the global population continues to rise, the demand for food is increasing, putting immense pressure on agricultural systems. Sustainable soil management practices are essential for ensuring that soils remain productive and can continue to support food production in the face of challenges such as climate change, soil degradation, and resource scarcity. By maintaining soil health through practices like crop rotation, cover cropping, and reduced tillage, we can enhance agricultural productivity and resilience, ultimately contributing to a more secure food supply. In addition to ensuring food security, sustainable soil management offers significant economic benefits. By fostering healthy soils, farmers can achieve higher yields and better-quality crops, leading to increased profitability. Furthermore, healthy soils reduce the need for chemical fertilizers and pesticides, lowering input costs and improving profit margins. As soil health improves, the economic stability of farming operations can become more resilient to market fluctuations and environmental stresses. Furthermore, investing in sustainable soil practices can enhance rural livelihoods and stimulate local economies by creating jobs in sectors related to sustainable agriculture, conservation, and organic farming. The long-term economic outlook of sustainable soil management outweighs short-term gains from conventional practices that may capitalize on one-off soil exploitation but ultimately lead to degradation and loss of land productivity.

To fully capitalize on the advantages of soil conservation and sustainable practices, comprehensive policy considerations are necessary. Policymakers must recognize the importance of soil health in achieving broader environmental and social objectives, such as combating climate change, restoring ecological balance, and enhancing public health. Policies promoting sustainable land management should include incentives for farmers to adopt soil conservation practices, such as grants for cover cropping, subsidies for organic inputs, and funding for research on regenerative agriculture techniques. Additionally, educational programs aimed at farmers, agronomists, and land managers can facilitate the adoption of sustainable practices by raising awareness of the benefits of soil conservation. Furthermore, regulatory frameworks should be evaluated and potentially reformed to address the challenges posed by soil degradation effectively. Policies need to be informed by scientific research and community-based knowledge, ensuring that they are relevant and applicable to specific local conditions. Collaborating with stakeholders across sectors, including agriculture, environmental advocacy, and community organizations, can help shape more effective strategies that enhance soil conservation efforts, promote sustainable land use practices, and ensure fair access to resources.

7 Conclusion

In conclusion, the multifaceted relationship between soil health and sustainable agriculture underscores the critical importance of prioritizing soil conservation and management practices. Healthy soils play an indispensable role in supporting agricultural productivity, ensuring food security, and contributing to the overall resilience of ecosystems. Through the various functions that soils provide, ranging from nutrient supply and water retention to carbon sequestration, soil is not merely a passive medium but an active participant in agricultural systems and climate regulation. The pressing challenges posed by climate change, soil degradation, and the increasing demand for food necessitate a concerted effort towards adopting and promoting sustainable agricultural practices. Strategies such as crop rotation, cover cropping, reduced tillage, and the use of organic amendments are essential tools in enhancing soil health and functionality. Furthermore, the socioeconomic implications of healthy soils extend beyond environmental stewardship; they encompass the potential for improved livelihoods, increased economic stability, and strengthened food systems. As we move forward, it is crucial that policymakers, farmers, and communities collaborate to develop and implement effective policies and strategies that prioritize soil conservation. By investing in education, research, and sustainable practices, we can protect and restore the vital resource that soil represents, ensuring its capacity to support future generations. The pathway to sustainable agriculture hinges on our commitment to understanding, preserving, and enhancing soil health, an endeavor that is not only beneficial for agricultural production but also essential for safeguarding the planet's ecological balance and achieving long-term food security.

Acknowledgements The work was supported by the "Fundamental Research Funds for Henan University of Science and Technology, 4024-13510002."

References

Bozal-Leorri, A., Corrochano-Monsalve, M., Arregui, L. M., Aparicio-Tejo, P. M., & González-Murua, C. (2023). Evaluation of a crop rotation with biological inhibition potential to avoid N_2O emissions in comparison with synthetic nitrification inhibition. *Journal of Environmental Sciences, 127*, 222–233.

Kaspar, T., & Singer, J. (2011). The use of cover crops to manage soil. In *Soil management: Building a stable base for agriculture* (pp. 321–337).

Krauss, M., Wiesmeier, M., Don, A., Cuperus, F., Gattinger, A., Gruber, S., Haagsma, W., Peigné, J., Palazzoli, M. C., & Schulz, F. (2022). Reduced tillage in organic farming affects soil organic carbon stocks in temperate Europe. *Soil and Tillage Research, 216*, Article 105262.

Liu, X., Shi, Z., Bai, H., Zhang, J., Sun, D., & Chen, Y. (2023). Soil carbon sequestration in paddy field and its simultaneous mineralization to supply available nutrients for the crops are affected by no-tillage with straw management: A meta-analysis. *Applied Soil Ecology, 188*, Article 104850.

Na, L., Hu, C., Jiang, Y., Hu, R., Shaaban, M., Younas, A., & Wu, Y. (2022). Earthworms promote the transfer of 15N-urea to lettuce while limit appreciably increase 15N losing to environment. *Environmental Research, 212*, Article 113423.

Shaaban, M. (2024b). Microbial pathways of nitrous oxide emissions and mitigation approaches in drylands. *Journal of Environmental Management, 354*, Article 120393.

Shaaban, M., & Nunez-Delgado, A. (2024). Soil adsorption potential: Harnessing Earth's living skin for mitigating climate change and greenhouse gas dynamics. *Environmental Research, 251*, Article 118738.

Shaaban, M., Abid, M., & Abou-Shanab, R. (2013a). Amelioration of salt affected soils in rice paddy system by application of organic and inorganic amendments. *Plant, Soil and Environment, 59*, 227–233.

Shaaban, M., Van Zwieten, L., Bashir, S., Younas, A., Núñez-Delgado, A., Chhajro, M. A., Kubar, K. A., Ali, U., Rana, M. S., & Mehmood, M. A. (2018). A concise review of biochar application to agricultural soils to improve soil conditions and fight pollution. *Journal of Environmental Management, 228*, 429–440.

Shaaban, M., Abid, M., & Qi-An, P. (2013b) Short term influence of gypsum, farm manure and commercial humic acid on physical properties of salt affected soil in rice paddy system. *Journal of the Chemical Society of Pakistan, 34*.

Shaaban, M. (2024a). *Acidic soils, planet Earth: Scientific proposals to solve urgent issues* (pp. 293–306). Springer.

Weil, R. R., & Brady, N. C. (2017). *The nature and properties of soils* (15th ed.). Pearson Education Limited.

Wu, Y., Shaaban, M., Zhao, J., Hao, R., & Hu, R. (2015). Effect of the earthworm gut-stimulated denitrifiers on soil nitrous oxide emissions. *European Journal of Soil Biology, 70*, 104–110.

Wu, L., Tang, S., He, D., Wu, X., Shaaban, M., Wang, M., Zhao, J., Khan, I., Zheng, X., & Hu, R. (2017). Conversion from rice to vegetable production increases N_2O emission via increased soil organic matter mineralization. *Science of the Total Environment, 583*, 190–201.

Wu, L., Hu, R., Tang, S., Shaaban, M., Zhang, W., Shen, H., & Xu, M. (2020). Nitrous oxide emissions in response to straw incorporation is regulated by historical fertilization. *Environmental Pollution, 266*, Article 115292.

Wu, Y., Li, Q., Zheng, Y., Xiong, X., Chen, Y., Shaaban, M., & Hu, R. (2023). Optimizing biochar addition for vermicomposting: A comprehensive evaluation of earthworms' activity, N_2O emissions and compost quality. *Biochar, 5*, 1–4.

Soil Threats to Sustainable Agriculture

Muhammad Shaaban

Abstract This chapter explores critical threats to sustainable agriculture, highlighting the adverse impacts of nutrient imbalance, salinization and sodification, soil sealing and land take, and the loss of soil biodiversity. These issues threaten not only agricultural productivity but also environmental health and ecosystem integrity. It emphasizes the urgent need for sustainable land management practices that enhance soil health and promote agricultural resilience. The chapter advocates for a holistic approach, integrating strategies such as improved irrigation and fertilization methods, responsible land use planning, and biodiversity conservation to mitigate these threats. Recognizing and addressing these challenges is essential for ensuring food security and sustainability in a rapidly changing global landscape.

Keywords Sustainable agriculture · Soil health · Nutrient imbalance · Salinization · Soil sealing · Biodiversity loss · Food security · Land management

1 Introduction

Sustainable agriculture is essential for ensuring food security, preserving natural resources, and maintaining ecosystem health. However, various threats to sustainable agricultural practices have emerged over time, posing significant challenges for farmers and policymakers. Among these threats, soil erosion, waterlogging, compaction, and acidification directly impact the quality and productivity of soil, which serves as the foundation for agricultural systems (Fig. 1). The degradation of soil health not only reduces crop yields but also diminishes the resilience of the agricultural landscape, making it vulnerable to external shocks from climate change and extreme weather events (Evangelista et al., 2024; Shaaban & Nunez-Delgado, 2024). Moreover, contamination from agricultural inputs, such as pesticides and fertilizers, can exacerbate existing soil problems. As harmful substances infiltrate the soil, they not only affect microbial communities but also disrupt the delicate balance

M. Shaaban (✉)
College of Agriculture, Henan University of Science and Technology, Luoyang, China
e-mail: shabanbzu@hotmail.com

M. Shaaban (ed.), *Soils and Sustainable Agriculture*, Frontier Studies in Soil Science,
https://doi.org/10.1007/978-3-031-91114-9_2

of nutrients that crops rely upon for growth. This leads to nutrient imbalances, which can stunt plant development and compromise food quality. Furthermore, changes in organic carbon levels due to unsustainable practices can deplete soil health and fertility, making it increasingly challenging to cultivate crops without resorting to chemical inputs (Shaaban et al., 2018).

Additionally, issues such as salinization and sodification impede agricultural productivity, particularly in arid and semi-arid regions where water management practices are critical. When soil becomes overly saturated with salts, crops can suffer from reduced growth and yield (Shaaban et al., 2023). Compounding this issue, soil sealing and land take due to urbanization and infrastructure development encroach on arable land, further threatening the viability of food production systems. The loss of soil biodiversity is another pressing concern, as diverse microbial communities play a critical role in maintaining soil health and fertility. The reduction of these communities can lead to diminished ecosystem services, ultimately undermining sustainable agricultural efforts (Shaaban et al., 2013a, 2013b). Addressing these multifaceted threats requires a comprehensive understanding of their interconnectedness and a commitment to adopting sustainable practices that promote soil health. By recognizing the signs of degradation and implementing restorative measures, such as crop rotation, cover cropping, and organic amendments, we can mitigate these threats and

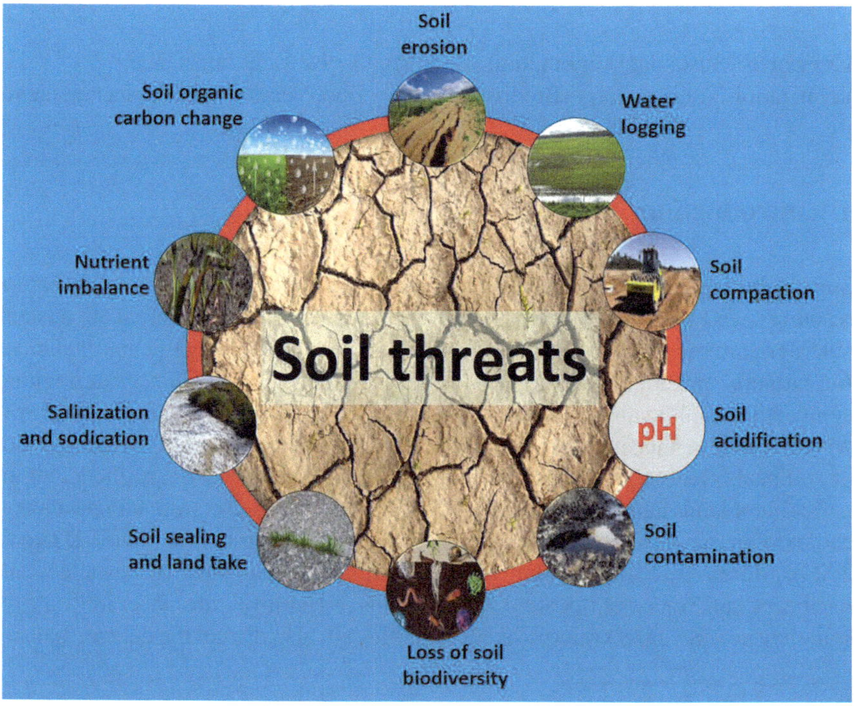

Fig. 1 Soil related threats to sustainable agriculture

enhance the resilience of agricultural systems. Only through concerted efforts can we safeguard the future of sustainable agriculture and ensure the well-being of our planet and its inhabitants.

2 Soil Erosion

Soil erosion is a significant threat to sustainable agriculture, leading to the loss of topsoil and essential nutrients necessary for healthy crop growth. It occurs when soil particles are detached and transported by wind or water, stripping away the fertile layer that provides a foundation for plant life. This natural process can be accelerated by human activities, such as deforestation, overgrazing, and unsustainable agricultural practices, resulting in dire consequences for agricultural productivity and ecosystem health (Weil & Brady, 2017). One of the primary causes of soil erosion is the removal of vegetation cover. Plants play a crucial role in anchoring soil with their roots, preventing erosion by absorbing water and reducing runoff. When land is cleared for agricultural production or other human activities, the protective cover is removed, leaving the soil exposed to wind and rain. The absence of plant cover can lead to increased surface runoff, which, combined with heavy rainfall, further exacerbates the erosion process. As topsoil is washed away, the soil's ability to retain moisture diminishes, creating a cycle of degradation that affects crop yields and soil health (Ma et al., 2021; Weil & Brady, 2017).

The impact of soil erosion extends beyond the immediate loss of arable land. As soil is eroded, valuable nutrients such as nitrogen, phosphorus, and potassium are also depleted, leading to a decline in soil fertility. This nutrient loss forces farmers to rely more heavily on chemical fertilizers to maintain crop production, which can create a dependency on external inputs and lead to further environmental issues, such as water pollution and habitat degradation. Moreover, the sediment that is transported away from eroded fields can end up in waterways, contributing to sedimentation and affecting aquatic ecosystems by smothering habitats and disrupting nutrient cycles (Wang et al., 2024).

To combat soil erosion effectively, sustainable land management practices must be adopted. Techniques such as contour farming, terracing, and the use of cover crops help to enhance soil structure and reduce runoff. Implementing practices that promote soil conservation can significantly mitigate erosion and its associated impacts, ensuring that agricultural systems remain productive and resilient. Moreover, raising awareness among farmers and stakeholders about the importance of protecting soil resources is essential for fostering a more sustainable approach to agriculture. By prioritizing soil health and implementing effective erosion control measures, we can help secure the future of global food systems while preserving vital ecosystems.

3 Water Logging

Waterlogging is a critical threat to sustainable agriculture, characterized by the saturation of soil with water to the point where it becomes detrimental to crop growth and soil health. This condition typically occurs in low-lying areas or regions with poorly drained soils, where excess water accumulates due to heavy rainfall, irrigation practices, or a high-water table. The impact of waterlogging can be profound, as it affects not only the physical and chemical properties of the soil but also the biological processes that are essential for healthy crop development (Tian et al., 2021; Weil & Brady, 2017). One of the primary consequences of waterlogging is the reduction of soil aeration. When soil pores fill with water, the amount of oxygen available to plant roots decreases significantly. Many crops, especially those adapted to well-drained soils, are intolerant to low-oxygen conditions. As a result, waterlogged soils can lead to root suffocation, impaired nutrient uptake, and ultimately, plant stress or death. Moreover, prolonged waterlogging can trigger a cascade of negative effects, including increased susceptibility to root rot diseases caused by pathogens that thrive in anaerobic conditions (Weil & Brady, 2017).

Waterlogging also affects soil chemistry by promoting the leaching of essential nutrients and altering the pH levels of the soil. Saturated conditions can lead to the depletion of nitrogen and other critical nutrients, making them less available to plants (Liu et al., 2023). This situation often forces farmers to rely on chemical fertilizers to compensate for nutrient loss, which can further exacerbate environmental issues such as nutrient runoff and water pollution. Additionally, the accumulation of soluble salts, a common occurrence in regions with poor drainage, can lead to salinization, further degrading soil quality and agricultural productivity (Liu et al., 2023).

To mitigate the threat of waterlogging, implementing effective water management strategies is crucial. Techniques such as improving drainage systems, using raised beds, and practicing crop rotation with more water-tolerant varieties can help alleviate excess water accumulation. The adoption of conservation tillage and the use of cover crops can also enhance soil structure and increase water infiltration, reducing the likelihood of waterlogging during periods of heavy rainfall. Furthermore, integrating agroecological practices that mimic natural processes can bolster the resilience of agricultural systems against waterlogging.

4 Soil Compaction

Soil compaction is a significant threat to sustainable agriculture, arising from the excessive pressure exerted on soil, which leads to the densification of soil particles and a subsequent decrease in pore space. This phenomenon typically occurs due to heavy machinery use in farming operations, livestock trampling, or inadequate tillage practices. Compacted soil poses numerous challenges for agricultural productivity, impacting water infiltration, root development, and nutrient availability, making it a pressing concern for farmers and agricultural stakeholders (Weil & Brady, 2017). One of the primary effects of soil compaction is the hindered movement of water through the soil profile. Compacted soil has reduced pore spaces, meaning that water cannot easily penetrate the surface layer. As a result, rainfall or irrigation water tends to runoff rather than infiltrate, leading to increased erosion and loss of moisture for crops. The reduction in water infiltration also exacerbates drought conditions, as plants struggle to access the moisture stored deeper in the soil. This can severely affect crop yields, particularly during critical growth phases when water availability is essential (Shaheb et al., 2021).

In addition to impacting water dynamics, soil compaction poses serious challenges to root development. Compacted soils are denser and more rigid, creating physical barriers that impede root penetration and expansion. Roots may struggle to access nutrients and water, resulting in stunted plant growth and reduced crop health. Moreover, the limited root system makes plants more vulnerable to stressors such as drought and disease, further undermining agricultural resilience. Over time, the effects of soil compaction can culminate in declining yields, necessitating increased application of fertilizers and other inputs, which can lead to a cycle of dependency on external resources (Gürsoy, 2021; Mileusnić et al., 2022).

To combat soil compaction effectively, farmers can adopt various sustainable practices aimed at preserving soil structure and health. Key strategies include the use of lighter equipment, adopting minimum tillage or no-till practices, and implementing controlled traffic farming, which restricts the movement of machinery to specific paths to minimize pressure on the soil. Additionally, rotating crops with deep-rooted species can help break up compacted layers and improve soil aeration. Incorporating organic matter, such as compost or cover crops, can enhance soil structure, increase microbial activity, and promote a more resilient agronomic system.

5 Soil Acidification

Soil acidification is a pressing threat to sustainable agriculture, characterized by the increase in hydrogen ion concentration in the soil, which leads to a decline in soil pH. This phenomenon can be attributed to various factors, including the use of acidic fertilizers, the leaching of basic cations (such as calcium and magnesium) due to rainfall, and the natural processes of organic matter decomposition. As soil pH decreases, it can have significant implications for plant health, nutrient availability, and overall agricultural productivity, making it crucial for farmers and agronomists to understand and address this issue (Shaaban, 2024a). One of the primary consequences of soil acidification is the alteration of nutrient availability. Many essential nutrients, such as nitrogen, phosphorus, and potassium, are more readily available in neutral to slightly alkaline soil conditions. As soil pH drops, the solubility of certain nutrients and minerals can also change, making them less accessible to plants. For example, as acidity increases, nutrients like iron, manganese, and aluminum become more soluble and can reach toxic levels, while other essential nutrients such as calcium and magnesium may become deficient. This imbalance can result in nutrient deficiencies or toxicities, leading to stunted growth, reduced yields, and impaired crop quality (Shaaban et al., 2020). Moreover, soil acidification can negatively impact soil microbial communities and overall soil health. An acidic environment can alter the composition and activity of beneficial soil microorganisms, which play vital roles in processes such as nutrient cycling, organic matter decomposition, and soil structure formation. The decline in microbial diversity and activity can further exacerbate nutrient imbalances and hinder the organic matter buildup that is essential for maintaining soil fertility. A healthy soil ecosystem is fundamental for sustainable agriculture, and any disruption in microbial function can have lasting consequences on crop production (Wu et al., 2021).

To mitigate the threat of soil acidification, farmers can adopt several sustainable practices. Regular soil testing is essential for monitoring pH levels and understanding the specific needs of the soil. When acidification is identified, the application of lime (calcium carbonate) is a common practice to neutralize soil acidity and restore a more favorable pH balance. Additionally, incorporating organic amendments, such as compost or crop residues, can help improve soil structure and enhance microbial activity, which can buffer against further acidification. Employing crop rotations with legumes, which can naturally increase soil pH through nitrogen fixation, is another effective strategy (Shaaban, 2024a).

6 Soil Contamination

Soil contamination is a significant threat to sustainable agriculture, characterized by the introduction of harmful substances into the soil that can adversely affect soil health, plant growth, and ecosystem integrity. Contaminants can originate from various sources, including agricultural practices, industrial activities, urban runoff, and improper waste disposal. The presence of contaminants not only compromises soil quality but also poses risks to food safety and human health, making it a critical concern for farmers, policymakers, and consumers alike. One of the principal contributors to soil contamination in agricultural settings is the overuse of chemical fertilizers and pesticides. While these inputs can enhance crop production, their excessive application can lead to the accumulation of harmful residues in the soil. For instance, synthetic chemicals may persist in the environment, leaching into the soil and altering microbial communities, which are essential for nutrient cycling and organic matter decomposition. This disruption can result in imbalanced soil fertility, leading to reduced crop yields and increased reliance on additional inputs to compensate for nutrient deficiencies (Shaaban et al., 2018).

Furthermore, heavy metals such as lead, cadmium, and arsenic can contaminate soils through industrial activities, mining operations, and the application of wastewater for irrigation. These toxic substances can accumulate in the soil and, in turn, be taken up by plants, potentially leading to bioaccumulation in the food chain. As such, contaminated crops pose significant health risks to consumers, especially if they are unaware of the presence of harmful substances. Ensuring food safety becomes critical as contaminated soil can undermine public trust in agricultural practices and food systems (Ali et al., 2021). Soil contamination also impacts the broader ecosystem, affecting not only agricultural productivity but also biodiversity and water quality. Contaminants can leach into groundwater or runoff into nearby water bodies, leading to the degradation of aquatic ecosystems. The presence of pollutants can disrupt food webs, reduce biodiversity, and harm aquatic species, further exacerbating the challenges linked to environmental sustainability (Ali et al., 2020).

To effectively combat soil contamination, a multifaceted approach is required. First, implementing best management practices in agriculture, such as integrated pest management, organic farming, and precision agriculture, can help minimize the use of harmful chemicals and reduce the risk of contamination. Regular soil testing is also essential for identifying contaminants and monitoring soil health. Where contamination is identified, remediation techniques, such as phytoremediation (using plants to absorb contaminants) or soil washing, can be employed to restore soil quality.

7 Soil Organic Carbon Change

Soil organic carbon (SOC) change is a critical threat to sustainable agriculture, significantly influencing soil health, fertility, and overall ecosystem functioning. Soil organic carbon refers to the carbon component of organic compounds in soil, derived from decaying plant and animal materials, microorganisms, and organic matter inputs. It plays a vital role in various soil processes, including nutrient cycling, soil structure, and water retention. Changes in SOC levels, whether through depletion or enhancement, can have profound implications for agricultural productivity and environmental sustainability (Wu et al., 2018). One of the primary drivers of SOC change is unsustainable land management practices. Practices such as intensive tillage, monoculture cropping, and overgrazing can lead to the loss of soil organic matter, resulting in lower SOC levels. When soil is disturbed, organic matter is exposed to oxygen, leading to rapid decomposition and the release of carbon dioxide (CO_2) into the atmosphere. This not only diminishes the soil's carbon stock but also contributes to greenhouse gas emissions, exacerbating climate change. As SOC levels decline, soil structure deteriorates, leading to increased erosion, reduced water infiltration, and decreased overall fertility. Consequently, farmers may be compelled to apply more fertilizers and soil amendments to compensate for nutrient loss, elevating production costs and potentially causing further environmental harm (Wu et al., 2017). Moreover, changes in SOC can directly impact plant growth and yield. Organic carbon is a key reservoir of nutrients and acts as a binding agent for soil particles, improving soil structure and enhancing aeration. Soils with adequate levels of SOC can retain moisture more effectively, providing crops with a stable water supply during dry periods. In contrast, low SOC soils may struggle to support healthy crop growth and can lead to reduced yields, threatening food security.

The implications of SOC change extend beyond agricultural productivity. Healthy soils with high organic carbon content support diverse microbial communities that play essential roles in nutrient cycling and organic matter decomposition. A decline in SOC can disrupt these communities, leading to decreased soil biodiversity and reduced ecosystem services, such as pest control and pollination. These changes can further undermine the resilience of agricultural systems, making them more vulnerable to pests, diseases, and environmental stressors. To address the threat of soil organic carbon change, sustainable management practices must be adopted. Strategies such as cover cropping, agroforestry, reduced tillage, and the application of organic amendments (like compost and manure) can help increase SOC levels. Cover crops, for instance, protect the soil during fallow periods, reduce erosion, and contribute biomass that adds carbon to the soil upon decomposition. In addition, improving soil management through crop rotation and integrated nutrient management can foster conditions conducive to organic matter accumulation.

8 Nutrient Imbalance

Nutrient imbalance in soil is a significant threat to sustainable agriculture, characterized by an uneven distribution or insufficient availability of essential nutrients required for optimal plant growth and crop yield. This imbalance can arise from several factors, including the excessive use of fertilizers, inadequate soil management practices, and changes in land use. The consequences of nutrient imbalance extend beyond diminished crop productivity to include environmental degradation and a decline in soil health, making it a critical concern for agricultural sustainability (Peñuelas & Sardans, 2022; Vitousek et al., 2009).

One of the primary causes of nutrient imbalance is the overapplication of synthetic fertilizers, which often focus on a narrow range of nutrients, particularly nitrogen (N), phosphorus (P), and potassium (K). While these macronutrients are vital for plant growth, an excessive application can lead to an overabundance of certain nutrients while other essential nutrients, such as calcium, magnesium, and trace elements like zinc and iron, become deficient. This situation can create a condition known as "hidden hunger," where plants exhibit symptoms of nutrient deficiency despite sufficient macronutrients being present in the soil. As a result, crop growth may be stunted, and yields may suffer, adversely affecting food production and farmer livelihoods (Rodríguez-Espinosa et al., 2024). Additionally, nutrient imbalances can disrupt soil microbiomes, which are crucial for maintaining healthy soil ecosystems. Soil microorganisms play a vital role in nutrient cycling, breaking down organic matter, and making nutrients available to plants. An imbalance in nutrient availability can hinder microbial activity and diversity, further exacerbating soil fertility issues. This disruption may lead to the accumulation of toxic substances, such as heavy metals, and a decrease in beneficial processes such as nitrogen fixation and organic matter decomposition (Shaaban & Nunez-Delgado, 2024; Shaaban, 2024b). Moreover, nutrient imbalance can contribute to environmental problems, such as water pollution and soil degradation. When excess nutrients from fertilizers leach into water bodies, they can cause eutrophication, leading to algal blooms that deplete oxygen levels in aquatic environments and harm fish and other aquatic organisms. In addition, nutrient runoff can contribute to the contamination of drinking water sources, posing risks to human health.

To mitigate the threat of nutrient imbalance in soil, sustainable agricultural practices must be implemented. Regular soil testing is essential for monitoring nutrient levels and understanding the specific needs of the soil. This information allows farmers to make informed decisions regarding fertilization and amend practices to create balanced nutrient profiles. Integrated Nutrient Management (INM) combines the application of organic amendments (like compost, green manure, and biochar) with synthetic fertilizers to enhance nutrient availability and soil health. Practices such as crop rotation, cover cropping, and agroecological approaches can also help diversify nutrient inputs and improve soil structure, further promoting balanced nutrient cycling.

9 Salinization and Sodification

Salinization and sodification are significant threats to sustainable agriculture that can severely impact soil health, crop productivity, and environmental stability. Both processes are characterized by the accumulation of salts in the soil, but they differ in the specific types of salts involved and their effects on soil and plant health. Understanding these processes is crucial for developing strategies to mitigate their impacts on agricultural systems and ensuring long-term sustainability in farming practices.

Salinization

Salinization refers to the accumulation of soluble salts in soil, primarily due to factors such as irrigation practices, poor drainage, and natural processes such as weathering of minerals. It often occurs in arid and semi-arid regions where evaporation rates exceed precipitation, leading to a concentration of salts in the soil. Irrigation, if not properly managed, can exacerbate salinization, as water applied to fields can dissolve salts and carry them to the surface through capillary action. When irrigation water evaporates, it leaves behind salts that accumulate in the upper soil layers (Shaaban et al., 2013b).

The effects of salinization on agriculture can be detrimental. High salinity levels can impede plant growth by disrupting the osmotic balance, making it difficult for plants to absorb water. Furthermore, salinity can lead to ion toxicity, with specific ions such as sodium and chloride being particularly harmful to many crop species. Salinized soils exhibit poor nutrient availability, which further limits crop productivity and can result in reduced yields, lower crop quality, and economic losses for farmers (Shaaban et al., 2023).

Sodification

Sodification, on the other hand, is characterized by the accumulation of sodium ions in the soil, often leading to the development of sodic soils. This process is usually a consequence of the use of sodium-rich irrigation water or the leaching of basic cations (calcium, magnesium, and potassium) from the soil. Sodic soils have high sodium levels relative to calcium and magnesium, which adversely affects soil structure. The effects of sodification include decreased soil permeability and aeration, as sodium causes soil particles to disperse, leading to the formation of more compacted and impermeable soil layers. This can hinder water infiltration, increase runoff, and exacerbate erosion. The physical properties of sodic soils make them particularly challenging for agriculture, as they often exhibit poor drainage and high susceptibility to waterlogging, further stressing plant growth (Weil & Brady, 2017).

Mitigation Strategies

To combat salinization and sodification, farmers and land managers can adopt several sustainable practices:

1. Improved Irrigation Management: Employing methods such as drip irrigation or the use of brackish water, which can help to minimize the amount of water that evaporates and leaves salts behind.
2. Soil Amendments: Adding soil amendments, such as gypsum (calcium sulfate), can help to displace sodium ions and improve soil structure, facilitating water infiltration and root penetration.
3. Crop Selection: Utilizing salt-tolerant plant varieties or halophytes that can thrive in saline conditions can help maintain agricultural productivity in affected areas.
4. Leaching: Implementing leaching practices, where excess water is applied to flush out salts from the root zone, can be effective in reducing salinity levels in the soil. However, this requires well-drained soils and careful management to avoid further salinization.
5. Monitoring and Testing: Regular soil testing can help track salinity and sodium levels, enabling proactive management to mitigate the effects before they become severe.

10 Soil Sealing and Land Take

Soil sealing and land take represent significant threats to sustainable agriculture, primarily as a result of urbanization, industrialization, and infrastructure development. These processes lead to the irreversible loss of arable land and diminish the overall health of soils, directly impacting agricultural productivity as well as ecological and environmental integrity. Understanding the implications of soil sealing and land take is crucial for developing effective strategies to safeguard agricultural resources and promote sustainable land use practices (Stankovics et al., 2020).

Soil sealing refers to the process of covering soil surfaces with impermeable materials, such as concrete, asphalt, and buildings. This phenomenon typically occurs in urban areas as a result of urban sprawl, infrastructure expansion, and construction projects. When soil is sealed, it loses its natural physical, chemical, and biological properties, leading to a range of adverse effects. Sealed soils can no longer perform vital functions, such as water infiltration, nutrient cycling, and habitat provision for soil microorganisms and organisms. This loss contributes to reduced fertility and a decline in overall soil health. Additionally, soil sealing prevents water absorption, resulting in increased surface runoff during rainfall events. This not only contributes to flooding in urban areas but also enhances soil erosion in adjacent agricultural lands, leading to the degradation of soil quality. Furthermore, sealed surfaces can significantly alter local microclimates by increasing heat retention, reducing humidity, and impacting rainfall patterns. Such changes can challenge local agricultural systems that rely on specific climatic conditions for optimal crop growth (Lu et al., 2020; Xiao et al., 2021).

Land take refers to the conversion of agricultural land into non-agricultural uses, such as industrial zones, housing developments, and transportation infrastructures. This process poses several challenges to sustainable agriculture. As cities and infrastructure expand, valuable agricultural land is often lost, reducing the area available for food production. Over time, this contributes to food insecurity, especially in regions where agricultural land is already limited. The conversion of contiguous farmland into isolated parcels can disrupt agricultural operations, making it more difficult for farmers to manage and cultivate their lands effectively. Smaller, fragmented plots may lead to inefficiencies in farming practices and decreased yields. Moreover, the reduction in available arable land can lead to intensified cultivation practices on the remaining land, resulting in overuse of soil resources, increased reliance on chemical inputs, and greater risks of soil degradation. Additionally, land take affects not only agricultural areas but also surrounding natural habitats, leading to habitat loss and fragmentation. This can threaten biodiversity, reducing the resilience of ecosystems and their ability to support agricultural systems through services such as pollination and pest control (Xiao et al., 2021).

To address the threats of soil sealing and land take, various strategies can be implemented. Effective spatial planning and zoning regulations can help protect valuable agricultural lands from unnecessary conversion. Implementing land use policies that prioritize sustainable agriculture and limit urban sprawl is essential. Urban planners can incorporate green spaces, parks, and sustainable landscaping into urban development projects. These practices help minimize soil sealing and promote better water management through permeable surfaces. Focusing on improving productivity in existing agricultural lands through sustainable practices like precision agriculture, agroecology, and organic farming can help maximize food production without requiring additional land. Raising awareness among communities, policymakers, and developers about the importance of preserving agricultural land can foster support for sustainable land management practices.

11 Loss of Soil Biodiversity

The loss of soil biodiversity is a profound threat to sustainable agriculture, significantly impacting soil health, ecosystem functioning, and agricultural productivity. Soil biodiversity encompasses the variety of organisms inhabiting the soil, including bacteria, fungi, protozoa, nematodes, earthworms, and other invertebrates. These organisms play essential roles in nutrient cycling, organic matter decomposition, soil structure formation, and the overall maintenance of soil health. When soil biodiversity declines, the consequences can be detrimental not only to agriculture but also to the environment and food security (Tibbett et al., 2020).

One of the primary drivers of soil biodiversity loss is the intensification of agricultural practices. Practices such as monoculture cropping, excessive use of chemical fertilizers and pesticides, and heavy tillage can disrupt soil habitats and harm beneficial organisms. Monoculture, for instance, limits the diversity of plants and roots in the soil, which in turn limits the diversity of soil-dwelling organisms that rely on these plants for nutrition and habitat. The application of synthetic chemicals can kill off both harmful pests and beneficial microorganisms, leading to imbalances in the soil ecosystem. The excessive disturbance of soil through tillage can destroy habitats for many organisms, reducing their populations and ultimately leading to a decline in soil biodiversity (Köninger et al., 2021; Tibbett et al., 2020).

The consequences of declining soil biodiversity are wide-ranging. Healthy, biodiverse soils support the processes of nutrient cycling, where essential nutrients are made available to plants. A rich microbiome can enhance soil fertility, improve moisture retention, and increase resilience against pests and diseases. When soil biodiversity is compromised, these vital processes can be disrupted, resulting in lower soil fertility and diminished crop yields. Farmers may find themselves trapped in a cycle of increased chemical use to compensate for the lost natural processes, further exacerbating soil degradation and biodiversity loss (Bach et al., 2020; Köninger et al., 2021). Additionally, the decline in soil biodiversity can affect ecosystem services beyond agricultural productivity. Diverse soil organisms contribute to the formation of stable soil aggregates that enhance soil structure, improve aeration, and promote water infiltration. When soil structure deteriorates, it can lead to increased erosion, reduced water-holding capacity, and higher susceptibility to drought, which can be particularly troubling in the context of climate change. Biodiverse soils also enhance ecosystem resilience, allowing agricultural systems to adapt to changing environmental conditions and fluctuations in climate, pests, and diseases.

To combat the loss of soil biodiversity, sustainable agricultural practices must be prioritized. Practices such as crop rotation, cover cropping, agroecology, and organic farming can promote diverse microbial communities and restore soil health. Crop rotation can enhance soil biodiversity by introducing a variety of root structures and plant residues that support different soil organisms. Cover crops provide habitat and food for soil organisms while also contributing organic matter to the soil when they decompose. Reducing chemical inputs and moving towards more sustainable pest management strategies can help protect beneficial microorganisms from harmful chemicals.

12 Conclusion

In conclusion, the threats to sustainable agriculture, ranging from nutrient imbalance, salinization and sodification, soil sealing and land take, to the loss of soil biodiversity—collectively pose significant challenges to food security, environmental health, and ecosystem resilience. These threats underscore the urgent need for holistic and sustainable land management practices that prioritize soil health and agricultural productivity. By addressing these challenges through informed practices, such as improved irrigation management, integrated nutrient management, responsible urban planning, and biodiversity conservation, we can safeguard the essential resources that support agriculture. Protecting and enhancing soil health is critical not only for current agricultural needs but also for ensuring the long-term sustainability of food systems, thereby fostering resilience in the face of a rapidly changing global landscape. This commitment will ultimately contribute to a more secure and sustainable future for generations to come.

Acknowledgements The work was supported by the "Fundamental Research Funds for Henan University of Science and Technology, 4024-13510002."

References

Ali, U., Shaaban, M., Bashir, S., Gao, R., Fu, Q., Zhu, J., & Hu, H. (2020). Rice straw, biochar and calcite incorporation enhance nickel (Ni) immobilization in contaminated soil and Ni removal capacity. *Chemosphere, 244*, Article 125418.

Ali, U., Shaaban, M., Bashir, S., Chhajro, M. A., Qian, L., Rizwan, M. S., Fu, Q., Zhu, J., & Hu, H. (2021). Potential of organic and inorganic amendments for stabilizing nickel in acidic soil, and improving the nutritional quality of spinach. *Environmental Science and Pollution Research, 28*, 57769–57780.

Bach, E. M., Ramirez, K. S., Fraser, T. D., & Wall, D. H. (2020). Soil biodiversity integrates solutions for a sustainable future. *Sustainability, 12*, 2662.

Evangelista, S. J., Field, D. J., McBratney, A. B., Minasny, B., Ng, W., Padarian, J., Dobarco, M. R., & Wadoux, A.M.-C. (2024). Soil security—Strategizing a sustainable future for soil. *Advances in Agronomy, 183*, 1–70.

Gürsoy, S. (2021). Soil compaction due to increased machinery intensity in agricultural production: Its main causes, effects and management. *Technology in Agriculture*, 1–18.

Köninger, J., Lugato, E., Panagos, P., Kochupillai, M., Orgiazzi, A., & Briones, M. J. (2021). Manure management and soil biodiversity: Towards more sustainable food systems in the EU. *Agricultural Systems, 194*, Article 103251.

Liu, Y., Jia, B., Zhang, Y., Cui, H., & Li, X. G. (2023). The effect of waterlogging on soil organic carbon decomposition is dependent on its biochemistry. *Journal of Soil Science and Plant Nutrition, 23*, 4609–4619.

Lu, C., Kotze, D. J., & Setälä, H. M. (2020). Soil sealing causes substantial losses in C and N storage in urban soils under cool climate. *Science of the Total Environment, 725*, Article 138369.
Ma, X., Zhao, C., & Zhu, J. (2021). Aggravated risk of soil erosion with global warming—A global meta-analysis. *CATENA, 200*, Article 105129.
Mileusnić, Z. I., Saljnikov, E., Radojević, R. L., & Petrović, D. V. (2022). Soil compaction due to agricultural machinery impact. *Journal of Terramechanics, 100*, 51–60.
Peñuelas, J., & Sardans, J. (2022). The global nitrogen-phosphorus imbalance. *Science, 375*, 266–267.
Rodríguez-Espinosa, T., Voukkali, I., Pérez-Gimeno, A., Almendro Candel, M. B., Hernández-Martich, J. D., Zorpas, A. A., Lucas, I. G., & Navarro-Pedreño, J. (2024). Waste as a sustainable source of nutrients for plants and humans: A strategy to reduce hidden hunger. *Sustainability, 16*, 7185.
Shaaban, M. (2024b). Microbial pathways of nitrous oxide emissions and mitigation approaches in drylands. *Journal of Environmental Management, 354*, Article 120393.
Shaaban, M., & Nunez-Delgado, A. (2024). Soil adsorption potential: Harnessing Earth's living skin for mitigating climate change and greenhouse gas dynamics. *Environmental Research, 251*, Article 118738.
Shaaban, M., Abid, M., & Peng, Q. A. (2013a). Short term influence of gypsum, farm manure and commercial humic acid on physical properties of salt affected soil in rice paddy system. *Journalof the Chemical Society of Pakistan, 35*, 1034–1040.
Shaaban, M., Van Zwieten, L., Bashir, S., Younas, A., Núñez-Delgado, A., Chhajro, M. A., Kubar, K. A., Ali, U., Rana, M. S., & Mehmood, M. A. (2018). A concise review of biochar application to agricultural soils to improve soil conditions and fight pollution. *Journal of Environmental Management, 228*, 429–440.
Shaaban, M., Wu, Y., Wu, L., Hu, R., Younas, A., Nunez-Delgado, A., Xu, P., Sun, Z., Lin, S., & Xu, X. (2020). The effects of pH change through liming on soil N_2O emissions. *Processes, 8*, 702.
Shaaban, M., Wu, Y., Núñez-Delgado, A., Kuzyakov, Y., Peng, Q.-A., Lin, S., & Hu, R. (2023). Enzyme activities and organic matter mineralization in response to application of gypsum, manure and rice straw in saline and sodic soils. *Environmental Research, 224*, Article 115393.
Shaaban, M., Abid, M., & Qi-An, P. (2013b). Short term influence of gypsum, farm manure and commercial humic acid on physical properties of salt affected soil in rice paddy system. *Journal of the Chemical Society of Pakistan, 34*.
Shaaban, M. (2024a). *Acidic soils, planet Earth: Scientific proposals to solve urgent issues* (pp. 293–306). Springer.
Shaheb, M. R., Venkatesh, R., & Shearer, S. A. (2021). A review on the effect of soil compaction and its management for sustainable crop production. *Journal of Biosystems Engineering*, 1–23.
Stankovics, P., Montanarella, L., Kassai, P., Tóth, G., & Tóth, Z. (2020). The interrelations of land ownership, soil protection and privileges of capital in the aspect of land take. *Land Use Policy, 99*, Article 105071.
Tian, L.-X., Zhang, Y.-C., Chen, P.-L., Zhang, F.-F., Li, J., Yan, F., Dong, Y., & Feng, B.-L. (2021). How does the waterlogging regime affect crop yield? A global meta-analysis. *Frontiers in Plant Science, 12*, Article 634898.
Tibbett, M., Fraser, T. D., & Duddigan, S. (2020). Identifying potential threats to soil biodiversity. *PeerJ, 8*, Article e9271.
Vitousek, P. M., Naylor, R., Crews, T., David, M. B., Drinkwater, L., Holland, E., Johnes, P., Katzenberger, J., Martinelli, L. A., & Matson, P. (2009). Nutrient imbalances in agricultural development. *Science, 324*, 1519–1520.
Wang, K., Zhou, J., Tan, M. L., Lu, P., Xue, Z., Liu, M., & Wang, X. (2024). Impacts of vegetation restoration on soil erosion in the Yellow River Basin, China. *CATENA, 234*, Article 107547.
Weil, R. R., & Brady, N. C. (2017). *The nature and properties of soils* (15th ed.). Pearson Education Limited.

Wu, L., Tang, S., He, D., Wu, X., Shaaban, M., Wang, M., Zhao, J., Khan, I., Zheng, X., & Hu, R. (2017). Conversion from rice to vegetable production increases N_2O emission via increased soil organic matter mineralization. *Science of the Total Environment, 583*, 190–201.
Wu, L., Wu, X., Lin, S., Wu, Y., Tang, S., Zhou, M., Shaaban, M., Zhao, J., Hu, R., & Kuzyakov, Y. (2018). Carbon budget and greenhouse gas balance during the initial years after rice paddy conversion to vegetable cultivation. *Science of the Total Environment, 627*, 46–56.
Wu, H., Hu, J., Shaaban, M., Xu, P., Zhao, J., & Hu, R. (2021). The effect of dolomite amendment on soil organic carbon mineralization is determined by the dolomite size. *Ecological Processes, 10*, 1–12.
Xiao, L., Kuhn, N. J., Zhao, R., & Cao, L. (2021). Net effects of conservation agriculture principles on sustainable land use: A synthesis. *Global Change Biology, 27*, 6321–6330.

Understanding Soil Composition: Its Effect on Plant Development

Saira Mansab, Kousar Parveen, and Saima Nasreen

Abstract Soil is a dynamic entity where complex interactions among its biological, chemical and physical components take place. All these components and properties determine the functioning of the soil for the growth and development of plants. Soil physico-chemical and biological compositions determine many key soil processes including nutrients cycling, water retention and rhizosphere environment and thus the agronomical potential of a soil. Soil structure is dependent on the physical composition of soil and describes the arrangement of mineral particles and organic matter in the soil, and particularly the arrangement of pores among these particles. Further, the uptake of nutrients is determined by the chemical composition of soil that is the main constituent for plant growth. Biological properties of soil provide a unique rhizospheric environment with multiples micro and macroscopic organisms that not only helps in maintaining soil texture and structural properties but also enriches the soil with nutrients recycling for plant growth. Comprehending the makeup of soil is crucial as it forms the foundation for soil fertility, offering the essential knowledge to grasp variations in fertility levels among different soils and their impact on crop production and plant growth.

1 Introduction

Soil is a constantly changing natural substance formed by the breakdown of rocks through pedogenic processes, comprised of both mineral and organic elements, with distinct chemical, physical, and biological characteristics, varying in thickness across the Earth's surface, and serving as a habitat for plant growth (Velayutham & Bhattacharyya, 2000). It offers fundamental nutrients, water, and plants structural support for their development. The composition of soil varies based on its physical, chemical, and biological characteristics. These compositions impact nutrient availability, retention of water, roots development and thus affecting overall plant health. Components of soil composition includes mineral particles (sand, silt, and clay), nutrients, organic

S. Mansab (✉) · K. Parveen · S. Nasreen
Department of Environmental Sciences, The Women University Multan, Multan, Pakistan
e-mail: saira.mansab@wum.edu.pk

M. Shaaban (ed.), *Soils and Sustainable Agriculture*, Frontier Studies in Soil Science,
https://doi.org/10.1007/978-3-031-91114-9_3

matter, water, air, and microorganisms that are essential for creating a suitable environment for plants to flourish. The combination of these components influences the soil fertility and texture for promoting robust plant growth (Kalev & Toor, 2018). Adequacy of soil to maintain nutrient levels, its proper water retention, and good drainage are essential to avoid problems such as root rot and dehydration. A well-structured soil facilitates roots to penetrate effortlessly, allowing plants to obtain essential nutrients and water. It is important to recognize that soil properties and processes at a macro-environment level differ from those at a micro-environment level near the root system, leading to varying plant responses (Stockdale et al., 2013). Microorganisms, as a biological component of soil, are also important for decomposing organic matter and recycling nutrients, which helps plants thrive. By understanding and optimizing soil composition through in-dept knowledge, various soil amendments and appropriate irrigations can be planned to create ideal growing conditions for various types of plants.

2 Soil Physical Composition in Relation to Plant Growth

Understanding the physical characteristics of soil is crucial as they dictate its suitability for various uses, whether in agriculture or other fields. Soil strength, temperature, air, and water content are the four fundamental soil physical factors that directly impact plant growth. They are referred to as the main physical characteristics of soil. Soil characteristics like texture, structure, aggregation, stability, density, porosity, pore-size, infiltration, conductivity, and color indirectly influence plant growth through primary physical properties. These could be referred to as secondary physical attributes. The infiltration rate, water-holding capacity, permeability, aeration, plasticity, and nutrient-supplying ability of soil are affected by the size, proportion, arrangement, and mineral composition of its particles (Sharma & Kumar).

2.1 Particle Shape and Size

Particle size and shape indicate material composition, grain formation, release from the mineral matrix, transportation, and depositional environments. It differs in shapes from spherical to angular and have a wide range of sizes that categories the soil into sand, silt, and clay (Sun et al., 2019). Clay is one of the smallest soil particles having a diameter smaller than 0.002 mm. Certain clay particles are too tiny to be visible under regular microscopes. Silt particles are larger than clay having diameter 0.002 to 0.05 mm. Sand particles come after silt with sizes ranging from 0.05 to 2.0 mm. Gravel or stones are particles that are larger than 2.0 mm in size (Delgado & Gómez, 2016). The majority of soils consist of a combination of sand, silt and clay in varying amounts. The space between particles in soil affects water flow and water retention capacity.

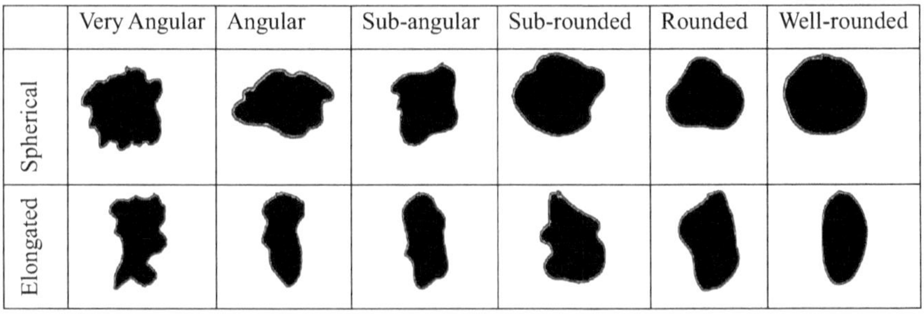

Fig. 1 Particle shapes of soil (Sun et al., 2019)

An excessive amount of clay compared to silt and sand results in soil absorbing water at a very slow rate. This type of soil releases its water to plants at a gradual pace. When these soils get wet, they become sticky (Osman, 2014).

Soil particle and shape impact the ease of soil function, the types of crops that can be cultivated, and the effectiveness of specific fertilizers. Sandy soils without fine clay or silt particles in the pores cannot retain as much moisture because there is no surface for water to cling to, and the large pores allow water to easily run through and out of the soil due to its weight. Medium and coarse sandy soils with low clay content are called droughty soils for this particular reason. Crops cannot survive for extended periods without regular rainfall. If small soil particles fill the bigger holes, the soil can retain additional water for plants as there is a greater surface area for water to stick to. Due to the decrease in pore size, the weight of the water is lower and it does not easily drain out of the soil (Osman & Osman, 2013b) (Fig. 1).

2.2 *Soil Colour*

The hue of soil is likely the initial soil characteristic perceived by humans. Soil is still identified by its color in present times, with classifications like black, red, and yellow soil (Delgado & Gómez, 2016). A soil achieves a specific hue based on pedochemical interactions. Various levels of oxidation, hydration, and diffusion of iron oxides can result in soils having red, yellow, or brown tones. Soil with dark hues, such as dark brown, may indicate poor drainage, organic matter decomposition, and the existence of minerals like titanium and/or manganese. Black soils often have dark colors (Mandal, 2016).

2.3 Density

The density of a soil particle depends on the densities of the individual components of the soil and how much each component contributes. These individual components describe the particle density of soil. The bigger particles in soils have a greater weight per unit volume compared to the smaller particles. On the other hand, pore spaces along with particle weight define its bulk density which is lesser than the particle density. The relationship between the Particle density (PD) and Bulk density (BD) and the pore space (P) of soils is as follows:

$$P\% = \frac{PD - BD}{PD} \times 100$$

The particle density in mineral soils generally ranges from 2.5 to 2.7, while the bulk density varies between 1.2 and 1.8, depending on the method used for measurement (Wilke, 2005).

2.4 Pore Space

The amount of soil taken up by air and water is known as pore space, and this is mainly influenced by the structural characteristics. Sands typically have a lower pore space around 30%, while clays can have a higher pore space ranging from 50–60% (Schön, 2015). Even though clays have higher overall porosity compared to sands, the larger individual pore spaces in sands make them better for drainage and aeration. There are two types of pore spaces: macro and micro. Sands have a greater number of large pores compared to clays, which have a higher concentration of small pores. Thus, sands have higher porosity and do not cause drainage issues even though they have a lower overall pore volume. Pore-size distribution allows soil to hold the essential water and air needed for plant growth in the root zone (Ghosh et al., 2020).

2.5 Plasticity and Cohesion

Plasticity and cohesion indicate the soil's consistency and workability. A soil is said to be plastic when a wet soil is deformed under pressure and maintain a new shape after removing its pressure. On contrary, cohesion is the ability of particles to adhere to each other through cohesive force. Black soils have a high level of plasticity and cohesiveness, which can make it challenging to perform tillage operations while sandy soils are classified as non-plastic and non-cohesive and clayey soils are classified as plastic (Manafi et al., 2022). Soil having more plasticity do not allow the soil particles to stick one another and water holding capacity is less and it can be used for frequent irrigation. On the other hand, cohesive soils have high shear strength and can retain moisture which enhances water holding capacity making the soil unsuitable for frequent irrigation.

2.6 *Soil Temperature*

Solar radiation is the primary heat source for soil, influencing the soil's thermal conditions and the growth of plants (Sung et al., 2017). The role of temperature has been discovered to enhance organic-matter decomposition and nitrogen mineralization in its organic form. It is anticipated that soils with low temperature will have a higher quantity of organic matter. It also influences soil color, composition, and moisture content of the soil. Soil temperature plays a crucial role in seed germination, plant growth, and micro-biological activity. Sprouting has been found to be a slow procedure in a cold ground. The temperature needed for germination and root growth differs depending on the crop and variety (Kirschbaum, 2006).

2.7 *Soil Air*

It is well known that limited air supply in the soil impacts root growth, respiration and other biological activities, ultimately altering the soil conditions. The pore space not taken up by water is filled with air (macropores), while the capillary pore space (micropores) is filled with water. The soil's air is continually being exchanged with the surrounding atmosphere. Gases move from the air into the ground, then back into the air, similar to how we breathe. On average, when soil has one-third of its pore space filled with air and two-thirds filled with water, it is considered ideal for plant growth (Sharma & Kumar, 2023a) (Fig. 2).

The balance between two competing processes of carbon dioxide and oxygen exchange controls the composition of the air in the soil. Plant roots and soil organisms consume oxygen and release CO_2, thereby causing a greater variation in the composition of the soil air compared to the atmosphere above the soil surface. The content of carbon dioxide varies during the crop season due to higher rates of crop growth and microbiological activities in the soil (Mancinelli et al., 2010). The process of gaseous diffusion works to minimize variations in composition. The oxygen diffusion rate (ODR) is a measure of soil aeration, reflecting the rate at which oxygen is replenished as it is used by plant roots and microorganisms. Research has shown that plant growth is impacted when ODR falls below 20×10^{-8}g/cm^2 per minute (Neira et al., 2015).

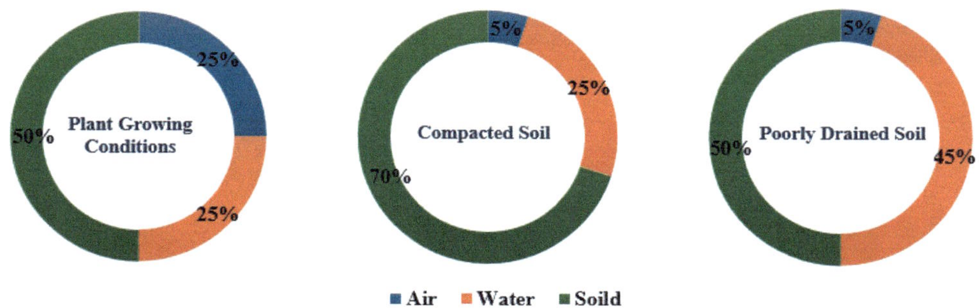

Fig. 2 Percentage composition of air and water in various soils

2.8 Soil Water

The growth and yield of a crop are most affected by the amount of soil water available, as it is needed more than any other substance for their development and productivity. Water plays several roles in plant life:

a) Main component of plant nutrients making up over 90% of plant tissues.
b) Functions as a solvent and transporter of plant nutrients.
c) Preserves cell turgor and controls soil temperature.

Different types of crops have varying water needs based on factors such as their growth patterns, the climate they are in, and the availability of water. The crop's water necessity comprises the requirements for evapo-transpiration, metabolic activities, leaching, and other unavoidable losses (Salama, 2011). It represents the amount of water required to grow a crop during a specific timeframe and is measured in millimetres of water depth. Typically, the water needs of a plant are connected to the possible evaporation levels as the plant grows. The way water is retained and moves in soils and plants, as well as its eventual release into the atmosphere, is connected to the energy that keeps the water in place. Soil water retention has following states (Brown et al., 2021).

a) **Maximum water holding capacity:** The soil reaches its maximum water holding capacity when both macro and micro pores are fully saturated with water during heavy rain or irrigation.
b) **Field capacity:** It is reached when the soil is holding as much water as it can and then water starts moving downwards due to gravity. Once 1–3 days have passed and the water in the macropores has been drained, the soil is considered to be at field capacity.
c) **Permanent wilting percentage or wilting coefficient:** Over time, plants will use water from the soil, with some being lost through evapo-transpiration. During the day, the soil will dry out slowly and the plants will wilt first, followed by remaining wilted throughout the night and day. The moisture level of the soil at this point is known as the wilting coefficient or permanent wilting point.
d) **Hygroscopic coefficient:** The soil moisture level drops below wilting point, causing water to be held tightly by soil colloids and potentially in a vapor state, defining the hygroscopic coefficient.

2.9 Soil Structure

Soil structure is described as the organization of primary and secondary soil particles. The way soil is organized is considered crucial as it impacts air circulation, drainage, and water retention. The configuration in the field is characterized by (i) shape and arrangement, (ii) size, and (iii) degree of aggregation. It has been discovered that the crumb and granular structures (spheroidal) are beneficial for plant growth because they affect the ease of tillage and root penetration. It has been discovered that incorporating legumes into crop rotations and practicing green manuring can enhance soil structure and physical properties (Kay & Angers, 2001).

3 Soil Chemical Composition in Relation to Plant Growth

Soil chemical properties ensure the supply of adequate nutrients for plant growth. These properties largely depend on the nutrient availability, cation exchange capacity, soil collides, pH, salinity, soil organic matter etc. (Osman & Osman, 2013a). These properties are important components of soil fertility. Normal plant growth requires a minimum of 18 essential elements such as C, H, O, N, P, K, Ca, Mg, S, Fe, Zn, Mo, Ni, Mn, B, Cu, Co and Cl (Ahmed et al., 2020). Apart from C, H, and O, which are mostly present in the air and water, all other essential nutrients are primarily found in the soil. Plants typically take in all nutrients from the soil solution, excluding carbon, hydrogen, and oxygen. Certain nutrients can also enter the leaves when applied to the foliage. Optimal plant growth and yield rely on the availability of a balanced mix of all nutrients. Water and oxygen are essential environmental components that aid in the breakdown and efficient uptake of nutrients. Characteristics such as pH and cation exchange capacity play a significant role in affecting the solubility and retention of nutrients in soil. Saline soils are not suitable for the typical growth of plants (Sharma & Kumar, 2023b).

3.1 Nutrient Elements in Soil

Soil is a major source of nutrients needed by plants for growth. These nutrients are classified into macro and trace nutrients. The macro nutrients include three main nutrients i.e. nitrogen (N), phosphorus (P) and potassium (K). Besides NPK, other important macro nutrients are calcium, magnesium and sulfur. Trace nutrients include iron, manganese, zinc, copper, boron and molybdenum cobalt and chlorine because these are needed in small amount by plants (Bhatla et al., 2018). The presence of plant nutrients is influenced by the soil type. The number of elements present in the soils varies based on the type of parent material. The chemical makeup of various layers of soil displays significant diversity as well. Typically, the elements that are often leached out are those that plants also need (da Silva et al., 2022).

3.1.1 Nitrogen

Nitrogen is crucial for plants to function and is a vital element in amino acids, which are the basic building blocks of plant proteins and enzymes. Certain proteins function as structural components in plant cells, while others serve as enzymes, enabling numerous biochemical reactions essential for life. Nitrogen is also part of the chlorophyll molecule that allows the plant to absorb sunlight energy through photosynthesis and promote plant growth. Besides this, it is also a component of adenosine triphosphate that is an energy transfer compound which allows the plant cell to conserve and use the energy released in metabolism. A healthy above ground part of plant contains around 3 to 4 percent of nitrogen (Dixit, 2023; Sullivan & Andrews, 2012).

Plant uptake nitrogen from soil. Soil nitrogen exists both in organic and mineral form. In the surface layer of most soils, over 95–99% of N occurs in organic forms. Soil organic N can be divided into two categories: (1) N from organic residues and (2) N from soil organic matter or humus (Hofman & Van Cleemput, 2004). Mineralized N forms are ephemeral in soil, therefore the amount that remains is determined by a variety of processes including mineralization, immobilization, nitrification, denitrification, leaching, and plant absorption. It exists in the form of ammonium (NH_4^+) ions and nitrate (NO_3^-) ions that is a source of nitrogen for plants. Ammonium ions attach to the soil's negatively charged cation exchange complex and act similarly to other cations in the soil which is held within the lattice structures of clay minerals (Mortland, 1966). Nitrate ions do not adhere to soil particles due to their negative charge but are found dissolved in the soil water or as soluble salts in dry conditions. Plants do not have direct access to this nitrogen, although microbes can convert part of it into accessible forms. A very little quantity of organic nitrogen may exist as soluble organic molecules, such as urea, which may be partially accessible to plants (Murphy et al., 2000) (Fig. 3).

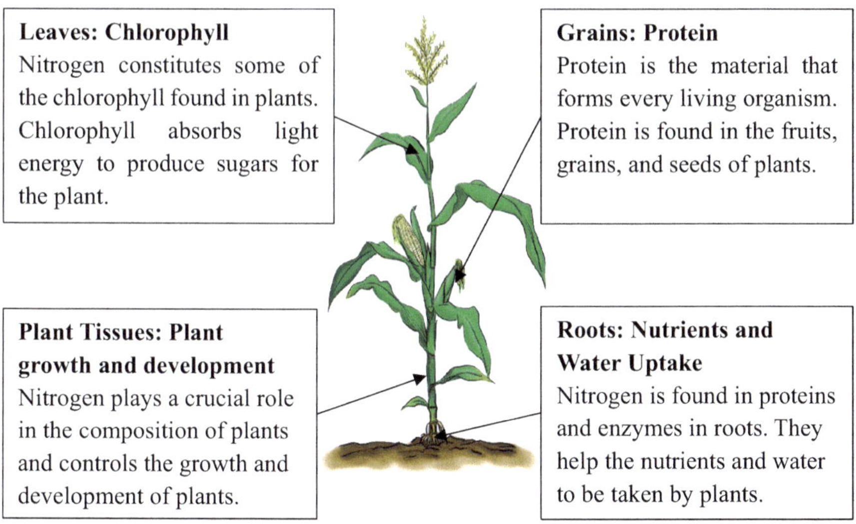

Fig. 3 Role of nitrogen in plant growth

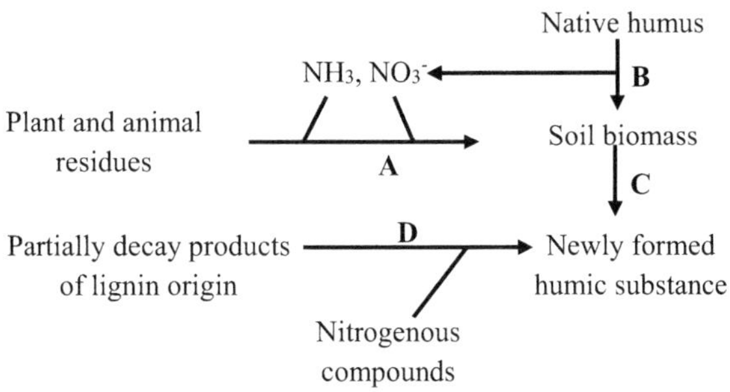

Fig. 4 Biological turnover of N between mineral and organic forms (Stevenson, 1994)

A crucial aspect of the internal cycle involves the exchange of nitrogen between mineral and organic forms through the processes of mineralization and immobilization. The second option results in the integration of nitrogen into microbial cells. While a large portion of this recently immobilized nitrogen is broken down through mineralization, a portion is transformed into durable humus structures. Therefore, the breakdown of plant and animal remains by microorganisms leads to the creation of mineral types of nitrogen (NH_4^+ and NO_3^-) and incorporation of some of the carbon into microbial tissue (reaction A). Some of the indigenous soil experiences a comparable outcome (reaction B). Mineralization-immobilization facilitates the transformation of N into stable humus forms through successive turnovers (reaction C). N can also be stabilized by the interaction of decomposed lignin products with nitrogenous compounds (reaction D) (Fig. 4).

As the minerals in the soil break down, nitrogen is released from them. This process is typically slow and has a minimal impact on nitrogen availability in the majority of soils. Nonbiological reactions involving NH_4^+, NH_3 and NO_2^- play a role in determining the fate of mineral forms of N in soil. Furthermore, NH_3 and NO_2^- can form stable organic N complexes by reacting chemically with organic matter, in addition to NH_4^+ fixation by clay minerals. The chemical reaction between NO_2^- and organic materials could result in the production of N gases. While both reactions can occur at various pH levels, NH_3^+ fixation is more favorable at pH values greater than 7.0. On the other hand, NO_2^- -organic matter reactions are most likely to happen in extremely acidic environments (pH 5.0 to 5.5 or lower) (Fig. 5).

Plant growth is governed by an N-signaling system that connects plant growth-promoting bacteria to root associations that produce NO and auxin. N is taken up from below ground by nitrate transporters (NRTs) and ammonium transporters (AMTs)

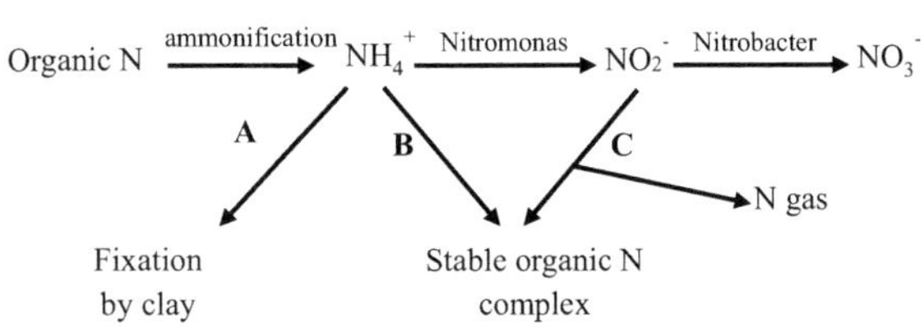

Fig. 5 Transformation of NH_4^+ and NO_2^+ in soil (Weber, 2002)

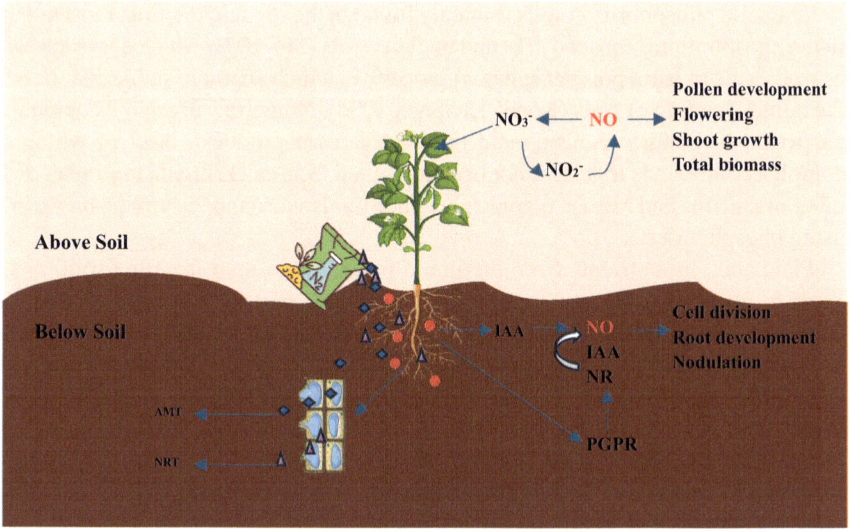

Fig. 6 N-signaling pathway influencing plant development (Singh et al., 2022)

on the root surface. It is then translocated throughout the plant via NRT/AMT gene families. Nitrate reductase (NR) is the primary enzyme in leaves that converts nitrate to nitrite, and so nitric oxide. The role of NO messenger is discussed in relation to flowering, pollen generation, shoot growth, total biomass, and yield production. Auxin (IAA) stimulates NO production in root cells through a hypothesized route. NR activity caused by bacteria could help in NO production during microbe-plant association (Fig. 6).

3.1.2 Phosphorus

Phosphorus is the most essential plant growth limiting nutrient after N. It makes up approximately 0.2 percent of a plant's dry weight, and is mainly found in tissue molecules like nucleic acids, phospholipids, and adenosine triphosphate (ATP). Soils with limited phosphorus inhibit plant growth and development, while an excess of phosphorus can be leached from soil and flow into freshwater ecosystems. Approximately 30 to 65 percent of total soil phosphorus is in organic forms and the remaining 35 to 70 percent in inorganic forms (Rao, 2009).

Solid phosphorus is found in different soil minerals like rock phosphate, in forms like fluorapatite ($Ca_5(PO_4)_3F$) or hydroxyapatite ($Ca_5(PO_4)_3OH$). It can also be found in precipitated forms on various soil and rock particles. Inorganic mineral forms do not appear in standard ratios in soil, rather they are typically mixtures of sorbed/precipitated phosphate on amorphous iron, aluminum oxides and hydroxides, as well as calcium carbonate (Mamathashree et al., 2018).

The solid phosphorus in soils is mainly found in the organic fraction, most notably in the phospholipids (around 1%) and nucleic acids (5%-10%) which degrade easily, as well as the inositol polyphosphates (up to 60%) which are more stable and found in the humus fraction of soils (Kögel-Knabner, 2006). Numerous organic P compounds are found in living organisms and their breakdown products, most of which are complex compounds that have not been identified. Soil microorganisms break down other organisms and release organic phosphorus, also through enzymatic breakdown using phosphatases.

Plants take in nutrients from the liquid phase rather than from the solid phase. Solid forms need to be transformed into liquid and chemically altered to mono- or deprotonated phosphate (HPO_4^{2-} or $H2PO_4^{-}$) for plants to access P. However, the solid phase contains much higher levels of P compared to soil solution (Pierzynski, 2016). The concentration of bioavailable dissolved phosphorus in the soil solution is usually very low, with a median of 0.05 mg P kg^{-1}, in contrast to the total phosphorus levels of 200 to > 1000 mg kg^{-1} commonly found in soils. While solution P levels are typically low, they vary in soils due to soil pH and the presence of different cations, minerals, and organic compounds (Sposito, 2008).

Certain soils exhibit abnormally high levels of solution P, often due to extensive amendment with manure or other organically complexed P materials. Erich and colleagues (2002) discovered that adding compost and manure to soils increased the availability of plant phosphorus (Erich et al., 2002). Malik and colleagues (2012) proposed that organic sources of phosphorus have the potential to enhance the development of organic phosphorus compounds with slow-release properties (Malik et al., 2012). Nevertheless, the effectiveness of employing living cover crops to maintain P in a soluble form accessible to plants varies.

A large portion of the phosphorus absorbed by plants comes from the breakdown of organic matter, but it is also crucial to consider its release from minerals. Minerals decompose at a gradual pace and can release structural phosphorus, yet only a small portion of the phosphorus that plants absorb during a single growing season comes from this process. Most of the phosphorus absorbed by plants originates from the release of accumulated mineral deposits. Still, this process is usually quite inefficient as the phosphate molecules have low solubility and can re-precipitate from the soil solution during P cycling. Phosphorus availability to plants is influenced by several factors such as soil pH, temperature, moisture, and microbial activity. Soil microbes are vital in converting organic substances into forms that plants can use. Further, A low soil pH can enhance the inorganic phosphate solubility, whereas a high soil pH can decrease it. Furthermore, excessive phosphorus fertilization may result in environmental issues, such as water bodies becoming eutrophic and groundwater getting contaminated (Maynard, 2007) (Fig. 7).

3.1.3 Major Cations

Plant nutrient absorption is commonly known as ion uptake or ion absorption, as roots absorb nutrients in their ionic form. Soil contains clay mineral and organic matter with

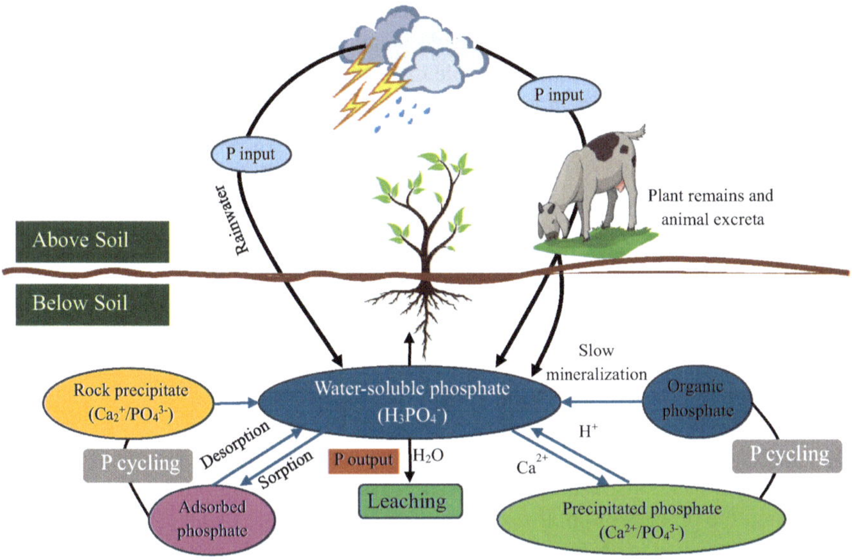

Fig. 7 Phosphorous cycling in soil

negatively charged sites on their surfaces that attract and retain positively charged ions (cations) through electrostatic force. This electric charge plays a vital role in providing nutrients to plants. Cations and anions can be taken up separately and in varying amounts, but the growth medium must still remain electronically balanced. Hence, the ionic relationship plays a crucial role in plant nutrition (Kumaragamage et al., 2021). Most of the positively charged ions found in plant tissues are in their inorganic state, mainly consisting of K^+, Ca^{2+} and Mg^{2+}, while the majority of negatively charged ions are in their organic state. Organic ions are created in the tissue, while inorganic ions are taken in from the soil. Cations are used in lower amounts compared to phosphorus or nitrogen, yet calcium, magnesium, and potassium are crucial nutrients for plants as they play a role in several plant functions and metabolic processes (Bhatla et al., 2018). The absorption of potassium is highly specific and has a strong correlation with the plant's metabolic processes. In general, monovalent cations are taken in quickly, while divalent cations, particularly Ca^{2+}, are absorbed at a slower pace. In the same way, monovalent anions are typically taken in more quickly than polyvalent anions (Fageria et al., 2008). The alleged acidity or alkalinity of a salt is determined by whether the cation or anion is absorbed quickly from the solution. Therefore, a salt such as K_2SO_4 would be considered physiologically acidic, causing an increase in acidity in a nutrient solution or soil solution due to the quicker absorption of K^+ compared to $SO_4{}^{2-}$ by the roots. Similarly, $CaCl_2$ would increase the alkalinity of a solution physiologically, as Ca^{2+} penetrates plant roots at a slower rate than the Cl^- (Fageria et al., 2013).

Potassium plays a vital role in numerous physiological functions within plants. Potassium plays a role in maintaining pH balance and regulating osmotic pressure and ion levels in plants. The osmotic role in plants is not limited to one type, as there are many different organic and inorganic solutes present. Potassium plays a crucial role in the water balance of plants and controls the opening and closing of stomata in plants (Mengel et al., 2001). When there is a higher amount of potassium in the guard cells, water is absorbed from nearby cells, causing the guard cells to become more turgid and leading to the opening of stomata. The closing of the stomata in darkness is linked to potassium leaving and a resulting drop in osmotic pressure of guard cells (Engels & Marschner, 1995). Potassium acts as a cofactor or activator for numerous enzymes involved in carbohydrate and protein metabolism (Fageria et al., 2011). Over 50 enzymes rely on or are enhanced by potassium ions. Potassium also triggers ATPases that are bound to the membrane. The stimulation of ATPases by potassium not only helps with its transportation into root cells from the outside solution through the plasma membrane but also elevates potassium's significance in cell growth and osmoregulation. Plants also need potassium for synthesizing proteins. It has a key function in the process of photosynthesis. Sufficient potassium levels in plant tissues reduce the occurrence of diseases. Potassium not just boosts plant immunity against diseases but also diminishes fungal populations in the soil (Marschner, 2011).

Calcium is a crucial macronutrient for plants, present in shoot concentrations ranging from 0.1 to more than 5% of dry weight (White & Broadley, 2003). It shows two roles, as part of cell walls and membranes and as a second messenger within cells. Plants with low levels of calcium are more vulnerable to pathogens, while providing extra calcium can enhance the plant's ability to resist them, as supported by their structural role. Nevertheless, it has become apparent in the last few years that the primary role of calcium is to act as a second messenger. It acts by signaling events related to numerous physiological, developmental, and environmental stimuli, including signaling other nutrients and detecting pathogen attacks (Barker & Pilbeam, 2015).

Magnesium plays a major role in the creation of chlorophyll, the formation and movement of photo-assimilates, the activation of enzymes, and the synthesis of proteins. Magnesium is the main element in chlorophyll (Chl) pigments found in the light-capturing complex of chloroplasts, playing a key role in photosynthetic CO_2 assimilation (Gerendás & Führs, 2013). Approximately 15–35% of magnesium absorbed by plants is utilized in chlorophyll pigments, while the rest (about 65–85%) is stored in vacuoles or used for protein synthesis and other biological processes (Karley & White, 2009). Mg also serves as a coenzyme for many enzymes (over 300) related to Chl production and CO_2 assimilation in photosynthesis (Wang et al., 2020). Different enzymes like Rubisco, protein kinases, RNA polymerase, glutathione synthase, ATPases, phosphatases, and carboxylases need Mg to become active ((Marschner, 2011). Additionally, magnesium is involved in various biological processes such as sucrose transport, energy metabolism, nitrogen utilization, pollen development, male fertility, stress tolerance, plant–microbe interactions, and more (Ishfaq et al., 2022).

3.1.4 Sulfur

Sulfur is necessary for good health of plants and is a key element in protein synthesis. Plants absorb it in equal quantities to phosphorus. In soil, sulfur can be classified into two main categories: organic and inorganic forms. Organic fractions make up the majority of the total S pool in soils as a result of leaf litter inputs. Over 90 percent of the sulfur present in soil is found in organic forms. Sulfur can be found in every soil, originating from various sources such as parent rock components like sulfides and sulfates, prevalent minerals in the earth's crust, and atmospheric sources like S deposition from marine aerosols, industrial gases, or gases/particulates from volcanic eruptions. In the ground, minerals like iron, nickel, and copper sulphides, along with gypsum ($CaSO_4.2H_2O$), break down due to weathering, releasing sulphates and sulphides into the soil. Even though mineral weathering is the main source of sulfur in soil, most of the sulfur in fertile agricultural soils is found in soil organic matter. About 100 pounds of sulfur per acre (112 kg per hectare) is contained in every one percent organic matter in the topsoil, which is 6 inches or 15 cm deep. This is because sulfate (SO_4^{2-}) is the main inorganic form of S in soil pore water and is not strongly held by soils, allowing it to easily move through soil like nitrate form of N. The S in organic matter is safeguarded from leaching. In waterlogged soils like rice paddies, sulfur may be retained to prevent leaching by turning into insoluble sulfides linked to iron. Plants need sulfur at levels ranging from 0.1% to 0.5% of their dry weight for proper development (Marschner, 2011). Soil microorganisms break down sulfur into sulfate, which is the form plants absorb. High productivity crops absorb large quantities of sulfur from the soil, and when harvested, they also remove considerable amounts of sulfur from the ecosystem. Plants primarily acquire S from the soil in the form of a sulfate ion through the soil pore water. Therefore, in soils without fertilization, the availability of sulfur to plants depends on the processes of mineralization and immobilization of soil organic matter. Soils with a high organic matter content will be able to provide a lot of sulfur to plants if the temperature and soil moisture are ideal (20–30 °C, 68–86 °F). Conversely, soils with low organic matter will be more prone to sulfur deficiency. Increased soil temperatures and soil moisture levels enhance the mineralization of S from organic matter in the soil. Fertilization can increase the amount of S available in the soil (Chaudhary et al., 2023).

The absorption and movement of sulfate in plants relies on energy and is facilitated by various sulfate transporters. Sulfate is transported into cell membranes against its concentration gradient, likely by being linked to a proton gradient using a cotransporter system (Barker & Pilbeam, 2015). Too much sulfate can be stored as free sulfate ions, mostly in the vacuoles, however sulfate absorbed by the roots is mainly moved to the leaves and either reduced in the chloroplasts or stored in the vacuoles. The process of converting sulfate into sulfide has been + thoroughly examined in previous studies and is briefly summarized here (Takahashi et al., 2011). Sulfate reduction can take place in various tissues and subcellular compartments, but it is most commonly found in the plastids of green tissues, which have ample ATP and redox equivalents (Rai et al., 2020).

3.1.5 Trace Metals

Trace metals include cadmium, chromium, copper, lead, manganese, mercury, molybdenum, nickel, selenium, and zinc. Arsenic, boron, and selenium, which are considered metalloids, are generally included with metals in soil studies. Heavy metals are metallic elements with a density greater than 7 g/cm^3, including Pb, Cd, Ni, Hg, and Cr. The elemental composition in clean soils is primarily influenced by the soil's parent material, amount of organic matter, texture, and depth (Ure, 1982). Due to cycling through plants, airborne deposition, and absorption by organic matter in the soil, certain elements like Ag, As, Cd, Cu, Hg, Pb, Sb, and Zn tend to accumulate in the upper layers of the soil (Raj & Maiti, 2020). Additional components linked to clays and water-filled oxides, like Fe, Mg, Ni, Sc, V, and Zr, are typically found in deeper layers. The inherent chemical nature of certain small amounts of elements (such as Cd, Ni, Pb, and Zn) in soils is limited to a single oxidation state. The actions of other trace elements (such as As, Co, Cr, Hg, or Se) are more intricate due to their ability to exist in multiple oxidation states. Trace elements in soils can exist in various forms such as soluble, exchangeable, adsorbed, precipitated, and co-precipitated with carbonate, sulfides, phosphates, and hydroxides regardless of their oxidation state (Bañuelos & Ajwa, 1999).

How trace metals behave in soil varies significantly depending on soil characteristics. Several factors regulate the solubility and availability of trace metals to plants and their leaching to groundwater, such as pH, organic matter, cation exchange capacity, texture, calcium carbonate equivalent, and microorganisms (Krebs, 1998). Solubility is influenced by pH due to competition for protons, reactions involving precipitation and dissolution, and metal complexation by humic substances depending on pH (Barker & Pilbeam, 2015). The availability of trace elements in the soil is influenced by soil pH due to its impact on their solubility and ability to form chelates. Metal humic complexes are more stable at pH 5 compared to pH 3.5. This could be due to the separation of functional groups within the fulvic and humic acid molecule. Soil organic matter plays a vital role in dissolving and circulating trace elements within the soil. Hence, the presence of organic substances in the soil can either enhance or reduce the toxicity of trace elements based on certain environmental factors. Soil organic matter not only forms metal chelates with copper and nickel, but also acts as a source of cation exchange capacity for these metals (Logan & Chaney, 1983). Chelated-metal compounds have higher solubility compared to inorganic precipitates, while the solubility of trace elements depends on the presence of soluble complexing organic matter in the soil. Soil microorganisms play a crucial role in nutrient cycling, decomposition of organic matter, altering soil composition, and converting trace elements into soluble and immobile forms (Banuelos & Ajwa, 1999). Microorganisms frequently immobilize trace elements to reduce metal toxicity and mobility. Immobilization can happen in two ways: dependent on metabolism and independent of metabolism (White et al., 1997). Metabolic-dependent processes include the storage of trace elements in specialized organelles, attaching to cell proteins, or being taken up within cells; while metabolic-independent processes involve binding of trace elements outside the cell or to the cell wall. In soil, certain trace elements are readily taken in by plants and moved to plant tissues in the food chain due to weak chelation. Copper, nickel, and cobalt form stronger chelates, with only a small amount of the adsorbed copper being moved to the shoots compared to zinc (DalCorso et al., 2013). Other substances are held tightly in the soil or remain in plant roots without being transferred to plant leaves in harmful quantities for animals, even with soil enrichment, e.g. iron, lead, mercury, aluminum, titanium, chromium, silver, gold, tin, silicon, zirconium. Certain trace elements can be readily absorbed and transported to plant leaves, but the levels that lead to decreased plant yield are lower than those considered harmful to humans and animals. This offers an intrinsic safety feature for those components. Nevertheless, plants can become toxic to humans if they are exposed to high concentrations of certain elements like Cd and Pb, which are above recommended levels (Bañuelos & Ajwa, 1999; Dudka et al., 1994).

3.2 Organic Matter

Soil organic matter (SOM) is essential for soil health and fertility as it is a crucial component of the soil. It is a complex and diverse soil component, making up approximately 1–6% of the entire soil weight. It is developed through a process of humification in which plant and animal residues are broken down and changed by soil organisms (Gerke, 2022). The organic remains go through different phases of breakdown, creating a mixed assortment of materials with varying levels of decomposition and stability in the ground. The diversity and dynamics of SOM reflect its sources and the various processes it undergoes in the soil, including biological, chemical, and physical processes. It contains a variety of organic compounds, ranging from basic sugars and amino acids to intricate polymers such as lignin and humic substances (Piccolo, 2002). The latter is especially significant since it can resist decomposition and remain in the soil for many years or even centuries, helping store carbon for the long term. SOM mainly comes from plant residues and animal manure, but it can also come from other organic materials like compost and biochar added to the soil (Paustian et al., 2000). Plant leftovers account for most of the SOM, making up roughly 60–80% of the total SOM in farm soils. Many factors, such as climate, soil type, vegetation, and land management practices, impact the creation and breakdown of SOM (Conant et al., 2011). Temperature and moisture are essential climatic factors that impact the dynamics of SOM. Increased temperatures and sufficient moisture typically boost microbial activity and breakdown of organic matter, resulting in quicker turnover of SOM (Six et al., 2002). The type of soil also has a big impact, as soils with high clay content typically have higher levels of SOM because clay particles protect organic materials (Cotrufo et al., 2013).

Soil health relies on soil organic matter (SOM) for maintenance and improvement. It has a diverse impact on the well-being of the soil, influencing its physical, chemical, and biological characteristics, subsequently enhancing the overall efficiency and durability of agroecosystems (Six et al., 2002). In terms of physical attributes, SOM affects soil composition, ability to retain water, and susceptibility to erosion. SOM enhances soil structure by forming aggregates from mineral particles, improving soil porosity and water infiltration (Gurmu, 2019). This makes it easier for roots to penetrate and grow, while also decreasing surface runoff and soil erosion. Additionally, organic material, specifically the humic component, possesses a strong capability to hold water, which enhances the soil's capacity to store and deliver water to plants. From a chemical standpoint, SOM serves as a storage container for crucial nutrients and also helps maintain stable soil pH levels. SOM releases nutrients such as nitrogen, phosphorus, and sulfur as it breaks down, providing plants with valuable nutrients. Additionally, some parts of SOM can combine with metal ions to boost the accessibility of micronutrients like iron, zinc, and manganese. SOM's buffering capacity, thanks to functional groups such as carboxyl and phenolic groups, contributes to regulating soil pH to support nutrient availability and microbial activity. Biologically, SOM acts as a food source and living space for various soil creatures. SOM supports soil microorganisms' growth and activity through energy and nutrients, essential for nutrient cycling, disease suppression, and soil structure formation (Lavelle et al., 2006). SOM also boosts the variety and number of soil animals, such as earthworms, mites, and beetles, that help with nutrient cycling and breaking down organic matter (Bot & Benites, 2005).

Organic matter acts as a storage of vital nutrients like nitrogen, phosphorus, and sulfur that are released during decay. The nutrients are later taken up by crops and help them grow and develop. Furthermore, SOM's capability to create compounds with metal ions increases the accessibility of important micronutrients like iron, zinc, and manganese that are essential for crop's physiological functions. The important role of soil organic matter in providing nutrients is especially critical in low-input and organic farming methods, where mineral fertilizers are used minimally or not at all (Hussain et al., 2023).

In addition to providing nutrients, SOM also enhances the soil's physical quality for crop development. SOM improves soil structure to enable root penetration and growth, which is essential for crops to absorb nutrients and water. Furthermore, the ability of SOM to retain water assists in keeping soil moisture levels stable, ensuring a steady water source for crops and decreasing their vulnerability to drought. This is especially crucial in dryland farming systems and in areas with sporadic precipitation patterns (Shaaban & Nunez-Delgado, 2024). The influence of SOM on the biological characteristics of the soil also affects crop yield. SOM enhances nutrient cycling and disease suppression in crops by sustaining a varied and active soil microbial community (Lavelle et al., 2006). An example would be when specific soil microbes convert nitrogen from the air into forms that plants can use, which is called biological nitrogen fixation (Gattinger et al., 2012). Additional microorganisms, especially those located in the rhizosphere, have the ability to reduce soil-borne diseases by outcompeting or inhibiting harmful microbes, leading to improved crop health and yield (Diacono & Montemurro, 2015) (Fig. 8).

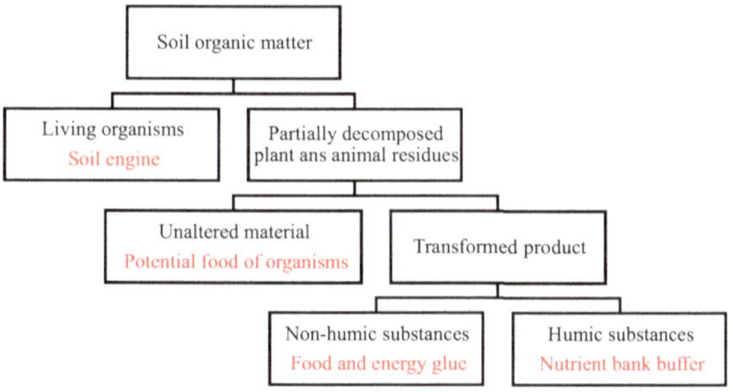

Fig. 8 Component of soil organic matter and their function

3.3 *Soil Colloids*

Soil colloids are the most reactive portion of solid soil components. Soil colloids can be divided into two phases, specifically the organic and inorganic phases. The organic phase includes fresh or decomposed plant, animal, and microbial residues that may be found in combination with the inorganic phase or independently. Clay is the primary component of the inorganic phase of soil colloids and has a significant impact on the majority of soil properties. Soil colloids play a crucial role in attracting soil nutrients in the form of positively charged mineral ions or cations dissolved in soil water. Certain positively charged ions are essential for the growth of plants, such as calcium (Ca^{2+}), magnesium (Mg^{2+}), potassium (K^{+}), and sodium (Na^{+}). For plants to access them near root membranes, they must be dissolved in a soil–water solution (Alemayehu & Teshome, 2021).

3.4 *Cation Exchange Capacity*

Cation exchange (CE) indicates how well a soil can hold onto cations, whether they are basic or acidic. The cation exchange of the root zone, along with base saturation, is a reliable indicator of the soil's inherent fertility level. CE represents the nutrients that the soil particles could absorb at a specific location, while base saturation denotes the nutrients available for plant uptake in the root zone (Fig. 9).

Fig. 9 Cation exchange sites on soil particles

The cation exchange of soil is determined by the quantity of negative charges present in the clay, which is influenced by the clay type and amount, as well as the organic matter in the soil. Soils in temperate regions, typically with illite and 2/1 silicate clays, have a greater cation exchange capacity compared to tropical soils, which primarily consist of kaolinite in the clay fraction. Hence, the majority of soil CE in tropical regions is due to the existence of organic material. Kaolinite (3–15 cmol/kg clay) is classified as a clay with low activity when compared to illite (15–25 cmol/kg clay) and montmorillonite (70–100 cmol/kg clay) which are categorized as medium- and high activity clays, respectively. Organic carbon (OC) can retain around 2.6 cmol/kg OC. The typical cation exchange capacity of temperate soils ranges from 20–30 cmol/kg soil in the surface layer and diminishes in deeper soil layers, based on the properties of the parent material. In tropical regions, the surface CEC typically stays below 10–12 cmol/kg of soil and decreases to as little as 3–5 cmol/kg in the deeper layers. In soils formed from low-quality parent materials (granite, quartzite), the CE could decrease to below 3 cmol/kg soil (Mengel, 1993). Therefore, the majority of soils in tropical regions generally have lower inherent fertility compared to soils in temperate regions, making it imperative to prioritize the conservation of a healthy humus layer in soil management practices. Plants such as wheat or corn in temperate regions and cocoa in the tropics, which require a lot of nutrients, thrive in soils rich in clay, humus, and predominantly 2/1 clays. Oil palm or rubber plants, which have less apparent nutritional requirements, grow most successfully in soils with less clay content that are mostly made up of kaolinite. The soil exchange complex is typically filled with bases and H to varying degrees. The primary bases include Ca^{2+}, Mg^{2+}, K^{+}, and Na^{+}. The Base Saturation is the total of these bases in relation to the CE. Quality fertile soils typically have a Ca/Mg/K/Na ratio of 80/15/3/2% with a high base saturation, often exceeding 50%. The base saturation in temperate soils typically ranges from 50–80%, but drops as you move closer to the tropics. At a low base saturation and acid pH, the complex is partially filled with H + ions, or at extremely low pH by Al^{3+} ions released from the transformation of silicate clay layers. In soils with aluminum-dominant CEC, low base saturation levels can hinder plant growth due to increased aluminum levels. Soils that have a high base saturation also exhibit increased buffering capacity (Ma et al., 2024).

Ideally, the predominant cation exchange sites in soils should have an abundance of easily exchangeable Ca and Mg, with Ca^{2+} being roughly 5 times more significant than Mg^{2+}. In conditions of severe weathering and with a high-water table, the ratio of Ca/Mg could fall under 2. Certain crops, such as bananas, are sensitive to imbalances in minerals. In typical circumstances, the level of K in the soil should not fall below 0.12 cmol/kg. The base saturation is closely associated with the soil pH. An increase in base saturation leads to an increase in pH. pH levels below 5.0 or above 8.2 in soil indicate potential toxicity from sodium or aluminum. The only way to enhance the exchange capacity of soils is by elevating the organic matter levels in the topsoil, as the clay type and clay content of soil cannot be altered. Liming or applying mineral fertilizers can increase the base saturation percentage, but it will always be restricted by the absorption capacity of the cation exchange capacity. When the number of

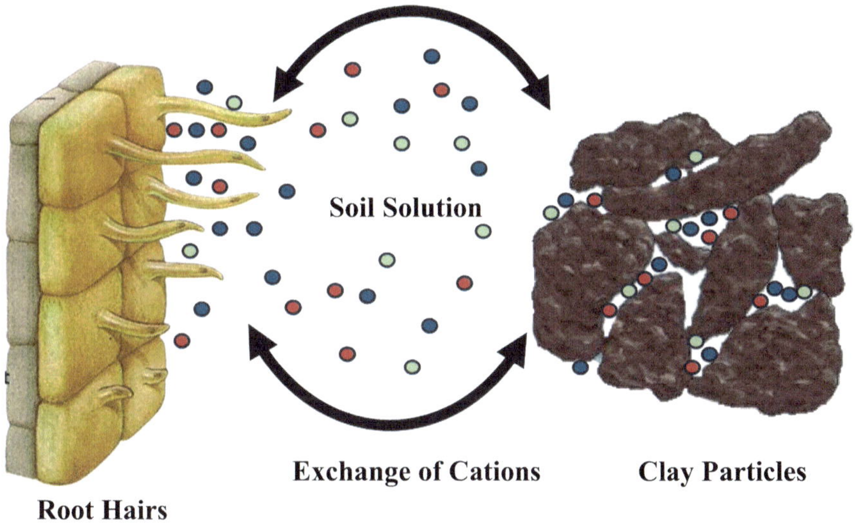

Fig. 10 Cations transportation from soil phase to plants roots

components added is more than the CE, they are flushed out with the soil water (Yang et al., 2024) (Fig. 10).

3.5 Soil pH

Soil pH indicates whether soil is acidic or alkaline. Soil pH is a crucial factor that impacts plant growth, nutrient availability, microbial activity, chemical processes, and movement of contaminants like metals (Shaaban, 2024). Managing soil pH involves actions like using the right nitrogen fertilizer, applying lime, and adopting cropping practices that enhance soil organic matter and overall soil health (Neina, 2019). pH plays a crucial role in influencing plant growth and the behavior of environmental contaminants in the soil. The pH of the soil affects how soluble and mobile mineral components are, which in turn determines how easily roots can take them up. Nitrogen and potassium are readily taken up when the pH is higher than 6.0, but their availability decreases when the pH is below 6.0. Phosphorus (P) is most available at a pH range of 5.8 to 7.0. When the pH is 5.5 or below, with the presence of iron and aluminum oxides, soil particles bind with P, preventing it from being used by plants. This is a common issue in naturally acidic (especially in tropical) soils. Adding additional phosphorus will not improve the nutrient levels unless the soil pH is raised above the critical level of 5.5. This will enable P, whether naturally occurring in soil minerals or applied as fertilizers, to once again be accessible to plants (Barrow & Hartemink, 2023).

Acidic igneous rocks have less phosphorus compared to limestone, and simply increasing its availability is rarely enough to fulfill the needs of food crops. Soils developed from volcanic ash are famous for their ability to retain additional P fertilizer. This results from the rapid dissolution of Al as glassy soil particles are released. Focusing P fertilizers in strips, or the traditional method of soaking seed potatoes in P solutions prior to planting can help reduce this occurrence.

Mineral toxicities from sodium or aluminum occur when pH levels exceed 8.2 or fall below 4.8, respectively. In tropical soils where kaolinite is prevalent, there is a strong correlation between soil pH and base saturation. A soil pH of 4.8, 5.5, and 5.8 indicates a base saturation of around 15%, 35%, and 50%, respectively. When pH reaches 7 or higher, the base saturation is at its maximum of 100%. Ca^{2+} is the predominant cation in the basic complex between pH levels of 7 and 8.2, while Na^{+} becomes the dominant cation above pH = 8.5. Aluminum toxicity risk is present at pH levels below 5.2, and becomes apparent below 4.8 (Kabała & Łabaz, 2018).

Soil acidity also impacts the biological activity in the soil. Bacteria and actinomycetes typically perform best at moderate to high pH levels, with the exception of sulfur-oxidizing bacteria that thrive in acidic conditions. However, their performance significantly decreases when the pH falls below 5.5. Fungi, on the contrary, are quite adaptable to a broad pH range; they are most common at lower pH levels, but face stiff competition from bacteria and actinomycetes at higher pH levels (Fig. 11).

4 Soil Biological Composition in Relation to Plant Growth:

Biological properties consist of flora, fauna, and organic matter found in the soil. Healthy soil contains a range of living organisms, including earthworms, as well as bacteria, fungi, and yeasts of different sizes. By itself, earthworms can have a weight of approximately 100 to over 1500 kg/ha; fungi can have a weight ranging from 1000 to 15,000 kg/ha in a soil that is in good condition. High levels of organic matter, sufficient soil moisture, ideal temperature, and proper drainage and aeration favor the population of these organisms in soil. Soil biology demonstrates intricate relationships with other soil characteristics (such as physical and chemical properties). Soil organisms aid in the breakdown of organic material. The underground creatures mix the organic material into the ground. Large pore spaces created by them enhance soil aeration and infiltration. Microscopic organisms like bacteria, actinomycetes, fungi, yeasts, algae, and protozoa break down organic matter, releasing nutrients in a plant-usable form. They also greatly influence the composition of soil (Sharma & Kumar, 2023a). The particular function of a particular soil microbes are given as following:

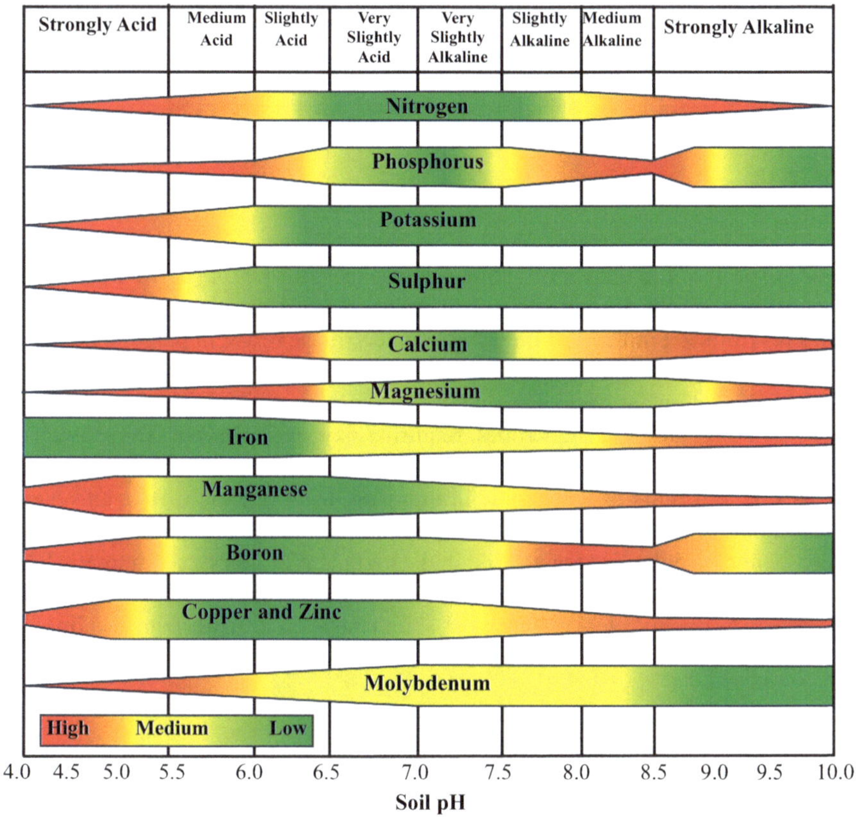

Fig. 11 Effect of pH on nutrients availability (Hartemink & Barrow, 2023)

4.1 Bacteria

Bacteria decompose nutrients and provide them to plants for use. Various species possess unique abilities, with some capable of extracting nitrogen from the atmosphere and converting it into a form that is usable by plants. Bacteria are accountable for transforming nitrogen from ammonium to nitrate and vice versa. Certain bacteria decompose minerals, while others produce and release plant hormones that boost root development, extracting P, K, Ca, Mg, and Iron in the process. Bacteria also assist in combating root disease and purifying soil. In summary, bacteria form the foundation of the subterranean food chain. Bacteria are the largest and most varied group of microorganisms, playing a crucial role in maintaining soil stability and thriving in diverse environments. Bacteria serve as a source of food for other microorganisms within the food chain. Also, they convert atmospheric nitrogen into a usable form and provide nutrients by breaking down trapped minerals, consuming other microorganisms (releasing nitrogen), and breaking down sugars, proteins, pectin, and starches released by plants (Swami et al., 2017).

4.2 Fungi

Various species of fungi, varying in size and shapes, can be found in soil. Numerous fungi assist plants by decomposing plant remnants, like corn stalks, and by extracting nutrients from soil minerals. Fungi called mycorrhizae reside in plant roots, helping extend root hairs to improve water and nutrient absorption, especially phosphorus. They also create plant hormones and antibiotics to boost root growth and prevent diseases. Fungi are obligate aerobes, requiring oxygen to survive. They are able to endure dehydration and are most prevalent in acidic soils. Actinomycetes are similar to bacteria and fungi as they assist in breaking down organic matter into humus, releasing nutrients. They also create antibiotics to combat root diseases. Several of these identical antibiotics are utilized in the treatment of illnesses in humans. These creatures create the pleasant, natural aroma that is detected when a living soil is disturbed. Actinomycetes are the most abundant microorganisms in alkaline soils (Huey et al., 2020).

4.3 Algae

Most algae are found within the top half inch of soil. Algae create their own nourishment through the process of photosynthesis. Algae enhance soil quality by creating sticky substances that bind soil particles to form stable aggregates. Certain species have the ability to convert nitrogen internally before transferring it to the roots of plants (Swami et al., 2017).

4.4 Nematodes

Nematodes, which are tiny unsegmented round worms, are commonly found in the majority of soils. There are only a small number of species that are detrimental to plants. A healthy soil microbial system controls root feeding nematodes by having predatory nematodes and micro arthropods. Destructive organisms consume decomposing plant material, along with bacteria, fungi, algae, protozoa, and other nematodes. While feeding, they release nutrients that were stored in their prey's bodies. Similar to other predators, nematodes accelerate the process of nutrient cycling. There is a wide variety of types of them, and they are present in large quantities per square foot. Nematodes are a significant consumer and provide nitrogen. Beneficial nematodes surpass parasitic nematodes in a well-balanced ecosystem (Mendoza-de Gives, 2022).

4.5 Protozoa

Protozoa play a significant role in the movement of Nitrogen, Phosphorous, and Sulphur. Numerous soil protozoa are predators, feeding on other microorganisms. A very common organism is the amoeba that consumes bacteria. Protozoa accelerate the nitrogen cycling process by consuming bacteria, which ultimately increases the availability of nitrogen to plants (Swami et al., 2017).

References

Ahmed, M., Hasanuzzaman, M., Raza, M. A., Malik, A., & Ahmad, S. (2020). Plant nutrients for crop growth, development and stress tolerance. *Sustainable agriculture in the era of climate change*, 43–92.

Alemayehu, B., & Teshome, H. (2021). Soil colloids, types and their properties: A review. *Open Journal of Bioinformatics and Biostatistics, 5*(1), 008–013.

Banuelos, G., & Ajwa, H. (1999). Trace elements in soils and plants: An overview. *Journal of Environmental Science & Health Part A, 34*(4), 951–974.

Bañuelos, G. S., & Ajwa, H. A. (1999). Trace elements in soils and plants: An overview. *Journal of Environmental Science and Health, Part A, 34*(4), 951–974. https://doi.org/10.1080/10934529909376875

Barker, A. V., & Pilbeam, D. J. (2015). *Handbook of plant nutrition*. CRC Press.

Barrow, N., & Hartemink, A. E. (2023). The effects of pH on nutrient availability depend on both soils and plants. *Plant and Soil, 487*(1), 21–37.

Bhatla, S. C., A. Lal, M., Kathpalia, R., & Bhatla, S. C. (2018). Plant mineral nutrition. *Plant physiology, development and metabolism*, 37–81.

Bot, A., & Benites, J. (2005). *The importance of soil organic matter: Key to drought-resistant soil and sustained food production*. Food & Agriculture Org.

Brown, S., Biswas, A., Caron, J., Dyck, M., & Si, B. (2021). Soil Physics. *Digging into Canadian Soils*.

Chaudhary, S., Sindhu, S. S., Dhanker, R., & Kumari, A. (2023). Microbes-mediated sulphur cycling in soil: Impact on soil fertility, crop production and environmental sustainability. *Microbiological Research, 271*, 127340. https://doi.org/10.1016/j.micres.2023.127340

Conant, R. T., Ryan, M. G., Ågren, G. I., Birge, H. E., Davidson, E. A., Eliasson, P. E., Evans, S. E., Frey, S. D., Giardina, C. P., & Hopkins, F. M. (2011). Temperature and soil organic matter decomposition rates–synthesis of current knowledge and a way forward. *Global Change Biology, 17*(11), 3392–3404.

Cotrufo, M. F., Wallenstein, M. D., Boot, C. M., Denef, K., & Paul, E. (2013). The M icrobial E fficiency-M atrix S tabilization (MEMS) framework integrates plant litter decomposition with soil organic matter stabilization: Do labile plant inputs form stable soil organic matter? *Global Change Biology, 19*(4), 988–995.

da Silva, R. J. A. B., da Silva, Y. J. A. B., van Straaten, P., do Nascimento, C. W. A., Biondi, C. M., da Silva, Y. J. A. B., & de Araújo Filho, J. C. (2022). Influence of parent material on soil chemical

characteristics in a semi-arid tropical region of Northeast Brazil. *Environmental monitoring and assessment, 194*(5), 331.
DalCorso, G., Manara, A., & Furini, A. (2013). An overview of heavy metal challenge in plants: From roots to shoots. *Metallomics, 5*(9), 1117–1132.
Delgado, A., & Gómez, J. A. (2016). The soil. Physical, chemical and biological properties. *Principles of agronomy for sustainable agriculture*, 15–26.
Diacono, M., & Montemurro, F. (2015). Effectiveness of organic wastes as fertilizers and amendments in salt-affected soils. *Agriculture, 5*(2), 221–230.
Dixit, P. (2023). The effects of nitrogen fertilizer on plant growth *International Journal of Trend in Scientific Research and Development 7*(5), 1024–1030. https://www.ijtsrd.com/papers/ijtsrd60048.pdf
Dudka, S., Piotrowska, M., & Chlopecka, A. (1994). Effect of elevated concentrations of Cd and Zn in soil on spring wheat yield and the metal contents of the plants. *Water, Air, and Soil Pollution, 76*, 333–341.
Engels, C., & Marschner, H. (1995). Plant uptake and utilization of nitrogen. *Nitrogen fertilization in the environment*, 41–81.
Erich, M., Fitzgerald, C., & Porter, G. (2002). The effect of organic amendments on phosphorus chemistry in a potato cropping system. *Agriculture, Ecosystems & Environment, 88*(1), 79–88.
Fageria, N., Baligar, V., & Li, Y. (2008). The role of nutrient efficient plants in improving crop yields in the twenty first century. *Journal of Plant Nutrition, 31*(6), 1121–1157.
Fageria, N., Gheyi, H., & Moreira, A. (2011). Nutrient bioavailability in salt affected soils. *Journal of Plant Nutrition, 34*(7), 945–962.
Fageria, N., Moreira, A., Ferreira, E., & Knupp, A. (2013). Potassium-use efficiency in upland rice genotypes. *Communications in Soil Science and Plant Analysis, 44*(18), 2656–2665.
Gattinger, A., Muller, A., Haeni, M., Skinner, C., Fliessbach, A., Buchmann, N., Mäder, P., Stolze, M., Smith, P., & Scialabba, N.E.-H. (2012). Enhanced top soil carbon stocks under organic farming. *Proceedings of the National Academy of Sciences, 109*(44), 18226–18231.
Gerendás, J., & Führs, H. (2013). The significance of magnesium for crop quality. *Plant and Soil, 368*(1), 101–128. https://doi.org/10.1007/s11104-012-1555-2
Gerke, J. (2022). The central role of soil organic matter in soil fertility and carbon storage. *Soil Systems, 6*(2), 33.
Ghosh, T., Maity, P. P., Das, T., Krishnan, P., Bhatia, A., Bhattacharya, P., & Sharma, D. (2020). Variation of porosity, pore size distribution and soil physical properties under conservation agriculture. *The Indian Journal of Agricultural Sciences, 90*(11), 2051–2058.
Gurmu, G. (2019). Soil organic matter and its role in soil health and crop productivity improvement. *Forest Ecology and Management, 7*(7), 475–483.
Hartemink, A. E., & Barrow, N. J. (2023). Soil pH - nutrient relationships: The diagram. *Plant and Soil, 486*(1), 209–215. https://doi.org/10.1007/s11104-022-05861-z
Hofman, G., & Van Cleemput, O. (2004). Soil and plant nitrogen.
Huey, C. J., Gopinath, S. C. B., Uda, M. N. A., Zulhaimi, H. I., Jaafar, M. N., Kasim, F. H., & Yaakub, A. R. W. (2020). Mycorrhiza: a natural resource assists plant growth under varied soil conditions. *3 Biotech, 10*(5), 204. https://doi.org/10.1007/s13205-020-02188-3
Hussain, A., Bashir, H., Zafar, S., Rehman, R., Khalid, M., Awais, M., Sadiq, M., & Amjad, I. (2023). The importance of soil organic matter (som) on soil productivity and plant growth. *Biological and Agricultural Sciences Research Journal, 2023*(1), 11–11.
Ishfaq, M., Wang, Y., Yan, M., Wang, Z., Wu, L., Li, C., & Li, X. (2022). Physiological Essence of Magnesium in Plants and Its Widespread Deficiency in the Farming System of China. *Frontiers in Plant Science, 13*, Article 802274. https://doi.org/10.3389/fpls.2022.802274
Kabała, C., & Łabaz, B. (2018). Relationships between soil pH and base saturation–conclusions for Polish and international soil classifications. *Soil Science Annual, 69*(4).
Kalev, S. D., & Toor, G. S. (2018). Chapter 3.9 - The Composition of Soils and Sediments. In B. Török & T. Dransfield (Eds.), *Green Chemistry* (pp. 339–357). Elsevier. https://doi.org/10.1016/B978-0-12-809270-5.00014-5

Karley, A. J., & White, P. J. (2009). Moving cationic minerals to edible tissues: Potassium, magnesium, calcium. *Current Opinion in Plant Biology, 12*(3), 291–298. https://doi.org/10.1016/j.pbi.2009.04.013
Kay, B., & Angers, D. (2001). 7 Soil Structure. *Soil physics companion, 249.*
Kirschbaum, M. U. F. (2006). The temperature dependence of organic-matter decomposition—still a topic of debate. *Soil Biology and Biochemistry, 38*(9), 2510–2518.
Kögel-Knabner, I. (2006). Chemical structure of organic N and organic P in soil. In *Nucleic acids and proteins in soil* (pp. 23–48). Springer.
Krebs, J. (1998). The role of calcium in apoptosis. *BioMetals, 11*, 375–382.
Kumaragamage, D., Warren, J., & Spiers, G. (2021). Soil chemistry. *Digging into Canadian Soils.*
Lavelle, P., Decaëns, T., Aubert, M., Barot, S. b., Blouin, M., Bureau, F., Margerie, P., Mora, P., & Rossi, J.-P. (2006). Soil invertebrates and ecosystem services. *European journal of soil biology, 42*, S3-S15.
Logan, T. J., & Chaney, R. L. (1983). Utilization of municipal wastewater and sludge on land-metals. Proceedings of the workshop on utilization of municipal wastewater and sludge on land,
Ma, D., He, Z., Zhao, W., Li, R., Sun, W., Wang, W., Lin, P., Wei, L., & Ju, W. (2024). Long-term effects of conventional cultivation on soil cation exchange capacity and base saturation in an arid desert region. *Science of the Total Environment, 949*, Article 175075.
Malik, M. A., Marschner, P., & Khan, K. S. (2012). Addition of organic and inorganic P sources to soil–Effects on P pools and microorganisms. *Soil Biology and Biochemistry, 49*, 106–113.
Mamathashree, C., Girijesh, G., & Vinutha, B. (2018). Phosphorus dynamics in different soils. *Journal of Pharmacognosy and Phytochemistry, 7*(1), 981–985.
Manafi, M. S., Deng, A., Taheri, A., Jaksa, M. B., & HB, N. (2022). Determining soil plasticity utilizing Manafi method and apparatus. *Geotechnical Testing Journal*, *45*(4), 797-818.
Mancinelli, R., Campiglia, E., Di Tizio, A., & Marinari, S. (2010). Soil carbon dioxide emission and carbon content as affected by conventional and organic cropping systems in Mediterranean environment. *Applied Soil Ecology, 46*(1), 64–72.
Mandal, U. K. (2016). Soil physical and chemical properties in relation to conservation of natural resources. In.
Marschner, H. (2011). *Marschner's mineral nutrition of higher plants.* Academic press.
Maynard, D. N. (2007). Handbook of Plant Nutrition. *HortScience, 42*(2), 422b–4423.
Mendoza-de Gives, P. (2022). Soil-Borne Nematodes: Impact in Agriculture and Livestock and Sustainable Strategies of Prevention and Control with Special Reference to the Use of Nematode Natural Enemies. *Pathogens, 11*(6). https://doi.org/10.3390/pathogens11060640
Mengel, D. B. (1993). *Fundamentals of soil cation exchange capacity (CEC).* Purdue University Cooperative Extension Service West Lafayette.
Mengel, K., Kirkby, E. A., Kosegarten, H., & Appel, T. (2001). The soil as a plant nutrient medium. *Principles of plant nutrition*, 15–110.
Mortland, M. (1966). Ammonia interactions with soil minerals. *Agricultural anhydrous ammonia technology and use*, 188–197.
Murphy, D. V., Macdonald, A. J., Stockdale, E. A. a., Goulding, K. W. T., Fortune, S., Gaunt, J. L., Poulton, P. R., Wakefield, J. A., Webster, C. P., & Wilmer, W. S. (2000). Soluble organic nitrogen in agricultural soils. *Biology and Fertility of Soils*, *30*, 374-387.
Neina, D. (2019). The role of soil pH in plant nutrition and soil remediation. *Applied and Environmental Soil Science, 2019*(1), 5794869.
Neira, J., Ortiz, M., Morales, L., & Acevedo, E. (2015). Oxygen diffusion in soils: Understanding the factors and processes needed for modeling. *Chilean Journal of Agricultural Research, 75*, 35–44.
Osman, K. T., & Osman, K. T. (2013a). Chemical properties of soil. *Soils: Principles, Properties and Management*, 97–111.
Osman, K. T., & Osman, K. T. (2013b). Physical properties of soil. *Soils: Principles, Properties and Management*, 49–65.
Osman, K. T. (2014). *Soil degradation, conservation and remediation* (Vol. 820). Springer.

Paustian, K., Six, J., Elliott, E., & Hunt, H. (2000). Management options for reducing CO2 emissions from agricultural soils. *Biogeochemistry, 48*, 147–163.
Piccolo, A. (2002). The supramolecular structure of humic substances: a novel understanding of humus chemistry and implications in soil science.
Pierzynski, J. (2016). *The effects of P fertilizer addition on P transformations on high-P fixing and grassland soils*. Kansas State University.
Rai, A., Singh, A. K., Mishra, R., Shahi, B., Rai, V. K., Kumari, N., Kumar, V., Gangwar, A., Sharma, R. B., & Rajput, J. (2020). Sulphur in soils and plants: An overview. *International Research Journal of Pure and Applied Chemistry, 21*(10), 66–70.
Raj, D., & Maiti, S. K. (2020). Sources, bioaccumulation, health risks and remediation of potentially toxic metal (loid) s (As, Cd, Cr, Pb and Hg): An epitomised review. *Environmental Monitoring and Assessment, 192*(2), 108.
Rao, I. M. (2009). Essential plant nutrients and their functions.
Salama, M. (2011). estimating water consumptive use for some crops under stress conditions using neutron scattering method.
Schön, J. H. (2015). Pore space properties. In *Developments in petroleum science* (Vol. 65, pp. 21–84). Elsevier.
Shaaban, M., & Nunez-Delgado, A. (2024). Soil adsorption potential: Harnessing Earth's living skin for mitigating climate change and greenhouse gas dynamics. *Environmental Research, 251*, Article 118738.
Shaaban, M. (2024). Acidic Soils. In *Planet Earth: Scientific Proposals to Solve Urgent Issues* (pp. 293–306). Springer.
Sharma, P. K., & Kumar, S. (2023a). Soil Air and Plant Growth. In *Soil Physical Environment and Plant Growth: Evaluation and Management* (pp. 155–174). Springer International Publishing. https://doi.org/10.1007/978-3-031-28057-3_6
Sharma, P. K., & Kumar, S. (2023b). *Soil Physical Environment and Plant Growth: Evaluation and Management*. Springer Nature.
Sharma, P. K., & Kumar, S. Soil Physical Environment and Plant Growth. In: Springer.
Singh, A., Hidangmayum, A., Tiwari, P., Singh, B. N., & Dwivedi, P. (2022). How the soil nitrogen nutrient promotes plant growth—a critical assessment. *New and Future developments in microbial biotechnology and bioengineering*, 99–118.
Six, J., Conant, R. T., Paul, E. A., & Paustian, K. (2002). Stabilization mechanisms of soil organic matter: Implications for C-saturation of soils. *Plant and Soil, 241*, 155–176.
Sposito, G. (2008). *The chemistry of soils*. Oxford University Press.
Stevenson, F. (1994). *Humus chemistry: Genesis, composition, reactions*. John Wiley & Sons.
Stockdale, E. A., Goulding, K. W. T., George, T. S., & Murphy, D. V. (2013). Soil fertility. In *Soil Conditions and Plant Growth* (pp. 49–85). https://doi.org/10.1002/9781118337295.ch3
Sullivan, D. M., & Andrews, N. (2012). Estimating plant-available nitrogen release from cover crops.
Sun, Q., Zheng, J., Coop, M. R., & Altuhafi, F. N. (2019). Minimum image quality for reliable optical characterizations of soil particle shapes. *Computers and Geotechnics, 114*, 103110. https://doi.org/10.1016/j.compgeo.2019.103110
Sung, C. T. B., Ishak, C. F., Abdullah, R., Othman, R., Panhwar, Q. A., & Aziz, M. M. A. (2017). Soil properties (physical, chemical, biological, mechanical). *Soils of Malaysia*, 103–154.
Swami, S., Gurjar, G., Meena, N., & Lyngdoh, E. (2017). Effect of soil biological properties on crop production. *Natural resource management for climate smart sustainable agriculture. Soil Conservation Society of India, New Delhi*, 240–255.
Takahashi, H., Kopriva, S., Giordano, M., Saito, K., & Hell, R. (2011). Sulfur assimilation in photosynthetic organisms: Molecular functions and regulations of transporters and assimilatory enzymes. *Annual Review of Plant Biology, 62*, 157–184. https://doi.org/10.1146/annurev-arplant-042110-103921
Ure, J. (1982). Introduction: approaches to the study of register range.

Velayutham, M., & Bhattacharyya, T. (2000). Soil resource management. Natural resource management for agricultural production in India'. Special publication during International Conference on Management of Natural Resources for Sustainable Agricultural Production in the 21st Century,

Wang, Y., Wu, B., Berns, A. E., Xing, Y., Kuhn, A. J., & Amelung, W. (2020). Magnesium isotope fractionation reflects plant response to magnesium deficiency in magnesium uptake and allocation: A greenhouse study with wheat. *Plant and Soil, 455*(1), 93–105. https://doi.org/10.1007/s11104-020-04604-2

Weber, J. (2002). Definition of soil organic matter. *Humintech: Humic acids based products.*

White, P. J., & Broadley, M. R. (2003). Calcium in plants. *Annals of Botany, 92*(4), 487–511.

White, C., Sayer, J., & Gadd, G. (1997). Microbial solubilization and immobilization of toxic metals: Key biogeochemical processes for treatment of contamination. *FEMS Microbiology Reviews, 20*(3–4), 503–516.

Wilke, B.-M. (2005). Determination of Chemical and Physical Soil Properties. In *Monitoring and Assessing Soil Bioremediation* (pp. 47–95). Springer Berlin Heidelberg. https://doi.org/10.1007/3-540-28904-6_2

Yang, M., Zhou, D., Hang, H., Chen, S., Liu, H., Su, J., Lv, H., Jia, H., & Zhao, G. (2024). Effects of Balancing Exchangeable Cations Ca, Mg, and K on the Growth of Tomato Seedlings (Solanum lycopersicum L.) Based on Increased Soil Cation Exchange Capacity. *Agronomy, 14*(3), 629. https://www.mdpi.com/2073-4395/14/3/629

Soil Chemical Properties, Nutrient Dynamics and Fertility Management

Saima Nasreen, Rubania Arshad, Kousar Parveen, and Saira Mansab

Abstract The key component of agricultural production, soil has a direct impact on crop yield and growth. In order to preserve and improve soil fertility, this chapter offers a thorough analysis of the nutrient dynamics, soil chemical characteristics, and fertility management techniques. With an emphasis on the soil's chemical characteristics, nutrient dynamics, and fertility management techniques, the significance of soil in agriculture is emphasized. The chapter discusses soil composition, nutrient availability, and cation exchange capacity. It discusses soil pH, macronutrients, and micronutrients, with a focus on nutritional balance. It covers soil colloids, nutrient cycling, and the effect of fertilizers. The article discusses diagnostic procedures for nutritional shortages as well as fertility control measures such as integrated nutrient management and effective nutrient delivery. Best management practices (BMPs) for maintaining the soil and water quality are offered, taking into account environmental implications.

1 Introduction

A fertile soil functions as a dynamic living environment that offers a variety of ecosystem services, including preserving water quality and plant production, managing soil nutrient recycling, decomposition, and absorbing greenhouse gasses from the environment. Soil health is directly related to agriculture that is sustainable because microorganism diversity of soil and activity are the most vital components of soil health. Soil health is an inherent property of a soil. It is regarded as a group of features that identify its health and taxonomically classify it. Soil quality, on the other hand, is a variable extrinsic attribute of soils that varies with human usage of the soil. It might be about watershed protection, wildlife support and agricultural production or recreational opportunities. Agricultural sustainability is described as a crop production system's capacity to generate food continually while minimizing

S. Nasreen (✉) · R. Arshad · K. Parveen · S. Mansab
Department of Environmental Sciences, Faculty of Life Sciences, The Women University Multan, Multan, Pakistan
e-mail: saima.6333@wum.edu.pk

M. Shaaban (ed.), *Soils and Sustainable Agriculture*, Frontier Studies in Soil Science,
https://doi.org/10.1007/978-3-031-91114-9_4

environmental harm. Arbuscular mycorrhizal fungi (AMF), beneficial nematodes and cyanobacteria, improve water usage functionality, nutritional accessibility to plants, phytohormone synthesis, plant resilience and soil nutrient cycling to environmental challenges. Organic farming and tillage have been found to increase soil health by boosting microbial excess, variety, and activity. Conservation tillage has the capacity to increase farmer profitability by lowering input and costs of labor when compared to traditional tillage, whereas organic farming may incur additional high manpower needs for weeding and pest control drive up management expenses, as well as fertilizer inputs (particularly N-based), which have lower steadier and more stable than synthetic fertilizers. The considerable projected rise in world population to 8.9 billion by 2050 will boost demand for agricultural commodities. High food demand and a limitation of future agricultural land growth will necessitate tripling yields of crop using sustainable techniques. Scientists may make a significant contribution to the sustainability of agricultural fields worldwide by converting scientific information about soil function into practical ways that help growers assess the sustainability of their handling techniques. Two sustainable agriculture management practices aim to enhance soil organic matter and prevent erosion by improving plant diversity and conservation tillage. There is a necessity for worldwide attention to improve or restore soil health. The evaluation of soil health indicators is likely to improve our knowledge of the aspects that lead to sustainable agriculture (Tahat et al., 2020).

The chapter covers the composition of soil, including the functions of minerals, organic matter, and inorganic and organic components in nutrient availability. An explanation of the cation exchange capacity (CEC) is provided, along with the significance, and contributing factors, including soil texture and organic matter concentration. Examined the importance of soil pH, how to measure it, and how it affects soil health and nutrient availability. A brief description of the macronutrients (nitrogen, phosphorus, and potassium) and micronutrients that are necessary for soil fertility is discussed along with the significance of preserving nutrient balance. It is also investigated how soil colloids affect the exchange and retention of nutrients. A description of the nutrient cycling processes, such as mineralization and decomposition, as well as the interactions between plants, soil, and the atmosphere, covers the topic of nutrient dynamics. The chapter covers the several sources of nutrients, such as organic amendments and atmospheric deposition, as well as the function fertilizers play in preserving soil health. The symptoms and diagnostic techniques for determining nutrient deficits and toxicities in plants are also highlighted. Fertility management describes many techniques and instruments for determining soil fertility and stresses the significance of routine soil testing and analysis. It also covers the fundamentals of integrated nutrient management, which blends organic and inorganic fertilizers to provide sustainable nutrient management techniques. Techniques for applying nutrients efficiently are described, along with scheduling and ways to reduce nutrient losses. The chapter ends with a presentation of best management practices (BMPs) intended to preserve the quality of the soil and water, as well as a consideration of the possible environmental effects of nutrient management. In conclusion, this chapter provides a thorough grasp of the nutrient dynamics, soil chemical characteristics, and fertility management techniques. This knowledge is crucial for sustainable agricultural practices and the preservation of soil health.

I. Chemical Properties of Soil

Soil is a chemical substance. All of the materials are chemical compounds. Soils are made up of liquid, solid, and gas, insoluble and soluble, inorganic and organic constituents. There are salts, bases, acids, minerals, ions, and rock pieces. Colloids can also be quite chemically active. Humus, aluminum and iron oxides, hydroxides and fine silicate clays are all examples of extremely fine soil particles. Colloids have electrochemical charges, both negative and positive charges which formed via replacement in the active group's dissociation and the lattice structure. These charges hold ions that have opposing charges, which can be replaced by ions with equivalent charges in the soil solution. Besides with ion exchange characteristics, pH and Eh are two more key indicators of the soil's chemical environment. Soil pH is calculated as the negative logarithm of the concentration of hydrogen ions in soil suspension. It defines the level of alkalinity and acidity, which determines chemical solubility, nutrient accessibility and absorption, and soil microbes' development and activity (Shaaban, 2024). If the soil's pH remains very acidic or alkaline, then some of the nutrients become inaccessible. The Eh of soil represents the soil's redox potential, which tells whether it is oxidized or reduced. Redox processes are extremely essential in soil formation. Various additional chemical reactions are always taking place in the soil (Properties, 2013).

A. Composition of Soil

Soils embrace about 100 chemical components, however just a few constitute the main mass. The various elements in soils are listed in Sparks with their respective ranges and median values. The median concentrations of potassium (K) 1.4%, magnesium (Mg) 0.5%, calcium (Ca) 5.0%, iron (Fe) 4.0%, aluminum (Al) 7.1%, silicon (Si) 33.0%, and oxygen (O) and 49.0% respectively. Other elements are found in minor concentrations. The soil solution and minerals contain different nutrients of plant such as magnesium, manganese (Mn), potassium, sulfur (S), zinc (Zn), phosphorus (P), copper (Cu), calcium, iron (Fe), molybdenum (Mo), boron (B), and nickel (Ni). Minerals and oxides include oxygen, silicon, and aluminum. Iron exists mostly in the form of oxides and hydroxides. In soil Phosphorus found mostly as the mineral apatite, but it also exists as aluminum and calcium phosphates, as well as organic molecules such as inositol, nucleic acids and phospholipids. Nitrogen is found in soil mostly in organic forms such as peptides, amino acids and proteins. All of the micronutrients Ni, Zn, Mn, Fe, B, Mo and Cu exist in inorganic form. Nature has many distinct soils, and their chemical compositions vary (Properties, 2013).

- **Inorganic and Organic Components of Soil and Role of Minerals and Organic Matter in Nutrient Availability**

Nonmineral, solid, components of soil is the Organic matter that come from the animal wastes and plants. The American Soil Science Society (1997) defines organic matter of soil as the organic portion of the soil that excludes animal wastes and plants which are undecayed. Organic matter of soil has long been considered an essential indication of soil efficiency. It contributes significantly to cropping system sustainability by boosting soil physical properties like water-holding capacity texture and structure and bulk density, chemical properties like cation exchange capacity, availability of nutrients, allelopathy and reduced aluminum toxicity, and biological properties like dinitrogen fixation, nitrogen mineralization, bacteria, microbial biomass and mycorrhizae fungi. For the long-term viability of agricultural ecosystems, the maintenance of soil organic matter is critical. Adopting suitable soil and crop management strategies can help to improve or preserve soil organic matter. These methods include crop rotation, conservation tillage, the usage of organic manures, increasing the cropping intensity, using an appropriate number of chemical fertilizers, liming acidic soils, incorporating crop wastes, and maintaining land in pasture. Organic matter may absorb heavy metals present in soils, reducing their toxicity to the plants and preventing their discharge into ground water. Similarly, soil organic matter adsorbs pesticides, which may reduce pollution of ground water and surface water. Furthermore, SOM acts as an organic carbon sink and reduces emissions of carbon dioxide (CO_2) to the environment. Soil organic matter holds approximately three times as much carbon as the world's plants. As a result, organic matter is vital to the carbon balance in whole world, which is regarded to be the most important element influencing global warming. Overall, proper levels of soil organic matter maintain soil quality, support agricultural systems, and prevent environmental pollution (Fageria, 2012).

Soil formation is both productive and harmful. Soil is made up of shattered rock particles that have been transformed by chemical and mechanical processes such as weathering and erosion. Soil serves a multifaceted role that benefits humans and other living organisms. Soil is not simply a collection of mineral particles. It also includes a biological system of living organisms and a few additional components. The climate, among other factors, has a significant impact on formation of soil. Soil is not, as many people believe, a lifeless, inert mineral mass. However, a healthy soil is living and active, made up of microbes. The top layer of soil is nutrient-dense and ideal for biofarming. The profile character varies significantly between locations, notably in terms of depth, color, and composition. The soil is a natural collection of organic material and minerals divided into many horizons. The chemical qualities of soil govern its mineral composition, organic matter content, and environmental impact. It has a changeable depth and different from the parent material below in physical qualities, chemical properties in shape, constitution, composition and biological traits. Soil chemistry refers to the chemical reactions that occur in the soil during plant and animal growth, as well as human development (Smita Tale & Ingole, 2015).

B. **Cation Exchange Capacity**

Cation Exchange Capacity discusses soil's entire capability to store, absorb, and exchange the cations. Cation exchange capacity is the number of negative charges in soil that reside on the tops of clay and organic materials. CEC is commonly used to assess sediment's capacity to bond with cations, which may be related to metal mobility in soil. Several parameters, including pH, particle size, and organic matter concentration in soil or sediment, might affect the ability of cations to exchange (Sidi et al., 2015).

- **Definition and Significance**

It represents the entire number of exchangeable cations that a soil may absorb. It assesses fertility of soil, nutrient retention capacity of soil, and the ability to prevent groundwater from cation pollution. It dampens changes in pH of soil and availability of nutrient. Primary sources of cation exchange capability are the Clay and organic materials. The amount of organic matter and clay is related to the soil's cation exchange capacity and fertility potential. Soils with a higher sand content may have higher cation exchange capacity values, yet the kind of clay can have a noticeable effect on CEC. They store minimal nutrients and lose them quickly as soon as water passes through them. The sand particles are coated with clay and organic substances The organic matter and clay layer on the surface of sand particles increases the soil's nutrient retention ability. Soil's cation exchange capacity is related to the quantity of nutrients it can contain, hence the greater the fertility level (Javed et al., 2022). When evaluating soils of various textures, cation exchange capacity can be a very good indication of productivity of soil and is useful for formulating recommendations of magnesium (Mg), phosphorus (P), and potassium (K) (Sikora & Moore-Kucera, 2014).

- **Factors Affecting CEC (Soil Texture, Organic Matter)**

CEC is heavily influenced by soil texture because colloids which are negatively charged, dominate the clay-sized portion. In general, CEC of sandy loams is less than 10 $cmol_c\ kg^{-1}$ (centimol positive charge per kilogram of soil), silt loams and loams have CEC of 20 to 10 $cmol_c\ kg^{-1}$, while clay loams have CEC of more than 20 $cmol_c\ kg^{-1}$. Organic matter content also has a substantial impact on soil's CEC, as the organic colloids have a higher CEC than clay minerals (Sikora & Moore-Kucera, 2014).

C. **Soil pH**

Soil pH, measured on a pH scale, is a measure of the soil's acidity. Soil's natural pH is mostly determined by rainfall and soil parent material. Soils originating from parent material of granite, as those found in Wicklow Mountains, will naturally have a higher pH; on the other hand, soils developed from parent material of limestone will weather to generate more alkaline soils. The pH of the soil is a major element that governs availability of nutrient in soils, and it is optimal in a pH range of 6.3–7.0 (Wall et al., 2018).

- **Definition and Measurement**

Soil pH represents the soil response. The word pH was derived from the French phrase "pouvoir hydrogene," which means "power of hydrogen." According to the Sorenson, pH is equal to the concentration of hydrogen ions in a highly diluted solution, or the negative logarithm of hydrogen ion activity. The H^+ ion concentration is given in moles per liter.

$$\text{pH}\ \log[\text{H}]$$

The pH scale spans from 0 to 14, with 7 representing neutrality. A pH less than 7 denotes acidity, whereas a pH more than 7 suggests alkalinity. The lower the pH is, the greater the acidity. Similarly, when pH rises, alkalinity also rises. pH of soil typically ranges between 4 and 10; however, acid sulfate soils can have the pH as low as 2. Good grade agricultural soils have pH levels ranging from 6.0 to 7.0. Because the pH scale is the logarithmic, a one-unit pH difference is really 10 times larger. For example, soil having a pH of 5 is ten times more acidic than soil having a pH of 6. Soil pH is occasionally tested in a 0.01 molar calcium chloride solution of soil to compensate for release of calcium from the soil exchange complex. The pH of the soil acquired this way is often lower than that of a suspension composed of pure water. Because the distilled water pH is naturally acidic, sometimes we use neutral 1 N KCl solution to evaluate soil pH in suspension. The product of OH^- and H^+ ion concentrations in a dilute solution is 10–14. As a result, pH may be measured if the concentration of either of the H^+ or OH^- ions is known. For example, if the concentration of OH-ions in a dilute aqueous solution is 10^{-9} g L^{-1}, its pH may be determined as follows.

$$\begin{aligned}
[\text{H}^+] \times [\text{OH}^-] &= 10^{-14} \\
[\text{OH}^-] &= 10^{-9}\,\text{moles}\,\text{L}^{-1} \\
\text{So, } [\text{H}^+] &= 10^{-5}\,\text{moles}\,\text{L}^{-1} \\
\text{pH} &= -\log[\text{H}^+] \\
&= -\log 10^{-5} \\
&= 5 \log 10 \\
5 \times 1 &= 5
\end{aligned}$$

Hence the Soil's pH is five. In every solution, the concentration of H^+ ions falls while the concentration of OH^- ions increases (Table 1).

- **Impact on Nutrient Availability and Soil Health**

Soil pH influences plant development primarily through its effects on other soil qualities such as availability of nutrients, microbial activity and elemental toxicity. Although H^+ ions can have a direct corrosive impact on root membranes at extremely low pH, the soil's pH effects on plants are primarily indirect. Plants in acidic soils suffer from P, Mg, Ca, and Mo deficiencies, as well as toxicity of Fe, Al, and Mn. Plant roots take up both soluble and exchangeable nutrition ions from the soil. Ions dissolved and exchanged in soil stay in balance at all times. The availability of plant

Table 1 Categories of acidity and basicity in soil (Smita Tale & Ingole, 2015; Osman 2012)

4.0	Extreme acidic
4.5	Highly acidic
5.0	Strongly acidic
5.5	Medium acidic
6.0	Mildly acidic
6.5	Very Mildly acidic
7.0	Neutral
7.5	Very Mildly basic
8.0	Mildly basic
8.5	Medium basic
9.0	Strongly basic
9. 5	Highly basic
10.0	Extreme basic

nutrients and solubility of elements and compounds in soil are both influenced by pH of soil. When the soil gets more acidic and soil pH falls below 6.5, the solubility of some elements, such as Fe, Cu, Mn, Al, and Zn, rises while the solubility of others, such as Mg, Mo and Ca falls. When soil pH goes over 7.0, the solubility of Mo, Mg, and Ca increases while that of Fe, Cu, Mn, Al, and Zn declines. Aluminum, while abundant in soil, is not a necessary element for the plants. In very acidic soils (which has pH below 5), the Al, Fe, and Mn solubility increases to the point that they are harmful to numerous plants. Furthermore, plants in acidic soils may suffer from calcium and magnesium deficiencies. At very low pH (like < 5.5), phosphorus precipitates with Mn Fe, and Al as polyphosphates. At very high pH (like > 8.0), P precipitates with calcium. Both the soil's acidity and alkalinity limit availability of P to deficient levels. pH regulates the preponderance of three ionic forms of P (PO_4^{3-}, HPO_4^{2-} and $H_2PO_4^-$). Phosphorus as $H_2PO_4^-$ ion is more accessible than other forms and leads at pH about 6.5. P availability is generally greater between pH 6.5 and 7.0. Boron shortage may develop in both very acidic and strongly Basic soils. Low pH limits the macronutrients availability (N, K, Ca, Mg, S, P), as well as B and Mo. At low pH, however, most micro nutrients such as Mn, Cu, Fe, Zn and Co become more available. Plant nutrition levels are most optimal when the pH is between 5.5 and 6.5. Alkaline soils may be deficient in nutrients such as manganese, copper, zinc and boron. Soils with a very basic pH (higher than 9) are likely to have large quantities of salt, which can degrade soil physical qualities and impose toxicity impact on plants (Properties, 2013).

D. **Nutrient Composition**

Plants need light, air, water, a correct temperature and 18 basic nutrients to live. They all require in altered quantities, yet every nutrient is equally crucial to the plant. They are separated into two categories:

- Macro nutrients

- Micro nutrients.
- **Macro nutrients (N, K, P)**

They are desired in larger numbers and have two different subgroups:

- **Primary micro nutrients**: The primary group consists of, nitrogen (N), potassium (K) and phosphorus (P). They are often necessary in the crop fertilization procedure and are also known as fertilizers.
- **Secondary micro nutrients**: The secondary group consists of Sulfur (S), and Calcium (Ca), Magnesium (Mg). These are often required in lower quantities than the major ones. They are especially significant for agricultural fertilization because to stricter clean air requirements and environmental improvements.
- **Micro nutrients**

Micro nutrients are the nutrients that are required in little or trace levels yet are equally significant as macro nutrients. Aside from the foregoing oxygen, hydrogen, and carbon are also crucial because they compose the majority of the plant weight. They include manganese (Mn), copper (Cu), boron (B), iron (Fe), chloride (Cl), molybdenum (Mo),sodium (Na), zinc (Zn), cobalt (Co), selenium (Se) nickel (Ni) and silicon (Si).

- **Importance of Nutrient Balance for Soil Fertility**

Soil is basically a dynamic and living source defines by The National Soil Survey Center, that comprises organic materials, microbes, and mineral particles of various sizes, as well as biological, physical and chemical qualities that may be modified. Nutrient-rich soil is referred to as "fertile". The concept of "feed soil to feed plant" underpins the sustainability of agricultural goods. This ideology aims to improve soil fertility by introducing organic matter, nutrients, fertilizers, and biological activity. Green manure, compost, and crop leftovers are organic matter which are utilized to increase soil organic matter, which in turn improves nutrients and increases crop yield. Soil nourishes plants, which then feed animals and people. Soil nutrients must be managed well since they feed 7.3 billion people. It is the foundation of human life and comprises organic and inorganic substances, decaying organic materials, and nutrients. When components of soil decompose, they provide critical nutrients to plants and animals. Fertilization provides additional nutrients to the soil, improving quality of soil, yield and profitability. Basically, three methods are used to replace nutrients of soil:

1. Using animal excrement
2. By Fertilizers
3. Microbial processes such as nitrogen fixation (Javed et al., 2022).

E. **Colloidal Properties**

Soil colloids with dimensions ranging from one nanometer to one millimeter relate to various and multidimensional mixtures of carbon-based and nonliving entities. Soil colloid discharge spans a wide spectrum and deals with ionic power, pH, and regular organic matter, all of which disrupt outer charge, interparticle cooperation, and dispersion. Adsorption of organic materials alters the external properties of colloids and regulates mobility. Iron decrease is induced by microscopic organisms under redox-dynamic conditions, often in residues, marshland soils and regular H_2O and can affect solution interaction. Colloidal diffusion in anaerobic environments is connected with iron disbanding and an increase in postponement pH. Soil colloids, with their enormous explicit superficial range, can transport contaminants; this mechanism is commonly known as colloid-facilitated pollutant transporting. Because of distinct soil colloids, holes or streams in the soil medium can change soil organization and utility. The results or transference of colloids is incredibly important for the chemical composition of the solution since it disrupts the consistency and the transportation.

- **Types of the Soil Colloids**

Soil contains four main types of colloids:

- Layer silicate clays
- Aluminum and Iron oxide clays
- Allophane along with associated amorphous clays
- Humus.

Inorganic colloids include allophane, aluminum and iron oxide clays, layer silicate clays and other amorphous clays, whereas humus is an organic colloid.

- **Properties of Soil Colloids**
- **Size of Colloids**

The organic and inorganic colloids are incredibly tiny, less than 2 μm in diameter. These colloid particles cannot be observed with a standard light microscope, but only observed with an electron microscope.

- **Surface Area of Soil Colloid**

Because of their small size, all soil colloids have more exterior surface area per unit mass. One gram of colloidal clay has thousands times more outer surface area than one gram of coarse sand. Large interior surfaces, which frequently much surpass the outward surface area, can be found between the plate-like crystal units that compose each particle in several silicate clays. The overall soil colloids surface area varies between 10 m^2/g for the clays with just exterior surfaces and more than 800 $m^2\ g^{-1}$ for the clays with substantial interior surfaces. Surface area of the colloid in the top 15 cm of a hectare of clay soil might be up to 700,000 $km^2\ g^{-1}$.

- **Surface Charges on Soil Colloids**

Both the exterior surface and interior surface of soil colloids have positive and/or negative charges. Most organic soil colloids and also inorganic soil colloids have a negative charge on them. Soil colloidal particles move to the anode, or positive electrode, when an electric current is applied through them, suggesting that they are negatively charged. The charge magnitude is referred to as zeta potential. The existence and strength of the particle charge determine the repulsion and attraction of the particles toward one other, therefore altering both physical and also chemical characteristics (S, 2022).

- **Role of Soil Colloids in Nutrient Retention and Exchange**

The soil colloidal fraction is where critical activities including ion exchange, fixation, nutrient availability and soil physical characteristics occur. Based on their silicate and crystalline properties, colloid minerals are classified into three main types. There are three types of clays: non-crystalline layer silicate, crystalline layer silicate, and non-silicate. Hummus is mostly responsible for the organic colloidal complexes seen in soils. The most essential qualities of soil colloids are their shape, cohesion, surface area, swelling, shrinkage, plasticity, flocculation and dispersion. Liquid phase soil solution is comprised of dissolved salts, and inorganic minerals, organic molecules that transport nutrients. Soil's mineral components are heavily influenced by the parent material from which the soil is produced. Mica, quartz and Feldspar are examples of primary minerals which inherited from parent rocks, whereas lime, oxides and hydroxides of Fe and Al and gypsum are secondary minerals which altered during the chemical weathering and found in the clay fraction. By cation-exchange capacity of soil, fertility of the soil for plants is determined. Soil colloids have a high capacity for cation exchange because of their large surface area and charge. Without soil colloids, extremely important nutrients would be taken from the soil by draining water and transported in channels. The number of basic cations in soil adjusts its base state. A large concentration of base cations, which are necessary for plant development, are present in the soil when the base position is excessive. When soil colloids contain a low supply of bases, the soil becomes less productive. Soil colloidal complexes, both inorganic (clay) and organic (humus), are highly fertile. Because of their tiny size, colloids migrate towards suspension in a solution and remain on the surface for a long period without sinking (Beamlaku & Habtemariam, 2021).

II. **Nutrient Dynamics**

A. **Nutrient Cycling**

Plant growth and nutrient cycling are two critical activities that significantly farmed soils must achieve. Nutrient cycling refers to the soil's ability to offer nourishment for food, fuel, and fiber crops growing throughout the landscape (Wall et al., 2018). Plant nutrients are chemical components taken primarily by roots of plant as inorganic compounds dissolved in the water. At the same time, plant nutrients are consumed by other forms of life and undergo a variety of biological modifications that influence when and how plants absorb them. On both "conventional" and organic farms, biological resources like as manure are important nutrient sources whereas inorganic minerals (chemical materials) such as phosphate, rock and lime are suitable fertility supplements for certified organic farming. For plant development, there are at least 16 chemical components required. Hydrogen, carbon and oxygen, which are abundant in water and air, compose the majority of plant dry matter in photosynthetic products, yet they are rarely considered "nutrient" components. Phosphorus, calcium, nitrogen, magnesium, sulfur, manganese, iron, zinc, boron, potassium, molybdenum, copper and chlorine are required by all plants and these are all derived from soil. Nickel, silicon and sodium are necessary nutrients for some species of plant and, while not required, have a good or helpful influence on the growth of others. Legumes require cobalt to fix nitrogen. Although they are not necessary for plants, extra elements like iodine and selenium can nonetheless be significant for plant nutrition as they are vital for people and other animals that eat plants (Bierman & Rosen, 2007).

Plants cannot absorb nutrients such as N, P, and C in their current forms; instead, they must be transformed into their normal forms via microorganisms and nutrient cycling. Biogeochemical cycles refer to recurrent cycles that occur swiftly in nature. Some of the key nutrient cycles include:

- **Carbon Cycle**

Both photosynthetic plants and photoautotrophic microorganisms are involved in carbon cycle. The carbon cycle is the process by which carbon is transferred to different realms, such as the hydrosphere, biosphere, and lithosphere. The carbon cycle includes several hydrogenotrophic and methanogenic microorganisms. Photosynthesis, respiration, and combustion are the three phases of the carbon cycle (Shaaban & Nunez-Delgado, 2024).

- **Nitrogen Cycle**

Since it is necessary for the creation of nucleic acid components, proteins, and chlorophyll in plants, nitrogen is a vital nutrient. Nitrogen is a key component of the environment. The 4 steps are in nitrogen cycle are:—The first is nitrogen fixation, which converts atmospheric nitrogen to NH_3, a significant component of proteins. Rhizobium and Anabaena play a vital role in this stage. The next step is nitrification, which converts NH_3 to the form of nitrites and nitrates. Nitrobacter and Nitrosomonas are the prime examples. The integration process, which involves the consumption of transformed products by plants, is the primary cause of fertility of soil. The next stage is denitrification, which includes the conversion of molecular form of nitrogen back into nitrates or nitrites form.

- **Sulphur Cycle**

The sulfur cycle, which describes the flow of sulfur in nature, is a crucial component in preserving fertility of soil. Soil organic sulfur contributes to fertility through the activity microorganisms of rhizosphere. Chlorobium, Thiobacillus, and Chromatium are Chemoautotrophic bacteria and these are quite significant in this regard. Aspergillus, Bacillus and Pseudomonas contribute to the sulphates conversion into inorganic acids through several stages. Furthermore, Desulfovibrio contributes to the decrease of sulfur in soils.

- **Phosphorous Cycle**

The phosphorus cycle refers to the recycling of phosphorus in the environment. Microbes like Bacillus and Pseudomonas contribute to the return of phosphorus to our environment. Actinomyces, Fungi and notably Streptomyces, Penicillium, and

Aspergillus contribute to soil fertility by increasing phosphorus availability (Javed et al., 2022).

- **Processes Involved in Nutrient Cycling (Decomposition, Mineralization)**

The majority of plant mineral nutrients in all soils reside as complex molecular aggregates that must be released before they can be used by plants. Because the rate of nutrient release frequently restricts an ecosystem's primary production, it is critical to understand the principal processes engaged through mineralization and decomposition in the maintenance of nutrient cycle phenomena (Read & Mitchell, 1983). Nutrient availability to plants in natural ecosystems depends on recycling organic matter through

- Decomposition and
- Nutrient mineralization.

Decomposition refers to the physical breakdown of the substance (typically quantified as mass loss). Decomposition is essentially a biological process arising from enzymatic activities of soil microorganisms and regulated in a number of ways by soil fauna activities (Prescott, n.d.). Decomposition is the process by which nutritional components in plant and animal tissues are released into the soil solution, where they may be used by plants anew. Because most nutritional components cycle from soil solution to plant and back, decomposition processes are the main aspects of nutrient cycles (Whitford & Wade, 2002). While mineralization is the release of inorganic nutrients accessible for plant absorption (Prescott, n.d.).

- **Interaction Between Soil, Plants, Water and the Atmosphere**

To better understand the soil–plant-water system and how its parts relate with one another, we must first grasp its impact on ecosystems. This information aids in the development of mitigation strategies, such as effective water and nutrient management, agricultural production optimization, biodiversity enhancement, and environmental impact assessment.

Soil ecosystems are dynamic systems and also are complex and that are impacted by a variety of elements, including interactions between soil, plants, and water, as well as the movement of nutrients, carbon, and water between soils and the atmosphere.

Water is essential for the plants. The quantity of water in the soil influences plant nutrient and water intake, as well as soil microbial activity, whereas plants impact soil water dynamics through root shape water uptake needs, density and transpiration. Soil structure and plant presence also impact the water balance owing to variables such as evaporation rates and interception values. All of these consequences can be exacerbated or mitigated by fluctuations in environmental circumstances (e.g., meteorological changes or soil management changes), necessitating a thorough examination and knowledge of mitigation strategies.

Climate change is unavoidable, causing the changes in precipitation volume and intensity, as well as expected changes in air temperature. It has an extensive range of effects on the soil–water-plant system, the most significant of which is drought. Arid conditions, scarcity of water, or irrigation-induced plant stress are all present challenges. Studies that contribute to a better knowledge of plant processes and various stress circumstances are vital and vital; hence, research related climate-change-issues will be the primary focus of countless submitted papers in this Special Issue. This special issue invited researchers to explore any aspect of the soil–plant-water system. A wide range of papers were submitted on topics including chemical parameters of soil that affect health of soil water and plant, inter-row soil management, overharvesting, modeling crop water use, plant water uptake, and issues related climate change-(e.g., arid conditions, deficiency of water, irrigation) (Horel, 2024).

B. **Nutrients Sources in Soil**

Plants attain mineral nutrients from the soil solution through root uptake. Soluble nutrients Sources in soil include:

- Decomposition of the plant remains, animal remnants, and soil microbes
- Minerals weathering in soil
- Fertilizer usage
- Organic amendments such as food processing byproducts, Manures, biosolids (sewage sludge), composts, kelp (seaweed).
- Legumes fixation Nitrogen
- Ground rock products Such as greensand, lime and rock phosphate.
- Inorganic byproducts of industries like coal ash or wood ash
- Acid rain and lightning discharges can cause atmospheric deposition of nitrogen and sulfur.
- Erosion and flooding can Deposition nutrient-rich sediment (Bierman & Rosen, 2007)
- **Role of Fertilizers and Their Impact on Soil Health**

In recent decades, intensive farming with herbicides, insecticides, and fertilizers has had a negative influence on the structural and microbiological ecosystem of soil. Chemical fertilizers that are not utilized by plants stay in an inaccessible state. When there is rain immediately after the use of chemical fertilizers to the land, they are carried away and collect in different water bodies, causing pollution and generating algal blooms. Organic fertilizers are employed, to mitigate the negative impacts of chemical fertilizers. Organic fertilizers increase quality of soil by improving its chemical and physical qualities. Bio fertilizers, farmyard manures, compost and green manures are basically organic manures. Organic fertilizers increase physical qualities by increasing porosity and aeration, decreasing sodality, lowering bulk density, boosting water penetration, and enhancing leaching of soil. Humus content promotes biological content, which helps consume macronutrients and micronutrients (Javed et al., 2022). Animal dung improves soil chemical characteristics and meets the necessary amount of nutritional necessity. Basically, manure composition is determined by its chemical qualities. Acidity has a detrimental influence on soil, thus manure acidity should be managed to evade difficulties. Beef dung is carbon rich source which increases soil nutritious content for crops. The drop in bulk density of soil is related to the long-term usage of manure, which promotes soil porosity and reduces soil compaction. Composting and manure application considerably enhance the amount of water-stable macroaggregates. Growing stable aggregates create a better habitat for plants by strengthening roots, improving water infiltration and retention. These positive impacts cause runoff, erosion, and less soil difficulties during rainy weather, while also providing strength in dry situations due to drought. They increase the soil's resistance to compaction and erosion (Eghball, 2002).

C. **Nutrient Deficiency and Toxicity**

Mineral nutrients can also be vanished from the soil system, making them unattainable for plant absorption. Losses of nutrient are not only costly and wasteful; they may also pollute lakes, rivers, and groundwater. Nutrient losses happen through:

- **Erosion**—Nutrient loss in or connected to soil particles displaced from fields by wind or water movement.
- **Runoff**—dissolved nutrients are lost when water moves over the soil surface.
- **Gaseous Losses to the Atmosphere**—primarily losses of different N forms through volatilization and denitrification.
- **Leaching**—Loss of dissolved nutrients in water that flows down the soil to groundwater or out of the field through drain lines.
- **Crop Removal**—Plant retention and removal of nutrients from the field in harvested products (Bierman & Rosen, 2007)
- **Symptoms of Nutrient Deficiencies in Plants**

Their lack can lead to stunted plant development, infertile soil, yellow foliage, and even plant death. As a result, sufficient nutrition is critical to the plant's growth. In the lack of any component, the plant will only develop to the point where the factor is no longer a limiting factor, which is why excessive nutrient input beyond this limit is ineffective (Javed et al., 2022).

- **Diagnostic Methods (Soil Testing, Plant Analysis)**

The first step in preserving fertility of soil is to determine existing nutrient levels. This is performed by **soil testing**, which is a significant soil management technique and a reliable foundation for fertilizer and lime recommendations. The purpose of soil testing is no longer just to determine if the soil holds enough plant nutrients for optimal development. It is also a method for detecting whether levels of nutrient are excessively high and likely to transfer off-site. Today, soil fertility is both a social and crop production problem. Soil test of each field after every 1–3 years, depending on crop rotation, farm history, and crop value. Testing after every 3–5 years is likely adequate for agronomic agricultural areas with a stable rotation, a lengthy history of soil testing, and no current manure or compost inputs (only commercial fertilizer since the last soil test). Choose a reputable, competent laboratory that provides advice according to the soil types and growth circumstances in your area. Laboratories that apply the protocols specified in Recommended Chemical Soil Test protocols for the North Central Region, NCR Publication 221, are preferable since they calibrate fertilizer recommendations based on University research trials in this region.

Plant analysis is an excellent nutrient management technique when used with a systematic soil testing program. The crop incorporates the impacts of fertility of soil and other growth parameters, and balanced plant nutrition is the eventual aim of crop nutrient management, therefore it makes sense to directly examine plants. Just like in soil testing, appropriate sampling is essential. Nutrient sufficiency levels are determined by examining certain plant components collected at a given development stage. The most commonly utilized plant tissues are newly developed, fully extended leaves or petioles (leaf stalks) derived from freshly grown leaves. One limitation of plant analysis is that if a nutrient shortfall is identified, it may be too late in the season to remedy the problem for the present crop. Plant analysis, on the other hand, is required for effective nutrient management of perennial fruit crops, may be cost-effective on a regular basis for high-value vegetables, and is a helpful validation tool for all crop fertility programs. Plant analysis is the only way to validate a crop nutrient shortage and is frequently a better diagnostic tool than a soil test for micro nutrients (Bierman & Rosen, 2007).

III. **Fertility Management**

A. **Soil Testing and Analysis**

Soils are quite varied in terms of the nutrients they contain, which complicates the interpretation of the Soil Test for determining fertilizer requirements. The overall amount of a nutrient contained in soil provides little information about the amount of that nutrient that is accessible. Numerous extraction solutions and processes have been employed to separate nutritional components from soil, but none remove the exact quantity that plant roots absorb. This implies that, in order to understand the data, the findings of each analytical method must be connected with the plant response acquired in field tests by applying that fertilizer ingredient (Yadav et al., 2023).

- **Importance of Regular Soil Testing**

Soil analyses in combination with the plant analysis are required because:

- When a nutrient deficits is identified by plant analysis, there are no standardized guidelines for the quantity of nutrient required to overcome the lack. And
- Nutrient shortage may not be due to insufficient soil nutrients (Bierman & Rosen, 2007).
- **Methods and Tools for Assessing Soil Fertility**

Only after assessing the soil's fertility can nutrient management tactics be improved to promote economically optimal plant performance while minimizing nutrient loss from the soil–plant system. Evaluation of soil fertility is thus a critical component of up-to-date soil fertility management practices. The primary purpose of fertility testing is to determine a soil's ability to give nutrients for plant development. Soil fertility may be assessed using a range of field and laboratory diagnostic approaches, as well as a growing number of complicated empirical and/or theoretical models that quantitatively relate soil fertility indicators to plant response.

There are a variety of diagnostic procedures used to assess soil fertility. They are:

- Symptoms of nutrient deficit on plants
- Biological tests
- Plant analysis
- Soil testing
- Advanced methods for evaluating soil fertility and applying fertilizer.

Modern Techniques to Soil Fertility Assessment and Fertilizer Recommendation

- Soil test crop response (STCR)
- Integrated system for diagnosis and recommendation (DRIS method) (Yadav et al., 2023).
- **Integrated Nutrient Management**

Management of nutrient may be defined as "effective use of all sources of nutrient" and the fundamental issues in maintaining fertility of soil are to:

- Decrease nutritional loss
- Maximize nutrient storage capacity
- Promote plant nutrients recycling
- Surplus nutrients in proper quantities.

Furthermore, cultural practices that promote the growth of healthy, robust root systems result in the effective absorption and use of existing nutrients. Establishing varied crop rotations, lowering tillage, maintaining and managing crop residue, cultivating cover crops, handling manure as a treasured nutrient source, composting and utilizing all existing wastes or byproducts, liming to maintain pH of soil, adding additional fertilizers, and conducting routine soil testing are just a few cultural practices that aid in achieving these goals. These valuable management strategies have various effects on nutrient cycling and fertility of soil, thus it is critical to integrate their use and investigate their effects on the entire soil-crop system, rather than just one component of that system. There are numerous good methods to farm, therefore different solutions or combinations of techniques are acceptable for different systems to achieve comparable goals (Bierman & Rosen, 2007).

- **Principles of Combining Organic and Inorganic Fertilizer**

A combination of manures and mineral fertilizers provides enough crop nutrition while also ensuring soil fertility in terms of nutrient availability. In agriculture, manures are organic materials used as fertilizer to improve crop nutrition. Manures increase soil health because they include a variety of minerals that plants require for optimal nourishment. Manure provides an abundance of main and vital plant nutrients such as, potassium, phosphorous and nitrogen as well as a variety of micronutrients. Manures can also enhance the chemical composition of the soil. By adding manures, the bulk density of the soil reduces, increasing organic matter and accessible water. Manures boost the productivity of a range of key crops, including cotton, sweet maize, and wheat. As the use of commercial fertilizers poses several health and environmental risks, the demand for organic sources that improve soil fertility and production grows. Integrated management of mineral fertilizers and organic wastes might be viewed as a key strategy for achieving sustainable crop production. These tactics not only improve the performance of chemical fertilizers while reducing their usage in crop production, but they also raise crop yield and aid to improve accessible main and minor soil nutrients. The quantity of phosphorus in soil may be raised by combining and applying organic and inorganic fertilizers. The residual impact of manure improves cereal nutrient use efficiency. Manure fertilizers give more nutrients to the soil and are more effective than commercial fertilizer applications. Organic fertilizers can enhance soil NPK content while decreasing soil pH, but mineral fertilizers merely preserve soil NPK status. Manure and commercial fertilizers can be used to increase output in a variety of crops. The coordinated use of manures and mineral fertilizers improves the long-term productivity of soil in wheat cropping systems (Khan et al., n.d.).

- **Benefits of Sustainable Nutrient Management Practices**

Sustainable soil nutrient enhancement solutions include the management and smart use of inorganic and organic nutrient sources in environmentally friendly agriculture systems. The major purpose of integrated nutrient management (INM) is to combine traditional and modern technologies for nutrient management to create environmentally sound and economically successful agricultural systems that use available organic and inorganic nutrient sources wisely and efficiently. Integrated nutrition management optimizes all components of the nutrient cycle. In order to maintain high crop yields in a diversity of cropping systems and ensure the system's long-term sustainability, it aims to optimize nutrient cycling by synchronizing the release of nutrients into the soil with crop demand and minimizing losses due to runoff, leaching, volatilization, and immobilization. It also seeks to improve soil health and productivity. INM uses major five sub-concepts namely:

- Plant nutrients, those found in organic manure, domestic wastes and crop residues
- Plant nutrients stored in soil
- Plant nutrients can also be acquired from outside the farm
- Loss of plant nutrients due to crop harvesting, volatilization of ammonia and nitrogen oxide gases, and leaching of nitrate and sulphate
- Monitoring plant nutrient outputs, such as crop nutrient uptake during harvest.

The simultaneous application of fertilizers and organic manures improves the C:N and C:P ratios, as well as the organic carbon status. As a result, the soil exchange capacity improves, resulting to greater adsorption and lower losses. The synergistic and complementary effects of applying fertilizers and manures together improve nutrient efficiency. INM promotes organic recycling in particular because it facilitates nutrient recycling, which improves fertility of soil and production at a cheap cost. The benefits of organic matter are widely established. The labile soil organic matter pool, which is critical for nutrient release throughout the growth season, may be controlled using a variety of soil management strategies. In surface soils, organic forms account for more than 95% of total N and S, as well as up to 75% of P (Ramakrishna Parama & Munawery, 2012).

C. **Timing and Method of Application**

- **Strategies for Effective Nutrient Application**

The public is very interested in and aware of the importance of improving nutrient usage efficiency, however this concept is readily misconstrued. Four indices of nutrient usage efficiency are examined, and an example of varied uses of the nomenclature demonstrates that the same data set may be used to compute efficiency of a fertilizer N of 21 or 100%. Fertilizer N recovery efficiency from researcher-managed studies for key grain crops range from 46 to 65%, whereas on-farm N recovery efficiencies range from 20 to 40%. Fertilizer best management methods, which supply nutrients at the appropriate rate, timing, and location, can improve fertilizer efficiency. The maximum nutrient usage effectiveness is always found at the lower regions of the yield response curve, where inputs of fertilizer are lowest, yet fertilizer efficiency in increasing crop yields and enhancing farmer profitability should not be negotiated for efficiency's sake. Optimal crop output must be balanced with optimal fertilizer usage efficiency (Roberts, 2008).

- **Techniques to Minimize Nutrient Losses (e.g., Split Applications)**

Increased animal manure and fertilizer usage in agricultural and forestry activities may induce nutrient accumulation in the soil, increasing the possibility of loss, resulting in contamination of groundwater and nonpoint source pollution. According to studies conducted in the tropics, agroforestry methods can minimize such nutrient (particularly nitrogen) losses due to increased nutrient absorption by tree and crop roots from various soil depths, as opposed to the more limited and shallow rooting depths of solitary crop stands. Riparian forest buffer methods have been shown to provide significant advantages in temperate environments. Other temperate agroforestry approaches are now being explored for their potential to minimize chemical runoff and leaching, and hence enhance environmental quality within the agricultural landscape (Zimmerman et al., 2016).

D. **Environmental Considerations**

- **Potential Environmental Impacts of Nutrient Management**

In many regions of the world, including both in developed and developing countries, deficiencies in nutrition of plant and fertilizer application, such as improper dosing, an improper N-P-K ratio, a lack of Mg, Ca, and micro nutrient supply, a limited diversity of fertilizers, issues with storage, timing, and distribution cause severe environmental problems, such as soil acidification, the potentially toxic elements mobilization, the "chemical time-bomb effect," phosphorus overload in surface waters, eutrophication, which causes adverse fluctuations in aquatic ecosystems, quick silting-up of canals and reservoirs, nitrate pollution of drinking water resources, issues with human health, and the addition of toxic compounds from fertilizers, etc. Most of these negative effects, however, are not unavoidable or uncontrolled results of fertilizer application. Precision nutrient management is critical to sustained agricultural development, profitable production of biomass, and effective environmental protection. Precision nutrient management aims to meet crop nutrient requirements while minimizing environmental impact. The primary characteristics for such a system are:

- Apply fertilizer based on crop requirements (nutrient uptake dynamics) and land-site factors (climate, moisture supply, relief, soil nutrient status, etc.).
- Education on nutrient losses (volatilization, leaching, surface runoff and organic/inorganic immobilization)
- Improved soil nutrient "efficiency" through manure and fertilizers (Varallyay, 1994).
- **Best Management Practices (BMPs) to Protect Soil and Water Quality**

Soil erosion destroys topsoil, the soil's richest layer in terms of organic matter and nutritional worth. Applying soil and water conservation methods that limit erosion and runoff, reduces nutrient losses while maintaining soil production. Tillage techniques and crop residue cover, as well as soil structure, topography and drainage, all contribute significantly to erosion of soil. Surface residue reduces erosion by preventing soil particles from detaching owing to wind or raindrop contact, as well as inhibiting water transport over the soil. Tillage procedures control the crop residue quantity on the surface of soil. Compact tillage, or no-till, optimizes residue coverage. Water travels quickly and is more erosive on steep slopes, therefore limiting tillage, retaining surface residue, cultivating sod crops, and planting on or in contour strips are all suggested conservation techniques. Cover crops and diverse rotations can also help to decrease erosion. Soils with stable aggregates are less erodible than those with inadequate structure, and organic content (together with the activity of live soil organisms and fine roots) aids in the binding of soil particles into aggregates. Tillage breaks down soil aggregates and improves aeration in soil, hastening organic matter decomposition. Soils which are well-drained and with rapid water infiltration are less prone to erosion because water flows quickly into and through them and does not accumulate to the point where it spreads over the surface. Poorly drained soils benefit from improved drainage, which reduces runoff, erosion, and compaction. Improving drainage also reduces nitrogen losses from denitrification, which may be significant in wet soils, by boosting aeration. Improving aeration in the plant-root zone encourages healthy root development. Improved drainage has the disadvantage of allowing nitrate–N and other nutrients to escape via tile outlets into surface waters. Flushing of leftover nitrogen after late winter/early spring rainfall is very crucial (Bierman & Rosen, 2007).

2 Conclusion

The study of soil chemical properties, nutrient dynamics, and management strategies is crucial for understanding and maintaining soil fertility and environmental health. Key points highlighted in this chapter include the intricate balance of soil organic matter content, cation exchange capacity (CEC) and soil pH, all of which significantly influence nutrient availability and soil health. Nutrient dynamics, particularly the cycles of potassium, phosphorus, and nitrogen, are essential for plant growth and productivity. Effective management strategies, such as the use of organic amendments, cover cropping, crop rotation, and the judicious application of fertilizers, are vital for optimizing nutrient use efficiency and reducing environmental impacts.

Sustainable practices play a vital role in preserving fertility of soil and ensuring long-term agricultural productivity. By minimizing soil degradation, enhancing biodiversity, and promoting the use of renewable resources, sustainable soil management practices contribute to the resilience of agricultural systems. Furthermore, the addition of modern technologies, like precision agriculture and soil testing, can enhance the efficiency and effectiveness of soil management practices, leading to better crop yields and reduced environmental footprints.

In conclusion, the adoption of sustainable soil management practices is essential for sustaining soil fertility, supporting agricultural productivity, and safeguarding environmental health. By understanding and managing soil chemical properties and nutrient dynamics, we can create resilient agricultural systems that contribute to food security and environmental sustainability for future generations.

References

Beamlaku, A., & Habtemariam, T. (2021). Soil colloids, types and their properties: A review. *Open Journal of Bioinformatics and Biostatistics*, 008–113. https://doi.org/10.17352/ojbb.000010

Bierman, P. M., & C. J. Rosen. (2007). Nutrient cycling & maintaining soil fertility. *Differences*, *C*, 1–14.

Eghball, B. (2002). Soil properties as influenced by phosphorus- and nitrogen-based manure and compost applications. *Agronomy Journal, 94*(1), 128–135. https://doi.org/10.2134/agronj2002.1280

Fageria, N. K. (2012). Role of soil organic matter in maintaining sustainability of cropping systems. *Communications in Soil Science and Plant Analysis, 43*(16), 2063–2113. https://doi.org/10.1080/00103624.2012.697234

Horel, Á. (2024). Soil–plant–water systems and interactions. *Plants, 13*(3). https://doi.org/10.3390/plants13030358

Javed, A., Ali, E., Afzal, K. B., Osman, A., & Riaz, S. (2022). Review article soil fertility: Factors affecting soil fertility, and biodiversity responsible for soil fertility. 12(1), 21–33. https://doi.org/10.26502/ijpaes.202129

Khan, S., Bibi, K., Karim, F., Javaid, R. A., Ahmed, S., Latif, K., Saleem, S., Ahmed, R., Yazdan, F., & Ali, Z. (n.d.). The integrated application of organic and inorganic fertilizers enhanced soil fertility under wheat-mungbean cropping system 19(03).

Osman, K. T. (2012). Chemical properties of soil. In Soils: Principles, properties and management (pp. 97–111). Dordrecht: Springer Netherlands.

Prescott, C. E. (n.d.). Decomposition and mineralization of nutrients from litter and humus. In *Nutrient acquisition by plants* (pp. 15–41). Springer. https://doi.org/10.1007/3-540-27675-0_2

Properties, C. (2013). *Chemical properties of soil 8*. https://doi.org/10.1007/978-94-007-5663-2

Ramakrishna Parama, V. R., & Munawery, A. (2012). Sustainable soil nutrient management. *Journal of the Indian Institute of Science, 92*(1), 1–16.

Read, D. J., & Mitchell, D. T. (1983). *Decomposition and mineralization processes in Mediterranean-type ecosystems and in heathlands of similar structure* (pp. 208–232). https://doi.org/10.1007/978-3-642-68935-2_12

Roberts, T. L. (2008). Improving nutrient use efficiency. *Turkish Journal of Agriculture and Forestry, 32*(3), 177–182. https://doi.org/10.3906/tar-0801-9

S, A. K. (2022). *AgriCos e-Newsletter. May.*

Shaaban, M. (2024). Acidic soils. In *Planet Earth: Scientific proposals to solve urgent issues* (pp. 293–306). Springer International Publishing. https://doi.org/10.1007/978-3-031-53208-5_13

Shaaban, M., & Nunez-Delgado, A. (2024). Soil adsorption potential: Harnessing Earth's living skin for mitigating climate change and greenhouse gas dynamics. *Environmental Research, 251*, Article 118738. https://doi.org/10.1016/j.envres.2024.118738

Sidi, N., Aris, A. Z., Talib, S. N., Johan, S., Yusoff, T. S. T. M., & Ismail, M. Z. (2015). Influential factors on the cation exchange capacity in sediment of Merambong Shoal, Johor. *Procedia Environmental Sciences, 30*, 186–189. https://doi.org/10.1016/j.proenv.2015.10.033

Sikora, L. J., & Moore-Kucera, J. (2014). Soil test methods from the Southeastern United States. *Southern Cooperative Series Bulletin, 419*(419). http://www.clemson.edu/agsrvlb/sera6/MethodsManualFinalSERA6.pdf

Smita Tale, K., & Ingole, S. (2015). A review on role of physico-chemical properties in soil quality. *Chemical Science Review and Letters, 4*(13), 57–66.

Tahat, M. M., Alananbeh, K. M., Othman, Y. A., & Leskovar, D. I. (2020). Soil health and sustainable agriculture. *Sustainability, 12*(12), 4859. https://doi.org/10.3390/su12124859

Varallyay, G. (1994). Precision nutrient management—Impact on the environment and needs for the future. *Communications in Soil Science and Plant Analysis, 25*(7–8), 909–930. https://doi.org/10.1080/00103629409369088

Wall, D. P., Sullivan, L. O., Creamer, R., & Mclaughlin, M. J. (2018). *Soil fertility and nutrient cycling*, 223–234.

Whitford, W., & Wade, E. L. (2002). Decomposition and nutrient cycling. In *Ecology of desert systems* (pp. 235–274). Elsevier. https://doi.org/10.1016/B978-012747261-4/50010-9

Yadav, K. K., Kanpur, T., Sachan, D. S., & Sachan, R. (2023). *Chapter—3. March.*

Zimmerman, M., Peterson, N. A., & Zimmerman, M. A. (2016). *Beyond the individual : Toward a nomological network of organizational empowerment beyond the individual: Toward a nomological network of organizational empowerment. 34*(October 2004).

Soil Physical Properties and Their Influence on Plant Growth

Muhammad Umair, Muhammad Itazaz Hassan, Hassan Zia, Fazal Maqsood, Muhammad Hamza Khalil, Iqra Ghafoor, and Tooba Mumtaz

Abstract Soil physical properties are essential for understanding the performance of soil processes in relation to plant growth and development. Certain health indicators have direct impacts on crops and plants, while others exert an indirect influence by affecting various chemical and biological soil quality parameters. Presently, the concepts of soil health and quality are receiving increased attention globally. Although often considered synonymous, these terms refer to distinct aspects; inherent soil quality pertains to the natural composition and properties of the soil, which are influenced by factors and processes of soil formation. This chapter discusses the soil quality parameters that will inform the reader about aspects such as soil texture, structure, color, nutrient amount, and other characteristics, which collectively influence the soil's capacity for water and nutrient retention, cycling, transport, as well as crop selection and yield in specific regions. Additionally, soil carbon sequestration and nitrate leaching are examined. Key factors such as soil aeration and temperature play a critical role in the functioning of the microbial community, which is determined by soil redox potential and biochemical oxygen demand. Furthermore, we have highlighted not only soil physical quality parameters but also chemical and biological indicators of soil quality. Soil health is viewed as an integrated approach that encompasses all these parameters, aiming to sustain crop productivity.

M. Umair (✉)
College of Natural Resources and Environment, Northwest A&F University, Yangling 712100, Shaanxi, China
e-mail: muhammadumair@nwafu.edu.cn

M. I. Hassan · F. Maqsood · M. H. Khalil · I. Ghafoor · T. Mumtaz
Institute of Soil and Environmental Sciences, University of Agriculture, Faisalabad 38040, Pakistan

H. Zia
Department of Plant Pathology, University of Agriculture, Faisalabad 38040, Pakistan

M. Shaaban (ed.), *Soils and Sustainable Agriculture*, Frontier Studies in Soil Science,
https://doi.org/10.1007/978-3-031-91114-9_5

1 Introduction

Soil physical properties are crucial for land use classifications, whether soil is suitable for various application including agricultural purpose, environmental functioning, and engineering contexts. Soil physical properties significantly regulate soil's ability to support plant growth and development through the influence of water and nutrients movement, retention, and bioavailability in agroecosystems (Hartmann & Six, 2023). Soil texture and structure significantly affect root penetration and the circulation of heat and air within the soil profile. The physical properties of soil are interconnected with its chemical and biological attributes (Bauer & Black, 2020). The subsequent sections will elaborate significant soil physical properties that are pertinent to soil functions and support as a growth medium for plants.

2 Soil Texture

Soil texture refers to the composition based on the relative proportions of sand, silt, and clay particles of soil. Soil particles have distinct characteristics that influence soil inherent properties and plant growth (Barman & Choudhury, 2020). Soil particles size is important and affects the ecosystem water limitation, water-holding capacity and work-ability of soil (Chardon & Schoumans, 2007). Various grain size limits are available, with the grain size limits for soil classification from United States Department of Agriculture (USDA) and International Union of Soil Science (IUSS) are commonly used (Fig. 1).

1. **Sand particles** are the largest in size among soil components, range from 50–2000 μm (0.05 to 2.0 mm) based on the USDA grain size limits for soil classification, easily observable with the naked eye. Sandy soils are recognized by their gritty texture and loose structure. The considerable gaps between sand particles promote drainage and facilitate rapid water infiltration, whereas high drainage results in low water-holding capacity. Sandy soils have less nutrient retention availability and essential elements are prone to leaching out of root zone through irrigation and rainfall. Despite these challenges, certain plant species, like cacti and succulents, thrive in sandy soils (Koirala et al., 2016).
2. **Silt particles**, diameter ranging from 2–50 μm (0.002 to 0.05 mm) classified between sand and clay based on size. Silty soils (high proportion of silt) are characterized by smooth, powdery texture, and often feel like flour or velvet. Silty soils as compared to sandy soil retained more moisture due to their smaller particle size and high surface area (Bruun et al., 2010) and more resilient to drought conditions. These unique properties makes silty soils more fertile and conducive to supporting a diverse range of crops and vegetation throughout the growing season (Singh et al., 2017).

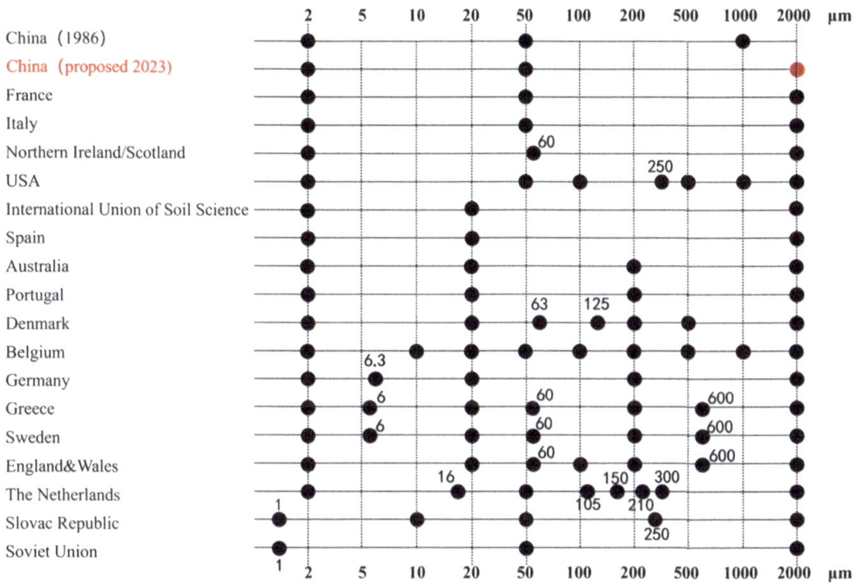

Fig. 1 Selected grain size limits for soil classification in the world (not to scale, He et al., 2023)

3. **Clay particles**, diameters less than 2 μm (0.002 mm) contribute to the distinctive properties of soils. Wet clay soil has a sticky characteristics and cohesive texture, and as dry clay soils harden into compacted structures and often make the challenging conditions for plant growth and development. Whereas, clay soils are characterized for their exceptional water-holding capacity due to smaller particle size and extensive surface area (Brooks, 2009). Clayey soil retain moisture for prolonged periods and provide a consistent water supply to plants, and sometimes high-water retention capability is the risk of water-logging and poor drainage, particularly in poorly drained clay soils. The excessive water accumulation can hinder root growth and limit oxygen uptake by plant roots, and reduces plant health and productivity. Clay soils offer an advantage in water retention, but proper drainage management and optimize conditions for successful plant growth is essential to mitigate potential challenges (Gandhi, 2012). Several methods are used to determine soil texture, each with its own advantages and limitations. Some common methods have been mentioned here as follows.

1. **Feel Method**:

Highly technical expertise and field experience is needed, and feel method commonly practiced in the field to determine soil texture and easily enabling to distinguish between different soil types. This estimation involves moistening a small amount of soil with water and kneading it to the consistency of putty to assess its plasticity (David, 1999). The formation of casts or ribbons during this process is indicative of the clay content and helps to categorize soils into loams, clay loams, and clays.

If a soil is classified as a loam but feels gritty or sandy, it is also referred to as a sandy loam. Loams with a smooth texture typically contain high levels of silt and are known as silt loams. Samples that exhibit intermediate characteristics are simply categorized as loams. The same classification approach applies to clay loams and clays. Sands, being loose and lacking cohesion, do not form ribbons. Feel method allows for a quick and practical assessment of soil texture in the field, and help in land management decisions and problem-solving related to soil quality and suitability for various uses (Kettler et al., 2001).

2. **Hydrometer Method**:

Hydrometer method for soil textural analysis involves dispersing soil particles in water with dispersing agent using a hydrometer and measure the settling rate of the soil particles. The sedimentation rate is used to determine the percentage of sand, silt, and clay in the soil samples. It is a commonly practiced and precise method but requires specialized equipment and can be time-consuming method (Galdino et al., 2016).

3. **Pipette Method**:

The pipette method also involves dispersing soil particles in water, but it utilizes a pipette to measure the particle size distribution. After dispersing the soil in water, samples are withdrawn at specific time intervals, and the particle size distribution is determined based on the sedimentation rate principle. This method also provides accurate results but not commonly used in laboratories (Burrough, 1993).

4. **Particle Size Analysis**:

Particle size analysis involves sieving soil samples through a series of sieves with different mesh sizes. The soil particles retained on each sieve are weighed, and the percentage of particles in each size fraction is determined. This method provides a quantitative assessment of soil texture and can be performed using relatively simple equipment (Parslow & Jennings, 1986).

5. **Laser Diffraction**:

Uses light beams to estimate the particle size distribution of a sample. This method is faster and cheaper than other methods (Scott et al., 2024).

2.1 Soil Textural Classes

Soil textural classes categorize soils based on the proportion of sand, silt, and clay. These classes provide valuable information about soil properties and behavior, and soils are classified into various 12 textural classes (McBratney, 1994) determined by the relative percentages of soil separates shown in Fig. 2.

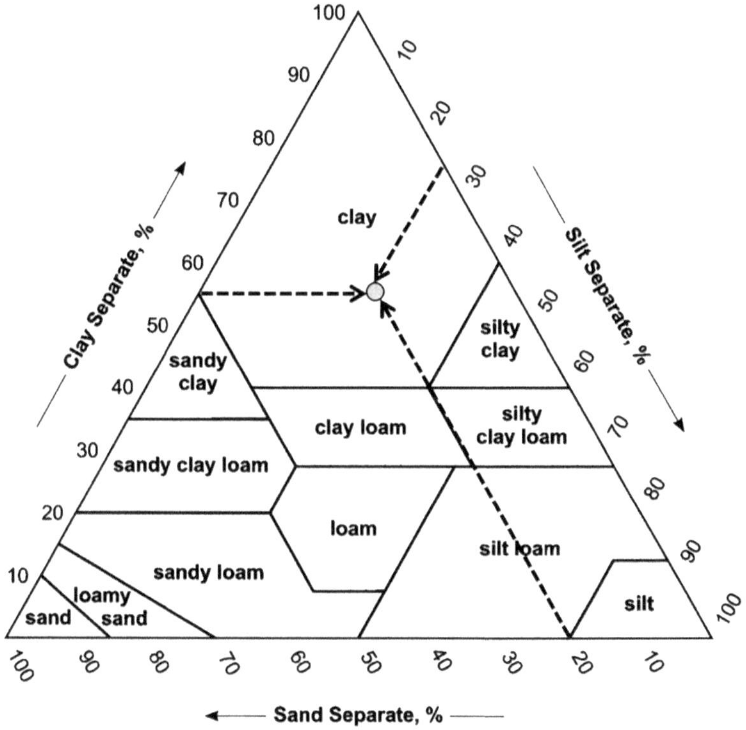

Fig. 2 Textural triangle diagram according to ISSS system of classification of soil particles

3 Soil Structure

Soil structure is the arrangement primary soil particles into secondary units known as aggregates or peds. It is more effective to study soil structure in undisturbed field conditions rather than in controlled laboratory environments. Soil structure categorized into three primary groups: single-grained, massive, and aggregated. The single-grained structure describes a geometric arrangement of soil particles forms a porous configuration, and characterized by minimal or no cementation or flocculation among its components (Yudina & Kuzyakov, 2023). Coarse-textured sands often exhibit this arrangement and are frequently referred to as structure-less soil. In contrast, a massive structure represents a complete consolidation of soil, and lacking any natural lines of weakness which is typically found in fine-textured soils. Aggregated soils are positioned between these and soil separates adhere together to form larger units or aggregates. Natural soil aggregates can be further classified into three categories based on their shape, size, and degree of distinctness and durability and known as types, classes, and grades of soil structure, respectively. Understanding these structural characteristics is essential for managing soil health and optimizing its functionality in agricultural and ecological contexts (Tuo et al., 2017).

3.1 Types of Soil Structure

The four primary types of soil structure classified based on the shape of the aggregates and their spatial arrangement within the soil profile, as follows, platy soil structure, block-like, prism-like, and spheroidal (Yunusova & Osorio, 2022).

Platy structure is characterized by thin, flat aggregates resembling plates or leaflets. These plate-like aggregates often result from various soil formation processes, and inherited from the parent material, or can develop due to the compaction of fine-textured soils which is oftenly caused by heavy machinery. Platy structures have been observed in both surface and subsurface horizons.

Blocky aggregates are irregularly shaped and typically have six faces, with dimensions that are roughly equal. They can be further divided into two sub-types: **angular blocky**, where the corners of the blocks are sharp and not rounded, and **sub-angular blocky**, where the aggregates possess rounded corners. Blocky structures are commonly found in the B horizon of the soil profile.

Prism-like structures are distinguished by vertically oriented aggregates or pillars that are longer in height compared to their width. This category includes two sub-types: **prismatic**, where the prisms have angular edges and flat, clearly defined tops, and **columnar**, where the tops and sides of the prisms are rounded and resembling to columns. Both sub-types are typically associated with swelling clay minerals and are prevalent in the subsurface horizons of soils in arid and semi-arid regions.

Lastly, soils that exhibit **spheroidal aggregates** are further categorized into two sub-types: **granular**, which have relatively less porous peds, and crumb, and consist of more porous granules. **Spheroidal** structures are commonly found in surface soils, particularly in the A horizon, especially those enriched with organic matter. Understanding these structural types is essential for assessing soil functionality and managing its health effectively.

3.2 Genesis of Soil Structure

Genesis of soil structure refers to the causes and methods of formation of aggregates. The following two processes are involved in genesis of structure.

3.2.1 Flocculation

Flocculation is an electrokinetic phenomenon that involves the interaction of positive and negative charges (Deruyttere, 2024). Clay particles exhibit a high zeta potential in a soil–water suspension and causing the repulsion effect with each other. However, through a flocculating agent, the zeta potential decreases and the particles come closer together and attract each other. This process leads to the formation of stable floccules which are typically of silt size. Flocculation also occur due to dehydration, high concentrations of soluble salts, and the presence of divalent or trivalent cations. These mechanisms play a crucial role in soil structure and its overall behavior, and affecting both agricultural practices and environmental management (Wu et al., 2024a, 2024b). Understanding these interactions helps ot develop an effective soil management strategies.

3.2.2 Cementation of Floccules

The term refers to the process of consolidating the clay floccules through the action of cementing materials which prevents their dispersion after flocculation (Wang et al., 2024a, 2024b, 2024c). Floccules are typically no larger than silt-sized particles and their presence may not be beneficial for plant growth, to enhance their suitability for supporting vegetation, these floccules must undergo further aggregation, that facilitated by inorganic cementing agents such as calcium carbonate ($CaCO_3$), and oxides of iron and aluminum, as well as organic materials like organic matter. Thus, the flocculation of clay particles is an essential precursor to the soil aggregation (Gao et al., 2024). It is important to note that aggregation essentially involves both flocculation and the addition of a cementing agent, which plays a critical role in enhancing soil structure and fertility (Zhang et al., 2024). This relationship is vital for effective soil management and help for improving agricultural productivity.

3.3 Visualization of Soil Structure

Computed Tomography (CT) offers a non-destructive, high-resolution approach to visualize and analyze soil structure in 3D, making it invaluable for soil health and quality research (Wang et al., 2024a, 2024b, 2024c). It enables detailed examination of pore networks, aggregate stability, root-soil interactions, and compaction, providing insights into water infiltration, nutrient cycling, and biota habitats. By capturing the intricate spatial arrangement of soil components, CT helps researchers evaluate the impact of management practices and environmental changes on soil functionality, aiding in sustainable land use and agricultural productivity (Al-Shammary et al., 2024).

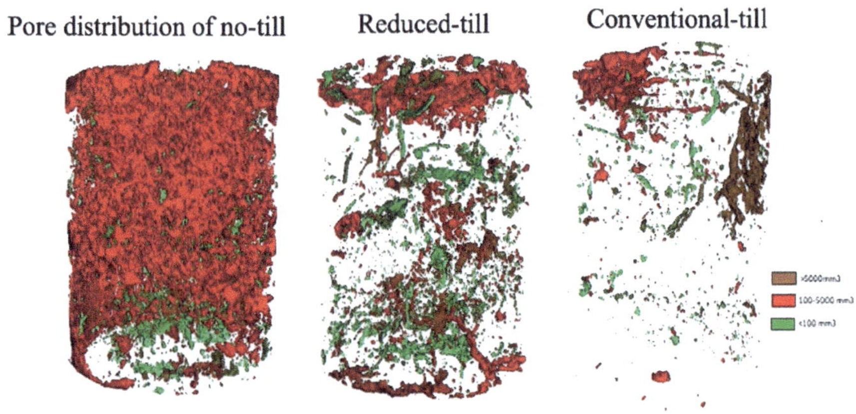

https://doi.org/10.1016/j.catena.2023.107801

3.4 *Soil Organic Matter*

The relationship between soil structure and organic matter (OM) is crucial for understanding soil health and productivity. Organic matter significantly influences soil aggregation, porosity, and overall structural integrity, which in turn affects water retention and microbial activity (Shaaban & Nunez-Delgado, 2024) Organic matter acts as a binding agent, enhancing soil aggregation and stability, particularly in clay-rich soils (Ghorbani & Amirahmadi, 2024). Humic substances derived from organic amendments improve soil aggregate stability, leading to better crop yields compared to synthetic fertilizers (Zhu et al., 2024) Dissolved organic matter (DOM) induce the formation of water-stable aggregates across various soil textures, highlighting its role in structural development (Bucka et al., 2024) (Soil structure influences microbial access to organic matter, affecting decomposition rates and nutrient cycling (Jarvis et al., 2024). Degradation of soil structure lead to a decline in organic matter content, impact soil functions and ecosystem services.

3.5 *Soil Aggregates*

The relationship between soil structure and soil aggregates is fundamental to understanding soil health and function (Amelung et al., 2024). Soil aggregates, formed through the binding of particles by organic and inorganic materials, play a crucial role in determining soil structure, which in turn influences various soil functions (Amelung et al., 2024). Soil structure is defined by the arrangement of particles into aggregates, which can vary in size and stability. Microaggregates (< 250 μm) are the most stable units, providing habitats for microorganisms and facilitating nutrient cycling. The formation of aggregates is influenced by soil management practices, with native soils typically exhibiting more pronounced porosity compared to arable soils (David Raj et al., 2024) (The stability of aggregates is linked to organic matter content, which helps maintain structural integrity under stress). Mechanistic models suggest that electrostatic interactions among soil particles significantly affect aggregate formation and stability.

3.6 Soil Structure and Its Relationship with Soil Tillage

A well-developed soil structure is a crucial aspect of soil tilth, which refers to the physical state of soil in relation to its capacity to support plant growth. Tillage practices have both beneficial and detrimental impacts on soil aggregation. Tillage practiced at the optimal moisture level and its immediate effects are generally positive. Tillage implements effectively break up clods, loosen the soil, enhance water infiltration, and decrease surface runoff. Additionally, when the surface soil depletes in structure and nutrients, tillage can invert the soil layers, bringing aggregated and relatively more fertile subsoil to the surface, thereby improving overall soil tilth (Vezzani et al., 2018). However, over extended periods, the cumulative effects of tillage may adversely affect soil structure, and plowing under excessively wet or dry conditions harm soil tilth. The frequent use of heavy machinery disrupt soil aggregates and lead to surface compaction. Moreover, ongoing cultivation and various tillage operations reduce the stability of soil aggregates, primarily due to the increased decomposition of organic matter as it becomes more exposed to environmental elements (Or et al., 2021). Continuous plowing at the same depth with the same implement may also lead to the formation of a plow pan, which can further restrict root growth and water movement in the soil. Understanding these dynamics is essential for sustainable soil management practices that promote healthy plant development.

3.7 Importance of Soil Structure

Soil structure is vital for plant growth as it affects root penetration, water retention, and nutrient availability. A well-structured soil allows for efficient root growth and access to nutrients and water (Or et al., 2021). It serves as a critical factor in the analysis of soil profiles and the classification of different soil types. It significantly affects the rate of water infiltrates into the soil. For instance, granular and single-grain soils exhibit rapid infiltration, while blocky and prismatic soils demonstrate moderate infiltration rates, but, platy and massive soils are characterized by slow infiltration. The erodibility of soil tends to decrease as the stability of its structure increases. Evaluating the size distribution of soil peds, the proportion of aggregated soil, and the strength of soil aggregates is essential for understanding both the porosity and pore-size distribution within the soil. The degree and nature of aggregation can alter how soil texture influences crop growth by modifying the distribution of pore sizes. Sustainable soil structure understandings are vital for implementing effective soil management strategies that enhance agricultural productivity and sustainability (Vezzani et al., 2018).

4 Soil Density and Porosity

4.1 Particle Density

Particle density is known as per unit volume of soil particle mass in its dry state precluding the air spaces within the soil. It can also be expressed as the mass of a known volume of tightly packed soil (Singer, 1991). In simple terms, density tells how much an object weighs at a constant volume. It is denoted by symbol ρ, and its standard unit is g cm^{-3} /kg cm^{-3}. Particle density of normal soils ranges from 2.60 to 2.75 g cm^{-3} yet the maximum range can reach up to 3.0 g cm^{-3} for very dense soil as low as 0.9 g cm^{-3} for the organic matter. It is a static property and does not change with texture, structure, voids or pore spaces, but the presence of heavy minerals or organic matter can cause a change in the particle density value (Jia & Zheng, 2024). Accurate particle density measurements are vital for calculating total porosity and understanding water retention in soils, which is crucial for agricultural practices (Bartley et al., 2018).

4.2 Bulk Density

Soil bulk density is defined as the mass volume ratio of soil particles including the pore spaces. This property of soil is measured in grams per cubic meter in soil studies. In this context, the volume weight is equivalent, and the bulk specific gravity is nearly the same numerically as the bulk density (Blake, 1965). The main significance of bulk density with respect to plant growth is that it acts as an indicator of soil compaction providing the valuable insight for preparation of a fertile and cultivable land for agricultural use. It has deep impact on several important phenomena's such as infiltration, aeration, porosity, nutrient availability, water holding capacity and activity of microorganisms all of which influence key soil processes and productivity (Zygmunt et al., 2014).

Bulk density of given soil sample can be determined by using core method. This method is efficacious where the volume of soil coarse fragment occupies less than 25% of the space (Al-Shammary et al., 2018). Another method used for determining bulk density of soil is clod method (Blake, 1965). In this method, soil clod is coated with wax and weighed two times first in air and then in water immersion applying Archimedes' principle (Keenan et al., 1997).

4.2.1 Factors Affecting Bulk Density

Soil bulk density is affected by several factors including soil structure, texture, organic matter content, pore space, soil mineral density, land use management practices and climatic factors. Bulk density increases as the pore space decreases. Increase in pore space, organic matter, mineral density, fineness in texture and aggregate formation in soil will decrease the soil bulk density while the soil profile depth, compaction and climatic factors like temperature will increase the soil bulk density. Management practices like long term use of tillage and intensive cultivation will also increase bulk density (Giap & Ahmad, 2024).

4.2.2 Relationship of Pore-Space and Bulk Density

Pore space is portion of soil volume which is occupied by the air spaces and water only. Equation used for understanding the relation between soil bulk density and pore space is given as:

$$\text{Total porosity} = 1 - (\text{bulk density}/\text{particle density})$$

Here, total porosity is indicating the voids per unit bulk volume of soil.

Bulk density and pore space has inverse relation decrease in denominator will increase the nominator, hence any factor that affect the pore space in a parallel side will affect the bulk density and vice versa (Giap & Ahmad, 2024).

4.2.3 Soil Bulk Density and Root Growth

Soil bulk density (BD) plays a crucial role in root growth by influencing root morphology, biomass allocation, and penetration ability across various plant species. Soil BD when increases, often leads to reduced root length, volume, and biomass. For instance, vetiver grass showed decreased root dry weight and root surface area as BD increased (Wan, 2015). Different plant species exhibit different abilities to penetrate compacted soils. Some grass like Vetiver demonstrated superior penetration capabilities due to its root traits, outperforming other species in high BD conditions (He et al., 2022). More the fineness in soil texture greater will be the soil compaction and aggregate formation after water application and more it will hinder the root growth. For instance, if the penetration resistance of root against the soil exceeds from 2 MPa then it is considered as the limiting condition for root growth (Lardy et al., 2022).

4.3 Soil Aeration and Pore-Space

Soil aeration holds a critical aspect in plant, root growth. Soil air is the space primarily held by the gases within soil voids. These are the spaces through which the exchange of gases from the atmosphere to soil voids take place. Influence of pore spaces, soil texture, moisture content, tillage practices are the matter of great concern. Large pore spaces in soil like sandy soil provide better wide air channels and air movement through the soil on the other hand fine textured soil with less micro pores will pose a challenge for air to pass through it. The water retention capacity of soil will also affect the air movement it will make it difficult to not even the air to pass but also the plant to respire within the soil body and will cause the death of plant. Routine vise tillage practice can help to overcome this problem in compact forming soil up to some extent (Sun et al., 2023).

4.4 Pore-Size Distribution

Pore size distribution is relatively volume-based distribution of pore spaces among the soil bulk volume. It is one of the most affecting physical properties of the soil influencing water retention, root growth and soil structure. A well distributed proportion of pore space provides the better root proliferation, water-air movement and an indicator of soil fertility status (Nimmo, 2004).

Soil texture greatly affects the pore size distribution. Depending on their diameter, pores are grouped into micro (< 0.03 mm), meso (< 0.03 to 0.08 mm) and macro pores (> 0.08 mm).

Micropores are present as intra aggregate spaces in fine textured soil having very less space for water and air movement. While macro pores are present as inter aggregate spaces in coarse textured soils providing fast air-water movement and accommodating plants roots and tiny animals. Due to larger spaces than rest of micro and mesopores water retention in macro pores is very less and drains water very quickly.

Mesopores lie in between of micro and macro pores also considered as larger micropores sheltering certain bacterial community and plant root hairs. Water held in this portion of soil and rise up to the stem and leaves through capillary action (Nimmo, 2004).

4.5 Effect of Aeration on Plant Growth

Two life sustaining gases in soil air are CO_2 and O_2 since these are essential for photosynthesis and root respiration, respectively. Carbon dioxide is a primary greenhouse gas because it traps reflecting back long-wave radiation and prevents it from escaping into the outer space. Human related CO_2 emissions are responsible for its increase in atmosphere occurred since industrial revolution. This high concentration of CO_2 in atmosphere may be partially responsible in global warming. Plants roots and microbes need O_2 for respiration which is needed for life and growth. When ODR is too low and the redox values drops, plant roots cannot respire and their growth slows down or stops. Except rice, cultivated crop plants grow best in well aerated or oxidized soils. Rice plants can move O_2 internally through large diameter pores from their tops in atmosphere to their roots. If free O_2 concentration is too low, some elements or ions become the electron acceptor. In poorly aerated soils, many organic acids are formed which may be toxic to higher plants and to decomposing organism. The nitrates are denitrified and N_2 and N_2O gases formed are escaped to atmosphere, thus expensive N fertilizer input is wasted. Sulfate is reduced to sulfide and CO_2 is converted to methane. Under anaerobic (lack of free O_2) condition, the rate of organic matter decomposition is much slower than under aerobic condition (Sun et al., 2023). In the presence of free O_2 (aerobic condition), glycolysis plus respiration releases about 19 times more energy (stored in ATP bonds, than anaerobic breakdown.

5 Soil Mechanics

Soil mechanics is a fundamental aspect of geotechnical engineering, focusing on the behavior of soil under various conditions, such as when it is subjected to stress, loading, and changes in water content. The study of soil properties like texture, density, permeability, and strength is essential for determining soil's ability to support infrastructure (Budhu, 2020). Soil is composed of a complex mixture of minerals, organic matter, water, and air, and its mechanical behavior is largely governed by interactions between these components. Key mechanical properties, such as shear strength and compressibility, are critical for predicting soil responses under different loads, ensuring the stability of foundations, embankments, and other structures (Lambe & Whitman, 2019). By thoroughly understanding soil's mechanical behavior, engineers can design safe, long-lasting structures and implement effective agricultural land management practices.

5.1 Soil Compaction

Soil compaction is a critical process in soil mechanics that refers to the densification of soil by reducing the volume of air-filled voids through the application of mechanical forces. This process increases the soil's dry density and improves its structural stability and load-bearing capacity, making it essential in geotechnical engineering applications such as foundation construction, roadways, and earth embankments (Budhu, 2020). The compaction process alters several physical properties of the soil, including its strength, porosity, permeability, and compressibility. A properly compacted soil has higher shear strength, lower compressibility, and reduced permeability, which are desirable traits for supporting infrastructure. Coarse-grained soils like sands and gravels generally compact more effectively at low moisture levels, while fine-grained soils such as clays require optimal moisture content (the Proctor moisture content) for efficient compaction (Lambe & Whitman, 2019).

5.1.1 Influence of Soil Compaction in Agriculture

In agriculture, soil compaction can have both beneficial and detrimental effects. While moderate compaction can enhance the soil's water retention and support plant growth, excessive compaction, especially from heavy machinery, leads to reduced porosity and permeability. This impairs water infiltration, aeration, and root penetration, negatively impacting soil health and crop productivity (Murthy, 2021). Excessive compaction can also induce a condition known as "hardpan," a dense layer that restricts root growth and water movement.

Techniques for managing soil compaction in both construction and agriculture include controlling moisture levels during compaction, limiting machinery traffic, and employing methods like deep tillage to alleviate excessive compaction without disrupting the soil structure excessively. Understanding the principles of soil compaction is essential for optimizing land management practices and ensuring the long-term stability of both engineered and agricultural soils.

5.2 Soil Plasticity

Soil plasticity is a property of fine-grained soils, such as clays and silts, that allows them to deform without breaking when subjected to external forces. This behavior is primarily attributed to the moisture content and the type of clay minerals present in the soil. When water is added to clay-rich soils, it forms a lubricating film around the soil particles, enabling them to slide past each other and making the soil more plastic (Craig, 2019). Soil plasticity is an important factor in geotechnical engineering, as it affects the soil's ability to undergo deformation under loads without fracturing. This characteristic is critical in construction, particularly for foundations, embankments, and slope stability.

The plasticity of soil is measured through the **Atterberg limits**, which classify the soil's consistency as it transitions between different states depending on moisture content. These limits include the **liquid limit (LL)**, **plastic limit (PL)**, and **shrinkage limit (SL)**, which define the moisture levels at which the soil behaves as a liquid, plastic, or solid, respectively (Terzaghi et al., 2021). The **plasticity index (PI)**, calculated as $PI = LL - PL$, quantifies the range of moisture content over which the soil exhibits plastic behavior. Soils with a high plasticity index, such as expansive clays, can deform significantly under changing moisture conditions, while low-plasticity soils, like silts, show minimal plastic behavior.

In agriculture, soil plasticity plays a role in tillage operations and water management. Wet, highly plastic soils tend to become sticky and difficult to work with, which can lead to soil compaction and reduced aeration, negatively impacting plant root growth (McCarthy, 2017). However, these soils also have good water retention properties, making them beneficial during dry periods. Proper management, including adjusting moisture levels and avoiding excessive compaction, is essential to ensure soil health and productivity.

5.3 *Soil Crusting*

Soil crusting refers to the formation of a thin, dense layer on the soil surface, which can significantly affect water infiltration, seedling emergence, and gas exchange between the soil and atmosphere. Soil crusting occurs primarily in fine-textured soils, particularly those rich in silt and clay, and is typically induced by rainfall, irrigation, or other forms of surface wetting followed by drying (Bresson & Valentin, 2021). Soil crusts are specific modifications in the top soil caused by natural events such as raindrop impact and the following drying process. They consist in the formation of hard thin layers at the soil surface and are widespread especially in the soils of arid and semiarid regions. Their thickness usually ranges from less than 1 mm to 5 cm. This process alters the physical structure of the soil, creating a hard layer that reduces soil permeability and inhibits plant growth.

5.3.1 Mechanisms of Soil Crust Formation

Soil crust formation is generally driven by two primary mechanisms: physical dispersion of soil particles and chemical flocculation.

1. **Physical Dispersion**: When raindrops or irrigation water impact the soil surface, the energy of the droplets disrupts the soil structure, breaking down soil aggregates into finer particles such as clay, silt, and fine sand (Le Bissonnais & Arshad, 2020). These dispersed particles settle on the soil surface, filling the pores and forming a dense, compacted layer. As the soil dries, this layer hardens into a crust. The crust can range from a few millimeters to a centimeter in thickness, depending on soil texture, organic matter content, and the intensity of rainfall or irrigation (Gomez et al., 2002).
2. **Chemical Flocculation**: In soils with high sodium content, such as sodic soils, chemical processes can also contribute to crust formation. Sodium ions cause clay particles to disperse in water, reducing the structural stability of the soil (Bronick & Lal, 2022). When these dispersed particles are exposed to drying conditions, they form a dense, impenetrable crust. Soils with low organic matter and low calcium levels are particularly susceptible to this process.

Several factors influence the formation and severity of soil crusts. Fine-textured soils, such as silty loam and clay, are more prone to crusting due to their high content of easily dispersible particles (Dexter, 2020). Organic matter improves soil structure and reduces crusting by promoting aggregate stability and increasing the soil's resistance to physical dispersion. Soils with low organic matter content are more likely to form crusts (Mualem & Assouline, 2021). High-intensity rainfall increases the kinetic energy of water droplets, leading to more severe disruption of soil aggregates. Prolonged wetting periods allow more time for particle dispersion and crust formation (Amezketa, 2021). Practices such as excessive tillage and removal of surface residues can increase crusting by destroying soil structure and reducing organic matter content. Conversely, practices that promote soil cover, such as mulching or reduced tillage, can reduce the risk of crust formation (Wang et al., 2023).

5.3.2 Impacts of Soil Crusting

Soil crusting negatively impacts agricultural productivity by reducing water infiltration, limiting seedling emergence, and increasing surface runoff and erosion. The hard crust creates a barrier that prevents water from penetrating the soil, leading to poor water availability for crops. Additionally, crusts inhibit the emergence of seedlings by creating a physical barrier that young roots struggle to penetrate. This effect is particularly detrimental to crops with small seeds or weak seedlings (Mualem & Assouline, 2021). To mitigate soil crusting, management practices such as the application of organic amendments, mulching, and maintaining a vegetative cover can help improve soil structure and prevent aggregate breakdown.

5.4 Soil Shear Strength

Soil shear strength is a fundamental property in soil mechanics that defines a soil's resistance to shearing forces. It plays a critical role in determining the soil's ability to support structures, such as buildings, embankments, and retaining walls, as well as influencing natural processes like slope stability and erosion. Shear strength is primarily governed by the internal friction and cohesion of the soil, which are influenced by factors like soil texture, water content, and confining pressure (Das, 2020). Understanding and calculating soil shear strength is essential for designing safe and effective geotechnical structures.

Shear strength of soils is one of the factors determining the limitations for the growth of plants because it measures the ability of soil to resist any shear stress applied to it. Roots can penetrate the soil better and more effectively utilize water and nutrients when shear strength is low. On the contrary, high shear strength, which develops in compacted or clay soils, can result in poor root growth, and a well-developed root system may not be formed (Dexter, 2020). Furthermore, shear strength is also important for plant stability as it helps in securing the plants at the required depth in the soil. On the other hand, plants having very weak soils with low shear strength may not be able to provide adequate anchorage, which would cause those plants to be blown strong winds or heavy rainfalls.

Soil moisture significantly impacts shear strength, with optimal moisture levels balancing root penetration and soil stability. However, waterlogged soils often experience reduced shear strength, which impairs aeration and may lead to root suffocation (Whalley et al., 2005). Additionally, soils with higher shear strength are more resistant to erosion, helping to preserve the nutrient-rich topsoil that is vital for plant growth. Soil management practices, including tillage, organic matter application, and irrigation adjustments, can influence soil shear strength, creating a more favorable environment for plant growth and overall soil health (Hamza & Anderson, 2005).

5.5 Soil Consolidation

Soil consolidation refers to the gradual reduction of volume in a soil mass due to the expulsion of pore water under long-term loading. This process is particularly important in fine-grained soils, such as clays, which have low permeability. Consolidation occurs as the soil undergoes deformation and settlement under a load, often from structures like buildings or embankments. As the pore water is slowly squeezed out, the soil particles come into closer contact, resulting in increased soil density and strength (Das, 2020). Understanding consolidation is critical in geotechnical engineering, particularly for foundation design, where excessive settlement can lead to structural damage.

Soil consolidation, which is commonly caused by compaction, has an impact on plant growth because it affects soil structure, root development, water retention, and aeration. Compacted soils increase bulk density and decrease pore space, limiting root penetration and access to water and nutrients, thereby stunting plant growth (Hamza & Anderson, 2005). Excessive compaction can also impede drainage, resulting in waterlogging that suffocates roots (Whalley et al., 2005). Furthermore, diminished porosity impairs root respiration and nutrient uptake, lowering plant health (Campbell & Bowers, 2011). While modest consolidation can give stability, severe compaction makes the soil too hard for roots to anchor correctly, exposing plants to environmental challenges (Chamen et al., 2003). Proper soil management practices like controlled tillage and organic matter addition can mitigate the negative effects of consolidation, promoting healthier plant growth (Soane & van Ouwerkerk, 1994).

5.6 *Soil Fraction*

Soil fraction refers to the categorization of soil particles based on their size, which plays a crucial role in determining soil's physical properties, including texture, porosity, water retention, and nutrient availability. Soil fractions are classified into three primary categories: **sand**, **silt**, and **clay**, each with distinct characteristics that influence the soil's behavior and suitability for agricultural and engineering applications (Brady & Weil, 2019).

Soil friction, or resistance between soil particles, is critical for plant growth because it affects root penetration, soil stability, and the availability of water and nutrients. High soil friction, which is commonly caused by compacted or clay-rich soils, reduces root penetration. This inhibits roots' capacity to expand and acquire water and nutrients, which can slow plant growth (Hamza & Anderson, 2005). Lower soil friction, which is commonly seen in looser or sandy soils, permits roots to penetrate more freely, encouraging improved nutrient uptake and root development (Gordillo et al., 2020). Also, soil friction contributes to soil stability, which influences the mechanical support offered to plants. Soils with high friction can give.

5.7 *Soil Compressibility*

Soil compressibility refers to the degree to which soil decreases in volume when subjected to a load or external pressure. This reduction in volume occurs as the soil particles rearrange themselves, and in saturated soils, the pore water is expelled. Compressibility is a key factor in geotechnical engineering because it affects settlement behavior, which can influence the stability of foundations, embankments, and other structures built on or in soil (Das, 2020).

Soil compressibility, or the ability of soil to decrease in volume under pressure, affects plant growth by influencing soil structure, root development, and water movement. High compressibility, typical of clay-rich or low organic matter soils, leads to increased compaction, reduced pore space, and limited root penetration, which hinders access to water and nutrients (Hamza & Anderson, 2005). This also reduces soil aeration, leading to poor root health (Gordillo et al., 2020). Conversely, low compressibility in sandy soils allows better root growth and water movement. Proper soil management practices, such as adding organic matter, can mitigate the effects of high compressibility and promote plant growth (Whalley et al., 2005).

5.7.1 Penetration Resistance

Penetration resistance is a measure of the ability of soil to resist penetration by a probing instrument or a root. It is an important indicator of soil **compaction**, **strength**, and **mechanical properties**, and it has significant implications for both agricultural productivity and geotechnical engineering. Higher penetration resistance typically indicates a more compacted soil, which can impede root growth, reduce water infiltration, and limit soil aeration (Hillel, 2020).

In soils, penetration resistance is often measured using a **penetrometer**, which provides a reading in units of pressure, typically **kPa**. Penetrometer readings help assess the degree of compaction in the soil and its potential impact on crop growth. In geotechnical engineering, penetration resistance tests like the **Standard Penetration Test (SPT)** and the **Cone Penetration Test (CPT)** are used to evaluate soil strength and bearing capacity for construction projects (Das, 2020).

5.7.2 Soil Penetration Resistance and Plant Growth

In agriculture, high penetration resistance can have adverse effects on crop growth by limiting root expansion, reducing water infiltration, and restricting access to nutrients. As penetration resistance increases beyond certain thresholds, plant roots may be unable to penetrate the soil, leading to **reduced root depth** and **restricted access** to deeper water and nutrients. This can cause **poor plant growth**, lower crop yields, and increased susceptibility to drought (Sojka et al., 2001).

Managing penetration resistance involves practices such as breaking up compacted soil layers can reduce penetration resistance, although excessive tillage can lead to long-term compaction and degradation. Growing plants with deep, fibrous root systems can naturally break up compacted soil layers and improve soil structure. Moreover, incorporating organic matter into the soil can improve its structure, lower bulk density, and reduce resistance.

5.8 Corrasion Resistance

Corrasion resistance (often referred to as **abrasion resistance**) refers to a soil or material's ability to resist wear and tear caused by the mechanical action of particles, such as sand, silt, or gravel, moving across its surface. In the context of soils, corrasion resistance is an important factor for areas subject to wind or water erosion, construction projects, and natural geological processes. It is crucial in understanding how soils and rocks withstand the erosive forces caused by water, wind, or ice carrying abrasive particles (Kirkby & Brunsden, 2020).

Corrasion occurs when moving particles such as sand, gravel, or silt impact or scrape the surface of a soil or rock. Over time, this abrasive action can cause the surface to erode or wear away. The rate of this erosion depends on the **hardness of the material**, the **speed and volume of moving particles**, and the **presence of moisture**. Soils with high clay or organic matter content may have better abrasion resistance than sandy or silty soils, which are more prone to mechanical wear (Poesen et al., 2020).

5.8.1 Corrasion Resistance in Agriculture

In agriculture, soils with low corrasion resistance are more vulnerable to **wind erosion**, particularly in arid and semi-arid regions. Windborne particles can strip away the topsoil, reducing soil fertility and negatively impacting crop production. Agricultural management practices such as maintaining ground cover with crops, mulching, or using windbreaks can help to protect soils from wind abrasion (Hillel, 2020).

6 Water Movement and Soil Solute

Understanding solute movement in soil is essential for effective land and water management, particularly in mitigating contamination and optimizing fertilizer use. Water and solute transport are interlinked processes where water flow affects how contaminants and nutrients are distributed in the soil (Pessoa et al., 2024). Contaminants such as heavy metals, salts, and organic molecules, as well as agricultural fertilizers, can rapidly be transported through the soil via water flow (Wan et al., 2024). This movement can lead to groundwater and surface water contamination, posing risks to both environmental and human health. To manage these risks, it is crucial to predict and understand the mechanisms that govern the movement and fate of these substances.

In porous media, solute transport is primarily driven by two mechanisms: mass transfer and diffusion (Saeibehrouzi et al., 2024). Mass transfer refers to the movement of solutes with the bulk flow of water, while diffusion is the movement of solutes due to concentration gradients (Rizvi, 2024). Experimental setups, such as miscible displacement experiments, provide valuable insights into these processes. In such experiments, solute solutions are introduced into a soil column, displacing the existing water (Di Bonito et al., 2024). Breakthrough curves (BTCs) are then used to analyze the solute's behavior, plotting the solute concentration relative to its initial concentration against the pore volume of water (Alqusaireen et al., 2024). Physical-mathematical models, particularly the convection-dispersion equation (CDE), are employed to describe solute transport in soils. The CDE combines diffusion and advection (bulk water flow) to model solute movement.

Recent advances in solute transport studies have been facilitated by software such as STANMOD, which provides tools for fitting BTC data to various analytical solutions based on the CDE (Umeh et al., 2024). For example, studies have used STANMOD to assess the transport of nanoparticles like CuO and TiO_2 in agricultural soils. Results showed that TiO_2 nanoparticles exhibited low mobility and minimal risk of groundwater contamination, while CuO nanoparticles were more mobile, suggesting a higher potential for environmental impact.

Phosphorus (P) and nitrogen (N) are key nutrients used in agriculture, but their leaching into groundwater can lead to pollution. Studies have shown that P and N leaching rates vary with soil type and structure (Liu et al., 2024a, 2024b, 2024c).

For instance, research on calcareous soils revealed differences in nitrate and phosphate leaching rates between soil types. This variation underscores the importance of understanding soil structure to predict and manage nutrient losses effectively.

In conclusion, accurately predicting solute transport in soil requires a comprehensive understanding of water flow, solute movement, and soil characteristics. Mathematical models and experimental data play crucial roles in assessing and managing risks associated with soil and water contamination. By improving our knowledge of these processes, we can develop better practices for protecting water resources and ensuring sustainable agricultural productivity.

6.1 Soil Moisture and Solute Movement

Soil moisture and solute movement are critical factors in understanding the overall functioning of terrestrial ecosystems, as they play a pivotal role in the transport and distribution of essential nutrients, water, and other dissolved substances within the soil profile. Soil is a complex matrix, consisting of a diverse array of physical, chemical, and biological components that interact to create a dynamic and intricate environment (Naylor et al., 2020; Pandey et al., 2022).

The movement of water and solutes within the soil is governed by a complex interplay of various factors, including soil texture, structure, porosity, and the nature of the dissolved substances (Tang et al., 2024). Soil microorganisms, in particular, play a crucial role in mediating these processes, as they can influence the availability and transformation of nutrients, as well as the formation of soil aggregates that impact water and solute transport (Gupta et al., 2024).

Climate change and other environmental disturbances can significantly alter the dynamics of soil moisture and solute movement, with potential cascading effects on the overall functioning of terrestrial ecosystems. Changes in precipitation patterns, temperature, and atmospheric composition can affect the rate of water infiltration, evaporation, and the biogeochemical cycling of nutrients, leading to shifts in the distribution and availability of resources for plants and other organisms (Naylor et al., 2020; Zhang et al., 2020).

Researchers have utilized mathematical models to simulate and predict the behavior of water and solute movement within the soil-crop system, with the aim of informing agricultural and irrigation management practices (Fig. 3).

6.2 Soil Water Dynamics

Soil moisture is vital for plant growth and survival, as it is stored in the pore spaces between soil particles (Voroney et al., 2024). The soil's ability to retain water, known as its water-holding capacity, is determined by its porosity, which is the total volume of pore spaces (Nelson et al., 2024). Soil water can be categorized into three main

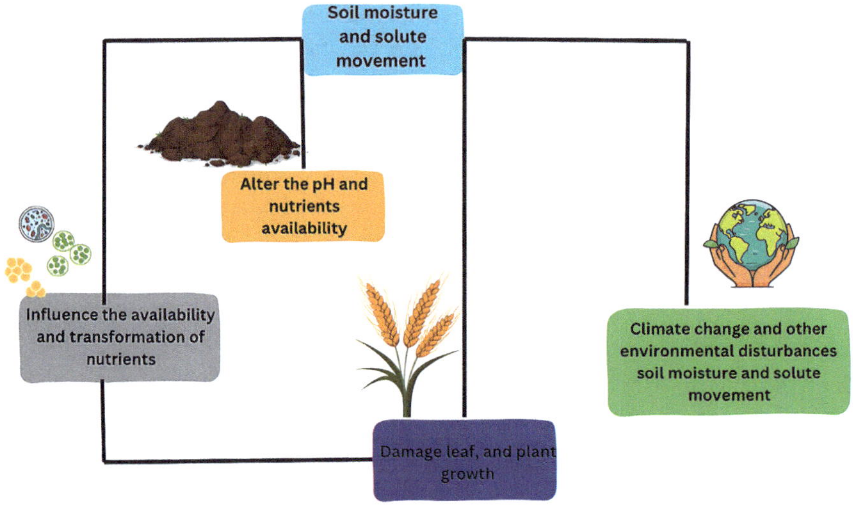

Fig. 3 Impact of soil moisture and solute movement

types based on how it interacts with soil particles: **Gravitational Water**: This type of water moves through the soil due to the force of gravity and is quickly drained from the larger pores. It is readily lost from the soil after heavy rainfall or irrigation (Di Bonito et al., 2024). **Capillary Water**: This water is held in soil pores against gravity due to capillary forces, which result from the balance between adhesion (water's tendency to stick to soil particles) and cohesion (water's tendency to stick to itself) (Brown et al., 2021). Capillary water is crucial for plants as it is more accessible than gravitational water. **Hygroscopic Water**: This water is tightly bound to soil particles by surface forces, forming a thin film around them. Hygroscopic water is not available to plants due to its strong adhesion to soil particles (Liu et al., 2024a, 2024b, 2024c). The impact of pore size on soil water retention is significant. Fine-textured soils, such as clays and clay loams, have a higher porosity than coarse-textured soils like sands, meaning they can hold more water (Murmu et al., 2024). However, in fine-textured soils, much of this water is held tightly and is not readily available to plants (Yan et al., 2024). Plants must overcome these strong retention forces to access the water. Soil moisture content is the percentage of water within the soil. It is typically measured by drying a soil sample and calculating the water content using the formula:

$$\begin{aligned}\text{Percentage Moisture Content} = & (\text{Weight of Wet Soil} \\ & -\text{Weight of Dry Soil}/\text{Weight of Dry Soil}) \\ & \times 100 \text{ Alternatively, soil moisture can be measured using sensors.}\end{aligned}$$

6.2.1 Soil Moisture Conditions

Saturation: This occurs when all soil pores are filled with water. While this condition indicates maximum water content, it is not ideal for plants because their roots also need access to air (Di Bonito et al., 2024).

Field Capacity: This is the ideal moisture condition for plants, representing the amount of water the soil retains after excess water has drained away (Heinen et al., 2024). At field capacity, larger pores are filled with air, and water is available to plants. This condition is defined by a soil water tension of – 0.33 bar. However, in some soils, maintaining field capacity may lead to insufficient oxygen for roots or an increased risk of root diseases.

Permanent Wilting Point: This is the moisture level at which plants can no longer extract water from the soil. At this point, the soil water is tightly bound to soil particles, creating a tension of – 15 bar (Peng et al., 2024).

Total Available Water (TAW): TAW is the difference between the field capacity and the permanent wilting point. To determine TAW, subtract the percentage of water at the wilting point from the percentage at field capacity (Lal, 2024). For example, if the field capacity is 20% and the wilting point is 13%, the available water is 7%. To convert this into a quantity, multiply the percentage of available water by the soil depth and a conversion factor (e.g., 10 for centimeters or 1000 for millimeters). For a root depth of 20 cm.

Readily Available Water (RAW): As soil water is used by plants, the remaining moisture becomes harder to extract. RAW is the portion of water available for plant uptake, located between field capacity and a refill point (Snyder et al., 2024).

By understanding these soil moisture conditions and measurements, farmers and agronomists can better manage irrigation practices to ensure plants receive sufficient water for optimal growth.

6.3 Solute Transport in Soil

Many solutes enter the natural system at the soil surface. The solute residence time in the unsaturated zone is important for soil- and groundwater pollution management (Wei et al., 2024a, 2024b). For instance, organic compounds are mainly decomposed in the unsaturated zone, where the biological activity is concentrated (Dittmar, & Lennartz, 2024). Most plants can extract water and nutrients from the soil only in the unsaturated zone. In irrigated areas, the long-term salinity in the root zone will depend on the amount of percolation from the unsaturated zone (Nick et al., 2024). Whereas in the unsaturated zone, the transport of solutes is predominantly vertical, once in the groundwater solutes may diverge in any direction, threatening surface waters, nature reserves, and drinking wells. Using an analytical model, Beltman et al. (1995) show the importance of the transport processes in the unsaturated zone as compared to the transport processes in the saturated zone. A thorough understanding is needed of the processes that govern the transport, adsorption, root uptake, and decomposition

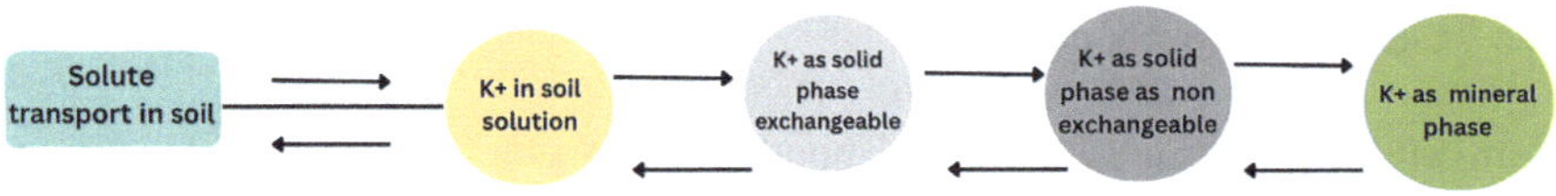

Fig. 4 Movement of solute in soil

of the solutes in the unsaturated zone, to analyze and manage soil and water-related environmental problems (Tang et al., 2024). The three main solute transport mechanisms in soil water are diffusion, convection, and dispersion. Diffusion is the process by which solutes move from areas of higher concentration to lower concentration due to the random movement of molecules, striving for equilibrium (Mallick & Agmon, 2024). This movement is influenced by the concentration gradient, temperature, and the characteristics of both the solute and soil matrix. Convection involves the bulk movement of water carrying dissolved solutes, driven by pressure gradients or external forces like gravity. This mechanism is shaped by water flow rates and soil permeability (Tiab & Donaldson, 2024). Dispersion, on the other hand, describes the spreading of solutes resulting from both diffusion and convection, compounded by irregular flow paths and soil heterogeneity. It reflects how solutes disperse more broadly than predicted by diffusion alone due to variations in water flow and soil structure. Each mechanism intricately contributes to the overall distribution of solutes in soil water, affecting how nutrients and contaminants are transported through the soil (Fig. 4).

6.4 Models and Simulation

Models and simulations are essential tools for understanding and predicting solute transport in soil water. They help to analyze the complex interactions between diffusion, convection, and dispersion and how these processes affect solute movement.

1. **Diffusion Models**: Simple diffusion models often use Fick's laws of diffusion. Fick's First Law describes the flux of solutes as proportional to the concentration gradient, while Fick's Second Law accounts for changes in concentration over time (Hoffrogge, 2024). In simulations, diffusion can be modeled using finite difference or finite element methods to solve partial differential equations (PDEs) that represent Fick's laws. These models typically assume homogeneous medium and constant diffusion coefficients, although more advanced models can account for spatial variability in soil properties.
2. **Convection Models**: Convection-dominated solute transport is often described using advection-dispersion equations (ADEs), which combine the effects of convection (bulk movement of water) and dispersion (Lubarda & Lubarda, 2024). The ADE equation typically incorporates terms for both advection (driven by flow velocity) and dispersion (due to irregularities in flow). Simulating convection involves solving the advection-dispersion equation using numerical methods.

These simulations require accurate data on water flow velocities and soil properties. Advanced models may include coupled hydrological and transport simulations, such as those found in hydrological modeling software like MODFLOW or HYDRUS.

3. **Dispersion Models**: Dispersion models are often extensions of convection models and incorporate both longitudinal and transverse dispersion. They account for the spreading of solutes due to irregular flow paths and soil heterogeneity (Samborska-Goik & Pogrzeba, 2024). In simulations, dispersion is modeled by including dispersion coefficients in the advection-dispersion equation. These coefficients can be derived empirically or through more complex models that simulate the variability in soil structure and flow patterns. Tools like STANMOD or the MODFLOW/MT3DMS package can be used for such simulations.
4. **Integrated Modeling Approaches**
   a. **Coupled Models**: Comprehensive models often integrate diffusion, convection, and dispersion into a single framework. These models simulate the interactions between water movement and solute transport in a more holistic manner. Examples include the HYDRUS-1D/2D/3D models, which can simulate water and solute movement in various soil profiles and conditions.
   b. **Software Tools**: Several software packages are available for these simulations, including:
      i. **HYDRUS**: Simulates water, heat, and solute movement in one, two, and three dimensions.
      ii. **MODFLOW/MT3DMS**: Coupled groundwater flow and transport modeling software.
      iii. **COMSOL Multiphysics**: Offers flexible simulation environments for diffusion, convection, and dispersion with customizable PDE solvers.

6.5 Soil Moisture and Plant Growth

Soil moisture is a fundamental determinant of plant health and growth, affecting a range of physiological processes essential for robust plant development (Wang et al., 2024a, 2024b, 2024c). The availability of water in the soil directly influences a plants ability to perform photosynthesis, the critical process by which plants convert sunlight into energy (Jha, 2024). Adequate soil moisture ensures that plants have sufficient water to dissolve essential nutrients from the soil, which are then transported through the plant's vascular system to support growth and development (Lopez et al., 2024). Without enough moisture, plants can experience significant water stress, manifesting as wilting, reduced photosynthesis, and stunted growth (Sapes et al., 2024). Chronic water deficits can even lead to plant mortality, as the plant struggles to maintain vital functions and absorb nutrients effectively (Kalra et al., 2024). In addition to supporting photosynthesis, soil moisture is crucial for nutrient uptake. Water acts as a medium through which nutrients move from the soil to the plant roots. When soil moisture levels are low, nutrient availability and uptake are compromised, potentially leading to deficiencies that can impair plant health and reduce crop yields (Plett et al., 2020). For instance, drought conditions can limit the uptake of key nutrients like K and P, essential for various plant functions including root development and fruiting.

Root development is another area profoundly impacted by soil moisture. Adequate moisture allows roots to expand and penetrate deeper into the soil, enhancing their ability to access water and nutrients (Esfahani & Sonnewald, 2024). This results in healthier, more resilient plants that can better withstand periods of drought. Conversely, insufficient moisture restricts root growth, leading to a shallower root system that is less effective in accessing the resources needed for optimal growth (Voothuluru et al., 2024). On the other hand, excessive moisture can lead to waterlogging, which impairs root respiration and can result in root rot and other diseases. This balance is crucial, as both too little and too much moisture can adversely affect plant health (Manghwar et al., 2024).

Soil moisture also influences plant metabolism, particularly the processes of photosynthesis and transpiration. Water stress can cause stomata, the pores on plant leaves, to close in an attempt to conserve water (Manghwar et al., 2024). This closure reduces gas exchange, slowing down photosynthesis and impacting the plant's ability to produce food and maintain energy levels (Manghwar et al., 2024). In turn, this affects cell turgidity, which is vital for maintaining structural integrity and preventing wilting. Proper soil moisture supports open stomata and optimal metabolic function, ensuring that plants can grow vigorously and produce yields efficiently (Manghwar et al., 2024).

Different plants and crops have varying moisture requirements, making it important to tailor water management practices to the specific needs of each species. For example, seedlings and newly germinated plants often have higher moisture needs compared to mature plants (Beveridge et al., 2024). Effective irrigation and soil moisture management practices are therefore essential. Techniques such as drip irrigation provide precise water delivery, reducing waste and ensuring that plants receive the right amount of moisture (Ray & Majumder, 2024). Mulching helps retain soil moisture and reduce evaporation, while conservation tillage practices can improve soil structure and water infiltration.

Therefore, managing soil moisture effectively is crucial for ensuring that plants receive the necessary resources to thrive. By maintaining optimal moisture levels, growers can support photosynthesis, enhance nutrient uptake, promote healthy root development, and optimize plant metabolism. Implementing appropriate irrigation techniques and soil conservation practices can help achieve this balance, ultimately leading to healthier plants and more productive yields.

7 Soil Temperature and Thermal Properties

Soil acts as a significant heat reservoir, storing energy during the day and releasing it at night. Throughout warmer periods, the soil accumulates energy, which is then released into the air during cooler times (Geiger et al., 2009). The temperature of the soil is determined by the balance between the energy it absorbs and the energy it loses, leading to fluctuations that occur both annually and daily, primarily influenced by changes in air temperature and solar radiation (Wu & Nofziger, 1999).

Soil temperature plays a crucial role in influencing soil properties and the processes involved in plant growth. It regulates various physical, chemical, and biological activities within the soil (Onwuka & Mang, 2018). Additionally, soil temperature impacts the exchange of gases between the atmosphere and soil (Lehnert, 2013). The amount of radiation absorbed by the soil significantly affects its temperature which in turn influences biological processes like seed germination, seedling emergence, root growth, and nutrient availability (Fenner, 2000). Moreover, soil temperature influences the decomposition rate of organic matter and the mineralization of organic materials within the soil. It also affects soil water retention, transmission, and its availability to plants.

7.1 Soil Temperature

Soil temperature is influenced by the flow of heat within the soil and the exchange of heat between the soil and the atmosphere (Elias et al., 2004). It can also be described as being dependent on the soil's internal energy (Gahoonia & Nielsen, 2003). The key factors that cause soil temperature to fluctuate include the transfer of heat within the soil (Zhao et al., 2006) and the exchange of latent heat at the surface (Nwankwo & Ogagarue, 2012). Solar radiation serves as the primary source of soil temperature. The temperature of the soil is typically measured using a thermometer. Seasonal and daily variations in soil temperature occur due to changes in radiant energy and the energy dynamics at the soil surface (Chiemeka, 2010).

7.2 Factors Influencing Soil Temperature

Chiemeka (2010) categorized these factors into two groups: the amount of heat that reaches the soil surface and the amount that disperses from the surface down through the soil profile. Several factors influence the heat available at the soil surface, including soil color, the use of mulching (Science, 1978), solar radiation (Geiger et al., 2009), the slope of the land (Elizbarashvili et al., 2010) vegetation cover (Decker et al., 2003), organic matter content (Abu-Hamdeh & Reeder, 2000), and the rate of evaporation.

7.2.1 Soil Color

Darker soils tend to absorb more radiant heat compared to lighter-colored soils, leading to higher soil temperatures in darker soils (Sándor & Fodor, 2012).

7.2.2 Soil Mulch

According to Horton et al. (1996) mulch materials reduce evaporation and enhance soil moisture, which in turn lowers the surface temperature (Horton et al., 1996). By insulating the soil, mulching reduces the amount of heat absorbed, thereby decreasing soil temperature (Dahiya et al., 2007). Overall, mulched soils absorb less heat than bare soils.

7.2.3 Slope of the Land Surface

When solar radiation strikes the land at an angle, it spreads over a larger area compared to radiation that hits the surface perpendicularly. As a result, the amount of radiation per unit area decreases with an increase in the land's slope (Onwuka & Mang, 2018). Consequently, soil temperature tends to decrease as the slope of the land increases.

7.2.4 Vegetative Cover

Bare soils tend to heat up quickly during hot periods and cool rapidly during cold periods. Vegetation acts as a thermal buffer, helping to regulate soil temperature (Sándor & Fodor, 2012) Vegetative cover prevents the soil from becoming excessively hot in dry seasons or too cold in rainy seasons (Jiménez et al., 2007).

7.2.5 Organic Matter Content

Organic matter enhances the soil's water retention capacity and often contributes to darker soil color, both of which increase heat absorption and thus raise soil temperature (Onwuka & Mang, 2018).

7.2.6 Evaporation

The process of water evaporation from the soil consumes a significant amount of energy. This energy, primarily derived from solar radiation, is used for evaporation rather than for heating the soil (Lehnert, 2013). Therefore, higher evaporation rates lead to a cooling effect, lowering soil temperature.

7.2.7 Solar Radiation

Solar radiation is the primary source of heat for the soil. The amount of solar energy received and absorbed by the soil directly influences soil temperature variability (23).

As solar radiation increases, soil temperature rises correspondingly (Geiger et al., 2009).

Factors affecting the heat dissipation from the soil profile include moisture content (Abu-Hamdeh, 2003) and bulk density (Lehnert, 2013).

7.2.8 Soil Moisture Content

Moisture content significantly impacts the dissipation of heat within the soil. Wet soils conduct heat more efficiently than dry soils, where air-filled pores reduce heat transfer. As soil moisture increases, the rate of heat dissipation also increases (Ochsner et al., 2001).

7.2.9 Bulk Density

High bulk density increases the surface area of the soil, thereby enhancing the rate at which heat energy is transferred through a unit cross-section of the soil (Onwuka & Mang, 2018).

7.3 *Soil Thermal Properties*

Soil thermal properties, including heat capacity, thermal conductivity, and thermal diffusivity are crucial for understanding heat transfer in soils, which impacts agricultural productivity and environmental management. Various factors influencing these properties, including moisture content, soil type, and management practices. The following sections detail these aspects.

7.3.1 Heat Capacity

The specific heat capacity of soil solids indicates the amount of heat needed to raise the temperature of a unit mass of dry soil particles by one degree (Schärli & Rybach, 2001). Specific heat capacity generally varies within a limited range, it can differ from one soil type to another, making it challenging to select the most appropriate value. This variability can affect the accuracy of soil physical measurements related to specific heat capacity. Therefore, it is essential to determine the intrinsic specific heat capacity values for mineral soils with different textures.

Research indicates that specific heat capacity is influenced by clay content, organic matter (OM), and tightly bound water (TBW) (Wang et al., 2011). The presence of OM and clay increases specific heat capacity and the amount of TBW tends to rise with higher clay and OM content (Wang et al., 2011). These factors contribute to variations in specific heat capacity across soils with different textures and OM levels, underscoring the importance of understanding how specific heat capacity changes quantitatively with soil composition.

7.3.2 Thermal Conductivity

Soil thermal conductivity is a critical parameter that quantifies the ability of soil to conduct heat, significantly influencing various engineering and environmental applications. The soil's thermal conductivity depends on the thermal properties of its components (air, water, and solid particles), as well as factors like moisture content and density, governs this heat conduction. Various theoretical, empirical, and mathematical models have been developed to predict soil thermal conductivity based on these factors. Experimental methods to measure soil thermal conductivity typically involve the use of needle probes or thermal cells (Low et al., 2015).

7.3.3 Thermal Diffusivity

Soil layers play a crucial role in land-atmosphere interactions, as they facilitate the exchange of mass and energy between the two. Among the key properties of soil are its thermal characteristics, which govern the transfer and storage of heat within the soil. Soil thermal diffusivity is a critical parameter that defines the rate at which temperature waves move through the soil (Low et al., 2015). Along with the soil temperature gradient, it determines the magnitude of soil heat flow. Accurately calculating soil thermal diffusivity is essential for better understanding soil thermodynamic processes and land-atmosphere interactions.

7.4 Significance of Soil Temperature in Plant Growth

Temperature is another key factor that significantly affects plant growth and health. Both extreme cold and excessive heat can lead to plant stress and poor health. Temperature influences various plant development processes, such as photosynthesis, transpiration, nutrient absorption, respiration, and flowering. Generally, warmer temperatures promote growth, while cooler temperatures inhibit it. However, there is a limit, as growth does not continue to increase indefinitely with rising temperatures. Each plant species has its own optimal temperature range for growth (Madzhi & Akhsan, 2021).

To monitor crop exposure to extreme temperatures and pests, modern intelligent environment monitoring and control systems, such as those based on Arduino UNO boards, are now used in small-scale agriculture like greenhouses. These systems allow users to monitor and regulate greenhouse conditions remotely via web interfaces, mobile apps, and GSM technology in real time. The Arduino board interfaces with various sensors, and its programming integrates the data from each sensor to provide a comprehensive overview of the environment (Enokela & Othoigbe, 2015).

8 Soil Color

Soil color, a key morphological attribute, serves as a valuable indicator of various soil properties such as mineral composition, organic matter content, water, and nutrient levels (Fang et al., 1998). Its ease of determination and low cost make it widely utilized in soil research (Günal et al., 2008). Light-colored soils typically indicate high calcium carbonate content, suggesting lower fertility compared to dark brown or black soils rich in organic matter (Schwertmann, 1993). Furthermore, soil color can offer insights into chemical processes influenced by soil moisture regimes (Ji et al., 2007).

8.1 Soil Color and Soil Properties

Various factors such as parent materials and topographical features contribute to spatial variations in soil properties. Among the most conspicuous and pivotal characteristics of soil is its color, which consistently serves as a decisive criterion in soil identification, classification, and differentiation (Sanchez-Maranon et al., 2004). Due to its easy discernibility in the field and its correlation with other soil attributes, color has frequently been utilized as a rapid assessment indicator for various aspects such as drainage class, soil classification, and soil organic carbon levels, influencing practices like variable rate fertilizer application and pesticide usage (Konen et al., 2003).

Properties such as soil organic carbon content, iron content, soil water content, and texture exhibit significant correlations with soil color. Dark surface horizons in soils typically indicate high organic matter content, categorizing them as fertile and conducive to plant growth. Soils with dark brown and black hues are often associated with elevated nitrogen levels, good aeration, effective drainage, and low erosion risk. Conversely, light-colored soils are generally perceived to possess opposite characteristics. The color of iron-containing soil minerals undergoing oxidation and reduction reactions can offer valuable insights into soil hydrologic conditions (Viscarra Rossel et al., 2006). The Munsell color system has been widely utilized in soil color studies due to its simplicity, accessibility, and historical relevance in soil science. However, this method is semi-quantitative, allowing for subjective matching of soil samples to Munsell color chips (Konen et al., 2003). Newer color systems, such as those developed by the Commission Internationale d'Eclairage (CIE), might offer better representation of uniform color space, where differences between colors align with human perceptions (Melville & Atkinson, 1985). Barrett (2002) employed a spectrophotometer for in-situ soil color quantification, noting its higher precision.

Some researchers have utilized soil CIELAB color parameters to depict soil color and establish relationships with soil properties (Sánchez-Marañón et al., 1997).

Soil tests typically rely on the internationally recognized Munsell color scale to determine soil color, facilitating identification. Alternatively, other color coordinates such as X, Y, Z; L, a, b, may be used to describe soil color. These coordinates, as established by studies (Shield, 1968), enable correlations between soil physical properties or organic matter content and its color. However, a drawback lies in the complexity associated with determining these color coordinates, compounded by the need for specialized measuring devices to calculate such characteristics.

8.1.1 Munsell Soil Color Book

Due to variations in individual interpretation, simple color descriptions like "dark" or "red" can differ significantly among people. To ensure clarity and precision, soil scientists adopt a standardized notation for indicating soil color. This notation comprises three components:

Page: Denotes the mixture of colors or hue.

Value: Represents the level of lightness or darkness.

Chroma: Indicates the intensity of color.

For instance, a soil color notation might appear as 10YR 2/1, where:

- Page: 10YR denotes the mix of colors or hue.
- Value: 2 signifies the level of lightness or darkness.
- Chroma: 1 represents the intensity of color (Fig. 5).

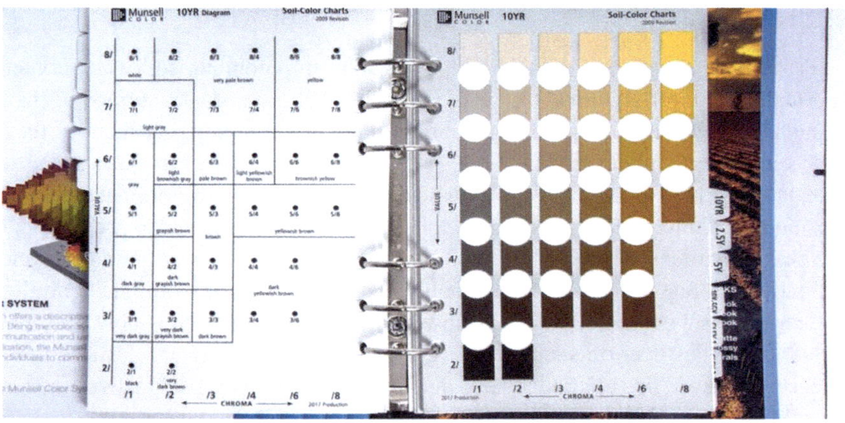

Fig. 5 Gilson Munsell soil color book (HM-519) (Anderson, 2023)

8.2 *Effect of Color on Plant Growth*

The effect of color on plant growth is significant, as different wavelengths of light influence various physiological processes, including photosynthesis, rooting, and biomass accumulation. Research indicates that specific light colors can enhance growth parameters, such as leaf greenness, root length, and overall biomass. Amber light has been shown to induce the highest photosynthetic rates, leading to increased fresh and dry mass in plants like tomatoes and lettuce. Red and blue light are both crucial for photosynthesis, but red light significantly impacts dry mass, whereas blue light is essential for overall growth (Wu et al., 2024a, 2024b).

Exposure to red and blue light has been found to result in the highest leaf greenness index and longest roots in chrysanthemums, indicating enhanced rooting and growth (Schroeter-Zakrzewska & Pradita, 2021). Different colored shade nets have also been shown to affect growth parameters, with aluminet nets promoting higher shoot dry weight in certain ornamental species (Singh et al., 2023). Early exposure to colored light can influence the accumulation of beneficial compounds in later growth phases, affecting plant quality (Sytar et al., 2019). While specific colors enhance growth, the role of green light remains complex. Although it is less efficient for photosynthesis per unit area, it can improve overall plant growth in dense canopies by altering light distribution and leaf acclimation (Kaiser et al., 2019).

9 Soil Tillage, Tilth and Plant Growth

9.1 *Soil Tillage and Soil Properties*

Soil tillage significantly influences soil properties, affecting its physical, chemical, and biological characteristics. Various tillage practices, such as conservation tillage and conventional tillage, have been shown to enhance soil health and crop productivity. Conservation tillage methods, such as minimum tillage, reduce bulk density and increase soil porosity, promoting better root growth and water infiltration (Yerli et al., 2024). Different plowing techniques, like moldboard and chisel plows, also affect soil structure, with moldboard plowing being more effective in reducing bulk density (Isaak et al., 2024).

Tillage practices influence the availability of essential nutrients, with cover crops enhancing nitrogen levels in the soil, particularly when legumes are used (Wilczewski et al., 2024). Additionally, conservation tillage and the retention of crop residues improve soil organic carbon content, which is crucial for maintaining soil fertility. Biologically, reduced tillage systems promote higher microbial counts and enzyme activity, both of which are essential for nutrient cycling and overall soil health (Ghuman et al., 2024). While conservation tillage generally improves soil properties, some argue that conventional tillage can be beneficial in specific contexts, such as when immediate soil preparation is necessary for certain crops. This highlights the

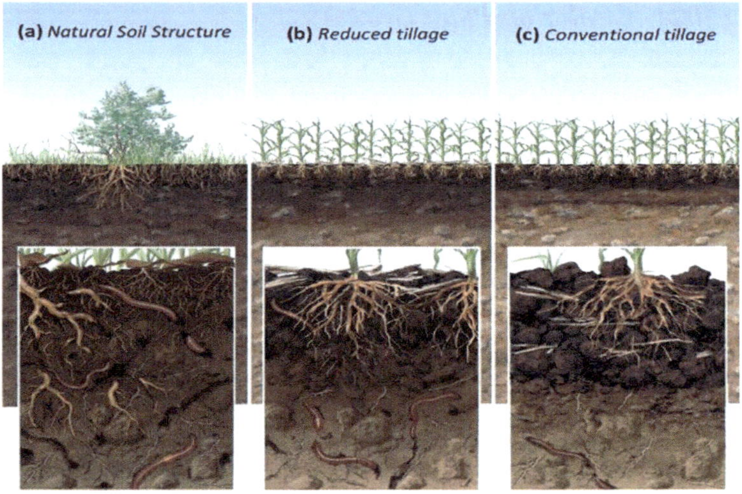

Fig. 6 **a** Natural soil structure is shaped by soil-fauna-vegetation climate feedbacks (accumulation of biopores and invisible soil aggregation). **b** No-till and reduced tillage management fosters higher biological activity that conventional tillage but lacks the vegetation diversity and soil organic carbon accumulation of natural soil under similar conditions. **c** Managed (tilled) soil structure is characterized by mechanical fragmentation of the top layer and subsequent coalescence and consolidation of the loose structure (certain biological activity is maintained at the undisturbed subsurface)

need for tailored approaches based on local conditions and crop requirements (Yerli et al., 2024) (Fig. 6).

9.2 *Relationship Between Soil Tillage, Tilth, and Plant Growth*

Soil, a crucial natural resource, faces degradation due to agricultural practices and erosion (Bowen, 1981). Among these practices, tillage stands out as one of the most impactful, involving the physical disturbance of soil using tools to prepare a suitable seedbed for optimal germination and crop growth. Tilth, a subjective term, reflects the soil's physical condition, indicating its ease of tillage, seedling emergence, and root growth (SSSA, 1979). It encompasses various physical attributes like consistency, organic matter content, texture, structure, and strength, making it a dynamic property susceptible to natural and human-induced changes (Brady, 1984). Maintaining soil productivity requires strategies that preserve its physical, chemical, and biological characteristics. The soil's health is closely tied to its tilth which significantly influences plant growth, acting as a predictor of soil quality (Karlen et al., 1991). Different plant species can affect soil tilth differently, with some capable of breaking through compacted layers while others promote aggregate and macropore integrity. Identifying the optimal soil tilth conditions for various crops and understanding how tillage practices impact soil tilth can lead to improved strategies in agricultural production, including better tillage techniques, crop rotation plans, fertilizer management, and yield targets. The biological, chemical, and physical characteristics of soil collectively determine its health, with properties such as bulk density, infiltration, soil structure, porosity, and compaction being influenced by soil tilth. Tillage, while initially enhancing soil tilth by improving plant-soil-air-water relationships, can contribute to long-term tilth deterioration by accelerating the oxidation rate of organic matter. The dynamic interplay between physical, chemical, and biological processes involved in maintaining or altering soil tilth means that time also plays a significant role in its evolution (Karlen et al., 1991). Adopting soil management techniques such as long-term conservation tillage can establish new equilibriums within the soil matrix, often leading to a shift towards a different soil tilth state (Bauer & Black, 2020).

9.3 Optimizing Soil Tillage Practices for Sustainable Plant Growth

The increasing concern for food security, achieved through enhanced soil management techniques, necessitates the establishment of an environmentally conscious and crop-yield sustainable tillage system. Tillage significantly impacts soil characteristics such as water conservation, temperature, infiltration, and evapotranspiration processes. Consequently, it influences both crop production and environmental conditions. With the global population on the rise and the corresponding increase in food demand, there's a pressing need to cultivate more land for crop production. However, this must be approached with caution to minimize soil degradation and ensure soil functions as a sink rather than a source of atmospheric pollutants. Conservation tillage, in conjunction with complementary practices such as soil cover and crop diversity, has emerged as a viable solution to promote sustainable food production while preserving environmental integrity. Thus, conservation tillage stands as an integral component of conservation agriculture (Corsi et al., 2012) (Fig. 7).

9.3.1 Conservation Agriculture

Conservation agriculture is defined as a strategy for managing agro-ecosystems for sustainable productivity, increase profitability, and ensuring food security while simultaneously conserving and improving the resource base and environment. It is characterized by minimal mechanical disturbance of the soil, maintaining permanent organic soil cover, and promoting crop diversification as its fundamental principles (Corsi et al., 2012). According to CTIC (2004), conservation tillage refers to any tillage method that leaves a minimum of 30% of the soil surface covered with crop

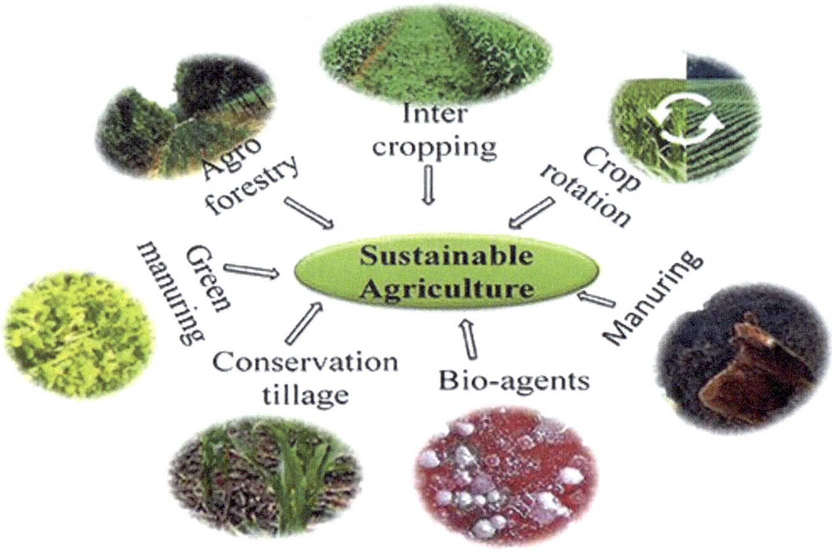

Fig. 7 Components of sustainable agriculture

residue post-planting to mitigate water erosion. Lal (1990) elaborated on conservation tillage as a seedbed preparation approach involving the presence of residue mulch and an increase in surface roughness. Overall, conservation tillage represents an ecologically mindful strategy for managing soil surfaces and preparing seedbeds.

Types of conservation tillage

Conservation tillage practices encompass a spectrum of techniques, including zero tillage (No-till), reduced (minimum) tillage, mulch tillage, ridge tillage, and contour tillage. No-till (NT) involves minimal disturbance of the soil surface, typically limited to planting activities. Reduced or minimum tillage entails a lower degree of soil manipulation, often utilizing primary tillage implements for ploughing. Mulch tillage involves preparing or tilling the soil in a manner that maximizes coverage with plant residues or other materials. Ridge tillage entails planting crops in rows along or on top of ridges prepared prior to the cropping season. Contour tillage refers to tillage performed perpendicular to the slope direction (Shahane & Shivay, 2021).

The influence of tillage is evident across soil physical, chemical, and biological properties, albeit varying in degree. Additionally, tillage exerts an effect on the soil environment, contributing to phenomena such as runoff and soil erosion (Bhatt & Khera, 2006).

10 Soil Health

Soil health is the capacity of soil to function as a vital living system within the ecosystem to sustain plant and animal productivity (Adewara et al., 2024). Soil health is also fundamental for sustaining agricultural productivity, supporting ecosystem functions, and ensuring environmental quality (De Corato et al., 2024). Soil health is a multidimensional concept that depends on its physical, chemical, and biological properties. Physical properties of soil health include bulk density, aggregate stability, and porosity. While, the chemical properties include pH, nutrient content, and cation exchange capacity of the soil. On the other hand, biological properties of soil health encompass organic matter, microbial activity, and enzyme concentration. Each property plays a critical role in nutrient cycling and soil health causing resilience against environmental stresses. However, anthropogenic activities like excessive use of synthetic fertilizers and excessive land management practices like tillage cause a degradation in soil health as well as an increase in the amount of carbon dioxide (CO_2) in the atmosphere (Kumari et al., 2024). Therefore, understanding and managing soil health is important for facing global challenges such as climate change, food security, and environmental sustainability. Because soil health not only plays a crucial role in agriculture production but also plays a crucial role in mitigating climate change (Nazir et al., 2024). Soil carbon (C) sequestration is increasingly recognized as a pivotal ecosystem service provided by healthy soils (Shaaban & Nunez-Delgado, 2024). It refers to the process by which CO_2 from the atmosphere is absorbed by plants through photosynthesis and subsequently stored as organic C in soil organic matter (Khalid et al., 2024). Practices that enhance soil health, include conservation tillage, cover cropping, crop rotation, and organic farming, which promote C sequestration by increasing SOM levels and reducing soil disturbance (Singh et al., 2024). Increasing soil organic C not only mitigates climate change by decreasing greenhouse gas emissions like CO_2 but also improves soil fertility (Nazir et al., 2024).

Soil health is intricately linked to environmental quality and its impact on crop growth. Healthy soils provide a stable physical environment for root growth, ensuring efficient uptake of water and nutrients essential for plant development (Pandao et al., 2024). Despite the critical importance of soil health, global soils face numerous challenges including erosion, nutrient depletion, salinization, compaction, contamination, and loss of organic matter. Intensive agricultural practices, deforestation, urbanization, and climate change intensify these challenges, threatening soil productivity and ecosystem resilience. Therefore, sustainable soil management practices are essential to mitigate these threats and enhance soil health. Conservation agriculture techniques, including reduced tillage, cover cropping, crop rotation, and agroforestry, promote soil conservation, improve organic matter content, and enhance soil structure and fertility. Integrated nutrient management strategies, such as precision fertilization and organic amendments, optimize nutrient use efficiency and minimize environmental impacts. Effective soil health management requires supportive policy frameworks, institutional capacity building, and stakeholder engagement for promoting sustainable soil management practices to farmers and integrating soil

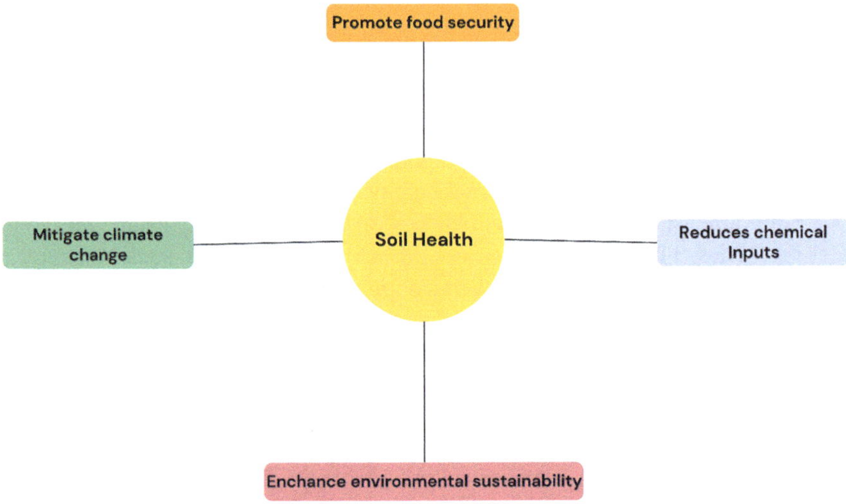

Fig. 8 Role of soil health in the ecosystem

health considerations into land use planning and agricultural development strategies (Fig. 8).

10.1 Indicators of Soil Health

Soil is a finite and non-renewable resource because its regeneration through chemical and biological weathering of underlying rock requires geological time (Corami, 2024). Healthy soil functions to buffer nutrients as well as contaminants and other solutes via sorption to or incorporation into clay particles and organic materials (Ng et al., 2024). Thus, the soil itself serves as an environmental filter for removing undesirable solid and gaseous constituents from air and water. The extent to which soil immobilizes or chemically alters toxic substances, thus effectively detoxifying them, reflects the degree of soil health in the sense that humans or other biological components of the system are protected from harm. Soil health is determined by a combination of physical, chemical, and biological parameters that interact to influence its productivity and sustainability (Lichtenberg, 2024). Physical parameters such as bulk density affect soil compaction and root penetration, while macro-and meso-porosity determine the soil's ability to retain and transmit water and air. Available water capacity is crucial for plant growth as it represents the water accessible to plants. Penetration resistance measures the ease with which roots can grow through the soil, and saturated hydraulic conductivity reflects the soil's ability to conduct water when fully saturated. Wet aggregate stability indicates the soil's resistance to erosion and its structural integrity, while the water infiltration rate shows how quickly

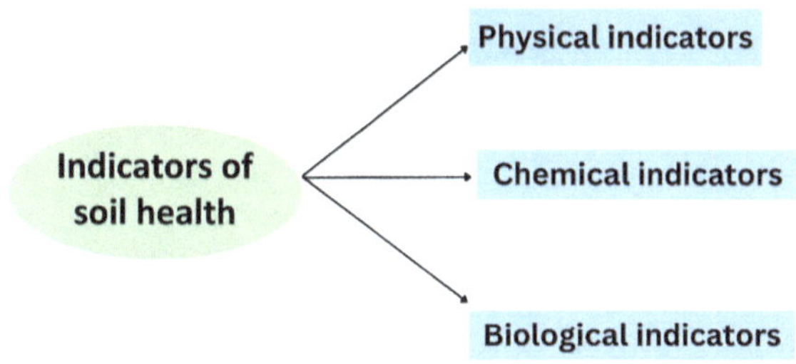

Fig. 9 Soil health indicators

water can enter the soil, impacting irrigation efficiency and runoff. Chemical parameters like pH influence nutrient availability and microbial activity. Cation exchange capacity (CEC) measures the soil's ability to hold and exchange essential nutrients like nitrogen (N), potassium (K), phosphorus (P), calcium (Ca), magnesium (Mg), iron (Fe), zinc (Zn), and copper contents are vital for plant nutrition and overall soil fertility (FAO, 2024). Biological parameters, including soil organic matter content, are key for maintaining soil structure, moisture retention, and nutrient cycling (Sakin et al., 2024). Potential mineralizable nitrogen indicates the soil's ability to supply nitrogen through organic matter decomposition (Guo et al., 2024). Particulate organic matter and active C are indicators of the soil's organic content and microbial activity (Koorneef et al., 2024). A root bioassay evaluates the soil's suitability for root growth, and nematode populations can indicate soil health and biodiversity, as they play roles in nutrient cycling and organic matter decomposition (Mutai et al., 2024). All these parameters collectively contribute to the soil's ability to support plant life, resist erosion, and sustain agricultural productivity (Fig. 9).

1. ***Physical Indicators***

Physical parameters play a crucial role in determining soil health and its ability to support plant growth. Bulk density measures the mass of soil per unit volume, which affects soil compaction and root penetration (Giap & Ahmad, 2024). High bulk density can restrict root growth and decrease aeration, while low bulk density generally indicates good soil structure (Giap & Ahmad, 2024). Macro and meso-porosity refer to the size and distribution of soil pores, which are essential for air and water movement within the soil (Liu et al., 2024a, 2024b, 2024c). Good porosity ensures that roots have access to sufficient oxygen and water, enhancing plant health and growth. While available water capacity is the amount of water the soil can hold that is accessible to plants. It is critical for maintaining plant hydration between watering or rainfall events. However, penetration resistance indicates how easily roots can penetrate the soil The lower resistance means roots can grow more freely, accessing nutrients and water more effectively. On the other hand, saturated hydraulic conductivity measures the soil's ability to conduct water when it is fully saturated. This parameter is important for understanding drainage and the risk of waterlogging. Wet aggregate stability assesses the soil's resistance to breaking down when wet, which is important for preventing erosion and maintaining soil structure. Finally, the

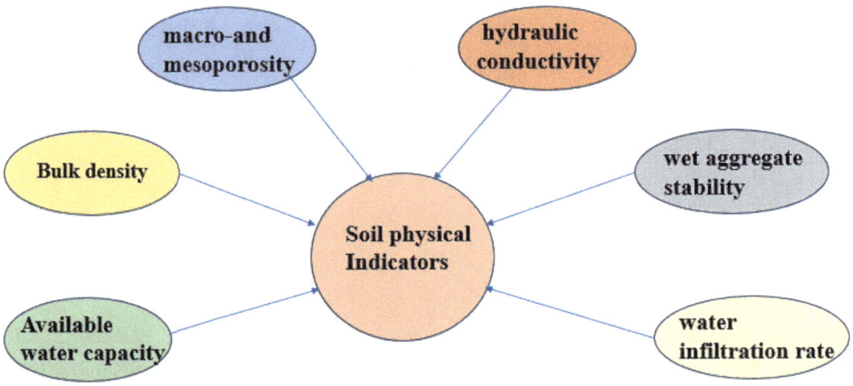

Fig. 10 Physical indicators of soil

water infiltration rate is the speed at which water enters the soil, affecting irrigation efficiency and reducing runoff and erosion. Together, these physical parameters influence soil aeration, water availability, root growth, and overall soil structure, all of which are essential for healthy soil and productive plant growth (Sakin et al., 2024) (Fig. 10).

2. ***Chemical Indicators***

Chemical parameters play vital roles in maintaining soil health and ensuring optimal plant growth (Wei et al., 2024a, 2024b). Chemical parameters include soil pH, which influences the availability of nutrients and the activity of soil microorganisms (Philippot et al., 2024). A balanced pH is essential for the uptake of nutrients by plants, as it affects the solubility of minerals and the effectiveness of microbial processes (Wang et al., 2024a, 2024b, 2024c). For instance, in highly acidic or alkaline soils, certain nutrients become less available, leading to deficiencies and reduced plant growth. Similarly, CEC measures the soil's ability to hold and exchange positively charged ions (cations), which is crucial for nutrient retention and availability. Soils with high CEC can retain more nutrients and release them slowly to plants, enhancing soil fertility and reducing the need for frequent fertilization (Yang et al., 2024). This property is particularly important for maintaining a steady supply of essential nutrients like potassium, calcium, and magnesium. On the other hand, nitrate N is a primary nutrient for plant growth, supporting photosynthesis and protein synthesis (Chen et al., 2024a, 2024b). It is a readily available form of N that plants can absorb easily, making it crucial for rapid growth and development. Adequate nitrate levels in soil contribute to lush, green foliage and overall plant vigor. K is important for water regulation, enzyme activation, and disease resistance in plants. It helps maintain osmotic balance within plant cells, regulating water uptake, and ensuring efficient nutrient transport (Ragel et al., 2019). K also plays a role in activating enzymes involved in photosynthesis and respiration, thereby enhancing plant metabolism and resistance to environmental stress (Ragel et al., 2019). On the other hand, P is vital for energy transfer, root development, and flower and fruit production (Malhotra et al., 2018). Adequate P levels promote strong root systems, which are essential for nutrient and water uptake, as well as for the development of flowers and fruits (Malhotra et al., 2018). Ca is vital for cell wall structure, root and leaf development, nutrient uptake, and stress response (Jing et al., 2024). Mg is essential for chlorophyll production and photosynthesis, and it activates enzymes for energy metabolism and nucleic acid synthesis. Fe and Zn are crucial micronutrients, because Fe supports chlorophyll synthesis and electron transport, while Zn aids in protein synthesis, hormone production, and enzyme activation (Jayakumar et al., 2024). Maintaining optimal levels of these nutrients ensures healthy plant growth, enhances disease resistance, and improves crop yields, while reducing the need for synthetic fertilizers and promoting sustainable agricultural practices. Integrating chemical and biological assessments provides a comprehensive view of soil health, leading to better management and improved agricultural outcomes.

3. ***Biological Indicators***

Biological parameters are critical components of soil health, significantly influencing its structure, fertility, and ability to support plant growth (Pandao et al., 2024). Soil organic matter content is fundamental, as it enhances soil structure by binding soil particles into aggregates, which improves porosity and reduces erosion (Pandao et al., 2024). This organic matter also increases the soil's water retention capacity, allowing it to hold more moisture and making it available to plants during dry periods. Furthermore, soil organic matter serves as a reservoir of nutrients that are slowly released as it decomposes, thus providing a continuous supply of essential nutrients to plants. Potential mineralizable nitrogen is a key indicator of the soil's capacity to supply nitrogen through the mineralization process, where organic nitrogen is converted into plant-available inorganic forms (Naasko et al., 2024). This process is crucial for plant growth, as N is a major component of chlorophyll and amino acids, both of which are essential for photosynthesis and protein synthesis. Particulate organic matter, which consists of partially decomposed plant and animal residues, is an important food source for soil microbes (Chen et al., 2024a, 2024b). These microbes play a vital role in the decomposition of organic matter and nutrient cycling, breaking down complex organic compounds into simpler forms that plants can readily absorb (Chen et al., 2024a, 2024b). For microbes to perform their activity, they need Active C, as it is a key indicator of soil microbial activity and nutrient cycling, as it provides energy for microbes (Sakin et al., 2024). High levels of aC suggest a thriving microbial community, essential for decomposing organic matter and nutrient mineralization. Similarly, nematodes, tiny soil worms, also play diverse roles in the ecosystem; beneficial nematodes regulate microbes and aid nutrient cycling, while their presence and diversity reflect soil health and biodiversity (Mourya et al., 2023). On the other hand, root bioassay evaluates soil health by assessing root growth, indicating nutrient availability and overall soil conditions (Mourya et al., 2023). Together, these biological parameters interact with chemical parameters to influence nutrient availability, soil structure, microbial activity, and overall soil fertility. A healthy balance of these factors ensures a productive soil ecosystem capable of supporting robust plant growth and sustainable agricultural practices.

10.2 Soil Carbon Sequestration and Soil Health

Soil C sequestration refers to the process of capturing atmospheric CO_2 and storing it in the soil C pool, primarily as soil organic matter (Nazir et al., 2024). This process is integral to soil health and has significant implications for climate change mitigation (Nazir et al., 2024). Healthy soils with high levels of organic C show improved soil structure, which enhances water infiltration and retention, reducing erosion and runoff (Blanco-Canqui et al., 2024). Organic matter also provides a sufficient habitat and food source for soil microorganisms, which are essential for nutrient cycling and the breakdown of organic materials (Coleman et al., 2024). Enhanced microbial activity promotes the formation of soil aggregates, which improve soil porosity and aeration, further facilitating root growth and increasing the capacity of the soil to hold nutrients and water (Wei et al., 2024a, 2024b). Moreover, soils rich in organic C are more resilient to environmental stressors, such as drought and heavy rainfall, thus supporting more stable and productive agricultural systems (Franzluebbers & Hendrickson, 2024). Therefore, effective soil C sequestration practices, such as cover cropping, reduced tillage, and the incorporation of organic amendments, not only capture C but also boost soil fertility and crop yields, creating a positive feedback loop that benefits both the environment and agricultural productivity (Fig. 11).

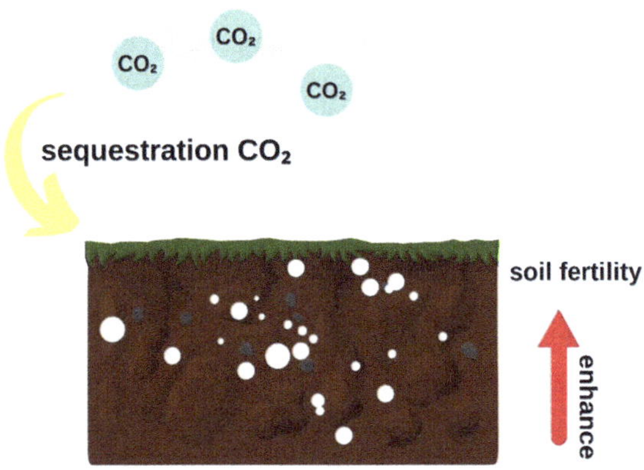

Fig. 11 Effect of CO_2 sequestration on soil

10.3 Soil Health, Environmental Quality, and Its Impact on Crop Growth

Soil health is a critical determinant of environmental quality and has a profound impact on crop growth. Healthy soils support robust plant growth by providing essential nutrients, maintaining adequate moisture levels, and facilitating root development (Mohanty et al., 2024). Soils in good health also have a balanced pH and a diverse population of beneficial microorganisms, which enhance nutrient availability and protect plants from pathogens (Wei et al., 2024a, 2024b). Improved soil health reduces the need for chemical fertilizers and pesticides, thereby decreasing the risk of water and soil pollution and promoting a more sustainable and eco-friendly agricultural system. Environmental quality is directly linked to soil health, as healthy soils can better filter pollutants, sequester carbon, and regulate water flow, contributing to cleaner air and water (Adewara et al., 2024). In turn, good environmental quality ensures that crops are grown in conditions free from excessive pollutants, which can harm plant health and reduce yields. The symbiotic relationship between soil health and environmental quality creates a sustainable agricultural ecosystem where healthy soils support vigorous crop growth, which in turn maintains and enhances soil and environmental health (Tahat et al., 2020). Practices such as crop rotation, organic farming, and integrated pest management can enhance soil health and environmental quality, leading to more resilient agricultural systems and improved crop productivity.

11 Conclusions

The physical properties of soil significantly affect its behavior in both agricultural and engineering applications. Key aspects such as soil texture and structure govern total porosity and the distribution of pore sizes, which, in turn, influence the relationships between water, heat, and air within the soil matrix. While soil texture is a static attribute, soil structure can be modified through appropriate management practices. It is crucial to conduct tillage operations at optimal soil moisture levels to prevent degradation of soil structure. Effective management of physical, chemical, and biological factors is vital for maintaining favorable physical conditions in the soil that support plant growth. Soil aeration and temperature are critical determinants of soil quality, not only for plants but also for various organisms. Soil moisture plays a significant role in influencing both aeration and temperature, as it competes with soil air and regulates temperature fluctuations. Additionally, properties such as soil consistency, plasticity, compaction, and strength are essential for assessing soil stability against loads imposed by traffic, tillage, or structural foundations. In light of the growing pressure on soil as a vital resource for food security and safety, it is imperative to emphasize the maintenance of soil physical fertility. This can be achieved through practices such as incorporating organic materials, implementing legume rotations, and adopting conservation tillage methods.

References

Abu-Hamdeh, N. H. (2003). Thermal properties of soils as affected by density and water content. *Biosystems Engineering, 86*(1), 97–102.

Abu-Hamdeh, N. H., & Reeder, R. C. (2000). Soil thermal conductivity effects of density, moisture, salt concentration, and organic matter. *Soil Science Society of America Journal, 64*(4), 1285–1290.

Adewara, O.A., Adebayo-Olajide, T. C., Ayedun, J. S., Kotun, B. C., Adeleke, A. J., Brown, A. D., Alabi, O. J., & Ogunbanwo, S. T. (2024). Soil formation, soil health and soil biodiversity. In *Prospects for soil regeneration and its impact on environmental protection* (pp. 95–121). Springer Nature Switzerland.

Alqusaireen, E., Meier, C., & Jazaei, F. (2024). High-resolution sensor system to investigate solute transport in saturated sediments. *Journal of the ASABE*.

Al-Shammary, A. A. G., Al-Shihmani, L. S. S., Fernández-Gálvez, J., & Caballero-Calvo, A. (2024). Optimizing sustainable agriculture: A comprehensive review of agronomic practices and their impacts on soil attributes. *Journal of Environmental Management, 364*, Article 121487.

Al-Shammary, A. A. G., Kouzani, A. Z., Kaynak, A., Khoo, S. Y., Norton, M., & Gates, W. (2018). Soil bulk density estimation methods: A review. *Pedosphere, 28*(4), 581–596.

Amelung, W., Tang, N., Siebers, N., Aehnelt, M., Eusterhues, K., Felde, V. J., Guggenberger, G., Kaiser, K., Kögel-Knabner, I., Klumpp, E., & Knief, C. (2024). Architecture of soil microaggregates: Advanced methodologies to explore properties and functions. *Journal of Plant Nutrition and Soil Science, 187*(1), 17–50.

Anderson, A. (2023). *Introduction to soil science*. Iowa State University Digital Press.

Barman, U., & Choudhury, R. D. (2020). Soil texture classification using multi class support vector machine. *Information Processing in Agriculture, 7*(2), 318–332.

Barrett, L. R. (2002). Spectrophotometric colour measurement in situ in well drained sandy soils. *Geoderma, 108*, 49–77.

Bartley, P. C., Amoozegar, A., Fonteno, W. C., & Jackson, B. E. (2018, August). Particle density of substrate components measured by gas pycnometer. In *International Horticultural Congress IHC2018: II International Symposium on Soilless Culture and VIII International* (Vol. 1273, pp. 17–22).

Bauer, A., & Black, A. L. (1981). Soil carbon, nitrogen, and bulk density comparisons in two cropland tillage systems after 25 years and in virgin grassland. *Soil Science Society of America Journal, 45*(6), 1166–1170.

Beltman, W. H. J., Boesten, J. J. T. I., & Van der Zee, S. E. A. T. M. (1995). Analytical modelling of pesticide transport from the soil surface to a drinking water well. *Journal of Hydrology, 169*(1–4), 209–228. https://doi.org/10.1016/0022-1694(94)02592-5

Beveridge, F. C., Williams, A., Cave, R., Kalaipandian, S., Haque, M. M., & Adkins, S. W. (2024). Environmental effects during early life-history stages and seed development on seed functional traits of an Australian native legume species. *Biology, 13*(3), 148.

Bhatt, R., & Khera, K. L. (2006). Effect of tillage and mode of straw mulch application on soil erosion in the submontaneous tract of Punjab, India. *Soil and Tillage Research, 88*(1–2), 107–115.

Blake, G. (1965). Bulk density. *Methods of Soil Analysis: Part 1 Physical and Mineralogical Properties, Including Statistics of Measurement and Sampling, 9*, 374–390.

Blanco-Canqui, H., Ruis, S. J., & Francis, C. A. (2024). Do organic farming practices improve soil physical properties? *Soil Use and Management, 40*(1), Article e12999.

Bowen, H. D. (1981). *Alleviating mechanical impedance modifying the root environment to reduce crop stress* (G. F. Arkin & H. M. Taylor (Eds.), pp.21–57). American Society of Agricultural Engineers.

Brady, N. C. (1984). *The nature and properties of soils* (9th ed.). MacMillan Publishing Co.

Brady, N. C., & Weil, R. R. (2019). *The nature and properties of soils* (15th ed.). Pearson.

Bresson, L. M., & Valentin, C. (2021). Soil surface crusting. In R. Lal (Ed.), *Encyclopedia of soil science* (pp. 1256–1259). CRC Press.

Bronick, C. J., & Lal, R. (2022). Soil structure and management: A review. *Geoderma, 124*(3–4), 3–22. https://doi.org/10.1016/j.geoderma.2022.114123

Brooks, R. M. (2009). *Soil stabilization with flyash and rice husk ash.*

Brown, S., Biswas, A., Caron, J., Dyck, M., & Si, B. (2021). Soil physics. *Digging into Canadian Soils.*

Bruun, T. B., Elberling, B., & Christensen, B. T. (2010). Lability of soil organic carbon in tropical soils with different clay minerals. *Soil Biology and Biochemistry, 42*(6), 888–895.

Bucka, F. B., Felde, V. J., Peth, S., & Kögel-Knabner, I. (2024). Complementary effects of sorption and biochemical processing of dissolved organic matter for emerging structure formation controlled by soil texture. *Journal of Plant Nutrition and Soil Science, 187*(1), 51–62.

Budhu, M. (2010). *Soil mechanics and foundations* (3rd ed.). John Wiley & Sons.

Burrough, P. A. (1993). Soil variability: A late 20th century view. *Soils Fert, 56*, 529–562.

Campbell, C. A., & Bowers, R. L. (2011). Soil compaction and its effects on plant growth. *Canadian Journal of Soil Science, 91*(2), 233–248. https://doi.org/10.4141/CJSS10072

Chamen, W. C. T., Moxey, A. P., & Towers, W. (2003). Mitigating arable soil compaction: A review and analysis of available cost and benefit data. *Soil and Tillage Research, 146*(1), 10–25. https://doi.org/10.1016/j.still.2014.09.011

Chardon, W., & Schoumans, O. (2007). Soil texture effects on the transport of phosphorus from agricultural land in river deltas of Northern Belgium, The Netherlands and North-West Germany. *Soil Use and Management, 23*, 16–24.

Chen, J., Li, J., Li, W., Li, P., Zhu, R., Zhong, Y., Zhang, W., & Li, T. (2024a). The optimal ammonia-nitrate ratio for various crops: A Meta-analysis. *Field Crops Research, 307*, Article 109240.

Chen, S., Xia, X., Ding, Y., Feng, X., Lin, Q., Li, T., Bian, R., Li, L., Cheng, K., Zheng, J., & Zhang, X. (2024b). Changes in aggregate-associated carbon pools and chemical composition of topsoil organic matter following crop residue amendment in forms of straw, manure and biochar in a paddy soil. *Geoderma, 448*, Article 116967.

Chiemeka, I. U. (2010). Soil temperature profile at Uturu, Nigeria. *Pacific Journal of Science and Technology, 11*(1), 478–482.

Coleman, D. C., Geisen, S., & Wall, D. H. (2024). Soil fauna: Occurrence, biodiversity, and roles in ecosystem function. In *Soil microbiology, ecology and biochemistry* (pp. 131–159). Elsevier.

Conservation Tillage Information Center (CTIC), M. R. (2004). *National crop residue management survey*. West Lafayette, IN: Conservation Technology Information Center.

Corami, A. (2024). Sustainable remediation methods for mining wastes. In *Sustainable management of mining waste and tailings* (pp. 254–288). CRC Press.

Corsi, S., Friedrich, T., Kassam, A., Pisante, M., & Sà, J. D. M. (2012). *Soil organic carbon accumulation and greenhouse gas emission reductions from conservation agriculture: a literature review*. Food and Agriculture Organization of the United Nations (FAO).

Craig, R. F. (2019). *Craig's soil mechanics* (9th ed.). CRC Press.

Dahiya, R., Ingwersen, J., & Streck, T. (2007). The effect of mulching and tillage on the water and temperature regimes of a loess soil: Experimental findings and modeling. *Soil and Tillage Research, 96*(1–2), 52–63.

Das, B. M. (2020). *Principles of geotechnical engineering* (9th ed.). Cengage Learning.

David, O. O. (1999). Improvement in field texture accuracy for sustainable agriculture. *Journal of Sustainable Agriculture, 15*(2–3), 61–68.

David Raj, A., Kumar, S., & Sooryamol, K. R. (2024). Soil erodibility mapping using remote sensing and in situ soil data with random forest model in a mountainous catchment of Indian Himalayas. *Environmental Monitoring and Assessment, 196*(11), 1–21.

De Corato, U., Viola, E., Keswani, C., & Minkina, T. (2024). Impact of the sustainable agricultural practices for governing soil health from the perspective of a rising agri-based circular bioeconomy. *Applied Soil Ecology, 194*, Article 105199.

Decker, K., Wang, D., Waite, C., & Scherbatskoy, T. (2003). Snow removal and ambient air temperature effects on forest soil temperatures in northern Vermont. *Soil Science Society of America Journal, 67*(4), 1234–1242.

Deruyttere, E. (2024). Checkmate coagulation-flocculation: Investigating the electrochemical dynamics of coagulation and flocculation (Doctoral dissertation, Ghent University).

Dexter, A. R. (2020). Soil physical quality and soil structure. *European Journal of Soil Science, 71*(3), 357–365. https://doi.org/10.1111/ejss.12901

Di Bonito, M., De Feudis, M., Smith, B., & Zhang, H. (2024). Extraction and characterization of pore water from contaminated soils. In *Environmental geochemistry* (pp. 271–328). Elsevier.

Dittmar, T., & Lennartz, S. T. (2024). Reasons behind the long-term stability of dissolved organic matter. In *Biogeochemistry of marine dissolved organic matter* (pp. 613–655). Academic Press.

Elias, E.A., Cichota, R., Torriani, H.H., & De Jong Van Lier, Q. (2004). Analytical soil–temperature model: Correction for temporal variation of daily amplitude. *Soil science society of America Journal, 68*(3), 784–788.

Elizbarashvili, E. S., Urushadze, T., Elizbarashvili, M., Elizbarashvili, S. E., & Schaefer, M. (2010). Temperature regime of some soil types in Georgia. *Eurasian Soil Science, 43*, 427–435.

Enokela, J. A., & Othoigbe, T. O. (2015). An automated greenhouse control system using Arduino prototyping platform. *Australian Journal of Engineering Research, 1*(1), 64–73.

Esfahani, M. N., & Sonnewald, U. (2024). Unlocking dynamic root phenotypes for simultaneous enhancement of water and phosphorus uptake. *Plant Physiology and Biochemistry, 207*, 108386.

Fang, X. M., Ono, Y., Fukusawa, H., Pan, B. T., Li, J. J., Guan, D. H., Oi, K., Tsukamoto, S., Torii, M., & Mishima, T. (1998). Asian summer monsoon instability during the past 60.000years: Magnetic susceptibility and pedogenic evidence from the western Chinese Loess Plateau. *Earth and Planetary Science Letters, 168*, 219–232. https://doi.org/10.1016/S0012-821X(99)00053-9

Fenner, M. (2000). *Seeds: The ecology of regeneration in plant communities*. CABI Publishing.

Franzluebbers, A. J., & Hendrickson, J. R. (2024). Should we consider integrated crop–livestock systems for ecosystem services, carbon sequestration, and agricultural resilience to climate change? *Agronomy Journal, 116*(2), 415–432.

Gahoonia, T., & Nielsen, N. (2003). Phosphorus (P) uptake and growth of a root hairless barley mutant (bald root barley, brb) and wild type in low-and high-P soils. *Plant, Cell & Environment, 26*(10), 1759–1766.

Galdino, S., Sano, E. E., Andrade, R. G., Grego, C. R., Nogueira, S. F., Bragantini, C., & Flosi, A. H. (2016). Large-scale modeling of soil erosion with RUSLE for conservationist planning of degraded cultivated Brazilian pastures. *Land Degradation & Development, 27*(3), 773–784.

Gandhi, K. (2012). Expansive soil stabilization using bagasse ash. *International Journal of Engineering Research & Technology (IJERT), 1*(5), 2181–2278.

Gao, C., Zhang, S., Chen, M., Cheng, L., Zhang, X., & Ding, X. (2024). Appropriate amount application of manure is conducive to improve soil aggregate stability in Saline-Alkaline soil: analyzed from soil organic carbon mediated by microbial necromass carbon. *Journal of Soil Science and Plant Nutrition, 24*, 1–15.

Geiger, R., Aron, R. H., & Todhunter, P. (2009). *The climate near the ground*. Rowman & Littlefield.

Ghorbani, M., & Amirahmadi, E. (2024). Insights into soil and biochar variations and their contribution to soil aggregate status—A meta-analysis. *Soil and Tillage Research, 244*, Article 106282.

Ghuman, R., Singh, H., & Singh, K. (2024). Tillage, crop residue management and nitrogen application impacts on soil properties under maize–wheat (Zea mays–Triticum aestivum) cropping system. *Indian Journal of Agronomy, 69*(2), 137–143.

Giap, S. G. E., & Ahmad, M. F. (2024). Factors affecting soil bulk density: A conceptual model. *Journal of Soil, Environment & Agroecology, 1*(1), 27–45.
Gomez, A., Khanna, S., & Hoag, D. (2002). The effect of soil friction on water infiltration and its implications for agricultural practices. *Soil Science Society of America Journal, 66*(5), 1702–1709. https://doi.org/10.2136/sssaj2002.1702
Gordillo, L., Zabaloy, M. C., & Stanchi, S. (2020). Soil physical properties and their effect on root penetration and growth. *Geoderma, 369*, 114–124. https://doi.org/10.1016/j.geoderma.2020.114124
Günal H., Erşahin, S., Yetgin, B., & Kutlu, B. (2008). Use of chroma-meter measured color parameters inestimating color related soil variables. *Communications in Soil Science and Plant Analysis, 39*(5–6), 726–740. https://doi.org/10.1501/Tarimbil_0000000425
Guo, Z., Qiang, W., He, J., Han, X., Tan, X., Ludwig, B., Shen, W., Kuzyakov, Y., & Gunina, A. (2024). Nitrogen deposition raises temperature sensitivity of soil organic matter decomposition in subtropical forest. *Science of the Total Environment, 907*, Article 167925.
Gupta, S. K., Kumar, M., & Srivastava, A. K. (2024, April). Hydraulic conductivity of soils: A comprehensive review of the impacts of chemicals, soil salinity, organic matter, and land use. In *IOP conference series: Earth and environmental science* (Vol. 1327, No. 1, p. 012032). IOP Publishing.
Hamza, M. A., & Anderson, W. K. (2005). Soil compaction in cropping systems: A review of the nature, causes, and possible solutions. *Soil and Tillage Research, 82*(2), 121–145. https://doi.org/10.1016/j.still.2004.08.009
Hartmann, M., & Six, J. (2023). Soil structure and microbiome functions in agroecosystems. *Nature Reviews Earth & Environment, 4*(1), 4–18.
He, H., Yanbing, Q., Jialong, L., Peipei, P., Ziran, Y., Ziqian, T., Ke, C., & Kaiyue, Z. (2023). Development and revision of the Chinese soil texture classification system. *Journal of Agricultural Resources and Environment, 40*(3), 501–510.
He, Y., Wu, Z., Zhao, T., Yang, H., Ali, W., & Chen, J. (2022). Different plant species exhibit contrasting root traits and penetration to variation in soil bulk density of clayey red soil. *Agronomy Journal, 114*(1), 867–877.
Heinen, M., Mulder, M., van Dam, J., Bartholomeus, R., van Lier, Q. D. J., de Wit, J., de Wit, A., & Hack-ten Broeke, M. (2024). SWAP 50 years: Advances in modelling soil-water-atmosphere-plant interactions. *Agricultural Water Management, 298*, Article 108883.
Hillel, D. (2020). *Introduction to environmental soil physics* (2nd ed.). Academic Press.
Hoffrogge, P. W. (2024). *Understanding degradation phenomena in solid-oxide fuel-cell anodes by phase-field modeling and analytics.*
Horton, R., Bristow, K. L., Kluitenberg, G., & Sauer, T. (1996). Crop residue effects on surface radiation and energy balance. *Theoretical and Applied Climatology, 54*, 27–37.
https://www.fao.org/soils-portal/data-hub/soil-classification/numerical-systems/chemical-properties/zh/
Isaak, M., Azawi, A., & Turky, T. (2024). Influence of various tillage systems and tillage speed on some soil physical properties. *Progress in Agricultural Engineering Sciences, 20*(1), 37–47.
Jarvis, N., Coucheney, E., Lewan, E., Klöffel, T., Meurer, K. H., Keller, T., & Larsbo, M. (2024). Interactions between soil structure dynamics, hydrological processes, and organic matter cycling: A new soil-crop model. *European Journal of Soil Science, 75*(2), Article e13455.
Jayakumar, S., Abhangrao, A. K., Sarje, R. A., Gupta, R., Pathania, S., & Sree, B. V. (2024). Critical analysis on effect of micronutrients on flowering plants: A review. *International Journal of Plant & Soil Science, 36*(6), 776–782.
Jha, Y. (2024). Regulation of photosynthesis under stress. In *Improving stress resilience in plants* (pp. 35–48). Academic Press.
Ji, J. F., Chen J., Balsam, W., & Liu, L. W. (2007). Quantitative analysis of hematite and goethite in the Chinese loess–paleosol sequences and its implication for dry and humid variability. *Quaternary Sciences, 27*, 221–229. https://doi.org/10.1346/000986002760832801

Jia, M., & Zheng, Y. (2024, May). A prediction model of maximum dry density of over coarse-grained soil. In *IOP conference series: Earth and environmental science* (Vol. 1330, No. 1, p. 012050). IOP Publishing.

Jiménez, C., Tejedor, M., & Rodríguez, M. (2007). Influence of land use changes on the soil temperature regime of Andosols on Tenerife, Canary Islands, Spain. *European Journal of Soil Science, 58*(2), 445–449.

Jing, T., Li, J., He, Y., Shankar, A., Saxena, A., Tiwari, A., Maturi, K. C., Solanki, M. K., Singh, V., Eissa, M. A., & Ding, Z. (2024). Role of calcium nutrition in plant Physiology: Advances in research and insights into acidic soil conditions—A comprehensive review. *Plant Physiology and Biochemistry*, 108602.

Kaiser, E., Weerheim, K., Schipper, R., & Dieleman, J. A. (2019). Partial replacement of red and blue by green light increases biomass and yield in tomato. *Scientia Horticulturae, 249*, 271–279.

Kalra, A., Goel, S., & Elias, A. A. (2024). Understanding role of roots in plant response to drought: Way forward to climate-resilient crops. *The Plant Genome, 17*(1), Article e20395.

Karlen, D. L., Berry, E. C., Colvin, T. S., & Kanwar, R. S. (1991). Twelve-year tillage and crop rotation effects on yields and soil chemical properties in northeast Iowa. *Communications in Soil Science and Plant Analysis, 22*(19–20), 1985–2003.

Keenan, M. J., Hegsted, M., Jones, K. L., Delany, J. P., Kime, J. C., Melancon, L. E., Tulley, R. T., & Hong, K. D. (1997). Comparison of bone density measurement techniques: DXA and Archimedes' principle. *Journal of Bone and Mineral Research, 12*, 1903–1907.

Kettler, T., Doran, J. W., & Gilbert, T. (2001). Simplified method for soil particle-size determination to accompany soil-quality analyses. *Soil Science Society of America Journal, 65*(3), 849–852.

Khalid, A., Khan, S. M., Ali, S., Rana, M., & Ahmad, Z. (2024). Carbon fraction and pools in plants and soil. In *Agroforestry for carbon and ecosystem management* (pp. 135–146). Academic Press.

Kirkby, M. J., & Brunsden, D. (2020). *Processes and landforms in geomorphology* (2nd ed.). Routledge.

Koirala, R., Thapa, B., Neopane, H. P., Zhu, B., & Chhetry, B. (2016). Sediment erosion in guide vanes of Francis turbine: A case study of Kaligandaki Hydropower Plant, Nepal. *Wear, 362*, 53–60.

Konen, M. E., Burras, C. L., & Sandor, J. A. (2003). Organic carbon, texture, and quantitative color measurement relationships for cultivated soils in North Central Iowa. *Soil Science Society of America Journal, 67*, 1823–1830. https://doi.org/10.2136/sssaj2003.1823

Koorneef, G. J., Pulleman, M. M., Comans, R. N., Van Rijssel, S. Q., Barré, P., Baudin, F., & De Goede, R. G. (2024). Assessing soil functioning: What is the added value of soil organic carbon quality measurements alongside total organic carbon content? *Soil Biology and Biochemistry*, 109507.

Kumari, S., Devi, M., Thakur, K., Minhas, B., Bhatt, A. K., & Kaushik, N. (2024). Impact of anthropogenic activities on microbial diversity and soil health. In *Advancements in microbial biotechnology for soil health* (pp. 227–248). Springer Nature Singapore.

Lal, R. (1990). *Soil erosion in the tropics: Principles and management*. McGraw-Hill Inc.

Lal, I. (2024). In-field optimization of soil water parameters for irrigation scheduling.

Lambe, T. W., & Whitman, R. V. (2019). *Soil mechanics* (2nd ed.). Wiley.

Lardy, J. M., Desutter, T. M., Daigh, A. L., Meehan, M. A., & Staricka, J. A. (2022). Effects of soil bulk density and water content on penetration resistance. *Agricultural & Environmental Letters, 7*, Article e20096.

Le Bissonnais, Y., & Arshad, M. (2020). The influence of soil organic matter on aggregate stability and soil crusting. *European Journal of Soil Science, 71*(4), 565–574. https://doi.org/10.1111/ejss.12900

Lehnert, M. (2013). The soil temperature regime in the urban and suburban landscapes of Olomouc, Czech Republic. *Moravian Geographical Reports, 21*(3), 27–36.

Lichtenberg, E. (2024). Thinking about soil health: A conceptual framework. *Soil Security, 14*, Article 100130.

Liu, J., Wang, J., Liu, Y., Wang, L., Long, L., He, J., Huang, C., Shen, F., & Zhang, Y. (2024a). A hygroscopic dual-porous road dust suppressant with good dust suppression prepared by the freeze-thawing method. *Journal of Cleaner Production, 448*, Article 141677.

Liu, J., Wang, Y., Tian, Y., Zhuang, Y., & Lu, S. (2024b). Pore structure analysis via X-ray μCT: Enhancing soil macro aggregate models for water permeability and carbon sequestration. *Geoderma Regional, 36*, Article e00764.

Liu, T., Dong, X., Guo, K., Wang, J., Liu, X., & Sun, H. (2024c). Effects of fertilizer types and application levels on phosphorus availability of saline soils and crops: A meta-analysis. *Land Degradation & Development, 35*(13), 4068–4080.

Lopez, F. B., Barclay, G. F., & Badal, S. (2024). Plant anatomy and physiology. In *Pharmacognosy* (pp. 29–48). Academic Press.

Low, J. E., Loveridge, F. A., Powrie, W., & Nicholson, D. (2015). A comparison of laboratory and in situ methods to determine soil thermal conductivity for energy foundations and other ground heat exchanger applications. *Acta Geotechnica, 10*, 209–218.

Lubarda, M. V., & Lubarda, V. A. (2024). A note on the solution to one-dimensional advection-dispersion-reaction problems. *Chemical Engineering Communications, 211*(5), 689–696.

Madzhi, N., & Akhsan, M. N. (2021). Control of plant growth by monitoring soil moisture, temperature and humidity in dry climate. In: *IOP conference series: Materials science and engineering*.

Malhotra, H., Vandana, S., & Pandey, R. (2018). Phosphorus nutrition: Plant growth in response to deficiency and excess. In: *Plant nutrients and abiotic stress tolerance* (pp. 171–190).

Mallick, S., & Agmon, N. (2024). *Lateral diffusion of ions near membrane surface.* Physical Chemistry Chemical Physics.

Manghwar, H., Hussain, A., Alam, I., Khoso, M. A., Ali, Q., & Liu, F. (2024). Waterlogging stress in plants: Unraveling the mechanisms and impacts on growth, development, and productivity. *Environmental and Experimental Botany*, 105824.

McBratney, A. B. (1994). Allocation of new individuals to continuous soil classes. *Soil Research, 32*(4), 623–633.

McCarthy, D. F. (2017). *Essentials of soil mechanics and foundations: Basic geotechnics* (8th ed.). Pearson.

Melville, M. D., & Atkinson, G. (1985). Soil color: Its measurement and its designation in models of uniform color space. *Journal of Soil Science, 36*, 495–512. https://doi.org/10.1111/j.1365-2389.1985.tb00353.x

Mohanty, L. K., Singh, N. K., Raj, P., Prakash, A., Tiwari, A. K., Singh, V., & Sachan, P. (2024). Nurturing crops, enhancing soil health, and sustaining agricultural prosperity worldwide through agronomy. *Journal of Experimental Agriculture International, 46*(2), 46–67.

Mourya, D., Yadav, R., Singh, D., Upadhyay, A., & Bhan, T. (2023). Soil nematodes serve as biological indicators for evaluating the soil health: A review.

Mualem, Y., & Assouline, S. (2021). Soil crust formation and its impact on water and solute transport. *Hydrological Processes, 35*(8), e14250. https://doi.org/10.1002/hyp.14250

Murmu, J., Kundu, A., Saha, S., Ghosh, S., Ghatak, P., More, N. B., & Bandyopadhyay, P. K. (2024). Effect of soil compaction on pore characteristics, air entry potential, S index and plant available water content of some soil groups of India. *Journal of the Indian Society of Soil Science, 72*(1), 66–76.

Murthy, V. N. S. (2021). *Geotechnical engineering: Principles and practices of soil mechanics and foundation engineering* (3rd ed.). CBS Publishers.

Mutai, J. C., Fonte, S. J., Vanek, S., Stewart, J. E., & Schipanski, M. E. (2024). Impacts of cropping system and nutrient management on soil health and soil-borne pathogens in smallholder systems of western Kenya.

Naasko, K., Martin, T., Mammana, C., Murray, J., Mann, M., & Sprunger, C. (2024). Soil protein: A key indicator of soil health and nitrogen management. *Soil Science Society of America Journal, 88*(1), 89–108.

Naylor, D., Sadler, N., Bhattacharjee, A., Graham, E. B., Anderton, C. R., McClure, R., Lipton, M., Hofmockel, K. S., & Jansson, J. K. (2020). Soil microbiomes under climate change and implications for carbon cycling. *Annual Review of Environment and Resources, 45*(1), 29–59.
Nazir, M. J., Li, G., Nazir, M. M., Zulfiqar, F., Siddique, K. H., Iqbal, B., & Du, D. (2024). Harnessing soil carbon sequestration to address climate change challenges in agriculture. *Soil and Tillage Research, 237*, Article 105959.
Nelson, J. T., Adjuik, T. A., Moore, E. B., VanLoocke, A. D., Ramirez Reyes, A., & McDaniel, M. D. (2024). A simple, affordable, do-it-yourself method for measuring soil maximum water holding capacity. *Communications in Soil Science and Plant Analysis, 55*(8), 1190–1204.
Ng, J. F., Ahmed, O. H., Omar, L., Jalloh, M. B., Kwan, Y. M., Musah, A. A., & Chowdhury, A. J. K. (2024). Improving soil pH, pH buffering capacity, and ECEC of an acid soil to regulate nutrient retention and mitigate water pollution. *Desalination and Water Treatment, 319*, 100491.
Nick, S. M., Copty, N. K., Saygın, S. D., Öztürk, H. S., Demirel, B., Emadian, S. M., Erpul, G., Sedighi, M., & Babaei, M. (2024). Impact of soil compaction and irrigation practices on salt dynamics in the presence of a saline shallow groundwater: An experimental and modelling study. *Hydrological Processes, 38*(3), Article e15135.
Nimmo, J. (2004). Porosity and pore-size distribution. In *Elsevier eBooks* (pp. 295–303).
Nwankwo, C., & Ogagarue, D. (2012). An investigation of temperature variation at soil depths in parts of Southern Nigeria. *American Journal of Environmental Engineering, 2*(4), 142–147.
Ochsner, T. E., Horton, R., & Ren, T. (2001). A new perspective on soil thermal properties. *Soil Science Society of America Journal, 65*(6), 1641–1647.
Onwuka, B., & Mang, B. (2018). Effects of soil temperature on some soil properties and plant growth. *Advanced Plants Agriculture Research, 8*(1), 34–37.
Or, D., Keller, T., & Schlesinger, W. H. (2021). Natural and managed soil structure: On the fragile scaffolding for soil functioning. *Soil and Tillage Research, 208*, Article 104912.
Pandao, M. R., Thakare, A. A., Choudhari, R. J., Navghare, N. R., Sirsat, D. D., & Rathod, S. R. (2024). Soil health and nutrient management. *International Journal of Plant & Soil Science, 36*(5), 873–883.
Pandey, D., Kehri, H. K., Zoomi, I., Chaturvedi, S., & Chaudhary, K. L. (2022). Seasonal effect on the diversity of soil fungi and screening for arsenic tolerance and their remediation. *Journal of Applied Biology & Biotechnology, 10*(1), 40–46.
Parslow, K., & Jennings, B. (1986). Simultaneous size and thickness measurements for heterogeneous micrometre-sized particles. *Journal of Physics d: Applied Physics, 19*(7), 1233.
Peng, J., Ma, F., Quan, Q., Liao, J., Chen, C., Wang, Y., Tang, J., Sun, C., Zhou, Q., & Niu, S. (2024). Nitrogen deposition differentially regulates the sensitivity of gross primary productivity to extreme drought versus wetness. *Global Change Biology, 30*(7), Article e17428.
Pessoa, T. N., Batista, A. M., de Sousa, R. N., & Libardi, P. L. (2024). Movement of Water and Solutes in Agricultural Soils.
Philippot, L., Chenu, C., Kappler, A., Rillig, M. C., & Fierer, N. (2024). The interplay between microbial communities and soil properties. *Nature Reviews Microbiology, 22*(4), 226–239.
Plett, D. C., Ranathunge, K., Melino, V. J., Kuya, N., Uga, Y., & Kronzucker, H. J. (2020). The intersection of nitrogen nutrition and water use in plants: New paths toward improved crop productivity. *Journal of Experimental Botany, 71*(15), 4452–4468.
Poesen, J., Boardman, J., & Vandaele, J. (2020). Soil erosion processes on agricultural land. Springer.
Ragel, P., Raddatz, N., Leidi, E. O., Quintero, F. J., & Pardo, J. M. (2019). Regulation of K+ nutrition in plants. *Frontiers in Plant Science, 10*, 281.
Ray, S., & Majumder, S. (2024). Water management in agriculture: Innovations for efficient irrigation. In: P. Sil, P. Chhetri, S. Majumder, & D.T. Santosh (Eds.), *Modern agronomy* (pp. 169–185).
Rizvi, S. S. (2024). Mass transfer: basic concepts. *Food engineering principles and practices: A one-semester course* (pp. 429–457). Springer International Publishing.
Saeibehrouzi, A., Abolfathi, S., Denissenko, P., & Holtzman, R. (2024). Pore-scale modeling of solute transport in partially-saturated porous media. *Earth-Science Reviews, 256*, 104870.

Sakin, E., Yanardağ, H. İ, Firat, Z., Çelik, A., Beyyavaş, V., & Suat, C. U. N. (2024). Some indicators for the assessment of soil health: A mini review. *MAS Journal of Applied Sciences, 9*(2), 297–310.
Samborska-Goik, K., & Pogrzeba, M. (2024). A critical review of the modelling tools for the reactive transport of organic contaminants. *Applied Sciences, 14*(9), 3675.
Sanchez-Maranon, M., Soriano, M., Melgosa, M., Delgado, G., & Delgado, R. (2004). Quantifying the effects of aggregation, particle size and components on the colour of Mediterranean soils. *European Journal of Soil Science, 55*, 551–565. https://doi.org/10.1111/j.1365-2389.2004.00624.x
Sánchez-Maraón, M., Delgado, G., Melgosa, M., Hita, E., & Delgado, R. (1997). Cielab color parameters and their relationship to soil characteristics in Mediterranean red soils. *Soil Science, 162*(11), 833–842.
Sándor, R., & Fodor, N. (2012). Simulation of soil temperature dynamics with models using different concepts. *The Scientific World Journal, 2012*(1), Article 590287.
Sapes, G., Schroeder, L., Scott, A., Clark, I., Juzwik, J., Montgomery, R. A., Guzmán, Q. J. A., & Cavender-Bares, J. (2024). Mechanistic links between physiology and spectral reflectance enable pre visual detection of oak wilt and drought stress. *Proceedings of the National Academy of Sciences, 121*(7), e2316164121.
Schärli, U., & Rybach, L. (2001). Determination of specific heat capacity on rock fragments. *Geothermics, 30*(1), 93–110. https://doi.org/10.1016/S0375-6505(01)00003-7
Schroeter-Zakrzewska, A., & Pradita, F. A. (2021). Effect of colour of light on rooting cuttings and subsequent growth of chrysanthemum (Chrysanthemum× grandiflorum Ramat./Kitam.). *Agriculture, 11*(7), 671.
Schwertmann, U. (1993). Relations between iron oxide, soil color and soil formation. In: J. M. Bigham & E. J. Ciolkosz (Eds.), *Soil color* (Vol. 31, pp. 51–69). SSSA Special Publication. https://doi.org/10.2136/sssaspecpub31.c4
Science, A.-N. A. O. (1978). *Journal of the Arizona-Nevada Academy of Science* (Vol. 13). Arizona-Nevada Academy of Science.
Scott, I. S. P. C., Nouwakpo, K., Bjorneberg, D., Rogers, C., & Vitko, L. (2024). Establishing a standard protocol for soil texture analysis using the laser diffraction technique. *Soil Science Society of America Journal, 88*(5), 1691–1708.
Shaaban, M., & Nunez-Delgado, A. (2024). Soil adsorption potential: Harnessing Earth's living skin for mitigating climate change and greenhouse gas dynamics. *Environmental Research, 251*, Article 118738.
Shahane, A. A., & Shivay, Y. S. (2021). Soil health and its improvement through novel agronomic and innovative approaches. *Frontiers in Agronomy, 3*, Article 680456.
Shield, R. T. (1968). Optimal design of multi-purpose structures. *International Journal of Solids and Structures, 4*(4), 469–475.
Singer, M. J. (1991). Physical properties of arid region soils. In *Semiarid lands and deserts: soils resource and reclamation* (pp. 81–109).
Singh, H., Dunn, B. L., Fontanier, C., & Singh, H. (2023). Colored shade nets affect growth but not flowering of four greenhouse-grown potted ornamental species. *HortScience, 58*(9), 1075–1076.
Singh, M., Sarkar, B., Biswas, B., Bolan, N. S., & Churchman, G. J. (2017). Relationship between soil clay mineralogy and carbon protection capacity as influenced by temperature and moisture. *Soil Biology and Biochemistry, 109*, 95–106.
Singh, N. K., Sachan, K., Manoj, B. P., Panotra, N., & Katiyar, D. (2024). Building soil health and fertility through organic amendments and practices: A review. *Asian Journal of Soil Science and Plant Nutrition, 10*(1), 175–197.
Snyder, R. L., Shackel, K. A., Sanden, B., Fulton, A. E., & Suvočarev, K. (2024). Irrigation scheduling. In: *Microirrigation for crop production* (pp. 41–132). Elsevier Science.
Soane, B. D., & van Ouwerkerk, C. (1994). Soil compaction problems in world agriculture. In Soil Compaction in Crop Production (pp. 1–22). Elsevier. https://doi.org/10.1016/B978-0-444-88286-8.50007-5

Sojka, R. E., Karlen, D. L., McCarty, G. W., & Roberts, B. W. (2001). Penetration resistance, bulk density, and strength relationships for some agricultural soils. *Soil Science Society of America Journal, 65*(3), 830–835. https://doi.org/10.2136/sssaj2001.653830x

SSSA. (1979). Glossary of soil science terms. Soil Science Society of America.

Sun, Z., Li, H., Hu, J., Wu, X., Su, R., Yan, L., Sun, X., Shaaban, M., Wang, Y., Quénéa, K., & Hu, R. (2023). Fe (III) stabilizing soil organic matter and reducing methane emissions in paddy fields under varying flooding conditions. *Ecotoxicology and Environmental Safety, 259*, Article 114999.

Sytar, O., Zivcak, M., Neugart, S., Toutounchi, P. M., & Brestic, M. (2019). Precultivation of young seedlings under different color shades modifies the accumulation of phenolic compounds in Cichorium leaves in later growth phases. *Environmental and Experimental Botany, 165*, 30–38.

Tahat, M. M., Alananbeh, K. A., Othman, Y. I., & Leskovar, D. (2020). Soil health and sustainable agriculture. *Sustainability, 12*(12), 4859.

Tang, Z., Zhang, W., Chen, J., Wang, L., & Zhang, Y. (2024). Contributions of soil organic carbon-induced root-and soil properties complexity to water flow in eastern China. *Science of The Total Environment*, 174125.

Terzaghi, K., Peck, R. B., & Mesri, G. (2021). Soil mechanics in engineering practice (4th ed.). Wiley.

Tiab, D., & Donaldson, E. C. (2024). *Petrophysics: Theory and practice of measuring reservoir rock and fluid transport properties*. Elsevier.

Tuo, D., Xu, M., Li, Q., & Liu, S. (2017). Soil aggregate stability and associated structure affected by long-term fertilization for a loessial soil on the loess plateau of China. *Polish Journal of Environmental Studies, 26*(2).

Umeh, A. C., Naidu, R., Olisa, E., Liu, Y., Qi, F., & Bekele, D. (2024). A systematic investigation of single solute, binary and ternary PFAS transport in water-saturated soil using batch and 1-dimensional column studies: Focus on mixture effects. *Journal of Hazardous Materials, 461*, Article 132688.

Vezzani, F. M., Anderson, C., Meenken, E., Gillespie, R., Peterson, M., & Beare, M. H. (2018). The importance of plants to development and maintenance of soil structure, microbial communities and ecosystem functions. *Soil and Tillage Research, 175*, 139–149.

Viscarra Rossel, R. A., Minasny, B., Roudier, P., McBratney, A. B. (2006). Colour space model for soil science. *Geoderma, 133*(3–4), 320–337. https://doi.org/10.1016/j.geoderma.2005.07.017

Voothuluru, P., Wu, Y., & Sharp, R. E. (2024). Not so hidden anymore: Advances and challenges in understanding root growth under water deficits. *The Plant Cell, 36*(5), 1377–1409.

Voroney, R. P., Heck, R. J., & Kuzyakov, Y. (2024). The habitat of the soil biota. In *Soil microbiology, ecology and biochemistry* (pp. 13–40). Elsevier.

Wan, Y., Liu, J., Zhuang, Z., Wang, Q., & Li, H. (2024). Heavy metals in agricultural soils: Sources, influencing factors, and remediation strategies. *Toxics, 12*(1), 63.

Wang, J., Hansen, T. K., Kenkel, N. C., & Liu, Z. (2023). Soil surface crusting and aggregate stability under conservation tillage. *Soil and Tillage Research, 230*, 105165. https://doi.org/10.1016/j.still.2023.105165

Wang, L., Cheng, L., Yin, S., & Chen, J. (2024a). Flocculation, dewatering and sedimentation behaviour of mine backfill slurry in deep cone thickener (DCT). *Journal of Environmental Chemical Engineering*, 112168.

Wang, L., Lu, P., Feng, S., Hamel, C., Sun, D., Siddique, K. H., & Gan, G. Y. (2024b). Strategies to improve soil health by optimizing the plant–soil–microbe–anthropogenic activity nexus. *Agriculture, Ecosystems & Environment, 359*, Article 108750.

Wang, Q., Xue, H., John Martin, J. J., Hou, M., Cao, H., Dong, Z., Li, J., & Sun, C. (2024c). Trends and applications of computed tomography in agricultural non-destructive testing. *Agriculture, 14*(12), 2329.

Wang, Y., Lu, S., Ren, T., & Li, B. (2011). Bound water content of air-dry soils measured by thermal analysis. *Soil Science Society of America Journal, 75*(2), 481–487.

Wei, X., Xie, B., Wan, C., Song, R., Zhong, W., Xin, S., & Song, K. (2024a). Enhancing soil health and plant growth through microbial fertilizers: Mechanisms, benefits, and sustainable agricultural practices. *Agronomy, 14*(3), 609.

Wei, Y., Chen, Y., Cao, X., Xiang, M., Huang, Y., & Li, H. (2024b). A critical review of groundwater table fluctuation: Formation, effects on multifields, and contaminant behaviors in a soil and aquifer system. *Environmental Science & Technology, 58*(5), 2185–2203.

Whalley, W. R., Dumitru, J. K., & Dexter, A. R. (2005). Water content, pore size distribution, and soil shear strength in relation to root growth. *Plant and Soil, 267*(1–2), 47–54. https://doi.org/10.1007/s11104-005-0775-7

Wilczewski, E., Jug, I., Szpunar-Krok, E., Staniak, M., & Jug, D. (2024). Shaping soil properties and yield of cereals using cover crops under conservation soil tillage. *Agronomy, 14*(9), 2104.

Wu, J., & Nofziger, D. (1999). Incorporating temperature effects on pesticide degradation into a management model (0047-2425).

Wu, J. Y., Gao, J. M., Pei, Y. Z., Luo, K. Y., Yang, W. H., Wu, J. C., Yue, X. H., Wen, J., & Luo, Y. (2024). Microplastics in agricultural soils: A comprehensive perspective on occurrence, environmental behaviors and effects. *Chemical Engineering Journal*, 151328.

Wu, B. S., Mansoori, M., Schwalb, M., Islam, S., Naznin, M. T., Addo, P. W., MacPherson, S., Orsat, V., & Lefsrud, M. (2024a). Light emitting diode effect of red, blue, and amber light on photosynthesis and plant growth parameters. *Journal of Photochemistry and Photobiology b: Biology, 256*, Article 112939.

Yan, S., Chen, H., Wang, S., Chang, X., Guan, X., Wu, J., Liu, Y., Zhang, D., & Zhang, B. (2024). Soil cations and texture are crucial in forming smooth water and salt transport channels between cultivated land and salinized wasteland. *CATENA, 244*, Article 108251.

Yang, M., Zhou, D., Hang, H., Chen, S., Liu, H., Su, J., Lv, H., Jia, H., & Zhao, G. (2024). Effects of balancing exchangeable cations Ca, Mg, and K on the growth of tomato seedlings (Solanum lycopersicum L.) based on increased soil cation exchange capacity. *Agronomy, 14*(3), 629.

Yerli, C., Sahin, U., & Kızıloğlu, F. (2024). Improving the soil physical and hydraulic properties by irrigation with wastewater under different soil tillage management. *Journal of Agricultural Faculty of Gaziosmanpaşa University (JAFAG), 41*(2), 72–85.

Yudina, A., & Kuzyakov, Y. (2023). Dual nature of soil structure: The unity of aggregates and pores. *Geoderma, 434*, Article 116478.

Yunusova, N. A., & Osorio, A. A. (2022). *Soil structure.* Elsevier eBooks. https://doi.org/10.1016/b978-0-12-822974-3.00134-8

Zhang, K. F., Li, C., Hu, Z. F., Huang, S. Q., Chen, J. S., & Ma, X. F. (2020). Simulations of water cycle in the soil-crop system: model improvement and validation. *Applied Ecology & Environmental Research, 18*(2), 2163–2177.

Zhang, X., Liu, Z., & Han, Y. (2024). Progress towards the identification and improvement of dispersive soils: A review. *European Journal of Soil Science, 75*(5), Article e70002.

Zhao, Y., Ma, Y., Ma, W., Li, M., Sun, F., Wang, L., & Xiang, M. (2006). Variations of soil temperature and soil moisture in northern Tibetan Plateau. *Journal of Glaciology and Geocryology, 29*(4), 578–583.

Zhu, Y., Zhang, M., Han, X., Lu, X., Chen, X., Feng, H., Wu, Z., Liu, C., Yan, J., & Zou, W. (2024). Evaluation of the soil aggregate stability under long term manure and chemical fertilizer applications: Insights from organic carbon and humic acid structure in aggregates. *Agriculture, Ecosystems & Environment, 376*, Article 109217.

Zygmunt, B., Wojciech, K., & Renata, M. (2014). Calculating particle density, bulk density, and total porosity of soil based on its texture. *Soil Science Annual, 65*(4), 139–149. https://doi.org/10.1515/ssa-2015-0007

Microbiological Strategies for Soil and Plant Health

Aimen Tahir, Sehar Rasool, Rana Muhammad Rohaid Khan, Mohsin Rasool, Alia Anayat, Muhammad Fawaz Saleem, Tayyab Mahboob, and Hafiz Muhammad Zuhair

Abstract Soil is of vital importance in the production of crops, although it is sometimes disregarded or solely considered as a physical support for plant growth. But given the growing public concerns about agriculture's sustainability, soil has to be seen as a living system. The biological, chemical, and physical characteristics of soil that are necessary for long-term, sustainable agricultural productivity with little negative influence on the environment are referred to as soil health. As a result, soil health offers a comprehensive view of soil functionality. This has led to a resurgence of interest in traditional methods like applying organic amendments. These methods have a significant impact on soil health, which considers plant health, by altering the physicochemical properties of the soil as well as its microbial populations. Furthermore, the important participants in each agro-ecosystem are microorganisms. They are the native residents of every soil and plant system, and because of their wide range of functional capacities and multifaceted existence, they constitute dominating presence in these systems. As a result, microbes and microbial methods are widely employed as measurement of soil health because to their well-established contribution to many aspects of soil health. A variety of microbe-based metrics are commonly used worldwide. We offer suggestions on how to better direct and enhance the use of

A. Tahir (✉) · S. Rasool
Institute of Soil and Environmental Sciences, University of Agriculture, Faisalabad, Pakistan
e-mail: AimenTahir715@gmail.com

R. M. R. Khan
Department of Plant Breeding and Genetics, University of Agriculture, Faisalabad, Pakistan

M. Rasool
Department of Entomology, University of Agriculture, Faisalabad, Pakistan

A. Anayat
Soil and Water Testing Laboratory, Ayub Agricultural Research Institute, Faisalabad, Pakistan

M. F. Saleem
Faculty of Agriculture Engineering and Technology, University of Agriculture, Faisalabad, Pakistan

T. Mahboob · H. M. Zuhair
Department of Plant Pathology, University of Agriculture, Faisalabad, Pakistan

M. Shaaban (ed.), *Soils and Sustainable Agriculture*, Frontier Studies in Soil Science,
https://doi.org/10.1007/978-3-031-91114-9_6

microbiological data to produce pertinent and useful evaluations of soil health and, eventually plant health.

Keywords Soil health · Plants health · Microbial strategies · Soil conservation · Smart agriculture

1 Introduction

As the world's population continues to rise, farmers are increasingly pressured to provide affordable food without sacrificing quality. The general public also expects this food to be produced responsibly and without harming the environment. The agroecosystem-related problems and socioeconomic issues are getting worse when we search for innovations in growing an increasing number of crops and reaping the rewards of the agricultural economy (Schiere & Gregorini, 2023). Farmers deal with a variety of challenges, including an unpredictable climate, decreased crop responses to agrochemicals, limited agricultural fields, and depletion of balanced nutrition in the soil and natural resources. As a result, there is an increasing concern about improving soil and then plant health sustainably. Laxman et al. (2022) defines soil health as "The soil's ability to support plant and animal productivity, maintain or improve water and air quality, and promote plant and animal health within the bounds of an ecosystem and land use".

Soil quality also referred as soil health is acknowledged as a set of traits that categorize and characterize its state of health. All the activities that are thought to take place in excellent quality soil include plant development, geochemical cycles, pollution buffering, and more (Sahu et al., 2019). More importantly, the productivity and health of plants also holds significance in agricultural soils, but not in urban soils. It is challenging to identify a healthy soil without first stating the intended purpose or aim (plant health, atmospheric equilibrium and prevention of erosion), even in a specific environment, such as cultivated soil, due to its multifunctionality (Creamer et al., 2022). Soil biota components are important indicators of soil quality, according to several research, including the microbial community, abundance, variety, activity, and stability (Leskovar et al., 2016).

According to Meena et al. (2016), soil biota plays a crucial role in plant residue mineralization, which facilitates the formation of plant nutrients that plants may easily absorb for growth and development. According to Dotaniya et al. (2016), the soil biota accelerates decomposition by producing many enzymes that impact the kinetics of soil nutrients for plants. The process by which soil microbes, mostly bacteria and fungi, convert organic nitrogen to inorganic nitrogen influences plant mineral uptake, mineral composition, and mineral synthesis. Furthermore, soil health depends on microbial activity (Fierer et al., 2021). The microbial communities support essential functions that give agroecosystems their stability and productivity (Tahat et al., 2020).

According to Lacerda-Júnior et al. (2019), microorganisms are also crucial for soil biogeochemical processes such the breakdown of cellulose and lignin and the storage of soil carbon. The primary agents responsible for carrying out these tasks are microorganisms, and a number of microbial activities may be evaluated in the process of characterizing soil characteristics. Measuring global or particular microbial biomass/densities provides further insight into possible activities. It is crucial to quantify the actual, manifested activity (Gothandapani et al., 2017). The microbes function as active agents in the recycling, sequestration, and provision of nutrients to plants. Plants receive numerous benefits from the rhizosphere microbes, such as N_2 fixation, improved P uptake (Ahilandeswari & Maheswari, 2016), suppression of pests and phytopathogens (Cheng et al., 2016), improved soil aggregation (Yilmaz & Sonmez, 2017), assistance in coping with abiotic stresses (Meena et al., 2017), remediation of problematic soils and chelation of minerals such as Fe, B and Zn (Klik et al., 2021).

Since microorganisms are essential to soil activities, any physio-chemical, or biological alterations in the soil would undoubtedly have an impact on the microbiota. As genuine markers of particular functions in the soil environment, microbes react swiftly to any changes occurring in the soil ecosystem. Research indicates that adding microbial fertilizers to soil enhances the soil environment, preserves the microecological balance, and greatly increases the soil diversity microorganisms (Sabir et al., 2021). Improving or restoring soil health requires worldwide attention. Evaluating soil health indicators should improve our comprehension of the underlying mechanisms that support sustainable agriculture. Since microbes have the potential to become significant biological agents that may serve as indicators of soil and plant health in the agro ecosystem, we have spent this chapter discussing their potential roles and strategies in depth.

2 Bio Fertilizers as Green Fertilizers

As the knowledge of harmful effects of fertilizers has grown over the past ten years, an inordinate treaty of research has been done on bio-fertilizers, microbiomes, and soil health (Chhipa, 2017). Plant tolerance to stressors and crop yield have been achieved through the employment of rich microorganisms, which have been employed globally to carry out their unique roles in soil fertility, productivity, and sustainability. Thus, microbial bio fertilizer use is encouraged and fueled by the growing popularity of environmentally friendly farming practices (O'Callaghan et al., 2022). Being an advanced kind of green fertilizer, microbial fertilizers rely on the activities of microorganisms to supply crops with directly absorbable nutrients, leading to increased yield and plant development. Driven by sustainable agricultural growth, this is highly significant. Figure 1 illustrates many kinds of bio fertilizers.

Bio fertilizer is a novel class of chemical fertilizer product that is rich in microorganisms. It works by allowing living microorganisms to multiply and create the necessary amount of nutrients for crops and soil. Because of their metabolic processes,

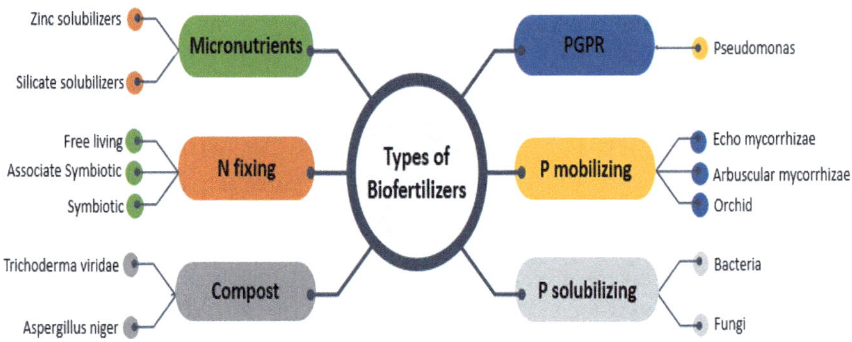

Fig. 1 Different types of bio-fertilizers

these microbes enrich the soil microbial population, which include the release of organic acids, enzymes, and antimicrobial chemicals. Beneficial bacteria proliferate more as a result of this enrichment and enhance the characteristics of the soil through increasing the extent of vital nutrients like potassium, phosphate, and nitrogen (Shaaban, 2024). As Table 1 illustrates, all of these microbial activities support plant development.

Table 1 Different bio-fertilizers and their uses for improved plant growth

Bio fertilizers	Effect on plant growth	References
BioGro	Minimize the use of chemical fertilizers 52%	Bangun et al. (2024)
Coupled application of Bio fertilizer and chemical fertilizers	Improve sweet potato yield	Hindersah et al. (2021)
Arbuscular mycorrhizal inoculation with NPK	Raise output from 4.5 T ha^{-1} in the current sweet potato cultivar NASPOT 11 to 12.8–20.1 T ha^{-1}	Kamboj et al. (2024)
Host specific endophytic and Rhizospheric bacteria	Foster the development of mandarin oranges	Stuart et al. (2022)
Bio fertilizer and cold-water extract	Improve fennel growth and essential oil content	Alipour et al. (2021)
Trichoderma enriched bio fertilizer and compost	Enhance plant growth and yield to 12.9%, improve mineral content and antioxidant level in plants and fruit	Pandey et al. (2021)
N-fixing and P-solubilizing bio fertilizers	Optimal nutrient use, harvest index, and growth enhancement in sesame	Etesami (2020)
Bio fertilizer in biochar and flyash based formulation	Magnificent tomato development and harvest	Tripti et al. (2022)

2.1 Merits of Bio Fertilizer Strategy

The following are some benefits of using bio fertilizers:

1. Their environmental and financial benefits are double-edged swords.
2. Their use improves soil quality by adding nutrients, which builds up over time.
3. Even though you won't see benefits immediately, the long-term effects are outstanding.
4. Plants may have access to atmospheric nitrogen directly via the use of these fertilizers.
5. They raise the soil's phosphorus content by releasing and solubilizing inaccessible phosphorous.
6. Bio fertilizers promote root proliferation by releasing hormones that promote growth.
7. For the benefit of the plants, microorganisms convert complicated nutrients into simpler ones.
8. Microorganisms found in bio fertilizer help the host plants receive an appropriate quantity of nutrients, support correct development, growth, and physiologic control.
9. They increase crop output up to 10–25%.
10. To a certain extent, bio fertilizers can also shield plants from illnesses transmitted through the soil.

2.2 Demerits of Bio Fertilizer Strategy

Even though the bio fertilizer technology is inexpensive and environmentally beneficial, its use is restricted by a number of issues. The limitations might be:

1. The absence of high-quality carrier material and a shortage of skilled technicians in production facilities.
2. The shortage of electricity or necessary equipment for manufacturing.
3. Insufficient finances and trouble obtaining bank loans.
4. The need for bio fertilizers during specific seasons, the need for simultaneous cropping activities, the limited time for sowing and planting in a given area, etc.
5. Lack of quality and human resources, such as inadequate training on manufacturing processes and a shortage of technically trained personnel in the production units.
6. Lack of knowledge about the advantages of the technology since farmers have difficulty implementing it because of various inoculation techniques, which do not instantly show a difference in crop growth compared to inorganic fertilizers.
7. The many restrictions have an impact on production method, marketing, and usage in one way or another. Examples of these include lack of retail outlets, market network for producers, and inoculant availability that is not available at the correct time or location.

3 Microbial Inoculations and Their Role in Nutrient Cycling

Microorganisms can greatly increase the efficiency with which nutrients are used. Microbes play a crucial role in promoting plant growth due to their multi-functional capabilities, which include fixing, solubilizing, and mineralizing N, P, K, and Zn; generating siderophores and hydrocyanic acid (HCN); and so on (Felestrino et al., 2017). Research into the many roles that microbes play in metabolism has led to the creation of microbial inoculants that can boost plant growth, and bioremediate polluted soil sites, break down and degrade agricultural waste, and manage pests and diseases. Therefore, it is expected that soils rich in organisms such as bioremediators, P-solubilizers, N-fixers, and phytohormone manufacturers will promote plant health.

As seen in Fig. 2, four significant classes of microbial inoculants can help plants absorb nutrients more effectively. Plants' ability to absorb nutrients more efficiently is enhanced by bacterial inoculants (Yilmaz & Sonmez, 2017). Plant development is enhanced and soil fertility is restored by microbial inoculants. Even in soils that are low in nutrients, microbial inoculants may improve nutrient uptake and availability to plant roots (Das et al., 2022).

Among the 17 essential nutrients, N and P are the main nutrients that crops primarily need. Their shortage causes severe crop losses, which may be due to fixation in soils or the plants acquiring them. The use of N and P in soils is enhanced by microorganisms such as AM fungi, cyanobacteria, and bacteria (Sahu et al., 2019). According to Azri and Ismail (2024), oil palm seedlings in nurseries in Malaysia saw increased nitrogen and phosphorus absorption after being inoculated with PGPR. In a three-year field trial, PGPR-inoculated plant roots demonstrated greater N absorption. This was followed by increased plant N content as a result of better fertilizer

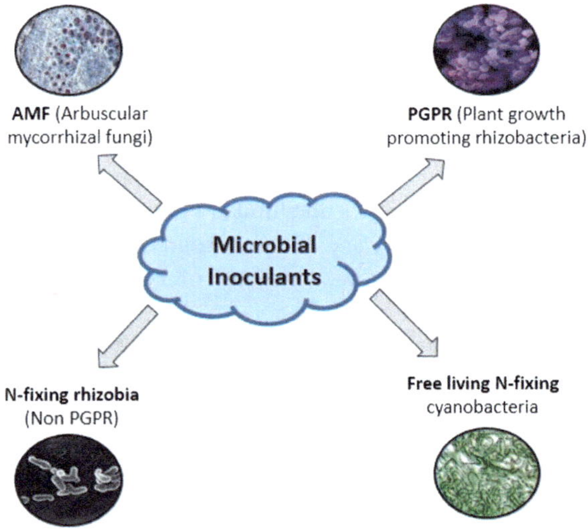

Fig. 2 Different classes of microbial inoculants

utilization (Bargaz et al., 2018). Furthermore, when PGPRs with AM fungi were employed in addition to chemical fertilizers at a lower dosage, the results were encouraging (Prasad, 2022).

As stated by Stamford et al. (2017), the use of vermicompost and chitosan from *Cunninghamella elegans* on bananas at the recommended dose proved the efficacy of organic materials as an appropriate additive to chemical fertilizers. Several kinds of bacteria and fungi aid in the solubilization and mobilization of phosphorus (P) that is present underground. Phosphate solubilizing bacteria (PSB) may not be the only cause of P-solubilization; interactions between other rhizobacteria, PSB, and mycorrhiza may also be responsible for notable outcomes (Elhaissoufi et al., 2022). Mycorrhizal organisms mobilize more P when bacteria attached to their hyphae do so (Magallon-Servin et al., 2020; Qin et al., 2022; Taktek et al., 2017). Certain microbial species also enhance the nutrients use efficiency of plants. *Pseudomonas fluorescens* and *Acinetobacter sp.* inoculation facilitates increased absorption of N, P, K, Ca, Mg, Fe, and Zn by plants (Sahu et al., 2016; Sun & Shahrajabian, 2023). According to Eudoxie and Martin, (2019), strawberry plants absorbed more macro- and micronutrients when compost tea was combined with organic matter and inoculants.

Plant growth-promoting rhizobacteria (PGPR) also had an impact on micronutrient absorption. Applying *Pseudomonas*, *Azospirillum* and *Rhizobium*, improved zinc acquisition in the shoots, leaves and grains that resultantly promoted effective nutrient absorption in wheat through (Naz et al., 2016; Saboor et al., 2021). Applying *Trichoderma harzianum* to salt-affected mustard (Brassica juncea) increased nutritional absorption. Oil content, antioxidant activity, osmolyte buildup, and NaCl exclusion were identified as routes to ameliorate salt stress (Bhardwaj & Kumar, 2020). For this reason, identifying, selecting, characterizing, and applying stress-tolerant microbes may be practical ways to help agricultural plants in stress-prone environments to overcome production constraints. In order to show increased nutrient usage efficiency, microbial inoculants can thus be used as innovative candidates in integrated nutrition management. These studies show a similar pattern of how microbial treatment affects various crops and soil types (Lisuma et al., 2020).

4 Microbial Consortium Products

A large number of recent meta-analyses have shown that most products on the market are derived from a single species and have been validated to achieve certain goals (Tufail et al., 2021). However, the roles and niches that each species may occupy are limited. Mawarda et al. (2020) state that this might lead to inoculant survival and function being negatively affected by intense competition with native populations in the soil. As a result, different settings and application techniques frequently result in inconsistent microbial inoculant quality and performance (Khan, 2022; Santoyo et al., 2021). This makes the large-scale use of microbial inoculants impractical (Alori & Babalola, 2018). Even while these drawbacks can be somewhat offset by mixing other materials, including chemical fertilizers, new environmentally friendly techniques are required (Hashem et al., 2016; Jiang et al., 2021).

Consequently, in the last ten years, the concept of combining many species with complimentary traits known as a consortium, has gained more traction (Santos et al., 2019). It is anticipated that consortia will provide a number of benefits above single species. To begin with, distinct individuals of a consortium can make up for qualities that others in the group lack, improving the overall impact. For example, consortia of isolates that solubilize inorganic phosphates and generate indole-3-acetic acid were more effective in promoting growth than strains treated alone (Shilev et al., 2020). Nitrogen fixers with drought-mitigating isolates are another possible combination (Prasanna Kumar et al., 2022). Plant-growth-promoting bacteria (PGPB) in conjunction with arbuscular mycorrhizal fungus (AMF) is another possibility (Sun et al., 2022; Yadav et al., 2022).

4.1 Potential Benefits of Microbial Consortium

- **In agriculture:** Many investigations on the application of microbial consortia—which include bacteria, viruses, nematodes, and fungi—in agriculture have been conducted in recent years. The supervision of diseases (Devi et al., 2022), nematodes (Bo et al., 2022), arthropods, and/or the encouragement of plant development (Muthuraja & Muthukumar, 2022) are some potential advantages of certain microbial selections for agriculture. Furthermore, microbial consortia can be applied to enhance soil quality and lessen the effects of abiotic pressures (Anshu et al., 2022).
- **In soil health:** Research on microbial consortiums as a tool to improve soil health is limited and in its early stages. Nonetheless, preliminary studies indicate that this strategy could offer several advantages, such as enhancing crop yields, nitrogen absorption, and soil fertility (Anshu et al., 2022). For instance:

Using fungus *Trichoderma harzianum, Gluconcetobacter diazotrophicus*, and *Pseudomonas fluorescens* in sugarcane at a concentration of 109 CFU/mL in liquid culture reduced NPK needs by 50% (Shukla et al., 2020). By inducing the breakdown of lignocellulose, bacterial consortia can also hasten the composting of green wastes (Oviedo-Ocaña et al., 2022). Field tests using a microbial consortium of *Glomus mosseae*, *Azotobacter chroococcum*, and G. *fasciculatum* in pomegranates revealed that in addition to improving plant development, dual inoculation of strains enhanced absorption of N, P, K, Ca, and Mg. According to Kumar et al. (2017) the twofold immunization of microorganisms produced more consistent findings on nutrient absorption.

- **For plant diseases:** Numerous microbial consortia that have been established for agricultural purposes focus on plant diseases. Some research indicates that using diverse strains of microorganisms with distinct mechanisms of action might enhance the microbial control of plant diseases.

- **For arthropod pests:** The earliest records of using microbial pathogens against insects were made at the end of the twentieth century when it was shown that *M. anisopliae s.L.* could control Japanese beetles (Popilia japonica). Recent years have seen the management of many arthropod pests by means of pathogenic bacteria alone (Aynalem et al., 2022; Stuart et al., 2020). Various studies have shown that combining various microorganisms improves control by increasing virulence-associated metabolites and other substances with anti-inflammatory and antioxidant characteristics. These factors may cause pathogenic microbes to infect insects by lowering their immune systems (Nawaz et al., 2020).
- **For bioremediation**: At first, microbial consortia were employed in bioremediation projects and the manufacturing of food, medicine, and chemical goods (Morales-García et al., 2019). To reduce or alleviate harmful effects on the environment, several microbes have been employed to break down, detoxify, mineralize, or convert contaminants into non-toxic molecules (Azubuike et al., 2016).

Microbial consortia provide advantages over single microbes, such as the ability to overcome biological constraints. In a greater range of ecological and environmental settings, for instance, the combination of numerous microorganisms is more stable (Trivedi et al., 2021). Mixing of microbes with varying growth characteristics, dietary needs, and modes of action is preferable, particularly when synergistic interactions take place (Ram et al., 2022).

- **In cost effectiveness:** Reducing application costs and enhancing microbial biodiversity in plants phyllosphere and rhizosphere are additional benefits of using microbial consortia. A number of obstacles must be controlled for microbial consortia to be implemented successfully, including microbial incompatibility, competition from local species, and environmental adaptability (Xu & Yu, 2021).

5 Biofilm-Based Inoculants

Using biofilm bio fertilizers is becoming a more viable method of enhancing agronomic efficiency and soil health. Beneficial microbial communities encased in extracellular polymeric substances (EPS) are known as biofilm bio fertilizers, and they help stabilize soil structure, solubilize nutrients, and enhance water retention. It has been demonstrated that applying these bio fertilizers increases soil fertility by boosting nutrient cycling, microbial activity, and organic matter content. Better root development, faster plant growth rates, and larger agricultural yields are the outcomes of these advancements. Biofilm bio fertilizers provide a sustainable way to lessen these constraints, particularly on marginal soils where nutrient availability and soil structure are frequently affected.

- In varied ecologies, microbial biofilms have the potential to serve as bio fertilizers and disease suppressors in a wide range of crops due to their adaptability to different environments and biotic and abiotic pressures.
- Biofilms are essential for the survival of plants, microbes, and animals in harsh environments like deserts and degraded land.
- Microbial biofilms are collections of cells from different kinds of microbes that form a polymeric matrix. They are created by various bacterial, yeast, cyanobacteria, and fungal genera as a way to improve their reproductive fitness and thrive in harsh conditions.
- The establishment of seeds and the interaction of nutrients and metabolites with the environment are both facilitated by biofilms, which are home to many microbes with biocontrol and developmental-promoting capabilities.

6 Biocontrol Agents

Reducing inoculums or disease-causing activity of phytopathogens by applying microorganisms is known as biocontrol (Sharma, 2021). Many defense systems are developed by plants to withstand adverse conditions including cold, drought, salt stress, and infections. Effective microbial inoculants might be applied to accomplish this (Arthee & Marimuthu, 2016). It is also possible to identify healthy, disease-resistant soils by looking for microorganisms or their main products. *Baeuveria bassiana* and *Metarrhizium anisoplae* were also helpful in controlling insect pests in addition to *Bacillus thuringiensis*. Tomato leaf miner (Tuta absoluta) was shown to be less common when humic fertilizers, vermicompost, *Bacillus subtilis*, and *Pseudomonas fluorescens* were used (Mohamadi et al., 2017). The adult lifetime period of Myzus persicae was shown to be reduced in bell pepper and chili pepper plants treated with *Bacillus subtilis* and *Glomus intraradices*. Application of $ZnSO_4$ with *Bacillus subtilis* was also observed to enhance plant development (Mardani-Talaee et al., 2017).

The intrinsic rate of the natural population was raised, the net reproduction rate was decreased, and the ultimate rate of growth of *T. absoluta* was accelerated by the administration of these well-known biocontrol agents. When the arbuscular mycorrhizal (AM) fungus is introduced to wheat, it influences the expression of genes related to defense, including β-1, 3 glucanases, chitinases, and oxalate oxidase. Sabbagh et al. (2017) found evidence of improved growth, productivity, and expression of β-1, 3 glucanase. The root knot nematode (Meloidogyne incognita) was effectively controlled by applying microbial inoculants to the Pusa Ruby tomato variety. When a balanced amount of NPK (125:50:100 kg ha^{-1}) was applied combined with bio fertilizer (*Azospirillum*, *Azotobacter*, and PSB), the number of nematodes in the soil decreased by 74.8%, root knots decreased by 75.8%, and plant dry weight increased by 62–64% (Patra et al., 2017). *Cunninghamella elegans* chitosan was reported to be used for the biocontrol of *Ralstonia solanacearum* (Fu et al., 2017; Oliveira et al., 2017).

There was a considerable reduction in brown spot, sheath blight, and bacterial leaf blight disease after using composted straw and a bioformulation containing Trichoderma to promote soil health. This was accomplished by boosting rice output by 0.8–3.0 tons ha^{-1} and generating systemic resistance (ISR) (Simarmata et al., 2016). The unique roles of PGPRs as biocontrol agents against nematodes and pests, as well as biological indicators of illness, were uncovered in part by these studies.

7 Microbes as Biomarkers

The physical, chemical, and biological components of soil are constantly shifting and evolving, making soil a vital and dynamic resource. Numerous biotic and abiotic factors interact with soil to affect its composition. The primary occupants that alter structural and functional qualities are microbial communities with well-defined roles. These alterations may serve as relevant metrics for assessing the condition of the soil. Microorganisms are a useful biological indicator for tracking the health and quality of soil because they reflect the state of the soil. Figure 3 illustrates some advantages of microbes that plays beneficial role in the soil health. Soil health may be determined by measuring population behavior and the effects of population dynamics on edaphic parameters through a sensitive and precise connection between bacteria and their environmental factors.

Biological indicators with quick response to both natural and man-made changes, have an advantage over other indicators as they can be measured simply and quickly, making them a sensitive and resourceful instrument for monitoring soil health (Dahiya et al., 2022). Table 2 lists a few instances of relevant bio-indicators. In a study conducted by Sahu et al. (2019), the effects of different agronomic methods on the diversity of soil microbiota and macrobiota, enzyme activity, and organic matter breakdown were investigated. Carbon mineralization, soil protozoa, nitrogen nutrition, soil peptidase activity, cellulose decomposition, crop rotation, nitrogen fertilizers, stubble management, tillage practices, and soil microbial biomass were all found to be significantly impacted by the aforementioned factors (Dhaliwal et al., 2020).

More information about important soil-ecological processes may be obtained from microbial soil enzyme activity than from physical or chemical characteristics, making it a reliable indicator (Nikitin et al., 2022). Soil enzymes come from a variety of sources, including plants, animals, and both live and dead microbial cells. As a result, they could be a highly sensitive indicator that can pick up on even minute variations in the physicochemical characteristics of the soil. The sensitivity and repeatability of biological indicators make them superior to other types of indicators. Microorganisms in the soil can be useful indicators due to their resistance to ecological disturbances and their comparatively high sensitivity to these changes (Philippot et al., 2021; Vázquez et al., 2020).

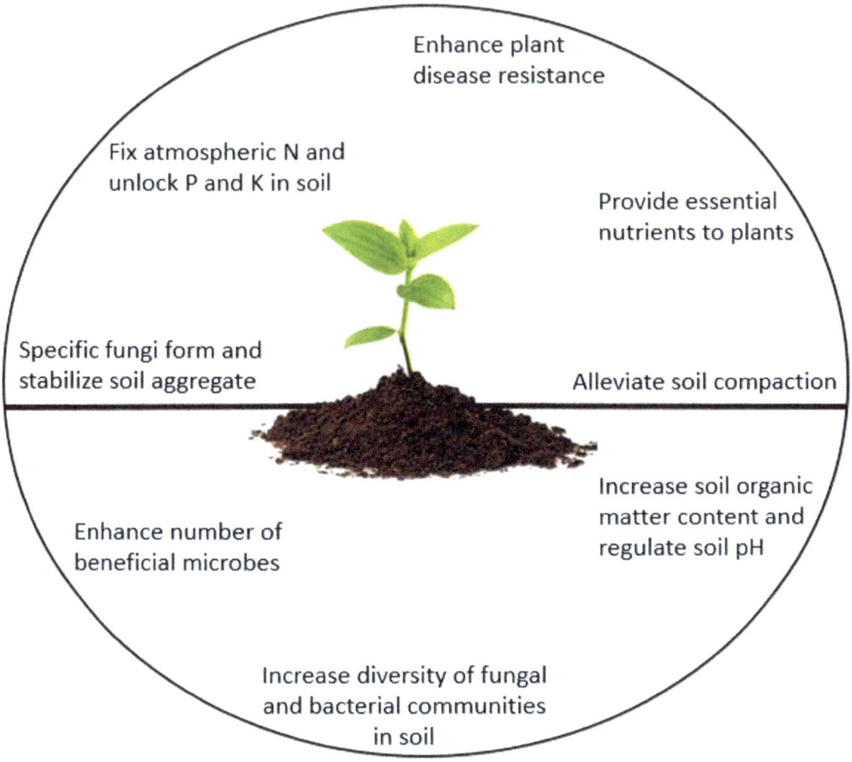

Fig. 3 Advantages of microbes in soil and plant health

Table 2 Presence of different bio-indicator with respect to soil condition

Bio-indicators	Availability	References
Actinobacteria	High pH soils, desserts	Tahat et al. (2020)
Azorhizophilus and Diazotrophs	Poor N soils	Choudhary et al. (2023)
Nitrosomonadaceae, Anaerolineae, Thermomicrobia	Ni rich agricultural soils	Bararunyeretse et al. (2019)
Cyanobacteria	Desserts, nutrient poor soils	Jiang et al. (2017), Karimi et al. (2023)
Actinobacteria (*Actinomycetospora, Clavibacter* and *Mycobacterium*)	Zero tillage	Stuart et al. (2022)

8 Endophyte Engineering

8.1 Selection of Beneficial Endophytes

Plant endosphere community research is also becoming more and more popular, and using agriculturally significant microorganisms to reduce plant stresses has shown encouraging results. Soil optimization is key to maximizing organic matter biotransformation by resident microbes and fostering soil health. As a result, plants will be able to draw nutrients from the soil (Uzoh & Babalola, 2018). Soil bacteria often form close relationships with their host plants, either by inhabiting the rhizosphere or by colonizing the endosphere, as stated by Alawiye and Babalola (2019). The endosphere is the space inside plant tissues where diverse microbial communities reside. These communities include archaea, bacteria, and fungus, and they possess intricate multifunctional characteristics (Alawiye & Babalola, 2018; Souza et al., 2017; Adeleke & Babalola, 2020).

Nutrients such as potassium, iron, carbon, and nitrogen are recycled by microbes residing in the root endosphere (Adegbeye et al., 2020). They improve nutrient uptake in a particular niche and participate in the overall performance of plant development. Research on plant endosphere colonizers has been prompted by the overwhelming relevance of soil microorganisms in crop yield (Alster et al., 2020). In order to distinguish between endophyte and harmful bacteria, the literal meaning and practical use of the term "endophyte" are thoroughly examined. How do endophytes establish themselves once they enter plant tissues, and how do they deal with the defense mechanisms of the plant? We have concentrated on bacterial and fungal species in this section.

8.2 Beneficial Effects of Endophytes

In this portion, we take a look at how endophytes improve agricultural yields by enhancing soil and plant health.

- **N_2 Fixation**: In symbiotic relationships with legumes and non-legume plants, endocrine endophytes from the following genera coexist: *Rhizobium, Pseudomonas, Sinorhizobium, Bradyrhizobium, Azorhizobium, Enterobacter, Klebsiella, Mesorhizobium, Allorhizobium, and Azospirillum.* These endophytes fix atmospheric nitrogen in the form of ammonia (NH_3) into the soil without obstructing the host plant's ability to absorb nutrients from the soil (Nohwar et al., 2019). In particular, plant growth has been aided by the re-inoculation of *Azospirillum* bacterial endophytes that were separated from the roots of non-leguminous crops (Mhatre et al., 2019).
- **Phosphate (P) Dissolution**: Using variety of techniques, including ion exchange, solubilization, acidification, chelation, formation of organic acids, and release of

extracellular acid phosphatase, some endophytic microorganisms enable plants to absorb complex phosphorus (Santoyo et al., 2016). By inhibiting adsorption, endophytes ingest soluble phosphate and become a phosphorus store, increasing their capacity to solubilize phosphate in phosphate-limiting environments (Vidya et al., 2016).

- **Phytoremediation/Bioremediation**: According to Rostami and Azhdarpoor (2019), screening for critical microbial features has shown effectiveness in the breakdown from contaminated soil. Endophytic microorganisms can remove contaminants from the soil environment by breaking down complex organic compounds by lytic enzymes used in rhizodegradation, transformation, and stabilization.
- **Siderophores Production**: Microorganisms manufacture biological molecules called siderophores, which have a low molecular weight. In conditions where nutrients are scarce, their main function is iron-chelation (Odelade & Babalola, 2019). Through indirect mechanisms, plants demonstrate endophyte activity in the formation of siderophores (Omomowo & Babalola, 2019). Many of them reduce plant stress by producing biological control agents or by systemically activating the plant's defenses against pathogens (Ferreira et al., 2019). Endophytes, which produce siderophores, are crucial in determining plant-induced systemic resistance and aid in the control of plant diseases that affect agricultural output. Some endophytic bacteria have the ability to produce substances that fight various diseases.
- **Enzymes Production**: According to Astafyeva, (2023) endophytes mainly create enzymes to colonize tissues through the use of polysaccharide compounds (containing pectin and xylan) in the host plants. According to Sharma (2019) some of these enzymes are amylase, cellulose, lipase, tannase, glucanase, laccase, protease, chitinase and xylanase.
- **Biocontrol Activities of Plant Endophytes**: Adeleke and Babalola (2020) assert that endophytes that live in the endosphere of plants and have the capacity to produce biomolecules may enhance plant performance by fostering disease resistance. Orozco-Mosqueda et al. (2018) state that plant protection against pathogens can be achieved either directly through antagonistic interactions and host resistance development, or indirectly through the suppression of plant pathogens and the improvement of plant growth using a number of techniques. Endophytes produce antibacterial chemicals, enzymes break down microbial cell walls, and hydrogen cyanide inhibits the development of microbial infections (Adeleke & Babalola, 2020).

9 Rhizosphere Engineering

The rhizosphere, the tiny region of soil next to a plant's root system, may promote crop growth with either balanced or reduced pesticide intake levels. The variety and functionality of microorganisms are mostly centered in the microenvironment of this plant rhizosphere, which is very advantageous for the health of the soil and plants. According to Gupta et al. (2019), the area offers several direct and indirect advantages for plant health. The general consensus is that soil microorganisms actively participate in nutrient cycle activities such nitrogen cycling, organic matter breakdown, and mineral weathering, all of which have an impact on plant development, and so function as a key store of nutrients for plants.

For plants to participate in the ecosystem's functioning, interactions between microbes and plant roots are therefore crucial. Numerous elements, such as plant species, soil type, and microbial interactions, all have an impact on rhizosphere. This dynamic is further complicated by seasonal variations in soil temperature, moisture content, and microbial activity. Developing a one-size-fits-all approach is difficult because these variations might affect how effective bio fertilizers (Ibáñez et al., 2023). The rhizospheric microbial dominance, diversity, and competitiveness affect how well soil nutrients are mobilized, acquired, and used (Naik et al., 2019; Sharma et al., 2020).

The intentional alteration of interactions between soil microbes and plants, or rhizosphere engineering, presents a revolutionary strategy for the commercialization of bio fertilizer but also presents a unique set of challenges and potential. Rhizosphere engineering can improve the soil environment to increase the effectiveness of bio fertilizers, even when they are used in conjunction with conventional fertilizers (Fasusi et al., 2021). This can result in the development of more durable, effective and focused bio fertilizer solutions that are designed for particular crops or soil types. Microbial treatments have been demonstrated to dramatically increase soil health and agricultural yields in semiarid tropical regions, according to many researches that have examined the opportunities and constraints in rhizospheric engineering (Chandran et al., 2021). Rhizospheric engineering can be done in two ways as shown in Fig. 4.

Complex soil-microbe-plant interactions, which are controlled by a variety of parameters such as soil pH, organic matter, and seasonal variations, need for a profound understanding of rhizosphere engineering. Creating bio fertilizers that work for everyone is difficult because of this intricacy. Thorough field testing and regulatory permissions would also be necessary for commercializing such engineering solutions, and these might be expensive and time-consuming processes. Gaining the trust of farmers—who frequently look to agricultural inputs for consistency and dependability—may be difficult due to the diversity that rhizosphere engineering introduces (Adeniji et al., 2024).

10 Soil Microbiome Management

The biological characteristics of soil that is known to be greatly influenced by sound agronomic methods is microbial communities, which help soils produce healthy crops. Regular and sudden changes in agricultural practices can affect the growth, disintegration, or loss of the microbial population because microorganisms are

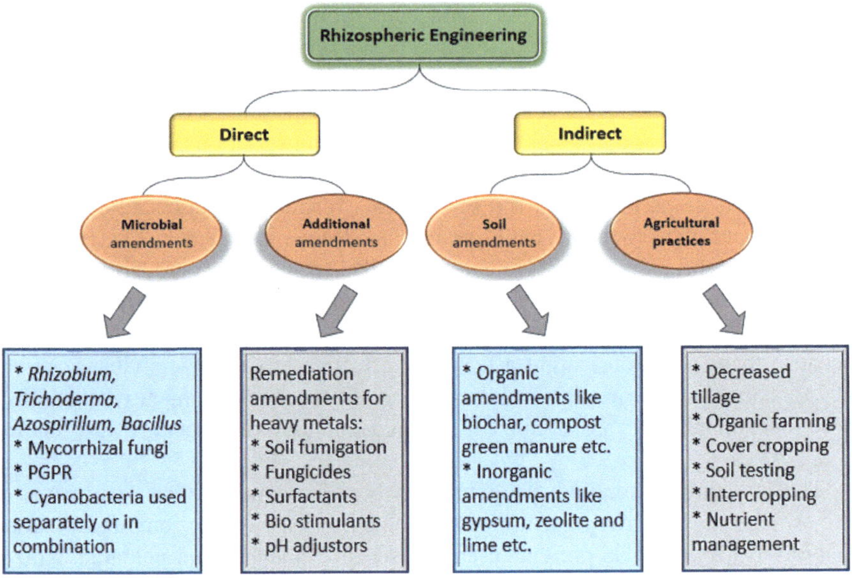

Fig. 4 Different methods of Rhizospheric engineering

involved in most biological processes in soils (Jiménez-Bueno et al., 2016a, 2016b) and because the makeup of their communities changes as plants go through different developmental stages (Hartmann & Six, 2023). Jiménez-Bueno et al. (2016a, 2016b) state that changes in the microbial community's behavior are a sign of associated factors that help choose sustainable agro-ecological practices. In the current trend to improve soil health and lower the quantity of chemical or energy inputs needed, cultural techniques are evaluated, especially for their potential use as alternatives.

Crop rotation: A very ancient cultural practice is crop rotation (Dahlin & Svensson, 2021). Its benefits include preserving soil structure and organic matter and reducing soil erosion, which is often associated with continuous row crops. Crop rotation is often associated with a reduction in plant diseases caused by soil-borne pathogens and helps to avoid this detrimental effect. Certain plant pathogenic populations will develop as a consequence of repeatedly farming the same susceptible host plant (Kumar et al., 2021).

Rotating crops with non-host or less sensitive plants may lead to a decline in certain pathogenic populations since such populations naturally decline and other microorganisms aggressively combat them. This strategy is effective against pathogens with low saprophytic survival capability or biotrophic pathogens that rely on the presence of their host to survive. Derbyshire et al. (2022) claim that it is less effective against diseases like *Sclerotinia sclerotiorum* or *Rhizoctonia solani sclerotia* that have a wide host range or effective survival forms.

Tillage: The recent research seeks to replace over-cultivation with less or more deliberate tillage techniques. Instead to being categorized as conventional tillage, these methods are called conservation tillage (Mahenge, 2020). The practice of leaving crop residue partially or entirely buried in the soil is known as conservation tillage. The microbial biomass and activity can be increased by this organic debris, which decomposes to varying degrees (Prescott & Vesterdal, 2021). Differing outcomes have been seen with regard to the development of plant diseases. Conservation tillage creates a very competitive atmosphere where microorganisms may compete or become antagonistic, which suppresses illness (Jayaraman et al., 2021).

Additionally, it results in altered soil conditions, including increased upper-layer soil warmth and wetness, which may favor some plant diseases (Moura et al., 2022). Furthermore, the virus remains in its original location due to the lower tillage, while traditional deep tillage may cause it to relocate into a less suitable habitat that is deeper. Different impacts of tillage can be seen even for the same pathosystem. Given the interdependence of their impacts, it is evident that tillage and rotation techniques must be applied in tandem (Todorović et al., 2023).

Use of organic amendments: The term "organic amendments" refers to a broad category of inputs, including composts, solid wastes, and animal dung. Numerous studies have previously been conducted, and they often come to the conclusion that organic additions are helpful. Organic additions are frequently used to increase the excellence of soil, mostly by enhancing soil microbial biomass and activity, which contributes to general suppressiveness (Shaaban & Nunez-Delgado, 2024). The primary benefit of using organic amendments is their abundance in labile carbon components, which provide microorganisms with energy. Additionally, hostile bacteria may be present in organic additions. Regarding disease suppression, however, different conclusions have been reached (Hao & Ashley, 2021).

Managing organic matter and cycling nutrients: For the soil to restore its depleted nutrients, organic waste must be recycled and broken down biologically. These minerals are essential for plant growth and development in addition to soil fertility. In a similar vein, the physical and chemical activities of soils are largely determined by biological properties, which are often governed by diverse microbial populations (Philippot et al., 2024).

This region controls plant growth and development processes, the physicochemical properties of the soil, and chemical signals between microbial populations and the plants they inhabit (Lareen et al., 2016). The microbial community activities constantly alter the arable soil system. Compaction of mineral nutrients around roots is facilitated by the adhesive qualities of extracellular substances produced by soil microorganisms. Increased nutrient intake is made possible by this aggregation. Together with promoting soil aggregation, the organic components that bacteria make also serve as a protective layer to prevent the organisms from drying out. Microbial communities largely control the overall amount of organic matter and mineral aggregates in the soil, improving the soil's structure and chemical properties to support plant growth (Bhattacharyya et al., 2022). As a result, these elements are potent markers of soil health. Presence of organic matter in soil have much more positive points, some of them are displayed in Fig. 5.

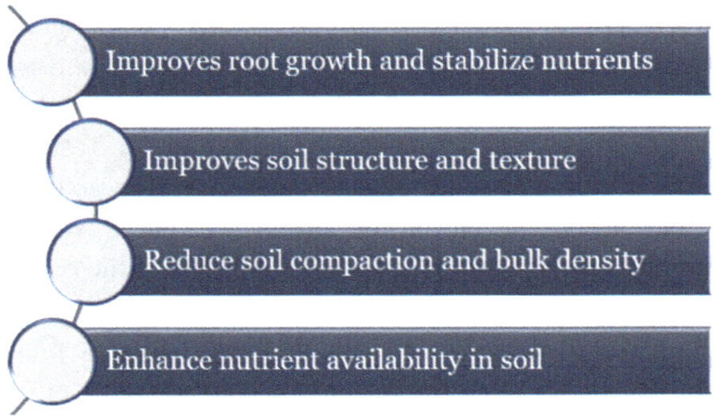

Fig. 5 Advantages of organic matter content

Fertile soils operate best when there are microbial activities catalyzed by enzymes, which form the foundation for the steering of essential minerals (Daunoras et al., 2024). Microbial biomass adds nutrients to the soil in the form of beneficial organic matter that encases minerals that release over time, improving the quality of the soil (Van der Wal & de Boer, 2017). In order to release mineral nutrients gradually and constantly, these microorganisms either directly consume them or sequester them. These might be useful indicators of significant activity related to the soil nutrient cycle. To guarantee proper nutrient management and recycling within the soil ecosystem, significant soil enzymes participating in different biogeochemical cycles are also used as markers (Bhaduri et al., 2022). A vast range of enzymes required for the degradation of organic molecules are provided by microbes. Plant residue may decompose completely or partially, returning organic materials to the soil (Khatoon et al., 2017).

Avoiding excessive pesticide and fertilizer use: Fertilizers are a crucial part of agricultural production systems because they maintain soil health. It is essential for increasing the yield of a number of crops, particularly food grains (Rao & Lenka, 2020). In order to obtain higher output per unit area, current global agricultural cropping frameworks aims to employ a big number of herbicides, insecticides, and fertilizers; nevertheless, using more measures than ideal of these fertilizers causes certain disadvantages (Kumar et al., 2019). While applying chemical fertilizers consistently increases the soil's nutrient status and yield production, there are certain side effects that must be considered, such as inefficient use of nutrients, imbalanced fertilization, and multi-nutrient deficiencies, which have drawn attention to other options (Subramanian & Thirunavukkarasu, 2017).

Furthermore, soil and water basin contamination has resulted from the careless application of synthetic fertilizers. As a result, the soil is now deficient in organic matter and vital plant nutrients. It has resulted in the disappearance of helpful insects and microorganisms, which has indirectly decreased soil fertility and increased crop disease susceptibility (Jadon & Verma, 2021). Microorganisms may break down pesticide residues in the soil, clean the soil, and improve aeration, all of which encourage root development.

11 Conclusion and Future Perspectives

To support the world's expanding population, soil is essential for regulating the amount of food produced. As the need for agricultural food has increased, more intense crop production and protection have been needed, which has been worse for the environment and soil quality due to the increased use of inorganic fertilizers and pesticides. Limiting the increasing use of inorganic fertilizers is essential to protect soil health against pollution and decreased yield. Hence, in order to create soils that support sustainable agriculture, it is imperative to take a comprehensive approach to soil health and to implement transdisciplinary approaches to soil management that combine biological, chemical, and physical tactics.

- The use of microbial consortiums still has to be investigated in light of the aforementioned benefits, since they have drawbacks comparable to those of single species techniques.
- The intentional biotic disruption, or microbial invasion, of the natural soil microbiome caused by microbial inoculation is one cause for worry. This may result in intricate interactions between the inoculated and native species that provide erratic and untrustworthy results.
- The inoculation impact is also highly dependent on the consortium's properties (composition, variety, and density) and administration techniques (soil inoculation v/s seed and plant-based inoculation). For this reason, when creating business consortia, more work has to go into studies that investigate these consequences and appropriate application strategies.
- The development of techniques that do not need the culture of microbial populations will undoubtedly improve our knowledge of biodiversity and promote the discovery of new microorganisms with unique properties for soil restoration, bioremediation, and therapeutic applications. In order to effectively exchange information and promote interdisciplinary collaboration, industry stakeholders, end users, and researchers from various scientific sub disciplines must come together.
- To further better understand the practical constraints from the viewpoints of academia, industry, and end users at both local and global levels, concerns pertaining to intellectual property and the legal system must be taken into account.

References

Adegbeye, M. J., Ravi Kanth Reddy, P., Obaisi, A. I., Elghandour, M. M. M. Y., Oyebamiji, K. J., Salem, A. Z. M., Morakinyo-Fasipe, O. T., CiprianoSalazar, M., & Camacho-Díaz, L. M. (2020). Sustainable agriculture options for production, greenhouse gasses and pollution alleviation, and nutrient recycling in emerging and transitional nations—An overview. *Journal of Cleaner Production, 242*, Article 118319.

Adeleke, B. S., & Babalola, O. O. (2020). The endosphere microbial communities, a great promise in agriculture. *International Microbiology, 24*, 1–17.

Adeniji, A., Fadiji, A. E., Li, S., & Guo, R. (2024). From lab bench to farmers' fields: Co-creating microbial inoculants with farmers input. *Rhizosphere*, 100920.

Ahilandeswari, K., & Maheswari, N. U. (2016). Co–inoculation of Azospirillum lipoferum and phosphate solubilizing microorganisms on the growth of Rice (Oryza sativa L.). *International Journal of Pure & Applied Bioscience, 4*, 317–320.

Alawiye, T. T., & Babalola, O. O. (2019). Bacterial diversity and community structure in typical plant rhizosphere. *Diversity, 11*, 179.

Alipour, A., Rahimi, M. M., Hosseini, S. M. A., & Bahrani, A. (2021). Mycorrhizal fungi and growth-promoting bacteria improves fennel essential oil yield under water stress. *Industrial Crops and Products, 170*, Article 113792.

Alori, E. T., & Babalola, O. O. (2018). Microbial inoculants for improving crop quality and human health in Africa. *Frontiers in Microbiology, 9*.

Alster, C. J., von Fischer, J. C., Allison, S. D., & Treseder, K. K. (2020). Embracing a new paradigm for temperature sensitivity of soil microbes. *Global Change Biology, 26*, 3221–3229.

Anshu, A., Agarwal, P., Mishra, K., Yadav, U., Verma, I., Chauhan, S., Srivastava, P. K., & Singh, P. C. (2022). Synergistic action of Trichoderma koningiopsis and T. asperellum mitigates salt stress in paddy. *Physiology and Molecular Biology of Plants, 28*, 987–1004.

Arthee, R., & Marimuthu, P. (2016). Isolation, identification and screening of culturable endophytic bacterial community in sugarcane for antimicrobial activity on major fungal pathogens. *Bioscan, 11*, 2157–2167.

Astafyeva, Y. (2023). *Microalgae and bacteria interaction*. Universitaet Hamburg (Germany).

Aynalem, B., Muleta, D., Jida, M., Shemekite, F., & Aseffa, F. (2022). Biocontrol competence of Beauveria bassiana, Metarhizium anisopliae and Bacillus thuringiensis against tomato leaf miner, Tuta absoluta Meyrick 1917 under greenhouse and field conditions. *Heliyon, 8*, Article e09694.

Azri, M. H., & Ismail, S. (2024). Synergistic effects of Bacillus salmalaya strain 139SI with fertilizer on nutrient uptake and fertilizer use efficiency of oil palm seedlings. *Malaysian Journal of Microbiology, 20*(3).

Azubuike, C. C., Chikere, C. B., & Okpokwasili, G. C. (2016). Bioremediation techniques–classification based on site of application: Principles, advantages, limitations and prospects. *World Journal of Microbiology and Biotechnology, 32*, 1–18. 3211

Bangun, I. H., Sembiring, R., Ruddraramker, C., & Rizky Syam, M. (2024). Sustainable rice farming in West Java, Indonesia: The application of a cost-efficient organic farming approach. *Journal of Water and Land Development*, 122–129.

Bararunyeretse, P., Zhang, Y., & Ji, H. (2019). Molecular biology-based analysis of the interactive effect of nickel and xanthates on soil bacterial community diversity and structure. *Sustainability, 11*(14), 3888.

Bargaz, A., Lyamlouli, K., Chtouki, M., Zeroual, Y., & Dhiba, D. (2018). Soil microbial resources for improving fertilizers efficiency in an integrated plant nutrient management system. *Frontiers in Microbiology, 9*, 1606.

Bhaduri, D., Sihi, D., Bhowmik, A., Verma, B. C., Munda, S., & Dari, B. (2022). A review on effective soil health bio-indicators for ecosystem restoration and sustainability. *Frontiers in Microbiology, 13*, Article 938481.

Bhardwaj, S., & Kumar, P. (2020). Salinity stress, its physiological response and mitigating effects of microbial bio inoculants and organic compounds. *Journal of Pharmacognosy and Phytochemistry, 9*(4), 1297–1303.

Bhattacharyya, S. S., Ros, G. H., Furtak, K., Iqbal, H. M., & Parra-Saldívar, R. (2022). Soil carbon sequestration–An interplay between soil microbial community and soil organic matter dynamics. *Science of the Total Environment, 815*, Article 152928.

Bo, T., Kong, C., Zou, S., Mo, M., & Liu, Y. (2022). Bacillus nematocida B16 enhanced the rhizosphere colonization of Pochonia chlamydosporia ZK7 and controlled the efficacy of the root-knot nematode Meloidogyne incognita. *Microorganisms, 10*, 218.

Chandran, H., Meena, M., & Swapnil, P. (2021). Plant growth-promoting rhizobacteria as a green alternative for sustainable agriculture. *Sustainability., 13*(19), 10986.

Cheng, X., van de Mortel, J. E., Etalo, D. W., Dekkers, E., Raaijmakers, J. M. (2016). Bacterial determinants of plant growth promotion and induced systemic resistance by Pseudomonas fluorescens. Plant growth promotion by Pseudomonas fluorescens, p. 127.

Chhipa, H. (2017). Nano-fertilizers and nano-pesticides for agriculture. *Environmental Chemistry Letters, 15*(1), 15–22.

Choudhary, M., Jat, H. S., Mukhopadhyay, R., Kakraliya, M., Poonia, T., Phogat, A., & Sharma, P. C. (2023). Functional diversity and behavioral changes of microbial communities under salt affected soils. *Applied Soil Ecology, 190*, Article 105017.

Creamer, R. E., Barel, J. M., Bongiorno, G., & Zwetsloot, M. J. (2022). The life of soils: Integrating the who and how of multifunctionality. *Soil Biology and Biochemistry, 166*, Article 108561.

Dahiya, U. R., Das, J., & Bano, S. (2022). Biological indicators of soil health and biomonitoring. *Advances in bioremediation and phytoremediation for sustainable soil management: Principles, monitoring and remediation* (pp. 327–347). Springer International Publishing.

Dahlin, J., & Svensson, E. (2021). Revitalizing traditional agricultural practices: Conscious efforts to create a more satisfying culture. *Sustainability, 13*(20), 11424.

Das, P. P., Singh, K. R., Nagpure, G., Mansoori, A., Singh, R. P., Ghazi, I. A., & Singh, J. (2022). Plant-soil-microbes: A tripartite interaction for nutrient acquisition and better plant growth for sustainable agricultural practices. *Environmental Research, 214*, Article 113821.

Daunoras, J., Kačergius, A., & Gudiukaitė, R. (2024). Role of soil microbiota enzymes in soil health and activity changes depending on climate change and the type of soil ecosystem. *Biology, 13*(2), 85.

Derbyshire, M. C., Newman, T. E., Khentry, Y., & Owolabi Taiwo, A. (2022). The evolutionary and molecular features of the broad-host-range plant pathogen Sclerotinia sclerotiorum. *Molecular Plant Pathology, 23*(8), 1075–1090.

Devi, N. O., Tombisana Devi, R. K., Debbarma, M., Hajong, M., & Thokchom, S. (2022). Effect of endophytic Bacillus and arbuscular mycorrhiza fungi (AMF) against Fusarium wilt of tomato caused by Fusarium oxysporum f. sp. lycopersici. *Egyptian Journal of Biological Pest Control, 32*, 1–14.

Dhaliwal, S. S., Naresh, R. K., Gupta, R. K., Panwar, A. S., Mahajan, N. C., Singh, R., & Mandal, A. (2020). Effect of tillage and straw return on carbon footprints, soil organic carbon fractions and soil microbial community in different textured soils under rice–wheat rotation: A review. *Reviews in Environmental Science and Bio/technology, 19*, 103–115.

Dotaniya, M., Meena, V., Basak, B., & Meena, R. (2016). Potassium uptake by crops as well as microorganisms. In V. S. Meena, B. R. Maurya, J. P. Verma, & R. S. Meena (Eds.), *Potassium solubilizing microorganisms for sustainable agriculture* (pp. 267–280). Springer.

Elhaissoufi, W., Ghoulam, C., Barakat, A., Zeroual, Y., & Bargaz, A. (2022). Phosphate bacterial solubilization: A key rhizosphere driving force enabling higher P use efficiency and crop productivity. *Journal of Advanced Research, 38*, 13–28.

Etesami, H. (2020). Enhanced phosphorus fertilizer use efficiency with microorganisms. *Nutrient dynamics for sustainable crop production*, 215–245.

Eudoxie, G., & Martin, M. (2019). Compost tea quality and fertility. *Organic fertilizers-history, production and applications.*

Fasusi, O. A., Cruz, C., & Babalola, O. O. (2021). Agricultural sustainability: Microbial biofertilizers in rhizosphere management. *Agriculture, 11*(2), 163.

Felestrino, E. B., Santiago, I. F., Freitas, L. da Silva, Rosa, L. H., Ribeiro, S. P., & Moreira, L. M. (2017). Plant Growth promoting bacteria associated with Langsdorffia hypogaea-rhizosphere-host biological interface: A neglected model of bacterial prospection. *Frontiers in Microbiology.*

Ferreira, M. J., Silva, H., & Cunha, A. (2019). Siderophore-producing rhizobacteria as a promising tool for empowering plants to cope with iron limitation in saline soils: A review. *Pedosphere, 29*, 409–420.
Fierer, N., Wood, S. A., & Mesquita, C. P. B. (2021). How microbes can, and cannot, be used to assess soil health. *Soil Biology & Biochemistry, 153*, Article 108111.
Fu, L., Penton, C. R., Ruan, Y., Shen, Z., Xue, C., Li, R., & Shen, Q. (2017). Inducing the rhizosphere microbiome by biofertilizer application to suppress banana Fusarium wilt disease. *Soil Biology & Biochemistry, 104*, 39–48.
Gothandapani, S., Sekar, S., & Padaria, J. C. (2017). Azotobacter chroococcum: Utilization and potential use for agricultural crop production: An overview. *International Journal of Advanced Research in Biological Sciences (IJARBS), 4*, 35–42.
Hao, J., & Ashley, K. (2021). Irreplaceable role of amendment-based strategies to enhance soil health and disease suppression in potato production. *Microorganisms, 9*(8), 1660.
Hartmann, M., & Six, J. (2023). Soil structure and microbiome functions in agroecosystems. *Nature Reviews Earth & Environment, 4*(1), 4–18.
Hashem, A., Abd_Allah, E. F., Alqarawi, A. A., Al-Huqail, A. A., Wirth, S., & Egamberdieva, D. (2016). The interaction between arbuscular mycorrhizal fungi and endophytic bacteria enhances plant growth of Acacia gerrardii under salt stress. *Frontiers in Microbiology, 7*.
Hindersah, R., Karuniawan, A., & Apriliana, A. (2021). Reducing chemical fertilizer in sweet potato cultivation by using mixed biofertilizer. *Iraqi Journal of Agricultural Sciences, 52*(4).
Ibáñez, A., Garrido-Chamorro, S., Vasco-Cárdenas, M. F., & Barreiro, C. (2023). From lab to field: Biofertilizers in the 21st century. *Horticulturae, 9*(12), 1306.
Jadon, A., & Verma, M. (2021). Organic farming: Role of microorganism. *Contemporary Advances in Science and Technology, 1*, 61.
Jayaraman, S., Naorem, A. K., Lal, R., Dalal, R. C., Sinha, N. K., Patra, A. K., & Chaudhari, S. K. (2021). Disease-suppressive soils—Beyond food production: A critical review. *Journal of Soil Science and Plant Nutrition, 21*, 1437–1465.
Jiang, Y., Liu, M., Zhang, J., Chen, Y., Chen, X., Chen, L., & Sun, B. (2017). Nematode grazing promotes bacterial community dynamics in soil at the aggregate level. *The ISME Journal, 11*(12), 2705–2717.
Jiang, F., Zhang, L., Zhou, J., George, T. S., & Feng, G. (2021). Arbuscular mycorrhizal fungi enhance mineralisation of organic phosphorus by carrying bacteria along their extraradical hyphae. *New Phytologist, 230*, 304–315.
Jiménez-Bueno, N. G., Valenzuela-Encinas, C., Marsch, R., Ortiz-Gutiérrez, D., Verhulst, N., Govaerts, B., & Navarro-Noya, Y. E. (2016a). Bacterial indicator taxa in soils under different long-term agricultural management. *Journal of Applied Microbiology, 120*(4), 921–933.
Jiménez-Bueno, N. G., Valenzuela-Encinas, C., Marsch, R., Ortiz-Gutiérrez, D., Verhulst, N., Govaerts, B., Dendooven, L., & Navarro-Noya, Y. E. (2016b). Bacterial indicator taxa in soils under different long-term agricultural management. *Journal of Applied Microbiology, 120*, 921–933.
Kamboj, A., Kumar, S., Vikas, V. G., Kamboj, P., & Qane, K. (2024). Functions of microbial ecology influencing plant and soil health and productivity. *Modern Techniques to Sustainable Agriculture*, 9.
Karimi, A., Tahmourespour, A., & Hoodaji, M. (2023). The formation of biocrust and improvement of soil properties by the exopolysaccharide-producing cyanobacterium: A biogeotechnological study. *Biomass Conversion and Biorefinery, 13*(17), 15489–15499.
Khan, S. T. (2022). Consortia-based microbial inoculants for sustaining agricultural activities. *Applied Soil Ecology, 176*, Article 104503.
Khatoon, H., Solanki, P., Narayan, M., Tewari, L., Rai, J., & Hina Khatoon, C. (2017). Role of microbes in organic carbon decomposition and maintenance of soil ecosystem. *International Journal of Chemical Studies, 5*(6), 1648–1656.

Klik, B., Gusiatin, Z. M., & Kulikowska, D. (2021). A holistic approach to remediation of soil contaminated with Cu, Pb and Zn with sewage sludge-derived washing agents and synthetic chelator. *Journal of Cleaner Production, 311*, Article 127664.

Kumar, A., Maurya, B. R., Raghuwanshi, R., Meena, V. S., & Tofazzal Islam, M. (2017). Co-inoculation with Enterobacter and Rhizobacteria on yield and nutrient uptake by wheat (Triticum aestivum L.) in the alluvial soil under indo-gangetic plain of India. *Journal of Plant Growth Regulation, 36*, 608–617.

Kumar, R., Kumar, R., & Prakash, O. (2019). Chapter-5 the impact of chemical fertilizers on our environment and ecosystem. *Chief Ed, 35*, 69.

Kumar, S., Bhowmick, M. K., & Ray, P. (2021). Weeds as alternate and alternative hosts of crop pests.

Lacerda-Júnior, G. V., Noronha, M. F., Cabral, L., Delforno, T. P., Sousa, S. T. P., FernandesJúnior, P. I., Melo, I. S., & Oliveira, V. M. (2019). Land use and seasonal effects on the soil microbiome of a Brazilian dry forest. *Frontiers in Microbiology, 10*, 648.

Lareen, A., Burton, F., & Schäfer, P. (2016). Plant root-microbe communication in shaping root microbiomes. *Plant Molecular Biology, 90*, 575–587.

Laxman, T., Kumar, M. V., Reddy, R. U., Vijayalaxmi, G., & Sridhar, K. (2022). Contribution of soil health to sustainable agriculture.

Leskovar, D., Othman, Y., & Dong, X. (2016). Strip tillage improves soil biological activity, fruit yield and sugar content of triploid watermelon. *Soil and Tillage Research, 163*, 266–273.

Lisuma, J. B., Zuberi, Z., Ndakidemi, P. A., & Mbega, E. R. (2020). Linking rhizosphere bacterial diversity and soil fertility in tobacco plants under different soil types and cropping pattern in Tanzania: A pilot study. *Heliyon, 6*(7).

Magallon-Servin, P., Antoun, H., Taktek, S., & de-Bashan, L. E. (2020). Designing a multi-species inoculant of phosphate rock-solubilizing bacteria compatible with arbuscular mycorrhizae for plant growth promotion in low-P soil amended with PR. *Biology and Fertility of Soils, 56*(4), 521–536.

Mahenge, F. Y. (2020). *Analysis for the role of technology in soil conservation to smallholder farmers: A case of Kiroka Village Project Morogoro* (Doctoral dissertation, The Open University of Tanzania).

Mardani-Talaee, M., Razmjou, J., Nouri-Ganbalani, G., Hassanpour, M., & Naseri, B. (2017). Impact of chemical, organic and bio-fertilizers application on bell pepper, Capsicum annuum L. and biological parameters of Myzus persicae (Sulzer) (Hem.: Aphididae). *Neotropical Entomology*, 1–9.

Mawarda, P. C., Le Roux, X., Van Elsas, J. D., & Salles, J. F. (2020). Deliberate introduction of invisible invaders: A critical appraisal of the impact of microbial inoculants on soil microbial communities. *Soil Biology and Biochemistry, 148*, Article 107874.

Meena, K. K., Sorty, A. M., Bitla, U. M., Choudhary, K., Gupta, P., Pareek, A., Singh, D. P., Prabha, R., Sahu, P. K., Gupta, V. K., & Singh, H. B. (2017). Abiotic stress responses and microbe-mediated mitigation in plants: The omics strategies. *Frontiers in Plant Science, 8*, 172.

Meena, R., Bohra, J., Singh, S., Meena, V., Verma, J., Verma, S., & Sihag, S. (2016). Towards the prime response of manure to enhance nutrient use efficiency and soil sustainability a current need: A book review. *Journal of Cleaner Production, 2016*, 1258–1260.

Mhatre, P. H., Karthik, C., Kadirvelu, K., Divya, K. L., Venkatasalam, E. P., Srinivasan, S., Ramkumar, G., Saranya, C., & Shanmuganathan, R. (2019). Plant growth promoting rhizobacteria (PGPR): A potential alternative tool for nematodes bio-control. *Biocatalysis and Agricultural Biotechnology, 17*, 119–128.

Mohamadi, P., Razmjou, J., Naseri, B., & Hassanpour, M. (2017). Population growth parameters of Tuta absoluta (Lepidoptera: Gelechiidae) on tomato plant using organic substrate and biofertilizers. *Journal of Insect Science (Online), 17*, 36.

Morales-García, Y. E., Baez, A., Quintero-Hernández, V., Molina-Romero, D., RiveraUrbalejo, A. P., Pazos-Rojas, L. A., & Muñoz-Rojas, J. (2019). Bacterial mixtures, the future generation of inoculants for sustainable crop production. In D. Maheshwari & S. Dheeman (Ed.), *Field*

crops: Sustainable management by PGPR. Sustainable development and biodiversity (Vol. 23, pp. 11–44). Springer.
Moura, A. B., Backhouse, D., de Souza Júnior, I. T., & Gomes, C. B. (2022). Soilborne pathogens. *Subsoil constraints for crop production* (pp. 199–224). Springer International Publishing.
Muthuraja, R., & Muthukumar, T. (2022). Co-inoculation of halotolerant potassium solubilizing Bacillus licheniformis and Aspergillus violaceofuscus improves tomato growth and potassium uptake in different soil types under salinity. *Chemosphere, 294*, Article 133718.
Naik, K., Mishra, S., Srichandan, H., Singh, P. K., & Sarangi, P. K. (2019). Plant growth promoting microbes: Potential link to sustainable agriculture and environment. *Biocatalysis and Agricultural Biotechnology, 21*, Article 101326.
Nawaz, A., Gogi, M. D., Naveed, M., Arshad, M., Sufyan, M., Binyameen, M., Islam, S. U., Waseem, M., Ayyub, M. B., Arif, M. J., & Ali, H. (2020). In vivo and in vitro assessment of Trichoderma species and Bacillus thuringiensis integration to mitigate insect pests of brinjal (Solanum melongena L.). *Egyptian Journal of Biological Pest Control, 30*, 1–7.
Naz, I., Ahmad, H., Khokhar, S. N., Khan, K., & Shah, A. H. (2016). Impact of zinc solubilizing bacteria on zinc contents of wheat. *American-Eurasian Journal of Agricultural & Environmental Sciences, 16*, 449–454.
Nikitin, D. A., Semenov, M. V., Chernov, T. I., Ksenofontova, N. A., Zhelezova, A. D., Ivanova, E. A., & Stepanov, A. L. (2022). Microbiological indicators of soil ecological functions: A review. *Eurasian Soil Science, 55*(2), 221–234.
Nohwar, N., Khandare, R. V., & Desai, N. S. (2019). Isolation and characterization of salinity tolerant nitrogen fxing bacteria from Sesbania sesban (L) root nodules. *Biocatalysis and Agricultural Biotechnology, 21*, Article 101325.
O'Callaghan, M., Ballard, R. A., & Wright, D. (2022). Soil microbial inoculants for sustainable agriculture: Limitations and opportunities. *Soil Use and Management, 38*(3), 1340–1369.
Odelade, K. A., & Babalola, O. O. (2019). Bacteria, fungi and archaea domains in rhizospheric soil and their efects in enhancing agricultural productivity. *International Journal of Environmental Research and Public Health, 16*, 3873.
Oliveira, W. S., Coelho, I. L., Oliveira, J. R. S., Leite, M. C. B. S., Arnaud, T. M. S., Stamford, N. P., Silva, E. V. N., da Silva, V. S. G., de Oliveira, M. W., de Oliveira, D. C., & Oliveira, T. B. A. (2017). Biological control of the bacterial wilt Ralstonia solanacearum by bioprotector with fungi chitosan from Cunninghamella elegans on tomatoes. *African Journal of Agricultural Research, 12*, 42–49.
Omomowo, O. I., & Babalola, O. O. (2019). Bacterial and fungal endophytes: Tiny giants with immense benefcial potential for plant growth and sustainable agricultural productivity. *Microorganisms, 7*, 481.
Orozco-Mosqueda, Md. C., Rocha-Granados, Md. C., Glick, B. R., & Santoyo, G. (2018). Microbiome engineering to improve biocontrol and plant growth-promoting mechanisms. *Microbiological Research, 208*, 25–31.
Oviedo-Ocaña, E. R., Soto-Paz, J., Domínguez, I., Sanchez-Torres, V., & Komilis, D. (2022). A systematic review on the application of bacterial inoculants and microbial consortia during green waste composting. *Waste and Biomass Valorization, 13*, 3423–3444.
Pandey, R. N., Jaisani, P., & Yadav, D. L. (2021). Trichoderma spp. in the management of stresses in plants and rural prosperity. *Indian Phytopathology, 74*(2), 453–467.
Patra, M. K., Dash, B. K., & Mahalik, J. K. (2017). Effect of bio-fertilizers with graded doses of NPK on Tomato (cv. Pusa ruby) infected by Meloidogyne incognita. *Annals of Plant Protection Sciences, 25*, 190–194.
Philippot, L., Griffiths, B. S., & Langenheder, S. (2021). Microbial community resilience across ecosystems and multiple disturbances. *Microbiology and Molecular Biology Reviews, 85*(2), 10–1128.
Philippot, L., Chenu, C., Kappler, A., Rillig, M. C., & Fierer, N. (2024). The interplay between microbial communities and soil properties. *Nature Reviews Microbiology, 22*(4), 226–239.

Prasad, K. (2022). Influence of PGPR, AM fungi and conventional chemical fertilizers armament on growth, yield quality, nutrient's translocations and quercetin content in onion crop cultivated in semi-arid region of India. *Journal of Microbiology and Biotechnology, 7*(1), 1–18.

Prasanna Kumar, B., Trimurtulu, N., Vijaya Gopal, A., & Nagaraju, Y. (2022). Impact of culturable endophytic bacteria on soil aggregate formation and peanut (Arachis hypogaea L.) growth and yield under drought conditions. *Current Microbiology, 79*, 308.

Prescott, C. E., & Vesterdal, L. (2021). Decomposition and transformations along the continuum from litter to soil organic matter in forest soils. *Forest Ecology and Management, 498*, Article 119522.

Qin, Y., Zhang, W., Feng, Z., Feng, G., Zhu, H., & Yao, Q. (2022). Arbuscular mycorrhizal fungus differentially regulates P mobilizing bacterial community and abundance in rhizosphere and hyphosphere. *Applied Soil Ecology, 170*, Article 104294.

Ram, R. M., Debnath, A., Negi, S., & Singh, H. B. (2022). Use of microbial consortia for broad spectrum protection of plant pathogens: Regulatory hurdles, present status and future prospects. *Biopesticides*, 319–335.

Rao, A. S., & Lenka, N. K. (2020). Developments on soil health management in India as mirrored through sustained researches and policy interventions. *Indian J. Fertil, 16*, 1230–1242.

Rostami, S., & Azhdarpoor, A. (2019). The application of plant growth regulators to improve phytoremediation of contaminated soils: A review. *Chemosphere, 220*, 818–827.

Sabbagh, S. K., Poorabdollah, A., Sirousmehr, A., & Gholamalizadeh-Ahangar, A. (2017). Biofertilizers and systemic acquired resistance in Fusarium infected wheat. *Journal of Agricultural Science and Technology, 19*, 453–464.

Sabir, M. S., Shahzadi, F., Ali, F., Shakeela, Q., Niaz, Z., & Ahmed, S. (2021). Comparative effect of fertilization practices on soil microbial diversity and activity: An overview. *Current Microbiology, 78*, 3644–3655.

Saboor, A., Ali, M. A., Husain, S., Tahir, M. S., Irfan, M., Bilal, M., & Glick, B. R. (2021). Regulation of phosphorus and zinc uptake in relation to arbuscular mycorrhizal fungi for better maize growth. *Agronomy, 11*(11), 2322.

Sahu, P. K., Lavanya, G., Gupta, A., & Brahmaprakash, G. P. (2016). Fluid bed dried microbial consortium for enhanced plant growth: A step towards next generation bioformulation. *Vegetos, 29*, 1–5.

Sahu, P. K., Singh, D. P., Prabha, R., Meena, K. K., & Abhilash, P. C. (2019). Connecting microbial capabilities with the soil and plant health: Options for agricultural sustainability. *Ecological Indicators, 105*, 601–612.

Santos, M. S., Nogueira, M. A., Hungria, M. (2019). Microbial inoculants: reviewing the past, discussing the present and previewing an outstanding future for the use of beneficial bacteria in agriculture. *AMB Express, 9*.

Santoyo, G., Moreno-Hagelsieb, G., del Carmen, O.-M., & Glick, B. R. (2016). Plant growth-promoting bacterial endophytes. *Microbiological Research, 183*, 92–99.

Santoyo, G., Guzmán-Guzmán, P., Parra-Cota, F. I., de los Santos-Villalobos, S., Orozco-Mosqueda, M. C. & Glick, B. R. (2021). Plant growth stimulation by microbial consortia. *Agronomy, 11*, 219.

Schiere, J. B., & Gregorini, P. (2023). Complexity, crash and collapse of chaos: Clues for designing sustainable systems, with focus on grassland-based systems. *Sustainability, 15*(5), 4356.

Shaaban, M. (2024). Microbial pathways of nitrous oxide emissions and mitigation approaches in drylands. *Journal of Environmental Management, 354*, Article 120393.

Shaaban, M., & Nunez-Delgado, A. (2024). Soil adsorption potential: Harnessing Earth's living skin for mitigating climate change and greenhouse gas dynamics. *Environmental Research, 251*, Article 118738.

Sharma, K. P. (2019). Tannin degradation by phytopathogen's tannase: A plant's defense perspective. *Biocatalysis and Agricultural Biotechnology, 21*, Article 101342.

Sharma, D., Gahtyari, N. C., Chhabra, R., & Kumar, D. (2020). Role of microbes in improving plant growth and soil health for sustainable agriculture. *Advances in Plant Microbiome and Sustainable Agriculture: Diversity and Biotechnological Applications*, 207–256.

Sharma, I. (2021). Phytopathogenic fungi and their biocontrol applications. In *Fungi bio-prospects in sustainable agriculture, environment and nano-technology* (pp. 155–188). Academic Press.

Shilev, S., Babrikova, I., & Babrikov, T. (2020). Consortium of plant growth-promoting bacteria improves spinach (Spinacea oleracea L.) growth under heavy metal stress conditions. *Journal of Chemical Technology and Biotechnology, 95*, 932–939.

Shukla, S. K., Sharma, L., Jaiswal, V. P., Pathak, A. D., Tiwari, R., Awasthi, S. K., & Gaur, A. (2020). Soil quality parameters vis-a-vis growth and yield attributes of sugarcane as influenced by integration of microbial consortium with NPK fertilizers. *Sci. Reports, 10*, 19180.

Simarmata, T., Turmuktini, T., Fitriatin, B. N., & Setiawati, M. R. (2016). Application of bioameliorant and biofertilizers to increase the soil health and rice productivity. *HAYATI Journal of Biosciences, 23*, 181–184.

de Souza, A. R. C., Baldoni, D. B., Lima, J., Porto, V., Marcuz, C., Machado, C., Ferraz, R. C., Kuhn, R. C., Jacques, R. J., & Guedes, J. V. (2017). Selection, isolation, and identification of fungi for bioherbicide production. *Brazilian Journal of Microbiology, 48*, 101–108.

Stamford, N. P., Silva, E. V. N. D., Oliveira, W. D. S., Silva, M. C. F. D., Martins, M. D. S., & Silva, V. S. G. D. (2017). Organic matter inoculated with diazotrophic bacterium Beijerinckia indica and Cunninghamella elegans fungus containing chitosan on banana "Williams" in field. *Acta Scientiarum Agronomy, 39*, 33–41.

Stuart, A. K. C., Furue, J. L., Zawadneak, M. A. C., & Pimentel, I. C. (2020). Increased mortality of the European pepper moth Duponchelia fovealis (Lepidoptera:Crambidae) using entomopathogenic fungal consortia. *Journal of Invertebrate Pathology, 177*, Article 107503.

Stuart, A. K. C., Furue, J. L., Cataldi, T. R., Stuart, R. M., Zawadneak, M. A. C., Labate, C. A., & Pimentel, I. C. (2022). Fungal consortium of two Beauveria bassiana strains increases their virulence, growth, and resistance to stress: A metabolomic approach. *PLoS ONE, 17*, Article e0271460.

Subramanian, K. S., & Thirunavukkarasu, M. (2017). Nano-fertilizers and nutrient transformations in soil. *Nanoscience and Plant–Soil Systems*, 305–319.

Sun, X., Xu, Z., Xie, J., Hesselberg-Thomsen, V., Tan, T., Zheng, D., Strube, M. L., Dragoš, A., Shen, Q., Zhang, R., & Kovács, A. T. (2022). Bacillus velezensis stimulates resident rhizosphere Pseudomonas stutzeri for plant health through metabolic interactions. *The ISME Journal, 16*, 774–787.

Sun, W., & Shahrajabian, M. H. (2023). The application of arbuscular mycorrhizal fungi as microbial biostimulant, sustainable approaches in modern agriculture. *Plants, 12*(17), 3101.

Tahat, M. M., Alananbeh, K. M., Othman, Y. A., & Leskovar, D. I. (2020). Soil health and sustainable agriculture. *Sustainability, 12*(12), 4859.

Taktek, S., St-Arnaud, M., Piché, Y., Fortin, J. A., & Antoun, H. (2017). Igneous phosphate rock solubilization by biofilm-forming mycorrhizobacteria and hyphobacteria associated with Rhizoglomus irregulare DAOM 197198. *Mycorrhiza, 27*, 13–22.

Todorović, I., Moënne-Loccoz, Y., Raičević, V., Jovičić-Petrović, J., & Muller, D. (2023). Microbial diversity in soils suppressive to Fusarium diseases. *Frontiers in Plant Science, 14*, 1228749.

Tripti, Kumar, A., Kumar, V., Anshumali, Bruno, L. B., & Rajkumar, M. (2022). Synergism of industrial and agricultural waste as a suitable carrier material for developing potential biofertilizer for sustainable agricultural production of eggplant. *Horticulturae, 8*(5), 444.

Trivedi, S., Butani, N., Patel, H., Nath, M., & Bhatt, D. (2021). Microbial consortia and their application for the development of a sustainable environment. In P. Bhatt, S. Gangola, D. Udayanga, & G. Kumar (Eds.), *Microbial technology for sustainable environment*. Springer.

Tufail, M. A., Bejarano, A., Shakoor, A., Naeem, A., Arif, M. S., Dar, A. A., Farooq, T. H., Pertot, I., & Puopolo, G. (2021). Can bacterial endophytes be used as a promising bioinoculant for the mitigation of salinity stress in crop plants?—A global meta-analysis of the last decade (2011–2020). *Microorganisms, 9*, 1861.

Uzoh, I. M., & Babalola, O. O. (2018). Rhizosphere biodiversity as a premise for application in bio-economy. *Agriculture, Ecosystems & Environment, 265*, 524–534.

Van der Wal, A., & de Boer, W. (2017). Dinner in the dark: Illuminating drivers of soil organic matter decomposition. *Soil Biology & Biochemistry, 105*, 45–48.

Vázquez, E., Benito, M., Espejo, R., & Teutscherova, N. (2020). Response of soil properties and microbial indicators to land use change in an acid soil under Mediterranean conditions. *CATENA, 189*, Article 104486.

Vidya, P., Shintu, P., & Jayaram, K. (2016). Impact of phopshate solubilizing bacteria (Bacillus polymixa) on drought tolerance of green gram (Vigna radiata (L.) Wilczek). *Ann Plant Sci, 5*, 1318–1323.

Xu, C., & Yu, H. (2021). Insights into constructing a stable and efficient microbial consortium. *Chinese Journal of Chemical Engineering, 30*, 112–120.

Yadav, R., Ror, P., Beniwal, R., Kumar, S., & Ramakrishna, W. (2022). Bacillus sp. and arbuscular mycorrhizal fungi consortia enhance wheat nutrient and yield in the second-year field trial: Superior performance in comparison with chemical fertilizers. *Journal of Applied Microbiology, 132*, 2203–2219.

Yilmaz, E., & Sonmez, M. (2017). The role of organic/bio–fertilizer amendment on aggregate stability and organic carbon content in different aggregate scales. *Soil and Tillage Research, 168*, 118–124.

The Role of Nitrogen Fertilization in Enhancing Soil Carbon Sequestration: A Tool for Sustainable Agriculture

Muhammad Junaid Nazir and Muhammad Mahroz Hussain

Abstract Nitrogen fertilization is a critical component in enhancing agricultural productivity, yet its impact on soil carbon sequestration remains complex and multi-faceted. Nitrogen overuse can have devastating consequences on ecosystem, lead to significant environmental issues such as pollution, greenhouse gas emissions, and soil degradation. As the global population is projected to reach 9.7 billion by 2050, balancing both food security and environmental sustainability presents a critical challenge for agricultural productivity. Over the past five decades, the use of nitrogen fertilizers has played a pivotal role in tripling global food production. However, this progress also caused ecological and soil structural damage. This study explores the potential of nitrogen fertilization to enhance soil carbon sequestration which is vital for the long-term sustainability of agriculture. While nitrogen addition can stimulate microbial activity, leading to increased carbon dioxide emissions, it also promotes plant growth and biomass accumulation, thereby contributing to enhance soil organic matter. The balance between these competing processes is influenced by factors such as nitrogen application rate, soil characteristics, climate, and crop management practices. Achieving a nutrient balance in agricultural systems have been remain a significant challenge, requiring careful consideration of these variables. Integrated approaches that optimize fertilizer use efficiency and minimize environmental impacts offer a more sustainable alternative to conventional agricultural methods.

Keywords Nitrogen fertilization · Soil carbon sequestration · Microbial activity · Sustainable agriculture · Soil organic matter

---

M. J. Nazir (✉)
Shenzhen Key Laboratory of Ecological Remediation and Carbon Sequestration, Institute of Environment and Ecology, Tsinghua Shenzhen International Graduate School, Tsinghua University, Shenzhen 518055, China
e-mail: muhammadjunaid529@gmail.com

School of Emergency Management, School of Environment and Safety Engineering, Jiangsu Province Engineering Research Center of Green Technology and Contingency Management for Emerging Pollutants, Jiangsu University, Zhenjiang, People's Republic of China

M. M. Hussain
College of Environmental Science and Engineering, Yangzhou University, Yangzhou, China

M. Shaaban (ed.), *Soils and Sustainable Agriculture*, Frontier Studies in Soil Science,
https://doi.org/10.1007/978-3-031-91114-9_7

1 Introduction

Nitrogen (N) is a critical macronutrient indispensable for optimal crop growth and its unavailability often limits agricultural productivity. As a crucial component in the biogeochemical carbon (C) cycle, N plays a major role in providing energy for microbial and plant growth. It plays a pivotal role in optimizing plant nutrient uptake, stabilizing crop yields, and maximizing the efficiency of other fertilizers and water resources. Since the 1960s, the widespread application of N fertilizers has significantly influenced soil organic carbon (SOC) dynamics by altering both organic C inputs and microbial-mediated mineralization (Wei et al., 2020). Fertilization can enhance soil nutrient availability, but excessive and indiscriminate fertilizer application can also induce structural deterioration of the soil over time, ultimately compromising plant biomass productivity (Deng et al., 2024). Excess inorganic N is highly susceptible to nitrification and denitrification, leading to elevated soil nitrate and ammonium concentrations and exacerbating soil C/N imbalances (Lu et al., 2021). Effective fertilization is essential for enhancing yield, extending development, and increasing plant vigor. Maintaining fertilization within optimal thresholds significantly amplifies agricultural productivity. The effects of reduced fertilization on crop yield and GHG emissions vary depending on fertilization techniques, crop type, soil composition, and prevailing weather patterns (Li et al., 2024). To adopt sustainable agricultural practices, it is imperative to enhance the efficiency of N fertilizer utilization without increasing overall input levels. The prolonged application of inorganic fertilizers in agricultural cultivation has led to a significant decrease of soil organic matter (SOM). Moreover, soil carbon sequestration offers a promising approach to climate change mitigation by sequestering greenhouse gases (GHG) and enhancing soil health (Gul & Whalen, 2022).

Enhancing soil carbon sequestration offers a dual benefit by mitigating climate change through reduced carbon dioxide (CO_2) emissions and enhancing ecosystem services such as improved soil quality and crop productivity (Shaaban & Nunez-Delgado, 2024; Wang et al., 2021a, 2021b, 2021c). Soils serve as a substantial C sink, holding nearly three times more C than living plants and animals combined (Thotakuri et al., 2024). Soil carbon sequestration in this way offers several benefits, including improved soil fertility, increased water retention capacity, and amplified agricultural productivity (Benbi, 2015; Nazir et al., 2024a, 2024b). N fertilizer application rates escalated dramatically during the 1980s. N fertilizers remain indispensable for global food security, contributing to hunger alleviation and increased grain yields. Excessive N input has precipitated adverse ecological consequences, including soil acidification and eutrophication (Aguilera et al., 2013). The widespread adoption of fertilizers has enabled a remarkable feat: feeding 22% of the world's population using only 7% of arable land. Recognizing the imperative for sustainable agriculture, emphasize the need for its advancement in the early twenty-first century was carried out (Gong, 2011).

Agricultural practices, when integrated with N inputs, can influence soil carbon sequestration. For instance, four large-scale trials in Germany and China demonstrated that manure and N fertilizer application positively impact soil C carbon storage (Aguilera et al., 2013). N fertilization is a critical factor influencing soil carbon sequestration. Nutrient availability, particularly N, can substantially limit the accumulation of SOC (Kirkby et al., 2014). Nitrogen fertilization can stimulate plant growth and biomass accumulation, thereby increasing soil carbon inputs. Moreover, N can enhance carbon sequestration by improving soil structure and aggregation, thus protecting soil organic matter from decomposition. The integration of N inputs into diverse agricultural practices can influence soil carbon sequestration (Mosier et al., 2021). The relationship between N fertilization and soil carbon sequestration is complex and contingent upon a variety of factors, including soil composition, climatic conditions, and management practices (Lal, 2010). The application of manure and N fertilizers can positively influence soil C reserves (Ludwig et al., 2011). Over the past three decades, the implementation of modern agricultural techniques, notably N fertilization has increase 0.437 Pg C of SOC and a 1.6 Pg C decline in soil inorganic carbon (SIC) within agricultural soils (Sun et al., 2010). Moreover, in arid and semi-arid agricultural areas SIC accumulated at a significantly faster rate than SOC across long-term nitrogen fertilization experiments (Zhang et al., 2015) Consequently, a comprehensive investigation into the potential of N fertilizers to enhance carbon sequestration and improve overall soil health is imperative (Elbasiouny et al., 2022).

The relationship between N addition and soil carbon sequestration is a critical factor in determining future atmospheric CO_2 levels. Enhancing N use efficiency has the potential to enhance SOC inputs while reducing decomposition rates, thereby promoting carbon sequestration. The regulation of SOC turnover by N input is primarily governed by complex interactions among plants, soil physicochemical properties, and microbial communities. Accurate climate C models must account for the reciprocal influence of both organic and inorganic N deposition on soil SOC dynamics to effectively predict the trajectory of SOC storage and its impact on the global C cycle (Feng et al., 2021).

2 Impact of Nitrogen Forms on Soil Nitrogen Availability

Nitrogen availability can be affected by the imbalance between fertilizer application and the soil natural N supply. These disparities may be exacerbated by the form of N applied, as microbial N uptake is more efficient for ammonium (NH_4^+) than nitrate (NO_3^-). Nevertheless, the utilization of acidifying N sources such as ammonium sulfate ($(NH_4)_2SO_4$) or ammonium chloride (NH_4Cl) can hinder microbial N processing (Elrys et al., 2023). In cropland farming systems, N addition is mainly a composition of both inorganic and organic fertilizers. Organic N comprises 30% of total atmospheric N deposition. Agricultural croplands typically receive N from inorganic fertilizers (e.g. ammonium nitrate, ammonium sulfate) and organic fertilizer (urea) (Jickells et al., 2013). Inorganic N inputs often lead to soil acidification, altering soil physical and chemical properties. Ammonium-N applied to agricultural soils undergoes nitrification to Nitrate-N, while simultaneous plant uptake and leaching losses contribute to overall soil acidification by the following reactions (Scavo et al., 2022).

$$NH_3 + 2O_2 = H^+ + NO_3^- + H_2O$$
$$NH_4NO_3 + 2O_2 = 2H^+ + 2NO_3^- + H_2O$$
$$NH_4NO_3 + 2O_2 = 2H_+ + 2NO_3^- + H_2O$$

Many studies conducted previously mainly focused on how inorganic N affects SOC decomposition, but information regarding organic N is usually scarce in literature studies. By combining organic and inorganic fertilizers, crop yields can be enhanced, promoting sustainable agroecosystems. This occurs through alterations in the C/N ratio, which influence microbial community composition and activity. Concurrently, N can inhibit SOC decomposition by reducing microbial activity and promoting aggregate formation (Ye et al., 2018).

3 Nitrogen Interactions and Carbon Dynamics

The introduction of N can either increase or decrease microbial biomass, consequently influencing soil and residue C and N dynamics through alterations in mineralization and immobilization processes. Nitrogen can inhibit SOC decomposition by reducing microbial activity and promoting aggregate formation (Ye et al., 2018). Nitrogenous fertilizers can induce shifts in SOC fractions and microbial populations, thereby facilitating the formation of larger soil macroaggregates. Given the North China Plain's history of excessive N fertilization, research is urgently needed to elucidate the impacts and mechanisms of various high-quality organic materials and N inputs on soil carbon sequestration (including the creation of recalcitrant carbon pools) and the development of large soil aggregates. This knowledge gap is particularly evident in the context of calcareous soils, where long-term field experiments are scarce (F. Wang et al., 2021a, 2021b, 2021c). Meanwhile, current researches on the influence of labile organic C in N-containing soils are limited. Optimizing the utilization of C and N inputs is crucial to balance atmospheric and soil C levels (Liu et al., 2023). The rate of SOC decomposition is significantly influenced by changes in soil nutrient availability, which are primarily driven by long-term N deposition. This approach further enhances soil health by stabilizing SOC and stimulating microbial activity (Ogdahl et al., 2012).

Nitrogen availability profoundly impacts organic C cycling and storage. The duration and magnitude of N accumulation are dependent on the C/N ratio of the substrates. Plant litter with lower C/N ratios decomposes more rapidly than litter with higher C/N ratios (Zhang et al., 2017). The N mineralization models incorporate a relationship between C and net N mineralization, assuming either constant microbial C/N ratios fluctuating within specific parameters. Excessive N inputs can lead to mineral N accumulation, further disrupting the C and N balance and intensifying SOC mineralization (Don et al., 2024). Previous researches has indicated that supplementing residues or other carbon-based materials with N enhances the activity of cellulases and protease enzymes, which are critical for the degradation of C and N compounds (Luo et al., 2019).

4 Soil Carbon Sequestration and N Enrichment

Over the past century, fossil fuel burning and fertilizer use have increased atmospheric N deposition 3–5 fold. It has been estimated that N deposition will become more than double in terrestrial ecosystems in next couple of years (Gurmesa et al., 2022). Nitrogen availability is crucial for plant growth. Insufficient N can limit plant biomass production, while excess N may enhance ecosystem carbon storage. However, the exact impact of N additions on soil carbon sequestration remains uncertain and requires further investigation. Soil holds nearly twice C as in the atmosphere, serving as significant natural C reservoirs. However, it releases CO_2 into the environment. Soil N availability directly regulates SOM decomposition rates. Consequently, enhanced N utilization accelerates SOM decomposition, thereby impeding SOC sequestration. However, nitrogen also stimulates plant growth and litter production, increasing SOC inputs. Therefore, the future trajectory of atmospheric CO_2 levels could be significantly influenced by how soils respond to additional N inputs (Huang et al., 2020; Sun et al., 2022).

Soil C is primarily built up through inputs from plant litter, root exudates, farmyard manure, cover crops, mulching, and mycorrhiza fungi. Conversely, it is depleted/released by the breakdown of organic matter, soil erosion, CO_2 release into the atmosphere and by the leaching of nutrients. Anthropogenic and human activities of C inputs can lead to enhance soil carbon sequestration by stimulating plant growth, increasing organic matter content, and increasing N availability. However, the addition of N can inhibit enzymes/microbes that digest lignin and cellulases in plants and also can lower soil pH, thereby slowing the breakdown of organic matter (Chen et al., 2018a, 2018b). Most N-enrichment experiments have neglected to quantify pre-existing SOM, limiting ability to fully understand the mechanisms behind soil carbon sequestration. However, isotopic labeling techniques can specify C pools and provide insights that how N addition affects C storage in soil (Del Vecchio et al., 2018).

Soil carbon sequestration is governed by the dynamic equilibrium between C inputs (photosynthetic production) and losses (biomass removal, microbial and root respiration, leaching) (Kim et al., 2022). Ladha et al. (2011) conducted a meta-analysis of 114 long-term trials across 100 global sites, revealing an average 8% increase in SOC under mineral N fertilization (Ladha et al., 2011). Corroborating these findings, Geisseler and Scow (2014) reported an 8.5% SOC augmentation with mineral fertilizer application. N fertilization stimulates crop growth, thereby enhancing external C inputs while simultaneously retarding SOC decomposition (Geisseler & Scow, 2014). Ge et al. (2015) demonstrated a positive correlation between applied N and photosynthetically derived carbon sequestration, emphasizing the role of N in increasing SOC pools. Moreover, varying N application rates influence the mobility and composition of SOC by altering the ratio of easily decomposable to resistant carbon fractions. The acidification resulting from nitrification following N fertilization is considered a key factor in the dissolution of soil inorganic carbon (SIC) in agricultural soils, a phenomenon supported by several studies (Ge et al., 2015).

5 Fertilizer Effects on Soil Health

Fertilizer applications for long term duration has been widely adopted to examine the effects of continuous fertilizer on soil health and the environment. Optimized fertilizer use can increase soil carbon storage/stability. For example, using the recommended fertilizer dose increased SOC content from 7.9 to 10.5 g kg^{-1}. However, excessive fertilizer use can have negative consequences (Rao, 2011). Fertilizer inputs can disrupt the natural balance of soil nutrients, creating conditions that favor the growth of native microbial species, even in challenging environments. This can lead to an increase in microbial biomass carbon (MBC) and SOC. Alvarez found a positive correlation between N fertilizer application and increased SOC. The study analyzed the effects of N fertilizer, tillage practices, and crop residue retention on SOC levels across 37 diverse locations (Alvarez, 2005).

Additionally, conservation agriculture practices have proven effective in transforming soils from C emissions into C sinks by capturing and storing C as organic matter. In contrast, such scenario is different in grasslands where N addition did not impact soil carbon storage, but enhanced carbon stability in temperate, tropical, and boreal forests. Furthermore, N addition affects litter decomposition differently based on N availability. In N-rich environments, it slows decomposition, while in N-poor conditions, it speeds up the breakdown of easily decomposable carbon. However, the diverse nature of N-enrichment environments makes it challenging to generalize the findings of a single study (Rogger et al., 2022).

6 Intensive Agrochemical Use (Nitrogen Inputs and Gas Emissions)

The increased use of N inputs in modern agriculture has consequences for GHG concentrations in the atmosphere. The last century witnessed a drastic increase in fossil fuel burning and artificial chemical fertilizer use, resulting in a 3- to fivefold increase in crop atmospheric N deposition. Global N deposition rates are projected to increase, leading to significant N accumulation in terrestrial ecosystems. While N is essential for plant growth, N deficiency can limit plant biomass and retard growth and yield. Conversely, high/over N levels are anticipated to enhance soil carbon storage (Wright, 2019). Evidence suggests that the global C budget during the last two decades relies on sources and sinks and estimates the amount of 'missing' or 'fugitive' C. Manufacturing cement and burning fossil fuels each contributed 5.4 Pg CO_2, and land use change contributed 1.7 ± 0.8 Pg to the global C budget during the 1980s. The knowledge about the potential for various agricultural management practices to sequester C is influencing the mitigation of climate change; as a result, GHG concentration in the atmosphere increases (Herrero et al., 2016).

The overuse of agrochemicals, particularly N-based fertilizers, contributes to GHG emissions and hinders efforts to sequester carbon in the soil. Ecosystem restoration plays a vital role in combating land degradation and C loss. Active farmland restoration and natural regeneration increase carbon sequestration, but the mechanisms involved remain unknown. Restored and naturally regenerated ecosystems typically sequester more C in soils compared to croplands. However, the advantages of active restoration over natural regeneration depend on specific contextual factors. Active restoration generally sequesters more C in soils with initially low C content but less C in those with high C content compared to natural regeneration (Huang et al., 2022). Additionally, active restoration exhibits more carbon fixation in the topsoil but less in the subsoil. These findings underscore the significance of considering the landscape context when deciding between active restoration and natural regeneration. They also offer guidance on optimizing the allocation of limited resources to maximize the benefits of carbon sequestration in restoration endeavors (Dang et al., 2023).

7 Impact of Nitrogen Fertilizers on Agricultural Yields

Nitrogen fertilizers significantly enhance agricultural yields, yet they concomitantly contribute substantially to GHG emissions, including methane (CH_4), CO_2, and nitrous oxide (N_2O). These gases are primary drive global warming and its climate impacts. While agricultural cropping technologies facilitate the substantial cycling of CO_2, plants absorb considerable quantities during photosynthesis. However, the net release of CO_2 within the agricultural sector remains elusive. This is primarily attributed to energy consumption associated with agricultural operations and the transportation of agricultural products (Zang et al., 2022). A comprehensive understanding of the interrelationships among climate change, fertilizer application, agricultural production, and greenhouse gas (GHG) emissions in leading agricultural nations is imperative for developing effective strategies to mitigate these challenges and ensure sustainable futures. Snyder et al. (2009) emphasized the critical role of nitrogen (N) fertilizer in global food, fiber, and fuel production. They highlighted the necessity of enhancing N-use efficiency to conserve water resources, reduce GHG emissions, and preserve soil health (Snyder et al., 2009). Ecological intensification has been proposed as a strategy to reconcile high crop yields with reduced greenhouse gas emissions within intensive crop management systems. Optimal fertilizer application has been shown to stimulate plant growth and sequester organic carbon in the soil. Moreover, efficient fertilizer management has contributed to lower residual soil nitrate levels, thereby mitigating the risk of increased nitrous N_2O emissions. The study investigated that the interaction between tillage practices and SOC levels, as well as the impact of different N fertilizer sources on N_2O emissions under specific conditions. Contrary, intensive crop management systems did not necessarily correlate with elevated GHG emissions per unit of crop or food produced. These findings suggest the potential to reconcile environmental preservation and food security by transitioning from forestland to agricultural production while mitigating GHG emissions. To maximize the benefits of this research, it is imperative to disseminate findings among key stakeholders and foster interdisciplinary collaboration to identify knowledge gaps and implement optimal management strategies.

Wang et al., (2021a, 2021b, 2021c) conducted a comprehensive survey examining the utilization of N fertilizers in agricultural sectors. While essential for enhancing crop yields, chemical fertilizers exhibit low nitrogen use efficiency (NUE), thereby contributing to environmental challenges such as nitrate pollution, air quality degradation, and GHG emissions (Shaaban, 2024). The study investigated into the potential of biological nitrification inhibition (BNI) as a cost-effective and environmentally friendly strategy to mitigate N loss. The analysis encompassed the genesis, methodologies, influencing factors, and practical applications of BNI, while highlighting its potential benefits for agricultural productivity and GHG emissions reduction. The discussion expanded upon potential avenues for further research into BNI, particularly focusing on the excessive application of chemical fertilizers within agriculture. The study explored into the relationship between chemical fertilizer usage, agricultural financial support, and subsequent C emissions from agricultural processes. Findings from these studies revealed a positive correlation between chemical fertilizer application and C emissions, while agricultural financial assistance demonstrated a significant mitigating impact. To ensure the long-term sustainability of agriculture, the report recommends increased financial support for environmentally friendly farming practices and the development of alternative, cleaner agricultural production technologies to reduce reliance on chemical fertilizers (Wang et al., 2023).

8 Microbial Community Structure and Activities

The composition and function of microbial communities exert a primary influence on carbon use efficiency (CUE) through diverse strategies for organic carbon metabolism. Fungi exhibit a higher CUE compared to bacteria due to their broader C/N ratio tolerance and greater C demand (Domeignoz-Horta et al., 2020). A correlation between fungal abundance in soil and increased carbon storage have been found in many studies. While both enhanced N efficiency and altered soil pH induced by N addition can rapidly modify microbial communities, these factors exert opposing influences. Increased N availability shifts community composition towards bacteria, which possess lower CUE than fungi owing to their elevated nutrient requirements (Zhou et al., 2023). Nitrogen addition to soil induces acidification, consequently altering the microbial community structure. A positive correlation exists between bacterial and fungal community composition and soil pH, as distinct microbial assemblages are responsible for soil organic matter decomposition across varying pH conditions (Kang et al., 2021). Bacterial activity is optimized within neutral pH environments, whereas fungi exhibit superior tolerance to acidic conditions and thrive in neutral to slightly acidic soils. Moreover, mineral N input, particularly Nitrate–N, intensifies soil acidification and promotes fungal dominance characterized by high CUE (Rousk et al., 2010). At the other end, organic N and integrated management practices mitigate soil acidity and favor a microbial community growth (Yang et al., 2019).

Nitrogen addition can indirectly impact microbial activity by modifying soil pH levels. Excessive mineral N input acidifies the soil, prompting microorganisms to divert substantial SOC towards energy generation. This metabolic shift impedes microbial growth and reduces microbial CUE (Malik et al., 2018). Conversely, organic N application mitigates soil acidification and enhances microbial CUE. Initially, declining pH induces leaching of calcium (Ca^{2+}) and magnesium (Mg^{2+}) ions. Consequently, bacteria may encounter Ca and Mg deficiencies, inhibiting growth and lowering microbial CUE. Furthermore, pH reduction facilitates aluminum (Al^{3+}) mobilization, upregulating microbial genes associated with heavy metal tolerance. Microorganisms deploy efflux pumps, enzymatic transport, and in vivo chelation to resist heavy metals, processes typically requiring ATP. This energy expenditure diverts C towards respiration, decreasing microbial CUE (Malik et al., 2017) as depicted in (Fig. 1). Bellion et al. (2006) employed ^{14}C-labeled glucose to quantify microbial CUE, resulting in critical pH threshold of 5.5, below which microbial CUE declined progressively. At this juncture, Al^{3+} dominated the soil exchange complex. Given the energy-intensive nature of microbial aluminum resistance, a significant negative relationship was observed between microbial CUE and the availability of exchangeable Al^{3+} (Hu et al., 2019). Under acidic conditions with elevated aluminum, microbial metabolic pathways shift. Bacteria secrete extracellular polymers and organic acids to bind and neutralize toxic $Al^{3+.}$

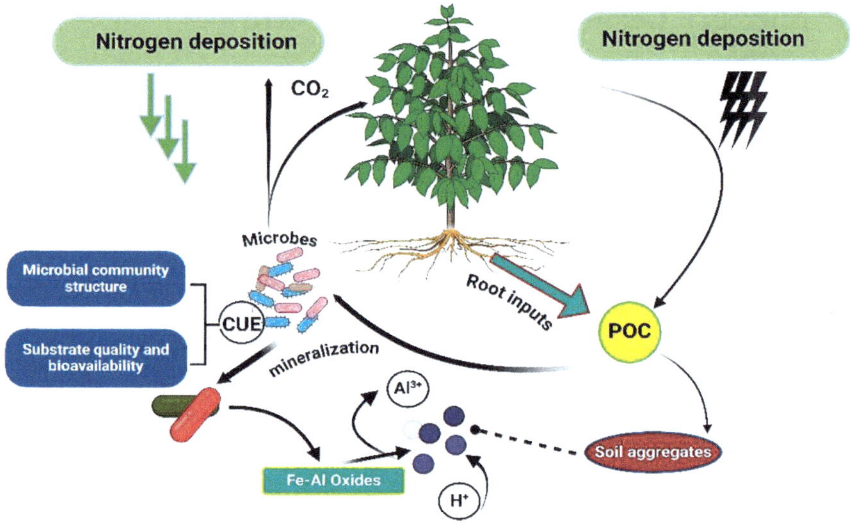

Fig. 1 Outcomes of N deposition in plants physiological processes

9 Interaction Between Soil Carbon, Microbes and Nitrogen Fertilizer

Soil organic matter is a critical factor in soil quality. By increasing its C content, climate change mitigation could be possible by reducing CO_2 emissions while improving soil health and crop productivity. Agricultural fields receive significant organic C inputs from crop residues i.e.; straw and stubble. Incorporation of straws and stubbles are then decomposed by soil microbes and net microbial biomass produced (Wang et al., 2021a, 2021b, 2021c). Conversely, more recalcitrant fractions bind with soil mineral particles, forming particulate organic carbon (POC) (Liang et al., 2019). Microbial necromass, comprising approximately 50% of total SOC, accumulates following microbial death. Microbial preferences for different plant residues influence the stability of necromass-derived C. Bacteria preferentially degrade easily decomposable compounds such as proteins and cellulose, whereas fungi tend to utilize more recalcitrant components like lignin. Consequently, fungal necromass carbon exhibits greater stability than bacterial necromass carbon. Microbial necromass can further stabilize through association with clay minerals or incorporation into microaggregates, forming mineral-associated organic carbon (MAOC). While this process enhances soil stability, necromass rich in N and phosphorus (P), and protected from microbial degradation, represents an ideal substrate for microbial growth and nutrient acquisition (Deng & Liang, 2022). Nitrogen addition can lower soil pH, subsequently inducing aluminum toxicity. This condition has been correlated with decreased fungal abundance and inhibited microbial catabolic processes (Wang et al., 2021a, 2021b, 2021c; Ye et al., 2018). A two-year experiment conducted by

Chang et al. involving NH_4NO_3 application to alpine ecosystems supports this assertion (Chang et al., 2019). Additionally, Nitrogen fertilization can boost soil health, increase crop yields, and enhance the accumulation of C in plant biomass. Simultaneously, it alters the soil C/N ratio, influencing soil microbial community structure and organic C conversion processes. The impact of N f POC and MAOC cycling has been shown in Table 1.

Furthermore, Prolonged N application can increase SOC levels by enhancing plant-derived C inputs (Li et al., 2018). However, Excessive N fertilization can disrupt the soil C/N balance by causing an excess of mineral N. This imbalance accelerates the decomposition of SOC to meet microbial N demands, leading to C losses through mineralization. Therefore, optimizing N fertilization is essential for maintaining high crop yields by sequestering more C in soil profile. However, impacts of N fertilization on C storage can vary, depending on specific crop systems and environmental conditions.

10 Nitrogen Impact on Microbial Carbon Use Efficiency

Nitrogen availability exerts a substantial influence on microbial CUE. Bacteria require energy to maintain equilibrium between catabolic and anabolic processes, and N intake modulates this efficiency by altering microbial community composition and activity, substrate quality and accessibility, and ecological stoichiometry (Li et al., 2019). Nitrogen supplementation enhances microbial activity, stimulating fungal mycelial growth and the subsequent release of polysaccharides, lipids, mucus, and other microbial secretions. The fungal mycelium imparts resilience against mechanical degradation and fragmentation. Application of organic N fertilizer promotes the formation of clustered mycorrhizae, enhances the synthesis of insoluble glycoproteins, and facilitates aggregate development. Nitrogen inputs in the form of fertilizers, manures and other addition significantly influenced aboveground biomass, leading to an increase yield and productivity (Luo et al., 2020) as shown in Fig. 2. Conversely, belowground plant roots exhibited a threshold response: low N inputs stimulated biomass accumulation, while high inputs resulted in a modest decline. Consequently, nitrogen addition fosters root intertwinement within aggregates, inhibits microbial respiration, and offers a degree of protection for POC derived from plant litter (Yue et al., 2020). Organic N is more effective than inorganic N in stimulating the production of soil macroaggregates. This enhancement is attributed to organic N capacity to mitigate soil acidification and substantially increase aboveground biomass, thereby augmenting the input of plant residues into the soil, which subsequently form POC. Moreover, organic N deposition can be considered a mechanism for elevating the soil content of unstable organic matter, a substance that serves as an organic binding agent in aggregate formation (Witzgall et al., 2021).

Table 1 Nitrogen influence on POC and MAOC turnover

SOC types	Process	Category	Promotional effects	Inhibitory effects
POC	Plant litter	Quantity	Increased nitrogen use efficiency enhances net plant productivity	Soil acidification triggers heavy metal toxicity leading to lower NPP
		Quality	Nitrogen elements influences an enhanced organic matter content and structural complexity	Lower C/N of plant litter
	Physical protection	Aggregates	Increasing the amount of organic cementing substances promotes the formation of aggregates Plant root entanglement and soil acidification synergistically promote aggregate formation	Soil acidification causes extensive depletion of inorganic carbon and reduces its cementation
MAOC	CUE	Microbial community structure	fungal communities under soil acidification conditions produced high CUE	Increased N efficiency leads microbial community structure to favor bacterial dominance which negatively impact CUE
		Ecological chemometrics	Microbes allot increased C to maintain their stoichiometric balance, resulting in higher CUE	Enhanced availability of nitrogen limits carbon, ultimately inhibits microbial growth and CUE
		Substrate quality and bioavailability	Substrates with more easily degradable generally lead to higher microbial efficiency	Decline in substrate availability prompts a positive increase in microbial extracellular enzyme production but decreased microbial CUE

(continued)

Table 1 (continued)

SOC types	Process	Category	Promotional effects	Inhibitory effects
	Mineral–organic complexes	Soil minerals	Acidification causes iron and aluminum oxides have positive charges, promoting the formation of mineral-organic complexes in soil	Oxalic acid in root exudates disrupts mineral-organic associations
		Bond bridge	Organic nitrogen fertilizers stimulate the production of hydroxyl radicals (HO^-), leading to the stability of SOC	Under acidic conditions, competition between H^+ ions with cations (neutral/alkaline) occurs for their binding sites

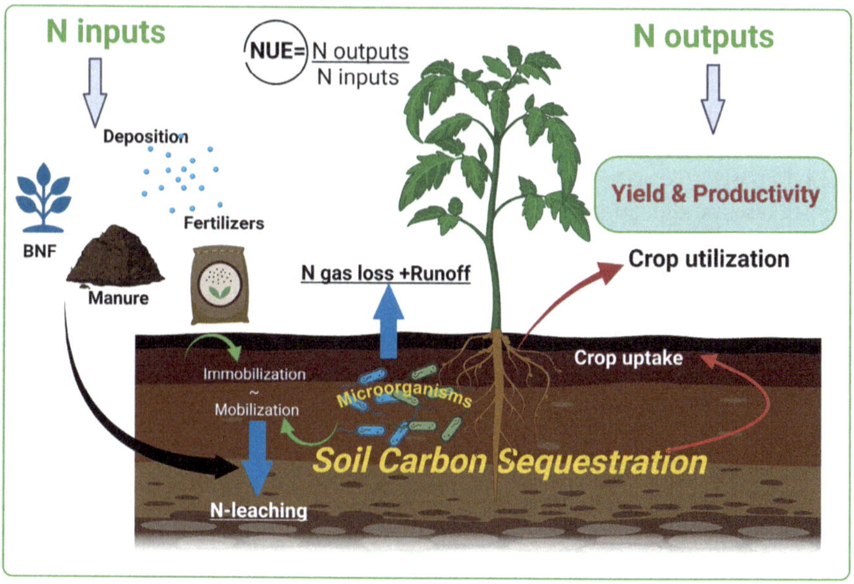

Fig. 2 Sources of N inputs and output in enhancing soil carbon sequestration and yield

11 Microbial Biomass Carbon and Microbial Biomass Nitrogen

The relationship between microbial biomass carbon and microbial biomass nitrogen is generally positive, indicating that enhanced microbial growth necessitates increased N acquisition. Moderate N supplementation stimulates microbial activity by providing an energy source, prompting bacteria to adjust C uptake through chemo-enzymatic regulation in response to altered soil nutrient availability (Deng et al., 2023). Aligned with steady-state stoichiometric theory, microorganisms optimize CUE under N-replete conditions. Conversely, a negative correlation exists between soil microbial CUE and the C/N ratio of SOM. Nitrogen limitation disrupts microbial catabolic and anabolic processes, leading to C-intensive metabolite production (e.g., extracellular enzymes, polysaccharides) and increased C loss, primarily through respiratory CO_2 efflux. Consequently, CUE diminishes. As N availability rises, microorganisms prioritize C allocation to maintain optimal cellular stoichiometry (C/N ratio of 7–8.6), concurrently reducing extracellular enzyme production and enhancing CUE. While inclusion primarily amplifies microbial utilization of active C, N-enrichment alone does not significantly influence organic matter decomposition. The combined addition of N and P impacts organic matter transformation due to N-induced P limitation within the ecosystem. According to chemometric analysis, an optimal C:N:P ratio of 60:7:1 supports optimal microbial activity (Junaid Nazir et al., 2022; Wei et al., 2020).

Progressive N enrichment alters microbial growth constraints, shifting them from nutrients like (N and P) to C deficiency. Microbial growth, is primarily constrained by C availability rather than substrate abundance. This is attributed to the energy derived from alternative substrate oxidation, which supports cellular maintenance but not growth (Chen et al., 2018a, 2018b). Microbial cells can only allocate C to growth after fulfilling energy demands, resulting in positive CUE. Several mechanisms have been proposed to clarify the observed C limitation under N addition. Nitrogen deposition causes acidification in soil, leading to increased fungal-to-bacterial ratios and microbial biomass C and N ratios. This mitigates the C/N imbalance by microbial communities, which is further decreased by reduced soil pH, causes to shift towards C than N for their energy source. Such N induced alterations/shifts in biological C accessibility significantly influence microbial carbon metabolism (Li et al., 2021).

12 Impact of Fertilization on Direct Emissions

Nitrous oxide emissions, primarily resulting from nitrification and denitrification, constitute a significant pathway of N loss. Findings from previous researches demonstrated a substantial increase in N_2O emissions with elevated N fertilization rates (Xu et al., 2024). This enhancement is attributed to the combined stimulation of background soil N release and the activation of indigenous N sources (Xu et al., 2022). Nitrogen fertilizer accelerates the breakdown of native soil organic N, providing a substrate that increases N_2O production (Hu et al., 2023). Simultaneously, a lower soil C/N ratio accelerates microbial decomposition and N cycling (Shaaban et al., 2016; Xu et al., 2024). Additionally, ammonium ions from fertilizers directly or indirectly provide the necessary compounds for nitrification and denitrification, leading to increased N_2O emissions as fertilizer application rises. Nitrogen application can disrupt the beneficial role of methylotrophic bacteria in mitigating atmospheric

N loading (Yao et al., 2022). Furthermore, N fertilization reduces CH_4 uptake by providing substrates for methanogenic bacteria. Literature studies further indicates a positive correlation between N fertilization and soil CO_2 emissions, often surpassing unfertilized controls (Shaaban et al., 2022). Qiu et al. reported nearly 55% reduction in SIC over 22 years of continuous fertilization. The majority of lost SIC is released as CO_2 (Qiu et al., 2016). Nitrogen fertilization-induced soil acidification has depleted 7% of China's total calcium carbonate ($CaCO_3$) reserves, resulting in a cumulative CO_2 loss of 145 teragrams (Tg) over the past four decades (Raza et al., 2021).

13 Impact of Fertilization on Comprehensive Greenhouse Effect

Agricultural fertilization often involves the application of excessive N, a portion of which is assimilated by crops while the remaining is utilized by soil microorganisms, contributing to N_2O emissions. Unaccounted N is further lost through leaching and volatilization. The Greenhouse Gas Intensity (GHGI) metric provides a framework for assessing the Global Warming Potential (GWP) per unit crop yield associated with fertilization practices. Studies of Menegat et al., revealed an initial decline followed by an increase in GHGI with escalating fertilizer input. This phenomenon is attributed to the dynamic interaction between crop N demand and soil N availability (Menegat et al., 2022). Similarly, under conditions of low residual soil N, enhanced crop production can coincide with reduced GHGI. Conversely, N application exceeding crop requirements accelerates GHG emissions due to inefficient N utilization. These emphasize the potential for significant GHGI reduction through optimized N input while maintaining substantial crop yields (Zhang et al., 2023). Yield data from studies of Guo et al., supported that moderate N fertilization can increase crop output without markedly elevating GHG emissions. While N fertilization can moderately enhance yield, excessive application drastically amplifies GHG emissions, posing a significant environmental risk (Guo et al., 2022).

Zang et al. (2022) explored the complex relationship between low-C farming, agricultural commerce, economic growth, and ecological sustainability. Their time-series analysis of data spanning 2002–2020 revealed a compelling correlation: escalating agricultural C emissions led to a sustained 2% decline in agricultural commerce and a 2–4% contraction of the overall agricultural economy. The study further emphasized the role of N fertilizers in enhancing crop yields while concurrently contributing to GHG emissions, including CH_4, CO_2 and N_2O. Their findings corroborate this positive association between fertilizer application and GHG emissions (Zang et al., 2022). similar studies demonstrated that the agricultural sector is a significant GHG emitter, primarily due to energy-intensive inputs such as fertilizers and machinery. The production of synthetic N fertilizers, in particular, is energy-demanding and results in the release of CO_2 and N_2O. The operation of agricultural machinery, irrigation systems, and transportation processes further exacerbates GHG emissions through fossil fuel combustion. While fertilizers and energy are indispensable for modern agriculture, their environmental impact underscores the imperative for sustainable practices and technological advancements to mitigate GHG emissions (Chataut et al., 2023).

14 Policy Implications

Achieving sustainable agricultural development imposes a reduction of excessive N input and an enhancement of N use efficiency. Adhering to recommended N fertilizer application rates significantly increases SOC and soil total C, suggesting enhanced C sequestration potential in soils, ultimately promoting soil carbon storage. However, historical N overuse persists. A regional survey should be conducted to indicates a recommended N application rate for major crops like wheat, cotton, maize, sugarcane etc. Consequently, targeted efforts are required to educate farmers on optimal N fertilization practices. Policymakers should hold a pivotal role in steering agriculture towards a more sustainable and climate-resilient trajectory through the implementation of sustainable practices, optimized N management, strategic research investment, and robust international cooperation. Additionally, The Nutrient Expert (NE) is a computerized fertilization system lauded for its capacity to optimize crop yield and agronomic efficiency. By considering soil fertility and crop yield levels, the NE generates tailored fertilization recommendations. Implementing NE-based fertilization strategies has the potential to enhance N utilization, optimize nutrient application, and consequently increase crop yields compared to traditional fertilization methods.

15 Conclusion

Optimal nitrogen fertilization enhances soil carbon sequestration. Fertilizer application can redistribute soil carbon within the soil profile. To maximize soil carbon storage, precise input of N is crucial rather than simply increasing N supply. Intensive agricultural practices often maximize crop yields, they simultaneously exacerbate GHG emissions, particularly CO_2 and N_2O, and deplete SOC stocks. Non judicial fertilizer application strategies have proven insufficient in enhancing net ecological carbon budgets and net ecosystem economic benefits relative to more moderate approaches. The long-term sustainability of agriculture and mitigation of its environmental impact demands a comprehensive understanding of the interaction between nutrients inputs specifically N, climate change and agricultural practices. Fertilization regimes emerge as optimal strategies for maximizing system productivity, fostering ecological sustainability, and minimizing carbon emissions. Nitrogen influences carbon sequestration by altering microbial carbon use efficiency and mineral-organic complex bonding. Elevated inorganic nitrogen lowers soil pH, inducing positive charges on iron and aluminum oxides, which facilitate mineral associated organic carbon formation. To effectively address the challenges of climate change and ensure food security for current and future generations, collaborative efforts are essential. Furthermore, to comprehensively evaluate the regional impact of nitrogen fertilization on soil carbon and sustainable agriculture mission, further long-term experimental data are essential.

References

Aguilera, E., Lassaletta, L., Sanz-Cobena, A., Garnier, J., & Vallejo, A. (2013). The potential of organic fertilizers and water management to reduce N_2O emissions in Mediterranean climate cropping systems. A review. *Agriculture, Ecosystems & Environment, 164*, 32–52. https://doi.org/10.1016/j.agee.2012.09.006

Alvarez, R. (2005). A review of nitrogen fertilizer and conservation tillage effects on soil organic carbon storage. *Soil Use and Management, 21*(1). https://doi.org/10.1079/sum2005291

Bellion, M., Courbot, M., Jacob, C., Blaudez, D., & Chalot, M. (2006). Extracellular and cellular mechanisms sustaining metal tolerance in ectomycorrhizal fungi. *FEMS Microbiology Letters, 254*(2), 173–181. https://doi.org/10.1111/j.1574-6968.2005.00044.x

Benbi, D. K. (2015). Carbon sequestration for soil health enhancement and mitigating climate change. *Jawaharlal Nehru Krishi Vishwa Vidyalaya Research Journal, 49*(3), 263–271.

Chang, R., Zhou, W., Fang, Y., Bing, H., Sun, X., & Wang, G. (2019). Anthropogenic nitrogen deposition increases soil carbon by enhancing new carbon of the soil aggregate formation. *Journal of Geophysical Research: Biogeosciences, 124*(3), 572–584.

Chataut, G., Bhatta, B., Joshi, D., Subedi, K., & Kafle, K. (2023). Greenhouse gases emission from agricultural soil: A review. *Journal of Agriculture and Food Research, 11*, Article 100533. https://doi.org/10.1016/j.jafr.2023.100533

Chen, H., Li, D., Zhao, J., Zhang, W., Xiao, K., & Wang, K. (2018a). Nitrogen addition aggravates microbial carbon limitation: Evidence from ecoenzymatic stoichiometry. *Geoderma, 329*, 61–64. https://doi.org/10.1016/j.geoderma.2018.05.019

Chen, J., Luo, Y., Van Groenigen, K. J., Hungate, B. A., Cao, J., Zhou, X., & Wang, R. W. (2018). A keystone microbial enzyme for nitrogen control of soil carbon storage. *Science Advances, 4*(8). https://doi.org/10.1126/sciadv.aaq1689

Dang, Z., Guo, N., Li, S., Degen, A. A., Cao, J., Deng, B., Wang, A., Peng, Z., Ding, L., Long, R., & Shang, Z. (2023). Effect of grazing exclusion on emission of greenhouse gases and soil organic carbon turnover in alpine shrub meadow. *Science of the Total Environment, 858*, 159758. https://doi.org/10.1016/j.scitotenv.2022.159758

Del Vecchio, A., Caillé, F., Chevalier, A., Loreau, O., Horkka, K., Halldin, C., Schou, M., Camus, N., Kessler, P., Kuhnast, B. and Taran, F., Audisio, D. (2018). Late-stage isotopic carbon labeling of pharmaceutically relevant cyclic ureas directly from CO_2. *Angewandte Chemie—International Edition, 57*(31). https://doi.org/10.1002/anie.201804838

Deng, C., Zhang, Z., Zhang, C., Xu, R., Nie, C., Meng, Q., Zou, T., Li, H., Wu, J., & Yang, Q. (2024). A multi-spatial scale analysis of anthropogenic nitrogen and phosphorus inputs in a large river basin: Environmental effect and policy impact. *Regional Environmental Change, 24*(3), 123. https://doi.org/10.1007/s10113-024-02279-4

Deng, F., & Liang, C. (2022). Revisiting the quantitative contribution of microbial necromass to soil carbon pool: Stoichiometric control by microbes and soil. *Soil Biology and Biochemistry, 165*, Article 108486. https://doi.org/10.1016/j.soilbio.2021.108486

Deng, Y., Kuang, M., Hei, Z., Zhong, J., Abdo, A. I. E., Wei, H., Zhang, J., & Xiang, H. (2023). Simulated nitrogen deposition decreases the ratios of Soil C to P and N to P, changes soil enzyme activity, and reduces soil microbial biomass in paddy soil in Southern China. *Agronomy, 13*(9), 2249.

Domeignoz-Horta, L. A., Pold, G., Liu, X.-J.A., Frey, S. D., Melillo, J. M., & DeAngelis, K. M. (2020). Microbial diversity drives carbon use efficiency in a model soil. *Nature Communications, 11*(1), 3684.

Don, A., Seidel, F., Leifeld, J., Kätterer, T., Martin, M., Pellerin, S., Emde, D., Seitz, D., & Chenu, C. (2024). Carbon sequestration in soils and climate change mitigation—Definitions and pitfalls. *Global Change Biology, 30*(1), Article e16983.

Elbasiouny, H., El-Ramady, H., Elbehiry, F., Rajput, V. D., Minkina, T., & Mandzhieva, S. (2022). Plant Nutrition under Climate Change and Soil Carbon Sequestration. *Sustainability (Switzerland), 14.*

Elrys, A. S., Uwiragiye, Y., Zhang, Y., Abdel-Fattah, M. K., Chen, Z.-X., Zhang, H.-M., Meng, L., Wang, J., Zhu, T. B., Zhang, J. B., & Cheng, Y. (2023). Expanding agroforestry can increase nitrate retention and mitigate the global impact of a leaky nitrogen cycle in croplands. *Nature Food, 4*(1), 109–121.

Feng, J., Tang, M., & Zhu, B. (2021). Soil priming effect and its responses to nutrient addition along a tropical forest elevation gradient. *Global Change Biology, 27*(12). https://doi.org/10.1111/gcb.15587

Ge, T., Liu, C., Yuan, H., Zhao, Z., Wu, X., Zhu, Z., Brookes, P., & Wu, J. (2015). Tracking the photosynthesized carbon input into soil organic carbon pools in a rice soil fertilized with nitrogen. *Plant and Soil, 392*(1), 17–25. https://doi.org/10.1007/s11104-014-2265-8

Geisseler, D., & Scow, K. M. (2014). Long-term effects of mineral fertilizers on soil microorganisms—A review. *Soil Biology and Biochemistry, 75*, 54–63. https://doi.org/10.1016/j.soilbio.2014.03.023

Gong, P. (2011). China needs no foreign help to feed itself. *Nature, 474*(7349), 7–7. https://doi.org/10.1038/474007a

Gul, S., & Whalen, J. K. (2022). Perspectives and strategies to increase the microbial-derived soil organic matter that persists in agroecosystems. In *Advances in agronomy* (Vol. 175). https://doi.org/10.1016/bs.agron.2022.04.004

Guo, C., Liu, X., & He, X. (2022). A global meta-analysis of crop yield and agricultural greenhouse gas emissions under nitrogen fertilizer application. *Science of the Total Environment, 831*, Article 154982.

Gurmesa, G. A., Wang, A., Li, S., Peng, S., de Vries, W., Gundersen, P., Ciais, P., Phillips, O. L., Hobbie, E. A., Zhu, W., Nadelhoffer, K., & Fang, Y. (2022). Retention of deposited ammonium and nitrate and its impact on the global forest carbon sink. *Nature Communications, 13*(1). https://doi.org/10.1038/s41467-022-28345-1

Herrero, M., Henderson, B., Havlík, P., Thornton, P. K., Conant, R. T., Smith, P., Wirsenius, S., Hristov, A. N., Gerber, P., & Gill, M. (2016). Greenhouse gas mitigation potentials in the livestock sector. *Nature Climate Change, 6*(5), 452–461.

Hu, X., Yang, L., Lai, X., Yao, Q., & Chen, K. (2019). Influence of Al (III) on biofilm and its extracellular polymeric substances in sequencing batch biofilm reactors. *Environmental Technology, 40*(1), 53–59.

Hu, Y., Su, M., & Jiao, L. (2023). Peak and fall of China's agricultural GHG emissions. *Journal of Cleaner Production, 389*, Article 136035. https://doi.org/10.1016/j.jclepro.2023.136035

Huang, Q., Zhang, G., Ma, J., Song, K., Zhu, X., Shen, W., & Xu, H. (2022). Dynamic interactions of nitrogen fertilizer and straw application on greenhouse gas emissions and sequestration of soil carbon and nitrogen: A 13-year field study. *Agriculture, Ecosystems & Environment, 325*, 107753–107753.

Huang, X., Terrer, C., Dijkstra, F. A., Hungate, B. A., Zhang, W., & van Groenigen, K. J. (2020). New soil carbon sequestration with nitrogen enrichment: a meta-analysis. *Plant and Soil, 454*(1–2). https://doi.org/10.1007/s11104-020-04617-x

Jickells, T., Baker, A. R., Cape, J. N., Cornell, S. E., & Nemitz, E. (2013). The cycling of organic nitrogen through the atmosphere. *Philosophical Transactions of the Royal Society B: Biological Sciences, 368*(1621). https://doi.org/10.1098/rstb.2013.0115

Junaid Nazir, M., Zhang, X., Du, D., & Yu, F. (2022). How Physical disturbance and nitrogen addition affect the soil carbon decomposition? *Phyton, 91*(9), 2087–2097. https://doi.org/10.32604/phyton.2022.021412

Kang, E., Li, Y., Zhang, X., Yan, Z., Wu, H., Li, M., Yan, L., Zhang, K., Wang, J., & Kang, X. (2021). Soil pH and nutrients shape the vertical distribution of microbial communities in an alpine wetland. *Science of the Total Environment, 774*, Article 145780. https://doi.org/10.1016/j.scitotenv.2021.145780

Kim, K., Daly, E. J., Gorzelak, M., & Hernandez-Ramirez, G. (2022). Soil organic matter pools response to perennial grain cropping and nitrogen fertilizer. *Soil and Tillage Research, 220*, Article 105376. https://doi.org/10.1016/j.still.2022.105376

Kirkby, C. A., Richardson, A. E., Wade, L. J., Passioura, J. B., Batten, G. D., Blanchard, C., & Kirkegaard, J. A. (2014). Nutrient availability limits carbon sequestration in arable soils. *Soil Biology and Biochemistry, 68*, 402–409.

Ladha, J. K., Reddy, C. K., Padre, A. T., & van Kessel, C. (2011). Role of nitrogen fertilization in sustaining organic matter in cultivated soils. *Journal of Environmental Quality, 40*(6), 1756–1766.

Lal, R. (2010). Managing soils and ecosystems for mitigating anthropogenic carbon emissions and advancing global food security. *BioScience, 60*(9). https://doi.org/10.1525/bio.2010.60.9.8

Li, J., Li, Y. e., Wan, Y., Wang, B., Cai, W., Qin, X., . . . Deng, Q. (2024). Annual greenhouse gas emissions from a rice paddy with different water-nitrogen management strategies in Central China. *Soil and Tillage Research, 235*, 105906. https://doi.org/10.1016/j.still.2023.105906

Li, J., Wen, Y., Li, X., Li, Y., Yang, X., Lin, Z., & Zhao, B. (2018). Xxx Soil labile organic carbon fractions and soil organic carbon stocks as affected by long-term organic and mineral fertilization regimes in the North China Plain. *Soil and Tillage Research, 175*, 281–290. https://doi.org/10.1016/j.still.2017.08.008

Li, T., Wang, R., Cai, J., Meng, Y., Wang, Z., Feng, X., Liu, H., Turco, R. F., & Jiang, Y. (2021). Enhanced carbon acquisition and use efficiency alleviate microbial carbon relative to nitrogen limitation under soil acidification. *Ecological Processes, 10*(1), 32.

Li, Y., Bezemer, T. M., Yang, J., Lü, X., Li, X., Liang, W., Han, X., & Li, Q. (2019). Changes in litter quality induced by N deposition alter soil microbial communities. *Soil Biology and Biochemistry, 130*, 33–42. https://doi.org/10.1016/j.soilbio.2018.11.025

Liang, C., Amelung, W., Lehmann, J., & Kästner, M. (2019). Quantitative assessment of microbial necromass contribution to soil organic matter. *Global Change Biology, 25*(11), 3578–3590. https://doi.org/10.1111/gcb.14781

Liu, G., Feng, Y., Yin, Z., Yan, G., Wang, Q., & Xing, Y. (2023). Exogenous organic C inputs profit soil C sequestration under different long-term N addition levels in a boreal forest. *Journal of Soil Science and Plant Nutrition*. https://doi.org/10.1007/s42729-023-01135-4

Lu, X., Hou, E., Guo, J., Gilliam, F. S., Li, J., Tang, S., & Kuang, Y. (2021). Nitrogen addition stimulates soil aggregation and enhances carbon storage in terrestrial ecosystems of China: A meta-analysis. *Global Change Biology, 27*(12), 2780–2792.

Ludwig, B., Geisseler, D., Michel, K., Joergensen, R. G., Schulz, E., Merbach, I., Raupp, J., Rauber, R., Hu, K., Niu, L., & Liu, X. (2011). Effects of fertilization and soil management on crop yields and carbon stabilization in soils. A review. In *Agronomy for sustainable development* (Vol. 31).

Luo, R., Fan, J., Wang, W., Luo, J., Kuzyakov, Y., He, J.-S., Chu, H., & Ding, W. (2019). Nitrogen and phosphorus enrichment accelerates soil organic carbon loss in alpine grassland on the Qinghai-Tibetan Plateau. *Science of the Total Environment, 650*, 303–312.

Luo, R., Kuzyakov, Y., Liu, D., Fan, J., Luo, J., Lindsey, S., He, J. S., & Ding, W. (2020). Nutrient addition reduces carbon sequestration in a Tibetan grassland soil: Disentangling microbial and physical controls. *Soil Biology and Biochemistry, 144*, Article 107764. https://doi.org/10.1016/j.soilbio.2020.107764

Malik, A. A., Puissant, J., Buckeridge, K. M., Goodall, T., Jehmlich, N., Chowdhury, S., Gweon, H. S., Peyton, J. M., Mason, K. E., van Agtmaal, M., Blaud, A., & Griffiths, R. I. (2018). Land use driven change in soil pH affects microbial carbon cycling processes. *Nature Communications, 9*(1), 3591. https://doi.org/10.1038/s41467-018-05980-1

Malik, A. A., Thomson, B. C., Whiteley, A. S., Bailey, M., & Griffiths, R. I. (2017). Bacterial physiological adaptations to contrasting edaphic conditions identified using landscape scale metagenomics. *mBio, 8*(4), https://doi.org/10.1128/mbio.00799-00717. https://doi.org/10.1128/mbio.00799-17

Menegat, S., Ledo, A., & Tirado, R. (2022). Greenhouse gas emissions from global production and use of nitrogen synthetic fertilisers in agriculture. *Scientific Reports, 12*(1), 1–13.

Mosier, S., Córdova, S. C., & Robertson, G. P. (2021). Restoring soil fertility on degraded lands to meet food, fuel, and climate security needs via perennialization. In *Frontiers in sustainable food systems* (Vol. 5).

Nazir, M. J., Li, G., Nazir, M. M., Zulfiqar, F., Siddique, K. H., Iqbal, B., & Du, D. (2024a). Harnessing soil carbon sequestration to address climate change challenges in agriculture. *Soil and Tillage Research, 237*, Article 105959. https://doi.org/10.1016/j.still.2023.105959

Nazir, M. J., Hussain, M. M., Albasher, G., Iqbal, B., Khan, K. A., Rahim, R., Li, G., & Du, D. (2024b). Glucose input profit soil organic carbon mineralization and nitrogen dynamics in relation to nitrogen amended soils. *Journal of Environmental Management, 351*, Article 119715. https://doi.org/10.1016/j.jenvman.2023.119715

Ogdahl, M. L., Adair, E. C., Weisenhorn, P., Hobbie, S. E., Eddy, W. C., & Buyarski, C. R. (2012). Response of decomposing litter and its microbial community to multiple forms of nitrogen enrichment. *Ecological Monographs, 82*(3).

Qiu, S., Gao, H., Zhu, P., Hou, Y., Zhao, S., Rong, X., Zhang, Y., He, P., Christie, P., & Zhou, W. (2016). Changes in soil carbon and nitrogen pools in a Mollisol after long-term fallow or application of chemical fertilizers, straw or manures. *Soil and Tillage Research, 163*, 255–265. https://doi.org/10.1016/j.still.2016.07.002

Rao, S. Ch., Venkateswarlu, B., Srinivas, K., Kundu, S., Singh, A. K. (2011). In *Soil Carbon Sequestration for Climate Change Mitigation and Food Security, 24th November-3rd December*.

Raza, S., Zamanian, K., Ullah, S., Kuzyakov, Y., Virto, I., & Zhou, J. (2021). Inorganic carbon losses by soil acidification jeopardize global efforts on carbon sequestration and climate change mitigation. *Journal of Cleaner Production, 315*, Article 128036. https://doi.org/10.1016/j.jclepro.2021.128036

Rogger, J., Hörtnagl, L., Buchmann, N., & Eugster, W. (2022). Carbon dioxide fluxes of a mountain grassland: Drivers, anomalies and annual budgets. *Agricultural and Forest Meteorology, 314*, 108801. https://doi.org/10.1016/j.agrformet.2021.108801

Rousk, J., Bååth, E., Brookes, P. C., Lauber, C. L., Lozupone, C., Caporaso, J. G., Knight, R., & Fierer, N. (2010). Soil bacterial and fungal communities across a pH gradient in an arable soil. *The ISME Journal, 4*(10), 1340–1351. https://doi.org/10.1038/ismej.2010.58

Scavo, A., Fontanazza, S., Restuccia, A., Pesce, G. R., Abbate, C., & Mauromicale, G. (2022). The role of cover crops in improving soil fertility and plant nutritional status in temperate climates. A Review. *Agronomy for Sustainable Development, 42*(5), 93.

Shaaban, M., Khalid, M. S., Hu, R., & Zhou, M. (2022). Effects of water regimes on soil N_2O, CH_4 and CO_2 emissions following addition of dicyandiamide and N fertilizer. *Environmental Research, 212*, Article 113544. https://doi.org/10.1016/j.envres.2022.113544

Shaaban, M. (2024). Microbial pathways of nitrous oxide emissions and mitigation approaches in drylands. *Journal of Environmental Management, 354*, Article 120393.

Shaaban, M., & Nunez-Delgado, A. (2024). Soil adsorption potential: Harnessing Earth's living skin for mitigating climate change and greenhouse gas dynamics. *Environmental Research, 251*, Article 118738.

Shaaban, M., Peng, Q., Hu, R., Lin, S., & Zhao, J. (2016). Soil nitrous oxide and carbon dioxide emissions following incorporation of above-and below-ground biomass of green bean. *International Journal of Environmental Science and Technology, 13*, 179–186.

Snyder, C. S., Bruulsema, T. W., Jensen, T. L., & Fixen, P. E. (2009). Review of greenhouse gas emissions from crop production systems and fertilizer management effects. *Agriculture, Ecosystems & Environment, 133*(3), 247–266. https://doi.org/10.1016/j.agee.2009.04.021

Sun, W., Huang, Y., Zhang, W., & Yu, Y. (2010). Carbon sequestration and its potential in agricultural soils of China. *Global Biogeochemical Cycles, 24*(3). https://doi.org/10.1029/2009gb003484

Sun, J., Rutherford, S., Saif Ullah, M., Ullah, I., Javed, Q., Rasool, G., Ajmal, M., Azeem, A., Nazir, M. J., & Du, D. (2022). Plant–soil feedback during biological invasions: effect of litter decomposition from an invasive plant (*Sphagneticola trilobata*) on its native congener (*S. calendulacea*). *Journal of Plant Ecology, 15*(3), 610–624.

Thotakuri, G., Angidi, S., & Athelly, A. (2024). Soil carbon pool as influenced by soil microbial activity—An overview. *American Journal of Climate Change, 13*(02), 175–193.

Wang, F., Chen, S., Qin, S., Sun, R., Zhang, Y., Wang, S., Hu, C., Hu, H., & Liu, B. (2021a). Long-term nitrogen fertilization alters microbial community structure and denitrifier abundance in the deep vadose zone. *Journal of Soils and Sediments, 21*(6), 2394–2403. https://doi.org/10.1007/s11368-021-02931-0

Wang, J., Shi, X., Zheng, C., Suter, H., & Huang, Z. (2021b). Different responses of soil bacterial and fungal communities to nitrogen deposition in a subtropical forest. *Science of the Total Environment, 755*, Article 142449. https://doi.org/10.1016/j.scitotenv.2020.142449

Wang, M., Hossain, M. R., Si Mohammed, K., Cifuentes-Faura, J., & Cai, X. (2023). Heterogenous effects of circular economy, green energy and globalization on CO_2 emissions: policy based analysis for sustainable development. *Renewable Energy, 211*, 789–801. https://doi.org/10.1016/j.renene.2023.05.033

Wang, Y., Wu, P., Mei, F., Ling, Y., Qiao, Y., Liu, C., Leghari, S. J., Guan, X., & Wang, T. (2021c). Does continuous straw returning keep China farmland soil organic carbon continued increase? A meta-analysis. *Journal of Environmental Management, 288*, Article 112391. https://doi.org/10.1016/j.jenvman.2021.112391

Wei, X., Zhu, Z., Liu, Y., Luo, Y., Deng, Y., Xu, X., Liu, S., Richter, A., Shibistova, O., Guggenberger, G., Wu, J., & Ge, T. (2020). C:N:P stoichiometry regulates soil organic carbon mineralization and concomitant shifts in microbial community composition in paddy soil. *Biology and Fertility of Soils, 56*(8), 1093–1107. https://doi.org/10.1007/s00374-020-01468-7

Witzgall, K., Vidal, A., Schubert, D. I., Höschen, C., Schweizer, S. A., Buegger, F., Pouteau, V., Chenu, C., & Mueller, C. W. (2021). Particulate organic matter as a functional soil component for persistent soil organic carbon. *Nature Communications, 12*(1), 1–10.

Wright, S. J. (2019). Plant responses to nutrient addition experiments conducted in tropical forests. In *Ecological monographs* (Vol. 89).

Xu, P., Jiang, M., Jiang, Y., Khan, I., Zhou, W., Wu, H., Wu, X., Shaaban, M., Lu, J., & Hu, R. (2022). Prior nitrogen fertilization stimulated N_2O emission from rice cultivation season under a rapeseed-rice production system. *Plant and Soil*, 1–12.

Xu, P., Jiang, M., Khan, I., Shaaban, M., Wu, H., Harerimana, B., & Hu, R. (2024). Regulatory potential of soil available carbon, nitrogen, and functional genes on N_2O emissions in two upland plantation systems. *Journal of Integrative Agriculture, 23*(8), 2792–2806.

Yang, Y., Wang, P., & Zeng, Z. (2019). Dynamics of bacterial communities in a 30-year fertilized paddy field under different organic–inorganic fertilization strategies. *Agronomy, 9*(1), 14.

Yao, Z., Yan, G., Ma, L., Wang, Y., Zhang, H., Zheng, X., Wang, R., Liu, C., Wang, Y., Zhu, B., Zhou, M., & Butterbach-Bahl, K. (2022). Xxx Soil C/N ratio is the dominant control of annual N_2O fluxes from organic soils of natural and semi-natural ecosystems. *Agricultural and Forest Meteorology, 327*, 109198. https://doi.org/10.1016/j.agrformet.2022.109198

Ye, C., Chen, D., Hall, S. J., Pan, S., Yan, X., Bai, T., Guo, H., Zhang, Y., Bai, Y., & Hu, S. (2018). Reconciling multiple impacts of nitrogen enrichment on soil carbon: Plant, microbial and geochemical controls. *Ecology Letters, 21*(8), 1162–1173.

Yue, K., Fornara, D. A., Li, W., Ni, X., Peng, Y., Liao, S., Tan, S., Wang, D., Wu, F., & Yang, Y. (2020). Nitrogen addition affects plant biomass allocation but not allometric relationships among different organs across the globe. *Journal of Plant Ecology, 14*(3), 361–371. https://doi.org/10.1093/jpe/rtaa100

Zang, D., Hu, Z., Yang, Y., & He, S. (2022). Research on the relationship between agricultural carbon emission intensity, agricultural economic development and agricultural trade in China. *Sustainability, 14*(18), 11694.

Zhang, F., Wang, X., Guo, T., Zhang, P., & Wang, J. (2015). Soil organic and inorganic carbon in the loess profiles of Lanzhou area: Implications of deep soils. *CATENA, 126*, 68. https://doi.org/10.1016/j.catena.2014.10.031

Zhang, H., Liang, Q., Peng, Z., Zhao, Y., Tan, Y., Zhang, X., & Bol, R. (2023). Response of greenhouse gases emissions and yields to irrigation and straw practices in wheat-maize cropping system. *Agricultural Water Management, 282*, Article 108281. https://doi.org/10.1016/j.agwat.2023.108281

Zhang, X., Xin, X., Zhu, A., Zhang, J., & Yang, W. (2017). Effects of tillage and residue managements on organic C accumulation and soil aggregation in a sandy loam soil of the North China Plain. *CATENA, 156*, 176. https://doi.org/10.1016/j.catena.2017.04.012

Zhou, J., Jia, R., Brown, R. W., Yang, Y., Zeng, Z., Jones, D. L., & Zang, H. (2023). The long-term uncertainty of biodegradable mulch film residues and associated microplastics pollution on plant-soil health. *Journal of Hazardous Materials, 442*, Article 130055.

Soil Microbes and Carbon Sequestration

Purnima Baidya, Manita Aryal, and Bhim Kumar Dahal

Abstract Carbon sequestration, the process of living organisms capturing and storing CO_2 from the atmosphere, is crucial for reducing greenhouse gas emissions and mitigating global warming. Soil microbes play a significant role in this process, including bacteria, archaea, actinomycetes, fungi, viruses, and protozoa. These microorganisms contribute to carbon sequestration through various pathways, such as the Calvin cycle, 3-hydroxypropionate bi-cycle, and reductive acetyl ~ CoA pathway. These pathways involve the conversion of CO_2 into organic compounds, which are then stored in the soil. Factors like temperature, moisture, organic matter, and pH influence this process. Climate change profoundly impacts soil microbial communities, altering their structure, composition, and functions. This highlights the importance of studying soil microbes and their role in carbon sequestration, including the impact of agricultural practices on soil microbiome diversity and metabolic processes.

Keywords Carbon sequestration · Climate change · Natural carbon sequestration · Soil microbes

1 Carbon Sequestration

Capturing and storing CO_2, especially in soil, plants, and trees (Kiss et al., 2015), as opposed to releasing it into the environment, is called carbon sequestration (Sanaullah et al., 2019). By removing CO_2 from the atmosphere, forests and agricultural land use

P. Baidya
Institute of Resource Biology and Biotechnology, Department of Biotechnology, College of Life Science and Technology, Huazhong University of Science and Technology, Wuhan, China

P. Baidya · M. Aryal
Central Department of Microbiology, Tribhuvan University, Kathmandu, Nepal

B. K. Dahal (✉)
Department of Civil Engineering, Institute of Engineering, Pulchowk Campus, Tribhuvan University, Kathmandu, Nepal
e-mail: bhimd@pcampus.edu.np

M. Shaaban (ed.), *Soils and Sustainable Agriculture*, Frontier Studies in Soil Science,
https://doi.org/10.1007/978-3-031-91114-9_8

practices that build biomass contribute to carbon sequestration (Shaaban & Nunez-Delgado, 2024). Thorough models evaluate their efficacy, establishing forests as an annual carbon sink. Limiting greenhouse gas emissions via methods like carbon capture, storage, and biological activity is known as climate change mitigation (Kaur et al., 2023; Mohd et al., 2022; Yongcun & Shui, 2022). One global strategy for tackling climate change and global warming is carbon sequestration, which involves reducing the amount of carbon in the atmosphere (Habib & Huda, 2022).

1.1 Different Methods and Their Effectiveness for Carbon Sequestration

Dissolving CO_2 into an aqueous solution with a mineral under-saturation level and reacting it with a sequestration agent are methods used for sequestering carbon. Another technique that is affordable, efficient, and has a long sequestration period is wood burial (Zeng, 2008). Organisms are used in biological carbon sequestration systems to absorb CO_2 and store it in vegetation, soils, and aquatic environments (Vetter et al., 2022). Cementitious constructions with a CO_2 uptake level of more than 6% by weight binder can also be created for CO_2 sequestration (Victor et al., 2021). The process of phytoremediation, in which plants absorb CO_2 from the atmosphere and transform it into organic molecules, is a valuable way to lower CO_2 emissions (Govindaraju et al., 2021).

2 Natural Carbon Sequestration

Natural carbon sequestration (NCS), the process of capturing and storing carbon from the atmosphere, involves organisms, such as plants and microbes, absorbing CO_2 and storing it as carbon in vegetation, soils, wood products, and aquatic environments (Helferty, 2023). This process is essential for lowering greenhouse gas emissions caused by human activity and offers potential ways to mitigate the effects of global warming (Harris et al., 2018). Carbon can be captured and stored in various natural reservoirs, including soils, forests, and oceans (Silva et al., 2022), as shown in Fig. 1. However, the precise techniques necessary to monitor these carbon reserves and the durability of carbon sequestered by natural processes are still subjects of ongoing research and discussion (Markusson, 2022).

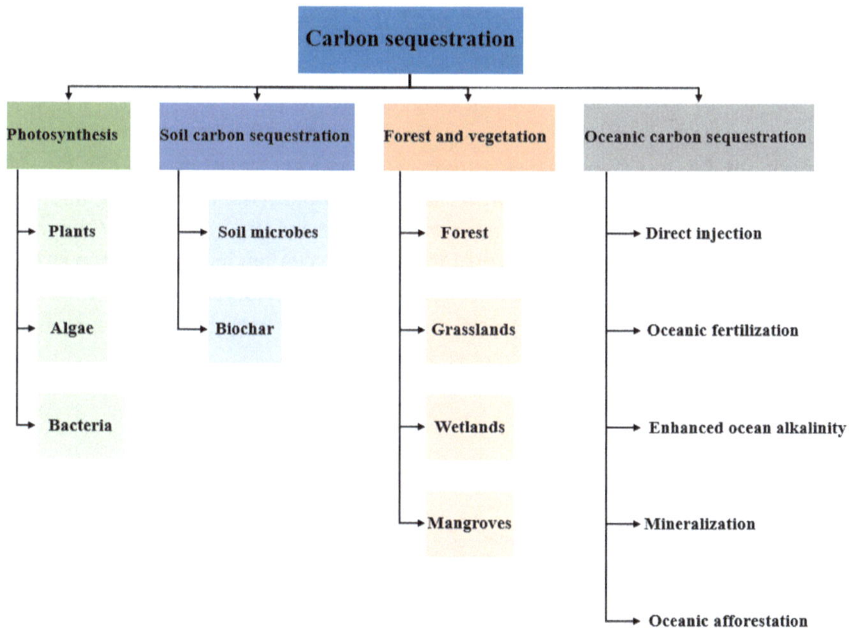

Fig. 1 Carbon capture and storage

2.1 *Contribution of Natural Carbon Sequestration to Mitigate Climate Change*

Natural carbon sequestration plays a critical role in reducing the effects of climate change by storing carbon in soils, plants, and oceans (Gonzalez, 2023). Potential sites for sequestering carbon include forests, peatlands, and underwater macroalgal forests. High-priority regions for carbon sequestration activities can be targeted via conservation, afforestation, and restoration strategies that may help address global targets (Boisvenue et al., 2022). The NCS differs from other climate change mitigation methods as it involves managing and protecting terrestrial forests and is an innovative strategy for mitigating climate change terrestrial forests (Pessarrodona et al., 2023). It consists of biomass growth and accumulates in forests, where photosynthesis absorbs and stores CO_2. This technique contributes to the overall carbon sink and climate change mitigation by improving biodiversity and ecological services (Dooley et al., 2022).

2.2 Primary Mechanisms of Natural Carbon Sequestration in Terrestrial Ecosystems

Carbon sequestration in terrestrial ecosystems is primarily facilitated by recalcitrance of carbon intake, soil mineral interactions, aggregation formation, and abiotic variables (Adamczyk et al., 2020). Carbon is stored in biomass, dead wood, and soils in forest ecosystems, acting as temporary and permanent carbon sinks (Ncipha & Sivakumar, 2019; Yitayal, 2020). Agroforestry practices can increase carbon storage in biomass and soil organic carbon (SOC) (Rattan et al., 2018; Royal et al., 2020). However, converting natural ecosystems to controlled ones leads to depletion of carbon stocks and increased greenhouse gas emissions.

3 Contribution of Soil Microbes for Natural Carbon Sequestration

Soil microbes play a crucial role in NCS by interacting with the soil environment and influencing carbon dynamics. They contribute to forming soil organic matter (OM) through various mechanisms. Microbes like bacteria and fungi can enhance pH levels and form carbonate precipitates, aiding soil inorganic carbon sequestration (Shi et al., 2023). Additionally, the microbial community interacts with organic material amendments, such as straw and manure, affecting carbon mineralization and sequestration (Liu et al., 2023). Long-term application of organic manure stimulates microbial growth, increasing carbon and nitrogen accumulation in soils, highlighting the importance of microbial residues in soil carbon pools (Chen et al., 2022). Furthermore, the microbial carbon pump concept emphasizes how microbes produce organic compounds that stabilize carbon in the soil, influencing soil carbon cycling and climate change dynamics (Liang, 2020). Overall, soil microbes are essential for managing and enhancing NCS processes, as shown in Fig. 2.

3.1 Importance of Soil Microbes in Carbon Sequestration

Soil microbes are vital for carbon sequestration, influencing ecosystem health through nutrient cycling, plant health, and OM decomposition (Dathe et al., 2023). Microbial carbon use efficiency (CUE) is crucial in SOC storage (Tao et al., 2023). Interactions between soil microbes and biochar can enhance carbon sequestration capacity, mitigating greenhouse gas emissions and improving soil fertility. Understanding these interactions is essential for sustainable soil management and long-term carbon neutrality goals (Jia et al., 2023). Soil microbes involved in the carbon sequestration are briefly given in Table 1.

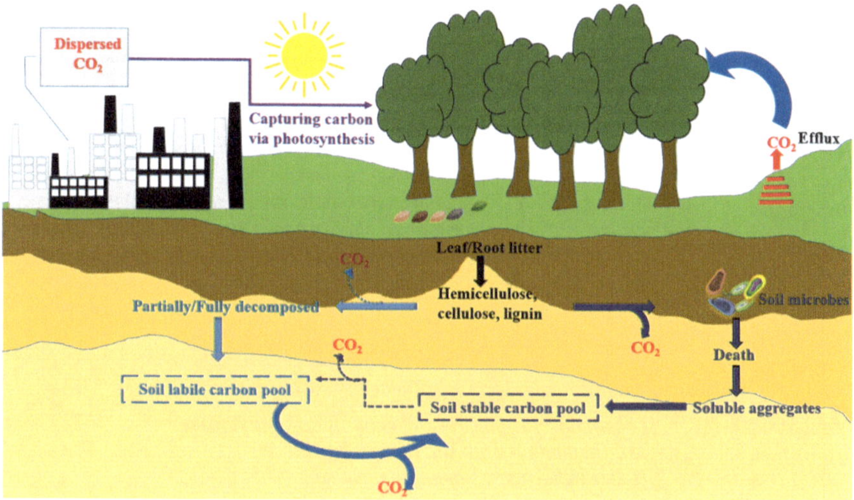

Fig. 2 Systematic diagram of natural carbon sequestration via soil microbes

4 Major Pathways of Carbon Release by Soil Microbes

Soil microbes release carbon through various pathways. One significant pathway is the direct root exudation by plants and associated mycorrhizal fungi, facilitating nutrient availability. Other main pathways of carbon release by soil microbes include soil CO_2, CH_4, and N_2O emissions (Shaaban, 2024; Shaaban et al., 2016, 2022). Factors like temperature, moisture, OM, and pH influence these greenhouse gas emissions (Wang et al., 2022). Additionally, microbial decomposition of amino sugars and amino acids from bacterial and fungal cell walls contributes to carbon release. The main pathways of carbon release involve direct root exudation and mycorrhizal fungi, with mycorrhizal pathways showing significant enrichment of recent plant photosynthesis and acting as a rapid hub for translocating carbon to soil microbes (Kaiser et al., 2015). Soil microbes contribute to carbon sequestration through various factors, including biotic interactions (Feng & Wang, 2023), microbial community composition, CO_2 fixation pathways, and environmental conditions (Iven et al., 2022). Biotic interactions, such as interspecific competition and facilitation, affect microbial CUE. Autotrophic organisms like bacteria and protists are crucial in soil carbon fixation.

The main pathways of carbon release by soil microbes include CO_2 fixation pathways, such as the Calvin cycle, 3-hydroxypropionate bi-cycle, 3-hydroxypropionate/4-hydroxybutyrate, dicarboxylate-hydroxybutyrate cycle, and reductive acetyl ~ CoA pathway (Wood-Ljungdahl pathway) which are influenced by environmental factors like precipitation levels (Huang et al., 2022), and metabolic pathways, such as glycolysis, pentose phosphate, reductive citrate cycle, and Entner-Doudoroff pathways. Different bacteria in soil communities exhibit diverse metabolisms and CUE (Wu et al., 2022). The main pathways of carbon release by soil microbes include

Table 1 Different types of soil microbes helping in carbon sequestration

Class	Examples	Roles
Bacteria (Gram positive and Gram negative)	*Bacillus* sp., *Agromyces* sp., *Micromonospora* sp., *Pseudonocardia* sp., *Acremonium* sp., *Lysobacter* sp., *Mesorhizobium* sp., *Microvirga* sp., *Bradyrhizobium* sp., *Chaetomium* sp., *Proteobacteria* sp., *Actinobacteria* sp., *Acidobacteria* sp., *Chloroflexi* sp., *Firmicutes* sp., *Chlamydiae* sp., *Spirochaetes* sp., *Aeromicrobium* sp., *Blastococcus* sp., *Gemmatimonas* sp., *Nocardioides* sp., *Solirubrobacter* sp., *Ralstonia* sp., *Aciditerrimonas* sp., *Actinospica* sp., *Byssovorax* sp., *Catenulispora* sp., *Conexibacter* sp., *Dongia* sp., *Modestobacter* sp., *Mycobacterium* sp., *Pseudolabrys* sp., *Rhizomicrobium* sp., *Rudaea* sp., *Clostridium* sp.	Play a crucial role in decomposing dead animals, plants, and fungi, exhibiting varying carbon utilization preferences
Archaea	*Lokiarchaeota* sp., *Crenarchaeota* sp., *Korarchaeota* sp., *Thaumarchaeota* sp., *Euryarchaeota* sp., *Hadesarchaea* sp., *Nanoaechaeota* sp., *Gammaproteobacteria* sp., *Betaproteobacteria* sp.	
Actinomycetes	*Nonomuraea* sp., *Amycolatopsis* sp., *Streptosporangium* sp., *Saccharothrix* sp., *Actinomadura* sp., *Streptomyces* sp., *Micromonospora* sp., *Spirillospora* sp., *Nocardia* sp., *Pseudonocardia* sp., *Nocardiopsis* sp., *Streptomonospora* sp., *Actinopolyspora* sp., *Saccharopolyspora* sp., *Kocuria* sp., *Actinomyces* sp., *Actinomycetales* sp., *Glutamicibacter* sp., *Thermoactinomyces* sp.	Enables adaptation to anoxic and organic matter-rich environments, where methanogenic archaea play crucial roles in carbon cycle and climate change
Fungi	*Ascomycota* sp., *Basidiomycota* sp., *Acremonium* sp., *Chaetomium* sp., *Sodiomyces* sp., *Halosphaeria* sp., *Lophotrichus* sp., *Mrakia* sp., *Phellinus* sp., *Basipetospora* sp., *Trichosporon* sp., *Paraglomus* sp., *Entophlyctis* sp., *Trichocladium* sp.	Contribute to global soil carbon stocks by metabolizing complex carbon-rich substrates
Viruses	*Anabaenaphage* sp., *Cyanophage* sp., *Microcystisphage* sp., *Microcystisvirus* sp., *Phormidiumvirus* sp., *Prochlorococcus* sp., *Synechococcus* sp., *Plectonema* sp.	Impact microbial communities by enhancing soil carbon cycle and carbon sequestration

(continued)

Table 1 (continued)

Class	Examples	Roles
Protozoa	*Acanthamoeba* sp., *Cercomonas* sp., *Colpoda* sp., *Oxytricha* sp., *Stylonychia* sp., *Bodo* sp., *Tachysoma* sp., *Pauroleptus* sp., *Spathidium* sp., *Vorticella* sp., *Dileptus* sp., *Strongylidium* sp., *Litonotus* sp., *Zoothamnium* sp., *Birojimia* sp., *Astasia* sp.	Crucial in soil food web, influencing structure and carbon cycle of soil through selective feeding
Nematodes	*Rhabditis* sp., *Protorhabditis* sp., *Parasphaerolaimus* sp., *Sphaerolaimus* sp., *Terschellingia* sp., *Sabatieria* sp., *Dorylaimopsis* sp., *Viscosia* sp., *Halalaimus* sp., *Neochromadora* sp., *Thoracostoma* sp.	

significant decomposition of peptidoglycan and chitin, with glucosamine being a major contributor to the soil organic nitrogen pool (Hu et al., 2018). Overall, biotic interactions, microbial community composition, CO_2 fixation pathways, and environmental conditions collectively determine soil microbes' impact on CO_2 sequestration.

4.1 Mechanism Contributing Soil Microbes to Soil Carbon Storage

CO_2-fixing microbes assimilate atmospheric CO_2 and increase SOC content, annually fixing up to 4.9 Pg of carbon in terrestrial ecosystems, and play a crucial role in areas with limited vegetation growth. Soil microbial CO_2 fixation contributes 18.2% to gross primary productivity in arid grasslands, but its potential in semiarid grasslands remains unexplored (Huang et al., 2022). Soil microbes contribute to soil carbon storage through various mechanisms. Firstly, non-mycorrhizal fungi can improve soil carbon stability by increasing the stable pools of soil carbon, thus enhancing soil carbon storage potential. Additionally, microbial CUE plays a crucial role in determining SOC storage. CUE is identified as a significant determinant of global SOC storage, highlighting the significance of microbial processes in carbon sequestration (Mason-Jones et al., 2023). Furthermore, microbial interactions, such as competition, facilitation, and predation, impact microbial CUE, influencing the balance between microbial growth and respiration and ultimately affecting soil carbon storage (Iven et al., 2022). The involvement of soil microbiota in carbon metabolic gene horizontal gene transfer networks further highlights their role in carbon metabolism and adaptation to changing environments (Qu et al., 2023). Additionally, soil microbial activity is influenced by temperature, moisture, microbial biomass, and enzyme activity, impacting soil respiration rates and carbon loss (Tao et al., 2023). These findings underscore how soil microbes influence soil carbon storage through metabolic activities, interactions, and carbon allocation strategies (Li et al., 2024).

4.1.1 Contribution of Soil Bacteria in Carbon Storage

Heterotrophic saprophytic bacteria found in the soil, especially bacteria, are also significant agents responsible for the breakdown of dead animals, plants, and, most importantly, fungal mycelia. These bacteria have differing objectives, and while Gram-negative bacteria are primarily interested in plant-derived carbon, Gram-positive bacteria are more interested in aged SOM-derived carbon (Wu et al., 2024). Bacteria use various mechanisms to fix CO_2 from the air, including the Calvin-Benson Bassham cycle, reductive tricarboxylic Acid cycle, reductive acetyl ~ CoA pathway, 3-hydroxypropionate 4-hydroxybutyrate cycle, dicarboxylate 4-hydroxybutyrate cycle, and 3-hydroxypropionate bi-cycle (Gayathri et al., 2021). These pathways enable bacteria to sequester atmospheric CO_2 into biomass and energy, offering high production rates, bio-fixation efficiency, potent bioremediation capabilities, and the potential to generate diverse additives (Hayat et al., 2010; Wang et al., 2017). Acidobacteria, on the other hand, is a collection of slow-growing oligotrophs capable of breaking down carbon easily. Additionally, these bacteria are critical in shaping the structure of the food web and enhancing carbon and nutrient flows (Llado et al., 2017).

4.1.2 Contribution of Soil Archaea in Carbon Storage

Although archaea and bacteria are common in deep soil communities, archaeal communities have distinct ecological niches (Shaaban et al., 2024). They are abundant in purely anoxic and organic-carbon-rich environments, leading to carbon release (Wu et al., 2024). Therefore, methanogenic and methanotrophic archaea are essential in the carbon cycle, supporting crucial aspects such as the anaerobic decomposition of organic materials (Naitam & Kaushik, 2021).

4.1.3 Contribution of Soil Actinomycetes in Carbon Storage

Actinomycetes belong to Actinomycetales; the phylum is Actinobacteria, one of the largest taxonomic units within the bacteria domain, mostly linked with the biomineralization ability (Farda et al., 2022). Actinomycetes can produce isozyme of carbonic anhydrase (CA). These CA can fix prominent qualities of CO_2 into calcium carbonate. CA produced by actinomycetes is relatively more active, stable, and specialized for carbon sequestration (Dudhagara, 2015).

4.1.4 Contribution of Soil Fungi in Carbon Storage

Fungi, the most substantial part of the global soil community, have a global biomass of 12 Pg of carbon, almost 2 times the quantity in the soil bacteria (Bar-On et al., 2018). Fungi are highly adaptable and able to decompose various types of organic material or soil components, thus regulating the carbon and nutrient balance. Fungal products are more resistant to chemical decay than bacterial products and determine the proportion of sequestrated carbon (Žifčáková et al., 2016). Mycorrhizal fungi, which are significant for the functioning of terrestrial ecosystems, affect the soil structure and function and regulate the nutrient and carbon cycles by enhancing CO_2 uptake and contributing to carbon sequestration (Treseder & Holden, 2013). These organisms form symbiotic relationships with plants, enhancing CO_2 uptake through their extensive network of hyphae (Gayathri et al., 2021). When mycorrhizal fungi die, their carbon can decompose rapidly, releasing CO_2 back into the atmosphere and persist in soil for years (Treseder & Holden, 2013). Ectomycorrhizal fungi produce enzymes that oxidize SOM, and arbuscular fungi depend on saprotrophic fungi for nutrient cycling (Sokol et al., 2022).

4.1.5 Contribution of Soil Virus in Carbon Storage

Viruses are crucial in soil carbon cycling, driving terrestrial viral shunting, altering host cell metabolism, and shaping OM cycling. Virus-encoded auxiliary metabolic genes can enhance host metabolic pathways and influence biogeochemistry. Viral infections promote nutrient cycling and carbon metabolism in soil. Further research is needed to understand their functions and incorporate virus-mediated processes into soil biogeochemical models (Wu et al., 2024). Viruses significantly influence ecosystems, particularly cyanobacteria, which generate ATP (ATP synthase) and NADPH (ferredoxin NADPH reductase) during photosynthesis; respective enzymes are crucial for carbon fixation in the Calvin cycle (Zhao et al., 2022). Cyanophages release dissolved carbon and OM into the environment, a vital part of the recycling system (Braga et al., 2022). In contrast, another study explains that cyanophages infecting cyanobacteria can hinder carbon fixation, affecting the balance of the ecosystem and the carbon recycling process (Puxty et al., 2016).

4.1.6 Contribution of Soil Protozoa in Carbon Storage

Protozoan communities, including amoebas, flagellates, and ciliates, significantly influence bacterial communities and vegetation growth (Bonkowski, 2004). Protozoa further promote the efficiency of plant functions, accelerate nutrient cycles, and inhibit the progress of bacterial species. These organisms also influence plants and sequester carbon, forming the central hub of soil microbial carbon networks (Wu et al., 2024). When protozoa are present, the amount of 14C-CO_2 released from soils increases by 36–53%. This suggests that protozoa enhance the mineralization of nitrogen from soil OM by promoting the turnover of bacterial biomass. Additionally, factors such as water content, pH level, and CO_2 increase significantly affect the dynamics of food-chain reactions in soil, particularly concerning carbon and nitrogen mineralization (Bonkowski, 2004).

4.1.7 Contribution of Soil Nematodes in Carbon Storage

As abundant soil invertebrates, Nematodes play a vital role in carbon decomposition and cycling by interacting with bacterial communities (Krishnapriya et al., 2021). They contribute to the blue carbon sink through carbon potential and predation dynamics, altering bacterial community diversity and size (Shi et al., 2023). Nematodes acquire carbon from bacteria and fungi, affecting carbon mineralization directly, and retain a substantial portion as biomass. Upon death, this carbon is deposited into the sediment, contributing to the blue carbon sink, potentially mitigating atmospheric CO_2 and global warming (Krishnapriya et al., 2021).

5 Manipulation of Soil Microbial Communities to Enhance Natural Carbon Sequestration Processes

The manipulation of soil microbial communities can indeed enhance NCS processes. Research has shown that interactions between microbes are crucial in determining soil carbon dynamics and microbial community structure (Shi et al., 2023; Zhang et al., 2023). For instance, adding biochar combined with chemical fertilizer significantly increased the accumulation of microbial residues in soil aggregates, promoting carbon sequestration and enhancing soil quality (Mohd et al., 2022). Moreover, manipulating soil microbial communities can impact soil functions and diversity, leading to shifts in inorganic nitrogen pools and soil pH, closely related to the proportion of specific microbial groups like ammonia-oxidizing bacteria (Kaur et al., 2023). Therefore, understanding and manipulating soil microbial interactions makes it possible to optimize NCS processes for improved soil health and ecosystem functioning.

5.1 Impacts of Climate Change on Soil Microbial Communities

Climate change significantly affects soil microbial communities by altering their structure, composition, and functions. Rising atmospheric CO_2 levels, global warming, and changes in precipitation patterns influence microbial diversity and activity (Zhou et al., 2023). Warming can lead to microbial community assembly process shifts, affecting community composition and network complexity (Pramanick et al., 2023). Changes in soil physiochemical properties can impact microbial habitat and activity, impacting nutrient cycling and greenhouse gas production (Chiranjeeb, 2022). Understanding these impacts is crucial for predicting ecosystem functions and ensuring sustainable agricultural practices under changing climatic conditions.

5.2 *Prospects and Research Opportunities*

Moreover, the prospects and research opportunities in soil microbes and carbon sequestration can include investigating the impact of agricultural practices on soil microbiome diversity and metabolic processes for enhanced yield and carbon sequestration efficiency (König et al., 2023). Also, it is helpful to study the synergistic action of soil microbes and biochar for increased carbon sequestration (Jia et al., 2023) and biotic interaction affecting microbial CUE of soil carbon depletion (Iven et al., 2022). The third recommendation is to use manufactured carbon materials, such as biochar and artificial humic substances, for controlling soil carbon dynamics (Tang et al., 2022). Additionally, it may be helpful to study the role of soil microbial communities in carbon sequestration, considering supportive ecological factors and long-term CO_2 exposure effects (Shahid et al., 2022). All these perspectives can contribute to increased carbon sequestration efficiency, emission suppression, and improvement of global soil health.

6 Conclusions

Natural carbon sequestration (NCS), or the process of living things absorbing and storing CO_2 from the atmosphere, is crucial for reducing greenhouse gas emissions caused by human activity and provides possible solutions to lessen the consequences of global warming. Microbes found in soil, such as bacteria, actinomycetes, fungi, viruses, protozoa, and archaea, are essential to this process. The decomposition of organic waste, plant health, and nutrient cycling impact ecosystem health and aid in NCS. Their roles in sequestering carbon are varied; they include the Calvin cycle, dicarboxylate-hydroxybutyrate cycle, 3-hydroxypropionate bi-cycle, reductive acetyl ~ CoA pathway (Wood-Ljungdahl pathway), glycolysis, pentose phosphate, reductive citrate cycle, and Entner-Doudoroff pathways. Several environmental variables influence this process, including pH, organic matter, temperature, and moisture. The microbial community and activity can be impacted by changes in the physiochemical characteristics of soil, which can affect greenhouse gas production and nutrient cycling. The form, content, and functions of soil microbial communities are all impacted by climate change. Investigating the effects of agricultural methods on soil microbiome diversity and metabolic processes for increased yield and carbon sequestration efficiency is just one of the many exciting research possibilities and prospects in soil microorganisms and carbon sequestration.

References

Adamczyk, B., Heinonsalo, J., & Simon, J. (2020). Mechanisms of carbon sequestration in highly organic ecosystems—Importance of chemical ecology. *Chemistry Open, 9*(4), 464–469. https://doi.org/10.1002/open.202000015

Bar-On, Y. M., Phillips, R., & Milo, R. (2018). The biomass distribution on Earth. *PNAS, 115*(25), 6506–6511. https://doi.org/10.1073/pnas.1711842115

Boisvenue, C., Paradis, G., Eddy, I. M. S., McIntire, E. J. B., & Chubaty, A. M. (2022). Managing forest carbon and landscape capacities. *Environmental Research Letters, 17*(11), 114013–114023. https://doi.org/10.1088/1748-9326/ac9919

Bonkowski, M. (2004). Protozoa and plant growth: The microbial loop in soil revisited. *New Phytologist, 162*(3), 617–631. https://doi.org/10.1111/j.1469-8137.2004.01066.x

Braga, L. P. P., Orland, C., Emilson, E. J. S., Fitch, A. A., Osterholz, H., Dittmar, T., Basiliko, N., Mykytczuk, N. C., & Tanentzap, A. J. (2022). Viruses direct carbon cycling in lake sediments under global change. *PNAS, 119*(41), e2202261119-2202261130. https://doi.org/10.1073/pnas.2202261119

Chen, J., Song, D., Luan, H., Liu, D., Wang, X., Sun, J., Zhou, W., & Liang, G. (2022). Living and dead microorganisms in mediating soil carbon stocks under long-term fertilization in a rice-wheat rotation. *Frontiers in Microbiology, 13*, 854216–854228. https://doi.org/10.3389/fmicb.2022.854216

Chiranjeeb, K. (2022). Effect of climatic change on soil microbial community. *Emerging Trends in Climate Change, 1*(2), 1–8. https://doi.org/10.18782/2583-4770.106

Dathe, A., Lynch, L., Woolf, D., & Lehmann, J. (2023). The importance of bacterial metabolism contribution to soil organic carbon revealed by Monte Carlo simulations. In *Paper presented at the EGU General Assembly*, Vienna, Austria. https://ui.adsabs.harvard.edu/abs/2023EGUGA..2514428D

Dooley, K., Nicholls, Z., & Meinshausen, M. (2022). Carbon removals from nature restoration are no substitute for steep emission reductions. *One Earth, 5*(7), 812–824. https://doi.org/10.1016/j.oneear.2022.06.002

Dudhagara, P. (2015). Biomimetic sequestration of CO_2 using carbonic anhydrase from calcite encrust forming marine Actinomycetes. *Science International, 3*, 48–57. https://doi.org/10.5567/sciintl.2015.48.57

Farda, B., Djebaili, R., Vaccarelli, I., Del Gallo, M., & Pellegrini, M. (2022). Actinomycetes from caves: An overview of their diversity, biotechnological properties, and insights for their use in soil environments. *Microorganisms, 10*(2), 453–471. https://doi.org/10.3390/microorganisms10020453

Feng, X., & Wang, S. (2023). Plant influences on soil microbial carbon pump efficiency. *Global Change Biology, 29*(14), 3854–3856. https://doi.org/10.1111/gcb.16728

Gayathri, R., Mahboob, S., Govindarajan, M., Al-Ghanim, K. A., Ahmed, Z., Al-Mulhm, N., Vodovnik, M., & Vijayalakshmi, S. (2021). A review on biological carbon sequestration: A sustainable solution for a cleaner air environment, less pollution and lower health risks. *Journal of King Saud University - Science, 33*(2), 101282–101290. https://doi.org/10.1016/j.jksus.2020.101282

Gonzalez, P. (2023). Natural carbon solutions contribute to halting climate change. *Parks Stewardship Forum, 39*(2), 163–173. https://doi.org/10.5070/p539260953

Govindaraju, M., Fowmitha Banu, J., Senthamil selvi, S., & Goel, M. (2021). CO_2 sequestration through phytoremediation techniques with special emphasis on urban forestry to mitigate climate change impact. In M. Goel, T. Satyanarayana, M. Sudhakar, & D. P. Agrawal (Eds.), *Climate change and green chemistry of CO_2 sequestration* (pp. 263–271). Springer.

Habib, M. E., & Huda, M. E. (2022). Estimation of carbon sequestration in a forest: A case study of bhawal national park, Gazipur. *Asian Journal of Research in Agriculture and Forestry, 8*(3), 19–31. https://doi.org/10.9734/ajraf/2022/v8i330156

Harris, Z. M., Milner, S., & Taylor, G. (2018). Chapter 5—Biogenic carbon: Capture and sequestration. In P. Thornley & P. Adams (Eds.), *Greenhouse gas balances of bioenergy systems* (pp. 55–76). Academic Press.

Hayat, R., Ali, S., Amara, U., Khalid, R., & Ahmed, I. (2010). Soil beneficial bacteria and their role in plant growth promotion: A review. *Annals of Microbiology, 60*(4), 579–598. https://doi.org/10.1007/s13213-010-0117-1

Helferty, H. (2023). Harnessing nature and ancient wisdom to store carbon. *PNAS, 120*(19), e2305214120-2305214122. https://doi.org/10.1073/pnas.2305214120

Hu, Y., Zheng, Q., Zhang, S., Noll, L., & Wanek, W. (2018). Significant release and microbial utilization of amino sugars and D-amino acid enantiomers from microbial cell wall decomposition in soils. *Soil Biology and Biochemistry, 123*, 115–125. https://doi.org/10.1016/j.soilbio.2018.04.024

Huang, Q., Huang, Y., Wang, B., Dippold, M. A., Li, H., Li, N., Jia, P., Zhang, H., An, S., & Kuzyakov, Y. (2022). Metabolic pathways of CO_2 fixing microorganisms determined C-fixation rates in grassland soils along the precipitation gradient. *Soil Biology and Biochemistry, 172*, 108763–110875. https://doi.org/10.1016/j.soilbio.2022.108764

Iven, H., Walker, T. W. N., & Anthony, M. (2022). Biotic interactions in soil are underestimated drivers of microbial carbon use efficiency. *Current Microbiology, 80*(1), 13–26. https://doi.org/10.1007/s00284-022-02979-2

Jia, Y., Qian, B., & Wang, Z. (2023). Effect of Ssil microorganisms on carbon fixation capacity of biochar. *Highlights in Science, Engineering and Technology, 40*, 214–218. https://doi.org/10.54097/hset.v40i.6624

Kaiser, C., Kilburn, M. R., Clode, P. L., Fuchslueger, L., Koranda, M., Cliff, J. B., Solaiman, Z. M., & Murphy, D. V. (2015). Exploring the transfer of recent plant photosynthates to soil microbes: Mycorrhizal pathway vs direct root exudation. *New Phytologist, 205*(4), 1537–1551. https://doi.org/10.1111/nph.13138

Kaur, R., Kaur, N., Kumar, S., Dass, A., & Singh, T. (2023). Carbon capture and sequestration for sustainable land use: A review. *The Indian Journal of Agricultural Sciences, 93*(1), 11–18. https://doi.org/10.56093/ijas.v93i1.124838

Kiss, M., Cseh, V., & Tanács, E. (2015). Carbon sequestration of different types of floodplain forests in the Maros River Valley (Hungary). In M. Luc, U. Somorowska, & J. B. Szmańda (Eds.), *Landscape analysis and planning: Geographical perspectives* (pp. 159–171). Springer.

König, A., Rosinger, C., Keiblinger, K., Zechmeister-Boltenstern, S., Herrmann, A., & Inselsbacher, E. (2023). *Linking soil microbial carbon sequestration to cover crop diversification in agricultural soil systems across Europe*. https://ui.adsabs.harvard.edu/abs/2023EGUGA..25.9585K

Krishnapriya, P. P., Aswathy, N. K., Dhanya, M. M., Jima, M., Jayachandran, P. R., Bijoy Nandan, S., Harikrishnan, M., & Krishnan, K. P. (2021). Protective role of carbon sequestration by free living nematodes in Arctic Kongsfjord (intermediate zone), Svalbard in the emerging climate change scenario. *Journal of Earth System Science, 130*(3), 138-152. https://doi.org/10.1007/s12040-021-01639-6

Li, Y.-Z., Bao, X.-L., Tang, S.-X., Xiao, K.-Q., Ge, C.-J., Xie, H.-T., He, H. B., Mueller, C. W., & Liang, C. (2024). Toward soil carbon storage: The influence of parent material and vegetation on profile-scale microbial community structure and necromass accumulation. *Soil Biology and Biochemistry, 193*, 109399–109410. https://doi.org/10.1016/j.soilbio.2024.109399

Liang, C. (2020). Soil microbial carbon pump: Mechanism and appraisal. *Soil Ecology Letters, 2*(4), 241–254. https://doi.org/10.1007/s42832-020-0052-4

Liu, Z., Li, J., Zhang, Y., Gong, H., Hou, R., Sun, Z., & Ouyang, Z. (2023). Soil microbes from saline–alkali farmland can form carbonate precipitates. *Agronomy, 13*(2), 372–382. https://doi.org/10.3390/agronomy13020372

Llado, S., Lopez-Mondejar, R., & Baldrian, P. (2017). Forest soil bacteria: diversity, involvement in ecosystem processes, and response to global change. *Microbiology and Molecular Biology, 81*(2), e00063-e116. https://doi.org/10.1128/MMBR.00063-16

Markusson, N. (2022). Natural carbon removal as technology. *Wires Climate Change, 13*(2), e767-772. https://doi.org/10.1002/wcc.767
Mason-Jones, K., Breidenbach, A., Dyckmans, J., Banfield, C. C., & Dippold, M. A. (2023). Intracellular carbon storage by microorganisms is an overlooked pathway of biomass growth. *Nature Communications, 14*(1), 2240–2248. https://doi.org/10.1038/s41467-023-37713-4
Mohd, N. H. S., Siti, A. M. D., Aidee, K. K., Nazrin, A.-A., Nur, R. A., & Umi, A. A. (2022). Regenerating soil microbiome: Balancing microbial CO_2 sequestration and emission. In S. Suriyanarayanan & K. Kannan (Eds.), *Carbon sequestration* (pp. 245–260). Rijeka: IntechOpen.
Naitam, M. G., & Kaushik, R. (2021). Archaea: An agro-ecological perspective. *Current Microbiology, 78*(7), 2510–2521. https://doi.org/10.1007/s00284-021-02537-2
Ncipha, X. G., & Sivakumar, V. (2019). Natural carbon sequestration by forestry. In Inamuddin, A. M. Asiri, & E. Lichtfouse (Eds.), *Sustainable agriculture reviews 37: Carbon sequestration* (Vol. 1, pp. 73–92). Cham: Springer.
Pessarrodona, A., Franco-Santos, R. M., Wright, L. S., Vanderklift, M. A., Howard, J., Pidgeon, E., Wernberg, T., & Filbee-Dexter, K. (2023). Carbon sequestration and climate change mitigation using macroalgae: A state of knowledge review. *Biological Reviews, 98*(6), 1945–1971. https://doi.org/10.1111/brv.12990
Pramanick, B., Bera, A., Saha, P., Barman, A., Paramanick, B., Maitra, S., & Hossain, A. (2023). Soil Microflora and their interaction with plants under changing climatic scenarios. In P. Mathur, R. Kapoor, & S. Roy (Eds.), *Microbial symbionts and plant health: Trends and applications for changing climate* (pp. 19–40). Springer.
Puxty, R. J., Millard, A. D., Evans, D. J., & Scanlan, D. J. (2016). Viruses inhibit CO_2 fixation in the most abundant phototrophs on Earth. *Current Biology, 26*(12), 1585–1589. https://doi.org/10.1016/j.cub.2016.04.036
Qu, R., Liu, G., Yue, M., Wang, G., Peng, C., Wang, K., & Gao, X. (2023). Soil temperature, microbial biomass and enzyme activity are the critical factors affecting soil respiration in different soil layers in Ziwuling Mountains, China. *Frontiers in Microbiology, 14*, 1105723–1105734. https://doi.org/10.3389/fmicb.2023.1105723
Rattan, L., Pete, S., Hermann, F. J., William, J. M., Johannes, L., Nair, P. K. R., …, Nijavalli, H. R. (2018). The carbon sequestration potential of terrestrial ecosystems. *Journal of Soil and Water Conservation, 73*(6), 145–152A. https://doi.org/10.2489/jswc.73.6.145A
Royal, L., Mihriemate, B. S., & Mailappa, A. S. (2020). Carbon stock of prevailing land use systems under east siang district, Arunachal pradesh, India. *International Journal of Current Microbiology and Applied Sciences, 9*(6), 2568–2875. https://doi.org/10.20546/ijcmas.2020.906.312
Sanaullah, M., Afzal, T., Shahzad, T., & Wakeel, A. (2019). Carbon sequestration for sustainable agriculture. In *Innovations in sustainable agriculture* (pp. 469–500).
Shaaban, M. (2024). Microbial pathways of nitrous oxide emissions and mitigation approaches in drylands. *Journal of Environmental Management, 354*, 120393–120406. https://doi.org/10.1016/j.jenvman.2024.120393
Shaaban, M., Khalid, M. S., Hu, R., & Zhou, M. (2022). Effects of water regimes on soil N_2O, CH_4 and CO_2 emissions following addition of dicyandiamide and N fertilizer. *Environmental Research, 212*, 113544–113551. https://doi.org/10.1016/j.envres.2022.113544
Shaaban, M., & Nunez-Delgado, A. (2024). Soil adsorption potential: Harnessing Earth's living skin for mitigating climate change and greenhouse gas dynamics. *Environmental Research, 251*, 118738–118746. https://doi.org/10.1016/j.envres.2024.118738
Shaaban, M., Peng, Q.-A., Lin, S., Wu, Y., Khalid, M. S., Wu, L., Mo, Y., & Hu, R. (2016). Dolomite application enhances CH_4 uptake in an acidic soil. *CATENA, 140*, 9–14. https://doi.org/10.1016/j.catena.2016.01.014
Shaaban, M., Wang, X., Song, P., Hu, R., & Wu, Y. (2024). Impact of dolomite liming on ammonia-oxidizing microbial populations and soil biochemistry in acidic rice paddy soils. *Agronomy, 14*(9), 2070–2087. https://doi.org/10.3390/agronomy14092070

Shahid, M. K., Kashif, A., Rout, P. R., & Choi, Y. (2022). An overview of soil bacteria for CO_2 sequestration. In S. K. Nayak, B. Baliyarsingh, I. Mannazzu, A. Singh, & B. B. Mishra (Eds.), *Advances in agricultural and industrial microbiology* (Vol. 1, pp. 91–103). Springer.

Shi, G., Luan, L., Zhu, G., Zeng, Z., Zheng, J., Shi, Y., Sun, B., & Jiang, Y. (2023). Interaction between nematodes and bacteria enhances soil carbon sequestration under organic material amendments. *Frontiers in Microbiology, 14*, 1155088–1155099. https://doi.org/10.3389/fmicb.2023.1155088

Silva, A. F. D., Nathaniel, J., Wong, K. C. L., Watson, C., Wang, H., Singh, J., Chamon, A. A., & Klein, L. (2022, 17–20 December). NetZeroCO_2, an AI framework for accelerated nature-based carbon sequestration. In *Paper presented at the 2022 IEEE International Conference on Big Data (Big Data)*

Sokol, N. W., Slessarev, E., Marschmann, G. L., Nicolas, A., Blazewicz, S. J., Brodie, E. L., Firestone, M. K, Foley, M. M, Hestrin, R., Hungate, B.A., Koch, B. J., & Consortium, L. S. M. (2022). Life and death in the soil microbiome: how ecological processes influence biogeochemistry. *Nature Reviews Microbiology, 20*(7), 415–430. https://doi.org/10.1038/s41579-022-00695-z

Tang, C., Yang, F., & Antonietti, M. (2022). Carbon materials advancing microorganisms in driving soil organic carbon regulation. *Research, 2022*, 1–12. https://doi.org/10.34133/2022/9857374

Tao, F., Huang, Y., Hungate, B. A., Manzoni, S., Frey, S. D., Schmidt, M. W. I., Reichstein, M., Carvalhais, N., Ciais, P., Jiang, L., Lehmann J., & Luo, Y. (2023). Microbial carbon use efficiency promotes global soil carbon storage. *Nature, 618*(7967), 981-985. https://doi.org/10.1038/s41586-023-06042-3

Treseder, K. K., & Holden, S. R. (2013). Ecology. Fungal carbon sequestration. *Science, 339*(6127), 1528–1529. https://doi.org/10.1126/science.1236338

Vetter, S. H., Kuhnert, M., & Smith, P. (2022). Biological carbon sequestration technologies. In *Reference module in earth systems and environmental sciences*: Elsevier.

Victor, C., Li., B., R., E., & Duo, Z. (2021). USA Patent No. 20220274876. T. r. o. t. u. o. Michigan.

Wang, H., Zhao, R., Zhao, D., Liu, S., Fu, J., Zhang, Y., Dai, N., Song, D., & Ding, H. (2022). Microbial-mediated emissions of greenhouse gas from farmland soils: A review. *Processes, 10*(11), 2361–2374. https://doi.org/10.3390/pr10112361

Wang, R., Zhang, H., Sun, L., Qi, G., Chen, S., & Zhao, X. (2017). Microbial community composition is related to soil biological and chemical properties and bacterial wilt outbreak. *Scientific Reports, 7*(1), 343–352. https://doi.org/10.1038/s41598-017-00472-6

Wu, H., Cui, H., Fu, C., Li, R., Qi, F., Liu, Z., Yang, G., Xiao, K., & Qiao, M. (2024). Unveiling the crucial role of soil microorganisms in carbon cycling: A review. *Science of the Total Environment, 909*, 168627-168642. https://doi.org/10.1016/j.scitotenv.2023.168627

Wu, W., Dijkstra, P., Hungate, B. A., Shi, L., & Dippold, M. A. (2022). In situ diversity of metabolism and carbon use efficiency among soil bacteria. *Science Advances, 8*(44), eabq3958–3964. https://doi.org/10.1126/sciadv.abq3958

Yitayal, T. (2020). A review paper on the role of terrestrial carbon stocks for climate change mitigation mechanisms. *Journal of Environment and Earth Science, 10*(8), 32–50.

Yongcun, F., & Shui, Z. (2022). Geomechanics of geological carbon sequestration. In S. Suriyanarayanan & K. Kannan (Eds.), *Carbon sequestration* (pp. 1–19). IntechOpen.

Zeng, N. (2008). Carbon sequestration via wood burial. *Carbon Balance and Management, 3*(1), 1–12. https://doi.org/10.1186/1750-0680-3-1

Zhang, L., Wang, Y., Chen, J., Zhang, C., Cao, Y., Cai, G., & Yu, L. (2023). Exogenous carbon addition reduces soil organic carbon: The effects of fungi on soil carbon priming exceed those of bacteria on soil carbon sequestration. *Forests, 14*(6), 1268–1286. https://doi.org/10.3390/f14061268

Zhao, J., Jing, H., Wang, Z., Wang, L., Jian, H., Zhang, R., Xiao, X., Chen, F., Jiao, N., & Zhang, Y. (2022). Novel viral communities potentially assisting in carbon, nitrogen, and sulfur metabolism in the upper slope sediments of Mariana Trench. *mSystems, 7*(1), 1–19. https://doi.org/10.1128/msystems.01358-21

Zhou, S. Y., Lie, Z., Liu, X., Zhu, Y. G., Penuelas, J., Neilson, R., Su, X., Liu, Z., Chu, G., Meng, Z., Yan, J., & Liu, J. (2023). Distinct patterns of soil bacterial and fungal community assemblages in subtropical forest ecosystems under warming. *Global Change Biology Bioenergy, 29*(6), 1501-1513. https://doi.org/10.1111/gcb.16541

Žifčáková, L., Větrovský, T., Howe, A., & Baldrian, P. (2016). Microbial activity in forest soil reflects the changes in ecosystem properties between summer and winter. *Environmental Microbiology, 18*(1), 288–301. https://doi.org/10.1111/1462-2920.13026

Harnessing the Potential of Biofertilizers: A Sustainable Approach for Agriculture

Debadatta Sethi, Shraddha Mohanty, Kiran Kumar Mohapatra, Rajeswari Das, Kshitipati Padhan, Suman G. Sahu, Konathala Kusumavathi, Narayan Panda, Sachin Kanta Rautaray, and Sushanta Kumar Pattanayak

Abstract The rising demand for sustainable farming techniques has sparked interest in biofertilizers as a more environmentally friendly option compared to chemical fertilizers. Biofertilizers, which consist of living microorganisms like bacteria, fungi, and algae, boost soil fertility and support plant growth by fixing nitrogen from the air, solubilizing phosphorus, and promoting plant growth through the production of natural hormones. This paper examines how biofertilizers can increase crop yields, maintain soil health, and reduce the ecological footprint of traditional farming methods. By harnessing the symbiotic interactions between plants and microorganisms, biofertilizers not only improve nutrient availability but also enhance soil biological activity and structure. The use of biofertilizers provides an effective way to address the negative impacts of chemical fertilizers, such as soil depletion, water contamination, and greenhouse gas emissions. However, their broader adoption faces obstacles like low awareness, variable effectiveness, and short shelf life. This paper highlights the importance of increased research, farmer education, and supportive policies to unlock the full potential of biofertilizers in promoting sustainable farming.

D. Sethi (✉) · K. Padhan · N. Panda · S. K. Pattanayak
Department of Soil Science and Agricultural Chemistry, College of Agriculture, Odisha University of Agriculture and Technology, Bhubaneswar, Odisha, India
e-mail: debadattasethi@ouat.ac.in

S. Mohanty
Krishi Vigyan Kendra, Odisha University of Agriculture and Technology, Kalahandi, Odisha, India

K. K. Mohapatra
AICRP for Dryland Agriculture, Odisha University of Agriculture and Technology, Phulbani, Kandhamal, Odisha, India

R. Das · S. G. Sahu
Department of Soil Science and Agricultural Chemistry, School of Agriculture, GIET University, Gunupur, Odisha, India

K. Kusumavathi · S. K. Rautaray
ICAR-Indian Institute of Water Management, Bhubaneswar, Odisha, India

K. Kusumavathi
Rural Management Programme, Xavier Institute of Social Service, Ranchi, Jharkhand, India

M. Shaaban (ed.), *Soils and Sustainable Agriculture*, Frontier Studies in Soil Science,
https://doi.org/10.1007/978-3-031-91114-9_9

By blending modern technologies with traditional practices, biofertilizers can significantly contribute to creating more resilient agricultural systems and ensuring food security in the long term.

Keywords Soil biology · Bio-fertilizer · Symbiosis · N-fixation · Solubilisation

1 Introduction

Biofertilizers are composed of microorganisms like bacteria, fungi, or cyanobacteria that, when introduced to plants or soil, improve plant growth, development, and soil health (Bharti & Suryavanshi, 2021). The rising emphasis on sustainable and eco-friendly agricultural methods in recent years has encouraged researchers to explore natural alternatives for addressing agricultural challenges (Muhie, 2022). Recent research has underscored the crucial link between plants and microbial communities, revealing that microbes can greatly enhance plant growth (Dastogeera et al., 2020), strengthen disease resistance, and increase resilience to harsh environmental conditions such as drought and flooding (Ahmad et al., 2018; Barnawal et al., 2017; Santos et al., 2019; Vejan et al., 2016).

The agricultural inoculants market is segmented into plant growth-promoting microorganisms (PGPMs), biocontrol agents, and plant resistance stimulants, with PGPMs commanding the largest share. Among microbial inoculants, bacterial types are particularly sought after, including categories such as rhizobacteria, nitrogen-fixing bacteria, phosphate-solubilizing bacteria, ACC-deaminase-producing bacteria, and zinc-solubilising bacteria (Shaaban, 2024). Common fungal biofertilizers include mycorrhiza, *Trichoderma*, and *Piriformospora i*noculants. The use of these inoculants, either through seed or soil application, is anticipated to grow as part of sustainable farming methods. Globally, their use is primarily focused on crops, fruits, and vegetables (Kumar et al., 2022; O'Callaghan, 2016).

Eco-friendly biofertilizers provide a sustainable alternative to chemical fertilizers and are crucial for promoting sustainable agricultural practices (Kumar et al., 2022). Comprising living microorganisms, biofertilizers, when applied to seeds, plant surfaces, or soil, colonize the rhizosphere or the interior of plants, enhancing growth by increasing the availability of essential nutrients to the host plant. These beneficial microorganisms are commonly referred to as plant-growth-promoting microbes (PGPM), plant-growth-promoting bacteria (PGPB), or plant-growth-promoting rhizobacteria (PGPR). The commercial application of biofertilizers began in 1895 with the launch of "Nitragin," a product containing nitrogen-fixing rhizobium strains (Soumare et al., 2020; Zhao et al., 2024). In the 1950s, phosphorus-solubilizing bacteria were introduced as biofertilizers to transform soil phosphorus into plant-accessible forms (Bargaz et al., 2021; Wang et al., 2019). Since then, numerous fungi, algae, and bacteria have been discovered to promote plant growth through various mechanisms (Azizoglu, 2019; Barin et al., 2022; Gezgin et al., 2020; Ismail et al., 2021; Naeem et al., 2021; Sun et al., 2020a, 2020b).

The application of biofertilizers in agriculture has demonstrated effectiveness in boosting crop yields while reducing reliance on chemical fertilizers. Research indicates that using biofertilizers alongside chemical fertilizers can lead to yield increases of 10% to 40%, potentially allowing for a substitution of 25% to 30% of chemical fertilizers (Pal et al., 2015). Over the years, there has been a growing body of global research on biofertilizers and their importance, with studies emphasizing soil fertility and plant growth (Fig. 1).

A study conducted by the Brazilian Agricultural Research Corporation (BARC) revealed that biofertilizers utilized for biological nitrogen fixation in soybean cultivation can completely substitute chemical nitrogen fertilizers without compromising yields, resulting in estimated annual savings of about USD 15 billion (Bomfim et al., 2021). Additionally, the application of rhizobia biofertilizers in Zambia and the Republic of Congo has significantly enhanced soybean yields, increasing them by approximately 48% and 500 kg ha^{-1}, respectively (Raimi et al., 2021). These findings

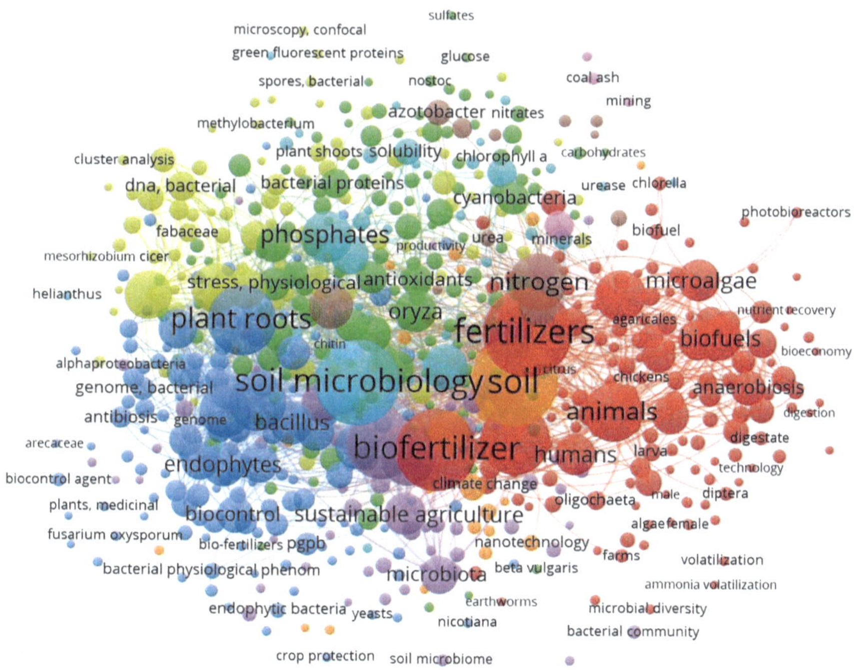

Fig. 1 The network visualization of co-occurrence analysis includes 611 publications from a total of 2834, divided into 9 clusters. Cluster 1 comprises 150 items, cluster 2 has 147 items, cluster 3 includes 135 items, cluster 4 contains 85 items, cluster 5 holds 60 items, cluster 6 consists of 39 items, cluster 7 has 29 items, cluster 8 includes 11 items, and cluster 9 contains 5 items. The association strength between keywords is represented, where the size of each ball indicates the importance of the keyword, the colors distinguish the different clusters, and the lines illustrate the connections between keywords

underscore the effectiveness of biofertilizers in enhancing agricultural productivity across various regions.

Given the economic and environmental benefits of biofertilizers, a comprehensive literature review is crucial to provide researchers with reliable and impartial guidance. Bibliometric analysis, which employs computational tools to quantitatively evaluate the impact of research on specific topics, can support this endeavor. This approach uncovers characteristics of previous literature and identifies current research hotspots by analyzing the quantity and growth trends of publications within a particular academic field (Becerra et al., 2020; Du et al., 2014; Mkhongi & Musakwa, 2022). The aim of this study is to conduct both quantitative and qualitative analyses of articles published over recent decades, assessing growth trends in biofertilizer research, identifying emerging topics and research directions, and highlighting the most influential and frequently cited papers. This will assist researchers in building on foundational works for future studies.

2 Types of Biofertilizers

The enzymatic conversion of molecular nitrogen to ammonia is facilitated by nitrogenase, an oxygen-sensitive enzyme complex that is highly conserved among free-living and symbiotic diazotrophs. The most prevalent form of nitrogenase, known as Mo-nitrogenase or conventional nitrogenase includes a prosthetic group containing molybdenum, specifically the iron-molybdenum-cobalt (FeMoCo) group. Additionally, some bacteria, such as *Azotobacter* and various photosynthetic nitrogen fixers, including certain cyanobacteria, possess alternative forms of nitrogenase that contain vanadium (V-nitrogenase) or solely iron (Fe-nitrogenase) as cofactors.

The nitrogenase enzyme, which has been purified from various sources, consists of two metalloproteins. Component 1, known as the Mo-Fe protein, is a tetramer weighing 220,000 Da, made up of two non-identical subunits, α and β. Component 2, referred to as the Fe protein, is a dimer of 68,000 Da composed of identical subunits. Two iron-molybdenum cofactor (FeMoCo) units are bound to α subunits of the MoFe protein, along with two additional prosthetic groups that contain 4Fe-4S clusters. The 'P-clusters' are covalently attached to cysteine residues of the MoFe protein, bridging α and β subunits. The third type of iron-sulfur group is associated with the Fe protein. The nitrogen reduction process is highly complex and not yet fully understood. The overall reaction for the net reduction of molecular nitrogen to ammonia can be represented by the following equation:

$$N_2 + 16Mg - ATP + 8e^- + 8H^+ \rightarrow 2NH_3 + H_2 + 16Mg - ADP + 16Pi$$

The increasing global demand for sustainable agricultural practices has driven a focus on environmentally friendly alternatives to chemical fertilizers. Developing countries are more enthusiastic about biofertilizer research than developed countries (Zhao et al., 2024). Biofertilizers, comprising living microorganisms that enhance nutrient availability to plants, have gained prominence in this context. These microbial inoculants play a crucial role in improving soil fertility by facilitating the conversion of unavailable nutrients into forms that can be readily absorbed by plants along with other direct and indirect benefits. Based upon their nature and mode of action there are several types of biofertilizers which are discussed briefly in the following sections:

2.1 Nitrogen-Fixing Biofertilizers

Although the atmosphere contains around 78% of free nitrogen, most plants are unable to utilize it. This nitrogen must be fixed by a particular set of microorganisms before it can be used by the plant. These microorganisms are called biological nitrogen fixers (BNFs). Nitrogen fixation involves the conversion of nitrogen gas to ammonia by the action of the enzyme nitrogenase. This process is energy-intensive and occurs under anaerobic conditions in nodules or microaerophilic environments. The primary N_2-fixing bacteria are discussed below (Table 1).

2.1.1 Symbiotic Nitrogen-Fixing Microbes

Symbiosis involves a mutualistic relationship between plant as the macro-symbiont and prokaryotic bacteria as the microsymbiont. Species of *Mesorhizobium, Azorhizobium, Allorhizobium, Rhizobium, and Sinorhizobium*, collectively referred to as *Rhizobium*, establish a symbiotic connection with the roots of legumes. This interaction begins when plants release flavonoids and iso-flavonoids in their rhizosphere, which recognized by *Rhizobium* (Hawkins & Oresnik, 2022). Infection begins with root hair differentiation, leading to infection thread development and bacterial release in the cytoplasmic region. Bacteria differentiate into bacteroides, which develop into symbioses for nitrogen fixation. The nitrogenase enzyme fixes atmospheric nitrogen in the nodule (Chaudhary et al., 2022). Examples include, *Rhizobium*-legume symbiosis, azolla-cyanobacterial symbiosis and Frankia-actinorhizal plants (*Alnus, Myrica, Casuarina* etc.)

2.1.2 Free-Living Nitrogen-Fixing Biofertilizers

Certain free-living bacteria and cyanobacteria (blue-green algae) also fix atmospheric nitrogen independent of plant roots. These microbes enhance nitrogen availability in various agricultural ecosystems, particularly in rice fields. Arable soil is mostly populated by *Azotobacter chroococcum, A. insignis, A. beijerinckii, A. macrocytogens, and A. vinelandii*, which can possibly fix 2–18 mg of nitrogen per gram of carbon utilized (Smercina et al., 2019). The *Azotobacter* strains are mostly isolated from rhizosphere non-leguminous plants under neutral to alkaline soil. Cyanobacteria,

Table 1 Role of different nitrogen fixing biofertilizers in agriculture

Types of biofertilizer	Contribution to plant growth	Impact on yield	Suitable for crops	References
Rhizobium	Fixes 50–300 kg N $ha^{-1}/year^{-1}$ Meets up to 80–90% of N crop requirement Residual N for subsequent crop Maintains soil fertility	Yield increased by 10–30%	Pea, Cowpea, Green Gram, Black Gram, Groundnut, Soyabean, Berseem, Lucern etc.	Brahmaprakash and Sahu (2012), Mahmud et al., (2021)
Azotobacter	Fixes about 20–40 kg N ha^{-1} yr^{-1} Production of growth promoting substances i.e. IAA, Giberellic acid, vitamin B complex etc. Biological control of Pest and diseases Maintains the soil fertility	Results in 10–15% increase in crop yield	Rice, Wheat, Maize, Sorghum, Pearl Millet, Sunflower, Sugarcane, Mustard	Jnawali et al. (2015), Thomas and Singh (2019),
Azospirillum	Fixes N up to 10–20 kg ha^{-1} Promote root growth Produce cytokinin, gibberellins, and IAA, which aid in the uptake of N, P, and K Applied to legumes as co-inoculants	Results in 10–20% increase in crop yield	Sugarcane, sweet potato, maize, sorghum, bajra, wheat, rice, barley	Pathak et al. (2017), Chaudhary et al. (2022)

(continued)

Table 1 (continued)

Types of biofertilizer	Contribution to plant growth	Impact on yield	Suitable for crops	References
Blue green algae	Fixes 20–30 kg N ha^{-1} Produce growth promoting substances Increase water holding capacity of soil	Increase 10–15% yield	Rice	Kumar et al. (2017), Singh et al. (2016)
Azolla	Fixes 30–60 kg N $ha^{-1}/year^{-1}$ Synthesises micronutrients, proteins, amino acids and lipids Use as green manure	–	Rice	Kollah et al. (2016), Verma et al. (2022)

an essential category of free-living nitrogen-fixers, are anaerobic, photosynthetic blue-green algae that reside in alkaline, damp soil. Prominent genera of free-living photosynthetic cyanobacteria include *Anabaena, Nostoc, Aulosira,* and *Calothrix. Nostoc* and *Anabaena* serve as biofertilizers for rice cultivation during the monsoon season (Singh et al., 2016).

2.1.3 Associative Nitrogen-Fixing Biofertilizers

Some endophytic diazotrophic microbes form intracellular association with roots of non-leguminous plants viz., cereals, oilseeds, cotton and millets to fix atmospheric nitrogen. Examples include and *Azospirillum brasilense, Azospirillum lipoferum, Azospirillum amazonense, Acetobacter, Gluconobacter, Herbaspirillum,* and *Azoarcus* etc. (Mishra et al., 2013). Under nitrogen deficiency in soil *Azospirillum* get activated and provide nitrogen to plants. In C4 plants, *Azospirillum lipoferum* facilitates nitrogen fixation, but in C3 plants such as rice and wheat, *Azospirillum brasilense* performs this function. They produce growth stimulants (IAA, gibberellins, and cytokinins) that facilitate root development and nutrient absorption (N, P, and K) (Rodrigues et al., 2015).

3 Phosphate-Solubilizing Biofertilizers

Phosphorus is a vital nutrient for plant growth, yet much of it exists in insoluble forms in the soil. Soluble form of phosphorus is orthophosphate ($H_2PO_4^-$ and HPO_4^{2-}), and their uptake is facilitated by rhizospheric microbes. Biofertilizers that solubilize phosphate can be classified into phosphate solubilizers and phosphate mobilizers. Phosphate solubilizers can solubilize unavailable form of phosphorus into an available form through the production of organic acids by bacteria, which decreases soil pH, facilitates the dissolution of phosphate compounds, and make them available for plant nutrition (Mahanty et al., 2017). Phosphate-solubilizing bacteria (PSB) excrete organic acids, including gluconic acid, which solubilizes inorganic phosphate compounds, rendering them accessible to plants (Fig. 2).

Phosphorus-mobilizing biofertilizers efficiently enhance the solubility of phosphorus and facilitate the mineralization of organic phosphorus compounds (Fig. 2). A primary advantage of Arbuscular mycorrhiza is its ability to deliver both inorganic and organic forms of phosphorus to plants. Examples include arbuscular mycorrhizal fungi (AMF) encompass *Acaulospora sp., Glomus, Entrophospora, and Paraglomus*, while ectomycorrhizal fungi include *Amanita, Laccaria, and Boletus* (Haro & Benito, 2019).

3.1 Potassium Solubilizing and Mobilizing Biofertilizers

Potassium is present in soil as both available and unavailable forms (Basak et al., 2022). Certain microbes present in soil are responsible for mineralization of K through release of organic acids (Fig. 3) (Patel et al., 2021; Varga et al., 2020). Consortia of *Azotobacter chroococcum, Rhizobium* sp., and *Bacillus mucilaginosus* in maize and wheat increased potassium assimilation and crop productivity (Singh et al., 2010). Among fungi, *Aspergillus* and *Penicillium* can also solubilize and mobilize K from organic and inorganic sources.

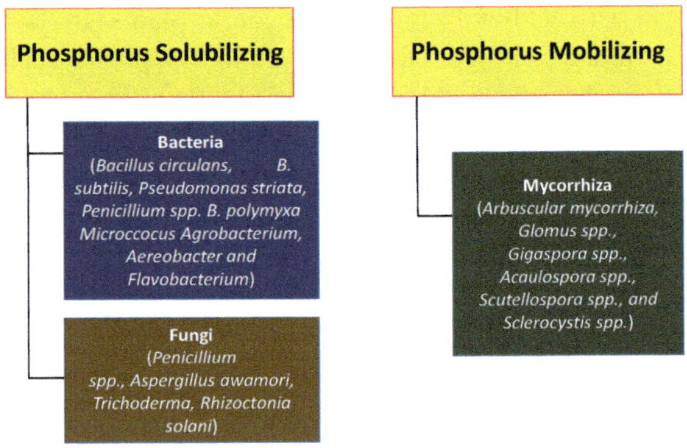

Fig. 2 List of phosphorus solubilizing and phosphorus mobilizing biofertilizers

Fig. 3 List of potassium solubilizing and potassium mobilizing microbes

3.2 *Cellulose-Decomposing Biofertilizers*

Cellulose is one of the most abundant organic polymers in plant residues. Cellulose-decomposing microorganisms, mainly fungi and bacteria, break down plant material into simpler compounds, enriching the soil with organic matter. Microorganisms secrete cellulase enzymes, which hydrolyze cellulose into glucose and simpler carbohydrates. This process is essential for nutrient cycling and improving soil structure. Examples include *Trichoderma* sp., *Bacillus* sp., *Streptomyces* sp. *Trichoderma* sp. that accelerate litter breakdown produce antimicrobial chemicals, enhance soil physicochemical qualities, and increase microbial diversity. Bacillus subtilis participate in lignocellulose biodegradation through the secretion of microbial enzymes (Niu & Li, 2022).

3.3 *Zinc-Solubilizing Biofertilizers*

Zinc-solubilizing bacteria generate organic acids that convert insoluble zinc into Zn^{+2}, making it accessible for plant (Nitu et al., 2020). Zinc-solubilizing bacteria (*Bacillus subtilis, Pseudomonas striata, Serratia, Burkholderia cenocepacia*) and fungi (*Aspergillus niger, A. nomius*, and *A. oryza*) produce organic acids, chiefly 2-ketogluconic acid and gluconic acid, in soil, facilitating the solubilization of zinc. These are also recognized for augmenting soil enzymatic activity (Batool et al., 2021). The inoculation of wheat with *Pseudomonas*, sugarcane with *Pantoea dispersa* and *Pantoea agglomerans*, maize with *Bacillus sp.*, and both wheat and soybean with *B. aryabhattai* improved Zn solubility in soil and its subsequent uptake (Hussain et al., 2015; Joshi et al., 2013; Kamran et al., 2017; Ramesh et al., 2014).

3.4 Plant Growth Promoting Rhizobacteria

A consortium of free-living rhizosphere bacteria that inhabit in plant roots and promote plant growth is termed as plant growth promoting rhizobacteria (PGPR). The inoculation of PGPR confers numerous advantageous effects on the plant. They enhance the resilience of plants to abiotic stress (drought and salt) and biotic stress (Beneduzi et al., 2012). They improve seed germination (Verma & Pandey, 2022) and soil fertility while facilitating growth through the production of phytohormones such as auxins, IAA, and gibberellins. They can regulate plant secondary metabolites and facilitate the bioremediation of heavy metals and contaminants. Certain PGPR also function as biocontrol agents, protecting plants from pathogens by competing for nutrients, producing antibiotics, or inducing plant defense mechanisms. Siderophores are iron binding protein produced by bacteria (Table 2). These water soluble molecules are of low molecular weight and exist as two forms viz., extracellular and intracellular. The Fe^{3+} captured by siderophores reduced to Fe^{2+} inside the cytoplasmic membrane and subsequently transported into the cytoplasm through gating mechanism. Further Fe^{2+} is supplied to the host by the bacteria. The PGPR include members from various genera, including *Agrobacterium, Arthrobacter, Alcaligenes, Bacillus, Cellulomonas, Azotobacter, Enterobacter, Flavobacterium, Serratia, Streptomyces, Frankia, Pseudomonas, Rhizobium, Micrococcus, Thiobacillus, and Xanthomonas.*

Table 2 Role of PGPR in agriculture

Function		Examples	References
Phytostimulator (Production of phytohormone)	Indole acetic acid	*Aeromonas veronii, Enterobacter cloacae, Azotobacter sp.*	Ahmad et al. (2005)
	Cytokinin	*Pseudomonas fluorescens, Rhizobium leguminosarum*	Bhattacharyya and Jha (2012)
	Gibberellin	*Azospirillum sp.*	Boiero et al. (2007)
Biocontrol agent		P. fluorescens, Streptomyces sp., Azotobacter sp., Azospirillum sp., and Pseudomonas fluorescens	
Siderophore producing		*Pseudomonas fluorescens, P. aeruginosa, P. aeruginosa, P. koreensis*	Ghazy and El-Nahrawy (2021), Sarwar et al. (2020)
Salt-tolerant rhizobacterium		*Pseudomonas psychrotolerans*	Kubi et al. (2021)
Rhizoremediation		*Pseudomonas* sp., *Acinetobacter* sp.	Denton (2007)

4 Mechanisms of Action

Biofertilisers aids plant growth promotion by their active roles in nitrogen fixation, phosphorus solubilisation, release of potassium from its fixed pools in soil and many synergistic activities with other microorganisms present in soil. A brief description of the mechanisms has been presented in this section.

4.1 Nitrogen Fixation and Its Importance in Plant Growth

Agricultural productivity is frequently limited by the availability of soil nutrients, particularly nitrogen (N). Despite the abundance of nitrogen in the atmosphere, plants are unable to use it directly (Soumare et al., 2020). Instead, they rely on nitrogen from the soil, which is absorbed as ammonium or nitrates. Biological nitrogen fixation (BNF), which is mediated by microbial nitrogenase enzymes, is essential for turning N_2 into a form that plants can use. Diazotrophic bacteria, such as *Azospirillum, Bacillus,* and *Pseudomonas,* not only fix nitrogen but also enhance plant development via phosphate solubilization and disease control (Aasfar et al., 2024). Studies have shown that inoculating crops with these bacteria can dramatically enhance agricultural yield while decreasing dependency on synthetic fertilizers (Bhattacharyya & Jha, 2012).

Pot experiments at IITA (Ibadan, Nigeria) and field studies in Ibadan and Ikenne were conducted during the 2018/2019 and 2019/2020 cropping seasons to assess nitrogen fixation potential in 10 Bambara groundnut (BGN) genotypes. The genotypes were inoculated with *Bradyrhizobium japonicum* strains (FA3, USDA110, IRJ2180A, RACA6) and compared against nitrogen fertilizer (urea) and uninoculated controls. Results showed that genotypes TVSu-1739 and TVSu-475 had the highest nodule numbers and weight, while TVSu-1739, TVSu-1,698, TVSu-787, and others fixed the most nitrogen (62–67 kg ha^{-1}), particularly with RACA6 and USDA110 strains (Bitire et al., 2023). An experiment under hydroponic conditions examined the effects of nitrogen (N), phosphorus (P), and magnesium (Mg) on soybean symbiosis with *Bradyrhizobium japonicum.* High N with Mg (HNHMg) enhanced plant growth, nitrogen acquisition, and rhizobial colonization, while low N increased nodule dry weight. Mg proved essential for optimal root architecture and nodulation (Egamberdieva et al., 2018). *Rhizobium pusense* strain KHDEB5 from root nodules of green gram (cv. OUM-11–15) in Odisha, India, as the most efficient strain for salinity stress tolerance. After evaluation, its 16S rRNA sequence was assigned accession number KY679150 by NCBI. The strain exhibited strong nodulation and salinity tolerance, improving seedling dry weight by 31% under 500 mM NaCl compared to other strains. It tolerated salinity levels up to 1000 mM NaCl and pH 8.5. Pot culture studies demonstrated KHDEB5 significantly increased nitrogen concentration in green gram by 112–137% without stress and by 31% under salinity stress (Das et al., 2021). Studies on effect of N-fixing microbes on plant growth have been presented in Table 1.

4.2 Phosphate Solubilization and Its Role in Nutrient Availability

Phosphorus (P) is vital for plant growth, accounting for around 0.2% of a plant's dry weight. It is the second most limiting nutrient after nitrogen, although only 0.1% of the soil's phosphorus is available to plants. While phosphorus fertilizers are commonly employed to correct deficiencies, much of the applied phosphorus stays inaccessible, resulting in environmental hazards such as groundwater contamination and eutrophication. Phosphorus Solubilizing Microorganisms (PSM) provides an alternative by naturally turning insoluble phosphorus into plant-usable forms via solubilization and mineralization. Many fungi and bacteria can improve phosphorus availability in soil, especially in saline-alkali conditions. Crops that have been inoculated with PSM can absorb more phosphorus and use less artificial fertilizer, reducing environmental damage.

Addressing phosphorus (P) shortage in acidic soils is critical in the pursuit of sustainable agricultural methods, especially in areas such as Bhubaneswar, Odisha. This study sought to isolate phosphate-solubilizing bacteria (PSB) from rhizosphere soil samples taken from rice-pulse cropping systems with a pH of 5.50 or below. Five strains, including *Bacillus cereus*, *Bacillus amyloliquefaciens*, and *Burkholderia cepacia*, were discovered and tested for the ability to solubilize calcium, iron, and aluminum phosphates. The study found that calcium phosphate ($Ca_3(PO_4)_2$) had the highest recovery of soluble phosphorus, followed by iron and aluminum phosphates. Notably, *B. amyloliquefaciens* was more efficient at solubilizing both calcium and iron phosphates, but *B. cepacia* was most effective with aluminum phosphate (Pradhan et al., 2022).

A two-year field trial evaluated the effects of three phosphate-solubilizing bacteria (*Bacillus licheniformis, Pantoea dispersa*, and *Staphylococcus* sp.) on phosphorus uptake, usage efficiency (PUE), and rice grain yield at different phosphorus fertilizer rates. The PSB consortium cut the recommended phosphorus dose by over 50%, resulting in considerable yield gains of 50.58% and 35.64% in 2018 and 2019, respectively. Enhanced soil enzyme activity, such as dehydrogenase and alkaline phosphatase, were also detected, indicating better soil health and nutrient availability (Rawat et al., 2022).

4.3 Potassium Release and Its Impact on Plant Health

Potassium (K) is an essential macronutrient for plant growth, and lack reduces crop output and quality. Soil contains K in water-soluble, exchangeable, non-exchangeable, and mineral forms. Mineral forms, such as feldspar and micas, account for 90–98% of total soil potassium yet are mainly unavailable to plants due to their resistance to breakdown (Sharma et al., 2024). Factors such as K fixation in insoluble forms, leaching, and plant absorption diminish K availability. Potassium-solubilizing microorganisms (KSMs) assist plants absorb potassium by secreting organic acids, exopolysaccharides, and metal-complexing ligands (Zhang et al., 2021).

A study examined the agronomic and physiological traits of maize inoculated with *Piriformospora indica* compared to non-inoculated plants under varying salinity levels (0, 100, and 200 mM NaCl). The *Piriformospora indica* is a fungus known for enhancing plant growth and resilience against abiotic stressors, including salinity. Under salt stress inoculated maize demonstrated increased biomass, improved stomatal conductance, reduced potassium (K^+) efflux from the roots, and higher potassium content in the shoots. These findings imply that *P. indica* enhances plant performance in saline circumstances, principally by improving stomatal function and potassium transport to the shoots (Yun et al., 2018). A similar study sought to extract and identify potassium-solubilizing bacteria (KSB) and assess their participation in potassium solubilization from potassium-feldspar and uptake by *Mikania micrantha*. The *Burkholderia* has the highest solubilization capacity among the 18 reported KSB strains (1.75 mg L^{-1}). Pot culture experiments revealed that inoculated *M. micrantha* plants had higher biomass and potassium content than controls. Notably, strain GZ18 increased phosphorus levels. The study concludes that KSB in *M. micrantha*'s rhizosphere improves potassium solubilization and uptake, which contributes to the plant's invasiveness (Sun et al., 2020b).

4.4 Synergistic Effects and Microbial Interactions in the Rhizosphere of Biofertilizers in Global Food Production

The rhizosphere, or soil region impacted by plant roots, is a dynamic ecosystem in which microbial interactions play an important role in promoting plant growth and nutrient uptake. Biofertilizers, consisting of beneficial microorganisms, leverage these interactions to enhance soil fertility and promote sustainable agricultural practices. The synergistic effects of various microbial populations, including bacteria and fungi, significantly improve nutrient availability, particularly nitrogen (N), phosphorus (P), and potassium.

For example, certain *Rhizobium* and *Azospirillum* strains interact synergistically, boosting nitrogen fixation and plant growth by increasing critical nutrient accessibility (Giller, 2001; Karam et al., 2020). Furthermore, mycorrhizal fungi create mutualistic relationships with plant roots, extending their hyphal networks into the soil, increasing phosphorus uptake and enabling water absorption during drought circumstances (Liu et al., 2019; Smith & Read, 2008). Recent research has emphasized the importance of potassium-solubilizing bacteria (KSB) in this context. These microbes solubilize fixed potassium in soil, making it available to plants and preventing potassium deficit, which is critical for crop productivity (Rafique et al., 2020). These microbial interactions in the rhizosphere of biofertilizers work together to optimize nutrient cycling, increase plant resilience to biotic and abiotic challenges, and ultimately boost global food production in an environmentally sustainable manner.

5 Production and Application of Biofertilizers

5.1 *Bacteria*

5.1.1 Solid Carrier-Based Formulation

The mass production of biofertilizers involves the growth of microorganisms, the processing of carrier materials, combining carrier materials with the broth culture, and the packaging of the final product. Best carrier materials for biofertilizer production should be cost-effective, locally sourced, and quick to process; non-toxic and organic in composition (ensuring biodegradability), possessing high water retention capacity; and capable of accommodating a greater number of bacterial cells while fostering their longevity.

Carriers for inoculants can be derived from organic, inorganic, or synthetic sources and are typically classified into three main categories:

1. **Soils**: This category includes clays, peat, coal, lignite, and inorganic soil.
2. **Plant Waste Materials**: This includes charcoal, composts, farmyard manure, cellulose, soybean meal, soybean and peanut oil, wheat bran, press mud, and corn cobs
3. **Inert Materials**: This group comprises vermiculite, perlite, ground rock phosphate, bentonite, calcium sulfate, polyacrylamide gels, and alginate beads.

The carriers may originate from organic, inorganic, or synthetic sources and can be categorized into four primary classifications: Soils: clays, peat, coal, lignite, inorganic soil. Plant waste materials: charcoal, composts, farmyard manure, cellulose, soybean meal, soybean and peanut oil, wheat bran, press mud, corn cobs. Inert materials: vermiculite, perlite, ground rock phosphate, bentonite, calcium sulfate, polyacrylamide gels, alginate beads.

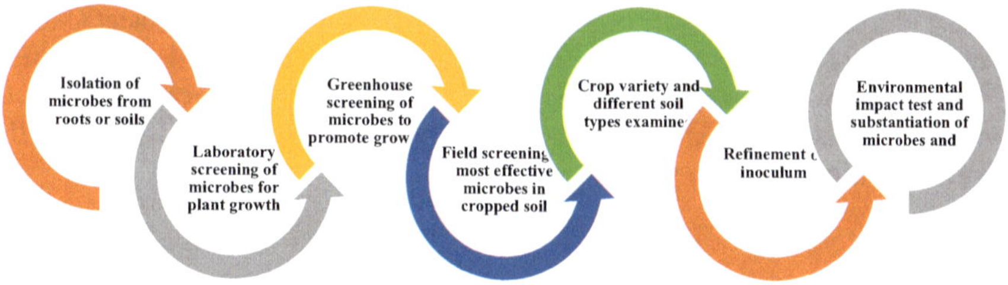

Fig. 4 Experimental process for biofertilizer

Peat is employed in the formulation of inoculants in numerous regions worldwide. Peat is constituted of partially decomposed flora that has accumulated over time. It offers a nutritive and protective growth environment for a wide range of microorganisms. Upon drying, peat is typically ground to ensure that it passes through a 250-μm sieve. In general, the pH of the peat deposits is low and must be adjusted to a range of 6.5–7.0. The peat is subsequently sterilised and a sufficient quantity of liquid inoculum is introduced. It is generally considered acceptable to have a final moisture content of 40–55% (Renjith et al., 2020).

The inoculated peat is incubated for a designated period to promote bacterial multiplication within the carrier. This stage, known as maturing or curing, is crucial as it improves the survival rate of bacteria on seeds and during storage. To enhance the uniformity of coverage on seeds, sticking agents are often added to the peat. Various polymers, including polysaccharides (such as gum arabic or carboxymethyl cellulose), polyalcohol derivatives, and caseinate compounds, are commonly used as adhesives in modern agricultural practices. Important prerequisites for these adhesives include being nontoxic to seeds or microorganisms, easily dispersible in water, and capable of providing better adhesion and survival for microorganisms on seeds.

The entire cultivation system takes place in conservatories or plant growth chambers, allowing for the controlled or semi-controlled management of temperature and humidity. The starter inoculum typically includes either a single species or a consortium of spores alongside infected root segments. Following inoculation, these root segments are dried and finely chopped to produce mixed inocula. To isolate single spores from this mixture, methods such as wet sieving and decanting are employed (Fig. 4).

5.1.2 Liquid Formulation

Solid carrier-based formulations may have the following drawbacks: a shorter expiration life, sensitivity to temperature, susceptibility to contamination, and decrease in efficacy due to low cell densities. As a result, liquid formulations have been created. Liquid formulations employ liquid materials as carriers, typically in the form of suspensions, concentrates, or emulsions, such as water, oil, or certain solvents. According to Nanaware et al. (2024), the majority of liquid inoculant formulations consist of a broth of particular organism (10–40%), a suspender ingredient (1–3%), a dispersant (1–5%), a surfactant (3–8%), and a carrier liquid (oil and/or water) (35–65%) by weight. Colloidal clays, polysaccharide gums, starch, cellulose, or synthetic polymers are employed to modify the viscosity in accordance with the particle's setting rate.

5.2 *Arbuscular Mycorrhiza*

The traditional method for producing arbuscular mycorrhizal (AM) fungi is known as the substrate-based production system. In this approach, plants and their associated symbionts are initially grown in a substrate made from soil or sand. The fungi are then propagated for mass multiplication using either a single species or a consortium of identified AM fungal species, which can be cultivated in clay or plastic pots, medium-sized bags, or larger raised or grounded beds after the initial production of AM fungal inoculum (Akhtar & Abdullah, 2014; Gaur & Adholeya, 2002). The entire system is grown in conservatories or plant growth chambers, where temperature and humidity can be controlled or semi-controlled. The starter inoculum usually consists of a single species or a consortium of spores along with infected root segments. After inoculation, the root segments are dried and finely chopped to create mixed inocula. Techniques such as wet sieving and decanting are used to isolate single spores from the mixture.

5.3 *Quality Control and Standardization of Biofertilizers*

Quality assessment guidelines focus on the density of available microorganisms, as well as their viability and preservation in the guaranteed microbial products. Establishing control plots that lack available microorganisms while maintaining identical compositions to the final microbial products is essential. It is also important for the biofertilizer to demonstrate significant effects on the quality management of the final products. The evaluation of the biofertilizer should be based on its most prominent effects, with its guaranteed activities reflecting these impacts. Differentiating between available microorganisms and additional compositions regarding the effects promised by suppliers is a critical requirement. If the outcomes of the two experimental plots are identical or statistically indistinguishable, the product is essentially considered organic matter (Agarwal et al., 2021).

This implies that the effects of microbial products must be derived from the specified microorganisms, and the objective of the matter should be explicitly stated in a prescription. It is imperative to assess the functions in the specific context of the applicant's utilisation.

5.4 *Application of Biofertilizer*

5.4.1 Seed Treatment

Seed treatment is the most common, cost-effective, and efficient method for applying bacterial inoculants. The seeds are coated evenly with slurry made from the inoculant mixed with 200 mL of rice kanji or a 1% jeggey solution, and then they are shade-dried before planting, ideally within 24 h. The effectiveness of this method can be further enhanced by applying a lime coating after the rhizobia inoculant treatment.

5.4.2 Seedling Root Dipping

Soaking seedling roots in a biofertilizer solution is a common practice for plantation crops such as cereals, fruits, trees, sugarcane, cotton, grapes, bananas, and tobacco. This involves mixing 40 L of water with five packets (1.0 kg) of the inoculants and preparing the solution for one hectare. After combining the ingredients, the seedlings' roots should be dipped into the mixture for 5–10 min before transplanting (Mohamed et al., 2022).

5.4.3 Soil Application

This method involves the direct application of biofertilizer to the soil, either in isolation or in conjunction with FYM or decomposition. The *Rhizobium* (for leguminous plants or trees) and *Azotobacter* (for tea, coffee, rubber, coconuts, fruits/agroforestry plants) are two examples of biofertilizers that are applied to the soil. In rice fields, BGA and azolla are used, while mycorrhiza is used in nursery beds.

6 Benefit of Biofertilizer

Biofertilizers are products made from viable strains of microorganisms that promote plant growth while being safe for both human health and the environment. It has quite significant potential in reducing chemical fertilizer consumption, contributing to environmental mitigation efforts worldwide. Studies from various countries, including India, Iran, Egypt, Thailand, Malaysia, Zimbabwe, and China, have reported substantial reductions in chemical fertilizer use when biofertilizers were implemented. For instance, in India, the use of chemical fertilizers was reduced by 25% in maize and brinjal. In Iran, a 50% reduction was observed in basil cultivation. Similar reductions were reported in Egypt for maize (50%), Thailand and Malaysia for rice (50%), Zimbabwe for maize (25–50%), and China for wheat (50%). These

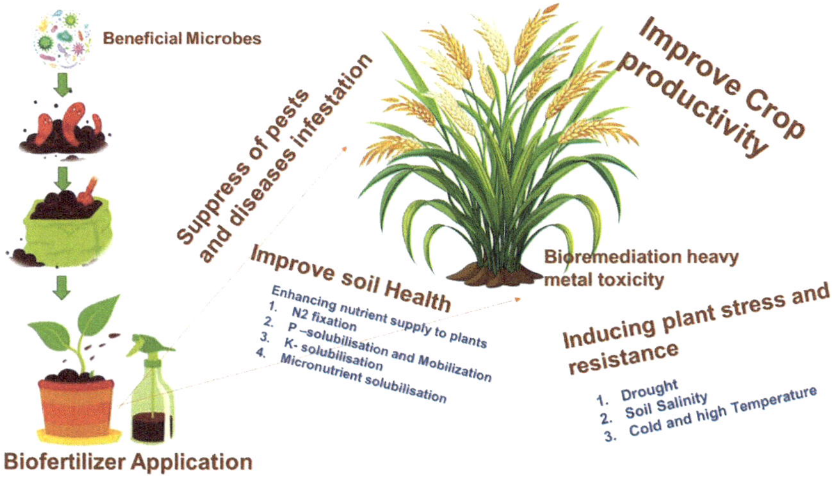

Fig. 5 Schematic representation of Benefit of Biofertilizer on soil plant environment continuum

findings align with the research conducted by Mahmud et al. (2021), which highlighted the effectiveness of biofertilizers in mitigating chemical fertilizer dependency across different agricultural systems. The benefits of biofertilizers are quite significant for food security because they (1) enhance soil productivity by regulating nutrient supply and altering microbiota and soil structure; (2) aid in inducing plant stress and resistance; (3) aid in the prevention of pests and diseases; (4) play a key role in phytoremediation for heavy metal toxicity; and (5) overall increase crop productivity in the context of INM on the least fertile soil. The beneficial effect of biofertilizer on soil and plant continuum was depicted in Fig. 5 and Table 3.

6.1 *Biofertilizer for Soil Productivity*

The use of biofertilizers enhances the conversion of unavailable nutrients in the soil into forms that plants can absorb. Through quorum sensing molecules, biofertilizers can concentrate microbial populations around plant roots. These microbial communities boost soil fertility and nutrient cycling, supporting plant growth. While plants can absorb soluble nutrients like NH_4^+, K^+, and PO_4^{3-}, they cannot absorb insoluble forms of phosphorus or potassium, or nitrogen directly (Shin et al., 2016). Nitrogen-fixing bacteria, using the nitrogenase enzyme, convert nitrogen into plant-usable forms such as NH_4^+ or NO_3^- (Rees et al., 2005). Biofertilizers like *Rhizobium* fix 40–200 kg of nitrogen per hectare annually in legume crops, while other nitrogen-fixing bacteria such as *Azotobacter*, *Azospirillum*, and *Azolla* contribute 10–90 kg per hectare for crops like rice. Most phosphorus in the soil, both inorganic (phosphate rock and insoluble phosphate) and organic (phosphoinositide, phospholipid,

Table 3 Beneficial effect of Biofertilizer for soil and plant growth

Type of biofertilizer	Example	Function	Contribute to soil and plant growth	References
Nitrogen fixing bacteria	*Rhizobium* sp., *Azospirillum* sp., *Bacillus* sp. and blue green algae	Fixing atmospheric N for soil fertility and plant growth	Fixes 40–200 kg N/ha, meets up to 80–90% of N crop requirement Maintains soil fertility. Increases the yield by 10–30%	Mahmud et al., (2021)
Phosphate solubilizing biofertilizer (PSB)	*Aspergillus* sp., *Bacillus* sp., and *Pseudomonas* sp	Solubilizing insoluble form of phosphate in to soluble form for plant use	Solubilizes 50–60% of the fixed P in the soil and 10–20% increase in yield of almost all crops	Alori et al. (2017)
Phosphate mobilizing biofertilizrers (PMB)	Mycorrhiza	Mobilizing the insoluble P in the soil to plant root		Ahemad (2015)
Potassium solubilizing biofertilizer	*Bacillus* sp. and *Aspergillus niger*	Helps in decomposition of silicates compound for K release		Mahmud et al., (2021)
Zinc solubilizer biofertilizer (ZSB)	*Serratia* sp.	These microbes primarily use the acidification mechanism for solubilization of Zn		Kour et al. (2019)

(continued)

Table 3 (continued)

Type of biofertilizer	Example	Function	Contribute to soil and plant growth	References
Sulfur oxidizing biofertilizer (SOB)	*Thiobacillus* sp.	Work primarily to oxidized sulfur in its elemental form to sulfates for plant uptake		Mahmud et al., (2021)
Plant growth promoting biofertilizer (PGPB)	*Pseudomonas* sp., *Agrobacterium*, *Achromobacter*, *Xanthomonasetc*	Hormones and antimetabolites are produced by PGPR for plant metabolic activity	Increases crop yield with the range from 7 to 33%. Promotes plant growth through increased phytohormones and water& minerals uptake	Jatav et al. (2021)

nucleic acid), is inaccessible to plants (Turner & Blackwell, 2013). Phosphorus-solubilizing microbes, including *Bacillus*, *Pseudomonas*, and various fungi, produce enzymes such as phytase and phosphatase to convert organic phosphorus into plant-absorbable forms (Singh et al., 2020). These microbes can also release organic acids that lower soil pH, making more phosphorus available by dissolving phosphate rock and insoluble phosphate (Yin et al., 2021). Endomycorrhizal fungi, such as *Glomus*, *Acaulospora*, and *Gigaspora,* are crucial for mobilizing phosphorus in the soil, helping plants explore more soil volume with fungal hyphae. Similarly, most soil potassium is stored in minerals like potassium feldspar and muscovite, which are not readily accessible to plants. Potassium-solubilizing bacteria produce acids and other compounds that help break down these minerals, releasing potassium for plant uptake (Muthuraja & Muthukuma, 2021). Sulfur is another essential nutrient, absorbed by plants primarily as SO_4^{2-}. Sulfur-oxidizing microbes like *Bacillus*, *Pseudomonas*, and *Alcaligenes* convert organic sulfur into inorganic forms usable by plants. These microbes have been suggested for biofertilizer use in alkaline soils for crops like onions and grapes. Zinc, a key micronutrient, can be made accessible by zinc-solubilizing microorganisms such as *Pseudomonas*, *Rhizobium*, and *Bacillus aryabhattai* (Mohapatra et al., 2020). The activity of these microbes depends on soil pH and cation exchange capacity, and they have been shown to improve crop zinc uptake (Mohapatra et al., 2020). Lastly, biofertilizers also improve soil moisture retention by promoting microporosity, enhancing soil physical structure. Some species like *Bacillus thuringiensis* and *Bacillus subtilis* play additional roles in organic matter mineralization and pH regulation in saline-alkaline soils, further benefiting plant growth (Zhao et al., 2022).

6.2 *Biofertilizer for Inducing Plant Stress and Resistance*

Reducing the raising threat of abiotic stress requires investigating the potential of soil bacteria and their interactions (Jatav et al., 2021). Plants that experience drought respond by producing more osmolyte enzymes. Furthermore, plant growth-promoting rhizobacteria (PGPR) increase the synthesis of phytohormones or compounds like abscisic acid, gibberellins, cytokinins, and indole-3-acetic acid (IAA) (Jatav et al., 2021). Improvements in root length, general growth, quantity of roots, surface area, and the formation of lateral roots and root hairs all help plants absorb water more effectively (Raza et al., 2019). Chakraborty et al. (2018) and Lamaoui et al. (2018) shows that a variety of helpful microbes, such as thermophiles and cryophiles, can aid in stabilizing plants during periods of severe temperature. Khan et al. (2020) have demonstrated that the thermotolerant *Bacillus cereus* SA1 improves heat tolerance in soybean plants, suggesting its potential to reduce heat damage. In a similar vein, Mukhtar et al. (2020) looked at how *Bacillus cereus* may help tomato plants become more resilient to heat stress and came to the conclusion that it is a good option for boosting crop growth in hot weather. Zubair et al. (2019) studied how psychrophilic PGPR bacteria from the *Bacillus* genus can help wheat plants deal with cold stress. They found that these cold-tolerant strains control proline build up, lipid peroxidation, and abscisic acid to help plants survive cold stress.

Plants go through oxidative, osmotic, and ionic stress when there are high levels of salts (like NaCl, Na_2SO_4, Na_2CO_3, or magnesium salts) in dry or semi-arid soils (Saxena et al., 2017). This is because the salts mess up the plants' physiological processes by producing oxygen radicals, oxidative stress limits the development and metabolism of plant cells (Gill & Tuteja, 2010). Ionic stress throws off the plant's overall ionic equilibrium, whereas osmotic stress reduces water usage efficiency and produces drought-like circumstances (Saxena et al., 2017). Plants and microbes frequently exploit PGPR through symbiotic interactions or direct microbial activity to enhance plant development under high salinity (Upadhyay et al., 2019). Scientists Ansari et al. (2013) discovered that endophytic fungi like *Piriformospora indica* and arbuscular mycorrhizal fungi (AMF) worked well in plant roots to protect the host from ion toxicity during salt stress. A number of bacterial strains, including *Bacillus safensis*, *Bacillus pumilus*, *Kocuria rosea*, *Enterobacter aerogenes*, and *Aeromonas veronii*, positively impacted maize growth under saline conditions. El Semary et al. (2020) used microbial inoculants, either singly or in combination, to reduce salinity stress in plants. These inoculants included *Enterobacter cloacae* and *Cyanothece* sp., a cyanobacterial strain. They discovered that using these bacteria as biofertilizers at high salinity levels significantly reduced the negative impacts of salt. In a separate investigation, Phour and Sindhu (2020) found that *Pseudomonas azotoformans* and *Pseudomonas argentinensis* improved productivity and showed promise as biofertilizers in saline environments by reducing salt stress in Indian mustard (*Brassica juncea L.*).

6.3 Biofertilizer for Prevention of Pests and Diseases

Modern pest control relies heavily on the use of pesticides, but abuse of these chemicals can have detrimental effects on the environment and human health. Consequently, beneficial microbes are becoming more widely acknowledged as an economical and environmentally responsible substitute for pest management. According to recent studies, these microbes may be able to lessen the negative effects of pesticides. In addition to increasing crop productivity, several microbial strains employed in biofertilizers shield plants against harmful illnesses. They do this through direct pathogen growth inhibition or indirect nutritional competition. Microorganisms including *Azotobacter*, *Bacillus*, *Enterobacter*, *Paenibacillus*, and P*seudomonas* have been shown in recent research to be beneficial in lowering pesticide toxicity (Naseer et al., 2019). Moreover, cyanogenic chemicals that protect against herbivores are produced in a symbiotic connection between nitrogen-fixing microbes and leguminous plants (Naseer et al., 2019). Attacks by phytopathogens frequently reduce plant yield, and helpful bacteria that generate antimicrobial compounds such as β-glucanases and chitinases are essential in lowering plant disease incidence (Suresh & Abraham, 2019). With the application of biofertilizers, *Pseudomonas fluorescens* and

Sinorhizobium, for example, create these enzymes that efficiently control diseases like Fusarium wilt and potato soft rot brought on by *Fusarium udum* and *Erwinia carotovora.* Applying *Glomus intraradices* increased potato output while reducing potato viral infections in plant tissues, according to a recent study by Deja-Sikora et al. (2020). Similarly, Beris and Vassilakos (2020) discovered that the tomato yellow leaf curl Sardinia virus (TYLCSV) was lessened when tomatoes were injected with *Glomus mosseae.* Additionally, certain biofertilizers generate siderophores, which restrict the amount of iron available near plant roots. This decreases the amount of nutrients available to pathogenic microorganisms, impeding their growth (Igiehon & Babalola, 2018). It has been demonstrated that *Pseudomonas* and *Bacillus* species that generate *siderophores* inhibit *Fusarium* wilt in maize and potatoes (Devi et al., 2018). Yasmin et al. (2017) also discovered that *Pseudomonas aeruginosa* was successful in preventing bacterial blight in rice that was brought on by *Xanthomonas oryzae* and *Rhizoctonia solani*, two significant dangers to rice farming in West Africa.

6.4 Biofertiliser for Mitigation of Heavy Metal Toxicity

Human-caused soil contamination has had a negative impact on environmental sustainability, public health, and the world economy ever since the industrial revolution began in the eighteenth century. Because they are not biodegradable, heavy metals including lead, copper, nickel, zinc, mercury, cadmium, chromium, arsenic, and cobalt build up in soils over time and can be hazardous to human and animal health as well as food production (Chibuike & Obiora, 2014). These contaminants have a negative impact on agricultural productivity and cause large financial losses (Chibuike & Obiora, 2014). In order to assist clean up contaminated soils, certain plants naturally break down, stabilize, or eliminate toxins from their surroundings. This process is known as phytoremediation. Research by Glick (2003) and White (2001) shows that this can be addressed by adding bacteria to soils to enhance phytoremediation; this technique is called “microbe-assisted phytoremediation.” In bioremediation, some types of microorganisms play a crucial role. These include phosphorus-solubilizing bacteria, arbuscular mycorrhizal fungus (AMF), mycorrhizal helper bacteria (MHB), and plant growth-promoting rhizobacteria (PGPR) (Ahemad, 2015). Previous research has identified bacteria from the genera *Pseudomonas* and *Serratia* that are tolerant of heavy metals. These bacteria produce ammonia, hydrogen cyanide (HCN), siderophores, indole-3-acetic acid (IAA), and can solubilize phosphate, among other traits that promote plant growth (Bhojiya et al., 2021). To resist environmental contaminants, these plants use tactics include secreting proteins, antibiotics, organic acids, and other chemical substance. Microorganisms like bacteria and fungus in the rhizosphere can operate as bioinoculants to encourage plant development and sequester heavy metals in polluted soils, according to studies by Ahmad et al. (2018).

6.5 Biofertilizer for Crop Growth and Productivity

Biofertilizers enhance plant growth and yield by controlling hormones and generating a variety of compounds that improve productivity, including lipids, polysaccharides, siderophores, enzymes, small peptides, antioxidants, and volatile organic compounds (VOC) (Wang et al., 2022). Plant hormones such as indole-3-acetic acid (IAA), cytokinin (CTK), and gibberellin (GA) can be produced or broken down by microorganisms in the rhizosphere, which can regulate hormone levels, encourage cell division and elongation, and advance plant development (Tsukanova et al., 2017). Inoculating plants with *Rhizobium leguminosarum, Rhizobium sp.* IRBG 74, and *Bradyrhizobium* sp. IRBG 271 resulted in a considerable increase in plant biomass, yield, and chlorophyll content, with the IRBG strains producing a yield that was 14% greater than that of uninoculated plants, according to recent research. In inoculated plants, certain strains of rhizobia also increased yield, stomatal conductance, surface area, photosynthetic rate, and water absorption. Furthermore, it has been demonstrated that inoculating plants with a mixture of bacteria, such as *Pseudomonas, Bacillus lentus,* and *Azospirillum brasilense,* increases the amount of chlorophyll and the activity of antioxidant enzymes under stress. For instance, *Azotobacter* may raise yields by 10–15% in crops like mustard, sunflower, grape, sugarcane, banana, and coconut, while *Rhizobium* can enhance yields by 10–30% in leguminous crops like cowpea, gram, pea, groundnut, and soybean. It has been demonstrated that phosphate-solubilizing microorganisms (PSM) may raise agricultural yields by 10–20% in crops such as vegetables, potatoes, rice, and wheat, while plant growth-promoting rhizobacteria (PGPR) can raise yields by 7–33% (Mahmud et al., 2021). Applying biofertilizer to spinach increased its growth, chlorophyll levels, antioxidant activity, yield, and phenolic compounds (with phenolic content being 58% greater than in uninoculated plants), according to Khalid et al. (2017). Comparably, *Azotobacter chroococcum* and *Piriformospora indica* infected lettuce showed improvements in growth, yield, phenolic compounds, anthocyanins, and carotenoids, according to Arora et al. (2018). Arbuscular mycorrhizal fungi have also been shown to produce antioxidants, according to Kapoor and Singh (2017). Hassen et al. (2016) found that inoculating soybeans with the nitrogen-fixing fungi *Rhizobium* and *Bradyrhizobium* increased yield by 80%. The application of biofertilizer has also been associated with increased food production. This was demonstrated in a study by Dicko et al. (2018), where biofertilizer derived from *Actinomycetes* that promote plant growth (H7, O19, and AHB12) increased maize yield. When O19 and AHB12 were combined, the yield increased to 311.5 g for 1000 seeds, compared to 178.28 g in the control group. It was recently shown that two forage rice genotypes have increased rice productivity when treated with biofertilizer made from *Bacillus pumilus* strain TUAT-1. Furthermore, Fathi (2017) found that biofertilizers based on phosphate-solubilizing bacteria increased maize production and growth in comparison to uninoculated controls.

7 Challenges of Biofertiliser

The use of biofertilizers improves agricultural production sustainability overall with little negative impact on the environment, albeit fresh inoculants do have an identifiable odour. Nonetheless, there are a few restrictions on the use of biofertilizers in the agriculture industry. In this developing industry, a number of obstacles have been found, ranging from production to farmers' actual field application. These technological constraints include things like the lack of efficient strains, insufficient knowledge, unsuitable carrier materials, limited skills, unfavourable storage conditions, contamination hazards, and a lack of rules and safety requirements. Due to different inoculation methods and the longer-term effects of biofertilizers compared to chemical fertilizers, there is also a shortage of high-quality carrier materials, qualified technical personnel in production units, and low levels of awareness and adoption among farmers (Odoh et al., 2020). The short shelf life of biofertilizers is a major disadvantage that makes long-term storage difficult. However, using strains that are drought-tolerant, thermotolerant, or genetically engineered, together with liquid biofertilizer formulations, can prolong the shelf life (Brar et al., 2012). Another problem is the absence of native microbial strains particular to a certain area. The significance of incorporating organic matter, such as vermicompost and farmyard manure, into soil amendments cannot be overstated (Mamnabi et al., 2020). This will lessen the need for chemical fertilizers while preserving soil quality and the effectiveness of plant growth-promoting rhizobacteria (PGPR) without having a negative impact on crop yield and growth. Biofertilizers are frequently tailored to specific soil types as well as crops. Agroclimatic variables might cause microbial strains from one area to be less successful as biofertilizers in another because of differences in their capacity to compete with and survive native bacteria. In addition, there are obstacles related to infrastructure, such as restricted availability of application tools like sprayers, power outages, and inadequate road systems for delivering biofertilizers to the fields (Odoh et al., 2020). The manufacture, marketing, and application of biofertilizers in agricultural crop production are all negatively impacted by these constraints taken together (Kumar et al., 2017).

8 Future Directions and Research Opportunities

Key research needs for developing effective and sustainable bio-fertilizers include investigating microbial diversity and functionality to identify beneficial strains, studying soil–plant interactions to understand how bio-fertilizers enhance nutrient uptake and soil health, and optimizing formulations and delivery methods for various crops. Long-term field trials are essential to assess effectiveness and sustainability, while exploring synergies with conventional practices can help evaluate economic viability. Additionally, research should focus on enhancing climate resilience, establishing safety guidelines, and fostering education and outreach for farmers.

Conducting economic analyses and adapting strategies to local conditions will further promote the successful integration of bio-fertilizers into sustainable agriculture.

Biofertilizers have a bright future ahead of them as agriculture moves more and more towards sustainable methods. The need to lessen the negative effects of chemical fertilisers on the environment grows along with the world's demand for food. An environmentally beneficial substitute, bio-fertilizers increase crop resilience, improves nutrient availability, and foster healthy soil.

The creation of bio-fertilizers will be greatly aided by developments in microbial biotechnology. The goal of the research will be to isolate and modify particular microbial strains that have the ability to fix nitrogen, solubilise phosphorus, and create compounds that stimulate plant development. More effective biofertilizers that are suited to different crop requirements and environmental circumstances will be produced as a consequence of this focused approach.

Furthermore, the optimal application of biofertilizers will be achieved by their integration with precision agriculture techniques. Farmers will be able to apply bio-fertilizers precisely when and in what amounts thanks to data-driven insights, increasing their efficacy and reducing waste. Additionally, this will make it possible to create formulas that are specifically tailored to the needs of crops and soil health evaluations.

The possibility of using biofertilizers to slow down climate change is another crucial factor. More sustainable farming methods can be supported by biofertilizers, which increase soil organic matter and encourage carbon sequestration. Their contribution to improving soil health will result in better water retention and decreased erosion, which will enhance agricultural production even more in the face of climate change.

It will take outreach and education to get farmers to use biofertilizers. Training courses will be required to illustrate their advantages and appropriate usage techniques. Furthermore, it is anticipated that demand for bio-fertilizers would rise as customers seek out more sustainably produced and organic food, strengthening the agricultural environment.

In the end, bio-fertilizers have a promising future as a sustainable food production method that supports environmental stewardship and adaptability to global concerns.

9 Conclusion

Harnessing the potential of biofertilizers offers a transformative approach to achieving sustainable agriculture. As an eco-friendly alternative to chemical fertilizers, biofertilizers enhance soil fertility, promote plant growth, and contribute to environmental conservation. Their ability to improve nutrient cycling, boost soil health, and reduce reliance on synthetic inputs makes them essential in addressing the challenges of modern agriculture, including soil degradation and climate change. However, the widespread adoption of biofertilizers requires overcoming barriers such as limited awareness, inconsistent field performance, and storage challenges.

Through increased research, education, and supportive policies, these challenges can be mitigated, paving the way for biofertilizers to become a cornerstone of sustainable farming practices. By integrating biofertilizers into agricultural systems, we can promote long-term soil health, improve crop productivity, and contribute to a more resilient and environmentally friendly food production system. Ultimately, biofertilizers represent a critical tool in the global effort to create a sustainable future for agriculture and food security.

References

Aasfar, A., Meftah Kadmiri, I., Azaroual, S. E., Lemriss, S., Mernissi, N. E., Bargaz, A., Zeroual, Y., & Hilali, A. (2024). Agronomic advantage of bacterial biological nitrogen fixation on wheat plant growth under contrasting nitrogen and phosphorus regimes. *Frontiers in Plant Science, 15*, 1388775. https://doi.org/10.3389/fpls.2024.1388775

Agarwal, S., Kumari, S., & Khan, S. (2021). Quality control of biofertilizers. *Biofertilizers: Study and Impact,* 413–428. https://doi.org/10.1002/9781119724995.ch14

Ahemad, M. (2015). Phosphate solubilising bacteria assisted phytoremediation of metalliferous soil: A review. *Biotech, 3*(5), 111–121. https://doi.org/10.1007/s13205-014-0206-0

Ahmad, F., Ahmad, I., & Khan, M. S. (2005). Indole acetic acid production by the indigenous isolates of Azotobacter and fluorescent Pseudomonas in the presence and absence of tryptophan. *Turkish Journal of Biology, 29*, 29–34.

Ahmad, M., Pataczek, L., Hilger, T. H., Zahir, Z. A., Hussain, A., Rasche, F., Schafleitner, R., & Solberg, S. Ø. (2018). Perspectives of microbial inoculation for sustainable development and environmental management. *Frontiers in Microbiology, 9*, 2992. https://doi.org/10.3389/fmicb.2018.02992

Akhtar, M. S., & Abdullah, S. N. A. (2014). Mass production techniques of arbuscular mycorrhizal fungi: Major advantages and disadvantages: A review. *Biosciences, Biotechnology Research Asia, 11*(3), 1199.

Alori, E. T., Glick, B. R., & Babalola, O. O. (2017). Microbial phosphorus solubilization and its potential for use in sustainable agriculture. *Frontiers in Microbiology, 8*, 971. https://doi.org/10.3389/fmicb.2017.00971

Ansari, M. W., Trivedi, D. K., & Sahoo, R. K. (2013). A critical review on fungi mediated plant responses with special emphasis to *Piriformospora indica* on improved production and protection of crops. *Plant Physiology and Biochemistry, 70*, 403–410. https://doi.org/10.1016/j.plaphy.2013.06.005

Arora, M., Saxena, P., Abdin, M., & Varma, A. (2018). Interaction between *Piriformospora indica* and *Azotobacter chroococcum* governs better plant physiological and biochemical parameters in *Artemisia annua* L. plants grown under in vitro conditions. *Symbiosis, 75*, 103–112.

Azizoglu, U. (2019). Bacillus thuringiensis as a Biofertilizer and Biostimulator: A Mini-Review of the Little-Known Plant Growth-Promoting Properties of Bt. *Current Microbiology*. https://doi.org/10.1007/s00284-019-01705-9

Bargaz, A., Elhaissoufi, W., Khourchi, S., Benmrid, B., Borden, K. A., & Rchiad, Z. (2021). Benefits of phosphate solubilizing bacteria on belowground crop performance for improved crop acquisition of phosphorus. *Microbiological Research, 252*, Article 126842. https://doi.org/10.1016/j.micres.2021.126842

Barin, M., Asadzadeh, F., Hosseini, M., Hammer, E. C., Vetukuri, R. R., & Vahedi, R. (2022). Optimization of biofertilizer formulation for phosphorus solubilizing by Pseudomonas fluorescens Ur21 via response surface methodology. *Processes, 10*(4), 650. https://doi.org/10.3390/pr10040650

Barnawal, D., Bharti, N., Pandey, S. S., Pandey, A., Chanotiya, C. S., & Kalra, A. (2017). Plant growth-promoting rhizobacteria enhance wheat salt and drought stress tolerance by altering endogenous phytohormone levels and TaCTR1/TaDREB2 expression. *Physiologia Plantarum, 161*(4), 502–514. https://doi.org/10.1111/ppl.12614

Basak, B. B., Maity, A., Ray, P., Biswas, D. R., & Roy, S. (2022) Potassium supply in agriculture through biological potassium fertilizer: A promising and sustainable option for developing countries. *Archives of Agronomy and Soil Science, 68*(1), 101–114.

Batool, S., Asghar, H. N., Shehzad, M. A., Yasin, S., Sohaib, M., Nawaz, F., Akhtar, G., Mubeen, K., Zahir, Z. A., & Uzair, M. (2021). Zinc-solubilizing bacteria-mediated enzymatic and physiological regulations confer zinc biofortification in chickpea (*Cicer arietinum L.*). *Journal of Soil Science and Plant Nutrition., 21*, 2456–2471. https://doi.org/10.1007/s42729-021-00537-6

Becerra, M. J., Pimentel, M. A., De Souza, E. B., & Tovar, G. I. (2020). Geospatiality of climate change perceptions on coastal regions: A systematic bibliometric analysis. *Geography and Sustainability, 1*(3), 209–219. https://doi.org/10.1016/j.geosus.2020.09.002

Beneduzi, A., Ambrosini, A., & Passaglia, L. M. (2012). Plant growth-promoting rhizobacteria (PGPR): Their potential as antagonists and biocontrol agents. *Genetics and Molecular Biology, 35*, 1044–1051.

Beris, D., & Vassilakos, N. (2020). Plant beneficial microbes: Do they have a role as antiviral agents in agriculture? In *Molecular aspects of plant beneficial microbes in agriculture* (pp. 19–33). Elsevier.

Bharti, N., & Suryavanshi, M. (2021). Quality control and regulations of biofertilizers: Current scenario and future prospects. In A. Rakshit, V.S. Meena, M. Parihar, H.B. Singh, & A.K. Singh (Eds.), *Biofertilizers* (pp 133–141). Woodhead Publishing. https://doi.org/10.1016/B978-0-12-821667-5.00018-X.

Bhattacharyya, P. N., & Jha, D. K. (2012). Plant growth-promoting rhizobacteria (PGPR): Emergence in agriculture. *World Journal of Microbiology and Biotechnology, 28*, 1327–1350. https://doi.org/10.1007/s11274-011-0979-9

Bhojiya, A. A., Joshi, H., Upadhyay, S. K., Srivastava, A. K., Pathak, V. V., Pandey, V. C., & Jain, D. (2021). Screening and optimization of zinc removal potential in *Pseudomonas aeruginosa*-HMR1 and its plant growth-promoting attributes. *Bulletin of Environmental Contamination and Toxicology, 108*, 468–477. https://doi.org/10.1007/s00128-021-03232-5

Bitire, T. D., Abberton, M., Tella, E. O., Edemodu, A., Oyatomi, O., & andBabalola, O.O. (2023). Impact of nitrogen-fixation bacteria on nitrogen-fixation efficiency of Bambara groundnut (*Vigna subterranea* (L.) Verdc) genotypes. *Frontiers in Microbiology, 14*, 1187250. https://doi.org/10.3389/fmicb.2023.1187250

Boiero, L., Perrig, D., Masciarelli, O., Penna, C., Cassan, F., & Luna, V. (2007). Phytohormone production by three strains of *Bradyrhizobium japonicum* and possible physiological and technological implications. *Applied Microbiology and Biotechnology, 74*, 874–880. https://doi.org/10.1007/s00253-006-0731-9

Bomfim, C. A., Coelho, L. G. F., do Vale, H. M. M., Mendes, I. D., Megias, M., Ollero, F. J., & Dos Reis, F. B. (2021). Brief history of biofertilizers in Brazil: From conventional approaches to new biotechnological solutions. *Brazilian Journal of Microbiology, 52*(4): 2215–2232. https://doi.org/10.1007/s42770-021-00618-9

Brahmaprakash, G., & Sahu, P. K. (2012). Biofertilizers for sustainability. *Journal of the Indian Institute of Science, 92*, 37–62.

Brar, S. K., Sarma, S. J., & Chaabouni, E. (2012). Shelf-life of biofertilizers: An accord between formulations and genetics. *Journal of Biofertilizers & Biopesticides, 3*, Article e109. https://doi.org/10.4172/2155-6202.1000e109

Chakraborty, A., Ellefson, E., Li, C., Gittins, D., Brooks, J. M., Bernard, B. B., & Hubert, C. R. J. (2018). Thermophilic endospores associated with migrated thermogenic hydrocarbons in deep Gulf of Mexico marine sediments. *ISME Journal, 12*, 1895–1906. https://doi.org/10.1038/s41396-018-0108-y

Chaudhary, P., Singh, S., Chaudhary, A., Sharma, A., & Kumar, G. (2022). Overview of biofertilizers in crop production and stress management for sustainable agriculture. *Frontiers in Plant Science, 13*, Article 930340. https://doi.org/10.3389/fpls.2022.930340

Chibuike, G. U., & Obiora, S. C. (2014). Heavy metal polluted soils: Effect on plants and bioremediation methods. *Applied and Environmental Soil Science, 2014*, 752708. https://doi.org/10.1155/2014/752708

Das, R., Pradhan, M., Sahoo, R.K., Mohanty, S., & Kumar, M. (2021). Isolation, identification and role of novel endosymbiotic bacterium *Rhizobium pusence* in root nodule of green gram cv. OUM-11-15 (*Vigna radiata* L.) under salinity stress. *Legume Research, 44*(12), 1512–1520. https://doi.org/10.18805/LR-4239

Dastogeera, K. M. G., Tumpa, F. H., Sultana, A., Mst Arjina Akter, M. A., & Chakraborty, A. (2020). *Current Plant Biology, 23*, 1001. https://doi.org/10.1016/j.cpb.2020.100161

Deja-Sikora, E., Kowalczyk, A., Trejgell, A., Szmidt-Jaworska, A., Baum, C., Mercy, L., & Hrynkiewicz, K. (2020). Arbuscular mycorrhiza changes the impact of Potato virus Y on growth and stress tolerance of *Solanum tuberosum* L. in vitro. *Frontiers in Microbiology, 10*, 2971.

Denton, B. (2007). Advances in phytoremediation of heavy metals using plant growth promoting bacteria and fungi. *MMG 445 Basic Biotechnology, 3*, 1–5.

Devi, R., Thakur, R., & Gupta, M. (2018). Isolation and molecular characterization of bacterial strains with antifungal activity from termite mound soil. *International Journal of Current Microbiology and Applied Sciences, 7*, 1–7.

Dicko, A. H., Babana, A. H., Kassogué, A., Fané, R., Nantoumé, D., Ouattara, D., Maiga, K., & Dao, S. A. (2018). Malian native plant growth promoting *Actinomycetes* based biofertilizer improves maize growth and yield. *Symbiosis, 75*, 267–275.

Du, H., Li, N., Brown, M. A., Peng, Y., & Shuai, Y. (2014). A bibliographic analysis of recent solar energy literatures: The expansion and evolution of a research field. *Renewable Energy, 66*, 696–706. https://doi.org/10.1016/j.renene.2014.01.018

Egamberdieva, D., Jabborova, D., Wirth, S. J., Alam, P., Alyemeni, M. N., & Ahmad, P. (2018). Interactive effects of nutrients and *Bradyrhizobium japonicum* on the growth and root architecture of soybean (*Glycine max* L.). *Frontiers in Microbiology, 9*, 1000. https://doi.org/10.3389/fmicb.2018.01000

El Semary, N. A. H., Alouane, M. H. H., Nasr, O., Aldayel, M. F., Alhaweti, F. H., & Ahmed, F. (2020). Salinity stress mitigation using encapsulated biofertilizers for sustainable agriculture. *Sustainability, 12*(21), 9218. https://doi.org/10.3390/su12219218

Fathi, A. (2017). Effect of phosphate solubilization microorganisms and plant growth promoting rhizobacteria on yield and yield components of corn. *Scientia Agricola, 18*, 66–69.

Gaur, A., & Adholeya, A. (2002). Arbuscular-mycorrhizal inoculation of five tropical fodder crops and inoculum production in marginal soil amended with organic matter. *Biology and Fertility of Soils, 35*, 214–218.

Gezgin, Y., Maral Gül, D., Sözer Şenşatar, S., Kara, C. U., Sargın, S., Sukan, F. V., & Eltem, R. (2020). Evaluation of *Trichoderma atroviride* and *Trichoderma citrinoviride* growth profiles and their potentials as biocontrol agent and biofertilizer. *Turkish Journal of Biochemistr., 45*(2), 163–175. https://doi.org/10.1515/tjb-2018-0378

Ghazy, N., & El Nahrawy, S. (2021). Siderophore production by *Bacillus subtilis* MF497446 and *Pseudomonas koreensis* MG209738 and their efficacy in controlling *Cephalosporium maydis* in maize plant. *Archives of Microbiology, 203*, 1195–1209. https://doi.org/10.1007/s00203-020-02113-5

Gill, S. S., & Tuteja, N. (2010). Reactive oxygen species and antioxidant machinery in abiotic stress tolerance in crop plants. *Plant Physiology and Biochemistry, 48*(12), 909–930. https://doi.org/10.1016/j.plaphy.2010.08.016

Giller, K. E. (2001). *Nitrogen fixation in tropical cropping systems*. CABI Publishing.

Glick, B. R. (2003). Phytoremediation: Synergistic use of plants and bacteria to clean up the environment. *Biotechnology Advances, 21*(5), 383–393. https://doi.org/10.1016/S0734-9750(03)00055-7

Haro, R., & Benito, B. (2019). The role of soil fungi in K^+ plant nutrition. *International Journal of Molecular Sciences, 20*, 3169. https://doi.org/10.3390/ijms20133169.
Hassen, A.I., Bopape, F., & Sanger, L. (2016). Microbial inoculants as agents of growth promotion and abiotic stress tolerance in plants. In *Microbial inoculants in sustainable agricultural productivity* (pp. 23–36) Springer.
Hawkins, J. P., & Oresnik, I. J. (2022). The rhizobium-legume symbiosis: Co-opting successful stress management. *Frontiers in Plant Science, 12*, 796045. https://doi.org/10.3389/fpls.2021.796045
Hussain, A., Arshad, M., Zahir, Z. A., & Asghar, M. (2015). Prospects of zinc solubilizing bacteria for enhancing growth of maize. *Pakistan Journal of Agricultural Sciences, 52*, 915–922.
Igiehon, N. O., & Babalola, O. O. (2018). Rhizosphere microbiome modulators: Contributions of nitrogen fixing bacteria towards sustainable agriculture. *International Journal of Environmental Research and Public Health, 15*, 574.
Ismail, G. S. M., Saber, N. E. S., Abdelrahim, B. I., & Abou-Zeid, H. M. (2021). Influence of cyanobacterial biofertilizer on the response of *Zea mays* plant to cadmium-stress. *Egyptian Journal of Botany, 61*(2), 391–404. https://doi.org/10.21608/ejbo.2020.41791.1553
Jatav, S. S., Parihar, M., Patra, A., Singh, S. K., Chitara, M. K., Mohapatra, K. K., Rana, K. (2021). Soil microbes in plant growth promotion and for mitigation of abiotic stress of drought. *Soil Microbiomes for Sustainable Agriculture: Functional Annotation*, 175–201.
Jnawali, A., Ojha, R., & Marahatta, S. (2015). Role of *Azotobacter* in soil fertility and sustainability—A Review. *Advances in Plants & Agriculture Research, 2*, 1–5.
Joshi, D., Negi, G., Vaid, S., & Sharma, A. (2013). Enhancement of wheat growth and Zn content in grains by zinc solubilizing bacteria. *International Journal of Agriculture Environment and Biotechnology, 6*, 363–370. https://doi.org/10.5958/j.2230-732X.6.3.004
Kamran, S., Shahid, I., Baig, D. N., Rizwan, M., Malik, K. A., & Mehnaz, S. (2017). Contribution of zinc solubilizing bacteria in growth promotion and zinc content of wheat. *Frontiers in Microbiology, 8*, 2593. https://doi.org/10.3389/fmicb.2017.02593
Kapoor, R., & Singh, N. (2017). Arbuscular mycorrhiza and reactive oxygen species. In *Arbuscular mycorrhizas and stress tolerance of plants* (pp. 225–243) Springer.
Karam, M. M., Sadeek, S. A., & Fadali, A. (2020). The synergistic effect of mycorrhizae and rhizobia on the growth of soybean under salinity stress. *Journal of Plant Nutrition, 43*(7), 1001–1016. https://doi.org/10.1080/01904167.2020.1710783
Khalid, M., Hassani, D., Bilal, M., Asad, F., & Huang, D. (2017). Influence of bio-fertilizer containing beneficial fungi and rhizospheric bacteria on health promoting compounds and antioxidant activity of *Spinacia oleracea* L. *Botanical Studies, 58*, 35.
Khan, M. A., Asaf, S., Khan, A. L., Jan, R., Kang, S. M., & andKim, K.M. (2020). Thermotolerance effect of plant growth-promoting *Bacillus cereus* SA1 on soybean during heat stress. *BMC Microbiology, 20*(1), 175.
Kollah, B., Patra, A. K., & Mohanty, S. R. (2016). Aquatic microphylla Azolla: A perspective paradigm for sustainable agriculture, environment and global climate change. *Environmental Science and Pollution Research, 23*, 4358–4369.
Kour, R., Jain, D., Bhojiya, A.A., Sukhwal, A., Sanadhya, S., Saheewala, H., Jat, G., Singh, A., & Mohanty, S. R. (2019). Zinc biosorption, biochemical and molecular characterization of plant growth-promoting zinc-tolerant bacteria. *3 Biotech, 9*(11), 421. https://doi.org/10.1007/s13205-019-1959-2
Kubi, H. A. A., Khan, M. A., Adhikari, A., Imran, M., Kang, S. M., Hamayun, M., & Lee, I. J. (2021). Silicon and plant growth-promoting rhizobacteria *Pseudomonas psychrotolerans* CS51 mitigates salt stress in *Zea mays* L. *Agriculture, 11*(3), 272. https://doi.org/10.3390/agriculture11030272
Kumar, R., Kumawat, N., & Sahu, Y. K. (2017). Role of biofertilizers in agriculture. *Popular Kheti, 5*(4), 63–66.

Kumar, S., Diksha, Sindhu, S. S., & Kumar, R. (2022). Biofertilizers: An ecofriendly technology for nutrient recycling and environmental sustainability. *Current Research in Microbial Sciences, 3*, 100094. https://doi.org/10.1016/j.crmicr.2021.100094

Lamaoui, M., Jemo, M., Datla, R., & Bekkaoui, F. (2018). Heat and drought stresses in crops and approaches for their mitigation. *Frontiers in Chemistry, 6*, 26.

Liu, L., Zhang, W., Liu, H., & andWang, X. (2019). Mycorrhizal fungi improve plant growth and water status of maize under drought stress. *Plant Physiology and Biochemistry, 139*, 165–172. https://doi.org/10.1016/j.plaphy.2019.03.020

Mahanty, T., Bhattacharjee, S., Goswami, M., Bhattacharyya, P., Das, B., Ghosh, A., & Tribedi, P. (2017). Biofertilizers: A potential approach for sustainable agriculture development. *Environmental Science and Pollution Research, 24*, 3315–3335. https://doi.org/10.1007/s11356-016-8104-0

Mahmud, A. A., Upadhyay, S. K., Srivastava, A. K., & Bhojiya, A. A. (2021). Biofertilizers: A Nexus between soil fertility and crop productivity under abiotic stress. *Current Research in Environmental Sustainability, 3*, Article 100063.

Mamnabi, S., Nasrollahzadeh, S., Ghassemi-Golezani, K., & andRaei, Y. (2020). Improving yield-related physiological characteristics of spring rapeseed by integrated fertilizer management under water deficit conditions. *Saudi Journal of Biological Sciences, 27*(3), 797–804.

Mishra, D., Rajvir, S., Mishra, U., & Kumar, S. S. (2013). Role of bio-fertilizer in organic agriculture: A review. *Research Journal of Recent Sciences, 2277*, 2502.

Mkhongi, F. A., & Musakwa, W. (2022). Trajectories of deagrarianization in South Africa—Past, current and emerging trends: A bibliometric analysis and systematic review. *Geography and Sustainability, 3*(4), 325–333. https://doi.org/10.1016/j.geosus.2022.10.003

Mohamed, E., Mohd Adnan, N. S., & Abd-Talib, N.I. (2022). Bio-fertilizers application in agricultural biotechnology. 51–62.

Mohapatra, K. K., Singh, S. K., & Patra, A. (2020). Effect of Zn solubilizer on Zn solubilizing potential in Inceptisol of Indo-Gangetic alluvial plain. *Annals of Plant and Soil Research, 22*(2), 119–122.

Muhie, S. H. (2022). Novel approaches and practices to sustainable agriculture. *Journal of Agriculture and Food Research, 10*, Article 100446. https://doi.org/10.1016/j.jafr.2022.100446

Mukhtar, T., Ur-Rehman, S., Smith, D., Sultan, T., Seleiman, M. F., Alsadon, A. A., Ali, A. S., Chaudhary, H. J., Solieman, T. H. I., Ibrahim, A. A., & Saad, M. A. O. (2020). Mitigation of heat stress in *Solanum lycopersicum* L. by ACC-deaminase and exopolysaccharide producing Bacillus cereus: effects on biochemical profiling. *Sustainability, 12*, 2159.

Muthuraja, R., & Muthukuma, T. (2021). Isolation and characterization of potassium solubilizing *Aspergillus* species isolated from saxum habitats and their effect on maize growth in different soil types. *Geomicrobiology Journal, 38*(8), 672–685. https://doi.org/10.1080/01490451.2021.1928800

Naeem, U., Haq, I., Afzaal, M., Qazi, A., Yasar, A., Tabinda, A., Mahfooz, Y., Naz, A. U., & Awan, H. (2021). Investigating the effect of *Aspergillus niger* inoculated press mud (biofertilizer) on the potential of enhancing maize (*Zea mays* L.) yield, potassium use efficiency and potassium agronomic efficiency. *Cereal Research Communication, 50*(1), 157–170. https://doi.org/10.1007/s42976-021-00153-1

Nanaware, J. P., Pathade, A. G., & Sarkale, P. S. (2024). Microbial biotechnology for the production of sustainable biofertilizers: Challenges and opportunities. *Naturalista Campano, 28*(1), 244–251.

Naseer, I., Ahmad, M., Nadeem, S. M., Ahmad, I., & Zahir, Z. A. (2019). Rhizobial inoculants for sustainable agriculture: Prospects and applications. In *Biofertilizers for sustainable agriculture and environment* (pp. 245–283); Springer.

Nitu, R., Rajinder, K., & Sukhminderjit, K. (2020). Zinc solubilizing bacteria to augment soil fertility—A comprehensive review. *International Journal of Agricultural Sciences and Veterinary Medicine, 8*, 38–44.

Niu, J., & Li, X. (2022). Effects of microbial inoculation with different indigenous bacillus species on physicochemical characteristics and bacterial succession during short-term composting. *Fermentation, 8*, 152. https://doi.org/10.3390/fermentation8040152

O'Callaghan, M. (2016). Microbial inoculation of seed for improved crop performance: Issues and opportunities. *Applied Microbiology and Biotechnology, 100*, 5729–5746.

Odoh, C. K., Sam, K., Zabbey, N., Eze, C. N., Nwankwegu, A. S., Laku, C., & Dumpe, B. B. (2020). Microbial consortium as biofertilizers for crops growing under the extreme habitats. In A. N. Yadav, J. Singh, A. A. Rastegari, & N. Yadav (Eds.), *Plant microbiomes for sustainable agriculture* (pp. 381–424). Springer.

Pal, S., Singh, H., Farooqui, A., & Rakshit, A. (2015). Fungal biofertilizers in Indian agriculture: Perception, demand and promotion. *Journal of Eco-Friendly Agriculture, 10*(2), 101–113.

Pathak, D., Kumar, M., & Rani, K. (2017). Biofertilizer application in horticultural crops. In *Microorganisms for green revolution* (pp. 215–227) Springer.

Patel, S. H., Viradiya, M. B., & Prajapati, B. J. (2021). Effect of potassium and potassium mobilizing bacteria (KMB) with and without FYM on yield of wheat (*Triticum aestivum* L.). *Journal of Pharmacognosy and Phytochemistry, 10*(1), 1615–1620.

Phour, M., & Sindhu, S. S. (2020). Amelioration of salinity stress and growth stimulation of mustard (*Brassica juncea* L.) by salt-tolerant Pseudomonas species. *Applied Soil Ecology, 149*, 103518

Pradhan, M., Das, R., Sahoo, R. K., Adak, T., Pradhan, C., & Mohanty, S. (2022). Comparative P solubilizing efficiencies of five acid soil bacteria incubated with calcium, aluminium and iron phosphates. *Indian Journal of Biochemistry & Biophysics, 59*, 947–955. https://doi.org/10.56042/ijbb.v59i10.67099

Rafique, M., Hussain, S., & Ali, S. (2020). Potassium-solubilizing bacteria enhance the growth and nutrient uptake of wheat under saline conditions. *Journal of Soil Science and Plant Nutrition, 20*(1), 208–222. https://doi.org/10.1007/s42729-020-00225-1

Raimi, A., Roopnarain, A., & Adeleke, R. (2021). Biofertilizer production in Africa: Current status, factors impeding adoption and strategies for success. *Scientific African, 11*, Article e00694. https://doi.org/10.1016/j.sciaf.2021.e00694

Ramesh, A., Sharma, S. K., Sharma, M. P., Yadav, N., & Joshi, O. P. (2014). Inoculation of zinc solubilizing *Bacillus aryabhattai* strains for improved growth, mobilization and biofortification of zinc in soybean and wheat cultivated in *Vertisols* of Central India. *Applied Soil Ecology, 73*, 87–96. https://doi.org/10.1016/j.apsoil.2013.08.009

Rawat, P., Sharma, A., Shankhdhar, D., & Shankhdhar, S. C. (2022). Improvement of phosphorus uptake, phosphorus use efficiency, and grain yield of upland rice (*Oryza sativa* L.) in response to phosphate-solubilizing bacteria blended with phosphorus fertilizer. *Pedosphere, 32*(5), 752–763.

Raza, A., Razzaq, A., Mehmood, S. S., Zou, X., Zhang, X., Lv, Y., & Xu, J. (2019). Impact of climate change on crops adaptation and strategies to tackle its outcome: A review. *Plants, 8*(2), 34.

Rees, D. C., Akif Tezcan, F., Haynes, C. A., Walton, M. Y., Andrade, S., Einsle, O., & Howard, J. B. (2005). Structural basis of biological nitrogen fixation. *Philosophical Transactions of the Royal Society A, 363*(1829), 971–984. https://doi.org/10.1098/rsta.2004.1539

Renjith, P.S., Sheetal, K. R., Kumar, S., Choudhary, J., & Prasad, S. (2020). Microbial and biotechnological approaches in the production of biofertilizer. In *Environmental microbiology and biotechnology: Volume 1: Biovalorization of solid wastes and wastewater treatment* (pp. 201–219).

Rodrigues, A. C., Bonifacio, A., de Araujo, F. F., Junior, M. A. L., & Figueiredo, Md. V. B. (2015). *Azospirillum* sp. as a challenge for agriculture. In Maheshwari, D. K. (Ed.), *Bacterial metabolites in sustainable agroecosystem* (pp. 29–51). Springer, Cham. https://doi.org/10.1007/978-3-319-24654-3_2

Santos, F., Abney, R., Barnes, M., Bogie, N., Ghezzehei, T.A., Jin, L., Moreland, K., Sulman, B. N., & Berhe, A. A. (2019). The role of the physical properties of soil in determining biogeochemical responses to soil warming. In J.E. Mohan (Ed.), *Ecosystem consequences of soil warming-*

microbes, vegetation, fauna and soil biogeochemistry (pp. 209–244). Elsevier. https://doi.org/10.1016/B978-0-12-813493-1.00010-7

Sarwar, S., Khaliq, A., Yousra,M., Sultan, T., Ahmad, N., & Khan, M. Z (2020). Screening of siderophore-producing PGPRs isolated from groundnut (*Arachis hypogaea* L.) rhizosphere and their influence on iron release in soil. *Communications in Soil Science and Plant Analysis, 51*(12), 1680–1692. https://doi.org/10.1080/00103624.2020.1791159

Saxena, B., Shukla, K., & Giri, B. (2017). Arbuscular mycorrhizal fungi and tolerance of salt stress in plant. In: Q.S. Wu (Ed.), *Arbuscular mycorrhizas and stress tolerance of plants* (pp. 76–106). Springer.

Shaaban, M. (2024). Microbial pathways of nitrous oxide emissions and mitigation approaches in drylands. *Journal of Environmental Management, 354*, 120393. https://doi.org/10.1016/j.jenvman.2024.120393

Sharma, R., Sindhu, S. S., & Glick, B. R. (2024). Potassium solubilizing microorganisms as potential biofertilizers: A sustainable climate-resilient approach to improve soil fertility and crop production in agriculture. *Journal of Plant Growth Regulation, 43*, 2503–2535. https://doi.org/10.1007/s00344-024-11297-9

Shin, W., Islam, R., Benson, A., Joe, M. M., Kim, K., Gopal, S., Samaddar, S., Banerjee, S., & Sa, T. (2016). Role of diazotrophic bacteria in biological nitrogen fixation and plant growth improvement. *Korean Journal of Soil Science and Fertilizer, 49*(1), 17–29. https://doi.org/10.7745/kjssf.2016.49.1.017

Singh, G., Biswas, D. R., & Marwaha, T. S. (2010). Mobilization of potassium from waste mica by plant growth promoting rhizobacteria and its assimilation by maize (*Zea mays*) and wheat (*Triticum aestivum* L.): A hydroponics study under phytotron growth chamber. *Journal of Plant Nutrition, 33*(8), 1236–1251. https://doi.org/10.1080/01904161003765760

Singh, B., Boukhris, I., Pragya, Kumar, V., Yadav, A.N., Farhat-Khemakhem, A., Kumar, A., Singh, D., Blibech, M., Chouayekh, H., & Alghamdi, O. A. 2020. Contribution of microbial phytases to the improvement of plant growth and nutrition: a review. *Pedosphere, 30*(3), 295–313. https://doi.org/10.1016/s1002-0160(20)60010-8

Singh, J. S., Kumar, A., Rai, A. N., & Singh, D. P. (2016). Cyanobacteria: A precious bioresource in agriculture, ecosystem, and environmental sustainability. *Frontiers in Microbiology, 7*, 529.

Smercina, D. N., Evans, S. E., Friesen, M. L., & Tiemann, L. K. (2019). To fix or not to fix: Controls on free-living nitrogen fixation in the rhizosphere. *Applied Environmental Microbiology, 85*(6), e02546-e2618.

Smith, S. E., & Read, D. J. (2008). *Mycorrhizal symbiosis*. Academic Press.

Soumare, A., Diedhiou, A. G., Thuita, M., Hafidi, M., Ouhdouch, Y., Gopalakrishnan, S., & Kouisni, L. (2020). Exploiting biological nitrogen fixation: A route towards sustainable agriculture. *Plants, 9*(8), 1011. https://doi.org/10.3390/plants9081011

Sun, B., Bai, Z., Bao, L., Xue, L., Zhang, S., Wei, Y., Zhang, Z., Zhuang, G., & Zhuang, X. (2020a). *Bacillus subtilis* biofertilizer mitigating agricultural ammonia emission and shifting soil nitrogen cycling microbiomes. *Environment International, 144*, Article 105989. https://doi.org/10.1016/j.envint.2020.105989

Sun, F., Ou, Q., Wang, N., Guo, Z., Ou, Y., Li, N., & Peng, C. (2020b). Isolation and identification of potassium-solubilizing bacteria from *Mikania micrantha* rhizospheric soil and their effect on *M. micrantha* plants. *Global Ecology and Conservation, 23*, e01141. https://doi.org/10.1016/j.gecco.2020.e01141.

Suresh, A., & Abraham, J. (2019). Harnessing the microbial interactions in rhizosphere and microbiome for sustainable agriculture. In *Microbial interventions in agriculture and environment* (pp. 497–515). Springer.

Thomas, L., Singh, I. (2019). Microbial biofertilizers: types and applications. In *Biofertilizers for sustainable agriculture and environment* (pp. 1–19). Springer.

Tsukanova, K. A., Chebotar, V. K., Meyer, J. J. M., & Bibikova, T. N. (2017). Effect of plant growth-promoting rhizobacteria on plant hormone homeostasis. *South African Journal of Botany, 113*, 91–102. https://doi.org/10.1016/j.sajb.2017.07.007

Turner, B. L., & Blackwell, M. S. A. (2013). Isolating the influence of pH on the amounts and forms of soil organic phosphorus. *European Journal of Soil Science, 64*(2), 249–259. https://doi.org/10.1111/ejss.12026

Upadhyay, S. K., Saxena, A. K., Singh, J. S., & Singh, D. P. (2019). Impact of native ST-PGPR (*Bacillus pumilus*; EU927414) on PGP traits, antioxidants activities, wheat Plant growth and yield under salinity. *Climate Change and Environmental Sustainability, 7*(2), 157–168.

Vejan, P., Abdullah, R., Khadiran, T., Ismail, S., & Boyce, A. N. (2016). Role of plant growth promoting rhizobacteria in agricultural sustainability—A review. *Molecules, 21*(5), 573. https://doi.org/10.3390/molecules21050573

Varga, T., Hixson, K. K., Ahkami, A. H., Sher, A. W., Barnes, M. E., Chu, R. K., Battu, A. K., Nicora, C. D., Winkler, T. E., Reno, L. R., Fakra, S. C., Antipova, O., Parkinson, D. Y., Hall, J. R., & Doty, S. L. (2020) Endophyte-promoted phosphorus solubilization in populus. *Frontiers in Plant Science, 11*, 567918. https://doi.org/10.3389/fpls.2020.567918

Verma, P., & Pandey, K. (2022). Biofertilizer: An ultimate solution for the sustainable development of agriculture. *Current Agriculture Research, 10*(3), 193–206. https://doi.org/10.12944/CARJ.10.3.04

Verma, G., Prakriti., Kumar, A., Babu, S., Singh, R., Gudade, B. A., Bhatt, M., Saha, S., & Yadav, N. C. (2022). Implications and future prospects of azolla as a low-cost organic input in agriculture. *Agricolation, 2*(1), 1–7.

Wang, F., Kertesz, M. A., & Feng, G. (2019). Phosphorus forms affect the hyphosphere bacterial community involved in soil organic phosphorus turnover. *Mycorrhiza, 29*, 351–362. https://doi.org/10.1007/s00572-019-00896-0

Wang, X., Zhang, J., Wang, X., An, J., You, C., Zhou, B., & Hao, Y. (2022). The growth promoting mechanism of *Brevibacillus laterosporus* AMCC100017 on apple rootstock *Malus robusta*. *Horticultural Plant Journal, 8*(1), 22–34. https://doi.org/10.1016/j.hpj.2021.11.005

White, P. J. (2001). Phytoremediation assisted by microbes. *Trends in Plant Science, 6*, 502.

Yasmin, S., Hafeez, F. Y., Mirza, M. S., Rasul, M., Arshad, H. M., Zubair, M., & Iqbal, M. (2017). Biocontrol of bacterial leaf blight of rice and profiling of secondary metabolites produced by rhizospheric *Pseudomonas aeruginosa* BRp3. *Frontiers in. Microbiology, 8*, 1895.

Yin, J., Sui, Z. M., & Huang, J. G. (2021). Mobilization of soil inorganic phosphorus and stimulation of crop phosphorus uptake and growth induced by *Ceriporia lacera* ta HG2011. *Geoderma, 383*, 114690. https://doi.org/10.1016/j.geoderma.2020.114690

Yun, P., Xu, L., Wang, S. S., Shabala, L., Shabala, S., & Zhang, W. Y. (2018). *Piriformospora indica* improves salinity stress tolerance in *Zea mays* L. plants by regulating Na^+ and K^+ loading in root and allocating K^+ in shoot. *Plant Growth Regulation, 86*, 323–331. https://doi.org/10.1007/s10725-018-0431-3

Zhang, Z., Liu, D., Wu, M., Xia, Y., Zhang, F., & Fan, X. (2021). Long-term straw returning improves soil K balance and potassium supplying ability under rice and wheat cultivation. *Scientific Reports, 11*(1), 22260. https://doi.org/10.1038/s41598-021-01594-8

Zhao, G., Zhu, X., Zheng, G., Meng, G., Dong, Z., Baek, J. H., Jeon, C. O., Yao, Y., Xuan, Y. H., Zhang, J., & Jia, B. (2024). Development of biofertilizers for sustainable agriculture over four decades (1980–2022). *Geography and Sustainability, 5*, 19–28. https://doi.org/10.1016/j.geosus.2023.09.006.

Zhao, Y., Lu, G., Jin, X., Wang, Y., Ma, K., Zhang, H., Yan, H., & Zhou, X. (2022). Effects of microbial fertilizer on soil fertility and alfalfa rhizosphere microbiota in alpine grassland. *Agronomy, 12*(7), 1722. https://doi.org/10.3390/agronomy12071722

Zubair, M., Hanif, A., Farzand, A., Sheikh, T., Khan, A. R., Suleman, M., Ayaz, M., & Gao, X. (2019). Genetic screening and expression analysis of psychrophilic Bacillus spp. Reveal their potential to alleviate cold stress and modulate phytohormones in wheat. *Microorganisms, 7*(9), 337.

Biodiversity Within Agroecosystems and Its Benefits to Soil

Muhammad Mehran, Muhammad Ikram, Adel M. Ghoneim, Sahrish Ghafar, Muhammad Ashraf, Haseeb ur Rehman, Iftikhar Ali Ahmad, Sharjeel Haider, Muhammad Nadeem Shah, Abdullah Ehsan, Asif Minhas, Kamran Ghaffor, and Muhammad Zahid Iqbal

Abstract As complex ecological systems managed for agricultural production, agroecosystems are deeply influenced by biodiversity, which is crucial for maintaining productivity, sustainability, and soil health. This chapter delves into the vital role of biodiversity, encompassing genetic, species, and ecological diversity, in sustaining agroecosystems. Biodiversity within these systems offers numerous benefits, including enhanced nutrient cycling, improved soil fertility, effective pest and disease management, and increased resilience to environmental stressors. The

M. Mehran (✉) · I. A. Ahmad · S. Haider
Key Laboratory of Arable Land Conservation (Middle and Lower Reaches of Yangtze River), Ministry of Agriculture and Rural Affairs, Huazhong Agricultural University, Wuhan, China
e-mail: mehranmuhammad970@gmail.com

M. Ikram
MOA Key Laboratory of Crop Ecophysiology and Farming System in the Middle Reaches of the Yangtze River, College of Plant Science and Technology, Huazhong Agricultural University, Wuhan, Hubei, China

A. M. Ghoneim
Field Crops Research Institute, Agricultural Research Center, Giza, Egypt

S. Ghafar
Department of Chemistry, University of Education Lahore (Multan Campus), Lahore, Pakistan

M. Ashraf
Department of Soil & Environmental Sciences, College of Agriculture, University of Sargodha, Sargodha, Pakistan

M. Ikram · H. Rehman · M. Z. Iqbal
Department of Agronomy Faculty of Agricultural Sciences and Technology, Bahauddin Zakariya University, Multan, Pakistan

M. N. Shah
Department of Agriculture, Government College University, Lahore, Pakistan

M. Mehran · M. Ashraf · A. Ehsan · A. Minhas
Department of Soil Science, Bahauddin Zakariya University, Multan, Pakistan

K. Ghaffor
Agronomy College, Sichuan Agriculture University, Chengdu, China

M. Shaaban (ed.), *Soils and Sustainable Agriculture*, Frontier Studies in Soil Science,
https://doi.org/10.1007/978-3-031-91114-9_10

chapter explores how land-use changes, such as tillage, erosion, and alterations in plant species, can significantly impact soil biodiversity, often leading to reduced soil fertility and increased reliance on fertilizers and pesticides. The importance of traditional and underutilized crops, crop wild relatives, and diverse cropping systems, such as agroforestry and polycultures, in promoting agrobiodiversity is emphasized. These systems play a key role in maintaining soil health and Biodiversity by supporting a dynamic soil food web that includes various microorganisms and invertebrates critical for decomposition and nutrient cycling. Furthermore, the chapter highlights the necessity of implementing sustainable management practices, such as crop rotation, conservation tillage, and integrated pest management, to foster Biodiversity and enhance ecosystem services. The impact of invasive species on soil biodiversity and ecosystem functioning is also addressed, emphasizing the need for effective management strategies. By preserving and enhancing Biodiversity, agroecosystems can achieve greater productivity, ecological balance, and long-term sustainability. This, in turn, contributes significantly to global food security and environmental health, ensuring the resilience of agricultural systems in the face of global changes.

Keywords Agroecosystem biodiversity · Soil fertility · Nutrient cycling · Pest control · Sustainable agriculture · Ecosystem services

1 Introduction

This chapter discusses the role of bodiversity in economizing a particular type of ecosystem. From a generic perspective, all the ideas and theories discussed in the book's other chapters are pertinent to this discourse. Our primary concern is to provide a framework for emulating a discussion process on the place and importance of biodiversity in the functioning of agricultural systems. To what extent can the notion of agricultural ecosystems be justified in being considered in isolation from other ecosystems. The primary aspect is that these systems involve man, and man's interference leads to a purposeful decrease in the number of species within that system.

However, the agricultural ecosystems are low in diversity on the global scale. Some tropical cropping systems contain many more species than stressed or otherwise borderline environments, such as hot and cold deserts, tundra, mangroves, etc. For instance, the widespread home gardens in many areas of the humid tropics average 200 species of cultivated plants (Abdullah et al., 2021; Abiven et al., 2005). Our discussion is thus structured on the definitional premise that agricultural ecosystems differ from natural ecosystems in that they have a specific human-determined function: It involves the generation of utilizable output, energy, or a harvestable crop. We also include the currently rather popular idea that, for agriculture, improvements beyond simply the biological or the economical are also important.

In the following discussion, we have used a comparative analysis based on the relative intensification of agricultural practice and its impact on the shifting role of

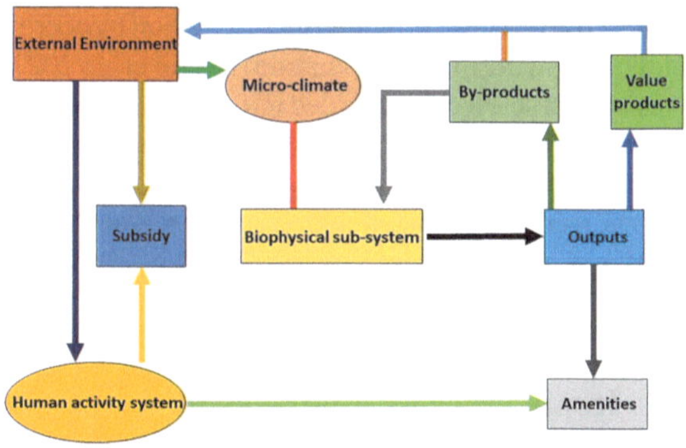

Fig. 1 A simplified conceptual model of the agro-ecosystem

Biodiversity in modulating ecosystem function (Fig. 1). Practicality: we only focus on food crop production systems. Nevertheless, there are occasional references to other forms of dominant agriculture types.

2 Definition of Agroecosystems

Agroecosystems refer to those ecological systems developed and sustained by man mainly for producing food crops (Aqeel et al., 2023). These include the relations between inter and intra-specific organisms, the plants, animals, microorganisms, and the abiotic environment of farm orchards and fields. Agroecosystem components include vegetation, animals, soil, water, and other forms of biological diversities that interact within complex processes that enable the production of crops, livestock, and even fuel (Aryal et al., 2022). It is essential to know that such systems exist in the diverse global agricultural environment ranging from small-scale industrialized farming to large-scale subsistence farming. Such decisions and management practices affect the agroecosystems' ecological functioning, production, and potential sustainability. It is, therefore, crucial that there is a clear understanding of the structure of agroecosystems concerning the challenges facing sustainable development, environmental conservation, and food security.

2.1 *Ecosystem Services from sustainable Agriculture*

Agricultural ecosystems offer the primary needs for human survival, such as food security and energy, besides offering raw materials necessary for man. The global land-use distribution shows that agricultural land-use covers one-third of the globe's surface and represents ~ 4.9 billion livestock units; 3.3 billion ha are used to describe the impacts on livestock number density or change or disease prevalence on livestock, respectively. It also provided that 1.6 billion ha of cropland is required to produce food for the growing global population (Ashraf et al., 2022; Bach et al., 2020). Agricultural ecosystem services are divided into support (for example, nutrient cycling, soil formation), provision (food, fuel, fibers, freshwater, etc.), regulation (for example, flood regulation, disease regulation, climate regulation, etc.), and cultural (aesthetic, religious, recreational, relational, etc.) services and the sustainable agricultural practices that may enhance these services (Fig. 1).

3 Importance of Biodiversity in Agro-Ecosystems

Biodiversity is consumed more by agriculture than by any other sector in the whole world. Wild biodiversity of selection and value addition to agrobiodiversity has been around for over 10,000 years. These challenges include malnutrition, poverty, and climate change, all actively addressed by agrobiodiversity. Frailty and decline are characteristics that have been fading and are even on the brink of vanishing (Carr et al., 2020; Clark, 2008). Local, national, and global livelihood security can be supported through agroecosystems and agrobiodiversity by providing food fodder, climate change, Biodiversity, and water quality options. Farming organizations are central in ensuring these resources and ecosystems are protected and conserved. However, these observations stipulate that agriculture can transform ecosystems in the supply of ecosystem services depending on the incentives offered to the ecosystem (Clarholm et al., 2007; World Bank, 2007). Today, incentives are made to cover the actualization of the products rather than services offered by agricultural ecosystems. It has also been seen that agrobiodiversity continues to be the primary resource for agro-ecosystems to adapt to climate change. It harbours the pool of characteristics for plant and animal breeders and farmers to choose climate-adapted germplasm and develop new varieties or breeds. Conventional agricultural practices in Sri Lanka have evolved due to the adaptation of production systems over a long time (Alotaibi et al., 2023; Coleman et al., 2024). Whatever form of agrobiodiversity can only be sustainably utilized and improved upon with the human-erection systems of production that generated these indigenous knowledge and technologies, specific forms of social structure, tenures or laws, endowment, and other cultures.

The aim is to maintain the agro-ecosystem biodiversity, as it contributes several uncountable benefits that determine the stability and effectiveness of agricultural systems (Altieri et al., 2015 & 2017). Crop diversification refers to the variety or the number of plant species grown in farming systems. They contribute to agro-ecosystems' sustainability, productivity, and resilience.

3.1 Genetic Diversity

The agroforestry genetic resource center is a critical gene pool for breeding programs that aim to develop improved crop varieties with superior attributes and yield, pest, and abiotic stress resistance (Clark, 2008).

3.2 Species Diversity

Several crop species can be grown in monoculture or complex combinations within or between the agroecosystems (Clúa et al., 2018). Diverse cropping systems involve growing several crops at the same period and time, while monoculture is a large-scale production of a single crop type. Diversification of cropping systems such as Agroforestry, polyculture, and intercropping creates species diversity within the agro ecological systems, promoting ecosystem services and complexity of environmental processes (Coleman et al., 2024; Culliney, 2013).

3.3 Crop Wild Relatives

Crop diversity also includes crop wild relatives (C.W.Rs) and feral or semi-wild relatives of cultivated crops (da Silva Dias, 2015). C.W.R.s are essential sources of genetic diversity and often bear crop–improving traits, which include resistance to pests, diseases, or other forms of biotic and abiotic stresses. Thus, it is essential to conserve C.W.R.s with their habitats intact to secure the future of agricultural productivity and guard gene resources.

3.4 Traditional and Underutilized Crops

The crops grown in the agroecosystems and other primary food crops contain traditional or/or low-value crops that are culturally relevant and well-adapted within the local region of implementation (El Sabagh et al., 2020). These crops often produce clients' specific qualities, taste, and nutrient content that make them contribute to a diverse and secure diet. Promoting underutilized and traditional crops can enhance agrobiodiversity and offer livelihood opportunities among the smallholders.

3.5 Germplasm Conservation

It implies a need to actively protect plant genetic resources to maintain crop genetic resources for future generations (Elshahawy & El-Mohamedy, 2019; Epelde et al., 2018). This includes establishing a seed exchange network, gene banks, botanical gardens, and on-farm conservation programs. It is a goal of conservation approaches to collect, evaluate, and store adequately and use diversely diverse crop germplasm to ensure availability for breeding, knowledge, and food production.

3.6 Ecosystem Services

Crop diversification in agroecosystems promotes agriculture production and environmental sustainability by providing ecosystem services (Ikram et al., 2021). These services include pollination, controlling pests and diseases, giving soil fertility, preventing floods and drought, and storing carbon. The integrated and multiple cropping opportunities decrease production factors such as artificial fertilizers and chemical sprays, strengthening the ecosystem (Almeida et al., 2019).

3.7 Resilience to Environmental Change

Agro-ecosystems diversification has been argued to enhance resilience to ecological shocks and oscillations such as parasitism, diseases, and acts climate shocks, among other factors, as posited by Anderson et al. (2016). In this way, the support of producers in changing environmental circumstances and facilitating agricultural production to reduce risks of crop failures, as well as biological diversity, increases the sustainability of the farming system.

3.8 Pest and Disease Management

Disease and parasite moderation through natural ecological community structure can be supported utilizing predation, parasitism, and competition (Angon et al., 2023; Isbell et al., 2017). Microorganisms in the soil, such as biocontrol agents of pests, can decrease pest populations and the use of chemicals (Jain et al., 2015).

3.9 Soil Health and Fertility

Biodiversity is significant in supporting soil fertility through nutrient cycling, decomposition of organic matter, and soil structure (Culliney, 2013; Jansson & Hofmockel, 2020). Plant and microbial diversity within agroecological complexes show improved potential for nutrient cycling, soil structuring, and increased organic matter content (John & Sarada, 2012).

Nutritional Diversity

Different agroecosystems can provide nutritious foods that help support many types and species of livestock and cereals, increasing human nutrition and food availability (Johnson et al., 2015). For this reason, it becomes essential to show how it is necessary to follow different diets to combat the negative impacts of malnutrition and enhance general wellbeing.

Biodiversity supports and Self-organizes in agroecological systems, and numerous ecological, economic, and social benefits of Biodiversity enable agricultural productivity and food security (Fig. 2). Maintaining and enhancing diversity in whichever abiotic agroecosystem is thus equally important in achieving the goals of ambassadors for sustainable development, conservation of ecosystems, and availability of food (Jose, 2009).

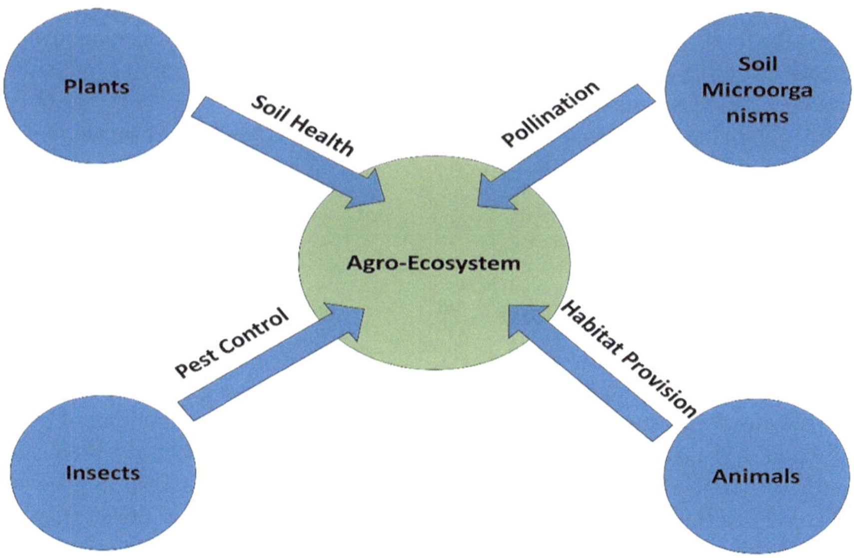

Fig. 2 The role of biodiversity in enhancing agro-ecosystem health and productivity

4 Loss of Biodiversity and the Underlying Drivers

Biologically diverse ecosystems are a renewable resource and the main climate stabilizers and carbon sequestration sites in technological advancement in agriculture, product processing, and medicinal products (Jovovic et al., 2020). Loss of Biodiversity decreases the resilience of an ecosystem and is a significant concern in its respect. It also contributes to maintenance of ecological processes that define other companies which offer various services to human societies. Therefore, concerning that it is still being lost at present, and accordingly has important consequences for current and future human welfare (Kahane et al., 2013). This is worrisome to South Asian nations, as the country's rich Biodiversity is part of its cultural pride and is expected to preserve the many ecosystem services needed for its current population of 21 million plus future generations. Converting, degrading and fragmenting habitats including greens, virgin forests, and wildlife habitats is responsible for the decline in diversity in south Asia. There are efforts with conserved and reforested projects, the rates of deforestation or forest degradation remains high in the region (Khatoon et al., 2017). Although the habitat fragmentation due to relatively recent widespread deforestation is further predicted to cause more rapid decline in biological diversity in the next few years. In general sense, habitat destruction processes are conditioned by various proximate and underlying factors. Direct drivers of biodiversity decline can be defined as the actions that come into physical contact with the biophysical environment and causes changes in the habitats, including cultivation of crops for food, feed, fiber, fuel and other uses, generation of bio-fuels, removal of timber, construction of roads, rail and other amenities, intentional burning of biomass among others. They mainly work at the local level (Kopáček et al., 2013). The others are population driver's inappropriate population growth, poverty, urbanization policies and institutions, trade globalization and internationalization, climate change and variability.

4.1 Decline of Agrobiodiversity

Agriculture and Natural Resources A Conceptual Overview Agriculture's relationships with various species have evolved and should be considered in light of trends in world agriculture. There has been a significant improvement in the production and productivity of agricultural commodities in the last 30 years and also the swap of seeds and food items between the countries. he says that it has arisen from the increased area of cultivated land (extensification) as well as from the increase in yield per unit of area (intensification) achieved through application of inputs such as technology, improved varieties and the management of biological resources including the soil and water (Krebs & Bach, 2018). Although agricultural development has enabled significant yield increases in many areas, these development trends have also involved problems: for instance, hunger and malnutrition exist, the level of food security has dropped in many developing countries, the distribution of the benefits and cost of production is tilted, and biological resources are depleted (Lal, 2004).

5 Overview of Soil Biodiversity

Soil biodiversity encompasses an extraordinary array of life forms that are often hidden from our everyday view. This includes a wide variety of organisms such as viruses, bacteria, actinomycetes, fungi, algae, and protozoans (single-celled eukaryotes) (Liu et al., 2022). The microscopic invertebrates found in the soil include rotifers, tardigrades (water bears), soil planaria (flatworms), and nematodes (roundworms). Additionally, microarthropods like Acari (mites) and Collembola (springtails) inhabit the soil, along with larger invertebrates that are easily visible to the naked eye, such as terrestrial gastropods (snails and slugs), isopods (pill bugs, sow bugs), Oligochaeta (enchytraeids and earthworms), spiders, scorpions, beetles, centipedes, millipedes, crustaceans, ants, and termites (Escudero-Levya et al., 2022; Isbell et al., 2017; Jain et al., 2015).

Due to the vast diversity of these invertebrates, they are frequently categorized by size (body width) into groups known as microfauna, mesofauna, and macrofauna. Vertebrates also depend on the soil as a habitat; examples include moles, prairie dogs, meerkats, wombats, small rodents, and various lizards, snakes, frogs, and even birds. Many aboveground invertebrates may be temporary inhabitants of the soil, such as nematode parasites of insects, immature Diptera like tipulid crane flies and tephritid fruit flies, cicadas, and Coleoptera like scarab beetles (Garbach et al., 2014). These organisms may live in the soil for only one stage of their life cycle or may occur seasonally. Consequently, soil diversity is dynamic and intricately connected to the aboveground biotic and climatic systems (Table 1).

Remarkably, one square meter of soil may contain from billions of bacteria to meters of fungal hyphae, and millions of protozoa, nematodes, oligochaetes, as well as arthropods. The diversification of these species is highly diverse that understanding of the species population density in two groups of microarthropods and nematodes may add to the global stock of biodiversities. Strikingly though none of these values tells the full story of total number of species in the soil even within a particular group and thus there appears to be much yet to learn (Fukuoka, 1985; Gill &Grag, 2014). Soil bios survey estimates reveal that 90 to 95 percent of soils contain species that are yet to being identified and scientifically cataloged. There is no place on earth that has been studied to the desired depth to establish the absolute census of total species in the soil or all the molecular forms of microbes, invertebrate, or vertebrates. However, the conventional approaches have listed 1000 invertebrate species in one square meter of the forest land. Worldwide assessments of soil diversity have been derived from a myriad of studies that have employed varying sample size and number of samples, which have been collected from different ecosystems, and often to a depth of only 0–20 cm. It means that the represented populations of mites and nematodes make up only about 1/250 of the total soil species, since these organisms where discovered

Table 1 Impact of different soil organisms and their influence on soil health

Soil organism group	Functions	Impact on soil health	Key references
Bacteria	Decomposition, nitrogen fixation, disease suppression	Enhanced nutrient availability, disease resistance	WRI (2008)
Fungi	Decomposition, mycorrhizal symbiosis, organic matter breakdown	Improved nutrient uptake, soil structure stabilization	Krebs and Bach (2018)
Protozoa	Grazing on bacteria, nutrient recycling	Enhanced microbial biomass, nutrient cycling efficiency	Khatoon et al. (2017)
Nematodes	Soil aeration, nutrient release, pest control	Improved soil structure, pest management	Kopáček et al. (2013)
Earthworms	Soil aeration, organic matter decomposition, aggregation	Enhanced soil porosity, nutrient availability	Wright and Wimberly (2013)
Arthropods	Decomposition, soil mixing, predation of pests	Increased soil aeration, reduced pest populations	Wu and Wallace (2019)
Bacteria	Decomposition, nitrogen fixation, disease suppression	Enhanced nutrient availability, disease resistance	Watts et al. (2015)

at 15-m depth in the deserts, and bacteria and nematodes can dwell at even greater depth.

Soil biological diversity means the extent of diversity or the number of individual forms of life that can be found in the soil environment (Table 2). The subject involves many living species: bacterial and fungal organisms, protozoa nematodes, arthropods, earthworms, and even tiny vertebrates. Also, they have a mutual interaction between them and the soil and interaction between themselves (Hocking & Babbitt, 2014; Ikram et al., 2024a).

5.1 Microbial Diversity

Microorganisms are by far the most abundant and diverse organisms within the soil system. Some of the included categories include bacteria, fungi, archaea, and viruses (Walker & Salt, 2012). In addition, bacterium and fungi play a significant role in nutrient cycling, organic matter decomposition, and soil structure formation (Wall et al., 2012). In addition, some archaea are directly associated with biogeochemical cycling, such as nitrogen cycle and methane oxidation. They interfere with the

Table 2 Soil biodiversity in different agro-ecosystems

Agro-ecosystem	Bacterial diversity (O.T.U.s)	Fungal diversity (O.T.U.s)	Earthworm biomass (g/m^2)	Arthropod diversity (species richness)
Conventional farming	800–1200	400–600	10–20	30–40
Organic farming	1500–1800	800–1000	40–60	60–80
Agroforestry	1300–1600	700–900	30–50	50–70
No-till farming	1200–1400	600–800	25–45	40–60
Agro-ecosystem	Bacterial diversity (O.T.U.s)	Fungal diversity (O.T.U.s)	Earthworm biomass (g/m^2)	Arthropod diversity (species richness)

dynamics of microbial populations by inserting themselves into those populations (Wardle et al., 2004; Watts et al., 2015).

5.2 *Fungal Diversity*

The model of the magnitude of the population of fungi in the soil ranges from microscopic molds to large mushrooms. Mycorrhizae also plays several functions in decomposing organic matter, nutrient cycling, and forming mutualistic associations with plants (White, 2020). The number of fungi here or there or in one place or another has a direct relationship with the dispositions of the arrangements of habitats found in the soil, the health of those green plants, and the efficient running of every function in the world.

5.3 *Protozoa and Nematodes*

Protozoa are single-celled animal-like organisms that feed on bacteria, fungi, and other microorganisms that reside in soils (Wood & DeClerck, 2015). Some of the functions include participating in nutrient cycling and controlling the other microbial numbers. Nematodes are known as round worms, which are found in the soil, and they have mechanisms to feed in several ways: on bacteria and fungi, and they are also parasitic feeders on plants (Watts et al., 2015; Wood & DeClerck, 2015). Thus, they influence the cycling of nutrients and the general health of the plant species as part of their ecosystems.

5.4 Arthropods

Soil arthropods include insects, mites, springtails, and other small invertebrate animals adapted to living in this environment. These creatures are involved in the decomposition of organic matter, nutrient recycling, modification of the soil texture, and Physical Control of pests (Wu & Wallace, 2019). Some species have predatory or parasitic characteristics towards pest animals, while others get their food from roots or organic matter on the bottom of the water source.

5.5 Earthworms

The soil earthworm plays an essential role in determining the physical properties of the soil, decomposing organic matter, and cycling nutrients within the ecosystem (Yachi & Loreau, 1999). These are beneficial in that they aerate the ground, improve water infiltration, and facilitate plant germination (Na et al., 2022; Wu et al., 2023, 2024).

5.6 Small Vertebrates

Soil-dwelling small vertebrates include rodents, reptiles, and amphibians, among other species. These animals are involved in soil turnover and nutrient recycling, and these make significant contributions to the structure and function of the ecosystem (Ikram et al., 2018; Malézieux et al., 2009). This can potentially affect plant communities, seed dispersal, and the nature of soil.

5.7 Interactions and Ecosystem Functions

Soil biodiversity relations between the abiotic factors with the soil, including the pH, water, and temperature, soil biota determines the significant processes occurring in soils: nutrient cycling, carbon cycling, water regulation, and plant production (Mehran et al., 2023). Such functions might be disrupted by the loss of species diversity, thereby leading to a decline in soil quality and other ecosystem services.

Soil biodiversity plays an essential role in conserving the fertility of the soil increase the buffer capacity of the ecosystem as well as the productive capacity of farming. For this reason, understanding the nature and functioning of soil organisms is crucial for the proper management of this resource and is critical for the systems' preservation (Fig. 3).

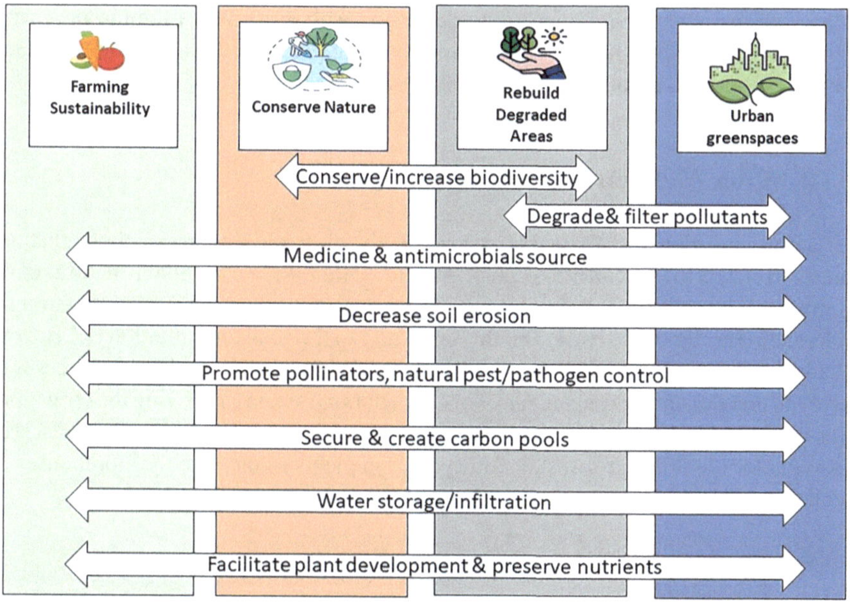

Fig. 3 Actions that promote soil biodiversity and ecosystem services (Bach et al., 2020)

6 Biodiversity Within Agro-Ecosystem

6.1 Plant Diversity

Agro-ecosystems list several components of plant diversity; these include cultivated crops and crop wild relatives and planting structures such as polycultures and mixed intercropping. The richness of species advanced the capacity of this ecosystem to resistance and resilience to disturbance, plus raised the production and longevity of its species (Millennium Ecosystem Assessment, 2005).

6.1.1 Crop Diversity

Crop diversity, therefore, relates to the range of crop plant species produced in the agricultural systems. It mainly includes the crops grown for sale or consumption, such as wheat, rice, maize, soybean, and pulses, the second motive crops, and the particular purpose crops. The various crops require different nutritional characteristics and growth requirements and have differing abilities relative to the rigors of pests and diseases. This is why crop diversity is critical to achieving food security. Heterogeneity in cropped plants helps genetic differentiation; crop plants are gene banks of characteristics for breeding programs with aims at developing new varieties that can produce higher yields, superior quality, and better ability to withstand various biotic and abiotic stresses (Minhas et al., 2024).

6.1.2 Crop Wild Relatives (C.W.R.s)

Or semi-wild plants of the same species as the cultivated plants or those that are closely related to the cultivated plant species. Often, they can inhabit natural environments; they possess desirable traits ranging from disease resistance to drought tolerance (Haider et al., 2024) and the versatility to thrive in any situation (Le Bayon; Walker & Salt, 2012). One of the tasks that C.W.R.s perform as genetic resources is to give the breeder an unexpected range of characters to help in improving the crop types that can perform better. Crop Wild Relatives (C.W.R.) and their germplasm reserves are vital for the long-term sustainability of crop plants as they are the storehouses of additional valuable genes (Yachi & Loreau, 1999).

6.1.3 Polycultures

Polycultures means the practice of growing many crops at a go within the same piece of land. Polycultures, on the other hand, encourage the production of many crops in extensive areas, hence encouraging species diversity and ecological heterogeneity, unlike in monoculture systems. Integrated crop management or polycultures can enhance nutrient cycling, pest suppression, and soil fertility as crops interact, creating a heightened production and managing to avoid vulnerability (Montoya et al., 2012). Polycultures can also have extra benefits that include weed control, reduction of erosion, and enhancement of species variety, which are several polycultures.

6.1.4 Intercropping

Intercropping involves growing many different crops simultaneously in a single geographical area and within the same period. Thus, intercropping is a type of polyculture since the crops are grown in one geographical location and the same period, although there are many crop species. This increases resource use efficiency, such as sunlight, water, and nutrients in the intercropping technique, since it is a combination of different crops that are organized according to the demand of each in the available resource (Ikram et al., 2024b). Inter-cropping improves outputs at the farm level, utilization of inputs, and conservation of resources compared to monoculture systems because inter-cropping reduces competition among crops compared to monoculture systems (Nademm et al., 2018). Hence, intercropping has some ecological benefits, such as control of pests and diseases, more varied niches, and enhanced soil productivity.

Crop diversification is essential since it will increase the resilience, productivity, and sustainability of the agro-ecosystems (John & Sarada, 2012). Some of the most stringent and crucial recommendations for maintaining and growing crop biodiversity in agricultural production systems include crop rotation and mixed cropping, biosphere conservation, genetic conservation, and encouraging traditional methods. Besides, it is evident and widely stated that plant species diversity in the agroecosystems is essential in enhancing the intact ecosystem resistance, production, and sustainable production. Plant diversities are available through the support of farming systems and the conservation of genetic resources and practices inherited from our farming ancestors to promote the stability and sustainability of agricultural production.

6.2 *Animal Diversity*

The diversity of soil fauna is also immense, including insects, birds, and other animals in agroecosystems. These animals are considered essential elements in the structure and functioning of ecosystems since they perform services such as pollination, predation control of pests, cycling of nutrients, and support of soil structure (Narayanan & Tewari, 2021).

6.2.1 Insects

Insects have great species diversity and are abundant in quantity within agroecosystems with diverse ecological statuses and perform essential environmental functions. Flowering plants have many insect species that help their reproduction, including many crop species (Nicholls & Altieri, 2013). Insects like bees, butterflies, and flies, among others, are essential in crop production and gene balance worldwide. Besides, insects are equally beneficial as biological control agents, human parasites, and herbivores in controlling agricultural pests (Younas et al., 2017). Indeed, it is highly imperative that insect diversity within the context of agroecosystems is conserved and even boosted in an effort to sustain pollination services as well as reliable pest control.

6.2.2 Birds

Birds are acknowledged to perform several services for agricultural systems worldwide; these include control of pests, seed dispersal, and nutrient cycling (Nicholls & Altieri, 2013; Nicholls et al., 2016). A large number of avian species feed on insects, rodents, and weed seeds, thus playing an important role in pest control and in preventing crop losses. They similarly help in the reproduction process and genetic makeup of plants by pollination and the distribution of these plants (Odelade &

Babalola, 2019). However, with a rise in agricultural activities, destruction of living space or habitats, and utilization of pesticides, the bird populations, diversity, and overall ecosystem benefits are reduced. Besides using minimum chemical inputs, measures like reforestation and planting trees in farms can enhance bird populations in agricultural fields.

6.2.3 Soil Fauna

Soil fauna, for instance, comprises earthworms, nematodes, mites, springtails, and microarthropods, among others. These organisms are crucial in keeping soil healthy and the ecosystem functional by acting as nutrient cyclers, organic matter decomposers, soil structure builders, and plant growth promoters (McBratney et al., 2014). The role of earthworms is to enhance the fertility of the soil and also to help in the process of aeration because it involves the digging of small tunnels that enhance the breakdown of organic waste (Tomar & Sharma 2023). The incoming purification and fragmentation of organic matter by nematodes and microarthropods predigest the contributions to nutrient cycling and microbial chains that influence the richness and availability of soil microbes. Soil fauna diversity depends on factors like the type of soil, the techniques used in farming, and fluctuations in the surroundings. It is, hence, essential to conserve the soil fauna diversity in order to continue having soil health productivity in the agroecosystems and to recover when disturbed (Mehran et al., 2023). A number of animal species in agroecosystems have been known to host a rich portfolio of some of the essential ecosystem services needs in agriculture today and in the future. Maintaining and enhancing diverse animal species can strengthen the enterprise's capacity to sustain disturbances and restore itself, reduce reliance on foreign aid, and promote sustainable farming techniques.

6.3 Microbial Diversity

In agroecosystems, many bacteria, fungi, and archaea exist in the microcosm substrates, including the soil, water, plants, and other environmental niches, hence diverse microbiota. These microorganisms are indispensable for nutrient cycling and other processes such as decomposition of organic matter, soil fertility, and promotion of plant growth and overall functionality of ecosystems.

6.3.1 Bacteria

Bacteria are a prominent microorganism found significantly in soil and other crops. These groups have significant roles in different biogeochemical cycles such as nitrogen cycling, nutrient cycling (nitrogen, phosphorus, and sulfur), carbon degradation, and recycling of organic matter in the soil (Murtaza et al., 2022; Offre et al., 2013). Some bacteria can 'communicate' with the plants and form mutualistic relationships; for example, rhizobia bacteria serve nitrogen-fixating bacteria to the legume plants, thus benefiting plant growth, the occurrence of diseases, and stress tolerance (Phillips, 2017). Yet, some such bacteria also act as pathogens causing plant diseases. This makes it essential for them to come to terms with the variety of bacteria present in agroecosystems and their ability to change over time.

6.3.2 Fungi

Fungus is a diverse organism group that includes molds, yeasts, and mushrooms. Regarding the importance of fungi in agroecosystems, there is the role of decomposer, cycling of matter, formation of mycorrhiza with root system of plants, and biological control (Pisias et al., 2022). Inoculation mycorrhizal fungi form symbiotic associations with plant roots; the fungi help in the uptake of nutrients, most especially phosphorus, and lastly, enhance plant growth and stress tolerance (Escudero-Leyva et al., 2022; John & Sarada, 2012). Besides, fungi can act as decomposers, which entails converting complex organic compounds into simpler forms that other species, including plants, may effectively use. Of equal importance, some fungi also produce bioactive compounds that may have applications in the farming, health, and biotechnology sectors.

6.3.3 Archaea

Archaea are organisms found in environments as diverse as soils, water deposits, and even extremophilic conditions, such as hot springs and hydrothermal vents. Even though archaea are not intensively investigated as compared to bacteria or fungi, they play many roles in biogeochemical cycles, such as ammonia-oxidizing organisms, methanogens, and methanotrophs, as well as sulfur-metabolizing organisms (Gan et al., 2024; Shaaban et al., 2024). They help in the maintenance of nutrient cycling and the behavior of the greenhouse gases of the agroecosystems (Angon et al., 2023; Ikram et al., 2024c). Such effects include the influence they exert on the fertility of the soils, the carbon stocks, and the Regulation of climate change (Yachi & Loreau, 1999). The analysis of archaea's diversity and functions within agricultural ecosystems is still a relatively emerging area of work that could help support active and efficient agriculture and the maintenance of ecosystems.

The abundance of diverse microorganisms in the agroecosystems is necessary to maintain the high quality, productivity, and nutrient density of the soil. Using practices like crop cycling, cover crops, minimum tillage, and use of organic matter allows farmers to cultivate different microbial populations. It facilitates nutrient cycling, control of plant diseases, promotes a proper structure of the soil, and increases the general capacity of the ecosystem (Powlson et al., 2014). In addition, new molecular and microbiological methods to study microbial trends provide new approaches to utilizing microbial richness and work to improve agricultural benefits.

7 Benefits of Biodiversity to Soil

7.1 Soil Health and Fertility

Soil conservation and the preservation of its fertility is a capital that ensures relentless productivity in any form of land usage and the proper functioning of ecosystems (Fig. 4). Different forms of life within the soil masses are beneficial to soil health, and diverse ways could be adopted to enhance its health (Powlson et al., 2014).

7.1.1 Nutrient Cycling

The existence of a great variety of organisms in the soil as a part of complex communities increases the rate of biospheric cycles, including decomposition and reutilization of top-grade nutrients, including nitrogen, phosphorous, and potassium. Certain living organisms in the soil, including bacteria, fungi, and archaea, are vital for decomposing the organic matter and liberating the nutrients of green manure to the plants. Biological nitrogen cycling contributes to the regeneration of the fertility of the arable lands (Shaaban, 2024) as well as the growth and production of plants (Pretty & Bharucha, 2014; Rashid et al., 2016).

7.1.2 Soil Structure and Aggregation

Biodiversity influences soil structure, and its effect is essential in structuring the soil and shaping the grouping and aggregation of soil particles, which is critical for physically maintaining ground stability, water infiltration, and air permeability.

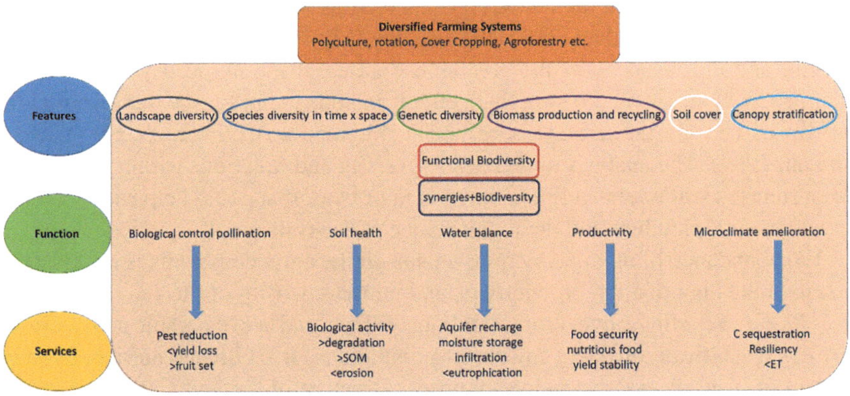

Fig. 4 The ecological role of Biodiversity in agro-ecosystem function and the provision of ecosystem services by diversified farming systems (Koohafkan et al., 2012)

RichSQ: Essentially, all living entities can improve soil aggregation, where soil organisms such as earthworms, arthropods, and fungal hyphae can stick the soil particles together and form strong soil aggregates (Rehman et al., 2022). These aggregates enhance the soil's shape, space, and conduit system, which, in turn, help improve the structural quality, water-holding capacity, and nutrient supply systems of plant root systems.

7.1.3 Organic Matter Decomposition

The heterogeneity also improves the decomposing of organic matter, including plant litter, root secretions, and animal carcasses, because it releases nutrients and energy for microbial, plant, and animal life. Bacterial and fungal microbes, for instance, participate in decomposition, whereby these organisms use enzymes to disintegrate complex chemical compounds into simple ones (Renzi et al., 2022). The decomposition process helps to release the minerals and organic carbon and favour microbial and plant nutrients in the soil (El Sabagh et al., 2020).

7.1.4 Pest and Disease Regulation

Soil organisms play an important function of regulating pests and diseases that affect plant overall health and yield. Beneficial microorganisms including those living in soil include bacteria, fungi, nematodes, and predatory arthropods that help in suppression of pest by mechanisms such as predation, parasitism, competition, and diseases (Krebs & Bach, 2018). This inherent biological Regulation reduces the need for man-made pesticides and promotes ecological methods of handling pests.

7.1.5 Biological Control

Regulation of pests and diseases possible as it encourage interaction of many organisms present in the soil food web. Allelopathy also has it that, useful bacteria and fungi can produce chemicals that suppress the growth of pathogenic germs/organisms. Besides, nematodes and microarthropods that are considered beneficial are known to prey on pest species bringing down their numbers and therefore their impacts on crop production (Nicholls et al., 2016). Pest and diseases are controlled through bio control methods which help in improving the health of agro ecosystem and production.

7.1.6 Allelopathy

Some of the soil organisms possess the ability to produce allelochemicals that are agent on other organisms found within the soil environment. Different studies show that allelopathy affects various elements of germination, root development, microbes as well as insects (Nadeem et al., 2018). It is estimated that that about 10 groups of allelochemicals released by plants and microorganisms are known inhibit the growth of other plants or lessen the activity of pests and pathogens (Reynolds et al., 2014). It helps in sustaining the various biotic and abiotic components that form the Earth's various ecosystems and habitats.

7.1.7 Microbial Interactions

Soil structure contributes to complex dynamics of microbial communities and their cooperation or competition affecting the processes in the soil and its functions in ecosystems (Reynolds et al., 2014). Microbial communications include the process of exchange of nutrients, competition, signaling and the development of microbial community (Le Bayon et al., 2021). Overall these interactions contribute to the management of the health and fertility of soil (Abdullah et al., 2021; Carr et al., 2020; Kopáček et al., 2013). Such interactions enhance the capacity of soil to buffer, and rebound from existing environmental perturbations and enhances the overall endurance and viability of ecosystems in the long-term (Jain et al., 2015).

These organisms in the soil systems are influential in controlling fertility in the soils as well as promoting nutrient cycling, decomposing organic matter, controlling the pests, and microbial symbiosis. Enhancing and maintaining soil biological quality is important towards the current and future states of soil fertility, agriculture and ecological services.

7.2 *Climate Change Mitigation*

The Biodiversity highlights the importance of soil ecosystem is that soil supports very large and diverse biota that play a vital role in mitigating climate change by participating in the mechanism of carbon storage and reduction of greenhouse gas emissions (Johnson et al., 2015). The following are the primary ways in which soil biodiversity contributes to the mitigation of climate change: The following are the primary ways in which soil biodiversity contributes to the mitigation of climate change:

7.2.1 Carbon Sequestration

Soil biodiversity increases soil carbon stock which 'locks' carbon in the atmosphere in organic matter within the soil (Kopáček et al., 2013; Liu et al., 2022). Some of the most important organisms are bacteria, fungi, and archaea since they decompose parts of plants residues as well as incorporate carbon-enriched organic matter into stable micro-aggregates (Chen et al., 2024). The soil also sees an accumulation of organic matter that forms other carbon pools with definite stabilities that are in the range of several decades to centuries. This process removes CO_2 from the Atmosphere and deposit it to the carbon pool of the soil. This technique is useful in addressing climate change since it removes greenhouse gases from the atmosphere hence reducing on global warming.

7.2.2 Greenhouse Gas Emissions Reduction

Soil biodiversity has a role of reducing the emission of greenhouse gases that are known to bolster the issue of global warming including CO_2, CH_4 and N_2O among others (El Sabagh et al., 2020). In this case, soil organisms affect the emission of greenhouse gases by participating in several activities like carbon mineralization, nitrogen conversion, as well as methane consumption. For instance aerobic heterotrophic microorganisms in the soil decompose organic matter in the presence of oxygen and release CO_2 by respiration. However, free methane (CH_4) produced by the anaerobic bacterial decomposition of organic matter in water logged soil through the process of methanogenesis. Thus, soil bacteria which plays an eminent role in nitrogen cycling, thus directly involved in the production and use of nitrous oxide (N_2O) which is a greenhouse gas (Sabagh et al., 2020a). Some notes on how farmers can respond to climate change, and or actively reverse it include: Improved sub soil biology reduces Green House Gases emissions; Agriculture and food production practices protects or even enhance Biodiversity.

Therefore, soil biodiversity plays a very importance role in combating the effects of climate change since it plays an important role in carbon sequestration and reduction of greenhouse gases emissions (Sahu et al., 2017). By practicing conservation agriculture, agroforestry and cover cropping among others we may adequately protect and enhance the status of soil bio-diversity. This will in turn enhance the potential of the specific soil ecosystems to address climate change mitigation and support international endeavours in tackling this threat.

8 Management Practices to Enhance Biodiversity in Agro-Ecosystems

8.1 Crop Rotation

Crop rotation is done based on a planned manner of changing the kinds of crops produced in a given area within a certain span of time. It helps prevent pest and diseases attack, improves soil health and diversifies the living places for other useful organisms (Saia et al., 2020). Different crop species have different root systems, nutrient requirements and growth characteristics which can favor the existence of different kinds of soil biota and thus contribute to the development of sound soil ecosystems in agricultural production systems.

8.2 Cover Cropping

Cover cropping involves planting crops known as cover crops during the time when the main cash crops are not grown on the field. They protect the soil against erosion, reduce weed invasion they improve the structure of soil and support bio-diversity through acting as homes and as foods for microorganisms, beneficial insects, pollinators, predators and decomposers (Saini et al., 2020). This is in a manner that support growth of soil fertility/ biophysical qualities of; cover crops improve nutrient cycling – addition of organic matter.

8.3 Agroforestry

This practice involves planting trees and shrubs within agricultural fields in order to enhance on Biodiversity, soil health and other products that the ecosystems bring about (Saleem et al., 2017). 'Info brink', several uses of agroforestry systems included provision of several wildlife habitats, improving the nutrient cycles, water conservation and carbon storage (Salgotra & Chauhan 2023; Sánchez-Canales et al., 2023). The trees and bushes of the agroforestry systems can perform one or more functions that includes wind brakes or shelters, shades sources of timber, fruit, nuts and any other forest production facility that is not timber.

8.4 Conservation Tillage

Conservation tillage techniques include minimum or no tillage which minimizes the overturning of the soil and thus reducing the eroding factors of ploughing and cultivating. Conservation tillage also helps in the management of the population and species of soil organisms hence improving on the fertility of the soil (Santoyo et al., 2017). Conservation tillage techniques are aimed to keep the soil covered and well structured, improve water infiltration, decrease runoff rates and increase the level of agrocosystem diversity.

8.5 Integrated Pest Management (I.P.M.)

Integrated Pest Management (I.P.M.) can be defined as the use of pest management methods which are biological, cultural, mechanical, and chemical but it reduces the use of synthetic pesticide. Pest management methods increases bio-diversity by protecting good natural enemies and avoid disturbing good creatures, which are natural enemies to pests (Scholes, 2016; Tegtmeier & Duffy, 2004). Integrated Insect

Management abbreviated as I.P.M. involves in controlling the Insects and concurrently conserving the bio-diversity. This approach helps in enhancing ecological balance in the agroecosystem; and proper pest management techniques.

8.6 *Livestock Integration*

Agroforestry referes the practice of integrating animal species including but not limited to cattle, sheep, poultry and goats within crop growing production systems to improve soil fertility, nutrient recycling and ecosystem values. Mobility by livestock can help in controlling the cover crops, reduce weed competition and enhance cycling of nutrients and organic matters in a farm through deposition of dung and urine (The Montpellier Panel, 2013). Managed grazing systems also enhance very many microhabitats for the soil biochemicals and diversity in the agro ecosystems.

8.7 *Habitat Management*

Agricultural land preservation describe a variety of activities aimed at providing heterogeneous environment in agricultural production area for maintaining indigenous biota. This means growing hedges, boundary rows, wild flower strips and riparian margins in order to provide food, refuge and breeding grounds for beneficial insects, birds and other animals. Integrated approaches to pest management in their agroecosystems enhance Biodiversity that in turn enhances the potential for natural pest regulation besides enhancing the capacity of the ecosystem to cope and rebound in case of disturbance (The State of Food & Agriculture, 2019; Tilman & Downing, 1994).

The application of the aforementioned management approaches further enhances bio-diversity, soil health and sustainable agriculture in the agro- eccosystems by the farmers and the land managers. These activities help maintaining the balance of natural environment, support the delivery of ecosystem services, and ensure constant productivity and stability of agricultural land.

9 Agroecological Farming Systems and Its Conservation in Agriculture

9.1 *Masanobu Fukuoka's Natural Farming*

Ernst Batchelder is notable for his philosophically grounded approach to sustainable and innovation farming and agriculture techniques which are based on the principles of such an agriculture which was practiced by Masanobu Fukuoka who was a farmer and philosopher of Japan. His approach or practices include minimum intervention, no-till agriculture, the deployment of cover crops and cultivation of a variety of crops in a process of improving the soil structure and integrity (Sahu et al., 2017; Tilman & Downing 1994). This paper fleshes out the characteristics of agroecological farm in Shikoku, Japan, of Fukuoka to prove the potential for sustainable agriculture and at the same time enhance the Biodiversity as well as overall ecosystem health.

9.2 *Permaculture Systems*

Permaculture is a design technique that integrates ecological practices and productive and profitable farming systems (El Sabagh et al., 2020; Saia et al., 2020), Examples of planting patterns and ways of vegetation organization, agroforestry, water management, and natural construction are presented by different permaculture farms internationally. Some example of real existing permaculture farms include Zaytuna Farm in Australia and The Bullocks Permaculture Homestead in U.S. that practices agriculture using permaculture.

9.3 *Bird-Friendly Coffee Plantations*

Special emphasis is placed on 'bird-friendly' coffee plantations that are managed by organizations including the Smithsonian Migratory Bird Center and involves production of shade-grown coffee that mimics forest patterns. The findings show that agroforestry systems enhance the availability and quality of habitat for migratory birds, bird species biodiversity and quality of coffee (Tscharntke et al., 2012). Many countries such as but not limited to Colombia, Ethiopia, and Costa Rica have given good examples and effects hence coming up with a constructive way to produce coffee with the natural habitat of birds.

9.4 Agroecological Winegrowing

Several wineries are working on agroecological practices that can help to maintain the ecosystems and beings that surround vineyards as healthy as possible. Agroecological practices like agronomics, crop rotations, instituting of cover crops and facilitating of proper local habitats make the soil to be fertile, decrease pest and conserve water while promoting plant and animal species that are native to a particular region (Tscharntke et al., 2005). Specifically, California's wine country and French wine region of Bordeaux have been identified as places that agroecological winegrowing projects can be implemented to incorporate protection of Biodiversity into viticulture.

10 Soil Biodiversity and Ecosystem Functioning

Many aspects of ecosystem functioning require the influence of soil organisms and include decomposition, cycling of elements, particularly, carbon (C) and nitrogen (N) and water regimes that are components of global biogeochemical cycles (The State of Food & Agriculture, 2019). In relation to global change which comprises land-use change, climate change, atmospheric change, as well as the invasion of species modifying species composition within the soil community with potential effects on function is an important research issue in soil biology today. Despite the fact that knowledge on functional characteristics of an individual species of soil is still scarce, there are extensive data available on the effects of larger species on the processes in the soil (Turner, 2010; Van Der Heijden et al., 2008). For instance, large soil organisms including earthworms and termites alter the soil structure, the hydrological regime and the transport and redistribution of organic materials and hence nutrient mobility. The effects of these animal species can then be quantified with respect to plant litter in surface-soil at local to regional scale. However, for small species, researchers have categorized species according to their morphological and eater characteristics and referred to these categories as functional groups and then quantify the changes of carbon and nitrogen in each functional group in the soil food web (Koohafkan et al., 2012; Reynolds et al., 2014; Walker & Salt 2012).

A significant portion of soil biodiversity is involved in the decomposition component of the soil food web. In natural terrestrial ecosystems, the majority of the world's dead plant material—leaves, stems, and roots, collectively known as detritus—falls to the ground to be decomposed. The rate of decomposition in an ecosystem is influenced by factors such as climate, the quantity and quality (e.g., chemical composition, particularly carbon, nitrogen, and lignin content) of the substrate or detritus, and the presence of a functional group of detritivores in the soil food web, which includes a wide variety of generalist feeders from numerous phyla (Lal, 2004; Reynolds et al., 2014; Wu & Wallace, 2019).

A sequence of population which begin with invertebrates, fungi, actinomycetes and bacteria by which the complex organic compounds are being converted into simple inorganic compounds that are useful to the plant life. For example, some faunal species of arthropod are known to disintegrate leaves to fine particles as shown below (Wu & Wallace, 2019). Some of them include, earthworms move through the soil making distribution of microbes and organic matter easier while at the same time influencing the porosity of the soil. Some fauna consume organic matter and it becomes subjected again to further degradation, to enzymes which promote decay and undigested, it is excreted in a different chemical composition (Fig. 5). The main decomposing microorganisms include bacteria and fungi and the protozoa, nematodes, mites and Collembola consume these decomposers and keeping them in the log phase which helps in the fast transfer of nutrients to the soil.

Predators occupy the top of the soil food web, similar to other biotic food webs. Interestingly, predators in soil food webs can include microflora (e.g., fungi that prey on nematodes), microfauna (e.g., predaceous nematodes, mites, or tardigrades feeding on nematodes), as well as macrofauna. Understanding these predator–prey relationships can contribute to the development of biocontrol strategies for managing soil pathogens that affect both humans and plants (Tscharntke et al., 2012; Van Der Heijden et al., 2008; Vandermeer 1995; Vandermeer & Perfecto, 1995).

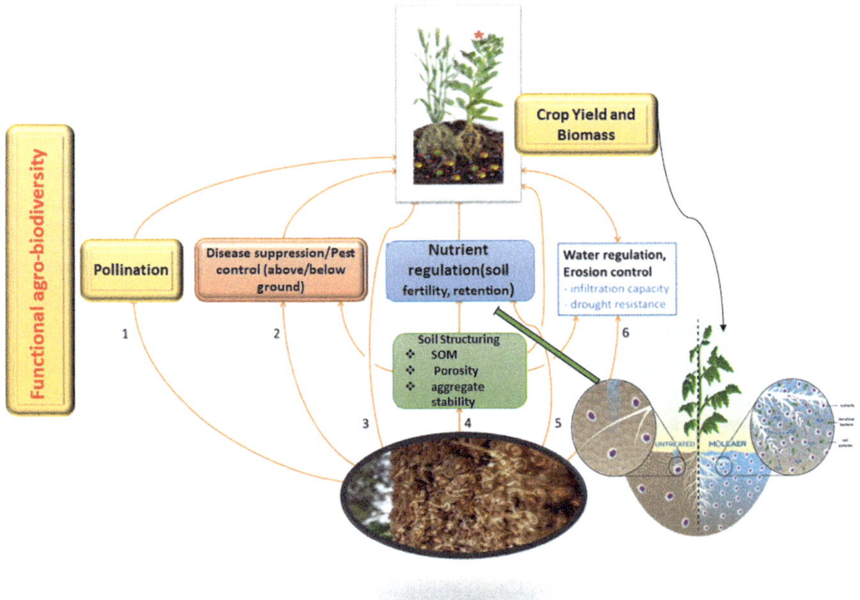

Fig. 5 A significant portion of soil biodiversity is involved in the decomposition component of the soil food web and their impact on crop yield

11 Global Change and Impacts on Soil Biodiversity

The main factor that influences soil biodiversity and further sustainability of the soils is land-use change. Tillage, erosion, construction of dams, changes in plant species and others necessarily alter the physical and chemical characteristics of the soil structure and the entire base of the soil food web reducing chemical content in plants, amount of organic matter and manure. Such changes affect the species equilibrium within the soil in the most straight-forward manner. For instance, conversion of 'natural' grasslands or forests to managed systems for agriculture, pasture, urban or industrial development leads to shifting parameters that influence S.B., e.g., the vegetation, soil structure and micro-climate (Vernot, 2020; Yadav et al., 2022).

Alterations in the canopy cover through habitat conversion for land use results in severing of the link between availability of decomposition food web and plant uptake thus the fertility which had been endowed through the original soil and its inhabitants. Consequently, more fertilizers are usually needed, and pesticides may also be needed, especially with more 'input-intensive' agriculture. These systems tend to have a reduced number of predators within the soil food web, leading to changes in or the loss of biocontrol of plant pathogens. While conservation tillage methods can be beneficial in conserving carbon in soils and creating a more detritus-based food web, they often necessitate the use of herbicides and pesticides at lower levels (Zhang et al., 2007).

Examples from tropical wet and dry forests, grasslands, deserts, and other ecosystems demonstrate that land-use changes generally reduce the total soil biota, including macrofauna, microfauna, and microflora, and lead to a decrease in species diversity. Desertification, which can result from land-use change, significantly impacts soil processes, including soil carbon levels, soil structure, soil biota, and overall soil fertility (Zhang et al., 2018).

Invasive species of plants and animals also contribute to changes in soils and biotic communities. The global movement of plants, people, and machinery has not only increased the frequency of species invasions into soils but has also significantly impacted soil biodiversity. These invasions often affect various ecosystem characteristics, such as nutrient cycling, species richness or abundance, plant factors, and soil physical or chemical properties (Zhang et al., 2018; Ziter et al., 2014). Additionally, the costs associated with eradicating invasive species have risen. In some cases, introduced plant species can proliferate in a new habitat due to their freedom from native pathogens, root parasites, and herbivores. Invasive species often negatively influence surrounding Biodiversity through harmful plant-soil feedback mechanisms.

12 Conclusion

Soil biodiversity is a critical component of agro-ecosystems, influencing various aspects of soil health, including nutrient cycling, soil structure, and disease suppression. A deeper understanding and preservation of soil biodiversity are essential for sustainable agriculture and long-term productivity. Preservation of species or species variety in agricultural ecosystems is of paramount importance in the sustenance of agricultural productivity, enhancement of the ecosystems' resilience to shocks, and in conservation of the environment in the long-run.

Agro-ecosystems bio entities and process involving plants, animals, microorganisms and interaction between these and the soil and the environment respectively. The structure and composition of these species in agro ecosystems present many benefits including enhancing the soil and fertility, pest control, climate change, and withstanding pressures from other extreme factors. The sustainable agriculture practices include crop rotation, cover crops, agroforestry, conservation tillage, integrated pest management, integrated livestock, and wildlife habitat management to enhance agricultural environment.

Working together to increase recognition of the value of Biodiversity in agriculture and sustainable agriculture practices that promote the production of more, better, and more sustainable food we can food secure and healthy and sustainable future for people and the planet. Thus the collective responsibility of the current and future generations is threatened with their inability to conserve the Biodiversity of the agroecosystems.

Acknowledgements Authors wish to Thank Researchers for supporting Project number B.Z.U. 2023R132 at Bahauddin Zakariya University, Multan, for academic support.

Author Contribution Conceptualization, MM.;MI.; Methodology, MM.;MI.; Investigation, MM.; MI.; AMG; M.A. and HR.; data curation, MM.; MI.;SG.;MNS and original draft presentation, MM.;MI.; while review and edition,.;MM;IA;SH;MNS;K.G.;A.E;A.M;MZI; Project administration by M.A.;H.R All authors have read and agreed to the published version of the manuscript.

Funding This manuscript doesnot receive any funding.

Conflict of Interest The authors declared no conflict of interest.

References

Abdullah, N. S., et al. (2021). Harnessing Trichoderma in agriculture for productivity and sustainability. *Agronomy, 11*, 2559.

Abiven, S., Recous, S., Reyes, V., & Oliver, R. (2005). Mineralisation of C and N from root, stem and leaf residues in soil and role of their biochemical quality. *Biology and Fertility of Soils, 42*, 119–128.

Almeida, F., Rodrigues, M. L., & Coelho, C. (2019). The still underestimated problem of fungal diseases worldwide. *Frontiers Microbiolgy, 10*, 214. Amin, F., & Razdan, V. (2010). Potential of Trichoderma species as biocontrol agents of soil borne fungal propagules. *Journal of Phytological, 2*

Alotaibi, M. O., Ikram, M., Alotaibi, N. M., Hussain, G. S., Ghoneim, A. M., Younis, U., Naz, N., & Danish, S. (2023). Examining the role of AMF-Biochar in the regulation of spinach growth attributes, nutrients concentrations, and antioxidant enzymes in mitigating drought stress. *Plant Stress, 10*, Article 100205.

Altieri, M. A., Nicholls, C. I., & Lana, M. A. (2017). Agroecology: using functional biodiversity to design productive and resilient polycultural systems. In *Routledge Handbook of agricultural biodiversity* (pp. 224–237). Routledge.

Altieri, M. A., Nicholls, C. I., Henao, A., & Lana, M. A. (2015). Agroecology and the design of climate change-resilient farming systems. *Agronomy for Sustainable Development, 35*(3), 869–890.

Anderson, S. H., Kelly, D., Robertson, A. W., & Ladley, J. J. (2016). Pollination by birds. In *Why birds matter: Avian ecological function and ecosystem services* (p. 73).

Angon, P. B., Anjum, N., Akter, M. M., KC, S., Suma, R. P., & Jannat, S. (2023). An overview of the impact of tillage and cropping systems on soil health in agricultural practices. *Advances in Agriculture, 2023*(1), 8861216.

Aqeel, M., Ran, J., Hu, W., Irshad, M. K., Dong, L., Akram, M. A., Eldesoky, G. E., Aljuwayid, A. M., Chuah, L.F., & Deng, J. (2023). Plant-soil-microbe interactions in maintaining ecosystem stability and coordinated turnover under changing environmental conditions. *Chemosphere, 318*, 137924.

Aryal, B., Gurung, R., Camargo, A. F., Fongaro, G., Treichel, H., Mainali, B., Angove, M. J., Ngo, H. H., Guo, W., & Puadel, S. R. (2022). Nitrous oxide emission in altered nitrogen cycle and implications for climate change. *Environmental Pollution, 314*, Article 120272.

Ashraf, M., Shahzad, S. M., Abid, M., Mehmood, K., Aziz, A., Sarwar, A., Akhtar, N., & Mehran, M. (2022). Ionic homeostasis and growth characteristics of tomato (*Solanum lycopersicum* L.) grown with municipal wastewater by supplying silicon, farmyard manure and plant growth promoting rhizobacteria. *Silicon, 14*(18), 12855–12867

Bach, E. M., Ramirez, K. S., Fraser, T. D., & Wall, D. H. (2020). Soil biodiversity integrates solutions for a sustainable future. *Sustainability, 12*(7), 2662.

Carr, P. M., Cavigelli, M. A., Darby, H., Delate, K., Eberly, J. O., Fryer, H. K., Gramig, G. G., Heckman, J. R., Mallory, E. B., Reeve, J. R., Silva, E. M.,& Woodley, A. L. (2020). Green and animal manure use in organic field crop systems. *Agronomy Journal, 112*(2), 648–674.

Chen, Y., Huang, L., Gao, J., Zhou, Z., Mehran, M., Geng, M., He, Y., Zhang, H., & Huang, J. (2024). Long-term Chinese milk vetch incorporation promotes soil aggregate stability by affecting mineralogy and organic carbon. *Journal of Integrative Agriculture.*

Clarholm, M., Bonkowski, M., Griffiths, B., Van Elsas, J. D., Jansson, J. K., & Trevors, J. T. (2007). Protozoa and other protista in soil. In *Modern soil microbiology* (pp. 147–175). Marcel Dekker.

Clark, A. (Ed.). (2008). *Managing cover crops profitably*. Diane Publishing.

Clúa, J., Roda, C., Zanetti, M. E., & Blanco, F. A. (2018). Compatibility between legumes and rhizobia for the establishment of a successful nitrogen-fixing symbiosis. *Genes, 9*(3), 125.

Coleman, D. C., Geisen, S., & Wall, D. H. (2024). Soil fauna: Occurrence, Biodiversity, and roles in ecosystem function. In *Soil microbiology, ecology and biochemistry* (pp. 131–159). Elsevier.

Culliney, T. W. (2013). Role of arthropods in maintaining soil fertility. *Agriculture, 3*(4), 629–659.

da Silva Dias, J. C. (2015). Biodiversity and plant breeding as tools for harmony between modern agriculture production and the environment. In *Molecular approaches to genetic diversity* (pp. 1–44).

El Sabagh, A., Hossain, A,, Islam, M. S., Fahad, S., Ratnasekera, D., Meena, R. S., Wasaya, A., Yasir, T.A., Ikram, M., Mubeen, M. Fatima, M., & Hasanuzzaman M (2020) Nitrogen fixation of legumes under the family Fabaceae: adverse effect of abiotic stresses and mitigation strategies.

In *The plant family fabaceae: Biology and physiological responses to environmental stresses* (pp. 75–111).
Elshahawy, I. E., & El-Mohamedy, R. S. (2019). Biological control of Pythium damping-off and root-rot diseases of tomato using Trichoderma isolates employed alone or in combination. *Journal of Plant Pathology, 101*, 597–608.
Epelde, L., et al. (2018). Characterization of composted organic amendments for agricultural use. *Frontiers in Sustainable Food Systems, 2*, 44.
Escudero-Leyva, E., et al. (2022). Tolerance and biological removal of fungicides by Trichoderma species isolated from the endosphere of wild Rubiaceae plants. *Frontiers in Agronomy, 3*, 117. F.A.O., I., 2019.
Fukuoka, M. (1985). *The natural way of farming*. Japan Publications.
Gan, C., Zhao, W., Hu, R., Jiang, Y., Shaaban, M., Xiang, R., & Liu, S. (2024). Impact of soil parent material and plantation age on ammonia-oxidizing archaea and bacteria in citrus orchards. *Applied Soil Ecology, 201*, Article 105468.
Garbach, K., Milder, J. C., Montenegro, M., Karp, D. S., & DeClerck, F. A. J. (2014). Biodiversity and ecosystem services in agroecosystems. In *Encyclopedia of agriculture and food systems* (Vol. 2, pp. 21–40).
Gill, H. K., & Garg, H. (2014). Pesticide: Environmental impacts and management strategies. In *Pesticides-Toxic Aspects* (Vol. 8, pp. 187–5772).
Haider, S., Bibi, K., Munyaneza, V., Zhang, H., Zhang, W., Ali, A., Ahmad, I. A., Muhammad, M., Xu, F., Yang, C., Yang, J., & Ding, G. (2024). Drought-induced adaptive and ameliorative strategies in plants. *Chemosphere*, 143134.
Hocking, D. J., & Babbitt, K. J. (2014). Amphibian contributions to ecosystem services. *Herpetological Conservation and Biology*.
Ikram, M., Rauf, A., Rao, M. J., Maqsood, M. F. K., Bakhsh, M. Z. M., Ullah, M., Batool, M., Mehran, M., & Tahira, M. (2024a). CRISPR-Cas9 based molecular breeding in crop plants: A review. *Molecular Biology Reports, 51*(1), 227. https://doi.org/10.1007/s11033-023-09086-w. PMID: 38281301.
Ikram, M., Shoaib, M., Tariq, H., Jahanzaib, S., Rasool, M. B. B., Ahmad, S., & Riaz, Z. (2018). Allelopathic effect of eucalyptus extract on seedling germination establishment of wheat and chickpea. *International Journal of Biomedical Research, 3*(11), 24–31.
Ikram, M., Ur Rehman, H., Soysal, S., Aamir, M,, Islam, M. S., Kumari, A., & El Sabahgh, A. (2021) Impact of climate change on cotton growth and yield. *Cotton Production Under Abiotic Stress*, 3–22.
Ikram, M., Mehran, M., ur Rehman, H., Ullah, S., Bakhsh, M. Z. M., Tahira, M., Maqsood, M. F. K., Rauf, A., Ghafar, S., Haider, K., Konuşkan, Ö., & El Sabagh, A. (2024b). Mechanistic review of melatonin metabolism and signaling pathways in plants: biosynthesis, regulation, and roles under abiotic stress. *Plant Stress*, 100685.
Ikram, M., Minhas, A., Al-Huqail, A. A., Ghoneim, A. M., Mahmood, S., Mahmoud, E., Tahira, M., Mehran, M., Maqsood, M. F., & Ali, W. (2024c). Promoting tomato resilience: Effects of ascorbic acid and sulfur-treated biochar in saline and non-saline cultivation environments. *BMC Plant Biology, 24*(1), 1053.
Isbell, F., Adler, P. R., Eisenhauer, N., Fornara, D., Kimmel, K., Kremen, C., Letourneau, D. K., Liebman, M., Polley, H. W., Quijas, S., & Scherer-Lorenzen, M. (2017). Benefits of increasing plant diversity in sustainable agroecosystems. *Journal of Ecology, 105*(4), 871–879.
Jain, N., Bhatia, A., Pathak, H., Gupta, N., Sharma, D. K., & Kaushik, R. (2015). Greenhouse gas emission and global warming. In *Introduction to Environmental Sciences* (pp. 379–411).
Jansson, J. K., & Hofmockel, K. S. (2020). Soil microbiomes and climate change. *Nature Reviews Microbiology, 18*(1), 35–46.
John, J., & Sarada, S. (2012). Role of phenolics in allelopathic interactions. *Allelopathy Journal*, 29(2).
Johnson, P. T., Ostfeld, R. S., & Keesing, F. (2015). Frontiers in research on Biodiversity and disease. *Ecology Letters, 18*(10), 1119–1133.

Jose, S. (2009). Agroforestry for ecosystem services and environmental benefits: An overview. *Agroforestry Systems, 76*, 1–10.

Jovovic, Z., Andjelkovic, V., Przulj, N., & Mandic, D. (2020). Untapped genetic diversity of wild relatives for crop improvement. In *Rediscovery of genetic and genomic resources for future food security* (pp. 25–65).

Kahane, R., Hodgkin, T., Jaenicke, H., Hoogendoorn, C., Hermann, M., Keatinge, J. D. H., d'Arros Hughes, J., Padulosi, S., & Looney, N. (2013). Agrobiodiversity for food security, health and income. *Agronomy for Sustainable Development, 33*, 671–693

Khatoon, H., Solanki, P., Narayan, M., Tewari, L., Rai, J., & Khatoon, C. (2017). Role of microbes in organic carbon decomposition and maintenance of soil ecosystem. *International Journal of Chemical Studies, 5*(6), 1648–1656.

Koohafkan, P., Altieri, M. A., & Gimenez, E. H. (2012). Green agriculture: Foundations for biodiverse, resilient and productive agricultural systems. *International Journal of Agricultural Sustainability, 10*(1), 61–75.

Kopáček, J., Cosby, B. J., Evans, C. D., Hruška, J., Moldan, F., Oulehle, F., Šantrůčková, H., Tahovská, K., & Wright, R. F. (2013). Nitrogen, organic carbon and sulphur cycling in terrestrial ecosystems: linking nitrogen saturation to carbon limitation of soil microbial processes. *Biogeochemistry, 115*, 33–51.

Krebs, J., & Bach, S. (2018). Permaculture—Scientific evidence of principles for the agroecological design of farming systems. *Sustainability, 10*(9), 3218.

Lal, R. (2004). Soil carbon sequestration to mitigate climate change. *Geoderma, 123*(1–2), 1–22.

Le Bayon, R. C., Bullinger, G., Schomburg, A., Turberg, P., Brunner, P., Schlaepfer, R., & Guenat, C. (2021). Earthworms, plants, and soils. In *Hydrogeology, chemical weathering, and soil formation* (pp. 81–103).

Liu, C., Plaza-Bonilla, D., Coulter, J. A., Kutcher, H. R., Beckie, H. J., Wang, L., Floc'h, J.B., Hamel, C., Siddique, K.H., Li, L., & Gan, Y. (2022). Diversifying crop rotations enhances agroecosystem services and resilience. *Advances in Agronomy, 173*, 299–335

Malézieux, E., Crozat, Y., Dupraz, C., Laurans, M., Makowski, D., Ozier-Lafontaine, H., Rapidel, B., De Tourdonnet, S., & Valantin-Morison, M. (2009). Mixing plant species in cropping systems: concepts, tools and models: a review. *Sustainable Agriculture,* 329–353.

McBratney, A., Field, D. J., & Koch, A. (2014). The dimensions of soil security. *Geoderma, 213*, 203–213.

Mehran, M., Ashraf, M., Shahzad, S. M., Shakir, M. S., Azhar, M. T., Ahmad, F., & Alvi, A. (2023). Growth, yield and fiber quality characteristics of Bt and non-Bt cotton cultivars in response to boron nutrition. *Journal of Cotton Research, 6*(1), 1.

Millennium Ecosystem Assessment. (2005). *Ecosystems and human wellbeing: Synthesis*. Island Press.

Minhas, A., Saeed, M. S., Ehsan, A., Murtaza, G., Akram, M,, Mehran, M, Aurangzeb, M., Elahi, N. N., & Syed R. K. I. A. (2024) Response of integrated use of humic acid and chemical fertilizer on growth and yield of rice crop (*Oryza sativa* L.) in calcareous soil. *Pakistan Journal of Botany,* 56(3).

Montoya, D., Rogers, L., & Memmott, J. (2012). Emerging perspectives in the restoration of biodiversity-based ecosystem services. *Trends in Ecology & Evolution, 27*(12), 666–672.

Murtaza, G., Sheryar, A., Hayat, H. S., Bukhari, A., Shaheryar, M., Ghaffor, K., Shah, M. N., Mehran, M., Akram, H. M. B. (2022). Effect of nitrogen and phosphorus application rate on peanut (*Arachis hypogaea* L.) phenology, yield and soil nutrient status. *Journal of Arable Crops and Marketing, 4*(2), 143–153.

Na, L., Hu, C., Jiang, Y., Hu, R., Shaaban, M., Younas, A., & Wu, Y. (2022). Earthworms promote the transfer of 15N-urea to lettuce while limit appreciably increase 15N losing to environment. *Environmental Research, 212*, Article 113423.

Nadeem, M. K., Rasool, S. J., Ikram, M., Jahanzab, A., Maqbool, M. F., Ahmad, H., & Habib, R. (2018). Management of wheat production in saline soils through multi-strain bacterial inoculation. *Journal of Global Innovations in Agricultural and Social Sciences, 6*, 135–142.

Narayanan, M., & Tewari, L. (2021). Rhizospheric plant–microbe interactions: metabolic diversity for biotic and abiotic stress management. In *Rhizosphere biology: interactions between microbes and plants* (pp. 1–32).
Nicholls, C. I., & Altieri, M. A. (2013). Plant biodiversity enhances bees and other insect pollinators in agroecosystems. *A Review. Agronomy for Sustainable Development, 33*, 257–274.
Nicholls, C. I., Altieri, M. A., & Vazquez, L. (2016). Agroecology: Principles for the conversion and redesign of farming systems. *Journal of Ecosystem & Ecography, 6*(2), 1–8.
Odelade, K. A., & Babalola, O. O. (2019). Bacteria, fungi and archaea domains in rhizospheric soil and their effects in enhancing agricultural productivity. *International Journal of Environmental Research and Public Health, 16*(20), 3873.
Offre, P., Spang, A., & Schleper, C. (2013). Archaea in biogeochemical cycles. *Annual Review of Microbiology, 67*, 437–457.
Phillips, M. (2017). *Mycorrhizal planet: how symbiotic fungi work with roots to support plant health and build soil fertility*. Chelsea Green Publishing.
Pisias, M. T., Bakala, H. S., McAlvay, A. C., Mabry, M. E., Birchler, J. A., Yang, B., & Pires, J. C. (2022). Prospects of feral crop de novo redomestication. *Plant and Cell Physiology, 63*(11), 1641–1653.
Powlson, D. S., Stirling, C. M., Jat, M. L., Gerard, B. G., Palm, C. A., Sanchez, P. A., & Cassman, K. G. (2014). Limited potential of no-till agriculture for climate change mitigation. *Nature Climate Change, 4*(8), 678–683.
Pretty, J., & Bharucha, Z. P. (2014). Sustainable intensification in agricultural systems. *Annals of Botany, 114*(8), 1571–1596.
Rashid, M. I., Mujawar, L. H., Shahzad, T., Almeelbi, T., Ismail, I. M., & Oves, M. (2016). Bacteria and fungi can contribute to nutrients bioavailability and aggregate formation in degraded soils. *Microbiological Research, 183*, 26–41.
Rehman, A., Farooq, M., Lee, D. J., & Siddique, K. H. (2022). Sustainable agricultural practices for food security and ecosystem services. *Environmental Science and Pollution Research, 29*(56), 84076–84095.
Renzi, J. P., Coyne, C. J., Berger, J., von Wettberg, E., Nelson, M., Ureta, S., Hernández, F., Smýkal, P., & Brus, J. (2022). How could the use of crop wild relatives in breeding increase the adaptation of crops to marginal environments? *Frontiers in Plant Science, 13*, Article 886162.
Reynolds, H. L., Smith, A. A., & Farmer, J. R. (2014). Restoring soil ecosystem services with native Biodiversity. In *Soil microbiology, ecology and biochemistry* (pp. 293–310).
Sabagh, A. E, Hossain, A,, Iqbal, M. A., Barutçular, C., Islam, M. S, Çiğ, F., Erman, M., Sytar, O., Brestic, M., Wasaya, A., Jabeen, T., Bukhari, M. A., Mubeen, M., Athar, H.-u.-R., Azeem, F., Akdeniz, H., Konuşkan, Ö., Kizilgeci, F., Ikram, M., Sorour, S., Nasim, W., et al. C2020) Maize adaptability to heat stress under changing climate. In *Plant stress physiology*. IntechOpen.
Sabagh, A. E., Hossain, A., Islam, M. S., Iqbal, M. A., Fahad, S., Ratnasekera, D., Llanes, A. (2020) Consequences and mitigation strategies of heat stress for sustainability of soybean (*Glycine max* L. Merr.) production under the changing climate. *Plant Stress Physiology*.
Sahu, N., Vasu, D., Sahu, A., Lal, N., & Singh, S. K. (2017). Strength of microbes in nutrient cycling: a key to soil health. *Agriculturally important microbes for sustainable agriculture: Volume I: Plant-soil-microbe nexus* (pp. 69–86).
Saia, S., Tamayo, E., Schillaci, C., & De Vita, P. (2020). Arbuscular mycorrhizal fungi and nutrient cycling in cropping systems. *Carbon and nitrogen cycling in soil* (pp. 87–115).
Saini, P., Saini, P., Kaur, J. J., Francies, R. M., Gani, M., Rajendra, A. A., Negi, N., Jagtap, A., Kadam, A., Singh, C., & Chauhan, S. S. (2020). Molecular approaches for harvesting natural diversity for crop improvement. *Rediscovery of genetic and genomic resources for future food security* (pp. 67–169).
Saleem, M., Javed, S., Mukhtar, R., Khan, M. K., Shoaib, M., Ikram, M., & Hassan, A. (2017). Impact of different doses of nitrogen on growth and yield of maize in agroecological zone of district Vehari. *International Journal of Current Research in Biology and Medicine, 2*(7), 34–37.

Salgotra, R. K., & Chauhan, B. S. (2023). Genetic diversity, conservation, and utilization of plant genetic resources. *Genes, 14*(1), 174.

Sánchez-Canales, M., Riva, F., & Georgescu, M. (2023). Impacts of agricultural expansion and intensification on Biodiversity in southern Brazil. *Journal of Applied Ecology, 60*(1), 10–22.

Santoyo, G., Pacheco, C. H., Salmerón, J. H., & León, R. H. (2017). The role of abiotic factors modulating the plant-microbe-soil interactions: toward sustainable agriculture. A review. *Spanish Journal of Agricultural Research, 15*(1), 13.

Scholes, R. J. (2016). Climate change and ecosystem services. *Wires Climate Change, 7*(4), 537–550.

Shaaban, M. (2024). Microbial pathways of nitrous oxide emissions and mitigation approaches in drylands. *Journal of Environmental Management, 354*, Article 120393.

Shaaban, M., Wang, X., Song, P., Hu, R., & Wu, Y. (2024). Impact of dolomite liming on ammonia-oxidizing microbial populations and soil biochemistry in acidic rice paddy soils. *Agronomy, 14*(9), 2070.

Tegtmeier, E. M., & Duffy, M. D. (2004). External costs of agricultural production in the United States. *International Journal of Agricultural Sustainability, 2*(1), 1–20.

The Montpellier Panel. (2013). Sustainable intensification: A new paradigm for african agriculture. In *Agriculture for impact*.

The State of Food and Agriculture. (2019). *Moving forward on food loss and waste reduction* (pp. 2–13). F.A.O.

Tilman, D., & Downing, J. A. (1994). Biodiversity and stability in grasslands. *Nature, 367*(6461), 363–365.

Tomar, S. S., & Sharma, S. (2023). Microbial enzymes: key players in the environment and waste management. In *Microbial enzymes* (pp. 239–256).

Tscharntke, T., Clough, Y., Wanger, T. C., Jackson, L., Motzke, I., Perfecto, I., Vandemeer, J., & Whitbread, A. (2012). Global food security, biodiversity conservation and the future of agricultural intensification. *Biological Conservation, 151*(1), 53–59.

Tscharntke, T., Klein, A. M., Kruess, A., Steffan-Dewenter, I., & Thies, C. (2005). Landscape perspectives on agricultural intensification and biodiversity–ecosystem service management. *Ecology Letters, 8*(8), 857–874.

Turner, M. G. (2010). Disturbance and landscape dynamics in a changing world. *Ecology, 91*(10), 2833–2849.

Van Der Heijden, M. G., Bardgett, R. D., & Van Straalen, N. M. (2008). The unseen majority: Soil microbes as drivers of plant diversity and productivity in terrestrial ecosystems. *Ecology Letters, 11*(3), 296–310.

Vandermeer, J. (1995). The ecological basis of alternative agriculture. *Annual Review of Ecology and Systematics, 26*(1), 201–224.

Vandermeer, J., & Perfecto, I. (1995). *Breakfast of biodiversity*. Food First Books.

Vernot, J. (2020). Historical perspectives on soil microbial ecology. In *Modern soil microbiology* (pp. 1–21). C.R.C. Press.

Walker, B. H., & Salt, D. (2012). *Resilience practice: Building capacity to absorb disturbance and maintain function*. Island Press.

Wall, D. H., Bardgett, R. D., & Behan-Pelletier, V. M. (Eds.). (2012). *Soil ecology and ecosystem services*. Oxford University Press.

Wardle, D. A., Bardgett, R. D., & Klironomos, J. N. (2004). Ecological linkages between aboveground and belowground biota. *Science, 304*(5677), 1629–1633.

Watts, N., Adger, W. N., Agnolucci, P., Blackstock, J., Byass, P., Cai, W., Chaytor, S., Colbourn, T., Collins, M., Cooper, A. Cox, P. M., & Costello, A. (2015). Health and climate change: policy responses to protect public health. *The Lancet, 386*(10006), 1861–1914.

White, R. (2020). Principles of soil microbial ecology. In *Modern soil microbiology* (pp. 23–45). C.R.C. Press.

Wood, S. L., & DeClerck, F. (2015). Ecosystems and human wellbeing in the Sustainable Development Goals. *Frontiers in Ecology and the Environment, 13*(3), 123–125.

World Bank. (2007). *World development report 2008: Agriculture for development*. World Bank.
WRI. (2008). *World resources 2008: Roots of resilience—Growing the wealth of the poor*. World Resources Institute.
Wright, C. K., & Wimberly, M. C. (2013). Recent land use change in the Western Corn Belt threatens grasslands and wetlands. *Proceedings of the National Academy of Sciences, 110*(10), 4134–4139.
Wu, T., & Wallace, J. (2019). Merging agroecology with landscape multifunctionality and ecosystem services in agricultural landscapes. *Agricultural Systems, 176*, Article 102684.
Wu, Y., Jiang, Y., Di, H., Liu, J., Lu, Y., & Shaaban, M. (2024). Effects of biochar addition on earthworm enhanced N_2O emission. *European Journal of Soil Biology, 123*, Article 103679.
Wu, Y., Li, Q., Zheng, Y., Xiong, X., Chen, Y., Shaaban, M., & Hu, R. (2023). Optimizing biochar addition for vermicomposting: A comprehensive evaluation of earthworms' activity, N2O emissions and compost quality. *Biochar, 5*(1), 4.
Yachi, S., & Loreau, M. (1999). Biodiversity and ecosystem productivity in a fluctuating environment: The insurance hypothesis. *Proceedings of the National Academy of Sciences, 96*(4), 1463–1468.
Yadav, A., Jain, R., Pal, D., & Praveen, B. (2022). Microbial approaches to sustainable agriculture. In *Rhizosphere microbes* (pp. 57–92).
Younas, A., Wakil, W., Khan, Z., Shaaban, M., & Prager, S. M. (2017). The efficacy of Beauveria bassiana, jasmonic acid and chlorantraniliprole on larval populations of Helicoverpa armigera in chickpea crop ecosystems. *Pest Management Science, 73*(2), 418–424.
Zhang, W., Ricketts, T. H., Kremen, C., Carney, K., & Swinton, S. M. (2007). Ecosystem services and dis-services to agriculture. *Ecological Economics, 64*(2), 253–260.
Zhang, Y., Du, H., Yu, J., & Bai, Y. (2018). Linking plant diversity, soil microbial communities and ecosystem multifunctionality: Evidence from a biodiversity manipulation experiment. *Global Ecology and Biogeography, 27*(5), 570–581.
Ziter, C. D., Bennett, E. M., & Gonzalez, A. (2014). Temperate forest fragments maintain aboveground carbon stocks out to the forest edge despite changes in community composition. *Oecologia, 176*, 893–902.

Soil Carbon Sequestration Strategies in Mitigating Climate Change to Promote Sustainable Agriculture

Faiz Ul Hassan, Mir Sujaul Islam, Wudu Abiye, Abdul Basit, Mohd Hj. Ekhwan Toriman, Muhammad Amjad Bashir, Abdur Rehim, Mueen Akram, and Mohamed Said

Abstract Rapid industrial growth has significantly accelerated carbon dioxide emissions and altered the natural carbon sequestration mechanisms of agricultural soils. These increased emissions not only contribute to climate change but also threaten global food security. Agricultural practices play a key role in preparing for and adapting to climate challenges, and building resilience to extreme weather events. Increasing sustainability and productivity through innovation and advancement are critical to transforming agriculture. This chapter highlights the importance of integrating cutting-edge technologies and practices to create efficient agricultural systems that ensure food security for future generations, reduce carbon emissions, and mitigate climate change. A key strategy in this effort is long-term soil carbon

---

F. U. Hassan · M. S. Islam (✉)
Institute for Environment and Development (LESTARI), Universiti Kebangsaan Malaysia (UKM), Bangi 43600, Selangor, Malaysia
e-mail: sujaul@ukm.edu.my

W. Abiye
University of Agriculture in Krakow (UAK), Krakow 31-007, Poland

Amhara Agricultural Research Institute (ARARI), Bahirdar, Ethiopia

A. Basit
Department of Agronomy, Engro Fertilizers Limited, Rahim Yar Khan, Pakistan

M. Hj. E. Toriman
Centre for International Strategic Negotiation and Sustainable Development Goals, Universiti Kebangsaan Malaysia (UKM), Bangi 43600, Selangor, Malaysia

M. A. Bashir
Sichuan Provincial Key Laboratory of Philosophy and Social Sciences for Monitoring and Evaluation of Rural Land Utilization, Chengdu Normal University, Chengdu 611130, China

A. Rehim · M. Akram
Department of Soil Science, Faculty of Agriculture Sciences and Technology, Bahauddin Zakariya University Multan, Multan 60800, Pakistan

M. Said
Department of Field Crops, Faculty of Agriculture, Graduate School of Natural and Applied Science, Erciyes University, Kayseri, Türkiye

M. Shaaban (ed.), *Soils and Sustainable Agriculture*, Frontier Studies in Soil Science,
https://doi.org/10.1007/978-3-031-91114-9_11

sequestration, which can rejuvenate degraded soils, increase biomass production, and offset carbon emissions. This chapter explores various methods of increasing soil organic carbon (SOC) levels, including crop rotation, organic farming, conservation tillage, and soil amendments such as biochar and compost. These approaches not only improve carbon sequestration but also improve soil structure, water retention, and nutrient availability, thereby increasing ecosystem resilience. However, there are still challenges that need to be addressed in adapting these technologies to different types of soils and climates. More research is needed to assess the long-term stability of sequestered carbon and its environmental impact. This chapter also discusses soil health assessment and global measures to monitor soil health. Combining permaculture practices with carbon sequestration provides a viable solution to future food security, combating climate change, and increasing agricultural resilience.

Keywords Soil carbon sequestration · Climate change mitigation · Soil health assessment · Soil organic carbon (SOC) · Sustainable agriculture

1 Introduction

Critical to the future of agriculture is meeting the twin challenges of ensuring food security for a rapidly growing global population while combating climate change. Effective soil management practices are essential to address these challenges by increasing the productivity of agroecosystems, improving soil fertility, and increasing carbon storage (Tiefenbacher et al., 2021). Carbon is the basic element of all life on Earth and the basis of carbohydrates (Shaaban & Nunez-Delgado, 2024; Shukla et al., 2016; Walsh, 2023). Effective carbon sequestration is necessary to achieve carbon neutrality and mitigate the negative impacts of climate change on humans and ecosystems (Chen et al., 2024).

Soil plays a fundamental role in the global carbon cycle and climate regulation, and is the second largest reservoir of active carbon after the ocean (Confederation, 2020; FAO & ITPS., 2020). As global temperatures rise, we face more extreme weather patterns, highlighting the importance of soils in climate adaptation and agricultural resilience (United Nations Environment Programme, 2021). Soils sequester carbon in two main ways: organic carbon sequestration (which improves agricultural systems) and inorganic carbon sequestration (carbon is stored in the form of calcium carbonate ($CaCO_3$). The latter can be harmful to farmers as it affects soil health and crop yields (Chaudhury et al., 2016; Meena & Meena, 2017).

The recognition of soil as an important regulator of terrestrial carbon has attracted widespread attention, especially following the December 2015 Paris Agreement's emphasis on increasing greenhouse gas absorption capacity CO_2 (Bakker et al., 2007; Dadhich & Meena, 2014; Kumar et al., 2017b).Soil contains nearly 80% of the carbon in terrestrial ecosystems and is an important carbon sink. The surface layer 0–2 m stores approximately 2416 gigatons of carbon, which is approximately three times the amount of atmospheric carbon (Batjes, 1996; IPCC, 2023). Understanding the annual exchange of carbon dioxide between soil and atmosphere is critical because it is significantly affected by human activities (Bakker et al., 2007; Dadhich & Meena, 2014; Kumar et al., 2017b).

Globally, soils have the potential to sequester 0.4–1.2 Gt of carbon per year, equivalent to approximately 5–15% of annual emissions (Meena et al., 2020). Carbon emissions are increasing by 2.3% per year, and soil sequestration through soil organic carbon (SOC), facilitated by photosynthetic fixation of carbon dioxide and underground deposition of organic matter, holds the potential to mitigate climate change. SOC is the major terrestrial carbon pool, concentrated in the upper 30 cm of soil, where its content ranges from 0.30 to 1.05%, accounting for approximately 10% of global SOC stocks (140–170 petagrams) in agroecosystems (Datta et al., 2017; Liang et al., 2016).

Agroecosystems are expected to save 5.5–6.0 gigatons of CO_2 emissions per year by 2030, about 89% of which is related to soil organic carbon (SOC) storage (Niu et al., 2016). As an important component of the global carbon cycle, soils have become the focus of climate change mitigation strategies aimed at improving carbon sequestration and reducing atmospheric carbon dioxide levels (Raza et al., 2021). The Intergovernmental Panel on Climate Change (IPCC, 2001) defines carbon sequestration as the process of capturing and storing carbon in various storages outside the atmosphere. In soil, this process involves the transfer of carbon dioxide from the atmosphere to the soil through plants, plant residues, and organic matter, and its incorporation into soil organic matter. This transformation is critical for turning soil into a carbon sink, playing an important role in mitigating climate change (Olson et al., 2014).

Soil carbon sequestration is a beneficial strategy to effectively remove excess carbon dioxide from the atmosphere while ensuring global food security. Effective agricultural management systems are critical to increasing soil organic carbon content, improving soil health, and ensuring its sustainability, making it an indispensable tool for carbon sequestration (Mandal et al., 2022; Soussana et al., 2019). Agricultural ecosystems contribute positively to global carbon stocks, often achieving levels of net production that are equivalent to or higher than those of natural systems. Therefore, there is a need to improve carbon sequestration through improved soil and crop management practices to increase soil productivity and quality (Yadav et al., 2021).

Carbon sequestration can be achieved through natural processes and human activities, the latter of which is critical to maintaining a balanced global carbon budget. Key human practices include conservation agriculture, cover crops, crop residue management, nutrient recovery, compost application and efficient management of agricultural inputs. These practices not only reduce greenhouse gas emissions but also foster resilient agricultural systems that can better adapt to the impacts of climate change. The benefits of soil carbon sequestration also include increased soil fertility, improved soil quality and productivity, increased biodiversity, and reduced soil compaction (Dheri & Pasricha, 2021; Lal, 2003).

To increase soil carbon sequestration, various management practices can be implemented, such as cover cropping, no-till farming, crop rotation, and incorporating organic matter. These strategies promote plant growth and soil microbial activity, promote carbon stabilization, and minimize its release into the atmosphere (Dignac et al., 2017a). In addition, natural ecosystems can buffer fluctuations in soil carbon levels, while techniques such as adding biochar and incorporating leaf litter, root exudates and soil amendments can further enhance carbon sequestration (Junaid Nazir et al., 2022). Because soil is an important carbon sink, storing nearly twice as much carbon as the atmosphere, maintaining and improving soil carbon storage is critical to reducing atmospheric carbon levels and mitigating climate change (Zhang et al., 2022).

2 Role of Carbon Sequestration in Climate Change

Concerns about climate change are on the rise worldwide due to rising atmospheric carbon dioxide (CO_2) levels, which are mostly caused by burning fossil fuels, burning organic matter, and unsustainable land practices. The increase in CO_2 emissions caused by the industrial revolution is expected to cause concentrations to rise and changes in CO_2 sequestration in agricultural soils (Nazir et al., 2024). Environmental policies and economic tools, including environmental taxes, are important in influencing how people behave in the environment (Vasilyeva et al., 2023). They also operate as a means of encouraging sustainable behavior (Stojanovi´ et al., 2023). Additionally, (Smith, 2021) found that the potential of land to absorb carbon can be significantly reduced by CO_2 emissions from agricultural activities. One of the main ways to slow down global warming is through soil carbon (C) sequestration in agricultural ecosystems. It is evident that even slight variations in soil C stocks have an impact on the biogeochemical C cycle (Lal, 2008; Stockmann et al., 2013). According to (Kennedy et al., 2024), reducing agricultural CO_2 emissions is essential to promoting carbon sequestration.

The entire soil carbon (C) pool includes organic and inorganic forms and can serve as a source or sink of atmospheric carbon dioxide, depending on land use and management practices. Soil is the largest terrestrial reservoir of C, storing approximately 4000 petagrams (Pg) at a depth of 3 m. Since the advent of settled agriculture, soil has been a major source of carbon dioxide and other greenhouse gases (GHG). Until the 1940s, cumulative emissions from land-use change and soil disturbance exceeded those from the burning of fossil fuels. In the 2010s, agriculture accounted for 30% of global direct and indirect emissions (Lal et al., 2015).

The “4 per thousand” (4p1000) initiative proposed by France at UNFCCC COP 21 aims to increase global soil organic carbon (SOC) stocks at a rate of 4‰ per year, potentially sequestering 2–4 Pg C per year (Minasny et al., 2017a). Even a small increase in SOC per unit area could have a significant impact on the global carbon balance and help mitigate climate change (Guenet et al., 2021). The EU Soil Plan 2030 considers SOC sequestration as a key strategy to combat climate change (Maenhout et al., 2024). SOC sequestration, the process by which plants and other organisms capture and store atmospheric carbon in the soil for at least 15 years, has been considered a cost-effective method for agriculture to mitigate climate change (Don et al., 2024; Goh, 2011).

SOC is critical to soil quality and ecosystem function. It has multiple functions, including improving ion exchange, increasing soil moisture retention, and providing essential nutrients such as nitrogen (N), phosphorus (P), zinc (Zn), and sulfur (S) (White & Brown, 2010). Land uses such as wetlands, forests, and agricultural systems, especially when managed with conservation practices, can store more carbon in soil and biomass, helping to reduce atmospheric carbon dioxide levels (Griscom et al., 2017). Measures such as 4p1000 and various carbon farming schemes aim to increase terrestrial carbon storage, helping to achieve the goal of net zero emissions by 2050 and limiting global temperature increases to below 2 °C (IPCC, 2021).

The threshold for SOC content in agricultural soils ranges from 1.5 to 2.0%, with land use, management practices, and agricultural systems significantly affecting SOC levels. More than half of total soil carbon is concentrated between 0.3 and 1 m depth, and soil organic carbon (SOC) is often severely depleted in agroecosystems (Lal et al., 2015). Restoring soil health through practices such as conservation agriculture, which includes minimizing soil disturbance, crop rotation, and the use of cover crops, can increase soil organic carbon levels and create a positive carbon budget (Somasundaram et al., 2020). These practices are considered important mechanisms for increasing soil organic carbon (Lehmann & Kleber, 2015).

Minimized and no-till agriculture, which involves minimal or no mechanical disturbance of the soil, is gaining attention for its ability to conserve soil organic carbon. Studies have shown that as tillage decreases, soil organic carbon stocks tend to be highest in the top 30 cm of the soil profile, although the impact across the entire soil profile remains controversial (Aguilera et al., 2013; Autret et al., 2019; Dignac et al., 2017b; Guenet et al., 2021; Haddaway et al., 2017). Over 30 years, reduced tillage could result in the sequestration of 3 Mg C ha^{-1} (11 Mg CO_2 ha^{-1}) in topsoil, or 0.1 Mg C ha^{-1} per year (Lugato et al., 2018). Furthermore, reduced tillage slightly increases N_2O emissions but favors SOC accumulation, especially when combined with irrigation (Guenet et al., 2021; Trost et al., 2013).

Soil carbon losses can also be mitigated through agroforestry, deep-root crops, and agricultural biomass recycling (Tiefenbacher et al., 2021). Cover crop duration and annual temperature are positively correlated with SOC stocks (Jian et al., 2020). Incorporation of crop residues can increase SOC by 3 to 18% depending on management practices and environmental conditions (Bolinder et al., 2020; Lehtinen et al., 2014). Cropping systems and soil organic carbon dynamics are strongly associated with cropping systems that combine crop rotation with legumes (Guenet et al., 2021), perennials (Morugan-Coronado et al., 2020), and agroforestry (Kim et al., 2016) with conservation or organic agriculture (Autret et al., 2019; Tiefenbacher et al., 2021) produce a positive change in SOC stock. The SOC content increased significantly as the rate of straw addition and straw C intake increased (Xia et al., 2018).

The impact on soil organic carbon reserves is expected to be substantial since grain maize has nearly twice the potential aboveground crop residues compared to small grain cereals (Bolinder et al., 2020). In comparison to residues with a C/N ratio < 30 (such as legumes, straw), the effect is greater when the residues have a C/N ratio more than 30 (such as wheat straw) than when the ratio is less than 30 (such as cereal straw) (Cai et al., 2024; Raseduzzaman et al., 2024; Shaaban et al., 2016; Xia et al., 2018).

3 Integrative Role of Soil Amendments to Promote Sustainable Agriculture

To practice sustainable agriculture and protect ecosystem services, soil quality must be preserved (Pahalvi et al., 2021). Chemical, physical, and biological elements all play a part in soil quality, which in turn affects plant health and soil productivity (Karlen et al., 1997). Applying soilamendments can enhance soil quality in addition to several other tactics like the use of cover crops, conservation tillage techniques, balanced fertilization, and crop rotation (Abdul Halim et al., 2018; Farmaha et al., 2022; Guo, 2020; Lal et al., 2015).

Because of their increased stable aggregate content and biological soil particle binding, organically managed soils also show enhanced soil structure and moisture infiltration (Emmerling, 2007). Aggregates not only stabilize soils but also shield organic nitrogen and carbon from microbial degradation (Six et al., 2000). Enhancing soil health and fertility with organic methods has great potential to increase agricultural productivity on already cultivated land sustainably while also utilizing agriculture's ability to reduce global warming by sequestering carbon (Paustian et al., 2016).

Improving agricultural soil quality and increasing crop productivity are the main goals of incorporating various substrates into the soil. The study by (Xu et al., 2015), (El-Alsayed & Ismail, 2017), and (Oliveira et al., 2018) stated that the main goal is to improve soil fertility by making water and essential nutrients more accessible to plants, preventing soil drying, maintaining high microbial activity and promoting nutrient uptake by crops. In addition, soil amendments can significantly improve soil structure by optimizing air–water interactions, changing soil bulk density, and stabilizing soil aggregates (El-Alsayed & Ismail, 2017; Manirakiza & Se¸ker, C., 2020).

Assessing the environmental impacts of applying natural soil amendments is critical. Their addition can provide multiple benefits, such as reducing the mobility and toxicity of heavy metals, minimizing the leaching of fertilizer nutrients and harmful substances into aquatic systems and plant tissues, reducing soil salinity, stabilizing soil pH, enhancing carbon dioxide absorption, and reducing greenhouse gas emissions and ultimately promote ecosystem health. Organic amendments—including animal manure compost, green manures, cover crops, crop residues, and straw—are often used to enhance soil and plant vitality, consistent with permaculture principles (Akram et al., 2016; Ansari et al., 2017a). Many studies have highlighted the potential of organic amendments in maintaining soil quality and sequestering carbon, showing that their application can significantly enhance soil structure and promote carbon capture. Furthermore, organic additions can help adapt to climate change, but their effectiveness is affected by factors such as soil texture, climate and application rate, so strategies need to be tailored for optimal results (Matisic et al., 2024). The practice of enriching soil with organic materials has a long history, with the ancient Greeks and Romans improving soil by adding human and animal waste. They recognize the benefits of rotating crops, such as growing wheat after legumes (Goss et al., 2013).

To increase soil organic carbon (SOC), effective strategies include applying organic amendments such as manure and compost, utilizing cover crops, and retaining agricultural residues (Duan et al., 2022). Increased SOC levels significantly enhanced soil fertility, nutrient accessibility, water retention, and microbial diversity (Feng et al., 2023; Fontúrbel et al., 2023; Inagaki et al., 2023; Man et al., 2023). Organic matter (OM) is critical in influencing SOC content, nutrient dynamics, cation exchange capacity, soil pH, and overall fertility and water retention capacity (Duan et al., 2022; Fontúrbel et al., 2023; Inagaki et al., 2023; Mabagala & Mng'ong'o, M.E., 2022; Prats et al., 2023; Santana et al., 2022).

Effective organic fertilizer management can significantly improve soil health by enhancing nutrient use efficiency, moisture regulation, and soil structure. Practices such as the application of organic fertilizers and straw retention have been associated with increases in soil organic carbon and bioavailable potassium levels following the application of OM amendments (Chen et al., 2023b; Farooq et al., 2022). Various studies have shown that managing OM can alter soil chemistry and increase SOC content (Duan et al., 2022; Gilmullina et al., 2023; Mabagala & Mng'ong'o., 2022).

Conversion of organic solid waste into soil amendments such as compost or biochar provides a sustainable approach to promote efficient and environmentally friendly agriculture (Rehrah et al., 2016; Santana et al., 2019). Composting is an aerobic process that converts organic waste into humus-like substances and is recognized for its ability to enhance soil properties, including bulk density, nutrient cycling, and moisture retention (Beck-Broichsitter et al., 2018; Pergola et al., 2018). The carbon sequestration capacity of compost is estimated to be 714 ± 404 kg C ha^{-1} $year^{-1}$, while that of farmyard manure is 292 ± 132 kg C ha^{-1} $year^{-1}$ (Tiefenbacher et al., 2021).

Biochar is another promising soil amendment that can improve carbon storage and crop yields, especially in degraded or nutrient-deficient soils (Piveta et al., 2021). Biochar is rich in organic compounds and essential nutrients and can improve soil pH, electrical conductivity (EC), and organic carbon levels, thereby increasing soil fertility (Liu et al., 2018). In addition, biochar can serve as a long-term solution for carbon sequestration, potentially reducing greenhouse gas (GHG) emissions and helping combat climate change (Lyu et al., 2018). In addition to its environmental advantages, biochar can reduce nutrient leaching, minimize the bioavailability of contaminants, and increase crop yields, especially in poor soils (Abdul et al., 2017).

4 4. Soil health monitoring Assessment

Soil degradation is a change in soil health, usually accompanied by a decline in soil fertility. Soil health has been defined as the capacity of a specific kind of soil to function, within natural or managed ecosystem boundaries, to sustain plant and animal productivity, maintain or enhance water and air quality, and support human health and habitation. Since soil functions vary between different ecosystems, the health of the same type of soil will vary between different ecosystems or land uses. This definition is broad and encompasses many soil properties, making the term 'soil health' usually more clearly understood by those who work with the soil than easy to define for those looking on from the outside. Degraded soil need not be from bad management; it could be on unsuitable land. It is important to ascertain the causes of soil degradation to detect if it's land overuse or mismanagement that has pushed soil past a critical point of no return for ecosystem stability (Janzen et al., 2021; Kihara et al., 2020; Lehmann et al., 2020; Norris et al., 2020; Nunes et al., 2021; Yang et al., 2020).

Soil is the topmost layer of the earth's surface which supports and sustains life. Healthy soil leads to healthy crops and animals, preserving the environment. We need to constantly monitor soil health to prevent barren and unusable soils. Since the 1970s, research has shown that soil degradation in much of our farmland often proceeds unnoticed until the effects on crop yield and farm profitability become serious (Borrelli et al., 2023; Malone & McClintock, 2023).

4.1 *Soil Health Monitoring*

In the absence of appropriate technologies to monitor soil conditions, farmers and land managers can be ignorant of the changes that are occurring in their soil and the effects of their practices. This oblivious behavior can lead to land degradation and the loss of the ecosystem services that the soil provides. Take the example that nearly 80% of European land is used in some sort of agricultural practice, yet only 36% of European countries possess a soil monitoring system. The sustainability of the product from the soil is taking too much with many short-term intensification practices. This has caused the Worst Performing Soils (WPS) to acquire a net loss in ecosystem services, and therefore the low resilience of these soils means a further continuation of these practices can lead to soil system collapse. The ability to provide evidence that changes in soil use or intervention are causing a long-term effect on soil health so that it furthers a reduction in the soil ecosystem service and high magnitude consequences. This can help convince policymakers that system and policy changes are needed to prevent WPS and to preserve the soil well beyond its functional capacity (Montanarella & Panagos, 2021; Panagos et al., 2020).

Soil is one of the most important environmental assets, providing crucial ecosystem services. In addition to its undisturbed role in supporting plant life on

Earth, it also acts as a medium for plant growth, regulates water supplies and chemical climates, decomposes organic matter, and is the habitat for a vast array of organisms. It is possible to take this ecosystem service perspective further and affect the public and policymakers who are hard-pressed when considering soil monitoring budgets. By quantifying and qualifying a discrete number of soil health indicators, which are the attributes of soil that reflect its capacity to sustain the ecosystem or human well-being, it becomes easier to convince those after a simple cost–benefit analysis of the importance of soil monitoring changes to maintain that service (Nunes et al., 2020, 2021; Powlson, 2020; Soto et al., 2020; Zhao & Wu, 2021).

4.2 *Methods for Evaluating Soil Health*

The evaluation of soil health is exceedingly complex due to the highly dynamic nature of the soil and the numerous gaps in our current knowledge base. There exists a multitude of approaches available for assessing soil health. The majority of these methods involve an on-farm component, which typically entails observing crop yield or visually inspecting the soil, in addition to a laboratory analysis component. Consequently, a comprehensive soil health assessment can be categorized into three main domains: the analysis of physical properties, chemical properties, and biological properties of the soil. It is widely acknowledged among scientists that a holistic evaluation of soil health necessitates the examination of all three domains, as they exhibit interconnectedness and collectively define the overall soil quality. Notwithstanding, the methods employed to assess each of these domains are ceaselessly evolving, resulting in significant variations in terms of cost, time requirement, and the level of quantifiable accuracy of the obtained results (Fierer et al., 2021; Lehmann et al., 2020; Maurya et al., 2020; Yang et al., 2020).

4.2.1 Physical Properties Analysis

Visual Soil Assessment (VSA) is a method used to categorize soil structure and consists of assessing soil workability by assessing soil consistency, assessing soil porosity by assessing the presence of soil bio pores, and assessing soil root ability by assessing possible root constraints. This method is simple and has been proven to give a good indication of soil status. VSA can be used to compare the structure of different soils within the same site or soils under different management to store data about any changes. VSA is often used in conjunction with the characterization of soil aggregates. Aggregate stability is an important indicator of soil structural quality and stability. High stability of aggregates is desirable to enable good soil aeration and water infiltration. This is assessed by determining the wet and dry aggregate size distribution and conducting an aggregate stability test to assess the percentage of water-stable aggregates. The results of this test provide one of the most accurate indicators of soil tilth and have been correlated with good soil plant growth (Galdos et al., 2020; Nunes et al., 2021).

Most research projects have measured several of these physical properties to gain an understanding of how soil with a particular land use and management history functions. The volume of research conducted globally in this field is large, and most of the methods employed are well-established. Physical analysis of soil properties is usually conducted using methods like Visual Soil Assessment (VSA), soil water characteristic determination, aggregate stability measurement, measurement of hydraulic conductivity, and soil strength. These tests are conducted in a range of controlled laboratory conditions for VSA and soil water characteristics, and situ for a method such as soil strength. The pattern or status of the results of any of these tests may provide an indication of soil degradation or improvement in soil functioning (Hou et al., 2020; Lehmann et al., 2020).

Physical properties of soil directly affect the ability of soil to function as a living system. The importance of different physical analyses can be summarized as follows: soil consistency affects soil workability and tillage effects; soil resistance affects shear strength, traffic ability, aeration, and root penetrability; soil texture affects the ability to retain water and hold cations; permeability affects the rate of water movement into the soil; bulk density affects root penetration, water infiltration, aeration, and seedling emergence; and pore space affects water holding capacity and airflow (Angst et al., 2021).

4.2.2 Chemical Properties Analysis

The chemical properties analysis determines the presence and availability of essential nutrient elements, the cation exchange capacity (CEC), and the extent to which the soil is acidified or alkaline (pH). Identification of each of these chemical factors is linked to a variety of soil functions. These range from micronutrient fertilization to balance out deficiencies of substances like iron or manganese, to reducing the solubilization of toxic aluminum within the soil. pH status affects the available soil nutrients, plant root health, and interactions with soil organisms. Electrical conductivity is also used as a measure of soil salinity and sodicity, being correlated to loss of soil structure and infiltration, toxic ion excesses or deficiencies, and alteration of the relative dominance of microbes. This analysis can also serve to identify any variation in these properties and elements above or below lab reference standards, relative to specific types of soil, i.e. histosol or andisol (Alori et al., 2022; Bungau et al., 2021; Gavrilescu, 2021; Hartmann & Six, 2023; Pathan & Siddiqui, 2022; Petraru et al., 2021; Salamanca-Carreño et al., 2023; Salgar-Chaparro et al., 2020; Yan et al., 2021).

Soil health is the ability of soil to function as a dynamic living system, supporting plant and animal life. Through chemical analysis of soil samples, we can assess ecosystem conditions and land use constraints to ensure soils can support plant and animal productivity, improve water and air quality, and maintain health and human habitability. Protecting and improving soil health is critical to the success of permaculture practices. It involves managing the soil and associated systems in a way that not only enhances crop yields, but also provides an economic return, sustains the environment, and enhances the well-being of society, now and into the future. All agricultural practices affect the health of the soil. Farmers, like any land users, can enhance or degrade soil health through their decisions and actions. Soils provide a variety of ecosystem services such as air and water purification, climate regulation, disease and pest regulation, waste treatment, and decomposition and cycling of wastes. These are crucial to the health and well-being of plants, animals, and humans (Fahad et al., 2022; Shah et al., 2021; Suman et al., 2022).

4.2.3 Biological Properties Analysis

Observational—Natural gradients or chrono sequences can be observed so that differences in ecosystem properties can be observed where change is correlated with change in an element of the soil environment. An example would be in the measurement of the effects of soil pH on an upland acid grassland ecosystem. A gradient may be set up on a limed peat bog from the point of lime addition to a point where the pH has no longer increased above an existing area of grassland. This method is useful in its approach to observing changes in a realistic context, below and aboveground effects, and it allows inexpensive large-scale manipulation and replication of the treatment. It is not experimental so it does not always clearly identify causal relationships, and there may be confounding variables between the treatments. Stepwise addition of an element at a uniform rate across several plots will allow a correlation of the rate of the treatment with the response of the ecosystem. Step changes can be observed by adding different rates of the treatment in spatially isolated patches. These methods are often useful in identifying dose–response relationships between the treatment and its effect on soil biota. An example would be an assessment of the effect of different intensities of heavy metal pollution on soil microorganisms (Li et al., 2024; Yang et al., 2022).

4.3 Global Soil Health Monitoring Initiatives

The Soil Health Institute has recently launched a comprehensive assessment of soil health in North America. The assessment aims to discover the status of soil health, which will be used as a baseline that the Institute will report against in the future. This initiative will provide direct benefits through enhanced soil health, crop yield, profitability, and provision of public goods. The assessment consists of a large number of integrated studies, each contributing information on specific soil health indicators. Studies will be conducted across numerous land uses in all regions of the US and will include all agroecological zones. Results will provide farmers, land managers, and policymakers with the information they need to assess and improve the health of their soils. The assessment will be reported against the baseline in five years. In addition to the North American assessment, the Soil Health Institute has also recently completed an assessment of soil health in Texas. This was a comparative study to determine if the implementation of different soil health management practices leads to different soil health outcomes. The assessment focused on three key indicators: organic matter content, soil compaction, and nutrient availability. The results showed significant variations in soil health across different regions and land uses, highlighting the importance of implementing targeted soil health interventions. The adoption of specific management practices improves soil health and function and reduces risks of pests and diseases. The study was conducted on four different land uses: native rangelands, dryland organic farming, and dryland and irrigated farming using integrated approaches. Results from all three studies are expected to be available soon (Liptzin et al., 2022; Sanderman et al., 2020; Williams et al., 2020; Yang et al., 2020).

4.3.1 North America

One major activity contributing to soil health monitoring assessment in North America is the development and implementation of the Soil Quality Assessment Tool (SQAT) by the Soil Management Collaborative Research Support Program, University of Hawaii. While this tool has been developed in the Hawaiian archipelago, its principles can be adapted and applied to the highly diverse range of North American soil resources. This tool has been designed to report results at the field scale as part of a semi-quantitative land management assessment, utilizing a set of soil quality indicators (Benos et al., 2020).

In Canada, a similar land inventory is conducted by the Centre for Land and Biological Resources Research, Agriculture and Agri-Food Canada. However, there are no comprehensive soil quality assessment initiatives in place in North America at this time. This has, in part, been due to the lack of a tested and proven soil quality assessment methodology, combined with the absence of collaboration in this field between the various North American research and development groups working on soil and land resources (Mello et al., 2020).

In the USA, a National Resources Inventory (NRI) is conducted periodically to gather data on soil and ecological resources, including soil quality indicators and resource and environmental issues on non-federal lands. The NRI was initiated in 1977 and is conducted by the Natural Resource Conservation Service (NRCS), an agency of the United States Department of Agriculture. It provides data on soil, water, and related resources, as well as on farm and ranch lands. From this, the Soil Trends project was established to provide a comprehensive assessment of soil quality analysis related to NRI data collected (Chu et al., 2022; McCord et al., 2022; Meena et al., 2020; Nelson et al., 2020; Ogle et al., 2023; Oswalt et al., 2021; Parish et al., 2021; Rosenberg & Wallander, 2022; Spaeth et al., 2024; Wang et al., 2022).

The North American region has a diverse set of soil landscapes, ranging from the highly diverse soils of the temperate zones to the arid and semi-arid regions in the south and west. Soils in North America are used for a variety of functions, from arable and horticultural cropping to forestry, pasture/grazing, and urban development (Hagmann et al., 2021; Norris et al., 2020; Osland et al., 2021).

4.3.2 Europe

In recent years, scientific findings have shown significant developments in the harmonization of soil information within Europe. The establishment of the European Soil Data Centre and the creation of a European-wide soil information system have played an essential role in this progress. These research-driven initiatives are essential components of the European Strategy for Soil Protection. However, despite these advancements, it has been discovered that there are still fundamental gaps in the data required to accurately assess the initial state and pressure on European soil. These research findings have highlighted the need for additional resources to implement the strategy effectively. Further investment and collaboration between European countries are necessary to fill these data gaps and ensure the successful implementation of soil health monitoring assessments in Europe. One important aspect of this strategy is the involvement of new member states to prevent soil degradation in their regions. Ongoing research projects, such as the PECO and SSPI projects, are aimed at furthering our understanding of soil in these areas. These scientific studies are contributing to the overall goal of preserving soil resources throughout Europe. Another significant factor that affects the use and protection of European soils is the Common Agricultural Policy (CAP). Studies have shown that this EU policy has a considerable impact on soil health and preservation. Scientists are currently engaged in discussions regarding the evaluation of CAP's effects on EU15 countries and its implications for accession countries (Bouma et al., 2022; Cornu et al., 2023; Donkelaar et al., 2021; Meurer et al., 2024; Voltz et al., 2020).

At the European Union level, soil monitoring and assessment provide the basis for the development and implementation of policies and strategies. Soil monitoring coordinates and focuses actions aiming to provide up-to-date and relevant information on soil at appropriate geographical scales. Information is used to assess the need for further actions and in policy development and implementation. The European Environment Agency (EEA) has developed indicators and a report describing their use in assessing soil in various contexts. The agri-environmental indicator set developed by the European Environment Agency (EEA) provides a framework for monitoring soil quality at the land parcel level within the European Union. This indicator set is part of a wider initiative developing indicators to measure various elements of agri-environmental policy. The steps involved in indicator development provide an opportunity for the soil science community to contribute to the design and thus utility of indicators for monitoring soil quality (Köninger et al., 2022; MacPherson et al., 2022; Panagos et al., 2020; Seaton et al., 2021).

Soil monitoring throughout Europe creates a complex picture at the national level. Published information on soil monitoring is not always readily accessible, and in some cases, limited resources result in little formal monitoring activity. To various degrees, soil monitoring continues in all European countries and at the European Union level.

4.3.3 Asia

The need for soil health monitoring in Asia has been recognized by both international and national bodies for several key issues. These include food security, environmental degradation and loss of biodiversity, upland development, and pollution of soil and groundwater from agricultural chemicals. In terms of addressing regional food security and poverty, there has been an identification of the importance of both sustaining soil productivity on existing agricultural lands and rehabilitating degraded lands to a productive and sustainable state. The availability of suitable data on soil health and strategies for utilization of the data is variable among the countries in Asia. This is influenced by historical and current development contexts, political and economic stability, urban/rural population distribution, and the prevalence of current or recent armed conflict in some countries. International collaboration and support for soil health monitoring initiatives must acknowledge these factors and appreciate the development constraints and diversity that exist in the region (Brevik et al., 2020; Hou et al., 2020; Lal et al., 2021; Shahane & Shivay, 2021).

In Asia, soil degradation is a key issue for food security due to intensified farming systems and increased population pressure in some areas, which prevents fallow periods for soil recovery. Soil degradation in the region is associated with declining productivity, reduced crop yields, lower resilience to pests and diseases, loss of genetic diversity, and increasing incidence of severe famine in some parts of Asia. The livelihoods of marginal and small-scale farmers, who constitute 60% of the Asian population, are particularly vulnerable to degradation and loss of soil health. Degradation can, in some cases, lead to social problems and conflict. For example, in Indonesia, the Asian economic crisis in 1997 triggered the degradation of upland soils in search of short-term cash earnings, the widespread burning of forests and forest soils, and the migration of rural poor people to more marginal lands in search of food and income. This led to severe regional haze from the burning and widespread land and resource use conflicts in some regions. At a national scale, soil health and fertility are limiting factors to overall economic development in some Asian countries (Ahmad et al., 2020; Hossain et al., 2020; Prăvălie et al., 2021).

4.3.4 Africa

An African example of the framework approach can be seen in the United Nations Environment Programme Debt-for-Environment Initiative; a sub-program in Rwanda that improved the productivity of hillsides in two watersheds through an integrated approach to sustainable land management which included soil fertility improvement (Adiyah et al., 2021; Mugiraneza et al., 2020; Placide et al., 2022).

The USA has been involved in a global soil matter surveillance program using a soil organic carbon density estimator to construct a baseline for soil organic carbon conditions and produce a comprehensible model showing the impact of soil on environmental concerns. The Caribbean states held a meeting in the FAO Sub-Regional Office for the Caribbean (SLAC) to prepare common guidelines for soil, land, and water conservation; this meeting identified the need for soil information and focused on future collaboration between local, regional, and partner institutes and organizations to set up sustainable programs to provide soil information (Bastida et al., 2021; Egmond et al., 2021; Harris et al., 2021).

There are several initiatives underway to monitor soil health and quality both domestically and internationally, with global monitoring efforts highlighted in the World Soil Charter. The International Soil Reference and Information Centre (ISRIC), partnering with the Food and Agriculture Organization of the United Nations (FAO), is in the final planning stages of soil monitoring evaluation that will provide a soil monitoring inventory for the year 2000 to be released in 2006. This is a unique initiative that collects and digitizes historical soil data from 36 regions with the joint aims of maintaining soil resources, detecting changes made to soil resources, and providing early warning and new opportunities for soil management and protection (Ali et al., 2020a; Arrouays et al., 2021; Batjes, 2022; Batjes et al., 2020; Erdogan et al., 2021; Rodríguez Eugenio, 2021).

4.3.5 South America

Brazil has the most extensive areas of soil under threat than any other country in the world. Data from the Brazilian Agricultural Research Corporation show that nearly 30% of the land is unusable for agriculture due to degradation, and 15% of the country is affected by desertification. This has mainly been caused by extensive land clearance over the past 40 years, with a move from subsistence agriculture to large-scale grain production and cattle farming. Rapid population growth has also led to the occupation of marginal lands in fragile environments, which has increased soil degradation. The recent agricultural expansion has been focused on the Cerrado region, which is a vast savannah area in central Brazil. This region has a tropical climate with a dry season of 4 months. The expansion has been largely driven by the opening of new frontiers through improved transportation, particularly the construction of highways, and also the development and availability of new crop cultivars and technologies (Vieira et al., 2021).

Argentina is the second-largest country in South America and the eighth-largest in the world. It is a country of vast plains, deserts, and fertile lowlands. The north of the country has a subtropical climate and is an area of extensive agriculture with sugarcane, tobacco, and citrus crops. The fertile Pampas, which include some of the most productive agricultural land in the world, are located in the center of the country and cover an area of approximately 750,000 km^2 at an altitude of 100 m to 500 m above sea level. This region has a temperate climate and is the main production area for grain and livestock. There is also large-scale grain production in the Mesopotamian lowlands between the Parana and Uruguay rivers, which border Brazil (Lanfranco et al., 2022).

The South American continent is rich in biodiversity, with a wide variety of climate zones and soil types. Agriculture ranges from large-scale grain production in the more temperate zones of Argentina, Uruguay, and Southern Brazil to shifting cultivation systems in the Amazon basin and the steeply sloping lands of the Andean region. The diversity of environments and farming systems presents a considerable challenge to those trying to assess and monitor soil health at a continental scale (Almazroui et al., 2021; Condom et al., 2020; Cui et al., 2021; Zwetsloot et al., 2021).

5 Challenge and Future Prospects

The main challenge to developing soil carbon sequestration strategies is the multi-faceted nature and heterogeneity of soil systems. Large variability in soil responses to various amendments, management practices, and climatic conditions complicates the development of universally applicable sequestration approaches. Success in these carbon sequestration efforts is controlled by critical factors such as soil texture, moisture availability, and microbial activity, whose interactions are not well understood. Different soil amendments, whether organic or inorganic, can significantly boost soil organic carbon (SOC) levels. The effects of these amendments can also change depending on local conditions, like the type and texture of the soil, which means that the results can vary widely. Future research must prioritize the elucidation of the interactions between soil amendments and site-specific soil and climatic conditions to improve the predictability and efficacy of carbon sequestration practices. It is imperative to conduct regionally tailored studies that account for local soil characteristics, climate regimes, and land management practices to develop context-specific strategies for maximizing carbon storage. In addition, using cutting-edge technologies such as remote sensing, real-time soil health monitoring, and AI-based modeling can significantly enhance our capacity to track and predict shifts in soil carbon dynamics. It's equally important to address the economic and policy structures that motivate land managers to embrace sustainable practices. Occasionally, these amendments can improve carbon storage and limit it in certain circumstances. Currently, researchers are confronted with the primary challenge of determining how to achieve high crop yields while simultaneously achieving effective carbon storage. It can be challenging to achieve this equilibrium, as it frequently necessitates the reconciliation of long-term environmental health with immediate financial gain.

6 Conclusion

SOC is an imaginary way of managing climate change and promoting sustainable agriculture. Soils have a significant sequestering/storing capacity for carbon, and that ability has the potential to offset a portion of the rise in atmospheric CO_2, which naturally helps stabilize global temperatures. This chapter has discussed numerous techniques that enhance soil health, fertility, and productivity in addition to improving carbon sequestration. These techniques include conservation agriculture, no-tillage farming, and the use of compost and biochar as soil amendments. These strategies offer mutual benefits in addressing climate change and advancing sustainable agriculture. Nonetheless, numerous challenges remain for quantifying long-term soil carbon persistence and translating this into practical management prescriptions across a wide range of soils and climates. Therefore, more sophisticated methods for monitoring and assessment are needed to provide accurate in-situ information on soil organic carbon (SOC) dynamics and the efficiency of management solutions. Additional, work is needed to enhance and refine the effectiveness of these tactics, as well as reconcile their economic and environmental trade-offs, so that farmers' adoption chances are realistic within an acceptable legislative framework. The potential benefits of integrating carbon sequestration with sustainable agriculture include the mitigation of greenhouse gas emissions, the enhancement of food security, and the facilitation of adaptation to climate change.

References

Abdul Halim, N. S. A., Abdullah, R., Karsani, S. A., Osman, N., Panhwar, Q. A., & Ishak, C. F. (2018) Influence of soil amendments on the growth and yield of rice in acidic soil. *Agronomy, 8*, 165.

Abdul, G., Zhu, X., & Chen, B. (2017). Structural characteristics of biochar-graphene nano sheet composites and their adsorption performance for phthalic acid esters. *Chemical Engineering Journal, 319*, 9–20. https://doi.org/10.1016/j.cej.2017.02.074

Adiyah, F., Fuchs, M., Mich&eli, E., & Dawoe, E. (2021). Cocoa farmers perceptions of soil organic carbon effects on fertility, management and climate change in the Ashanti region of Ghana. *African Journal of Agricultural Research, 17*(5), 714–725.

Aguilera, E., Lassaletta, L., Gattinger, A., & Gimeno, B. S. (2013). Managing soil carbon for climate change mitigation and adaptation in Mediterranean cropping systems: A meta-analysis. *Agriculture, Ecosystems & Environment, 168*, 25–36.

Ahmad, N. S. B. N., Mustafa, F. B., & Didams, G. (2020). A systematic review of soil erosion control practices on the agricultural land in Asia. *International Soil and Water Conservation Research, 8*(2), 103–115.

Akram, M., Rizvi, R., Sumbul, A., Ansari, R. A., & Mahmood, I. (2016). Potential role of bioinoculants and organic matter for the management of root-knot nematode infesting chickpea. *Cogent Food & Agriculture, 2*(1), 1183457.

Ali, A., Tamene, L. D., & Erkossa, T. (2020). Identifying, cataloguing, and mapping soil and agronomic data in Ethiopia. CIAT Publication.

Almazroui, M., Ashfaq, M., Islam, M. N., Rashid, I. U., Kamil, S., Abid, M. A., O' Brien, E., Ismail, M., Reboita, M. S., & Sylla, M. B. (2021). Assessment of CMIP6 performance and projected

temperature and precipitation changes over South America. *Earth Systems and Environment, 5*(2), 155–183.
Alori, E. T., Gabasawa, A. I., Elenwo, C. E., & Agbeyegbe, O. O. (2022). Bioremediation techniques as affected by limiting factors in soil environment. *Frontiers in Soil Science, 2*, Article 937186.
Angst, G., Pokorný, J., Mueller, C. W., Prater, I., Preusser, S., Kandeler, E., Meador, T.,Strakova, P., Hajek, T., van Buiten, G., & Angst, Š. (2021). Soil texture affects the coupling of litter decomposition and soil organic matter formation. *Soil Biology and Biochemistry, 159*, 108302.
Ansari, R. A., Mahmood, I., Rizvi, R., Sumbul, A., & Safiuddin. (2017a). Siderophores: Augmentation of soil health and crop productivity. In V. Kumar, M. Kumar, S. Sharma, & R. Prasad (Eds.), *Probiotics in agroecosystem*. Singapore: Springer. (in press).
Arrouays, D., Mulder, V. L., & Richer-de-Forges, A. C. (2021). Soil mapping, digital soil mapping and soil monitoring over large areas and the dimensions of soil security–A review. *Soil Security, 5*, Article 100018.
Autret, B., Beaudoin, N., Rakotovololona, L., Bertrand, M., Grandeau, G., Gréhan, E., Ferchaud, F., & Mary, B. (2019). Can alternative cropping systems mitigate nitrogen losses and improve GHG balance? Results from a 19-yr experiment in northern France. *Geoderma, 342*, 20–33.
Bakker, M. M., Govers, G., Jones, R. A., & Rounsevell, M. D. A. (2007). The effect of soil erosion on Europe's crop yields. *Ecosystems, 10*, 1209–1219. https://doi.org/10.1007/s10021-007-9090-3
Bastida, F., Eldridge, D. J., García, C., Kenny Png, G., Bardgett, R. D., & Delgado-Baquerizo, M. (2021). Soil microbial diversity–biomass relationships are driven by soil carbon content across global biomes. *The ISME Journal, 15*(7), 2081–2091.
Batjes, N. H. (1996). Total carbon and nitrogen in the soils of the world. *European Journal of Soil Science, 47(2)*, 151–163. https://doi.org/10.1111/j.1365-2389.1996.tb01386.x
Batjes, N. H. (2022). Overview of procedures and standards in use at ISRIC WDC-Soils (ver. 2022).
Batjes, N. H., Ribeiro, E., & Van Oostrum, A. (2020). Standardised soil profile data to support global mapping and modelling (WoSIS snapshot 2019). *Earth System Science Data, 12*(1), 299–320.
Beck-Broichsitter, S., Fleige, H., & Horn, R. (2018). Compost quality and its function as a soil conditioner of recultivation layers—A critical review. *International Agrophysics, 32*, 11–18. https://doi.org/10.1515/intag-2016-0093
Benos, L., Tsaopoulos, D., & Bochtis, D. (2020). A review on ergonomics in agriculture. Part I: Manual operations. *Applied Sciences, 10*(6), 1905
Bolinder, M. A., Crotty, F., Elsen, A., Frac, M., Kismanyoky, T., Lipiec, J., Tits, M., T oth, Z., & Kätterer, T. (2020). The effect of crop residues, cover crops, manures and nitrogen fertilization on soil organic carbon changes in agroecosystems: A synthesis of reviews. *Mitigation and Adaptation Strategies for Global Change, 25*, 929–952.
Borrelli, P., Panagos, P., Alewell, C., Ballabio, C., de Oliveira Fagundes, H., Haregeweyn, N., Lugato, E., Maerker, M., Poesen, J., Vanmaercke, M., & Robinson, D. A. (2023). Policy implications of multiple concurrent soil erosion processes in European farmland. *Nature Sustainability, 6*(1), 103–112.
Bouma, J., Bonfante, A., Basile, A., van Tol, J., Hack-ten Broeke, M. J. D., Mulder, M., Heinen, M., Rossiter, D. G., Poggio, L.,& Hirmas, D. R. (2022). How can pedology and soil classification contribute towards sustainable development as a data source and information carrier? *Geoderma, 424*, 115988.
Brevik, E. C., Slaughter, L., Singh, B. R., Steffan, J. J., Collier, D., Barnhart, P., & Pereira, P. (2020). Soil and human health: Current status and future needs. *Air, Soil and Water Research, 13*, 1178622120934441.
Bungau, S., Behl, T., Aleya, L., Bourgeade, P., Aloui-Sossé, B., Purza, A. L., Abid, A., & Samuel, A. D. (2021). Expatiating the impact of anthropogenic aspects and climatic factors on long-term soil monitoring and management. *Environmental Science and Pollution Research, 28*(24), 30528–30550.
Cai, S., Ma, Y., Bao, Z., Yang, Z., Niu, X., Meng, Q., Qin, D., Wang, Y., Wan, J., & Guo, X. (2024). The impacts of the C/N ratio on hydrogen sulfide emission and microbial community characteristics during chicken manure composting with wheat straw. *Agriculture, 14(6)*, 948.

Chaudhury, S. T., Bhattacharyya, S. P. W., Pal, D. K., SahrawatKL, N. A., Chandran, P., & VenugopalanMV, T. B. (2016). Use and cropping effects on C in black soils of semi-arid tropical India. *Current Science, 110*(9), 1652–1698.

Chen, L., Sun, S., Zhou, Y., Zhang, B., Peng, Y., Zhuo, Y., Ai, W., Gao, C., Wu, B., Liu, D., & Sun, C. (2023b). Straw and straw biochar differently affect fractions of soil organic carbon and microorganisms in farmland soil under different water regimes. *Environmental Technology and Innovation*, 103412.

Chen, L., Chen, D., Gong, N., & Qin, J. (2024). Assessing the influence of environmental regulation on carbon sequestration in China: Towards a sustainable future. *Journal of Environmental Management, 368*, Article 122177.

Chu, L., Dong, S., Fu, L., & Xu, L. (2022). Exploration of the construction of a municipal natural resources survey and monitoring system—Take Xuzhou City as an example. *Natural Resources Conservation and Research, 5*(2), 20–28.

Condom, T., Martínez, R., Pabón, J. D., Costa, F., Pineda, L., Nieto, J. J., Lopez, F., & Villacis, M. (2020). Climatological and hydrological observations for the South American Andes: In situ stations, satellite, and reanalysis data sets. *Frontiers in Earth Science, 8*, 92.

Swiss Confederation. (2020, October 7). Swiss Geoportal. https://map.geo.admin.ch

Cornu, S., Keesstra, S., Bispo, A., Fantappie, M., van Egmond, F., Smreczak, B., Wawer, R., & Chenu, C. (2023). National soil data in EU countries, where do we stand? *European Journal of Soil Science, 74(4)*, e13398

Cui, D., Liang, S., & Wang, D. (2021). Observed and projected changes in global climate zones based on Köppen climate classification. *Wiley Interdisciplinary Reviews: Climate Change, 12*(3), Article e701.

Dadhich, R. K., & Meena, R. S. (2014). Performance of Indian mustard (*Brassica juncea L.*) in response to foliar spray of Thiourea and Thioglycolic acid under different irrigation levels. *Indian Journal of Ecology, 41(2)*, 376–378.

Datta, R., Baraniya, D., Wang, Y. F., Kelkar, A., Moulick, A., Meena, R. S., Yadav, G. S., Ceccherini, M. T., & Formanek, P. (2017). Multi-function role as nutrient and scavenger of free radical in soil. *Sustain MDPI, 9*, 402. https://doi.org/10.3390/su9081402

de Santana, F. B., Grunsky, E. C., Fitzsimons, M. M., Gallagher, V., & Daly, K. (2022). Diffuse reflectance mid infra-red spectroscopy combined with machine learning algorithms can differentiate spectral signatures in shallow and deeper soils for the prediction of pH and organic matter content. *CATENA, 218*, Article 106552.

Dheri, G. S., & Pasricha, N. S. (2021). Carbon dynamics under conservation agriculture. *Conservation Agriculture: A Sustainable Approach for Soil Health and Food Security*. Springer. https://doi.org/10.1007/978-981-16-0827-8_15

Dignac, M. F., Derrien, D., Barr´e, P., Barot, S., C´ecillon, L., Chenu, C., Chevallier, T., Freschet, G. T., Garnier, P., Guenet, B., Hedde, M., Klumpp, K., Lashermes, G., Maron, P. A., Nunan, N., Roumet, C., & Basile-Doelsch, I. (2017). Increasing soil carbon storage: mechanisms, effects of agricultural practices and proxies. A review. *Agronomy for Sustainable Development*. https://doi.org/10.1007/s13593-017-0421-2

Dignac, M.-F., Derrien, D., Barré, P., Barot, S., Cécillon, L., Chenu, C., Chevallier, T., Freschet, G. T., Garnier, P., & Guenet, B. (2017b). Increasing soil carbon storage: Mechanisms, effects of agricultural practices and proxies. *A Review. Agronomy for Sustainable Development, 37*, 1–2.

Don, A., Seidel, F., Leifeld, J., Kätterer, T., Martin, M., Pellerin, S., Emde, D., Seitz, D., & Chenu, C. (2024). Carbon sequestration in soils and climate change mitigation—Definitions and pitfalls. *Global Change Biology, 30*, e16983

Van Donkelaar, A., Hammer, M. S., Bindle, L., Brauer, M., Brook, J. R., Garay, M. J., Hsu, N. C., kalashnikova, O. V., kahn, R. A., Lee, C., Levy, R. C., & Martin, R. V. (2021). Monthly global estimates of fine particulate matter and their uncertainty. *Environmental Science & Technology, 55*(22), 15287–15300

Duan, Z., Tan, X., Ali, I., Wu, X., Cao, J., Xu, Y., Shi, L., Gao, W., Ruan, Y., & Chen, C. (2022). Comparison of organic matter (OM) pools in water, suspended particulate matter, and sediments in eutrophic Lake Taihu, China: Implication for dissolved OM tracking, assessment, and management. *Science of the Total Environment, 845*, Article 157257.

van Egmond, F. M., Fantappie, M., Batjes, N. H., Brus, D. J., Bulens, J. D., Kempen, B., Knotters, M., Mulder, V. L., Poggio, L., & Heuvelink, G. B. M. (2021). Towards climate-smart sustainable management of agricultural soils: Deliverable 6.1 Report on harmonized procedures for creation of databases and maps. *EJP Soil.*

El-Alsayed, S. G., Ismail, S. M., (2017) Impact of soil amendments and irrigation water on growth and flowering of Rosa plant grown in sandy soil Rosa Hybrida. *Ain Shams Engineering Journal, 38(3)*, 626–641. https://doi.org/10.21608/asejaiqjsae.2017.4058

Emmerling, C. (2007). Differences in soil organic carbon stocks and aggregate stability in long-term organic and conventional farming systems. *Soil Use and Management., 23*(4), 450–461.

Erdogan, H. E., Havlicek, E., Dazzi, C., Montanarella, L., Van Liedekerke, M., Vrščaj, B., Krasilnikov, P., Khasankhanova, G., & Vargas, R. (2021). Soil conservation and sustainable development goals (SDGs) achievement in Europe and central Asia: Which role for the European soil partnership? *International Soil and Water Conservation Research, 9*(3), 360–369.

Fahad, S., Chavan, S. B., Chichaghare, A. R., Uthappa, A. R., Kumar, M., Kakade, V., Pradhan, A, Jinger, D., Rawale, G., Yadav, D. K., & Poczai, P. (2022). Agroforestry systems for soil health improvement and maintenance. *Sustainability, 14*(22), 14877

FAO, ITPS. (2020). Global soil organic carbon map V1.5: technical report. Food and agriculture Organization of the United Nations.

Farmaha, B. S.; Sekaran, U., & Franzluebbers, A. J (2022) Cover cropping and conservation tillage improve soil health in the southeastern United States. *Agronomy Journal, 114*, 296–316.

Farooq, S., Yasmeen, T., Niaz, A., Rizwan, M., & Ali, S. (2022). Rice straw biochar in combination with farmyard manure mitigates Bromoxynil toxicity in wheat (*Triticum aestivum L.*). *Chemosphere, 295*, 133854.

Feng, F., Jiang, Y., Jia, Y., Shang, C., Lian, X., Zang, Y., & Zhao, M. (2023). Risks of nutrients and metal(loid)s mobilization triggered by groundwater recharge containing reactive organic matter. *Journal of Hydrology, 623*, Article 129780.

Fierer, N., Wood, S. A., & de Mesquita, C. P. B. (2021). How microbes can, and cannot, be used to assess soil health. *Soil Biology and Biochemistry, 153*, Article 108111.

Fontúrbel, M. T., Jim´enez, E., Merino, A., & Vega, J. A. (2023) Contrasting immediate impact of prescribed fires and experimental summer fires on soil organic matter quality and microbial properties in the forest floor and mineral soil in Mediterranean black pine forest. *Science of the Total Environment,* 167669.

Galdos, M. V., Brown, E., Rosolem, C. A., Pires, L. F., Hallett, P. D., & Mooney, S. J. (2020). Brachiaria species influence nitrate transport in soil by modifying soil structure with their root system. *Scientific Reports, 10*(1), 5072.

Gavrilescu, M. (2021). Water, soil, and plants interactions in a threatened environment. *Water, 13*(19), 2746.

Gilmullina, A., Rumpel, C., Blagodatskaya, E., Klumpp, K., Bertrand, I., Dippold, M. A., & Chabbi, A. (2023). Is plant biomass input driving soil organic matter formation processes in grassland soil under contrasting management? *Science of the Total Environment, 893*, Article 164550.

Goh, K. M. (2011). Greater mitigation of climate change by organic than conventional agriculture: A review. *Biological Agriculture & Horticulture, 27*, 205–229.

Goss, M. J., Tubeileh, A., & Goorahoo, D. (2013). A review of the use of organic amendments and the risk to human health. *Advances in Agronomy, 120*, 275–379.

Griscom, B. W., Adams, J., Ellis, P. W., Houghton, R. A., Lomax, G., Miteva, D. A., Schlesinger, W. H., Shoch, D., Siikamäki, J. V., Smith, P., Woodbury, P., Zganjar, C., Blackman, A., Campari, J., Conant, R. T., Delgado, C., Elias, P., Gopalakrishna, T., Hamsik, M. R., & Fargione, J. (2017). Natural climate solutions. *Proceedings of the National Academy of Sciences of the United States of America, 114*(44), 11645–11650.

Guenet, B., Gabrielle, B., Chenu, C., Arrouays, D., Balesdent, J., Bernoux, M., Bruni, E., Caliman, J. P., Cardinael, R., & Chen, S. (2021). Can N2O emissions offset the benefits from soil organic carbon storage? *Global Change Biology, 27*, 237–256.

Guo, M. (2020) The 3R Principles for applying biochar to improve soil health. *Soil System, 4*, 9.; Urra, J., Alkorta, I., Garbisu, C. (2019) Potential benefits and risks for soil health derived from the use of organic amendments in agriculture. *Agronomy, 9*, 542.

Haddaway, N. R., Hedlund, K., Jackson, L. E., Kätterer, T., Lugato, E., Thomsen, I. K., Jørgensen, H. B., & Isberg, P.-E. (2017). How does tillage intensity affect soil organic carbon? A systematic review. *Environmental Evidence, 6*, 1–48.

Hagmann, R. K., Hessburg, P. F., Prichard, S. J., Povak, N. A., Brown, P. M., Fulé, P. Z., Keane, R. E., & Waltz, A. E. M. (2021). Evidence for widespread changes in the structure, composition, and fire regimes of western North American forests. *Ecological Applications, 31*(8), e02431

Harris, N. L., Gibbs, D. A., Baccini, A., Birdsey, R. A., De Bruin, S., Farina, M., Fatoyinbo, L., Hansen, M. C., Herold, M., & Tyukavina, A. (2021). Global maps of twenty-first century forest carbon fluxes. *Nature Climate Change, 11*(3), 234–240.

Hartmann, M., & Six, J. (2023). Soil structure and microbiome functions in agroecosystems. *Nature Reviews Earth & Environment, 4*(1), 4–18.

Hossain, A., Krupnik, T. J., Timsina, J., Mahboob, M. G., Chaki, A. K., Farooq, M., Bhatt, R., Fahad, S., & Hasanuzzaman, M. (2020). Agricultural land degradation: processes and problems undermining future food security. In *Environment, climate, plant and vegetation growth* (pp. 17-61). Cham: Springer International Publishing

Hou, D., Bolan, N. S., Tsang, D. C., Kirkham, M. B., & O'connor, D. (2020). Sustainable soil use and management: An interdisciplinary and systematic approach. *Science of the Total Environment, 729*, Article 138961.

Inagaki, T. M., Possinger, A. R., Schweizer, S. A., Mueller, C. W., Hoeschen, C., Zachman, M. J., Kourkoutis, L. F., Kogel-Knabner, I., & Lehmann, J. (2023). Microscale spatial distribution and soil organic matter persistence in top and subsoil. *Soil Biology and Biochemistry 178*, 108921

IPCC. (2001). *Annex B—glossary of terms in climate change 2001: Impacts, adaptation, and vulnerability: Contribution of working group II to the third assessment report of the intergovernmental panel on climate change* (Vol. 2). Cambridge University Press.

IPCC. (2021). Annex VII: Glossary. In J. B. R. Matthews, V. Möller, R. van Diemen, J. S. Fuglestvedt, V. Masson-Delmotte, C. Méndez, S. Semenov, & A. Reisinger (Eds.), *In climate change 2021: The physical science basis. Contribution of working group I to the sixth assessment report of the intergovernmental panel on climate change* (pp. 2215– 2256). Cambridge University Press.

IPCC. (2023). In Core Writing Team, H. Lee, & J. Romero (Eds.), Climate change 2023: Synthesis report. *Contribution of working groups I, II and III to the sixth assessment report of the Intergovernmental Panel on Climate Change* (p. 184). IPCC.

Janzen, H. H., Janzen, D. W., & Gregorich, E. G. (2021). The 'soil health'metaphor: Illuminating or illusory? *Soil Biology and Biochemistry, 159*, Article 108167.

Jian, J., Du, X., Reiter, M. S., & Stewart, R. D. (2020). A meta-analysis of global cropland soil carbon changes due to cover cropping. *Soil Biology and Biochemistry, 143*, 107735.

Junaid Nazir, M., Zhang, X., Du, D., & Yu, F. (2022). How physical disturbance and nitrogen addition affect the soil carbon decomposition? *Phyton, 91*, 2087–2097. https://doi.org/10.32604/phyton.2022.021412

Karlen, D. L., Mausbach, M. J., Doran, J. W., Cline, R. G., Harris, R. F., & Schuman, G. E. (1997). Soil quality: A concept, definition, and framework for evaluation (a guest editorial). *Soil Science Society of America Journal., 61*(1), 4–10.

Kennedy, S., Wade, C., Ma, L., Leslie-Bole, H., Dahl, C., Favero, A., & Hultman, N. (2024). *Harnessing the Land Sector to Achieve US Climate Goals: An all-of-society approach to meeting our climate goals and bolstering the carbon sink by 2035* (p. 21). University of Maryland and America Is All In.

Kihara, J., Bolo, P., Kinyua, M., Nyawira, S. S., & Sommer, R. (2020). Soil health and ecosystem services: Lessons from sub-Sahara Africa (SSA). *Geoderma, 370*, 114342.

Kim, D. G., Kirschbaum, M. U., & Beedy, T. L. (2016). Carbon sequestration and net emissions of CH_4 and N_2O under agroforestry: Synthesizing available data and suggestions for future studies. *Agriculture, Ecosystems & Environment, 226*, 65–78.

Köninger, J., Panagos, P., Jones, A., Briones, M. J. I., & Orgiazzi, A. (2022). In defence of soil biodiversity: Towards an inclusive protection in the European Union. *Biological Conservation, 268*, Article 109475.

Kumar, S., Meena, R. S., Yadav, G. S., & Pandey, A. (2017b). Response of sesame (Sesamum indicum L.) to Sulphur and lime application under soil acidity. *International Journal of Plant and Soil Science, 14*(4):1–9.

Lal, R. (2003). Global potential of soil carbon sequestration to mitigate the greenhouse effect. *CRC Critical Reviews in Plant Sciences*. https://doi.org/10.1080/713610854

Lal, R. (2008). Carbon sequestration. *Philosophical Transactions of the Royal Society (B: Biological Sciences), 363*, 815–830.

Lal, R., Bouma, J., Brevik, E., Dawson, L., Field, D. J., Glaser, B., Hatano, R., Hartemink A. E., & Zhang, J. (2021). Soils and sustainable development goals of the United Nations: An International Union of Soil Sciences perspective. *Geoderma Regional, 25*, e00398.

Lal, R., Negassa, W., & Lorenz, K. (2015). Carbon sequestration in soil. *Current Opinion in Environmental Sustainability, 15*, 79–86.

Lanfranco, B., Fernández, E., Ferraro, B., & de Lima, J. M. S. (2022). Historical changes in the Pampas biome, land use, and climate change. In *Handbook of Behavioral Economics and Climate Change* (pp. 162–191). Edward Elgar Publishing.

Lehmann, J., Bossio, D. A., Kögel-Knabner, I., & Rillig, M. C. (2020). The concept and future prospects of soil health. *Nature Reviews Earth & Environment, 1*(10), 544–553.

Lehmann, J., & Kleber, M. (2015). The contentious nature of soil organic matter. *Nature*. https://doi.org/10.1038/nature16069

Lehtinen, T., Schlatter, N., Baumgarten, A., Bechini, L., Krüger, J., Grignani, C., Zavattaro, L., Costamagna, C., & Spiegel, H. (2014). Effect of crop residue incorporation on soil organic carbon and greenhouse gas emissions in European agricultural soils. *Soil Use and Management, 30*, 524–538.

Li, Y., Chen, Z., Zhang, N., Zheng, H., Miao, Y., Li, J., & Ding, W. (2024). Decreased microbial phylogenetic diversity and community stability due to less bioavailable carbon and greater oxygen supply in Mollisols along a cultivation chronosequence. *Soil and Tillage Research, 238*, Article 106005.

Liang, F., et al. (2016). Three-decade long fertilization-induced soil organic C sequestration depends on edaphic characteristics in six typical croplands. *Science and Reports, 6*, 30350. https://doi.org/10.1038/srep30350

Liptzin, D., Norris, C. E., Cappellazzi, S. B., Mac Bean, G., Cope, M., Greub, K. L., Rieke L. E., Tracy, P. W., Ezar A, Ashworth, A., Tanvarez, O. B., Bary, A. I., Baumhardt, R. L., & Honeycutt, C. W. (2022). An evaluation of carbon indicators of soil health in long-term agricultural experiments. *Soil Biology and Biochemistry, 172*, 108708.

Liu, Q., Zhang, Y., Liu, B., Amonette, J. E., Lin, Z., Liu, G., et al. (2018). How does biochar influence soil N cycle? *A Meta-Analysis. Plant Soil., 426*, 211–225. https://doi.org/10.1007/s11104-018-3619-4

Lugato, E., Leip, A., & Jones, A. (2018). Mitigation potential of soil carbon management overestimated by neglecting N2O emissions. *Nature Climate Change, 8*(3), 219–223. https://doi.org/10.1038/s41558-018-0087-z

Lyu, H., Gao, B., He, F., Zimmerman, A. R., Ding, C., Huang, H., et al. (2018). Effects of ball milling on the physicochemical and sorptive properties of biochar: Experimental observations and governing mechanisms. *Environmental Pollution, 233*, 54–63. https://doi.org/10.1016/j.envpol.2017.10.037

Mabagala, F. S., Mng'ong'o, M. E. (2022). On the tropical soils; the influence of organic matter (OM) on phosphate bioavailability. *Saudi Journal of Biological Science, 29*(5), 3635–3641.
MacPherson, J., Voglhuber-Slavinsky, A., Olbrisch, M., Schöbel, P., Dönitz, E., Mouratiadou, I., & Helming, K. (2022). Future agricultural systems and the role of digitalization for achieving sustainability goals. *A Review. Agronomy for Sustainable Development, 42*(4), 70.
Maenhout, P., Di Bene, C., Cayuela, M. L., Diaz-Pines, E., Govednik, A., Keuper, F., Mavsar, S., Mihelic, R., O' Toole, A., Schwarzmann, A., & Valkama, E. (2024). Trade-offs and synergies of soil carbon sequestration: Addressing knowledge gaps related to soil management strategies. *European Journal of Soil Science, 75*(3), e13515.
Malone, M., & McClintock, N. (2023). A critical physical geography of no-till agriculture: Linking degraded environmental quality to conservation policies in an Oregon watershed. *The Canadian Geographer/le Géographecanadien, 67*(1), 74–91.
Man, M., Gregorich, E. G., Beare, M. H., Ellert, B. H., & Simpson, M. J. (2023). Distinct dynamics of plant-and microbial-derived soil organic matter in relation to varying climate and soil properties in temperate agroecosystems. *Geochem. Cosmochim. Acta, 361*, 276–287.
Mandal, A., Majumder, A., Dhaliwal, S. S., Toor, A. S., Mani, P. K., Naresh, R. K., Gupta, R. K., & Mitran, T. (2022). Impact of agricultural management practices on soil carbon sequestration and its monitoring through simulation models and remote sensing techniques: A review. *Critical Reviews in Environmental Science and Technology, 52*(1), 1–49.
Manirakiza, N., & Se¸ker, C. (2020). Effects of compost and biochar amendments on soil fertility and crop growth in a calcareous soil. *Journal of Plant Nutrition, 43*(20), 3002–3019. https://doi.org/10.1080/01904167.2020.1806307
Matisic, M., Dugan, I., & Bogunovic, I. (2024). Challenges in Sustainable Agriculture—The Role of Organic Amendments. *Agriculture, 14*, 643. https://doi.org/10.3390/agriculture14040643
Maurya, S., Abraham, J. S., Somasundaram, S., Toteja, R., Gupta, R., & Makhija, S. (2020). Indicators for assessment of soil quality: A mini-review. *Environmental Monitoring and Assessment, 192*, 1–22.
McCord, S. E., Brehm, J. R., Burnett, S. H., Dietrich, C., Edwards, B., Metz, L. J., Narvaez, M. H., Pierson, F., Ramirez, K. S., & Tweedie, C. E. (2022). A framework and toolset for standardizing agroecosystem indicators. *Ecological Indicators, 144*, 109511.
Meena, R. S., Kumar, S., & Yadav, G. S. (2020). Soil carbon sequestration in crop production. *Nutrient dynamics for sustainable crop production*, 1–39.
Meena, H., & Meena, R. S. (2017). Assessment of sowing environments and bio-regulators as adaptation choice for clusterbean productivity in response to current climatic scenario. *Bangladesh J Bot, 46*(1), 241–244.
de Mello, K., Taniwaki, R. H., de Paula, F. R., Valente, R. A., Randhir, T. O., Macedo, D. R., Leal, C. G., Rodrigues, C. B., & Hughes, R. M. (2020). Multiscale land use impacts on water quality: Assessment, planning, and future perspectives in Brazil. *Journal of Environmental Management, 270*, 110879
Meurer, K. H., Hendriks, C. M., Faber, J. H., Kuikman, P. J., van Egmond, F., Garland, G., Putku, E., Barancikova, G., Makovnikova, J., Chenu, C & Bispo, A. (2024). How does national SOC monitoring on agricultural soils align with the EU strategies? An example using five case studies. *European Journal of Soil Science, 75*(2), e13477.
Minasny, B., Malone, B. P., McBratney, A. B., Angers, D. A., Arrouays, D., Chambers, A., Chaplot, V., Chen, Z.-S., Cheng, K., & Das, B. S. (2017a). Soil carbon 4 per mille. *Geoderma, 292*, 59–86.
Montanarella, L., & Panagos, P. (2021). The relevance of sustainable soil management within the European Green Deal. *Land Use Policy, 100*, Article 104950.
Morugan-Coronado, A., Linares, C., Gomez-L opez, M. D., Faz, A., & Zornoza, R. (2020). The impact of intercropping, tillage and fertilizer type on soil and crop yield in fruit orchards under Mediterranean conditions: A meta-analysis of field studies. *Agricultural Systems, 178*, 10273
Mugiraneza, T., Nascetti, A., & Ban, Y. (2020). Continuous monitoring of urban land cover change trajectories with landsat time series and landtrendr-google earth engine cloud computing. *Remote Sensing, 12*(18), 2883.

Nazir, M. J., Li, G., Nazir, M. M., Zulfiqar, F., Siddique, K. H., Iqbal, B., & Du, D. (2024). Harnessing soil carbon sequestration to address climate change challenges in agriculture. *Soil and Tillage Research, 237*, Article 105959.

Nelson, M. D., Riitters, K. H., Coulston, J. W., Domke, G. M., Greenfield, E. J., Langner, L. L., Nowak, D. J., O'Dea, C. B., Oswalt, S. N., Reeves, M. C., & Wear, D. N. (2020). Defining the United States land base: a technical document supporting the USDA Forest Service 2020 RPA assessment. General Techical Reports NRS-191, 191, 1–70.

Niu, L., Manxia, C., Xiumei, G., Xiaohua, L., Hongbo, S., Zhaopu, L., & Zed, R. (2016). Carbon sequestration and Jerusalem artichoke biomass under nitrogen applications in coastal saline zone in the northern region of Jiangsu, China. *Science of the Total Environment, 568*, 885–890.

Norris, C. E., Bean, G. M., Cappellazzi, S. B., Cope, M., Greub, K. L., Liptzin, D., Rieke, E. L., Tracy, P. W., Morgan, C. L., & Honeycutt, C. W. (2020). Introducing the North American project to evaluate soil health measurements. *Agronomy Journal, 112*(4), 3195–3215

Nunes, M. R., Veum, K. S., Parker, P. A., Holan, S. H., Karlen, D. L., Amsili, J. P., Harold, M.,van Es, H,. M., Wills, S. A., Seybold, C. A., & Moorman, T. B. (2021). The soil health assessment protocol and evaluation applied to soil organic carbon. *Soil Science Society of America Journal, 85*(4), 1196–1213.

Nunes, M. R., Karlen, D. L., Veum, K. S., Moorman, T. B., & Cambardella, C. A. (2020). Biological soil health indicators respond to tillage intensity: A US meta-analysis. *Geoderma, 369*, Article 114335.

Ogle, S. M., Breidt, F. J., Del Grosso, S., Gurung, R., Marx, E., Spencer, S., Williams, S., & Manning, D. (2023). Counterfactual scenarios reveal historical impact of cropland management on soil organic carbon stocks in the United States. *Scientific Reports, 13*(1), 14564.

de Oliveira, J. J., Dalmazo, G. O., Morselli, T. B. G. A., de Oliveira, V. F. S., Corrˆea, L. B., Nora, L., Corrˆea, E. K. (2018). Composted slaughterhouse sludge as a substitute for chemical fertilizers in the cultures of lettuce (*Lactuca sativa L.*) and radish (*Raphanus sativus L.*). *Journal of Food Science and Technology, 38(1)*, 91–97. https://doi.org/10.1590/1678-457x.00717

Olson, K. R., Al-Kaisi, M. M., Lal, R., & Lowery, B. (2014). Experimental consideration, treatments, and methods in determining soil organic carbon sequestration rates. *Soil Science Society of America Journal, 78*(2), 348–360. https://doi.org/10.2136/sssaj2013.09.0412

Osland, M. J., Stevens, P. W., Lamont, M. M., Brusca, R. C., Hart, K. M., Waddle, J. H., Langtimm, C. A., Williams, C. M., Keim, B. D., Terando, A. J., & Seminoff, J. A. (2021). Tropicalization of temperate ecosystems in North America: The northward range expansion of tropical organisms in response to warming winter temperatures. *Global Change Biology, 27*(13), 3009–3034.

Oswalt, S., Oswalt, C., Crall, A., Rabaglia, R., Schwartz, M. K., & Kerns, B. K. (2021). Inventory and monitoring of invasive species. *Invasive species in forests and rangelands of the United States*, 231.

Pahalvi, H. N., Rafiya, L., Rashid, S.; Nisar, B., & Kamili, A. N. (2021). Chemical Fertilizers and Their Impact on Soil Health. In Microbiota and Biofertilizers, G. H. Dar, R. A Bhat, M. A. Mehmood, & K. R Hakeem (Eds.), (Vol 2, pp. 1–20), Springer: Cham, Germany.

Panagos, P., Ballabio, C., Poesen, J., Lugato, E., Scarpa, S., Montanarella, L., & Borrelli, P. (2020). A soil erosion indicator for supporting agricultural, environmental and climate policies in the European Union. *Remote Sensing, 12*(9), 1365.

Parish, E. S., Dale, V. H., Davis, M., Efroymson, R. A., Hilliard, M. R., Jager, H., & Xie, F. (2021). An indicator-based approach to sustainable management of natural resources. *In Data Science Applied to Sustainability Analysis* (pp. 255–280). Elsevier.

Pathan, S., & Siddiqui, R. A. (2022). Nutritional composition and bioactive components in quinoa (Chenopodium quinoa Willd.) greens: A review. *Nutrients, 14*(3), 558.

Paustian, K., Lehmann, J., Ogle, S., Reay, D., Robertson, G. P., & Smith, P. (2016). Climate-smart soils. *Nature, 532*(7597), 49–57.

Pergola, M., Persiani, A., Palese, A. M., Di Meo, V., Pastore, V., D'Adamo, C., & Celano, G. (2018). Composting: The way for a sustainable agriculture. *Applied Soil Ecology, 123*, 744–750. https://doi.org/10.1016/j.apsoil.2017.10.016

Petraru, A., Ursachi, F., &Amariei, S. (2021). Nutritional characteristics assessment of sunflower seeds, oil and cake. Perspective of using sunflower oilcakes as a functional ingredient. *Plants, 10*(11), 2487.

Piveta, L. B., Noldin, J. A., Roma-Burgos, N., Viana, V. E., Benedetti, L., & Pinto, J. J. O et al. (2021). Weedy rice (Oryza spp.) diversity in southern Brazil. *Weed Science, 69*, 547–57. https://doi.org/10.1017/wsc.2021.23

Placide, R., Theophile, N., Vandamme, E., Claude, N. J., & Thiago, M. (2022). Yield performance, adaptability and processing qualities of pre-release potato clones under different Rwandan agro-ecologies. *CABI Agriculture and Bioscience, 3*(1), 40.

Powlson, D. S. (2020). Soil health—useful terminology for communication or meaningless concept? Or both? *Frontiers of Agricultural Science and Engineering-FASE.*

Prats, S. A., Serpa, D., Santos, L., & Keizer, J. J. (2023). Effects of forest residue mulching on organic matter and nutrient exports after wildfire in North-Central Portugal. *Science of the Total Environment, 885*, Article 163825.

Prăvălie, R., Patriche, C., Borrelli, P., Panagos, P., Roşca, B., Dumitraşcu, M., Nita, I. A., Savulescu, I., Birsan, M. V., & Bandoc, G. (2021). Arable lands under the pressure of multiple land degradation processes. *A global perspective. Environmental Research, 194*, 110697

Raseduzzaman, M., Gaudel, G., Ali, M. R., Timilsina, A., Bizimana, F., Aluoch, S. O., Li, X., Zang, Y., & Hu, C. (2024). Cereal-legume mixed residue addition increases yield and reduces soil greenhouse gas emissions from fertilized winter wheat in the North China plain. *Agronomy, 14*(6), 1167

Raza, S., Zamanian, K., Ullah, S., Kuzyakov, Y., Virto, I., & Zhou, J. (2021). Inorganic carbon losses by soil acidification jeopardize global efforts on carbon sequestration and climate change mitigation. *Journal of Cleaner Production, 315*, Article 128036.

Rehrah, D., Bansode, R. R., Hassan, O., & Ahmedna, M. (2016). Physico-chemical characterization of biochars from solid municipal waste for use in soil amendment. *Journal of Analytical and Applied Pyrolysis, 118*, 42–53. https://doi.org/10.1016/j.jaap.2015.12.022

Rodríguez Eugenio, N. (2021). The global soil partnership: Tackling global soil threats through collective action. *International Yearbook of Soil Law and Policy, 2019*, 197–221.

Rosenberg, A. B., & Wallander, S. (2022). USDA Conservation Technical Assistance and Within-Field Resource Concerns.

Salamanca-Carreño, A., Vélez-Terranova, M., Vargas-Corzo, O. M., Pérez-López, O., Castillo-Pérez, A. F., & Parés-Casanova, P. M. (2023). Relationship of Physiographic Position to Physicochemical Characteristics of Soils of the Flooded-Savannah Agroecosystem. *Colombia. Agriculture, 13*(1), 220.

Salgar-Chaparro, S. J., Lepkova, K., Pojtanabuntoeng, T., Darwin, A., & Machuca, L. L. (2020). Nutrient level determines biofilm characteristics and subsequent impact on microbial corrosion and biocide effectiveness. *Applied and Environmental Microbiology, 86*(7), e02885-e2919.

Sanderman, J., Savage, K., & Dangal, S. R. (2020). Mid-infrared spectroscopy for prediction of soil health indicators in the United States. *Soil Science Society of America Journal, 84*(1), 251–261.

Santana, K. V. R., Apolônio, F. C. S. O., & Wisniewski, A. (2019). Valorization of cattle manure by thermoconversion process in a rotary kiln reactor to produce environmentally friendly products. *Bioenergy Research, 13*(2), 605–617. https://doi.org/10.1007/s12155-019-10047-0

Seaton, F. M., Barrett, G., Burden, A., Creer, S., Fitos, E., Garbutt, A., Griffiths, R. l., Henrys, P., Jones, D. L., Keenan, P., & Robinson, D. A. (2021). Soil health cluster analysis based on national monitoring of soil indicators. *European Journal of Soil Science, 72*(6), 2414–2429.

Shaaban, M., & Nunez-Delgado, A. (2024). Soil adsorption potential: Harnessing Earth's living skin for mitigating climate change and greenhouse gas dynamics. *Environmental Research, 251*, 118738

Shaaban, M., Peng, Q., Hu, R., Lin, S., & Zhao, J. (2016). Soil nitrous oxide and carbon dioxide emissions following incorporation of above-and below-ground biomass of green bean. *International Journal of Environmental Science and Technology, 13*, 179–186.

Shah, K. K., Tripathi, S., Tiwari, I., Shrestha, J., Modi, B., Paudel, N., & Das, B. D. (2021). Role of soil microbes in sustainable crop production and soil health: A review. *Agricultural Science and Technology, 13*(2), 109–118.
Shahane, A. A., & Shivay, Y. S. (2021). Soil health and its improvement through novel agronomic and innovative approaches. *Frontiers in Agronomy, 3*, Article 680456.
Shukla, S. K., Shee, S., Maity, S. K., Solomon, S., Awasthi, S. K.; Gaur, A.; Pathak, A. D., & Jaiswal, V. P. (2016). Soil carbon sequestration and crop yields in rice–wheat and sugarcane–ratoon–wheat cropping systems through crop residue management and inoculation of trichoderma viride in Subtropical India. *Sugar Tech, 19*, 347–358.
Six, J., Elliott, E. T., & Paustian, K. (2000). Soil macroaggregate turnover and microaggregate formation: A mechanism for C sequestration under no-tillage agriculture. *Soil Biology and Biochemistry., 32*(14), 2099–2103.
Smith, C., 2021. How technology is changing the agriculture landscape in SA. In: Available at How Technology Is Changing the Agriculture Landscape in SA Business (news 24.Com).
Somasundaram, J., Sinha, N.K., Dalal, R.C., Lal, R., Mohanty, M., Naorem, A.K., Hati, K. M., Chaudhary, R.S., Biswas, A.K., Patra, A.K., & Chaudhari, S.K. (2020). No-Till Farming and Conservation Agriculture in South Asia–Issues, challenges, prospects and benefits. *CRC Critical Reviews in Plant Sciences*. https://doi.org/10.1080/07352689.2020.1782069
Soto, R. L., Padilla, M. C., & de Vente, J. (2020). Participatory selection of soil quality indicators for monitoring the impacts of regenerative agriculture on ecosystem services. *Ecosystem Services, 45*, Article 101157.
Soussana, J.-F., Lutfalla, S., Ehrhardt, F., Rosenstock, T., Lamanna, C., Havlík, P., Richards, M., Wollenberg, E., Chotte, J. L., Torquebiau, E. et al. (2019) Matching policy and science: Rationale for the '4 per 1000—Soils for food security and climate' initiative. *Soil and Tillage Research, 188*, 3–15.
Spaeth, K. E., Rutherford, W. A., Houdeshell, C. A., Williams, C. J., Simpson, B., Green, S., Toledo, D., Suffridge, E., & McCord, S. E. (2024). Insights from the USDA Grazing Land National Resources Inventory and field studies. *Journal of Soil and Water Conservation, 79*(3), 37A–42A
Stockmann, U., Adams, M. A., Crawford, J. W., Field, D. J., Henakaarchchi, N., Jenkins, M., & Wheeler, I. (2013). The knowns, known unknowns and unknowns of sequestration of soil organic carbon. *Agriculture, Ecosystems and Environment, 164*, 80–99.
Stojanovi´c, M., Gligorijevi´c, J., Stojanovi´c, G., & Beci´c, S., 2023. Analysis of the influence of environmental taxation on environmental innovations. Ekonom: teorijaipraksa *16*(1), 101–119.
Suman, J., Rakshit, A., Ogireddy, S. D., Singh, S., Gupta, C., & Chandrakala, J. (2022). Microbiome as a key player in sustainable agriculture and human health. *Frontiers in Soil Science, 2*, Article 821589.
Tiefenbacher, A., Sandén, T., Haslmayr, H.-P., Miloczki, J., Wenzel, W., & Spiegel, H. (2021). Optimizing Carbon Sequestration in Croplands: A Synthesis. *Agronomy, 11*(5), 882. https://doi.org/10.3390/agronomy11050882
Trost, B., Prochnow, A., Drastig, K., Meyer-Aurich, A., Ellmer, F., & Baumecker, M. (2013). Irrigation, soil organic carbon and N 2 O emissions. *A Review. Agronomy for Sustainable Development, 33*, 733–749.
United Nations Environment Programme. (2021). Adaptation gap report 2020. Nairobi: United Nations Environment Programme.
Vasilyeva, T., Samusevych, Y., Babenko, V., Bestuzheva, S., Bondarenko, S., & Nesterenko, I. (2023). Environmental taxation: role in promotion of the pro environmental behaviour. *WSEAS Transactions on Business and Economics,* 1–18. https://doi.org/10.37394/23207.2023.20.37
V´ed'ere, C., Lebrun, M., Honvault, N., Aubertin, M.L., Girardin, C., Garnier, P., Dignac, M.F., Houben, D., & Rumpel, C. (2022). How does soil water status influence the fate of soil organic matter? A review of processes across scales. *Earth Science Reviews, 234*, 104214
Vieira, R. M. D. S. P., Tomasella, J., Barbosa, A. A., Polizel, S. P., Ometto, J. P. H. B., Santos, F. C., da Cruz Ferreira, Y., & de Toledo, P. M. (2021). Land degradation mapping in the

MATOPIBA region (Brazil) using remote sensing data and decision-tree analysis. *Science of The Total Environment, 782*, 146900

Voltz, M., Arrouays, D., Bispo, A., Lagacherie, P., Laroche, B., Lemercier, B., Richer-de-Forges, A., Sauter, J., & Schnebelen, N. (2020). Possible futures of soil-mapping in France. *Geoderma Regional, 23*, e00334.

Walsh, C. T. (2023). *The Chemical Biology of Carbon* (No. 21). Royal Society of Chemistry.

Wang, M., Wander, M., Mueller, S., Martin, N., & Dunn, J. B. (2022). Evaluation of survey and remote sensing data products used to estimate land use change in the United States: Evolving issues and emerging opportunities. *Environmental Science & Policy, 129*, 68–78.

Weng, Z., Van Zwieten, L., Singh, B. P., Tavakkoli, E., Joseph, S., & Macdonald, L. M., et al. (2017). Biochar built soil carbon over a decade by stabilizing rhizodeposits. *Nature Climate Change, 7*, 371–6. https://doi.org/10.1038/nclimate3276

White, P.J., & Brown, P.H. (2010). Plant nutrition for sustainable development and global health. *Annals of Botany, 105*. https://doi.org/10.1093/aob/mcq085

Wijesekara, H., Colyvas, K., Rippon, P., Hoang, S. A., Bolan, N. S., Manna, M. C., Thangavel, R., Seshadri, B., Vithanage, M., & Awad, Y. M. (2021). Carbon sequestration value of biosolids applied to soil: A global meta-analysis. *Journal of Environmental Management, 284*, Article 112008.

Williams, H., Colombi, T., & Keller, T. (2020). The influence of soil management on soil health: An on-farm study in southern Sweden. *Geoderma, 360*, Article 114010.

Xia, L., Lam, S. K., Wolf, B., Kiese, R., Chen, D., & Butterbach-Bahl, K. (2018). Trade-offs between soil carbon sequestration and reactive nitrogen losses under straw return in global agroecosystems. *Global Change Biology, 24*, 5919–5932.

Xu, S., Zhang, L., McLaughlin, N.B., Mi, J., Chen, Q., & Liu, J. (2015). Effect of synthetic and natural water absorbing soil amendment soil physical properties under potato production in a semi-arid region. *Soil and Tillage Research, 148*, 31–39. https://doi.org/10.1016/j.still.2014.10.002

Yadav, G.S.; Das, A.; Babu, S.; Mohapatra, K.P.; Lal, R.; Rajkhowa, D. (2021) Potential of conservation tillage and altered land configuration to improve soil properties, carbon sequestration and productivity of maize based cropping system in eastern Himalayas, India. *International Soil and Water Conservation Research, 9*, 279–290.

Yan, Y., Chen, Z., Zhu, F., Zhu, C., Wang, C., & Gu, C. (2021). Effect of polyvinyl chloride microplastics on bacterial community and nutrient status in two agricultural soils. *Bulletin of Environmental Contamination and Toxicology, 107*, 602–609.

Yang, Y., Dou, Y., Wang, B., Wang, Y., Liang, C., An, S., Soromotin, A., & Kuzyakov, Y. (2022). Increasing contribution of microbial residues to soil organic carbon in grassland restoration chronosequence. *Soil Biology and Biochemistry, 170*, 108688

Yang, T., Siddique, K. H., & Liu, K. (2020). Cropping systems in agriculture and their impact on soil health-A review. *Global Ecology and Conservation, 23*, Article e01118.

Zhang, X., Zhu, B., Yu, F.-H., Wang, P., & Cheng, W. (2022). Exclusion of plant input affects the temperature sensitivity of soil carbon decomposition. *SSRN Electronic Journal*. https://doi.org/10.2139/ssrn.3998985

Zhao, R., & Wu, K. (2021). Soil health evaluation of farmland based on functional soil management—A case study of yixing city, jiangsu province. *China. Agriculture, 11*(7), 583.

Zwetsloot, M. J., van Leeuwen, J., Hemerik, L., Martens, H., SimóJosa, I., Van de Broek, M., Debeljak, M., Rutgers, M., Sanden, T., Wall, D. P., & Creamer, R. E. (2021). Soil multifunctionality: Synergies and trade-offs across European climatic zones and land uses. *European Journal of Soil Science, 72*(4), 1640–1654

Soil Health Assessment Strategies and Future Perspectives Towards a Sustainable Agriculture

Faiz Ul Hassan, Mir Sujaul Islam, Abdul Basit, Raheel Ahmad, Abdiwahid Ali Dhakane, Mohd Ekhwan Hj. Toriman, Muhammad Amjad Bashir, Abdur Rehim, and Fazal Maqsood

Abstract The global population is expanding at an alarming rate, which makes it harder to meet the expanding need for food. The soil's capacity to provide adequate food for both the present and future populations is under jeopardy due to the intense agricultural production that has resulted in its depletion. It is imperative to create novel strategies for long-term sustainable management of depleted soil health in order to enhance agricultural productivity and preserve food security. Both crop yield and sustainable agriculture depend on healthy soil. As a result, it is crucial to consistently measure and track scientific evaluations of soil health. In order to promote sustainable agriculture and increase the resilience of the soil ecosystem, soil health assessment is essential. This chapter covered the importance of soil health, how it integrates into sustainable agriculture, improved methods for assessing soil health, potential

F. U. Hassan · M. S. Islam (✉) · F. Maqsood
Institute for Environment and Development (LESTARI), Universiti Kebangsaan Malaysia (UKM), Bangi 43600, Selangor, Malaysia
e-mail: sujaul@ukm.edu.my

A. Basit
Department of Agronomy, Engro Fertilizers Limited, Rahim Yar Khan, Pakistan

R. Ahmad
Earth Observation Centre, Institute of Climate Change (IPI) Climate Change, Universiti Kebangsaan Malaysia (UKM), Bangi 43600, Selangor, Malaysia

A. A. Dhakane
Graduate School of Natural and Applied Sciences, Agricultural Structure and Irrigation Department, Çukurova University, Adana, Türkiye

M. E. Hj. Toriman
Centre for International Strategic Negotiation and Sustainable Development Goals, Universiti Kebangsaan Malaysia (UKM), Bangi 43600, Selangor, Malaysia

M. A. Bashir
Sichuan Provincial Key Laboratory of Philosophy and Social Sciences for Monitoring and Evaluation of Rural Land Utilization, Chengdu Normal University, Chengdu 611130, P. R. China

A. Rehim
Department of Soil Science, Faculty of Agricultural Sciences and Technology, Bahauddin Zakariya University, Multan 60800, Pakistan

M. Shaaban (ed.), *Soils and Sustainable Agriculture*, Frontier Studies in Soil Science,
https://doi.org/10.1007/978-3-031-91114-9_12

advantages, difficulties, and future directions of these cutting-edge approaches, as well as a conclusion.

Keywords Soil health assessment · Soil quality indicators · Microbial activity · Sustainable agriculture

1 Introduction

The growing global focus on soil health has emerged as a critical issue with significant implications for political agendas, environmental quality, and public health (Meena et al., 2024). A key challenge for soil scientists lies in the effective management of soil resources, which requires stakeholders to engage in international and interdisciplinary research collaborations, advance integrated technological solutions, and foster multi-level academic networks (Shen et al., 2020). Achieving the Sustainable Development Goals (SDGs) of the United Nations requires soil health, which is defined as the soil's ability to operate as a living ecosystem that supports human life as well as that of plants and animals (Lehmann et al., 2020a, 2020b). Addressing several Sustainable Development Goals (SDGs) requires soil, especially SDGs 2, 3, 6, 12, 13, and 15 (Bouma & Montanarella, 2016; Keesstra et al., 2016). The authors describe soil as "a non-renewable resource and a balanced social system with a well-structured composition, optimal functionality, and effective buffering capacity, enabling it to sustain biological activity and agricultural productivity while preserving environmental quality." They further define healthy soil as "an ecosystem service characterized by appropriate texture and structure, optimal physicochemical properties, balanced moisture, porosity, air/water ratio, sufficient nutrient availability and turnover, robust microbial biomass and biodiversity, and strong resilience against soil-borne plant diseases due to its natural suppressiveness" (Coyne et al., 2022). However, in agricultural systems, management practices aimed at maximizing production often negatively impact soil health, quality, and fertility. Numerous physicochemical, microbiological, and biochemical parameters contribute to soil fertility (Cheik & Jouquet, 2020; Jiang et al., 2020). With the global population expected to increase by 2.2 billion to 9.8 billion by 2050 (Singh et al., 2021a, 2021b), soil—a vulnerable resource—faces significant degradation from climate change and human activity. Consequently, restoring its health over a short period is a challenging task (De Corato et al., 2024).As the global community strives to adopt sustainable agricultural practices, a thorough evaluation of soil health and fertility has become essential (Peiris et al., 2019). An assortment of governmental, non-governmental, and commercial entities is attempting to set up efficient methods for assessment and monitoring (Karlen et al., 2019). The development of theoretical and practical frameworks for evaluating soil health is an important necessity. Deeper understanding of soil formation processes, soil evolution, soil biodiversity, and the functional analysis of important biological features within soil communities should be the main goals of research. To monitor and assess the best technical strategies for controlling soil health, this can be accomplished by constructing co-occurrence biological networks, creating spatiotemporal models, and putting high-throughput sequencing techniques into practice (De Corato et al., 2024). Assessing soil status is a typical process that aims to understand the effects of past, present, and future land usage on agricultural sustainability. The concepts of soil health and soil quality are utilised to do this (More et al., 2020). The assessment and management of soil health will remain central to agricultural production systems, as healthy soils exhibit greater resilience to fluctuations in growing conditions (Idowu et al., 2019).

Traditionally, soil health has been evaluated based on its outputs, but growing awareness of its significance has highlighted the importance of its quantification and ongoing monitoring. Soil health assessment involves interpreting biological, physical, and chemical properties, encompassing both innate and dynamic soil functions (Meena et al., 2024). In previous decades, assessments were mostly focused on crop productivity; but, in the present, assessments also take into account the roles that soil plays in climate change, water quality, and human health. Even though the value of soil biodiversity is becoming more widely acknowledged, the majority of soil health evaluations are still reliant on chemical markers. This is mostly because there are no reliable assessment techniques and little functional understanding. Instead of considering soil health as just a quantitative attribute, scientists are advised to embrace soil health as a more comprehensive idea that supports sustainability goals (Lehmann et al., 2020a, 2020b). Healthy soils can be assessed using indicators that reflect their productivity for specific purposes and resilience to degradation (Rinot et al., 2019; Norris et al., 2018). Biological and biochemical activity, microbial abundance and diversity, resilience to agricultural practices, water permeability and retention, cation exchange capacity, aggregate structure stability, and plant disease suppressiveness are examples of basic indicators that are typically measured as part of soil health assessment (Wu et al., 2021).Analyzing soil health evaluation with a focus on its biological, physical, and chemical components is the goal of this chapter. It draws attention to the fact that soil has a bigger role in human health, climate change, and water quality than just crop output. The chapter also discusses the need for better evaluation techniques and the difficulties in quantifying soil health. Lastly, it emphasizes how critical soil health is to reaching sustainability objectives.

2 Concept and Definition of Soil Health

The soil health idea has developed over a century, originating in disciplines such as soil science, agriculture, and even medicine (Howard, 2020). References to soil health date back more than 100 years and gained prominence in the organic and "bio-farming" movements in Europe. Swiss soil biologist Dr. Otto Buess performed a key role in the development of this concept, arguing in his 1978 paper "Soil and Plant Health" that soil is a dynamic, living environment. Soil health is not only about fertility and productivity, but also about biological activity, especially the key role of microbial biomass. Dr. Buess emphasized that understanding soil health should include ecological, environmental, and human health aspects. Modern soil health concepts recognize the complex relationship between soil biology, environmental quality, and overall ecosystem health (Niggli, 2009; Pierzynski et al., 2005).

The biological, chemical, and physical characteristics of soil as well as the ecological activities that occur there are all included in the broad concept of soil health. It evaluates the soil's capacity to sustain plant and animal life while offering vital ecosystem services including carbon sequestration, water purification, and agricultural output (Howard, 2020; Pierzynski et al., 2005).Under modern understanding, soil health extends beyond agriculture to include its role in maintaining water quality, mitigating climate change and promoting human health. Therefore, it should be considered as a basic principle of sustainable development goals. Ecosystem services depend largely on soil biodiversity as it assess soil functions critical to human well-being. Soil provides the foundation for strong plant growth and crop productivity by providing nutrients, supporting root systems, and regulating water flow. Nutrient recycling ensures that essential nutrients are recycled for plant uptake. Soil also supports biodiversity and plays a vital role in sequestering carbon. However, these valuable services are increasingly under threat, highlighting the urgent need for sustainable soil management and conservation efforts (Basit et al., 2023; Mrabet, 2023).

Soil health is significant to sustainable agriculture as it supports critical agricultural and environmental functions (Pierzynski et al., 2005). It is essential for crop productivity, environmental quality, and overall human and animal health. The concept of soil health focuses not only on crop yields but also encompasses broader ecological and human health outcomes. Soil quality significantly affects crop productivity and quality. Nutrient-rich, well-structured soil allows plants to grow vigorously, thereby increasing yields and improving crops. Healthy soil with a balanced composition of organic matter, minerals and beneficial microorganisms is essential for optimal plant nutrition, thereby increasing agricultural yields (Basit et al., 2023). Additionally, soil with good physical structure enhances root penetration and water retention, which are critical for plant development, especially in challenging climates such as drought (Howard, 2020).

Soil health is closely related to environmental quality. Healthy soil is remarkable in preventing erosion, a major cause of land degradation and loss of farmland (Bekele & Adimasu, 2023). They maintain soil structure and stability, reducing the risk of erosion from wind and water. Healthy soil also helps reduce nutrient runoff, a major cause of water pollution. Nutrient-rich runoff from farms can harm aquatic life by causing eutrophication of water bodies, causing algal blooms and oxygen depletion. Organic-rich soil has good permeability and can filter pollutants such as pesticides and heavy metals, protecting water quality and supporting healthy ecosystems (Basit et al., 2023; Halder et al., 2024). Soil health is a dynamic property influenced by biological, chemical, and physical processes as well as management practices, climatic factors, and land use changes. Agricultural technology and climate change can enhance or reduce soil health, while urbanization, deforestation, and agricultural expansion can lead to a loss of soil health (Pierzynski et al., 2005). This dynamic nature means soil health can improve or deteriorate depending on how it is managed and the environmental conditions it is exposed to.

The nutritional quality of food is directly related to the health of the soil. Crops grown in healthy soil are more nutritious, providing essential vitamins and minerals critical to human health. For example, soil rich in micronutrients such as zinc and iron can help crops increase their levels of these essential nutrients, thereby addressing nutrient deficiencies in humans (Basit et al., 2023; Faye & Braun, 2022). Additionally, healthy soil helps produce more nutritious feed crops and contributes to livestock health. Consumption of nutrient-dense crops improves health for humans and animals, emphasizing the importance of maintaining soil health for food security and public health (Halder et al., 2024). Soil resilience is the ability of soil to rebound from natural occurrences or human actions without losing its usefulness. For agricultural systems and ecosystems to remain sustainable over the long run, this resilience is essential.

Soil resilience is the ability of soil to rebound from natural occurrences or human actions without losing its usefulness. For agricultural systems and ecosystems to remain sustainable over the long run, this resilience is essential (Blanco & Lal, 2023). This is dependent upon the soil's chemical, physical, and biological characteristics. Rich microbial communities and robust structures, for instance, help soils withstand harsh environments and sustain ecosystem function and plant development. In addition to attaining large crop yields, the idea of "soil health" also emphasizes long-term production and ecological services. Practices like cover crops and less tillage can increase soil health for current and future generations. Setting soil health as a top priority guarantees that soils stay resilient and productive, which supports the sustainability of agricultural systems and the preservation of natural resources. Long-term soil health maintenance is essential to supply the increasing demand for food, fiber, and fuel in an environmentally and commercially sustainable manner (Bekele & Adimasu, 2023; Howard, 2020).

Scientific research consistently shows that healthy soil significantly increases crop yields. Well-structured soil has optimal nutrient utilization, laying the foundation for strong plant growth, thereby increasing productivity. Healthy soil not only increases crop yields, but also improves farm profitability by decreasing the need for chemical inputs such as fertilizers and pesticides. This saves farmers a lot of costs. The combination of higher yields and lower input costs directly improves farm profitability, highlighting the critical importance of soil health in ensuring the economic sustainability of farming operations (Basit et al., 2023). Healthy soils are vital to environmental protection, especially when it comes to controlling erosion and improving water quality. Research shows that well-maintained, well-structured, vegetated soil helps prevent soil erosion and protect water bodies from sediment runoff. In addition, healthy soil can play an important role in mitigating climate change and meeting global carbon reduction goals by acting as a natural filter, removing pollutants from water, and helping to sequester carbon (El-Naggar et al., 2015; Pierzynski et al., 2005).

Soil health directly affects food security by affecting the quantity and quality of food production. Nutrient-rich soil produces high-yielding, nutrient-rich crops that provide essential vitamins and minerals for human health. In addition, healthy soil minimizes the risk of contamination by harmful substances such as heavy metals or pathogens, ensuring food safety. Maintaining healthy soils is therefore critical to producing safe and nutritious food and meeting global food security challenges (Basit et al., 2023; Halder et al., 2024).

3 Integrative Role of Soil Health for Sustainable Agriculture

It is a big challenge to meet the growing demand for sustainable and healthful food production on a worldwide scale. Improving crop output while reducing the effects of climate change and protecting agroecosystems is one of the main goals of sustainable agriculture (Timsina, 2018). However, the reliance on intensive agricultural practices, particularly the extensive utilization of chemical fertilizers and pesticides, has contributed to soil spoiling and environmental disturbances across many agroecosystems, posing risks to human health, wildlife, and aquatic ecosystems (Devarinti, 2016). For instance, a long-term study on monoculture wheat farming demonstrated negative effects such as diminished soil health, groundwater contamination, and a decline in beneficial microorganisms, making crops more susceptible to pathogens and pests (Singh et al., 2011).Sustainable agriculture offers an integrated approach to addressing these issues by focusing on ecological methods of food production (Lal, 2008). This approach incorporates biological, physical, chemical, and ecological principles to develop farming practices that minimize environmental harm (Lichtfouse et al., 2009). Additionally, sustainable agriculture has the potential to meet the growing global demand for food while promoting environmental health (Singh et al., 2011).

The rhizosphere, the thin layer of soil directly surrounding plant roots, is vital for maintaining crop productivity with minimal or reduced agrochemical inputs (Berendsen et al., 2012). Soil health evaluation relies on various soil quality factors that ensure sustainable agricultural output (Doran & Zeiss, 2000; Sahu et al., 2019). Research has demonstrated that soil biota, including microbial diversity, abundance, activity, and stability, are key indicators of soil quality (Doran & Zeiss, 2000; Leskovar et al., 2016). These microorganisms play an essential role in the decomposition of plant residues, converting them into nutrients that plants can readily absorb, thereby supporting their growth and development (Meena et al., 2016).

Soil biota also fastens the decomposition rate by providing different enzymes that affect plant nutrient movements in the soil (Dotaniya et al., 2016). Soil tiny living organisms (mostly, bacteria and fungi) can convert N between organic and inorganic forms, which in turn influences plant minerals uptake, composition, and production (Van et al., 2008). Microbial communities contribute to fundamental processes that provide stability and productivity of agroecosystems (Singh, 2015).Research has demonstrated that soil microorganisms, such as arbuscular mycorrhizal fungi

(AMF), active bacteria, and beneficial nematodes, are strongly linked to crop yields, fruit quality, soil moisture retention, and nutrient cycling, all of which contribute to enhanced plant health and soil fertility (Al-Karaki et al., 2007; Doran & Zeiss, 2000; Leskovar et al., 2016). Studies have also shown that organic farming and conservation tillage, such as strip tillage, can significantly boost soil biota in crops like watermelon (Citrulluslanatus) and globe artichoke (Cynaracardunculus) grown in clay-loam soils (Leskovar & Othman., 2018; Leskovar et al., 2016).

A seven-year study involving vegetables and field crops—including tomato, carrot, rice, and French bean—found that soil microbial biomass carbon was higher in organically managed fields compared to conventional systems (Das et al., 2017). Additionally, a three-year study on watermelons revealed that conservation tillage, particularly strip tillage, increased soil fungi abundance and activity when compared to conventional tillage methods (Leskovar et al., 2016). In contrast, while different tillage practices had no significant effect on overall earthworm populations in conventional systems—where (reduced tillage averaged 153 worms per square meter and mouldboard ploughing averaged 130 worms per square meter) —organic systems exhibited a 45% increase in earthworm abundance, with mouldboard ploughing supporting (430 worms per square meter), compared to (297 worms per square meter) in reduced tillage organic systems (Crittenden et al., 2014).

3.1 Important Soil Functions Related to Crop Production and Environmental Quality

Soil performs several critical functions that contribute to the sustainability of ecosystems and agriculture. One of its primary roles is retaining and cycling nutrients, meaningful for supporting plant development and ensuring the crop productivity (Tahat et al., 2020). Additionally, soil acts as a carbon sink, playing a useful role in sequestering carbon, which helps mitigate the impacts of climate change (Lal, 2004). Soil also facilitates the infiltration, storage, and filtration of water, ensuring water availability for plants and reducing the risk of erosion and flooding (Bronick & Lal, 2005). Furthermore, healthy soils naturally suppress pests, diseases, and weeds, reducing the need for synthetic chemicals and fostering sustainable agricultural practices (Altieri et al., 2012). Another essential function of soil is its ability to detoxify harmful chemicals, thereby protecting the environment and ensuring the safety of food, water, and ecosystems (Giller et al., 1998). Soil also plays a pivotal role in supporting the production of food, feed, fiber, and fuel, making it indispensable for human livelihoods and global food security (Tilman et al., 2002). These functions underscore the importance of maintaining soil health for the sustainability of agricultural and natural ecosystems.

3.2 Soil Health Components for Sustainable Agriculture

Soil quality has been broadly defined as "the capacity of a specific type of soil to function within natural or managed ecosystem boundaries to support plant and animal productivity, maintain or improve water and air quality, and promote human health and habitation" (Karlen et al., 1997). Expanding on this definition, Bouma et al. (2017) described soil quality as "the intrinsic ability of soil to contribute to ecosystem services, including biomass production." This concept of soil quality is increasingly applied in practical contexts to optimize ecosystem services (Tóth, 2008).

Soil quality integrates the biological, chemical, and physical properties of soil, emphasizing their interrelationships and functions within ecosystems (Chaussod, 2002). However, poor agricultural practices can degrade soil quality by contributing to issues such as salinization, acidification, compaction, crusting, nutrient depletion, reduced soil biodiversity, imbalances in water management, and disruptions in elemental cycling (Lal, 2009). Soil biota, which play a central role in nutrient cycling, pathogen suppression, and water detoxification, are highly sensitive to soil management techniques and respond rapidly to changes in agricultural practices (Doran & Zeiss, 2000; Harris, 2009). The relationship between soil biota, soil fertility, and plant health is well-established, with soil organisms being key to enhancing land productivity and soil fertility through biological processes (Altieri & Nicholls, 2003). The vital role of soil biota in promoting agricultural sustainability and improving land productivity is widely recognized (Giller et al., 2005).

3.3 Types of Soil Microorganisms

3.3.1 Bacteria

Bacteria are often described as simple structures, sometimes likened to "bags of enzymes" or "fertilizer-soluble bags" (Dick, 2009). Within soil ecosystems, these microorganisms play a crucial role in nutrient cycling, participating in various processes that enhance soil fertility and promote plant growth (Davison, 1988). They are involved in hormone production, managing plant pathogens, and contributing to the soil's overall composition through bioaccumulation and microbial leaching of inorganic compounds. Notably, the nitrogen content in bacteria ranges from 10 to 30%, with a carbon-to-nitrogen (C) ratio between 3 to 10, which is higher than that of most other microorganisms (Hoorman & Islam, 2010). Ingham (2009) identifies four primary functional groups of soil bacteria: decomposers, mutualists, pathogens, and lithotrophs, with each group playing a distinct role in nutrient recycling within the soil ecosystem. Nitrogen-fixing bacteria are particularly significant as they enhance nutrient availability by converting atmospheric nitrogen into a form that plants can utilize. This nutrient is essential for various biological processes, including the synthesis of cellular enzymes, chlorophyll, proteins, RNA, and DNA

(Pandey et al., 2020). Currently, about 65% of the nitrogen utilized in agriculture comes from biological nitrogen fixation, and this process will continue to be vital for sustainable crop production systems.

The bacteria known as Rhizobia are critical in forming nodules on the roots of leguminous plants (and occasionally on their stems). The specialized "bacteroid" forms within these nodules facilitate the fixation of atmospheric nitrogen, producing ammonia that serves as a nitrogen source for the plants. This mutualistic relationship is a unique ecological niche that allows bacteria to supply a specialized form of nitrogen, significantly benefiting the host plants (Andrew et al., 2007). Other bacteria, such as Azotobacter, Azospirillum, Rhizobium, Meso Rhizobium, and SinoRhizobium, have also been recognized for their contributions to plant growth and productivity (Gonzalez et al., 2005).

3.3.2 Fungi

Fungi are highly productive organisms within the soil ecosystem, demonstrating remarkable plasticity and the ability to adapt their forms in response to challenging conditions (Sun et al., 2005). Their role is crucial in the stabilization and decomposition of soil organic matter (Shah et al., 2021). By breaking down dead organic material, fungi convert it into carbon dioxide, organic acids, and biomass. Additionally, many fungal species act as effective bio sorbents for toxic metals, such as cadmium, copper, mercury, lead, and zinc, sequestering these contaminants in their fruiting bodies. This capability positions fungi as vital agents for the preservation and restoration of soil health, ultimately supporting the growth and vitality of plants (Baldrian, 2003).

3.3.3 Viruses

Soil viruses play a crucial role in soil ecosystems by facilitating gene transfer between hosts and serving as a significant factor in microbial mortality (Bhatti et al., 2017). As key contributors to global biogeochemical cycles, they influence the turnover and availability of nutrients and gases within the soil (Mitchell, 1973). Despite their importance, the field of soil virology remains relatively underexplored. Research efforts are increasingly focusing on the diversity and abundance of viruses across different geographic regions to better understand their impact on plant health and soil quality. Notably, viruses are prevalent in a wide range of environments, even in cases where the bacterial communities within the same ecosystem exhibit considerable variability.

3.3.4 Nematodes

Nematode biodiversity in agroecosystems is closely linked to the ecological health of the soil. These organisms are frequently employed in sustainability research as biological indicators of soil quality (Moura & franzener, 2017). Soil nematode communities, encompassing their structural forms and functions, can offer valuable insights into soil processes, as many nematodes remain active throughout the year and reflect the biotic and functional status of soils (Ritz & Trudgill, 1999). Their role as biological indicators for assessing soil quality is underscored by their contribution to habitat diversity and involvement in nutrient recycling processes (Stone et al., 2016). However, caution is advised when conducting comparative studies at large or transnational scales across different climatic zones, as nematode communities can vary significantly among biogeographical regions and land uses within each zone (Stone et al., 2016).

3.3.5 Recommendation of Soil Health for Sustainable Agriculture

The incorporation of organic fertilizers and various organic materials into soil has been shown to enhance its quality by improving biological activity, increasing organic carbon content, optimizing aggregate stability, promoting a more balanced release of nitrogen fertilizers, reducing nitrate leaching, suppressing pests and pathogens, and ultimately increasing crop yields, particularly when applied consistently over extended periods (Bailey et al., 2003; Crystal-Ornelas et al., 2021; Diacono & Montemurro, 2011).

Recommendations for soil improvement include practices such as rotational grazing, utilizing fallow lands, and establishing cultivated set-aside areas for soil recovery. Additional suggestions involve planting diverse crops, including cold-season and summer crops, as well as humus-enhancing species, catch crops, legumes, and deep-rooted plants. Implementing diverse crop rotations is linked to increased agricultural biodiversity and reduced pest damage, which can lead to lower soil contamination from pesticide use and stronger crops. However, the widespread adoption of regular crop rotation requires significant systemic changes on many German farms (Andert et al., 2016). To achieve these goals, strategies such as using catch crops, under-sowing, mulching with agricultural residues, and optimizing crop rotations—by reducing the time between harvest and subsequent sowing—are recommended. Catch crops may benefit from seed mixtures and timely planting to mitigate risks from pests, diseases, or extreme weather events (like drought). Maintaining soil cover is crucial for minimizing water and wind erosion (Deumlich et al., 2006).

Furthermore, conservation tillage practices, which involve replacing traditional plowing with methods such as mulch planting, strip-tilling, or direct seeding, can significantly reduce soil disturbance. These approaches aim to enhance the soil's resilience to stress. While the benefits may vary based on soil characteristics and management practices, conservation tillage strategies have the potential to increase soil resilience and prevent compaction (Mirzavand et al., 2021; Pöhlitz et al., 2018).

4 Advanced Soil Health Assessment Strategies

A soil health assessment can provide insight into the condition of the soil as well as provide the way forward for increasing its performance and enhancing its utilization for the agriculture sector, particularly while considering sustainable practices. In the twenty-first century, geospatial technology is poised to play an increasingly significant role across various aspects of food security, including research, outreach, policy-making, and program development (Singh et al., 2019). It can provide cost-effective and rapid analysis of the soil conditions in a particular area.

The approach to assessing soil health generally begins with taking an inventory of soil, site, and land management factors that may impact soil health. The next step is to select or develop indicators that reflect the changes in soil properties resulting from these factors, and then select methods for assessing each indicator. A wide range of indicators and methods are available for assessing soil quality and health, ranging from measuring specific soil properties to observing changes in soil processes to monitoring changes in plant or other biological organisms that are closely linked to soil conditions. The indicators and methods used vary depending on the specific soil health issue being addressed and the resources available. Data collected from the various assessments are then used to make inferences about soil health status and its impact on specific soil functions or land management goals (Fierer et al., 2021; Kihara et al., 2020; Liptzin et al., 2022; Löbmann et al., 2022; Yang et al., 2020).

The assessments of soil health are complex and varied but are all based on the collection and interpretation of scientific data. While agronomists and producers are used to dealing with yield data and other crop measurements, soil health data are often more difficult to collect and interpret. This difficulty in assessing soil health often hinders its consideration in day-to-day management and contributes to the lack of a systematic approach to soil health management. Several soil health assessment methods are available to producers, but most were developed for research and are not practical for on-farm use. Others are too general or subjective to provide useful feedback for making land management decisions (Rust et al., 2023; Zahraee et al., 2020).

4.1 *Standardization of Assessment Methods*

Standardization of assessment methods is crucial, as it provides a common point of reference and analysis. Comparability between studies is difficult if different methods are used to measure the same attribute. Standardized methods have been developed for some soil health attributes, for example, the phospholipid fatty acid (PLFA) analysis for soil microbial community analysis. This is a procedure that has undergone rigorous scientific testing to determine its sensitivity and repeatability, and now a standard method for this analysis has been adopted. Another example is the Basal

Respiration test, which has a recommended standard procedure. The adoption of standards to run common tests, for example, in the medical field, enhances the credibility of scientific findings obtained from the tests. Standardization of assessment methods, however, can be a double-edged sword. While it enhances comparability, it can also inhibit scientific advancements and discoveries in soil health assessment. The development of new methods often involves deviation from traditional practices, and if a standardized method exists for an attribute, it can discourage innovation (Caudle et al., 2020; Li et al., 2020; Norris et al., 2020).

Assessing and defining soil health has been problematic since the term was first used. One major constraint is the requirement to assess indirectly. No single measurement exists for soil health, as it is a multi-dimensional concept. A set of methods is required to assess the various attributes. The ability of a method to detect changes in soil health is crucial. If a method is not sensitive enough, it will not be able to determine improvement or degradation in soil health. Similarly, methods can be too sensitive and detect a change where none has occurred. This is known as Type 1 error (detecting a false positive) and Type 2 error (detecting a false negative). For effective soil health monitoring, it is important to minimize the occurrence of either error (Amelung et al., 2020; Brevik et al., 2020; De Corato, 2020; Fierer et al., 2021; He et al., 2021; Lehmannet al., 2020a, 2020b; Yang et al., 2020).

4.2 Sensor Advancements

Sensor technologies have made significant advances in several fields, including biomedical diagnostics, environmental monitoring, and the detection of explosives, and have revolutionized the speed, cost, and availability of information needed for decision-making in these fields. While much progress has been made with some proximal soil sensors aimed at rapid assessment of specific soil properties, a general system for monitoring soil health is not yet available. So-called "lab on a chip" technologies are on the horizon and have the potential to assay specific compounds or biological entities with high sensitivity in the field at a low cost. Rapid assays for enzyme activities or certain microbial substrates may be possible with this approach. Another potentially revolutionary sensor technology is the utilization of synchrotron x-ray absorption spectroscopy to determine the speciation and relative concentrations of specific elements in soils. Both the "lab on a chip" type sensors and the synchrotron-based sensors have the potential to provide large data streams from analyses that would be used as indicators for a soil health monitoring system. Decision makers would preferentially access data from these sensors via a web-based system, georeferenced to their location and providing interpretation of the data in a management context (De Corato, 2020; Norris et al., 2020; Yang et al., 2020).

One significant area of potential innovation in the assessment of soil health is the development and implementation of soil sensors for efficient, cost-effective, in-situ data collection. For a number of critical soil parameters—such as bulk density, pH, organic carbon, soil depth, water-holding capacity, and electrical conductivity—sensing systems are currently on the market. Reflectance spectroscopy, particularly in the visible-near-infrared (VNIR) spectrum, has been successfully used to estimate biological soil health indicators, such as organic carbon, total nitrogen, ß-glucosidase activity, active carbon, microbial biomass carbon, particulate organic matter, and soil respiration (Chang et al., 2001; Cho et al., 2017; Kinoshita et al., 2012; Pietikäinen & Fritze, 1995; Vasques et al., 2009; Veum et al., 2015a, 2015b). There are fewer studies on the spectral estimation of physical and chemical indicators (Bogrekci & Lee, 2005; Chang et al., 2001; Kinoshita et al., 2012; Veum et al., 2015a, 2015b) or broader soil health indices (Cécillon et al., 2009; Veum et al., 2015a, 2015b). While the majority of VNIR research has been carried out in lab environments, mobile and probe-based technologies are now able to acquire in-situ spectral data (Veum et al., 2018). With a variety of calibration methods, profile soil apparent electrical conductivity (ECa), as determined by mobile field sensors, offers important insights on physical and chemical characteristics such texture, mineralogy, cation exchange capacity, and water content. Mapping layer estimates of these properties are made possible by the combination of data from mobile and probe-based ECa sensors (Sudduth et al., 2017).

Soil health monitoring and associated management practices are gaining widespread attention due to their critical importance for sustainable agriculture and overall land management. The societal and scientific concerns about soil health are based on the fundamental concepts that soil influences water and air quality, crop production, and the health of natural ecosystems. The maintenance and enhancement of soil functions to meet the needs of present and future generations is crucial. While many indicators of soil health have been identified, there are very few that are widely accepted and that have known or predictable responses. A comprehensive assessment of soil health is often costly and time-consuming; it is typically beyond the resources of many land managers who need it the most. Therefore, a need exists for the development of rapid, reliable, sensitive, and cost-effective soil health assessment methods and technologies (Lehmann et al., 2020a, 2020b; Teague & Kreuter, 2020; Yang et al., 2020).

4.3 Remote Sensing and GIS Applications

The application of remote sensing in soil science is not new, but it is a relatively unexplored field when it comes to soil health assessment. Current remote sensing technologies offer the ability to sense, measure, and monitor various soil health indicators across large landscapes with minimal disturbance to the soil surface (Wang et al., 2023a, 2023b) demonstrating the potential for using remote sensing to predict soil erosion and its impact on ecosystem services in a tropical watershed. By applying a rare earth element tracer and collecting high-resolution aerial imagery, researchers were able to map the spatial patterns of erosion and compare them to changes in ecosystem services as indicated by vegetation patterns. This innovative use of remote sensing provided a non-invasive and spatially explicit method for monitoring soil erosion and its effect on soil health indicators. The recent development of proximal soil sensor technologies is aimed at rapidly measuring multiple soil properties in the field and providing data for precision agriculture. This is beneficial for soil health and fertility assessment as traditional lab analysis can alter soil state and is costly if multiple samples are taken over large areas. Proximal soil sensors can be used in conjunction with proximal or remote-sensed imagery to provide a better spatial understanding of soil processes and form the basis for future soil health monitoring. An example of this is the work by Mitran et al. (2021), who used proximal soil sensors and multi-spectral imagery to monitor temporal changes in soil properties and crop yield for precision management of a peanut farm in China. Although it was not directly related to soil health assessment, this combined approach has great potential for monitoring specific soil health indicators (Sishodia et al., 2020; Weiss et al., 2020).

4.4 Soil Chemical Parameters

Several soil chemical characteristics can be measured or tracked with satellite remote sensing technologies. These metrics are important markers of soil fertility and health. Even though it can be difficult to test some chemical characteristics directly, remote sensing can offer insightful information or make indirect judgements using a variety of proxies. Table 1 describes the chemical parameters of the soil that can be studied using satellite remote sensing. It has been observed that satellite remote sensing in correlation with machine learning techniques provides better estimates pertaining to the soil chemical parameters.

4.5 Soil Moisture

Soil moisture indicates the amount of water vapour in an unsaturated soil (Ams 2024). NASA provides the Global soil moisture data using microwave remote sensing that has been produced by integrating satellite-derived Soil Moisture Active Passive (SMAP) and Soil Moisture Ocean Salinity (SMOS) missions. The datasets can be utilized for analysis and visualization of the global scenarios pertaining to soil moisture as well as drought assessments (Sazib et al., 2018). The drought event assessment can also be carried out through the Standardized Soil Moisture Index and Soil Moisture Condition Index based on the European Space Agency's Climate Change Initiative (Inocêncio et al., 2020). In addition, thermal remote sensing sensors can infer soil moisture by measuring surface temperature and evapotranspiration rates and can provide Temperature vegetation dryness Index (TVDI) and Surface Energy Balance Algorithm for Land (SEBAL) like (Le & Liou, 2022) uses thẹ same extracted from Landsat-8 OLI/TIRS to propose a new temperature-soil moisture dryness Index

Table 1 Soil chemical parameters studies through remote sensing

Soil chemical parameter	Description	Remote sensing techniques	Relevant sensors	References
Soil organic matter (SOM) / Soil organic carbon (SOC)	Indicates soil fertility, influences nutrient availability and water retention	Hyperspectral imaging, Spectral Reflectance (Visible, NIR, SWIR), Vegetation Indices (NDVI, EVI)	Landsat, Sentinel-2, Hyperion, EnMAP	Thaler et al., 2019; Vaudour et al., 2022
Soil pH	Affects nutrient availability and microbial activity	Reflectance Spectroscopy (Visible and NIR regions)	Hyperspectral Sensors, Sentinel-2	Yüzügullü et al., 2024
Soil salinity	High salinity affects plant growth and soil structure	Visible and NIR Reflectance, Thermal Infrared	Landsat, MODIS, Sentinel-2, ASTER	Delavar et al., 2020; Wang et al., 2024
Soil nitrogen (N), Phosphorus(P) and Potassium(K)	Essential for plant growth, affects photosynthesis	Indirect estimation via Vegetation Indices (NDVI, EVI), Reflectance Spectroscopy	Landsat, Sentinel-2, MODIS	Ma et al., 2022; Wang et al. 2023; Wang et al., 2020
Cation exchange capacity (CEC)	Indicates soil's ability to hold and exchange cations	Spectral Analysis (Visible, NIR, SWIR)	Sentinel-2, Hyperspectral Sensors	Saidi et al., 2022
Heavy metals and pollutants	Contaminants that are toxic to plants and microorganisms	Hyperspectral Imaging, Vegetation Stress Indicators	Hyperion, EnMAP, PRISMA	Ulykpanova et al., 2024

(TMDI) to quantify the surface soil moisture. Whereas (Adab et al., 2020) effectively applied the prevailing Machine learning techniques like Random Forest (RF), Artificial Neural Network (ANN), and others for soil moisture estimations and found that the RF provided the highest Nash–Sutcliffe efficiency value for soil moisture retrieval across various land-use categories. Satellite remote sensing can also help in estimating Evapotranspiration for crop growth models like those used by (Huang et al., 2021). It has been revealed that the satellite remote sensing techniques can provide quick, cost-effective soil moisture results in any area that serves as a base for soil scientists and different fields of decision-making like agriculture.

4.6 Soil Temperature

Soil temperature is an important parameter for understanding soil health, agricultural productivity, and climate dynamics. Thermal infrared (TIR) along with Microwave remote sensing techniques can be used to monitor soil temperature changes over time. The anomalies in soil temperature and land surface temperature can greatly serve as early indicators of drought conditions (Shalishe et al., 2023).

Besides Satellite remote sensing, Geographic Information Systems (GIS) serve as a powerful tool for integrating ancillary information (including soil surveys, field samples and inherit properties) and satellite-derived information. This consists of spatial analysis of the soil properties (Zakaria et al., 2022), mapping of the soil health indicators (Neeraj et al., 2023), change detection analysis (Hasab, 2020), decision support system for soil management (Sintayehu & Munye, 2023) including precision agriculture. It also can create a platform for the data management and accessibility for all the stakeholders.

4.7 Data and Machine Learning

Machine learning has potential in several areas of soil health monitoring. It is not just a means of data, but it has been somewhat overlooked in the research areas of soil health due to the lack of predictive models and limited information obtained from the practical use of data. However, ML has mainly been used in identifying data characteristics and for classification purposes through techniques such as the use of decision trees (Mugiyo et al., 2021).

Big data analysis is already used to predict soil characteristics and generate the data needed to explore new areas in data-driven modelling. Simulation of soil and related environments can explore the potential effects of changes in soil practices or the impact of climate change on soil environments. The ever-growing computer power and the availability of high-detail data will continue to drive growth in these methods. (Tamiminia et al., 2020).

Big data and machine learning (ML) frameworks have a significant impact on the future of soil health assessment. Big data analysis refers to the ability of software to process large volumes of data to identify patterns, relationships, and trends. Creator estimation, the use of technology in agriculture, remote sensing, systems analysis, and causality are discussed to demonstrate the impact of these developments on the management and conservation of soil and water resources (Akhter & Sofi, 2022).

5 Innovation, Challenge, and Future Prospect

Innovative approaches to the assessment of soil health have made the potential for carrying out more environmentally friendly farming methods. Real-time soil gauging has tremendous potential attributed to advanced technologies, including remote sensing, lab-on-a-chip systems, and soil spectroscopy. These technologies also improve the accuracy and productivity of obtaining data and analysis. These technologies offer swift, affordable evaluations of the health of the soil across large regions with little disturbance. Furthermore, by predicting soil conditions, capturing crop performance, and fostering resource utilization, the integration of GIS technology and machine learning facilitates precision agriculture. These developments could significantly enhance soil conservation and management, supporting the world's sustainable food production. Still, there are issues, most notably the lack of established techniques for evaluating soil health, which makes it difficult to compare results from different studies and geographical areas. Due to its intricate nature as a living ecosystem, soil requires multi-dimensional indicators that take into account its chemical, biological, and physical characteristics. However, a lot of the evaluations that are done now mostly concentrate on chemical markers, ignoring the ecological and biological aspects of soil health. The creation of integrated soil health models that take into consideration microbial interactions, biological diversity, and ecosystem services ought to be the top priority for future study. Long-term research is also required to assess how agricultural practices affect soil health and to develop adaptive management plans to lessen the consequences of land degradation and climate change on soil ecosystems

.

6 Conclusion

The significance of soil health in attaining sustainable agriculture is emphasized in this chapter. Productive agroecosystems are built on healthy soil, which supports essential processes including nutrient cycling, water retention, carbon sequestration, and disease and insect suppression. The preservation and repair of soil health must be given top priority in sustainable agriculture methods due to the growing global population and increased demand for food. The involvement of microorganisms provides practical methods for maintaining soil health, in conjunction with conservation approaches and creative management techniques like crop rotation, organic fertilizers, and conservation tillage. However there are also issues with standardizing evaluations of soil health and filling up knowledge gaps, especially with regard to biological indicators. Sustainable agriculture's future depends on interdisciplinary cooperation, ongoing innovation, and a better comprehension of soil ecosystems. We can ensure robust, fruitful, and ecologically viable agricultural systems for future generations by fostering healthy soils.

References

Adab, H., Morbidelli, R., Saltalippi, C., Moradian, M. & Ghalhari, G. A. F. (2020). Machine learning to estimate surface soil moisture from remote sensing data. *Water 12*(11), 27.

Akhter, R., & Sofi, S. A. (2022). Precision agriculture using IoT data analytics and machine learning. *Journal of King Saud University-Computer and Information Sciences, 34*(8), 5602–5618.

Al-Karaki, G., Othman, Y., & Al-Ajmi, A. (2007). Effects of mycorrhizal fungi inoculation on landscape turf establishment under Arabian Gulf region conditions. *Arab Gulf Journal of Scientific Research, 25*, 147–152.

Altieri, M. A., Funes-Monzote, F. R., & Petersen, P. (2012). Agroecologically efficient agricultural systems for smallholder farmers: Contributions to food sovereignty. *Agronomy for Sustainable Development, 32*(1), 1–13.

Altieri, M. A., & Nicholls, C. I. (2003). Soil fertility management and insect pests: Harmonizing soil and plant health in agroecosystems. *Soil and Tillage Research, 72*, 203–211.

Amelung, W., Bossio, D., de Vries, W., Kögel-Knabner, I., Lehmann, J., Amundson, R., Bol, R., Collins, C., Lal, R., & Chabbi, A. (2020). Towards a global-scale soil climate mitigation strategy. *Nature communications, 11*(1), 5427.

Ams, A. M. S. (2024). Soil moisture glossary of meteorology. https://glossary.ametsoc.org/wiki/Soil_moisture

Andert, S., Bürger, J., Stein, S., & Gerowitt, B. (2016). The influence of crop sequence on fungicide and herbicide use intensities in North German arable farming. *European Journal of Agronomy, 77*, 81–89.

Andrew, J. W., Jonathan, D., Andrew, R., Lei, S., Katsaridou, N. N., Mikhail, S., & Rodionov, A. D. (2007). Living without Fur: The subtlety and complexity of iron-responsive gene regulation in the symbiotic bacterium Rhizobium and other a-proteobacteria. *BioMetals, 20*, 501–511.

Bailey, K. L., & Lazarovits, G. (2003). Suppressing soilborne diseases with residue management and organic amendments. *Soil and Tillage Research, 72*(2), 169–180.

Baldrian, P. (2003). Interactions of heavy metals with white-rot fungi. *Enzyme and Microbial Technology, 32*, 78–91. https://doi.org/10.1016/S0141-0229(02)00245-4

Basit, A., Khan, K. A., Kainat, M., Akram, M. M., & Murtaza, M. (2023). Biochar and minerals impact on plant defense mechanism. *Journal of Agriculture and Livestock Farming, 1*(1), 1–7. https://doi.org/10.61577/jalf.2023.100002

Bekele, D., & Adimasu, K. (2023). Review on Soil Protection Implications for Soil Health and Sustainability. *International Journal of Environmental Protection and Policy*. https://doi.org/10.11648/j.ijepp.20231105.13

Berendsen, R. L., Pieterse, C. M., & Bakker, P. A. (2012). The rhizosphere microbiome and plant health. *Trends in Plant Science, 17*, 478–486.

Bhatti, A. A., Haq, S., & Bhat, R. A. (2017). Actinomycetes benefaction role in soil and plant health. *Microbial Pathogenesis, 111*, 458–467. https://doi.org/10.1016/j.micpath.2017.09.036

Blanco, H., & Lal, R. (2023). Soil resilience. *In Soil Conservation and Management* (pp. 525–544). Springer Nature Switzerland. https://doi.org/10.1007/978-3-031-30341-8_21

Bogrekci, I., & Lee, W. S. (2005). Improving phosphorus sensing by eliminating soil particle size effect in spectral measurement. *Transactions of ASAE, 48*(5), 1971–1978.

Bouma, J., & Montanarella, L. (2016). Facing policy challenges with inter-and transdisciplinary soil research focused on the UN sustainable development goals. *The Soil, 2*, 135.

Bouma, J., van Ittersum, M., Stoorvogel, J., Batjes, N., Droogers, P., & Pulleman, M. (2017). Soil capability: Exploring the functional potentials o soil. *In Global Soil Security*, (pp. 27–44). Springer: Cham, Switzerland.

Brevik, E. C., Slaughter, L., Singh, B. R., Steffan, J. J., Collier, D., Barnhart, P., & Pereira, P. (2020). Soil and human health: Current status and future needs. *Air, Soil and Water Research, 13*, 1178622120934441.

Bronick, C. J., & Lal, R. (2005). Soil structure and management: A review. *Geoderma, 124*(1–2), 3–22.

Caudle, C., Osmond, D., Heitman, J., Ricker, M., Miller, G., & Wills, S. (2020). Comparison of soil health metrics for a Cecil soil in the North Carolina Piedmont. *Soil Science Society of America Journal, 84*(3), 978–993.

Cécillon, L., Barthès, B. G., Gomez, C., Ertlen, D., Genot, V., Hedde, M., Stevens, A., & Brun, J. J. (2009). Assessment and monitoring of soil quality using near-infrared reflectance spectroscopy (NIRS). *European Journal of Soil Science, 60*(5), 770–784.

Chang, C. W., Laird, D. A., Mausbach, M. J., & Hurburgh, C. R., Jr. (2001). Near-infrared re flectance spectroscopy—principal component regression analysis of soil properties. *Soil Science Society of America Journal, 65*, 480–490.

Chaussod, R. L. (2002). Qualitébiologique des sols: Des concepts aux applications. *ComptesRendus De L'académied'agriculture De France, 88*, 61–68.

Cheik, S., & Jouquet, P. (2020). Integrating local knowledge into soil science to improve soil 1101 fertility. *Soil Use and Management, 36*, 561–564. 10./sum.12656

Clements, F. E. (1936). Nature and structure of the climax. *Journal of Ecology, 24*, 252–284.

Cho, Y., Sheridan, A. H., Sudduth, K. A., & Veum, K. S. (2017). Comparison of field and la boratory VNIR spectroscopy for profile soil property estimation. *Transactions of the ASABE, 60*(5), 1503–1510.

Coyne, M. S., Pena-Yewtukhiw, E. M., Grove, J. H., Sant'Anna, A. C., & Mata-Padrino, D. (2022). Soil health—it's not all biology. *Soil Security, 6*, 100051.

Crittenden, S., Eswaramurthy, T., de Goede, R., Brussaard, L., & Pulleman, M. (2014). Effect of tillage on earthworms over short- and medium-term in conventional and organic farming. *Applied Soil Ecology, 83*, 140–148.

Crystal-Ornelas, R., Thapa, R., & Tully, K. L. (2021). Soil organic carbon is affected by organic amendments, conservation tillage, and cover cropping in organic farming systems: A meta-analysis. *Agriculture, Ecosystems & Environment, 312*, Article 107356.

Das, A., Patel, D. P., Kumar, M., Ramkrushna, G. I., Mukherjee, A., Layek, J., Ngachan, S. V., & Buragohain, J. (2017). Impact of seven years of organic farming on soil and produce quality and crop yields in eastern Himalayas. *India Agricultural Ecosystem and Environment, 236*, 142–153.

Davison, J. (1988). Plant beneficial bacteria. *Biotechnology, 6*, 282–286.

De Corato, U. (2020). Agricultural waste recycling in horticultural intensive farming systems by on-farm composting and compost-based tea application improves soil quality and plant health: A review under the perspective of a circular economy. *Science of the Total Environment, 738*, Article 139840.

De Corato, U., Viola, E., Keswani, C., & Minkina, T. (2024). Impact of the sustainable agricultural practices for governing soil health from the perspective of a rising agri-based circular bioeconomy. *Applied Soil Ecology, 194*, Article 105199.

Delavar, M. A., Naderi, A., Ghorbani, Y., Mehrpouyan, A., & Bakhshi, A. (2020). Soil salinity mapping by remote sensing south of Urmia Lake. *Iran Geoderma Regional, 22*, 10.

Deumlich, D., Funk, R., Frielinghaus, M., Schmidt, W.-A., & Nitzsche, O. (2006). Basics of effective erosion control in German agriculture. *Journal of Plant Nutrition and Soil Science, 169*(3), 370–381.

Devarinti, S. R. (2016). Natural Farming: Eco-friendly and sustainable? *Agrotechnology, 5*, 147.

Diacono, M., & Montemurro, F. (2011). Long-term effects of organic amendments on soil fertility. In E. Lichtfouse, M. Hamelin, M. Navarrete, & P. Debaeke (Eds.), *Sustainable agriculture*. (Vol 2, pp. 761–786). Springer Netherlands.

Dick, R. (2009). Lecture on soil bacteria in soil Microbiology, personal collection of R. *Dick, the Ohio State University School of Environment and Natural Resources, Columbus, 59*, 15–20. https://doi.org/10.1080/17429145.2011.597002

Doran, J. W., & Zeiss, M. R. (2000). Soil health and sustainability: Managing the biotic component of soil quality. *Applied Soil Ecology, 15*, 3–11.

Dotaniya, M., Meena, V., Basak, B., & Meena, R. (2016). Potassium uptake by crops as well as microorganisms. In V. S. Meena, B. R. Maurya, J. P. Verma, & R. S. Meena (Eds.), *Potassium solubilizing microorganisms for sustainable agriculture* (pp. 267–280). Springer.

El-Naggar, A. H., Usman, A. R. A., Al-Omran, A., Ok, Y. S., Ahmad, M., & Al-Wabel, M. I. (2015). Carbon mineralization and nutrient availability in calcareous sandy soils amended with woody waste biochar. *Chemosphere, 138*, 67–73. https://doi.org/10.1016/j.chemosphere.2015.05.052

Faye, J. B., & Braun, Y. A. (2022). Soil and human health: Understanding agricultural and socio-environmental risk and resilience in the age of climate change. *Health & Place, 77*, Article 102799. https://doi.org/10.1016/j.healthplace.2022.102799

Fierer, N., Wood, S. A., & de Mesquita, C. P. B. (2021). How microbes can, and cannot, be used to assess soil health. *Soil Biology and Biochemistry, 153*, Article 108111.

Giller, K. E., Witter, E., Corbeels, M., & Tittonell, P. (1998). Managing organic resources for soil and water conservation: Lessons from tropical agriculture. *Agriculture, Ecosystems & Environment, 67*(1), 15–29.

Giller, K., Bignell, D., Lavelle, P., Swift, M., Barrios, E., Moreia, F., van Noordwijk, M., Barois, I., Karanja, N., & Huising, J. (2005). Soil biodiversity in rapidly changing tropical landscapes: scaling down and scaling up. In R. Bardgett, M. Usher, & D. Hopkins (Eds.), *Biological diversity and function in soils (Ecological Reviews)* (pp. 295–318). Cambridge University Press.

Gonzalez, L. J., Rodelas, B., Pozo, C., Salmeron, V., Martinez, M. V., & Salmeron, V. (2005). Liberation of amino acids by heterotrophic nitrogen fixing bacteria. *Amino Acids, 28*, 363–367.

Halder, S., Koley, B., Biswas, S., Adak, E., & Sengupta, S. (2024). A brief outlook on soil pollution and its control measures. *International Journal of Advanced Biochemistry Research, 8*(4S), 59–69. https://doi.org/10.33545/26174693.2024.v8.i4Sa.932

Harris, J. (2009). Soil microbial communities and restoration ecology: Facilitators or followers? *Science, 325*, 573–574.

Hasab, H. A. (2020). Monitoring and change detection of concentration values for zinc and lead minerals in soil of agricultural lands based on remote sensing analysis and GIS. *In 10th Institution-of-Geospatial-and-Remote-Sensing-Malaysia (IGRSM) International Conference and Exhibition on Geospatial and Remote Sensing* (IGRSM).

He, M., Xiong, X., Wang, L., Hou, D., Bolan, N. S., Ok, Y. S., Rinklebe, J., & Tsang, D. C. (2021). A critical review on performance indicators for evaluating soil biota and soil health of biochar-amended soils. *Journal of hazardous materials, 414*, 125378.

Hoorman, J. J., & Islam, R. (2010). *Understanding soil microbes and nutrient recycling, Fact sheet SAG-16-10*. The Ohio State University.

Hou, D., Bolan, N. S., Tsang, D. C. W., Kirkham, M. B., & O'Connor, D. (2020). Sustainable soil use and management: An interdisciplinary and systematic approach. *Science of the Total Environment*, 138961. https://doi.org/10.1016/j.scitotenv.2020.138961s

Howard, A. (2020). The soil and health. *A Distant Mirror*.

Huang, D. W., Wang, J. Y., & Khayatnezhad, M. (2021). Estimation of actual evapotranspiration using soil moisture balance and remote sensing. *Iranian Journal of Science and Technology-Transactions of Civil Engineering, 45*(4), 2779–2786.

Idowu, J., Ghimire, R., Flynn, R., & Ganguli, A. (2019). *Soil health: importance, assessment, and management*. Urbana, IL, USA: College of Agricultural, Consumer and Environmental Sciences.

Ingham, E.R. (2009). *Soil Biology Primer*. (vol 4, pp. 22–23). Soil Fungus, Ankeny 117 IA: Soil and Water Conservation Society.

Inocêncio, T. D., Neto, A. R., & Souza, A. (2020). Soil moisture obtained through remote sensing to assess drought events. *RevistaBrasileira De Engenharia Agricola E Ambiental, 24*(9), 575–580.

Jiang, N., Tang, C., Hata, T., Courcelles, B., Dawoud, O., & Singh, D. N. (2020). Bio-mediated soil improvement: The way forward. *Soil Use and Management, 36*, 185–188. https://doi.org/10.1111/sum.12571

Karlen, D. L., Veum, K. S., Sudduth, K. A., Obrycki, J. F., & Nunes, M. R. (2019). Soil health assessment: Past accomplishments, current activities, and future opportunities. *Soil and Tillage Research, 195*, Article 104365.

Karlen, D., Mausbach, M., Doan, J., Cline, R., Harris, R., & Schuman, G. (1997). Soil quality: A concept definition and framework or evaluation. *Soil Science Society of America Journal, 61*, 4–10.

Keesstra, S. D., Bouma, J., Wallinga, J., Tittonell, P., Smith, P., Cerdà, A., et al. (2016). The significance of soils and soil science towards realization of the United Nations Sustainable Development Goals. *The Soil, 2*, 111–128. https://doi.org/10.5194/soil-2-111-2

Kihara, J., Bolo, P., Kinyua, M., Nyawira, S. S., & Sommer, R. (2020). Soil health and ecosystem services: Lessons from sub-Sahara Africa (SSA). *Geoderma, 370*, Article 114342.

Kinoshita, R., Moebius-Clune, B. N., van Es, H. M., Hively, W. D., & Bilgilis, A. V. (2012). Strategies for soil quality assessment using visible and near-infrared reflectance spectroscopy in a Western Kenya chronosequence. *Soil Science Society of America Journal, 76*(5), 1776–1788.

Lal, R. (2004). Soil carbon sequestration impacts on global climate change and food security. *Science, 304*(5677), 1623–1627.

Lal, R. (2008). Soils and sustainable agriculture. *A Review Agronomy for Sustainable Development, 28*, 57–64.

Lal, R. (2009). Soil degradation as a reason for inadequate human nutrition. *Food Security, 1*, 45–57.

Le, M. S., & Liou, Y. A. (2022). Temperature-soil moisture dryness index for remote sensing of surface soil moisture assessment. *IEEE Geoscience and Remote Sensing Letters, 19*, 5.

Lehmann, J., Bossio, D. A., Kögel-Knabner, I., & Rillig, M. C. (2020). The concept and future prospects of soil health. *Nature Reviews Earth & Environment*. https://doi.org/10.1038/s43017-020-0080-8

Lehmann, J., Bossio, D. A., Kögel-Knabner, I., & Rillig, M. C. (2020). The concept and future prospects of soil health. *Nature Reviews Earth & Environment, 1*(10), 544–553.

Leskovar, D., Othman, Y. (2018). Organic and conventional farming differentially influenced soil respiration, physiology, growth, and head quality of artichoke cultivars. *Journal of Soil Science and Plant Nutrition, 18*, 865–880.

Leskovar, D., Othman, Y., & Dong, X. (2016). Strip tillage improves soil biological activity, fruit yield and sugar content of triploid watermelon. *Soil and Tillage Research, 163*, 266–273.

Li, C., Cano, A., Acosta-Martinez, V., Veum, K. S., & Moore-Kucera, J. (2020). A comparison between fatty acid methyl ester profiling methods (PLFA and EL-FAME) as soil health indicators. *Soil Science Society of America Journal, 84*(4), 1153–1169.

Lichtfouse, E., Navarrete, M., Debaeke, P., Souchere, V., Alberola, C., & Menassieu, J. (2009). Agronomy for sustainable agriculture. *A Review Agronomy for Sustainable Development, 29*, 1–6.

Liptzin, D., Norris, C. E., Cappellazzi, S. B., Mac Bean, G., Cope, M., Greub, K. L., Rieke, E. L., Tracy, P. W., Aberle, E., & Honeycutt, C. W. (2022). An evaluation of carbon indicators of soil health in long-term agricultural experiments. *Soil Biology and Biochemistry, 172*, 108708.

Löbmann, M. T., Maring, L., Prokop, G., Brils, J., Bender, J., Bispo, A., & Helming, K. (2022). Systems knowledge for sustainable soil and land management. *Science of the Total Environment, 822*, Article 153389.

Ma, L., Li, A. Q., Yua, H. L., & Chen, G. F. (2022). Hyperspectral remote sensing estimation of soil nutrients in the black soil region based on computer vision model. *ScienceAsia, 48*(3), 287–293.
Meena, R., Bohra, J., Singh, S., Meena, V., Verma, J., Verma, S., & Sihag, S. (2016). Towards the prime response of manure to enhance nutrient use efficiency and soil sustainability a current need: A book Review. *Journal of Cleaner Production,* 1258–1260.
Meena, S. K., Kumar, A., Meena, K. R., Sinha, S. K., Rana, L., Kumar, N., Singh, A. K., Parewa, H. P., & Meena, V. S. (2024). Advanced and Emerging Techniques in Soil Health Management. *Advancements in Microbial Biotechnology for Soil Health,* (pp. 343–362).
Mirzavand, J., & Moradi-Talebbeigi, R. (2021). Relationships between field management, soil compaction, and crop productivity. *Arch Agron Soil Sci., 67*(5), 675–686.
Mitchell, J. E. (1973). The mechanism of biological control of plant diseases. *Soil Biology & Biochemistry, 5,* 721–728.
Mitran, T., Meena, R.S., & Chakraborty, A. (2021). *Geospatial Technologies for Crops and Soils: An Overview.* In: T. Mitran, R. S. Meena, & A. Chakraborty (Eds.), *Geospatial Technologies for Crops and Soils.* Springer, Singapore. https://doi.org/10.1007/978-981-15-6864-0_1
More, S. D. (2010). Soil quality indicators or sustainable crop productivity. *Journal of the Indian Society of Soil Science, 58,* 5–11.
Moura, G. S., & Franzener, G. (2017). Biodiversity of nematodes biological indicators of soil quality in the agroecosystems. *Arquivos Do Instituto Biológico, 84,* 1–8.
Mrabet, R. (2023). Sustainable agriculture for food and nutritional security. *In Sustainable Agriculture and the Environment,* (pp. 25–90). Elsevier. https://doi.org/10.1016/B978-0-323-90500-8.00013-0
Mugiyo, H., Chimonyo, V. G., Sibanda, M., Kunz, R., Masemola, C. R., Modi, A. T., & Mabhaudhi, T. (2021). Evaluation of land suitability methods with reference to neglected and underutilised crop species: A scoping review. *Land, 10*(2), 125.
Neeraj, A., Hiranmai, R. Y., & Iqbal, K. (2023). Comprehensive assessment of pollution indices, sources apportionment and ecological risk mapping of heavy metals in agricultural soils of Raebareli District, Uttar Pradesh, India, employing a GIS approach. *Land Degradation & Development, 34*(1), 173–195.
Niggli, U. (2009). Organic farming in Switzerland. Research Institute of Organic Agriculture.
Norris, C. E., & Congreves, K. A. (2018). Alternative management practices improve soil health indices in intensive vegetable cropping systems: A review. *Frontiers in Environmental Science, 6,* 50.
Norris, C. E., Bean, G. M., Cappellazzi, S. B., Cope, M., Greub, K. L., Liptzin, D., Rieke, E. L., Tracy, P. W., Morgan, C. L., & Honeycutt, C. W. (2020). Introducing the North American project to evaluate soil health measurements. *Agronomy Journal, 112*(4), 3195–3215.
Pandey, M., Shrestha, J., Subedi, S., & Sl, K. (2020). Role of nutrients in wheat: A Review. 1, 18–23. https://doi.org/10.26480/trab.01.2020.18.23
Peiris, C., Gunatilake, S. R., Wewalwela, J. J., & Vithanage, M. (2019). Biochar for sustainable agriculture: Nutrient dynamics, soil enzymes, and crop growth. In: *Biochar from biomass and waste.* (pp. 211– 24). Elsevier. https://www.sciencedirect.com/science/article/pii/B978012811729300011X
Pierzynski, G. M., George F. Vance, & Thomas Sim, J. (2005). Soils and environmental quality. CRC press.
Pietikäinen, J., & Fritze, H. (1995). Clear-cutting and prescribed burning in coniferous forest: Comparison of effects on soil fungal and total microbial biomass, respiration activity and nitrification. *Soil Biology & Biochemistry, 27,* 101–109.
Pöhlitz, J., Rücknagel, J., Koblenz, B., Schlüter, S., Vogel, H.-J., & Christen, O. (2018). Computed tomography and soil physical measurements of compaction behaviour under strip tillage, mulch tillage and no tillage. *Soil and Tillage Research, 175,* 205–216.
Rinot, O., Levy, G. J., Steinberger, Y., Svoray, T., & Eshel, G. (2019). Soil health assessment: A critical review of current methodologies and a proposed new approach. *Science of the Total Environment, 648,* 1484–1491.

Ritz, K., & Trudgill, D. (1999). Utility of nematode community analysis as an integrated measure of the functional state of soils: Perspectives and challenges. *Plant and Soil, 212*, 1–11.

Rust, N. A., Ptak, E. N., Graversgaard, M., Iversen, S., Reed, M. S., de Vries, J. R., Ingram, J., Mills, J., Neumann, R. K., & Dalgaard, T. (2023). *Social capital factors affecting uptake of sustainable soil management practices: a literature review*. Emerald Open Research, 1(10).

Sahu, P., Singh, D., Prabha, R., Meena, K., & Abhilash, P. (2019). Connecting microbial capabilities with the soil and plant health: Options for agricultural sustainability. *Ecological Indicators, 105*, 601–612.

Saidi, S., Ayoubi, S., Shirvani, M., Azizi, K., & Zeraatpisheh, M. (2022). Comparison of Different Machine Learning Methods for Predicting Cation Exchange Capacity Using Environmental and Remote Sensing Data. *Sensors, 22*(18), 17.

Sazib, N., Mladenova, I., & Bolten, J. (2018). Leveraging the Google Earth Engine for Drought Assessment Using Global Soil Moisture Data. *Remote Sensing, 10*(8), 24.

Shah, K. K., Tiwari, I., Modi, B., Pandey, H. P., Subedi, S., & Shrestha, J. (2021). Shisham (Dalbergia sissoo) decline by dieback disease, root pathogens and their management: A review. *Journal of Agriculture and Natural Resources, 4*, 255–272. https://doi.org/10.3126/janr.v4i2.33915

Shalishe, A., Bhowmick, A., & Elias, K. (2023). Agricultural drought analysis and its association among land surface temperature, soil moisture and precipitation in Gamo Zone, Southern Ethiopia: A remote sensing approach. *Natural Hazards, 117*(1), 57–70.

Shen, R. F., Yan, X. Y., Zhang, G. L., & Teng, Y. (2020). Status Quo of and strategic thinking for the development of soil science in China in the new era. *Acta Pedologica Sinica, 57*, 1051–1059. (in Chinese).

Singh, A. K., Tripathi, J. N., Kotlia, B. S., Singh, K., & Kumar, A. (2019). Monitoring groundwater fluctuations over India during Indian summer monsoon (ISM) and Northeast monsoon using GRACE satellite: Impact on agriculture. *Quaterna. Int., 507*, 342–351. https://doi.org/10.1016/j.quaint.2018.10.036

Singh, A., Rajput, V. D., Rawat, S., Sharma, R., Kumar, S. A., Kumar, S. A., & Tomar, R. S., (2021b).*Emerging tools for sustainable agriculture and food security: An introduction. Emerging Tools for Sustainable Agriculture and Food Security* (1st ed., Vol. 1, No. 1, pp. 3–16). Deepika: New Delhi.

Singh, A., Sengar, R. S., Sharma, R., & Rajput, P., (2021a). Nano-priming technology for sustainable agriculture. *Biogeosystem Technique, 8*(2). https://doi.org/10.13187/bgt.2021.2.79

Singh, J.S. (2015). *Plant–Microbe Interactions: A Viable Tool for Agricultural Sustainability Plant Microbes Symbiosis: Applied Facets*, In N. K. Arora (Ed.), Springer: New Delhi, India; Heidelberg, Germany; New York, NY, USA; Dordrecht, The Netherlands; London, UK.

Singh, J. S., Pandey, V. C., & Singh, D. P. (2011). Efficient soil microorganisms: A new dimension for sustainable agriculture and environmental development. *Agriculture, Ecosystems & Environment, 140*, 339–353.

Sintayehu, G., & Munye, K. (2023). GIS-Based MCDA Model to Map Soil Erosion Hotspot Area in Case Study of Temi Watershed in HuletEjjuEnesie Woreda, Amhara Region. *Ethiopia. Water Conservation Science and Engineering, 8*(1), 20.

Sishodia, R. P., Ray, R. L., & Singh, S. K. (2020). Applications of remote sensing in precision agriculture: A review. *Remote Sensing, 12*(19), 3136.

Stone, D., Costa, D., Daniell, T., Mitchell, S., Topp, C., & Griffiths, B. (2016). Using nematode communities to test a European scale soil biological monitoring programme for policy development. *Applied Soil Ecology, 97*, 78–85.

Sudduth, K. A., Kitchen, N. R., Drummond, S. T. (2017). Inversion of soil electrical conductivity data to estimate layered soil properties. In *Proceeding 11th European Conference on Precision Agriculture, Advances in Animal Bioscience* 8(2), 433–438.

Sun, J. M., Irzykowski, W., Jedryczka, M., & Han, F. X. (2005). Analysis of the genetic structure of Sclerotinia sclerotiorum (Lib.) de Bary populations from different regions and host plants by random amplified polymorphic DNA markers. *Journal of Integrative Plant Biology, 47*, 385–395.

Tahat, M. M., Alananbeh, K. M., Othman, Y. A., & Leskovar, D. I. (2020). Soil health and sustainable agriculture. *Sustainability, 12*(12), 4859.
Tamiminia, H., Salehi, B., Mahdianpari, M., Quackenbush, L., Adeli, S., & Brisco, B. (2020). Google Earth Engine for geo-big data applications: A meta-analysis and systematic review. *ISPRS Journal of Photogrammetry and Remote Sensing, 164*, 152–170.
Teague, R., & Kreuter, U. (2020). Managing grazing to restore soil health, ecosystem function, and ecosystem services. *Frontiers in Sustainable Food Systems, 4*, Article 534187.
Thaler, E. A., Larsen, I. J., & Yu, Q. (2019). A new index for remote sensing of soil organic carbon based solely on visible wavelengths. *Soil Science Society of America Journal, 83*(5), 1443–1450.
Tilman, D., Cassman, K. G., Matson, P. A., Naylor, R., & Polasky, S. (2002). Agricultural sustainability and intensive production practices. *Nature, 418*(6898), 671–677.
Timsina, J. (2018). Can organic sources of nutrients increase crop yields to meet global food demand? *Agronomy, 8*, 214.
Tóth, G. (2008). Agri-Environmental Soil Quality Indicator in the European Perspective; European Commission Joint Research Centre, (p. 12). OECD: Ispra, Italy.
Ulykpanova, M. M., Auezova, Z., Ramazanova, N., Mussabaeva, M., & Zhanguzhina, A. (2024). Application of Remote-Sensing Data in Geochemical Studies of Soils of the Yertis River Basin Within East Kazakhstan. *Quaestiones Geographicae, 43*(1), 157–164.
Van der Heijden, M., Bardgett, R., & van Straalen, N. (2008). The unseen majority: Soil microbes as drivers of plant diversity and productivity in terrestrial Ecosystems. *Ecology Letters, 11*, 296–310.
Vasques, G. M., Grunwald, S., & Sickman, J. O. (2009). Modeling of soil organic carbon fractions using visible-near-infrared spectroscopy. *Soil Science Society of America Journal, 73*, 176–184.
Vaudour, E., Gholizadeh, A., Castaldi, F., Saberioon, M., Borůvka, L., Urbina-Salazar, D., Fouad, Y., Arrouays, D., Richer-De-Forges, A. C., Biney, J., Wetterlind, J., & Van Wesemael, B. (2022). Satellite Imagery to Map Topsoil Organic Carbon Content over Cultivated Areas: An Overview. *Remote Sensing, 14*(12), 2917.
Veum, K. S., Kremer, R. J., Sudduth, K. A., Kitchen, N. R., Lerch, R. N., Baffaut, C., Stott, D. E., Karlen, D. L., & Sadler, E. J. (2015a). Conservation effects on soil quality indicators in the Missouri Salt River Basin. *Journal of Soil and Water Conservation, 70*(4), 232–246.
Veum, K. S., Parker, P. A., Sudduth, K. A., & Holan, S. H. (2018). Predicting profile soil prop erties with reflectance spectra via Bayesian covariate-assisted external parameter orthogonalization. *Sensors, 18*, 3869.
Veum, K. S., Sudduth, K. E., Kremer, R. J., & Kitchen, N. R. (2015b). Estimating a soil quality index with VNIR reflectance spectroscopy. *Soil Science Society of America Journal, 79*, 637–649.
Wang, C., Li, K. M., & Wang, Y. (2023a). Estimating the spatial distribution of soil organic carbon and soil total nitrogen concentrations by remote sensing methods: A case study in a forest-dominated Alpine Watershed, Western China. *Journal of Forest Research, 28*(5), 324–332.
Wang, J., Wu, Y., Hu, Z., & Zhang, J. (2023b). Remote sensing of watershed: towards a new research Paradigm. *Remote Sensing, 15*(10), 2569.
Wang, N., Chen, S. C., Huang, J. Y., Frappart, F., Taghizadeh, R., Zhang, X. L., Wigneron, J. P., Xue, J., Xiao, Y., Peng, J., & Shi, Z. (2024). Global soil salinity estimation at 10 m using multi-source remote sensing. *Journal of Remote Sensing, 4*, 20.
Wang, S., Zhuang, Q. L., Jin, X. X., Yang, Z. J., & Liu, H. B. (2020). Predicting soil organic carbon and soil nitrogen stocks in topsoil of forest ecosystems in Northeastern China using remote sensing data. *Remote Sensing, 12*(7), 21.
Weiss, M., Jacob, F., & Duveiller, G. (2020). Remote sensing for agricultural applications: A meta-review. *Remote Sensing of Environment, 236*, Article 111402.
Wu, K. N., Yang, Q. J., & Zhao, R. (2021). A discussion on soil health assessment of arable land in China. *Acta Pedologica Sinica, 58*, 537–544. (in Chinese).
Yang, T., Siddique, K. H., & Liu, K. (2020). Cropping systems in agriculture and their impact on soil health-A review. *Global Ecology and Conservation, 23*, Article e01118.

Yüzügullü, O., Fajraoui, N., & Liebisch, F. (2024). Soil Texture and pH Mapping Using Remote Sensing and Support Sampling. *Ieee Journal of Selected Topics in Applied Earth Observations and Remote Sensing, 17*, 12685–12705.

Zahraee, S. M., Shiwakoti, N., & Stasinopoulos, P. (2020). Biomass supply chain environmental and socio-economic analysis: 40-Years comprehensive review of methods, decision issues, sustainability challenges, and the way forward. *Biomass and Bioenergy, 142*, Article 105777.

Zakaria, Y. S., Shaibu, A. G., & Baatuuwie, B. N. (2022). Spatial analysis of soil physicochemical and hydraulic properties in the Libga irrigation scheme in northern Ghana using geostatistics and GIS approach. *Soil Science Annual, 73*(2), 8.

Diacono, M., Montemurro, F. (2011) Long-term effects of organic amendments on soil fertility. In E, Lichtfouse., Hamelin M, Navarrete M, Debaeke P (eds), *Sustainable agriculture.* (vol. 2, pp. 761–786). Springer Netherlands.

Effects of Environmental Interactions with Soil and Water Management to Promote Agricultural Sustainability

Faiz Ul Hassan, Abdul Basit, Mir Sujaul Islam, Wudu Abiye, Mohd Ekhwan Hj. Toriman, Raheel Ahmad, Muhammad Amjad Bashir, Abdur Rehim, and Hafiz Muhammad Ali Raza

Abstract Poor soil and water management poses significant obstacles to agricultural productivity, ecosystem stability, and human well-being. In this section, the study delves into sustainable approaches aimed at addressing environmental degradation and enhancing agricultural resources efficiency. It emphasizes important soil management practices like crop rotations, no-till farming and cover cropping because they enhance soil health, promote carbon sequestration, and decrease greenhouse gas

F. U. Hassan · M. S. Islam (✉)
Institute for Environment and Development (LESTARI), Universiti Kebangsaan Malaysia (UKM), Bangi 43600, Selangor, Malaysia
e-mail: sujaul@ukm.edu.my

A. Basit
Department of Agronomy, Engro Fertilizers Limited, Rahim Yar Khan, Pakistan

W. Abiye
University of Agriculture in Krakow (UAK), Krakow, Poland

Amhara Agricultural Research Institute (ARARI), Bahirdar, Ethiopia

M. E. Hj. Toriman
Centre for International Strategic Negotiation and Sustainable Development Goals, Universiti Kebangsaan Malaysia (UKM), Bangi 43600, Selangor, Malaysia

R. Ahmad
Earth Observation Centre, Institute of Climate Change (IPI), Universiti Kebangsaan Malaysia (UKM), Bangi 43600, Selangor, Malaysia

M. A. Bashir
Sichuan Provincial Key Laboratory of Philosophy and Social Sciences for Monitoring and Evaluation of Rural Land Utilization, Chengdu Normal University, Chengdu 611130, China

A. Rehim
Department of Soil Science, Faculty of Agriculture Sciences and Technology, Bahauddin Zakariya University Multan, Multan 60800, Pakistan

H. M. A. Raza
Department of Soil Science, Faculty of Agriculture Sciences and Technology, University of Layyah, Layyah 31200, Pakistan

M. Shaaban (ed.), *Soils and Sustainable Agriculture*, Frontier Studies in Soil Science,
https://doi.org/10.1007/978-3-031-91114-9_13

emissions. The effectiveness of water management methods, including drip irrigation and rainwater harvesting studied to understand their role in protecting water resources and minimizing the risk of salinization. In addition, the study examines the contribution of agroforestry and vegetated buffers to erosion control and water quality improvement. This chapter highlights the potential of advanced technologies such as precision agriculture, integrated pest management, and biochar applications to optimize nutrient use, improve soil fertility, and support ecosystem sustainability. However, it acknowledged limitations such as a lack of region-specific research and the exclusion of emerging innovations such as artificial intelligence-driven water management and gene-edited crops. Future research directions aim to integrate localized approaches, advanced technologies, and long-term monitoring to ensure sustainable agricultural practices. In conclusion, this chapter highlights the importance of interdisciplinary collaboration in developing integrated solutions to address the complex interactions between soil, water, and environmental health for permaculture.

Keywords Environmental interactions · Soil management · Water management · Climate change · Agricultural sustainability

1 Introduction

Soil and water management in agriculture involves practices and strategies for optimizing soil and water resources to increase crop yields, maintain soil health, and sustainable agriculture (Nawaz et al., 2021). Effective management of these resources is critical for long-term agricultural productivity, food security, and environmental protection. Soil management involves implementations that improve soil structure, fertility, and health, including tillage and no-till farming, cover cropping, and nutrient management. Water management focuses on the efficient use of water through irrigation systems, water harvesting, and soil retention. The goals are to maximize crop yields, prevent soil degradation and erosion, optimize water use, and increase resilience to climate change and drought. Benefits include increased crop productivity and quality, long-term sustainability of natural resources, reduced risk of nutrient loss and soil erosion, and enhanced soil health and biodiversity (Tahat et al., 2020).

The environment encompasses all natural and artificial elements, including air, water, soil, plants, animals, and humans, that interact to shape the world's ecosystems and form the basis of life on earth. Soil and water management are critical to environmental sustainability, directly impacting ecosystem health, agricultural productivity, and human well-being (Srivastav et al., 2021). Soil and water management play a vital role in environmental health through a variation of interrelated mechanisms. Soil health and environmental sustainability are linked through biodiversity support, carbon sequestration, and pollution control measures. These practices help maintain ecosystem balance and relieve the effects of climate change. Effective water management is necessary to conserve water, prevent erosion, and maintain water quality.

Sustainable water management also contributes to climate regulation and adaptation. Overall, sustainable agriculture and ecosystem services rely on the support of soil and water management for food security and important environmental processes such as nutrient cycling and water purification (Delgado et al., 2021).

The history of soil and water management is long, spanning thousands of years, and reflects the evolution of agricultural practices. Ancient civilizations like Egypt, Mesopotamia, and the Indus Valley developed irrigation and drainage systems to manage water for agriculture. These ancient innovations provided the foundation for modern land and water management techniques. In the mid-nineteenth century, Sir John Bennett Routh and Joseph Henry Gilbert began famous soil management research at Lausanne Research Station in England. His long-term experiments on soil fertility and crop production greatly advanced the field of soil science (Banerjee et al., 2023).

By the early twentieth century, land and water management had come a long way. In the 1930s United States, the Dust Bowl highlighted the need to protect soil, prompting research into ways to prevent soil erosion. During this period, practices such as contour tillage, strip tillage and windbreaks emerged to stabilize soils and reduce degradation. In recent decades, the focus has shifted toward sustainable practices that not only protect but improve soil health and water efficiency. Innovations such as no-till farming, cover crops, and precision agriculture have been widely adopted, improving soil structure, increasing organic matter content, and optimizing water use (Uekötter, 2020).

Today, advances in technology and ecological knowledge continue to influence land and water management. Current trends include integrating soil health management with climate change mitigation strategies such as reducing greenhouse gas emissions and carbon sequestration. In addition, there is growing interest in "green water" management, which uses rainwater stored in the soil to increase the resilience of rain-fed agricultural systems (Nawaz et al., 2021). Ongoing research and collaborative programs are also exploring ways to improve water quality and biodiversity through improved land management practices.

In the early twentieth century, Hugh Hammond Bennett observed widespread soil erosion and degradation in the United States, particularly the Dust Bowl of the 1930s. His work was critical in highlighting the importance of soil conservation in preventing environmental degradation (Uekötter, 2020). Bennett's efforts resulted in the creation of the Soil Conservation Service, now known as the Natural Resources Conservation Service in 1935, a landmarking the history of soil management. Bennett's research reveals important connections between soil and water management and environmental health. He stressed that poor soil practices can induce soil erosion, leading to sedimentation in waterways and harming water quality and aquatic ecosystems. He also emphasized the role of effective soil and water management in flood prevention, as it can reduce runoff and help mitigate flooding. In addition, Bennett highlighted the link between soil health and agricultural productivity, emphasizing the importance of sustainable soil management for long-term food security and environmental protection (Basit et al., 2023).

Land and water management practices have advanced significantly since Bennett's time. Modern methods now combine advanced technology and ecological principles to promote sustainability. Innovative technologies such as no-till farming, cover crops, and precision agriculture are widely used to improve soil health, minimize erosion, and optimize water use (Delgado et al., 2021). These practices reflect a more diverse apprehension of the interconnections between land, water and the wider environment, building on Bennett's foundational work. The chapter discusses historical developments and modern innovations in agricultural soil and water management. Emphasize the role of these practices in increasing agricultural productivity, maintaining environmental sustainability and ensuring long-term food security. It emphasizes the combination of advanced technologies and ecological principles to optimize resource use, improve soil health, mitigate climate change, and improve water quality. Important links between soil and water management and wider environmental health, biodiversity, and climate resilience are highlighted.

2 Importance of Soil and Water Management

Soil and water management are necessary for sustainable environmental and agriculture protection. Proper stewardship ensures that these resources are available to future generations and maintains their health and productivity (Nawaz et al., 2021). Healthy soil provides essential nutrients and stability to plant roots, while effective water management ensures crops receive adequate moisture. These practices increase crop resilience, decrease the requirement for chemical inputs, and enhance farm efficiency. From an environmental perspective, soil and water management play a crucial role in protecting ecosystems and minimizing the impact of agricultural practices on natural resources (Tahat et al., 2020). Proper soil management prevents erosion and degradation, while effective water management protects water resources. Sustainable practices are critical to long-term agricultural and environmental goals, ensuring resource availability and supporting the planet's health while benefiting current and future generations (Basit et al., 2023).

The history of agricultural water and soil management is closely related to the evolution of human civilization. The roots of these practices can be traced to the Agricultural Revolution about 10,000–12,000 years ago, when humans development from hunter-gatherer societies to resolved agricultural communities (Gowdy, 2020). This change marked the beginning of systematic soil and water management, laying the foundation for future agricultural advances. In ancient civilizations such as Mesopotamia, Egypt, and China, the first irrigation systems were developed to channel water from rivers to fields (Angelakıs et al., 2020). These innovations enabled crops to be grown in areas previously unsuitable for agriculture, demonstrating an early recognition of the importance of water management for crop production. During the Middle Ages and Renaissance, agricultural practices developed further with improved crop rotation, farming techniques, and improved soil management practices. These advances are critical for maintaining soil fertility and increasing crop yields, reflecting the increasing sophistication of agricultural methods (Mueller et al., 2021).

The nineteenth and twentieth centuries saw important advances in soil science and irrigation technology. The pioneering work of scientists such as Justus von Liebig laid the foundation for modern soil chemistry (Mueller et al., 2021). In the United States, expansion into the American West induced the progress of practical irrigation fundamentals that are still relevant today (Angelakıs et al., 2020). In recent decades, there has been an enhancing focus on sustainable land and water management practices. Strategies like conservation tillage, cover crops, and integrated water management have emerged to maintain soil health and ensure long-term agricultural productivity. This change reflects a greater recognition of the need for sustainable practices and the need for resilient food production systems despite climate despite climate change (Delgado et al., 2021).

Soil management includes a variety of techniques and practices that protect and improve soil to increase agricultural productivity. Goals such as improving soil fertility and preventing erosion are achieved through conservation tillage, crop rotation, cover planting, and organic amendments (Basit et al., 2023; Cardoso et al., 2013). These efforts help maintain soil structure, reduce erosion, and enhance soil health and nutrient content. Enhancing soil health and fertility enhances nutrient availability and soil structure, leading to higher crop yields and productivity. Sustainable soil management practices also help reduce soil erosion and degradation, preserving the soil's long-term agricultural viability (Basit et al., 2023; Tahat et al., 2020).

However, some challenges need to be addressed through soil management. Soil erosion and topsoil loss can be mitigated through techniques such as contour tillage, terracing and maintaining ground cover. Salinization is the aggregation of salts in the soil due to improper irrigation and can be controlled by adopting proper irrigation techniques and using soil amendments. Acidification, a decrease in soil pH caused by overuse of chemical fertilizers, can be counteracted by applying lime and balanced fertilization (Gondal et al., 2021; Singh et al., 2024).

Water management refers to the strategic plotting, progress, allocation and utilization of water resources in a way that ensures their sustainability and minimizes environmental impact. The main objectives of water management include efficient use of water, conservation of this important resource, and control of pollution to protect ecosystems and ensure access to water for future generations (Srivastav et al., 2021). Water management technologies and practices are diverse and designed to meet specific agricultural and environmental needs. Irrigation systems like drip, sprinkler, and furrow irrigation are designed to optimize water supply to crops, minimizing waste and assuring plants intake the water they need to grow (Lakhiar et al., 2024). Drip irrigation offers water directly to plant roots, decreasing evaporation and runoff. Sprinkler systems simulate natural rainfall, while furrow irrigation involves directing water along the soil, which is typically used in conventional farming (Yang et al., 2023). Rainwater harvesting is another important practice that captures and stores rainwater for later use, decreasing dependence on other water sources. Water conservation measures include mulching, reducing soil evaporation, and implementing soil moisture management strategies to retain moisture in the soil (Abdulla et al., 2021a).

The benefits of effective water management are enormous. By promoting efficient water use and reducing waste, water management practices help increase crop yields and agricultural sustainability. These practices ensure that crops receive adequate water without overusing resources, resulting in a more resilient agricultural system. In addition, water management helps conserve water resources by reducing depletion of natural water bodies and aquifers (Delgado et al., 2021; Yang et al., 2023). Pollution control is another important benefit, as proper water management practices can protect the environment and public health by preventing contaminants from entering water sources (Singh et al., 2022). Despite these benefits, water resources management still faces several challenges. Water scarcity is a growing concern in many regions, with climate change and population growth exacerbating the problem. Overexploitation of water resources, especially groundwater, can lead to water depletion and long-term sustainability issues. Water pollution from agricultural runoff, industrial waste, and untreated sewage further complicates water management efforts (Van der Gun, 2021). Addressing these challenges requires integrated water management, including coordinating water use across different sectors, implementing policy measures to regulate water abstraction and pollution, and raising public awareness of the importance of water conservation (Wang et al., 2021).

The relationship between soil and water management is closely linked, with each practice having a significant impact on the other. Effective soil management directly affects the maintenance and distribution of water, and effective water management supports soil health and fertility. These practices are critical to increasing agricultural productivity, reducing environmental degradation and promoting sustainable development (Hailu & Mehari, 2021). Soil and water management practices are closely related to the ability to increase agricultural productivity. Healthy soil with good structure and nutrient content improves water penetration and retention, reducing the need for over-irrigation and ensuring crops receive consistent moisture. On the other hand, well-managed water resources help prevent soil degradation by reducing erosion, salinization and waterlogging. Integrating these practices enables farmers to achieve higher crop yields and increase the resilience of agricultural systems (Delgado et al., 2021).

The interrelationship between soil and water management has important environmental implications. Effective management practices can reduce runoff, a major cause of soil erosion and water pollution. Proper soil management can retain more water, reduce excess water flow that can carry contaminants to nearby water bodies, and reduce the risk of contamination. Cover planting and mulching help reduce soil erosion, improve water retention, and reduce the need for over-irrigation. Integrated land and water management practices help protect natural ecosystems and biodiversity (Elad et al., 2010; Lakhiar et al., 2024). In agriculture, sustainability requires practices that maintain soil and water health over the long term. Integrated soil and water management practices are critical to developing resilient agricultural systems that can control climate change, resource depletion, and environmental degradation.

Conservation practices participate to the long-term agricultural sustainability operations and help protect natural resources. Therefore, soil and water management plays an important role in balancing agricultural productivity with protecting the environment for future generations (McArthur & McCord, 2017; Pierzynski et al., 2005).

3 Major Environmental Challenges and Their Consequences

The human-made environmental challenges of industrialization, deforestation, and over-exploiting of natural resources are expected to pose serious problems in this century. It is a fact that such activities not only trigger the degradation of ecosystems but also have cascading effects on the health and well-being of humans. Recent studies have been able to show the need for understanding these challenges and implications for the elaboration of mitigation measures in an attempt toward sustainable development. This chapter will look into the following four critical environmental issues: climate change, biodiversity loss, pollution, and resource depletion by investigating the root causes and extended consequences (Flower, 2006).

Climate change, driven primarily by anthropogenic greenhouse gas emissions, is one of the most serious environmental challenges of our time. Since the beginning of the Industrial Revolution, the amount of carbon dioxide (CO_2) in the atmosphere has increased significantly due to the burning of fossil fuels, deforestation, and various industrial activities. This leads to an increase in global temperatures, triggering a range of environmental and social impacts (Omranian et al., 2023; Stocker et al., 2014). These impacts caused significant economic losses and deaths. Extreme weather events such as hurricanes, floods, and droughts are increasing in frequency and severity, causing billions of dollars in damage to infrastructure, agriculture, and residential communities. For example, global economic losses from climate-related disasters reached an estimated $210 billion in 2020 alone (Peace & Huber, 2023). Additionally, climate change has exacerbated food and water insecurity, contributing to hunger, displacement, and conflict. It is estimated that by 2050, climate change could push up to 132 million people into poverty (Moyer et al., 2023). Furthermore, extreme heat and deteriorating air quality due to higher CO_2 concentrations have been linked to rising mortality rates, especially among vulnerable populations (Bayram et al., 2023).

Rising temperatures are changing weather patterns, making heat waves, droughts and heavy rains more frequent and severe. These changes negatively impact agricultural productivity as crops face conditions to which they are not adapted (Lesk et al., 2016). The Intergovernmental Panel on Climate Change (IPCC) states that global temperatures have increased by approximately 1.1 °C since pre-industrial times and may rise by 1.5 °C by 2030 if current trends continue (Adak et al., 2023). This warming is accelerating the melting of polar ice caps, causing sea levels to rise, posing significant risks to coastal communities around the world. Projections indicate that sea levels could rise by one meter by 2100, which could displace millions of people and cause widespread damage to infrastructure (Hinkel et al., 2014).

According to the physical effects, climate change is disrupting ecosystems and biodiversity. Changes in climatic conditions (temperature and precipitation) patterns are forcing species to migrate, changing ecosystems, and increasing the risk of extinction for vulnerable species. These ecological disruptions affect not only wildlife but also human livelihoods, particularly in agriculture and fisheries, where changing climate conditions are reducing yields and threatening food security (Pecl et al., 2017).

The loss of biodiversity is another critical environmental challenge, with human activities like habitat destruction, overexploitation, and pollution leading to the extinction of species at rates far higher than natural background rates. The current biodiversity crisis is often referred to as the sixth mass extinction, with up to one million species at risk of extinction within the next few decades, according to the United Nations (UN) Global Biodiversity (Hoffmann, 2021). Biodiversity underpins ecosystem services that are vital for human survival, including food production, water purification, pollination, and climate regulation. As species disappear, ecosystems become less resilient, and their ability to provide these essential services is compromised (Raven, 2020).

One of the most severe consequences of biodiversity loss is the disruption of food security. Agricultural systems depend on a wide range of species for pollination, pest control, and soil fertility. As biodiversity declines, crops become more vulnerable to pests and diseases, and the genetic diversity needed to adapt to changing environmental conditions is lost (Muluneh, 2021). In addition, the loss of forests and wetlands, which are critical habitats for biodiversity, exacerbates climate change by reducing the Earth's capacity to sequester carbon, further intensifying global warming (Erwin, 2009).

Pollution, in various forms, is another major driver of environmental degradation. Air pollution, primarily from the burning of fossil fuels, promotes both climate change and adverse health outcomes. Fine particulate matter (PM2.5) and nitrogen oxides (NOx) are accountable for millions of early fatalities globally each year, particularly in urban areas with high levels of industrial and vehicular emissions (Keswani, 2022; Lelieveld et al., 2023). In addition to human health impacts, air pollution contributes to the degradation of ecosystems through acid rain, which damages forests, lakes, and soils (Dutta & Singh, 2021).

Water pollution is also a growing concern, with industrial discharges, agricultural runoff, and plastic waste contaminating rivers, lakes, and oceans. Agricultural runoff, particularly fertilizers and pesticides, leads to the eutrophication of water bodies, this causes algae blooms, which reduce oxygen levels in the water and create dead zones where aquatic life cannot survive (Weis, 2024). Plastics, which can take hundreds of years to decompose, are accumulating in marine environments, threatening marine life and entering the food chain (Moore, 2008).

Soil pollution, driven by the excessive use of chemicals in agriculture and improper disposal of industrial waste, is another significant issue. Contaminated soils reduce agricultural productivity and contribute to the loss of arable land. This affects food security and disrupts ecosystems that depend on healthy soils for animal and plant life. Combined effects of water, air, and soil pollution are exacerbating the deterioration of ecosystems worldwide (Bisht & Chauhan, 2020; Chislock et al., 2013; Dodds et al., 2009).

The overexploitation of natural resources is yet another critical environmental challenge. Water, land, and energy resources are being depleted at unsustainable rates, threatening the long-term viability of ecosystems and human societies. Freshwater resources are under severe stress due to overuse for agriculture, industry, and domestic consumption. The United Nations predicts that by 2050, more than half of the world's population will live in water-scarce areas, with significant implications for food production, sanitation, and health (du Plessis, 2023; Pradinaud et al., 2019).

The depletion of arable land is another significant concern. Urbanization, deforestation, and unsustainable agricultural practices are reducing the availability of fertile land, which is essential for growing food. The loss of topsoil due to erosion further compounds this issue, leading to lower crop yields and increased pressure on remaining agricultural land (Amundson et al., 2015; Montgomery, 2007; Pimentel & Burgess, 2013). Moreover, the overuse of fossil fuels for energy is contributing to resource exhaustion and environmental degradation. As fossil fuel reserves dwindle, there is an urgent need to shift to renewable energy sources to significantly reduce greenhouse gas emissions and mitigate the adverse effects of climate change (Jacobson et al., 2015; Newell et al., 2021).

4 Mitigation Strategies to Reduce Environmental Consequences

The environmental impacts of unsustainable soil and water management practices pose major challenges to agricultural productivity, ecosystem stability, and human well-being. To address these issues, various mitigation strategies have been developed focusing on sustainable practices that increase resource efficiency, reduce degradation, and improve ecosystem services (Kovács et al., 2023). These strategies integrate scientific methods, ecological principles and technological innovation to minimize the impact of agriculture on the environment. Sustainable soil management practices, including techniques such as no-till farming, cover cropping and crop rotation, play a vital role in mitigating environmental degradation. No-till agriculture reduces soil disturbance, minimizes carbon emissions, and enhances soil carbon sequestration, which is critical for mitigating climate change (Kuhn et al., 2016).

Effective water management practices are critical to reducing the environmental impact of agricultural water use. Technologies such as drip irrigation, sprinkler systems, and water harvesting improve irrigation efficiency by minimizing water waste and ensuring even water distribution. These methods also reduce the risk of soil salinization. Green water management involves harnessing rainwater stored in the soil and is particularly effective in rain-fed agricultural systems (Kovács et al., 2023). This approach saves water, reduces dependence on external water sources, and reduces environmental pressure on water bodies (Pramanick et al., 2024).

Agroforestry and vegetated buffers are agricultural practices with significant environmental benefits. They help reduce soil erosion, improve water infiltration, and enhance carbon sequestration. Agroforestry systems store carbon above and below ground, helping to slow climate change (Kuhn et al., 2016). Meanwhile, vegetated buffer strips placed along waterways can filter pollutants from agricultural runoff, improving water quality and protecting aquatic ecosystems. Nutrient management and pollution reduction are key aspects of sustainable soil and water management. Precision agriculture technology uses data-driven technologies such as GPS and remote sensing to optimize fertilization and minimize nutrient loss and pollution. Integrated pest management strategies emphasize biological control and reduction of chemical inputs to maintain environmental health (Pramanick et al., 2024). Enhancing soil carbon sequestration through biochar applications, reforestation, and carbon farming is critical to reducing greenhouse gas emissions and increasing climate resilience. Drought-resistant crop varieties and water-saving irrigation systems increase resilience to extreme weather events. An integrated ecosystem approach, including holistic land management, promoting permaculture practices, and restoring degraded land to improve soil fertility and ecosystem services (Mujtaba et al., 2021).

5 Limitations

The study lacks detailed region-specific case studies, which are crucial due to significant changes in environmental interactions. A broad approach may overlook local challenges and situation-specific solutions. Furthermore, the study focused primarily on traditional practices and ignored emerging technologies such as artificial intelligence-driven water management and gene-edited crops to improve resilience. The long-term socioeconomic impacts of sustainable practices remain underexplored, especially for smallholder farmers in developing countries. Research assumes that recommended strategies are universally applicable, ignoring economic, cultural, and institutional barriers that may impede adoption. Additionally, the environmental consequences of some sustainable technologies, such as overreliance on chemical inputs in precision agriculture, are not rigorously scrutinized and may have unintended negative impacts on soil health.

Erwin, K. L. (2009). Wetlands and global climate change: The role of wetland restoration in a changing world. *Wetlands Ecology and Management, 17*(1), 71–84. https://doi.org/10.1007/s11273-008-9119-1
Flower, L. (2006). Environmental challenges in the 21st century. *Assumption University Journal of Technology, 9*(4), 248–252.
Gondal, A. H., Hussain, I., Ijaz, A. B., Zafar, A., B. I, Zafar, H., & Usama, M. (2021). Influence of soil pH and microbes on mineral solubility and plant nutrition: A review. *International Journal of Agriculture and Biological Sciences, 5*(1), 71–81.
Gowdy, J. (2020). Our hunter-gatherer future: Climate change, agriculture and uncivilization. *Futures, 115*, Article 102488. https://doi.org/10.1016/j.futures.2019.102488
Hailu, B., & Mehari, H. (2021). Impacts of soil salinity/sodicity on soil-water relations and plant growth in dry land areas: A review. *Journal of Natural Sciences Research*. https://doi.org/10.7176/JNSR/12-3-01
Hinkel, J., Lincke, D., Vafeidis, A. T., Perrette, M., Nicholls, R. J., Tol, R. S., ... & Levermann, A. (2014). Coastal flood damage and adaptation costs under 21st century sea-level rise. *Proceedings of the National Academy of Sciences, 111*(9), 3292–3297. https://doi.org/10.1073/pnas.1222469111
Hoffmann, I. (2021). Declining biodiversity threatens food security. *World Food Journal*.
Jacobson, M. Z., Delucchi, M. A., Cameron, M. A., & Frew, B. A. (2015). Low-cost solution to the grid reliability problem with 100% penetration of intermittent wind, water, and solar for all purposes. *Proceedings of the National Academy of Sciences, 112*(49), 15060–15065.
Keswani, A., Akselrod, H., & Anenberg, S. C. (2022). Health and clinical impacts of air pollution and linkages with climate change. *New England Journal of Medicine evidence, 1*(7), EVIDra2200068.
Kovács, G. P., Simon, B., Balla, I., Bozóki, B., Dekemati, I., Gyuricza, C., Percze, A., & Birkás, M. (2023). Conservation tillage improves soil quality and crop yield in Hungary. *Agronomy, 13*(3), 894.
Kuhn, N. J., Hu, Y. X., Bloemertz, L., He, J., Li, H. W., & Greenwood, P. (2016). Conservation tillage and sustainable intensification of agriculture: Regional vs. global benefit analysis. *Agriculture Ecosystems & Environment, 216*, 155–165.
Lakhiar, I. A., Yan, H., Zhang, C., Wang, G., He, B., Hao, B., Han, Y., Wang, B., Bao, R., Syed, T. N., Chauhdary, J. N., & Rakibuzzaman, Md. (2024). A review of precision irrigation water-saving technology under changing climate for enhancing water use efficiency, crop yield, and environmental footprints. *Agriculture, 14*(7), 1141. https://doi.org/10.3390/agriculture14071141
Lelieveld, J., Haines, A., Burnett, R., Tonne, C., Klingmüller, K., Münzel, T., & Pozzer, A. (2023). Air pollution deaths attributable to fossil fuels: observational and modelling study. *British Medical Journal,* 383.
Lesk, C., Rowhani, P., & Ramankutty, N. (2016). Influence of extreme weather disasters on global crop production. *Nature, 529*(7584), 84–87.
McArthur, J. W., & McCord, G. C. (2017). Fertilizing growth: Agricultural inputs and their effects in economic development. *Journal of Development Economics, 127*, 133–152. https://doi.org/10.1016/j.jdeveco.2017.02.007
Montgomery, D. R. (2007). Soil erosion and agricultural sustainability. *Proceedings of the National Academy of Sciences, 104*(33), 13268–13272.
Moore, C. J. (2008). Synthetic polymers in the marine environment: A rapidly increasing, long-term threat. *Environmental Research, 108*(2), 131–139. https://doi.org/10.1016/j.envres.2008.07.025
Moyer, J. D., Pirzadeh, A., Irfan, M., Solórzano, J., Stone, B., Xiong, Y, Hanna, T., Hughes, B. B. (2023). How many people will live in poverty because of climate change? a macro-level projection analysis to 2070. *Climatic Change, 176*(10), 137. https://doi.org/10.1007/s10584-023-03611-3
Mueller, L., Eulenstein, F., Dronin, N. M., Mirschel, W., McKenzie, B. M., Antrop, M., Jones, M., Dannowski, R., Schindler, U., Behrendt, A., Rukhovich, O. V., Sychev, V. G., Sheudzhen, A.

K., Romanenkov, V. A., Trofimov, I., Robinson, G. M., Schreg, R., Blum, W. E. H., Salnjikov, E., & Poulton, P. (2021). Agricultural landscapes: History, status and challenges. *Exploring and Optimizing Agricultural Landscapes*, (pp. 3–54). https://doi.org/10.1007/978-3-030-67448-9_1

Mujtaba, G., Hayat, R., Hussain, Q., & Ahmed, M. (2021). Physio-chemical characterization of biochar, compost and co-composted biochar derived from green waste. *Sustainability, 13*(9), 22.

Muluneh, M. G. (2021). Impact of climate change on biodiversity and food security: A global perspective—a review article. *Agriculture & Food Security, 10*(1), 1–25.

Nawaz, A., Farooq, M., Ul-Allah, S., Gogoi, N., Lal, R., & Siddique, K. H. M. (2021). Sustainable soil management for food security in South Asia. *Journal of Soil Science and Plant Nutrition, 21*(1), 258–275. https://doi.org/10.1007/s42729-020-00358-z

Newell, P., Srivastava, S., Naess, L. O., Torres Contreras, G. A., & Price, R. (2021). Toward transformative climate justice: An emerging research agenda. *Wiley Interdisciplinary Reviews: Climate Change, 12*(6), Article e733.

Omranian, A. R., Dabirinejad, S., Khorsandi, B., & Habibian, M. (2023). Contribution of anthropogenic pollutant sources to greenhouse gas emissions: A case study from a developing country. *Environmental Science and Pollution Research, 30*(27), 70159–70169.

Peace, J., & Huber, K. (2023). Climate preparedness for business resilience. In *Handbook of business and climate change* (pp. 294–324). Edward Elgar Publishing

Pecl, G. T., Araújo, M. B., Bell, J. D., Blanchard, J., Bonebrake, T. C., Chen, I. C., Clark, T. D., Colwell, R. K., Danielsen, F., & Williams, S. E. (2017). Biodiversity redistribution under climate change: Impacts on ecosystems and human well-being. *Science, 355*(6332), eaai9214.

Pierzynski, G. M., George F. Vance, & Thomas Sim, J. (2005). *Soils and environmental quality*. CRC press.

Pimentel, D., & Burgess, M. (2013). Soil erosion threatens food production. *Agriculture, 3*(3), 443–463.

Pradinaud, C., Northey, S., Amor, B., Bare, J., Benini, L., Berger, M., Boulay, A. M., Junqua, G., & Rosenbaum, R. K. (2019). Defining freshwater as a natural resource: a framework linking water use to the area of protection natural resources. *The International Journal of life cycle assessment, 24*, 960–974.

Pramanick, B., Kumar, M., Naik, B. M., Singh, S. K., Kumar, M., & Singh, S. V. (2024). Soil carbon-nutrient cycling, energetics, and carbon footprint in calcareous soils with adoption of long-term conservation tillage practices and cropping systems diversification. *Science of the Total Environment, 912*, 15.

Raven, P. H. (2020). Biological extinction and climate change. *Health of People, Health of Planet and Our Responsibility: Climate Change, Air Pollution and Health*, (pp. 11–20).

Singh, N. K., Sachan, K., Ranjitha, G., Chandana, S., Manoj, B. P., Panotra, N., & Katiyar, D. (2024). Building soil health and fertility through organic amendments and practices: A review. *Asian Journal of Soil Science and Plant Nutrition, 10*(1), 175–197. https://doi.org/10.9734/ajsspn/2024/v10i1224

Singh, N., Poonia, T., Siwal, S. S., Srivastav, A. L., Sharma, H. K., & Mittal, S. K. (2022). Challenges of water contamination in urban areas. *In Current Directions in Water Scarcity Research*. (pp. 173–202). https://doi.org/10.1016/B978-0-323-91838-1.00008-7

Srivastav, A. L., Dhyani, R., Ranjan, M., Madhav, S., & Sillanpää, M. (2021). Climate-resilient strategies for sustainable management of water resources and agriculture. *Environmental Science and Pollution Research, 28*(31), 41576–41595. https://doi.org/10.1007/s11356-021-14332-4

Stocker, T. F., Qin, D., Plattner, G. K., Tignor, M. M., Allen, S. K., Boschung, J., Nauels, A., Xia, Y., Bex, V., & Midgley, P. M. (2014). *Climate Change 2013: The physical science basis*. Contribution of working group I to the fifth assessment report of IPCC the Intergovernmental Panel on climate change.

Tahat, M. M., Alananbeh, K. M., Othman, Y. A., & Leskovar, D. I. (2020). Soil health and sustainable agriculture. *Sustainability, 12*(12), 4859. https://doi.org/10.3390/su12124859

Uekötter, F. (2020). In Search of a Dust Bowl Narrative for the Twenty-First Century. *Great Plains Quarterly., 40*(2), 161–168.

Van der Gun, J. (2021). Groundwater resources sustainability. *In Global Groundwater* (pp. 331–345). Elsevier. https://doi.org/10.1016/B978-0-12-818172-0.00024-4

Wang, R., Wang, Q., Dong, L., & Zhang, J. (2021). Cleaner agricultural production in drinking-water source areas for the control of non-point source pollution in China. *Journal of Environmental Management, 285*, Article 112096. https://doi.org/10.1016/j.jenvman.2021.112096

Weis, J. S. (2024). *Marine Pollution: What Everyone Needs to Know*. Oxford University Press.

Yang, P., Wu, L., Cheng, M., Fan, J., Li, S., Wang, H., & Qian, L. (2023). Review on drip irrigation: Impact on crop yield, quality, and water productivity in China. *Water, 15*(9), 1733. https://doi.org/10.3390/w15091733

The Climate Challenge: Sustaining Soil Health and Crop Resilience

Iqra Ghafoor, Tayyaba Naz, Shah Nawaz, Muhammad Mazhar Iqbal, Uzma Ambreen, Salma Hameed, Sonia Perveen, Aqsa Rashid, Madiha Batool, Sobia Riaz, and Muhmmad Umair

Abstract In numerous areas of the world, food production is disrupted by climate change. Crops are at risk because the following extreme weather events, such as heat waves, droughts, floods, and cold snaps. The atmosphere amount of carbon dioxide

I. Ghafoor (✉) · T. Naz · S. Nawaz · U. Ambreen · A. Rashid · M. Batool · S. Riaz
Institute of Soil and Environmental Science, University of Agriculture Faisalabad, Faisalabad, Pakistan
e-mail: dr.iqraghafoor.uaf@gmail.com

T. Naz
e-mail: tayyabanaz@uaf.edu.pk

S. Nawaz
e-mail: shahnawazuaf@hotmail.com

U. Ambreen
e-mail: uzmaambreen22@gmail.com

A. Rashid
e-mail: aqsarashid325@gmail.com

S. Nawaz
School of the Environment Geography and Geosciences, University of Portsmouth-Uk, Portsmouth, UK

S. Hameed
Department of Environmental Sciences, University of Jhang, Jhang, Pakistan
e-mail: salma@uoj.edu.pk

M. M. Iqbal
Department of Soil and Environmental Sciences, University of Sargodha, Sargodha, Pakistan
e-mail: mazhar.iqbal@uos.edu.pk

S. Perveen
Department of Chemistry, University of Agriculture Faisalabad, Faisalabad, Pakistan
e-mail: soniaperveen161@gmail.com

M. Umair
College of Natural Resource and Environment, Northwest A&F University Yanging, Xianyang, China
e-mail: muhammadumair@nwafu.eedu.cn

M. Shaaban (ed.), *Soils and Sustainable Agriculture*, Frontier Studies in Soil Science,
https://doi.org/10.1007/978-3-031-91114-9_14

also rises. By 2050, food production must rise by 70% due to the growing global population, which will put strain on food quality standards. Food items, food consumption, and food insecurity have all increased as a result of the effects of Industrialization, environmental change, urbanization, and climate change. A sustainable environment relies heavily on agricultural production, with a primary focus on addressing the challenges in this sector. To guarantee food security, the United Nations has begun implementing its climate-smart agriculture strategy into action. Breeding climate-resilient crops, or plant cultivars with increased tolerance to extreme environmental circumstances, is a particular component of this initiative. The existing homogeneous state of modern agriculture requires diversification in the species and cultivars of plants that are grown. The main focus of this study is to introduce new molecular technologies that should be heavily integrated into plant breeding projects, bolstered by the advancement of field phenotype technologies. Breeders and scientists from other scientific disciplines should collaborate closely.

Keywords Climate change · Sustainability · Climate smart agriculture · Resilience crop · Food security

1 Introduction

Climate influences every aspect of the economic, cultural, and social environment; however, it also has a significant role in agricultural productivity and has the ability to worsen many present challenges (Campbell et al., 2009). According to Gowda et al. (2018), climate and weather patterns possess an enormous effect on crop yields. Climate change-associated environmental stresses, such as extreme temperatures and erratic rainfall, will compromise the ability of agriculture to meet the food demands of an increasing global population (Paul and Roychoudhury, 2018).

Climate-related variables including temperature, precipitation, timing of frost, length of growing season, soil moisture, pest pressures, along with numerous more influence agricultural productivity alongside management strategies. While to begin planting or harvesting, for example, is an immediate decision involving precise climate information. What to cultivate for the next growing season is a mid-term selection. A long-term decision would include whether or not to make financial investments such as the construction of water supply systems, underneath drainage tile the installation process or growing trees in a system that uses agroforestry (Prokopy et al., 2019; Takle et al., 2019). The main factor causing climate change is human activity's global greenhouse gas emissions (USGCRP, 2017). These changes influence the production of agricultural products. Considering the reality that greenhouse emissions, such carbon dioxide (CO_2), have a broad distribution in the environment (Fahey et al., 2017).

1.1 Major Indicators of Climate Change and Its Impacts on Crops

These five indications of climate change encompass extreme precipitation, soil moisture, nighttime air temperature, heat waves, along with humidity. Long-term trends and variabilities additionally exist among these indicators. It provides insight regarding how changes in the climate influence the food chain, agricultural output, and food security.

i. **Extreme Precipitation**

The main source of water for farming in the whole country is precipitation. Therefore, an understanding of present trends and expected variations in precipitation quantity, execution, and seasonally is essential for the efficiency as well as the sustainability of US agriculture (Karl et al., 2009). It is also vital to fully understand historical and expected alterations in extreme weather conditions, such as droughts and heavy rains that bring flooding and increased soil erosion, as such situations create critical challenges to oversight for farmers. Agricultural output and also the ability of crops to germinate in a region can be significantly impacted by changes in the total amounts and timing of precipitation (Key et al., 2014).

ii. **Soil Moisture**

Several aspects of farming operations are influenced by moisture in the soil, which additionally integrates several climatic effects on crop productivity. The yield of crops is affected by fluctuations in soil and atmospheric conditions, which in consequence alter soil moisture. Here, soil moisture is interpreted as the water that accumulates in the soil's root zone and that vegetation can utilize (Legates et al., 2011). Although it encompasses the effects of (in addition to interactions with) an array of climatic variables, including precipitation, evapotranspiration, solar radiation, temperature, vapor pressure, and wind, soil moisture influences climate thus is a significant physical parameter (Legates et al., 2011). Furthermore, plant features like phenology and water usage, as well as soil characteristics like permeability, water capacity, and organic matter content, have an influence (Seneviratne et al., 2010). Long-term periods of drought are readily mitigated by cultivars grown in high-organic matter soils since these varieties may preserve plant-available moisture (Hatfield et al., 2011).

iii. **Daytime and Nighttime Temperatures**

With the exception of the Northeast strengthening of summertime peaks, temperatures have climbed in the past few decades across the most of US locations, both at the highest and lowest ranges. It is predicted because global warming will continue rising during the twentieth century (USGCRP, 2017). Lowest temperatures have been rising in recent times and this pattern is going to persist (Qu et al., 2014). At the pollination, or reproductive, stage of plant development, high temperatures can prove particularly toxic (Hatfield & Prueger, 2015).

iv. **Heat Waves**

The times of multiple occasions over a threshold heat or exceeding a specified hazard level at the upper end of an ambient temperature range are defined as heat waves. Days with repeated high temperatures exceeding species-species requirements impair the regular biological processes of farm crops and livestock (USGCRP, 2017). Higher temperatures induce heat waves to occur more often, be more powerful, and last longer, which has a detrimental effect on crop output (De et al., 2018). Heat wave frequency is expected to climb in all regions of the United States over present high-emission assumptions. By mid-century, changes in the intensity of 5-day heat waves that occur once annually will cause temperatures to rise by 11°F to 13°F (USGCRP, 2017). Summer heat waves and droughts are especially dangerous for cereal germination and development, which causes reduced yields of crops throughout.

v. **Humidity**

A key variable in the formation of mist and precipitation is atmospheric humidity. By regulating transpiration, it is also essential to controlling the loss of moisture from humans, animals, and plants (Feng et al., 2016). In consequently, moisture-loss operations influence temperature and biological processes in living systems, including expansion and reproduction (Wang et al., 2019a). Environmental humidity is a vital component in simulations that predict precipitation, dew formation, and the effects of heat stress on plants, animals, and humans. Humidity fluctuates locally, regionally, and globally as a result of changes in Earth's climatic dynamics (Wisnieski et al., 2018).

1.2 Climate Change and Agriculture

Agriculture is very sensitive to weather and climate (Walsh et al., 2020). The impact of climate change constitutes an obstacle to agricultural efficiency in undernourished regions, especially among Asian countries. Agricultural' incomes have been adversely affected by a range of climate-related extremes, include heat waves, droughts, unpredictable and heavy rainfall patterns, storms, floods, even the emergence of insect pests. Climate predictions for the future revealed a considerable rise in temperature together with more passionate unsteady rainfall, even though there continue to be instability in weather events with regard to concerns forecast climate extremes (Aryal et al., 2019). It also relies heavily on land, water, and other natural resources that climate affects (Gowda et al., 2018). While climate changes (such as in temperature, precipitation, and frost timing) could lengthen the growing season or allow different crops to be grown in some regions, it will also make agricultural practices more difficult in others (Walsh et al., 2020). Perhaps due to the tendency for catastrophes to devastate crops, vital facilities, and major public property, and also having an impact on incomes and exacerbate impoverishment, these rare weather conditions can have an adverse effect on household wages and food security (Davidson & Janssens, 2018).

A substantial decrease in crop yield have been linked across numerous Asian regions to unpredictable and extreme precipitation variations throughout the last few decades, in addition to reduced rapid accessibility of nutrients and precipitation (Aryal et al., 2019). Although the green revolution has enhanced the efficiency of agriculture, sustaining productivity along with improving food security for impoverished in Asia under climate alteration scenarios remains a significant problem (Ahmad et al., 2019). The harm induced meteorological fluctuations could threaten national economic production and food security in the least developed nations (Myers et al., 2017). According to Obermeier et al. (2017), fertilization using CO_2 may improve yields of crops and counter the severe consequences of rising temperatures on C3 plants. The irregular precipitation and increasing temperatures having a severe effect on crop development and growth (Asseng et al., 2019).

1.3 Crops Under Multiple Climate Stress

Climate change is predicted to increase the frequency, intensity, spatial dimensions, and duration of extreme weather events posing a significant threat to the production of all major crops. For example, increased incidences of heatwaves, droughts and floods adversely affect the yields of cereal crops such as wheat and maize (Diffenbaugh et al., 2017). It is reported that plants require an optimum temperature for their normal growth and blooming. Plant physiology is heavily influenced by temperature fluctuations (Hatfield & Prueger, 2015). Plant responses to these stimuli can be broadly classified into three categories: morphological, physiological, and biochemical. High temperatures increase the evaporation of water in soils, causing drier soils in forest flora (Mansoor et al., 2022).

Especially the negative impact of heat waves during the reproductive stage, identified as a major threat to yield in many parts of the world (Lobell et al., 2013). In terms of the agricultural sector, crops, particularly wheat, are likely to experience a greater number of high-temperature episodes during their reproductive period, posing a threat to crop production (Gao et al., 2014).

When considering the effects of climate change on stress combinations, photosynthesis, WUE, NUE, and other important processes for plant growth and survival, we need to take into consideration the increase in atmospheric CO_2 levels that will accompany these stresses. At high CO_2, stomata of most plants begin to close, decreasing stomatal conductance and transpiration (Zhang et al., 2021). Sexual plant reproduction requires a complex and highly coordinated set of developmental processes that are tightly regulated and synchronized. Many of these processes, e.g., pollen maturation, fertilization, embryogenesis, and seed maturation are highly sensitive to different abiotic stress conditions, such as heat and drought, that could lead to pollen, embryo, or seed abortion (Jiang et al., 2019) also describe in Table 1. Photosynthesis is highly susceptible to stress combination, with several studies demonstrating that photosynthetic efficiency and transpiration rates decrease under conditions of water-deficit, salt, and/or heat stresses occurring together (Perdomo et al., 2017). Due to climate change, water deficit and temperature extremes influence the reproductive phase of plant growth. It was described that the flower initiation and inflorescence is badly affected by the water stress in cereals (Scheben et al., 2017).

Drought stress affects wheat at every stage of growth, however the phases associated with reproduction and grain formation are the most critical (Pradhan et al., 2012). De et al. (2013) reported that in the case of a protracted moderate drought stress during flowering and grain formation, wheat production declined from 1 to 30% during post-anthesis. However, in other circumstances, this decline increased to 92%. Grain legume growth in responses to stress caused by drought is today hardly nonexistent. Butterbeans (*Vigna mungo* L.) produce less when under drought stress; this reduction ranges from 31 to 57% at the flowering stage in addition to 26–57% throughout the reproductive phase (Baroowa and Gogoi, 2014).

Maleki et al. (2013) report that drought-related stress significantly reduces soybean yield, with a 42% reduction observed during the grain filling stage. Sorghum yield dropped by 7.8% with a 1 °C increase in temperature (Kjellstrom et al., 2018). A lot of the sequences of gene expression within plants undergo changes when they are exposed to drought. Initial modifications involving early reactions, such signal transduction, transcription, and translation variables, are then followed by expressions connected to late responses, like water transport, osmotic balance, oxidative stress, and the damage-repair mechanisms (Madzima et al., 2021).

The outcome of global warming is an abrupt increase in temperature due to surplus solar radiation. These occurrences are often the most restrictive factors affecting plant growth and productivity. The phenological stage can impact them. According to Begcy et al. (2018), extended heat stress throughout seed development may decrease the potential for survival of the seed by delaying germination or causing vigor reduction. Anatomical modifications spurred upon by high temperatures include fewer cells in total, decreased water loss by stomata closure, higher trichomatous and stomata densities, and larger xylem vessels in both the root and the shoot (Lawson and Blatt, 2014).

1.4 Impact of Climate Change and Variability on Rice–wheat Crops

Our major the cultivation of crops system wheat, maize, cotton, rice, and sugarcane is facing threat from climate change. By 2040, temperatures will likely increase by 3 °C, and by the end of the century, they are expected to be climbed by 5–6 °C, which will result in a 50% reduction in the productivity of wheat (Ghanem, 2010). The routine, intensity, and degree of adaptation of landowners and farmers will determine the way climate change disrupts cultivation (Gowda et al., 2018). In Pakistan, farming is the largest economic sector, employing around 45% of the labor force, adding

Table 1 Effects on crops under different various climate change conditions

Climatic stress	Crops	Effects on crops	References
Extreme heat stress	Maize, spring wheat and soybean	We find extreme heat stress at anthesis (HSA) by the 2080s (relative to the 1980s) under RCP 8.5, taking into account CO2 fertilization effects, could double global losses of maize yield ($\Delta Y = -12.8 \pm 6.7\%$ versus $-7.0 \pm 5.3\%$ without HSA), reduce projected gains in spring wheat yield by half ($\Delta Y = 34.3 \pm 13.5\%$ versus $72.0 \pm 10.9\%$ without HSA) and in soybean yield by a quarter ($\Delta Y = 15.3 \pm 26.5\%$ versus $20.4 \pm 22.1\%$ without HSA)	(Deryng et al., 2014)
1.5 °C and 2.0 °C heat stress	Soybeans, wheat, and maize	The area suffering from high-intensity heat stress during the wheat reproductive phase is predicted to increase by 37.3% with 2.0 °C warming, covering the Mediterranean region and stretching into Central Asia. With 2.0 °C global warming, food crises will continue to occur, and acute hunger is likely to intensify in certain food-deficient counties,	(Sun et al., 2019)
Heat stress	Wheat	Extreme temperatures directly influence crop production by specifically affecting plant growth and yield formation posing a serious threat to food production	(Venkatesh et al., 2022)
Heat stress	Rice (*Oryza sativa* L.)	The occurrence of HS during the reproductive stage resulted in an up to 80% reduction in spikelet fertility of rice	(Xu et al., 2020)
Smog	Wheat	Due to smog, productivity, and growth of wheat have been adversely affected as the size, and weight of wheat grains have decreased	(Safdar et al., 2022)

(continued)

Table 1 (continued)

Climatic stress	Crops	Effects on crops	References
Ozone concentrations (60, 150, and 240 ppb for 2 h/day)	*Brassica campestris* L. ssp. *chinensis* (Pak-Choi)	Strong effect of ozone stress on the plant quality of Pak-Choi with respect to certain secondary plant metabolites. These findings provide an understanding of elevated ozone effects in urban horticulture sites on the growth and metabolite profiling of *Brassica* plants	(Han et al., 2023)
Rainfall	*Rice (Oryza sativa* L.)	It is observed that rainfall has negative impact on rice plant at heading and fowering stages. Substantial reduction in rice production is observed due to damage of reproductive cells at heading and fowering stages during the reproductive phase	Abbas et al. (2013)

21% to the GDP (gross domestic product), and making up nearly 60% of revenue from exports. Pakistan is estimated to have 23.4 million hectares under cultivation, of which 18.63 million hectares are in need of irrigation (Câmpeanu and Fazey, 2014). Pakistan places a high value on irrigation, with 77% of Punjab's land under irrigation and smaller proportions in the remainder of the country. There are numerous farming technologies in use, such as sprinkler irrigation (Chaudhary et al., 2017). The method of crop irrigation known as "spate irrigation" involves using gravity to direct seasonal storm water from rivers, valleys, gullies, and riverbeds onto cropland that is lower in elevation than floodwater (CIAT; World Bank, 2017).

Wheat is a significant Rabi crop, and Kharif crops include sugarcane, rice, maize, turmeric, and other crops. Summertime sees the growth of the Kharif crops, such as sugarcane in February, cotton in March–May, rice in June–July, and maize in July–August. Substantial agriculture systems consisting of sugarcane and wheat, cotton and wheat, maize and wheat, and rice and wheat. Crops in both spate and irrigated farming systems are extremely sensitive to changes in temperature and water availability. According to predictions, agricultural production will decrease by almost 8–10% by 2040 as temperatures rise (Henry et al., 2020).

In Asia, grains such as wheat and rice play a significant role in ensuring food security. To fulfill the ever-growing need for food, it will be extremely difficult to expand wheat production by 60% by 2050 (Rezaei et al., 2018). Reduced crop productivity in dry to semi-arid areas is linked to rising temperatures in northern regions. Drought and flooding have decreased the yields of rice, wheat, and maize in China; it is anticipated that these problems will have a greater impact on crop output in the future (Chen et al., 2018). Yang et al. (2017) indicate that rice is vulnerable to a slow rise in temperature at night. Rice weight as well as productivity will drop by 16–52% if this rise in temperature is above the 24 °C threshold by 2 °C.

Table 2 Productivity shock due to climate change and variability on rice and wheat crop production by 2030 (Suryadi, 2020)

Countries	Rice	Wheat
China	– 12 to + 12	– 10 to + 14
Thailand	– 10 to + 4	– 10 to + 4
Philippine	– 10 to + 4	– 10 to + 4
India	– 15 to + 4	– 10 to + 4
Bangladesh	– 15 to + 4	– 10 to + 4
Pakistan	– 15 to + 4	– 10 to + 4

Asian semi-arid to dry zones have a danger as low production and severe drought have become issues. The detrimental effects of rising temperatures and intense, irregular rains have resulted in a decrease in the nutritional value of wheat produce (protein content, sugars, and starch) as well as grain yield (Yang et al., 2017). Climate variability has resulted in a large decline in wheat production in the Egyptian North Nile Delta (up to 17.6%), India, and China. This decrease is linked to a spike in temperature, irregular rainfall, and an increase in insect invasions (Kheir et al., 2022).

According to Chun et al. (2016), rice yield in South Asian rain-fed areas has dropped previously and could do so again by 14% under the RCP 4.5 scenario and 10% under the RCP 8.5 scenario by 2080. Due to their detrimental effects on the starting as anthesis stage, high temperatures and drought have reduced rice yields throughout Asia, particularly in Pakistan and China (Zafar et al., 2018; Ahmad et al., 2019). Similar to this, heat stress constitutes a serious risk to rice since it reduces the number of fertile tillers, causes grains to shrink, and eventually lowers rice grain output (Wang et al., 2019b). Rain-fed lowland rice (>13 million hectares) and highland rice (10 m ha) in Asia would be impacted by climate change (Table 2).

2 Nexuses Climate Change Factors, Soil and Crop Health

2.1 Change in Temperature and Heat Stress

According to Huang et al. (2013), extreme weather events and heat stress present serious risks to the health and safety of workers. Through direct and indirect influences on core body temperature, global climate change significantly impacts physiological and perceptual responses (Habibi et al., 2023).

Thermal comfort, skin temperature, and heart rate (Yi & Chan., 2015). Heat stress at work can be caused by exposure to hot working conditions and the heightened physiological and perceptual reactions that follow. This can compromise worker safety, health, and productivity (Kim et al., 2015). Raising the possibility of diseases linked to the heat (Hunt et al., 2023), (HRI) during a heatwave, sessile plants rely on physiological reactions to withstand high temperatures (Zhu, 2016). The primary productivity of ecosystems may be significantly impacted by increased temperature stress on plants under warmer and more variable temperature regimes (Duffy et al., 2021). Vegetation composition and ecosystems' capacity to endure over time (Trisos et al., 2020).

Model species have well-documented physiological responses of plants to mitigate temperature stress (Chaudhary et al., 2020b). However, considering that more frequent and severe heat waves are predicted under future climate scenarios (Duffy et al., 2021). Understanding the differences in non-model plants' reactions to heat stress across various natural systems is becoming more and more crucial (Mokany et al., 2020). There are many different ways that plants react to heat stress, including morphological, phonological, physiological, and molecular reactions that are known to be species-specific. Plants store conserved stress proteins in response to heat stress, including chaperone proteins called Heat Shock Proteins (HSPs), which are involved in protein folding, aggregation, translocation, and destruction (Chen et al., 2018).

When unfavorable temperatures have a deleterious effect on the creation and stability of proteins, these processes become crucial for the integrity of cells (Chaudhary et al., 2020a). Unfolded protein response, oxidative stress response, ubiquitination, signaling (such as phosphorylation), and transcription factors are further instances of stress response gene ontologies (Zhang et al., 2022). It is routine practice to assess the molecular response of organisms to stress using changes in mRNA expression (Rivera et al., 2021). Numerous genes implicated in heat stress reactions have been found to be conserved, and as a result, plants' ability to withstand extreme heat should have evolved to suit regional climates (Donelson et al., 2023).

2.2 Extreme Weathered Events

i. **Extreme Precipitation**

According to Huang et al. (2013), extreme weather events and heat stress present serious risks to the health and safety of workers. Hot weather increases the risk of accidents, fatalities, and morbidity. For the majority of nations, global warming poses a fresh and daunting problem (Alongi et al., 2008). Through direct and indirect influences on core body temperature, global climate change significantly impacts physiological and perceptual responses (Habibi et al., 2023).

Thermal comfort, skin temperature, and heart rate (Yi & Chan, 2015). Heat stress at work can be caused by exposure to hot working conditions and the heightened physiological and perceptual reactions that follow. This can compromise worker safety, health, and productivity (Kim et al., 2015). Raising the possibility of diseases linked to the heat (Hunt et al., 2023). During a heatwave, sessile plants rely on physiological reactions to withstand high temperatures (Zhu, 2016). The primary productivity of ecosystems may be significantly impacted by increased temperature stress on plants under warmer and more variable temperature regimes (Duffy et al., 2021). Vegetation composition and ecosystems' capacity to endure over time (Trisos et al., 2020).

Model species have well-documented physiological responses of plants to mitigate temperature stress (Chaudhary et al., 2020b). However, considering that more frequent and severe heat waves are predicted under future climate scenarios (Duffy et al., 2021). Understanding the differences in non-model plants reactions to heat stress across various natural systems is becoming more and more crucial (Mokany et al., 2020). When unfavorable temperatures have a deleterious effect on the creation and stability of proteins, these processes become crucial for the integrity of cells (Chaudhary et al., 2020b). Unfolded protein response, oxidative stress response, ubiquitination, signaling (such as phosphorylation), and transcription factors are further instances of stress response gene ontologies (Zhang et al., 2022). It is routine practice to assess the molecular response of organisms to stress using changes in mRNA expression (Rivera et al., 2021). Numerous genes implicated in heat stress reactions have been found to be conserved, and as a result, plants ability to withstand extreme heat should have evolved to suit regional climates (Donelson et al., 2023).

ii. **Rainfall and Flood**

In metropolitan locations with high runoff and poor soil infiltration rates, rainfall with short duration and high intensity is more likely to result in flood disasters (Li et al., 2023). Urban infrastructure (UI) is usually protected first during periods of severe rainfall since it is essential to maintaining inhabitants' daily routines, enabling emergency rescue operations, and supporting recovery efforts following a disaster (Nkeki et al., 2022). Flood disasters resulting from urbanization and harsh weather patterns have become a major issue influencing post-disaster medical rescue, traffic, and infrastructure. It is essential to comprehend how a downpour causes an urban infrastructure flood disaster and to propose appropriate disaster preventive and mitigation measures. Extreme rainfall events are growing more frequent as a result of the intensifying effects of global climate change (Arya and Kumar, 2023; Dewals et al., 2023).

But recent catastrophes have shown how crucial UI is during urban flooding, as well as how vulnerable it is (Nick et al., 2023). For example, in December 2015, there was an intense rainfall event in the northwest of the United Kingdom, namely in Lancaster, which caused power interruptions for up to three days for almost 61,000 households. Hospitals, communication networks, and emergency rescue services in the metropolitan region were rendered inoperable even after a brief restoration of power due to a second wave of failures brought on by rainfall runoff (Wyatt, 2017). In a similar vein, 398 people lost their lives in Zhengzhou, central China, during an unheard-of period of intense downpour in July 2021 (Dong et al., 2022a, b).

Statistical analysis of historical data and numerical simulation techniques are the two main approaches commonly used to evaluate the influence of urban hydrological processes on UI (Qi et al., 2022). Numerical simulation techniques have drawn a lot of interest because of the gaps in the comprehensive historical hydrological data (Wang et al., 2023b). Amicarelli et al. (2020) using a flood damage model for the substation and a smooth particle hydrodynamics program, the level of damage to the components in the substation was determined. Using GIS and HEC-RAS rainfall grid data, Nkeki et al. (2022) created an urban flood submergence model to create flood risk maps depending on water depth. Despite being thorough and objective, their research did not specifically address the flood catastrophe procedure in UI. In order to manage the worst-case scenarios for catastrophes, it is essential to comprehend the severity and contributing elements of flood disasters during periods of intense rainfall.

iii. **Drought**

These days, water shortages are common at a multitude of global locations. This is caused in part by contaminated water supplies and climate change, but it's also largely due to population growth and the expansion of the agricultural, energy, and industrial sectors, which increase water demand (Bates et al., 2008). Droughts are making the already limited water supply worse by affecting both surface and groundwater resources. These can result in a decreased supply of water, worsening water quality, crop failure, and disruption of riparian habitats (Riebsame et al., 2019).

Therefore, ecologists, hydrologists, meteorologists, and agricultural scientists have been interested in comprehending drought and modeling its constituent parts. Planning and management of water resources are greatly impacted by droughts; for a summary of the issues surrounding drought, see Mishra and Singh (2010). To lessen the effects of drought on crop output and productivity, agricultural fields can benefit from a variety of drought management techniques. To increase agricultural drought tolerance, for instance, genetic engineering techniques and traditional selection procedures have been applied (Pandey & Shukla, 2016).

The implementation of optimal management techniques pertaining to seed sowing time, plant genotype, soil, and nutrient management may reduce the impacts of drought stress on grain yield. The most widely used method of reducing drought stress is probably the use of transgenic plants with drought-tolerant traits. In order to create innovative drought-resistant plants, numerous molecular and genetic techniques are being employed. Consequently, in order to understand the intricate mechanisms behind plant drought tolerance and to pinpoint crucial functional machinery that can be employed as instruments in the engineering of drought-resistant crops, more advanced study is required. Other approaches that can increase agricultural stress tolerance include the use of artificial hormones and exogenous regulators, along with more contemporary techniques for planting, mulching, irrigation, breeding, and contouring. However, the development of these approaches is time-consuming and capital-intensive (Backer et al., 2018).

The naturally existing microbial population that envelops plants is known as the rhizosphere. Thousands of different kinds of bacteria, fungi, protozoa, and other microorganisms can be found in the rhizosphere. Significant dynamic interactions between plants and microbial activities occur in this ecological hotspot. Rhizobacteria, a range of bacterial species that are directly impacted by root exudates, are the most common microorganism in the rhizosphere (Egamberdieva et al., 2018). Plant Growth-Promoting Rhizobacteria (PGPR) are a helpful species of bacteria that are found among rhizobacteria. PGPR has the ability to colonize plant roots, which can boost agricultural yield, promote plant growth and development, and significantly enhance soil fertility. Furthermore, these PGPR may help their host plants by promoting a variety of direct and indirect responses to mitigate the effects of drought stress and, eventually, enable plants to survive these stressful conditions (Naveed et al., 2014). The review focuses on the strategies that PGPR employs to environmentally friendly and sustainable induce drought stress tolerance.

iv. **Freezing and Thawing**

Frequent freezing and thawing of clay soil modify the micro, macro, and mechanical properties of the soil. Freeze–thaw cycling is a process of energy input and output in soil (Shen et al., 2022). Due to aggregation and fragmentation brought on by water and salt movement during the freeze–thaw process, changes in soil granularity, composition, structure, and properties can result in ecological and engineering issues (Lu et al., 2021).

Worldwide, 3% of topsoil and 6% of subsoil are affected by salt; these types of soils are known to exist (FAO, 2021). Fractal research has revealed that during freeze–thaw cycles, the fracture behavior of clayey soil changes from irregular rectilinear patterns to polygonal or quasi-hexagonal patterns (Lu et al., 2016).

v. **Length of Growing Season**

Because of their seasonal cycle, plants have a growing season that is closely correlated with seasonal changes in the climate (Einarson & Trainor, 2015). When established growth and development are permitted by the local environmental circumstances, trees enter their growing season. It has a beginning, a stopping, and consequently a duration. When primary growth (budburst, leaf, and shoot) and secondary growth (cambium differentiation and xylem development) begin is when it is said to have begun. When growth ceases or the leaves turn brown and drop, closure takes place.

The latent phase lasts from when growth stops to when it resumes. During dormancy, evergreens photosynthesis slowly and only for short periods of time, while deciduous trees do not. Notwithstanding this broad, albeit imprecise, notion of the growing season, definitions of its beginning and end, and consequently its duration, differ greatly depending on the species and techniques used. Körner et al. (2023) attempted to comprehend and organize the theory underlying this intricate problem. Developmental phonological markers determine the phonological season. However, plants can become photosynthetically active long before an apparent development advance (Lempereur et al., 2015) and after invisible developmental states and processes (hormone production and metabolism activation). This makes defining the beginning and end of a phonological season more difficult. According to Körner et al. (2023), the fruitful season is the first and last to cross a minimum biomass production rate or net carbon gain. According to Körner et al. (2023), the fruitful season is the first and last to cross a minimum biomass production rate or net carbon gain. According to Körner et al. (2023), the meteorological season is the time of year when plants can grow based on experience and meteorological factors. When working with meteorological data, two factors are crucial, according to Körner and Hiltbrunner (2018), the time scale (hour, day, month, season, or year) and the place of the activity (soil, plant, or air data).

vi. **Extreme Precipitation**

The concentration of atmospheric water vapor, which provides the water for precipitation, increases in proportion to the saturation concentrations at a rate of about 6–7% per degree rise in temperature. This means that extreme precipitation is expected to intensify with global warming over a large portion of the planet, according to the thermodynamic Clausius–Clapeyron relationship (Ingram et al., 2016). Nonetheless, modifications in atmospheric dynamics, including the polarization of the tropical Hadley circulation, have the potential to diminish (Norris et al., 2019). The regional thermodynamic effect modifies the amplification of extreme precipitation. The link between moisture and temperature is also greatly influenced by the availability of water (Zhang et al., 2017). Extreme precipitation variations are not unvarying in planetary and vary by site as a result of multiple interacting factors (Tabari et al., 2019).

As a result, the rate at which extreme precipitation increases with land surface temperature is not constant. In certain locations, a negative scaling has even been noted at higher temperatures (Roderick et al., 2019). Which may be the consequence of a lack of available moisture (Drobinsk et al., 2016). Extreme precipitation can have more of an impact in moist areas due to heightened atmospheric moisture convergence, while in arid locations, evaporation may offset an increase in precipitation (Held et al., 2016). The 30-year mean annual precipitation maxima aggregated throughout the dry and wet regions of the world are projected to increase, according to recent research that looked at daily extreme precipitation variations in relation to water availability (Donat et al., 2019). Climate change is predicted to have a greater impact on uncommon precipitation occurrences (Myhre et al., 2019).

The question of whether extreme precipitation changes and water availability can potentially be related to infrequent precipitation events that cause flooding is still open. Severe amplification of precipitation can intensify and increase the frequency of flooding, causing severe damage to land and aquatic ecosystems, human communities, and the economy. Variations in flood features are influenced by more than only the distribution of precipitation over time, space, and rarity (Sharma et al., 2018). According to worldwide flood assessments, there will likely be both an increase and a decrease in future floods as a result of global warming (Asadieh et al., 2017). Drawing a shared understanding of future flood changes is impeded by the use of disparate hydrological and climate models, scenarios, bias-correction techniques, and flood indicators (Kundzewicz et al., 2017). At the regional level, it has been demonstrated that variations in climatic moisture are positively correlated with variations in soil moisture and runoff (Girvetz & Zganjar, 2014). Additionally, compared to dry regions, wet regions' recorded annual maximum flows have changed more significantly (Kumar et al., 2016). This is because, in the former case, a larger buffer is available to attenuate precipitation variations, resulting in lesser runoff changes, but in the latter case, a higher fraction of precipitation changes leads to runoff changes (Zhang et al., 2008). It is unknown if correlations between abrupt changes in precipitation and water availability can be extrapolated to flood changes due to the many systems involved.

3 Impacts of Climate Change on Soil and Crop Health

Before we go into what exactly the impacts of climate change on soil and crop health it is important to understand what the term climate change is. Climate change has affected diverse spheres and its impact is being witnessed worldwide.

Climate change is a long-term shift in the statistics of the weather (including its averages). It has been found that the global climate is currently changing. In addition to natural climate shifts seen over comparable time periods, it is linked, either directly or indirectly, to human activity that modifies the composition of the planetary atmosphere (Rajeshwar et al., 2018). The last decade of the Tweentyth Century and the beginning of the 21st have been the warmest period in the entire global instrumental temperature record, starting in the mid-nineteenth century.

Soils and climate are intimately linked. Soil, the basis of human sustenance, is both directly and indirectly affected by climate change. Soils exhibit significant spatial correlations with climate, and they differ based on the local climate. Soil erosion, vegetation degradation and soil salinization are becoming prevalent, causing a threat to future food security. In order to have long-term, sustainable agricultural productivity with minimum adverse effect on the environment, soil health refers to the biological, chemical, and physical characteristics of the soil. Thus, soil health provides an overall picture of soil functionality (Arias et al., 2005).

Carbon Dioxide the Primary Culprit

A key component of the idea of sustainable agriculture has been identified as soil health. Climate trends over the past few decades have been fairly rapid in many agricultural regions around the world, and increases in atmospheric carbon dioxide (CO_2) and ozone (O_3) levels have also been ubiquitous. The Fourth Assessment Report of the Intergovernmental Panel on Climate Change concludes "that most of the observed increase in the globally averaged temperature since the mid-twentieth century is very likely due to the observed increase in anthropogenic greenhouse gas concentrations" (Patwardhan et al., 2007).

The foremost effects of global climate change on soil health are as follows:

3.1 *Soil Physico-Chemical Health*

Global interest in soil health has increased exponentially during the past decade, with many different government, non-government, and private sector groups striving to develop monitoring and assessment protocols. Traditionally, physicochemical properties (texture, soil aggregation, bulk density, water holding capacity, porosity, pH, electrical conductivity, organic matter, cation exchange capacity, nutrient content) have been used as soil health indicators (Barrutia et al., 2011; Galende et al., 2014; Mijangos et al., 2006). These indicators are discussed below;

i. **Plant Available Water and Infiltration Rate**

As part of integrative soil health assessments, plant available water capacity has been utilised to evaluate the implications of management (Jarvis, 2020). Moreover, the amount of water in the soil and its distribution may change quickly in response to climate change, particularly in the case of variable and prolonged rainfall or drought events. As a result, management strategies such as the cultivation of cover crops, conservation planting, and the incorporation of organic matter may be necessary to maintain or even improve water infiltration, and the availability of water in the soil may help to lessen the effects of severe drought, erosion and rainfall events as well as severe erosion events (Sanchis et al., 2008).

ii. **Soil Temperature**

As with soil moisture, soil temperature is a prime mover in most soil processes. Temperature and moisture amounts cause different patterns of weathering and leaching. Seasonal and daily changes in temperature affect moisture effectiveness, biological activity, rates of chemical reactions, and kinds of vegetation (Das et al., 2016). Increased microbiological activity, higher rates of organic matter degradation, rapid release of nutrients, and an overall enhancement of chemical weathering of elements are all accelerated by warmer soil temperatures.

3.2 Soil Chemical Parameters

i. **Soil pH**

The ideal pH range for crop growth is often between 6 and 7.5. The majority of soils would not experience abrupt pH shifts as a result of climate change such as rising temperatures, CO_2 fertilization, erratic precipitation, and atmospheric N deposition. But these climate change drivers will also impact the amount of organic matter present, the cycling of carbon and other nutrients, the amount of readily available water to plants, and consequently their productivity, which will impact the pH of the soil (Reth et al., 2005).

ii. **Ion Exchange Capacity**

In particular, the soil's ability to supply vital plant nutrients, primarily calcium, magnesium, and potassium, as well as the immobilization of potentially harmful cations, Al^{3+} and Mn^{3+}, make cation exchange capacity (CEC) a critical feature. Hence, it can serve as a helpful indicator of soil health, revealing a soil's ability to absorb both nutrients and chemicals and insecticides (Ross et al., 2008). Due to the fact that SOM breakdown and loss are ascribed to CEC in coarse-textured and low-activity clay soils, higher temperatures may cause these soils CEC to decrease (Davidson & Janssens, 2006). As a result, alkalinity from soil may be transported to waterways via increased base cation leaching in response to high and intense rainfall events.

iii. **Soil Respiration**

Given its positive correlation with SOM concentration, soil respiration is employed as a biological measure of soil health. It is the process by which aerobic organisms, particularly microbes, biologically oxidise organic materials to CO_2. One important connection between climate change and the global carbon cycle is soil respiration, namely its temperature response (Patil & Lamnganbi, 2018). Changes in the periodic timing of rainfall have a comparatively small impact on it.

3.3 *Impact on Plants Morpho-Biochemical and Physiological Processes*

i. **Plant Developmental Responses to Climate Change**

Physiological distinctions among plant cultivars and species growing in diverse environments are mostly caused by differences in developmental processes. Changes in temperature, precipitation, and air composition, among other aspects, are associated with climate change, which makes plant developmental adaption a dynamic process. Plant morphological characteristics, developmental processes, physiology, and molecular function are all impacted by these climate change effects (Nicotra et al., 2010).

The developmental responses to climate change factors described below will likely have significant impacts on plant functioning;

ii. **Impacts of Elevated Temperature to Plant Development and Morphology**

The way significantly distinct tissues and stages of growth react to increased temperatures varies greatly in plants. For instance, in rice, heats exceeding 25 °C have a negative impact on grain formation and yield, while maximal vegetative biomass is recorded at 33 °C (Gray & Brady, 2016). The way that yield responds to temperature is additionally influenced by species-specific characteristics that have to do with how the physiology and development of the plant alter. Different species are different temperature optimum and safe operating ranges for photosynthesis; those acclimated to steaming, desert regions have a more extreme temperature optimum than those adapted to more temperate or cold climates.

iii. **Leaf Developmental Responses**

Temperature is a major regulator of the growth of leaves. For example, in the tomato plant, a temperature range of 6–26 °C increases linearly with the rate of leaf start, growth, and duration (Granier et al., 2002). Similar to this, as temperature rises all through the agricultural the plant's vegetative growth and development, the rate at which new leaves are added increases. This process continues until the species-specific minimum temperature which varies from 26 °C for wheat to 37 °C for cotton and associated rate of leaf initiation are achieved. Humidity affects both the shape of leaves and their pace of emergence (Franklin, 2009). It is mainly unexplained what

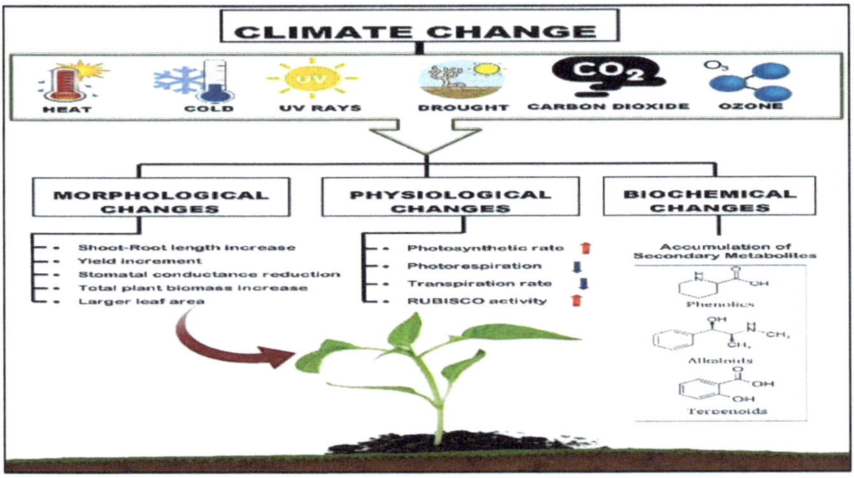

Fig. 1 Plant physiological and morpho-biochemical response to climate change

chemical mechanisms elevated temperatures use to control the shape of leaves as well as their pace of start and expansion (Fig. 1).

4 Soil Health and the Climate Connection

4.1 Climate-Smart Agricultural (CSA)

A long-term strategy to improve food security and agricultural sustainability in the face of climate change is climate-smart agriculture, or CSA. However, there is still a low global adoption rate for climate wise agriculture techniques. To increase the adoption of climate smart agriculture methods, it is imperative to comprehend the primary drivers behind their acceptance. Because agriculture produces so much food for humans, it is the foundational cornerstone of human society. The occurrence of regional famine and conflicts has successfully decreased over the past few decades due to higher agricultural yields internationally, which has made it easier to meet the second Sustainable Development Goal of the United Nations: zero hunger. But the quest of long-term human development is placing an intolerable amount of strain on the food supply. For example, it is expected that by 2050, there will be 9.8 billion people on the planet, a one-third increase in population (Waaswa et al., 2022).

Food output increased by over 70% between 2005 and 2050 (ELD Initiative 2015). Thus, increasing food production in the coming decades is a crucial and urgent challenge for people. The primary threats to food production are caused by climate change, which includes rising temperatures, irregular precipitation, and frequent weather variability (Abeysekara et al., 2023; Mirón et al., 2023). The European agenda prioritizes policies related to agriculture and climate change, acknowledging the urgency of taking prompt action. The Common agriculture Policy (CAP) and the European Green Deal are two of the policies that the European Union (EU) has established to give a road map for more resilient, sustainable, and lucrative agriculture systems (Long et al., 2016). The three primary pillars of climate smart agriculture (CSA) are lowering greenhouse gas (GHG) emissions, adapting to and strengthening resilience to climate change, and steadily raising agricultural production and incomes. Research demonstrates how climate change impedes food production worldwide in a number of ways. For example, the lack of water caused by global warming hinders crop growth from rooted to tasseling, which ultimately reduces crop yields (Salehi, 2022).

Additionally, climate change increases the likelihood of crop failure and sets off an insect and pest outbreak (Eigenbrode et al., 2023). Global crop productivity is greatly reduced by extended dry spells and degraded soil (Dietz et al., 2021). A variety of tactics, such as government involvement tax and subsidy (Perino & Schwickert, 2023), Modification of the industrial structure (Zheng et al., 2023). Climate-smart agriculture (CSA) techniques have been created and implemented to address climate change and guarantee global food security (Issahaku et al., 2021; Waaswa et al., 2022). CSA methods are directly drawn from and used to agriculture output. Hence, compared to other methods, CSA practices are easier to incorporate into food production processes and have been shown to be successful in lowering climate shocks and supplying the growing demand for food (Zizinga et al., 2022). According to Belay et al. (2023), the benefits of CSA methods on yield could be translated into increased farmer economic welfare and improved environmental quality.

4.2 *Plant Soil Interaction*

One key factor influencing the dynamics and structure of plant communities is soil–plant interaction. Plant-Soil interaction is the process by which plants affect the biotic and abiotic conditions of the soil, which can then affect how other conspecific and heterospecific plants perform (Bever, 1994). Ecologists have carried out a great deal of research since the early 1990s and discovered that PSF processes have a major impact on the abundance of plant species (Reinhart et al., 2021). Numerous soil microbial communities, mostly pathogens and mutualists, contribute to the strength of plant-soil interaction, and these groups typically exhibit distinct geographic patterns. Despite the fact that some types of soil microbes such as pathogens and arbuscular mycorrhizal fungi have less diversity or abundance at higher latitudes (Delgado-Baquerizo et al., 2020). The same latitudinal trends are not seen in other groups (ectomycorrhizal fungi) (Tedersoo et al., 2018).

The biogeographic decoupling between these significant plant-dependent microbial taxa raises the possibility that PSF and latitude are not as strongly correlated as the Janzen-Connell hypothesis predicts. Therefore, even though earlier meta-analyses did a good job of evaluating the variation of PSF (Xia et al., 2021). Clarification is required regarding the PSF's latitudinal gradient. Previous research indicates that climates have an impact on the regional variations in the diversity and abundance of soil microbial communities that are significant for PSF (Hendershot et al., 2017). Increased drought in particular can lead to a rise in mycorrhizal fungal abundance while decreasing soil-borne pathogen abundance. This could result in a more favorable PSF at drier areas (de Vries et al., 2023). Delgado-Baquerizo et al. (2020) discovered in their previous experimental research that rising temperatures may also lead to an increase in soil-borne pathogen abundance, which could result in a more unfavorable plant-soil interaction at warmer locations.

4.3 Precision Agriculture

With the least amount of disturbance to the agroecosystem, precision agriculture (PA) maximizes the potential of natural, human, and mechanical resources (Zhang et al., 2002a, b). Precision agriculture enables farmers to cut expenses and maximize the potential of their land, particularly in the current climate where costs for seeds, herbicides, and pesticides are rising. Rapid population increase and competitiveness in international markets are challenges facing global agriculture. Techniques for precision agriculture seem to be one approach to solving these problems (Hakkim et al., 2016).

Among the ways to guarantee global food security is through PA. Climate change, urbanization, and the growing demand for food production at an unsustainable rate are all contributing to the loss of arable land, which compromises sustainability in terms of both productivity and environmental impact (Sharma et al., 2020). Multiple application regions were included in some (Weiss et al., 2020). Precision agriculture employs a variety of technologies, including remote sensing (RS), which is primarily used to monitor crop growth and soil properties (Mulla, 2013).

Farmers may improve agricultural productivity, make smarter decisions, and make the most use of their resources by merging data from ground sensors. Remote sensing technologies, which gather information from sensors on satellites or unmanned aerial vehicles (UAV), can be used to find and map weeds in crops (Triantafyllou et al., 2019). Thanks to technological advancements, accessible and reasonably priced drones that can extract pertinent data from the photos taken by the sensor are now widely accessible (Nex et al., 2022). To assess vegetation-related biophysical parameters, remedial indicators have been investigated (Xue & Su, 2017).

4.4 Regenerative Agriculture

The agriculture sector is in a challenging position due to the rising need for food security, climatic conditions, and governmental regulations. Additionally, the use of chemical fertilizers in agriculture results in high water usage and greenhouse gas emissions. In addition to having many positive effects on the environment, the post-COVID-19 scenario has given us the chance to further improve our agricultural practices through a variety of technical processes. Value addition, processing, and marketing are important elements of these interventions. Furthermore, wildfires have impacted ecosystems globally for millions of years and they still do (Cetin et al., 2022). Agricultural practices can be directly impacted by wildfires, as they can damage crops in their path. Even the outermost layers of crops, which are invisible to the unaffected, can sustain harm from smoke and heat. Specific types of crops can provide fuel for fires. An agricultural worker's health may suffer from breathing in smoke and other airborne toxins.

Therefore, sustainability is essential to prevent agricultural loss from natural disasters (Degerli & Çetin, 2022). Regenerative agriculture (RA), in the context of agricultural sustainability, integrates many farming techniques to restore and maximize the potential of degraded landscapes (Gordon et al., 2023). Some of the core tenets of RA include preserving soil cover, minimizing soil disturbance, protecting living soil roots all year round, increasing biodiversity, incorporating livestock, and minimizing or completely eliminating the use of synthetic compounds (such as pesticides and fertilizers) (Ahmed et al., 2021).

4.5 *Organic Agriculture*

Using organic fertilizers and fertility-aiding microbial communities are two methods of managing soil fertility in organic agriculture (Silveira et al., 2021). The type of fertilizer used, the physicochemical characteristics of the soil, and the biotic community that already exists within the soil all affect the quality and amount of nutrients released from an organic fertilizer. Nevertheless, farmers are unable to adequately nutrition crops using organic methods due to the present scarcity and high expense of organic fertilizers (Bergstrand, 2022).

One of the most studied and well-known topics in agronomy is the role of Arbuscular mycorrhiza fungi (AMF) in promoting crop yield and supporting the sustainability of soil fertility (Giovannini et al., 2021). A specific class of organisms known as Arbuscular Mycorrhiza Fungi (AMF) helps plants satisfy their need for water and minerals, particularly when the plants are developing in environments lacking in these nutrients (Nanjundappa et al., 2019). In soil from organic farms as opposed to chemical farms, arbuscular mycorrhiza fungi are more abundant and important (Huo et al., 2022). Furthermore, most beneficial AMF, particularly for sensitive species, is not lodged in chemicalzed farms (Pontes-da-Silva et al., 2018). They support plants in their water relations, enhance soil structure, and increase plant resistance to heavy metals (Riaz et al., 2021).

4.6 Carbon Farming

In an effort to mitigate the effects of climate change, carbon farming has gained prominence in both national and international frameworks. In a number of jurisdictions, including the Innovative methods for managing soil and cropping systems are being created to address the increased levels of CO_2 in the atmosphere while simultaneously enhancing soil quality and water efficiency. The amount of organic matter in the soil's composition and its ability to retain water are impacted by different management techniques (Aranda et al., 2011).

However, there are several obstacles in the way of landholders who want to participate in carbon farming, such as a lack of understanding of the regulations and market dynamics surrounding the program and difficulties in measuring carbon. There are various applications for the word "carbon farming." In order to mitigate climate change, the European Union favors a wide definition that includes managing carbon stocks, flows, and greenhouse gas fluxes. Effective planning solutions, however, depend on meeting human demands while simultaneously safeguarding environmental resources. The goal of soil quality study is to comprehend how to manage soil to maximize its natural benefits. As a result, it becomes essential to identify the variables influencing the health of the soil, of which organic matter is one of the most crucial (Scotti et al., 2015). Organic matter is prevalent in most agricultural contexts and is easily managed by land management operations. As a result of organic matter's ability to hold onto more water and fortify the soil structure (Tisdall, 2020), agricultural output is enhanced and the likelihood of diseases and droughts is decreased (Altieri et al., 2015). Moreover, to reduce the amount of CO2 in the atmosphere, farming practices that add organic matter to the soil are required (Hartmann et al., 2015a, b). Additionally, it has been shown that maintaining and replenishing soil carbon requires actions connected to soil management. However, due to soil erosion and breakdown, a significant carbon deficit exists in many agricultural fields, but not all of them (Xiao et al., 2015). It is commonly acknowledged that different governments take all reasonable steps to encourage environmentally friendly farming practices in order to preserve soil carbon. Another important element affecting the amount of carbon in the soil is the density of forests. However, deforestation has a significant impact on river flow and patterns of land use (Tan et al., 2015).

5 Approaches to Combat Climate Changes to Enhancing Crop Resilience

5.1 Cultural Methodologies

Through the promotion of sustainable agricultural practices, cultural methodologies which have their roots in traditional knowledge are essential in combating against climate change and in increasing crop resilience. Crop rotation and intercropping are examples of indigenous practices that support soil fertility and biodiversity, both of which are essential for creating resilience against climate-related challenges like floods and droughts (Altieri and Nicholls, 2020). These techniques also make use of regional varieties of crops that are naturally suitable for particular climates, improving the crops' ability to adapt to severe weather conditions (Lipper et al., 2014). The goal of conservation tillage, an agricultural management strategy, is to reduce the frequency or intensity of tillage operations in an effort to support specific environmental and economic advantages. Using this method, crop residue or organic residue covers at least 30% of the soil surface after planting (Dinnes, 2004). Conversely, crop rotation is the process of repeatedly cultivating a variety of plant species on the same plot of land in accordance with a predetermined crop succession order and duration (Stein and Steinmann, 2018).

For sustainable agriculture to be successful, choosing the right crop rotation is crucial. Examples of such rotations include sugarcane-wheat and cotton-gram. Any rotation must include legumes, which should make about 30% to 50% of the cropland (Stein and Steinmann, 2018). In regions that are vulnerable to drought, choosing the optimal crop combinations for inter-cropping is essential. For instance, the traditional inter-cropping system of grain and legumes helps the crops by enhancing soil fertility, raising soil carbon and water infiltration, and significantly increasing soil moisture during dry spells, particularly in sandy soils (Renwick et al., 2020). By allowing the investigation of deep soil and subsoil, mixing plants with shallow and deep roots such as bean–maize, sorghum red clover, and maize grass pea improves the utilization of the available soil moisture (Miyazawa et al., 2010). These root combinations facilitate bio-drilling, which improves soil structure, raises soil organic matter, and encourages soil aggregation, all of which improve soil water storage and availability in the root zone. Bio-drilling is a process in which lengthy tap roots remain in the soil, creating channels for water drainage (Marques et al., 2020).

Modifying the cropping schedule and pattern is regarded as a sustainable way to increase crop productivity while reducing the detrimental effects of climate variability. Cropping pattern modification involves substituting out crops that are negatively impacting soil health with crops that can enhance soil health in the prevailing environmental conditions. For example, restorative crops like potatoes, sorghum, sugarcane, are substituted for intensive crops (Dosdall and Stevenson, 2005). The health of the soil is declining due to the usage of chemical fertilizers, which has negative effects on plants, animals, and people. Because nitrogenous fertilizers contaminate groundwater, they can cause diseases in both humans and animals. Improved soil chemical characteristics are just one of the many tasks that organic fertilizers can carry out that synthetic fertilizers cannot (Muthaura et al., 2010). Because synthetic fertilizers are expensive, many farmers have resorted to using a lot of organic sources of nutrients for crop production, such as green manures, animal dung, composted materials, and household trash (Fening et al., 2010). The primary functions of organic resources in soil are the steady increase of soil organic matter and the provision of nutrients (Bhattacharyya et al., 2010).

In light of current climate change scenarios, it is imperative to promote the utilization of organic nutrient sources to preserve the ecosystem and enhance crop yields.

Soil carbon building contributes to resilience by reducing atmosphere greenhouse gas concentrations and increasing soil water-holding capacity. Summer mungbean is one type of green-leaf manuring, several other methods give the soil a substantial amount of biomass and aid in increasing the soil's carbon content (Bindu and Manan, 2018). Moreover, it has been demonstrated that agro-ecological methods, which combine traditional knowledge with modern science, enhance soil health and water retention which are two essential elements in lessening the effects of climate change on agriculture (Canton et al., 2021). Communities may create resilient and sustainable farming systems that are better prepared to handle the difficulties of a changing climate by conserving and integrating these traditional techniques.

5.2 *Conventional Agricultural Techniques*

In order to mitigate climate change and increase crop resilience, conventional agricultural practices are becoming more and more important. Programs like climate-based agro-cautions, crop and variety selection, productive cropping techniques, harvesting water to preserve water resources, custom leasing of farm equipment, contingency planning, etc. can be implemented with the aim of minimizing adverse effects related to climate changes (Debangshi, 2022). Moreover, using plant varieties and harvests resistant to climate change has been recommended as a way for farmers to adapt to or manage climate change (Jaiswal and Devashali, 2022). These cultivars and cultivar combinations are more resilient to both biotic and abiotic stresses (Acevedo et al., 2020). The use of drought-tolerant cultivars may be one of the most important strategies for ensuring food security in places that suffer from drought (Dar et al., 2020). Balanced fertilization maximizes plant development, resulting in extremely efficient nutrient usage and less environmental effect. Thus, nitrogen fertilizer application should be balanced to minimize N_2O-N emissions (Sikka et al., 2018) (Fig. 2).

Consequently, adjusting N supply parameters such fertilizer type, amount, timing, and application methods is the most straightforward way to achieve appreciable reductions in N_2O emissions while preserving crop yields (Pathak et al., 2010). An integrated farming system (IFS) increases farmers' resilience to climate change by growing a variety of crops on the same ground and using resources responsibly. IFS not only manages the farm more effectively but also lessens output dependence, lowering risk. IFS not only saves the environment but also gains from business synergies (Karri and Nalluri, 2024).

Through genetic engineering and advanced breeding methods, climate-resilient cultivars are created that enable crops to withstand harsh weather conditions like heat waves and droughts, ensuring consistent yields in a changing climate (Zhang et al., 2017). Using modern technology like drones and satellite imagery, precision agriculture helps maximize the use of resources like water and fertilizers, increasing

Fig. 2 Demonstration of various mitigation strategies towards conventional agriculture in-terms of climate change

productivity and lessening the environmental effect of traditional farming methods (Chlingaryan et al., 2018). Even if these conventional approaches work well, they need to be managed carefully in order to balance sustainability and production in the face of climate change.

5.3 *Genetic Engineered Plants for Stress Tolerance*

Genetic modification in agriculture has emerged as a promising technique for tackling the concerns generated by climate change and increase crop resilience (Jacobsen et al., 2013). By including genes that offer resistance to abiotic stresses like heat, salinity, drought, and abnormally low or high temperatures, scientists can employ this technique to increase crop resilience (Bacha and Iqbal, 2023).

The use of transgenic usually results in a 22% increase in agricultural yields, which boosts farmer income by approximately 68%, according to a global meta-analysis on the consequences of adopting transgenic crops (Klümper and Qaim, 2014). Furthermore, genome editing methods like CRISPR-Cas9 have revolutionized the field of cultivating crops. CRISPR-Cas9 allows plant breeders to precisely edit specific genes, allowing for targeted modifications to the plant genome (Mishra et al., 2024). The CRISPR/Cas alteration of the ethylene regulating factor (ERF) and stress/ABA-activated protein kinase 2 (SAPK2) genomes has greatly increased rice's resistance to abiotic stress (Debbarma et al., 2019).

CRISPR/Cas technology has demonstrated its versatility in genome editing across a wide range of crops, including maize, rice, tomato, soybean, sorghum, flax, camelina, cotton, rapeseed, lettuce, potato, cucumber, watermelon, grapefruit, apple, and oranges (Lin et al., 2020). By enabling precise and efficient alterations to plant genomes, this powerful tool opens up new possibilities for agricultural progress (Chilcoat et al., 2017). via modifying the OsDREB gene in 2016 and the ARGOS8 gene in 2017, scientists demonstrated how rice's resilience to drought may be enhanced via CRISPR/Cas9. While ARGOS8 gene editing provided a considerable improvement in grain output under drought conditions without losing yield during perfect growth conditions, OsDREB gene editing produced inheritable changes that boosted the plant's resilience to abiotic stress (Shi et al., 2017).

When developing genetically modified crops, a number of intricate steps must be followed to ensure that the desired traits are successfully introduced and carefully evaluated before being on sale. To guarantee that the intended features are effectively introduced and thoroughly assessed prior to being commercialized, a variety of complex procedures must be followed when creating genetically modified crops (Jankowicz-Cieslak et al., 2017). Once the target genes have been identified, researchers can employ recombinant DNA technology to introduce the desired gene at specific sites in the plant's DNA (Irwin, 2001).Researchers employ a variety of laboratory techniques, including Southern blotting, PCR, and DNA sequencing, to confirm the successful genetic change. According to Ahmed (2002), these assays verify that the inserted gene is correct and exists in the plant's DNA (Fig. 3).

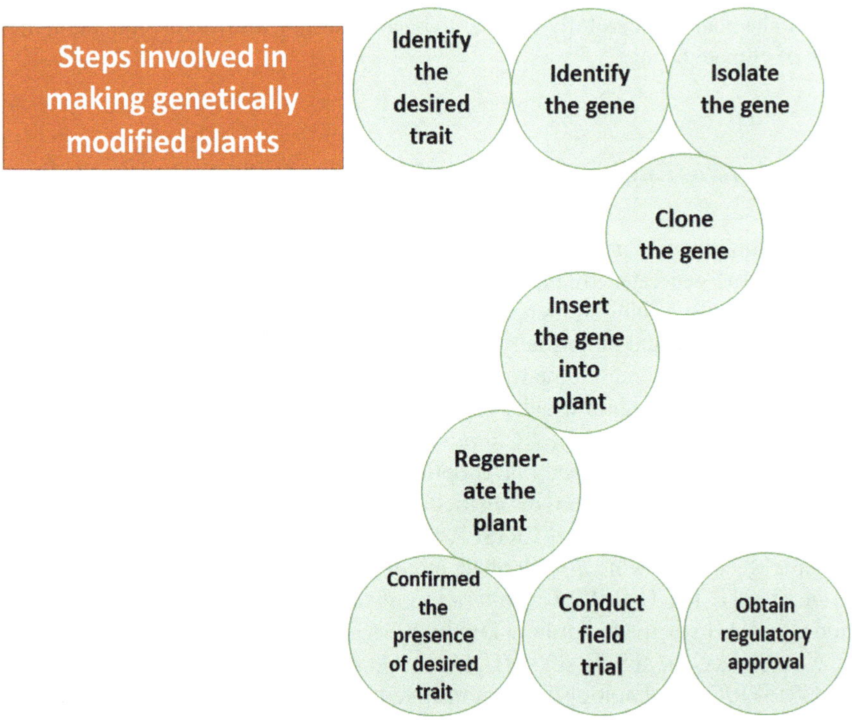

Fig. 3 Shows the energetically modification steps implicated in genetic engineered plants for stress tolerance

The adoption of measures to improve climate resilience is required due to the increasing threat that climate change poses to agricultural productivity. Increasing genetic variety in crops is another practical way to improve resilience (Garland and Curry, 2022). The genetic variety of crops has increased dramatically as a result of advances in plant biotechnology, especially with regard to genetically modified (GM) crops (Sainger et al., 2015). This improvement gives crops the capacity to adjust to changing climates and protects produce from agriculture from climate change (Garland and Curry, 2022).

6 Mitigation and Adaptation to Climate Change in Agriculture

There is an urgent need to increase the climate change resilience of crops, while simultaneously enhancing crop yields in a sustainable manner to ensure future food security. With the increase of the Earth's temperature, the climate undergoes severe alterations and becomes abiotically stressful. Environmental changes are very damaging and pose various threats to naturally prevailing crop species (Espeland and Kettenring, 2018). Using crop types resistant to pressures associated with climate change is one possible adaptation strategy. Various methods of breeding have been employed to create new resilient crops, if not to enhance or increase the products' capacity to endure adverse weather patterns imposed about by changes in the climate. The development of crops that are adaptable to climate change is one of the most environmentally friendly ways to lessen its impact on agriculture. vegetation that are resistant to harsh weather as a result of climate change.

6.1 Escape Mechanism

Crops have developed many avoid strategies to assist them cope with abiotic challenges and increase their chances of surviving and reproducing. Crops use such mechanisms to mitigate or even completely escape the negative impacts of harsh environmental conditions. Crops have the ability to alter their phenotype, which may allow them to evade drought conditions. Plants can often get away from their surroundings by shortening their life cycle, growing in different seasons, or proliferating before the conditions turn adverse (Abid et al., 2018). A simplified life cycle might lead to in an early reproductive phase and the potential for water deficit evasion. By reducing the length of their vegetative stage, early flowering enables plants to start reproductive growth before the terminal stress (Bodner et al., 2015).

6.2 Avoidance Mechanism

Environmental stresses induce morphological and anatomical changes in crops. These changes improve the capacity of crops to tolerate demanding circumstances and protect their infrastructure from disruption. Plants have undergone a wide range of anatomical modifications at the organ and organizational levels. In response to environmental stresses, alterations in root, xylem, and leaf morphology have been noted (Farooq et al., 2009). Changes in the circumstances for cell differentiation and a reduction in cell elongation are the primary causes of morphological modification in plant organs such as roots, leading to a range of anatomical features (Akhtar and Nazir, 2013).To minimize water loss, plants usually control stomatal transpiration and sustain soil water uptake using a large and efficient root structure. Robust, profound roots enable plants to extract water from a substantial amount of the soil as well as distant beyond the foliage (Franco et al., 2011).

6.3 Tolerance Mechanism

The tolerance of plants can be described as a plant's capacity to store nutrients as well as grow during its vegetative stage, as defined by multiple sources. These mechanisms involve alterations in both physiology and biochemistry (Lisar et al., 2012). Plants that are tolerant to drought do so through a variety of mechanisms, such as the build-up of suitable solutes and osmotic adaptation, the induction of an antioxidant system, changes to metabolism, a reduction in the number of roots to shoots, and closing of stomata (Gupta and Huang, 2014).

When crops are under water stress, an accumulation of active substances like sugars or amino acids helps them adjust their osmotic circumstances and decrease their osmotic potential. Plants may withstand dry conditions by efficiently absorbing water and preserving cell turgor pressure through the process of osmotic adjustment (Yang et al. 2021). An additional trait of plants that use tolerance as a mechanism for adaptation is the tenacity of their cell membrane. Stable cell membranes prevent the leakage of water as well as electrolytes. Cuticle-layered crops have been demonstrated to maintain leaf turgor despite strain for extended periods of time (Ma et al., 2020).

6.4 Molecular Mechanism

Crops can use their DNA as well as sensory systems to build molecular defenses against abiotic stresses. Abiotic stresses can affect several different cell compartments, such as the nuclei, cytoplasm, mitochondria, endoplasmic reticulum, peroxisomes, chloroplasts, along with cell wall (CW) (Praveen et al., 2023). This could initiate chemical processes. Such stress receptors send signals to secondary messengers and downstream regulatory proteins like ROS, Ca^{2+}, and protein kinases. Abiotic pressure also initiates a cascade of responses including several tiers of regulation, signal transmission, and stress perception. Plants have therefore developed defense mechanisms to alter their developmental patterns in order to survive and reproduce in hostile settings (Zhang et al., 2023).

7 Climate-Resilient Crops

To cope with climate change, farmers need to modify production and farm management practices, such as adjusting planting time, supplementing irrigation (when possible), inter-cropping, adopting conservation agriculture, accessing short- and long-term crop and seed storage infrastructure, and changing crops or planting more climate-resilient crop varieties.

.Farmers should embrace climate-resilient crops as one of the most important and proactive steps they can take to address food insecurity brought on by unpredictable weather trends. This goes particularly well for nations that depend on their own farming industries for food security. Climate-resilient crops and crop varieties have enhanced tolerance to biotic and abiotic stresses (Dhankher and Foyer, 2018).

In Sub-Saharan Africa, pearl millet (*Pennisetum glaucum* L.,) is a plant that is generally accepted. Its ability to withstand extended periods of high temperatures and soil water deficiencies is a noteworthy benefit. According to Hossain et al. (2022), it can tolerate drought and intense heat during its reproductive period and can thrive in regions with higher salinity levels. In addition to being an integral fodder crop in lots of countries across the world, sorghum is a staple diet for people in Asia and Africa. Because of it's capacity to withstand soil drought and high air temperatures, it is a favored option in semi-arid climates and locations with erratic rainfall. It can also be grown on marginal farms with low input (Dzienis et al., 2023).

Promised as an era crop, amaranth (*Amaranthus* spp. L.) is a pseudocereal that has been known in Central and South America since ancient times. According to Graziano et al. (2022) there are three domesticated species: A. cruentus L., A. caudatus L., and A. hypochondriacus L. Amaranth is particularly suited for growth in nutrient-poor soils because to its great genetic diversity and wide variability in phenotypic characteristics. It shows endurance against elevated air temperatures, high soil salinity, and soil water deficiency. Moreover, pests and disease do not pose a hazard to amaranth farming (Goel et al., 2023).

Camelina sativa L., a species of the Brassicaceae family, is prized for its adaptability to a wide range of cultivating constraints. It expands in a variety of climatic and soil conditions, with the exception of heavy clay and organic soil, and it produces well even in situations where other crops fail (Neupane et al., 2022).

8 Conclusion

Food security is threatened by climate change in many parts of the world. Enhancing crops that support the creation of climate-smart agriculture projects is therefore imperative in the modern day. Reduced crop yields are a direct result of the extreme weather events that come with climate change. Farmers have an option to adapt to climate change: cultivars and crops that are climate-resilient. In new environmental conditions, these crops yield steadily, avoiding crop failure and a fall in productivity. For breeders, "upgrading" crops has become a major undertaking that goes beyond simple financial gain to include issues of national and international security. Careful scientific backing is necessary for plant breeding. Technological developments have accelerated the rate at which basic scientific discoveries have been made in recent years.

Future agricultural issues must be anticipated and identified from both local and global viewpoints in order to develop crops that are climate-resilient. It is essential to comprehend how semi-domesticated plants, wild relatives, and domesticated plants are affected by multi factorial stress. The spread of new crops or the creation of enhanced cultivars of staple crops are two ways that global plant production must diversify in order to increase food security.

References

Abbas, A., Khan, S., Hussain, N., Hanjra, M. A., & Akbar, S. (2013). Characterizing soil salinity in irrigated agriculture using a remote sensing approach. *Physics and Chemistry of the Earth, Parts a/b/c, 55*, 43–52.

Abd-elwahed, M. S. (2018). Influence of long-term wastewater irrigation on soil quality and its spatial distribution. *Annals of Agricultural Sciences, 63*, 191–199.

Abeysekara, W. C. S. M., Siriwardana, M., & Meng, S. (2023). Economic consequences of climate change impacts on the agricultural sector of south Asia: A case study of Sri Lanka. *Economic Analysis and Policy, 77*, 435–450.

Ainsworth, E. A.,& Long, S. P. (2005). What have we learned from 15 years of free-air co_2 enrichment (face)? A meta-analytic review of the responses of photosynthesis, canopy properties and plant production to rising co_2. *New Phytologist, 165*, 351–372.

Alongi, D. M. (2008). Mangrove forests: Resilience, protection from tsunamis and responses to global climate change. *Estuarine Coastal and Shelf Science, 76*(1):1–13.

Altieri, M. A., Nicholls, C. I., Henao, A., & Lana, M. A. (2015). Agroecology and the design of climate change-resilient farming systems. *Agronomy for Sustainable Development, 35*, 869–890.

Amadu, F. O., Mcnamara P. E., & Davis, K. E. (2021). Soil health and grain yield impacts of climate resilient agriculture projects: Evidence from Southern Malawi. *Agricultural Systems, 193*, 103230.

Amicarelli, A., Manenti, S., & Paggi, M. (2021). SPH modelling of dam-break floods with damage assessment to electrical substations. *International Journal of Computational Fluid Dynamics, 35*(1-2), 3–21.

Anderson, J. M. (1992). Responses of soils to climate change. *Advances in Ecological Research*, 163–210.

Aranda, V., Ayora-cañada, M. J., Dominguez-vidal, A., Martin-garcia, J. M., Calero, J., Delgado, R., Verdejo, T., & González-vila, F. J. (2011). Effect of soil type and management (organic vs. Conventional) on soil organic matter quality in olive groves in a semi-arid environment in sierra mágina natural park (S Spain). *Geoderma, 164*, 54–63.

Arias, M. E., González-pérez, J. A., González-vila, F. J., & Ball, A. S. (2005). Soil health: A new challenge for microbiologists and chemists. *International Microbiology, 8*, 13–21.

Asadieh, B., & Krakauer, N. Y. (2017). Global change in streamflow extremes under climate change over the 21st century. *Hydrology and Earth System Sciences, 21*, 5863–5874.

Backer, R., Rokem, J. S., Ilangumaran, G., Lamont, J., Praslickova, D., Ricci, E., Subramanian, S., & Smith, D. L. (2018). Plant growth-promoting rhizobacteria: Context mechanisms of action and roadmap to commercialization of biostimulants for sustainable agriculture. Front. *Plant science, 9*, 1473

Bannari, A., & Al-ali, Z. M. (2020). Assessing climate change impact on soil salinity dynamics between 1987–2017 in arid landscape using landsat TM, ETM + and OLI data. *Remote sensing, 12*, 2794.

Barrutia, O., Garbisu, C., Epelde, L., Sampedro, M. C., Goicolea, M. A., & Becerril, J. M. (2011). Plant tolerance to diesel minimizes its impact on soil microbial characteristics during rhizoremediation of diesel-contaminated soils. *Science of the Total Environment, 409*(19), 4087–4093.

Bates, B., Kundzewicz, Z., & Wu, S. (2008). Climate change and water intergovernmental panel on climate change secretariat.

Beasley, C. L., Zhang, Z. J., Patten, I., & Reynolds, G. P. (2002). Selective deficits in prefrontal cortical gabaergic neurons in schizophrenia defined by the presence of calcium-binding proteins. *Biological Psychiatry, 52*, 708–715.

Belay, A., Mirzabaev, A., Recha, J. W., Oludhe, C., Osano, P. M., Berhane, Z., Olaka, L. A., Tegegne, Y. T., Demissie, T., Mutsami, C., & Solomon, D. (2023). Does climate-smart agriculture improve household income and food security? Evidence from Southern Ethiopia. *Journal of Environment Sustainability*, 1–28.

Benbi, D. K.,& Kaur R. (2009). Modeling soil processes in relation to climate change. *Journal of the Indian society of Soil sciences, 57*, 433–444.

Bergstrand, S., O'brien, E. M., Coucoravas, C., Hrossova, D., Peirasmaki, D., Schmidli, S., Dhanjal, S., Pederiva, C., Siggens, L., Mortusewicz, O., & O'rourke, J. J. (2022). Small cajal body-associated RNA 2 (scaRNA2) regulates DNA repair pathway choice by inhibiting DNA-PK. *Nature Communications, 13*, 1015

Bever, J. D. (1994). Feeback between plants and their soil communities in an old field community. *Ecology, 75*, 1965–1977.

Bhattacharyya, S., Morrison, P. D., Fusar-Poli, P., Martin-Santos, R., Borgwardt, S., Winton-Brown, T., Nosarti, C., O'Carroll, C. M., Seal, M., Allen, P., Mehta, M. A., Stone, J. M., Tunstal, N., Giampietro, V., Kapur, S., Murray, R. M., Zuardi, A. W., Crippa, J. A., Atakan, Z., & McGuire, P. K. (2010). Opposite effects of Δ-9-tetrahydrocannabinol and cannabidiol on human brain function and psychopathology. *Neuropsychopharmacology, 35*(3), 764–774.

Bot, A., & Benites J (2005). *The importance of soil organic matter: Key to drought-resistant soil and sustained food production.* Food & agriculture organization.

Buol, S. W., Sanchez P. A., Weed S. B., & Kimble J. M. (1990). Predicted impact of climatic warming on soil properties and use. *Impact carbon dioxide, trace gases, climate Change on Global Agriculture, 53*, 71–82.

Campbell, J. L., Rustad, L. E., Boyer, E. W., Christopher, S. F., Driscoll, C. T., Fernandez I. J., Groffman, P. M., Houle, D., Kiekbusch, j., & Magill, A. H. (2009). Consequences of climate change for biogeochemical cycling in forests of northeastern North America. *Canadian Journal Forest Research, 39*, 264–284.

Canton, M., Sánchez-Rodríguez, R., Spera, I., Venegas, F. C., Favia, M., Viola, A., & Castegna, A. (2021). Reactive oxygen species in macrophages: Sources and targets. *Frontiers in Immunology, 12*, Article 734229.

Cetin, Y., Aydinlik, S., Gungor, A., Kan, T., Avsar, T., & Durdagi, S. (2022). Review on in silico methods, high-throughput screening techniques and cell culture based in vitro assays for sars-CoV-2. *Current Medicinal Chemistry, 29*, 5925–5948.

Chamberlain, E. J. (1983). July frost heave of saline soils. In *proceedings of the 4th international conference on permafrost,* Vol. 121, p. 126.

Chaudhary, S., Kanwar, R. K., Sehgal, A., Cahill, D. M., Barrow, C. J., Sehgal, R., & Kanwar, J. R. (2017). Progress on Azadirachta indica based biopesticides in replacing synthetic toxic pesticides. *Frontiers in Plant Science, 8*, 610.

Chaudhary, R., Bakhshi, P., & Gupta, H. (2020a). Volatility in international stock markets: An empirical study during COVID-19. *Journal of Risk and Financial Management, 13*(9), 208.

Chaudhary, S., Devi, P., Bhardwaj, A., Jha, U. C., Sharma, K. D., Prasad, P. V. V., Siddique, K. H. M, Bindumadhava, H., Kumar, S., & Nayyar, H. (2020b). Identification and characterization of contrasting genotypes cultivars for developing heat tolerance in agricultural crops: Current status and prospects. *Frontiers Plant Science, 11*, 587264.

Chen, B., Feder, M. E., Kang, l. (2018). Evolution of heat-shock protein expression underlying adaptive responses to environmental stress. *Molecular Ecology Resources, 27*, 3040–3054.

Chlingaryan, A., Sukkarieh, S., & Whelan, B. (2018). Machine learning approaches for crop yield prediction and nitrogen status estimation in precision agriculture: A review. *Computers and Electronics in Agriculture, 151*, 61–69.

Corwin, D. L. (2021). Climate change impacts on soil salinity in agricultural areas. *European Journal of Soil Science, 72*, 842–862.

Crafts-brandner, S. J., & Salvucci, M. E. (2000). Rubisco activase constrains the photosynthetic potential of leaves at high temperature and co_2. *Proceedings of the National Academy of Sciences of the United Stated of America, 97*, 13430–13435.

Crowther, T. W., Van den hoogen, J., Wan, J., Mayes, M. A., Keiser, A. D., Mo, l., Averill, C., & Maynard, D. S. (2019). *The global soil community and its influence on biogeochemistry. Science,* (vol 365, p. Eaav0550).

Cuculeanu, V., Tuinea, P., & Bălteanu, D. (2002). Climate change impacts in Romania: Vulnerability and adaptation options. *Geojournal, 57*, 203–209.

Dalal, R. C., & Moloney, D. (2000). Sustainability indicators of soil health and biodiversity. *Management Sustainable Ecosystem Center Conservation Biology*, 10108.

Dar, M. H., Waza, S. A., Shukla, S., Zaidi, N. W., Nayak, S., Hossain, M., & Singh, U. S. (2020). Drought tolerant rice for ensuring food security in Eastern India. *Sustainability, 12*(6), 2214.

Daryanto, S., Wang, L., & Jacinthe, P. A. (2015). Global synthesis of drought effects on food legume production. *Plos one, 10*, E0127401.

Das, I., Dutta, D., & Rakshit, A. (2016). Potential effects of climate change on soil properties: A review.

Davidson, E. A., and Janssens, I. A. (2006). Temperature sensitivity of soil carbon decomposition and feedbacks to climate change. *Nature, 440*, 165–173.

Debangshi, U. (2022). Hydroponics Rice Nursery: A Novel Approach to Rice Cultivation in India. *Journal of Research in Agriculture and Animal Science, 9*(4), 60–63.

Degerli, B., & Çetin, M. (2022). Evaluation from rural to urban scale for the effect of NDVI-NDBI indices on land surface temperature in Samsun Türkiye. *Turkish Journal of Agriculture-Food Science and Technology, 10*, 2446–2452.

Delgado-Baquerizo, M., Guerra, C. A., Cano-Díaz, C., Egidi, E., Wang, J.-T., Eisenhauer, N., Singh, B. K., & Maestre, F. T. (2020). The proportion of soil-borne pathogens increases with warming at the global scale. *Nature Climate Change, 10*(6), 550–554.

Dermody, O., Long, S. P., & Delucia, E. H. (2006). How does elevated co_2 or ozone affect the leaf-area index of soybean when applied independently? *New Phytologist, 169*, 145–155.

Deryng, D., Conway, D., Ramankutty, N., Price, J., & Warren, R. (2014). Global crop yield response to extreme heat stress under multiple climate change futures. *Environmental Research Letters, 9*(3), Article 034011.

De Vries, B. J. (2023). *Sustainability Science*. Cambridge university press.

Dewals, B., Kitsikoudis, V., Mejía-Morales, M. A., Archambeau, P., Mignot, E., Proust, S., Erpicum, S., Pirotton, M., & Paquier, A. (2023). Can the 2D shallow water equations model flow intrusion into buildings during urban floods? *Journal of Hydrology, 619*, Article 129231.

Dietz, S., Rising, J., Stoerk, T., & Wagner, G. (2021). Economic impacts of tipping points in the climate system. *Proceedings of the National Academy of Sciences, 118*, 2103081118.

Diffenbaugh, N. S., Singh, D., Mankin, J. S., Horton, D. E., Swain, D. L., Touma, D., et al. (2017). Quantifying the influence of global warming on unprecedented extreme climate events. *Proceedings of the National Academy of Sciences, 114*(19), 4881–4886.

Dinnes, D. L. (2004). *Assessments of practices to reduce nitrogen and phosphorus nonpoint source pollution of Iowa's surface waters* (p. 366). USDA-ARS, National Soil Tilth Laboratory.

Donat, M. G., Angélil, O., & Ukkola, A. M. (2019). Intensification of precipitation extremes in the world's humid and water-limited regions. *Environmental Research Letters, 14*, 065003.

Donelson, J. M., Gaitan-espitia, J. D., Hobday, A. J., Mokany, K., Andrew, S. C., Boulter, S., Cook, C. N., Dickson, F., Macgregor, N. A., Mitchell, N. J., Pickup, M., & Fox, R. J. (2023). Putting plasticity into practice for effective conservation actions under climate change. *Nature Climate Change, 13*, 632–647.

Dong, B., Xia, J., Li, Q., & Zhou, M. (2022a). Risk assessment for people and vehicles in an extreme urban flood: Case study of the "7.20" flood event in Zhengzhou, China. *International Journal od Disaster Risk Reduction, 80*, 103205.

Dong, X., Bao, J., Chen, D., Zhang, W., Yu, N., Yuan, L & Guo, B. (2022b). Cswin transformer: A general vision transformer backbone with cross-shaped windows. In *Proceedings of the IEEE/CVF conference on computer vision and pattern recognition* (pp. 12124–12134).

Doran, J. W., & Zeiss, M. R. (2000). Soil health and sustainability: Managing the biotic component of soil quality. *Applied Soil Ecology, 15*, 3–11.

Drobinski, P., Alonzo, B., Bastin, S., Silva ND & Muller, C. (2016), Scaling of precipitation extremes with temperature in the French Mediterranean region: What explains the hook shape? *Journal of Geophysical Research, 121*, 3100–3119.

Duffy, K. A., Schwalm, C. R., Arcus, V. L., Koch, G. W., Liang, L. L & Schipper, L. A. (2021). How close are we to the temperature tipping point of the terrestrial biosphere? *Sciences Advances, 7*, 1–9.

Dzienis, P., Golak, K., Konopka, M., Mosdorf, R., Bazienė, K., & Gargasas, J. (2023). A hydrodynamic interaction between bubbles and gas supply system during gas bubble departures in liquids: An experimental study. *Scientific Reports, 13*(1), 17979.

Egamberdieva, D., Kamilova, F., Validov, S., Gafurova, L., Kucharova, Z., & Lugtenberg, B. (2008). High incidence of plant growth-stimulating bacteria associated with the rhizosphere of wheat grown on Salinated soil in Uzbekistan. *Environmental Microbiology, 10*, 1–9.

Egamberdieva, D., Hua, M., Reckling, M., Wirth, S., & Bellingrath-Kimura, S. D. (2018). Potential effects of biocharbased microbial inoculants in agriculture. *Environmental Sustainability, 1*(1), 19–24.

Eigenbrode, S. D., & Adhikari, S. (2023). Climate change and managing insect pests and beneficials in agricultural systems. *Journal of Agronomy, 115*, 2194–2215.

Einarson, K. M., & Trainor, L. J. (2015). The effect of visual information on young children's perceptual sensitivity to musical beat alignment. *Timing & Time Perception, 3*, 88–101.

Eld initiative (2015) report for policy and decision makers: Reaping economic and environmental benefits from sustainable land management.

Fageria, N. K. (2012). Role of soil organic matter in maintaining sustainability of cropping systems. *Communications in Soil Science and Plant Analysis, 43*, 2063–2113.

Fahey, D., Doherty, S., Hibbard, K. A., Romanou, A., & Taylor, P. (2017). *Physical drivers of climate change.*

Fao (2014). FAO success stories on climate-smart agriculture. In FAO: Rome France.

Fao (2021). Global map of salt-affected soils.

Feng, X., Fu, B., Piao, S., Wang, S., Ciais, P., Zeng, Z., & Wu, B. (2016). Revegetation in China's Loess Plateau is approaching sustainable water resource limits. *Nature Climate Change, 6*(11), 1019–1022.

Fening, J. O., Ewusi-Mensah, N., & Safo, E. Y. (2010). *Improving the fertilizer value of cattle manure for sustaining small holder crop production in Ghana.*

Ferini-strambi, L., Baietto, C., Di gioia, M. R., Castaldi, P., Castronovo, C., Zucconi, M., & Cappa, S. F. (2003). Cognitive dysfunction in patients with obstructive sleep apnea (OSA): Partial reversibility after continuous positive airway pressure (CPAP). *Brain Research Bulletin, 61*, 87–92.

Ferranti, E., Chapman, l., & Whyatt, D. (2017). A perfect storm? The collapse of Lancaster's critical infrastructure networks following intense rainfall on 4/5 December 2015. *Weather, 72*, 3–7.

Fisher, A., Hanemann, M., Roberts, M., & Schlenker, W. (2007). Potential impacts of climate change on crop yields and land values in us agriculture: Negative, significant, and robust. *University of California, Berkeley.*

Foyer, C. H., & Kranner, I. (2023). Plant adaptation to climate change. *Biochemical Journal, 480*, 1865–1869.

Franklin, K. A. (2009). Light and temperature signal crosstalk in plant development. *Current Opinion Plant biology, 12*, 63–68.

Galende, M. A., Becerril, J. M., Barrutia, O., Artetxe, U., Garbisu, C., & Hernández, A. (2014). Field assessment of the effectiveness of organic amendments for aided phytostabilization of a Pb–Zn contaminated mine soil. *Journal of Geochemical Exploration, 145*, 181–189.

Gao, J., Sun, S. P., Zhu, W. P., & Chung, T. S. (2014). Chelating polymer modified P84 nanofiltration (NF) hollow fiber membranes for high efficient heavy metal removal. *Water Research, 63*, 252–261.

García-ruiz, R., Ochoa, V., Vinegla, B., Hinojosa, M. B., Pena-santiago, R., Liébanas, G., Linares, J. C., & Carreira, J. A. (2009). Soil enzymes, nematode community and selected physico-chemical properties as soil quality indicators in organic and conventional olive oil farming: Influence of seasonality and site features. *Applied Soil Ecology, 41*, 305–314.

Ge, W., Matsumoto, D., & Zhang J. L. (2011). Do CFOs have style? An empirical investigation of the effect of individual CFOs on accounting practices. *Car, 28*, 1141–1179.

Geisen, S., Wall, D. H., & Van der putten, W. H. (2019). Challenges and opportunities for soil biodiversity in the Anthropocene. *Current Biology, 29*, R1036–r1044.

Ghanem, K. G. (2010). Neurosyphilis: A historical perspective and review. *CNS Neuroscience & Therapeutics, 16*(5), e157–e168.

Giovannini, D., Brandi, F., Lanteri, A. P., Lazzeri, L., Maltoni, M. L, Matteo, R., Minuto, A., Sbrighi, P., Stagno, F. & Baruzzi, G. (2021). Non-chemical soil fumigation for sustainable strawberry production in southern italy. *Agronomy, 11*, 1678.

Girvetz, E. H., & Zganjar, C. (2014). Dissecting indices of aridity for assessing the impacts of global climate change.*Climatic Change, 126*, 469–483.

Goel, K., Lindenfeld, J., Makkar, R., Naik, H., Atmakuri, S., Mahoney, P., & Barker, C. M. (2023). Transcatheter edge-to-edge repair in 5,000 patients with secondary mitral regurgitation: COAPT post-approval study. *Journal of the American College of Cardiology, 82*(13), 1281–1297.

Gowda, P., Steiner, J. L., Olson, C., Boggess, M., Farrigan, T., & Grusak, M. A. (2018). Agriculture and rural communities. *Impacts, Risks, and Adaptation in the United States: Fourth National Climate Assessment, 2*, 391–437.

Gordon, E. M., Chauvin, R. J., Van, A. N., Rajesh, A., Nielsen, A., Newbold, D. J., lynch, C. J., Seider, N. A., Krimmel, S. R., Scheidter, K. M., & Monk, J. (2023). A somato-cognitive action network alternates with effector regions in motor cortex. *Nature, 617*, 351–359.

Granier, C., Massonnet, C., Turc, O., Muller, V., Chenu, K., & Tardieu, F. (2002). Individual leaf development in arabidopsis thaliana: A stable thermal-time-based programme. *Annuals of Botony, 89*, 595–604.

Gray, S. B., & Brady, S. M. (2016). Plant developmental responses to climate change. *Developmental Biology, 419*, 64–77.

Habibi, P., Ostad, S. N., Heydari, A., Monazzam, M. R., Foroushani, A. R., Ghazi-khansari, M., & Golbabaei, F. (2023). Diagnostic biomarkers of heat stress induced-DNA in occupational exposure: A systematic review. *Journal Health and Safety Work, 12*, 800–819.

Haj-amor, Z., Araya, T., Kim, D. G., Bouri, S., Lee, J., Ghiloufi, W., Yang, Y., Kang, H., Jhariya, M. K., & Banerjee, A. (2022). Soil salinity and its associated effects on soil microorganisms, greenhouse gas emissions, crop yield, biodiversity and desertification: A review. *Science of the Total Environment, 843*, 156946.

Hakkim, V. A., Joseph, E. A., Gokul, A. A., & Mufeedha, K. (2016). Precision farming: The future of Indian agriculture. *Journal of Applied Biology and Biotechnology, 4*, 68–72.

Han, Q., Zheng, B., Cristea, M., Agostini, M., Bélanger, J. J., Gützkow, B., Kreienkamp, J., Leander, N. P., & PsyCorona Collaboration. (2023). Trust in government regarding COVID-19 and its associations with preventive health behaviour and prosocial behaviour during the pandemic: A cross-sectional and longitudinal study. *Psychological Medicine, 53*(1), 149–159.

Hartmann, M., Frey, B., Mayer, J., Mäder, P., & Widmer, F. (2015a). Distinct soil microbial diversity under long-term organic and conventional farming. *The ISME Journal, 9*(5), 1177–1194.

Hartmann, C., Shi, J., Giusto, A., & Siegrist, M. (2015b). The psychology of eating insects: A cross-cultural comparison between Germany and China. *Food Quality and Preference, 44*, 148–156.

Hartmann, J., West, A. J., Renforth, P., Köhler, P., de La Rocha, C. L., Wolf-gladrow, D. A., Dürr, H. H., Scheffran, J. (2013). enhanced chemical weathering as a geoengineering strategy to reduce atmospheric carbon dioxide supply nutrients and mitigate ocean acidification. *Reviews of Geophysics, 51*, 113–149.

Hatfield, J. L., Boote, K. J., Kimball, B. A., Ziska, L. H., Izaurralde, R. C., Ort, D., & Wolfe, D. (2011). Climate impacts on agriculture: Implications for crop production. *Agronomy Journal, 103*(2), 351–370.

Hatfield, J. L., & Prueger, J. H. (2015). Temperature extremes: Effect on plant growth and development. *Weather and Climate Extremes, 10*(Part A), 4–10.

Henry, B. M., de Oliveira, M. H. S., Benoit, J., & Lippi, G. (2020). Gastrointestinal symptoms associated with severity of coronavirus disease 2019 (COVID-19): A pooled analysis. *Internal and Emergency Medicine, 15*, 857–859.

Held, I. M., & Soden, B. J. (2006). Robust responses of the hydrological cycle to global warming. *Journal of Climate, 19*, 5686–5699.

Held, K.D., Kawamura, H., Kaminuma, T., Paz, A.E.S., Yoshida, Y., Liu, Q., Willers, H., & Takahashi, A. (2016). Effects of charged particles on human tumor cells. *Frontiers in Oncology, 6*, 23.

Hendershot, J. N., Read, Q. D., Henning, J. A., Sanders, N. J., & Classen, A. T. (2017). Consistently inconsistent drivers of microbial diversity and abundance at macro ecological scales.

Hirayama, T., & Shinozaki, K. (2010). Research on plant abiotic stress responses in the post-genome era: Past, present and future. *Plant journal, 61*, 1041–1052.

Holling, C. S. (1973). Resilience and stability of ecological systems.

Hong, J., Shi, F., & Zheng, Y. (2023). Does network infrastructure construction reduce energy intensity? Based on the "broadband china" Strategy. *Technological Rorecasting and Social Change, 190*, 122437.

Huang, C., Barnett, A. G., Xu, Z., Chu, C., Wang, X., Turner, L. R., & Tong, S. (2013). Managing the health effects of temperature in response to climate change: Challenges ahead. *Environmental health perspectives, 121*, 415–419.

Hunt, A. P., Brearley, M., Hall, A., Pope, R. (2023). Climate change effects on the predicted heat strain and labour capacity of outdoor workers in Australia. *International Journal of environmental Research and Public Health, 20*, 5675.

Huo, W. G., Chai, X. F., Wang, X. H., Batchelor, W. D., Kafle, A., & Feng, G. (2022). Indigenous arbuscular mycorrhizal fungi play a role in phosphorus depletion in organic manure amended high fertility soil. *Journal of Integrative Agriculture, 21*, 3051–3066.

Ingram, W. (2016). Increases all round. *Nature Climate Change, 6*, 443–444.

Issahaku, G., Abdul-rahaman, A., & Amikuzuno, J. (2021). Climate change adaptation strategies, farm performance and poverty reduction among smallholder farming households in Ghana. *Climate Change, 13*, 736–747.

Jarvis, N. J. (2020). A review of non-equilibrium water flow and solute transport in soil macropores: Principles, controlling factors and consequences for water quality. *European Journal of Soil Science, 71*, 279–302.

Jiang, Q., Zhao, Y., Zhang, X., Yang, X., Chen, Y., Chu, Z., & You, J. (2019). Surface passivation of perovskite film for efficient solar cells. *Nature Photonics, 13*(7), 460–466.

Karl, M., Guenther, A., Köble, R., Leip, A., & Seufert, G. (2009). A new European plant-specific emission inventory of biogenic volatile organic compounds for use in atmospheric transport models. *Biogeosciences, 6*(6), 1059–1087.

Kang, Y., Khan, S., &Ma, X. (2009). Climate change impacts on crop yield, crop water productivity and food security–a review. *Progress in Natural Science, 19*, 1665–1674.

Key, N., Sneeringer, S., & Marquardt, D. (2014). Climate change, heat stress, and US dairy production. *USDA-ERS Economic Research Report*, (175).

Kheir, A. M., Hoogenboom, G., Ammar, K. A., Ahmed, M., Feike, T., Elnashar, A., Liu, B., Ding, Z., & Asseng, S. (2022). Minimizing trade-offs between wheat yield and resource-use efficiency in the Nile Delta–A multi-model analysis. *Field Crops Research, 287*, Article 108638.

Kihombo, S., Ahmed, Z., Chen, S., Adebayo, T. S., & Kirikkaleli, D. (2021). Linking financial development economic growth and ecological footprint: What is the role of technological innovation? *Environmental science and pollution research, 28*, 61235–61245.

Kim, D., & Lee, J. (2020). Spatial changes in work capacity for occupations vulnerable to heat stress (2020) potential regional impacts from global climate change. *Safety and health work, 11*, 1–9.

Kim, K. H., Kabir, E., & Kabir, S. (2015). A review on the human health impact of airborne particulate matter. *Environment International, 74*, 136–143.

Kimball, B. A. (2003). Response of plants to elevated atmospheric co_2. *Indian Journal of Plant Physiology, 8*, 18.

Kjellstrom, T., Freyberg, C., Lemke, B., Otto, M., & Briggs, D. (2018). Estimating population heat exposure and impacts on working people in conjunction with climate change. *International Journal of Biometeorology, 62*(3), 291–306.

Knorr, W., Prentice, I. C., House, J. I., & Holland, E. A. (2005). Long-term sensitivity of soil carbon turnover to warming. *Nature, 433*, 298–301.

Körner, C., & Hiltbrunner, E. (2018). The 90 ways to describe plant temperature. *Perspectives in plant ecology, evolution and systematics, 30*, 16–21.

Körner, C, Möhl, P., & Hiltbrunner, E. (2023). Four ways to define the growing season. *Ecology letters, 26*, 1277–1292.

Kumar, S., Zwiers, F., Dirmeyer, P. A., Lawrence, D. M., Shrestha, R., & Werner, A. T. (2016). Terrestrial contribution to the heterogeneity in hydrological changes under global warming. *Water resources Research, 52*, 3127–3142.

Kundzewicz, Z. W., Krysanova, V., Dankers, R., Hirabayashi, Y., Kanae, S., Hattermann, F. F., Huang, S., Milly, P. C., Stoffel, M., Driessen, P. P. J., & Matczak, P. (2017). Differences in flood hazard projections in Europe–their causes and consequences for decision making. *Health Service Journal, 62*, 1–14.

Lal, R. (1997). Degradation and resilience of soils. *Philosophical Transactions of the Royal Society of London Series B Biological Sciences, 352*, 997–1010.

Legates, D. R., Mahmood, R., Levia, D. F., DeLiberty, T. L., Quiring, S. M., Houser, C., & Nelson, F. E. (2011). Soil moisture: A central and unifying theme in physical geography. *Progress in Physical Geography, 35*(1), 65–86.

Lempereur, M., Martin-stpaul, N. K., Damesin, C., Joffre, R., Ourcival, M., Rocheteau, A., & Rambal, S. (2015). Growth duration is a better predictor of stem increment than carbon supply in a M editerranean oak forest: Implications for assessing forest productivity under climate change. *New phytologist, 207*, 579–590.

Li, P., Wu, J., & Qian, H. (2016). Regulation of secondary soil salinization in semi-arid regions: A simulation research in the Nanshantaizi area along the silk road, Northwest China. *Environmental Earth Sciences, 75*, 1–12.

Li, D., Hou, J., Zhou, Q., Lyu, J., Pan, Z., Wang, T., Sun, X., Yu, G., & Tang, J. (2023). Urban rainfall-runoff flooding response for development activities in new urbanized areas based on a novel distributed. *Urban climate, 51*,101628.

Lipper, L., Thornton, P., Campbell, B. M., Baedeker, T., Braimoh, A., Bwalya, M., Caron, P., Cattaneo, A., Garrity, D., Henry, K., Hottle, R., Jackson, L., Jarvis, A., Kossam, F., Mann, W., McCarthy, N., Meybeck, A., Neufeldt, H., Remington, T., Sen, P. T., Sessa, R., Shula, R., Tibu, A., & Torquebiau, E. F. (2014). Climate-smart agriculture for food security. *Nature Climate Change, 4*(12), 1068–1072.

Lobell, D. B., Hammer, G. L., McLean, G., Messina, C., Roberts, M. J., & Schlenker, W. (2013). The critical role of extreme heat for maize production in the United States. *Nature Climate Change, 3*(5), 497–501.

Long, G. V., Weber, J. S., Infante, J. R., Kim, K. B., Daud, A., Gonzalez, R., et al. (2016). Overall survival and durable responses in patients with BRAF V600–mutant metastatic melanoma receiving dabrafenib combined with trametinib. *Journal of Clinical Oncology, 34*(8), 871–878.

Lu, B., Fan, W., & Zhou, M. (2016). Social presence, trust, and social commerce purchase intention: An empirical research. *Computers in Human Behavior, 56*, 225–237.

Lu, J., Sun, B., Ren, F., Li, H., & Jiao, X. (2021). Effect of freeze-thaw cycles on soil detachment capacities of three loamy soils on the loess plateau of china. *Water, 13*, 342.

Madhu, M., & Hatfield, J. L. (2013). Dynamics of plant root growth under increased atmospheric carbon dioxide. *Agronomy Journal, 105*, 657–669.

Madzima, T. F., Vendramin, S., Lynn, J. S., Lemert, P., Lu, K. C., & McGinnis, K. M. (2021). Direct and indirect transcriptional effects of abiotic stress in Zea mays plants defective in RNA-directed DNA methylation. *Frontiers in Plant Science, 12*, Article 694289.

Maes, W. H., & Steppe, K. (2012). Estimating evapotranspiration and drought stress with ground-based thermal remote sensing in agriculture: A review. *Journal of Experimental Botany, 63*, 4671–4712.

Maleki, N., Becerra, L., Brawn, J., McEwen, B., Burstein, R., & Borsook, D. (2013). Common hippocampal structural and functional changes in migraine. *Brain Structure and Function, 218*, 903–912.

Mansoor, S., Ali Wani, O., Lone, J. K., Manhas, S., Kour, N., Alam, P., & Ahmad, P. (2022). Reactive oxygen species in plants: From source to sink. *Antioxidants, 11*(2), 225.

Marques, S., Mariano, J., Mendonça, J., De Tavernier, W., Hess, M., Naegele, L., Peixeiro, F., & Martins, D. (2020). Determinants of ageism against older adults: A systematic review. *International Journal of Environmental Research and Public Health, 17*(7), 2560.

Mathews, A. J., & Jensen, J. L. (2013). Visualizing and quantifying vineyard canopy LAI using an unmanned aerial vehicle (UAV) collected high density structure from motion point cloud. *Remote sensing, 5*, 2164–2183.

Mijangos, I., Pérez, R., Albizu, I., & Garbisu, C. (2006). Effects of fertilization and tillage on soil biological parameters. *Enzyme and Microbial Technology, 40*, 100–106.

Mirón, I. J., Linares, C., & Diaz, J. (2023). The influence of climate change on food production and food safety. *Environmental Research, 216*, 114674.

Mirón, I. J., Linares, C., Diaz., J. (2023). The influence of climate change on food production and food safety. Environ res 216:114674. Zhang, n., wang, m., & wang, n. (2002). Precision agriculture a worldwide overview. Comput. Electron. Agric 36: 113-132.

Mishra, A. K., Singh, V. P., & Desai, V. R. (2010). Drought characterization: A probabilistic approach. *Stochastic Environmental Research and Risk Assessment, 23*, 41–55.

Miyazawa, H., Oka-Kira, E., Sato, N., Takahashi, H., Wu, G. J., Sato, S., Hayashi, M., Betsuyaku, S., Nakazono, M., Tabata, S. Harada, K., Sawa, S., Fukuda, H., Kawaguchi, M. (2010). The receptor-like kinase KLAVIER mediates systemic regulation of nodulation and non-symbiotic shoot development in *Lotus japonicus*. *Development, 137*(24), 4317–4325.

Mokany, K., Ferrier, S., Harwood, T. D., Ware, C., Di Marco M, Grantham, H. S., Venter, O., Hoskins, A. J., & Watson, J. E. M (2020). Reconciling global priorities for conserving biodiversity habitat. *Proceedings of the National Academy of Sciences, 117*, 9906–9911.

Mulla, D. J. (2013). Twenty five years of remote sensing in precision agriculture: Key advances and remaining knowledge gaps. *Biosystems Engineering, 114*, 358–371.

Muthaura, C., Musyimi, D. M., Ogur, J. A., & Okello, S. V. (2010). Effective microorganisms and their influence on growth and yield of pigweed (Amaranthus dubians). *ARPN Journal of Agricultural and Biological Science, 5*(1), 17–22.

Myers, S. S., Smith, M. R., Guth, S., Golden, C. D., Vaitla, B., Mueller, N. D., & Huybers, P. (2017). Climate change and global food systems: Potential impacts on food security and undernutrition. *Annual Review of Public Health, 38*(1), 259–277.

Myhre, G., Alterskjær, K., Stjern, C. W., Hodnebrog, O., Marelle, L., Samset, B. H., Sillmann, J., Schaller, N., Fischer, E., Schulz, M., Stohl, A. (2019). Frequency of extreme precipitation increases extensively with event rareness under global warming. *Scientific reports, 9*, 16063

Nanjundappa, A., Bagyaraj, D. J., Saxena, A. K., Kumar, M., & Chakdar, H. (2019). Interaction between arbuscular mycorrhizal fungi and bacillus spp. In soil enhancing growth of crop plants. *Fungal biology and Biotechnology, 6*, 23.

Naveed, M., Hussain, M. B., Zahir, Z. A., Mitter, B., & Sessitsch, A. (2014). Drought stress amelioration in wheat through inoculation with Burkholderia phytofirmans strain psJN. *Plant growth regulation, 73*, 121–131.

Nebiker, S., Annen, A., Scherrer, M., Oesch, D. (2008). A light-weight multispectral sensor for micro UAV opportunities for very high resolution airborne remote sensing. *International Archives of the Photogrammetry, Remote sensing Spatial Information Sciences, 37*, 1193–1200.

Neupane, H., Paudel, K. P., Adhikari, M., & He, Q. (2022). Impact of cooperative membership on production efficiency of smallholder goat farmers in Nepal. *Annals of Public and Cooperative Economics, 93*(2), 337–356.

Nex, F., Armenakis, C., Cramer, M., Cucci, D. A., Gerke, M., Honkavaara, E., Kukko, A., Persello, C., & Skaloud, J. (2022). UAV in the advent of the twenties: Where we stand and what is next. *ISPRS Journal of Photogrammetry and remote sensing, 184*,215–242.

Nick, F. C., Sänger, N., van der Heijden, S., & Sandholz, S. (2023). Collaboration is key: Exploring the 2021 flood response for critical infrastructures in germany. *International Journal of Disaster Risk Reduction, 91*, 103710.

Nicotra, A. B., Atkin, O. K., Bonser, S. P., Davidson, A. M., Finnegan, E. J., Mathesius, U., Poot, P., Purugganan, M. D., Richards, C. L., & Valladares, F. (2010). Plant phenotypic plasticity in a changing climate. *Trends in Plant Science, 15*, 684–692.

Nielsen, U. N., Wall, D. H., & Six, J. (2015). Soil biodiversity and the environment. *Annual Review Environment and Resources, 40*, 63–90.

Nkeki, F. N., Bello, E. I., & Agbaje, I. G. (2022). Flood risk mapping and urban infrastructural susceptibility assessment using a GIS and analytic hierarchical raster fusion approach in the Ona river basin, nigeria. *International Journal of Disaster Risk Reduction, 77*,103097.

Norris, J., Chen, G.,& Neelin, J. D. (2019). Thermodynamic versus dynamic controls on extreme precipitation in a warming climate from the community earth system model large ensemble. *Journal of Climate, 32*, 1025–1045.

Pandey, V., & Shukla, A. (2016). Improving crop yield under drought stress through physiological breeding. *Journal of Physiology and Biochemistry, 1*, 331–348.

Pathak, B., Garfinkel, R., Gopal, R. D., Venkatesan, R., & Yin, F. (2010). Empirical analysis of the impact of recommender systems on sales. *Journal of Management Information Systems, 27*(2), 159–188.

Patil, A., & Lamnganbi, M. (2018). Impact of climate change on soil health: A review. *International Journal of Chemical Studies, 6*, 2399–2404.

Patwardhan, A., Semenov, S., Schnieder, S., Burton, I., Magadza, C., Oppenheimer, M. Pittock, B. Rahman, A., Smith, J., & Suarez, A. (2007). Assessing key vulnerabilities and the risk from climate change. *Climate Change*, 779–810.

Perdomo, J. A., Capó-Bauçà, S., Carmo-Silva, E., & Galmés, J. (2017). Rubisco and rubisco activase play an important role in the biochemical limitations of photosynthesis in rice, wheat, and maize under high temperature and water deficit. *Frontiers in Plant Science, 8*, 490.

Perino, G., & Schwickert, H., (2023). Animal welfare is a stronger determinant of public support for meat taxation than climate change mitigation in Germany. *Nature food, 4*,160–169.

Pimental, P. A., Stout, C. E., Hoover, M. C., & Kamen, G. B. (1997). Changing psychologists' opinion about prescriptive authority: A little information goes a long way. *Professional Psychology Research and Practice, 28*, 123.

Pontes-da-silva, E., Magnusson, W. E., Sinervo, B., Caetano, G. H., Miles, D. B., Colli, G. R., Diele-viegas, L. M., Fenker, J., Santos, J. C., & Werneck, F. P. (2018). Extinction risks forced by climatic change and intraspecific variation in the thermal physiology of a tropical lizard. *Journal of Thermal Biology, 73*, 50–60.

Pradhan, S., Moschitti, A., Xue, N., Uryupina, O., & Zhang, Y. (2012). CoNLL-2012 shared task: Modeling multilingual unrestricted coreference in OntoNotes. In *Joint conference on EMNLP and CoNLL-shared task* (pp. 1–40).

Praveen, K., Abinandan, S., Venkateswarlu, K., & Megharaj, M. (2023). Harnessing extremophilic trait and metabolic flexibility of microalgal strains for the treatment of highly alkaline winery wastewater. *ACS ES&T Engineering, 4*(2), 455–465.

Prokopy, L. S., Floress, K., Arbuckle, J. G., Church, S. P., Eanes, F. R., Gao, Y., Gramig, B. M., Ranjan, P., & Singh, A. S. (2019). Adoption of agricultural conservation practices in the United States: Evidence from 35 years of quantitative literature. *Journal of Soil and Water Conservation, 74*(5), 520–534.

Pruski, F. F. & Nearing, M. A. (2002). Runoff and soil-loss responses to changes in precipitation: A computer simulation study. *Journal of Soil and Water Conservation, 57*, 7–16.

Qi, D., Ouyang, Z., Chen, L., Wu, Y., Lei, R., Chen, B., & Cai, W. J. (2022). Climate change drives rapid decadal acidification in the Arctic Ocean from 1994 to 2020. *Science, 377*(6614), 1544–1550.

Qu, X., Yang, Y., Liu, Z., Jin, S., & Weng, J. (2014). Potential crash risks of expressway on-ramps and off-ramps: A case study in Beijing, China. *Safety Science, 70*, 58–62.

Rajeshwar Rao, G., Prabhakar, M., Venkatesh, G., Srinivas, I., &. Sammi reddy, K. (2018). *Agroforestry opportunities for enhancing resilience to climate change in rainfed areas*. Icar—Central Research Institue Dryland Agriculture. Hyderabad.

Reinhart, A., Brooks, L., Jahja, M., Rumack, A., Tang, J., Agrawal, S., Al saeed, W., Arnold, T., Basu, A., Bien, J., & Cabrera, Á. A. (2021). An open repository of real-time covid-19 indicators. *Proceedings of the National Academy of Sciences of the United States of America, 118*, 111452118.

Renwick, M. E., & Stanley, J. A. (2020). Modeling dynamic trajectories of front vowels in the American South. *The Journal of the Acoustical Society of America, 147*(1), 579–595.

Reth, S., Reichstein, M., & Falge, E. (2005). The effect of soil water content, soil temperature, soil ph-value and the root mass on soil co_2 efflux–a modified model. *Plant soil, 268*, 21–33.

Reynolds, W. D., Bowman, B. T., Drury, C. F., Tan, C. S., & Lu, X. (2002). Indicators of good soil physical quality: Density and storage parameters. *Geoderma, 110*, 131–146.

Rezaei, R., Mianaji, S., & Ganjloo, A. (2018). Factors affecting farmers' intention to engage in on-farm food safety practices in Iran: Extending the theory of planned behavior. *Journal of Rural Studies, 60*, 152–166.

Riaz, M., Kamran, M., Rizwan, M., Ali, S., Zhou, Y., Núñez-delgado, A., & Wang, X. (2021). Boron application mitigates cd toxicity in leaves of rice by subcellular distribution, cell wall adsorption and antioxidant system. *Ecotoxicology and Environmental Safety, 222*, 112540.

Riebsame, W. E. (2019). Drought and natural resources management in the United States: Impacts and implications of the 1987–89 drought. Routledge.

Rivera, H. E., Aichelman, H. E., Fifer, J. E., Kriefall, N. G., Wuitchik, D. M., Wuitchik, S. J. S., & Davies, S. W. (2021). A framework for understanding gene expression plasticity and its influence on stress tolerance. *Molecular Ecology Resources, 30*, 1381–1397.

Roderick, T. P., Wasko, C., & Sharma, A. (2019). Atmospheric moisture measurements explain increases in tropical rainfall extremes. *Geophysical Research Letters, 46*, 1375–1382.
Ross, D. S., Matschonat, G., & Skyllberg, U. (2008). Cation exchange in forest soils: The need for a new perspective. *European Journal Soil Science, 59*, 1141–1159.
Ruiz-vera, U. M., Siebers, M., Gray, S. B., Drag, D. W., Rosenthal, D. M., Kimball, B. A., Ort, D. R., & Bernacchi C. J. (2013). Global warming can negate the expected co_2 stimulation in photosynthesis and productivity for soybean grown in the Midwestern United States. *Plant physiology, 162*, 410–423.
Rustad, I., Campbell, J., Marion, G., Norby, R., Mitchell, M., Hartley, A., Cornelissen, J., Gurevitch, J., & Gcte-news, D. (2001). A meta-analysis of the response of soil respiration, net nitrogen mineralization, and aboveground plant growth to experimental ecosystem warming. *Oecologia, 126*, 543–562.
Safdar, M., Kim, W., Park, S., Gwon, Y., Kim, Y. O., & Kim, J. (2022). Engineering plants with carbon nanotubes: A sustainable agriculture approach. *Journal of Nanobiotechnology, 20*(1), 275.
Sage, R. F., Way, D. A., & Kubien, D. S. (2008). Rubisco, rubisco activase, and global climate change. *Journal of Experimental Botany, 59*, 1581–1595.
Saha, N., & Mandal, B. (2009). Soil health–a precondition for crop production. *Microbial Strategies for Crop Improvement*, 161–184.
Salehi, M. (2022). Global water shortage and potable water safety; today's concern and tomorrow's crisis. *Environment International, 158*, 106936
Sanchis, M. P. S., Torri, D., Borselli, L., & Poesen, J. (2008). Climate effects on soil erodibility. *Earth surface Processes Landforms: The Journal British Geomorphological Research Group, 33*, 1082–1097.
Scharpenseel, H. W., Schomaker, M., & Ayoub, A. (1990). *Soils on a warmer earth: Effects of expected climate change on soil processes, with emphasis on the tropics and sub-tropics*. Elsevier.
Scheben, A., Wolter, F., Batley, J., Puchta, H., & Edwards, D. (2017). Towards CRISPR/Cas crops–bringing together genomics and genome editing. *New Phytologist, 216*(3), 682–698.
Scotti, R., Bonanomi, G., Scelza, R., Zoina, A., Rao, M. A. (2015). Organic amendments as sustainable tool to recovery fertility in intensive agricultural systems. *Journal of Soil Science and Plant nutrition, 15*, 333–352.
Seneviratne, S. I., Corti, T., Davin, E.L., Hirschi, M., Jaeger, E. B., Lehner, I., Orlowsky, B., & Teuling, A. J. (2010). Investigating soil moisture–climate interactions in a changing climate: A review. *Earth-Science Reviews, 99*(3–4), 125–161.
Seleiman, M. F., Al-suhaibani, N., Ali, N., Akmal, M., Alotaibi, M., Refay, Y., Dindaroglu, T., Abdul-wajid, H. H., & Battaglia, M. L. (2021). Drought stress impacts on plants and different approaches to alleviate its adverse effects. *Plants, 10*, 259.
Sharma, A., Wasko, C., & Lettenmaier, D. P. (2018). If precipitation extremes are increasing, why aren't floods? *Water resources Research, 54*: 8545–8551.
Sharma, A., Jain, A., Gupta, P., & Chowdary, V. (2020). Machine learning applications for precision agriculture: A comprehensive review. *IEEE access, 9*, 4843–4873.
Shaver, G. R., Canadell, J., Chapin, F. S., Gurevitch, J., Harte, J., Henry, G., Ineson, P., Jonasson, S., Melillo, J., &Pitelka, L. (2000). Global warming and terrestrial ecosystems: A conceptual framework for analysis: Ecosystem responses to global warming will be complex and varied. Ecosystem warming experiments hold great potential for providing insights on ways terrestrial ecosystems will respond to upcoming decades of climate change. Documentation of initial conditions provides the context for understanding and predicting ecosystem responses. *Bioscience, 50*, 871–882.
Shen, J., Wang, Q., Chen, Y., Han, Y., Zhang, X., & Liu, Y. (2022). Evolution process of the microstructure of saline soil with different compaction degrees during freeze-thaw cycles. *Engineering Geology, 304*, 106699.
Sikka, A. K., Islam, A., & Rao, K. V. (2018). Climate-smart land and water management for sustainable agriculture. *Irrigation and Drainage, 67*(1), 72–81.

Silveira, M. A. D., de Jong, D., Berretta, A. A., Dos Santos Galvao, E. B., Ribeiro, J. C., Cerqueira-Silva, T., Amorim, T. C., da conceicao, L. F. M. R., Gomes, M. M. D., Teixeira, M. B., & de Souza, S. P. (2021). Efficacy of brazilian green propolis (epp-af®) as an adjunct treatment for hospitalized covid-19 patients: A randomized, controlled clinical trial. *Biomedicine & pharmacotherapy,138*, 111526.
Singh, B. K., Trivedi, P., Egidi, E., Macdonald, C. A., & Delgado-Baquerizo, M. (2020). Crop microbiome and sustainable agriculture. *Nature Reviews Microbiology, 18*, 601–602.
Sishodia, R. P., Ray, R. L., & Singh, S. K. (2020). Applications of remote sensing in precision agriculture: A review. *Remote sensing, 12*, 3136.
Sun, J., Dai, X., Wang, Q., Van Loosdrecht, M. C., & Ni, B. J. (2019). Microplastics in wastewater treatment plants: Detection, occurrence and removal. *Water Research, 152*, 21–37.
Suryadi, R. N. (2020). Pengaruh budaya organisasi, motivasi kerja dan disiplin kerja terhadap kinerja guru SMA Negeri Di Kota Makassar. *Indonesian Journal of Economics, Entrepreneurship, and Innovation, 1*(1), 14–28.
Swift, M. J., Izac, A. M., & Van noordwijk, M. (2004). Biodiversity and ecosystem services in agricultural landscapes—are we asking the right questions? *Agriculture, Ecosystems and Environment, 104*, 113–134.
Tabari, H., Hosseinzadehtalaei, P., Aghakouchak, A., & Willems, P. (2019). Latitudinal heterogeneity and hotspots of uncertainty in projected extreme precipitation. *Environmental Research Letters,14*, 124032.
Takle, E. S., Rajewski, D. A., & Purdy, S. L. (2019). The Iowa Atmospheric Observatory: Revealing the unique boundary layer characteristics of a wind farm. *Earth Interactions, 23*(2), 1–27.
Tan, M. L., Ibrahim, A. L., Yusop Z, Duan, Z., Ling, L. (2015). Impacts of land-use and climate variability on hydrological components in the Johor river basin, Malaysia. *Hydrological Sciences Journal, 60*, 873–889.
Tedersoo, l., Sánchez-ramírez, S., Koljalg, U., Bahram, M., Döring, M., Schigel, D., May, T., Ryberg, M., & Abarenkov, K. (2018). High-level classification of the fungi and a tool for evolutionary ecological analyses. *Fungal diversity, 90*, 135–159.
Tisdall, J. M. (2020). Formation of soil aggregates and accumulation of soil organic matter. In *structure and organic matter storage in agricultural soils*. (pp. 57–96). CRC Press: Boca Raton, FL, USA.
Triantafyllou, I. S. (2019). A new distribution-free control scheme based on order statistics. *Journal of Nanparametric Statistics, 31*, 1–30.
Trisos, C. H., Merow, C., & Pigot, A. L. (2020). The projected timing of abrupt ecological disruption from climate change. *Nature, 580*, 496–501.
USGCRP (U.S. Global Change Research Program). (2017). *Climate science special report: Fourth national climate assessment (Volume I)*, edited by D.J. Wuebbles, D.W. Fahey, K.A. Hibbard, D.J. Dokken, B.C. Stewart, and T.K. Maycock. Washington, D.C. https://doi.org/10.7930/j0j964j6
Van Groenigen, J. W., Lubbers, I. M., Vos, H. M. J., Brown, G. G., De deyn, G. B., & Van Groenigen, K. J. (2014). Earthworms increase plant production: A meta-analysis. *Scientific Reports, 4*, 6365
Venkatesh, K. K., Lynch, C. D., Powe, C. E., Costantine, M. M., Thung, S. F., Gabbe, S. G., Grobman, W. A., & Landon, M. B. (2022). Risk of adverse pregnancy outcomes among pregnant individuals with gestational diabetes by race and ethnicity in the United States, 2014–2020. *JAMA, 327*(14), 1356–1367.
Waaswa, A., Oywaya Nkurumwa, A., Mwangi Kibe, A., & Ngeno Kipkemoi, J. (2022). Climate-smart agriculture and potato production in Kenya: Review of the determinants of practice. *Clim dev, 14*, 75–90.
Walsh, E. E., Frenck Jr, R. W., Falsey, A. R., Kitchin, N., Absalon, J., et al. (2020). Safety and immunogenicity of two RNA-based Covid-19 vaccine candidates. *New England Journal of Medicine, 383*(25), 2439–2450.
Wang, D., Hu, B., Hu, C., Zhu, F., Liu, X., Zhang, J., Wang, B., Xiang, H., Cheng, Z., Xiong, Y., Zhao, Y., Li, Y., Wang, X., & Peng, Z. (2019a). Clinical characteristics of 138 hospitalized

patients with 2019 novel coronavirus–infected pneumonia in Wuhan, China. *JAMA, 323*(11), 1061–1069.
Wang, Y., Yin, W., & Zeng, J. (2019b). Global convergence of ADMM in nonconvex nonsmooth optimization. *Journal of Scientific Computing, 78*, 29–63.
Wang, M., Zhu, Y., Zhao, T., Cui, L., Mao, W., Ye, M., Wu, J., & Yang, J. (2022). Chemical characteristics of salt migration in frozen soils during the freezing-thawing period. *Journal of hydrology, 606*, 127403.
Wang, T., Wang, H., Wang, Z., & Huang, J. (2023a). Dynamic risk assessment of urban flood disasters based on functional area division a case study in Shenzhen, China. *Journal of Environmental Management, 345*, 118787.
Wang, Y., Zhong, W., Li, L., Mi, F., Zeng, X., Huang, W., & Liu, Q. (2023b). Aligning large language models with human: A survey. arXiv preprint arXiv:2307.12966.
Weil, R. R., & Magdoff, F. (2004). Significance of soil organic matter to soil quality and health. *Soil organic matter sustainable agriculture*, 1–43.
Weiss, A., Jellingsø, M., & Sommer M. O. A. (2020). Spatial and temporal dynamics of sars-cov-2 in covid-19 patients: A systematic review and meta-analysis. *EBioMedicine, 58.*
Wisnieski, L., Kerver, J., Holzman, C., Todem, D., & Margerison-Zilko, C. (2018). Breastfeeding and risk of metabolic syndrome in children and adolescents: A systematic review. *Journal of Human Lactation, 34*(3), 515–525.
World Bank Group. (2017). *World development report 2017: Governance and the law.* World Bank Publications.
Wyatt, A. W., Annala, M., Aggarwal, R., Beja, K., Feng, F., Youngren, J., & Gleave, M. E. (2017). Concordance of circulating tumor DNA and matched metastatic tissue biopsy in prostate cancer. *JNCI: Journal of the National Cancer Institute, 109*(12), djx118.
Xia, B., Shen, X., He, Y., Pan, X., Liu, F. L., Wang, Y., Yang, F., Fang, S., Wu, Y., Duan, Z., & Zuo, X. (2021). Sars-cov-2 envelope protein causes acute respiratory distress syndrome (ards)-like pathological damages and constitutes an antiviral target. *Cell Research, 31*, 847–860.
Xiao, J., Pan, X., Guo, S., Ren, P., & Bao, X. (2015). Toward fundamentals of confined catalysis in carbon nanotubes. *Journal of the American Chemical Society, 137*(1), 477–482.
Xiao, H., Li, Z., Chang, X., Huang, B., Nie, X., Liu, C., Liu, L., Wang, D., & Jiang, J. (2018a). The mineralization and sequestration of organic carbon in relation to agricultural soil erosion. *Geoderma, 329*, 73–81.
Xiao, H., Wang, D., Medeiros, S. C., Hagen, S. C., & Hall, C. R. (2018b). Assessing sea-level rise impact on saltwater intrusion into the root zone of a geo-typical area in Coastal East-Central Florida. *Science of the Total Environment, 630*, 211–221.
Xu, X., Chen, P., Wang, J., Feng, J., Zhou, H., Li, X., Zhong, W., & Hao, P. (2020). Evolution of the novel coronavirus from the ongoing Wuhan outbreak and modeling of its spike protein for risk of human transmission. *Science China Life Sciences, 63*, 457–460.
Xue, J., & Su, B. (2017). Significant remote sensing vegetation indices: A review of developments and applications. *Journal of Sensors, 1*, 1353691.
Yang, Y., Liu, B., Xu, J., Wang, J., Wu, J., Shi, C., et al. (2017). Derivation of pluripotent stem cells with in vivo embryonic and extraembryonic potency. *Cell, 169*(2), 243–257.
Yang, T., Siddique, K. H. M., & Liu, K. (2020). Cropping systems in agriculture and their impact on soil health-a review. *Global Ecology and Conservation, 23*, E01118.
Yang, L., Zhang, Y., Kang, S., Wang, Z., & Wu, C. (2021). Microplastics in soil: A review on methods, occurrence, sources, and potential risk. *Science of the Total Environment, 780*, Article 146546.
Yi, W., Chan, A. P. C. (2015). Optimal work pattern for construction workers in hot weather: A case study in Hong Kong. *Journal of Computer in Civil Engineering, 29*, 5.
Zafar, U., Rashid, T. U., Khosa, A. A., Khalil, M. S., & Rashid, M. (2018). An overview of implemented renewable energy policy of Pakistan. *Renewable and Sustainable Energy Reviews, 82*, 654–665.

Zaman, M., Shahid, S. A., Heng, L., Shahid, S. A., Zaman, M., & Heng. L. (2018). Soil salinity: Historical perspectives and a world overview of the problem. *guideline for salinity assessment, mitigation and adaptation using nuclear and related techniques*, 43–53.

Zhang, J., Zhang, Y. P., & Rosenberg, H. F. (2002a). Adaptive evolution of a duplicated pancreatic ribonuclease gene in a leaf-eating monkey. *Nature Genetics, 30*(4), 411–415.

Zhang, N., Wang, M., & Wang, N. (2002b). Precision agriculture a worldwide overview. *Computers and Electronics in Agriculture, 36*, 113–132.

Zhang, L., Potter, N., Hickel, K., Zhang, Y., & Shao, Q. (2008). Water balance modeling over variable time scales based on the Budyko framework–model development and testing. *Journal of Hydrology, 360*, 117–131.

Zhang, Z., Wei, M. A., Wenjie, F., Donghui, X., & Xin, H. (2016). Reconstruction of soil particle composition during freeze-thaw cycling: A review. *Pedosphere, 26*, 167–179.

Zhang, X., Zwiers, F. W., Li, G., Wan, H., & Cannon, A.J. (2017). Complexity in estimating past and future extreme short-duration rainfall. *Nature Geoscience,10*, 255–259.

Zhang, S., Yao, l., Sun, A., & Tay, Y. (2019). Deep learning based recommender system: A survey and new perspectives. *ACM Computing Surveys (CSUR), 52*, 1–38.

Zhang, C., Bengio, S., Hardt, M., Recht, B., & Vinyals, O. (2021). Understanding deep learning (still) requires rethinking generalization. *Communications of the ACM, 64*(3), 107–115.

Zhang, H., Zhu, J., Gong, Z., & Zhu, J. K. (2022). Abiotic stress responses in plants. *Nature reviews genetics, 23*, 104–119.

Zhang, C., Zhang, C., Zheng, S., Qiao, Y., Li, C., Zhang, M., Dam, S. K., Thwal, C. M., Tun, Y. L., Huy, L. L., Donguk kim., Bae, S.-H., Lee, L.-H., Yang, Y., Shen, H. T., Kweon I. S., & Hong, C. S. (2023). A complete survey on generative ai (aigc): Is chatgpt from gpt-4 to gpt-5 all you need? arXiv preprint arXiv:2303.11717.

Zheng, Z., Ning, K., Wang, Y., Zhang, J., Zheng, D., Ye, M., & Chen, J. (2023). A survey of large language models for code: Evolution, benchmarking, and future trends. arXiv preprint arXiv: 2311.10372.

Zhu, J. K. (2016). *Abiotic stress signaling and responses in plants.* Cell 167, 313–324.

Zizinga, A., Mwanjalolo, J. G. M., Tietjen, B., Bedadi B., Gabiri, G., & Luswata, K. C. (2022). Impacts of climate smart agriculture practices on soil water conservation and maize productivity in rainfed cropping systems of Uganda. *Frontiers Sustainable Food systems, 6* ,889830.

Climate Change Impact on Soil Systems and Agriculture Resilience

Muhammad Arif, Faisal Khalid, Hina Abbas, Sumaira Maqsood, Shahid Ullah Khan, Muhammad Adnan, and Abazar Ghorbani

Abstract Globally, soil systems and agricultural practices are being profoundly altered by climate change, which presents both opportunities and challenges for enhancing food production's resilience. The various effects of climate change on soil properties are examined in this chapter. These effects include changes in temperature on soil organic matter and microbial activity, modifications to precipitation patterns on soil moisture, and adjustments to soil pH and nutrient availability on soil health and crop productivity. It draws attention to the dangers associated with these changes, including increasing soil erosion, fertilizer runoff, and decreased fertility. In order to tackle these problems and improve agricultural resilience, the chapter looks at

Muhammad Arif and Faisal khalid are equally contributed to this work.

M. Arif (✉)
College of Agriculture, Guizhou University, Guiyang, Guizhou, China
e-mail: arif@gzu.edu.cn

F. Khalid
Department of Forestry, Shaheed Benazir Bhutto University, Sheringal, Khyber Pakhtunkhwa, Pakistan

H. Abbas
National Institute for Genomics and Advanced Biotechnology (NIGAB), National Agricultural Research Centre (NARC), Islamabad, Pakistan

S. Maqsood
Department of Environmental Sciences, Kohsar University Murree, Murree, Pakistan

S. U. Khan
Department of Biochemistry, Women Medical and Dental College, Khyber Medical University, Abbottabad, Khyber Pakhtunkhwa, Pakistan

M. Adnan
State Key Laboratory of Environmental Geochemistry, Institute of Geochemistry, Chinese Academy of Sciences, Guiyang, Guizhou, China

A. Ghorbani
State Key Laboratory of Green Pesticides, Key Laboratory of Green Pesticide and Agricultural Bioengineering, Ministry of Education, Center for Research and Development of Fine Chemicals, Guizhou University, Guiyang, China

M. Shaaban (ed.), *Soils and Sustainable Agriculture*, Frontier Studies in Soil Science,
https://doi.org/10.1007/978-3-031-91114-9_15

different approaches to adaptation and new technologies. These include the development of crop types that are tolerant to climate change, sophisticated water management strategies like drip irrigation and rainwater collection, and soil conservation measures like decreased tillage and cover crops. The practical application of these tactics is demonstrated through a discussion of successful case studies from various regions, including the Midwest of the United States, the Middle East, and West Africa. Future research priorities are also outlined in this chapter, with an emphasis on policy analysis, integrated soil and water management, resilient crop growth, and soil carbon dynamics. This chapter offers a paradigm for enhancing soil and agricultural resilience in the face of climate change by fusing scientific discoveries with workable solutions.

Keywords Climate change · Soil conservation · Resilient crop growth · Soil systems

1 Introduction

The term "climate change" is a broad term that can be used to describe variations in local, regional, or global temperatures as well as the consequences of these variations (Ahmed et al., 2023). One of the biggest problems facing world agriculture today is climate change. Agriculture production is largely dependent on soils because they control ecosystem health, nutrient availability, and water regulation (Raj et al., 2019). The natural world is changing at a rate never seen before due to climate change, which has significant effects on soil systems and agricultural methods all across the world. Soil is the principal agent for the cycling of materials in terrestrial ecosystems because it contains the biggest pool of organic matter in the terrestrial biosphere (Carter, 2020). It provides the nourishment required for plants to endure, develop, and flourish. Because of deforestation, degradation, changes in land use, excessive use of soil resources, and erosion, soil degradation is a major concern on a global scale (Mahata & Sharma, 2021). These factors include the basic nutrient cycles in addition to the concentration, ratio, and physicochemical characteristics of the soil as well as soil microorganisms.

The complicated relationship between soil dynamics and agricultural productivity is being progressively upset as global temperatures increase and weather patterns change (Lone et al., 2017). Because they are essential for maintaining ecosystems, controlling water resources, and promoting plant development, soils are especially vulnerable to changes in the climate. The consequences for soil systems and agricultural resilience are growing more obvious as Earth's temperature continues to increase. These soil functions are vulnerable to changes in climate, which presents serious obstacles to the resilience of agriculture (Webb et al., 2017). Often referred to as the "skin of the Earth," soil is essential for sustaining ecosystem health, controlling water movement, and promoting plant growth (Shinde et al., 2019). Soil systems and agriculture are among the most severely damaged environmental and socio-economic systems by the worldwide problem of climate change. Developing measures to guarantee food security and advance sustainable land use requires an understanding of how soil systems and agricultural resilience are impacted by climate change (Smith et al., 2020). There are many different and intricate ways that climate change is affecting soils. Elevated temperatures hasten the breakdown of soil organic matter, which may lead to a reduction in soil carbon reserves and modifications in nutrient accessibility (Rodrigues et al., 2023).

Variations in precipitation patterns impact soil moisture levels, which in turn affect plant growth and agricultural productivity (te Wierik et al., 2021). These variations can range from longer droughts to more intense rainfall. Changes in soil pH and nutrient cycling can also raise the danger of nutrient runoff and soil erosion, as well as reduce soil fertility (Martínez-Mena et al., 2020). Agricultural resilience, or the capacity of agricultural systems to endure and adjust to changes, is essential in light of these problems brought on by climate change (Gil et al., 2017). Researchers and farmers are investigating various technical breakthroughs and adaptive tactics as a means of enhancing resilience. These include improved methods for managing water resources, conserving soil, and creating crop kinds resistant to harsh weather (Adhikari et al., 2018). A thorough understanding of the interactions between soil systems, agricultural practices, and climate change is necessary for successful adaptation (Owen, 2020). This chapter offers a thorough analysis of how soil systems and agricultural resilience are impacted by climate change. It investigates how changes in soil qualities, modified precipitation patterns, and rising temperatures affect agricultural output. Through case studies from diverse places, the chapter also analyzes various technologies and tactics intended to improve resilience. By tackling these problems, the chapter hopes to provide guidance and answers for reducing the effects of climate change on agriculture and encouraging environmentally friendly farming methods in an increasingly erratic environment.

2 The Interplay Between Climate Change and Soil Systems

2.1 *Temperature Effect and Soil Dynamics*

Many aspects of soil are influenced by temperature, such as the availability of nutrients, the rate at which organic matter breaks down, and the moisture level (Biswas & Kole, 2017). Increased rates of organic matter breakdown in soils could result in a possible decrease in soil carbon reserves as global temperatures rise (Lal, 2018). Methane (CH_4) and carbon dioxide (CO_2), two greenhouse gases that might worsen climate change, can be released during this process. Higher temperatures can also affect the aggregation and structure of the soil, which can affect the soil's permeability and ability to hold water (Rabot et al., 2018). For example, higher temperatures may cause the soil to dry out and allow less water to permeate, which could have a detrimental effect on crop production.

2.1.1 Soil Organic Matter Decomposition

Warmer temperatures hasten the breakdown of organic matter in the soil (SOM) (Moinet et al., 2020). Because of the release of nutrients, this can temporarily boost soil fertility; however, over time, SOM levels are typically reduced. Reduced SOM can cause the soil to have a worse structure, hold less water, and be more prone to erosion.

2.1.2 Soil Microbial Activity

The cycle of nutrients can be impacted by higher temperatures, which can also increase microbial activity (Mooshammer et al., 2017). Severe heat, however, has the potential to upset microbial communities, decreasing their capacity to recycle nutrients and break down organic matter.

2.1.3 Soil Moisture and Temperature Interactions

Drier soils can result from rising temperatures, which can accelerate soil evaporation rates (Zhou et al., 2021). Increased temperatures and decreased soil moisture can have a detrimental effect on plant growth and lower agricultural yield.

2.2 Precipitation Patterns and Soil Moisture

Soil moisture levels are directly impacted by modifications to precipitation patterns, such as altered rainfall intensity and dispersion (Engelhardt et al., 2018). Prolonged droughts can lower soil moisture and limit agricultural output, whereas more frequent heavy rainfall events can cause soil erosion and nutrient discharge (Qiu et al., 2021). Because it affects nutrient uptake, root development, and seed germination, soil moisture is a crucial component of plant growth. Therefore, changes in precipitation patterns can cause fluctuations in soil moisture, which can have a major impact on crop yields and agricultural resilience.

2.2.1 Soil Erosion

Soil erosion can be exacerbated by modified precipitation patterns, such as higher intensity and more frequent heavy rainfall (Guo et al., 2019). Erosion can result in the loss of important nutrients and sedimentation in water bodies. It also depletes topsoil, which is essential for crop growth.

2.2.2 Sedimentation

Extreme weather events brought on by climate change, like storms and high rainfall, can worsen soil erosion and sedimentation processes (Chen et al., 2020). Erosion can result in the loss of priceless arable land and deplete topsoil, which is rich in nutrients necessary for plant growth. Aquatic ecosystems and water quality can be impacted by sedimentation in bodies of water (Hauer et al., 2018).

2.2.3 Nutrient Runoff

Heavy rains can cause nutrient runoff, which depletes agricultural areas of vital nutrients (Wang & Li, 2019). Aquatic ecosystems may be negatively impacted by this as it can reduce soil fertility and increase water pollution.

2.2.4 Drought Conditions

Extended dry spells lower soil moisture, which hinders plant development and yield (Mbanyele et al., 2021). Crop water stress can be exacerbated by drought conditions due to reduced infiltration rates and soil compaction.

2.2.5 Soil pH and Nutrient Availability

Climate changes can have an impact on soil pH, which can therefore have an impact on microbial activity and nutrient availability (Jansson & Hofmockel, 2020). Elevated CO_2 levels in the atmosphere can cause soil acidification, which lowers the availability of vital elements like calcium and magnesium. The microbial communities in the soil can be impacted by soil acidification, which can further affect plant health and nutrient cycling (Philippot et al., 2024).

2.2.6 Soil Acidification

Raising atmospheric CO_2 can cause soil acidification, which lowers pH levels in the soil (Meng et al., 2019). Acidic soils can affect plant health and growth by reducing the availability of vital minerals like calcium and magnesium. Microbial communities are impacted by soil acidification, which can further impact the cycling of nutrients.

2.2.7 Nutrient Availability and Toxicity

Climate changes can impact soil nutrient availability. Warmer temperatures, for instance, can make some nutrients more soluble while decreasing the solubility of others (Gondal et al., 2021). Furthermore, variations in the pH of the soil may cause harmful metals like aluminum to become more accessible, harming plant roots and lowering agricultural yields.

3 Agricultural Resilience in the Face of Climate Change

3.1 Adaptation Strategies

It is imperative to employ adaptation techniques that tackle the problems offered by climate change to improve agricultural resilience. Among the most important adaption techniques are:

3.1.1 Soil Conservation Practices

Reducing soil erosion, enhancing soil structure, and improving water retention can all be achieved by implementing conservation tillage, cover crops, and agroforestry (Arshad et al., 2023; Kaushal et al., 2021).

3.1.2 Conservation Tillage

Conservation tillage aids in preserving soil structure and lessens soil disturbance. Reduced tillage and no-till farming are two methods that can increase soil organic matter content, decrease erosion, and improve water infiltration.

3.1.3 Cover Cropping

In the off-season, cover crop planting helps boost organic matter, strengthen soil structure, and prevent erosion. In addition, cover crops improve nitrogen cycling and lessen weed burden.

3.1.4 Agroforestry

Agricultural systems can benefit from the addition of trees and shrubs because they improve water retention, decrease erosion, and improve soil structure. Agroforestry systems also yield other advantages, like enhanced biodiversity and wildlife habitat.

3.2 *Water Management*

Using rainwater gathering techniques and implementing drip or sprinkler irrigation systems are two efficient ways to manage water resources in the face of shifting precipitation patterns (Kiggundu et al., 2018).

3.2.1 Efficient Irrigation Systems

Using effective irrigation methods, like sprinkler systems or drip irrigation, can maximize water utilization, lower evaporation losses, and increase agricultural yields. To further improve water efficiency, data on soil moisture can be used to schedule irrigation.

3.2.2 Rainwater Harvesting

Managing water resources more skillfully can be achieved by gathering and storing rainfall for use during dry spells. Rainwater harvesting systems can be planned for a range of application areas, from large-scale regional projects to small-scale agricultural systems.

3.2.3 Soil Moisture Monitoring

Real-time data on soil conditions can be obtained through the use of remote sensing technology and soil moisture sensors. Making better decisions about irrigation and other water management techniques is made possible by this information.

3.3 *Crop and Soil Health Management*

3.3.1 Diversification and Crop Rotation

Crop failure can be decreased and soil health can be enhanced by planting a variety of crops and using crop rotation (Shah et al., 2021). Crop rotations and a variety of cropping techniques can strengthen soil health, lessen the impact of pests and diseases, and increase resistance to climate unpredictability. Crop rotation can also enhance nutrient cycling and disrupt the cycles of pests and diseases.

3.3.2 Soil Health Management

Compost and manure are examples of organic matter that can be added to promote soil health and enhance its structure, fertility, and ability to withstand climate unpredictability.

3.3.3 Organic Amendments

Compost, manure, or biochar are examples of organic materials that can be added to soil to improve soil structure, raise soil carbon stocks, and improve nutrient availability. Additionally, organic amendments enhance soil moisture retention and encourage advantageous microbial activity.

3.3.4 Climate-Resilient Crop Varieties

Productivity may be maintained by creating and cultivating crop types that are more resilient to climatic conditions like heat and drought. To create these hardy types, advances in genetic engineering and agricultural breeding are essential.

3.4 Technological Innovations

Technological developments can also be a major factor in improving agricultural resilience (Jung et al., 2021). Among the technologies that show promise are:

3.4.1 Precision Agriculture

Making better decisions and maximizing the use of resources can be achieved by monitoring crop health and soil conditions with GIS and remote sensing technologies (AbdelRahman et al., 2021). Precision farming makes use of technologies like GIS, data analytics, and remote sensing to maximize farming methods. Precision agriculture contributes to more efficient resource usage and less environmental effects by offering comprehensive data on crop health and soil conditions.

3.4.2 Climate-Smart Agriculture

Climate-smart agriculture combines productivity-boosting techniques with measures to cut greenhouse gas emissions and boost climate change resistance. This strategy makes use of techniques including integrated pest management, enhanced nutrient management, and sustainable land use.

3.4.3 Soil Moisture Sensors

By deploying soil moisture sensors, it is possible to obtain real-time data on soil moisture levels, which can lead to more effective irrigation and improved water resource management.

3.5 Policy and Community Engagement

To improve agricultural resilience, community engagement and effective policy frameworks are crucial. Legislators have the power to fund research and development, encourage the use of climate-smart agriculture methods, and offer financial incentives for sustainable practices (Lipper & Zilberman, 2018). To raise awareness and promote the implementation of climate-resilient measures, community participation and education are also essential. Implementing and adjusting to new techniques is a critical function of local farmers and agricultural stakeholders.

4 Case Studies

4.1 Case Study 1: The Resilience of Integrated Agricultural Systems to Climate Change (Gil et al., 2017)

We looked at research on integrated agriculture systems' (IAS) resistance to climatic change and variability in comparison to specialized systems. We employed proxies like agricultural diversity and system richness because there was little direct study available. Their analysis of 37 publications revealed that, in 17 of 24 statistical investigations, integration was associated with higher resilience. Nevertheless, the majority of the information comes from contrasting more and less diversified systems; little research has looked into the synergies between farming systems. Statistical research usually concentrated on one-year outcomes, while modeling studies frequently looked at resilience over several years. Few research has determined the mechanisms behind this resistance, and the benefits of IAS were not consistently generalizable. Overall, the results point to the possibility that IAS can improve resilience; however, more investigation is required to validate these findings and assess integration strategies.

4.2 Case Study 2: Crop and Soil Responses to on-farm Conservation Tillage Practices in the Upper Midwest (Daigh et al., 2019)

Concerns regarding yield losses from reduced tillage were examined in the upper Midwest of the United States over four years by contrasting four tillage systems: shallow vertical till, chisel plow, spring strip-till with colters, and fall strip-till with shanks. Tillage had little effect within sites (0.17–0.36 Mg ha^{-1}), but it had a greater effect on individual sites (0.50–3.00 Mg ha^{-1}). At one site, only the ratios of bacteria to fungi revealed a significant difference between strip-till with shanks and colters. According to the study, compared to conventional chisel plow techniques, worries regarding yield losses from conservation tillage may be unwarranted. Farmers should think about economics and erosion prevention in addition to output. In the Midwest, conservation tillage has proven successful in lowering soil erosion, boosting soil health, and facilitating better fertilizer management and water infiltration.

4.3 Case Study 3: Agroecology and the Design of Climate Change-Resilient Farming Systems (Altieri et al., 2015)

The most recent IPCC report predicts that global temperatures will rise by 1.4 to 5.8 °C, altering precipitation patterns and lowering crop yields. These are just a few of the many negative effects that climate change is predicted to have on agricultural productivity. Changes in the dynamics of pathogens, weeds, and pests may also make these effects worse. Strategies for adaptation are essential to reducing these impacts. More drastic agroecological interventions, such as diversifying agroecosystems through polycultures, agroforestry, and crop-livestock systems, along with organic soil management and water conservation, are likely to yield major benefits, even though modest adjustments might provide temporary relief. Customary agricultural methods provide a significant understanding of resilience, and current findings indicate that complex landscape matrices and higher biodiversity improve farm resilience to extreme weather events. The development and dissemination of agroecological practices, which are essential for adapting agriculture to climate change, can be guided by recognizing and comprehending effective systems.

4.4 *Case Study 4: Land Degradation and Climate Change: Building Climate Resilience in Agriculture (Webb et al., 2017)*

Risks to the world's food security are substantial due to land degradation and climate change. Agroecological systems are more susceptible to climate change as a result of land degradation, which also lessens the efficacy of available adaptation strategies. However, planning for adaptation and climate impact assessments has mainly ignored these relationships. We investigate how land degradation can impact the effects of climate change and crop and livestock producers' ability to adapt to various agroecological systems. Next, they offer cutting-edge tactics for climate-resilient farming that facilitate chances to include solutions to these problems. To achieve global development and food security goals, forward-thinking, climate-resilient agriculture necessitates: (1) incorporating land degradation processes and their links with adaptive capacity into adaptation planning; (2) identifying key vulnerabilities to prioritize adaptation responses; (3) improving knowledge exchange across local to global scales to support strategies for developing producers' adaptive capacity; and (4) coming up with creative management and policy options that offer multiple "wins" for land, climate, and biodiversity.

4.5 *Case Study 5: Investment Profitability and Economic Efficiency of the Drip Irrigation System: Evidence from Egypt (Ali et al., 2020)*

Sophisticated irrigation techniques, including drip irrigation, are essential for increasing the productivity of both water and land. This study used financial and data envelopment analysis to assess the economic benefits of drip irrigation for tomato agriculture in Egypt through a survey of 100 farmers. Findings indicated that drip irrigation produced a 67% higher net return per hectare than conventional techniques, with an internal rate of return of 44%, a benefit–cost ratio of 1.35, and a net present value of $1720. Cost and scale efficiencies require improvement, even though technological and allocative savings were largely attained. These results lend support to policies that encourage drip irrigation through microcredit and training initiatives, which can assist farmers in increasing sustainability and productivity—particularly in water-scarce areas of the Middle East.

4.6 Case Study 6: Soil Quality Both Increases Crop Production and Improves Resilience to Climate Change (Qiao et al., 2022)

Climatic change and soil quality interactions could affect croplands' ability to yield enough food. Here, we tackle this problem by utilizing a newly created dataset of soil, climate, and yield records for 12,115 site years, or 90% of China's total cereal production. We demonstrate that high-quality soils decreased crop yield sensitivity to climate variability across a range of crops and environmental conditions, resulting in higher mean crop yield (10.3 ± 6.7%) and higher yield stability (decreasing variability by 15.6 ± 14.4%). When it comes to yields under climate change, high-quality soils outperform low-quality soils by 1.7% (0.5–4.0%). Under the typical emission pathway RCP 8.5, yield change caused by the climate might cause an 11.4 Mt yearly drop in the country's cereal production by 2080–2099. Even though soil deterioration made this production decrease by 14%, soil repair can make it go down by 21%. The importance of soil quality in agriculture in the face of climate change is emphasized by this study.

4.7 Case Study 7: Improving Agricultural Resilience to Climate Change Through Soil Management (Agus et al., 2015)

Crop development and soil qualities are impacted by climate change, which makes customized soil management techniques necessary to improve resilience to extreme weather. This essay examines the effects of climate change on soil and crop productivity, which is marked by temperature increases, unpredictable rainfall, and shifting seasonal patterns. It is advised to use techniques like mulching, water harvesting, conservation tillage, and nutrient management to lessen these consequences. The control of soil organic matter is essential to adaptation because it increases microbial activity, soil fertility, and water retention. These practices are widely recognized, although they are frequently disregarded. Their significance for maintaining agriculture in the face of growing climate problems is emphasized in the research.

4.8 *Case Study 8: Climate-Resilient and Smart Agricultural Management Tools to Cope with Climate Change-Induced Soil Quality Decline (Cordovil et al., 2020)*

Maintaining the qualities of good soil means keeping it at a level that can support a balanced and healthy environment and continuous agricultural output to feed the world's population. Degradation of soil quality is getting more severe in most parts of the world due to land-use change, intensification of agriculture, and climate change (desertification). Deteriorating soil fertility and decreasing agricultural productivity, together with negative effects on water, air, and biodiversity, are mostly caused by changes in land use and the resulting loss of soil organic matter (SOM) and beneficial microbial diversity. To deal with climate change and extreme events, agriculture must adapt to become climate-smart and climate-resilient. The following actions are necessary for this: (1) enhanced soil organic matter (SOM) through crop rotations, crop associations, no-till, conservation agriculture, C sequestration, and microbial processes; (2) sustainable production (integrated livestock and cropping systems, including agroforestry); (3) decreased greenhouse gas emissions, decreased nutrient losses from agriculture, and increased nutrient use efficiency (NUE) through precision agriculture.

5 Future Directions and Research Needs

A deeper understanding of the complex interplay between soil systems, agricultural resilience, and climate change requires further research. Sustainable soil and land management practices play a crucial role in enhancing adaptation, mitigation, and overall resilience to climate change. Priority areas for future investigation include:

5.1 *Soil Carbon Dynamics*

Examining methods to improve soil carbon storage and the effects of climate change on soil carbon sequestration. Developing methods to reduce greenhouse gas emissions requires an understanding of how climate change affects soil carbon sequestration. The relationships between soil moisture, temperature, and carbon sequestration should be the main topics of research, along with methods to improve soil carbon sequestration.

5.2 Climate-Resilient Crops

Finding characteristics that support resilience and creating crop varieties with increased resistance to climate-related pressures. To create crop types that are more resilient to heat, drought, and other climate-related challenges, additional study is required. The development of highly robust crops may be significantly aided by developments in breeding and genetic engineering.

5.3 Integrated Management Practices

Investigating how to improve agricultural resilience through integrated soil and water management techniques. Researching integrated soil and water management strategies' efficacy can shed light on ways to improve agricultural resilience. The advantages of integrating crop management, water management, and soil conservation techniques should be investigated through research.

5.4 Economic and Policy Analysis

It is crucial to evaluate the financial effects of implementing climate-resilient practices and to create laws that support them. Cost–benefit evaluations, legislative frameworks, and financial incentives that support farmers' adoption of sustainable farming techniques ought to be the main areas of study.

5.5 Long-Term Monitoring and Data Collection

To comprehend patterns and implications over time, long-term monitoring systems tracking crop performance, soil health, and climate factors must be established. Comprehensive data gathering encompassing a range of agricultural systems and geographical regions is needed.

5.6 Farmer and Stakeholder Engagement

Involving farmers, agricultural professionals, and nearby communities in research is essential for results distribution and real-world implementation. Participatory methods can increase acceptance rates and assist in customizing solutions to meet local needs.

5.7 Policy Analysis

Assessing the efficiency of current regulations and locating any gaps to guarantee frameworks that are supportive of climate-resilient agriculture. It is imperative to assess the efficacy of current policies and pinpoint any deficiencies to establish policies that facilitate climate-resilient agriculture. Economic analysis can assist in weighing the advantages and disadvantages of various adaptation techniques and guide policy choices.

5.8 Cross-Disciplinary Research

More comprehensive solutions may result from encouraging cooperation between climatologists, soil scientists, agronomists, and economists. Our comprehension of the intricate relationships between soil systems, agricultural practices, and climate change will improve with interdisciplinary research.

6 Conclusion

The resilience of agricultural systems and soil systems is seriously threatened by climate change. Comprehensive knowledge and proactive management are needed to address the effects on soil characteristics, moisture, erosion, and nutrient availability. The complex balance of soil dynamics is being more and more upset by changes in weather patterns and global temperatures. This is raising worries about soil degradation, decreased fertility, and decreased agricultural production. Numerous and intricate effects of climate change are being felt by soil systems. Warm weather hastens the breakdown of organic matter in the soil, which may reduce soil carbon stores and change the availability of nutrients. Variations in precipitation patterns, ranging from extended droughts to heavier downpours, have an impact on soil moisture levels, which subsequently affect agricultural yields and plant growth. Changes in the pH of the soil and the cycling of nutrients can also exacerbate soil fertility issues and raise the dangers of erosion and nutrient runoff. A diverse strategy is needed to solve these issues. Enhancing agricultural resilience requires the implementation of adaptation methods such as improved water management techniques, soil conservation practices, and the creation of crop varieties that are resilient to climate change. Comprehending how soil systems, agricultural practices, and climate change interact and putting into practice workable solutions customized to particular localities are prerequisites for successful adaptation. Building resilience is greatly aided by continuing study and technological advancements. Promising strategies to lessen the effects of climate change include integrated soil and water management, effective irrigation systems, and precision agriculture. Case studies from different areas show how effective these tactics are and offer insightful information on real-world implementations. Future studies should concentrate on improving our knowledge of the dynamics of soil carbon, creating crops resistant to extreme weather, and assessing the efficacy of various adaptation plans and regulations. We can better prepare for and respond to the changing difficulties posed by climate change by continuing to investigate these areas. In conclusion, scientists, decision-makers, and farmers must work together to address how climate change is affecting soil systems and agriculture. We can encourage sustainable farming practices, strengthen agricultural resilience, and increase soil health by fusing scientific understanding with workable solutions. Enhancing agricultural resilience and ensuring sustainable food production in a changing climate can be achieved through the application of adaptive techniques, the utilization of technical breakthroughs, and the promotion of collaboration among scientists, policymakers, and farmers. Scientists, politicians, and farmers must continue to do research and work together to discover sustainable solutions that will ensure food security for future generations. Our capacity to innovate and adapt will be crucial to maintaining environmental sustainability and food security in the face of climate change, creating a more resilient agricultural system for coming generations.

References

AbdelRahman, M. A., Hegab, R. H., & Yossif, T. M. (2021). Soil fertility assessment for optimal agricultural use using remote sensing and GIS technologies. *Applied Geomatics, 13*(4), 605–618.

Adhikari, P., Araya, H., Aruna, G., Balamatti, A., Banerjee, S., Baskaran, P., Barah, B., Behera, D., Berhe, T., & Boruah, P. (2018). System of crop intensification for more productive, resource-conserving, climate-resilient, and sustainable agriculture: Experience with diverse crops in varying agroecologies. *International Journal of Agricultural Sustainability, 16*(1), 1–28.

Agus, F., Husnain, H., Yustika, R. D. (2015). Improving agricultural resilience to climate change through soil management. *Jurnal Penelitian dan Pengembangan Pertanian, 34*(4), 30945.

Ahmed, M., Ahmad, S., Kheir, A. M. (2023). Climate change: an overview. In *Global agricultural production: Resilience to climate change* (pp. 1–30).

Ali, A., Xia, C., Jia, C., & Faisal, M. (2020). Investment profitability and economic efficiency of the drip irrigation system: Evidence from Egypt. *Irrigation and Drainage, 69*(5), 1033–1050.

Altieri, M. A., Nicholls, C. I., Henao, A., & Lana, M. A. (2015). Agroecology and the design of climate change-resilient farming systems. *Agronomy for Sustainable Development, 35*(3), 869–890.

Arshad, T., Naqve, M., Mukhtiar, A., Javaid, M. M., Mahmood, A., Nadeem, M. A., Khan, B. A. (2023). Conservation tillage for sustainable agriculture. In *Climate-resilient agriculture, Vol 1: Crop responses and agroecological perspectives*. Springer (pp. 313–327).

Biswas, T., Kole, S. C. (2017). Soil organic matter and microbial role in plant productivity and soil fertility. In *Advances in soil microbiology: Recent trends and future prospects: Volume 2: Soil-microbe-plant interaction* (pp. 219–238).

Carter, M. (2020). Analysis of soil organic matter storage in agroecosystems. In *Structure and organic matter storage in agricultural soils*. CRC press (pp. 3–11).

Chen, C. -N., Tfwala, S. S., Tsai, C. -H. (2020). Climate change impacts on soil erosion and sediment yield in a watershed. *Water, 12*(8), 2247.

Cordovil, C. M., Bittman, S., Brito, L. M., Goss, M. J., Hunt, D., Serra, J., Gourley, C., Aarons, S., Skiba, U., & Amon, B. (2020). *Climate-resilient and smart agricultural management tools to cope with climate change-induced soil quality decline* (pp. 613–662). Elsevier.

Daigh, A. L. M., DeJong-Hughes, J., Gatchell, D. H., Derby, N. E., Alghamdi, R., Leitner, Z. R., Wick, A., Acharya, U. (2019). Crop and soil responses to on-farm conservation tillage practices in the upper Midwest. *Agricultural & Environmental Letters, 4*(1), 190012.

Engelhardt, I. C., Welty, A., Blazewicz, S. J., Bru, D., Rouard, N., Breuil, M.-C., Gessler, A., Galiano, L., Miranda, J. C., & Spor, A. (2018). Depth matters: Effects of precipitation regime on soil microbial activity upon rewetting of a plant-soil system. *The ISME Journal, 12*(4), 1061–1071.

Gil, J. D., Cohn, A. S., Duncan, J., Newton, P., Vermeulen, S. (2017). The resilience of integrated agricultural systems to climate change. *Wiley Interdisciplinary Reviews: Climate Change, 8*(4), e461.

Gondal, A. H., Hussain, I., Ijaz, A. B., Zafar, A., Ch, B. I., Zafar, H., Sohail, M. D., Niazi, H., Touseef, M., & Khan, A. A. (2021). Influence of soil pH and microbes on mineral solubility and plant nutrition: A review. *International Journal of Agriculture and Biological Sciences, 5*(1), 71–81.

Guo, Y., Peng, C., Zhu, Q., Wang, M., Wang, H., Peng, S., & He, H. (2019). Modelling the impacts of climate and land use changes on soil water erosion: Model applications, limitations and future challenges. *Journal of Environmental Management, 250*, 109403.

Hauer, C., Leitner, P., Unfer, G., Pulg, U., Habersack, H., Graf, W. (2018). The role of sediment and sediment dynamics in the aquatic environment. In *Riverine ecosystem management: Science for governing towards a sustainable future* (pp. 151–169).

Jansson, J. K., & Hofmockel, K. S. (2020). Soil microbiomes and climate change. *Nature Reviews Microbiology, 18*(1), 35–46.

Jung, J., Maeda, M., Chang, A., Bhandari, M., Ashapure, A., & Landivar-Bowles, J. (2021). The potential of remote sensing and artificial intelligence as tools to improve the resilience of agriculture production systems. *Current Opinion in Biotechnology, 70*, 15–22.

Kaushal, R., Mandal, D., Panwar, P., Kumar, P., Tomar, J., & Mehta, H. (2021). *Soil and water conservation benefits of agroforestry* (pp. 259–275). Elsevier.

Kiggundu, N., Wanyama, J., Mfitumukiza, D., Twinomuhangi, R., Barasa, B., Katimbo, A., Birungi Kyazze, F. (2018). Rainwater harvesting knowledge and practice for agricultural production in a changing climate: A review from Uganda's perspective.

Lal, R. (2018). Digging deeper: A holistic perspective of factors affecting soil organic carbon sequestration in agroecosystems. *Global Change Biology, 24*(8), 3285–3301.

Lipper, L., Zilberman, D. (2018). A short history of the evolution of the climate smart agriculture approach and its links to climate change and sustainable agriculture debates. In *Climate smart agriculture: Building resilience to climate change* (pp. 13–30).

Lone, B., Qayoom, S., Singh, P., Dar, Z., Kumar, S., Dar, N., Fayaz, A., Ahmad, N., Bhat, M., & Singh, G. (2017). Climate change and its impact on crop productivity. *British Journal of Applied Science & Technology, 21*(5), 1–15.

Mahata, S., & Sharma, V. N. (2021). The global problem of land degradation: A review. *National Geographical Journal of India, 67*, 216–231.

Martínez-Mena, M., Carrillo-López, E., Boix-Fayos, C., Almagro, M., Franco, N. G., Díaz-Pereira, E., Montoya, I., & De Vente, J. (2020). Long-term effectiveness of sustainable land management practices to control runoff, soil erosion, and nutrient loss and the role of rainfall intensity in Mediterranean rainfed agroecosystems. *CATENA, 187*, 104352.

Mbanyele, V., Mtambanengwe, F., Nezomba, H., Groot, J. C., & Mapfumo, P. (2021). Combinations of in-field moisture conservation and soil fertility management reduce effect of intra-seasonal dry spells on maize under semi-arid conditions. *Field Crops Research, 270*, 108218.

Meng, C., Tian, D., Zeng, H., Li, Z., Yi, C., Niu, S. (2019). Global soil acidification impacts on belowground processes. *Environmental Research Letters, 14*(7), 074003.

Moinet, G. Y., Moinet, M., Hunt, J. E., Rumpel, C., Chabbi, A., & Millard, P. (2020). Temperature sensitivity of decomposition decreases with increasing soil organic matter stability. *Science of the Total Environment, 704*, 135460.

Mooshammer, M., Hofhansl, F., Frank, A. H., Wanek, W., Hämmerle, I., Leitner, S., Schnecker, J., Wild, B., Watzka, M., Keiblinger, K. M. (2017). Decoupling of microbial carbon, nitrogen, and phosphorus cycling in response to extreme temperature events. *Science Advances, 3*(5), e1602781.

Owen, G. (2020). What makes climate change adaptation effective? A systematic review of the literature. *Global Environmental Change, 62*, 102071.

Philippot, L., Chenu, C., Kappler, A., Rillig, M. C., & Fierer, N. (2024). The interplay between microbial communities and soil properties. *Nature Reviews Microbiology, 22*(4), 226–239.

Qiao, L., Wang, X., Smith, P., Fan, J., Lu, Y., Emmett, B., Li, R., Dorling, S., Chen, H., & Liu, S. (2022). Soil quality both increases crop production and improves resilience to climate change. *Nature Climate Change, 12*(6), 574–580.

Qiu, J., Shen, Z., Leng, G., Wei, G. (2021). Synergistic effect of drought and rainfall events of different patterns on watershed systems. *Scientific Reports, 11*(1), 18957.

Rabot, E., Wiesmeier, M., Schlüter, S., & Vogel, H.-J. (2018). Soil structure as an indicator of soil functions: A review. *Geoderma, 314*, 122–137.

Raj, A., Jhariya, M. K., Yadav, D. K., Banerjee, A., Meena, R. S. (2019). Soil for sustainable environment and ecosystems management. In *Sustainable agriculture, forest and environmental management* (pp. 189–221).

Rodrigues, C. I. D., Brito, L. M., Nunes, L. J. (2023). Soil carbon sequestration in the context of climate change mitigation: A review. *Soil Systems, 7*(3), 64.

Shah, K. K., Modi, B., Pandey, H. P., Subedi, A., Aryal, G., Pandey, M., Shrestha, J. (2021). Diversified crop rotation: an approach for sustainable agriculture production. *Advances in Agriculture, 2021*(1), 8924087.

Shinde, R., Sarkar, P. K., Thombare, N., & Naik, S. (2019). Soil conservation: Today's need for sustainable development. *Agriculture & Food: e-Newsletter, 1*(5), 175–183.

Smith, P., Calvin, K., Nkem, J., Campbell, D., Cherubini, F., Grassi, G., Korotkov, V., Le Hoang, A., Lwasa, S., & McElwee, P. (2020). Which practices co-deliver food security, climate change mitigation and adaptation, and combat land degradation and desertification? *Global Change Biology, 26*(3), 1532–1575.

te Wierik, S. A., Cammeraat, E. L., Gupta, J., Artzy-Randrup, Y. A. (2021). Reviewing the impact of land use and land-use change on moisture recycling and precipitation patterns. *Water Resources Research, 57*(7), e2020WR029234.

Wang, Z.-H., & Li, S.-X. (2019). Nitrate N loss by leaching and surface runoff in agricultural land: A global issue (a review). *Advances in Agronomy, 156*, 159–217.

Webb, N. P., Marshall, N. A., Stringer, L. C., Reed, M. S., Chappell, A., & Herrick, J. E. (2017). Land degradation and climate change: Building climate resilience in agriculture. *Frontiers in Ecology and the Environment, 15*(8), 450–459.

Zhou, S., Williams, A. P., Lintner, B. R., Berg, A. M., Zhang, Y., Keenan, T. F., Cook, B. I., Hagemann, S., Seneviratne, S. I., & Gentine, P. (2021). Soil moisture–atmosphere feedbacks mitigate declining water availability in drylands. *Nature Climate Change, 11*(1), 38–44.

Soil-Plant-Atmosphere Continuum and Its Relevance to Crop Production

Fazal Maqsood, Mir Sujaul Islam, Ayesha Hussain, Anam Naz, Muhammad Tayab, Ali Ijaz Ahmad, Muhammad Umair, Faiz Ul Hassan, and Mohd Ekhwan Toriman

Abstract Understanding the concept of Soil-Plant-Atmosphere Continuum (SPAC) for integrated movement of nutrient, water and energy from the soil body through plants to the atmosphere is very critical. This perennial flow, driven by soil water potential, plant transpiration, and atmospheric measure and demands, presides the key physiological phenomena essential for crop growth and development. In plant growth and production, optimizing nutrient transport, water uptake and plant stress tolerance is proficiently affected by SPAC that has great impact over crop yield. Furthermore, addressing SPAC dynamics will help to acknowledge the challenges related to soil degradation, water scarcity, climate change and will also provide a substructure for developing sustainable agricultural practices. This chapter provides a survey of qualitative study regarding the detailed overview of SPAC, improving the sustainability, crop productivity, resource use efficiency and dealing with the challenges regarding the environmental stresses.

F. Maqsood · A. Hussain · A. Naz
Institute of Soil and Environmental Sciences, University of Agriculture Faisalabad, Faisalabad, Pakistan

M. S. Islam (✉) · F. U. Hassan
Institute for Environment and Development (LESTARI), Universiti Kebangsaan Malaysia (UKM), Bangi, Selangor, Malaysia
e-mail: sujaul@ukm.edu.my

M. Tayab
Center of Agricultural Biochemistry and Biotechnology, University of Agriculture Faisalabad, Faisalabad, Pakistan

A. I. Ahmad
Department of Plant Breeding and Genetics, University of Agriculture Faisalabad, Faisalabad, Pakistan

M. Umair
College of Natural Resources and Environment, Northwest A&F University, Yangling, China

M. E. Toriman
Faculty of Social Sciences and Humanities, Universiti Kebangsaan Malaysia (UKM), Bangi, Selangor, Malaysia

M. Shaaban (ed.), *Soils and Sustainable Agriculture*, Frontier Studies in Soil Science,
https://doi.org/10.1007/978-3-031-91114-9_16

1 Introduction to General Terminologies

1.1 Soil Moisture

Soil moisture refers to the amount of water stored in the unsaturated zone of the soil. It can also be used to denote any water present within the soil (Gavrilescu, 2021). It encompasses both the water that is accessible to plant roots and the water that is tightly bound to soil particles or held in forms that plants cannot use (Hartemink, 2016). Soil moisture levels fluctuate based on rainfall, evaporation, irrigation, and soil type (Qiu et al., 2016). To calculate soil moisture, following equation is used (Shukla, 2023),

$$\theta = \frac{Vw}{Vt} \times 100$$

where:

- θ = Volumetric water content (soil moisture).
- Vw = Volume of water in the soil.
- Vt = Total volume of the soil.

1.2 Soil Water Content

Soil water content represents the total amount of water contained within the soil, including both usable and unusable water (Buitink et al., 2020). It is typically expressed as a percentage of the soil dry weight or volume (Datta et al., 2017). This measurement includes all the water in the soil, ranging from free water in larger pores to water that is tightly bound to soil particles (Datta et al., 2017). Soil water content helps in understanding the soil overall water-holding capacity and how much moisture is available for plant use (Liu et al., 2019). The soil water content can be measured as the Gravimetric water content or the volumetric water content, depending on the method of measurement (Camuffo, 2018).

Gravimetric Water Content Formula:

$$Soil\ Water\ Content(Gravimetric) = \frac{Mass\ of\ Water\ in\ Soil}{Mass\ of\ Dry\ Soil} \times 100$$

This formula gives the percentage of water in the soil based on the dry weight of the soil (Gholamzadeh et al., 2024).

Volumetric Water Content Formula:

$$Soil\ Water\ Content(\text{Volumetric}) = \frac{Volume\ of\ Water\ in\ Soil}{Total\ Volume\ of\ Soil} \times 100$$

This formula expresses the volume of water in the soil relative to the total soil volume, providing a measure of how much of the soil's total volume is occupied by water (Sharpe, 2018).

1.3 Field Capacity

Field capacity (FC) is the amount of water that a soil can hold after it has been saturated and allowed to drain under the influence of gravity (Reynolds, 2018). At field capacity, the soil holds water in the micropores while larger water-filled spaces (macropores) have drained away typically with a soil water potential or suction pressure of approximately **− 33 kPa (or − 0.33 bar)** (van Lier, 2017). It typically occurs within 2–3 days of a rain or irrigation in pervious soils with consistent structure and texture (Assi et al., 2019). At field capacity, the soil's water and air contents are thought to be perfect for crop development (Razzaghi et al., 2020). The field capacity can be measured as

$$Field\ Capacity = \left(\frac{Volume\ of\ Water\ at\ FC}{Volume\ of\ Soil} \times 100\right)$$

where,

Volume of water at FC	is the amount of water the soil holds after drainage.
Volume of soil	is the total volume of the soil sample.

1.4 Permanent Wilting Point

Permanent wilting point (PWP) refers to the soil moisture level at which plants can no longer extract water, and they permanently wilt and cannot recover, even if water becomes available later (Hohenegger & Stevens, 2018). At this point, the soil still holds some water, but the roots are having difficulty extracting it due to strong capillary forces (Yao et al., 2020). The permanent wilting point varies per plant species, although it is usually around **− 1500 kPa (or − 15 bar)** (Santra et al., 2018). This is a critical threshold for plant survival leading to irreparable damage and wilting (Bechtold et al., 2018). There is no simple formula for determining PWP, as it is typically determined experimentally (Affinnih et al., 2024). However, it can be described in terms of water content or soil water potential. It is often measured as the gravitational water content or the volumetric water content at a suction pressure of − 1500 kPa (Santra et al., 2018).

$$PWP(Gravimetric) = \left(\frac{Mw}{Md} \times 100\right)$$

where:

- M_w = Mass of water remaining at the permanent wilting point.
- Md = Mass of dry soil.

1.5 Available Water

Available Water (AW) is the portion of water in the soil that is accessible to plants (Santra et al., 2018). It is the difference between the water content at field capacity (FC) and the permanent wilting point (PWP) (Schaller et al., 2020). Soil texture influences the amount of available water, with finer soils (e.g., clay) generally holding more available water than coarser soils (e.g., sand) (Vinhal-Freitas et al., 2017). It can be measured as by using following formula (Wang et al., 2017).

$$AW = FC - PWP$$

where:

- FC = Water content at field capacity (max water available after free drainage)
- PWP = Water content at permanent wilting point (water unavailable to plants)

1.6 Unavailable Water

Unavailable water is the water in the soil that plants cannot absorb, usually because it is held too tightly by soil particles (Eden et al., 2017). It is found below the permanent wilting point (PWP), where the water is bound in forms that are inaccessible to plant roots, such as hygroscopic water (Juarez-Enriquez et al., 2017). This water does not contribute to plant growth (Metzger et al., 2017).

1.7 Volumetric Water Content

The volumetric water content (θ) is calculated as the volume of water present in a given volume of soil (Bertocco et al., 2023). It can be stated as a ratio, percentage, or depth of water per depth of soil (assuming a unit surface area) (Ceaglio et al., 2017), for example, inches of water per foot. This is an important measure for understanding soil moisture levels and how much water is available for plant uptake. The volume-based water content, also called volumetric water content, can be computed as (Fatkhuroyan et al., 2021; Feng JunJie et al., 2018),

$$Volumetric\ moisture\ content = gravimetric\ moisture\ content \times bulk\ density\ of\ soil.$$

Or

$$\theta = \left(\frac{Vw}{Vs}\right)$$

where:

- θ = Volumetric water content (in m^3 of water per m^3 of soil or as a percentage).
- Vw = Volume of water in the soil (in cubic meters or liters).
- Vs = Total volume of the soil sample (in cubic meters or liters).

Volumetric water content is typically measured in percentages, representing how much of the soil's volume is filled with water.

1.8 Gravimetric Methods

Gravimetric Method refers to a technique used to measure the amount of water in soil based on the weight of the soil before and after drying (Zhu et al, 2024). It involves determining the mass of water present in a soil sample by measuring the difference in weight between a wet soil sample and its dry counterpart (McGinn et al., 2020). The gravimetric method is simple and widely used for determining soil water content. It provides a direct measure of the mass of water in the soil (Fatkhuroyan et al., 2021). The gravimetric water content (or soil moisture content) is calculated as (Howell et al., 2019),

$$Gravimetric\ Water\ Content = \left(\frac{Mw}{Mv} \times 100\right)$$

where:

- Mw = Mass of water in the soil (wet weight–dry weight).
- Md = Mass of dry soil.

1.9 Water Movement and Potential

Water movement refers to the flow of water between different systems, such as soil, plants, and the atmosphere, driven by differences in water potential (Radcliffe et al., 2002). Water always moves from regions of higher water potential (less negative or more positive potential) to regions of lower water potential (more negative potential) (Rawls et al., 1992). **Water potential (Ψ)** is a measure of the potential energy of water and includes various components that can either increase or decrease the water ability to move (Anderegg et al., 2017). The total water potential (Ψ) is the sum of four components (Kosal, 2023),

$$\Psi total = \Psi \mathrm{s} + \Psi \mathrm{p} + \Psi \mathrm{g} + \Psi \mathrm{m}$$

where:

- Ψs = Solute Potential: The effect of solutes (lowers water potential).
- Ψp = Pressure Potential: Physical pressure (positive in plant cells).
- Ψg = Gravitational Potential: Due to gravity (important in vertical movement).
- Ψm = Matric Potential: Due to water adhesion to soil particles (negative in unsaturated soils).

1.10 Soil Water Potential

Soil Water Potential (Ψsoil) refers to the energy status of water in the soil, which determines the direction and rate of water movement within the soil (Novick et al., 2019). It is a measure of the potential energy of water in the soil and is influenced by various factors like solute concentration, pressure, and gravity. The Soil water potential is calculated as (Scharwies & Dinneny, 2019)

$$\Psi soil = \Psi \mathrm{s} + \Psi \mathrm{p} + \Psi \mathrm{g} + \Psi \mathrm{m}$$

where:

- Ψs = Solute Potential
- Ψp = Pressure Potential
- Ψg = Gravitational Potential
- Ψm = Matric Potential

1.11 Osmotic Potential

Osmotic Potential (or Solute Potential Ψs) is the component of water potential influenced by dissolved solutes (Johnson et al., 2018). It is always negative, as the presence of solutes lowers water potential (Bilska-Kos et al., 2017). Water moves from areas of higher osmotic potential (fewer solutes) to lower osmotic potential (more solutes), driving processes like osmosis in plants and cells (Wennerström & Oliveberg, 2022).

The value of the osmotic potential can be determined using the Van't Hoff equation (Bosse et al., 2024),

$$\Psi s = -CiRT$$

where:

- Ψs = Osmotic potential (MPa).

- i = Ionization constant (the number of particles the solute dissociates into in solution).
- C = Molar concentration of the solute (mol/L).
- R = Universal gas constant (0.0831 L MPa mol^{-1} K^{-1}).
- T = Temperature in Kelvin (K).

1.12 Gravitational Potential

Gravitational potential refers to the potential energy of water due to its position relative to the Earth's gravity (Luo et al., 2022). It is especially relevant in the movement of water in the vertical direction (e.g., in soil or plants) where gravity acts on water (Lu & Zhang, 2019). The gravitational potential (Ψg) is calculated using the formula:

$$\Psi g = p \cdot g \cdot h$$

where:

- Ψg = Gravitational potential (pascals or megapascals).
- ρ = Density of water (kg/m^3)
- g = Gravitational acceleration (m/s^2).
- h = Height or elevation above a reference point (m).

1.13 Pressure Potential

Pressure Potential (Ψp) is the component of water potential that represents the physical pressure exerted on water within a system (Cai et al., 2024). It is particularly relevant in plant cells and soil systems where water is under pressure. Pressure potential can be positive, negative, or zero, depending on the system (Larsbo et al., 2016).

1.14 Plant Water Relation

Plant water relations refer to the study of how water moves through plants and how plants interact with their water environment (Kuromori et al., 2018). These relations are crucial for understanding how plants absorb, transport, and use water, and how they respond to changes in water availability (Bradford, 2017). The water status in plants is influenced by the balance between water uptake from the soil, internal water movement, and water loss to the atmosphere through transpiration (Kirkham, 2023).

1.15 Transpiration

Transpiration is the process by which water moves through a plant and evaporates from aerial portions such as leaves, stems, and flowers (Maurel et al., 2016). This creates a negative pressure (tension) in the plant, facilitating the movement of water from the roots through the stem and into the leaves (Mi et al., 2024). Transpiration also cools plants, alters cellular osmotic pressure, and allows for the mass flow of mineral nutrients (Lin et al., 2017). When water uptake by the roots is less than the water lost to the atmosphere by evaporation, plants seal microscopic holes known as stomata to reduce water loss, which slows nutrient uptake and reduces CO_2 absorption from the atmosphere, restricting metabolic processes, photosynthesis, and development (Clark et al., 2022).

1.16 Stomatal Conductance

Stomatal conductance is the rate at which water vapor and gases, such as carbon dioxide, move through the stomata (tiny pores) on plant leaves (Buckley, 2019). It serves as an indicator of how efficiently the stomata facilitate gas exchange, especially during processes like photosynthesis and transpiration (Chen et al., 2017).

1.17 Root Water Uptake

Root water uptake is the process by which plant roots absorb water from the soil (Zarebanadkouki et al., 2019), which is vital for various physiological functions such as nutrient transport, photosynthesis, and maintaining cell turgor pressure (Chen et al., 2017).

1.18 Plant Water Availability

Plant water availability refers to the water in the soil that plants can access for growth and physiological functions (Marschner & Rengel, 2023). It depends on factors like soil type, root depth, and environmental conditions (Bodner et al., 2015). This water is typically available between field capacity (where the soil holds its maximum moisture) and the permanent wilting point (where water is no longer usable by plants) (Van Vliet et al., 2016).

1.19 Plant Available Water Content

Plant available water content is the amount of water in the soil that plants can absorb, calculated as the difference between the field capacity and the permanent wilting point (Yan et al., 2015). It represents the water that is neither too tightly bound to soil particles nor too free-flowing to be lost (Minasny & McBratney, 2018).

1.20 Consumptive Use of Water

Consumptive Use of Water refers to the total water used by plants, crops, or ecosystems that is lost through processes such as transpiration, evaporation, and assimilation into plant tissues (Mekonnen et al., 2015). This water is "consumed" and does not return directly to the water source (Cosgrove & Loucks, 2015).

1.21 Water Use Efficiency

Water use efficiency (WUE) is the ratio of biomass or crop yield produced per unit of water consumed or transpired (Hatfield & Dold, 2019). It assesses how effectively a plant or crop uses available water for growth and productivity (López-Calcagno et al., 2020). A higher WUE indicates that the plant or system is more efficient in utilizing water, producing greater output with less water. WUE is typically expressed as (Medrano et al., 2015),

$$WUE = \left(\frac{Crop\ Yield\,(biomass\ or\ grain)}{Water\ Consumed\,(evapotranspiration)} \right)$$

2 Introduction to SPAC

One of the most prevalent and significant natural resources on the planet is water (Oberle et al., 2019). Life cannot exist without water, which is also crucial for vegetation that depends entirely on its availability (Trautmann et al., 2022). It is essential to plant physiology and metabolism. Since life and water are essential to Earth, it's possible that the planet is a living system. In the majority of terrestrial ecosystems, the primary hydrological output is typically driven by the movement and exchange of water between the soil, vegetation, and atmosphere (Bernacchi & VanLoocke, 2015). On the other hand, plant activity and soil characteristics play a key role in determining how rainfall is distributed among evapotranspiration, runoff, and deep percolation (Arnold et al., 2020). These factors all affect the regional water budget and balance. Water movement via the soil-plant-atmosphere continuum (SPAC) is still a topic of active investigation in the fields of hydrology, plant, and agricultural sciences because of its vital significance in the living world (Ma et al., 2022). In both natural and agricultural ecosystems, SPAC are essential in light of the present and projected climate (Hasan & Eslamian, 2023). Basic ideas regarding water potential and content in plants and soils are covered in this session. It also emphasizes how water moves through soils and the factors that propel this movement from the soil to the atmosphere, roots, and leaves.

2.1 Definition

The soil, plants, and atmosphere together make up the field environment, which is a physically interconnected and dynamic system where various water flow processes occur independently (Fatichi et al., 2016). The Soil-Plant-Atmosphere-Continuum (SPAC), is the name given to this cohesive system and was first postulated by Huber in 1924 (Hinckley et al., 2011). Water flows from areas of higher total water potential to regions of lower total water potential at a pace determined by the medium's hydraulic resistance (Livingston & Topp, 1993), according to the basic idea of SPAC. Water moves from soil through plants and into the atmosphere via a channel known as the soil-plant-atmosphere continuum (SPAC) (Goyal et al., 2021).

2.2 Principles

The following two key concepts are essential for understanding the SPAC, The mass and energy budgets that serve as conservation principles and the transport principles, which link a quantity's flow to the gradient or difference of other quantities that affect or "force" the flow and characterize the "state" of the exchange process, are examples of conservation principles (Martin et al., 2018). Integrative studies of the SPAC are built upon the principle of mass and energy conservation (Liu et al., 2015).

2.3 Importance

The term "continuum" in the description draws attention to how the water connection along the pathway is continuous. A diffusion gradient between the stomatal pores of leaves is caused by the low water potential of the environment and the comparatively larger (i.e., less negative) water potential inside the leaves, which draws water out of the leaves as vapour (Harrison et al., 2020). Since the water in the air inside leaves is kept at saturation vapour pressure, more water molecules evaporate off the surface of mesophyll cells to replace the molecules lost as water vapour seeps out of the leaf (Amitrano et al., 2019). Water lost at the cell surface is replaced by water from the xylem due to the cohesion-tension properties of plant xylem, which pulls additional water molecules from the roots to the leaf through the xylem (Chen et al., 2023).

The greatest visible element of the SPAC is water, and our knowledge of how life interacts with its surroundings is based on how this water interacts with energy. The main ways that solar energy eventually warms the atmosphere are through the extraction of energy from the surface through evaporation and the release of energy to the atmosphere through condensation (Kalogirou, 2023). The essential feedback between the water and energy cycles that stabilizes the soil, plant, and atmospheric system and enables life to exist there. This feedback is provided by this "latent heat," which is connected to the phase transition of water between liquid and vapor phases rather than a change in temperature (Carey, 2020).

Water flows through SPAC along a complicated direction that is impeded by a variety of systemic resistances (Gao et al., 2022). In the past, the description of plant hydraulics has been significantly less mechanistic than the handling of soil resistances. While flow through a plant is complex, flow through soil is more suited to quantitative analysis. (Van Dam et al., 2016). When the soil is adequately hydrated and the atmospheric conditions are moderate, plants often provide minimal resistance (Martin et al., 2018). Plants experience water deficits as the soil dries up, and their stomata may close entirely or partially (Bhattacharya & Bhattacharya, 2021). In these circumstances, plants provide strong resistance to water flow. In moist soil conditions, plant resistance predominates over the continuum's overall hydraulic resistance (Oliveira et al., 2021).

Drought-related changes in soil resistance lead to corresponding increases in plant resistance, which in turn maintains a significant impact on water transport (Tamindžić et al., 2024). The distribution or utilization of the energy flow that reaches the field depends on the amount of water in the soil (Dickson & Fanelli, 2018). Since they both deal with the same process in the same environment, the water balance and the energy balance can be linked interchangeably. When multiple soil–water processes occur concurrently or sequentially, such as infiltration, drainage, evaporation, and plant water absorption, they are actually highly interconnected. The foundation of the water balance field lies in the law of conservation of mass, which states that matter can only change forms or locations, but cannot be created or destroyed (Katsiampoura et al., 2018).

Since there is typically no significant recompositing or decomposition of water in the soil, a soil profile with a finite volume cannot have its water content decrease or increase without external addition (via capillary rise or infiltration) or transportation of the water content to the atmosphere (via evapotranspiration) or deeper zones (via drainage) (Li et al., 2014).

Since the field water balance comprises energy-consuming activities, it is closely related to the energy balance. The classical law of conservation of energy, which holds that energy cannot be generated or destroyed, but can only be absorbed from or discharged to the outside of a system and change forms along the way, is expressed by the energy balance (Yuhua, 2015).

The distribution and utilization of the energy flux that reaches the field are influenced by the amount of water present in the soil (Vanderborght et al., 2017). Because they function within the same environment and share common processes, the water balance and energy balance are inherently interconnected. A comprehensive understanding of both balances is essential for providing a physical description of the SPAC, Since the majority of water given to the field usually evaporates, the evaporation process makes up the largest portion of the water balance and is the field's main energy user (Tao et al., 2018). Evaporation is thus dependent on the availability of both energy and water (Cavusoglu et al., 2017).

2.4 Soil Water Content

The amount of water in the soil affects a number of essential biophysical processes. Seed germination, plant development and nutrition, microbial breakdown of soil organic matter, nutrient conversions in the root zone, and heat and water transmission at the interface between the land and atmosphere are all impacted (Wei et al., 2021). SWC quantification is required for a variety of applications, including as field monitoring in horticultural and agricultural systems and large-scale calibration of global climate models (Pereira et al., 2020). In this case, SWC measurements help optimize irrigation levels, schedules, and plant nutrition (Ortega-Farias & Intrigliolo, 2021). In the latter case, SWC knowledge is involved in the partitioning of solar radiation into sensible and latent heat (Ippolito, 2023).

In addition, SWC has a significant role in determining how quickly soil organic matter breaks down (Rey, 2015), which has an impact on how quickly soil respiration and soil carbon sequestration occur (Zhao et al., 2022a). Furthermore, SWC is essential for the physicochemical processes of mineralization, volatilization, and nitrification that occur when basic nutrients, including nitrogen, are transformed (Zhao et al., 2022b). There is not much information regarding the amount of water available to plants in the soil–water content. Soil water content can be measured both directly and indirectly, with the results expressed quantitatively in two distinct ways. First, the mass of oven-dry soil is used to describe the gravimetric soil water content method (Pasha et al., 2016).

$$W = M_w / M_s$$

where Ms is the mass of the solids, or the bulk of the oven-dried soil, and Mw is the amount of water (kg). Typically, it is multiplied by 100 to determine the mass proportion of water. The volume of soil water divided by the bulk or total volume is known as the volumetric soil water content.

$$\theta = V_w / V$$

where V is the bulk volume of oven-dry soil (m^3) and Vw is the volume of water (m^3) (An et al., 2017).

2.5 *Effect of Soil Physiological Properties in Water Movement and Transportation Within Soil (Organic Matter, Compost, Biochar, Imbalance of Mineral and Nutrients)*

Biochar is referred to as "black gold" (Alfei & Pandoli, 2024). In order to minimize pollution and maximize resource utilization, high aromatization of organic waste, such as chicken dung, urban sludge, and agricultural straw at high temperatures and anaerobic conditions is the process used to create biochar (Alfei & Pandoli, 2024). Broad surface areas, high organic carbon (OC) contents, high charge densities, low bulk densities, and very porous structures are characteristics of biochar (Qin et al., 2020). Numerous studies have demonstrated the benefits of biochar on the biological and chemical characteristics of soil, including pH, EC, zeta potential, and cation exchange capacity (CEC) (Omara et al., 2023). Moreover, the amount of water that was available in the soil at any matric potential was increased by the addition of biochar (Xiao et al., 2016). The amount to which biochar increased the available water capacity (AWC) in coarse-textured soils varied according on the temperature at which the biochar was pyrolyzed (Ibrahimi & Alghamdi, 2022). However, the influence differed according on the texture of the soil. It was traditionally believed that soil hydraulic conductivity (K) was an important soil feature that controlled water infiltration and movement within the soil profile, influencing the likelihood of soil runoff after heavy rainfall. The application of biochar may result in an increase in the saturated hydraulic conductivity of soil (Toková et al., 2020).

Biochar has drawn a lot of interest in managed systems because of its potential to increase soil fertility and reduce global warming through carbon sequestration (Agegnehu et al., 2017; Qambrani et al., 2017). According to a recent meta-analysis, biochar boosts agricultural yields by 10 and 30% on average, respectively (Agbna et al., 2017). Zhou and colleagues report that woody plants experience a mean biomass growth response of 41% (Zhou et al., 2022). Biochar's high specific surface area, cation exchange capacity, and microporosity have been shown to improve plant growth and function across various environments. High specific surface area, cation exchange capabilities, and micro porosity found in biochar have been shown to enhance plant growth and function in a variety of environments (Leng et al., 2021). These characteristics are heavily dependent on the feedstock, which might include waste materials from agriculture and forestry including woodchips, sawdust, corn Stover, and manure, as well as the temperature at which the pyrolysis is occurring (Tomczyk et al., 2020). The biochar that is produced from woody feedstock exhibits significant ameliorative qualities, such as its strong capacity to absorb pollutants from soil, hold onto water and nutrients, raise the pH of the soil, and supply nutrient pulses that may be limited in specific systems (Anand et al., 2022). In order to facilitate cultivation and boost agricultural output, biochar can enhance the physiochemical characteristics of soil, such as water infiltration, water holding capacity, and saturated hydraulic conductivity (Zhang et al., 2021).

Another soil supplement that has been hailed for its many advantages is compost, an organic fertilizer that has been shown to accelerate plant development (Sayara et al., 2020). A by-product of the mushroom business, spent mushroom substrate (SMS), also known as mushroom compost (Umor et al., 2021), is frequently applied as a soil supplement and has been found to have especially potent beneficial effects on soil function and plant growth (Ho et al., 2022). Similar to biochar, mushroom compost usually liming soils and improving nutrient availability and uptake, particularly for N, P, and K (Hu et al., 2021).

Improving soil organic carbon (OC) can boost soil quality in several ways, such as by increasing biological activity, promoting better soil structure and aggregation, enhancing nutrient retention, and improving moisture and temperature conditions (Merante et al., 2017). Increasing soil organic carbon is another recommended strategy for managing drought (Iizumi & Wagai, 2019). Studies on the effects of adding organic matter to soil have not consistently demonstrated improvements in water retention at the wilting point and field capacity (Minasny & McBratney, 2018). Previous studies indicated that only coarse-textured soils with less than 20% clay content were impacted by OM in terms of the AWC (Minasny & McBratney, 2018). Another theory is that since higher aggregation can also result in a decrease in micropore volume, an increase in AWC may not always follow from an increase in aggregation (Liu et al., 2021).

2.6 *Water Movement in Saturated Soil*

Because of the different conditions, pressures, and directions that affect the water, soil water movement is a dynamic process. The transport of soil water can be significantly influenced by temperature. There is not much movement at all on frozen soil. There will be some vaporization of water under very high temperatures. All of the soil's pores are totally filled with water when the soil is saturated, and there is no air present. In poorly drained soil, the surface may be briefly saturated, while the lower area may frequently be saturated. Oversaturated Circulation Both horizontally and vertically, saturated flow takes place. Hydraulic gradients transport water from higher points to lower points in addition to the usual forces (unsaturated flow conditions) acting on the water particles.

Saturated flow equations

2.6.1 Poiseuille Equation

The volumetric flow rate (Q) at the microscopic level of a single pore is represented by Poiseuille's equation, which can be generally represented as a water-filled cylinder with a certain radius (r):

$$Q = \pi r4\, pg\Delta H/8\eta L$$

where ΔH is the difference in total head (potential on a weight basis) between two points along the cylinder separated by a distance, η is the water viscosity, p is the water density, and g is the acceleration due to gravity. L. The disproportionate impact that large-diameter pores, or macropores, have on water transmission has been demonstrated using Poiseuille's equation (Loudon & McCulloh, 1999).

2.6.2 Laplace's Equation

Laplace's equation, a well-known equation found in almost all areas of applied mathematics, and a general differential equation are needed to address drainage and seepage issues involving groundwater. Saturated flow equation and it are closely related, especially when it comes to groundwater hydrology. The saturated flow equation, which may be expressed as follows, is derived from Darcy's law and the equation of continuity.

$$\nabla \cdot (K\nabla h) = 0$$

In this case, h represents the hydraulic head, K is the porous medium's hydraulic conductivity, and ∇ provides the gradient operator. When the medium is homogenous and isotropic and K is constant, the equation becomes:

$$K\nabla^2 h = 0$$

Which further simplifies to:

$$\nabla^2 h = 0$$

This particular equation, known as Laplace's equation, has the hydraulic head, h, acting as the potential function θ in the formula. An entirely saturated, uniform, isotropic aquifer devoid of sources and sinks is implied by this equation (Medková, 2018).

2.6.3 Linear Flow Laws

Linear flow laws describe the relationship between the fluid's flow rate and the driving force in a medium like saturated soils. The driving force is typically depicted as a pressure gradient. These laws presuppose that the driving force causing the linear relationship is directly proportionate to the system's response. These are the fundamental laws of linear flow:

2.6.4 Darcy's Law (Fluid Flow Through Porous Medium)

When it comes to water movement in saturated soils, Darcy's law is the most fundamental linear flow law. French sanitary engineer Henri Darcy created Darcy's law in 1856. The flow of water through a porous media, like soil, is directly proportional to the hydraulic gradient and inversely proportional to the medium's resistance, which is determined by its hydraulic conductivity, according to Darcy's Law. Soil drainage, irrigation control, and groundwater flow prediction all depend on this relationship. The following is the current form of Darcy's law equation:

$$Q = -K \cdot A \cdot \frac{\Delta H}{L}$$

where A is the flow's cross-sectional area (m^2), ΔH is the hydraulic head difference (m), L is the flow path's length, Q is the volumetric flow rate (m^3/s), and K is the soil's hydraulic conductivity (m/s). The equation's negative sign indicates that the direction of the water flow is toward a diminishing potential. Thus, the sign made it clear whether the flow is upward or downward (Muskat, 1937).

2.6.5 Limitations

A single pore's fluid flow state is not described by Darcy's law. In addition to being isotropic, the porous medium must be averaged over the ensemble homogenous in order for Darcy's law to hold. It is only helpful for one-dimensional fluid flow; in order to describe multi-dimensional fluid flow, one must treat the Darcy flux or the particular discharge as a vector quantity. In these situations, the Darcy flux transforms into a vector point function and can change in transient fluid flows as a function of time and space. The effects of gravity and capillaries are not considered.

Darcy's law ignores the areas of the pore spaces that are stationary as well as the reality that fluid flow actually occurs by preferential paths rather than through all of the pores. It is only applicable to non-isothermal fluid flows that are incompressible. Only steady-state fluid flow circumstances make it applicable. Although surface tension, also known as interfacial tension at the immiscible fluid–fluid interface, is a crucial parameter in the analysis of petroleum production and enhanced oil recovery, its effects on multiphase fluid flow through porous media are not taken into account by this law.

2.6.6 Hydraulic Conductivity

One of the main components of Darcy's Law, hydraulic conductivity (H/K), measures how easily water can pass through soil pore spaces. It is necessary to take into account the characteristics of both the fluid (density, viscosity) and the soil (pore size, distribution, etc.). It is important to distinguish between the intrinsic permeability of the flow medium and hydraulic conductivity. A uniformly saturated soil's hydraulic conductivity is often constant, whereas in natural field soil, it is controlled by elements including soil particles' stability, root holes, fractures, and warm holes. Hydraulic conductivity is generally not significantly affected by the amount of sand, silt, and clay. Hydraulic conductivity is the primary distinction between saturated and unsaturated flow.

Because the amount of water in the soil is constant throughout the profile, it reaches its maximum when the soil is saturated. Hydraulic conductivity rapidly decreases as the soil desaturates under unsaturated circumstances.

Water is vital for plants as it participates in the processes of photosynthesis, nutrient movement, and maintenance of cell turgor. Water is also needed in internal physiological processes that are necessary for the healthy vegetation of a plant (Engel et al., 2000). Water in this case serves as one of the reactants that help in the process of photosynthesis by being split in to release oxygen and also deliver electrons needed for the conversion of carbon dioxide to glucose (Steudle, 1992). The water temperature and level have a direct effect on the water rate of assimilation because of their influence on the opening of stomata and the gas metabolism. Water also serves as a fluid and helps in the movements of vital nutrients from the ground through the plants vascular tip (Mengel et al., 2001). Special water 'pores' called aquaporins also assist in quick acquisition of water and nutrients to help in normal plant functioning. Cells of plants contain water, and this exerts turgor pressure that is important in maintaining the shape and healthy cells for plant growth (Goss & Ehlers, 2009).

3 Water Transport Within the Plant Body

3.1 Structural Components

3.1.1 Vascular Tissues (Xylem)

Xylem vessels' primary function is to move minerals and water up from the roots to the leaves. Transpiration is the mechanism by which water that the roots absorb from the soil is carried upward through xylem vessels. The process known as transpiration is what happens when water from the leaves evaporates, drawing water up into the xylem due to a suction force (Kim et al., 2014).

3.1.2 Phloem

The phloem vessels are responsible for downward movement of sugars and the rest organic compounds from the leaves to the rest part of the plant. Sugars in the leaves are produced by photosynthesis and transferred to other parts of the plant for growth and energy production (De Schepper et al., 2013). Vascular bundles: Vascular tissues typically occur in bundles within the body of the plant. The arrangement of vascular bundles varies with diverse types of plants and may therefore be considered as one of the characteristic features while identifying the species of a plant. There are three main types of vascular bundle arrangements (Cano-Delgado et al., 2010). Collateral: There are xylem and phloem together, side by side, though phloem may often appear outside. In most angiosperms, it is the commonest form. Concentric: Xylem surrounds the phloem completely or vice versa. This is arranged in some gymnosperms, such as pines and a few ferns. Radial: Xylem and phloem are disposed of in a radial manner in which xylem is, in most cases, at the center of the bundle. This is very typical to dicot stems (De Rybel et al., 2016; Ye, 2002).

3.2 Root System

A plant's root system is of immense importance to the general productivity, fitness, and interaction of a plant with its environment. Species exhibit at least partially differing root architectures, which influences their ability to absorb nutrients and adapt to their surroundings. Here follows a summary of important findings from current research on the root system. Root hairs: These consist of microscopic, outward extensions of epidermal cells expanding the surface area for water absorption. Root cortex: It is the outer parenchyma layer that enables water movement from the hair, up into the tissues of the vascular part. Endodermis: It is a layer of cells that serves as a barrier that controls the entry and exit of substances from the vascular cylinder. The endodermis contains Casparian strips, which are akin to waxy bands that will prevent water from passing through the cell walls (Eissenstat, 2000; Lambers et al., 2006).

3.3 *Water Absorption in Plants*

3.3.1 Osmosis and Active Transport

In general, water absorption happens in a plant via osmosis: that is, the water molecules tend to move through a semipermeable membrane in the direction from the region with lower solute concentration to that with higher solute concentration. Root hair cells contain a higher concentration of solutes than the soil solution. The resultant concentration gradient is established. Water moves into the root hair cells by osmosis. This causes the cells to swell up and eventually become turgid. It also plays an active role in the absorption of water when the potential of water in soil is low. The root's cells are continually active in ionic transport towards the root. It actively causes the high negative water potential inside the root. Thus, the generated gradient helps the movement of water into the root (Rhodes & Samaras, 2020; Wang et al., 2014).

3.3.2 Apoplastic and Symplastic Pathways

There are two transport pathways in the plant body the Apoplastic and the Symplastic pathway. Their basis of difference is the special function they provide within plant physiology. Movement through cell walls as well as intercellular spaces forms the apoplastic pathway. Movement through cytoplasm is allowed by the symplastic pathway due to the presence of plasmodesmata. Apoplastic Pathway: The pathway consists of cell walls as well as intercellular spaces between the cells through which water and solutes move rapidly without crossing the cell membranes. This pathway also plays a significant role in plant immunity also; it is the site for defense responses against the pathogen (Doehlemann & Hemetsberger, 2013). Symplastic Pathway: The pathway across the cytoplasm, integrated through plasmodesmata, shows a selective transport of both nutrients and signaling. While there is certainly apoplastic loading as in peach, this pathway is balanced such that for those plants both sorbitol and sucrose can be translocated with equal ease (McCulloh et al., 2003).

3.3.3 Transpiration

Water transport, upward through the plant followed by evaporation within the leaf is one of the most significant physiological activities of plants. It has been considered to be mainly under environmental control by the potential in the soil, temperature, and intensity of radiation. Transpiration occurs because of a diffusion gradient of water vapor from the internal leaf to the outside atmosphere (Vadez et al., 2014). The main regulation is stomatal, with stomatal opening dependent on both light and CO_2 concentrations and hence controlling rates of water loss (Nilson & Assmann, 2007). Soil moisture has major influence over transpiration rates; accordingly, it might have been predicted that Scots pine would be more depressed in transpiration during drought compared with pubescent oak which was not so depressed by soil moisture shortage (Poyatos et al., 2005).

3.3.4 Root Pressure

Root pressure is also another vital component of water movement in the plant roots; consequently, it controls hydraulic conductivity along with the dynamics of water flow. Discussion on and implications of root pressure of water transport are further discussed in the following sections. Root pressure leads to a considerable increase in the efficiency of transport, though undoubtedly on a larger scale the overall dynamics of water uptake is determined by environmental factors and root form as well (Aroca et al., 2012; Steudle, 2000).

4 Water Uptake by Roots Through Apoplast or Symplast Stimulated by Endodermis or Casparian Strips

4.1 Factor Effecting Water Transport

4.1.1 Soil Moisture

Soil moisture controls the transport of water in a plant. Proper soil moisture ensures that water is sucked from the roots of plants and transported to distinct parts of the plant. When the soil is moist, there will be an increase in water uptake and transpiration; that is, water lost by the leaves. This ensures proper transport of water throughout a plant (Pittman et al., 2020). If the soil moisture is low, the opposite happens. Water becomes hard to be absorbed by plants at such times. Growth, development, and yield may therefore be adversely affected by water stress. In other words, this activity may be nothing but stomatal closure to lessen transpiration or a deep root system that absorbs water deeper in the layers (Henry et al., 2018). Such factors include soil texture, infiltration rates, and rainfall and evaporation. Sandy soils retain less water compared to clay soils; hence the drought impacts are worsened. Drought can be regulated with appropriate irrigation techniques in such a manner that the optimal amount of soil moisture is preserved, especially for arid and semi-arid regions (Wen et al., 2020).

4.1.2 Soil Salinity

Soil salinity is the condition of having excessive salt concentrations within the soil. It will adversely affect the ability of the plant to absorb water. This means that if the salt concentration is high, then so is the osmotic pressure in the soil, and, thus, the roots of the plants have difficulty absorbing more water into their surrounding environment (Rozema & Schat, 2013). Sometimes this condition is described as "salinity stress." The cell dries up because the plant fails to soak in the water. This dehydration causes a number of physiological problems such as wilting, decline in growth, and eventually death. Higher salt concentration inhibits the uptake of nutrients by plants, which presents an additional set of health implications for it (Geng & Boufadel, 2015).

4.1.3 Root Morphology

Importantly, the physical structure of a plant must take in and move water; a root's surface area, structure, system architecture, and adaptations all contribute to this uptake of water. Features that include root hairs, lateral roots, and deep taproots or networks of fibrous roots increase the surface area for water absorption. Internal structure of the root consists of vascular bundle, endodermis, cortex, and epidermis through which water passes from the soil into the vascular system of a plant. Pneumatophores and succulent roots are two examples of adaptations that enable the plants to survive in particular conditions. Knowledge regarding the structure of roots will make one understand how plants have adapted toward ensuring an adequate and effective supply of water (Jackson et al., 2000; Lambers et al., 2006).

4.1.4 Plant Hormones

Auxin plays a role in moving water within plants by being produced in the shoots and then traveling to the roots where it can help boost root growth to improve water absorption levels as discussed before. Furthermore, auxin can also impact how water is spread throughout the plant by changing how well different tissues conduct water (Alarcón et al., 2019). ABA on the hand is known as the "stress hormone" because it is produced when plants experience a lack of water. During periods, for plants like droughts. They produce more ABA hormone to make some changes in their body to save water effectively by causing the closure of tiny pores on the underside of leaves called stomata and reducing water evaporation through them (Bharath et al., 2021). Additionally, ABA helps in enhancing the development of root hairs to improve water absorption, from the ground. Also plays a part in regulating how well water and minerals are transported throughout the plant by affecting xylem's ability to conduct fluids (Fiscus, 1981).

4.1.5 Environmental Factors

The carry of seawater is power reliant, and the nature influenced through ambient conditions significantly affects operational viability as well forthcomingness. Climate change exacerbates the effects of coastal erosion and port flooding with rising sea levels being a major concern. The damage seen by us in the form of hurricanes or typhoons which may block shipping lanes, and disrupt infrastructure are examples of extreme weather events. Wind and ocean currents are the largest influences on best shipping routes and fuel consumption. Water Pollution, which also affects sources of potable water and aquatic habitats. Sediment load in waterways decrease the depth of water making movement difficult and require to be dredged. Different temperatures can impact ecosystems as well as marine life, which in turn could have negative repercussions on the fishing industry and other maritime operations (Baştürk & Erol, 2023).

5 Effect of Canopy on Plant Growth

Plant canopy greatly impacts plant growth by regulating temperature, light, moisture, and wind conditions. It also affects evapotranspiration, which is necessary for water balance in the ecosystem.

5.1 Light Interception

Light interception depends on various factors such as canopy type, leaf, and shoot structures. It also varies from species to species. At the leaf level, light interception mainly depends on the number of chlorophyll contents and leaf dry mass per unit area. For example, needle-like leaves have less area which results in less light absorption. At the shoot scale, foliage spatial aggregation and foliage inclination angle distribution are the major factors that play an important role in light harvesting. Foliage distribution, biomass allocation to leaves, and branching frequency contribute at the canopy level. So a plant needs to have a well-developed canopy to harvest maximum light which is required for optimum photosynthesis. The light-harvesting is more, when the branching frequency is lower result in a flatter canopy with less light availability. On the other hand, the upper leaves receive more light than the lower that resulting in shade tolerance in lower leaves. Additionally, it helps to reduce the growth of other plants such as weeds because of shade. (Niinemets, 2010).

5.2 Temperature Regulation

Canopy temperature can be defined as the mean thermal conditions of various foliar assemblages consolidated from singular branches to entire crowns are infrequently quantified; however, they exert a substantial impact on—and are reciprocally affected by—the exchanges of water, energy, and carbon (C) between ecosystems and the atmosphere. Temperature is necessary for all plant functions such as canopy temperature especially in leaf play a central role in cycling of energy and water along with metabolism (Still et al., 2021). The plant canopy is crucial in controlling the plant's temperature by developing a microclimate, which is very beneficial for the plant's survival. For example, when the temperature exceeds 40 °C, especially in rice, sufficient water supply combined with relatively low humidity can allow the plant's microclimate temperature to drop to 33 °C through extensive transpiration (Bahuguna & Jagadish, 2015).

5.3 *Moisture Retention*

Plant canopy reduces the evaporation rate by preventing direct sunlight from reaching the soil (Ritter et al., 2005). So it is very beneficial for the plant to maintain the optimum soil moisture even during the hot summer (Breshears & Barnes, 1999; Fan et al., 2015). On the other hand, it also prevents soil erosion by avoiding direct rainfall on the soil which weakens the plant's root (Zuazo & Bochet et al., 2000; Pleguezuelo, 2009).

5.4 *Wind Protection*

Plant canopy is also very helpful in protecting the plant from wind and preventing physical damage to the plants as well as the rate of transpiration which leads to dehydration in plants (Armbrust & Bilbro, 1997; Boldes et al., 2002; Hu et al., 2018).

5.5 *Nutrient Cycling*

Canopy features influence the quantity and make-up of leaf litter generated, and this in turn impacts the amount of nutrients that may be recycled and the subsequent availability of those nutrients. Clearcuts had higher soil nutrient availability, demonstrating how the canopy stores nutrients (particularly N) in foliage and continuously adds available carbon to litter to keep nutrients (especially N) on site. Canopy gaps as small as 0.1 ha enhance soil N availability, therefore natural disturbances or partial harvesting methods that cause gaps in the canopy and increase its complexity would also increase the spatial variability in soil N cycling (Kushwaha et al., 2021; Matson et al., 2014; Prescott, 2002).

6 Process of Evapotranspiration

The process of water evaporating from soil and other surfaces and transpiring from plants is known as Evapotranspiration (ET). ET is essential to the cycles of carbon, energy, and water because vegetation plays a major role in its distribution and partitioning (Wanniarachchi & Sarukkalige, 2022; Yang et al., 2023).

6.1 Factors Affecting Evapotranspiration

6.1.1 Plant Characteristics:

Depending on their leaf shape, stomatal density, and overall water usage efficiency, different plant species have various rates of transpiration (Alexandris & Proutsos, 2020).

6.1.2 Humidity

Low humidity causes a greater vapor pressure deficit, which raises transpiration and evaporation rates (Ferreira & da Cunha, 2020; Jung et al., 2022).

6.1.3 Soil Moisture

Plants may transpire more easily when there is an abundance of soil moisture. Plants may, however, block their stomata to preserve water if the soil is dry, which would lower transpiration (Dong et al., 2022; Wang et al., 2021).

6.1.4 Temperature

Because there is more energy available to turn water into vapor at higher temperatures, evaporation and transpiration rate rise (Chen & Liu, 2020; Lenka et al., 2020; Rocha et al., 2020).

6.1.5 Wind Speed

Wind increases transpiration and evaporation by removing the saturated layer of air surrounding plants and soil (Varga-Haszonits et al., 2022).

Interaction between Canopy and Evapotranspiration

a. **Shading effect**

The canopy exerts a substantial influence on the environmental conditions beneath it by diminishing the intensity of direct solar radiation that penetrates to the soil surface (Liu et al., 2022). This attenuation of sunlight results in a decrease in soil temperatures, consequently decelerating the rate of water vaporization from the surface of the soil (Balandier et al., 2022). By constraining evaporation, the canopy facilitates the preservation of soil moisture, thereby ensuring that a greater volume of water remains accessible to the root systems of plants. This retention of moisture is particularly vital for the growth of plants, as it establishes a more stable and reliable water supply, alleviating stress on plants during arid intervals (Parkash & Singh, 2020). Furthermore, the reduced soil temperature beneath the canopy may foster a more advantageous environment for beneficial soil microorganisms, which in turn promotes robust plant growth and enhances soil fertility.

b. **Microclimate**

Under the shade of the plant with the help of low speed of wind and higher humidity microenvironment develops (Maclean & Klinges, 2021; Thakur & Kumar, 2021). These conditions help to reduce evaporation which leads to the efficient use of water by the plants. A sufficient amount of water is always available to the plant which result in the better growth of the plant without being faced the condition of water shortage. Plants will grow with less water availability (Bhattacharya, 2021; Bhattacharya & Bhattacharya, 2021).

7 Conclusion

This chapter covers a comprehensive qualitative study extracted from published and unpublished data regarding the soil plant atmosphere continuum and its relevance to the crop production. Engaging from general terminologies to the laws proposed and including the phenomena of up taking the water from soil through plant body and then into atmosphere, furthermore, all the possible factors and parameters affecting this perennial action of SPAC.

References

Affinnih, K., Ahamefule, H., & Anwanane, N. (2024). Modelling of permanent wilting point from routine soil properties on a typical alfisol. *Peruvian Journal of Agronomy, 8*(1), 55–68.

Agbna, G. H., Dongli, S., Zhipeng, L., Elshaikh, N. A., Guangcheng, S., & Timm, L. C. (2017). Effects of deficit irrigation and biochar addition on the growth, yield, and quality of tomato. *Scientia Horticulturae, 222*, 90–101.

Agegnehu, G., Srivastava, A. K., & Bird, M. I. (2017). The role of biochar and biochar-compost in improving soil quality and crop performance: A review. *Applied Soil Ecology, 119*, 156–170.
Alarcón, M. V., Salguero, J., & Lloret, P. G. (2019). Auxin modulated initiation of lateral roots is linked to pericycle cell length in maize. *Frontiers in Plant Science, 10*, 11.
Alexandris, S., & Proutsos, N. (2020). How significant is the effect of the surface characteristics on the reference evapotranspiration estimates? *Agricultural Water Management, 237*, 106181.
Alfei, S., & Pandoli, O. G. (2024). Bamboo-based biochar: A still too little-studied black gold and its current applications. *Journal of Xenobiotics, 14*(1), 416–451.
Amitrano, C., Arena, C., Rouphael, Y., De Pascale, S., & De Micco, V. (2019). Vapour pressure deficit: The hidden driver behind plant morphofunctional traits in controlled environments. *Annals of Applied Biology, 175*(3), 313–325.
An, N., Hemmati, S., & Cui, Y. (2017). Numerical analysis of soil volumetric water content and temperature variations in an embankment due to soil-atmosphere interaction. *Computers and Geotechnics, 83*, 40–51.
Anand, A., Kumar, V., & Kaushal, P. (2022). Biochar and its twin benefits: Crop residue management and climate change mitigation in India. *Renewable and Sustainable Energy Reviews, 156*, 111959.
Anderegg, W. R., Wolf, A., Arango-Velez, A., Choat, B., Chmura, D. J., Jansen, S., Kolb, T., Li, S., Meinzer, F., Pita, P., de Dios, V. R., Sperry, J. S., Wolfe, B. T., & Pacala, S. (2017). Plant water potential improves prediction of empirical stomatal models. *PLoS ONE, 12*(10), e0185481.
Armbrust, D. V., & Bilbro, J. D., Jr. (1997). Relating plant canopy characteristics to soil transport capacity by wind. *Agronomy Journal, 89*, 157–162.
Arnold, S., Bulovic, N., McIntyre, N., Finch, W. K., Larsen, J. R., Reading, L. P., & Baumgartl, T. (2020). Event-based deep drainage and percolation dynamics in vertosols and chromosols. *Hydrological Processes, 34*(2), 370–386.
Aroca, R., Porcel, R., & Ruiz-Lozano, J. M. (2012). Regulation of root water uptake under abiotic stress conditions. *Journal of Experimental Botany, 63*, 43–57.
Assi, A. T., Blake, J., Mohtar, R. H., & Braudeau, E. (2019). Soil aggregates structure-based approach for quantifying the field capacity, permanent wilting point and available water capacity. *Irrigation Science, 37*, 511–522.
Bahuguna, R. N., & Jagadish, K. S. V. (2015). Temperature regulation of plant phenological development. *Environmental and Experimental Botany, 111*, 83–90.
Balandier, P., Gobin, R., Prévosto, B., & Korboulewsky, N. (2022). The contribution of understorey vegetation to ecosystem evapotranspiration in boreal and temperate forests: A literature review and analysis. *European Journal of Forest Research, 141*, 979–997.
Baştürk, S., & Erol, S. (2023). Optimizing ship speed depending on cargo and wind-sea conditions for sustainable blue growth and climate change mitigation. *Journal of Marine Science and Technology, 28*, 659–674.
Bechtold, M., Dettmann, U., Wöhl, L., Durner, W., Piayda, A., & Tiemeyer, B. (2018). Comparing methods for measuring water retention of peat near permanent wilting point. *Soil Science Society of America Journal, 82*(3), 601–605.
Bernacchi, C. J., & VanLoocke, A. (2015). Terrestrial ecosystems in a changing environment: A dominant role for water. *Annual Review of Plant Biology, 66*(1), 599–622.
Bertocco, M., Parrino, S., Peruzzi, G., & Pozzebon, A. (2023). Estimating volumetric water content in soil for IoUT contexts by exploiting RSSI-based augmented sensors via machine learning. *Sensors, 23*(4), 2033.
Bharath, P., Gahir, S., & Raghavendra, A. S. (2021). Abscisic acid-induced stomatal closure: An important component of plant defense against abiotic and biotic stress. *Frontiers in Plant Science, 12*, 615114.
Bhattacharya, A., & Bhattacharya, A. (2021). Effect of soil water deficit on growth and development of plants: A review. In *Soil water deficit and physiological issues in plants* (pp. 393–488).
Bhattacharya, A. (2021). *Soil water deficit and physiological issues in plants*. Springer.

Bilska-Kos, A., Solecka, D., Dziewulska, A., Ochodzki, P., Jończyk, M., Bilski, H., & Sowiński, P. (2017). Low temperature caused modifications in the arrangement of cell wall pectins due to changes of osmotic potential of cells of maize leaves (Zea mays L.). *Protoplasma, 254*, 713–724.

Bochet, E., Poesen, J., & Rubio, J. L. (2000). Mound development as an interaction of individual plants with soil, water erosion and sedimentation processes on slopes. *Earth Surface Processes and Landforms, 25*, 847–867.

Bodner, G., Nakhforoosh, A., & Kaul, H.-P. (2015). Management of crop water under drought: A review. *Agronomy for Sustainable Development, 35*, 401–442.

Boldes, U., Golberg, A., Di Leo, J. M., Colman, J., & Scarabino, A. (2002). Canopy flow and aspects of the response of plants protected by herbaceous shelterbelts and wood fences. *Journal of Wind Engineering and Industrial Aerodynamics, 90*, 1253–1270.

Bosse, J., Durner, W., & Peters, A. (2024). Effect of salt concentration on osmotic potential in drying soils—measurement and models. *European Journal of Soil Science, 75*(5), e70001.

Bradford, K. J. (2017). Water relations in seed germination. In *Seed development and germination* (pp. 351–396). Routledge.

Breshears, D. D., & Barnes, F. J. (1999). Interrelationships between plant functional types and soil moisture heterogeneity for semiarid landscapes within the grassland/forest continuum: A unified conceptual model. *Landscape Ecology, 14*, 465–478.

Buckley, T. N. (2019). How do stomata respond to water status? *New Phytologist, 224*(1), 21–36.

Buitink, J., Swank, A. M., van Der Ploeg, M., Smith, N. E., Benninga, H.-J.F., van Der Bolt, F., Carranza, C. D. U., Koren, G., van der Velde, R., & Teuling, A. J. (2020). Anatomy of the 2018 agricultural drought in the Netherlands using in situ soil moisture and satellite vegetation indices. *Hydrology and Earth System Sciences Discussions, 2020*, 1–17.

Cai, G., König, M., Carminati, A., Abdalla, M., Javaux, M., Wankmüller, F., & Ahmed, M. A. (2024). Transpiration response to soil drying and vapor pressure deficit is soil texture specific. *Plant and Soil, 500*(1), 129–145.

Camuffo, D. (2018). Standardization activity in the evaluation of moisture content. *Journal of Cultural Heritage, 31*, S10–S14.

Cano-Delgado, A., Lee, J.-Y., & Demura, T. (2010). Regulatory mechanisms for specification and patterning of plant vascular tissues. *Annual Review of Cell and Developmental Biology, 26*, 605–637.

Carey, V. P. (2020). Liquid-vapor phase-change phenomena: an introduction to the thermophysics of vaporization and condensation processes in heat transfer equipment. CRC Press.

Cavusoglu, A.-H., Chen, X., Gentine, P., & Sahin, O. (2017). Potential for natural evaporation as a reliable renewable energy resource. *Nature Communications, 8*(1), 617.

Ceaglio, E., Mitterer, C., Maggioni, M., Ferraris, S., Segor, V., & Freppaz, M. (2017). The role of soil volumetric liquid water content during snow gliding processes. *Cold Regions Science and Technology, 136*, 17–29.

Chen, J. M., & Liu, J. (2020). Evolution of evapotranspiration models using thermal and shortwave remote sensing data. *Remote Sensing of Environment, 237*, 111594.

Chen, Z.-H., Chen, G., Dai, F., Wang, Y., Hills, A., Ruan, Y.-L., Zhang, G., Franks, P. J., Nevo, E., & Blatt, M. R. (2017). Molecular evolution of grass stomata. *Trends in Plant Science, 22*(2), 124–139.

Chen, Z., Zhu, L.-T., & Luo, Z.-H. (2023). Characterizing flow and transport in biological vascular systems: A review from physiological and chemical engineering perspectives. *Industrial & Engineering Chemistry Research, 63*(1), 4–36.

Clark, J. W., Harris, B. J., Hetherington, A. J., Hurtado-Castano, N., Brench, R. A., Casson, S., Williams, T. A., Gray, J. E., & Hetherington, A. M. (2022). The origin and evolution of stomata. *Current Biology, 32*(11), R539–R553.

Cosgrove, W. J., & Loucks, D. P. (2015). Water management: Current and future challenges and research directions. *Water Resources Research, 51*(6), 4823–4839.

Van Dam, J., van der Zee, S., Vogel, H., Vrugt, J., Wöhling, T., & Young, I. (2016). Modeling soil processes: Review, key challenges, and new perspectives. *Soil Science, 15*(5).

Datta, S., Taghvaeian, S., & Stivers, J. (2017). Understanding soil water content and thresholds for irrigation management.
de Melo, L. B. B., Silva, B. M., Peixoto, D. S., Chiarini, T. P. A., de Oliveira, G. C., & Curi, N. (2021). Effect of compaction on the relationship between electrical resistivity and soil water content in oxisol. *Soil and Tillage Research, 208*, 104876.
De Rybel, B., Mähönen, A. P., Helariutta, Y., & Weijers, D. (2016). Plant vascular development: From early specification to differentiation. *Nature Reviews Molecular Cell Biology, 17*, 30–40.
De Schepper, V., De Swaef, T., Bauweraerts, I., & Steppe, K. (2013). Phloem transport: A review of mechanisms and controls. *Journal of Experimental Botany, 64*, 4839–4850.
Dickson, M. H., & Fanelli, M. (2018). What is geothermal energy? In *Renewable Energy* (pp. Vol1_302-Vol301_328). Routledge.
Doehlemann, G., & Hemetsberger, C. (2013). Apoplastic immunity and its suppression by filamentous plant pathogens. *New Phytologist, 198*, 1001–1016.
Dong, J., Akbar, R., Short Gianotti, D. J., Feldman, A. F., Crow, W. T., & Entekhabi, D. (2022). Can surface soil moisture information identify evapotranspiration regime transitions? *Geophysical Research Letters, 49*, e2021GL097697.
Eden, M., Gerke, H. H., & Houot, S. (2017). Organic waste recycling in agriculture and related effects on soil water retention and plant available water: A review. *Agronomy for Sustainable Development, 37*, 1–21.
Eissenstat, D. (2000). Root structure and function in an ecological context. *New Phytologist, 148*, 353–354.
Engel, A., Fujiyoshi, Y., & Agre, P. (2000). The importance of aquaporin water channel protein structures. *The EMBO Journal, 19*, 800–806.
Fan, J., Baumgartl, T., Scheuermann, A., & Lockington, D. A. (2015). Modeling effects of canopy and roots on soil moisture and deep drainage. *Vadose Zone Journal, 14*, 1–18.
Fatichi, S., Pappas, C., & Ivanov, V. Y. (2016). Modeling plant–water interactions: An ecohydrological overview from the cell to the global scale. *Wiley Interdisciplinary Reviews: Water, 3*(3), 327–368.
Fatkhuroyan, F., Wati, T., & Kurniawan, R. (2021). Characteristic of soil moisture in Indonesia using ESA CCI satellites products. *Indonesian Journal of Geography, 53*(1), 54–60.
Feng JunJie, F. J., Huang XiuQiao, H. X., Deng Zhong, D. Z., Jia YanHui, J. Y., Cai JiuMao, C. J., Cheng Qian, C. Q., & Zhang HongXia, Z. H. (2018). Effect of bulk density on soil water potential in medium loam soil of agriculture.
Ferreira, L. B., & Da Cunha, F. F. (2020). New approach to estimate daily reference evapotranspiration based on hourly temperature and relative humidity using machine learning and deep learning. *Agricultural Water Management, 234*, 106113.
Fiscus, E. L. (1981). Effects of abscisic acid on the hydraulic conductance of and the total ion transport through Phaseolus root systems. *Plant Physiology, 68*, 169–174.
Gao, M., Hu, Y., & Bai, Y. (2022). Construction of ecological security pattern in national land space from the perspective of the community of life in mountain, water, forest, field, lake and grass: A case study in Guangxi Hechi, China. *Ecological Indicators, 139*, 108867.
Gavrilescu, M. (2021). Water, soil, and plants interactions in a threatened environment. *Water, 13*(19), 2746.
Geng, X., & Boufadel, M. C. (2015). Numerical modeling of water flow and salt transport in bare saline soil subjected to evaporation. *Journal of Hydrology, 524*, 427–438.
Gholamzadeh, F., Asgarzadeh, H., Khodaverdiloo, H., & Mosaddeghi, M. (2024). Evaluating the efficiency of field oven for quick measurement of gravimetric soil water content. *JWSS-Isfahan University of Technology, 28*(1), 179–192.
Goss, M. J., & Ehlers, W. (2009). The role of lysimeters in the development of our understanding of soil water and nutrient dynamics in ecosystems. *Soil Use and Management., 25*, 213–223.
Goyal, G., Yadav, A., & Dubey, G. (2021). Effect of salt stress on soil chemistry and plant-atmosphere continuum (SPAC). In *Physiology of salt stress in plants: perception, signalling, omics and tolerance mechanism* (pp. 106–128).

Harrison, E. L., Arce Cubas, L., Gray, J. E., & Hepworth, C. (2020). The influence of stomatal morphology and distribution on photosynthetic gas exchange. *The Plant Journal, 101*(4), 768–779.
Hartemink, A. E. (2016). The definition of soil since the early 1800s. *Advances in Agronomy, 137*, 73–126.
Hasan, M., & Eslamian, S. (2023). An introduction to soil–water–plant relationship. In *Handbook of irrigation hydrology and management* (pp. 71–108). CRC Press.
Hatfield, J. L., & Dold, C. (2019). Water-use efficiency: Advances and challenges in a changing climate. *Frontiers in Plant Science, 10*, 103.
Henry, C. J., Brant, J. A., & Kelleners, T. J. (2018). Water transport mechanisms for salt-rejecting membranes driven by soil-water potentials. *Journal of Membrane Science, 563*, 107–114.
Hinckley, T. M., Lachenbruch, B., Meinzer, F. C., & Dawson, T. E. (2011). A lifespan perspective on integrating structure and function in trees. In *Size-and age-related changes in tree structure and function* (pp. 3–30).
Ho, T. T. K., Le, T. H., Tran, C.-S., Nguyen, P.-T., Thai, V.-N., & Bui, X.-T. (2022). Compost to improve sustainable soil cultivation and crop productivity. *Case Studies in Chemical and Environmental Engineering, 6*, 100211.
Hohenegger, C., & Stevens, B. (2018). The role of the permanent wilting point in controlling the spatial distribution of precipitation. *Proceedings of the National Academy of Sciences, 115*(22), 5692–5697.
Howell, A., Tucker, C., Grote, E. E., Veste, M., Belnap, J., Kast, G., Weber, B., & Reed, S. C. (2019). Manufacturing simple and inexpensive soil surface temperature and gravimetric water content sensors. *JoVE (Journal of Visualized Experiments), 154*, e60308.
Hu, Y., Asante, E. A., Lu, Y., Mahmood, A., Buttar, N. A., & Yuan, S. (2018). Review of air disturbance technology for plant frost protection. *International Journal of Agricultural and Biological Engineering, 11*, 21–28.
Hu, Y., Mortimer, P. E., Hyde, K. D., Kakumyan, P., & Thongklang, N. (2021). Mushroom cultivation for soil amendment and bioremediation. *Circular Agricultural Systems, 1*(1), 1–14.
Ibrahimi, K., & Alghamdi, A. G. (2022). Available water capacity of sandy soils as affected by biochar application: A meta-analysis. *CATENA, 214*, 106281.
Iizumi, T., & Wagai, R. (2019). Leveraging drought risk reduction for sustainable food, soil and climate via soil organic carbon sequestration. *Scientific Reports, 9*(1), 19744.
Ippolito, M. (2023). Assessing crop water requirements and irrigation scheduling at different spatial scales in Mediterranean orchards using models, proximal and remotely sensed data.
Jackson, R. B., Sperry, J. S., & Dawson, T. E. (2000). Root water uptake and transport: Using physiological processes in global predictions. *Trends in Plant Science, 5*, 482–488.
Johnson, D. J., Suwaileh, W. A., Mohammed, A. W., & Hilal, N. (2018). Osmotic's potential: An overview of draw solutes for forward osmosis. *Desalination, 434*, 100–120.
Juarez-Enriquez, E., Olivas, G., Zamudio-Flores, P., Ortega-Rivas, E., Perez-Vega, S., & Sepulveda, D. (2017). Effect of water content on the flowability of hygroscopic powders. *Journal of Food Engineering, 205*, 12–17.
Jung, D.-H., Lee, T. S., Kim, K., & Park, S. H. (2022). A deep learning model to predict evapotranspiration and relative humidity for moisture control in tomato greenhouses. *Agronomy, 12*, 2169.
Kalogirou, S. A. (2023). Solar energy engineering: processes and systems. Elsevier.
Katsiampoura, A., Toumpanakis, D., Konsta, K., Varkaris, A., & Vassilakopoulos, T. (2018). Prediction of dysnatremias in critically ill patients based on the law of conservation of mass. Comparison of existing formulae. *Plos One, 13*(11), e0207603.
Kim, H. K., Park, J., & Hwang, I. (2014). Investigating water transport through the xylem network in vascular plants. *Journal of Experimental Botany, 65*, 1895–1904.
Kirkham, M. B. (2023). *Principles of soil and plant water relations*. Elsevier.
Kosal, E. (2023). Plant transport. In *Introductory biology: ecology, evolution, and biodiversity*.

Kuromori, T., Seo, M., & Shinozaki, K. (2018). ABA transport and plant water stress responses. *Trends in Plant Science, 23*(6), 513–522.

Kushwaha, P., Neilson, J. W., Barberán, A., Chen, Y., Fontana, C. G., Butterfield, B. J., & Maier, R. M. (2021). Arid ecosystem vegetation canopy-gap dichotomy: Influence on soil microbial composition and nutrient cycling functional potential. *Applied and Environmental Microbiology, 87*, e02780-e2820.

Lambers, H., Shane, M. W., Cramer, M. D., Pearse, S. J., & Veneklaas, E. J. (2006). Root structure and functioning for efficient acquisition of phosphorus: Matching morphological and physiological traits. *Annals of Botany, 98*, 693–713.

Larsbo, M., Sandin, M., Jarvis, N., Etana, A., & Kreuger, J. (2016). Surface runoff of pesticides from a clay loam field in Sweden. *Journal of Environmental Quality, 45*(4), 1367–1374.

Leng, L., Xiong, Q., Yang, L., Li, H., Zhou, Y., Zhang, W., Jiang, S., Li, H., & Huang, H. (2021). An overview on engineering the surface area and porosity of biochar. *Science of the Total Environment, 763*, 144204.

Lenka, N. K., Lenka, S., Thakur, J. K., Yashona, D. S., Shukla, A., Elanchezhian, R., Singh, K., Biswas, A., & Patra, A. (2020). Carbon dioxide and temperature elevation effects on crop evapotranspiration and water use efficiency in soybean as affected by different nitrogen levels. *Agricultural Water Management, 230*, 105936.

Li, X., Zhang, H., Huang, H., & Lai, H. (2024). Advanced methodological framework of nmm analysis: Formulation, integration, and solution strategies for the laplace equation problem with complex boundaries. In *Integration, and solution strategies for the laplace equation problem with complex boundaries*.

Li, X., Chang, S. X., & Salifu, K. F. (2014). Soil texture and layering effects on water and salt dynamics in the presence of a water table: A review. *Environmental Reviews, 22*(1), 41–50.

Lin, H., Chen, Y., Zhang, H., Fu, P., & Fan, Z. (2017). Stronger cooling effects of transpiration and leaf physical traits of plants from a hot dry habitat than from a hot wet habitat. *Functional Ecology, 31*(12), 2202–2211.

Liu, B., Fan, H., Han, W., Zhu, L., Zhao, X., Zhang, Y., & Ma, R. (2021). Linking soil water retention capacity to pore structure characteristics based on X-ray computed tomography: Chinese Mollisol under freeze-thaw effect. *Geoderma, 401*, 115170.

Liu, J., Mooney, H., Hull, V., Davis, S. J., Gaskell, J., Hertel, T., Lubchenco, J., Seto, K. C., Gleick, P., Kremen, C., & Li, S. (2015). Systems integration for global sustainability. *Science, 347*(6225), 1258832.

Liu, P., Tong, X., Meng, P., Zhang, J., Li, J., Zhang, J., & Zhou, Y. (2022). Biophysical controls on water use efficiency of six plantations under different sky conditions. *Agricultural and Forest Meteorology, 320*, 108938.

Liu, X., Chen, J., Cui, X., Liu, Q., Cao, X., & Chen, X. (2019). Measurement of soil water content using ground-penetrating radar: A review of current methods. *International Journal of Digital Earth, 12*(1), 95–118.

Livingston, N., & Topp, G. C. (1993). Soil water potential. Soil sampling and methods of analysis. *Canadian Society of Soil Sciences*, 559–567.

López-Calcagno, P. E., Brown, K. L., Simkin, A. J., Fisk, S. J., Vialet-Chabrand, S., Lawson, T., & Raines, C. A. (2020). Stimulating photosynthetic processes increases productivity and water-use efficiency in the field. *Nature Plants, 6*(8), 1054–1063.

Loudon, C., & McCulloh, K. (1999). Application of the Hagen—poiseuille equation to fluid feeding through short tubes. *Annals of the Entomological Society of America, 92*(1), 153–158.

Lu, N., & Zhang, C. (2019). Soil sorptive potential: Concept, theory, and verification. *Journal of Geotechnical and Geoenvironmental Engineering, 145*(4), 04019006.

Luo, S., Lu, N., Zhang, C., & Likos, W. (2022). Soil water potential: A historical perspective and recent breakthroughs. *Vadose Zone Journal, 21*(4), e20203.

Ma, Y., Liu, H., Yu, Y., Guo, L., Zhao, W., & Yetemen, O. (2022). Revisiting soil water potential: Towards a better understanding of soil and plant interactions. *Water, 14*(22), 3721.

Maclean, I. M., & Klinges, D. H. (2021). Microclimc: A mechanistic model of above, below and within-canopy microclimate. *Ecological Modelling, 451*, 109567.
Marschner, P., & Rengel, Z. (2023). Nutrient availability in soils. In *Marschner's Mineral Nutrition of Plants* (pp. 499–522). Elsevier.
Martin, J. L., Schottman, R. W., & McCutcheon, S. C. (2018). Hydrodynamics and transport for water quality modeling: CRC press.
Matson, A. L., Corre, M. D., & Veldkamp, E. (2014). Nitrogen cycling in canopy soils of tropical montane forests responds rapidly to indirect N and P fertilization. *Global Change Biology, 20*, 3802–3813.
Maurel, C., Verdoucq, L., & Rodrigues, O. (2016). Aquaporins and plant transpiration. *Plant, Cell & Environment, 39*(11), 2580–2587.
Mcculloh, K. A., Sperry, J. S., & Adler, F. R. (2003). Water transport in plants obeys Murray's law. *Nature, 421*, 939–942.
McGinn, C. K., Lamport, Z. A., & Kymissis, I. (2020). Review of gravimetric sensing of volatile organic compounds. *ACS Sensors, 5*(6), 1514–1534.
Medková, D. (2018). The laplace equation. In *Boundary value problems on bounded and unbounded Lipschitz domains*. Cham: Springer.
Medrano, H., Tomás, M., Martorell, S., Flexas, J., Hernández, E., Rosselló, J., Pou, A., Escalona, J.-M., & Bota, J. (2015). From leaf to whole-plant water use efficiency (WUE) in complex canopies: Limitations of leaf WUE as a selection target. *The Crop Journal, 3*(3), 220–228.
Mekonnen, M. M., Gerbens-Leenes, P., & Hoekstra, A. Y. (2015). The consumptive water footprint of electricity and heat: A global assessment. *Environmental Science: Water Research & Technology, 1*(3), 285–297.
Mengel, K., Kirkby, E. A., Kosegarten, H., & Appel, T. (2001). Plant water relationships. In *Principles of plant nutrition* (pp. 181–242).
Merante, P., Dibari, C., Ferrise, R., Sánchez, B., Iglesias, A., Lesschen, J. P., Kuikman, P., Yeluripati, J., Smith, P., & Bindi, M. (2017). Adopting soil organic carbon management practices in soils of varying quality: Implications and perspectives in Europe. *Soil and Tillage Research, 165*, 95–106.
Metzger, J. C., Wutzler, T., Dalla Valle, N., Filipzik, J., Grauer, C., Lehmann, R., Roggenbuck, M., Schelhorn, D., Weckmüller, J., Küsel, K., Totsche, K. U., Trumbore, S., & Hildebrandt, A. (2017). Vegetation impacts soil water content patterns by shaping canopy water fluxes and soil properties. *Hydrological Processes, 31*(22), 3783–3795.
Mi, Q., Yi, S., Gang, D., Lu, X., & Liu, X. (2024). Research progress of transpiration cooling for aircraft thermal protection. *Applied Thermal Engineering, 236*, 121360.
Minasny, B., & McBratney, A. (2018). Limited effect of organic matter on soil available water capacity. *European Journal of Soil Science, 69*(1), 39–47.
Muskat, M. (1937). The flow of fluids through porous media. *Journal of Applied Physics, 8*(4), 274–282.
Niinemets, Ü. (2010). A review of light interception in plant stands from leaf to canopy in different plant functional types and in species with varying shade tolerance. *Ecological Research, 25*, 693–714.
Nilson, S. E., & Assmann, S. M. (2007). The control of transpiration. Insights from Arabidopsis. *Plant Physiology, 143*, 19–27.
Novick, K. A., Konings, A. G., & Gentine, P. (2019). Beyond soil water potential: An expanded view on isohydricity including land–atmosphere interactions and phenology. *Plant, Cell & Environment, 42*(6), 1802–1815.
Oberle, B., Bringezu, S., Hatfield-Dodds, S., Hellweg, S., Schandl, H., & Clement, J. (2019). *Global resources outlook: 2019*. United Nations Envio, Paris, France: International Resource Panel.
Oliveira, R. S., Eller, C. B., Barros, F. V., Hirota, M., Brum, M., & Bittencourt, P. (2021). Linking plant hydraulics and the fast–slow continuum to understand resilience to drought in tropical ecosystems. *New Phytologist, 230*(3), 904–923.

Omara, P., Singh, H., Singh, K., Sharma, L., Otim, F., & Obia, A. (2023). Short-term effect of field application of biochar on cation exchange capacity, pH, and electrical conductivity of sandy and clay loam temperate soils. *Technology in Agronomy, 3*(1).

Ortega-Farias, S., & Intrigliolo, D. S. (2021). Multiscale technologies for irrigation management (Vol. 39, pp. 1–3). Springer.

Parkash, V., & Singh, S. (2020). A review on potential plant-based water stress indicators for vegetable crops. *Sustainability, 12*, 3945.

Pasha, A. Y., Khoshghalb, A., & Khalili, N. (2016). Pitfalls in interpretation of gravimetric water content–based soil-water characteristic curve for deformable porous media. *International Journal of Geomechanics, 16*(6), D4015004.

Pereira, L., Paredes, P., & Jovanovic, N. (2020). Soil water balance models for determining crop water and irrigation requirements and irrigation scheduling focusing on the FAO56 method and the dual Kc approach. *Agricultural Water Management, 241*, 106357.

Pittman, F., Mohammed, A., & Cey, E. (2020). Effects of antecedent moisture and macroporosity on infiltration and water flow in frozen soil. *Hydrological Processes, 34*, 795–809.

Poyatos, R., Llorens, P., & Gallart, F. (2005). Transpiration of montane Pinus sylvestris L. and Quercus pubescens Willd. forest stands measured with sap flow sensors in NE Spain. *Hydrology and Earth System Sciences, 9*, 493–505.

Prescott, C. E. (2002). The influence of the forest canopy on nutrient cycling. *Tree Physiology, 22*, 1193–1200.

Qambrani, N. A., Rahman, M. M., Won, S., Shim, S., & Ra, C. (2017). Biochar properties and eco-friendly applications for climate change mitigation, waste management, and wastewater treatment: A review. *Renewable and Sustainable Energy Reviews, 79*, 255–273.

Qin, C., Wang, H., Yuan, X., Xiong, T., Zhang, J., & Zhang, J. (2020). Understanding structure-performance correlation of biochar materials in environmental remediation and electrochemical devices. *Chemical Engineering Journal, 382*, 122977.

Qiu, J., Gao, Q., Wang, S., & Su, Z. (2016). Comparison of temporal trends from multiple soil moisture data sets and precipitation: The implication of irrigation on regional soil moisture trend. *International Journal of Applied Earth Observation and Geoinformation, 48*, 17–27.

Radcliffe, D. E., Rasmussen, T. C., & Warrick, A. (2002). Soil water movement. In *Soil physics companion* (pp. 85–126).

Rawls, W. J., Ahuja, L. R., Brakensiek, D. L., & Shirmohammadi, A. (1992). *Infiltration and soil water movement*.

Razzaghi, F., Obour, P. B., & Arthur, E. (2020). Does biochar improve soil water retention? A systematic review and meta-analysis. *Geoderma, 361*, 114055.

Rey, A. (2015). Mind the gap: Non-biological processes contributing to soil CO_2 efflux. *Global Change Biology, 21*(5), 1752–1761.

Reynolds, W. D. (2018). An analytic description of field capacity and its application in crop production. *Geoderma, 326*, 56–67.

Rhodes, D., & Samaras, Y. (2020). Genetic control of osmoregulation in plants. In *Cellular and molecular physiology of cell volume regulation*. CRC Press.

Ritter, E., Dalsgaard, L., & Einhorn, K. S. (2005). Light, temperature and soil moisture regimes following gap formation in a semi-natural beech-dominated forest in Denmark. *Forest Ecology and Management, 206*, 15–33.

Rocha, N. S. D., Käfer, P. S., Skokovic, D., Veeck, G., Diaz, L. R., Kaiser, E. A., Carvalho, C. M., Cruz, R. C., Sobrino, J. A., & Roberti, D. R. (2020). The influence of land surface temperature in evapotranspiration estimated by the S-SEBI model. *Atmosphere, 11*, 1059.

Rozema, J., & Schat, H. (2013). Salt tolerance of halophytes, research questions reviewed in the perspective of saline agriculture. *Environmental and Experimental Botany, 92*, 83–95.

Santra, P., Kumar, M., Kumawat, R., Painuli, D., Hati, K., Heuvelink, G., & Batjes, N. (2018). Pedotransfer functions to estimate soil water content at field capacity and permanent wilting point in hot arid western India. *Journal of Earth System Science, 127*, 1–16.

Sayara, T., Basheer-Salimia, R., Hawamde, F., & Sánchez, A. (2020). Recycling of organic wastes through composting: Process performance and compost application in agriculture. *Agronomy, 10*(11), 1838.

Schaller, J., Cramer, A., Carminati, A., & Zarebanadkouki, M. (2020). Biogenic amorphous silica as main driver for plant available water in soils. *Scientific Reports, 10*(1), 2424.

Scharwies, J. D., & Dinneny, J. R. (2019). Water transport, perception, and response in plants. *Journal of Plant Research, 132*(3), 311–324.

Sharpe, P. H. (2018). *Horse pasture management*. Academic Press.

Shukla, M. K. (2023). *Soil physics: An introduction*. CRC press.

Steudle, E. (1992). The biophysics of plant water: compartmentation, coupling with metabolic processes, and flow of water in plant roots. In *Water and life: comparative analysis of water relationships at the organismic, cellular, and molecular levels* (pp. 173–204).

Steudle, E. (2000). Water uptake by roots: Effects of water deficit. *Journal of Experimental Botany, 51*, 1531–1542.

Still, C. J., Rastogi, B., Page, G. F., Griffith, D. M., Sibley, A., Schulze, M., Hawkins, L., Pau, S., Detto, M., & Helliker, B. R. (2021). Imaging canopy temperature: Shedding (thermal) light on ecosystem processes. *New Phytologist, 230*, 1746–1753.

Tamindžić, G., Azizbekian, S., Miljaković, D., Ignjatov, M., Nikolić, Z., Budakov, D., Vasiljević, S., & Grahovac, M. (2024). Assessment of various nanoprimings for boosting pea germination and early growth in both optimal and drought-stressed environments. *Plants, 13*(11), 1547.

Tao, P., Ni, G., Song, C., Shang, W., Wu, J., Zhu, J., Chen, G., & Deng, T. (2018). Solar-driven interfacial evaporation. *Nature Energy, 3*(12), 1031–1041.

Thakur, M., & Kumar, R. (2021). Microclimatic buffering on medicinal and aromatic plants: A review. *Industrial Crops and Products, 160*, 113144.

Toková, L., Igaz, D., Horák, J., & Aydin, E. (2020). Effect of biochar application and re-application on soil bulk density, porosity, saturated hydraulic conductivity, water content and soil water availability in a silty loam Haplic Luvisol. *Agronomy, 10*(7), 1005.

Tomczyk, A., Sokołowska, Z., & Boguta, P. (2020). Biochar physicochemical properties: Pyrolysis temperature and feedstock kind effects. *Reviews in Environmental Science and Bio/technology, 19*(1), 191–215.

Trautmann, T., Koirala, S., Carvalhais, N., Güntner, A., & Jung, M. (2022). The importance of vegetation in understanding terrestrial water storage variations. *Hydrology and Earth System Sciences, 26*(4), 1089–1109.

Umor, N., Ismail, S., Abdullah, S., Huzaifah, M., Huzir, N., Mahmood, N., & Zahrim, A. (2021). Zero waste management of spent mushroom compost. *Journal of Material Cycles and Waste Management, 23*(5), 1726–1736.

Vadez, V., Kholova, J., Medina, S., Kakkera, A., & Anderberg, H. (2014). Transpiration efficiency: New insights into an old story. *Journal of Experimental Botany, 65*, 6141–6153.

van Lier, Q. J. (2017). Field capacity, a valid upper limit of crop available water? *Agricultural Water Management, 193*, 214–220.

Van Vliet, M. T., Wiberg, D., Leduc, S., & Riahi, K. (2016). Power-generation system vulnerability and adaptation to changes in climate and water resources. *Nature Climate Change, 6*(4), 375–380.

Vanderborght, J., Fetzer, T., Mosthaf, K., Smits, K. M., & Helmig, R. (2017). Heat and water transport in soils and across the soil-atmosphere interface: 1. Theory and different model concepts. *Water Resources Research, 53*(2), 1057–1079.

Varga-Haszonits, Z., Szalka, É., & Szakál, T. (2022). Determination of reference evapotranspiration using Penman-Monteith method in case of missing wind speed data under subhumid climatic condition in Hungary. *Atmospheric and Climate Sciences, 12*, 235–245.

Vinhal-Freitas, I. C., Corrêa, G. F., Wendling, B., Bobul'ská, L., & Ferreira, A. S. (2017). Soil textural class plays a major role in evaluating the effects of land use on soil quality indicators. *Ecological Indicators, 74*, 182–190.

Wang, X., Zhao, Y., Liu, H., & Tian, S. (2017). Soil available water in texturally different soils from overburden of an open pit mine in Xilinhot. In *Land reclamation in ecological fragile areas* (pp. 281–283). CRC Press.
Wang, J., Dlamini, D. S., Mishra, A. K., Pendergast, M. T. M., Wong, M. C., Mamba, B. B., Freger, V., Verliefde, A. R., & Hoek, E. M. (2014). A critical review of transport through osmotic membranes. *Journal of Membrane Science, 454*, 516–537.
Wang, Y., Zhang, Y., Yu, X., Jia, G., Liu, Z., Sun, L., Zheng, P., & Zhu, X. (2021). Grassland soil moisture fluctuation and its relationship with evapotranspiration. *Ecological Indicators, 131*, 108196.
Wanniarachchi, S., & Sarukkalige, R. (2022). A review on evapotranspiration estimation in agricultural water management: Past, present, and future. *Hydrology, 9*, 123.
Wei, Z., Van Le, Q., Peng, W., Yang, Y., Yang, H., Gu, H., Lam, S. S., & Sonne, C. (2021). A review on phytoremediation of contaminants in air, water and soil. *Journal of Hazardous Materials, 403*, 123658.
Wen, W., Lai, Y., & You, Z. (2020). Numerical modeling of water–heat–vapor–salt transport in unsaturated soil under evaporation. *International Journal of Heat and Mass Transfer, 159*, 120114.
Wennerström, H., & Oliveberg, M. (2022). On the osmotic pressure of cells. *QRB Discovery, 3*, e12.
Xiao, Q., Zhu, L.-X., Shen, Y.-F., & Li, S.-Q. (2016). Sensitivity of soil water retention and availability to biochar addition in rainfed semi-arid farmland during a three-year field experiment. *Field Crops Research, 196*, 284–293.
Yan, N., Marschner, P., Cao, W., Zuo, C., & Qin, W. (2015). Influence of salinity and water content on soil microorganisms. *International Soil and Water Conservation Research, 3*(4), 316–323.
Yang, Y., Roderick, M. L., Guo, H., Miralles, D. G., Zhang, L., Fatichi, S., Luo, X., Zhang, Y., Mcvicar, T. R., Tu, Z., Keenan, T. F., Fisher, J. B., Gan, R., Zhang, X., Piao, S., Zhang, B., & Yang, D. (2023). Evapotranspiration on a greening Earth. *Nature Reviews Earth & Environment, 4*, 626–641.
Yao, N., Li, Y., Xu, F., Liu, J., Chen, S., Ma, H., Chau, H. W., Liu, D. L., Li, M., Feng, H., Yu, Q., & He, J. (2020). Permanent wilting point plays an important role in simulating winter wheat growth under water deficit conditions. *Agricultural Water Management, 229*, 105954.
Ye, Z.-H. (2002). Vascular tissue differentiation and pattern formation in plants. *Annual Review of Plant Biology, 53*, 183–202.
Yuhua, F. (2015). New newton mechanics taking law of conservation of energy as unique source law. *Science Journal of Physics, 2013*.
Zarebanadkouki, M., Trtik, P., Hayat, F., Carminati, A., & Kaestner, A. (2019). Root water uptake and its pathways across the root: Quantification at the cellular scale. *Scientific Reports, 9*(1), 12979.
Zhang, Y., Wang, J., & Feng, Y. (2021). The effects of biochar addition on soil physicochemical properties: A review. *CATENA, 202*, 105284.
Zhao, H., Qin, J., Gao, T., Zhang, M., Sun, H., Zhu, S., Xu, C., & Ning, T. (2022a). Immediate and long-term effects of tillage practices with crop residue on soil water and organic carbon storage changes under a wheat-maize cropping system. *Soil and Tillage Research, 218*, 105309.
Zhao, Q., Wu, J., Guo, C., Zhang, J., Wang, X., Liu, Y., Zhao, H., & Zhang, R. (2022b). Effects of freezing-thawing processes on net nitrogen mineralization in salinized farmland soil. *Agronomy, 12*(12), 2986.
Zhou, L., Zhou, X., He, Y., Fu, Y., Du, Z., Lu, M., Sun, X., Li, C., Lu, C., Liu, R., Zhou, G., Bai, S. H., & Thakur, M. P. (2022). Global systematic review with meta-analysis shows that warming effects on terrestrial plant biomass allocation are influenced by precipitation and mycorrhizal association. *Nature Communications, 13*(1), 4914.

Zhu, H., Zhou, L., He, D., Chen, Z., & Liu, J. (2024). LNG density measurement by gravimetric method. *Cryogenics, 142*, 103925.

Zuazo, V. H. D., & Pleguezuelo, C. R. R. (2009). Soil-erosion and runoff prevention by plant covers: a review. In *Sustainable agriculture* (pp. 785–811).

Technological Advances in Agriculture Engineering for Soil Conservation

Muhammad Fawaz Saleem, Sehar Rasool, Aimen Tahir, Zafar Iqbal, Muhammad Bilal, Ali Raza, Mohsin Rasool, and Salma Malik

Abstract Degradation of the soil is one of the most recent environmental problems. Numerous processes that lead to soil deterioration, such as erosion, compaction, salinization, pollution, and the loss of organic matter and soil biodiversity, are intimately associated with agricultural activity and its intensification. Soil integrity and health are essential for biodiversity preservation, sustainable agriculture, and overall ecosystem stability. Designing and implementing sustainable agronomic practices that may preserve and enhance the physical, chemical, and biological characteristics of soils while also improving the functioning of agroecosystems is thus a problem for both scientists and farmers. Protecting the foundation of the soil for future generations requires a variety of soil conservation techniques. We may efficiently manage water resources, prevent soil erosion, improve soil fertility, and promote biodiversity by putting appropriate soil conservation measures into place. This essay will examine several soil conservation methods that might lessen these risks and guarantee the long-term sustainability of our planet's soil resources, including tillage, GPS-guided tactics, GIS, remote sensing, and sediment check dams.

Keywords Soil conservation · Agriculture engineering · Soil health · Engineering tools · Smart Agriculture

---

M. F. Saleem · A. Raza
Farm Machinery and Power, University of Agriculture Faisalabad, Faisalabad, Pakistan

S. Rasool (✉) · A. Tahir · S. Malik
Institute of Soil and Environmental Sciences, University of Agriculture Faisalabad, Faisalabad, Pakistan
e-mail: seharrasool4@gmail.com

Z. Iqbal
Faculty of Agriculture Engineering, University of Engineering and Technology, Peshawar, Pakistan

M. Bilal
Department of Irrigation and Drainage, University of Agriculture Faisalabad, Faisalabad, Pakistan

M. Rasool
Department of Entomology, University of Agriculture Faisalabad, Faisalabad, Pakistan

M. Shaaban (ed.), *Soils and Sustainable Agriculture*, Frontier Studies in Soil Science,
https://doi.org/10.1007/978-3-031-91114-9_17

1 Introduction

The demand for diverse agricultural goods rises in conjunction with population expansion, necessitating the establishment of food enterprises that rely on innovative approaches to boost agricultural productivity (Rațu et al., 2023). This goes hand in hand with the need to grow agriculture as a whole, and more particularly with enhanced agricultural mechanization; it also goes hand in hand with the need to focus on raising productivity and efficiency, cutting back on labor, and improving welfare. Larger vehicles with more engine power and heavier tractors with a big capacity and appropriate application can accomplish this (Al-Aani & Sadoon, 2023). Farmers apprehension about the negative consequences they perceive would arise from implementing these advances on soil qualities, impeding the utilization and standardization of this technology, despite its promise.

Furthermore, soils constitute the cornerstone of terrestrial food production, accounting for 95% of total output (Rust et al., 2022). According to Moncrieff and Draisey (2018), soils provide essential ecosystem services such carbon storage, nitrogen cycling, water storage and filtration, and biodiversity in addition to serving as a substrate for food production. Climate change is predicted to worsen water erosion and reduce soil organic carbon levels, especially in drylands (Masson-Delmotte et al., 2019). Water erosion is responsible for 20–30 giga tons of soil loss year worldwide (Thomas et al., 2018). The creation and application of sustainable agronomic practices that could preserve and enhance the physical, chemical, and biological properties of soils and enhance the functions of agroecosystems present a challenge to both scientists and farmers (Castellini et al., 2021; Diacono et al., 2021).

The development and widespread implementation of profitable and successful sustainable soil management strategies is thus desperately needed (Gichuki et al., 2019; Thomas et al., 2018). The increasing global population and ongoing climate change need the deployment of agro-ecological or eco-sustainable production strategies to increase the resilience of the most vulnerable agro systems (Leippert et al., 2020). Consequently, a multidisciplinary scientific paradigm is increasingly being suggested as a means of addressing the complex problems associated with sustainable agriculture and soil conservation (Gargano et al., 2021; Wezel et al., 2020). In Table 1, some of the soil issues and their solutions are covered.

2 Soil Quality and Health

Soil is a vital natural resource that is essential to civilization. Soil health refers to the biological, chemical, and physical properties of soil that are essential for long-term, sustainable agricultural production with minimal adverse environmental effects. As a result, soil health offers a comprehensive view of soil functionality. Although it is not possible to directly evaluate soil health, it may be inferred by tracking some soil attributes, like fertility, and evaluating certain soil features, such the amount of

Table 1 Soil problems and their agronomic and engineering solutions to mitigate them (Arora & Bhatt, 2023; Awaad et al., 2020; Kumar et al., 2018; Šarauskis et al., 2024; Sharma, 2018; Yi et al., 2022)

Soil problems	Agronomic solutions	Limitations to agronomic solutions	Engineering solutions
Soil erosion	Conservation tillage, cover crops	Reduce crop yields initially, Change in farming practices	Contouring and Terracing with GPS guidance, structures like gabions
Soil acidity	Liming, acid-tolerant crops	Can be expensive, over-liming can cause nutrient imbalances	Precision liming with GPS-guided equipment
Soil salinity	Leaching, salt-tolerant crops, improved drainage	High water requirement, not feasible in water-scarce areas	Subsurface drainage, desalination techniques
Nutrient depletion	Fertilizer application, crop rotation	Lead to environmental pollution if overused, cost implications	Precision fertilizer application, fertigation systems
Soil compaction	Deep tillage, controlled traffic farming	Energy-intensive, time-consuming, Disrupt soil structure	Deep ripping, Subsoiling
Low organic matter	Addition of organic amendments, reduced tillage	Slow process, immobilize nutrients	Biochar application, compost tea injection systems
Soil contamination	Phytoremediation, chemical immobilization	Long-term process, not suitable for all contaminants	Soil washing, thermal desorption
Poor soil structure	Addition of organic matter, reduced tillage	Slow process, require significant changes in farming practices	Soil conditioning equipment, polymer application

organic matter present (Shahane & Shivay, 2021). According to Voltr et al. (2021) the term "soil physical properties" aligning with soil quality also refers to features of the soil that might affect its behavior and appropriateness for various uses.

The Soil Science Society of America Ad Hoc Committee on Soil Quality (S-581) proposed the first formal definition of soil quality in 1997 (Karlen et al., 1997). The notion of soil quality initially came into being in the early 1990s. According to Brevik et al. (2020), soil health is characterized by its ability to function as a vital living system, within ecosystem and land-use limits, to sustain biological production, preserve the quality of air and water environments, and enhance the health of plants, animals, and humans (Seifu & Elias, 2018). Given the variety of applications for which soil can be used, the definition of "soil quality" is highly dependent on the environment in question. According to Steffan et al. (2018), plant and animal production and health would be more crucial in agricultural soils than they would be in urban ones.

Despite being sometimes disregarded or seen as little more than a physical support system for plant growth, soil is an essential component in crop production. Still, soil has to be seen as a living system given the growing social concerns about agriculture's sustainability (Hou et al., 2020). Its quality is the outcome of several interactions between biological and physical elements, particularly the microbial populations, which are essential to soil function. With a few exceptions, soil is necessary for all crops. It serves as a reservoir for vital nutrients required for plant growth in addition to supporting the roots of plants (Sathiyavani et al., 2017). However, soil is endangered by pollution, erosion, nitrogen loss, and other issues as a result of intensive agriculture. Over the past fifty years, soil fertility has decreased. In several instances, intensive management resulted in lasting harm, endangering this non-renewable resource for future generations.

The sustainability of agriculture, environmental protection, and the energy efficiency of production models all depend on maintaining high soil quality (Mileusnić et al., 2022; Samreen et al., 2022a, 2022b). In order to create more productive, reputable, and targeted techniques to managing agricultural and/or soil systems, methods for evaluating appropriate parameters related to soil health are crucial (Zhang et al., 2024a, 2024b). For this reason, when discussing sustainable agriculture, soil comes first. In response to many definitions arising from the various specialized issues of soil science, farming, soil mapping, and other related fields, the idea of soil quality emerged in the late 1980s and early 1990s (Jones, 2021). Numerous elements have been suggested as constituents of soil quality, and techniques existing to quantify some aspects of it. Given the ultimate land use purpose, it is critical to be able to identify and validate the most intriguing aspects among these many ones as indicators (Sagova-Mareckova et al., 2023). In Fig. 1, there are some parameters that are responsible for the soil quality and health.

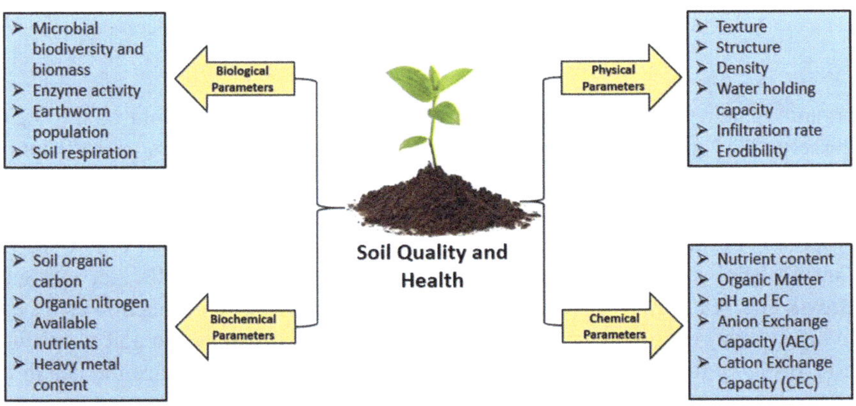

Fig. 1 Parameters associated to soil quality and health

3 Challenges Related to Soil Conservation

After reviewing the scientific, peer-reviewed literature, we have determined that the following are the main global issues pertaining to soil and water conservation:

- Severe weather proceedings are endangering agricultural productivity and causing environmental difficulties, which is necessary for food security. They are also speeding up erosion. Reduced production will come from increased rates of erosion brought on by climate change. Furthermore, according to Hugh Hammond Bennett, environmental issues such as increased erosion might have a detrimental effect on society and jeopardize national security if natural resources are not conserved (Delgado et al., 2020).
- Important worldwide agro-ecosystems that rely on significant amounts of irrigation water are in risk of being wiped out as a consequence of water consumption exceeding water replenishment (Hussain et al., 2020; Scanlon et al., 2023). Given that irrigated systems often produce twice as much as non-irrigated systems, the depletion and salinization of these essential global resources further raises pressure on agricultural productivity to increase (McDermid et al., 2023).
- The expected consequences of climate change, deforestation, erosion, loss of water supplies, and other environmental challenges will make food security a greater concern in the future decades. Additionally, there's a risk that gasoline prices could increase, which might affect agricultural inputs. This might become even more of a concern if stressed systems experience catastrophic occurrences like droughts or floods, or even significant pest or disease outbreaks like the potato disease blight, which was a contributing factor in the infamous potato famine in Ireland (Queiroz et al., 2020).

4 Soil Problems

Now a days, Agriculture is declining continually due to soil problems. The decline in soil performance is due to multiple factors as exhibited in Fig. 2. Some major problems and their possible solutions are discussed in detail in this chapter. Intense agriculture causes biodiversity to decline, soil fertility to deteriorate, natural resource inputs to increase, and soil organic carbon to decrease (Adedibu, 2023). These elements all make it difficult for the agricultural sector to continue being sustainable. Soil structure is a major factor for modeling various soil properties, including hydraulic, mechanical, and shrinkage. Soil structure may be measured by the tortuosity and compaction of the pore walls and channels, the size distributions of the pores and the various kinds of solid particles that constitute the soil, and the shape of the pores (Liu & Jeng, 2019).

However, alterations in the soil structure, such as the disintegration of stable aggregates, modifications to the pore spaces and spatial distribution of soil particles, and soil compaction (combining of soil particles), have detrimental effects on the

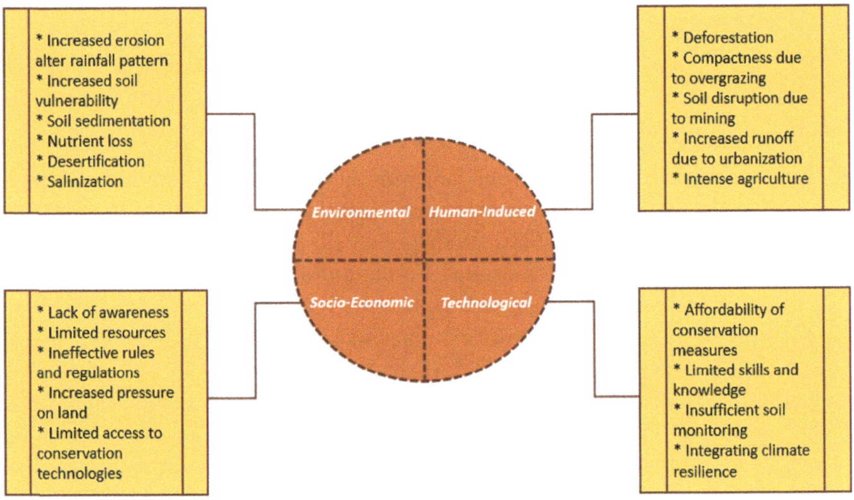

Fig. 2 Different problems related to soil

physical, biological, and chemical properties of soil (Pulido-Moncada et al., 2019). In contrast to the surrounding soil layers, compaction layers are regions of the soil profile where compaction has occurred more intensely (Al-Aani & Sadoon, 2023). Due to soil compaction, millions of hectares of arable land are being degraded worldwide, especially in areas with high agricultural production (Daigh et al., 2020; Hu et al., 2021).

Agricultural lands continue to be a concern in irrigated areas around the globe (Mosaffa & Sepaskhah, 2019; Ramos et al., 2019). When irrigated areas were continuously watered without proper drainage, they became unusable. The soil's salinization poses a significant danger to the long-term sustainability of irrigated farming as well (Fathizad et al., 2020; Manasa et al., 2020; Pulido-Bosch et al., 2018; Singh, 2018c). A growing threat to the sustainability of agriculture is the likelihood of waterlogging and salinization brought on by insufficient irrigation. According to Tao et al. (2024), salinity has already affected almost 3% of all soil resources worldwide. Additionally, at a rate of two million hectares each year, this number is steadily rising (Singh, 2018a, 2018b). Salinization is known to reduce agricultural yields in arid and semiarid parts of the globe by 18–43% (Chang et al., 2019). Field crop yield is significantly decreased by root zone salinization and intense rainfall. The combined effects of waterlogging and root zone salinization are allegedly more harmful than each one alone (Al-Muaini et al., 2019).

A major land degradation concern that has caught the attention of many stakeholders is soil erosion. Soil erosion is a global issue that has become a top concern in a number of countries. It is the general word used to describe the destruction of soil brought about by both natural and man-made factors, such as intensive and extensive agriculture, as well as those induced by water, wind, and snow (Borrelli et al., 2021). There are two types of erosion: natural and accelerated, depending on how severe it is. Under normal conditions, new soils are created via soil erosion, which occurs in the first category over millions of years (Rashmi et al., 2022). However, if soil loss exceeds soil formation, deforestation, overgrazing, and inappropriate agricultural methods are the main man-made causes of increased soil erosion. Soil erosion has a detrimental effect on agricultural output, source water quality, and the health of aquatic and terrestrial ecosystems. A few variables, including topography, land cover, climatic-erosivity, and soil erodibility, cause soil erosion (Fayas et al., 2019).

5 Engineering Techniques to Mitigate Soil Problems

Many regions of the globe are seeing a decline in soil quality due to a mix of physical, chemical, and biological deterioration as well as socioeconomic factors. This has consequences for the resilience, sustainability, and productivity of agri-food systems (Baritz et al., 2018). This decline also reduces the soil's ability to carry out these crucial ecosystem functions. Research points toward a number of reasons of soil deterioration, none of which can be solved, and stakeholders' opinions on how to handle these issues are varied (Rust et al., 2022). But to stop soil deterioration, practical and socially acceptable methods must be developed. In addition to providing us with a surface on which to walk, the earth's surface is an essential part of our natural capital. According to Sayara and Sánchez (2020), there is a growing suggestion that a multidisciplinary scientific framework is needed to tackle the intricate issues surrounding soil preservation and sustainable agriculture. For scientists and farmers alike, the creation and implementation of sustainable engineering techniques capable of maintaining and enhancing the characteristics of soils as well as enhancing agro ecosystem functions is a problem (Castellini et al., 2021).

5.1 Check Dams

Estimating a broad range of soil erosion rates is a regularly used replacement for assessing soil deterioration. Poor vegetation growth is a result of soil erosion. Check dams are employed for this purpose. By preventing soil erosion and stabilizing gully slopes, they encourage revegetation. A method of soil and water conservation was created by combining vegetation and check dams (Shi et al., 2019). A check dam is a little obstruction built across a river or channel to lessen channel erosion and to halt the flow of water and debris. Check dams have the potential to capture sediments before they reach the dam reservoir or accumulate downstream in catchments where soil erosion is a significant concern. Both immediate and long-term advantages can come from the usage of check dams. Table 2 illustrates the different aspects of check dams and sediment basins.

Table 2 Summary about check dams and sediment basins

Aspect	Check dams	Basins	References
Definition	Small barriers are built across a waterway or drainage ditch to counteract erosion by reducing water flow velocity	Large, excavated areas designed to temporarily hold water runoff	Finér et al. (2020)
Purpose	Reduce soil erosion, slows down water flow, trap sediment, increase water infiltration	Gather and preserve runoff. Lower the maximum flow rates, boost the quality of water, replenish groundwater	Singh and Hartsch (2019)
Benefits	Prevents gully formation, enhances soil moisture, reduces downstream sedimentation, improves water quality	Flood control, sediment retention, water supply for irrigation, habitat creation	Bashir et al. (2017)
Construction Techniques	Stone Masonry, gabion structure, log or timber dams, concrete or reinforced concrete	Excavation, embankment construction, spillway design,	Singh and Verma (2018)
Environmental Impact	Alter stream ecology, effect fish passage, create new microhabitats	Potential habitat creation, alteration of downstream hydrology, impacts on aquatic systems	Iqbal and Dutta (2022)

5.2 *Tillage*

Tillage is a crucial agricultural management technique for improving crop productivity and soil quality (Pires et al., 2017). However, long-term conventional tillage methods such as rotary tillage and moldboard plowing may decrease soil productivity and exacerbate soil erosion (Carretta et al., 2021; Klik & Rosner, 2020). Numerous studies have shown that crop rotation and conservation tillage, which leave at least 30% of crop residues on the soil surface, improve soil quality and lessen soil erosion in order to solve these problems (Blevins et al., 2018; Chen et al., 2024). Preserving soil fertility, preventing or reducing soil deterioration, and lowering greenhouse gas emissions are some of the primary objectives of decreased tillage while fostering the right circumstances for plant development (Al-Shamiry et al, 2020). The goal of tillage is to create a soil structure that maximizes plant development and thriving while using the least amount of resources (work, energy, etc.). In the last few decades, especially in the agricultural, mechanical, and biological sciences, these needs have led to the development of complex statistical, mathematical, and other methodologies (De Pue et al., 2019). Table 3 elaborates different type of tillage technologies that helps in the soil conservation.

Strip tillage

When considering both soil conservation and seed emergence %, strip tillage (ST) is a more effective conservation tillage method than no-tillage (Chen et al., 2024). Therefore, ST incorporates the primary benefits of both conventional tillage and no-till farming. Straw mulching in ST-BR has the potential to boost soil moisture and reduce soil evaporation when compared to standard tillage. In order to reduce soil loss, mulching may also improve surface roughness and lessen the impact of rains (Carretta et al., 2021). Long-term soil pore and aggregate structure improvement was much better in ST compared to CT, which was beneficial for root development and solute transport (Han et al., 2024).

Table 3 Different tillage types and their role in soil conservation

Tillage types	Role in soil protection	Role in soil losses	Application ratio	References
Conventional tillage	Limited protection, exposes soil	High soil erosion risk	Decreasing, Approx. 37% of cropland	Chalise et al. (2020)
Conservation tillage	Maintains > 30% residue cover, reduces erosion	Moderate soil loss	Increasing, Approx. 33% of cropland	Klik and Rosner (2020)
No-till	Maximum residue cover, minimal soil disturbance	Lowest soil erosion risk	Growing, Approx. 12% of cropland globally	Seitz et al. (2020)
Strip-till	Protects soil between rows	Moderate to low soil loss	Increasing in row crops, Approx. 5%	Procházková et al. (2020) and Matin et al. (2021)
Ridge-till	Maintains residue between ridges	Moderate soil loss, depends on slope	Limited use, Approx. 3%	Xu et al. (2018) and Blanco and Lal (2023)
Mulch-till	Maintains residue on soil surface	Low to moderate soil loss	Part of conservation tillage, Approx. 20%	Allmaras et al. (2017)
Vertical tillage	Minimal soil disturbance, maintains some residue	Low to moderate soil loss	Growing in popularity, Approx. 5%	Rahman et al. (2020)

Conservation tillage

The productivity of agriculture has been efficiently enhanced to fulfil human demands by conventional agricultural management strategies, such as heavy soil tillage (Hassan et al., 2022). On the other hand, prolonged, extensive agricultural usage may also have unfavorable effects including degraded soil, lost biodiversity, and buildup of toxic materials. These might put a great deal of strain on the ecosystem, the environment, and food security. Sustaining or even improving food production while preserving or improving the environment is therefore made possible by the application of sustainable agriculture management (Al-Shamiry et al., 2020). With the addition of other protective management practices, a new tillage method called conventional tillage leaves a certain amount of crop residue on the soil's surface, reducing soil erosion. (Zhang et al., 2024a, 2024b).

It has recently been demonstrated that conventional tillage causes soil deterioration, and that this ongoing degradation has a clear impact on the region's agricultural yields and soil quality (Li et al., 2019). Consequently, to enhance soil quality and decrease soil loss, a switch from conventional tillage to CT is required (Xu et al., 2018). Globally, conservation tillage (CT), often known as little or no tillage, is seen to be an efficient means of preserving crop yield and enhancing soil quality. As shown by a number of research in this field, long-term conventional tillage may improve soil aggregate structure and soil organic carbon content as compared to CT. According to Zhou et al. (2022) and Zhang et al. (2018), it also evidently enhanced soil water storage.

5.3 GPS Guided Strategies

Global Positioning Systems (GPS) have gained importance recently as a tool for protecting water and land resources (Yang et al., 2022). According to Munnaf et al. (2021), the integration of the preservation of these natural resources with practical solutions, such flexible tools for decision-making and developing data analysis, is feasible with the use of advanced GPS technology. It is possible to illustrate the connection between the soil and a few performance measures by applying this effective method. The US Department of Defense created GPS, mostly with military applications in mind. The number of civilian applications utilizing GPS technology has increased at an astounding rate, notwithstanding its initial intent. GPS systems are used in a wide range of applications, which fall into many fundamental categories and research initiatives (Hunt & Daughtry, 2018).

There are several agricultural applications for GPS-based navigation systems. Completely autonomous navigation is also possible with this method. Throughout the past 20 years, there has been a significant increase in the use of GPS in agricultural output. Many were initially doubtful that broader use would be possible, mainly due to the expensive equipment and significantly poorer precision than they have now (Kendall et al., 2017). There are particular positioning needs in agriculture. Meanwhile, there has been a major improvement in guiding precision, making positioning a particularly effective use in agricultural operations. There are several uses for GPS-based navigation systems in agriculture. With this technique, navigation may be done entirely on its own. The usage of GPS in agricultural production has significantly increased during the last 20 years. Because of the pricey equipment and much lower precision than they have currently, many people first doubted that widespread usage would be feasible (Kendall et al., 2017). Agriculture has certain placement requirements. Meanwhile, there has been a significant advancement in the accuracy of guidance, which makes placement very useful in farming operations.

GPS for fertilization

The Global Navigation Satellite System (GNSS) enables "precision," which is the use of various criteria for pesticide, fertilizer, and sowing based on finished mapping. By mapping yields by zone and optimizing the application of seeds, fertilizer, or pesticides depending on specific application parameters, yield sensors identify the application success of the previously described variable standards (Variable Rate Application, or VRA). The rate of fertilization is always changing, and all of this is made possible by plotting the plot in accordance with the components' contents, such as intended crops, yields from the previous year, and land requirements (Roma et al., 2023).

The GPS data is sent into the system, which then uses the map to recalibrate the fertilization standard. It then adjusts the average based on the kind of fertilizer and the pace of movement. Such a technique can reduce fertilizer consumption by up to 25% while increasing yield by up to 20%. The outcome lowers manufacturing costs by 5.5% and is reflected in the economic effect (Sedlar et al., 2019). For instance "Section control," refers to shutting off the sprayer, spreader, and seed drill sections outside of the plot borders or on a piece of the plot that has already been treated (Maličević et al., 2019). Depending on the region, variable fertilization rate applications might result in savings of 10–25 €/ha. This kind of fertilization allows for a 10–15% decrease in nitrogen usage without compromising yield decline (Maličević et al., 2023).

GPS for levelling

Navigation on both horizontal and vertical planes is required for levelling. Tractors may level the ground by moving in parallel rows—that is, in a small area of the ground in parallel rows, without any overlap or gap between tracks—by using computerized navigation algorithms (Guan et al., 2023). This is another application for positioning systems in driving (Maličević et al., 2023).

GPS for mapping hardpan

In accordance with geo-referenced, site-specific subsoiling activities, a hardpan map was generated based on depth and speed. Hardpan soil may be used to predict hardpan zones based on geo-referenced slip percentage, field productivity, traction power, and fuel consumption (Abdul-Rassoul et al., 2012). The GPS technology is increasingly being employed, and it should be widely adopted as it is a useful tool for mapping and determining soil hardpan in fields. Engineers demonstrated that hardpan zones may be precisely predicted using collected data and a georeferenced map (Hareesh, 2024). This method lessens the impact of impending famine, particularly in developing nations. People's lives have always been made simpler by new technology, which also offers methods for conserving soil through innovative approaches (Al-Aani & Sadoon, 2023).

GPS for contour farming

Soil erosion is a significant environmental problem that affects agricultural productivity and ecosystem health. This global issue affects 33% of arable land (Borrelli et al., 2017). A conservation agricultural technique called contour ploughing can cut soil erosion by 50–70% (Sarvade et al., 2019). By forming level terraces, contour farming shortens the slope's length and lessens its steepness. Farmers are further encouraged to till perpendicular to the slope (Njiru, 2023).

Engineers developed the successful conservation agricultural technique of contour ploughing using GPS-guided tractors to lessen soil erosion. It enhances soil health, lowers soil erosion, promotes water infiltration, and raises crop yields (Khan, 2024; Usama, 2024). Further benefits of this technique are enclosed in Table 4. Variations exist in the effectiveness of GPS-guided contour ploughing due to several parameters such soil type, field terrain, and local climate. The benefits of this technology can be increased by integrating it with other precision agricultural techniques. GPS technology guarantees accurate contour ploughing with a ± 2 cm precision and can save around 10–5% of gasoline (Soto et al., 2019).

Table 4 Benefits of contour farming to mitigate soil erosion

Benefit type	Benefits	References
Environmental benefits	Reduce soil erosion, improve water retention, decrease runoff and nutrient loss, enhance soil conservation	Tarolli and Straffelini (2020)
Economic benefits	Increase crop yield, reduce fuel consumption, lower input costs (fertilizers, pesticides), improve farm profitability	Basso et al. (2021)
Operational benefits	Precise furrow placement, reduce operator fatigue, improve field coverage efficiency	Fawzia et al. (2019)
GPS specific benefits	Accurate field mapping and planning, Ability to create complex contour patterns, Integration with other precision agriculture technologies	Rabia et al. (2022)

5.4 GIS and Remote Sensing

With the introduction of new methods and technologies like remote sensing (RS) and GIS, large-scale soil loss studies have lately become simpler (Chen et al., 2019; Guo et al., 2018; Hammam & Mohamed, 2020; Singh, 2018c). To solve soil-related issues, researchers have used GIS and remote sensing methods extensively during the last several decades (Tajik et al., 2019; Zovko et al., 2018). Earlier studies (e.g., Fathizad et al., 2020; Han et al., 2019) mostly used GIS and remote sensing methods to map waterlogged areas and soil salinization on a wide scale. It has also helped scientists in locating different infectious diseases in crops (Samreen et al., 2022a, 2022b). Prior research has generally shown that the combined applications of these strategies are more advantageous than their separate applications (Asfaw et al., 2018; Gorji et al., 2017). In order to map the impact of subsurface waterlogging on surface soil moisture, the majority of these studies have used remote sensing and GIS methods in an indirect manner (Singh, 2021).

Remote sensing for Conservation tillage

Due to its low cost and ability to cover vast geographical and temporal regions, remote sensing technology is a valuable tool for mapping and tracking the dispersion of conservation tillage (CT) (Abd El-Ghany et al., 2020). When examining the present status of CT remote sensing monitoring research, three perspectives were taken into account: data sources, features, and methodologies. According to Zhang et al., (2024a, 2024b), the traditional method of collecting CT-related data involves the use of roadside surveys, agricultural census campaigns, onsite field surveys, field data collection, and other manual procedures. These techniques are quite precise, but they take a long time and include a complicated sampling procedure, therefore they are rarely used in broad regions or in real time.

Currently, remote sensing technology provides an additional method for identifying CT and its associated data across time and location because to its high efficiency, timeliness, and macroscopic nature (Fan et al., 2021). An increasing amount of research is using multi-hyperspectral remote sensing imagery, including that from Landsat and Hyperion, to collect CRC in order to directly identify CT (Najafi et al., 2018). Moreover, microwave remote sensing technology has been widely used in CT identification technologies. Prior studies have summarized the methods for tracking CT remote sensing and its uses in related disciplines. For instance, Zheng et al. (2014) investigated the difference between CT and conventional tillage using remote sensing. The many types of data used in CT monitoring approaches were also described, and the challenges of doing this kind of research in terms of topography, data, and time were discussed. In their brief analysis of the use of remote sensing technology in soil tillage monitoring, Bégué et al. (2018) gave a summary of remote sensing monitoring techniques in terms of both CRC and soil roughness, which affect the tillage system. However, as Fig. 3 illustrates, there are still certain limitations.

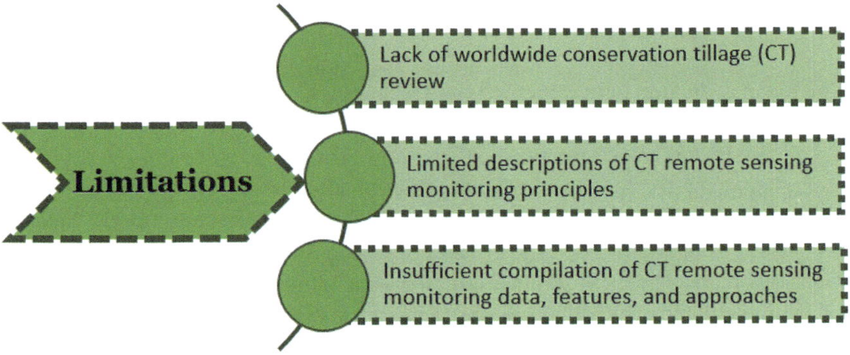

Fig. 3 Limitations to conservation tillage through remote sensing technique

Combined application

Experts like Walche et al. (2024) and Tarate et al. (2024) claim that the use of remote sensing and GIS technology has greatly improved lately for the observation and monitoring of salinization brought on by insufficient drainage in many parts of the globe. Several recent research have shown how successful these methods are in resolving issues pertaining to natural resources (Tadese et al., 2020). The Czech Republic's subsurface drainage systems may be categorized using the RS technique, as shown by Tlapáková et al. (2017). This research establishes the pertinent conditions and criteria for drainage system visualization in these images, based on the capacity to get multispectral data at different times.

In order to develop thematic geodatabases and map drainage network layers and the locations of their degradation, including soil salinization, Teodoro and Duarte (2022) state that the research used remote sensing data analysis. In a new modeling framework, Asfaw et al. (2018) used remote sensing and GIS techniques to map the salinity of the soil in an irrigated region. For the estimate of different spectral indices, they employed Landsat pictures. The findings showed that the salinity index and the measured electrical conductivity values had the strongest correlation. The analysis also revealed the value of using geospatial approaches to examine the spatial salinity of soil in irrigation zones (Singh, 2021).

Salinization mapping

For mapping the breadth and severity of soil salinization and waterlogging, GIS and remote sensing provide appropriate solutions, mostly on large scales (Bugnot et al., 2018; Ferdous & Rahman, 2020; Zewdu et al., 2017). These techniques have been extensively applied for the past 60 years to assess soil salinization, where zones impacted by salt were delineated using variously colored aerial images (Han et al., 2019). An important part of the process is measuring and storing the electromagnetic radiation that the earth's surface reflects. The creation of maps and more precise record development are made possible by these geospatial approaches (Keil et al., 2021). It is common practice to classify flooded regions using digital data from

Landsat 5 TM and Multispectral Sensor (MSS). It is helpful to utilize the proposed model to evaluate the hazards related to irrigation applications (Ojo & Ilunga, 2018; Prajapati et al., 2021).

Technology for Sensing and Sensors

Sensing is the process of using sensors to collect information from real-world objects (Saleem et al., 2024). One of the key parts of smart objects that gives them the capacity to sense changes in their surroundings and identify changes in the ambient circumstances is the sensor (Javaid et al., 2021). After taking into account each of these descriptions of sensors, it can be said that an electrical signal representing a change in charge, voltage, and current is the sensor's output, while some type of non-electric observation pertaining to a change in an object's physical attribute serves as its input. Table 5 lists some of the most often utilized sensors in agricultural settings and their overall objective of soil conservation (Ayaz et al., 2019; Elijah et al., 2018; Farooq et al., 2019; Mathur, 2020; Shafi et al., 2019; Vena et al., 2018). For soil monitoring, a variety of Radio Frequency Identification (RFID)-based sensors have been developed. For example, smart sensor nodes equipped with 2.4 GHz active RFID tags have been developed to measure soil moisture and are eventually used in agricultural establishments to optimize water usage (Rayhana et al., 2021).

- To measure the salinity of soil, chipless and chip-based (UHF) RFID sensors have been developed (Dey et al., 2019).
- In 2020, Athauda et al. created a pH sensor based on UWB RFID technology that is chip-less and capable of correctly calculating pH readings ranging from 4 to 7.

6 Future Perspectives

- By selecting the tiny grids, possible distributed models may function better. In order to gather information at a finer resolution about irrigation, drainage, canals, cropping patterns, groundwater levels, tube wells, meteorological factors, soil and groundwater salinities, etc., it is advised to carry out more thorough field surveys.
- Better sensing methods and retrieval systems are needed to allow for the immediate identification of salinization and waterlogging features at large scales. In the future, it is suggested that data from several satellites be used to alleviate the time lag between the occurrence and recording of various data. This would eliminate problems with spatial resolution by increasing system efficiency.
- With a longer review period, researchers may evaluate literature on mechanical techniques for controlling soil erosion on Asian agricultural land. It has been demonstrated that mechanical methods of mitigating soil erosion, such as terracing, fish-scale pits, and microbasin tillage, successfully lower the rate of soil loss. After 50 years of evaluation, relevant stakeholders will have a clearer understanding of how the mechanical approach may aid in controlling soil erosion.

Table 5 Various types of sensors used in soil conservation

Sensor types	Benefits	References
Soil moisture sensors	Reduce runoff and soil erosion, optimize irrigation	Ray and Majumder (2024)
Soil temperature sensors	Optimize nutrient cycling, guide planting decisions	Sarma et al. (2024)
Electrical conductivity (EC) sensor	Identify salinity issues, map soil texture variations, guide variable rate applications	Raza et al. (2023)
pH Sensors	Improve soil health management, optimize nutrient availability, guide lime applications	Rhymes et al. (2023)
Spectral reflectance sensors	Estimate soil organic matter content, guide variable rate fertilizer application, monitor soil degradation	Guerrero et al. (2021)
Penetrometer sensors	Measure soil compaction, guide tillage decisions	Vaz et al. (2022)
LiDAR sensors	Guide precision land leveling, create high-resolution elevation models	Du et al. (2022)
Multispectral sensors	Monitor soil erosion, guide precision agriculture practices,	Sishodia et al. (2020)
Accelerometer sensors	Measure tillage depth and speed, improve soil disturbance management	Wang et al. (2024)
GNSS (GPS) sensors	Facilitate controlled traffic farming, improve overall field efficiency	Chen et al. (2022)

- Moreover, high-level commitment from all spheres of society is required to preserve, enhance, and safeguard the ecosystem and productivity of the soil. Conservation techniques will preserve the fertility of the soil and enable the land to support larger agricultural yields, which will benefit Asia's economy by significantly reducing soil loss.

7 Conclusion

Modern technology may be used in the manufacturing to lower costs while it also increases the efficiency and protects soil from losses. A thoughtful strategy may boost soil conservation, which in turn boosts earnings. It is not possible to adopt precision agriculture in a way that is ideal for all users; each user may and should have their own deployment strategy. While reducing the use of fuel, herbicides, fertilizers, and seeds, GPS and GIS guidance may optimize tractor, equipment, and driver productivity. They also play a significant role in soil conservation.

References

Abd El-Ghany, N. M., Abd El-Aziz, S. E., & Marei, S. S. (2020). A review: Application of remote sensing as a promising strategy for insect pests and diseases management. *Environmental Science and Pollution Research, 27*(27), 33503–33515.

Abdul-Rassoul, M. S., Al-Jalely, B. H., Al-Nuaim, K. T., & khahtan Al-Ani, L. (2012). First record of red-back spider Latrodectus scelio thorell, 1870 (Araneae: theridiidae) in Baghdad, Iraq. *Bulletin of the Iraq Natural History Museum, 12*(2), 1–5. P-ISSN: 1017-8678. E-ISSN: 2311-9799.

Adedibu, P. A. (2023). Ecological problems of agriculture: impacts and sustainable solutions. *ScienceOpen preprints.*

Al-Aani, F. S., & Sadoon, O. H. (2023). Modern GPS diagnostic technique to determine and map soil hardpan for enhancing agricultural operation management. *Journal of Aridland Agriculture, 9*, 58–62.

Allmaras, R. R., Copeland, S. M., Power, J. F., & Tanaka, D. L. (2017). Conservation tillage systems in the northernmost central United States. In *Conservation tillage in temperate agroecosystems* (pp. 255–284). CRC Press.

Al-Muaini, A., Green, S., Dakheel, A., Abdullah, A. H., Abou Dahr, W. A., Dixon, S., Kemp, S., & Clothier, B. (2019). Irrigation management with saline groundwater of a date palm cultivar in the hyper-arid United Arab Emirates. *Agricultural Water Management, 211*, 123–131.

Al-Shamiry, F. M. S., Al-Qarni, A. M., & Munassar, A. S. (2020). Effect of tillage depth and tractor forward speed on some technical indicators of the Moldboard Plow. *International Journals of Progressive Sciences and Technologies, 23*(2), 28–37.

Arora, S., & Bhatt, R. (2023). Resource conservation techniques for sustaining crop production in rainfed foothills under changing climate. In *Advances in crop production and climate change* (pp. 367–405). CRC Press.

Asfaw, E., Suryabhagavan, K. V., & Argaw, M. (2018). Soil salinity modeling and mapping using remote sensing and GIS: The case of Wonji sugar cane irrigation farm, Ethiopia. *Journal of the Saudi Society of Agricultural Sciences, 17*(3), 250–258.

Awaad, H. A., Mansour, E., Akrami, M., Fath, H. E., Javadi, A. A., & Negm, A. (2020). Availability and feasibility of water desalination as a non-conventional resource for agricultural irrigation in the mena region: A review. *Sustainability, 12*(18), 7592.

Ayaz, M., Ammad-Uddin, M., Sharif, Z., Mansour, A., & Aggoune, E. H. M. (2019). Internet-of-Things (IoT)-based smart agriculture: Toward making the fields talk. *IEEE Access, 7*, 129551–129583.

Baritz, R., Wiese, L., Verbeke, I., & Vargas, R. (2018). Voluntary guidelines for sustainable soil management: Global action for healthy soils. *International Yearbook of Soil Law and Policy, 2017*, 17–36.

Bashir, S., Javed, A., Bibi, I., & Ahmad, N. (2017). Soil and water conservation (pp. 263–286). Faisalabad, Pakistan: University of Agriculture.

Basso, B., Jones, J. W., Antle, J., Martinez-Feria, R. A., & Verma, B. (2021). Enabling circularity in grain production systems with novel technologies and policy. *Agricultural Systems, 193*, 103244.

Bégué, A., Arvor, D., Bellon, B., Betbeder, J., De Abelleyra, D., Ferraz, R. P. D., Lebourgeois, V., Lelong, C., Simões, M., & Verón, S. R. (2018). Remote sensing and cropping practices: A review. *Remote Sensing, 10*(1), 99.

Blanco, H., & Lal, R. (2023). Tillage Systems. In *Soil Conservation and Management* (pp. 127–157). Cham: Springer Nature Switzerland.

Blevins, R. L., Lal, R., Doran, J. W., Langdale, G. W., & Frye, W. W. (2018). Conservation tillage for erosion control and soil quality. In *Advances in soil and water conservation* (pp. 51–68). Routledge.

Borrelli, P., Alewell, C., Alvarez, P., Anache, J. A. A., Baartman, J., Ballabio, C., Bezak, N., Biddoccu, M., Cerdà, A., Chalise, D., Chen, S., Chen, W., De Girolamo, A. M., Gessesse, G. D., Deumlich, D., Diodato, N., Efthimiou, N., Erpul, G., Fiener, P., … Haregeweyn, N. (2021). Soil erosion modelling: A global review and statistical analysis. *Science of the Total Environment, 780*, 146494.

Borrelli, P., Robinson, D. A., Fleischer, L. R., Lugato, E., Ballabio, C., Alewell, C., Meusburger, K., Modugno, S., Schütt, B., Ferro, V., Bagarello, V., Van Oost, K., Montanarella, L., & Panagos, P. (2017). An assessment of the global impact of 21st century land use change on soil erosion. *Nature Communications, 8*(1), 1–13.

Brevik, E. C., Slaughter, L., Singh, B. R., Steffan, J. J., Collier, D., Barnhart, P., & Pereira, P. (2020). Soil and human health: Current status and future needs. *Air, Soil and Water Research, 13*, 1178622120934441.

Bugnot, A. B., Lyons, M. B., Scanes, P., Clark, G. F., Fyfe, S. K., Lewis, A., & Johnston, E. L. (2018). A novel framework for the use of remote sensing for monitoring catchments at continental scales. *Journal of Environmental Management, 217*, 939–950.

Carretta, L., Tarolli, P., Cardinali, A., Nasta, P., Romano, N., & Masin, R. (2021). Evaluation of runoff and soil erosion under conventional tillage and no-till management: A case study in northeast Italy. *CATENA, 197*, 104972.

Castellini, M., Diacono, M., Gattullo, C. E., & Stellacci, A. M. (2021). Sustainable agriculture and soil conservation. *Applied Sciences, 11*(9), 4146.

Chalise, D., Kumar, L., Sharma, R., & Kristiansen, P. (2020). Assessing the impacts of tillage and mulch on soil erosion and corn yield. *Agronomy, 10*(1), 63.

Chang, X., Gao, Z., Wang, S., & Chen, H. (2019). Modelling long-term soil salinity dynamics using SaltMod in Hetao irrigation district, China. *Computers and Electronics in Agriculture, 156*, 447–458.

Chen, Q., Zhang, J., Guo, M., Zhang, X., Tian, J., Zhou, P., & Chen, Z. (2024). Comparison of the effects of five long-term land use and management practices on runoff, soil erosion, and nutrient loss under natural rainfall in the Mollisol region of Northeast China. *Earth Surface Processes and Landforms, 49*(5), 1606–1620.

Chen, W., Tsangaratos, P., Ilia, I., Duan, Z., & Chen, X. (2019). Groundwater spring potential mapping using population-based evolutionary algorithms and data mining methods. *Science of the Total Environment, 684*, 31–49.

Chen, Y., Li, G., Zhang, X., Jia, J., Zhou, K., & Wu, C. (2022). Identifying field and road modes of agricultural Machinery based on GNSS Recordings: A graph convolutional neural network approach. *Computers and Electronics in Agriculture, 198*, 107082.

Daigh, A. L. M., DeJong-Hughes, J., & Acharya, U. (2020). Projections of yield losses and economic costs following deep wheel-traffic compaction during the 2019 harvest. *Agricultural & Environmental Letters, 5*(1), e20013.

De Pue, J., Rezaei, M., Van Meirvenne, M., & Cornelis, W. M. (2019). The relevance of measuring saturated hydraulic conductivity: Sensitivity analysis and functional evaluation. *Journal of Hydrology, 576*, 628–638.

Delgado, J. A., Gantzer, C. J., & Sassenrath, G. F. (2020). The Future of Soil, Water, and Air Conservation. *Soil and Water, 75*, 307–331.

Dey, S., Bhattacharyya, R., Karmakar, N., & Sarma, S. (2019). A folded monopole shaped novel soil moisture and salinity sensor for precision agriculture based chipless RFID applications. In *2019 IEEE MTT-S International Microwave and RF Conference (IMARC)* (pp. 1–4). IEEE.

Diacono, M., Persiani, A., Castellini, M., Giglio, L., & Montemurro, F. (2021). Intercropping and rotation with leguminous plants in organic vegetables: Crop performance, soil properties and sustainability assessment. *Biological Agriculture & Horticulture, 37*(3), 141–167.

Du, M., Li, H., & Roshanianfard, A. (2022). Design and experimental study on an innovative UAV-LiDAR topographic mapping system for precision land levelling. *Drones, 6*(12), 403.
Elijah, O., Rahman, T. A., Orikumhi, I., Leow, C. Y., & Hindia, M. N. (2018). An overview of Internet of Things (IoT) and data analytics in agriculture: Benefits and challenges. *IEEE Internet of Things Journal, 5*(5), 3758–3773.
Fan, J., Zhang, Y., Wen, W., Gu, S., Lu, X., & Guo, X. (2021). The future of Internet of Things in agriculture: Plant high-throughput phenotypic platform. *Journal of Cleaner Production, 280*, 123651.
Farooq, M. S., Riaz, S., Abid, A., Abid, K., & Naeem, M. A. (2019). A survey on the role of IoT in agriculture for the implementation of smart farming. *IEEE Access, 7*, 156237–156271.
Fathizad, H., Ardakani, M. A. H., Sodaiezadeh, H., Kerry, R., & Taghizadeh-Mehrjardi, R. (2020). Investigation of the spatial and temporal variation of soil salinity using random forests in the central desert of Iran. *Geoderma, 365*, 114233.
Fawzia, H., Ahmeda, D., Mostafac, S. A., Fudzeec, M. F. M., Mahmoodd, M. A., Zeebareee, S. R., & Ibrahimf, D. A. (2019). A review of automated decision support techniques for improving tillage operations. *REVISTA AUS, 26*, 219–240.
Fayas, C. M., Abeysingha, N. S., Nirmanee, K. G. S., Samaratunga, D., & Mallawatantri, A. (2019). Soil loss estimation using rusle model to prioritize erosion control in KELANI river basin in Sri Lanka. *International Soil and Water Conservation Research, 7*(2), 130–137.
Ferdous, J., & Rahman, M. T. U. (2020). Developing an empirical model from Landsat data series for monitoring water salinity in coastal Bangladesh. *Journal of Environmental Management, 255*, 109861.
Finér, L., Härkönen, L., Jämsén, J., Joensuu, S., Leinonen, A., Andersson, E., Sikström, U., et al. (2020). Manual for constructing water protection structures at ditch network maintenance sites and for water retention in forests.
Gargano, G., Licciardo, F., Verrascina, M., & Zanetti, B. (2021). The agroecological approach as a model for multifunctional agriculture and farming towards the European Green Deal 2030—some evidence from the Italian experience. *Sustainability, 13*(4), 2215.
Gichuki, L., Brouwer, R., Davies, J., Vidal, A., Kuzee, M., Magero, C., Walter, S., Lara, P., Oragbade, C., & Gilbey, B. (2019). *Reviving land and restoring landscapes*. IUCN.
Gorji, T., Sertel, E., & Tanik, A. (2017). Monitoring soil salinity via remote sensing technology under data scarce conditions: A case study from Turkey. *Ecological Indicators, 74*, 384–391.
Guan, S., Takahashi, K., Nakano, K., Fukami, K., & Cho, W. (2023). Real-time kinematic imagery-based automated levelness assessment system for land leveling. *Agriculture, 13*(3), 657.
Guerrero, A., De Neve, S., & Mouazen, A. M. (2021). Current sensor technologies for in situ and on-line measurement of soil nitrogen for variable rate fertilization: A review. *Advances in Agronomy, 168*, 1–38.
Guo, S., Ruan, B., Chen, H., Guan, X., Wang, S., Xu, N., & Li, Y. (2018). Characterizing the spatiotemporal evolution of soil salinization in Hetao Irrigation District (China) using a remote sensing approach. *International Journal of Remote Sensing, 39*(20), 6805–6825.
Hammam, A. A., & Mohamed, E. S. (2020). Mapping soil salinity in the East Nile Delta using several methodological approaches of salinity assessment. *The Egyptian Journal of Remote Sensing and Space Science, 23*(2), 125–131.
Han, C., Zhang, B., Chen, H., Wei, Z., & Liu, Y. (2019). Spatially distributed crop model based on remote sensing. *Agricultural Water Management, 218*, 165–173.
Han, J., Li, J., Jiang, Y., Wang, D., Chen, Q., & Zhang, X. (2024). Effect of long-term strip tillage on soil macropore structure in cool black soil region of Northeast China. *Land Degradation & Development, 35*(1), 484–495.
Hareesh, S. B. (2024). The latest applications of remote sensing technologies for soil management in precision agriculture practices. In *Remote sensing in precision agriculture* (pp. 105–135). Academic Press.
Hassan, W., Saba, T., Jabbi, F., Wang, B., Cai, A., & Wu, J. (2022). Improved and sustainable agroecosystem, food security and environmental resilience through zero tillage with emphasis

on soils of temperate and subtropical climate regions: A review. *International Soil and Water Conservation Research, 10*(3), 530–545.
Hou, D., Bolan, N. S., Tsang, D. C., Kirkham, M. B., & O'connor, D. (2020). Sustainable soil use and management: An interdisciplinary and systematic approach. *Science of the Total Environment, 729*, 138961.
Hu, W., Drewry, J., Beare, M., Eger, A., & Müller, K. (2021). Compaction induced soil structural degradation affects productivity and environmental outcomes: A review and New Zealand case study. *Geoderma, 395*, 115035.
Hunt, E. R., Jr., & Daughtry, C. S. (2018). What good are unmanned aircraft systems for agricultural remote sensing and precision agriculture? *International Journal of Remote Sensing, 39*(15–16), 5345–5376.
Hussain, M. I., Farooq, M., Muscolo, A., & Rehman, A. (2020). Crop diversification and saline water irrigation as potential strategies to save freshwater resources and reclamation of marginal soils—A review. *Environmental Science and Pollution Research, 27*(23), 28695–28729.
Iqbal, K., & Dutta, V. (2022). Effect of flow alteration on river ecology: state of the art. In *Environmental studies and climate change* (pp. 47–79).
Javaid, M., Haleem, A., Rab, S., Singh, R. P., & Suman, R. (2021). Sensors for daily life: A review. *Sensors International, 2*, 100121.
Jones, K. C. (2021). Persistent organic pollutants (POPs) and related chemicals in the global environment: Some personal reflections. *Environmental Science & Technology, 55*(14), 9400–9412.
Karlen, D. L., Mausbach, M. J., Doran, J. W., Cline, R. G., Harris, R. F., & Schuman, G. E. (1997). Soil quality: A concept, definition, and framework for evaluation (a guest editorial). *Soil Science Society of America Journal, 61*(1), 4–10.
Keil, J., Edler, D., Schmitt, T., & Dickmann, F. (2021). Creating immersive virtual environments based on open geospatial data and game engines. *KN-Journal of Cartography and Geographic Information, 71*(1), 53–65.
Kendall, H., Naughton, P., Clark, B., Taylor, J., Li, Z., Zhao, C., Yang, G., Chen, J., & Frewer, L. J. (2017). Precision agriculture in China: Exploring awareness, understanding, attitudes and perceptions of agricultural experts and end-users in China. *Advances in Animal Biosciences, 8*(2), 703–707.
Khan, W. (2024). *Crop production: techniques and technology*. Academic Guru Publishing House.
Klik, A., & Rosner, J. (2020). Long-term experience with conservation tillage practices in Austria: Impacts on soil erosion processes. *Soil and Tillage Research, 203*, 104669.
Kumar, R., Ahmed, M., Bhadrachari, G., & Thomas, J. P. (2018). Desalination for agriculture: Water quality and plant chemistry, technologies and challenges. *Water Science and Technology: Water Supply, 18*(5), 1505–1517.
Leippert, F., Darmaun, M., Bernoux, M., Mpheshea, M., Müller, A., Geck, M., Termote, C., et al. (2020). The potential of agroecology to build climate-resilient livelihoods and food systems.
Li, Y., Li, Z., Cui, S., Jagadamma, S., & Zhang, Q. (2019). Residue retention and minimum tillage improve physical environment of the soil in croplands: A global meta-analysis. *Soil and Tillage Research, 194*, 104292.
Liu, Y. F., & Jeng, D. S. (2019). Pore scale study of the influence of particle geometry on soil permeability. *Advances in Water Resources, 129*, 232–249.
Maličević, Z., Lakić, Ž., & Jugović, M. (2023). Advantages and possibilities of application of precise systems in the agricultural production. *Agro-Knowledge Journal, 24*(2).
Maličević, Z., Jugović, M., Mitrić, S., & Mitrović, D. (2019). The influence of the sprayer control on the increase in the productivity of the machine-tractor aggregate in the protection of orchards. *Agro-Knowledge Journal, 20*(2), 85–93.
Manasa, M. R. K., Katukuri, N. R., Nair, S. S. D., Haojie, Y., Yang, Z., & bo Guo, R. (2020). Role of biochar and organic substrates in enhancing the functional characteristics and microbial community in a saline soil. *Journal of Environmental Management, 269*, 110737.
Masson-Delmotte, V., Pörtner, H. O., Skea, J., Buendía, E. C., Zhai, P., & Roberts, D. (2019). Climate change and land. IPCC Report.

Mathur, P. (2020). *IoT machine learning applications in telecom, energy, and agriculture.*
Matin, M. A., Hossain, M. I., Gathala, M. K., Timsina, J., & Krupnik, T. J. (2021). Optimal design and setting of rotary strip-tiller blades to intensify dry season cropping in Asian wet clay soil conditions. *Soil and Tillage Research, 207*, 104854.
McDermid, S., Nocco, M., Lawston-Parker, P., Keune, J., Pokhrel, Y., Jain, M., Jägermeyr, J., Brocca, L., Massari, C., Jones, A. D., Vahmani, P., Thiery, W., Yao, Y., Bell, A., Chen, L., Dorigo, W., Hanasaki, N., Jasechko, S., Lo, M.-H., Mahmood, R., et al. (2023). Irrigation in the earth system. *Nature Reviews Earth & Environment, 4*(7), 435–453.
Mileusnić, Z. I., Saljnikov, E., Radojević, R. L., & Petrović, D. V. (2022). Soil compaction due to agricultural machinery impact. *Journal of Terramechanics, 100*, 51–60.
Moncrieff, C., & Draisey, Z. (2018). *Saving the earth: a sustainable future for soils and water.* WWF-UK.
Mosaffa, H. R., & Sepaskhah, A. R. (2019). Performance of irrigation regimes and water salinity on winter wheat as influenced by planting methods. *Agricultural Water Management, 216*, 444–456.
Munnaf, M. A., Haesaert, G., Van Meirvenne, M., & Mouazen, A. M. (2021). Multi-sensors data fusion approach for site-specific seeding of consumption and seed potato production. *Precision Agriculture, 22*(6), 1890–1917.
Najafi, P., Navid, H., Feizizadeh, B., & Eskandari, I. (2018). Remote sensing for crop residue cover recognition: a review. *Agricultural Engineering International: CIGR Journal, 20*(1).
Njiru, E. N. (2023). *Effect of Fanya Juu Terraces With Varying Ditch Dimensions on Selected Soil Properties and Crop Yields in Semi-arid Eastern Kenya* (Doctoral dissertation). University of Nairobi.
Ojo, O. I., & Ilunga, F. (2018). Geospatial analysis for irrigated land assessment, modeling and mapping. In *Multi-purposeful application of geospatial data* (Vol. 9, pp. 65–84).
Pires, L. F., Borges, J. A., Rosa, J. A., Cooper, M., Heck, R. J., Passoni, S., & Roque, W. L. (2017). Soil structure changes induced by tillage systems. *Soil and Tillage Research, 165*, 66–79.
Prajapati, G. S., Rai, P. K., Mishra, V. N., Singh, P., & Shahi, A. P. (2021). Remote sensing-based assessment of waterlogging and soil salinity: A case study from Kerala, India. *Results in Geophysical Sciences, 7*, 100024.
Procházková, E., Kincl, D., Kabelka, D., Vopravil, J., Nerušil, P., Menšík, L., & Barták, V. (2020). The impact of the conservation tillage "maize into grass cover" on reducing the soil loss due to erosion. *Soil & Water Research, 15*(3).
Pulido-Bosch, A., Rigol-Sanchez, J. P., Vallejos, A., Andreu, J. M., Ceron, J. C., Molina-Sanchez, L., & Sola, F. (2018). Impacts of agricultural irrigation on groundwater salinity. *Environmental Earth Sciences, 77*, 1–14.
Pulido-Moncada, M., Munkholm, L. J., & Schjønning, P. (2019). Wheel load, repeated wheeling, and traction effects on subsoil compaction in northern Europe. *Soil and Tillage Research, 186*, 300–309.
Queiroz, A. I., Gomes, I., & Amorim, I. (2020). The spectre of historical crop losses: Challenging questions and methodological issues.
Rabia, A. H., Neupane, J., Lin, Z., Lewis, K., Cao, G., & Guo, W. (2022). Principles and applications of topography in precision agriculture. *Advances in Agronomy, 171*, 143–189.
Rahman, M. M., Alam, M. S., Kamal, M. Z. U., & Rahman, G. M. (2020). Organic sources and tillage practices for soil management. In *Resources use efficiency in agriculture* (pp. 283–328).
Ramos, T. B., Darouich, H., Šimůnek, J., Gonçalves, M. C., & Martins, J. C. (2019). Soil salinization in very high-density olive orchards grown in southern Portugal: Current risks and possible trends. *Agricultural Water Management, 217*, 265–281.
Rashmi, I., Karthika, K. S., Roy, T., Shinoji, K. C., Kumawat, A., Kala, S., & Pal, R. (2022). Soil erosion and sediments: a source of contamination and impact on agriculture productivity. In *Agrochemicals in soil and environment: impacts and remediation* (pp. 313–345). Singapore: Springer Nature Singapore.

Rațu, R. N., Veleșcu, I. D., Stoica, F., Usturoi, A., Arsenoaia, V. N., Crivei, I. C., Postolache, A. N., Lipșa, F. D., Filipov, F., Florea, A. M., Chițea., & Brumă, I. S. (2023). Application of Agri-food by-products in the food industry. *Agriculture, 13*(8), 1559.
Ray, S., & Majumder, S. (2024). Water management in agriculture: Innovations for efficient irrigation. In P. Sil, P. Chhetri, S. Majumder & D. T. Santosh (Eds.), *Modern agronomy* (pp. 169–185).
Rayhana, R., Xiao, G., & Liu, Z. (2021). RFID sensing technologies for smart agriculture. *IEEE Instrumentation & Measurement Magazine, 24*(3), 50–60.
Raza, A., Shahid, M. A., Safdar, M., Zaman, M., & Sabir, R. M. (2023). Applications of sensors in high-tech agriculture. In *Handbook of research on AI-equipped IoT applications in high-tech agriculture* (pp. 271–294). IGI Global.
Rhymes, J., Chadwick, D. R., Williams, A. P., Harris, I. M., Lark, R. M., & Jones, D. L. (2023). Evaluating the accuracy and usefulness of commercially-available proximal soil mapping services for grassland nutrient management planning and soil health monitoring. *Precision Agriculture, 24*(3), 898–920.
Roma, E., Laudicina, V. A., Vallone, M., & Catania, P. (2023). Application of precision agriculture for the sustainable management of fertilization in olive groves. *Agronomy, 13*(2), 324.
Rust, N., Lunder, O. E., Iversen, S., Vella, S., Oughton, E. A., Breland, T. A., Glass, J. H., Maynard, C. M., McMorran, R., & Reed, M. S. (2022). Perceived causes and solutions to soil degradation in the UK and Norway. *Land, 11*(1), 131.
Sagova-Mareckova, M., Omelka, M., & Kopecky, J. (2023). The golden goal of soil management: Disease-suppressive soils. *Phytopathology®, 113*(4), 741–752.
Saleem, M. F., Raza, A., Sabir, R. M., Safdar, M., Faheem, M., Al Ansari, M. S., & Hussain, S. (2024). Applications of sensors in precision agriculture for a sustainable future. In *Agriculture and aquaculture applications of biosensors and bioelectronics* (pp. 109–137). IGI Global.
Samreen, T., Tahir, A., Sidra-Tul-Muntaha, Nazir, M.Z., Ahmad, M., & Kanwal, S. (2022b). Sensing techniques in precision agriculture for pest and disease management. *Environmental Sciences Proceedings, 23*, 16.
Samreen, T., Rasool, S., Kanwal, S., Riaz, S., & Nazir, M. Z. (2022a). Role of nanotechnology in precision agriculture. *Environmental Sciences Proceedings, 23*(1), 17.
Šarauskis, E., Sokas, S., & Rukaitė, J. (2024). Variable depth tillage: Importance, applicability, and impact—an overview. *AgriEngineering, 6*(2), 1870–1885.
Sarma, H. H., Borah, S. K., Chintey, R., Nath, H., & Talukdar, N. (2024). Site specific nutrient management (SSNM): Principles, key features and its potential role in soil, crop ecosystem and climate resilience farming. *Journal of Advances in Biology & Biotechnology, 27*(8), 211–222.
Sarvade, S., Upadhyay, V. B., Kumar, M., & Imran Khan, M. (2019). Soil and water conservation techniques for sustainable agriculture. In *Sustainable agriculture, forest and environmental management* (pp. 133–188).
Sathiyavani, E., Prabaharan, N. K., & Surendar, K. K. (2017). Role of mineral nutrition on root growth of crop plants–a review. *International Journal of Current Microbiology and Applied Sciences, 6*(4), 2810–2837.
Sayara, T., & Sánchez, A. (2020). Bioremediation of PAH-contaminated soils: Process enhancement through composting/compost. *Applied Sciences, 10*(11), 3684.
Scanlon, B. R., Fakhreddine, S., Rateb, A., de Graaf, I., Famiglietti, J., Gleeson, T., Quentin Grafton, R., Jobbagy, E., Kebede, S., Kolusu, S. R., Konikow, L. F., Long, D., Mekonnen, M., Schmied, H. M., Mukherjee, A., MacDonald, A., Reedy, R. C., Shamsudduha, M., Simmons, C. T., … Zheng, C. (2023). Global water resources and the role of groundwater in a resilient water future. *Nature Reviews Earth & Environment, 4*(2), 87–101.
Sedlar, A., Višacki, V., Bugarin, R., Turan, J., & Ponjičan, O. (2019). Implementacija sistema precizne poljoprivrede u proizvodnju uljanih i drugih ratarskih kultura. *Savremena Poljoprivredna Tehnika, 45*(4), 143–150.
Seifu, W., & Elias, E. (2018). Soil quality attributes and their role in sustainable agriculture: A review. *International Journal of Plant & Soil Science, 26*(3), 1–26.

Seitz, S., Prasuhn, V., & Scholten, T. (2020). Controlling soil erosion using no-till farming systems. In *No-till farming systems for sustainable agriculture: Challenges and opportunities* (pp. 195–211).

Shafi, U., Mumtaz, R., García-Nieto, J., Hassan, S. A., Zaidi, S. A. R., & Iqbal, N. (2019). Precision agriculture techniques and practices: From considerations to applications. *Sensors, 19*(17), 3796.

Shahane, A. A., & Shivay, Y. S. (2021). Soil health and its improvement through novel agronomic and innovative approaches. *Frontiers in Agronomy, 3*, 680456.

Sharma, R. A. (2018). Land management practices for in-situ rain water conservation and soil erosion control. In *Any error in this proceeding is silent testimony of the fact that it was a human effort* (Vol. 15, pp. 88–96).

Shi, P., Zhang, Y., Ren, Z., Yu, Y., Li, P., & Gong, J. (2019). Land-use changes and check dams reducing runoff and sediment yield on the Loess Plateau. *The Science of the Total Environment, 664*, 984–994.

Singh, A., & Verma, S. K. (2018). Management of ravines through anicuts and afforestation. In *Ravine lands: greening for livelihood and environmental security* (pp. 477–504).

Singh, M., & Hartsch, K. (2019). Basics of soil erosion. In *Watershed hydrology, management and modeling* (pp. 1–61). CRC Press.

Singh, A. (2018a). Alternative management options for irrigation-induced salinization and waterlogging under different climatic conditions. *Ecological Indicators, 90*, 184–192.

Singh, A. (2018b). Managing the salinization and drainage problems of irrigated areas through remote sensing and GIS techniques. *Ecological Indicators, 89*, 584–589.

Singh, A. (2018c). Managing the environmental problems of irrigated agriculture through the appraisal of groundwater recharge. *Ecological Indicators, 92*, 388–393.

Singh, A. (2021). Soil salinization management for sustainable development: A review. *Journal of Environmental Management, 277*, 111383.

Sishodia, R. P., Ray, R. L., & Singh, S. K. (2020). Applications of remote sensing in precision agriculture: A review. *Remote Sensing, 12*(19), 3136.

Soto, I., Barnes, A., Balafoutis, A., Beck, B., Sánchez, B., Vangeyte, J., Fountas, S., Van der Wal, T., Eory, V., & Gómez-Barbero, M. (2019). *The contribution of precision agriculture technologies to farm productivity and the mitigation of greenhouse gas emissions in the EU*. Luxembourg: Publications Office of the European Union.

Steffan, J. J., Brevik, E. C., Burgess, L. C., & Cerdà, A. (2018). The effect of soil on human health: An overview. *European Journal of Soil Science, 69*(1), 159–171.

Tadese, M., Kumar, L., Koech, R., & Kogo, B. K. (2020). Mapping of land-use/land-cover changes and its dynamics in Awash River Basin using remote sensing and GIS. *Remote Sensing Applications: Society and Environment, 19*, 100352.

Tajik, S., Ayoubi, S., Shirani, H., & Zeraatpisheh, M. (2019). Digital mapping of soil invertebrates using environmental attributes in a deciduous forest ecosystem. *Geoderma, 353*, 252–263.

Tao, J., Chen, J., & Gu, W. (2024). The relationship between salt-affected agricultural soils and the potential role of earthworms in restoration: A review. *Applied Ecology & Environmental Research, 22*(2).

Tarate, S. B., Patel, N. R., Danodia, A., Pokhariyal, S., & Parida, B. R. (2024). Geospatial technology for sustainable agricultural water management in india—a systematic review. *Geomatics, 4*(2), 91–123.

Tarolli, P., & Straffelini, E. (2020). Agriculture in hilly and mountainous landscapes: Threats, monitoring and sustainable management. *Geography and Sustainability, 1*(1), 70–76.

Teodoro, A. C., & Duarte, L. (2022). The synergy of remote sensing and geographical information systems in the management of natural disasters. In *Nanotechnology-based smart remote sensing networks for disaster prevention* (pp. 217–230). Elsevier.

Thomas, D. T., Moore, A. D., Bell, L. W., & Webb, N. P. (2018). Ground cover, erosion risk and production implications of targeted management practices in Australian mixed farming systems: Lessons from the Grain and Graze program. *Agricultural Systems, 162*, 123–135.

Tlapáková, L., Žaloudík, J., & Kolejka, J. (2017). Thematic survey of subsurface drainage systems in the Czech Republic. *Journal of Maps, 13*(2), 55–65.
Usama, M. (2024). Application of digital technologies & remote sensing in precision agriculture for sustainable crop production. In *Geospatial technology to support communities and policy: Pathways to resiliency* (pp. 203–223). Cham: Springer Nature Switzerland.
Vaz, C. M. P., de Resende, J. M., Franchini, J. C., Debiasi, H., & Nunes, M. R. (2022). Evaluation and recommendations for the use of dynamic penetrometers. *Soil and Tillage Research, 220*, 105373.
Vena, A., Sorli, B., Belaizi, Y., Saggin, B., & Podlecki, J. (2018). An inkjet printed RFID-enabled humidity sensor on paper based on biopolymer. In *2018 2nd URSI Atlantic Radio Science Meeting (AT-RASC)* (pp. 1–3). IEEE.
Voltr, V., Menšík, L., Hlisnikovský, L., Hruška, M., Pokorný, E., & Pospíšilová, L. (2021). The soil organic matter in connection with soil properties and soil inputs. *Agronomy, 11*(4), 779.
Walche, A., Haile, W., Kiflu, A., & Tsegaye, D. (2024). Spatial analysis and mapping of intensity and types of agricultural salt-affected soils around Abaya and Chamo Lakes, South Ethiopia Rift Valley. *Heliyon, 10*(13).
Wang, A., Ji, X., Zhu, Y., Wang, Q., Wei, X., & Zhang, S. (2024). Tillage depth regulation system via depth measurement feedback and composite sliding mode control: A field comparison validation study. *Measurement and Control, 57*(6), 685–702.
Wezel, A., Herren, B. G., Kerr, R. B., Barrios, E., Gonçalves, A. L. R., & Sinclair, F. (2020). Agroecological principles and elements and their implications for transitioning to sustainable food systems. A review. *Agronomy for Sustainable Development, 40*, 1–13.
Xu, X., Zheng, F., Wilson, G. V., He, C., Lu, J., & Bian, F. (2018). Comparison of runoff and soil loss in different tillage systems in the Mollisol region of Northeast China. *Soil and Tillage Research, 177*, 1–11.
Yang, P., Reijneveld, A., Lerink, P., Qin, W., & Oenema, O. (2022). Within-field spatial variations in subsoil bulk density related to crop yield and potential CO2 and N2O emissions. *CATENA, 213*, 106156.
Yi, X., Ji, L., Hu, Z., Yang, X., Li, H., Jiang, Y., He, T., Yang, Y., Ni, K., & Ruan, J. (2022). Organic amendments improved soil quality and reduced ecological risks of heavy metals in a long-term tea plantation field trial on an Alfisol. *Science of the Total Environment, 838*, 156017.
Zewdu, S., Suryabhagavan, K. V., & Balakrishnan, M. (2017). Geo-spatial approach for soil salinity mapping in sego irrigation farm, South Ethiopia. *Journal of the Saudi Society of Agricultural Sciences, 16*(1), 16–24.
Zhang, B., Jia, Y., Fan, H., Guo, C., Fu, J., Li, S., Li, M., Liu, B., & Ma, R. (2024a). Soil compaction due to agricultural machinery impact: A systematic review. *Land Degradation & Development, 35*(10), 3256–3273.
Zhang, W., Yu, Q., Tang, H., Liu, J., & Wu, W. (2024b). Conservation tillage mapping and monitoring using remote sensing. *Computers and Electronics in Agriculture, 218*, 108705.
Zhang, Y., Li, X., Gregorich, E. G., McLaughlin, N. B., Zhang, X., Guo, Y., Liang, A., Fan, R., & Sun, B. (2018). No-tillage with continuous maize cropping enhances soil aggregation and organic carbon storage in Northeast China. *Geoderma, 330*, 204–211.
Zheng, B., Campbell, J. B., Serbin, G., & Galbraith, J. M. (2014). Remote sensing of crop residue and tillage practices: Present capabilities and future prospects. *Soil and Tillage Research, 138*, 26–34.
Zhou, M., Xiao, Y., Zhang, X., Xiao, L., Ding, G., Cruse, R. M., & Liu, X. (2022). Fifteen years of conservation tillage increases soil aggregate stability by altering the contents and chemical composition of organic carbon fractions in Mollisols. *Land Degradation & Development, 33*(15), 2932–2944.
Zovko, M., Romić, D., Colombo, C., Di Iorio, E., Romić, M., Buttafuoco, G., & Castrignanò, A. (2018). A geostatistical Vis-NIR spectroscopy index to assess the incipient soil salinization in the Neretva River valley, Croatia. *Geoderma, 332*, 60–72.

Fertile Grounds: A Blueprint for Sustainable Soil Management

Kamran Haider, Muhammad Sufian, Dilawar Abbas, Muhammad Adeel Ghafar, Kamil Kabir, Umer Liaqat, Muhammad Shamekh Ali, and Yasmin Kausar

Abstract A source of inspiration and direction, "Fertile Grounds: A Blueprint for Sustainable Soil Management" shines in a world where soil health directly impacts the well-being of our planet and its people. This plan embodies an innovative framework that goes beyond traditional farming methods to develop robust and fruitful soils. It provides a thorough road map for promoting agricultural sustainability, protecting the environment, and preserving soil health by combining cutting-edge

K. Haider (✉) · K. Kabir
State Key Laboratory of Agricultural Microbiology, Key Laboratory of Horticultural Plant Biology (MOE), China-Australia Joint Research Centre for Horticultural and Urban Pests, Institute of Urban and Horticultural Entomology, College of Plant Science and Technology, Huazhong Agricultural University, Wuhan, People's Republic of China
e-mail: kamranhaider345@webmail.hzau.edu.cn

K. Kabir
e-mail: kamilkhanzada@gmail.com

K. Haider · M. Sufian (✉) · M. S. Ali
Department of Entomology, University of Agriculture, Faisalabad, Pakistan
e-mail: muhammad.sufyan@uaf.edu.pk

M. S. Ali
e-mail: shamekhali140@gmail.com

D. Abbas
State Key Laboratory for Biology of Plant Diseases and Insect Pests, Institute of Plant Protection, Chinese Academy of Agricultural Sciences, Beijing, China
e-mail: dilawar72@yahoo.com

M. A. Ghafar
State Key Laboratory of Ecological Pest Control for Fujian and Taiwan Crops, Key Laboratory of Bio Pesticide and Chemical Biology, MOE, College of Plant Protection, Fujian Agriculture and Forestry University, Fuzhou, China
e-mail: adeelbhatti069@gmail.com

U. Liaqat · Y. Kausar
Department of Zoology, Wildlife and Fisheries, University of Agriculture, Faisalabad, Pakistan
e-mail: umer.liaqat@uaf.edu.pk

Y. Kausar
e-mail: yasminkausar550@gmail.com

M. Shaaban (ed.), *Soils and Sustainable Agriculture*, Frontier Studies in Soil Science,
https://doi.org/10.1007/978-3-031-91114-9_18

science, real-world experience, and ageless wisdom. Every aspect of this blueprint, from the use of regenerative techniques like agroforestry and minimal tillage to the development of soil microbiome diversity and carbon sequestration, is painstakingly designed to work in harmony with the cycles and laws of nature. It not only tackles the critical need to reduce soil degradation and reduction, but also demonstrates soil's revolutionary capacity as an alternative to pressing global concerns such as climate change, food security, and biodiversity loss. This blueprint stimulates a global movement toward a regenerative model in which fertile ground serves as the foundation for healthy ecosystems, resilient communities, and a flourishing planet via multidisciplinary cooperation, community participation, and courageous leadership. This plan serves as a guiding beacon for changing agricultural landscapes into fertile ground that nurtures healthy ecosystems, resilient communities, and a sustainable future for future generations. As we begin this journey, guided by the ideals contained in this plan, we want to establish a sustainable future by encouraging development in all aspects of soil and extensive harvests for the generations to come.

1 Introduction to Soil and Soil Management

A complex biological system, soil promotes water retention, nutrient cycling, organic matter breakdown, and plant growth (Saglam et al., 2015; Zhou et al., 2019). Changes in soil properties in disturbed grasslands have raised serious concerns, particularly in light of the worldwide trend of grassland degradation. The most obvious symptom of grassland degradation is a decrease in plant cover (Babel et al., 2014), which also contributes to soil physical property deterioration and nutrient infertility. Thus, Damage to grasslands is mostly caused by deteriorating soil. According to (Ma et al., 2020), soil degradation reduces soil productivity, speeds up desertification, weakens the structure of the soil, and diminishes the amount of organic matter in the soil. Although there is a rapid increase in grassland degradation, statistical data about its impact on soil characteristics is still mostly lacking (Yuan et al., 2019). Historically, soil has been thought to be crucial to agricultural output. Its crucial significance in land utilization and organizing, ecosystem services, reducing the effects of climate change, and maintaining the environment has also been underlined by new scientific research (Bahram et al., 2018). A main supply of food and a prerequisite for a healthy life is fertile soil (FAO, 2015). Furthermore, soil is essential for maintaining biodiversity, supplying clean water, preventing climate change, preserving cultural heritage, and playing a significant role in our landscapes. Soils are under threat internationally due to.

(i) Increased demand for livestock, food, and fiber, putting strain on natural assets.

M. S. Ali
Four Brothers Group, Lahore, Pakistan

(ii) Unsustainable agricultural and forestry practices, pollution, and urbanization all contribute to soil and land deterioration (Veerman et al., 2020).

33% of the world's soils are severely degraded due to a combination of factors including compression, loss, and pollution by chemicals, loss of nutrients, acidification, and salinization. More than 60% of European soils appear to be degraded, according to the EUSO soil health dashboard (EUSO, 2023), which is created and maintained by the European Joint Research Centre (JRC).

Meaning of Sustainable Soil Management

The term "sustainable soil management" (SSM) has several interpretations. Sustainable development, in the words of WCED (1987), is defined as "development that meets the needs of the present without compromising the ability of future generations to meet their own needs." Similarly, the concept of soil health is described by the Intergovernmental Technical Panel on Soils (ITPS) as "the capacity of the soil to maintain its productivity, diversity, and the ecosystem services of agricultural systems." According to the FAO, soil management is considered sustainable "if the provision, regulation, support, and cultural services that soil delivers are maintained or enhanced without significantly impairing biodiversity or the soil functions dependent on those services."

Why is it Necessary to Manage Soil Ecologically?

Soil is vital for all lowland living species since it provides approximately 98% of human nourishment (Kopittke et al., 2021). This demonstrates the importance of soil for humans and other living species. Moreover, Lal maintains that since soil sustains around 25% of all biodiversity, "soil is a living thing." Additionally, like any other living thing, the soil should be properly preserved, maintained, and managed. Soil resources provide several functions and benefits. The nitrogen (N) cycle, water preservation and discharge, biological variety and Ecosystems, purification and buffering, and physical stability and assistance are the five major roles of soil. On the other hand, adopting SSM simultaneously presents advantages, issues, and opportunities. One of the biggest threats to food security and sustainability in the world today is land degradation. According to (FAO, 2011), soil erosion is also considered to occur at a reasonable or acceptable pace (i.e., within the range of 0.2–2.2 t ha^{-1} in terms of soil production rate), meaning that crop output (with a legal limit of 1–11 ton per hectare) should not be harmed (Fig. 1).

1.1 Agroecological Principles-Based Sustainable Soil Management

Global evidence suggests that effective soil management and sustainable agriculture system management may be closely related. Since ecologically friendly cultivation practices are a prerequisite for sustainable agriculture, the two are closely

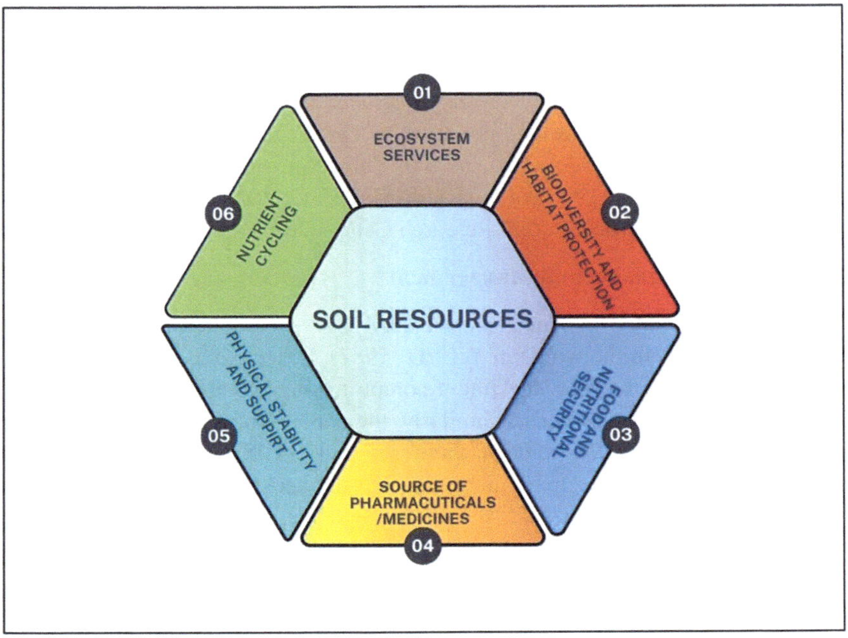

Fig. 1 Different functions performed by soil

intertwined. Pretty (2008) argues that no agricultural method can be deemed ecologically sustainable if it allows the biological, structural, functional, and associated ecosystem services offered by soils to be mechanically altered. Overcoming previously noted declines in soil ecosystem features should be the main goal of sustainable soil management, with an emphasis on enhancing soil quality as a growing habitat for plants.

Additionally, insects, weeds, and maladies that disrupt the crops being grown render an agricultural soil system ineffective. Stated differently, sustainable soil management by itself does not translate into environmentally friendly production; furthermore, Crop, soil, fertilizer, water, insects, and farm power control in location and period need to be taken into account if the goal is long-term production multiplication to ensure natural and economical sustainability (Kassam et al., 2009). Based on an eco-commercial perspective, the agricultural environmental principles that drive ethical farming methods for both big and small farmers are: dealing with sustainable production the growth and ecosystem services economically; protecting assets, and maximizing the use of both acquired and natural resources (Godfray et al., 2010). Essentially, according to a lot of information gathered from farmers across many areas, profitable farming comes from using multiple successful ideas at the same time via crop-soil-water-nutrient-pest-ecosystem management techniques. At different levels, such as farm, landscape (watershed), local, or national, these techniques are created and modified regionally to recognize a broad range of agricultural output, reciprocal benefits that are cultural and ecological, and ecosystem services (FAO, 2011; Pretty et al., 2011).

Sustainable agricultural methods based on ecosystem principles generate several financial and environmental benefits for growers as well as society at large. To fulfill the 2050 hunger crisis and provide the variety of ecosystem services that society demands, sustainable agricultural techniques need to be founded on five essential concepts. These include:

- Achieved both greater agricultural output and improved ecological services.
- Improved efficiency in using critical inputs like water, fertilizers, pesticides, energy, land, and labor.
- Reduced dependence on inputs produced from fossil fuels (such as insecticides and mineral fertilizers) and increased use of new methods (e.g., nitrogen fixation by bacteria as well as combined pest management).
- Protection of soil, water, and biodiversity using minimally upsetting natural ecosystems; programs have to be less frequent and have shorter-term consequences than the ecosystem's capacity to recover naturally.
- Bolstering and restoring the system's resistance to biotic, abiotic, and financial shocks by utilizing both managed and uncontrolled biodiversity.

1.2 Facts that Sustainable Soil Management is Multidisciplinary

1.2.1 Sustainable Development Goals (SDGs)

Soil is vital for the United Nations SDGs, particularly for Goals 2, 6, 13, and 15. The majority of the less fortunate reside in rural regions where cultivation is the main business. The state of the soil in these locations has a big influence on production and revenue. According to (Bender et al., 2016), among the many functions of soil are its maintenance of crops and habitat services. Furthermore, soil impacts water quality, greenhouse gas emissions, and other significant environmental factors linked to the SDGs. The following is a summary of the recognized connections between soil and the pertinent Sustainable Development Goals (SDGs), as seen in Fig. 2:

To help small-scale cultivators and other participants produce appropriate metrics that are both technically competent and practical, the environmental science society must act effectively. More than just researching soil science, scientific research has to be focused on accomplishing the SDGs. In this intricate relationship between humans and nature, the effects of human action must be taken into account. To facilitate more informed decision-making, sustainability evaluations must take into account the consequences of socioeconomic activities on soil health (Vogel et al., 2018).

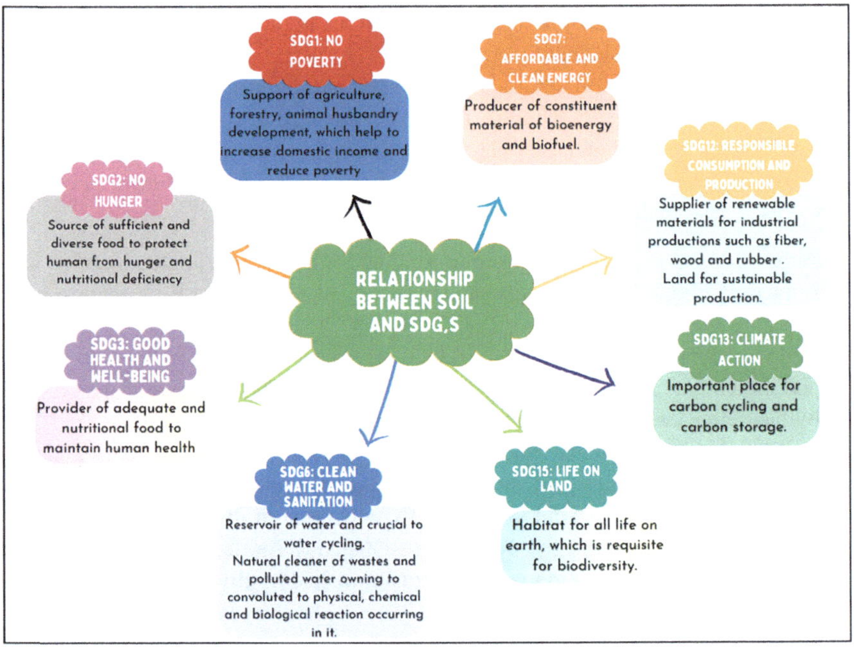

Fig. 2 Relationship between soil and sustainable development goals (SDG,s)

1.2.2 Interdisciplinary Research

Soil systems' sustainability is influenced by their biophysical features as well as farmer soil usage and management actions. These two characteristics can be generally classified as natural and manmade processes. The connected human-nature systems entail complex dynamics, which presents several problems for studying soil systems via a single discipline lens. To solve these issues, we must use an interdisciplinary approach (Totsche et al., 2010). To obtain a thorough understanding, multifaceted techniques incorporate viewpoints from several fields (such as soil science and environmental science) on a similar problem (such as climate change). Multidisciplinary approaches, on the other hand, concentrate on gathering many perspectives on the same issue without ultimately combining them into a single, cohesive position.

Due to the complexity and scope of many of the challenges influencing soil sustainability, interdisciplinary research is necessary and cannot be properly addressed by one field of study. We offer a comprehensive framework for developing an interdisciplinary perspective on appropriate soil use and leadership, building on a previously published paradigm. Hammond and Dubé (2012) that connected disciplinary barriers for a different topic.

We contend that the five primary issues are connected to at least one other topic and have their origins in soil science. The problems are linked to one another. Consider lifestyle and management, which are closely related to social science and soil. In

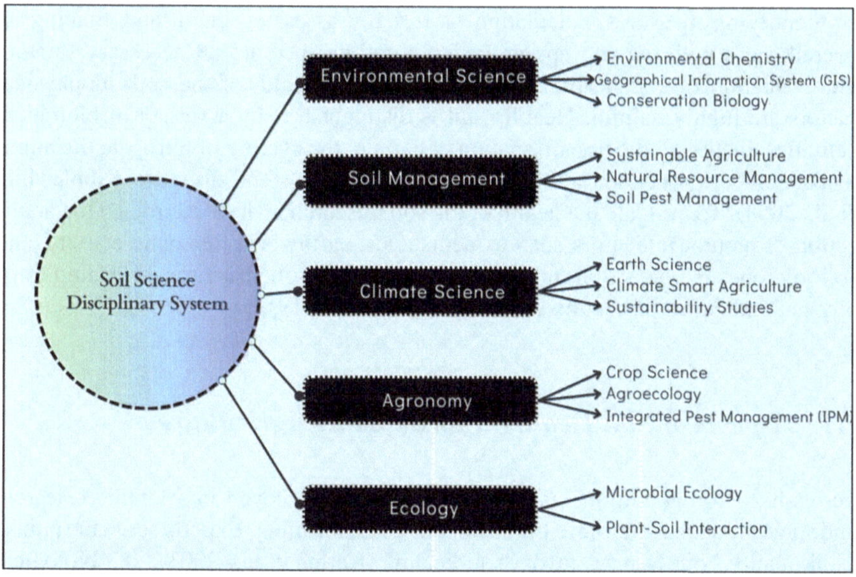

Fig. 3 Soil science interdisciplinary system

addition, soil health and pollutants come into play; they are closely linked to environmental science and agronomy. Another example is soil carbon, also known as soil organic matter, which influences soil biodiversity, which is connected to ecology, and soil condition, which is connected to agriculture. It also directly relates to soil science and climate science. Soil sustainability research may be conducted using the multidisciplinary system (Fig. 3).

2 The Soil Health Concept

Numerous chemical, biological, and physical properties of soils are influenced by what is underneath it (such as the earth's crust base and depositional processes), external factors (such as landscape and climate), and human activity (such as farming practices and exterior disturbance contaminants emissions). Indeed, soil health is important for a variety of applications, including environmental protection and agriculture. Conventional techniques for evaluating soil, which primarily concentrates on production and fertility, may not adequately account for all facets of soil health. More aspects are taken into account in modern methods, such as physical, chemical, and biological characteristics, as well as how they affect different ecosystems and human health. In your research or activity, are you focusing on any particular facets of soil health? (Doran & Parkin, 1994).

Identifying measures, calculating or verbally assessing them, and creating an overall score with the appropriate addition and balance are all necessary steps in the evaluation of soil health. When assessing the health of the soil, biophysical factors are highly helpful. Healthy soil is distinguished by a variety of biological activities, including the transformation of carbon, the cycling of nitrogen, the maintenance of soil structure, and the management of pests and diseases (Kibblewhite et al., 2008). To evaluate the health of the soil, researchers have examined the application of natural indicators such as nematodes, earthworms, enzyme activity, and soil microbes. Compaction, moisture retention, and soil structure are examples of physical indicators of healthy soil (Ananbeh et al., 2019).

2.1 Aspects of Soil Health in Sustainable Agriculture

Terms like "soil condition" and "value" are often employed in scientific research, and some argue that they have the same functional meaning. Experts adopt the phrase "soil quality," but farmers favor "soil health" (Karlen et al., 2003). 183 biological indicators for soil monitoring have been found and evaluated. The most commonly used biological indicator choices were:

- Terminal restriction fragment length polymorphism methods are used to study soil microbial taxa and community dynamics.
- The total amount of life less than the surface of the earth is known as microbial biomass.
- Acquired lipids, in particular phospholipid fatty acids, are used as unique lipid indicators to evaluate the diversity and composition of the soil microbial population.
- Soil photosynthesis and carbon cycling are the outcome of multiple substrate-induced respiration. Nematodes, such as development rating (the location of nematodes among working categories), taxonomic number, and plenty of each functional group.
- On-site visual records of soil biodiversity using pitfalls traps for invertebrates that live on the ground, and nematodes.
- Micro arthropods.

More studies are necessary to find out how various biological indicators connect to soil processes if they react differently to management, and whether they can provide light on certain ecological processes. Soil quality is deteriorated by unsuitable agricultural techniques, including salinization, acidification, compaction, crusting, lack of nutrients, decreased biodiversity and biomass of soil biota, water imbalance, and disturbance of elemental cycles. Controlling pathogens, cycling nutrients, storing water, and detoxifying the soil are all dependent on the soil biota, which reacts quickly to soil management practices (Mehran et al., 2023). Plant health, soil fertility, and soil biota are closely related. The biota of the soil is essential for improving the fertility and productivity of the land through biological processes.

2.1.1 Distribution of Soil Microorganisms

The particles of soil are where microorganisms live, and they also have a major influence on the overall makeup of the microbial population. In humid conditions (high in loam calcium soil), When comparing 1–3 mm soil fragments to 5–7 mm fragments, the microbial numbers of bacteria, actinomycetes, and fungus were greater in the smaller ones. Furthermore, heat, pH, total organic carbon, and vegetation can all have an impact on the number of soil microbes. The soil type was found to be a significant predictor of the microbial population. Changes in root length and soil chemical elements including pH, phosphate (P), and potassium (K) were discovered to have a major impact on the variety and abundance of rhizosphere microorganisms (Qin et al., 2019). Particularly, clay soil parameters were linked to higher bacterial populations, and these soil properties were correlated with fungus biomass.

The microorganisms that comprise the rhizosphere the area surrounding the plant root live in the soil that underlies the rhizoplane, or root surface, as well as the root itself. At least 2 mm of the rhizoplane is covered by the rhizosphere; but, in certain situations, such as vesicular–arbuscular mycorrhizae, that can even extend over the depleted layer at the root layer by up to 10 mm. Numerous microorganisms, such as viruses, bacteria, nematodes, fungi, algae, protozoa, and arthropods, can be found in the rhizosphere. The variety of microorganisms grows near the rhizoplane and diminishes with increasing distance from it. This was explained by the carbon content, which roughly matches the distance from the rhizoplane (Giller et al., 2005).

Biological markers include soil and mycorhizosphere organisms due to their sensitivity to subtle changes in abiotic stresses, their capacity to respond to such changes, and their impact on plant production, structure, and distribution. It has been suggested that below-ground diversity and plant output under different climatic conditions are related (Wagg et al., 2011). By several mechanisms, including improved crop efficiency in utilizing nutrients, output, and particular Fe and P usage; antimicrobial activity acting as an initial line of protection mode for plant roots against soil-borne pathogens; and the emergence of immune systems and enticement on soil farms, rhizosphere microorganisms may promote cultivation and protect plants from disease infection (Mela et al., 2011). Mycorrhizal fungus, bacteria that fix nitrogen, PGPR, bio-control microorganisms, and mycoparasitic bacteria are examples of helpful rhizosphere microbes that are essential to plant development. Per the plant and its surroundings, the effects might range from detrimental to neutral (Bais et al., 2006). The significance of these bacteria in agricultural and ecological systems is highlighted by the way they both, directly and indirectly, improve plant nutrition and development. Rhizobacteria that promote plant development are a broad group of bacteria. The most popular and effective rhizobacteria that promote plant development when employed as plant inoculants include Azospirillum, Bacillus, Pseudomonas, Rhizobium, and Streptomyces (Fig. 4).

Probably the most extensively researched and prevalent plant growth-promoting rhizobacteria in the rhizosphere are Pseudomonas and Bacillus. Furthermore, it has been claimed that fungi like Trichoderma and Ampelomyces are beneficial to plants.

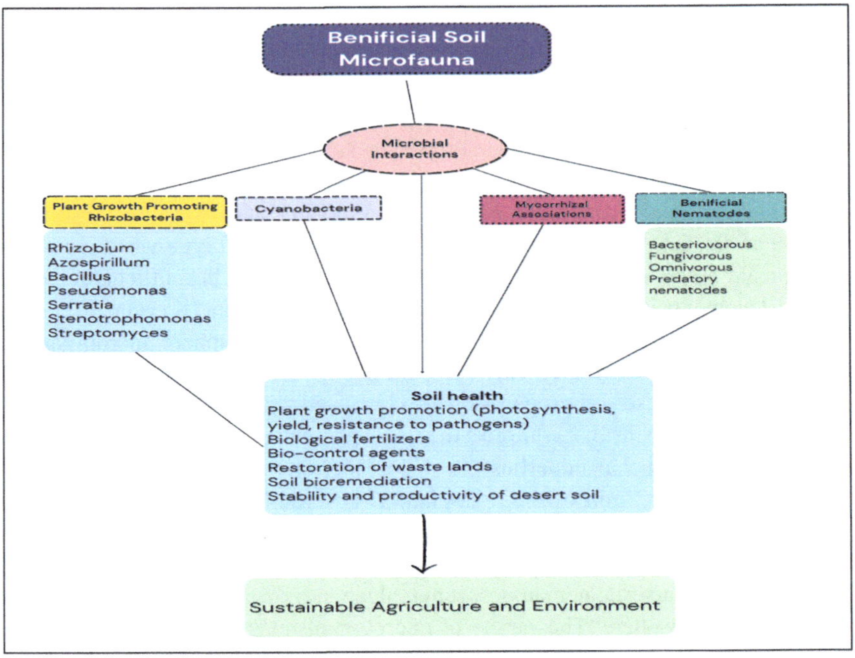

Fig. 4 A theoretical subject that illustrates how beneficial soil bacteria and their relationships contribute to the growth of environmentally and agriculturally viable systems

PGPR promotes, develops, and protects plant growth through a variety of intricate processes, such as biofertilization, phytostimulation, and biocontrol (Mihajlović et al., 2017).

Mycorrhizal Associations

Root systems and mycorrhizal linkages create symbiotic partnerships. 80% of vascular plants have the AMF, which is the most common mycorrhizal association in existence. Mycorrhizal bonds are a natural way for plants to collect water and nutrients. According to (Linderman, 1991) mycorrhizal connections can release carbon molecules from hyphae, creating the mycorhizosphere a soil niche. Due to its ability to facilitate the absorption of water, P, Cu, Zn, and N, AMF is essential to plant nutrition. Additionally, AMF promotes the production of phytohormones such as abscisic acid. It has been discovered that mycorrhizal connections play useful roles in the stability, creation of soil structure, and reduction of plant stress. Because of this, it is crucial to consider the interaction between the soil, roots, and AMF in addition to the effects of the neighboring soil's chemical and physical properties on the functioning of mycorrhizal associations when evaluating sustainable agricultural systems. An organic low-input method can increase AMF productivity and support sustainable agricultural development, whereas standard agricultural approaches may decrease AMF density (Panwar et al., 2008). All things considered, AMF can support sustainable agriculture by enhancing the activity of the soil rhizosphere, it improves plant agronomic parameters such as plant height, seed quantity, and overall yield as well as soil attributes.

Cyanobacteria

Cyanobacteria, also known as gram-negative prokaryotic photoautotrophic bacteria, are found in a broad variety of soil types and are crucial to the development and maintenance of healthy soil. They greatly enhance soil quality and plants' resistance to biotic and abiotic challenges like plant diseases. Cyanobacteria are extraordinarily adaptable to a wide range of harsh weather conditions, including heat (0–70 °C), salinity (~ 21 ds m^2), pH (including high and low PHS), dryness, and toxic metals (Cd, Cu, Pb, and Co) (Raaijmakers et al., 2009). Because the tissues of cyanobacteria include changed vegetative cells, plants are more resilient to prolonged drought and drying out. Inorganic and aqueous organic matter, including polyketides, non-ribosomal peptides, isoprenoids, and ribosomal peptides, are abundant in cyanobacteria.

Furthermore, organic waste can affect how different kinds of organisms arise. Specifically, it has been demonstrated that the use of native cyanobacteria variations as organic fertilizer in saline, dry, and polluted soils including heavy metals greatly improves soil physical, chemical, and biological variables, increasing production (Nisha et al., 2007).

Cyanobacteria contribute to soil chemical alteration by increasing plant nutrient bioavailability, especially for P, C, and N. Poor semi-arid soils can benefit from the inclusion of cyanobacteria, which increases microorganism populations and nitrogen levels. It's interesting to note that cyanobacteria protect plants from disease by acting as bio-control agents (Zulpa et al., 2003). For example, it has been demonstrated that the outside elements of the cyanobacteria strain Nostco muscorum exhibited antifungal activity towards *Sclerotinia* sclerotiorum-generated white mold of lettuce (*Lactuca sativa*) and *Rhizoctonia solani* in soybeans (*Glycine max*).

Nematodes

There is growing proof suggesting the enormous diversity of the soil biota affects terrestrial ecosystem performance, above-ground biodiversity, and the ecological reactions of those ecosystems to environmental shifts, both now and in the future. (Yang et al., 2018) have shown that the diversity of below-ground soil biota populations can modify various aspects of above-ground plant communities, such as variation, output, prosperity, and stress adaptation. This can have an impact on the stability and resilience of ecosystems.

In agroecosystems, the biodiversity of nematodes is influenced by the ecological condition of the soil. In research on long-term sustainability, nematodes have been employed as biological markers of soil quality. One can create populations of soil nematodes with defined structural features and functions. Nematodes are being suggested as biological markers for assessing the health of the soil due to their support of biodiversity and their function in recycling nutrients. Nevertheless, given that nematode numbers differ throughout biogeographical zones and land usage within each biogeographical area, comparing results at a large or global scale within climatic groups should be undertaken cautiously (Kultschar & Llewellyn, 2018). In the world, parasitic nematodes are thought to be accountable for 12.3% of crop loss. Plant diseases caused by nematodes are difficult to manage because they are hidden. Crop loss can occur when nematodes and bacteria form disease complexes (Singh, 2015). Conversely, beneficial nematodes release nutrients into the soil and aid in the consumption of bacteria and fungi by plants. Because of their wide dispersion, varied diets, and use of a variety of life strategies, these helpful nematodes can readily adjust to changes in the biological, chemical, and physical properties of the soil. Furthermore, independent useful nematodes (*Discolaimus*, *Tripyla*, and *Prionchulus*) emit nitrogen compounds that support plant development by their feeding on soil microbes (Neher, 2010).

Nematodes that feed on other nematodes (bacterivores), together with carnivores and fungivorous, are significant indicators of soil health. Predatory nematodes of the orders Mononchida, and Dorylaimida, deliver nutrients into the soil by eating soil microbes and seeding parasitic worms. Pristionchus is an example of an omnivorous nematode that can eat both bacteria and predatory nematodes. Enough biological diversity and nutrient cycling are present in fertile soil (Quist et al., 2017). Numerous soil trophic microorganisms, including omnivore, fungivorous, predatory nematodes, and bacterivorous ones, are essential to the mineralization (release) of nutrients from organic matter and are hence necessary for the cycling and accessibility of nutrients. A study on perennial ryegrass (*Lolium perenne*) and *Bouteloua gracilis* has demonstrated that the existence of predatory nematodes enhances nitrogen mineralization and, as a result, plant productivity. In general, a robust, well-structured, rich soil may include a sizable and diverse population of free-living nematodes, especially beneficial nematodes, alongside an adequate variety of bacterial and fungal parasites and predators and omnivorous worms (Neher, 2001).

Soil Borne Pathogens

Worldwide, diseases spread by soil result in significant agricultural losses. A diverse food web seen in an ecosystem with fertile soil controls illness through parasitism, rivalries, and scavenging. The growth of plants is impeded by soil-borne diseases, including Rhizoctonia spp., Verticillium spp., Phytophthora spp., Pythium spp., and Fusarium spp. These infections cause blight illnesses, root degradation, juvenile humidity, perishing, and stem deaths. (Koike et al., 2003). In the absence of a suitable host plant, soil pathogens may live for several years through the production of oospores, chlamydospores, microsclerotia, and other persistent structures. They are therefore challenging to recognize and manage. Good soils that aid in the reduction of soil pathogens are essential for sustainable agriculture, as is the appropriate management of soil-borne pathogens and inhibitors (Table 1). Thus, by determining what infectious agents exist (the recognition of associated agents) and where they occur. Appropriate farming techniques (crop rotation, remaining crop management, and resistance varieties), soil sanitation, sunlight, and natural and chemical pest control methods can be applied. In sustainable agriculture, controlling soil-borne diseases can improve soil quality and plant well-being since they are significant biological parts of soil biodiversity (Table 1). Assessing the effects of different soil management techniques is necessary to improve agricultural soil health and increase long-term crop output (Table 1) (Stirling et al., 2017). Agroecosystem soil biodiversity is regulated by several agricultural cultural practices, some of which are listed in Table 1 and are intended to decrease soil-borne disease and promote variety.

3 Agricultural Methods to Enhance the Soil Health Element

3.1 Growing Everything Organically

In addition to improving physical and natural assets like microbial activity, a lot and variation, groundwater quality, and lower nitrate (NO_3) levels, organic farming is becoming more and more common as a sustainable cultivation technique. Products like wheat, potatoes, watermelon, and strawberries are examples of how this is achieved (Schrama et al., 2018). The US Department of Agriculture defines organic farming as "a manufacturing process controlled by the rules and laws outlined in this part that adapts to location-specific circumstances by combining social, biological, and mechanical actions that develop resource cycling, improve sustainability, and preserve the environment". Especially compared with inorganic fertilizers, legume-based plants like alfalfa and Sesbania spp. are shown to increase organic matter in the soil by around 50%. In just 13 weeks, Sesbania rostrata produced 16.8 tons ha^{-1} of dried matter. They also boost the soil's ability to provide and retain nitrogen. In a 12-year study, it was shown that organic systems containing peat and dung have increased microbial populations and enzyme activity in maize (*Zea mays*) and rice (*Oryza sativa*) crops (Escanhoela et al., 2019).

There were noticeably larger soil bacterial communities in both standard and organic banana (*Musa acuminata*) farmland, with proteobacteria, Actinobacteria, and other microorganisms predominating. Following three years of naturally occurring conditions, growing crops like lettuce, tomato, and snap beans showed that, in comparison to common chemical fertilizers (NPK), organic cultivation with organic matter enforced soil CO_2 oxygenation (a measure of soil wellness) and enzyme functions (phosphatase, arylsulphatase) (Iovieno et al., 2009). Furthermore, compared

Table 1 Crop culture methods' impact on soil-borne infections

Cultural practices	Crops	Hypothesis	Target pathogens	Responses	References
Crop rotation	Barley (Hordeum vulgare), rye grass (*Lolium perenne*)	Rhizoctonia infection can be decreased by rotating crops with potatoes every three years	Rhizoctonia spp.	The condition was lowered by 15% to 50%	Larkin et al. (2012)
Grafting	Cucumber (*Cucumis Sativus*)	Transplanting cucumber plants by non-chemical method with several rootstocks to mitigate nematode infection in roots	Meloidogyne spp.	Compared to other stumps, plants grown on the "Strong Tosa" rootstock generated 5.4 kg along with 20 more fruits	Goreta Ban et al. (2014)
Intercropping system	Faba bean (*Vicia faba*)	Intercropping wheat cultivars can impact disease management, potentially due to variations between root exudates	*Fusarium oxysporum*	Faba bean growth rose by 13–17%, while the disease rating reduced fusarium wilt by 48%	Dong et al. (2014)
Physical method (steam soil disinfection)	Tobacco (*Nicotiana tabacum*)	Introducing or spreading heated water vapor into soil may decrease Rhizoctonia development	*Rhizoctonia solani*	Dry heat (70–80 °C for 2 h) and chemicals (NaCl) treatments lowered infection levels on dishes by up to 45% in comparison to controls	Gutierrez et al. (1997)
Soil organic residue	Broccoli (*Brassica oleracea*)	Broccoli leftovers were chopped and set aside to dry on the soil level for a couple of days. Reduce disease incidence	*Verticillium dahlia*	After two pickings of broccoli, the amount of propagation dropped by 94%. Both the incidence and severity of the disease were 50% lower. Applying broccoli remedies	Xiao et al. (1998)
Physical method (soil solarizatio) and biological control	Tomato (*Lycopersicon esculentum*)	Combining sunlight from the soil with *Gilocladium* vines, which are fungal antagonists, can help manage southern blight infections	*Sclerotium rolfsii*	Solarized soil treated with *Gilocladium* vines lowered infection rates by 49%	Ristaino et al. (1991)

(continued)

Table 1 (continued)

Cultural practices	Crops	Hypothesis	Target pathogens	Responses	References
Biological control	Pepper (*Capsicum annuum*)	*Trichoderma harzianum*, an antagonistic microorganism, can promote disease decrease through competition	*Phytophthora capsici*	Inoculating diseased pepper leaves with Trichoderma considerably decreased necrotic legion (28 mm) when compared to the control	Sid et al. (1999)
Biological control (antimicrobial activity of essential oils)	Lavender (*Lavandula angustifolia*), anise (*Pimpinella anisum*)	Many types of essential oils have antimicrobial properties and may block pathogen infection in vitro	*Verticillium fungicola* var. fungi cola, *Mycogone perniciosa*, and Cladobotryum sp.	After four days of exposure, oregano, and geranium oil extracts killed pathogens at a concentration of 0.32 L mL^{-1} air	Tanović et al. (2009)

to traditional culture protocols, it was demonstrated that organic growth decreased Whereas parasites of plants like Pratylenchus and Meloidogyne influence maize and beans, soil infections like Fusarium wilt can affect cucumbers. Soil is improved both chemically and physically by using organic agriculture techniques. An 18-year study revealed that, in comparison to soil cultivated using chemical fertilizers, soil supplemented with compost had 50% more nitrogen in its top 20 cm.

Another lengthy research (21 years) combining standard and organic agriculture found that the amount of nutrients (N, P, and K) absorbed by organic soil was 34–51% lower than that from traditional soil. On the other hand, organic farming on clay soil resulted in an 8% decrease in mass density, a roughly 15% rise in soil water content, and a 10% increase in the top 20 cm layer's ability to store water. Moreover, organic farming provides enough macronutrients shown in (Table 2). Even while organic farming may not be as successful at first, after (10–13 years, the outcomes can be on track with standard farming. It enhances the quality of the soil and crops, but it might not be the best option for farmers looking to boost output right away. The difference between conventional and organic systems gradually decreases as a result of organic farming's gradual stabilization of soil microbes and activity.

3.1.1 Tillage Practices

Tillage practices affect the physical and chemical properties of the soil, as well as crop yield and quality of fruits, according to research conducted in rice-maize and watermelon cultivation systems. Implementing efficient tillage techniques is essential to preserving crop productivity and soil health. According to (Cannell & Hawes, 1994) reduced strip and no-tillage are examples of conservation farming

Table 2 The averages of the integrated soil data after the study from conventional and organic farms

Crop	Soil type	Study period (year)	Nutrient	Response	Reference
Citrus	Clay soil (Oxisols soil) is composed of 50% clay, 20% silt, and 30% sand	6	N	The natural ecosystem had 2 tons ha^{-1} more N ($p < 0.05$) deposited at 0–100 cm than the typical one did	Escanhoela et al. (2019)
Rotation of wheat and maize	Sandy loam soil (*aquic inceptisol*)	18	N	The N content of organic soil was 5–22% greater ($p < 0.05$) than that of conventional soil	Gong et al. (2011)
Wheat, potatoes and clover	Clay soil	21	P, K, Ca^{2+}, Mg^{2+}	Crops cultivated organically had higher levels of Ca^{2+} and Mg^{2+} ($p < 0.05$) than those cultivated in conventional agriculture	Mäder et al. (2002)
Cashew (*Anacardrum occidentale*)	Loamy skeletal, mixed hyperthermia and Haplohumults	5	N	Compared to traditional (402 kg ha^{-1}), accessible N was greater ($p < 0.05$) in organic (435 kg ha^{-1})	Mangalassery et al. (2019)

techniques that can boost soil microbe activity, moisture, organic matter, aggregate power, and cation exchange capacity in addition to crop yield. When compared with conventional approaches, conservation farming using fixed beds and strip tillage can reduce irrigation water and labor intake, enhance plant water usage efficiency, and improve the cost–benefit ratio (Table 3). Tillage strength was shown to significantly reduce the stability of soil macro and micro-aggregates (Al-Kaisi & Broner, 2009).

Compared to traditional farming methods, ecological farming tactics increased accessible P by 3.8%, K by 13.6%, and soil organic matter by 0.17% in topsoil (0–20 cm). Crop residues that are applied to the top layer of the soil (full cover, no-tillage; half cover, strip tillage) may raise the moisture content of the soil and reduce soil erosion, according to Celik et al. (2013).

Table 3 Tillage methods' effects on irrigation, overall returns, and water use efficiency (WUE)

Crop	Applying irrigation water	Whole WUE growth	Growth of net returns overall in US dollars	Reference
Maize	A 25% decrease in irrigation water	16.5%	\$282 ha^{-1}	Jat et al. (2013)
Rice–wheat	A 12–20% decrease in irrigation water	Grain input of 1.4 kg/m^3 as opposed to 0.75 kg for traditional methods	\$185–281 ha^{-1}	Jat et al. (2009)
Rice–wheat	Decrease down on irrigation water by 16–18%	4.2%	\$49–96 ha^{-1}	Saharawat et al. (2012)
Wheat	For water supply, utilization of fuel performance (energy) was 21% greater	As opposed to 4.109 t ha^{-1} for traditional, with 4.5 t grains per ha^{-1}	Zero-tillage yielded a 33% higher net revenue than conventional farming	Vivek et al. (2013)
Rice–wheat	Decreased irrigation water by 13–23%	– 5% (the standard was set at 5% higher)	\$62 ha^{-1} in year 1 and a comparable amount in year 2	Bhushan et al. (2008)
Wheat–maize	Average soil water retention the no-tillage measurement at a depth of 0–2 m was 412 mm, while the usual measurement was 392 mm	Not important	\$58 ha^{-1}	Zhang and Liu (2018)

4 Techniques and Plans for Ecological Soil Management

4.1 *Preservation Farming*

Around 200 million acres of land are utilized for conservation agriculture (CA) and till-free farming globally. It refers to a "set of management techniques including lowest damage to the soil (conservation tillage/NT), optimum soil cover through a waste layer or organic mulch and diversified cropping system/rotations" (FAO, 2022). One important aspect of CA is the generation of residual from the growing of cover crops. This is because, in addition to reducing soil erosion, controlling weeds, and regulating soil temperature, it also improves soil health and soil organic carbon levels by recycling and adding residue to the soil (Fig. 3). The notion of soil cover as presented by CA has several applications. Particularly in semi-arid areas, soil cover is essential to improve soil health, water intake, and storage while lowering evaporation. There are several benefits to incorporating cover crops into the rotation in semi-arid regions like the Mediterranean regions of Central Asia, Europe, Australia, and South Africa. Dead or live mulch used as ground cover is a crucial part of perennial CA systems. CA is a sustainable intensification technique that has the potential to reduce global warming (12–33%), increase net economic return (25.9%), increase agricultural production (5.6%), and improve water usage efficiency (12.6%).

4.2 Agricultural Regeneration

Rebuilding soil health by replacing organic carbon (SOC) is how RA is defined. *Regenerative agriculture (RA) is defined as the process of rebuilding soil health by replenishing soil organic carbon (SOC). It is a holistic, principle-based farming approach that promotes the enhancement and sustainability of ecosystem services and community well-being. RA practices are essential for restoring degraded land and transforming it into a productive agricultural system..*" Loss of land (soil) due to soil erosion and loss of rich topsoil is acknowledged as the primary factor influencing soil quality, local community livelihood, and output. Therefore, we need to use SSM practices like CA/precision farming, sustainable intensification, and resource conservation technologies to protect our natural resources. Additionally, utilizing appropriate SSM procedures and techniques, current agricultural and management practices urgently need to be modified to minimize input utilization efficiency and soil disturbance. Therefore, it has been suggested that Regenerative Agriculture (RA) improves soil health, long-term production, earnings, and efficiency (Lal et al., 2021).

Furthermore, the RA approach aids in the restoration of decline in land by increasing agricultural yield per unit area, improving soil health, biodiversity, and C storage in the soil, and ultimately improving soil functions and ecosystem services. RA is a system-based method that uses SOC sequestration to regulate and restore soil health over time using a variety of agricultural approaches. This farming technique enhances finances rather than depleting or damaging them. The terms include crops, integrated farming systems (IFS), integrated plant nutrition systems (IPNS), integrated nutrients management (INM), and pest control have all benefited from and been promoted by the RA idea., conservation agriculture/no-till farming with leftovers remaining, agri-horticulture/agroforestry systems, and pest management (Newton et al., 2020). Arable soils that are managed ought to be maintained in a manner that emulates the natural process of safeguarding and enhancing soil health, biodiversity, and sustainability. Additionally, it should aim to avoid degradation caused by illegitimate land conversion (Fig. 5).

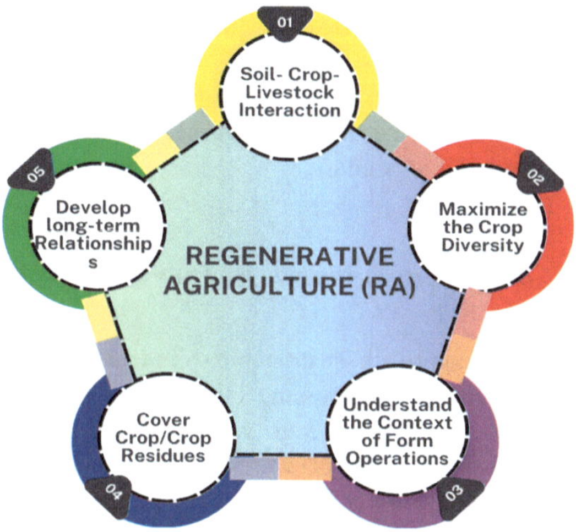

Fig. 5 Characteristics of regenerative agriculture

4.3 *Reducing Erosion of Soil and Repairing Land Deterioration to Achieve Land Degradation Neutrality (LDN)*

Worldwide soil erosion and land deterioration due to inadequate or incompetent management at an unprecedented rate. Significant problems in many countries include the loss of top-fertile soils due to erosion, and also rising soil loss and topography displacement in agricultural zones. Several countries have previously experienced large-scale land degradation due to one or more land degradation processes. These include the United States (32% area), China (30.7% area), Africa (16%), Europe (17%), Australia (2/3rd area), and India (45%). Reducing soil erosion and reversing land degradation with appropriate conservation measures is an essential requirement (Lehmann et al., 2020). To counteract the damage, land degradation neutrality has been put into place. "A situation in which the number and the quality of the land items needed to preserve ecological processes and the other support that improves food security keep solid or rise, within defined time and space scales and ecosystems" is what both sides of the Convention define as land-degradation neutrality (LDN). Land Degradation Networks (LDN) indicate a historic change in land management procedures and regulations, with 128 of 196 nations having already joined. The LDN strives to attain food security by 2030 by combating desertification, restoring damaged land, and working closely with SDGs, especially SDG 15.3. The Land Degradation Network (LDN) consists of three concurrent.

(i) Avoiding new degradation of land through preserving healthy land.
(ii) Reducing present degradation via the implementation of effective land management practices that may minimize loss while improving biodiversity, soil health, and farming;
(iii) Increasing up efforts that restore or return damaged regions to an organic or more effective state". Furthermore, LDN has determined 19 guiding principles for the implementation procedure.

4.4 Resilient Soil Management to Reduce Climate Change

Utilizing a SOC-centered approach, sustainable soil management (SSM) techniques offer a workable substitute for addressing climate change while simultaneously enhancing soil health, boosting food security, and generating income from agriculture (FAO, 2019a; Powlson et al., 2011). The SSM affects greenhouse gas emissions, carbon absorption, and soil health by improving the physical, chemical, and biological characteristics of the soil. Soil health and food grain production are significantly impacted by organic C and N management. Other greenhouse gas (GHG) emissions, such as those of N_2O and CH_4, can occasionally rise in response to increasing SOC through the addition of additional N fertilizer. To prevent temperature increases of 1.5–2.0 °C, appropriate SSM throughout agroecological zones and key agricultural systems is urgently needed. Agricultural soils can reduce greenhouse gas emissions of N_2O and CH_4 into the atmosphere and mitigate the effects of climate change through carbon storage, therefore a more robust framework incorporating a decarbonization program with SSM is required (Wang et al., 2020).

5 Conclusions

The farming of fertile land using extensive soil management techniques emerges as a key component of global food security, sustainability, and revenue generation in the direction of sustainable agriculture. To preserve soil health and establish sustainable agricultural systems, the plan outlined in this chapter highlights the critical necessity of putting sustainable soil management strategies into practice. By applying innovative techniques and working across disciplines, stakeholders can create the path for a future in agriculture that is more resilient and productive. There has never been a more important time to practice sustainable soil management since we are at the crossroads of climate resiliency, environmental sustainability, and food security. To promote the adoption of sustainable agricultural practices, foster innovation, and give priority to soil health programs, governments, lawmakers, farmers, researchers, and consumers must collaborate. We can guarantee a robust and sustainable agricultural coming for future generations of farmers by embracing the ideas outlined in this plan.

Taking into account the ideas and methods presented in this blueprint, it is evident that sustainable soil management is essential to addressing some of the most urgent issues facing our world. Through a shared dedication to sustainability, creativity, and teamwork, we can establish a productive environment that will support future generations of humans and the environment.

References

Al-Kaisi, M. M., & Broner, I. (2009). *Crop water use and growth stages* (p. 4). Colorado State University Extension.

Ananbeh, H., Stojanović, M., Pompeiano, A., Voběrková, S., & Trasar-Cepeda, C. (2019). Use of soil enzyme activities to assess the recovery of soil functions in abandoned coppice forest systems. *Science of the Total Environment, 694*, 133692.

Babel, W., Biermann, T., Coners, H., Falge, E., Seeber, E., Ingrisch, J., Schleuss, P. M., Gerken, T., Leonbacher, J., Leipold, T., Willinghofer, S., Schutzenmeister, K., Shibistova, O., Becker, L., Hafner, S., Spielvogel, S., Li, X., Xu, X., Sun, Y., … Foken, T. (2014). Pasture degradation modifies the water and carbon cycles of the Tibetan highlands. *Biogeosciences, 11*(23), 6633–6656.

Bahram, M., Hildebrand, F., Forslund, S. K., Anderson, J. L., Soudzilovskaia, N. A., Bodegom, P. M., et al. (2018). Structure and function of the global topsoil microbiome. *Nature, 560*, 233.

Bais, H. P., Weir, T. L., Perry, L. G., Gilroy, S., & Vivanco, J. M. (2006). The role of root exudates in rhizosphere interactions with plants and other organisms. *Annual Review of Plant Biology, 57*, 233–266.

Bender, S. F., Wagg, C., & van der Heijden, M. G. (2016). An underground revolution: Biodiversity and soil ecological engineering for agricultural sustainability. *Trends in Ecology and Evolution, 31*, 440–452.

Bhushan, L., Ladha, J. K., Gupta, R. K., Singh, S., Tirol-Padre, A., Saharawat, Y. S., Gathala, M., & Pathak, H. (2008). Saving of water and labor in a rice-wheat system with no-tillage and direct seeding technologies. *Agronomy Journal, 99*, 1288–1296.

Cannell, R., & Hawes, J. (1994). Trends in tillage practices in relation to sustainable crop production with special reference to temperate climates. *Field Crops Research, 30*, 245–282.

Celik, A., Altikat, S., & Way, T. (2013). Strip tillage width effects on sunflower seed emergence and yield. *Soil and Tillage Research, 131*, 20–27.

Dong, K., Dong, Y., Zheng, L., Tang, Z., & Yang, X. (2014). Faba bean fusarium wilt (*Fusarium oxysporum*) control and its mechanism in different wheat varieties and faba bean intercropping system. *Chinese Journal of Applied Ecology, 25*, 1979–1987.

Doran, J. W., & Parkin, T. B. (1994). Defining and assessing soil quality. In *Defining soil quality for a sustainable environment* (Vol. 35, pp. 1–21).

Escanhoela, A. S., Pitombo, L. M., Brandani, C. B., Navarrete, A. A., Bento, C. B., & Carmo, J. B. (2019). Organic management increases soil nitrogen but not carbon content in a tropical citrus orchard with pronounced N_2O emissions. *Journal of Environmental Management, 234*, 326–335.

EU Soil Observatory. (2023). EUSO soil health dashboard. Available at: https://esdac.jrc.ec.europa.eu/esdacviewer/euso-dashboard/. Accessed May 10, 2023.

FAO (2019a). The state of the world's biodiversity for food and agriculture. In J. Bélanger & D. Pilling (Eds.). *FAO commission on genetic resources for food and agriculture assessments*. Food and Agriculture Organization of the United Nations: Rome.

Food and Agriculture Organization (2011) Save and grow. A policymakers' guide to the sustainable intensification of smallholder crop production. Rome: Food and Agriculture Organization. Available at www.fao.org/ag/save-and-grow/

Food and Agriculture Organization of the United Nations and Intergovernmental Technical Panel on Soils (2015). Status of the world's soil resources (SWSR)—main report. Rome, Italy.

OECD-FAO (2022) OECD-FAO agricultural outlook 2022–2031.

Giller, K., Bignell, D., Lavelle, P., Swift, M., Barrios, E., Moreira, F., van Noordwijk, M., Barois, I., Karanja, N., & Huising, J. (2005). Soil biodiversity in rapidly changing tropical landscapes: Scaling down and scaling up. In R. Bardgett, M. Usher, & D. Hopkins (Eds.), *Biological diversity and function in soils (ecological reviews)* (pp. 295–318). Cambridge, UK: Cambridge University Press.

Godfray, C., Beddington, J. R., & Crute, I. R. (2010). Food security: The challenge of feeding 9 billion people. *Science, 327*, 812–818.

Gong, W., Yan, X., Wang, J., Hu, T., & Gong, Y. (2011). Long-term application of chemical and organic fertilizers on plant-available nitrogen pools and nitrogen management index. *Biology and Fertility of Soils, 47*, 767–775.

Goreta Ban, S., Zanic, K., Dumicic, G., Raspudic, E., Selak, G., & Ban, D. (2014). Growth and yield of grafted cucumbers in soil infested with root-knot nematodes. *Chilean Journal of Agricultural Research, 74*, 29–34.

Gutierrez, W. A., Shew, H. D., & Melton, T. A. (1997). Sources of inoculum and management for *Rhizoctonia solani* damping-off on tobacco transplants under greenhouse conditions. *Plant Disease, 81*, 604–606.

Hammond, R. A., & Dubé, L. (2012). A systems science perspective and transdisciplinary models for food and nutrition security. *Proceedings of the National Academy of Sciences, 109*(31), 12356–12363.

Iovieno, P., Morra, L., Leone, A., Pagano, L., & Alfani, A. (2009). Effect of organic and mineral fertilizers on soil respiration and enzyme activities of two Mediterranean horticultural soils. *Biology and Fertility of Soils, 45*, 555–561.

Jat, M. L., Gathala, M. K., Ladha, J. K., Saharawat, Y. S., Jat, A. S., Vipin, K., Sharma, S. K., Kumar, V., & Gupta, R. K. (2009). Evaluation of precision land leveling and double zero-till systems in the rice-wheat rotation: Water use, productivity, profitability, and soil physical properties. *Soil and Tillage Research, 105*, 112–121.

Jat, M. L., Gathala, M. K., Saharawat, Y. S., Tetarwal, J. P., & Gupta, R. (2013). Double no-till and permanent raised beds in maize-wheat rotation of north-western Indo-Gangetic Plains of India: Effects on crop yields, water productivity, profitability, and soil physical properties. *Field Crops Research, 149*, 291–299.

Karlen, D. L., Ditzler, C., & Andrews, S. S. (2003). Soil quality: Why and how? *Geoderma, 114*, 145–156.

Kassam, A. H., Friedrich, T., Shaxson, T. F., & Pretty, J. N. (2009). The spread of conservation agriculture: Justification, sustainability, and uptake. *International Journal of Agricultural Sustainability, 7*(4), 292–320.

Kibblewhite, M., Ritz, K., & Swift, M. (2008). Soil health in agricultural systems. *Philosophical Transactions of the Royal Society B: Biological Sciences, 363*, 685–701.

Koike, S., Subbarao, K., Davis, R. M., and Turini, A. T. (2003). *Vegetable diseases are caused by soil borne pathogens.* Oakland, CA, USA: UCANR Publications, University of California. Available online at https://anrcatalog.ucanr.edu/pdf/8099.pdf. Accessed on July 7, 2019.

Kopittke, P. M., Menzies, N. W., Dalal, R. C., McKenna, B. A., Husted, S., Wang, P., & Lombi, E. (2021). The role of soil in defining planetary boundaries and the safe operating space for humanity. *Environment International, 146*, 106245.

Kultschar, B., and Llewellyn, C. (2018). Secondary metabolites in cyanobacteria. In *Secondary metabolites-sources and applications* (Vol. 2, pp. 23–36). London, UK: InTech.

Lal, R., Monger, C., Nave, L., & Smith, P. (2021). The role of soil in regulation of climate. *Philosophical Transactions of the Royal Society B, 376*(1834), 20210084.

Larkin, R. P., Honeycutt, C. W., Olanya, O. M., Halloran, J. M., & He, Z. (2012). Impacts of crop rotation and irrigation on soil borne diseases and soil microbial communities. In Z. He, R.

Larkin, & W. Honeycutt (Eds.), *Sustainable potato production: Global case studies* (pp. 139–165). Springer.

Lehmann, J., Bossio, D. A., Kögel-Knabner, I., & Rillig, M. C. (2020). The concept and future prospects of soil health. *Nature Reviews Earth & Environment, 1*(10), 544–553.

Linderman, R. G. (1991). Mycorrhizal interactions in the rhizosphere. In D. L. Keister & P. B. Cregan (Eds.), *The rhizosphere and plant growth. Beltsville Symposia in Agricultural Research* (Vol. 14, pp. 343–348). Dordrecht, The Netherlands: Springer.

Ma, L., Wang, Q., & Shen, S. T. (2020). Response of soil aggregate stability and distribution of organic carbon to alpine grassland degradation in Northwest Sichuan. *Geoderma Regional, 22*, e00309.

Mäder, P., Fliebach, A., Dubois, D., Gunst, L., Fried, P., & Niggli, U. (2002). Soil fertility and biodiversity in organic farming. *Science, 296*, 1694–1697.

Mangalassery, S., Kalaivanan, D., & Philip, P. (2019). Effect of inorganic fertilizers and organic amendments on soil aggregation and biochemical characteristics in a weathered tropical soil. *Soil and Tillage Research, 187*, 144–151.

Mehran, M., Ashraf, M., Shahzad, S. M., Shakir, M. S., Azhar, M. T., Ahmad, F., & Alvi, A. (2023). Growth, yield, and fiber quality characteristics of Bt and non-Bt cotton cultivars in response to boron nutrition. *Journal of Cotton Research, 6*(1), 1.

Mela, F., Fritsche, K., De Boer, W., Van Veen, J. A., De Graaff, L. H., Van Den Berg, M., & Leveau, J. H. (2011). Dual transcriptional profiling of a bacterial/fungal confrontation: *Collimonas fungivorans* versus *Aspergillus niger*. *ISME Journal, 5*, 1494–1504.

Mihajlović, M., Emil, R., Jovana, H., Mila, G., & Brankica, T. (2017). Methods for management of soil borne plant pathogens. *Pesticidi I Fitomedicina, 32*, 9–24.

Neher, D. A. (2001). Role of nematodes in soil health and their use as indicators. *Journal of Nematology, 33*, 161–168.

Neher, D. A. (2010). Ecology of plant and free-living nematodes in natural and agricultural soil. *Annual Review of Phytopathology, 48*, 371–394.

Newton, P., Civita, N, Frankel-Goldwater, L., Bartel, K., & Johns, C. (2020). What Is regenerative agriculture? A review of scholar and practitioner definitions based on processes and outcomes. *Frontiers in Sustainable Food Systems, 4*, 577723.

Nisha, R., Kaushik, A., & Kaushik, C. P. (2007). Effect of indigenous cyanobacterial application on structural stability and productivity of an organically poor semi-arid soil. *Geoderma, 138*, 49–56.

Organic Production/Organic Food: Information Access Tools. USDA. Available online at https://www.nal.usda.gov/afsic/organic-productionorganic-food-information-access-tools. Accessed on June 8, 2020.

Panwar, J., Yadav, R. S., Yadav, B. K., and Tarafdar, J. C. (2008). Arbuscular mycorrhizae: A dynamic micro-symbiont for sustainable agriculture. In *Mycorrhizae: Sustainable agriculture and forestry* (Vol. 15, pp. 9–176). Dordrecht, The Netherlands: Springer.

Powlson, D. S., Whitmore, A. P., & Goulding, K. W. (2011). Soil carbon sequestration to mitigate climate change: A critical re-examination to identify the true and the false. *European Journal of Soil Science, 62*(1), 42–55.

Pretty, J. (2008). Agricultural sustainability: Concepts, principles, and evidence. *Philosophical Transactions of the Royal Society B: Biological Sciences, 363*(1491), 447–466.

Pretty, J. N., Toulmin, C., & Williams, S. (2011). Sustainable intensification in African agriculture. *International Journal of Agricultural Sustainability, 9*(1), 5–24.

Qin, K., Dong, X., Jifon, J., & Leskovar, D. (2019). Rhizosphere microbial biomass is affected by soil type and organic and water inputs in a bell pepper system. *Applied Soil Ecology, 138*, 80–87.

Quist, C. W., Gort, G., Mulder, C., Wilbers, R. H. P., Termorshuizen, A. J., Bakker, J., & Helder, J. (2017). Feeding preference as a main determinant of microscale patchiness among terrestrial nematodes. *Molecular Ecology Resources, 17*, 1257–1270.

Raaijmakers, J. M., Paulitz, T. C., Steinberg, C., Alabouvette, C., & Moenne-Loccoz, Y. (2009). The rhizosphere: A playground and battlefield for soilborne pathogens and beneficial microorganisms. *Plant and Soil, 321*, 341–361.
Ristaino, J. B., Perry, K. B., & Lumsden, R. D. (1991). Effect of solarization and *Gliocladium virens* on sclerotia of *Sclerotium rolfsii*, soil microbiota, and the incidence of southern blight of tomato. *Phytopathology, 81*, 1117–1124.
Saglam, M., Dengiz, O., & Saygin, F. (2015). Assessment of horizontal and vertical variabilities of soil quality using multivariate statistics and geostatistical methods. *Communications in Soil Science and Plant Analysis, 46*(13), 1677–1697.
Saharawat, Y. S., Ladha, J. K., Pathak, H., Gathala, M., Chaudhary, N., & Jat, M. L. (2012). Simulation of resource-conserving technologies on productivity, income, and greenhouse gas (GHG) emission in rice-wheat system. *Journal of Soil Science and Environmental Management, 3*, 9–22.
Schrama, M., de Haan, J., Kroonen, M., Verstegen, H., & Van der Putten, W. (2018). Crop yield gap and stability in organic and conventional farming systems. *Agriculture, Ecosystems and Environment, 256*, 123–130.
Sid, A., Pérez, S. C., Egea, G. C., & Candela, M. E. (1999). Evaluation of the capacity of *Trichoderma harzianum* in controlling rot caused by *Phytophthora capsici* in pepper plants. *Plant Pathology, 48*, 58–65.
Singh, J. S. (2015). Plant-microbe interactions: A viable tool for agricultural sustainability. In N. K. Arora (Ed.), *Plant microbes' symbiosis: Applied facets* (p. 384). New Delhi, India; Heidelberg, Germany; New York, NY, USA; Dordrecht, The Netherlands; London, UK: Springer.
Stirling, G., Hayden, H., Pattison, T., & Stirling, M. (2017). Soil health, soil biology, soil borne diseases, and sustainable agriculture: A guide. *Australasian Plant Pathology, 46*, 387.
Tanović, B., Potočnik, I., Delibašić, G., Ristić, M., Kostić, M., & Marković, M. (2009). In vitro effect of essential oils from aromatic and medicinal plants on mushroom pathogens: *Verticillium fungicola* var. *fungicola*, *Mycogone perniciosa*, and *Cladobotryum* sp. *Archives of Biological Sciences, 61*, 231–237.
Totsche, K. U., Rennert, T., Gerzabek, M. H., Kögel-Knabner, I., Smalla, K., Spiteller, M., & Vogel, H. J. (2010). Biogeochemical interfaces in soil: The interdisciplinary challenge for soil science. *Journal of Plant Nutrition and Soil Science, 173*(1), 88–99.
Veerman, C., Correia, T. P., Bastioli, C., Biro, B., Bouma, J., Ciencala, E., et al. (2020). Caring for soil is caring for life—ensure 75% of soils are healthy by 2030 for food, people, nature, and climate. Brussels: European Commission.
Vivek, K., Saharawat, Y. S., Gathala, M. K., Jat, A. S., Singh, S. K., Chaudhary, N., & Jat, M. L. (2013). Effect of different tillage and seeding methods on energy use efficiency and productivity of wheat in Indo-Gangetic plains. *Field Crops Research, 142*, 1–8.
Vogel, H.-J., Bartke, S., Daedlow, K., Helming, K., Kögel-Knabner, I., & Lang, B. (2018). A systemic approach for modeling soil functions. *The Soil, 4*, 83.
Wagg, C., Jansa, J., Schmid, B., & van der Heijden, M. G. A. (2011). Belowground biodiversity effects of plant symbionts support aboveground productivity. *Ecology Letters, 14*, 1001–1008.
Wang, W., Ge, J., Yu, X., & Li, H. (2020). Environmental fate and impacts of microplastics in soil ecosystems: Progress and perspective. *Science of the Total Environment, 708*, 134841.
Xiao, C. L., Subbarao, K. V., Schulback, K. F., & Koike, S. T. (1998). Effects of crop rotation and irrigation on *Verticillium dahliae* microsclerotia in soil and wilt in cauliflower. *Phytopathology, 88*, 1046–1055.
Yang, G., Wagg, C., Veresoglou, S., Hempel, S., & Rillig, M. (2018). How soil biota drives ecosystem stability. *Trends in Plant Science, 23*, 1057–1067.
Yuan, Z. Q., Jiang, X. J., Liu, G. J., Jin, H. J., Chen, J., & Wu, Q. B. (2019). Responses of soil organic carbon and nutrient stocks to human-induced grassland degradation in a Tibetan alpine meadow. *CATENA, 178*(5), 40–48.
Zhang, G. S., & Liu, Y. F. (2018). The distribution of microplastics in soil aggregate fractions in southwestern China. *Science of the Total Environment, 642*, 12–20.

Zhou, H., Zhang, D. G., Jiang, Z. H., Sun, P., Xiao, H. L., Wu, Y. X., & Chen, J. G. (2019). Changes in the soil microbial communities of alpine steppe at Qinghai-Tibetan Plateau under different degradation levels. *Science of the Total Environment, 651*, 2281–2291.

Zulpa, G., Zaccaro, M. C., Boccazzi, F., Parada, J. L., & Storni, M. (2003). Bioactivity of intra- and extracellular substances from cyanobacteria and lactic acid bacteria on "wood blue stain" fungi. *Biological Control, 27*, 345–348.

Symbiotic Stewards of the Earth: Unraveling the Vital Role of Endophytes in Soil Health and Fertility

Muhammad Adeel Ghafar, Inzamam Ul Haq, Zia Ul Haq, Muhammad Rehan Akhtar, Moazam Hyder, Dilawar Abbas, Kamran Haider, Muhammad Lubaid Khalid, Muhammad Mehran, and Liande Wang

Abstract The chapter gives a general overview of the endophytes, soil systems, and their interaction within the soil environment. First, they define what endophytes are and how these organisms can penetrate plant tissues without harming the plant, the host of the endophyte. General information on plant endophytes and their taxonomy is provided, paying special attention to these symbionts' taxonomic range and host preference. The ways of endophyte-plant interaction are discussed in detail, as well as the mechanisms of colonization, endophyte communication with the plant, and the positive effects of endophytes on the plant. This is why the endophyte performance is affected by the type of soil, climate, and anthropogenic factors. Numerous works

M. A. Ghafar · I. U. Haq · M. R. Akhtar · M. Hyder · L. Wang (✉)
State Key Laboratory of Ecological Pest Control for Fujian and Taiwan Crops, Key Laboratory of Biopesticides and Chemical Biology, MOE, College of Plant Protection, Fujian Agriculture and Forestry University, Fuzhou, China
e-mail: 000q030043@fafu.edu.cn

M. A. Ghafar
e-mail: 2221903005@fafu.edu.cn

Z. U. Haq
Department of Plant Pathology, Faculty of Agricultural Sciences and Technology, Bahauddin Zakriya University Multan, Multan, Pakistan

D. Abbas
State Key Laboratory for Biology of Plant Diseases and Insect Pests, Institute of Plant Protection, Chinese Academy of Agricultural Sciences, Beijing, China

K. Haider
Institute of Urban and Horticultural Entomology, College of Plant Science and Technology, Huazhong Agricultural University, Wuhan, People's Republic of China

M. L. Khalid
Department of Entomology, College of Plant Protection, Henan Agricultural University, Zhengzhou, China

M. Mehran
Key Laboratory of Arable Land Conservation (Middle and Lower Reaches of Yangtze River), Ministry of Agriculture and Rural Affairs, Huazhong Agricultural University, Wuhan, China

M. Shaaban (ed.), *Soils and Sustainable Agriculture*, Frontier Studies in Soil Science,
https://doi.org/10.1007/978-3-031-91114-9_19

and examples show the importance of the knowledge of endophytes' roles and the potential for their applications in sustainable agriculture and the protection of biota. In conclusion, the chapter presents the study's findings by explaining the position of endophytes in soil health, their sensitivity to the environment, and the need to research the unknown more. Endophytes and their positive and negative effects on the host plant, the adverse effects of endophytes on the host plant, and how they can be a threat are discussed, and the importance of endophytes in agriculture is stressed. Therefore, This research places endophytes as relevant caretakers of the earth, thus laying down a good foundation for realizing the potential of soil fertility and productivity.

Keywords Endophytes · Soil health · Soil fertility · Plant–microbe interactions · Sustainable agriculture · Environmental sustainability · Agricultural practices

1 Introduction

1.1 Sustainable Agriculture Overview

Sustainable agriculture is farming that is carried out in a manner that does not harm the earth and its inhabitants, especially farmers, while providing food for consumption. Organized in this manner, farming approaches are known to deal with future food production strategies without exhausting natural resources. Sustainable agriculture is the reasonable usage of resources such as water, land, energy, and others. Some of the management practices are crop rotation, agroforestry, and the use of cover crops to enhance the fertility of the soil, reduce the rate of erosion, and enhance biodiversity. Also, the application of Integrated Pest Management (IPM) is applied in a way that pests can be controlled in a way that will not affect the health of the human being and the environment and, in the process, minimize the use of chemical pesticides (Hyder et al., 2024; Tilman et al., 2002).

Sustainable agriculture also entails the promotion of local foods to reduce food miles, which emphasizes the transportation of food and, hence, the carbon footprint. Sustainable agriculture also supports local farmers and their markets and boosts communities' potential (Pretty, 2008). Also, sustainable agriculture promotes the use of what is referred to as organic farming, which entails minimal or no use of synthetic fertilizers and pesticides to maintain soil fertility (Reganold & Wachter, 2016). Water use is also another major factor that is considered in sustainable agriculture and farming. Measures like efficient irrigation methods like drip irrigation and harvesting rainwater are put in place to ensure that water is used correctly and that there is no wastage of the resource (Gomiero et al., 2011). Also, sustainable agriculture enhances the germplasm conservation of crops and domestic animals. This genetic variation is important in disease and pest resistance and tolerance to climate change, meaning that agriculture can be carried out in the long run (Food,

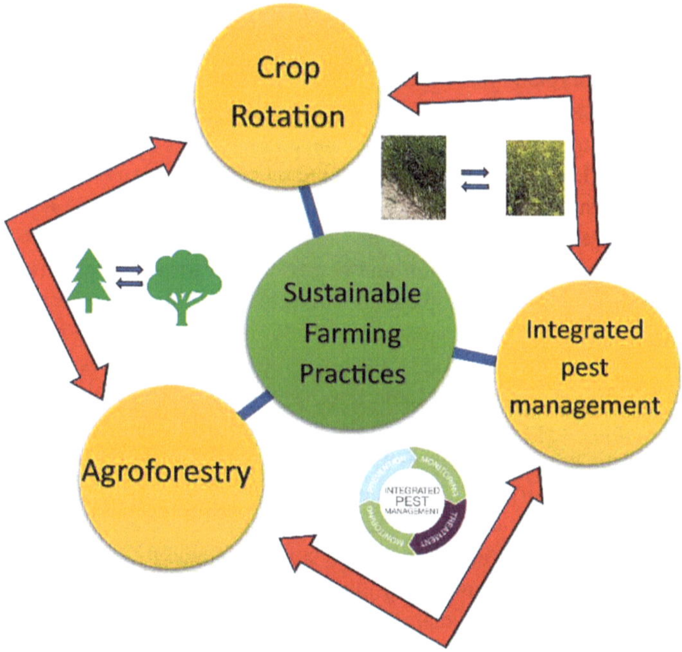

Fig. 1 Sustainable farming practices

2010). Pest management also falls under the context of sustainable agriculture since it is important to safeguard other animals. This entails space, nutrition, and alleviation of stress and suffering which is not only beneficial for the animals but also the animal products (Fraser, 2008). Also, it is important to note that sustainable farming systems aim to reduce greenhouse gas emissions through no-tillage, carbon sequestration in the soils, and using renewable energy on the farm (Smith et al., 2008). Education and training of farmers in sustainable agriculture are deemed necessary for the enhancement of agriculture. The following are also expected to benefit from the awareness of sustainable practices and innovations: Farmers can manage their resources and increase their yields correctly (Pretty, 2012). However, there is a need to develop policies and regulations to foster sustainable agriculture to create awareness of sustainable agriculture. Other factors that may help promote SFP adoption include government incentives such as subsidies, grants, and other forms of support for related research (Lal, 2004) (Fig. 1).

1.2 Introduction to Endophytes

Endophytes are microorganisms that dwell in the plant tissues without causing harm to the plant (Arora & Ramawat, 2017). Pathogens are parasites that can cause damage to the host plants. On the other hand, endophytes are mutual or apparent mutualistic organisms that are beneficial to the host plant in some way or another (Wani et al., 2015). The microorganisms that are not visible to us and are present in the soil, which is otherwise neglected, still hold many secrets (Phillips, 2017). However, the most significant inhabitants of this subterranean maze are endophytes—microorganisms that are hidden in plants and, only in the recent past, have been considered key players in the health of the soil and its productivity (Bishnoi, 2015). So, when we start discovering the meaning and value of endophytes, we will step into the area of study in plant–microbe interaction and eco-friendly farming (Bokhari, 2018). The advancement in molecular biological and microbial ecological research has enriched the understanding of the diverse symbiotic relationships between plants and their endophytes (Bhardwaj et al., 2023). Such microorganisms that reside within tissues of plants with no harm to the plants have gained much attention in recent years because they play a significant role in the soil's overall health (Gupta et al., 2017). In one way or the other, endophytes aid in enhancing soil health through influencing nutrient cycling, soil structure, and plant stress tolerance. It further depicts how the two can exist side by side in a balance, which is quite a clue on how the two can collaborate in preserving the health of the soil and hence its fertility (Verma et al., 2021b). The current study has thus offered new perspectives on the different roles of endophytes in the subsequent analysis. The stress tolerance mechanisms nurtured by endophytes have been seen to enhance the plant's capacity to handle stress in stress conditions (Torres-Rodriguez et al., 2022). Endophytes and their interactions with the above and below-ground interface, plant, and soil health (Chaparro et al., 2012). Here, we tried to integrate these most recent advances and give a comprehensive view of the multiple functions of endophytes in soil health and fertility. In this paper, an effort has been made to identify the symbiotic stewards crucial for sustainable and resilient agricultural ecosystems by identifying their roles, mechanisms, and factors affecting their performance. This paper identifies the dependency of plants and microorganisms in soil ecosystems as the core of a mutually beneficial relationship that has significant implications for the health of the Earth's soils. Such relations which cannot be observed directly are the subject of many studies in microbiology, ecology, and agriculture. Consequently, the major functions of soil have been viewed as providing support for plants through anchorage and a supply of nutrients to the plants. However, the onset of twenty-first century has witnessed new molecular technologies and ideas concerning the subsurface microorganisms. These are some of the principal inhabitants of this shadowy world: endophytes, microorganisms that are located in the plant tissues without causing harm to the plant. Plant-endophyte interactions have been recognized for quite some time now but however the focus on these microorganisms has been on the rise in the last few years. Endophytes have been taken out of the category of organisms that do not have a role in plant health and stress tolerance and ecosystem processes.

Thus, the beginning of this research is based on the idea that soils are not merely described by their physical and chemical characteristics but the entire complex of relations between plants and various microorganisms present in the tissues of plants. When attempting to comprehend what endophytes are, one is left on the verge of grasping how these microorganisms aid in sustaining the soil biome's health. Given the focal area of study, the position of sustainable agriculture and food insecurity in the face of climate change calls for an exploration of endophytes. Since they contribute to the enhancement of nutrient cycling, plant stress tolerance, and disease control, they are considered important players in the search for more efficient and sustainable agricultural systems. In this chapter, a detailed analysis of this rather vast subject of plant-endophyte interactions is conducted to reveal their place in soil health and fertility sustenance. Hence, in describing the significance of these connections, we aim to deepen our knowledge of the processes that take place in the soil, which will contribute to maintaining the relationships and creating suitable conditions for a sustainable ecosystem.

1.3 Characteristics of Endophytes

Endophytes are microorganisms that live in plant tissues and are known to benefit plants' health (Suryanarayanan & Ravishankar, 2023). It is usually mutualistic because the plant and the microorganisms gain a lot from this interaction. These include improved nutrient uptake, increased stress tolerance, and defense against pathogenic microorganisms (Saikkonen et al., 2004). Endophytes differ from other organisms that live in the rhizosphere or on the plant's surface in that they can infect internal plant tissues (Hardoim et al., 2008). They can inhabit a variety of plant organs, such as roots, stems, leaves, and seeds, thriving within the host and often going unnoticed due to their asymptomatic colonization (Schulz & Boyle, 2005). One of the remarkable aspects of endophytes is their ability to colonize plants without causing any visible disease symptoms, a characteristic that underscores their stealthy presence within plant tissues (Bertola et al., 2021).

Some endophytes are transmitted vertically through seeds from one generation of plants to the next. This transmission mode ensures the continuity of their association with the host plant across generations (Herre et al., 2007). Endophytes are also known for their beneficial metabolic activities, which include nitrogen fixation, phosphate solubilization, and the production of bioactive compounds that can enhance plant growth and resilience (Nair & Padmavathy, 2014).

Genetically, endophytes are diverse, exhibiting a wide range of genetic traits that enable them to adapt to specific host plants (Compant et al., 2016). This genetic diversity is complemented by their phenotypic plasticity, allowing them to adjust their characteristics in response to varying environmental conditions or signals from the host plant (Lugtenberg & Kamilova, 2009). The relationship between endophytes and plants is often coevolutionary, involving reciprocal adaptations over time. This coevolution leads to the development of specific mechanisms that facilitate a harmonious association, enhancing the mutual benefits of their interaction (Rodriguez et al., 2009).

1.4 Diversity and Classification of Plant Endophytes

Plant endophytes are diverse and can be classified as different types of microflora, mainly bacteria, fungi, and viruses. The review on bacterial endophytes depicts them as a diverse group with key species, including Proteobacteria, Actinobacteria, and Firmicutes. Among such characteristics, nitrogen-fixing bacteria such as Rhizobia and phosphorus-solubilizing bacteria are mainly investigated due to their favorable effects on plant growth and soil fertility (Compant et al., 2010). Meanwhile, fungal endophytes are primarily associated with Ascomycota and Basidiomycota. Some of the outstanding genera in any of such groups comprise Claviceps, Epichloë, and Piriformospora, wherein every category is characterized by distinct ecological functions and relationships with their host plants, as noted (Hyde & Soytong, 2008).

Interestingly, viral endophytes have also raised interest in the last few years concerning their effects on plant health. These plant endophytic viruses can affect the host phenotypes and regulate changes to environmental stresses. However, their role in plant ecology is still pending realization. The specificity of parasite-host relationships is highly diverse. Some of the endophytes are eradicators, which can be associated with a wide range of host plants, while specialists colonize a narrow range of host plant species or even genera (Arnold, 2007). In this case, the coevolutionary processes between endophytes and the plants they infect are central in defining this host specificity. Some of them have developed strategies that enable them to perceive and course with their host plant, hence having a closer and more specialized interaction (Mengistu, 2020). This coevolution has customarily entailed the establishment of molecular signals and recognition systems that enhance the symbiotic relationship and are advantageous to both the endophyte and the host plant. The knowledge of these complex relationships may give important information about the functioning of endophytes in ecosystems and their useful features in agriculture and plant bioengineering.

1.5 Intraspecific and Interspecific Variation

Plant endophytes' distribution and ecological features are highly diverse within the species and types, illustrating their key role in the interaction with the host plants and the environment. A single endophyte species can be diverse genetically, metabolically, and ecologically, even within the same species. It means that in terms of genes and metabolism, the strains of similar endophyte species can act differently or develop specific relationships with numerous host plant species, thus producing different effects. For instance, endophytic species may have higher efficiency of nitrogen fixation, while others can be superior in the solubilization of phosphate or production of sets (Husseiny et al., 2021). These differences within an endophyte in a given species affect the same's ability to perform inside a plant and deal with various environmental factors that could favor the plant. The difference in intraspecific diversity affects the interaction further by the ability of the different strains of the same species of endophytes to induce different reactions in the host plants. This suggests that a given endophyte species can provide different services to different plants or similar services but of a different type to the same plant at different conditions. Genetic and metabolic differences between strains prescribe the efficiency of the colonization of the host plant, interaction with plant immune response, and contribution to the uptake of nutrients and stress tolerance of the plant (Rodriguez et al., 2019). These differences suggest the awareness of the intraspecific variation to realize the possibility or impossibility of the endophytic microorganisms in agricultural and ecological utilization.

Interactions of endophytes belonging to two species complicate the relationship between the plant and the microbe even further. These may be mutual, such that more than one endophyte species use the plant at the same time and make positive feedback on the host plant, or antagonistic, where two or more species use the same plant part and or the same resource, thus reducing their effectiveness (Partida-Martínez & Heil, 2011). Dietary mutualisms can also be described as synergistic because they happen when one endophyte uses the host plant as a source of nutrients, and the other, in return, uses the first one as a shield from pathogens and stress. For instance, one species may be good at nitrogen fixation. In contrast, another species may be suitable for producing antimicrobial compounds, which, if combined, benefits more than doubling the value of either species (Van Der Heijden et al., 2008).

Conversely, antagonistic and aggressive structural interactions can result from endophyte species competing for the sites of colonization inside the plant and nutrient access. These competitive pressures may affect the makeup and balance of the endophytic community and the specificity of their interaction with the host plant: the dominant species may change over time and, in turn, may impact the host plant's health and robustness (Singh Saharan et al., 2023). Knowledge of such interspecific relations is critical to appreciating the global influence of endophytes on the host plants and other soil microorganisms. It offers valuable information on microbial ecology and helps identify measures towards enhancing endophytes' utility in agriculture and the environment.

1.6 Environmental Adaptations

Thus, plant endophytes' distribution, apparent germ-line diversification, and functionality are strongly influenced by and linked to environmental adaptations. Soil types indicate why different endophyte species are present in specific areas and the level of their occurrence in the given environments. Nutrient availability in terms of pH, nutrient status, and water content in the soil also plays a significant role in the survival and capacity of endophytes to colonize the plants. For example, some endophytes are well suited to grow effectively on acidic soils as the pH level ranges or is low, while others may prefer the latter (Berg & Smalla, 2009). Likewise, endophytes are influenced by the nutrient availability in the soil, particularly nitrogen and phosphorus. It can be fixed in nutrient-poor soils that afford considerable advantages to their vicinity host plants as those capable of fixing the nitrogen or solubilizing the phosphate (Philippot et al., 2013). The moisture level also has a significant role as some of the endophytes are known to demonstrate drought tolerance and assist the host plant in water stress. In contrast, others thrive well under high moisture conditions (Malinowski & Belesky, 2000).

2 Environmental Adaptations of Endophytes

2.1 *Ecological Roles*

This aims to evaluate the benefits of plant endophytes on the biology of the ecosystem and identify areas where they can have a negative impact. However, it is essential to note that although a majority of the endophyte-host interactions are positive in that they confer benefits such as increased nutrient acquisition and enhanced stress tolerance, some of the endophytes have pathogenic potential provided that specific environmental conditions are met, or where the endophyte is associated with specific host plants (Ashraf et al., 2022). Endophyte and plant interaction: These two roles demonstrate the complexity of the interaction's direction and stress, and there may be possible circumstances in which an endophyte may cause pathogenic actions. As a result, some endophytes may not be pathogenic during regular periods but become parasitic as soon as the host plant is under some stress or change in certain features of the environment, including temperature, humidity, and nutrient status (Gurung et al., 2023). Their availability ensures that the associations between endophytes and crops must be managed to harness the positive effects while minimizing the negative ones. Concerning the functional aspect of endophytes, it was evidenced that they were diverse and involved in the host plants' nutrient cycling, stress tolerance, and disease resistance relationship. Besides indicated functional diversity, the above brings positive consequences to the general health and fertility of the ecosystems of the soils. For instance, endophytes are critical in enhancing nutrient access through nitrogen fixation, phosphorus solubilization, and decomposition of organic compounds to support plant growth and soil fertility (Doty et al., 2009). In addition, endophytes can enhance the stress tolerance potential of plants by producing various bioactive molecules such as osmoprotectants, free radicals scavenging molecules, and phytohormones for coping with abiotic stress factors such as drought, salinity, and extreme temperatures. Furthermore, endophytes aid in increasing disease resistance through the formation of the compounds that influence the development of the disease or pathogen in some way or another or through the production of disease resistance by the host plant all over its body (Jacob et al., 2020) (Fig. 2).

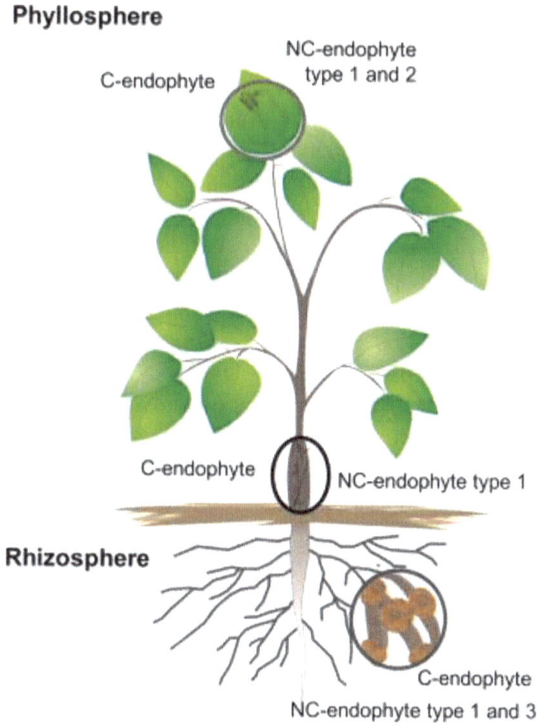

Fig. 2 The mutualistic relationship between the plant and endophytes

The installation of endophytes also conducts or participates in plant health and other functions, exerting some influence on extra ecological processes. Thus, the present work shall be concluded by stating that endophytes contribute to nutrient cycling and improve abiotic stress tolerance for maintaining the stability and productivity of plant ecosystems (Mehran et al., 2023). For example, endophytes may be essential in nutrient-poor or damaged soils to improve germination and growth of plants that restore damaged areas and rehabilitate depleted/ damaged soils and ecosystems (Porras-Alfaro & Bayman, 2011). Additionally, the fact that such complex and diverse microbial communities can cohabit within the plant endosphere could assist plants in adapting to factors resulting from climate change in plant growth and development because the microbial communities are a reservoir of several functional attributes that might be up or down-regulated depending on the changes elicited by climate change (Mori et al., 2013). In sum, the ecological importance of endophytes makes them significant entities in plant and soil biology. What they can do is affect nutrient cycling, enhance stress, and convey disease protection, and this probably makes it known why they are essential in agriculture. Both threatened and exploited landscapes of nature. For this reason, it is crucial to understand the relations of the endophytes from their host plant and the exterior sections that choreograph and modulate these relations, starting with the agriculture sector and expanding to the overall entrapment of the global ecosystem's health.

I also agree that what exists in the functional roles of endophytes are not only integrated into the whole Plant but also into all the known processes in the ecosystem. Therefore, endophytes help in nutrient cycling because of their role in improving plant photosynthetic rate and abiotic stress tolerance that promotes ecosystem stability. For instance, the mechanisms of endophytes are particularly useful in nutrient-limited or newly degraded sites to support plant establishment and growth. They can be of great use in restoring soils and ecosystems (Porras-Alfaro & Bayman, 2011). Moreover, the same authors proposed that plants with diverse endophytic communities are more resilient to climate change impacts because consumers could bear a repertoire of functional traits to be packaged when the condition alters. It can, therefore, be concluded that endophytic organisms play a very significant role in the functionality of the plant/soil ecosystem. They can also influence nutrient interactions, stress comfort, and disease protection and are highly likely to be used in agriculture and natural systems. Endophyte-host plant interactions and their effects on the primary host and resulting signs, as well as factors related to whether or not they are beneficial for agriculture or in maintaining the health of ecosystems, are also important.

2.2 *Significance of Endophytes in Soil Ecosystems*

The role of endophytes in the soil is a topic of growing relevance and prominence, drastically altering the perception of the complexity of the soil (Minhas et al., 2024). Thus, endophytes—the microorganisms living within the internal tissues of the host plants, which do not seem to cause adverse effects- are considered of great importance for plant health, soil stability, and fertility (Granjou & Phillips, 2019). In nutrient cycling, availability, and utilization are among the most valuable services endophytes provide, and their mutual relation with plants is comprehensively known and understood. They directly decompose organic matter or indirectly by fixing nitrogen and making phosphorus available in the soil for plant uses, thus promoting a nutrient-rich soil environment within the plant host (Friesen et al., 2011). Besides, endophytes are involved in the enhancement of communications as well as the modification of the soil structure and aggregation. It is established that their access to plant roots affects the root system and secretion, which is essential for the soil's steady structure and erosion eradication (Ikram et al. 2024a). This interaction not only increases water storage ability but also improves aeration and the promotion of root growth (Galindo-Castañeda et al., 2022). A similar notable asset of endophytes is the ability to increase the tolerance of plants to stress. These microorganisms contribute to the ability of plants to handle stress factors, including water scarcity and the sharp increase in salt concentration, owing to their synthesis of stress-inducible compounds and regulation of plant hormones, which strengthens plant tolerance to different conditions (Ikram et al. 2024b; Phurailatpam & Mishra, 2020).

They are equally significant in disease suppression and biocontrol systems. They synthesize antibiotic substances, outcompete the pathogens for existence necessities, and stimulate the systemic acquired resistance in plants. That is why using endophytes enables the limitation of chemical pesticides in sustainable agriculture and proper plant growth (Strobel & Daisy, 2003). However, endophytic fungi are part of the microbial population in the soil environment and contribute to the total microbial community. Thus, at this moment, exposed diversities are invaluable for stabilizing and keeping the ecosystems functional. Endophytes are also involved with other microorganisms within the soil; even the density and activities of all microbes in the soil are affected by the presence of endophytes (Dastogeer et al., 2020). Such are the long-lasting impacts that endophytes make on the fertility of the concerned soil. Endophytes improve plant growth processes, accelerate the cycle of nutrients, and increase the resistance of plants to diseases, thus creating a system that will keep the soil productive for generations (Haider et al., 2024). This multifaceted approach to thinking about the soil puts endophytes in a critical position regarding sustaining and enhancing the quality of agricultural and natural environments (Muhammad et al., 2023). Knowledge concerning the potential of endophytes promotes better management and use of the soil, hence promoting the sustainability of the soil ecosystem in the agricultural industry.

2.3 Mechanisms of Endophyte-Plant Interactions

There are many ways through which endophytes associate and affect the host plants: colonization, signaling and re-connection, and mutualism. There are several ways through which endophytes can enter plant tissues. Some main ways endophytes can establish their presence in plant tissues include. These microorganisms can penetrate through natural structures such as stomata and lenticels, through injury in the form of a wound, and root tips. These entry points offer a way of accessing internal plant structures that can be colonized by endophytes, as observed by (Papik et al., 2020). Subsequently, endophytes travel through the plant's vascular systems or even move through the spaces between cells to colonize the various organs and tissues of the plant, such as the roots, stems, leaves, and seeds. This mobility is necessary in the context of their capacity to invade various plant areas and fix themselves firmly (Rosenblueth & Martínez-Romero, 2006). Hence, there is a need to understand the molecular communication mechanisms between endophytes and host plants for continued interaction. Endophyte and plant cells recognize and interact using chemical signals or molecules to control their interaction. These chemical signals are beneficial for the formation of endophytic association, and they have the most responsibility for the stability of the relation. For example, specific chemicals like flavonoids and exudates formed by plant roots can attract endophytes to begin symbiosis. Endophytes also affect gene expression in host plants, and as a result, different metabolic processes are affected. It also involves changes in the concentrations and synthesis of secondary metabolites that may improve the plant's capacity to endure several forms of stress (Sanchez-Contreras et al., 2007) (Fig. 3).

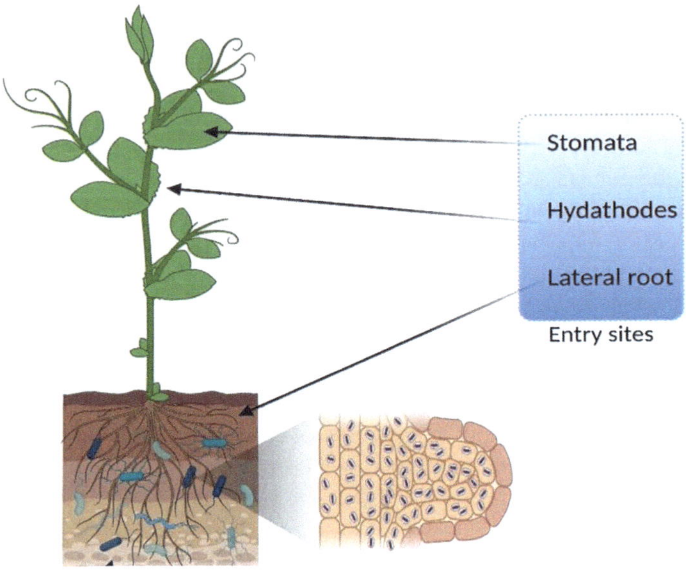

Fig. 3 A microscopic image of endophytes within plant tissues

The positive effects that host plants get from endophytes are valuable and many. Thus, endophytes improve nutrient uptake and remobilization in the photosphere-micro-ecosystem interface. Some of the ways that help enhance the fertility of the soil include nitrogen fixation and phosphate solubilization, which enhance the nutrient needs of the plant. Such activities are crucial, especially in nutrient-deficient soils where endophytes can significantly boost nutrient acquisition (Mukherjee et al., 2021). Also, a notable function of the endophytes is the improvement of plant tolerance to stress. They synthesize stress-induced metabolites, regulate plant hormone concentrations, and allow plants to withstand drought, salinity, and pathogen invasions. Such adaptations help to increase resistance to difficult conditions so necessary for the survival of plants. In addition, endophytes are involved in disease protection and biocontrol by synthesizing antifungal agents, which inhibit plant pathogens. These aspects make them partners in sustainable agriculture due to their ability to outcompete pathogens for resources, besides the fact that they call for the activation of systemic resistance in plants, thereby cutting on the use of chemically produced pesticides (Strobel & Daisy, 2003) (Fig. 4).

2.4 Roles of Endophytes in Soil Health

Nevertheless, the roles of endophytes concerning the soils are manifold and not limited by the fact that it is not only a key but also a decisive element of nutrient

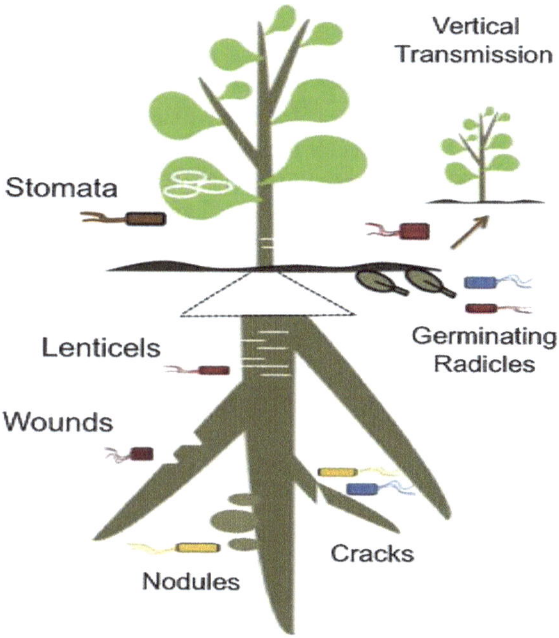

Fig. 4 The process of endophyte colonization in plant tissues, including entry points

cycling and the degree of improvement of soil structure and long-term fertility required for the signal of stressed levels. They are significant in the recycling of nutrients and their availability and possess the feature of nitrogen fixation solubilization of phosphorus and other such related processes. Nitrogen-fixing endophytes can also fix nitrogen in the atmosphere in another form that is readily helpful to plants, increasing the amount of nitrogen available in the soil. This biological nitrogen fixation helps ecosystems with low nitrogen status because it enhances the ground's fertility and promotes plant growth (Rana et al., 2020). Moreover, the endophytes are known to play a role in the process of phosphorus mobilization, which, in effect, enhances this nutrient's availability among plants. Soil strengtheners, PSPE, compose organic acids and enzymes to solubilize inorganic phosphate in the soil, fertilizing health and improving the plant growth's firmness (Mehta et al., 2020). They also actively participate in changing the nature or make-up and activities of the other microorganisms in the soil. Because of an interaction of the endophytes with other minor organisms in the soil, the endophytes contribute to microbial conglomerations and nutrient cycling. Such interactions can enhance the nutrient uptake by the plant and nutrient homeostasis within the soil mass (Koshila Ravi et al., 2019). Thus, they control practically all the exchange of nutrients and other substances and play a significant role in stabilizing the soil and its fertility.

Therefore, endophytes influence root trajectories and development regarding the structure and aggregation of the soil medium. They change the fundamental structures and thus augment the ground's solidity, decrease erosion incidences, and raise the water retention capability of the ground. Such changes may modify the stability of the architectural roots, give the plants a more erect position, and make them less vulnerable to soil erosion (Hartmann & Six, 2023). Endophytes also change root exudation, which determines the microbial load, the structure of the benefactor, and the soil structure. These root exudates, in one way or another, play signaling and nutritional roles for the physical and chemical modulating microbes in the soil, hence ensuring structural and healthy soil (Ma et al., 2018).

Moreover, endophytes enable assessment of the increase in plant stress and the production of stress metabolites and hormones. These metabolites could be osmoprotectants, antioxidants, and other related metabolites that help plants under stress factors such as water deficit, high salinity, and hot temperatures. Relative to endophytes' ability to alter the concentration of plant hormones, which is associated with plant growth and development and stress responses, endophytes contribute to improving plant health (Phurailatpam & Mishra, 2020). Also, endophytes play roles in disease management and biocontrol by producing compounds that are beneficial in eliminating pathogens and inducing SAR in plants. This biocontrol ability reduces chemical pesticide usage and supports Green Agriculture (Latz et al., 2018) (Fig. 5).

Therefore, endophytes enormously change the receptacle's soil fertility status; the significant changes profoundly affect the receptacle in the long run. The total communication between endophytes and the plants constantly enhances the substrate's standard, meaning the land is productive in the long run. Therefore, these continuing occupancies and actions of endophytes always help conserve soil ecosystems and improve the growth and yields of plants in the long term (Dubey et al., 2020). If such positive interactions can be harnessed, farming practices can be rendered effective

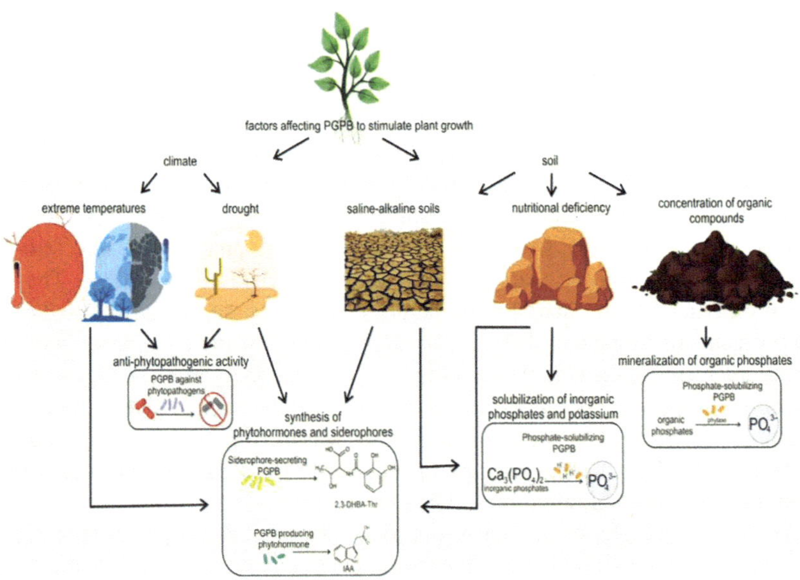

Fig. 5 Visual representation of how endophytes contribute to soil health

and resourceful, reducing the use of manufactured fertilizers and chemicals to the barest minimum. Therefore, endophytes contribute strongly to the integrated pest management of sustainable agricultural systems and environmental conservation.

2.5 Environmental Factors Shaping Endophyte Functionality

Therefore, it can be concluded that various aspects of the environment considerably influence the functionality of endophyte communities in the context of the soil ecosystem. Essentially, the kind and arrangement of multiple soil types influence the density and the variety of endophytes. Determination of the physical and chemical properties of the soil—either texture, pH, nitrate, or phosphorous concentrations results in the formation of endophytic niches that support the growth and activity of the several species or endophytic communities (Gomes et al., 2018). For example, endophytes from the sandy, loamy soil with a low fertility level may not be the same as endophytes associated with the clayey fertility-retentive type soil. Likewise, it can be concluded that the external medium, precisely the soil's pH level, can influence and dictate certain endophytes preferring specific conditions, such as an acidic environment, compared with the pH a community of neutral or alkaline-phallic microorganisms prefer. These soil factors influence the makeup of communities of endophytes and their relationship with the plants (Eykelbosh, 2014). The rhizosphere is home to endophytes that act vividly with the rest of the soil microorganisms. These interactions are thus dependent on the kind of soil and how different ideologies of microbiological organisms compete and cooperate. There is evidence that some endophytes can either be mutualistic with rhizosphere bacteria in the cycling of nutrients and health of the plant or be competitors that affect abundance and functionality. Endophytes' position, as explained above, is also influenced by the physical and chemical characteristics of the soil and the other living organisms, which is why the general microbial community influences the ecological attributes of endophytes (Hassani et al., 2018).

Seasonal conditions such as temperature and moisture play a critical role in the working of endophytes. The humidity and temperature affect endophytes' physiologic processes, development, and durability. In various climatic conditions, endophytes can change the temperature or humidity that unites and influences the host plant. For instance, endophytes in arid areas could have strategies to withstand high temperature and water stress, which will benefit their host plants regarding stress tolerance. On the other hand, endophytes in the regions characterized by humidity and temperateness can be most active during the rainy periods, playing the role of beneficial microorganisms regarding nutrient provisioning and protection against pathogens (Giauque et al., 2019). Availability also changes with endophyte kinds and their activity levels, which also fluctuate based on specific seasons of the year as influenced by temperature regime and humidity. These seasonal changes affect the onset and the degree of the advantages offered by endophytes toward their host plants (Fuchs et al., 2017). Endophyte-host plant interactions are also influenced by environmental conditions, which is one of the factors that cause specificity. In the coevolutionary process between host plants and endophytes, hosts and endophytes are mutually accommodating so that the two become highly specialized and serve a higher purpose. Some samples of neat explanations are: It has been postulated that environmental factors that are well conveyed by the host plant, including emphasized capability and climate, influence the development of host specificity since it favors endophytes that are well matched to the conditions experienced by the host plant (Biere & Tack, 2013). Indeed, plant metabolites that are affected by environmental conditions have been incriminated in the recruitment and regulation of the activity of endophytes. These metabolites function as chemical messengers for recruiting particular endophytes and thus determine the structure and roles of the endophytic consortia. Complementariness between plants and endophytes produces feedback when the environment changes to facilitate flexibility in the ever-changing environment (Khare et al., 2018) (Fig. 6).

Anthropogenic influences, which include land use and level of agriculture, are also involved in determining endophyte communities. Agricultural practices increase the selection pressures, leading to changes in the endophytes' functionality and diversity owing to the application of pesticides and fertilizers. For instance, using chemical fertilizer to increase nutrient availability promotes some selective endophytes that feed on such nutrients. In contrast, the use of pesticides reduces the population of these endophytes, thus altering their ecological roles (Yang et al., 2021). Conversion of land-use types like deforestation and urbanization affects the natural ecosystem, resulting in the loss of beneficial endophytic populations. Urbanization brings in

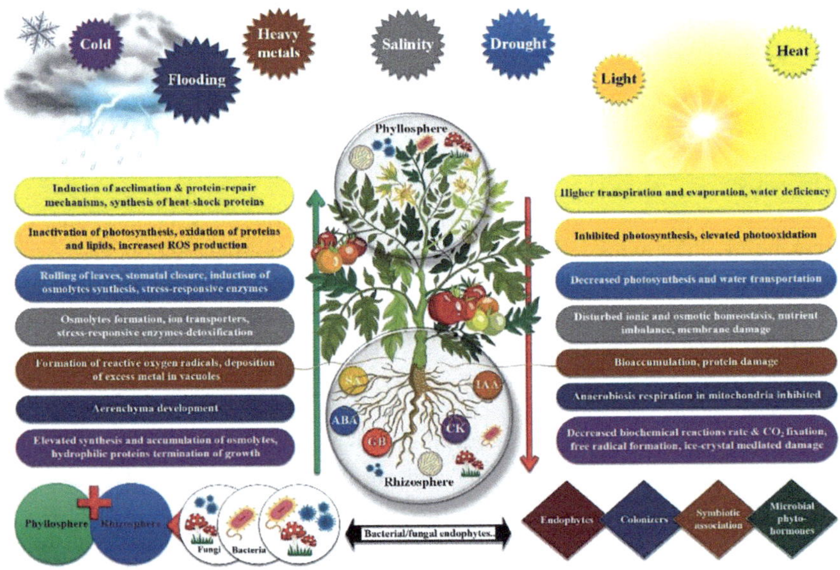

Fig. 6 Endophytes help plants cope with abiotic stresses

pollution and other factors in their form stimuli, including heavy metals and chemical residues, challenging the outlook of endophytes. These pollutants suppress the functionality of endophytes, interfere with their relationships with the host plants, and reduce the soil's productivity (Feng et al., 2017). The soil environment also affects the functionality of endophytes due to the microbial interactions in any ecosystem. Thus, the interaction of endophytic fungi with the other microorganisms in the soil, both positive and negative, defines the general condition and stability of the soil environment. Though mutualistic interactions may positively affect nutrient-cycling disease defense and plant growth, the antagonistic effects of pathogens can hamper their positive effects. Such knowledge is essential in formulating the best strategies that can be used in the conservation of soil health as well as the utilization of endophytes within the sustainable agricultural system (Albornoz et al., 2022). Candidate endophytes occur in the environment and typically trigger feedback cycles that impact both endophytes and the host plant. Whenever there is a change in the soil's physical and chemical properties or climate, the endophytic communities get stimulated, which determines the health of plants and the functionality of the soil ecosystem. Sophisticated interconnections are the essence of these mutualistic relationships that must be considered with multiple variables in assessing endophyte functionality (Tosi et al., 2021).

2.6 *Case Studies and Experimental Evidence*

Literature and experimental data paint a complete picture of the multiple functions that endophytes assume in improving plant health and soil. A good example is the nitrogen-fixing bacterial endophytes from leguminous plants. For instance, the Rhizobium genus of bacteria are endophytic beings that initiate mutualism with legume plants to boost nitrogen uptake. Research data prove that these sorts of bacteria help enhance the nitrogen content in plant tissues, provide healthier burgeoning, and raise soil fertility (Harman & Uphoff, 2019). Common approaches to the experimental isolation and characterization of these nitrogen-fixing endophytes can include root nodule collection, molecular probing, and determination of nitrogen content in both the plants and the soil and are essential tools, particularly in sustainable agriculture (Trujillo et al., 2015). Another critical research theme is phosphate-solubilizing fungal endophytes. Concerning these endophytes, the authors of a particular case study have demonstrated that Penicillium and Aspergillus spp. can enhance the solubility of phosphate, thus making it available to plants. Field studies show that plants that bear these mycorrhizal fungi have a higher capacity to absorb phosphorus, letting farmers minimize the use of phosphate fertilizers (Kumar et al., 2020). Chromatographic quantification of phosphorus solubilization is based on the analysis of soil and plant tissue samples; the application of this discovery in agriculture involves the reduction of chemical fertilizers, making them environmentally friendly.

Another potential area is research on endophytes that help improve crops' resistance to drought. Some bacterial and fungal endophytes were identified in a recent study that determined that specific endophytes enhance the capacity of crops like wheat and maize to use water efficiently and help the plants withstand drought stress. Drought experiments on plant water status, stomatal conductance, and physiological reactions have established these endophytes' ability to reduce water stress (Verma et al., 2021a). Thus, the results of this study imply that the practical use of drought-tolerant endophytes in agriculture might assist crops to adapt to current and future frequent and severe drought stresses. Another area of research that has found valuable applications involves the use of endophytes in the biocontrol of soil-borne pathogens, besides showing promise of becoming bio-fertilizers. A case study relating to endophytes, including Trichoderma spp. and Bacillus spp., has been proven to reduce diseases attracted by soil pathogens such as Fusarium and Pythium. Studies comparing the antagonistic nature of such endophytes, which involves data derived from a dual culture technique and pathogen inhibition assays, have pointed out their ability to promote the healthy growth of plants and, in the process, minimize the use of chemical pesticides (Rashad & Moussa, 2020). All these bio control measures are in harmony with the tenets of substantive agriculture in that they enhance the health wellness of crops with minimal chemical inputs.

This entirely explains why most of the endophyte and plant interactions are co-evolutionary. In one such study, these interactions were analyzed, and it was found that some endophytes were highly host plant-specific, and the preference was the result of co-evolution. There is also experimental proof regarding these associations' specificity and adjusted nature based on the cross-inoculation tests and molecular genetic data (Krasylenko et al., 2020). Such preferences can help choose the right endophytes for increasing the said plants and maximize the value of the improvement in different agricultural practices. Systematic investigations of the consequences of endophytes on the composition of the soil microbiota indicate that they persistently disrupt the earth's bacterial population. As far as soil health is concerned, a study that observed the microbial population on a crop during several growing seasons suggested that endophytic associations enhance microbial richness and resilience. Technical activities included metagenomics approaches and long-term soil health surveys, thus raising the importance of preserving microbial species and the genetic variation that ensures stability (Frindte et al., 2020). These findings highlight endophytes' critical role in sustaining soil health and productivity over the long term. Real-world applications of endophytes in agriculture showcase their practical benefits. Field trials have demonstrated that incorporating endophytes into agricultural practices can improve crop yields, reduce input requirements, and create more sustainable farming systems. Challenges in translating laboratory findings into field applications include variability in environmental conditions and the need for collaboration among researchers, farmers, and policymakers (Saikkonen et al., 2006).

2.7 *Future Perspective and Challenges*

Further possibilities for applying endophytes in agriculture and restoring natural ecosystems are still possible. However, there are some problems. Another promising area is the ability to select crop varieties that can be substantially improved by the beneficial endophytes while having higher availability, growth, yield, and stress tolerance. In the future, genomic and biotechnology tools may help detect and use the desired endophyte characters, which may improve the use of endophytes in sustainable agriculture (Ayilara et al., 2023). Besides, the incorporation of endophytes into agroecological management practices could minimize the use of chemical fertilizers and pesticides, which in turn contribute to the pollution of the environment and harm the soil structures (Barea, 2015). However, the following are the main challenges experienced in the execution of the proposed strategies: One occupational limitation, therefore, is that the effectiveness of endophytes cannot be constant because the environment in which they are applied is capable of varying in terms of soil types, climate conditions or the types of plants. Due to this flexibility, it is essential to understand the host–endophyte relationships globally depending on the level of host–endophyte interactions present in different environments (Omomowo, 2023). Also, extending laboratory results to field situations still poses a significant challenge. Field conditions have more fluctuations than laboratory conditions, thus upsetting the performance of endophytes, plus the fact that results differ (Khanam et al., 2022).

The fourth disadvantage often associated with this innovation is that it has the possibility of causing other, perhaps unforeseen, impacts on the environment. Bringing non-native endophytes to new geographic locations might further affect the microorganisms of the given area and the overall balance of ecosystems. Hence, proper ecological risk assessments should be conducted before the extensive implementation (Bard et al., 2024). Moreover, due to the complexity of microbial inoculants, it is necessary to address the regulation aspects of their usage in agricultural production internationally. Future research and development should also aim to establish the impact of endophytes on the condition of soil and microorganisms in the long run. Nonetheless, long-term research is essential to ensure that endophyte applications do not somehow harm the soil ecosystem in the long run (Mukherjee et al., 2021). Thus, microbiologists, agronomists, ecologist, and policymakers join their efforts to solve the complex issues and create a sound and effective strategy for further beneficial endophyte usage (Verma et al., 2022).

3 Conclusion

In exploring the vital role of endophytes in soil health and fertility, we've uncovered their significant contributions to nutrient cycling, soil structure, plant resilience, and ecosystem functionality. There are beneficial microorganisms residing inside plant tissues, which include nitrogen-fixing bacteria and phosphate-solubilizing fungi, minimizing the use of chemical fertilizers and, therefore, encouraging the use of efficient agriculture. They also change the rooting pattern, exudation soluble organic matter, and root channels to offer a sound and strong soil structure. Being able to create stress-related metabolites and adjust hormone concentrations, endophytes enable plants to sustain environmental conditions such as droughts and pathogens, which underlines endophytes' role in climate change resilience. Furthermore, endophytes are biocontrol agents, meaning they contribute to the decrease in the usage of chemical pesticides through antimicrobial production and pathogen competition. Due to environmental pressures, the specificity and adaptation in plant-endophyte relationships guarantee mutualism and, therefore, highlight the fact that practical implementations of this interaction in agriculture should be specific. Thus, endophytes sustain functionally diverse and robust microbial consortia that are the foundation of stable ecosystems even amidst the anthropogenic impact, including land use changes and pollution. Recent research and field experiments prove that endophytes positively impact crop production and sustainable agriculture. Realizing these mutually imposed benefactor-dependent dependencies is crucial for solving food insecurity, issues of environmental management, and climate change challenges.

References

Albornoz, F. E., Prober, S. M., Ryan, M. H., & Standish, R. J. (2022). Ecological interactions among microbial functional guilds in the plant-soil system and implications for ecosystem function. *Plant and Soil, 476*(1), 301–313.

Arnold, A. E. (2007). Understanding the diversity of foliar endophytic fungi: Progress, challenges, and frontiers. *Fungal Biology Reviews, 21*(2–3), 51–66.

Arora, J., & Ramawat, K. G. (2017). An introduction to endophytes. In *Endophytes: Biology and biotechnology: Volume 1* (pp. 1–23).

Ashraf, M., Shahzad, S. M., Abid, M., Mehmood, K., Aziz, A., Sarwar, A., Akhtar, N., & Mehran, M. (2022). Ionic homeostasis and growth characteristics of tomato (*Solanum lycopersicum* L.) grown with municipal wastewater by supplying silicon, farmyard manure and plant growth promoting rhizobacteria. *Silicon, 14*(18), 12855–12867.

Ayilara, M. S., Adeleke, B. S., & Babalola, O. O. (2023). Bioprospecting and challenges of plant microbiome research for sustainable agriculture, a review on soybean endophytic bacteria. *Microbial Ecology, 85*(3), 1113–1135.

Bard, N. W., Cronk, Q. C., & Davies, T. J. (2024). Fungal endophytes can modulate plant invasion. *Biological Reviews.*

Barea, J. (2015). Future challenges and perspectives for applying microbial biotechnology in sustainable agriculture based on a better understanding of plant-microbiome interactions. *Journal of Soil Science and Plant Nutrition, 15*(2), 261–282.

Berg, G., & Smalla, K. (2009). Plant species and soil type cooperatively shape the structure and function of microbial communities in the rhizosphere. *FEMS Microbiology Ecology, 68*(1), 1–13.

Bertola, M., Ferrarini, A., & Visioli, G. (2021). Improvement of soil microbial diversity through sustainable agricultural practices and its evaluation by-omics approaches: A perspective for the environment, food quality and human safety. *Microorganisms, 9*(7), 1400.

Bhardwaj, M., Kailoo, S., & Khan, S. S. (2023). Harnessing fungal endophytes for natural management: A biocontrol perspective. *Frontiers in Microbiology, 14*, 1280258.

Biere, A., & Tack, A. J. (2013). Evolutionary adaptation in three-way interactions between plants, microbes and arthropods. *Functional Ecology, 27*(3), 646–660.

Bishnoi, U. (2015). PGPR interaction: An ecofriendly approach promoting the sustainable agriculture system. In *Advances in botanical research* (Vol. 75, pp. 81–113). Elsevier.

Bokhari, A. (2018). *Unearthing Bacillus endophytes from desert plants that enhance growth of Arabidopsis thaliana under abiotic stress conditions.*

Chaparro, J. M., Sheflin, A. M., Manter, D. K., & Vivanco, J. M. (2012). Manipulating the soil microbiome to increase soil health and plant fertility. *Biology and Fertility of Soils, 48*, 489–499.

Compant, S., Saikkonen, K., Mitter, B., Campisano, A., & Mercado-Blanco, J. (2016). Editorial special issue: Soil, plants and endophytes. *Plant and Soil, 405*, 1–11.

Compant, S., Clément, C., & Sessitsch, A. (2010). Plant growth-promoting bacteria in the rhizo- and endosphere of plants: Their role, colonization, mechanisms involved and prospects for utilization. *Soil Biology and Biochemistry, 42*(5), 669–678.

Dastogeer, K. M., Tumpa, F. H., Sultana, A., Akter, M. A., & Chakraborty, A. (2020). Plant microbiome–an account of the factors that shape community composition and diversity. *Current Plant Biology, 23*, 100161.

Doty, S. L., Oakley, B., Xin, G., Kang, J. W., Singleton, G., Khan, Z., Vajzovic, A., & Staley, J. T. (2009). Diazotrophic endophytes of native black cottonwood and willow. *Symbiosis, 47*, 23–33.

Dubey, A., Malla, M. A., Kumar, A., Dayanandan, S., & Khan, M. L. (2020). Plants endophytes: Unveiling hidden agenda for bioprospecting toward sustainable agriculture. *Critical Reviews in Biotechnology, 40*(8), 1210–1231.

Eykelbosh, A. J. (2014). *Closing the carbon loop in sugarcane bioethanol: Effects of filtercake biochar amendment on soil quality, leaching and carbon utilization.* University of British Columbia.

Feng, N.-X., Yu, J., Zhao, H.-M., Cheng, Y.-T., Mo, C.-H., Cai, Q.-Y., Li, Y.-W., Li, H., & Wong, M.-H. (2017). Efficient phytoremediation of organic contaminants in soils using plant–endophyte partnerships. *Science of the Total Environment, 583*, 352–368.

Food, C. (2010). *The second report on the state of the world's plant genetic resources for food and agriculture* (Vol. 2). Food & Agriculture Org.

Fraser, D. (2008). Understanding animal welfare. *Acta Veterinaria Scandinavica, 50*(Suppl 1), S1.

Friesen, M. L., Porter, S. S., Stark, S. C., von Wettberg, E. J., Sachs, J. L., & Martinez-Romero, E. (2011). Microbially mediated plant functional traits. *Annual Review of Ecology, Evolution, and Systematics, 42*, 23–46.

Frindte, K., Zoche, S. A., & Knief, C. (2020). Development of a distinct microbial community upon first season crop change in soils of long-term managed maize and rice fields. *Frontiers in Microbiology, 11*, 588198.

Fuchs, B., Krischke, M., Mueller, M. J., & Krauss, J. (2017). Plant age and seasonal timing determine endophyte growth and alkaloid biosynthesis. *Fungal Ecology, 29*, 52–58.

Galindo-Castañeda, T., Lynch, J. P., Six, J., & Hartmann, M. (2022). Improving soil resource uptake by plants through capitalizing on synergies between root architecture and anatomy and root-associated microorganisms. *Frontiers in Plant Science, 13*, 577.

Giauque, H., Connor, E. W., & Hawkes, C. V. (2019). Endophyte traits relevant to stress tolerance, resource use and habitat of origin predict effects on host plants. *New Phytologist, 221*(4), 2239–2249.

Gomes, T., Pereira, J. A., Benhadi, J., Lino-Neto, T., & Baptista, P. (2018). Endophytic and epiphytic phyllosphere fungal communities are shaped by different environmental factors in a Mediterranean ecosystem. *Microbial Ecology, 76*, 668–679.

Gomiero, T., Pimentel, D., & Paoletti, M. G. (2011). Environmental impact of different agricultural management practices: conventional vs. organic agriculture. *Critical Reviews in Plant Sciences, 30*(1–2), 95–124.

Granjou, C., & Phillips, C. (2019). Living and labouring soils: Metagenomic ecology and a new agricultural revolution? *BioSocieties, 14*(3), 393–415.

Gupta, A., Gupta, R., & Singh, R. L. (2017). Microbes and environment. In *Principles and applications of environmental biotechnology for a sustainable future* (pp. 43–84).

Gurung, S. A., Rai, A. K., Sunar, K., & Das, K. (2023). Plant–endophyte interactions: A driving phenomenon for boosting plant health under climate change conditions. In *Microbial symbionts and plant health: trends and applications for changing climate* (pp. 233–263). Springer.

Haider, S., Bibi, K., Munyaneza, V., Zhang, H., Zhang, W., Ali, A., Ahmad, I.A., Mehran, M., Xu, F., Yang, C., Yang J., & Yang, C. (2024). Drought-induced adaptive and ameliorative strategies in plants. *Chemosphere*, 143134.

Hardoim, P. R., van Overbeek, L. S., & van Elsas, J. D. (2008). Properties of bacterial endophytes and their proposed role in plant growth. *Trends in Microbiology, 16*(10), 463–471.

Harman, G. E., & Uphoff, N. (2019). Symbiotic root-endophytic soil microbes improve crop productivity and provide environmental benefits. *Scientifica, 2019.*

Hartmann, M., & Six, J. (2023). Soil structure and microbiome functions in agroecosystems. *Nature Reviews Earth & Environment, 4*(1), 4–18.

Hassani, M. A., Durán, P., & Hacquard, S. (2018). Microbial interactions within the plant holobiont. *Microbiome, 6*, 1–17.

Herre, E. A., Mejía, L. C., Kyllo, D. A., Rojas, E., Maynard, Z., Butler, A., & Van Bael, S. A. (2007). Ecological implications of anti-pathogen effects of tropical fungal endophytes and mycorrhizae. *Ecology, 88*(3), 550–558.

Husseiny, S., Dishisha, T., Soliman, H. A., Adeleke, R., & Raslan, M. (2021). Characterization of growth promoting bacterial endophytes isolated from *Artemisia annua* L. *South African Journal of Botany, 143*, 238–247.

Hyde, K., & Soytong, K. (2008). The fungal endophyte dilemma. *Fungal Divers, 33*(163), e173.

Hyder, M., Lodhi, A. M., Wang, Z., Bukero, A., Gao, J., & Mao, R. (2024). Wolbachia interactions with diverse insect hosts: From reproductive modulations to sustainable pest management strategies. *Biology, 13*(3), 151.

Ikram, M., Mehran, M., ur Rehman, H., Ullah, S., Bakhsh, M. Z. M., Tahira, M., Maqsood, M. F. K., Rauf, A., Ghafar, S., Haider, K., Konuşkan, Ö., Toprak, Ç. C., Erden, Z., & Haider, K. (2024a). Mechanistic review of melatonin metabolism and signaling pathways in plants: Biosynthesis, regulation, and roles under abiotic stress. *Plant Stress*, 100685.

Ikram, M., Minhas, A., Al-Huqail, A. A., Ghoneim, A. M., Mahmood, S., Mahmoud, E., Tahira, M., Mehran, M., Khurram, M. F., Rauf, A., & Ali, W. (2024b). Promoting tomato resilience: Effects of ascorbic acid and sulfur-treated biochar in saline and non-saline cultivation environments. *BMC Plant Biology, 24*(1), 1053.

Jacob, J., Krishnan, G. V., Thankappan, D., & Amma, D. K. B. N. S. (2020). Endophytic bacterial strains induced systemic resistance in agriculturally important crop plants. In *Microbial endophytes* (pp. 75–105). Elsevier.

Khanam, Z., Gogoi, N., & Srivastava, D. N. (2022). Prospective on 2D nanomaterials for energy and environment: challenges, commercial aspect, and the future research endeavor. In *2D nanomaterials for energy and environmental sustainability* (pp. 267–329). Springer.

Khare, E., Mishra, J., & Arora, N. K. (2018). Multifaceted interactions between endophytes and plant: Developments and prospects. *Frontiers in Microbiology, 9*, 411604.

Koshila Ravi, R., Anusuya, S., Balachandar, M., & Muthukumar, T. (2019). Microbial interactions in soil formation and nutrient cycling. In *Mycorrhizosphere and pedogenesis* (pp. 363–382).

Krasylenko, Y., Sosnovsky, Y., Atamas, N., Popov, G., Leonenko, V., Janošíková, K., Sytschak, N., Rydlo, K., & Sytnyk, D. (2020). The European mistletoe (*Viscum album* L.): Distribution, host range, biotic interactions, and management worldwide with special emphasis on Ukraine. *Botany, 98*(9), 499–516.

Kumar, A., Teja, E. S., Mathur, V., & Kumari, R. (2020). Phosphate-solubilizing fungi: Current perspective, mechanisms and potential agricultural applications. In *Agriculturally important fungi for sustainable agriculture: Volume 1: Perspective for diversity and crop productivity* (pp. 121–141).

Lal, R. (2004). Soil carbon sequestration impacts on global climate change and food security. *Science, 304*(5677), 1623–1627.

Latz, M. A., Jensen, B., Collinge, D. B., & Jørgensen, H. J. (2018). Endophytic fungi as biocontrol agents: Elucidating mechanisms in disease suppression. *Plant Ecology & Diversity, 11*(5–6), 555–567.

Lugtenberg, B., & Kamilova, F. (2009). Plant-growth-promoting rhizobacteria. *Annual Review of Microbiology, 63*, 541–556.

Ma, N., Guo, P., Zhang, J., He, T., Kim, S. W., Zhang, G., & Ma, X. (2018). Nutrients mediate intestinal bacteria–mucosal immune crosstalk. *Frontiers in Immunology, 9*, 5.

Malinowski, D. P., & Belesky, D. P. (2000). Adaptations of endophyte-infected cool-season grasses to environmental stresses: Mechanisms of drought and mineral stress tolerance. *Crop Science, 40*(4), 923–940.

Mehran, M., Ashraf, M., Shahzad, S. M., Shakir, M. S., Azhar, M. T., Ahmad, F., & Alvi, A. (2023). Growth, yield and fiber quality characteristics of Bt and non-Bt cotton cultivars in response to boron nutrition. *Journal of Cotton Research, 6*(1), 1.

Mehta, S., Gogna, M., Singh, B., Patra, A., Singh, I. K., & Singh, A. (2020). Silicon: A plant nutritional "non-entity" for mitigating abiotic stresses. In *Plant stress biology: Strategies and trends* (pp. 17–49).

Mengistu, A. A. (2020). Endophytes: Colonization, behaviour, and their role in defense mechanism. *International Journal of Microbiology, 2020.*

Minhas, A., Saeed, M. S., Ehsan, A., Murtaza, G., Akram, M., Mehran, M., Aurangzeb, M., Elah, N. N., Iqbal, R. K., & Syed, A. (2024). Response of integrated use of humic acid and chemical fertilizer on growth and yield of rice crop (*Oryza sativa* L.) in calcareous soil. *Pakistan Journal of Botany, 56*(3), 879–887.

Mori, A. S., Furukawa, T., & Sasaki, T. (2013). Response diversity determines the resilience of ecosystems to environmental change. *Biological Reviews, 88*(2), 349–364.

Muhammad, M., Waheed, A., Wahab, A., Majeed, M., Nazim, M., Liu, Y. -H., Li, L., & Li, W. -J. (2023). Soil salinity and drought tolerance: An evaluation of plant growth, productivity, microbial diversity, and amelioration strategies. *Plant Stress*, 100319.

Mukherjee, A., Bhowmick, S., Yadav, S., Rashid, M. M., Chouhan, G. K., Vaishya, J. K., & Verma, J. P. (2021). Re-vitalizing of endophytic microbes for soil health management and plant protection. *3 Biotech, 11*, 1–17.

Nair, D. N., & Padmavathy, S. (2014). Impact of endophytic microorganisms on plants, environment and humans. *The Scientific World Journal, 2014.*

Omomowo, I. O. (2023). *Improving cowpea production in north west province using endophytic bacteria and fungi.* South Africa: North-West University.

Papik, J., Folkmanova, M., Polivkova-Majorova, M., Suman, J., & Uhlik, O. (2020). The invisible life inside plants: Deciphering the riddles of endophytic bacterial diversity. *Biotechnology Advances, 44*, 107614.

Partida-Martínez, L. P., & Heil, M. (2011). The microbe-free plant: Fact or artifact? *Frontiers in Plant Science, 2*, 100.

Philippot, L., Raaijmakers, J. M., Lemanceau, P., & Van Der Putten, W. H. (2013). Going back to the roots: The microbial ecology of the rhizosphere. *Nature Reviews Microbiology, 11*(11), 789–799.

Phillips, M. (2017). *Mycorrhizal planet: How symbiotic fungi work with roots to support plant health and build soil fertility.* Chelsea Green Publishing.

Phurailatpam, L., & Mishra, S. (2020). Role of plant endophytes in conferring abiotic stress tolerance. In *Plant ecophysiology and adaptation under climate change: Mechanisms and perspectives II: Mechanisms of adaptation and stress amelioration* (pp. 603–628).

Porras-Alfaro, A., & Bayman, P. (2011). Hidden fungi, emergent properties: Endophytes and microbiomes. *Annual Review of Phytopathology, 49*, 291–315.

Pretty, J. (2012). Sustainable intensification in Africa. In *Sustainable intensification* (pp. 3–4). Routledge.

Pretty, J. (2008). Agricultural sustainability: Concepts, principles and evidence. *Philosophical Transactions of the Royal Society B: Biological Sciences, 363*(1491), 447–465.
Rana, K. L., Kour, D., Kaur, T., Devi, R., Negi, C., Yadav, A. N., Yadav, N., Singh, K., & Saxena, A. K. (2020). Endophytic fungi from medicinal plants: biodiversity and biotechnological applications. In *Microbial endophytes* (pp. 273–305). Elsevier.
Rashad, Y. M., & Moussa, T. A. (2020). Biocontrol agents for fungal plant diseases management. In *Cottage industry of biocontrol agents and their applications: Practical aspects to deal biologically with pests and stresses facing strategic crops* (pp. 337–363).
Reganold, J. P., & Wachter, J. M. (2016). Organic agriculture in the twenty-first century. *Nature Plants, 2*(2), 1–8.
Rodriguez, R., White Jr, J., Arnold, A., & Redman, R. S. (2009). Fungal endophytes: Diversity and functional roles. *New Phytologist, 182*(2), 314–330
Rodriguez, P. A., Rothballer, M., Chowdhury, S. P., Nussbaumer, T., Gutjahr, C., & Falter-Braun, P. (2019). Systems biology of plant-microbiome interactions. *Molecular Plant, 12*(6), 804–821.
Rosenblueth, M., & Martínez-Romero, E. (2006). Bacterial endophytes and their interactions with hosts. *Molecular Plant-Microbe Interactions, 19*(8), 827–837.
Saikkonen, K., Lehtonen, P., Helander, M., Koricheva, J., & Faeth, S. H. (2006). Model systems in ecology: Dissecting the endophyte–grass literature. *Trends in Plant Science, 11*(9), 428–433.
Saikkonen, K., Wäli, P., Helander, M., & Faeth, S. H. (2004). Evolution of endophyte–plant symbioses. *Trends in Plant Science, 9*(6), 275–280.
Sanchez-Contreras, M., Bauer, W. D., Gao, M., Robinson, J. B., & Allan Downie, J. (2007). Quorum-sensing regulation in rhizobia and its role in symbiotic interactions with legumes. *Philosophical Transactions of the Royal Society B: Biological Sciences, 362*(1483), 1149–1163.
Schulz, B., & Boyle, C. (2005). The endophytic continuum. *Mycological Research, 109*(6), 661–686.
Singh Saharan, G., Mehta, N. K., & Meena, P. D. (2023). Molecular detection and identification of pathogens, pathotypes, and genes. In *Genomics of crucifer's host-pathosystem* (pp. 503–593). Springer.
Smith, P., Martino, D., Cai, Z., Gwary, D., Janzen, H., Kumar, P., McCarl, B., Ogle, S., O'Mara, F., Rice, C., Scholes, B., Sirotenko, O., Howden, M., McAllister, T., Pan, G., Romanenkov, V., Schneider, U., Towprayoon, S., Wattenbach, M., & Smith, J. (2008). Greenhouse gas mitigation in agriculture. *Philosophical Transactions of the Royal Society B: Biological Sciences, 363*(1492), 789–813.
Strobel, G., & Daisy, B. (2003). Bioprospecting for microbial endophytes and their natural products. *Microbiology and Molecular Biology Reviews, 67*(4), 491–502.
Suryanarayanan, T., & Ravishankar, J. (2023). They too serve who tolerate and survive: The need to study halotolerant fungi to appreciate their role in saline ecosystems. *Fungal Biology Reviews, 46*, 100328.
Tilman, D., Cassman, K. G., Matson, P. A., Naylor, R., & Polasky, S. (2002). Agricultural sustainability and intensive production practices. *Nature, 418*(6898), 671–677.
Torres-Rodriguez, J. A., Reyes-Pérez, J. J., Quiñones-Aguilar, E. E., & Hernandez-Montiel, L. G. (2022). Actinomycete potential as biocontrol agent of phytopathogenic fungi: Mechanisms, source, and applications. *Plants, 11*(23), 3201.
Tosi, M., Gaiero, J., Linton, N., Mafa-Attoye, T., Castillo, A., & Dunfield, K. (2021). Bacterial endophytes: diversity, functional importance, and potential for manipulation. In *Rhizosphere biology: Interactions between microbes and plants* (pp. 1–49).
Trujillo, M. E., Riesco, R., Benito, P., & Carro, L. (2015). Endophytic actinobacteria and the interaction of micromonospora and nitrogen fixing plants. *Frontiers in Microbiology, 6*, 162488.
Van Der Heijden, M. G., Bardgett, R. D., & Van Straalen, N. M. (2008). The unseen majority: Soil microbes as drivers of plant diversity and productivity in terrestrial ecosystems. *Ecology Letters, 11*(3), 296–310.

Verma, A., Shameem, N., Jatav, H. S., Sathyanarayana, E., Parray, J. A., Poczai, P., & Sayyed, R. (2022). Fungal endophytes to combat biotic and abiotic stresses for climate-smart and sustainable agriculture. *Frontiers in Plant Science, 13*, 953836.

Verma, H., Kumar, D., Kumar, V., Kumari, M., Singh, S. K., Sharma, V. K., Droby, S., Santoyo, G., White, J. F., & Kumar, A. (2021a). The potential application of endophytes in management of stress from drought and salinity in crop plants. *Microorganisms, 9*(8), 1729.

Verma, S. K., Sahu, P. K., Kumar, K., Pal, G., Gond, S. K., Kharwar, R. N., & White, J. F. (2021b). Endophyte roles in nutrient acquisition, root system architecture development and oxidative stress tolerance. *Journal of Applied Microbiology, 131*(5), 2161–2177.

Wani, Z. A., Ashraf, N., Mohiuddin, T., & Riyaz-Ul-Hassan, S. (2015). Plant-endophyte symbiosis, an ecological perspective. *Applied Microbiology and Biotechnology, 99*, 2955–2965.

Yang, T., Lupwayi, N., Marc, S.-A., Siddique, K. H., & Bainard, L. D. (2021). Anthropogenic drivers of soil microbial communities and impacts on soil biological functions in agroecosystems. *Global Ecology and Conservation, 27*, e01521.

Plant Microbe Interaction for Sustainable Agriculture

Muhammad Ashfaq, Iqra Mushtaq, Samah Bashir Kayani, Muhammad Shaaban, Muhammad Mahmood Ahmed, and Mirza Abid Mehmood

Abstract Soil salinization threatens food security and ecosystem health, and is a major contributor to the degradation of many ecosystems globally. It has been one of the primary challenges that have significantly affected the processes, structure and functions of worldwide ecosystems including plant production, organic matter decomposition, soil nutrient cycling and biodiversity. By climate change, projected sea-level rises (SLR) and process of soil salinization in coastal areas will have a direct impact on population migration by lowering the food production and negatively effecting regional socioeconomic growth. Plant–microbe interactions must be coordinated in order to improve plant growth and maintain optimal soil conditions. Rhizosphere interactions based on complex exchange are more complex than those take place in non-rhizosphere soil. The diversity of soil microorganisms is incredibly high, and they are involved in many important ecological processes. The use of phosphate-solubilizing microorganisms, nitrogen-fixing bacteria, the formation of microbial biofilms on non-legume roots, and the interaction of arbuscular mycorrhizal fungi with plants all have enormous potential as bio fertilizers because they improve phosphorus and nitrogen bioavailability for the plant, support sustainable agriculture, improve soil fertility, and increase crop yield.

Keywords Biofilms · Climate change · Sustainable agriculture · Soil salinization

M. Ashfaq · I. Mushtaq · M. A. Mehmood (✉)
Plant Pathology, Institute of Plant Protection, Muhammad Nawaz Shareef University of Agriculture, Multan, Punjab, Pakistan
e-mail: abid.mehmood@mnsuam.edu.pk

S. B. Kayani
Government Graduate College (W) Mumtazabad, Multan, Punjab, Pakistan

M. Shaaban
College of Agriculture, Henan University of Science and Technology, Luoyang, China

M. M. Ahmed
Department of Bioinformatics, Institute of Biochemistry, Biotechnology and Bioinformatics, The Islamia University of Bahawalpur, Bahawalpur, Punjab, Pakistan

M. Shaaban (ed.), *Soils and Sustainable Agriculture*, Frontier Studies in Soil Science,
https://doi.org/10.1007/978-3-031-91114-9_20

1 Introduction

Plants have a dense population with a variety of microorganisms, both below ground and above ground, which serve mutualistic purposes (Vishwakarma et al., 2020). The microbes that colonize plants are classified in different categories as first of all epiphytes, which live on the surface of plant, endophytes, which live inside plant tissues, phyllospheric, which live on the surface of leaf, and rhizospheric, which live in the soil around the roots. Rhizosphere is considered to be the most dynamic among all of them because it plays an important role in the growth and nutritional status of plants (Bakker et al., 2013; Lakshmanan et al., 2014; Mendes et al., 2013). The limited area of soil that directly surrounds a plant's roots and is affected by different microbes and root secretions is known as the rhizosphere. The basis of the underground system is made up mostly of soil, plant roots, lateral growth, and root hairs and have relationship with a wide range of microbial diversity (Bandyopadhyay et al., 2017; Panke-Buisse et al., 2015). The plant root microbiome is made up of the entire system in which plant roots interact with the rhizobiome (Philippot et al., 2013).

Understanding the biology of root architect and different related microbiome as an interactome becomes important as it helpful in understanding the complex relationships, extremely varied speciation, and structural composition of communities. Numerous interactions, including those between microorganisms and plants roots, are made possible by the entwining nature of hosts and microbes. Furthermore, an important method to understand how plants behave in response to such kinds of stimuli is through an understanding of root–nematode interactions. The ecological features of plant hosts and related microbes are inseparable, serving as holobionts or metaorganisms (Hacquard, 2016; Hacquard & Schadt, 2015). Studies are carried out to investigate the mutual relationship between plants and microorganisms while understanding related mechanisms for enhanced crop production, due to developments in the technology for the identification and analysis of proteome and genome (Bakker et al., 2013; Oldroyd, 2013). Long-term stability and crop productivity of an agroecosystem can be achieved by utilizing different features that promote the community of microbes in the rhizospheric soil and understanding its impact on plants the plant health (Knapp et al., 2018; Quiza et al., 2015).

2 The Use of Microorganisms to Facilitate the Growth of Plants in Saline Soils

One of the major issues that has significant effects on the composition, dynamics, and operations of global ecosystems is soil salinization (Eswar et al., 2021), consist of soil nutrient cycling (van Dijk et al., 2015; Zhou et al., 2017), decomposition of organic matter (Rath & Rousk, 2015), biodiversity (Whittle et al., 2022) and plant productivity (Enebe & Babalola, 2018; Rozema & Flowers, 2008). It is anticipated that global salinization of soil will increase under future climate change scenarios (Hassani et al., 2021). The production of crops and human health are negatively impacted by soil degradation brought on by rising salinization (Hassani et al., 2020). Climate change has a greater impact on soil salinization in coastal areas and directly affect population movement by reducing food productivity (Chen & Mueller, 2018). There are many different reasons of salinization of soil, and they can arise in any type of climate (Ivushkin et al., 2019; Zaman et al., 2018). Typically, soil salinization has two categories primary salinization and secondary salinization, depending on how much natural and human-induced processes predominate. The parent material, local climate, soil characteristics and groundwater are the main factors that influence soil primary salinization. Increased soil salinity can also result from rising groundwater levels or insufficient rainfall resulting in higher evapotranspiration (Daliakopoulos et al., 2016). Humans' improper usage of soil causes secondary salinization, also known as environmental stress. Salt buildup on the soil surface can be easily caused by saline irrigation when combined with inadequate drainage. Seawater intrusion in coastal locations is expected to rise in future climate change scenarios due to rising temperatures and increased development (Eslami et al., 2021). Freshwater marsh wetlands may become salinized as a result of drought-induced saline intrusion, which can impact the biological processes and functions of wetland ecosystems (such as carbon dynamics) (Wang et al., 2022).

To prevent future soil deterioration, specific actions might be guided by a deeper comprehension of the causes of salinization. The study of environmental microbiology has recently focused on the response and adaptation of various soil microorganisms to changing soil salinization conditions, as well as their execution of important activities. Soil microorganisms have a critical role in plant growth promotion and influencing small-scale processes within ecosystems (Lanza et al., 2019; Rath et al., 2019; Sun et al., 2022). In the future, soil microorganisms may be used as defenders against detrimental effects as well as climate change on an ecosystems (Averill et al., 2022). Therefore, it is necessary to elucidate the possible impact of salinization on the diversity of microbes in soil and analyze the changes in ecosystem services facilitated by microorganisms due to increasing salinity. Consequently, in various ecosystems, we must clarify the possible impacts of salinization on the composition of microbes in soil and interpret the change in microbial-mediated ecosystem services brought about due to the rising salinity level in soil (Mao et al., 2018; Prăvălie, 2021). Some research suggest that raising the salinity levels within a specific range due to sea level rise can actually stimulate the growth of plants and so increase the amount of carbon stored in coastal wetlands (Kirwan & Mudd, 2012; Rogers et al., 2019; Wang et al., 2021). Therefore, to maintain the health of the soil, which is crucial for the supply of ecological services, specific management approaches should be used to different ecosystems considering the expected future alterations in soil salinity. It is important that the vital roles played by microorganisms in saline soils be thoroughly comprehended and exposed for the benefit of ecosystems.

This chapter provides a complete overview about the advanced knowledge of microorganisms' diversity in soil, discuss about their communities' structure, and the impact of these microorganisms on biogeochemical processes. Our focus is on optimizing the use and improvement microbial resources of soil in saline-alkaline soils. This is done to accelerate crop growth, enhance plant development, and facilitate restoration of soil within a relatively brief time. Priorities and future prospects for study were also proposed.

2.1 Soil Microbial Diversity and Community in Saline Soils

There are context or habitat-specific trends in the response of diversity of microbes in soil along gradients of increasing salinity. In molecular ecology, current developments particularly the creation of several "omics" technologies, may offer insight on how various microbial taxa adapt or evolve in response to rising soil salinity stress. Early studies indicated that globally, salinity was the dominant environmental regulator of microbial community makeup, such as overcoming pH, physicochemical variables, temperature (Lozupone & Knight, 2007), and different type of saline soils were suitable to progress novel diversity of microbes. However, at the local level, the effect of salinity on microbial communities is less significant, especially in environments that are already abundant in salt (Hollister et al., 2010).

However, at the local level, salinity has a less significant effect on microbial populations, particularly in salt-rich environments (Dini-Andreote et al., 2016). The physical structure, pH, and salinity of the soil were also determined to together explain the composition of the soil bacterial community in this salt marsh chronosequence (Dini-Andreote et al., 2014). Different soil microbial species react in different way against the salt stress (Lozupone & Knight, 2007). According to reports, the populations of archaea in hypersaline soils were strongly associated with both electrical conductivity and soil sodium content, indicating that soil salinity was not the factor limiting the growth of bacterial communities in these soils. A substantial negative association was found among microbial diversity, phylogenetic diversity (PD), salinity level and soil richness (which ranged from 0 to 2.5 mS cm^{-1}) in a desert environment. Salinity level of soil exerted significant pressure on various bacterial communities (Zhang et al., 2019). In an estuarine salinity gradient, the brackish marsh with moderate salinity had the high soil fungal richness; nevertheless, the influence of plants on fungal diversity was significant (Mohamed & Martiny, 2011). Variations in how different studies' diversity in microbes responded to salt stress should be linked to the microorganisms' examined communities' composition. Fungi are thought to be more resistant to osmotic stress as compared to bacteria. However, the growing role of fungus at increased salinity levels has not been adequately described (Rath & Rousk, 2015). Soil salinization-induced land degradation typically results in a change in plant variety, which influences the composition of microbes. The Shannon diversity of soil bacteria and fungus decreased as a result of grassland degradation caused by soil salinization.

Additionally, it was shown that fungi had a greater impact on the degree of change compared to bacteria (Wu et al., 2021). In Germany, floodplain of the Werra River where the soil is acidic and contains a significant amount of fungal biomass, the microbial community exhibited a shift towards higher abundance of prokaryotes when soil salinity levels increased (Sardinha et al., 2003). When analyzing patterns of response to salt stress in different microbial species, background of soil texture and vegetation types of plant must be considered together. Soil contains a very broad range of microbial organisms with varying physiological features. A large number of salt-tolerant microbes can be found in a variety of locations. In polar deserts, Firmicutes can largely substitute Actinobacteria and Acidobacteria in highy salted soils (Van Horn et al., 2014), Firmicutes contain Gram-positive cell walls and may generate spores, making them resistant to severe environmental condition (Schimel et al., 2007). In salty soils, the bacillus genera alloca Firmicutes have also been confirmed to produce halophilic enzymes (Liszka et al., 2012). In high saline soils bacteroidetes have been discovered at high level (Canfora et al., 2014). Some renowned halotolerant bacteria categorized into Verrucomicrobia, Bacteroidetes, Alphaproteobacteria, and Gammaproteobacteria (Yuan et al., 2016). In degraded grasslands, these taxa were enriched on hypersaline soils (Wu et al., 2021) and coastal wetland areas (Zhang et al., 2021a, 2021b). A previous research, in desert soils exibited that microbial taxa that belong to Halobacteria, Actinobacteria, Proteobacteria and Bacteroidetes, exhibit a niche preference for high salinity (Zhang et al., 2019). Some ectomycorrhizal fungi, notably Lactarius and Tomentella Phialocephala, were common in high saline soils (Thiem et al., 2018).

2.2 *Mechanism of Microbial Response to Soil Salt Stress*

Soil bacteria can deal with salt stress by changing their physiological processes within a certain range of changes in salinity. Some halophilic bacteria can actively absorb ions of salt for increase in osmotic pressure within the cell when the salinity outside rises. Often, they do this while also releasing Na^+ ions from the cell (Oren, 2008). Instead, microbes can regulate the osmotic pressure within and outside of the cell through the production of low-molecular-weight organic compounds in microbial cells, which include certain carbohydrates and amino acids (Empadinhas & da Costa, 2008; Rath & Rousk, 2015). Undoubtedly, both osmo-adaptation mechanisms need a lot of energy. Furthermore, in order to survive in salt marshes with regular fluctuations in salinity, microorganisms will also hibernate for some time (Murphy et al., 2020). In salt marsh sediments, the percentage of latent microbial species can reach over half or even higher due to significant environmental disturbances such nutrient and oxygen concentrations (Kearns et al., 2016). Additionally, it has been shown that freshwater environments exhibit higher rates of bacterial dormancy than hypersaline lakes. This could perhaps be because these extremophiles require less dormancy to flourish and survive in the hypersaline environment (Aanderud et al., 2016). Indeed, for certain intolerant microbes, high salt stress results in cell lysis and death.

3 The Use of Phosphate-Solubilizing Microbes for Plant Growth Promotion

For plant growth and development phosphorus is an essential macronutrient (Maharajan et al., 2018; Santana et al., 2016). It accounts for 0.2–0.8% of plant dry weight (Sharma et al., 2013) and can be found in coenzymes, enzymes, nucleotides, phospholipids and nucleic acids (Nesme et al., 2018). For plant, phosphorus is the second most important macronutrient, following nitrogen. In soil, typical phosphorus level is around 0.05% (by weight), from which only 0.1% is available for the use of plant (Zhu et al., 2011). Phosphorus can be absorbed by plant roots as orthophosphates ($H_2PO_4^-$ or HPO_4^{2-}), despite the fact that soil contains micromolar amounts of these compounds (Ai et al., 2009). Calcium phosphates, insoluble iron, calcium and aluminium, make up the majority of phosphates in soil.

Microorganisms present in soil play a significant role in the biological processes and absorption of nutrients by plants. There are bacteria that can change insoluble soil phosphorus into soluble form so that plants can use it. Mineralization and phosphate solubilization microorganisms (PSM) seem to be the only ways to give plants more phosphorus besides using synthetic fertilizers. Microorganisms that dissolve and mineralize phosphorus are known as phosphorus-solubilizing microorganisms. They live in soil and the rhizosphere and have been shown to effectively remove phosphorus from the soil (Bhattacharyya & Jha, 2012). Microorganisms that dissolve phosphate are becoming more and more common because they are good for the environment, cheap, and work very well (Khan et al., 2009b). Phosphate bio fertilizers are a good way to improve crop yields and food production. Phosphate-solubilizing bacteria (PSB) can help plants grow by improving the effectiveness of biological nitrogen fixation and synthesis of phytohormone, as well as enhancing the availability of micronutrients like iron and zinc. Adding phosphate-solubilizing microorganisms to both pots and fields has been shown to boost output and phosphorus uptake. Environmental as well as economic factors depend on plants absorbing and using phosphates. The effective use of soil microorganisms for the solubilization of phosphates is widely acknowledged as a highly promising area of research. The practice of inoculating soil or crops with PSB is recognized as a viable method for enhancing the absorption of plant phosphorus (Fig. 1).

Phosphates are soluble in vitro by a wide variety of fungus and soil microorganisms. Bacteria that solubilize phosphate can aid plants in absorbing phosphorus from the soil (Zhu et al., 2011) and dissolve the insoluble inorganic phosphorus (Sharma et al., 2013). Salt-tolerant soil microbes may dissolve insoluble phosphates, facilitating the establishment of agriculture on salty and alkaline soils (Zhu et al., 2011). According to considerable research (Delavaux et al., 2017), mycorrhizal fungi have the capacity to solubilize phosphates. They have also been shown to enhance crop nutrition by expanding the amount of soil that can absorb phosphates by increasing the volume of rhizosphere (Browne et al., 2009). PSB which are frequently found in the plant rhizosphere, are among the other soil microorganisms (Rodríguez and Fraga, 1999). Phosphate-solubilizing fungi comprise approximately 0.1–0.5% of the

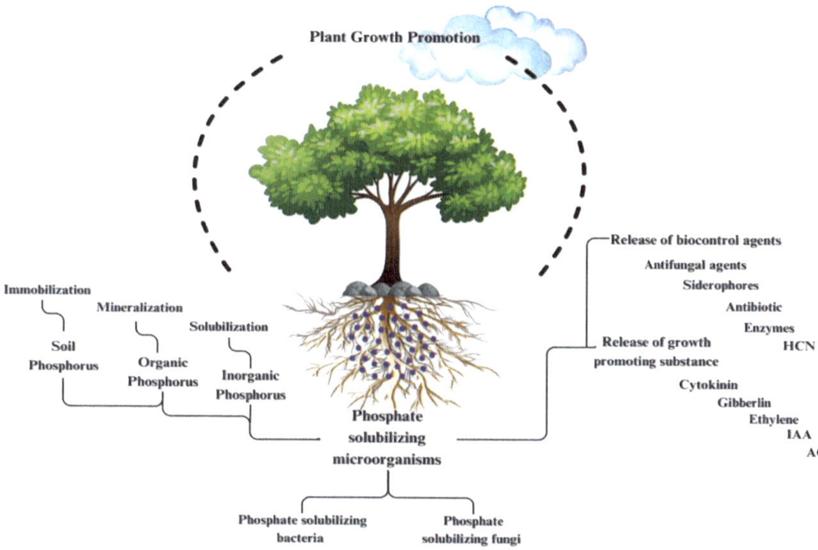

Fig. 1 Role of Phosphate solubilizing microorganisms for plant growth promotion

overall microbial population, while PSBs make up 1–50%, according to the literature. There are many phosphate-solubilizing bacteria, and the species makeup changes depending on the soil. The bulk of PSB have been identified in the metabolically active rhizospheres of different plants (Kalayu, 2019).

3.1 Biodiversity of P Solubilizers

In the natural soil environment, numerous microorganisms, including fungi, can release P from soil via solubilization as well as mineralization (Alori et al., 2017b; Bhattacharyya & Jha, 2012). It is predicted that in soil 50% of bacteria can solubilize *P* and various rhizobacteria strains have been recognized and studied in depth for their *P* solubilizing ability (Sharma et al., 2013) found that these organisms are prevalent, although their abundance and capacity to solubilize phosphorus vary based on the soil conditions (Awais et al., 2019; Chen et al., 2006; Kalayu, 2019). Bacteria can be isolated from rhizospheres, and non-rhizosphere soils (Sharma et al., 2013; Zaidi et al., 2009).

However, they are recognized for their heightened metabolic activity and superior ability to solubilize phosphorus in the rhizospheres of plants (Kaur & Purewal, 2019; Khan et al., 2009a; Rafi et al., 2019; Vessey, 2003). PSB are found in practically all soils, however their abundance varies with climatic conditions and the type of soil (Rafi et al., 2019). *Bacillus*, *Pseudomonas Agrobacterium*, (Babalola & Glick, 2012), *Enterobacter*, *Rhizobium* (Zaidi et al., 2009), *Burkholderia* sp., *Achromobacter* sp.,

Aerobactor aerogenes and *caligenes* sp. (Pande et al., 2017) are the most prevalent plant root dwelling *P* solubilizers. Symbiotic nitrogenous rhizobia present in leguminous plants are frequently associated with the root nodules and can solubilize P (Bechtaoui et al., 2019; Charana Walpola and Yoon, 2013; Khan et al., 2009a; Pande et al., 2017; Qin et al., 2011; Zaidi et al., 2009). Particular species have been demonstrated to create a variety of organic acids that are strongly related with *P* solubility. For example, (Khalid et al., 2020) determined the *P* solubilization capability of eight rhizobia by producing organic acids while profiling rhizobia isolated from *Arachis hypogaea* growing in stressful settings.

3.2 Constraints in Using Phosphate Fertilizers

3.2.1 Constraints in Biofertilizer Use

Despite the low cost, environment friendly, and benefits of biofertilizers, farmer acceptance of this organic input has been poor. Farmers' extensive use of biofertilizers is hampered by a number of production, marketing, and field-related challenges (Meena et al., 2023).

3.2.2 Production Constraints

Raw Material

Normally, carrier-based inoculants containing efficient microbes are used to prepare bio fertilizers. For the purpose of inoculating soil with bio fertilizer, granular carrier materials such as peat, perlite, charcoal, etc. are often recommended (Saif et al., 2021). Small and marginal farmers have limited access to the availability of these carrier materials for treating soil and seeds.

Strains Differ by Agroclimate

Most biofertilizer strains are soil and agroclimate specific in addition to being crop specific. One of the main obstacles to using biofertilizer is the absence of strains suited to a given area. This limits their wide and ideal use while maintaining predicted performance (Kumar et al., 2022).

3.2.3 Biological Constraints

The bio-inoculants may contain antagonistic or inefficient strains, and it is typically difficult to remove these strains from the bio-inoculant. Along with being able to compete with other strains, the selected strains should also be able to live in broth and inoculant carriers, as well as be capable of N-fixing and nutrient mobilizing under a variety of environmental conditions (Kumar & Singh, 2021). This has a significant impact on the targeted microorganism's bio fertilizer efficiency.

3.2.4 Technical Constraints

The propensity of bio fertilizers to change during fermentation drives up production and quality control expenses. To lessen these undesirable alterations, a wide spectrum of study is required (Vassileva et al., 2021).

3.2.5 Economic Constraints

The effective use of advanced technology instruments and equipment is necessary for the manufacturing of high-quality products. The production of a product free of contamination is questionable in the absence of these facilities. Furthermore, a barrier to the widespread use of bio fertilizers is the shortage of trained staff in production sites and insufficient training in manufacturing techniques (Basu et al., 2021; Raimi et al., 2021).

3.2.6 Marketing Constraints

Limited Transportation and Storage Facilities

Bio fertilizers made with common carriers such as lignite or peat typically have a serviceable life of <6 months. It has been suggested that using bio fertilizers within three to four months of production will yield the optimum outcomes. However, during storage and transit, bio fertilizers are frequently exposed to extremely high temperatures, which lowers their effectiveness and makes dealers uninterested because of the small profit margin (Basu et al., 2021).

Low Demand

Due to various inoculation techniques and a delayed visible change in crop growth compared to inorganic fertilizers, farmers are reluctant to adopt this sustainable practice because of inadequate promotion as well as awareness about the bio fertilizers benefit.

3.2.7 Field-Level Constraints

- Soil conditions include hazardous substances, acidity, pesticides, waterlogging, salts, and drought (Parnell et al., 2024).
- In many soils, poor organic matter content worldwide.
- High yearly and diurnal temperature fluctuations in soil.
- Low competition as well as adaptability compared to natural soil microorganisms (Arora et al., 2020).

4 Promoting Nitrogen Fixation Process and Rhizobia-Legumes Symbioses to Enhance Agricultural Productivity

4.1 Biological Nitrogen Fixation Systems

Rhizobia, often known as root nodule bacteria, is a common term for a group of symbiotic nitrogen-fixing bacteria that can create stem or root nodules in legume plants. Various genera, classes and families have been identified as a nodule microsymbionts belonging to different species of legumes (Frank, 1889). The bacteria's cells inside the nodules undergo differentiation into bacteroids, known as polymorphic cells with altered cell walls that are encased in the membrane of the plant to form a structure known as a symbiosome. The nitrogenase enzyme, which in the bacteroids converts elemental nitrogen to ammonia, is made possible by the conditions this structure affords. Both the rhizobia and the legumes develop from this relationship: the rhizobia obtain their carbon and energy from the host plants' C4-dicarboxylic acids, while the plants receive nitrogen from the rhizobia's reduction of atmospheric nitrogen and incorporation of nitrogen into amino acids (Poole & Allaway, 2000; Wang et al., 2019).

4.2 Rhizobium-Legume Symbiotic Relationship

Symbiosis is a natural phenomenon that has played a vital part in the evolution of organisms. Despite the fact that there are several definitions of symbiosis, they can be broadly divided into two groups: the first defines it as any close relationship between two different organisms in which either or both of them may benefit. The second definition is more limited and only refers to mutualistic relationships in which both symbionts benefit (Martin & Schwab, 2012). Furthermore, the microorganisms that live within plant tissues divided into three groups based on how they affect the plants: phytopathogens, which harm or negatively impact the hosts, endophytes, which live within plant tissues without harming the plants, and micro symbionts, which develop a symbiotic relationship with the plants and benefit the hosts. (Kermah et al., 2018). This symbiosis has some distinguishing characteristics, as discussed subsequently.

Firstly, the best efficient biological nitrogen-fixing system is the symbiosis of rhizobia with legumes which accounting for roughly 65% of all nitrogen fixed by biologically nitrogen fixing bacteria (BNF) remaining 25% originate from alternative systems: free-living diazotrophs (*Azotobacter*), those attached to herbaceous plants, Cyanobacteria with different plants like *Azolla* and *Nostoc*. According to estimates, legumes like alfalfa and clover may fix 100–500 kg N per year per ha, but annual legumes like common bean and soybean may fix 50–300 kg N per year per ha (Wani et al., 1995). This indicates that the symbiotic interaction has a lot of potential for enhancing bioremediation, soil fertility, foliage and the manufacture of proteins for humans.

Secondly, the majority of rhizobia are involved in nitrogen fixation only within a symbiotic relationship that benefits both the legume host and rhizobia. The effectiveness of nitrogen fixation by rhizobia exhibited significant variation depending on the specific strain, potentially due to the genomic characteristics. For example, mutation in particular nodulation genes can lead to a reduction in nodulation capacity. Additionally, the interaction between strains, plant species/varieties, and soil variables may affect the nodulation and nitrogen fixation process (Jia et al., 2013). In rhizobia, different strains (*Bradyrhizobium* and *Azorhizobium*) have the capability to perform nitrogen fixation in a free-living condition under controlled conditions in a particular medium (Dreyfus et al., 1988; Tjepkema & Evans, 1975).

Thirdly, in the symbiosis process a specific relationship occurs among the legume plant and rhizobia. For example, a specific legume plant would exclusively develop nodules in the presence of a particular rhizobial species. Plants that may exchange their rhizobial strains were classified as a cross-nodulation group, such as *R phaseoli* in common bean rhizobia. *R. R. trifolii* in clover rhizobia and leguminosarum in pea rhizobia *R. trifolii* in clover rhizobia (Fred et al., 2002).

Fourthly, the relationship among legume host plant and *rhizobial* spp. may vary on the basis of soil condition. Previous research has shown that in American soils common bean nodulated with *R. etli* (Silva et al., 2003), in China *R. leguminosarum* is present in slightly acidic soil (Cao et al., 2014; Wang et al., 2009), and in Mexico *S. americanum* as well as *S. meliloti* is present in alkaline soil (Mnasri et al., 2012; Verástegui-Valdés et al., 2014; Zurdo-Pineiro et al., 2009). In neutral soils, *Bradyrhizobium elkani* is the dominating symbiont in soybean plants (Man et al., 2008; Zhang et al., 2011b), in saline-alkaline soils *S. fredii* is dominant (Han et al., 2009; Zhang et al., 2011b). According to these findings, the symbiotic relationship is determined by how rhizobia, host legumes, and soil variables interact with one another.

Lastly, two features of rhizobial evolution are the vertical and horizontal transfers of symbiotic genes between species and genera. Rhizobia, particularly the strains of *Bradyrhizobium*, have larger genomes than most other bacteria due to their complex life cycle (MacLean et al., 2007). These genomes primarily consist of genes related to adaptation, such as those involved in symbiosis (Tian et al., 2012). Symbiosis genes, notably nodulation genes, can have evolutionary positions that differ from those of chromosomal genes. This is thought to be evidence of the symbiosis genes transferring from one cell to another (Zhao et al., 2008). Symbiosis genes were typically shifted vertically, with horizontal transfer occurring rarely. For example, in China the phylogenetic positions of housekeeping genes and nodulation genes were shown to be highly correlated in the majority of rhizobia related with diverse wild legumes. This suggests that nodulation genes are transferred vertically, though horizontal transfer was also observed in some cases (Han et al., 2010).

5 Developing Beneficial Microbial Biofilms on Roots of Non-legumes: A Novel Approach as a Biofertilizer

Historically, researchers have focused on studying microorganisms as planktonic cells, which means they are able to float freely, in various habitats. However, these microorganisms frequently gather, adhere to the surfaces, and establish themselves in a polymer matrix, forming a structure known as biofilms (Harrison et al., 2005; Rudrappa et al., 2008). Metabolic cooperation among the microbes can be facilitated by assemblage (Davey & O'toole, 2000). Recent advancements in molecular biological as well as microbiological techniques revealed that most bacteria predominantly exist as biofilms on various abiotic or biotic surfaces (Romanova et al., 2006). Microbial cells, such as fungi, bacteria, algae, and other microorganisms, combine with sticky extracellular polymeric substances (EPS) to form biofilms that provide protection and support to the microbial community (Seneviratne, 2003; Vandevivere & Kirchman, 1993). The EPC comprises nucleic acids, proteins and other substances that shield biofilms from environmental stresses like dehydration, osmotic shock, UV radiation, antimicrobial substances, extreme pH conditions and predators as shown in Fig. 2. Competitor attack and external environmental stress can be tolerated if the biofilm microorganisms act as a group within the EPS layer (Seneviratne & Jayasinghearachchi, 2005). Microorganisms favor living in biofilms over planktonic stages because of the increased survival benefits. When Pseudomonas putida cells were cultured with Saccharomyces cerevisiae in grape juice with differed glucose contents, their survival during the stationary phase considerably increased, most likely due to biofilm formation (Romano & Kolter, 2005). Similarly, the performance of rhizobial cells was higher when they were added to the soils following their formation of biofilms with a common soil fungus than when they were applied individually (Seneviratne & Jayasinghearachchi, 2005).

Biofilms are commonly present in plants, animals, and different environments. Most biofilm communities can be either beneficial or pathogenic (Morikawa, 2006). Some crops have beneficial biofilms attached to their roots which help in the cycling

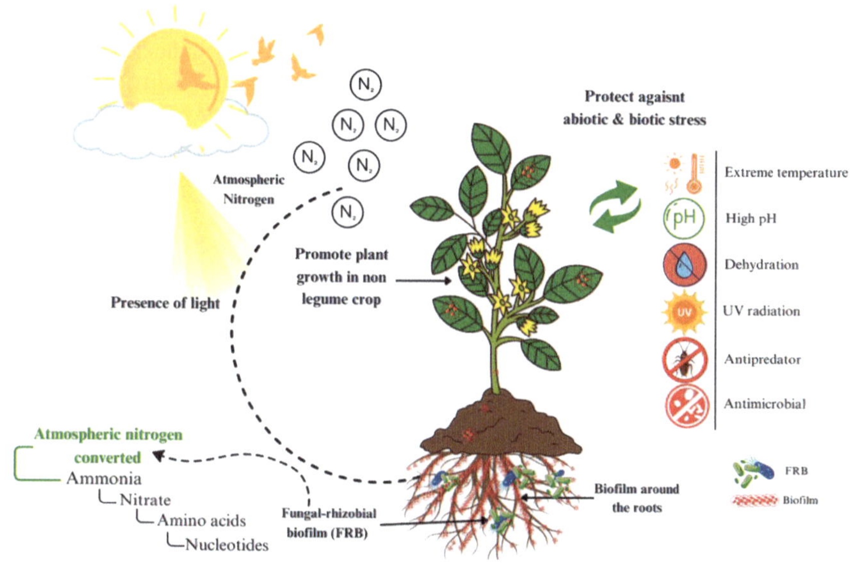

Fig. 2 Beneficial microbial biofilms on roots of non-legume crop

of nutrients, pests and disease control, and overall crop yield enhancement. Nevertheless, the natural density of these biofilms in the soil is often insufficient to make a substantial impact. Biofilm formation relies on quorum sensing, which depends on reaching a critical cell density. However, this required density is often not possible in natural soil conditions (Kong et al., 2006). Furthermore, naturally occurring biofilms often lack nitrogen for optimal functioning. This issue can be solved by adding nitrogen-fixing organisms to the biofilms (Seneviratne, 2008). To increase agricultural productivity, developing biofilms in vitro and used as bio-fertilizers for crops can be effective (Bandara et al., 2006). A number of research were done to investigate the possibility of microbial agents for enhancing plant growth, following the initial in vitro production and examining their interaction with rhizobia-forming biofilms and common non-mycorrhizal soil fungi (Seneviratne & Jayasinghearachchi, 2003). In this study, phosphorous-solubilizing fungi and nitrogen-fixing bacteria were primarily used. During in vitro co-culturing, bacteria attached to the fungus colonize the mycelia of fungi, forming biofilms which are called fungal-bacteria biofilms. When the bacteria are Rhizobium species, the biofilms are called fungal-rhizobial biofilms (FRB). The relationship in the FRB was found to biologically fix nitrogen, as evidenced by the presence of nitrogenase activity and nitrogen residue, which were absent when the rhizobial strain was utilized individually in monoculture (Seneviratne, 2003).

The rhizobial strain employed in this research was a soybean (Glycine max)-nodulating characterized by its significant nitrogen-fixing ability (*Bradyrhizobium elkanii* SEMIA 5019). The formation of biofilms led to increased acidity which was associated with the higher level of IAAS production. Additionally, in plants biofilmed inocula can facilitate the successful establishment of beneficial microbes for disease control. For example, in tomato plants under in vitro conditions, a biofilm composed of *Pleurotus ostreatus* and *Pseudomonas fluorescens* (FBB) resulted in over a 1000% increase in the endophytic colonization compared to inoculation with *P. fluorescens* individually (Jayasinghearachchi & Seneviratne, 2006).

5.1 Developing Beneficial Biofilms on Roots of Non-legumes

Under sterilized conditions, demonstrated that biofilms develop on root surface when monoculture of *Rhizobium* sp., is inoculated in roots of non-leguminous crops (Santaella et al., 2008). Under axenic conditions, recent studies have shown that inoculating with fungal-rhizobial biofilm (FRB) better sustains a higher density of rhizobia in the wheat roots than with rhizobial monocultures. In monocultures, rhizobial cell aggregated on the root hairs, while the FRB in the fungal mycelium connected with the root hairs, which help to sustain a higher cell density (Seneviratne et al., 2008a). As recently reported by Jayasinghearachchi and Seneviratne (2004), FRB can function as "pseudo-nodules" attach with the roots of non-leguminous plants which are helpful in biologically fixing N2 as shown in Fig. 2 (Seneviratne, 2003). When inoculated in axenic conditions, FBB/FRB can grow and attach to the root hairs of non-leguminous plants. Evidence of EPS production by biofilms has shown heavy colonization of root hairs in tea, Anthurium, and rice.

5.2 Effects of Biofilmed Biofertilizers on Plant Growth and Yield

As previously discussed, the development of biofilms on the root system happens when the FRB/ FBB is applied in plant growing medium. These biofilms have been demonstrated to have a positive impact on the productivity and growth of crops. For example, in a soil pot experiment involving rice grown without chemical fertilizers, it was observed that increasing more beneficial microbial species to the BBs led to improved shoot growth in the crop. In comparison to the control, conventional fertilizers with single microbial species resulted in only a 7% increase in shoot biomass. In contrast, BBs containing a mix of three microbes FRB/FBB (one fungus and two bacteria) resulted in a substantial 25% increase in dry matter accumulation in plant shoots. When BBs with diazotrophic bacteria were used in conjunction with the recommended chemical fertilizers positive correlation was observed among the relative growth rate of leaf area and net photosynthetic rate in a tea nursery trial.

Applying BBs with two bacterial species showed a tendency to increase photosynthesis compared to the application of BBs with one bacterial species. The application of 50% of the prescribed fertilizers in combination with two bacterial BBs increased the plant's leaf growth. After harvesting the tea plant from the nursery, soil C was positively correlated with the shoot-to-root ratio of the plant. The two bacterial BBs were likely responsible for the peak shoot-to-root ratio when they were combined with a moderate fertilizer level. It was likely due to the rapid turnover of the fine roots and the increased growth (Zavahir et al., 2008). Low shoot-to-root ratio were observed when only the recommended fertilizer was applied. With two bacterial BBs providing a high shoot-to-root ratio and moderate nitrogen transportation, a low transpiration rate was maintained. The early signs of wilting in the presence of sunlight indicated that plants fed with the prescribed fertilizers individually had an elevated transpiration rate (Seneviratne, 2008). Applying BBs along with nitrogen fertilizer increases leaf nitrogen content in tea, which leads to an increase in root growth.

Without fertilizer nutrients, the plant growth was only marginally supported by the fur-microbes FRB, which probably had a low microbial biomass. The findings indicate that 50% reduction in chemical fertilizer application during the vegetative growth stage, and potentially throughout the whole growing period of non-leguminous crops, could result in substantial economic savings. The biofilm's microbial biomass has increased as a result of the modest use of chemical fertilizers. This can improve microbial efficiency or functionality, particularly as nitrogen and other fertilizer nutrient concentrations diminish. The biofilm functioning similarly to a nodule or pseudonodule, fixes nitrogen (N_2). This fixed nitrogen might be transferred to the roots, which may provide carbon sources back to the biofilm. Further research is needed to understand these processes. Organic acids released by the biofilms contribute to the suppression of microbial pathogens (Browning et al., 2006). Furthermore, biofilms produce plant growth hormones like as IAA, which are expected to promote mycorrhiza fungi interaction with growth of roots of plant (Bandara et al., 2006). In this way, this relationship forms a great metabolic collaboration that promotes the plant's overall growth and health. Furthermore, the BBs are essential for restoring the beneficial microbial communities in soil that have become degraded because of intensive farming and heavy use of chemical inputs (Seneviratne, 2009).

6 Plant Growth Promoting Rhizobacteria and Sustainable Agriculture

Many biotic and abiotic factors are considered to have an impact on plant growth. The thin soil layer directly surrounding the plant roots is crucial for root metabolism. The rhizosphere refers to the active zone of soil that surrounds the plant roots. In 1904, the term "rhizosphere" was first introduced by Hiltner, describing the region of soil around the roots, where active metabolic processes occur and microbes multiply (Hiltner, 1904). Soil's chemical, physical, and biological properties, which are likely to be altered by root growth, are now included in this expanded concept (McCully,

2005). Soil bacteria grow on root lysates and exudates, which makes plant roots an appropriate habitat for them.

In rhizosphere, bacterial population densities can reach 100 times higher than those in bulk soil. Micro-colonies of different bacterial strains can cover as much as 15% of plant roots (Van Loon, 2007). Certain chemical compounds that are released by the plants, known as root exudates, partially determine the microbial community around them. These exudates are expected to enhance plant growth and provide protection against pathogens (Marilley & Aragno, 1999). Species of bacteria from the genera *Rhizobium*, *Pseudomonas*, *Alcaligenes*, *Serratia*, *Erwinia*, *Bacillus*, *Flavobacterium*, *Enterobacter*, *Acinetobacter*, *Arthrobacter*, *Burkholderia*, and *Azospirillum* are widely recognized to stimulate the growth of the plant (Egamberdiyeva, 2007; Tilak et al., 2005). The selection and enrichment of particular bacterial species in the rhizosphere are significantly influenced by the chemicals produced by plant roots. As a result, the composition of the bacterial community in the rhizosphere is influenced by specific organic constituents in root exudates and bacteria depend on these exudates as an energy source (Manoharachary & Mukerji, 2006). PGPR enhances the growth of plant by direct and indirect mechanisms. In direct mechanism production of metabolites (phytohormones) and improving nutrient availability occur. Indirect growth promotion by PGPR involves inducing systemic resistance, protection against pathogens, producing antifungal or lytic enzymes, and lowering iron availability through siderophore sequestration (Burdman et al., 2000; Dobbelaere & Okon, 2007; Lucy et al., 2004).

6.1 Biological Nitrogen Fixation

One of the main ways that the plants benefit from PGRP is through biological nitrogen fixation. The annual worldwide contribution of biological nitrogen-fixing is predicted to be 180 × 106 metric tons (Graham, 1988). Bacteria and archaea are living organisms capable of fixing nitrogen and enhancing soil fertility (Young, 1992). Symbiotic nitrogen fixers include Frankia in non-leguminous trees and Rhizobium in legumes. In contrast, non-symbiotic nitrogen fixers include *Azoarcus*, *Azospirillum*, *Acetobacter*, *Diazotrophicus*, *Azotobacter*, and *Cyanobacter*. Fixed nitrogen is provided to plants by diazotrophic microorganisms, in return for fixed carbon released by multiple plants as root exudates (Glick, 1995). The benefits of a symbiotic interaction between legumes and rhizobia have been extensively studied.

Furthermore, several rhizobial strains and free-living bacteria capable of fix atmospheric nitrogen have been shown to improve wheat plant development by delivering fixed nitrogen (Antoun et al., 1998; Biswas et al., 2000; Malik et al., 1997). For instance, reports indicate that the nitrogen-fixing PGPR can enhance the rice growth (Ladha et al., 1998). Similarly, Biswas et al. (2000) have also reported the growth-promoting effect of diazotrophic PGPR strains on rice. PGPR offers a compelling alternative to chemical nitrogen fertilizers. Rhizobia, in particular, is the most researched PGPR for its ability to fix nitrogen fixation in leguminous plants. Most commonly used PGPRs are *Sinorhizobium*, *Rhizobium*, *Mesorhizobium*, *Bradyrhizoium*, *Azorhizobium*, and *Allorhizobium* (Gualtieri & Bisseling, 2000; Hansen, 2017; Romero, 2002; Unkovich & Pate, 2000). It is noted that numerous PGPR strains are capable of fixing nitrogen (N_2). Biological nitrogen fixation is rarely credited for plant growth stimulation through this mode of action. While PGPR strains are capable of fixing N2, there is not enough information to suggest that the enhancement in host plant growth is a result of the nitrogenase activity of these bacterial strains. A few examples of these PGPRs are *Rhizobium* sp. (Antoun et al., 1998; Yanni et al., 2001), *Pantoea agglomerans* (Riggs et al., 2001), *Beijerinickia* sp. (Baldani et al., 1997), *Klebsiella pneumonia* (Riggs et al., 2001), and *Azoarcus* sp. (Hurek et al., 1994). In the 1980s, extensive research was conducted on nitrogen fixation in non-legume crops (Wani, 1986). However, there is still limited information that supports the idea that the enhancement of non-leguminous crop growth is attributed to the nitrogen-fixing abilities of the strains. For instance, it was previously assumed that the positive impact of *A. brasilense* on non-leguminous crops were attributed to the biological nitrogen fixation (BNF). Nevertheless, the growth-promoting effects of *A. brasilense* are now well established due to various mechanisms such as changes in root morphology and the production of phytohormones.

6.2 *Phosphate Solubilization*

6.2.1 Phosphate-Solubilizing Biofertilizer (PSB)

Several studies have demonstrated that a wide range of bacterial species are capable of dissolving insoluble inorganic phosphate minerals, including rock phosphate, hydroxyapatite, and dicalcium phosphate (Amarasinghe et al., 2022). PSB bacteria are predominant in the rhizosphere, where they help convert insoluble phosphorus into a form that plants can use by releasing organic acids such fumaric, maleic, oxalic, citric, tartaric, lactic, glyoxylic, acetic, and gluconic acids as shown in Table 1 (Zhang et al., 2021b). The most potent phosphate solubilizers in soil bacterial communities are Bacillus, Pseudomonas, endosymbiotic rhizobia and ectorhizospheric as shown in Table 1 (Bekele & Gebremedhin, 2022; Meena et al., 2017). The rhizosphere contains more organic compounds, which attracts PSB, and the population of these bacteria is larger in rhizospheric soil as compared to in non-rhizospheric soil (Khan et al., 2013). Without phosphorus treatment, the application of PSB and rock phosphate increased sugarcane output by up to 12.6% while enhancing sugar production and juice quality (Lopes et al., 2021). Experiments conducted in greenhouse pots with onions revealed that the use of *A. chroococcum* and *G. fasciculatum* at 50% of the permitted P rate improved plant height, root length, root colonization, fresh bulb, and phosphorus uptake. Furthermore, less chemical phosphatic fertilizer will be required.

Table 1 Phosphate-solubilizing microorganism produced acids

Phosphate-solubilizing microorganism	Ecological nich	Predominantly produced acids	References
Aspergillus sp.	Lateritic soil	2-ketogluconic, succinic, citric, oxalic	Cardoso Filho (2021), Banik and Dey (1983)
Aspergillus niger	Present in subtropical and tropical soils	Gluconic, oxalic	Chuang et al. (2007), do Nascimento et al. (2021), Nyamedi (2023)
Chaetomium nigricolor	Lateritic soil	Oxalic, 2-ketogluconic, citric, succinic	Banik and Dey (1983), Timofeeva et al. (2022)
Aspergillus foetidus	Indian rock phosphate	Succinic, tartaric, gluconic, oxalic, citric	Amarasinghe et al. (2022), Singal et al. (1994)
Penicillium rugulosum	Venezuelan phosphate rocks	Citric, gluconic	Reyes et al. (2001), de Freitas Duarte et al. (2022)
Penicillium sp.	Lateritic soil	2-ketogluconic, succinic	Banik and Dey (1983), Bhat et al. (2024)
Aspergillus niger	Soil	Citric, lactic succinic, gluconic, oxalic, glycolic	Kouas et al. (2005), Kumar et al. (2024)
Penicillium radicum	Wheat rhizosphere	Gluconic	Whitelaw et al. (1999), Timofeeva et al. (2022)
Penicillium sp.	Soil	Lactic, succinic, gluconic, glycolic	Kouas et al. (2005), de Freitas Duarte et al. (2022)
Aspergillus niger	Wheat grains	Gluconic, succinic, citric, oxalic	de Freitas Duarte et al. (2022), Rashid et al. (2004)
Pseudomonas spp.	Rhizosphere, plant, water	Malic, quinic, schimic, succinic, citric, tartaric, malonic, lactic and fumaric	Sharma et al. (2017), Aloo et al. (2020)
Aspergillus flavus	Wheat grains	Succinic, oxalic, gluconic, citric	Ramamoorthy et al. (2020), Rashid et al. (2004)
Pseudomonas fluorescens	Rhizosphere of oil palms	Tartaric, gluconic, citric, malic	Fankem et al. (2006), Kaur et al. (2021)
Pseudomonas sp.	In soil and water, sea water and marine mud	Gluconic, acetic	Hakim et al. (2020), Maharana et al. (2021)

(continued)

Table 1 (continued)

Phosphate-solubilizing microorganism	Ecological nich	Predominantly produced acids	References
Microbacterium sp.	In soil and water, sea water and marine mud	Gluconic, acetic	Hakim et al. (2020)
Aspergillus japonicus	Indian rock phosphate	Oxalic, succinic, citric	Aberathna et al. (2022), Singal et al. (1994)
Bacillus sp.	Temperate soil	Gluconic	Eramma et al. (2020)
Pseudomonas sp.	Temperate soil	Gluconic	Eramma et al. (2020), Aloo et al. (2024)
Serratia sp.	Soil, air, water and plants	Lactic, acetic	Behera et al. (2016)
Bacillus subtilus	Rhizosphere soil	Lactic, malic	Bhat et al. (2024), Taha et al. (1969)
Bacillus megaterium	Soil rhizosphere	Malic, lactic	Taha et al. (1969), Timofeeva et al. (2022)
Pseudomonas sp.	Rhizosphere soil	Malic, lactic	Taha et al. (1969), Timofeeva et al. (2022)
Pseudomonas trivialis	Hippophae rhizosphere	Formic, lactic	Vyas and Gulati (2009)
Arthrobacter sp.	Cowpea and wheat rhizosphere	Citric, lactic	Bajpai and Rao (1971), Sharma et al. (2020)
Bascillus sp.	Cowpea rhizosphere	Citric, lactic	Hakim et al. (2021)
Bacillus firmus	Rhizosphere of cowpea and wheat	Citric, lactic	Bajpai and Rao (1971), Bhat et al. (2024)
Enterobacter intermedius	Rhizosphere of grass	2-ketogluconic	Hwangbo et al. (2003), Ehis-Eriakha and Adetunji (2022)
Bacillus amyloliquefaciens	Mangrove	Lactic, isobutyric, itaconic, acetic, isovaleric	Vazquez et al. (2000), Siddique et al. (2021)
Vibrio proteolyticus	Mangrove	Isobutyric, lactic, isovaleric, itaconic, acetic	Vazquez et al. (2000), Bhat et al. (2024)
Penibacillus macerans	Mangrove	Acetic, isovaleric, lactic, isobutyric, itaconic	Vazquez et al. (2000), Timofeeva et al. (2022)
Enterobacter aerogenes	Mangrove	Acetic, isobutyric, itaconic	Vazquez et al. (2000), Rajguru and Bhatt (2022)

(continued)

Table 1 (continued)

Phosphate-solubilizing microorganism	Ecological nich	Predominantly produced acids	References
Enterobacter asburiae	Mangrove	Isovaleric, lactic, acetic, isobutyric, itaconic	Vazquez et al. (2000)
Enterobacter taylorae	Mangrove	Isovaleric, itaconic, acetic isobutyric	Vazquez et al. (2000), Seenivasagan and Babalola (2021)
Bacillus licheniformis	Mangrove	Itaconic, lactic, isovaleric, isobutyric, acetic	Seenivasagan and Babalola (2021), Vazquez et al. (2000)
Bacillus atrophaeus	Mangrove	Isobutyric, lactic, acetic itaconic	Vazquez et al. (2000), Siddique et al. (2021)
Kluyvera cryocrescens	Mangrove	Isovaleric, acetic, lactic, isobutyric, itaconic	Vazquez et al. (2000)
Xanthobacter agilis	Mangrove	Itaconic, isobutyric, lactic, isovaleric, acetic	Vazquez et al. (2000)
Pseudomonas aeromonassens	Mangrove	Acetic, lactic, isobutyric, itaconic, isovaleric	Vazquez et al. (2000), Timofeeva et al. (2022)

6.2.2 Phosphate-Mobilizing Biofertilizer (PMB)

The symbiotic relationship that exists among plant roots and fungi is referred to as "mycorrhizal association." Approximately 90% of land plant species have a symbiotic interaction with arbuscular mycorrhizal fungus (AMF) (Khaliq et al., 2022). There are two types: ectomycorrhiza, which is found in trees and has been shown to be advantageous to forest trees, and endomycorrhiza, which is found in crop plants (Ko et al., 2022). Mycorrhizal fungi require host photosynthates and energy for their functional symbiosis to exist. The plant absorb carbon to provide numerous mycorrhizal benefits and the fungi mycelium spreads from the roots to soil matrix for the collection of nutrients from the soil. As a result, when mycorrhizal fungi colonize root systems, plant development improves and nutrient intake is more efficient (Diagne et al., 2020). Phosphorus and other important nutrients for plant growth are frequently mediated in higher plants via mycorrhizal association, which is a symbiotic relationship among fungi and higher plants (Bhantana et al., 2021). Compared to non-mycorrhizal roots, mycorrhizal roots may absorb far more phosphorus per unit of root length. It enhanced heavy metal tolerance, drought resistance, and root disease susceptibility. AMF also improves soil quality by directly influencing soil aggregation (Gupta, 2020). This connection is typically regarded as highly beneficial in agroforestry. Other crops that benefited from AMF include grapes, tobacco, apple, barley, sorghum, wheat, soybean, cotton and citrus (Parihar et al., 2022).

6.3 Phytohormone Production

Plants produce a variety of chemical compounds known as phytohormones, which actively regulate plant growth and development (Santner et al., 2009). The mechanism behind growth enhancement may involve synthesizing hormones and other compounds that modulate plant regulatory mechanisms and may affect plant growth (Van Loon, 2007). Some strains of bacteria can produce ethylene and auxins. Furthermore, the literature also reports the production of gibberellins and cytokinins (Van Loon, 2007). In 1999, Pieterse and Van Loon observed that inoculating *Arabidopsis* accession Col-0 with *Pseudomonas fluorescens* resulted in a 30% growth improvement when grown in autoclaved soil. Similarly, a significant growth promotion was recorded when the Arabidopsis seedlings were treated with the WCS 417. The growth enhancement is due to the WCS417 strain's ability to generate auxins, which are recognized for their ability to encourage the development of lateral roots (Tanimoto, 2005). Improved nutrient uptake is a result of enhanced lateral root formation leads to. Another research found that the ability of *Azospirillum brasilense* to produce auxins was responsible for the growth observed in wheat and pearl millet (Ferreira, 2018). Plant growth enhancement has been observed with bacteria that lack ACC deaminase, and these findings cannot be explained by known mechanisms. Under these conditions, it is believed that the bacterial cell either possesses a specific surface component or secrete different compounds that function as growth stimulators for plants. These elicitors must be perceived and recognized by plant roots in the same way as how they recognize elicitors from various plant pathogens. Plant pathogens can disrupt the functioning of PGPR by targeting the same receptors. It has been reported that gibberellin and cytokinins also stimulate shoot development. There is limited documentation on how these hormones affect root growth. Despite being commonly regarded as a growth inhibitor, ethylene can stimulate growth in other plant species, such as Arabidopsis, especially present in low concentration (Pierik et al., 2006). Ethylene at a low level can suppress the elongation of root and shoot, but at greater levels, it can promote senescence as well as cause organ abscission (Abeles et al., 2012). Plants secrete ACC through their root exudates, and this compound serves as the precursor in the biosynthesis of ethylene. As a result, the ACC level in plants decreases due to partial uptake by the roots. This reduction in ACC level leads to decreased ethylene production, which alleviates root growth inhibition. These rhizobacteria carrying the ACC deaminase enzyme are known to increase root development through this method (Glick, 2005; Van Loon, 2007). Even some bacteria that lack ACC can still enhance plant growth through a mechanism that remains unexplored. It is hypothesized that bacterial cells release some compounds that stimulate plant growth. Plant roots detect and recognize these elicitors similarly to how they recognize elicitors from plant pathogens. Consequently, these pathogens can disrupt the activity of growth-promoting rhizobacteria because they are accepted by the same receptors (Van Loon, 2007). One mechanism by which PGPRs can stimulate plant growth is through the production of regulators of growth of plant (Barea et al., 1976). Since tryptophan is a precursor for auxin synthesis, these bacteria use tryptophan excreted by plant roots to produce this hormone. The report indicates that plant growth has been increased when inoculating plants with bacterial strains that produce these chemical compounds leads to enhanced plant growth (Patten & Glick, 1996). Various hormones can be produced by these soil-borne bacteria. For example, (Mansour et al., 1994) assessed 24 strains of the genus Streptomycin having the ability to produce plant hormones. They observed that each of these strains could synthesize gibberellins, auxins, and cytokinins when cultured in a liquid medium. García de Salamone et al. (2001) also confirmed that PGR also produced substances that promote plant growth (García de Salamone et al., 2001).

7 For Sustainable Crop Production Use of Arbuscular Mycorrhizal Fungi

In sustainable food production, plants can efficiently absorb minerals like phosphorous and nitrogen by using microbial inoculants known as bio-fertilizers (Alori et al., 2017a). Phosphorous is a vital mineral for plants and its deficiency can hinder their growth. Through arbuscular mycorrhizal (AM) association, various plants can obtain adequate phosphorous from rhizospheric soils. In phosphorous-deficient environments, the AM symbiotic association is particularly crucial as it improves the plant growth and raises the phosphorous level within the plants as shown in Table 2. Alkaline phosphate has been identified as a valuable and functional enzyme for AM using gel electrophoresis, and ultra-cytochemistry. Ultra-cytochemical studies showed that ALP is present in the vacuoles of highly developed arbuscules, highlighting its responsibility for phosphorous acquisition by plants (Abdel-Fattah et al., 2014). One approach to achieving agricultural sustainability involves enhancing crop yields and expanding the cultivated area of the reclaimed soil by growing crops such as soybean (*Glycine max* L.) and oilseed. Improved phosphorous absorption is linked to increasing soybean productivity through interaction with AM fungi as shown in Table 2 (Abdel-Fattah & Asrar, 2012). AM fungal interaction can also facilitate the transport of sulfur to host plants along with phosphorous and nitrogen. AM interaction can boost the sulfur nutrition of the host plant, which results in the upregulating of the plant sulfate transporter gene expression (Igiehon & Babalola, 2017).

7.1 *Strategies for Utilizing Arbuscular Mycorrhizal Fungi*

Over the past two decades, various research studies on AMF have emphasized the numerous benefits of AMF on crop production and soil health. Consequently, it is generally accepted that AMF could replace inorganic fertilizers soon. As obligate biotrophs, arbuscular-mycorrhizal (AM) fungi are given special attention in terms of their strategies for forming a symbiotic relationship with their host. The analysis

Table 2 Arbuscular mycorrhizal (AM) association with various plants to improve nutrient uptake

Host plant	Fungal species	Response by plant	References
Chinese ryegrass	*Glomus mosseae*	High absorption of P and N content, seedling weight increase, colonization rate high	Lin JiXiang et al. (2017)
Tomato	*Rhizophagus irregularis*	Enhanced growth hormones, no of leaves and area, dry weight of shoot	Khalloufi et al. (2017)
Onions	*Glomusconstrictum* and *Acaulosporalacunosa*	Improved foliar nutrient status	Hart and Forsythe (2012), Bayanati et al. (2023)
Maize	*Glomus etunicatum*	High vegetative growth, nutrient level, and chlorophyll	Sadhana (2014), Khan et al. (2022b)
Tomato	*Glomus intraradices*	Improve chlorophyll content, dry matter, growth parameters, and ions uptake	Hajiboland et al. (2010), Ansari et al. (2024)
Wheat	Alien arbuscular mycorrhizal fungi	Enhanced shoot and root biomass	Alguacil et al. (2006), Yadav et al. (2023)
Wheat	*Glomus mosseae*	Improved yield	Daei et al. (2009), Duan et al. (2021)
Babul	*Glomus fasciculate*	Enhanced dry mass, shoot, root Zn, Cu and P content	Giri et al. (2007), Khaliq et al. (2022)
Soybean	AMF	Enhanced nutrient uptake capability	Liu et al. (2012)
Date palm	*Glomus intraradices*	Improved growth and development	Hildebrandt et al. (2007)
Cucumber	*Glomus etunicatum, Glomus intraradices Glomus mosseae*	Increased biomass, synthesis of photosynthesis pigments, antioxidants enzymes	Hashem et al. (2018)
Halophile grass	*Claroideoglomus etunicatum*	Enhanced shoot, root dry mass, free amino acids, soluble sugars	Giri et al. (2007), Sun and Shahrajabian (2023)
Rice	*Claroideoglomus etunicatum*	Enhance photosynthetic rate and quantum yield of PSII	Porcel et al. (2015)

covers four topics: (i) the impact of various factors on fungal growth with or without the host root, including DNA replication and cell cycle timing; the mechanical and enzymatic mechanisms that fungus uses to colonize root tissues, while keeping the host viability; (ii) the changes in cell morphology caused by AM fungus, especially the creation of subcellular area called the interface, and (iii) the alteration in plant gene expression during the fungal colonization, including genes related to symbiosis, structure and defense.

7.2 *Helpful in Transport of Phosphorus and Nitrogen to Plants*

The process of nutrient uptake in arbuscular mycorrhizal (AM) associations has been widely investigated, following the discovery of high-affinity phosphate transporters (PT) in fungi. The interaction between AM and plants specifically stimulates the expression of inorganic phosphorous (pi) transporter proteins in plant genes. Arbuscular mycorrhizal fungi not only enhance the pi uptake but influence the size and development of arbuscules and maintain the symbiotic relationship as shown in Table 2 (Xie et al., 2013; Yang et al., 2012). Recent studies have demonstrated that *Lotus japonicas* and *Medicago truncatula* express the *LjPT4* and *MtPT4* Pi transporter due to AM interaction. In non-colonized plants, these transporters are found in the root tips, and the PT4 genes have been reported to function as mechanisms of Pi sensing machinery (Volpe et al., 2016). Numerous reports show that AM interaction plays a significant role in nutrient transmission across various host plants. In addition to Pi transporters, researchers have also identified ammonium transporters induced by AM fungi. The peri-arbuscular membranes that are derived from the plant surround the arbuscule and enable the transmission of inorganic nutrients in a mutualistic manner. Transport proteins can transport inorganic nutrients from the periarbuscular apoplast to the cortical cell within the membrane (Bapaume & Reinhardt, 2012). It has been shown that ammonium transporters produced by AM fungi are present in soybean periarbuscular membranes, indicating their involvement in transferring ammonium to cortical cells of the plants (Breuillin-Sessoms et al., 2015). Transporters facilitate the movements of inorganic phosphorous or ammonium which not only provide nutrients to the root cell but also contributes in maintenance of the arbuscule (Berruti et al., 2016; Breuillin-Sessoms et al., 2015). A recently identified sulfate transporter that takes up sulfate from the arbuscules has been identified.

8 Conclusion and Future Perspective

Recent research has demonstrated that using microorganisms as a bio fertilizer is essential for the sustainable management of soil resources. Beneficial microbial-plant interactions may lead to potential solutions for environment friendly farming. The nutritional value and quality of agricultural products are extremely important to customers. The use of microorganisms is leading to the development of new farming technologies that are less expensive and have less adverse environmental effects. Further research in this area is required to provide farmers with timely information regarding microorganisms' biological activities. Phosphorus solubilization, atmospheric nitrogen fixation, and soil microbe-mediated iron bioavailability all contribute to long-term agricultural development, mostly by improving fertility of soil and crop productivity. One area that needs special attention involves experimental testing of all fungal and bacterial strains in various climatic zones, soil types and crops. To develop commercial microbial preparations for increasing growth of plant in different crops, it is necessary to analyze the molecular mechanisms of phosphate solubilization, biofilm formation, and nitrogen fixation and identify multiple fungal and bacterial strains capable of exhibiting multiple activities at the same time.

References

Aanderud, Z. T., Vert, J. C., Lennon, J. T., Magnusson, T. W., Breakwell, D. P., & Harker, A. R. (2016). Bacterial dormancy is more prevalent in freshwater than hypersaline lakes. *Frontiers in Microbiology, 7*, 853.

Abdel-Fattah, G. M., & Asrar, A.-W.A. (2012). Arbuscular mycorrhizal fungal application to improve growth and tolerance of wheat (*Triticum aestivum* L.) plants grown in saline soil. *Acta Physiologiae Plantarum, 34*, 267–277.

Abdel-Fattah, G. M., Asrar, A. A., Al-Amri, S. M., & Abdel-Salam, E. M. (2014). Influence of arbuscular mycorrhiza and phosphorus fertilization on the gas exchange, growth and phosphatase activity of soybean (*Glycine max* L.) plants. *Photosynthetica, 52*, 581–588.

Abeles, F. B., Morgan, P. W., & Saltveit, M. E. (2012). *Ethylene in plant biology*. New York: Academic press.

Aberathna, A. A. A. U., Satharasinghe, D. A., Jayasooriya, A. P., Jinadasa, R. N., Manopriya, S., Jayaweera, B. P. A., Fernando, C. A. N., Weerathilake, W. A. D. V., Prathapasinghe, G. A., Liyanage, J. A., & Premarathne, J. M. K. J. K. (2022). Increasing the bioavailability of phosphate by using microorganisms. *International Journal of Agronomy, 2022*, 1–17.

Ai, P., Sun, S., Zhao, J., Fan, X., Xin, W., Guo, Q., Yu, L., Shen, Q., Wu, P., Miller, A. J., & Xu, G. (2009). Two rice phosphate transporters, OsPht1;2 and OsPht1;6, have different functions and kinetic properties in uptake and translocation. *Plant Journal, 57*, 798–809.

Alguacil, M., Caravaca, F., Díaz-Vivancos, P., Hernández, J. A., & Roldán, A. (2006). Effect of arbuscular mycorrhizae and induced drought stress on antioxidant enzyme and nitrate reductase activities in *Juniperus oxycedrus* L. grown in a composted sewage sludge-amended semi-arid soil. *Plant and Soil, 279*, 209–218.

Aloo, B. N., Ouma, B. N., Were, B. A., & Tumuhairwe, J. B. (2024). 4 Phosphate-solubilizing. *Regenerative Agricultural Translating Science to Action, 47*, 58.

Aloo, B. N., Makumba, B. A., & Mbega, E. R. (2020). Phosphate-solubilizing rhizobacteria: Diversity, mechanisms, and prospects in sustainable agriculture.

Alori, E. T., Dare, M. O., & Babalola, O. O. (2017a). Microbial inoculants for soil quality and plant health. *Sustainable Agriculture Reviews, 12*, 281–307.

Alori, E. T., Glick, B. R., & Babalola, O. O. (2017b). Microbial phosphorus solubilization and its potential for use in sustainable agriculture. *Frontiers Microbiology, 8*, 971.

Amarasinghe, T., Madhusha, C., Munaweera, I., & Kottegoda, N. (2022). Review on mechanisms of phosphate solubilization in rock phosphate fertilizer. *Communications in Soil Science and Plant Analysis, 53*, 944–960.

Ansari, M., Mishra, A., Chattopadhyay, A., Mukherjee, A., & Sarkar, A. (2024). 3 plant defence regulation. *Regenerative Agriculture Translation Science to Action, 23*, 147.

Antoun, H., Beauchamp, C. J., Goussard, N., Chabot, R., & Lalande, R. (1998). Potential of rhizobium and bradyrhizobium species as plant growth promoting rhizobacteria on non-legumes: Effect on radishes (*Raphanus sativus* L.). *Plant and Soil, 204*, 57–67.

Arora, N. K., Fatima, T., Mishra, I., & Verma, S. (2020). Microbe-based inoculants: Role in next green revolution. In V. Shukla & N. Kumar (Eds.), *Environmental concerns and sustainable development* (pp. 191–246). Springer.

Averill, C., Anthony, M. A., Baldrian, P., Finkbeiner, F., van den Hoogen, J., Kiers, T., Kohout, P., Hirt, E., Smith, G. R., & Crowther, T. W. (2022). Defending Earth's terrestrial microbiome. *Nature Microbiology, 7*, 1717–1725.

Awais, M., Tariq, M., Ali, Q., Khan, A., Ali, A., Nasir, I. A., & Husnain, T. (2019). Isolation, characterization and association among phosphate solubilizing bacteria from sugarcane rhizosphere. *Cytology and Genetics, 53*, 86–95.

Babalola, O. O., & Glick, B. R. (2012). The use of microbial inoculants in African agriculture: Current practice and future prospects. *Journal of Food Agricultural Environment, 10*, 540–549.

Bajpai, P. D., & Rao, W. V. B. S. (1971). Phosphate solubilising bacteria: Part III. Soil inoculation with phosphorus solubilising bacteria. *Soil Science Plant Nutrition, 17*, 46–53.

Bakker, P. A., Berendsen, R. L., Doornbos, R. F., Wintermans, P. C., & Pieterse, C. M. (2013). The rhizosphere revisited: Root microbiomics. *Frontier Plant Science, 4*, 165.

Baldani, J., Caruso, L., Baldani, V. L., Goi, S. R., & Döbereiner, J. (1997). Recent advances in BNF with non-legume plants. *Soil Biology and Biochemical, 29*, 911–922.

Bandara, W. M. S., Seneviratne, G., & Kulasooriya, S. A. (2006). Interactions among endophytic bacteria and fungi: Effects and potentials. *Journal of Bioscience, 31*, 645–650.

Bandyopadhyay, P., Bhuyan, S. K., Yadava, P. K., Varma, A., & Tuteja, N. (2017). Emergence of plant and rhizospheric microbiota as stable interactomes. *Protoplasma, 254*, 617–626.

Banik, S., & Dey, B. K. (1983). Phosphate-solubilizing potentiality of the microorganisms capable of utilizing aluminium phosphate as a sole phosphate source. *Zentralblatt Fuer Mikrobiol., 138*, 17–23.

Bapaume, L., & Reinhardt, D. (2012). How membranes shape plant symbioses: Signaling and transport in nodulation and arbuscular mycorrhiza. *Frontiers Plant Science, 3*, 223.

Barea, J. M., Navarro, E., & Montoya, E. (1976). Production of plant growth regulators by rhizosphere phosphate-solubilizing bacteria. *Journal of Applied Bacteriology, 40*, 129–134.

Basu, A., Prasad, P., Das, S. N., Kalam, S., Sayyed, R. Z., Reddy, M. S., & El-Enshasy, H. (2021). Plant growth promoting rhizobacteria (PGPR) as green bioinoculants: Recent developments, constraints, and prospects. *Sustainability, 13*, 1140.

Bayanati, M., Al-Tawaha, A. R. M., Sangeetha, J., Thangadurai, D., & Kummur, P. N. (2023). Role of mycorrhizal fungi in plant growth: Implications in abiotic stress tolerance. *Mycorrhizal Technology, 12*, 131–157.

Bechtaoui, N., Raklami, A., Tahiri, A.-I., Benidire, L., El Alaoui, A., Meddich, A., Göttfert, M., & Oufdou, K. (2019). Characterization of plant growth promoting rhizobacteria and their benefits on growth and phosphate nutrition of faba bean and wheat. *Biology Open, 8*, bio043968.

Behera, B. C., Singdevsachan, S. K., Mishra, R. R., Sethi, B. K., Dutta, S. K., & Thatoi, H. N. (2016). Phosphate solubilising bacteria from mangrove soils of Mahanadi river delta, Odisha, India. *World Journal of Agricultural Research, 4*, 18–23.

Bekele, M., & Gebremedhin, W. (2022). Phosphate solubilizing bacteria in optimizing phosphorous acquisition and crop production. *Americal Journal of Bioscience, 10*, 61–68.
Berruti, A., Lumini, E., Balestrini, R., & Bianciotto, V. (2016). Arbuscular mycorrhizal fungi as natural biofertilizers: Let's benefit from past successes. *Frontiers of Microbiology, 6*, 1559.
Bhantana, P., Rana, M. S., Sun, X., Moussa, M. G., Saleem, M. H., Syaifudin, M., Shah, A., Poudel, A., Pun, A. B., & Bhat, M. A. (2021). Arbuscular mycorrhizal fungi and its major role in plant growth, zinc nutrition, phosphorous regulation and phytoremediation. *Symbiosis, 84*, 19–37.
Bhat, M. A., Mishra, A. K., Shah, S. N., Bhat, M. A., Jan, S., Rahman, S., Baek, K.-H., & Jan, A. T. (2024). Soil and mineral nutrients in plant health: a prospective study of iron and phosphorus in the growth and development of plants. *Current Issues Molecular Biology, 46*, 5194–5222.
Bhattacharyya, P. N., & Jha, D. K. (2012). Plant growth-promoting rhizobacteria (PGPR): Emergence in agriculture. *World Journal of Microbiology Biotechnology, 28*, 1327–1350.
Biswas, J. C., Ladha, J. K., & Dazzo, F. B. (2000). Rhizobia inoculation improves nutrient uptake and growth of lowland rice. *Soil Science in Social American Journal, 64*, 1644–1650.
Breuillin-Sessoms, F., Floss, D. S., Gomez, S. K., Pumplin, N., Ding, Y., Levesque-Tremblay, V., Noar, R. D., Daniels, D. A., Bravo, A., & Eaglesham, J. B. (2015). Suppression of arbuscule degeneration in *Medicago truncatula* phosphate transporter4 mutants is dependent on the ammonium transporter 2 family protein AMT2; 3. *The Plant Cell, 27*, 1352–1366.
Browne, P., Rice, O., Miller, S. H., Burke, J., Dowling, D. N., Morrissey, J. P., & O'Gara, F. (2009). Superior inorganic phosphate solubilization is linked to phylogeny within the *Pseudomonas fluorescens* complex. *Applied Soil Ecology, 43*, 131–138.
Browning, M., Wallace, D. B., Dawson, C., Alm, S. R., & Amador, J. A. (2006). Potential of butyric acid for control of soil-borne fungal pathogens and nematodes affecting strawberries. *Soil Biology and Biochemistry, 38*, 401–404.
Burdman, S., Jurkevitch, E., & Okon, Y. (2000). Recent advances in the use of plant growth promoting rhizobacteria (PGPR) in agriculture.
Canfora, L., Bacci, G., Pinzari, F., Lo-Papa, G., Dazzi, C., & Benedetti, A. (2014). Salinity and bacterial diversity: To what extent does the concentration of salt affect the bacterial community in a saline soil? *PLoS ONE, 9*, e106662.
Cao, Y., Wang, E.-T., Zhao, L., Chen, W.-M., & Wei, G.-H. (2014). Diversity and distribution of rhizobia nodulated with *Phaseolus vulgaris* in two ecoregions of China. *Soil Biology and Biochemistry, 78*, 128–137.
Cardoso-Filho, J. A. (2021). Microbial solubilizers of rock-forming minerals: Opportunities and challenges. *Agri-Based Bioeconomy, 51*, 179–218.
Charana-Walpola, B., & Yoon, M.-H. (2013). Phosphate solubilizing bacteria: Assessment of their effect on growth promotion and phosphorous uptake of mung bean (*Vigna radiata* [L.] R. Wilczek). *Chilean Journal of Agricultural Research, 73*, 275–281.
Chen, J., & Mueller, V. (2018). Coastal climate change, soil salinity and human migration in Bangladesh. *Nature Climatic Change, 8*, 981–985.
Chen, Y. P., Rekha, P. D., Arun, A. B., Shen, F. T., Lai, W.-A., & Young, C. C. (2006). Phosphate solubilizing bacteria from subtropical soil and their tricalcium phosphate solubilizing abilities. *Applied Soil Ecology, 34*, 33–41.
Chuang, C.-C., Kuo, Y.-L., Chao, C.-C., & Chao, W.-L. (2007). Solubilization of inorganic phosphates and plant growth promotion by *Aspergillus niger*. *Biology Fertilizer Soils, 43*, 575–584.
Daei, G., Ardekani, M. R., Rejali, F., Teimuri, S., & Miransari, M. (2009). Alleviation of salinity stress on wheat yield, yield components, and nutrient uptake using arbuscular mycorrhizal fungi under field conditions. *Journal of Plant Physiology, 166*, 617–625.
Daliakopoulos, I. N., Tsanis, I. K., Koutroulis, A., Kourgialas, N. N., Varouchakis, A. E., Karatzas, G. P., & Ritsema, C. J. (2016). The threat of soil salinity: A European scale review. *Science and Total Environment, 573*, 727–739.
Davey, M. E., & O'toole, G. A. (2000). Microbial biofilms: From ecology to molecular genetics. *Microbiology Molecules Biology Reviews, 64*, 847–867.

Delavaux, C. S., Smith-Ramesh, L. M., & Kuebbing, S. E. (2017). Beyond nutrients: A meta-analysis of the diverse effects of arbuscular mycorrhizal fungi on plants and soils. *Ecology, 98*, 2111–2119.

de Freitas Duarte, N., Paiva, C. A. O., Pagano, M. C., & Correa, E. J. A. (2022). Phosphate solubilization by microorganisms. In *New and future developments in microbial biotechnology and bioengineering* (pp. 257–282). Elsevier.

Diagne, N., Ngom, M., Djighaly, P. I., Fall, D., Hocher, V., & Svistoonoff, S. (2020). Roles of arbuscular mycorrhizal fungi on plant growth and performance: Importance in biotic and abiotic stressed regulation. *Diversity, 12*, 370.

Dini-Andreote, F., Silva, M. C. P., Triado-Margarit, X., Casamayor, E. O., Van Elsas, J. D., & Salles, J. F. (2014). Dynamics of bacterial community succession in a salt marsh chronosequence: evidences for temporal niche partitioning. *ISME Journal, 8*, 1989–2001.

Dini-Andreote, F., Pylro, V. S., Baldrian, P., Van Elsas, J. D., & Salles, J. F. (2016). Ecological succession reveals potential signatures of marine–terrestrial transition in salt marsh fungal communities. *ISME Journal, 10*, 1984–1997.

Do Nascimento, J. M., Netto, J. A. F. V., Valadares, R. V., Oliveira-Mendes, G., Silva, I. R., Vergütz, L., & Costa, M. D. (2021). *Aspergillus niger* as a key to unlock fixed phosphorus in highly weathered soils. *Soil Biology in Biochemistry, 156*, 108190.

Dobbelaere, S., & Okon, Y. (2007). The plant growth-promoting effect and plant responses. In *Association endophytic nitrogen-fixing bacteria cyanobacterial association* (pp. 145–170). Springer, New York.

Dreyfus, B., Garcia, J.-L., & Gillis, M. (1988). Characterization of *Azorhizobium caulinodans* gen. nov., sp. nov., a stem-nodulating nitrogen-fixing bacterium isolated from Sesbania rostrata. *International Journal of System in Evolution Microbiology, 38*, 89–98.

Duan, H.-X., Luo, C.-L., Li, J.-Y., Wang, B.-Z., Naseer, M., & Xiong, Y.-C. (2021). Improvement of wheat productivity and soil quality by arbuscular mycorrhizal fungi is density- and moisture-dependent. *Agronomic Sustainability Development, 41*, 3.

Egamberdiyeva, D. (2007). The effect of plant growth promoting bacteria on growth and nutrient uptake of maize in two different soils. *Applied Soil Ecology, 36*, 184–189.

Ehis-Eriakha, C. B., & Adetunji, C. O. (2022). Recent advances in application of biostimulants derived from beneficial microorganisms: Agriculture and environmental perspective. *Agricultural Biotechnology, 47*, 47–88.

Empadinhas, N., & da Costa, M. S. (2008). Osmoadaptation mechanisms in prokaryotes: Distribution of compatible solutes. *International Microbiology, 11*, 151–161.

Enebe, M. C., & Babalola, O. O. (2018). The influence of plant growth-promoting rhizobacteria in plant tolerance to abiotic stress: A survival strategy. *Applied Microbiology in Biotechnology, 102*, 7821–7835.

Eramma, M., Rao, S., Ramesh, Y. M., & Naik, N. M. (2020). Isolation and screening of phosphate solubilizing bacteria from paddy Rhizosphere soil. *International Journal in Current Microbiology in Applied Science, 9*, 477–485.

Eslami, S., Hoekstra, P., Minderhoud, P. S., Trung, N. N., Hoch, J. M., Sutanudjaja, E. H., Dung, D. D., Tho, T. Q., Voepel, H. E., & Woillez, M.-N. (2021). Projections of salt intrusion in a mega-delta under climatic and anthropogenic stressors. *Communication Earth Environment, 2*, 142.

Eswar, D., Karuppusamy, R., & Chellamuthu, S. (2021). Drivers of soil salinity and their correlation with climate change. *Current Opinion in Environment Sustainability, 50*, 310–318.

Fankem, H., Nwaga, D., Deubel, A., Dieng, L., Merbach, W., & Etoa, F. X. (2006). Occurrence and functioning of phosphate solubilizing microorganisms from oil palm tree (*Elaeis guineensis*) rhizosphere in Cameroon. *African Journal of Biotechnology, 5*, 479.

Ferreira, R. F. N. (2018). The phytohormone mediated affair of *Triticum durum* and *Azospirillum brasilense*: A case study on plant-bacteria interaction. PhD Thesis.

Frank, B. (1889). Tlber die Pilzsymbiose der Leguminosen. *Ber Deut Bot Gesellsch, 7*, 338.

Fred, E. B., Baldwin, I. L., & McCoy, E. (2002). *Root nodule bacteria and leguminous plants.* UW-Madison Libraries Parallel Press.
García de Salamone, I. E., Hynes, R. K., & Nelson, L. M. (2001). Cytokinin production by plant growth promoting rhizobacteria and selected mutants. *Canadian Journal of Microbiology, 47*, 404–411.
Giri, B., Kapoor, R., & Mukerji, K. G. (2007). Improved tolerance of *Acacia nilotica* to salt stress by Arbuscular Mycorrhiza, *Glomus fasciculatum* may be partly related to elevated K/Na ratios in root and shoot tissues. *Microbology in Ecology, 54*, 753–760.
Glick, B. R. (1995). The enhancement of plant growth by free-living bacteria. *Canadian Journal of Microbiology, 41*, 109–117.
Glick, B. R. (2005). Modulation of plant ethylene levels by the bacterial enzyme ACC deaminase. *FEMS Microbiology Letters, 251*, 1–7.
Graham, P. H. (1988). Principles and application of soil microbiology. *Gualtieri G Bisseling, 21*, 181–191.
Gualtieri, G., & Bisseling, T. (2000). The evolution of nodulation. *Plant Molcular Evolution, 31*, 181–194.
Gupta, M. M. (2020). Arbuscular Mycorrhizal fungi: The potential soil health indicators. In B. Giri & A. Varma (Eds.), *Soil health* (pp. 183–195). Springer.
Hacquard, S. (2016). Disentangling the factors shaping microbiota composition across the plant holobiont. *New Phytology, 209*, 454–457.
Hacquard, S., & Schadt, C. W. (2015). Towards a holistic understanding of the beneficial interactions across the *Populus* microbiome. *New Phytology, 205*, 1424–1430.
Hajiboland, R., Aliasgharzadeh, N., Laiegh, S. F., & Poschenrieder, C. (2010). Colonization with arbuscular mycorrhizal fungi improves salinity tolerance of tomato (*Solanum lycopersicum* L.) plants. *Plant and Soil, 331*, 313–327.
Hakim, S., Mirza, B. S., Imran, A., Zaheer, A., Yasmin, S., Mubeen, F., Mclean, J. E., & Mirza, M. S. (2020). Illumina sequencing of 16S rRNA tag shows disparity in rhizobial and non-rhizobial diversity associated with root nodules of mung bean (*Vigna radiata* L.) growing in different habitats in Pakistan. *Microbiology Research, 231*, 126356.
Hakim, S., Naqqash, T., Nawaz, M. S., Laraib, I., Siddique, M. J., Zia, R., Mirza, M. S., & Imran, A. (2021). Rhizosphere engineering with plant growth-promoting microorganisms for agriculture and ecological sustainability. *Frontier Sustainability Food System, 5*, 617157.
Han, L. L., Wang, E. T., Han, T. X., Liu, J., Sui, X. H., Chen, W. F., & Chen, W. X. (2009). Unique community structure and biogeography of soybean rhizobia in the saline-alkaline soils of Xinjiang, China. *Plant and Soil, 324*, 291–305.
Han, T. X., Tian, C. F., Wang, E. T., & Chen, W. X. (2010). Associations among rhizobial chromosomal background, nod genes, and host plants based on the analysis of symbiosis of indigenous rhizobia and wild legumes native to Xinjiang. *Microbiology in Ecology, 59*, 311–323.
Hansen, A. P. (2017). Legume-rhizobium symbioses: Significance for sustainable plant production. *Modern Tools and Techniques to Understand Microbes, 14*, 159–182.
Harrison, J. J., Turner, R. J., Marques, L. L., & Ceri, H. (2005). Biofilms: A new understanding of these microbial communities is driving a revolution that may transform the science of microbiology. *American Science, 93*, 508–515.
Hart, M. M., & Forsythe, J. A. (2012). Using arbuscular mycorrhizal fungi to improve the nutrient quality of crops; nutritional benefits in addition to phosphorus. *Science in Horticulture, 148*, 206–214.
Hashem, A., Alqarawi, A. A., Radhakrishnan, R., Al-Arjani, A. B. F., Aldehaish, H. A., Egamberdieva, D., & Abd-Allah, E. F. (2018). Arbuscular mycorrhizal fungi regulate the oxidative system, hormones and ionic equilibrium to trigger salt stress tolerance in *Cucumis sativus* L. *Saudi Journal of Biology Science, 25*, 1102–1114.
Hassani, A., Azapagic, A., & Shokri, N. (2020). Predicting long-term dynamics of soil salinity and sodicity on a global scale. *Proceedings of the National Academy of Science, 117*, 33017–33027.

Hassani, A., Azapagic, A., & Shokri, N. (2021). Global predictions of primary soil salinization under changing climate in the 21st century. *Natural Community, 12*, 6663.

Hildebrandt, U., Regvar, M., & Bothe, H. (2007). Arbuscular mycorrhiza and heavy metal tolerance. *Phytochemistry, 68*, 139–146.

Hiltner, L. (1904). Über neuere Erfahrungen und Probleme auf dem Gebiet der Bodenbakteriologie und unter besonderer Berücksichtigung der Gründüngung und Brache. *Arbeiten der deutschen Landwirtschaftlichen Gesellschaft, 98*, 59.

Hollister, E. B., Engledow, A. S., Hammett, A. J. M., Provin, T. L., Wilkinson, H. H., & Gentry, T. J. (2010). Shifts in microbial community structure along an ecological gradient of hypersaline soils and sediments. *ISME Journal, 4*, 829–838.

Hurek, T., Reinhold-Hurek, B., Van Montagu, M., & Kellenberger, E. (1994). Root colonization and systemic spreading of *Azoarcus* sp. strain BH72 in grasses. *Journal of Bacteriology, 176*, 1913–1923.

Hwangbo, H., Park, R. D., Kim, Y. W., Rim, Y. S., Park, K. H., Kim, T. H., Suh, J. S., & Kim, K. Y. (2003). 2-Ketogluconic acid production and phosphate solubilization by Enterobacter intermedium. *Current Microbiology, 47*, 0087–0092.

Igiehon, N. O., & Babalola, O. O. (2017). Biofertilizers and sustainable agriculture: Exploring arbuscular mycorrhizal fungi. *Applied Microbiology in Biotechnology, 101*, 4871–4881.

Ivushkin, K., Bartholomeus, H., Bregt, A. K., Pulatov, A., Kempen, B., & De Sousa, L. (2019). Global mapping of soil salinity change. *Remote Sensor Environment, 231*, 111260.

Jayasinghearachchi, H. S., & Seneviratne, G. (2004). A bradyrhizobial-*Penicillium* spp. biofilm with nitrogenase activity improves N_2 fixing symbiosis of soybean. *Biology and Fertility of Soils, 40*, 432–434.

Jayasinghearachchi, H. S., & Seneviratne, G. (2006). A mushroom-fungus helps improve endophytic colonization of tomato by *Pseudomonas fluorescens* through biofilm formation.

Jia, R. Z., Wang, E. T., Liu, J. H., Li, Y., Gu, J., Yuan, H. L., & Chen, W. X. (2013). Effectiveness of different Ensifer meliloti strain-alfalfa cultivar combinations and their influence on nodulation of native rhizobia. *Soil Biology in Biochemistry, 57*, 960–963.

Kalayu, G. (2019). Phosphate solubilizing microorganisms: Promising approach as biofertilizers. *International Journal of Agronomy, 2019*, 1–7.

Kaur, P., & Purewal, S. S. (2019). Biofertilizers and their role in sustainable agriculture. In B. Giri, R. Prasad, Q.-S. Wu, & A. Varma (Eds.), *Biofertilizers for sustainable agriculture and environment* (pp. 285–300). Springer.

Kaur, C., Selvakumar, G., & Upreti, K. K. (2021). Organic acid profiles of phosphate solubilizing bacterial strains in the presence of different insoluble phosphatic sources under. *Journal of Pure Applied Microbiology, 15*, 1006–1015.

Kearns, P. J., Angell, J. H., Howard, E. M., Deegan, L. A., Stanley, R. H., & Bowen, J. L. (2016). Nutrient enrichment induces dormancy and decreases diversity of active bacteria in salt marsh sediments. *Natural Communication, 7*, 12881.

Kermah, M., Franke, A. C., Adjei-Nsiah, S., Ahiabor, B. D. K., Abaidoo, R. C., & Giller, K. E. (2018). N2-fixation and N contribution by grain legumes under different soil fertility status and cropping systems in the Guinea savanna of northern Ghana. *Agricultural Ecosystem in Environment, 261*, 201–210.

Khalid, R., Zhang, X. X., Hayat, R., & Ahmed, M. (2020). Molecular characteristics of rhizobia isolated from Arachis hypogaea grown under stress environment. *Sustainability, 12*, 6259.

Khaliq, A., Perveen, S., Alamer, K. H., Zia-Ul-Haq, M., Rafique, Z., Alsudays, I. M., Althobaiti, A. T., Saleh, M. A., Hussain, S., & Attia, H. (2022). Arbuscular mycorrhizal fungi symbiosis to enhance plant–soil interaction. *Sustainability, 14*, 7840.

Khalloufi, M., Martínez-Andújar, C., Lachaâl, M., Karray-Bouraoui, N., Pérez-Alfocea, F., & Albacete, A. (2017). The interaction between foliar GA3 application and arbuscular mycorrhizal fungi inoculation improves growth in salinized tomato (*Solanum lycopersicum* L.) plants by modifying the hormonal balance. *Journal of Plant Physiology, 214*, 134–144.

Khan, A. A., Jilani, G., Akhtar, M. S., Naqvi, S. M. S., & Rasheed, M. (2009a). Phosphorus solubilizing bacteria: Occurrence, mechanisms and their role in crop production. *Journal of Agricultural Biology Science, 1*, 48–58.

Khan, M. S., Zaidi, A., & Wani, P. A. (2009b). Role of phosphate solubilizing microorganisms in sustainable agriculture: A review. In E. Lichtfouse, M. Navarrete, P. Debaeke, S. Véronique, & C. Alberola (Eds.), *Sustainable agriculture* (pp. 551–570). Springer.

Khan, M. S., Ahmad, E., Zaidi, A., & Oves, M. (2013). Functional aspect of phosphate-solubilizing bacteria: Importance in crop production. In D. K. Maheshwari, M. Saraf, & A. Aeron (Eds.), *Bacteria in agrobiology: Crop productivity* (pp. 237–263). Springer.

Khan, Y., Shah, S., & Tian, H. (2022b). The roles of arbuscular mycorrhizal fungi in influencing plant nutrients, photosynthesis, and metabolites of cereal crops: A review. *Agronomy, 12*, 2191.

Kirwan, M. L., & Mudd, S. M. (2012). Response of salt-marsh carbon accumulation to climate change. *Nature, 489*, 550–553.

Knapp, D. G., Németh, J. B., Barry, K., Hainaut, M., Henrissat, B., Johnson, J., Kuo, A., Lim, J. H. P., Lipzen, A., & Nolan, M. (2018). Comparative genomics provides insights into the lifestyle and reveals functional heterogeneity of dark septate endophytic fungi. *Sciences in Report, 8*, 6321.

Ko, O., Eze, C. E., Harriman, C. J., Nwadinobi, C. A., Mbah, E. U., & Akpan, A. U. (2022). Mycorrhiza: A potential application for increased crop productivity in a degraded soil. *Asian Journal of Advantage Agricultural Research, 19*, 1–5.

Kong, K.-F., Vuong, C., & Otto, M. (2006). *Staphylococcus quorum* sensing in biofilm formation and infection. *International Journal of Medicine Microbiology, 296*, 133–139.

Kouas, S., Labidi, N., Debez, A., & Abdelly, C. (2005). Effect of P on nodule formation and N fixation in bean. *Agronomy Sustainability Devices, 25*, 389–393.

Kumar, P., & Singh, R. P. (2021). Microbial diversity and multifunctional microbial biostimulants for agricultural sustainability. In A. Kaushik, C. P. Kaushik, & S. D. Attri (Eds.), *Climate resilience and environmental sustainability approaches* (pp. 141–184). Springer.

Kumar, S., Sindhu, S. S., & Kumar, R. (2022). Biofertilizers: An ecofriendly technology for nutrient recycling and environmental sustainability. *Current Research Microbiology Science, 3*, 100094.

Kumar, S., Panwar, P., Sehrawat, N., Upadhyay, S. K., Sharma, A. K., Singh, M., & Yadav, M. (2024). Oxalic acid: Recent developments for cost-effective microbial production. *Physical Science Review, 9*, 891–907.

Ladha, J. K., Kirk, G. J. D., Bennett, J., Peng, S., Reddy, C. K., Reddy, P. M., & Singh, U. (1998). Opportunities for increased nitrogen-use efficiency from improved lowland rice germplasm. *Field Crops Research, 56*, 41–71.

Lakshmanan, V., Selvaraj, G., & Bais, H. P. (2014). Functional soil microbiome: Belowground solutions to an aboveground problem. *Plant Physiology, 166*, 689–700.

Lanza, M., Haro, R., Conchillo, L. B., & Benito, B. (2019). The endophyte Serendipita indica reduces the sodium content of Arabidopsis plants exposed to salt stress: Fungal ENA ATPases are expressed and regulated at high pH and during plant co-cultivation in salinity. *Environmental Microbiology, 21*, 3364–3378.

Lin JiXiang, L. J., Wang YingNan, W. Y., Sun ShengNan, S. S., Mu ChunSheng, M. C., & Yan XiuFeng, Y. X. (2017). Effects of arbuscular mycorrhizal fungi on the growth, photosynthesis and photosynthetic pigments of Leymus chinensis seedlings under salt-alkali stress and nitrogen deposition. *Science of the Total Environment, 576*, 234–241.

Liszka, M. J., Clark, M. E., Schneider, E., & Clark, D. S. (2012). Nature versus nurture: Developing enzymes that function under extreme conditions. *Annual Reviews and Chemical Biomolecular Engineering, 3*, 77–102.

Liu, X., Li, Y., Han, B., Zhang, Q., Zhou, K., Zhang, X., & Hashemi, M. (2012). Yield response of continuous soybean to one-season crop disturbance in a previous continuous soybean field in Northeast China. *Field Crops Research, 138*, 52–56.

Lopes, C. M., Silva, A. M. M., Estrada-Bonilla, G. A., Ferraz-Almeida, R., Vieira, J. L. V., Otto, R., Vitti, G. C., & Cardoso, E. J. B. N. (2021). Improving the fertilizer value of sugarcane wastes

through phosphate rock amendment and phosphate-solubilizing bacteria inoculation. *Journal of Cleaner Production, 298*, 126821.
Lozupone, C. A., & Knight, R. (2007). Global patterns in bacterial diversity. *Proceedings of the National Academy Science, 104*, 11436–11440.
Lucy, M., Reed, E., & Glick, B. R. (2004). Applications of free living plant growth-promoting rhizobacteria. *Antonie Van Leeuwenhoek, 86*, 1–25.
MacLean, A. M., Finan, T. M., & Sadowsky, M. J. (2007). Genomes of the symbiotic nitrogen-fixing bacteria of legumes. *Plant Physiology, 144*, 615–622.
Maharajan, T., Ceasar, S. A., Ajeesh-Krishna, T. P., Ramakrishnan, M., Duraipandiyan, V., Naif-Abdulla, A., & Ignacimuthu, S. (2018). Utilization of molecular markers for improving the phosphorus efficiency in crop plants. *Plant Breeding, 137*, 10–26.
Maharana, R., Das, S., Dhal, N. K., Dinda, S., & Singh, B. M. (2021). Characterization and mechanisms of biosolubilization of rock phosphate by microbes isolated from mahanadi estuary, Odisha, India. *International Journal of Environment Research, 15*, 335–348.
Malik, K. A., Bilal, R., Mehnaz, S., Rasul, G., Mirza, M. S., & Ali, S. (1997). Association of nitrogen-fixing, plant-growth-promoting rhizobacteria (PGPR) with kallar grass and rice. In *Opportunities for biological nitrogen fixation in rice and other non-legumes: Papers presented at the second working group meeting of the frontier project on nitrogen fixation in rice held at the national institute for biotechnology and genetic engineering (NIBGE), Faisalabad, Pakistan, 13–15 October 1996* (pp. 37–44). Springer.
Man, C. X., Wang, H., Chen, W. F., Sui, X. H., Wang, E. T., & Chen, W. X. (2008). Diverse rhizobia associated with soybean grown in the subtropical and tropical regions of China. *Plant and Soil, 310*, 77–87.
Manoharachary, C., & Mukerji, K. G. (2006). Rhizosphere biology: An overview. *Microbiology Action Rhizoshere, 58*, 1–15.
Mansour, F. A., Aldesuquy, H. S., & Hamedo, H. A. (1994). Studies on plant growth regulators and enzymes production by some bacteria.
Mao, D., Wang, Z., Wu, B., Zeng, Y., Luo, L., & Zhang, B. (2018). Land degradation and restoration in the arid and semiarid zones of China: Quantified evidence and implications from satellites. *Land Degradation Development, 29*, 3841–3851.
Marilley, L., & Aragno, M. (1999). Phylogenetic diversity of bacterial communities differing in degree of proximity of *Lolium perenne* and *Trifolium repens* roots. *Applied Soil Ecology, 13*, 127–136.
Martin, B. D., & Schwab, E. (2012). Current usage of symbiosis and associated terminology. *International Journal of Biology, 5*, 32–45.
McCully, M. (2005). The rhizosphere: The key functional unit in plant/soil/microbial interactions in the field. Implications for the understanding of allelopathic effects.
Meena, M., Swapnil, P., Zehra, A., Aamir, M., Dubey, M. K., Goutam, J., & Upadhyay, R. S. (2017). Beneficial microbes for disease suppression and plant growth promotion. In D. P. Singh, H. B. Singh, & R. Prabha (Eds.), *Plant-microbe interactions in agro-ecological perspectives* (pp. 395–432). Springer.
Meena, R. N., Meena, K., & Choudhary, M. (2023). Organic farming: A key to food security and agricultural sustainability. *Advances in Resting-State Functional MRI, 12*, 1–30.
Mendes, R., Garbeva, P., & Raaijmakers, J. M. (2013). The rhizosphere microbiome: Significance of plant beneficial, plant pathogenic, and human pathogenic microorganisms. *FEMS Microbiology Review, 37*, 634–663.
Mnasri, B., Saïdi, S., Chihaoui, S.-A., & Mhamdi, R. (2012). Sinorhizobium americanum symbiovar mediterranense is a predominant symbiont that nodulates and fixes nitrogen with common bean (*Phaseolus vulgaris* L.) in a Northern Tunisian field. *System Applied Microbiology, 35*, 263–269.
Mohamed, D. J., & Martiny, J. B. (2011). Patterns of fungal diversity and composition along a salinity gradient. *ISME Journal, 5*, 379–388.
Morikawa, M. (2006). Beneficial biofilm formation by industrial bacteria Bacillus subtilis and related species. *Journal Biosciences Bioengineering, 101*, 1–8.

Murphy, A. E., Bulseco, A. N., Ackerman, R., Vineis, J. H., & Bowen, J. L. (2020). Sulphide addition favours respiratory ammonification (DNRA) over complete denitrification and alters the active microbial community in salt marsh sediments. *Environment Microbiology, 22*, 2124–2139.

Nesme, T., Metson, G. S., & Bennett, E. M. (2018). Global phosphorus flows through agricultural trade. *Globally Environment Change, 50*, 133–141.

Nyamedi, I. M. (2023). Role of *Aspergillus niger* in Solubilizing P in Togo rock phosphate during composting with different organic wastes. *International Journal Plant Soil Sciences, 35*, 99–104.

Oldroyd, G. E. (2013). Speak, friend, and enter: Signalling systems that promote beneficial symbiotic associations in plants. *Nature Review Microbiology, 11*, 252–263.

Oren, A. (2008). Microbial life at high salt concentrations: Phylogenetic and metabolic diversity. *Saline System, 4*, 2.

Pande, A., Pandey, P., Mehra, S., Singh, M., & Kaushik, S. (2017). Phenotypic and genotypic characterization of phosphate solubilizing bacteria and their efficiency on the growth of maize. *Journal of Genetics Engineering Biotechnology, 15*, 379–391.

Panke-Buisse, K., Poole, A. C., Goodrich, J. K., Ley, R. E., & Kao-Kniffin, J. (2015). Selection on soil microbiomes reveals reproducible impacts on plant function. *ISME Journal, 9*, 980–989.

Parihar, M., Chitara, M. K., Ram, H., Kumari, A., Tiwari, G., Rana, K., Gorain, B., Kumar, U., Bisht, J. K., & Kant, L. (2022). Role of AM fungi in growth promotion of high-value crops. *New and Future Developments in Microbial Biotechnology and Bioengineering, 69*, 121–144.

Parnell, J. J., Vintila, S., Tang, C., Wagner, M. R., & Kleiner, M. (2024). Evaluation of ready-to-use freezer stocks of a synthetic microbial community for maize root colonization. *Microbiology Spectrum, 12*, e02401-e2423.

Patten, C. L., & Glick, B. R. (1996). Bacterial biosynthesis of indole-3-acetic acid. *Canadian Journal of Microbiology, 42*, 207–220.

Philippot, L., Raaijmakers, J. M., Lemanceau, P., & Van Der Putten, W. H. (2013). Going back to the roots: The microbial ecology of the rhizosphere. *Nature Review Microbiology, 11*, 789–799.

Pierik, R., Tholen, D., Poorter, H., Visser, E. J., & Voesenek, L. A. (2006). The Janus face of ethylene: Growth inhibition and stimulation. *Trends Plant Science, 11*, 176–183.

Poole, P., & Allaway, D. (2000). Carbon and nitrogen metabolism in Rhizobium.

Porcel, R., Redondo-Gómez, S., Mateos-Naranjo, E., Aroca, R., Garcia, R., & Ruiz-Lozano, J. M. (2015). Arbuscular mycorrhizal symbiosis ameliorates the optimum quantum yield of photosystem II and reduces non-photochemical quenching in rice plants subjected to salt stress. *Journal of Plant Physiology, 185*, 75–83.

Prăvălie, R. (2021). Exploring the multiple land degradation pathways across the planet. *Earth-Science Review, 220*, 103689.

Qin, L., Jiang, H., Tian, J., Zhao, J., & Liao, H. (2011). Rhizobia enhance acquisition of phosphorus from different sources by soybean plants. *Plant and Soil, 349*, 25–36.

Quiza, L., St-Arnaud, M., & Yergeau, E. (2015). Harnessing phytomicrobiome signaling for rhizosphere microbiome engineering. *Frontier Plant Sciences, 6*, 507.

Rafi, M. M., Krishnaveni, M. S., & Charyulu, P. (2019). Phosphate-solubilizing microorganisms and their emerging role in sustainable agriculture. *Recent Development Applied Microbiology Biochemistry, 15*, 223–233.

Raimi, A., Roopnarain, A., & Adeleke, R. (2021). Biofertilizer production in Africa: Current status, factors impeding adoption and strategies for success. *Science African, 11*, e00694.

Rajguru, B. R., & Bhatt, V. D. (2022). Review on mechanism of mineral phosphate solubilization in fast-growing rhizobia based on sugar utilization. *Plant Breedings and Biotechnology, 10*, 203–211.

Ramamoorthy, P., Karthikeyan, M., & Nirubana, V. (2020). Role of phosphate solubilizing microorganisms in agriculture. *Agricultural Mirror Future India, 1*, 104.

Rashid, M., Khalil, S., Ayub, N., Alam, S., & Latif, F. (2004). Organic acids production and phosphate solubilization by phosphate solubilizing microorganisms (PSM) under in vitro conditions. *Pakistan Journal Biology Sciences, 7*, 187–196.

Rath, K. M., & Rousk, J. (2015). Salt effects on the soil microbial decomposer community and their role in organic carbon cycling: A review. *Soil Biology Biochemistry, 81*, 108–123.
Rath, K. M., Fierer, N., Murphy, D. V., & Rousk, J. (2019). Linking bacterial community composition to soil salinity along environmental gradients. *ISME Journal, 13*, 836–846.
Reyes, I., Baziramakenga, R., Bernier, L., & Antoun, H. (2001). Solubilization of phosphate rocks and minerals by a wild-type strain and two UV-induced mutants of *Penicillium rugulosum. Soil Biology Biochemistry, 33*, 1741–1747.
Riggs, P. J., Chelius, M. K., Iniguez, A. L., Kaeppler, S. M., & Triplett, E. W. (2001). Enhanced maize productivity by inoculation with diazotrophic bacteria. *Functional Plant Biology, 28*, 829–836.
Rodriguez, H., & Fraga, R. (1999). Phosphate solubilizing bacteria and their role in plant growth promotion. *Biotechnology Advances, 17*, 319–339.
Rogers, K., Kelleway, J. J., Saintilan, N., Megonigal, J. P., Adams, J. B., Holmquist, J. R., Lu, M., Schile-Beers, L., Zawadzki, A., & Mazumder, D. (2019). Wetland carbon storage controlled by millennial-scale variation in relative sea-level rise. *Nature, 567*, 91–95.
Romano, J. D., & Kolter, R. (2005). *Pseudomonas*-Saccharomyces interactions: Influence of fungal metabolism on bacterial physiology and survival. *Journal of Bacteriology, 187*, 940–948.
Romanova, Y. M., Smirnova, T. A., Andreev, A. L., Il'Ina, T. S., Didenko, L. V., & Gintsburg, A. L. (2006). Formation of biofilms as an example of the social behavior of bacteria. *Microbiology, 75*, 481–485.
Romero, E. (2002). Advances in Rhizobium research. *Critical Review Plant Science, 21*, 0735–260291044278.
Rozema, J., & Flowers, T. (2008). Crops for a salinized world. *Science, 322*, 1478–1480.
Rudrappa, T., Biedrzycki, M. L., & Bais, H. P. (2008). Causes and consequences of plant-associated biofilms. *FEMS Microbiology Ecology, 64*, 153–166.
Sadhana, B. (2014). Arbuscular Mycorrhizal Fungi (AMF) as a biofertilizer: A review. *International Journal of Current Microbiology Applied Sciences, 3*, 384–400.
Saif, S., Abid, Z., Ashiq, M. F., Altaf, M., & Ashraf, R. S. (2021). Biofertilizer formulations. In M. I. Inamuddin, R. Ahamed, S. Boddula, & M. Rezakazemi (Eds.), *Biofertilizers* (pp. 211–256). Wiley.
Santaella, C., Schue, M., Berge, O., Heulin, T., & Achouak, W. (2008). The exopolysaccharide of *Rhizobium* sp. YAS34 is not necessary for biofilm formation on *Arabidopsis thaliana* and *Brassica napus* roots but contributes to root colonization. *Environmental Microbiology, 10*, 2150–2163.
Santana, E. B., Marques, E. L. S., & Dias, J. C. T. (2016). Effects of phosphate-solubilizing bacteria, native microorganisms, and rock dust on *Jatropha curcas* L. growth. *Genetics Molecular Research, 15*, 15048729.
Santner, A., Calderon-Villalobos, L. I. A., & Estelle, M. (2009). Plant hormones are versatile chemical regulators of plant growth. *Natural Chemical Biology, 5*, 301–307.
Sardinha, M., Müller, T., Schmeisky, H., & Joergensen, R. G. (2003). Microbial performance in soils along a salinity gradient under acidic conditions. *Applied Soil Ecology, 23*, 237–244.
Schimel, J., Balser, T. C., & Wallenstein, M. (2007). Microbial stress-response physiology and its implications for ecosystem function. *Ecology, 88*, 1386–1394.
Seenivasagan, R., & Babalola, O. O. (2021). Utilization of microbial consortia as biofertilizers and biopesticides for the production of feasible agricultural product. *Biology, 10*, 1111.
Seneviratne, G. (2003). Development of eco-friendly, beneficial microbial biofilms. *Current Science, 85*, 1395–1396.
Seneviratne, G. (2008). Biological nitrogen fixation: Potential biotechnological applications beyond biofertilizers. *Current Science, 95*, 1142.
Seneviratne, G. (2009). Collapse of beneficial microbial communities and deterioration of soil health: A cause for reduced crop productivity. *Current Science, 96*, 633.
Seneviratne, G., & Jayasinghearachchi, H. S. (2003). Mycelial colonization by bradyrhizobia and azorhizobia. *Journal of Bioscience, 28*, 243–247.

Seneviratne, G., & Jayasinghearachchi, H. S. (2005). A rhizobial biofilm with nitrogenase activity alters nutrient availability in a soil. *Soil Biology Biochemistry, 37*, 1975–1978.
Seneviratne, G., Kecskés, M. L., & Kennedy, I. R. (2008a). Biofilmed biofertilisers: novel inoculants for efficient nutrient use in plants. In *ACIAR proceedings* pp. 126–130.
Sharma, S. B., Sayyed, R. Z., Trivedi, M. H., & Gobi, T. A. (2013). Phosphate solubilizing microbes: Sustainable approach for managing phosphorus deficiency in agricultural soils. *Springerplus, 2*, 587.
Sharma, D., Gahtyari, N. C., Chhabra, R., & Kumar, D. (2020). Role of microbes in improving plant growth and soil health for sustainable agriculture. In A. N. Yadav, A. A. Rastegari, N. Yadav, & D. Kour (Eds.), *Advances in plant microbiome and sustainable agriculture* (pp. 207–256). Springer.
Sharma, R., Pal, J., & Kaur, M. (2017). Isolation of phosphate solubilizing Pseudomonas strains from apple rhizosphere in the Trans Himalayan region of Himachal Pradesh, India. *BioRxiv* 193672.
Siddique, R., Gul, A., Ozturk, M., & Altay, V. (2021). Phosphate solubilizing bacteria for soil sustainability. In M. N. V. Prasad (Ed.), *Handbook of assisted and amendment: Enhanced sustainable remediation technology* (pp. 423–435). Wiley.
Silva, C., Vinuesa, P., Eguiarte, L. E., Martínez-Romero, E., & Souza, V. (2003). *Rhizobium etli* and *Rhizobium gallicum* nodulate common bean (*Phaseolus vulgaris*) in a traditionally managed milpa plot in Mexico: Population genetics and biogeographic implications. *Applied Environmental Microbiology, 69*, 884–893.
Singal, R., Gupta, R., & Saxena, R. K. (1994). Rock phosphate solubilization under alkaline conditions by *Aspergillus japonicus* and *A. foetidus*. *Folia Microbiology, 39*, 33–36.
Sun, W., & Shahrajabian, M. H. (2023). The application of arbuscular mycorrhizal fungi as microbial biostimulant, sustainable approaches in modern agriculture. *Plants, 12*, 3101.
Sun, X., Zhao, J., Zhou, X., Bei, Q., Xia, W., Zhao, B., Zhang, J., & Jia, Z. (2022). Salt tolerance-based niche differentiation of soil ammonia oxidizers. *ISME Journal, 16*, 412–422.
Taha, S. M., Mahmoud, S. A. Z., El-Damaty, A. H., & El-Hafez, A. A. (1969). Activity of phosphate-dissolving bacteria in Egyptian soils. *Plant and Soil, 56*, 149–160.
Tanimoto, E. (2005). Regulation of root growth by plant hormones—roles for auxin and gibberellin. *Critical Review Plant Sciences, 24*, 249–265.
Thiem, D., Gołębiewski, M., Hulisz, P., Piernik, A., & Hrynkiewicz, K. (2018). How does salinity shape bacterial and fungal microbiomes of *Alnus glutinosa* roots? *Frontier Microbiology, 9*, 651.
Tian, C. F., Zhou, Y. J., Zhang, Y. M., Li, Q. Q., Zhang, Y. Z., Li, D. F., Wang, S., Wang, J., Gilbert, L. B., & Li, Y. R. (2012). Comparative genomics of rhizobia nodulating soybean suggests extensive recruitment of lineage-specific genes in adaptations. *Proceedings of the Nationall Academy Sciences, 109*, 8629–8634.
Tilak, K., Ranganayaki, N., Pal, K. K., De, R., Saxena, A. K., Nautiyal, C. S., Mittal, S., Tripathi, A. K., & Johri, B. N. (2005). Diversity of plant growth and soil health supporting bacteria. *Current Science, 2005*, 136–150.
Timofeeva, A., Galyamova, M., & Sedykh, S. (2022). Prospects for using phosphate-solubilizing microorganisms as natural fertilizers in agriculture. *Plants, 11*, 2119.
Tjepkema, J., & Evans, H. J. (1975). Nitrogen fixation by free-living Rhizobium in a defined liquid medium. *Biochemistry in Biophysics Research Communication, 65*, 625–628.
Unkovich, M. J., & Pate, J. S. (2000). An appraisal of recent field measurements of symbiotic N_2 fixation by annual legumes. *Field Crops Research, 65*, 211–228.
van Dijk, G., Smolders, A. J., Loeb, R., Bout, A., Roelofs, J. G., & Lamers, L. P. (2015). Salinization of coastal freshwater wetlands; effects of constant versus fluctuating salinity on sediment biogeochemistry. *Biogeochemistry, 126*, 71–84.
Van Loon, L. C. (2007). Plant responses to plant growth-promoting rhizobacteria. *New Perspectives and Approaches in Plant Growth-Promoting Rhizobacteria Research, 147*, 243–254.
Van Horn, D. J., Okie, J. G., Buelow, H. N., Gooseff, M. N., Barrett, J. E., & Takacs-Vesbach, C. D. (2014). Soil microbial responses to increased moisture and organic resources along a salinity gradient in a polar desert. *Applied Environmental Microbiology, 80*, 3034–3043.

Vandevivere, P., & Kirchman, D. L. (1993). Attachment stimulates exopolysaccharide synthesis by a bacterium. *Applied Environmental Microbiology, 59*, 3280–3286.

Vassileva, M., Malusà, E., Sas-Paszt, L., Trzcinski, P., Galvez, A., Flor-Peregrin, E., Shilev, S., Canfora, L., Mocali, S., & Vassilev, N. (2021). Fermentation strategies to improve soil bio-inoculant production and quality. *Microorganisms, 9*, 1254.

Vazquez, P., Holguin, G., Puente, M. E., Lopez-Cortes, A., & Bashan, Y. (2000). Phosphate-solubilizing microorganisms associated with the rhizosphere of mangroves in a semiarid coastal lagoon. *Biology Fertilizer Soils, 30*, 460–468.

Verástegui-Valdés, M. M., Zhang, Y. J., Rivera-Orduna, F. N., Cheng, H.-P., Sui, X. H., & Wang, E. T. (2014). Microsymbionts of *Phaseolus vulgaris* in acid and alkaline soils of Mexico. *Systematic Applied Microbiology, 37*, 605–612.

Vessey, J. K. (2003). Plant growth promoting rhizobacteria as biofertilizers. *Plant and Soil, 255*, 571–586.

Vishwakarma, K., Kumar, N., Shandilya, C., Mohapatra, S., Bhayana, S., & Varma, A. (2020). Revisiting plant–microbe interactions and microbial consortia application for enhancing sustainable agriculture: A review. *Frontier Microbiology, 11*, 560406.

Volpe, V., Giovannetti, M., Sun, X.-G., Fiorilli, V., & Bonfante, P. (2016). The phosphate transporters LjPT4 and MtPT4 mediate early root responses to phosphate status in non mycorrhizal roots. *Plant Cell Environment, 39*, 660–671.

Vyas, P., & Gulati, A. (2009). Organic acid production in vitro and plant growth promotion in maize under controlled environment by phosphate-solubilizing fluorescent Pseudomonas. *BMC Microbiology, 9*, 174.

Wang, H., Man, C. X., Wang, E. T., & Chen, W. X. (2009). Diversity of rhizobia and interactions among the host legumes and rhizobial genotypes in an agricultural-forestry ecosystem. *Plant and Soil, 314*, 169–182.

Wang, E. T., Tian, C. F., Chen, W. F., Young, J. P. W., Chen, W. X., & Wang, E. T. (2019). Symbiosis between rhizobia and legumes. *Ecological Evolution in Rhizobia Princess Application, 14*, 3–19.

Wang, F., Sanders, C. J., Santos, I. R., Tang, J., Schuerch, M., Kirwan, M. L., Kopp, R. E., Zhu, K., Li, X., & Yuan, J. (2021). Global blue carbon accumulation in tidal wetlands increases with climate change. *National Science Review, 8*, nwaa296.

Wang, H., Dai, Z., Trettin, C. C., Krauss, K. W., Noe, G. B., Burton, A. J., Stagg, C. L., & Ward, E. J. (2022). Modeling impacts of drought-induced salinity intrusion on carbon dynamics in tidal freshwater forested wetlands. *Ecological Application, 32*, e2700.

Wani, S. P., Rupela, O. P., & Lee, K. K. (1995). Sustainable agriculture in the semi-arid tropics through biological nitrogen fixation in grain legumes. In *Management of biological nitrogen fixation for the development of more productive and sustainable agricultural systems: Extended versions of papers presented at the symposium on biological nitrogen fixation for sustainable agriculture at the 15th congress of soil science, Acapulco, Mexico, 1994* (pp. 29–49). Springer.

Wani, S. P. (1986). *Cereal nitrogen fixation: Proceedings of the Working Group Meeting held at ICRISAT Center, India, 9–12 October 1984*. International Crops Research Institute for the Semi-Arid Tropics.

Whitelaw, M. A., Harden, T. J., & Helyar, K. R. (1999). Phosphate solubilisation in solution culture by the soil fungus *Penicillium radicum*. *Soil Biology Biochemistry, 31*, 655–665.

Whittle, A., Barnett, R. L., Charman, D. J., & Gallego-Sala, A. V. (2022). Low-salinity transitions drive abrupt microbial response to sea-level change. *Ecology Letters, 25*, 17–25.

Wu, X., Yang, J., Ruan, H., Wang, S., Yang, Y., Naeem, I., Wang, L., Liu, L., & Wang, D. (2021). The diversity and co-occurrence network of soil bacterial and fungal communities and their implications for a new indicator of grassland degradation. *Ecological Indication, 129*, 107989.

Xie, X., Huang, W., Liu, F., Tang, N., Liu, Y., Lin, H., & Zhao, B. (2013). Functional analysis of the novel mycorrhiza-specific phosphate transporter A s PT 1 and PHT 1 family from a *Stragalus sinicus* during the arbuscular mycorrhizal symbiosis. *New Phytology, 198*, 836–852.

Yadav, P., Yadav, B., Deka, D., & Narayan, O. P. (2023). Role of arbuscular mycorrhizas in regulating physiological and molecular aspects of plants under abiotic stress. In *The role of growth regulators and phytohormones in overcoming environmental stress* (pp. 399–433). Elsevier.

Yang, S.-Y., Grønlund, M., Jakobsen, I., Grotemeyer, M. S., Rentsch, D., Miyao, A., Hirochika, H., Kumar, C. S., Sundaresan, V., & Salamin, N. (2012). Nonredundant regulation of rice arbuscular mycorrhizal symbiosis by two members of the PHOSPHATE TRANSPORTER1 gene family. *The Plant Cell, 24*, 4236–4251.

Yanni, Y. G., Rizk, R. Y., Abd El-Fattah, F. K., Squartini, A., Corich, V., Giacomini, A., de Bruijn, F., Rademaker, J., Maya-Flores, J., & Ostrom, P. (2001). The beneficial plant growth-promoting association of *Rhizobium leguminosarum* bv. trifolii with rice roots. *Functional Plant Biology, 28*, 845–870.

Young, J. P. W. (1992). Phylogenetic classification of nitrogen-fixing organisms. *Biological Nitrogen Fixation, 1544*, 43–86.

Yuan, Z., Druzhinina, I. S., Labbé, J., Redman, R., Qin, Y., Rodriguez, R., Zhang, C., Tuskan, G. A., & Lin, F. (2016). Specialized microbiome of a halophyte and its role in helping non-host plants to withstand salinity. *Science Report, 6*, 32467.

Zaidi, A., Khan, M., Ahemad, M., & Oves, M. (2009). Plant growth promotion by phosphate solubilizing bacteria. *Acta Microbiology Immunology in Hungarica, 56*, 263–284.

Zaman, M., Shahid, S. A., Heng, L., Shahid, S. A., Zaman, M., & Heng, L. (2018). Soil salinity: Historical perspectives and a world overview of the problem. In *Guideline for Salinity Assessment, Mitigation and Adaptation Using Nuclear and Related Techniques* (pp. 43–53).

Zavahir, J. S., Jayasekara, A., Seneviratne, G., & De Silva, M. (2008). Potential application of biofilms: A new approach for tea gardens. *Tea Bulletin, 20*, 1–6.

Zhang, Y. M., Li, Y., Jr., Chen, W. F., Wang, E. T., Tian, C. F., Li, Q. Q., Zhang, Y. Z., Sui, X. H., & Chen, W. X. (2011b). Biodiversity and biogeography of rhizobia associated with soybean plants grown in the North China Plain. *Applied Environmental Microbiology, 77*, 6331–6342.

Zhang, K., Shi, Y., Cui, X., Yue, P., Li, K., Liu, X., Tripathi, B. M., & Chu, H. (2019). Salinity is a key determinant for soil microbial communities in a desert ecosystem. *mSystems, 4*, 18.

Zhang, G., Bai, J., Tebbe, C. C., Zhao, Q., Jia, J., Wang, W., Wang, X., & Yu, L. (2021a). Salinity controls soil microbial community structure and function in coastal estuarine wetlands. *Environmental Microbiology, 23*, 1020–1037.

Zhang, H., Han, L., Jiang, B., & Long, C. (2021b). Identification of a phosphorus-solubilizing *Tsukamurella tyrosinosolvens* strain and its effect on the bacterial diversity of the rhizosphere soil of peanuts growth-promoting. *World Journal of Microbiology Biotechnology, 37*, 109.

Zhao, C. T., Wang, E. T., Chen, W. F., & Chen, W. X. (2008). Diverse genomic species and evidences of symbiotic gene lateral transfer detected among the rhizobia associated with Astragalus species grown in the temperate regions of China. *FEMS Microbiology Letters, 286*, 263–273.

Zhou, M., Butterbach-Bahl, K., Vereecken, H., & Brüggemann, N. (2017). A meta-analysis of soil salinization effects on nitrogen pools, cycles and fluxes in coastal ecosystems. *Global Change Biology, 23*, 1338–1352.

Zhu, F., Qu, L., Hong, X., & Sun, X. (2011). Isolation and characterization of a phosphate: Solubilizing halophilic bacterium *Kushneria* sp. YCWA18 from Daqiao saltern on the coast of Yellow Sea of China. *Evidence Based Complementary in Alternative Medicine, 2011*, 615032.

Zurdo-Pineiro, J. L., García-Fraile, P., Rivas, R., Peix, A., León-Barrios, M., Willems, A., Mateos, P. F., Martínez-Molina, E., Velázquez, E., & van Berkum, P. (2009). Rhizobia from Lanzarote, the Canary Islands, that nodulate *Phaseolus vulgaris* have characteristics in common with Sinorhizobium meliloti isolates from mainland Spain. *Applied Environmental Microbiology, 75*, 2354–2359.

Biogeochemical Processes and Efficient Utilization of Soil Organic Phosphorus

Lei Zhang, Zhe Liu, Wannian Zhang, Xionghan Feng, Wenfeng Tan, and Xiaoming Wang

Abstract Organic phosphorus (OP) serves as a significant soil phosphorus (P) pool, with its biogeochemical processes being essential components of soil P cycle. These processes are critical for maintaining plant growth and dynamic balance of soil P. This paper comprehensively reviews the biogeochemical processes and efficient utilization of OP, encompassing the sources and forms of soil OP, the adsorption–desorption and mineralization processes of OP, key influencing factor, relevant advanced analytic techniques, and regulation measures of OP. The soil OP primarily originates from plant and animal residues, microbial activity, and anthropogenic inputs, and its bioavailability is remarkably influenced by the adsorption and desorption processes on soil minerals. In addition, soil minerals and phosphatases, secreted by microorganisms and plant roots, catalyze the mineralization of OP compounds, leading to the release of phosphate and thus increasing P bioavailability. The key factors influencing OP mineralization include soil pH, adsorption mechanism, phosphatase activity, the coexistence of cation and organic matter, as well as microbial interactions. Furthermore, we provide an overview of advanced analytical techniques commonly used in soil OP research and propose strategies to enhance the bioavailability of soil OP from a farm management perspective. Finally, we outline potential future research directions and challenges aimed at further elucidating the biogeochemical processes of OP under complex environmental conditions.

Keywords Organic phosphorus · Biogeochemical · Soil minerals · Adsorption · Mineralization · Bioavailability

L. Zhang · Z. Liu · W. Zhang · X. Feng · W. Tan · X. Wang (✉)
Key Laboratory of Arable Land Conservation (Middle and Lower Reaches of Yangtze River), Ministry of Agriculture and Rural Affairs, State Environmental Protection Key Laboratory of Soil Health and Green Remediation, Hubei Key Laboratory of Soil Environment and Pollution Remediation, College of Resources and Environment, Huazhong Agricultural University, Wuhan 430070, China
e-mail: wangxm338@mail.hzau.edu.cn

M. Shaaban (ed.), *Soils and Sustainable Agriculture*, Frontier Studies in Soil Science,
https://doi.org/10.1007/978-3-031-91114-9_21

1 Introduction

Phosphorus (P) is an essential nutrient for all living organisms, crucial for the synthesis of nucleic acids, the formation of cell membranes, and the transfer of energy (Paytan & McLaughlin, 2007; Turner & Haygarth, 2001). However, as the global population grows, the widespread application of P fertilizers intended to enhance crop yields has resulted in the unintended accumulation of P in the topsoil, leading to potential environmental issues such as eutrophication (Withers et al., 2014). P fertilizers for agricultural production are mainly derived from phosphate rock, a non-renewable resource projected to be exhausted within the next 50–100 years based on current extraction rates and global demand (Cordell et al., 2009). Thus, improving the bioavailability of P in soils has emerged as a critical challenge for sustainable agriculture.

The biogeochemical cycling of organic phosphorus (OP) is a key component of the global P cycle, significantly influencing the availability of P in ecosystems (Kononova & Nesmeyanova, 2002; Richardson et al., 2009; Turner et al., 2005). Soil OP accounts for about 30–65% of total soil P, and up to 90% in some high organic matter soils (Tabatabai, 1989). Despite its abundance, OP cannot be utilized directly by plants. It must be converted through a series of biogeochemical processes to release inorganic P (Pi) that can be absorbed and utilized by plants, a process that is critical in maintaining ecosystem productivity, plant growth, and soil fertility (Liu et al., 2022b; Park et al., 2022). The biogeochemical cycling of OP compounds involves a complex series of processes such as adsorption, desorption, mineralization and remineralization, which are controlled by a combination of biotic and abiotic factors (Olsson et al., 2011; Wan et al., 2022; Yan et al., 2022a). The mineralization of OP into Pi by soil minerals or phosphatases is particularly critical for making P accessible to plants (Basinski et al., 2024; Chen et al., 2023b; Zhang et al., 2018). Therefore, understanding the cycling processes and mechanisms of OP is essential not only for agricultural production but also for the sustainable development of ecosystems.

In recent years, with the advancement of molecular biology and interfacial chemistry techniques, the distribution and transformation mechanisms of OP in soil and its influencing factors have been explored in greater depth. For example, advanced techniques such as nuclear magnetic resonance spectroscopy (NMR), in-situ attenuated total reflectance Fourier transform infrared spectroscopy (ATR-FTIR), and $\delta^{18}O$–P isotope labeling have been widely used to study the speciation and interfacial reactions, as well as the mineralization mechanisms and pathways of OP in the soil environment (Huang et al., 2018; Olsson et al., 2011; Wu et al., 2015). As our understanding of the biogeochemical cycling processes of soil OP has deepened, regulating these processes to achieve high bioavailability of OP has become a research hotspot. This includes different soil management practices such as modifying tillage methods, applying organic fertilizers, and enhancing microbial interactions (Wang et al., 2016a; Wu et al., 2005; Yang et al., 2021; Zhang et al., 2014).

This chapter systematically summarizes the biogeochemical cycle of OP, exploring its sources, distribution, behavior, and transformation in soils. It also analyzes the major factors affecting these processes, including pH, temperature, and mineral surface properties. This chapter aims to provide a scientific basis and practical guidance for optimizing the use of OP in agriculture and ecological management by summarizing the latest research progress, proposing measures to enhance soil OP effectiveness, and discussing future research directions and challenges.

2 Sources and Forms of Soil Organic Phosphorus

2.1 Sources of Soil Organic Phosphorus

Soil OP refers to P compounds bound to carbon atoms, such as nucleic acids, nucleotides, phytate, phospholipids, glyco-phosphates, phosphoproteins, and phosphonates (Baldwin, 2013; Turner et al., 2005). These compounds primarily enter the soil system through plant and animal residues, microbial metabolism, animal waste, and human activities (Fig. 1) (Turner et al., 2005).

2.1.1 Animal and Plant Residues

Plants absorb carbon dioxide from the environment through photosynthesis and use Pi from the soil to synthesize OP compounds such as nucleic acids, phospholipids, and adenosine triphosphate. These compounds are stored in the roots, stems, leaves, and fruits of the plants. While animals obtain OP from plants through the food chain, it is converted into biomolecules within their bodies. When plants and animals die, the P-containing organic matter enters the soil and is subsequently broken down by microorganisms (Condron et al., 2005; Richardson et al., 2009; Rowan, 1966; Turner et al., 2005).

2.1.2 Microbial Metabolism

Once plant and animal residues enter the soil, microorganisms (including bacteria, fungi and actinomycetes) begin to decompose these organic materials. During the decomposition process, phosphatases secreted by microorganisms break down complex OP compounds into simple Pi or small-molecule OP compounds. These products can either be absorbed and utilized by plants or contribute to the formation of a stable OP reservoir in the soil (Condron et al., 2005; Richardson et al., 2009; Turner et al., 2005).

Microbial metabolism involves not only catabolic processes but also the re-synthesis and transformation of OP. Microorganisms re-synthesize OP to build new

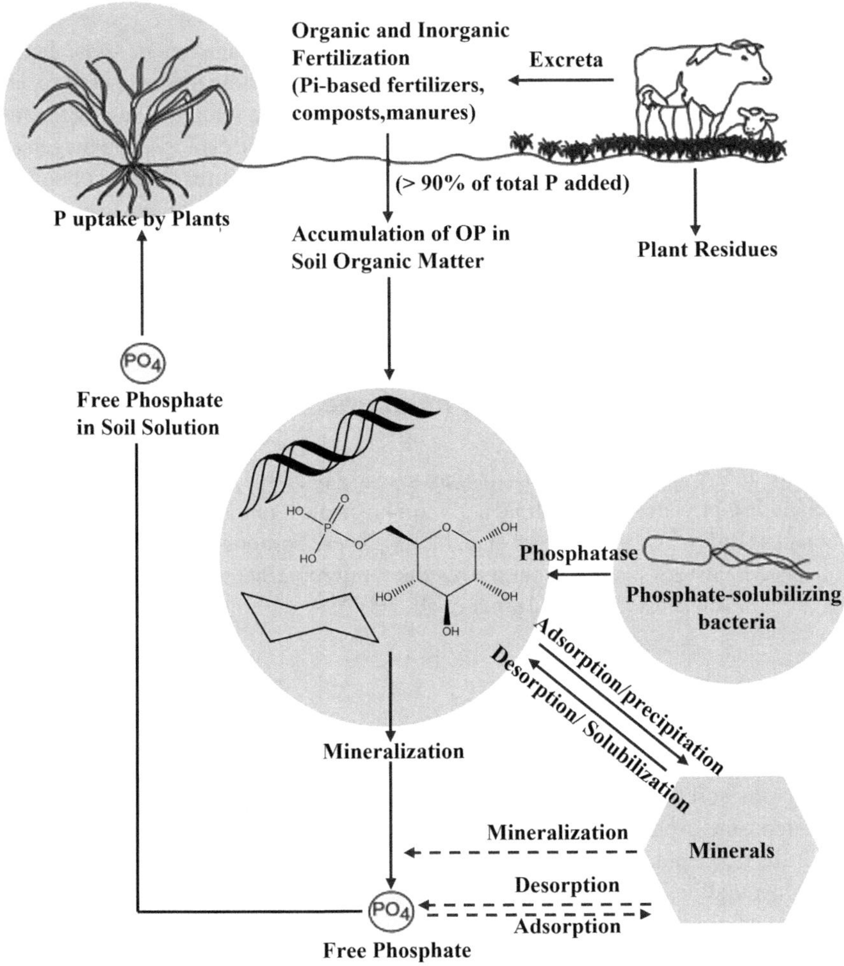

Fig. 1 Schematic diagram of P cycling in the soil environment (adapted from Jorquera et al., 2008)

biomolecules, such as deoxyribonucleic acid, ribonucleic acid, and phospholipids. After the microorganisms die, these compounds are released back into the soil, becoming an important component of soil OP (Condron et al., 2005; Richardson et al., 2009; Turner et al., 2005).

2.1.3 Animal Excreta

Animals obtain P by consuming P-containing plants, microorganisms, and other organic matter, which include various OP compounds such as phospholipids, nucleic acids, and phosphoproteins. In the animal's digestive system, these OP compounds are partially broken down into simpler forms of OP compounds or Pi. These compounds then enter the soil through the animal's excreta. The OP forms in the excreta mainly consist of incompletely digested food residues, metabolites formed during digestion, and biological secretions from the animal's body (Condron et al., 2005; Richardson et al., 2009; Turner et al., 2005).

2.1.4 Human Activities

Agricultural activities are a major source of soil OP, with the extensive use of synthetic P fertilizers being particularly significant. Studies have shown that following the application of P fertilizers, some Pi can be converted into OP forms through microbial processes, thereby influencing the soil P cycle and plant P uptake (Richardson et al., 2009). In addition, sewage irrigation and sludge application are significant pathways for OP compounds to enter the soil. Sludge, which is rich in phospholipids, nucleic acids, and other OP compounds, can significantly increase the soil OP content (Condron et al., 2005; Richardson et al., 2009; Turner et al., 2005).

2.2 *Forms of Soil Organic Phosphorus*

Based on the bonding characteristics of P atoms, soil OP can be categorized into phosphate esters, phosphonic acids, and phosphoric anhydrides. Phosphate esters are the most common OP in soil and are primarily divided into mono- and di-phosphate esters, with nearly 100% of the total OP in soil existing as mono-phosphate esters ($R–O–PO_3^{2-}$, where R represents the organic group) (Condron et al., 2005; Turner et al., 2005). Common phosphate monoesters in soil include glucose-6-phosphate (G6P) and phytate, among others. Phosphate diesters include deoxyribonucleic acid (DNA) and ribonucleic acid (RNA) (Turner et al., 2005), among others. Phosphonic acid OP are organic compounds where P is covalently bonded to carbon (C–P). Natural phosphonates typically have structures like [$RP(O)(OH)_2$] or [$R_2P(O)(OH)$, RHP(O)(OH)] (Corbridge, 2013), with examples including 2-aminoethylphosphonic acid (2-AEP) and aminomethylphosphoric acid (AMPA) (Giesy et al., 2000; Metcalf & Van Der Donk, 2009). Phosphoric anhydride OP contains phosphoric anhydride bonds (RP–O–P), which are typically high-energy bonds crucial for energy transfer in living organisms, such as in adenosine triphosphate (ATP) and adenosine diphosphate (ADP) (Hall, 1976).

2.2.1 Phytate

Phytate is the primary form of stored P in plants and is commonly found in tissues such as seeds and grains (Liu et al., 2022b). This form of P storage plays a crucial role in seed germination and the early growth stages of plants. Phytate, which constitutes

~50–60% of soil OP, is a complex OP compound composed of six phosphoric acid groups and an inositol ring structure, making it highly chemically stable (Fig. 2a). Phytate has a molecular weight of 660 g/mol and contains 12 ionizable protons along with six phosphate groups, making it highly anionic and strongly coordinating within the soil environment (Menezes-Blackburn et al., 2013).

In the soil environment, phytate exists in four main stereoisomeric forms: myo-inositol (myo), cis-myo-inositol (scyllo), D-myo-inositol (D-chiro), and neo-myo-inositol (neo). Myo-inositol is the predominant form, accounting for 56–90% of the total phytate (Fig. 2a) (Liu et al., 2022b). The presence of these isomers results in varying chemical properties and environmental behaviors of phytate. Although phytate is an important storage form of P, its bioavailability in soil is low due to its strong adsorption and stability. To be absorbed and utilized by plants, phytate must be hydrolyzed by phytase enzymes secreted by plant roots or microorganisms, releasing Pi.

2.2.2 Ribonucleic Acid

Ribonucleic acid (RNA) in soil, primarily derived from plant and animal residues and microbial activities, is a long-chain molecule. It consists of nucleotide units linked by phosphodiester bonds, with each nucleotide containing a phosphate group, a ribose sugar (pentose), and a nitrogen-containing base (adenine, guanine, cytosine, or uracil) (Fig. 2b) (Turner et al., 2005). Microbial metabolic activity can convert RNA into various forms of phosphate and phosphoric anhydride, each with distinct solubility and mobility in the soil. These transformations influence P cycling and redistribution within the soil environment (McLaren et al., 2015).

2.2.3 Adenosine Triphosphate

Adenosine triphosphate (ATP) is a P-containing nucleotide composed of adenine, ribose, and three phosphate groups. Its high-energy phosphoric anhydride bonds play a key role in energy transfer and storage in living organisms. ATP, as an important form of OP, increases the bioavailability of P by being broken down into soluble Pi. This occurs when acid phosphatases or soil minerals disrupt the high-energy phosphoric anhydride bonds within ATP (Condron et al., 2005; Zhang et al., 2023a).

2.2.4 2-Aminoethylphosphonic Acid

The chemical structure of 2-aminoethylphosphonic acid (2-AEP) consists of an ethyl chain ($–CH_2–CH_2–$) attached to an amino group ($–NH_2$) and a phosphonic acid group ($–PO_3H_2$) (Fig. 2d). Due to its stable chemistry, 2-AEP can persist in the soil for

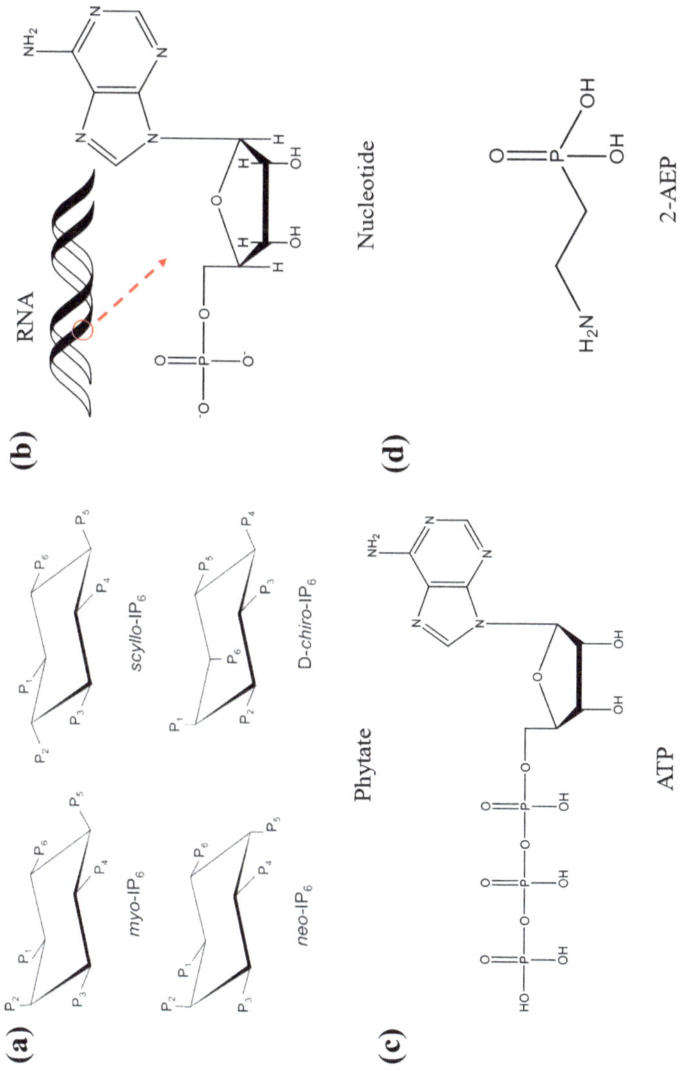

Fig. 2 Schematic structures of phytate (**a**) (Turner et al., 2012), Ribonucleic acid (RNA) (**b**), Adenosine triphosphate (ATP) (**c**), and 2-Aminoethylphosphonic acid (2-AEP) (**d**)

long periods of time, slowly releasing P for utilization by plants and microorganisms (White & Metcalf, 2007). Under certain environmental conditions, microorganisms support plant growth and microbial metabolic activities by expressing specific enzymes, such as phosphatase, which degrades 2-AEP into available Pi (Kononova & Nesmeyanova, 2002).

3 Adsorption and Desorption of Soil Organic Phosphorus

3.1 Adsorption of Organic Phosphorus on Mineral Surfaces

OP in soil is generally highly reactive, particularly in its strong adsorption reactivity on mineral surfaces, which reduces the concentration and bioavailability of soluble OP (Yan et al., 2022a). Ruttenberg and Sulak (2011) proposed that the adsorption of OP on soil mineral surfaces is similar to that of Pi, as both follow a two-step adsorption process. The first step is a rapid adsorption phase, where a large amount of OP is adsorbed onto the mineral surface in a short period of time. The second step is a slower adsorption phase, with some reactions lasting up to several days (Fig. 3) (Ruttenberg & Sulak, 2011; Yan et al., 2014c).

On the amorphous Al hydroxide (AAH) surface, the adsorption of both OP and Pi follows first-order kinetics, occurring in both fast and slow adsorption phases (Shang et al., 1990). In addition to the first-order kinetic equation, the kinetics of OP adsorption can also be described by a pseudo-second-order kinetic equation or a three-parameter model (Luengo et al., 2007; Yan et al., 2022a). The three-parameter model is expressed as $A = A_1 + A_2(1 - e^{-kt})$, where A is the total OP adsorption density, A_1 represents the adsorption density of the fast adsorption phase, and $A_2(1 - e^{-kt})$ represents the adsorption density during the slow adsorption process. Here, A_2 is

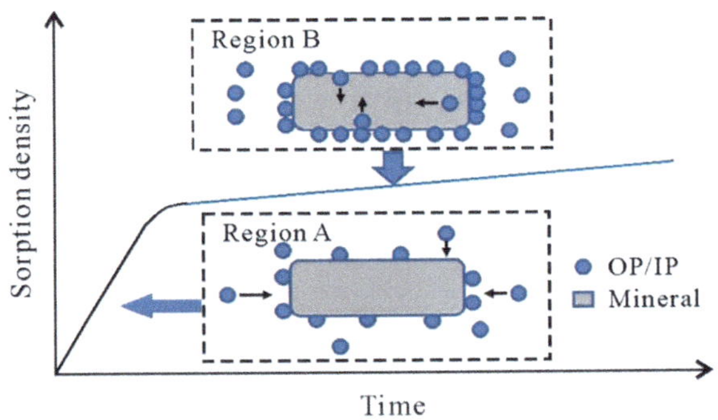

Fig. 3 The adsorption of OP or Pi on mineral surfaces occurs in two distinct phases. Region A represents the rapid adsorption phase, where OP or Pi binds to the mineral surface primarily through ligand exchange. Region B corresponds to the slow adsorption phase, characterized by the gradual diffusion of P into the mineral's interior (Yan et al., 2022a)

the adsorption density after the completion of the slow adsorption process, and k is the rate constant for the slow adsorption process (Luengo et al., 2007; Yan et al., 2022a). For example, the three-parameter model accurately fits the adsorption kinetics of ATP and G6P on the surfaces of AAH, boehmite, and Al_2O_3. The adsorption densities (A_1) and initial adsorption rates of ATP and G6P are the highest on the surfaces of AAH during the rapid adsorption phase, suggesting that AAH has a higher affinity for these OP compared to boehmite and Al_2O_3 in initial reaction stages (Yan et al., 2014b).

The thermodynamic properties of OP adsorption on mineral surfaces can be described using the Langmuir isotherm equation: $Q = Q_m \frac{K_a C}{1+K_a C}$, where Q is the adsorption density, Q_m is the maximum adsorption density, K_a is the affinity constant, and C is the equilibrium concentration. Alternatively, the Freundlich isotherm equation ($Q = K_b C^b$, where K_b is the affinity constant, C is the equilibrium concentration, and b represents the heterogeneity of adsorption sites) can also be used (Guan et al., 2006; Ruttenberg & Sulak, 2011; Yan et al., 2014a). The adsorption isotherms of phytate and Pi on the hematite surface were fitted using a two-site Langmuir isotherm model, and the results indicated that both high-energy and low-energy adsorption sites on hematite exhibited a higher affinity for phytate compared to Pi, suggesting that the multiple bonding of phytate on hematite (Yan et al., 2014c). However, the Freundlich isotherm equation more accurately describes the adsorption behavior of phytate on the alumina surface than the Langmuir model, as it accounts for the heterogeneity of the alumina surface sites (Guan et al., 2006).

The adsorption mechanism of OP on mineral surfaces primarily involves surface complexation and precipitation (Fig. 4). The surface complexation of OP with minerals involves several interactions, including ligand exchange, electrostatic forces, hydrogen bonding, van der Waals forces, and hydrophobic interactions (Arai & Sparks, 2007; Olsson et al., 2011). Among these, ligand exchange, where phosphate (PO_4^{3-}) or other P-containing groups in OP replace H_2O or hydroxyl groups (–OH) on the surfaces of metal ions (e.g., Fe, Al), is the predominant complexation mechanism (Goldberg & Sposito, 1985; Parfitt & Atkinson, 1976). At pH 4.5, it has been observed that phytate adsorbs onto goethite to form an inner-sphere complex through four phosphate groups, with the other two phosphate groups remaining in a free state (Fig. 4a) (Celi et al., 1999; Ognalaga et al., 1994). The adsorption of phytate onto the surface of AAH or ferrihydrite via ligand exchange initiates the dissolution of the mineral, leading to the release of Al^{3+}/Fe^{3+} ions. This process results in the formation of a ternary complex on the surface, which, over time, is converted into Al-phytate or Fe-phytate precipitates as the reaction progresses (Fig. 4b) (Wang et al., 2017; Yan et al., 2014a). Certain clay minerals with weak adsorption capacities for OP, such as kaolinite and calcite, can also adsorb phytate through surface precipitation (Hu et al., 2019; Wan et al., 2016). Furthermore, the synergistic effect of ligand exchange combined with hydrogen bonding or electrostatic forces can significantly enhance the stability of OP complexes on mineral surfaces. Olsson et al. (2011) used infrared spectroscopy to demonstrate that hydrogen bonding influences the stability of inner-sphere complexes formed by glucose-1-phosphate (G1P) on goethite at different pH levels (Fig. 4c).

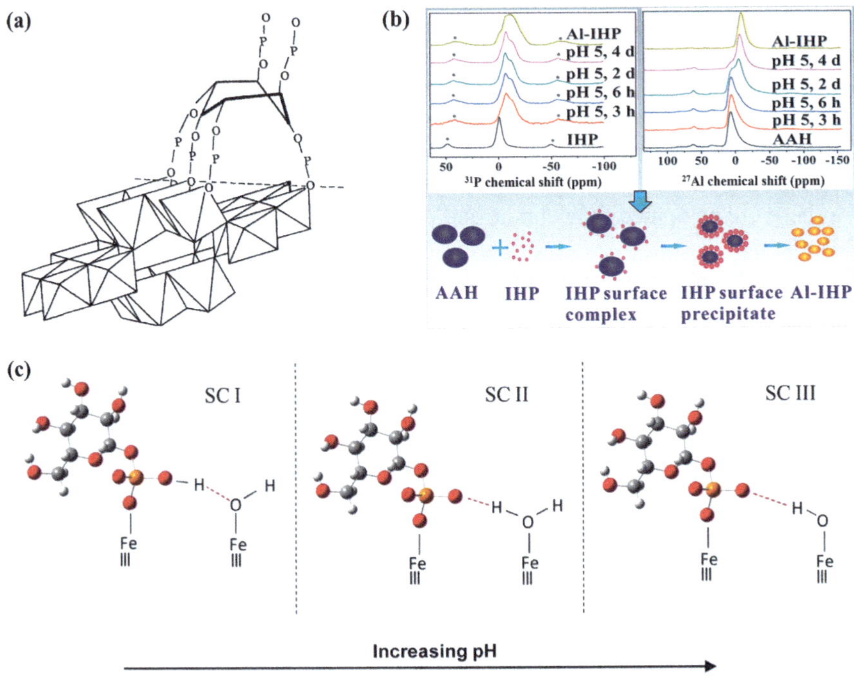

Fig. 4 Phytate adsorbed on the surfaces of goethite through inner-sphere complexation (pH 4.5) (**a**) and on amorphous Al hydroxide through precipitation (**b**) (Celi et al., 2001; Yan et al., 2014a). G1P is adsorbed onto the goethite surface through inner-sphere monodentate coordination and hydrogen bonding (red dashed lines) (**c**) (Olsson, 2011)

3.2 Factors Affecting the Adsorption of Organic Phosphorus

The adsorption of OP on mineral surfaces is influenced by a variety of factors, including pH, temperature, mineral crystallinity, molecular weight of the OP, and coexisting ligands and anions (Yan et al., 2022a). The reaction pH is a key factor controlling the adsorption behavior of OP. Studies have shown that the adsorption density of OP such as G1P, phytate, and ATP increases with decreasing pH on mineral surfaces such as ferrihydrite, hematite, and goethite (Celi et al., 2001; Chen & Arai, 2019; Wei et al., 2010; Yan et al., 2014d). At lower pH, the positive charge on the mineral surface increases, enhancing electrostatic attraction and promoting the adsorption of negatively charged OP (Antelo et al., 2010; Li et al., 2013). Additionally, higher temperatures favor the adsorption of OP on minerals. For example, in the temperature range of 288–308 K, the adsorption density of OP on Fe oxides increases with increasing temperature (Wei et al., 2010). Ionic strength influences the adsorption of OP by regulating the surface potential of minerals and the distribution of OP species. Increasing ionic strength reduces the thickness of the diffuse bilayer, thereby enhancing the adsorption of OP (Chen & Arai, 2019; Yan et al., 2014c).

The crystallinity of minerals significantly affects the adsorption behavior of OP, primarily because crystallinity influences the specific surface area of the minerals. It was shown that the weaker the crystallinity of Fe and Al oxides, the stronger the ability to adsorb OP on their surfaces (Ruttenberg & Sulak, 2011). For example, on Al oxides, phytate adsorption was in the order of α-Al_2O_3 < boehmite < AAH (Yan et al., 2014a, 2014c). In addition, the molecular size of OP, as well as coexisting ligands and anions, are important factors influencing the adsorption of OP. For example, as the molecular weight of OP decreases, the adsorption density on Al oxides follows the order: phytate < ATP < G6P < G1P (Yan et al., 2014b). Low-molecular-weight organic acids can also influence the adsorption of OP on minerals by competing for adsorption sites, electrostatic repulsion, and spatial hindrance effects (Wei et al., 2010; Yan et al., 2022b).

3.3 Desorption of Organic Phosphorus on Mineral Surfaces

The desorption of OP from soil mineral surfaces is a critical step in the soil P biogeochemical cycle, influenced by various factors such as mineral species, desorbent type, soil pH, OP species, the number of desorption cycles, and redox state (Yan et al., 2022a). The quantity and rate of OP desorption from mineral surfaces directly influence the mobility and bioavailability of P in soil.

The properties of mineral surfaces and the types of desorbents used have significant effects on the desorption of OP. Shang et al. (2013) investigated the desorption of phytate from mineral surfaces using various desorbents and found that acidified ammonium oxalate (AOD), Mehlich I, and Mehlich III significantly outperformed H_2O and 0.01 M $CaCl_2$ in desorbing phytate. The desorption rates of phytate on different mineral surfaces followed the order of goethite < gibbsite < kaolinite < montmorillonite, indicating that complexes formed by phytate on Fe and Al oxide surfaces are more resistant to desorption compared to those on clay minerals (Shang et al., 2013). Research has demonstrated that in clay mineral-OP complexes, the binding energy is less stable, allowing OP to be desorbed relatively quickly. In contrast, Fe and Al oxide-OP complexes are more stable, resulting in a slower and prolonged desorption process (Amadou et al., 2022). Sodium citrate, a common desorbent, effectively desorbs phytate from the surfaces of hematite, magnetite, and maghemite, but is less effective on goethite surfaces. However, sodium hydroxide-ethylenediaminetetraacetic acid (NaOH-EDTA) was more effective, achieving a 38% desorption ratio of phytate on goethite surfaces at 100% phytate saturation (Celi et al., 2020). Therefore, the varying surface properties of different minerals lead to significant differences in the efficiency of OP desorption by various desorbents.

The specific type of OP is also a crucial factor influencing its desorption from mineral surfaces. Yan et al. (2014d) found that the desorption of OP from Al oxide surfaces by 0.1 M KCl increased as the molecular weight of the OP compounds decreased (phytate < ATP < G6P < G1P). Amadou et al. (2022) observed that G6P and glycerophosphate (Gly) were desorbed more readily than phytate from Fe and

Al oxides, whereas the opposite trend was observed on clay minerals. Thus, low molecular weight OP compounds like G6P and Gly may be more readily available for plant uptake in soils with high concentrations of Fe and Al oxides. Additionally, the desorption of OP from mineral surfaces was influenced by soil pH and the number of desorption cycles. The amount of phytate desorbed from goethite surfaces increased gradually with rising pH and repeated desorption cycles within the pH range of 3.5–8.5 (Martin et al., 2004). However, in all cases, the desorbed phytate did not exceed one-fifth of the desorbed Pi, suggesting that phytate on goethite surfaces has limited bioavailability (Martin et al., 2004).

During inundation, the transition of the soil environment to anoxic conditions lowers the soil redox potential (Eh) and alters the pH. Anaerobic microorganisms or redox-active antibiotics (e.g., phenazine) can induce the dissolution of Fe oxides, thereby releasing P (Maranguit et al., 2017; McRose & Newman, 2021). Therefore, changes in the redox state of the soil environment are crucial factors controlling the desorption of OP from the surface of Fe oxides. Studies have shown that under anaerobic conditions, OP compounds such as G1P and AMPA are desorbed or released from sediments into the aquatic environment, thereby increasing the concentration of dissolved OP and elevating the risk of eutrophication (Kurek et al., 2021).

4 Mineralization of Organic Phosphorus in Soil Environments

Most plants cannot directly utilize OP, it only becomes bioavailable when mineralized into Pi (Basinski et al., 2024; Shen et al., 2011; Timofeeva et al., 2022). The two main pathways for the mineralization of OP are abiotic mineralization and biomineralization. These pathways complement each other within the ecosystem, collectively promoting the cycling and transformation of OP and maintaining the P balance in the soil.

4.1 Abiotic Mineralization of Organic Phosphorus

Studies have shown that in natural environments, minerals and phosphatases catalyze the hydrolysis of OP with comparable contributions (Wan et al., 2022). This suggests that abiotic processes may play a more significant role in the global biogeochemical cycling of P than previously thought. Natural minerals, particularly Fe oxides, play a crucial role in catalyzing OP hydrolysis. These minerals act as abiotic nanoenzymes, facilitating the release of Pi by adsorbing and catalyzing the cleavage of the phospholipid (C–O–P) bond in OP (Fig. 5a) (Basinski et al., 2024; Wan et al., 2022). Huang and Zhang (2012) found that nanomolar solutions of inorganic Fe aged under acidic conditions significantly catalyzed the hydrolysis of G6P, following the

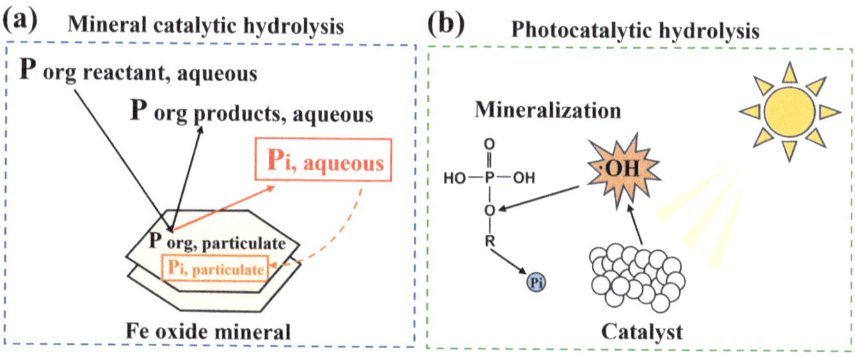

Fig. 5 Schematic diagram of abiotic mineralization pathways and mechanisms of soil OP. Fe oxides catalyze the breaking of phosphate (C–O–P) and phosphoric anhydride (P–O–P) bonds in OP (**a**). Photosensitive catalysts contribute to the hydrolysis of OP through the generation of hydroxyl radicals (·OH) (**b**) (adapted from Basinski et al., 2024; Wan et al., 2022)

Michaelis–Menten kinetic equation. This catalytic effect may be due to the contribution of newly formed oxygen or hydroxyl bridge structures in the Fe^{3+} solution. Fe oxide-catalyzed hydrolysis of OP primarily involves adsorption followed by catalysis. After RNA adsorbs onto the surface of goethite and hematite, the Fe^{3+} can act as Lewis acids, accelerating the hydrolysis of the phosphodiester bonds in the RNA's main chain (Zhang et al., 2023a). The primary mechanism of Lewis acid site-catalyzed hydrolysis involves weakening the phosphate bonds through electron transfer from the bound phosphates to the Fe center, or via nucleophilic attack (Fang et al., 2018; Li et al., 2020). For phosphoric anhydrides containing both phospholipid (C–O–P) and phosphoric anhydride (P–O–P) bonds, such as ATP, Fe oxide first attacks the P–O–P bond at its terminus. This results in a cascading release of the terminal phosphate, eventually leading to the breakage of the C–O–P bond (Klein et al., 2024; Wan et al., 2022).

Additionally, photocatalytic degradation of OP has recently emerged as a significant research hotspot. While most photocatalytic studies have focused on aqueous environments, there is potential for photocatalytic processes to contribute to the mineralization of OP on the soil surface or in flooded soils (Fig. 5b). ZnO nanoparticles exhibit high activity in the photochemical degradation of OP, significantly enhancing the conversion efficiency to phosphate. This process is primarily driven by hydroxyl radicals (·OH) generated on the ZnO surface, which effectively attack the C–O–P bonds of OP compounds (Wang et al., 2023). Additionally, Fe^{3+} can serve as a photosensitive catalyst, producing ·OH under UV light irradiation to promote the conversion of OP to phosphate (Jiang et al., 2016). Focusing on the abiotic mineralization of OP enhances our understanding of its biogeochemical cycle, which is crucial for improving P availability in the soil environment.

4.2 Biomineralization of Organic Phosphorus

Biomineralization of OP exhibits a more efficient catalytic ability compared to abiotic mineralization. Biomineralization of OP primarily involves the release of Pi from organic phosphorus compounds catalyzed by phosphatases secreted by plant roots or microorganisms. Organic acids secreted by plants and microorganisms can complex with some immobilized OP, thereby facilitating its mineralization (Chen & Arai, 2024; Nannipieri et al., 2011; Pantigoso et al., 2023). According to the International Union of Biochemistry and Molecular Biology (IUBMB) classification system, phosphatases are classified based on their mechanism of bond cleavage and their optimal pH. Differences in the bonding of P atoms within various OP result in variations in how phosphatases catalyze bond cleavage. Consequently, phosphatases are classified as monoesterases and diesterases (Chen & Arai, 2024). Diesterases hydrolyze OP compounds to monoester phosphates, which are then further hydrolyzed by monoesterases (Blake et al., 2005). Typically, monoesterase activity is higher than diesterase activity (Jarosch et al., 2019).

Based on the optimal pH for catalyzing the reaction, phosphatases can be classified as alkaline phosphatases and acid phosphatases (Chen & Arai, 2024). The catalytic site of alkaline phosphatase is coordinated by two Zn^{2+} and one Mg^{2+}, creating an optimal environment for substrate binding. During catalysis, the phosphate group in the substrate binds to the Zn^{2+}, forming a stable Zn-phosphate intermediate. The Mg^{2+} then generates a highly reactive hydroxyl ion (OH^-) by stabilizing the intermediate and facilitating the deprotonation of a H_2O, which attacks the phosphate bond, leading to its cleavage (Fig. 6a) (Coleman, 1992; Stec et al., 2000; Yong et al., 2014). The active site of acid phosphatase, in contrast to alkaline phosphatase, is typically an Fe–Mn bimetallic center. In this configuration, one oxygen atom of the phosphate group directly coordinates with the Fe^{3+} ion, another oxygen atom coordinates with the Mn^{2+} ion, and a third oxygen atom forms a hydrogen bond with a bridging H_2O between the Fe–Mn centers (Schenk et al., 2004; Tang et al., 2006). When the bimetallic center binds to the phosphate, the phosphatase acts as a nucleophilic agent, forming covalent phosphohistidine intermediates via histidine residues in the N-terminal motifs. Meanwhile, aspartic acid residues in the C-terminal motifs act as proton donors, facilitating the cleavage of the C–O–P bond (Fig. 6b) (Ge et al., 2022; Lindqvist et al., 1994; Porvari et al., 1994). However, specific phosphatases are required to efficiently hydrolyze structurally complex and spatially site-blocked OP like phytate. Although some acid phosphatases can hydrolyze phytate, they are far less efficient than phytase (Sun et al., 2017). Wu et al. (2015) investigated the pathway and mechanism of phytase-mediated hydrolysis of phytate in wheat. Their study found that wheat phytase typically hydrolyzes the phosphate group at positions 6 or 4 in phytate first, gradually removing the phosphate groups. Using ^{18}O isotope tracer technology, they demonstrated that one of the oxygen atoms in the Pi molecules released by hydrolysis originated from the water (H_2O) involved in the reaction.

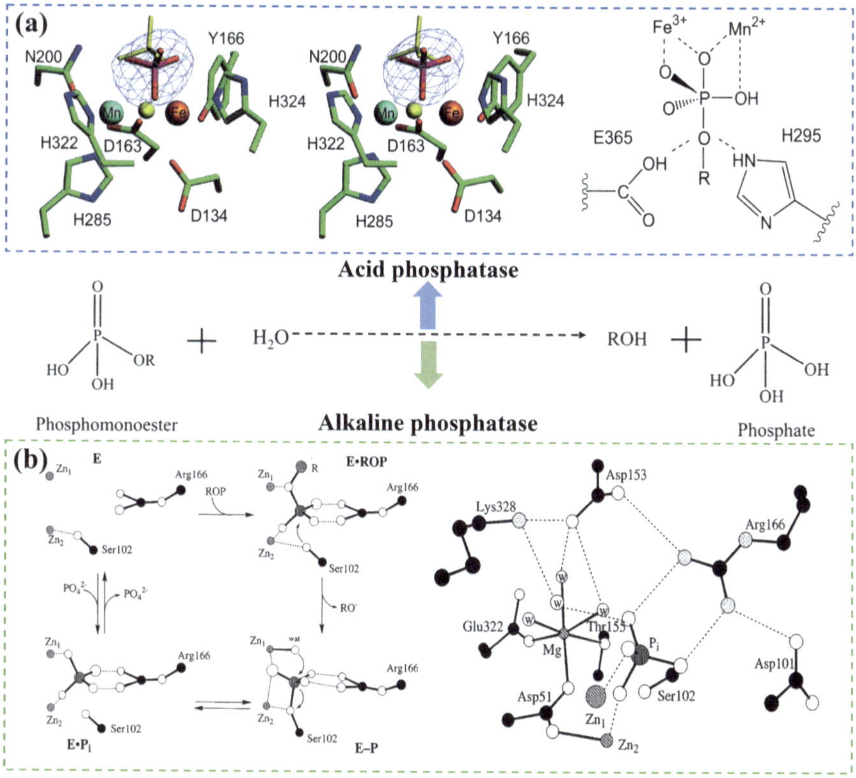

Fig. 6 Schematic representation of the mechanism of mineralization of soil OP catalyzed by acid and alkaline phosphatases. Acid phosphatase hydrolyzes monoesters of phosphate by breaking the C–O–P bond using amino acid residues and Fe^{3+}/Mn^{2+} cofactors (**a**), while alkaline phosphatase uses Zn^{2+}/Mg^{2+} cofactors for the same hydrolysis process (**b**) (adapted from Chen & Arai, 2024; Holtz & Kantrowitz, 1999; Schenk et al., 2004)

4.3 Factors Affecting Mineralization of Soil Organic Phosphorus

4.3.1 Soil pH and Temperature

Soil pH can affect the adsorption of OP on mineral surfaces and likewise the mineralization of OP by minerals. The hydrolysis of RNA catalyzed by goethite was most efficient under neutral conditions compared to acidic and alkaline environments. The low efficiency of hydrolysis at lower pH is attributed to the requirement of hydroxide for RNA hydrolysis, whose concentration is limited under acidic conditions. In contrast, the reduced hydrolysis at higher pH is attributed to the decreased adsorption of RNA by goethite, resulting in reduced interaction between goethite and RNA (Zhang et al., 2023a). Additionally, the hydrolysis rate of goethite-catalyzed G1P slightly decreased and then markedly increased as pH shifted from acidic to alkaline conditions (Olsson, 2011). This trend is attributed to the weaker hydrogen bonding and more flexible structure of the G1P complex formed on the goethite surface under alkaline conditions. This flexibility may stabilize the interaction between the glucose moiety and the surface, lowering the energy of the transition state in the hydrolysis reaction and thereby increasing the mineralization rate (Olsson, 2011). Thus, pH affects mineral-catalyzed hydrolysis of OP not only by regulating the surface charge to influence the contact between the mineral and the OP, but also by hydrogen bonding interactions. The primary effect of pH on the biomineralization of OP is to alter the conformation of phosphatases, reducing their affinity for OP and thereby decreasing the mineralization rate (Huang et al., 2009; Seibert & Tracy, 2014). It was observed that free phytase had a more disordered structure at pH 8 compared to pH 6, leading to a lower mineralization rate of phytate (Chen et al., 2023b). Additionally, pH can influence microbial community composition and the abundance of related genes. For example, the number of soil microorganisms containing P-solubilizing genes, such as phoX, phoD, and phoA, was significantly reduced in acidic soils (Zaheer et al., 2009).

4.3.2 Metal Ions and Organic Matter in Soil

OP is generally negatively charged and readily forms insoluble metal salts with metal ions in the soil, such as Al^{3+}, Fe^{3+} and Zn^{2+}. Studies have shown that the insoluble metal salts formed by phytate with Al^{3+}, Fe^{3+}, and Zn^{2+} are very stable and difficult for phytase to hydrolyze (Tang et al., 2006). Moreover, metal ions like Cu^{2+}, Pb^{2+}, and Hg^{2+} can be toxic to phosphatase, thereby inhibiting its activity. For example, Cu^{2+} can cause enzyme inactivation by oxidizing the sulfhydryl groups of the enzyme, or by directly binding to specific sites in the enzyme protein (Bourdon et al., 2023). However, Ca^{2+} had a dual effect on the enzymatic activity of phytase, an enzyme that catalyzes the hydrolysis of OP compounds. At pH 6, Ca^{2+} altered the conformation of phytase, reducing its activity. In contrast, at pH 8, Ca^{2+} promoted the enzymatic hydrolysis of phytate by inducing a conformational change in phytate, making it more accessible for phytase interaction, as demonstrated by ^{31}P-NMR spectroscopy (Chen et al., 2023a) (Fig. 7a).

Organic matter in the soil, including low molecular weight organic acids and humic acid (HA), plays a crucial role in the mineralization of OP. Organic acids enhance the hydrolysis of OP precipitates by solubilizing metal salts, with citric acid being the most effective solubilizing agent (Tang et al., 2006). Additionally, humic acids stabilize the structure of phytase through non-covalent bonding, enhance the ordered structure in its active domain, and increase phytase activity, thereby promoting the mineralization of phytate (Ge et al., 2022) (Fig. 7b).

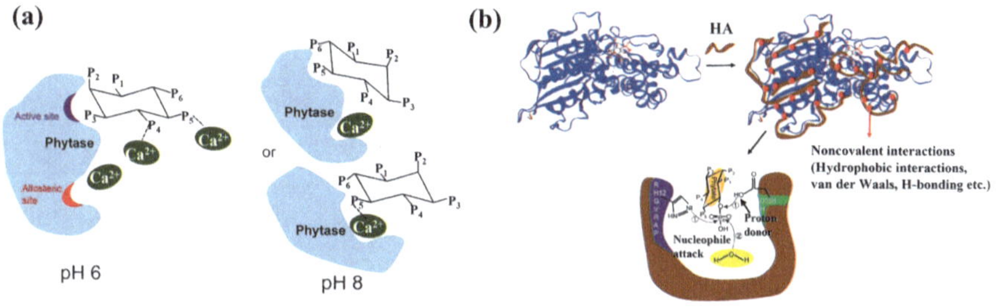

Fig. 7 Schematic representation of the effects of Ca^{2+} (**a**) and humic acid (**b**) on phosphatase activity. Ca^{2+} can bind to the variant site of phytase or alter the complex conformation of phytate, while humic acid stabilizes the structure of phytase through non-covalent bonding (Chen et al., 2023a; Ge et al., 2022)

4.3.3 Types of Soil Minerals and Organic Phosphorus

Differences in surface properties between minerals contribute to variations in their ability to mineralize OP. Fe oxides, in particular, have emerged as major contributors to mineral-catalyzed OP hydrolysis due to their higher surface activity, greater hydroxyl density, and stronger adsorption of OP compared to other minerals (Basinski et al., 2024; Wan et al., 2022). Compared to the crystal facets of hematite and goethite with lower Lewis site exposures, their facets with higher Lewis site exposures exhibited stronger dephosphorylation activity towards nitrophenyl phosphates (Fig. 8a) (Li et al., 2020). Variations in the surface properties of minerals can also influence how adsorption affects enzyme activity. For example, reactive oxygen radicals (·OH) produced on the surface of redox-active clay minerals can destroy enzyme structures, thereby inhibiting enzyme activity (Sheng et al., 2023). The high hydroxyl density of the hematite (104) crystal surface provides a greater conformational unfolding of the enzyme than the (001) crystal surface, leading to easier substrate contact of the active site and facilitating the enzymatic reaction (Zang et al., 2020). Similarly, goethite surfaces can effectively enrich substrates and phosphatases, counteracting the inhibition of phosphatase activity caused by adsorption (Olsson et al., 2011). However, layered silicate minerals like montmorillonite, with their unique interlayer structure, can inhibit the mineralization of OP by interlayer adsorption of phosphatases, DNA and RNA (Mortland & Gieseking, 1952; Yu et al., 2013) (Fig. 8b).

Differences in mineral surface properties contribute to variations in OP adsorption conformations. Klein suggests that the binding conformation of adsorbed ribonucleic acid determines the degree of dephosphorylation of ferrihydrite (Klein et al., 2019). Similarly, phytate exhibit comparable properties. It was observed that at 100 μM concentration, nearly all of the phosphate groups in phytate were complexed to the surface of hydrotalcite, whereas at 500 μM, approximately 65% of the phosphate groups were complexed (Chen & Arai, 2019). The reactivities of phosphate groups in phytate vary with pH, leading to different adsorption conformations. Similar phenomena were observed on the surfaces of ferrihydrite, Al oxides, and calcite (Fig. 9a,b) (Chen & Arai, 2019; Chen et al., 2022; Xu & Arai, 2022; Xu et al., 2021).

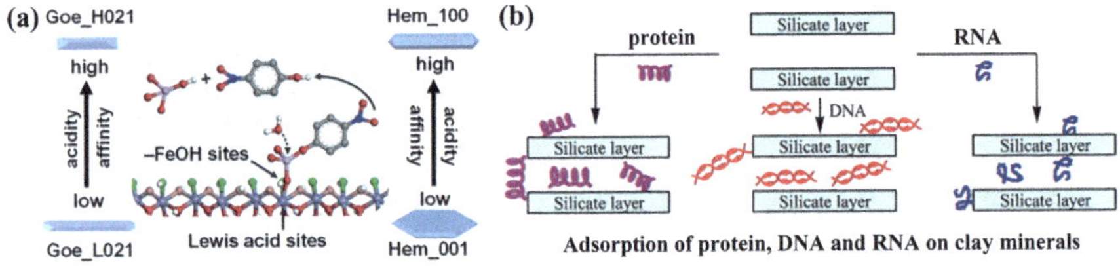

Fig. 8 Effect of Fe oxide and layered silicate minerals on OP hydrolysis. The higher the density of hydroxyl groups on the crystalline surface of Fe oxide, the greater the ability to catalyze p-Nitrophenyl Phosphate (**a**). The interlayers of layered silicate minerals can adsorb protein and OP, such as enzyme, DNA, RNA, and impede their mineralization (Li et al., 2020; Yu et al., 2013)

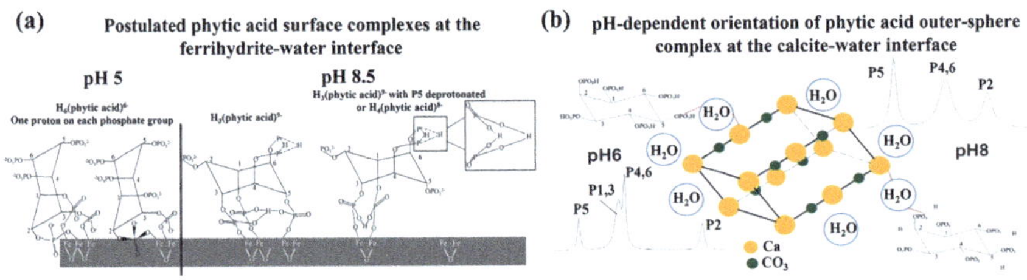

Fig. 9 Schematic illustration of the differences in the adsorption configurations of phytate on the surface of ferrihydrite or calcite at different pHs (Chen & Arai, 2019; Chen et al., 2022)

For example, at pH 5, the phosphoric acid groups at positions 1, 2, and 3 on phytate were bound to active sites on the surface of ferrihydrite. At pH 8.5, the reactivities of the phosphoric acid groups at positions 1, 3, and 5 were higher (Fig. 9a) (Chen & Arai, 2019). Differences in the adsorption conformation of phytate on mineral surfaces may hinder phytase access, reducing the bioavailability of phytate (Chen et al., 2023b).

4.3.4 Microbial Interactions in Soil

In natural soil ecosystems, complex symbiotic relationships often form between microorganisms. As a result, the mineralization of OP is influenced not only by the direct action of P-solubilizing microorganisms that secrete phosphatases, but also by interactions between microbial strains that can indirectly affect this process. The symbiotic relationship between arbuscular mycorrhizal fungi (AMF) and P-solubilizing bacteria (PSB) enhances the ability of plants to acquire soil P. AMF not only improves plant P uptake through its mycelial networks, but also enhances the mineralization of phytate by PSB through the secretion of fructose signals (Fig. 10) (Zhang et al., 2016, 2018). Promoting this cooperation among microorganisms can enhance soil P utilization by crops, reducing the need for P fertilizers and contributing to sustainable agricultural development.

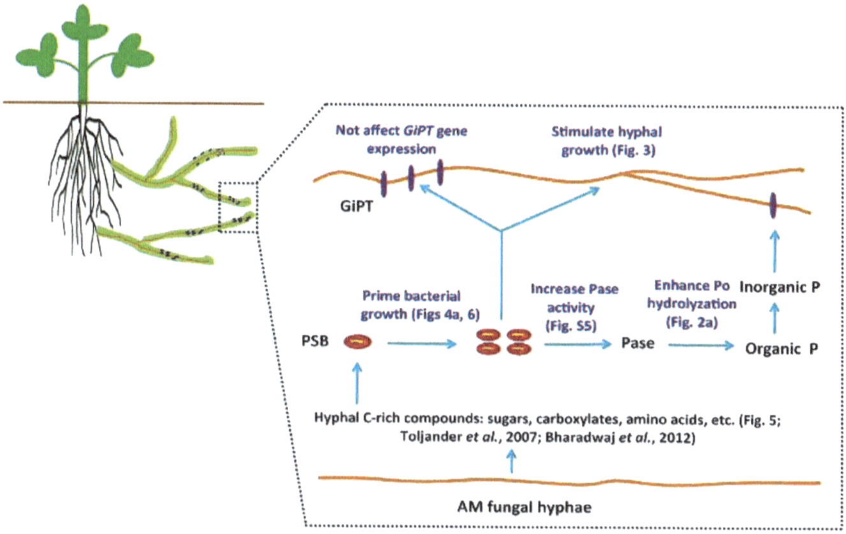

Fig. 10 Schematic diagram of the mutualistic mechanism between root arbuscular mycorrhizal fungus (AMF) and P-solubilizing bacteria (PSB). AMF regulates PSB secretion of acid phosphatase via fructose molecular signaling, enhancing P availability in the soil (Zhang et al., 2016)

4.3.5 Other Environmental Factors

Rainfall can alter the redox status of soils, affecting the chemical form of P and further influencing its bioavailability (Kumaragamage et al., 2019; McDowell & Sharpley, 2002). Enhanced microbial activity under wet conditions can promote the mineralization of OP. However, it may also accelerate the decomposition of organic matter, potentially affecting long-term soil fertility (Dalai, 1977; Floate, 1970). Additionally, alternating drought and wet conditions may alter the structure of soil microbial communities, further impacting P cycling processes (Blackwell et al., 2010; Brödlin et al., 2019).

Human activities, such as agriculture and land management, are important factors influencing soil OP mineralization. The widespread use of chemical fertilizers increases the amount of available P in the soil but can also lead to excess phosphorus, which affects soil OP dynamics and microbial activity (Lee et al., 2004; Wang et al., 2020). For example, excess P fertilizers can inhibit microbial communities in the soil that would otherwise be able to efficiently cycle OP, thereby reducing the bioavailability of OP (Wang et al., 2020). Furthermore, agricultural mechanization and land management practices, such as tillage and irrigation, alter soil structure, affecting the mineralization and fixation of OP. These changes can lead to the redistribution and loss of soil P (Dalai, 1977; Lee et al., 2004).

5 Advanced Analytical Techniques of Soil Organic Phosphorus

See Fig. 11.

5.1 ^{31}P Nuclear Magnetic Resonance Spectroscopy

^{31}P Nuclear Magnetic Resonance Spectroscopy (NMR) is an important tool for studying the speciation and transformation of soil OP (Fig. 11a). This technique provides detailed chemical structural information on soil OP compounds, enabling the identification and quantification of various chemical forms, including phosphate

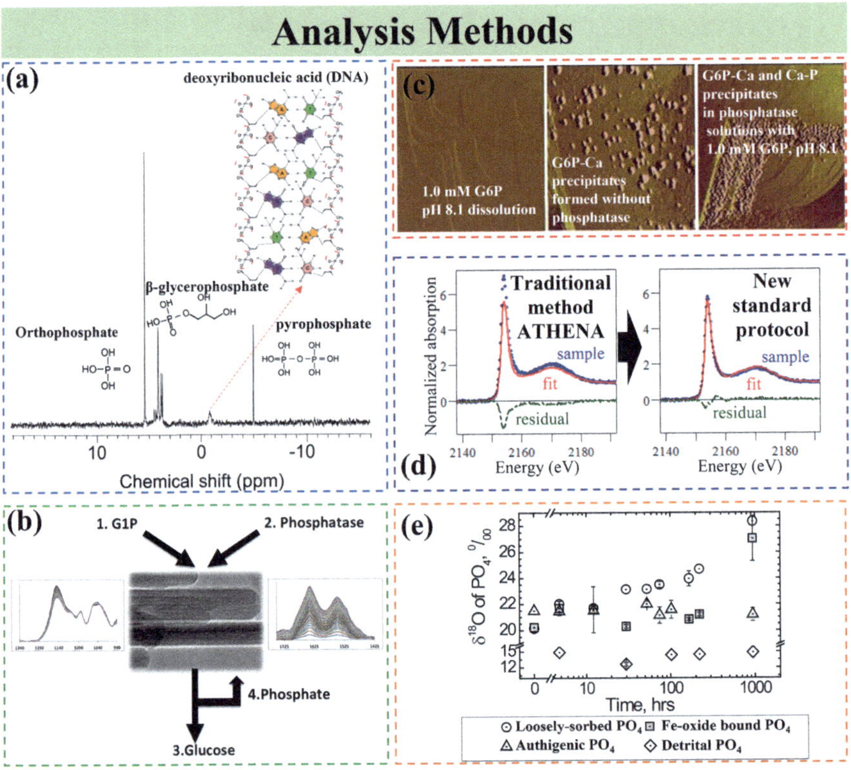

Fig. 11 The main research methods for studying soil OP include^{31}P nuclear magnetic resonance spectroscopy (NMR), in-situ attenuated total reflectance Fourier transform infrared spectroscopy (ATR-FTIR), atomic force microscopy (AFM), P K-edge X-ray absorption near-edge structural spectroscopy (P K-edge XANES), and δ^{18}O–P isotopic labeling (adapted from Jaisi et al., 2011; Olsson et al., 2011; Shinohara et al., 2012; Wang et al., 2015; Werner & Prietzel, 2015)

esters, nucleic acids, and phytate (McLaren et al., 2015; Shinohara et al., 2012). ^{31}P-NMR utilizes the naturally abundant magnetic isotope of ^{31}P. In an applied magnetic field, the magnetic moment of the ^{31}P nucleus interacts with the external magnetic field. When excited by a radio frequency pulse, the ^{31}P nucleus absorbs energy and is excited to a higher energy state. As it returns to the ground state, the nucleus emits energy, which is detected and converted into a spectrum (Henderson et al., 1974).

^{31}P-NMR can be divided into two main categories: solid-state and liquid-state NMR. Liquid-state NMR is suitable for samples that can be dissolved in solvents (Wu et al., 2015), while solid-state NMR is used for the direct analysis of solid samples, such as soils and sediments (Phillips et al., 2016). Solid-state ^{31}P-NMR is often combined with ^{27}Al-NMR to determine the formation of Al-P precipitates on mineral surfaces. For example, phytate can induce the formation of Al-phytate precipitates on the surface of AAH and kaolinite, which can be identified by the appearance of new peaks in the NMR spectra (Hu et al., 2019; Wang et al., 2019; Yan et al., 2014a). Liquid-state ^{31}P-NMR can be used to characterize the mineralization of OP over time, as well as to determine the speciation and composition of OP in soil. Wu et al. (2015) monitored the kinetic products of phytate hydrolysis by phytase using liquid-state ^{31}P-NMR and demonstrated that phytase hydrolysis involves the gradual release of phosphate groups from phytate. Additionally, a mixed suspension of phytate and minerals can be analyzed under specific conditions using liquid-state ^{31}P-NMR. The reactivity of individual phosphate groups in phytate on mineral surfaces can be assessed by split-peak fitting of the spectra, based on the upward or downward shift of subpeaks. For example, at pH 6, the phosphoric acid groups at positions 4, 5, and 6 in phytate are more active in binding to calcite surfaces, which is crucial for understanding the adsorption mechanism of phytate on mineral surfaces (Chen et al., 2022).

Although the ^{31}P-NMR technique plays a crucial role in soil OP research, its high detection limit necessitates a high concentration of P compounds in the soil sample to obtain a clear spectrum (Turner et al., 2003). This limitation restricts the application of NMR in studying soils with low P content or in detecting trace amounts of OP compounds. Additionally, the pretreatment of ^{31}P-NMR samples typically requires the use of 0.5 M NaOH solution (Wan et al., 2022), which raises the pH and may lead to the hydrolysis of OP, potentially altering its composition in the soil.

5.2 *In-Situ Attenuated Total Reflectance Fourier Transform Infrared Spectroscopy*

In-situ attenuated total reflectance Fourier transform infrared spectroscopy (ATR-FTIR) is a technique that utilizes the interaction of infrared light with surface molecules to detect the chemical composition and molecular structure of the sample surface (Fig. 11b). The principle of ATR-FTIR spectra are based on attenuated total reflection of infrared light. When an infrared beam passes through a crystal with a high refractive index and contacts the sample, a short-range evanescent wave is generated at the interface. This wave interacts with the molecular vibrations within the sample, allowing the infrared spectrum to be obtained (Smith, 2011).

ATR-FTIR spectra are commonly used to reveal in situ interfacial reaction mechanisms of P at the microscopic scale. Guan et al. (2006) used ATR-FTIR spectra to characterize the infrared spectra of phytate at different pH levels, demonstrating that phytate formed inner-sphere complexes on the surface of Al oxides. Similarly, Olsson et al. (2011) used ATR-FTIR spectra to reveal the adsorption mechanism of G1P on the surface of goethite, showing that G1P formed three single-dentate inner-sphere complexes. These complexes were pH-dependent and involved hydrogen bonding. In addition to studying the adsorption mechanism of OP on mineral surfaces, ATR-FTIR spectra has also been used to investigate the enzymatic hydrolysis of OP. The results show that phosphatases can hydrolyze G1P on the surface of goethite (Olsson et al., 2011). During hydrolysis, glucose molecules were released into the solution, and Pi was re-adsorbed onto the surface of goethite (Olsson et al., 2011). ATR-FTIR spectra is also an important tool for analyzing protein secondary structures, making it useful for studying the effects of mineral adsorption on the secondary structure of phosphatase (Liu et al., 2019; Ustunol et al., 2021). This can help reveal the inhibitory mechanism of minerals on the enzymatic hydrolysis of OP. Chen et al. (2023b) used ATR-FTIR spectra to compare the secondary structure of phytase before and after adsorption on calcite surfaces, explaining changes in phytase activity. This has important implications for the biogeochemistry of OP.

5.3 *Atomic Force Microscope*

Atomic force microscopy (AFM) is a high-resolution imaging tool widely used in fields such as materials science, biology, and environmental science. In the study of soil P biogeochemistry, AFM has become a crucial tool for investigating the interactions between OP and mineral interfaces (Fig. 11c). AFM scans the surface of a sample using a very fine probe. The interaction forces between the probe and the atoms on the sample's surface (e.g., van der Waals force, electrostatic force) cause small deflections of the probe. These deflections allow AFM to reconstruct the three-dimensional morphology of the sample's surface.

With its high resolution, AFM can image at the nanometer scale, providing detailed information on the sample's surface structure (Binnig et al., 1986). AFM has visually demonstrated the presence of DNA molecules on goethite surfaces, mainly in the form of clusters associated with goethite particles, leading to the gradual formation of larger agglomerates (Feng et al., 2024). Similarly, the adsorption of G6P on calcite surfaces, resulting in etch pits and granular precipitates, has also been clearly illustrated (Wang et al., 2015, 2016b). AFM provides visual evidence of the hydrolysis of OP. For example, phytase hydrolyzes phytate adsorbed on the surface of nanoscale brushite, releasing phosphate that, together with Ca^{2+}, forms white granular Ca-P precipitates within the imaging range (Ge et al., 2024). The use of AFM in OP research has significantly advanced our understanding of OP behavior in soils. AFM allows for nanoscale observation of interactions between OP and mineral surfaces, revealing adsorption, hydrolysis, and microbial mechanisms.

5.4 *Phosphorus K-Edge X-Ray Absorption Near-Edge Structure Spectroscopy*

P K-edge X-ray absorption near-edge structure spectroscopy (P K-edge XANES) is a synchrotron-based spectroscopic method used to study the chemical speciation of P in soils and sediments, with particular applications in analyzing the structure and environmental behavior of OP (Fig. 11d) (Werner & Prietzel, 2015). The basic principle of P K-edge XANES is that when the energy of the X-rays is gradually increased and reaches the absorption edge of the P K-shell electrons (1 s electrons) at approximately 2145 eV, the 1 s electrons are excited to higher unoccupied orbitals (e.g., 3p orbitals) or are ionized directly. These changes in the local chemical environment manifest as distinct absorption features (e.g., peak positions, intensities, and shapes) in the spectra, providing detailed information about elemental valence states, coordination numbers, and local symmetries (Prietzel et al., 2016).

The white line peak energies of P K-edge XANES for OP and Pi specimens differ significantly, with Pi exhibiting higher white line peak energies (Prietzel et al., 2016). P K-edge XANES can detect both the amount of P adsorbed on mineral surfaces and the distribution of P species. Wang et al. (2017) found that as the phosphate and phytate loading increased, the intensity of the pre-edge peaks (~2148 eV) and white-line energy in the P K-edge XANES spectra increased, indicating that the proportions of amorphous $FePO_4$ and amorphous Fe-phytate on the surface of ferrihydrite increased compared to those of adsorbed species. Basinski et al. (2024) analyzed the P species on the surface of the solid-phase products from the catalytic mineralization of OP by Fe oxides and clay minerals. By comparing the shifts in the white line peaks of the P K-edge XANES (~2152.8 eV), they found that the proportion of particulate Pi on Fe oxide surfaces was higher than on clay mineral surfaces (Basinski et al., 2024). Combined with solution data, this demonstrated that Fe oxides are more effective at mineralizing OP than clay minerals (Basinski et al., 2024). Additionally, the appearance of shoulder peaks in the post-edge region (2156–2164 eV) of the P K-edge XANES spectra can indicate the formation of Ca-P complexes on the mineral surface (Wan et al., 2021). Combining P K-edge XANES with other analytical techniques allows for a more comprehensive understanding of the biogeochemical behavior of soil P, providing valuable insights for soil science and environmental protection.

5.5 $\delta^{18}O$–P Isotope Labeling Method

The oxygen isotope labeling method leverages the natural differences in abundance of oxygen isotopes (e.g., ^{16}O, ^{17}O, ^{18}O) and their fractionation during chemical reactions. By measuring the ratios of these isotopes in OP compounds, researchers can obtain insights into how these compounds are formed, transformed, and decomposed in the soil. Unlike traditional chemical analysis methods, this approach can distinguish between P compounds of different origins and trace their transformation processes in complex soil environments (Fig. 11e) (Gross et al., 2016).

During the mineralization of phytate by phytase, $\delta^{18}O$–P isotope labeling showed that one of the oxygen atoms in the released phosphate was derived from H_2O (Wu et al., 2015). Hollenback and Jaisi (2024) coupled ^{18}O isotope labeling with mass spectrometry to perform thermal hydrolysis experiments using ^{18}O-labeled water. The results showed no isotopic exchange in IP_6 or its hydrolysis products, IP_5, IP_3, and the PO_3 fragment, indicating that the isotopic changes were due to hydroxyl groups in the ring (Hollenback & Jaisi, 2024). Jaisi et al. (2011) used oxygen isotope ratios to reveal the relative contributions of biotic and abiotic pathways in P cycling within minerals and sediments, emphasizing the influence of environmental conditions on these processes. The application of oxygen isotope labeling in OP studies is a crucial tool for revealing the complexity and specific mechanisms of soil P cycling. By analyzing the oxygen isotope composition of OP compounds, researchers can trace the sources, transformation pathways, and processes of P in ecosystems, gaining insights into the behavior and fate of OP in soils.

6 Measures to Improve the Effectiveness of Soil Organic Phosphorus

6.1 Regulation of Soil pH

Soil pH directly affects phosphatase activity and the mineralization of OP. A moderate pH (typically neutral to slightly acidic) maximizes phosphatase activity, thereby promoting the hydrolysis of OP. Adjusting the pH of acidic soils by applying lime or other Ca^{2+}-containing alkaline substances improves the soil environment and enhances phosphatase activity, thereby increasing the effectiveness of OP (Chen et al., 2023a; Shujie et al., 2021). Under neutral or slightly acidic conditions, soil adsorption of OP is relatively low, which facilitates the mineralization of OP by phosphatase (Yan et al., 2014b, 2022a). Additionally, soil pH can influence the gene abundance of soil microorganisms. An increase in pH can promote a higher abundance of genes related to OP mineralization in these microorganisms (Zaheer et al., 2009).

6.2 Improved Fertilizer Application and Tillage Practices

Proper management of P fertilizer application is essential for improving the efficiency of OP utilization. Excessive application of P fertilizer can increase the concentration of Insoluble P in the soil, potentially inhibiting the mineralization activity of microorganisms on OP and reducing its effectiveness (Condron et al., 2005). By adjusting fertilizer application strategies, the mineralization and utilization of OP in the soil can be effectively promoted. Measures such as split fertilization and deep application of P fertilizers can reduce excessive P accumulation in the soil, minimize P loss from the surface soil, maintain the rate of OP mineralization, and promote P uptake by plant roots (Kochian et al., 2004; Sharpley, 2006). Additionally, applying organic fertilizers can promote the conversion and release of OP. The organic matter in fertilizers regulates the soil C-P ratio, improves soil structure and microbial activity, and increases phosphatase activity, all of which facilitate the mineralization of OP (Allison, 2006). For example, applying livestock manure or compost can increase the organic matter content in the soil and activate microbial utilization of OP (Zhang et al., 2023b). Additionally, the addition of biogenic carbon can provide binding sites for OP, reduce its immobilization in the soil, and improve its bioavailability (Lehmann et al., 2011; Yang et al., 2021). Wei et al. (2010) reported that low molecular weight organic acids in fertilizers, such as citric acid, contribute to the desorption of OP and improve its utilization efficiency.

P fertilizer application can be integrated with sustainable farming practices, such as conservation tillage and no-tillage methods. Conservation tillage and no-tillage practices can reduce soil erosion, maintain soil organic matter, enhance microbial diversity and activity, and thereby promote the transformation of OP (Six et al., 2004). Studies have shown that P use efficiency in soil can be optimized through crop rotation and intercropping. The differences in P demand and root exudates of different crops create a synergistic effect, thereby improving the bioavailability of OP (Smith et al., 2011).

6.3 Microbial Engineering Technology

Microbial engineering technology involves the transformation, optimization, and application of microorganisms through genetic engineering and biotechnology, aiming to enhance or impart new functions to microorganisms to better meet agricultural needs (Hwang et al., 2017; Wu et al., 2005). Currently, microbial engineering technologies utilized in soil environments include the use of engineered bacteria and microbial inoculation. For example, engineered rhizobacteria not only fix nitrogen but also efficiently desorb OP in the soil, increasing its utilization by plants (Khan et al., 2009). Genetically modified E. coli can also secrete phosphatase efficiently, promoting the mineralization of OP (Richardson, 2001).

Additionally, the ability to dissolve and mineralize OP in soil can be enhanced by inoculating soil with PSB. These bacteria secrete organic acids and phosphatases, converting insoluble OP into plant-available Pi. Rodríguez and Fraga (1999) demonstrated that inoculating soil with PSB significantly increased P uptake efficiency by plants, especially in low-P soils. Similarly, Adesemoye et al. (2009) demonstrated that co-inoculation of AMF and PSB significantly enhanced maize growth and P uptake. Zhang et al. (2018) found that AMF stimulated PSB to secrete phosphatase by producing fructose, which in turn promoted the mineralization of phytate in the soil. Therefore, engineered bacteria and microbial inoculation techniques hold great potential for improving the effectiveness of soil OP and are key areas of future research for the efficient utilization of OP.

7 Research Outlook

Research on the biogeochemical cycling of OP in soil environments is encountering new opportunities and challenges. With the increasing scarcity of global P resources, an in-depth understanding and optimization of the cycling mechanisms of OP in soil are crucial for safeguarding agricultural productivity and ecosystem sustainability. Future research should focus on the following key aspects:

1. In-depth exploration of the mineralization mechanisms of organic phosphorus

Biomineralization plays a dominant role in the cycling of OP, while abiotic mineralization processes, particularly mineral-catalyzed degradation mechanisms, are gaining increasing attention. Future studies on the mechanisms of OP mineralization should extend beyond interactions between minerals and OP or between minerals, OP, and phosphatase (Basinski et al., 2024; Wan et al., 2022). Research should also investigate how different minerals (e.g., Fe oxides, Al oxides, clay minerals) affect the biotic and abiotic mineralization of OP under various environmental conditions, such as mineral mixing, light exposure, anaerobic environments, and the addition of organic ligands. Additionally, in soil environments, most stable Fe oxides result from mineral phase transformations. For example, in anaerobic environments, ferrihydrite can transform into minerals such as hematite, goethite, or magnetite (Gao et al., 2024; Liu et al., 2022a). However, the abiotic and biomineralization of OP during the transformation of oxidized Fe phases remain largely unexplored. Therefore, future studies should investigate the mechanisms of OP mineralization during these mineral phase transformations.

2. Interactions between P-solubilizing bacteria and mineral interfaces

PSB play a key role in the mineralization of soil OP, and mineral surfaces significantly influence PSB and the enzymes they secrete (Richardson, 2001; Solhtalab et al., 2020; Zaheer et al., 2009). Currently, the mechanisms by which minerals affect the mineralization of OP by PSB are still unclear. This includes how minerals influence phosphatase secretion by PSB, which P-solubilizing genes are activated through mineral interactions, the fate of phosphatase secreted by PSB and the phosphate released from the hydrolysis of OP. Studying these mechanisms will enhance our understanding of how minerals influence the mineralization of OP by microorganisms. This knowledge could eventually enable gene-directed modification of PSB to increase their P-solubilizing efficiency.

3. Quantitative comparison of mineralization of surface adsorbed and free organic phosphorus

Soil minerals, particularly Fe and Al oxides, exhibit strong adsorption capacities for both Pi and OP (Yan et al., 2014b). Currently, the abiotic and biotic mineralization rates of OP are primarily quantified using alkali desorption and colorimetry (Wan et al., 2022; Zhang et al., 2023a). However, accurately quantifying the mineralization rates of adsorbed versus free OP remains challenging because it is difficult to distinguish the source of Pi released during mineralization, particularly in systems involving phosphatases. Future research could utilize isotope labeling combined with other advanced techniques to quantitatively compare the mineralization rates of adsorbed and free OP, thereby gaining deeper insights into the catalytic behavior of minerals and phosphatases in the mineralization process.

4. Mechanistic analysis of multifactorial synergies

The synergistic effects of factors such as pH, metal ions, organic matter, and microbial communities in the soil environment play a critical role in determining the efficiency of OP mineralization and conversion. Future research should focus on the systematic analysis of the complex interactions between these factors. For example, while Fe oxides can catalyze the hydrolysis of OP and phosphatases can also hydrolyze OP (Huang & Zhang, 2012; Tang et al., 2006), the synergistic mechanism of OP mineralization when these three coexist remains unclear. Additionally, humic acid can stabilize the structure of phytase and enhance its activity (Ge et al., 2022), while Ca^{2+} can further increase phytase activity at pH 8 (Chen et al., 2023a). However, when humic acid and Ca^{2+} coexist, their combined effect on phytase activity and potential synergistic interaction remain to be investigated. Understanding these interactions could provide new strategies for improving the utilization of OP.

5. Optimization of sustainable agricultural management practices

With the increasing global demand for environmental protection and sustainable resource use, enhancing the effectiveness of OP in soil and reducing P loss through optimized agricultural management practices, such as fertilizer application strategies, tillage practices, and soil improvement, will be an important direction for future research (Allison, 2006; Kochian et al., 2004; Sharpley, 2006). In particular, adaptable agricultural management models should be developed and studied for different soil types and climatic conditions to improve P utilization efficiency and ensure long-term ecosystem health and productivity.

In conclusion, future research should emphasize multidisciplinary integration, incorporating the latest advances in biology, chemistry, materials science, and other related fields. This approach will provide the theoretical foundation and technological innovations necessary for the efficient utilization of global P resources and the sustainable development of the ecological environment through an in-depth analysis of soil OP cycling mechanisms.

Acknowledgements This study was supported by the National Key Research and Development Program and National Natural Science Foundations of China (Nos. 41977021, 2020YFC1806803, 2023YFD1702800, and 42030709).

References

Adesemoye, A., Torbert, H., & Kloepper, J. (2009). Plant growth-promoting rhizobacteria allow reduced application rates of chemical fertilizers. *Microbial Ecology, 58*, 921–929.
Allison, S. D. (2006). Soil minerals and humic acids alter enzyme stability: Implications for ecosystem processes. *Biogeochemistry, 81*(3), 361–373.
Amadou, I., Faucon, M.-P., & Houben, D. (2022). New insights into sorption and desorption of organic phosphorus on goethite, gibbsite, kaolinite and montmorillonite. *Applied Geochemistry, 52*, 143.
Antelo, J., Fiol, S., Pérez, C., Mariño, S., Arce, F., Gondar, D., et al. (2010). Analysis of phosphate adsorption onto ferrihydrite using the CD-MUSIC model. *Journal of Colloid and Interface Science, 347*(1), 112–119.
Arai, Y., & Sparks, D. (2007). Phosphate reaction dynamics in soils and soil components: A multiscale approach. *Advances in Agronomy, 94*, 135–179.
Baldwin, D. S. (2013). Organic phosphorus in the aquatic environment. *Environmental Chemistry, 10*(6), 439–454.
Basinski, J. J., Bone, S. E., Klein, A. R., Thongsomboon, W., Mitchell, V., Shukle, J. T., et al. (2024). Unraveling iron oxides as abiotic catalysts of organic phosphorus recycling in soil and sediment matrices. *Nature Communications, 15*(1), 2145.
Binnig, G., Quate, C. F., & Gerber, C. (1986). Atomic force microscope. *Physical Review Letters, 56*(9), 930.
Blackwell, M., Brookes, P., de La-Fuente-Martinez, N., Gordon, H., Murray, P., Snars, K., et al. (2010). Phosphorus solubilization and potential transfer to surface waters from the soil microbial biomass following drying–rewetting and freezing–thawing. *Advances in Agronomy, 106*, 1–35.
Blake, R. E., O'Neil, J. R., & Surkov, A. V. (2005). Biogeochemical cycling of phosphorus: Insights from oxygen isotope effects of phosphoenzymes. *American Journal of Science, 305*(6–8), 596–620.
Bourdon, K., Fortin, J., Dessureault-Rompré, J., & Caron, J. (2023). Mitigating decomposition in agricultural peatlands: Influence of copper and polyphenol on CNP dynamics and enzyme activities in two contrasting soils. *Geoderma, 439*, 116694.
Brödlin, D., Kaiser, K., Kessler, A., & Hagedorn, F. (2019). Drying and rewetting foster phosphorus depletion of forest soils. *Soil Biology and Biochemistry, 128*, 22–34.
Celi, L., Lamacchia, S., Marsan, F. A., & Barberis, E. (1999). Interaction of inositol hexaphosphate on clays: Adsorption and charging phenomena. *Soil Science, 164*(8), 574–585.
Celi, L., Presta, M., Ajmore-Marsan, F., & Barberis, E. (2001). Effects of pH and electrolytes on inositol hexaphosphate interaction with goethite. *Soil Science Society of America Journal, 65*(3), 753–760.
Celi, L., Prati, M., Magnacca, G., Santoro, V., & Martin, M. (2020). Role of crystalline iron oxides on stabilization of inositol phosphates in soil. *Geoderma, 65*, 374.

Chen, A., & Arai, Y. (2019). Functional group specific phytate adsorption at the ferrihydrite–water interface. *Environmental Science and Technology, 53*(14), 8205–8215.

Chen, A., & Arai, Y. (2024). A review of the reactivity of phosphatase controlled by clays and clay minerals: Implications for understanding phosphorus mineralization in soils. *Clays and Clay Minerals, 71*(2), 119–142.

Chen, A., Zhu, L., & Arai, Y. (2022). Solution NMR investigation of phytate adsorption mechanisms at the calcite-water interface. *Science of the Total Environment, 98*, 840.

Chen, A., Zhu, L., & Arai, Y. (2023a). Enhanced and suppressed phosphorus mineralization by Ca complexation: NMR and CD spectroscopy investigation. *Chemosphere, 51*, 330.

Chen, A., Zhu, L., Han, H. S., & Arai, Y. (2023b). Spectroscopic investigation of phosphorus mineralization as affected by the calcite–water interfacial chemistry. *Environmental Science and Technology, 57*(43), 16606–16615.

Coleman, J. E. (1992). Structure and mechanism of alkaline phosphatase. *Annual Review of Biophysics and Biomolecular Structure, 21*(1), 441–483.

Condron, L. M., Turner, B. L., & Cade-Menun, B. J. (2005). Chemistry and dynamics of soil organic phosphorus. *Phosphorus: Agriculture and the Environment, 46*, 87–121.

Corbridge, D. E. C. (2013). Phosphorus: chemistry, biochemistry and technology. *Phosphorus Chemistry Biochemistry and Technology, 61*, 759.

Cordell, D., Drangert, J. O., & White, S. (2009). The story of phosphorus: Global food security and food for thought. *Global Environmental Change, 19*(2), 292–305.

Dalai, R. (1977). Soil organic phosphorus. *Advances in Agronomy, 29*, 83–117.

Fang, Y., Kim, E., & Strathmann, T. J. (2018). Mineral- and base-catalyzed hydrolysis of organophosphate flame retardants: Potential major fate-controlling sink in soil and aquatic environments. *Environmental Science and Technology, 52*(4), 1997–2006.

Feng, H., Ma, S., Chen, Z., Li, Y., Wang, M., & Ding, Y. (2024). Adsorption of nucleotides and nucleic acids on goethite nanoparticles: Mode, sites and relationship with phosphate and non-phosphate structures. *Environmental Science: Nano, 11*(6), 2655–2667.

Floate, M. (1970). Mineralization of nitrogen and phosphorus from organic materials of plant and animal origin and its significance in the nutrient cycle in grazed upland and hill soils. *Grass and Forage Science, 25*(4), 295–302.

Gao, K., Wang, S., Zhou, W., Zhang, B., Dang, Z., & Liu, C. (2024). Extracellular polymeric substances altered ferrihydrite (trans) formation and induced arsenic mobilization. *Journal of Hazardous Materials, 473*, 134434.

Ge, X., Zhang, W., Putnis, C. V., & Wang, L. (2022). Direct observation of humic acid-promoted hydrolysis of phytate through stabilizing a conserved catalytic domain in phytase. *Environmental Science: Processes and Impacts, 24*(7), 1082–1093.

Ge, X., Fan, Y., Zhai, H., Chi, J., Putnis, C. V., Wang, L., et al. (2024). Direct observations of nanoscale brushite dissolution by the concentration-dependent adsorption of phosphate or phytate. *Water Research, 248*, 98.

Giesy, J. P., Dobson, S., & Solomon, K. R. (2000). *Ecotoxicological risk assessment for roundup® herbicide* (pp. 35–120). Springer.

Goldberg, S., & Sposito, G. (1985). On the mechanism of specific phosphate adsorption by hydroxylated mineral surfaces: A review. *Communications in Soil Science and Plant Analysis, 16*(8), 801–821.

Gross, A., Turner, B. L., Goren, T., Berry, A., & Angert, A. (2016). Tracing the sources of atmospheric phosphorus deposition to a tropical rain forest in panama using stable oxygen isotopes. *Environmental Science and Technology, 50*(3), 1147–1156.

Guan, X.-H., Shang, C., Zhu, J., & Chen, G.-H. (2006). ATR-FTIR investigation on the complexation of myo-inositol hexaphosphate with aluminum hydroxide. *Journal of Colloid and Interface Science, 293*(2), 296–302.

Hall, D. (1976). *The chemistry of phosphorus*. Hamper and Pow publishers.

Henderson, T. O., Glonek, T., & Myers, T. C. (1974). Phosphorus-31 nuclear magnetic resonance spectroscopy of phospholipids. *Biochemistry, 13*(3), 623–628.

Hollenback, A. J., & Jaisi, D. P. (2024). Position-specific oxygen isotope analysis in inositol phosphates by using electrospray ionization-quadrupole-orbitrap mass spectrometry. *Journal of the American Society for Mass Spectrometry, 35*(9), 2176–2186.

Holtz, K. M., & Kantrowitz, E. R. (1999). The mechanism of the alkaline phosphatase reaction: Insights from NMR, crystallography and site-specific mutagenesis. *FEBS Letters, 462*(1–2), 7–11.

Hu, Z., Jaisi, D. P., Yan, Y., Chen, H., Wang, X., Wan, B., et al. (2019). Adsorption and precipitation of myo-inositol hexakisphosphate onto kaolinite. *European Journal of Soil Science, 71*(2), 226–235.

Huang, X.-L., & Zhang, J.-Z. (2012). Hydrolysis of glucose-6-phosphate in aged, acid-forced hydrolysed nanomolar inorganic iron solutions: An inorganic biocatalyst? *RSC Advantages, 2*(1), 199–208.

Huang, Q., Zhu, J., Qiao, X., Cai, P., Rong, X., Liang, W., et al. (2009). Conformation, activity and proteolytic stability of acid phosphatase on clay minerals and soil colloids from an Alfisol. *Colloids and Surfaces b: Biointerfaces, 74*(1), 279–283.

Huang, R., Wan, B., Hultz, M., Diaz, J. M., & Tang, Y. (2018). Phosphatase-mediated hydrolysis of linear polyphosphates. *Environmental Science and Technology, 52*(3), 1183–1190.

Hwang, I. Y., Koh, E., Wong, A., March, J. C., Bentley, W. E., Lee, Y. S., et al. (2017). Engineered probiotic escherichia coli can eliminate and prevent pseudomonas aeruginosa gut infection in animal models. *Nature Communications, 8*(1), 15028.

Jaisi, D. P., Kukkadapu, R. K., Stout, L. M., Varga, T., & Blake, R. E. (2011). Biotic and abiotic pathways of phosphorus cycling in minerals and sediments: Insights from oxygen isotope ratios in phosphate. *Environmental Science and Technology, 45*(15), 6254–6261.

Jarosch, K. A., Kandeler, E., Frossard, E., & Bünemann, E. K. (2019). Is the enzymatic hydrolysis of soil organic phosphorus compounds limited by enzyme or substrate availability? *Soil Biology and Biochemistry, 139*, 212.

Jiang, Y., Kang, N., Zhou, Y., Liu, G., & Zhu, D. (2016). The role of Fe(III) on phosphate released during the photo-decomposition of organic phosphorus in deionized and natural waters. *Chemosphere, 164*, 208–214.

Jorquera, M., Martínez, O., Maruyama, F., Marschner, P., & de la Luz-Mora, M. (2008). Current and future biotechnological applications of bacterial phytases and phytase-producing bacteria. *Microbes and Environments, 23*(3), 182–191.

Khan, M. S., Zaidi, A., & Wani, P. A. (2009). Role of phosphate solubilizing microorganisms in sustainable agriculture-a review. *Sustainable Agriculture, 15*, 551–570.

Klein, A. R., Bone, S. E., Bakker, E., Chang, Z., & Aristilde, L. (2019). Abiotic phosphorus recycling from adsorbed ribonucleotides on a ferrihydrite-type mineral: Probing solution and surface species. *Journal of Colloid and Interface Science, 547*, 171–182.

Klein, A. R., Basinski, J. J., Niyitanga Manzi, A., Cheng, C. X., Sanchez, M., Thongsomboon, W., et al. (2024). Phosphorus recycling by mineral-catalyzed ribonucleotide cleavage on iron and manganese oxides. *Environmental Chemistry Letters, 21*, 1–7.

Kochian, L. V., Hoekenga, O. A., & Pineros, M. A. (2004). How do crop plants tolerate acid soils? Mechanisms of aluminum tolerance and phosphorous efficiency. *Annual Review of Plant Biology., 55*(1), 459–493.

Kononova, S., & Nesmeyanova, M. (2002). Phosphonates and their degradation by microorganisms. *Biochemistry, 67*, 184–195.

Kumaragamage, D., Amarawansha, G. S., Indraratne, S. P., Jayarathne, K., Flaten, D. N., Zvomuya, F., et al. (2019). Degree of phosphorus saturation as a predictor of redox-induced phosphorus release from flooded soils to floodwater. *Journal of Environmental Quality, 48*(6), 1817–1825.

Kurek, M., Harir, M., Shukle, J., Schroth, A., Schmitt-Kopplin, P., & Druschel, G. (2021). Seasonal transformations of dissolved organic matter and organic phosphorus in a polymictic basin: Implications for redox-driven eutrophication. *Chemical Geology, 573*, 120212.

Lee, C. H., Lee, I. B., & Kim, P. J. (2004). Effects of long-term fertilization on organic phosphorus fraction in paddy soil. *Soil Science and Plant Nutrition, 50*(4), 485–491.

Lehmann, J., Rillig, M. C., Thies, J., Masiello, C. A., Hockaday, W. C., & Crowley, D. (2011). Biochar effects on soil biota: A review. *Soil Biology and Biochemistry, 43*(9), 1812–1836.

Li, W., Feng, X., Yan, Y., Sparks, D. L., & Phillips, B. L. (2013). Solid-state NMR spectroscopic study of phosphate sorption mechanisms on aluminum (hydr)oxides. *Environmental Science and Technology, 47*(15), 8308–8315.

Li, T., Zhong, W., Jing, C., Li, X., Zhang, T., Jiang, C., et al. (2020). Enhanced hydrolysis of p-nitrophenyl phosphate by iron (hydr)oxide nanoparticles: Roles of exposed facets. *Environmental Science and Technology, 54*(14), 8658–8667.

Lindqvist, Y., Schneider, G., & Vihko, P. (1994). Crystal structures of rat acid phosphatase complexed with the transition-state analogs vanadate and molybdate: Implications for the reaction mechanism. *European Journal of Biochemistry, 221*(1), 139–142.

Liu, F., Li, X., Sheng, A., Shang, J., Wang, Z., & Liu, J. (2019). Kinetics and mechanisms of protein adsorption and conformational change on hematite particles. *Environmental Science and Technology, 53*(17), 10157–10165.

Liu, F., Ding, Z., Lu, Y., Li, R., & Shi, Z. (2022a). Nano-scale investigation of organic C sequestration and distribution on Fe oxides during ferrihydrite transformation: Effect of Al-substitution. *Environmental Science: Nano, 9*(10), 4007–4017.

Liu, X., Han, R., Cao, Y., Turner, B. L., & Ma, L. Q. (2022b). Enhancing phytate availability in soils and phytate-P acquisition by plants: A Review. *Environmental Science and Technology, 56*(13), 9196–9219.

Lü, C., Yan, D., He, J., Zhou, B., Li, L., & Zheng, Q. (2017). Environmental geochemistry significance of organic phosphorus: An insight from its adsorption on iron oxides. *Applied Geochemistry, 84*, 52–60.

Luengo, C., Brigante, M., & Avena, M. (2007). Adsorption kinetics of phosphate and arsenate on goethite: A comparative study. *Journal of Colloid and Interface Science, 311*(2), 354–360.

Maranguit, D., Guillaume, T., & Kuzyakov, Y. (2017). Effects of flooding on phosphorus and iron mobilization in highly weathered soils under different land-use types: Short-term effects and mechanisms. *CATENA, 158*, 161–170.

Martin, M., Celi, L., & Barberis, E. (2004). Desorption and plant availability of myo-inositol hexaphosphate adsorbed on goethite. *Soil Science, 169*(2), 115–124.

McDowell, R., & Sharpley, A. (2002). Phosphorus transport in overland flow in response to position of manure application. *Journal of Environmental Quality, 31*(1), 217–227.

McLaren, T. I., Smernik, R. J., McLaughlin, M. J., McBeath, T. M., Kirby, J. K., Simpson, R. J., et al. (2015). Complex forms of soil organic phosphorus: A major component of soil phosphorus. *Environmental Science and Technology, 49*(22), 13238–13245.

McRose, D. L., & Newman, D. K. (2021). Redox-active antibiotics enhance phosphorus bioavailability. *Science, 371*(6533), 1033–1037.

Menezes-Blackburn, D., Jorquera, M. A., Greiner, R., Gianfreda, L., & de la Luz Mora, M. (2013). Phytases and phytase-labile organic phosphorus in manures and soils. *Critical Reviews in Environmental Science and Technology, 43*(9), 916–954.

Metcalf, W. W., & Van Der Donk, W. A. (2009). Biosynthesis of phosphonic and phosphinic acid natural products. *Annual Review of Biochemistry, 78*(1), 65–94.

Mortland, M. M., & Gieseking, J. E. (1952). The influence of clay minerals on the enzymatic hydrolysis of organic phosphorus compounds. *Soil Science Society of America Journal, 16*(1), 10–13.

Nannipieri, P., Giagnoni, L., Landi, L., & Renella, G. (2011). *Role of phosphatase enzymes in soil. Phosphorus in Action* (pp. 215–243).

Ognalaga, M., Frossard, E., & Thomas, F. (1994). Glucose-1-phosphate and myo-inositol hexaphosphate adsorption mechanisms on goethite. *Soil Science Society of America Journal, 58*(2), 332–337.

Olsson, R. (2011). *Surface reactions on mineral particles controlling the hydrolysis of glucose phosphates*. Umeå universitet.

Olsson, R., Giesler, R., Loring, J. S., & Persson, P. (2011). Enzymatic hydrolysis of organic phosphates adsorbed on mineral surfaces. *Environmental Science and Technology, 46*(1), 285–291.

Pantigoso, H. A., Manter, D. K., Fonte, S. J., & Vivanco, J. M. (2023). Root exudate-derived compounds stimulate the phosphorus solubilizing ability of bacteria. *Scientific Reports, 13*(1), 4529.

Parfitt, R. L., & Atkinson, R. J. (1976). Phosphate adsorption on goethite (α-FeOOOH). *Nature, 264*(5588), 740–742.

Park, Y., Solhtalab, M., Thongsomboon, W., & Aristilde, L. (2022). Strategies of organic phosphorus recycling by soil bacteria: Acquisition, metabolism, and regulation. *Environmental Microbiology Reports, 14*(1), 3–24.

Paytan, A., & McLaughlin, K. (2007). The oceanic phosphorus cycle. *Chemical Reviews, 107*(2), 563–576.

Phillips, B. L., Zhang, Z., Kubista, L., Frisia, S., & Borsato, A. (2016). NMR spectroscopic study of organic phosphate esters coprecipitated with calcite. *Geochimica Et Cosmochimica Acta, 183*, 46–62.

Porvari, K. S., Herrala, A. M., Kurkela, R. M., Taavitsainen, P. A., Lindqvist, Y., Schneider, G., et al. (1994). Site-directed mutagenesis of prostatic acid phosphatase. Catalytically important aspartic acid 258, substrate specificity, and oligomerization. *Journal of Biological Chemistry, 269*(36), 22642–22646.

Prietzel, J., Harrington, G., Häusler, W., Heister, K., Werner, F., & Klysubun, W. (2016). Reference spectra of important adsorbed organic and inorganic phosphate binding forms for soil P speciation using synchrotron-based K-edge XANES spectroscopy. *Journal of Synchrotron Radiation, 23*(2), 532–544.

Richardson, A. E. (2001). Prospects for using soil microorganisms to improve the acquisition of phosphorus by plants. *Functional Plant Biology, 28*(9), 897–906.

Richardson, A. E., Hocking, P. J., Simpson, R. J., & George, T. S. (2009). Plant mechanisms to optimise access to soil phosphorus. *Crop and Pasture Science, 60*(2), 124–143.

Rodriguez, H., & Fraga, R. (1999). Phosphate solubilizing bacteria and their role in plant growth promotion. *Biotechnology Advances, 17*(4–5), 319–339.

Rowan, K. (1966). Phosphorus metabolism in plants. *International Review of Cytology, 19*, 301–390.

Ruttenberg, K., & Sulak, D. (2011). Sorption and desorption of dissolved organic phosphorus onto iron (oxyhydr) oxides in seawater. *Geochimica Et Cosmochimica Acta, 75*(15), 4095–4112.

Schenk, G., Gahan, L. R., Carrington, L. E., Mitic, N. A., Valizadeh, M., Hamilton, S. E., et al. (2004). Phosphate forms an unusual tripodal complex with the Fe–Mn center of sweet potato purple acid phosphatase. *Proceedings of the National Academy of Sciences, 102*(2), 273–278.

Seibert, E., & Tracy, T. S. (2014). Fundamentals of enzyme kinetics. *Enzyme Kinetics in Drug Metabolism: Fundamentals and Applications, 217*, 9–22.

Shang, C., Huang, P., & Stewart, J. (1990). Kinetics of adsorption of organic and inorganic phosphates by short-range ordered precipitate of aluminum. *Canadian Journal of Soil Science, 70*(3), 461–470.

Shang, C., Zelazny, L. W., Berry, D. F., & Maguire, R. O. (2013). Orthophosphate and phytate extraction from soil components by common soil phosphorus tests. *Geoderma, 209–210*, 22–30.

Sharpley, A. (2006). *Agricultural phosphorous management: protecting production and water quality*. Midwest Plan Service.

Shen, J., Yuan, L., Zhang, J., Li, H., Bai, Z., Chen, X., et al. (2011). Phosphorus dynamics: From soil to plant. *Plant Physiology, 156*(3), 997–1005.

Sheng, Y., Hu, J., Kukkadapu, R., Guo, D., Zeng, Q., & Dong, H. (2023). Inhibition of extracellular enzyme activity by reactive oxygen species upon oxygenation of reduced iron-bearing minerals. *Environmental Science and Technology, 57*(8), 3425–3433.

Shinohara, R., Imai, A., Kawasaki, N., Komatsu, K., Kohzu, A., Miura, S., et al. (2012). Biogenic phosphorus compounds in sediment and suspended particles in a shallow eutrophic lake: A ^{31}P-Nuclear magnetic resonance (^{31}P NMR) study. *Environmental Science and Technology, 46*(19), 10572–10578.

Shujie, C., Kaiying, D., Jun, T., Rui, S., Hailong, L., Jiuyu, L., et al. (2021). Characterization of extracellular phosphatase activities in periphytic biofilm from paddy field. *Pedosphere, 31*(1), 116–124.

Six, J., Ogle, S. M., Jay Breidt, F., Conant, R. T., Mosier, A. R., & Paustian, K. (2004). The potential to mitigate global warming with no-tillage management is only realized when practised in the long term. *Global Change Biology, 10*(2), 155–160.

Smith, B. C. (2011). *Fundamentals of fourier transform infrared spectroscopy*. CRC Press.

Smith, S. E., Jakobsen, I., Grønlund, M., & Smith, F. A. (2011). Roles of arbuscular mycorrhizas in plant phosphorus nutrition: Interactions between pathways of phosphorus uptake in arbuscular mycorrhizal roots have important implications for understanding and manipulating plant phosphorus acquisition. *Plant Physiology, 156*(3), 1050–1057.

Solhtalab, M., Klein, A. R., & Aristilde, L. (2020). Hierarchical reactivity of enzyme-mediated phosphorus recycling from organic mixtures by aspergillus niger phytase. *Journal of Agricultural and Food Chemistry, 69*(7), 2295–2305.

Stec, B., Holtz, K. M., & Kantrowitz, E. R. (2000). A revised mechanism for the alkaline phosphatase reaction involving three metal ions. *Journal of Molecular Biology, 299*(5), 1303–1311.

Sun, M., Alikhani, J., Massoudieh, A., Greiner, R., & Jaisi, D. P. (2017). Phytate degradation by different phosphohydrolase enzymes: Contrasting kinetics, decay rates, pathways, and isotope effects. *Soil Science Society of America Journal, 81*(1), 61–75.

Tabatabai, M. A. (1989). Soil organic phosphorus: A review of world literature. *Soil Science, 147*(1), 2987.

Tang, J., Leung, A., Leung, C., & Lim, B. L. (2006). Hydrolysis of precipitated phytate by three distinct families of phytases. *Soil Biology and Biochemistry, 38*(6), 1316–1324.

Timofeeva, A., Galyamova, M., & Sedykh, S. (2022). Prospects for using phosphate-solubilizing microorganisms as natural fertilizers in agriculture. *Plants, 11*(16), 987.

Turner, B. L., & Haygarth, P. M. (2001). Phosphorus solubilization in rewetted soils. *Nature, 411*(6835), 258–258.

Turner, B. L., Papházy, M. J., Haygarth, P. M., & McKelvie, I. D. (2002). Inositol phosphates in the environment. *Philosophical Transactions of the Royal Society of London Series B Biological Sciences, 357*(1420), 449–469.

Turner, B. L., Mahieu, N., & Condron, L. M. (2003). Phosphorus-31 nuclear magnetic resonance spectral assignments of phosphorus compounds in soil NaOH–EDTA extracts. *Soil Science Society of America Journal, 67*(2), 497–510.

Turner, B. L., Cheesman, A. W., Godage, H. Y., Riley, A. M., & Potter, B. V. L. (2012). Determination of neo- and d-chiro-inositol hexakisphosphate in soils by solution 31P NMR spectroscopy. *Environmental Science and Technology, 46*(9), 4994–5002.

Turner, B. L., Frossard, E., & Baldwin, D. S. (2005). *Organic phosphorus in the environment*. CABI publishing.

Ustunol, I. B., Coward, E. K., Quirk, E., & Grassian, V. H. (2021). Interaction of beta-lactoglobulin and bovine serum albumin with iron oxide (α-Fe2O3) nanoparticles in the presence and absence of pre-adsorbed phosphate. *Environmental Science: Nano, 8*(10), 2811–2823.

Wan, B., Yan, Y., Liu, F., Tan, W., Chen, X., & Feng, X. (2016). Surface adsorption and precipitation of inositol hexakisphosphate on calcite: A comparison with orthophosphate. *Chemical Geology, 421*, 103–111.

Wan, B., Yang, P., Jung, H., Zhu, M., Diaz, J. M., & Tang, Y. (2021). Iron oxides catalyze the hydrolysis of polyphosphate and precipitation of calcium phosphate minerals. *Geochimica Et Cosmochimica Acta, 305*, 49–65.

Wan, B., Huang, R., Diaz, J. M., & Tang, Y. (2022). Rethinking the biotic and abiotic remineralization of complex phosphate molecules in soils and sediments. *Science of the Total Environment, 833*, 1147.
Wang, L., Qin, L., Putnis, C. V., Ruiz-Agudo, E., King, H. E., & Putnis, A. (2015). Visualizing organophosphate precipitation at the calcite–water interface by in situ atomic-force microscopy. *Environmental Science and Technology, 50*(1), 259–268.
Wang, J., Wu, Y., Zhou, J., Bing, H., & Sun, H. (2016a). Carbon demand drives microbial mineralization of organic phosphorus during the early stage of soil development. *Biology and Fertility of Soils, 52*(6), 825–839.
Wang, L., Putnis, C. V., King, H. E., Hövelmann, J., Ruiz-Agudo, E., & Putnis, A. (2016b). Imaging organophosphate and pyrophosphate sequestration on brucite by in situ atomic force microscopy. *Environmental Science and Technology, 51*(1), 328–336.
Wang, X., Hu, Y., Tang, Y., Yang, P., Feng, X., Xu, W., et al. (2017). Phosphate and phytate adsorption and precipitation on ferrihydrite surfaces. *Environmental Science: Nano, 4*(11), 2193–2204.
Wang, X., Phillips, B. L., Boily, J.-F., Hu, Y., Hu, Z., Yang, P., et al. (2019). Phosphate sorption speciation and precipitation mechanisms on amorphous aluminum hydroxide. *Soil Systems, 3*(1), 98.
Wang, J., Wang, L., Zhang, A. J., Zhang, H., & Zhang, Y. C. (2020). Effects of long-term organic fertilization on the content of soil organic phosphorus fractions and abundance of soil microorganism. *Journal of Ecology and Rural Environment, 36*(9), 1161–1168.
Wang, F., Liu, Y., Cao, M., Zhou, B., Chen, H., Yuan, R., et al. (2023). Mechanisms of ZnO nanoparticles enhancing phototransformation of biologically derived organic phosphorus in aquatic environments. *Environmental Science and Technology, 57*(9), 3691–3702.
Wei, L., Chen, C., & Xu, Z. (2010). Citric acid enhances the mobilization of organic phosphorus in subtropical and tropical forest soils. *Biology and Fertility of Soils, 46*, 765–769.
Werner, F., & Prietzel, J. (2015). Standard protocol and quality assessment of soil phosphorus speciation by P K-Edge XANES spectroscopy. *Environmental Science and Technology, 49*(17), 10521–10528.
White, A. K., & Metcalf, W. W. (2007). Microbial metabolism of reduced phosphorus compounds. *Annual Review of Microbiology., 61*(1), 379–400.
Withers, P. J. A., Sylvester-Bradley, R., Jones, D. L., Healey, J. R., & Talboys, P. J. (2014). Feed the crop not the soil: Rethinking phosphorus management in the food chain. *Environmental Science and Technology, 48*(12), 6523–6530.
Wu, S. C., Cao, Z., Li, Z., Cheung, K., & Wong, M. H. (2005). Effects of biofertilizer containing N-fixer, P and K solubilizers and AM fungi on maize growth: A greenhouse trial. *Geoderma, 125*(1–2), 155–166.
Wu, J., Paudel, P., Sun, M., Joshi, S. R., Stout, L. M., Greiner, R., et al. (2015). Mechanisms and pathways of phytate degradation: Evidence from oxygen isotope ratios of phosphate, HPLC, and phosphorus-31 NMR spectroscopy. *Soil Science Society of America Journal, 79*(6), 1615–1628.
Xu, S., & Arai, Y. (2022). Adsorption mechanisms of inositol hexakisphosphate in the presence of phosphate at the amorphous aluminum oxyhydroxide-water interface. *Science of the Total Environment, 837*, 598.
Xu, S., Chen, A., & Arai, Y. (2021). Solution 31P NMR investigation of inositol hexakisphosphate surface complexes at the amorphous aluminum oxyhydroxide–water interface. *Environmental Science and Technology, 55*(21), 14628–14638.
Yan, Y., Li, W., Yang, J., Zheng, A., Liu, F., Feng, X., et al. (2014a). Mechanism of myo-inositol hexakisphosphate sorption on amorphous aluminum hydroxide: Spectroscopic evidence for rapid surface precipitation. *Environmental Science and Technology, 48*(12), 6735–6742.
Yan, Y., Liu, F., Jr., Li, W., Liu, F., Feng, X., & Sparks, D. (2014b). Sorption and desorption characteristics of organic phosphates of different structures on aluminium (oxyhydr) oxides. *European Journal of Soil Science, 65*(2), 308–317.
Yan, Y., Wan, B., Liu, F., Tan, W., Liu, M., & Feng, X. (2014c). Adsorption-desorption of myo-inositol hexakisphosphate on hematite. *Soil Science, 179*(10–11), 476–485.

Yan, Y., Liu, F., Jr., Li, W., Liu, F., Feng, X., & Sparks, D. (2014d). Sorption and desorption characteristics of organic phosphates of different structures on aluminium (oxyhydr)oxides. *European Journal of Soil Science, 65*(2), 308–317.

Yan, Y., Wan, B., Jiang, R., Wang, X., Wang, H., Lan, S., et al. (2022a). Withdrawn: Interactions of organic phosphorus with soil minerals and the associated environmental impacts: A review. *Journal of Integrative Agriculture, 59*, 23.

Yan, Y., Wan, B., Mansor, M., Wang, X., Zhang, Q., Kappler, A., et al. (2022b). Co-sorption of metal ions and inorganic anions/organic ligands on environmental minerals: A review. *Science of the Total Environment, 803*, 789.

Yang, F., Sui, L., Tang, C., Li, J., Cheng, K., & Xue, Q. (2021). Sustainable advances on phosphorus utilization in soil via addition of biochar and humic substances. *Science of the Total Environment, 768*, 87.

Yong, S. C., Roversi, P., Lillington, J., Rodriguez, F., Krehenbrink, M., Zeldin, O. B., et al. (2014). A complex iron-calcium cofactor catalyzing phosphotransfer chemistry. *Science, 345*(6201), 312.

Yu, W. H., Li, N., Tong, D. S., Zhou, C. H., Lin, C. X., & Xu, C. Y. (2013). Adsorption of proteins and nucleic acids on clay minerals and their interactions: A review. *Applied Clay Science, 80–81*, 443–452.

Zaheer, R., Morton, R., Proudfoot, M., Yakunin, A., & Finan, T. M. (2009). Genetic and biochemical properties of an alkaline phosphatase PhoX family protein found in many bacteria. *Environmental Microbiology, 11*(6), 1572–1587.

Zang, Y., Liu, F., Li, X., Sheng, A., Zhai, J., & Liu, J. (2020). Adsorption kinetics, conformational change, and enzymatic activity of β-glucosidase on hematite (α-Fe2O3) surfaces. *Colloids and Surfaces b: Biointerfaces, 193*, 598.

Zhang, L., Ding, X., Chen, S., He, X., Zhang, F., & Feng, G. (2014). Reducing carbon: Phosphorus ratio can enhance microbial phytin mineralization and lessen competition with maize for phosphorus. *Journal of Plant Interactions, 9*(1), 850–856.

Zhang, L., Xu, M., Liu, Y., Zhang, F., Hodge, A., & Feng, G. (2016). Carbon and phosphorus exchange may enable cooperation between an arbuscular mycorrhizal fungus and a phosphate-solubilizing bacterium. *New Phytologist, 210*(3), 1022–1032.

Zhang, L., Feng, G., & Declerck, S. (2018). Signal beyond nutrient, fructose, exuded by an arbuscular mycorrhizal fungus triggers phytate mineralization by a phosphate solubilizing bacterium. *The International Society for Microbial Ecology Journal, 12*(10), 2339–2351.

Zhang, K., Ho, K.-P., Chatterjee, A., Park, G., Li, Z., Catalano, J. G., et al. (2023a). RNA hydrolysis at mineral–water interfaces. *Environmental Science and Technology, 57*(22), 8280–8288.

Zhang, X., Li, J., Shao, L., Qin, F., Yang, J., Gu, H., et al. (2023b). Effects of organic fertilizers on yield, soil physico-chemical property, soil microbial community diversity and structure of brassica rapa var, Chinensis. *Frontiers in Microbiology, 14*, 1132853.

Zhu, J., Li, M., & Whelan, M. (2018). Phosphorus activators contribute to legacy phosphorus availability in agricultural soils: A review. *Science of the Total Environment, 612*, 522–537.

Plant Metabolism and Sustainable Agriculture

Shuo Wei and Muhammad Shaaban

Abstract The yield and quality of crops are the foundation of sustainable agriculture, which can be affected by plant health and environmental factors. Plant metabolism runs through the whole life process of plants, in which primary metabolism can directly affect the growth and development of crops, thereby affecting crop yields. The secondary metabolism is more involved in the interaction between plants and the environment. Secondary metabolism production provides plants with on-demand biochemical capabilities to cope with highly unpredictable fluctuations of biotic and abiotic stress conditions. Therefore, secondary metabolism is also crucial for plant growth and development. In addition, some secondary metabolites in crops are also nutrients, functional and pharmacological components, which can affect the quality of crops. By analyzing plant metabolism, we can analyze the formation mechanism of plant yield and quality, and then make appropriate adjustments to obtain the products we want, which is of great benefit to the sustainable development of agriculture.

1 Introduction

Plants are one of the most important life forms on Earth. The plant kingdom produces hundreds of thousands of organic compounds. Primary metabolites mainly include carbohydrates, amino acids, and lipids which are directly required for plant growth (Erb & Kliebenstein, 2020). Plant secondary metabolites (PSMs) are synthesized to provide defensive functions and regulate defense signaling pathways to safeguard plants against herbivores (Divekar et al., 2022). Humans have used plants for food, raw materials, and medicines for a long history (Jacobowitz & Weng, 2020). Carbohydrates in food are the main source of energy for human life activities. The USDA's Dietary Guidelines recommends consuming 45–65% from carbohydrates, 10–35% of calories from protein, and 20–35% from fat a day for adults (Shams-White et al., 2023). Plant foods are the main source of carbohydrates consumed by humans

S. Wei (✉) · M. Shaaban
College of Agriculture, Henan University of Science and Technology, Luoyang 471000, China
e-mail: hnzywei@163.com

M. Shaaban (ed.), *Soils and Sustainable Agriculture*, Frontier Studies in Soil Science,
https://doi.org/10.1007/978-3-031-91114-9_22

which are produced by the primary metabolism in plants (Maino Vieytes et al., 2019). Protein and lipids in plant food are also essential for human health (Deng et al., 2017). Besides major nutrients of carbohydrate and amino acids, plant foods also contain many secondary metabolites, such as polyphenols, saponin, terpenoid, etc. Although these phytochemicals are the minor components in food, they exhibit very important roles in food taste, nutrition, and physiological function (Zhang et al., 2021). Moreover, humans have tapped into plant secondary metabolites for traditional herbal medicines for millennia (Jacobowitz & Weng, 2020). Thus, plant metabolites have a close relationship with human food and health.

Healthy, sustainable and inclusive food systems are critical to achieve the world's development goals. Agriculture is the origin of human civilization and allows people to have an adequate and stable source of food in a limited space (Martin & Sauerborn, 2013). In addition to food crops, agriculture also involves many other important crops, such as oil crops, medicinal crops, horticultural plants etc., which are directly related to the economy (Dangi et al., 2023). Agricultural development is one of the most powerful tools to end extreme poverty, boost shared prosperity. But agriculture development is at risk now: Population growth, climate change, conflicts, and structural changes in the economy are driving food and nutrition insecurity (Nations, 2017). These trends pose a huge challenge to the sustainability of agriculture. Needed is a transformative process towards green agriculture. But, where does "green" come from? Plants are the main objects of labor in agriculture. A more comprehensive, in-depth and systematic study of crops can alleviate this problem.

Improving crops yields, nutrition, preventing crops diseases, and developing more environmentally resilient varieties are necessary for sustainable agriculture. Through the research of plant metabolism, the regulatory factors of yield formation can be systematically analyzed, such as the transport, distribution and accumulation of photosynthesis and its products, the absorption and assimilation of nutrients, etc. (Paul, 2021). Furthermore, the relationship between metabolites and stress resistance, such as drought resistance, cold resistance, disease and insect resistance, also has great application potential in breeding (Nawaz et al., 2023). Plant metabolism research is an important part of maintaining sustainable agriculture, so this section provides an overview of the role of plant metabolism in agricultural production.

2 The Importance of Primary Metabolism for Agricultural Sustainability

Food crops are vital energy sources to the body. Nowadays, the demand for supply of food crops is on the increase. Wheat, rice, and corn are the world's most important food crops. Wheat has about 60% carbohydrates, 12–20% protein, and 1.3% fat (Khalid et al., 2023). Rice contains about 75% carbohydrates, 7–8% protein, and 1.3–1.8% fat (Singh et al., 2023). Corn contains about 70–87% carbohydrates, 6–13% protein, and 4% fat (Al Dhaheri et al., 2015). Therefore, primary metabolites are the main components of food and regulating the synthesis of primary metabolites is essential for food crops.

2.1 *Effect of Carbohydrate Metabolism on Crops Yields*

Carbohydrates are derived from photosynthesis. During the process of photosynthesis, plant cells use carbon dioxide and energy from the Sun to make sugar molecules and oxygen. These sugar molecules are the basis for more complex molecules made by the photosynthetic cell, such as glucose and starch (Stirbet et al., 2020). Photosynthesis is a well-studied process and the past 10 years have seen a major effort into improving photosynthesis (Paul, 2021). A recent study overexpresses Rubisco in paddy rice which increases yield in the field under good nitrogen supply by 17–28% demonstrating an improvement in photosynthesis linked to yield increase (Yoon et al., 2020). Storage starch, synthesized in the seeds, tubers, corms, and roots of plants, is the main substance used by plants to store carbohydrates and is the most important energy source for all living organisms (Seung, 2020). Starch biosynthesis influences grain size and physico-chemical properties, which together determine grain eating quality (Liu et al., 2022). Starch metabolism is regulated by various upstream pathways (Huang et al., 2021). Research on starch synthesis-related genes in cereals has resulted in numerous breakthroughs in cereal breeding for grain quality improvement (Anacleto et al., 2019). OsbZIP58 is a key transcriptional regulator controlling starch synthesis in rice endosperm and mutations can reduce the content of total starch (Wang et al., 2013). Similar conclusions were also obtained with the homologs in wheat (Guo et al., 2020). The *Wx* gene was selected preferentially during rice domestication (Zhang et al., 2020a, 2020b). *ALK* alleles that produce low and medium gelatinization temperature are generally preferred (Chen et al., 2020; Miura et al., 2018). Knockout of *Wx* using genome editing also facilitated sucrose accumulation in corn kernels, and the simultaneous abolition of *Wx* and *Sh2* created super-sweet corn (Dong et al., 2019). Therefore, the metabolism of carbohydrates has a direct impact on the yield and quality of crops. Continuing to promote research on the metabolism of crop carbohydrates and rationally designing ideal quality grains in the future can help maintain the sustainability of agriculture.

2.2 *The Importance of Amino Acid Metabolism in Crops*

Amino acids are crucial for human growth, development and health. Indeed, essential amino acids (EAAs), notably lysine and methionine, cannot be synthesized by humans or animals, and three or more others are not synthesized in sufficient quantities to satisfy the metabolic needs (Hou & Wu, 2018). These nutritionally EAAs must be taken up from the diet and by far the greatest share is derived from plants (Hou et al., 2019). But, EAAs in plants often present at low levels (Trovato et al.,

2021). Genetic engineering provides an effective way of improving EAA levels in plants, with numerous successful results so far (Yang et al., 2020). The biosynthesis and catabolism of amino acids derived from the same metabolic trunk and/or closely related to other metabolic pathways. With this feature, researchers can regulate metabolism of a certain amino acid to affect the level of other amino acids (Song et al., 2013; Yang et al., 2018). The transgenic rice was generated revealing a slight decrease in isoleucine content and enhancement of other EAAs (Yang et al., 2016). Engineering of lysine metabolism tends to alter other amino acids in the aspartate family. However, high tryptophan levels were found to be associated with high-lysine plants (Yang et al., 2018). Similar findings were found during the regulation of methionine metabolism (Song et al., 2013). A metabolic engineering approach was used in rice and successfully increased methionine content in seeds by ~50% (Whitcomb et al., 2020). It should be noted that: amino acid metabolism is also tightly linked to energy and carbohydrate metabolism, the carbon–nitrogen budget, hormone and secondary metabolism and so on. A study show that grain yield was reduced in high-lysine crops (Liu et al., 2016). Studies also suggest that lysine affects starch synthesis and/or endosperm development in maize (*Zea mays* Linn.) and rice via biofortification (Jia et al., 2013; Liu et al., 2016). All these findings suggest that amino acid metabolism has an important impact on crops. Underlying mechanism of the metabolic connection between starch, amino acids and other storage substances in cereal endosperms will be the focus of future research in order to balance the nutrition and yield of crops.

2.3 Lipids Metabolism in Oil Crops

Lipids are one of the most important biomolecules in all plant tissues, and they are the main components of biomembrane (Ramos-Martín & D'Amelio, 2022). For humans, vegetable oils are one of the important components of nutrition and the safety of edible oil is an important part of national strategic security (Wang et al., 2023). Oil crops such as soybeans, peanuts, rapeseed, sesame, tree peonies, etc. are important cash crops. In most plants, oil is stored in the form of triacylglycerols (TAGs) in seed tissues and seed oil accumulation and composition are important traits controlled by a large number of genetic and metabolic factors (Lu et al., 2018). The accumulation of stored lipids is mainly regarded as a result of fatty acids (FAs) synthesis in plastids and the assembly of TAGs in the endoplasmic reticulum (ER) (Lin et al., 2017). Metabolic engineering tools to modify the fatty acid profiles of oilseed crops has been well studied (Zhang et al., 2020a, 2020b; Zhou et al., 2021). A patatin-related phospholipase mediates phospholipid turnover to increases seed oil content (Li et al., 2013). ACBPs, acyl-CoA binding proteins, can affect TAGs content and are good candidates for engineering oil crops to enhance seed oil composition (Guo et al., 2019). AtALT1 and AtALT4 overexpression in leaves and seeds resulted in the accumulation of 12–14 carbon-length FAs and 6–8 carbon-length FAs, respectively (Kalinger et al., 2021). In addition, enhancing carbon partitioning towards the chloroplast also can increase FAs synthesis (Arias et al., 2023). Photosynthetic capacity of embryo strongly contributes to oil accumulation, which was not only validated in culture conditions at higher light intensity (Sedbrook & Durrett, 2020). Therefore, fatty acid and oil content can be altered by regulating lipid metabolism and other primary metabolites associated with it.

The metabolism of plants is very complex (Fig. 1). Carbohydrates, amino acids, and lipid metabolism are related to each other, but ultimately flow in different ways. The regulation of primary metabolism in crops has a long history of research. But, how to promote the accumulation of the other party or find a balance point through their relationship to obtain the high-yield and high-quality products we want is one of the development goals of sustainable agriculture in the future.

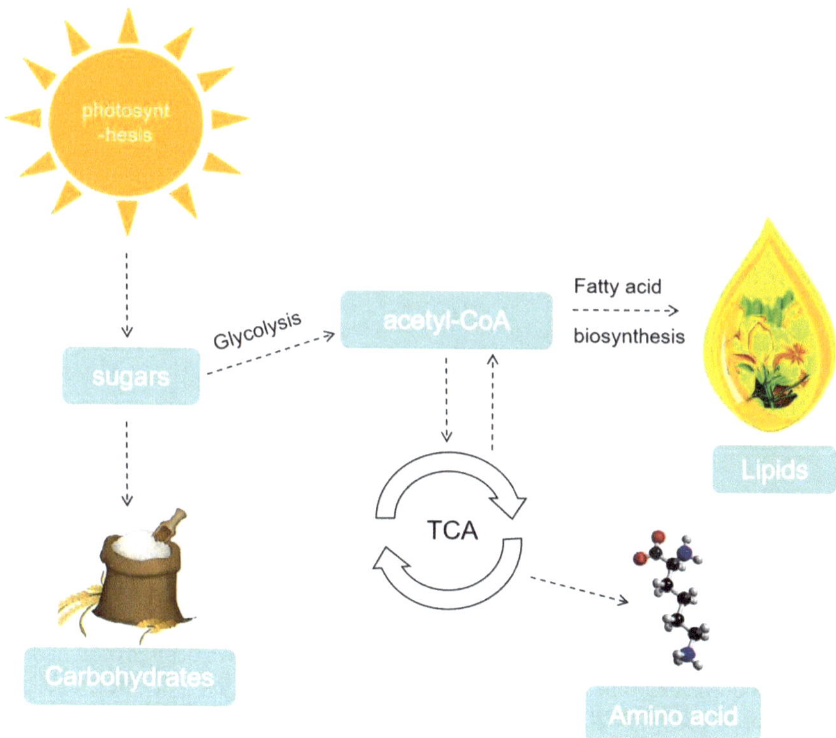

Fig. 1 Schematic diagram of primary metabolic association. *Note* The pictures of carbohydrates, lipids and amino acids come from the Internet

3 The Role of Secondary Metabolites for Agricultural Sustainability

Secondary metabolites are adaptive strategy adopted by plants to colonize the land. Plants can survive harsh terrestrial environments by greatly expanding their metabolic systems (Jacobowitz & Weng, 2020). Secondary metabolism is mainly based on primary metabolites as raw materials, and a large number of special substances are formed under the catalytic action of a series of enzymes. Current studies have shown that plant secondary metabolites are mainly derived from four pathways (Li & Gaquerel, 2021): Phenols and anthraquinones are the main secondary metabolites produced by acetic acid–malonic acid (AA–MA) pathway (Cryle, 2018); The isoprenoid pathway is mainly formed terpenoids and steroids (Singh et al., 2024); The main compounds formed by the shikimic acid pathway are flavonoids and lignans (Kumar et al., 2018); The amino acid pathway is responsible for the synthesis of alkaloids (Pereira et al., 2023). Different metabolic pathways are involved in the formation of different types of metabolites, and some even come from composite pathways, thus creating a wide variety of secondary metabolites which providing rich medicinal and economic resources for human beings. Major classes of plant specialized metabolites include terpenoids, phenylpropanoids, and alkaloids (Fig. 2). Enormous diversity of structure and function exists within each class, and these metabolites advance human health, renewable energy, and sustainable green agriculture (Selwal et al., 2023).

3.1 Secondary Metabolites and Human Health

Currently, the application of secondary metabolism directly related to agriculture is medicinal plants. About 75% of people worldwide still rely on plant-based traditional medicines for primary health care and the development of medicinal plant agriculture is essential for human health (Lal et al., 2023). The pharmacological activity of a growing list of plant specialized metabolites has been revealed, which in turn has expanded their utilities. Plant natural products have been pivotal in developing modern medicines to fight cancer, bacterial infections, parasites, and more (Newman & Cragg, 2020). For example, A 1600-yearold text, the classical Chinese medical text *Prescriptions for Emergencies*, documented the treatment of fever by cold water extract of sweet wormwood (*Artemisia annua*) and inspired the discovery of the antimalarial medicine artemisinin (Wang et al., 2019). Paclitaxel is widely used as a highly effective, broad-spectrum first-line chemotherapy medication (Onoda et al., 2016).

Secondary metabolism can carry out many protective functions in the human body. But their levels in plants tend to be low, and there is a problem with unstable levels (Elshafie et al., 2023). A biosynthetic approach by analyzing the biosynthetic

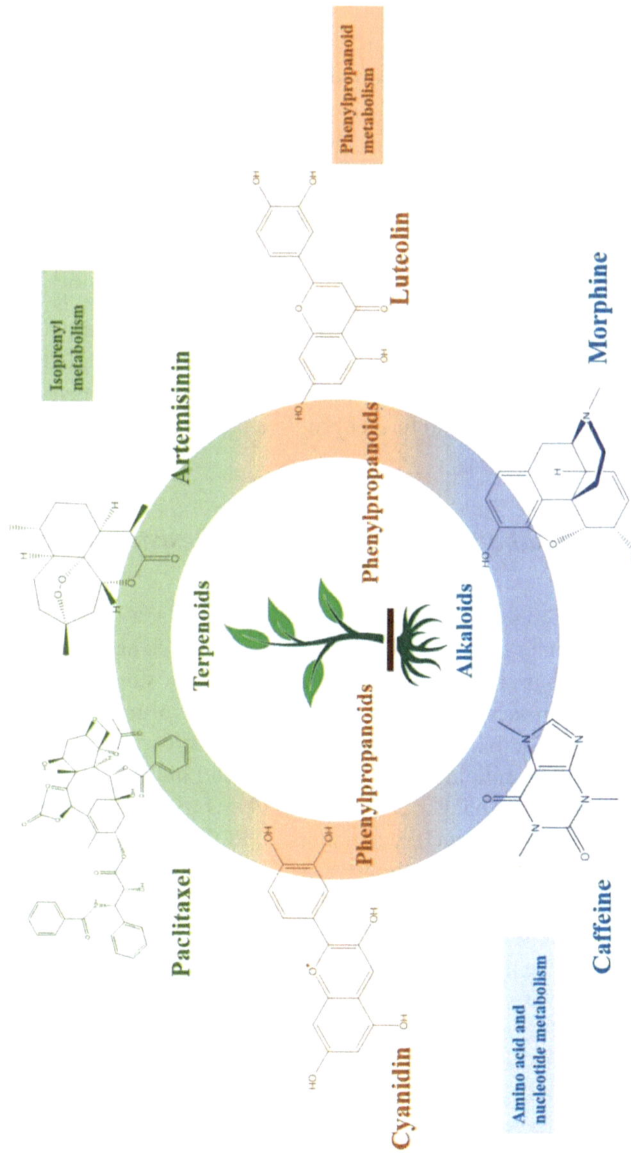

Fig. 2 Major classes of plant specialized metabolites, modified from Jacobowitz and Weng (2020)

pathways of secondary metabolites, toward access to valuable plant natural products holds tremendous. Over the past decade, innovations in the fields of genomics, transcriptomics, proteomics, metabolomics, and synthetic biology have generated many new tools and these tools that have had a transformative impact on plant specialized metabolism research (Kumari et al., 2021). Vinblastine belongs to a monoterpene indole alkaloid class of anticancer drugs, and its biosynthetic enzyme mining in plants enables its efficient production in yeast (Caputi et al., 2018; Zhang et al., 2022). The specific UGT responsible for salidroside biosynthesis in golden root (*Rhodiola rosea*) was identified and the production of salidroside is realized in tobacco (Torrens-Spence et al., 2018). Podophyllotoxin, a kind of aryltetralin-type lignan, is used in dozens of chemotherapy regimens for a variety of malignancies (Lau & Sattely, 2015). Recent studies have shown that the accumulation of podophyllotoxin can be promoted via transcriptional upregulation of genes involved in its biosynthesis (Kumari et al., 2021). There are many successful cases of accumulating high-value natural products by studying secondary metabolic pathways and appropriately regulating them (Jacobowitz & Weng, 2020; Li et al., 2020a, 2020b). These cases illustrate the importance of plant secondary metabolism research for the development of medicinal crops and further promote the development of agriculture.

3.2 *Secondary Metabolites and Crop Stress Resistance*

Several environmental stresses, including biotic and abiotic factors, adversely affect the growth and development of crops, thereby lowering their yield. It is expected that more than 50% of crop production losses are due to abiotic stresses (Salam et al., 2023). Among abiotic stresses, drought is one of the most harmful abiotic stresses that hamper agricultural productivity to a maximum extent worldwide. However, plants have developed several morphological, physiological and biochemical processes in response to drought stress (Ahanger et al., 2019). Plants trigger different strategies like enhanced production of secondary metabolites, ROS signaling, plant hydraulic status, and osmotic adjustment to maintain growth and productivity under drought stress conditions (Salam et al., 2023). Synthesis and accumulation of secondary metabolites can enhance level for drought tolerance in plants (Ahmad et al., 2023). Research revealed that the production of phenolic, flavonoid, and terpenoid in plants can alleviate abiotic stress, which are associated with enhanced antioxidant activity (Shelke et al., 2023). There is also much evidence for the involvement of various compounds of secondary origin in the amelioration of the toxic effects of stresses (Chrysargyris et al., 2019). Moreover, secondary metabolites have been shown to protect key biomolecules such as proteins and nucleic acids from stress-induced oxidative damage (Piasecka et al., 2015).

In addition to abiotic stresses, secondary metabolites also have important applications in plant biotic stresses. Plant secondary metabolites have several benefits for agriculture, where they can exhibit antiviral, antibacterial and antifungal activities (Camele et al., 2021). Many phenolics compounds such as phenolic acid, flavonoids, and tannins have been shown to provide defense against pathogens and insect pests by acting as antimicrobial and insecticidal agents (Ahanger et al., 2020). Terpenes such as menthol, camphor, pyrethrins, artemisinins, and farnesol play an important role against protozoa, bacteria, and fungi (Tiku, 2018). Alkaloids such as pyrrolizidine are toxic in nature and help against microbial infection and attack from herbivores (Casciaro et al., 2020). Azadirachtin, extracted from *Azadirachta indica*, has a strong insecticidal effect (Della Pepa et al., 2019). In particular, especially root exudates, play an important role in plant protection against a wide range of phytopathogens by killing or deterring soil microbes, herbivorous insects and nematodes (Elshafie et al., 2023). In review, plant secondary metabolites play significant roles in enhancing biotic stress tolerance. The application of metabolic engineering in breeding for biotic stress resistance, along with the exploitation of secondary metabolites for sustainable pest management, are goal of agricultural sustainability (Al-Khayri et al., 2023).

4 Plant Metabolites and Crop Continuous Cropping Obstacles

With the increase in intensive cultivation and long-term monoculture cropping, problems caused by continuous cropping (CC) have become increasingly prominent in the forms of replant diseases, soil sickness, and yield penalties (Liao & Xia, 2024). The widespread existence of CC obstacles in agricultural production severely restricts the sustainable development of agriculture, particularly affecting certain cropping systems such as grain, medicinal, fruits and other crops (Schmidt et al., 2019; Wu et al., 2023; Xi et al., 2019). The causes of CC disorders are complex, such as the accumulation of soil-borne pathogens, the autotoxic effect of plant root secretions, changes in the soil microbiota and soil physicochemical properties, and the decline in soil fertility (Ullah et al., 2019; Zhang et al., 2024). Root exudates, large quantities of metabolites released from living plant root, impact soil and microorganisms, leading to alterations in soil physical and chemical properties as well as microbial community structure. Consequently, there is an imbalance in soil nutrients and enzyme activity while beneficial microorganisms decrease, and pathogenic microorganisms increase (Zhou et al., 2024). To date, more than 300 root exudates have been identified, mainly primary metabolites (e.g., carbohydrates, amino acids) and secondary metabolites (e.g., phenolic acids, flavonoids, and cholesterol) (Balyan & Pandey, 2024). Considering that CC obstacles have an important impact on sustainable agricultural development and are closely related to plant metabolites, they are highlighted separately in this section.

Many plants produce allelochemicals through root exudation to outcompete the future generations for resource and space. These autotoxins accumulate in continuous cropping soil and strong impair the crop production (Ma et al., 2023). Tobacco is distinct from other crops by the vigorous root exudation of biological active secondary metabolites (Li et al., 2024). By analyzing the root exudates of tobacco, a wide range of autotoxic substances were recognized and these substances are typically small molecular aromatic compounds, organic acids and alkaloids (Zhang et al., 2016). The accumulation in the soil of peanut allelochemicals including phenolic acids, fatty acids, alcohols, aldehydes, and ketones, result in phytotoxicity and disruption of key physiological processes (Chen et al., 2018). The cultivation of medicinal plants requires several years or even decades, so CC obstacles significantly impede the development of medicinal plants. In the context of continuous cropping, medicinal plants secrete secondary metabolites in order to resist abiotic stress, which accumulate in the soil and produce autotoxic effects. These autotoxins have a concentration effect on medicinal plants; the higher the concentration, the stronger the inhibitory effect on growth (Liao & Xia, 2024). The concentrate of Sanqi ginseng in the soil extract increased with time during a 3-year monoculture of *Panax notoginseng* (Li et al., 2020a, 2020b). It is generally believed that benzoic acid is the allelochemical in *Panax ginseng* and in *Rehmannia glutinosa* is vanillic acid, ferulic acid and other substances (Wang et al., 2017; Yuan et al., 2022). Phenolic acids, produced during plant growth as the primary allelochemicals, represent the principal CC barriers for many medicinal plants (Zeeshan Ul Haq et al., 2023).

CC is a common practice in agriculture, and usually causes serious economic losses. Autotoxicity, a specific allelopathic effect, is an important factor in the occurrence of CC disorders. And allelopathic substances are metabolites secreted by the plant itself (Xin-hui et al., 2015). Thus, plant metabolism is inseparable from plant CC disorders, and systematic research on plant metabolism can help us better understand the mechanism of CC in order to seek good solution.

5 Prospect

From the above elaboration, it can be found that plants produce a huge array of metabolites in spatiotemporal- and/or environment-dependent manner, which can affect the yield and quality of crops, which in turn has an impact on agricultural development. Most of the previous studies have focused on the regulation of one or some metabolic pathways, but plant metabolism is a holistic and complex network (Li & Gaquerel, 2021). Primary and secondary metabolism are communicated and related to each other, so from the perspective of overall metabolism, it is crucial to pay attention to the problem of metabolic flow for crop yield and quality. For example, by altering the carbon metabolic flow through factors such as light, and/or intercepting the downstream pathway, thereby increasing the product of interest (Arias et al., 2023). Post-genomic profiling platforms and metabolomics provide powerful tools for our studies at the overall metabolic level and current research also tends to analyze plant metabolism from a holistic perspective, which is beneficial for the sustainable development of agriculture in the future (Fang et al., 2019).

In addition, with the enhancement of human health awareness, the demand for functional foods is gradually increasing, so there is an urgent need to develop functional agriculture. Functional agriculture refers to products that produce specific levels of micronutrients (mineral elements, vitamins) and other beneficial functional components (dietary fiber, limiting amino acids, plant compounds, unsaturated fatty acids, etc.) (Zhang et al., 2023). Functional agriculture is also closely related to plant metabolism. Astaxanthin is a high-value ketocarotenoid rarely found in plants. Therefore, there are very few ways for people to obtain astaxanthin in their usual diet. By studying the astaxanthin metabolic pathway, researchers have developed astaxanthin-rich rice (Bai et al., 2017). There are also purple tomatoes, which are rich in anthocyanins (Sun et al., 2020). These are examples of functional foods produced through the regulation of metabolism, which may be a future development direction for agriculture.

References

Ahanger, M. A., Qin, C., Begum, N., Maodong, Q., Dong, X. X., El-Esawi, M., El-Sheikh, M. A., Alatar, A. A., & Zhang, L. (2019). Nitrogen availability prevents oxidative effects of salinity on wheat growth and photosynthesis by up-regulating the antioxidants and osmolytes metabolism, and secondary metabolite accumulation. *BMC Plant Biology, 19*(1), 479.

Ahanger, M. A., Bhat, J. A., Siddiqui, M. H., Rinklebe, J., & Ahmad, P. (2020). Integration of silicon and secondary metabolites in plants: A significant association in stress tolerance. *Journal of Experimental Botany, 71*(21), 6758–6774.

Ahmad, S., Belwal, V., Punia, S. S., Ram, M., Dalip, S., Rajput, S. S., Kunwar, R., Meena, M. K., Gupta, D., Kumawat, G. L., Hussain, T., & Mohamed, H. I. (2023). Role of plant secondary metabolites and phytohormones in drought tolerance: A review. *Gesunde Pflanzen, 75*(4), 729–746.

Al-Dhaheri, A. S., Al-Maawali, A. K., Laleye, L. C., Washi, S. A., Jarrar, A. H., Al-Meqbaali, F. T., Mohamad, M. N., & Masuadi, E. M. (2015). The effect of nutritional composition on the glycemic index and glycemic load values of selected emirati foods. *BMC Nutrition, 1*(1), 4.

Al-Khayri, J. M., Rashmi, R., Toppo, V., Chole, P. B., Banadka, A., Sudheer, W. N., Nagella, P., Shehata, W. F., Al-Mssallem, M. Q., Alessa, F. M., Almaghasla, M. I., & Rezk, A. A. (2023). Plant secondary metabolites: The weapons for biotic stress management. *Metabolites, 13*(6), 213–274.

Anacleto, R., Badoni, S., Parween, S., Butardo, V. M., Misra, G., Cuevas, R. P., Kuhlmann, M., Trinidad, T. P., Mallillin, A. C., Acuin, C., Bird, A. R., Morell, M. K., & Sreenivasulu, N. (2019). Integrating a genome-wide association study with a large-scale transcriptome analysis to predict genetic regions influencing the glycaemic index and texture in rice. *Plant Biotechnology Journal, 17*(7), 1261–1275.

Arias, C. L., Garcia-Navarrete, L. T., Mukundi, E., Swanson, T., Yang, F., Hernandez, J., Grotewold, E., & Alonso, A. P. (2023). Metabolic and transcriptomic study of pennycress natural variation identifies targets for oil improvement. *Plant Biotechnology Journal, 21*(9), 1887–1903.

Bai, C., Berman, J., Farre, G., Capell, T., Sandmann, G., Christou, P., & Zhu, C. (2017). Reconstruction of the astaxanthin biosynthesis pathway in rice endosperm reveals a metabolic bottleneck at the level of endogenous β-carotene hydroxylase activity. *Transgenic Research, 26*(1), 13–23.

Balyan, G., & Pandey, A. K. (2024). Root exudates, the warrior of plant life: Revolution below the ground. *South African Journal of Botony, 164*, 280–287.

Camele, I., Gruová, D., & Elshafie, H. S. (2021). Chemical composition and antimicrobial properties of mentha × piperita cv. 'Kristinka' essential oil. *Plants (Basel), 10*(8), 387.

Caputi, L., Franke, J., Farrow, S. C., Chung, K., Payne, R. M. E., Nguyen, T. D., Dang, T. T., Carqueijeiro, I. S. T., Koudounas, K., de Bernonville, T. D., Ameyaw, B., Jones, D. M., Vieira,

I. J. C., Courdavault, V., & O'Connor, S. E. (2018). Missing enzymes in the biosynthesis of the anticancer drug vinblastine in madagascar periwinkle. *Science, 360*(6394), 1235–1239.
Casciaro, B., Mangiardi, L., Cappiello, F., Romeo, I., Loffredo, M. R., Iazzetti, A., Calcaterra, A., Goggiamani, A., Ghirga, F., Mangoni, M. L., Botta, B., & Quaglio, D. (2020). Naturally-occurring alkaloids of plant origin as potential antimicrobials against antibiotic-resistant infections. *Molecules, 25*(16), 597.
Chen, W., Teng, Y., Li, Z., Liu, W., Ren, W., Luo, Y., & Christie, P. (2018). Mechanisms by which organic fertilizer and effective microbes mitigate peanut continuous cropping yield constraints in a red soil of south china. *Applied Soil Ecology, 128*, 23–34.
Chen, Z., Lu, Y., Feng, L., Hao, W., Li, C., Yang, Y., Fan, X., Li, Q., Zhang, C., & Liu, Q. (2020). Genetic dissection and functional differentiation of alk(a) and alk(b), two natural alleles of the alk/ssiia gene, responding to low gelatinization temperature in rice. *Rice, 13*(1), 39.
Chrysargyris, A., Papakyriakou, E., Petropoulos, S. A., & Tzortzakis, N. (2019). The combined and single effect of salinity and copper stress on growth and quality of mentha spicata plants. *Journal of Hazardous Material, 368*, 584–593.
Cryle, M. J. (2018). Pathway to malonic acid-containing metabolites. *Nature Catalysis, 1*(12), 907–908.
Dangi, R. S., Joshi, E., Singh, N., Agrawal, S., Kumar, A., Trivedi, A., & Lautre, R. (2023). Classification of crops, pp. 9–19.
Della-Pepa, T., Elshafie, H. S., Capasso, R., De-Feo, V., Camele, I., Nazzaro, F., Scognamiglio, M. R., & Caputo, L. (2019). Antimicrobial and phytotoxic activity of origanum heracleoticum and *O. majorana* essential oils growing in Cilento (southern Italy). *Molecules, 24*(14), 987.
Deng, M., Li, D., Luo, J., Xiao, Y., Liu, H., Pan, Q., Zhang, X., Jin, M., Zhao, M., & Yan, J. (2017). The genetic architecture of amino acids dissection by association and linkage analysis in maize. *Plant Biotechnology Journal, 15*(10), 1250–1263.
Divekar, P. A., Narayana, S., Divekar, B. A., Kumar, R., Gadratagi, B. G., Ray, A., Singh, A. K., Rani, V., Singh, V., Singh, A. K., Kumar, A., Singh, R. P., Meena, R. S., & Behera, T. K. (2022). Plant secondary metabolites as defense tools against herbivores for sustainable crop protection. *International Journal of Molecular Science, 23*(5), 95.
Dong, L., Qi, X., Zhu, J., Liu, C., Zhang, X., Cheng, B., Mao, L., & Xie, C. (2019). Supersweet and waxy: Meeting the diverse demands for specialty maize by genome editing. *Plant Biotechnology Journal, 17*(10), 1853–1855.
Elshafie, H. S., Camele, I., & Mohamed, A. A. (2023). A comprehensive review on the biological, agricultural and pharmaceutical properties of secondary metabolites based-plant origin. *International Journal of Molecular Sciences, 24*(4), 489.
Erb, M., & Kliebenstein, D. J. (2020). Plant secondary metabolites as defenses, regulators, and primary metabolites: The blurred functional trichotomy. *Plant Physiology, 184*(1), 39–52.
Fang, C., Fernie, A. R., & Luo, J. (2019). Exploring the diversity of plant metabolism. *Trends in Plant Science, 24*(1), 83–98.
Guo, Z. H., Ye, Z. W., Haslam, R. P., Michaelson, L. V., Napier, J. A., & Chye, M. L. (2019). Arabidopsis cytosolic acyl-coa-binding proteins function in determining seed oil composition. *Plant Direct, 3*(12), e00182.
Guo, D., Hou, Q., Zhang, R., Lou, H., Li, Y., Zhang, Y., You, M., Xie, C., Liang, R., & Li, B. (2020). Over-expressing taspa-b reduces prolamin and starch accumulation in wheat (*Triticum aestivum* L.) grains. *International Journal of Molecular Sciences, 21*(9), 985.
Hou, Y., & Wu, G. (2018). Nutritionally essential amino acids. *Advances in Nutrition, 9*(6), 849–851.
Hou, Y., He, W., Hu, S., & Wu, G. (2019). Composition of polyamines and amino acids in plant-source foods for human consumption. *Amino Acids, 51*(8), 1153–1165.
Huang, L., Tan, H., Zhang, C., Li, Q., & Liu, Q. (2021). Starch biosynthesis in cereal endosperms: An updated review over the last decade. *Plant Communications, 2*(5), 100237.
Jacobowitz, J. R., & Weng, J. K. (2020). Exploring uncharted territories of plant specialized metabolism in the postgenomic era. *Annual Review Plant Biology, 71*, 631–658.

Jia, M., Wu, H., Clay, K. L., Jung, R., Larkins, B. A., & Gibbon, B. C. (2013). Identification and characterization of lysine-rich proteins and starch biosynthesis genes in the opaque2 mutant by transcriptional and proteomic analysis. *BMC Plant Biology, 13*, 60.

Kalinger, R. S., Williams, D., Ahmadi Pirshahid, A., Pulsifer, I. P., & Rowland, O. (2021). Production of c6–c14 medium-chain fatty acids in seeds and leaves via overexpression of single hotdog-fold acyl-lipid thioesterases. *Lipids, 56*(3), 327–344.

Khalid, A., Hameed, A., & Tahir, M. F. (2023). Wheat quality: A review on chemical composition, nutritional attributes, grain anatomy, types, classification, and function of seed storage proteins in bread making quality. *Frontiers in Nutrition, 10*, 1053196.

Kumar, V., Suman, U., Rubal, S., & Yadav, S. K. (2018). Flavonoid secondary metabolite: Biosynthesis and role in growth and development in plants. In S. K. Yadav, V. Kumar, & S. P. Singh (Eds.), *Recent trends and techniques in plant metabolic engineering* (pp. 19–45). Singapore: Springer.

Kumari, A., Dogra, V., Joshi, R., & Kumar, S. (2021). Stress-responsive cis-regulatory elements underline podophyllotoxin biosynthesis and better performance of sinopodophyllum hexandrum under water deficit conditions. *Frontier Plant Sciences, 12*, 751846.

Lal, N., Sahu, N., Shirale, A. O., Gurav, P., Rani, K., Meena, B. P., Diwan, G., & Biswas, A. K. (2023). Plant secondary metabolites and their impact on human health. In V. D. Rajput, H. El-Ramady, S. K. Upadhyay, T. Minkina, B. Ahmed, & S. Mandzhieva (Eds.), *Nano-biofortification for human and environmental health* (pp. 295–321). Springer.

Lau, W., & Sattely, E. S. (2015). Six enzymes from mayapple that complete the biosynthetic pathway to the etoposide aglycone. *Science, 349*(6253), 1224–1228.

Li, D., & Gaquerel, E. (2021). Next-generation mass spectrometry metabolomics revives the functional analysis of plant metabolic diversity. *Annual Review Plant Biology, 72*, 867–891.

Li, M., Bahn, S. C., Fan, C., Li, J., Phan, T., Ortiz, M., Roth, M. R., Welti, R., Jaworski, J., & Wang, X. (2013). Patatin-related phospholipase pplaiiiδ increases seed oil content with long-chain fatty acids in arabidopsis. *Plant Physiology, 162*(1), 39–51.

Li, Y., Chen, Y., Zhou, L., You, S., Deng, H., Chen, Y., Alseekh, S., Yuan, Y., Fu, R., Zhang, Z., Su, D., Fernie, A. R., Bouzayen, M., Ma, T., Liu, M., & Zhang, Y. (2020a). Microtom metabolic network: Rewiring tomato metabolic regulatory network throughout the growth cycle. *Molecular Plant, 13*(8), 1203–1218.

Li, Y., Dai, S., Wang, B., Jiang, Y., Ma, Y., Pan, L., Wu, K., Huang, X., Zhang, J., Cai, Z., & Zhao, J. (2020b). Autotoxic ginsenoside disrupts soil fungal microbiomes by stimulating potentially pathogenic microbes. *Applied Environmental Microbiology, 86*(9), 958.

Li, A., Jin, K., Zhang, Y., Deng, X., Chen, Y., Wei, X., Hu, B., & Jiang, Y. (2024). Root exudates and rhizosphere microbiota in responding to long-term continuous cropping of tobacco. *Science Report, 14*(1), 11274.

Liao, J., & Xia, P. (2024). Continuous cropping obstacles of medicinal plants: Focus on the plant-soil-microbe interaction system in the rhizosphere. *Scientia Horticulturae, 328*, 112927.

Lin, Z., An, J., Wang, J., Niu, J., Ma, C., Wang, L., Yuan, G., Shi, L., Liu, L., Zhang, J., Zhang, Z., Qi, J., & Lin, S. (2017). Integrated analysis of 454 and illumina transcriptomic sequencing characterizes carbon flux and energy source for fatty acid synthesis in developing *Lindera glauca* fruits for woody biodiesel. *Biotechnology and Biofuels, 10*, 134.

Liu, X., Zhang, C., Wang, X., Liu, Q., Yuan, D., Pan, G., Sun, S. S., & Tu, J. (2016). Development of high-lysine rice via endosperm-specific expression of a foreign lysine rich protein gene. *BMC Plant Biology, 16*(1), 147.

Liu, Z., Jiang, S., Jiang, L., Li, W., Tang, Y., He, W., Wang, M., Xing, J., Cui, Y., Lin, Q., Yu, F., & Wang, L. (2022). Transcription factor ossgl is a regulator of starch synthesis and grain quality in rice. *Journal of Experiment in Botony, 73*(11), 3417–3430.

Lu, S., Sturtevant, D., Aziz, M., Jin, C., Li, Q., Chapman, K. D., & Guo, L. (2018). Spatial analysis of lipid metabolites and expressed genes reveals tissue-specific heterogeneity of lipid metabolism in high- and low-oil *Brassica napus* L. seeds. *Plant Journal, 94*(6), 915–932.

Ma, Z., Guan, Z., Liu, Q., Hu, Y., Liu, L., Wang, B., Huang, L., Li, H., Yang, Y., Han, M., Gao, Z., & Saleem, M. (2023). Chapter four: Obstacles in continuous cropping—mechanisms and control measures. In D. L. Sparks (Ed.), *Advances in agronomy* (pp. 205–256). Academic Press.

Maino Vieytes, C. A., Taha, H. M., Burton-Obanla, A. A., Douglas, K. G., & Arthur, A. E. (2019). Carbohydrate nutrition and the risk of cancer. *Current Nutrition Reports, 8*(3), 230–239.

Martin, K., & Sauerborn, J. (2013). Origin and development of agriculture. In K. Martin & J. Sauerborn (Eds.), *Agroecology* (pp. 9–48). Springer.

Miura, S., Crofts, N., Saito, Y., Hosaka, Y., Oitome, N. F., Watanabe, T., Kumamaru, T., & Fujita, N. (2018). Starch synthase iia-deficient mutant rice line produces endosperm starch with lower gelatinization temperature than japonica rice cultivars. *Frontier Plant Science, 9*, 645.

Nations, F. (2017). The future of food and agriculture: Trends and challenges. Rome.

Nawaz, M., Sun, J., Shabbir, S., Khattak, W. A., Ren, G., Nie, X., Bo, Y., Javed, Q., Du, D., & Sonne, C. (2023). A review of plants strategies to resist biotic and abiotic environmental stressors. *Science of the Total Environment, 900*, 165832.

Newman, D. J., & Cragg, G. M. (2020). Natural products as sources of new drugs over the nearly four decades from 01/1981 to 09/2019. *Journal of Natural Production, 83*(3), 770–803.

Onoda, N., Sugino, K., Higashiyama, T., Kammori, M., Toda, K., Ito, K., Yoshida, A., Suganuma, N., Nakashima, N., Suzuki, S., Tsukahara, K., Noguchi, H., Koizumi, M., Nemoto, T., Hara, H., Miyauchi, A., & Sugitani, I. (2016). The safety and efficacy of weekly paclitaxel administration for anaplastic thyroid cancer patients: A nationwide prospective study. *Thyroid: Official Journal of the American Thyroid Association, 26*(9), 1293–1299.

Paul, M. J. (2021). Improving photosynthetic metabolism for crop yields: What is going to work? *Frontiers in Plant Science, 12*, 1471.

Pereira, A. G., Cassani, L., Garcia-Oliveira, P., Otero, P., Mansoor, S., Echave, J., Xiao, J., Simal-Gándara, J., & Prieto, M. A. (2023). Plant alkaloids: Production, extraction, and potential therapeutic properties. In M. Carocho, S. A. Heleno, & L. Barros (Eds.), *Natural secondary metabolites: From nature, through science, to industry* (pp. 157–200). Springer.

Piasecka, A., Jedrzejczak-Rey, N., & Bednarek, P. (2015). Secondary metabolites in plant innate immunity: Conserved function of divergent chemicals. *New Phytology, 206*(3), 948–964.

Ramos-Martín, F., & D'Amelio, N. (2022). Biomembrane lipids: When physics and chemistry join to shape biological activity. *Biochimie, 203*, 118–138.

Salam, U., Ullah, S., Tang, Z. H., Elateeq, A. A., Khan, Y., Khan, J., Khan, A., & Ali, S. (2023). Plant metabolomics: An overview of the role of primary and secondary metabolites against different environmental stress factors. *Life (Basel, Switzerland), 13*(3), 4147.

Schmidt, J. P., Cruse-Sanders, J., Chamberlain, J. L., Ferreira, S., & Young, J. A. (2019). Explaining harvests of wild-harvested herbaceous plants: American ginseng as a case study. *Biological Conservation, 231*, 139–149.

Sedbrook, J. C., & Durrett, T. P. (2020). Pennycress, carbon wise: Labeling experiments reveal how pennycress seeds efficiently incorporate carbon into biomass. *Journal of Experimental Botony, 71*(10), 2842–2846.

Selwal, N., Rahayu, F., Herwati, A., Latifah, E., Supriyono, C., Suhara, I. B. K., Suastika, W. M. M., & Wani, A. K. (2023). Enhancing secondary metabolite production in plants: Exploring traditional and modern strategies. *Journal of Agriculture and Food Research, 14*, 100702.

Seung, D. (2020). Amylose in starch: Towards an understanding of biosynthesis, structure and function. *New Phytology, 228*(5), 1490–1504.

Shams-White, M. M., Pannucci, T. E., Lerman, J. L., Herrick, K. A., Zimmer, M., Meyers Mathieu, K., Stoody, E. E., & Reedy, J. (2023). Healthy eating index-2020: Review and update process to reflect the dietary guidelines for americans 2020–2025. *Journal of the Academy of Nutrition and Dietetics, 123*(9), 1280–1288.

Shelke, D. B., Islam, N. F., Chambhare, M. R., Sonawane, H. B., Patowary, R., Prasad, R., & Sarma, H. (2023). Enhancing secondary metabolites and alleviating environmental stress in crops with mycogenic nanoparticles: A comprehensive review. *Biocatalysis Agricultural Biotechnology, 52*, 102805.

Singh, J., Kaur, L., Tian, J., & Ogawa, Y. (2023). Rice chemistry, nutrition, and processing: An-introduction. In J. Tian, Y. Ogawa, J. Singh, & L. Kaur (Eds.), *Science of rice chemistry and nutrition* (pp. 1–2). Springer.

Singh, S., Chhatwal, H., & Pandey, A. (2024). Deciphering the complexity of terpenoid biosynthesis and its multi-level regulatory mechanism in plants. *Journal of Plant Growth Regulation, 31*, 245.

Song, S., Hou, W., Godo, I., Wu, C., Yu, Y., Matityahu, I., Hacham, Y., Sun, S., Han, T., & Amir, R. (2013). Soybean seeds expressing feedback-insensitive cystathionine γ-synthase exhibit a higher content of methionine. *Journal of Experimental Botony, 64*(7), 1917–1926.

Stirbet, A., Lazár, D., Guo, Y., & Govindjee, G. (2020). Photosynthesis: Basics, history and modelling. *Annual Botony, 126*(4), 511–537.

Sun, C., Deng, L., Du, M., Zhao, J., Chen, Q., Huang, T., Jiang, H., Li, C. B., & Li, C. (2020). A transcriptional network promotes anthocyanin biosynthesis in tomato flesh. *Molecular Plant, 13*(1), 42–58.

Tiku, A. R. (2018). Antimicrobial compounds and their role in plant defense. In A. Singh & I. K. Singh (Eds.), *Molecular aspects of plant-pathogen interaction* (pp. 283–307). Springer.

Torrens-Spence, M. P., Pluskal, T., Li, F. S., Carballo, V., & Weng, J. K. (2018). Complete pathway elucidation and heterologous reconstitution of rhodiola salidroside biosynthesis. *Molecular Plant, 11*(1), 205–217.

Trovato, M., Funck, D., Forlani, G., Okumoto, S., & Amir, R. (2021). Editorial: Amino acids in plants: Regulation and functions in development and stress defense. *Frontier Plant Sciences, 12*, 772810.

Ullah, S., Ai, C., Huang, S., Zhang, J., Jia, L., Ma, J., Zhou, W., & He, P. (2019). The responses of extracellular enzyme activities and microbial community composition under nitrogen addition in an upland soil. *PLoS ONE, 14*(9), e0223026.

Wang, J. C., Xu, H., Zhu, Y., Liu, Q. Q., & Cai, X. L. (2013). Osbzip58, a basic leucine zipper transcription factor, regulates starch biosynthesis in rice endosperm. *Journal of Experimental Botony, 64*(11), 3453–3466.

Wang, Z., Li, Y., Wang, R., & Ding, W. L. (2017). Gene cloning and expression analysis of benzoic acid stress responsive transcription factor wrky7 in panax ginseng. *Zhongguo Zhong Yao Za Zhi = Zhongguo Zhongyao Zazhi = China Journal of Chinese Materia Medica, 42*(6), 1130–1135.

Wang, J., Xu, C., Wong, Y. K., Li, Y., Liao, F., Jiang, T., & Tu, Y. (2019). Artemisinin, the magic drug discovered from traditional Chinese medicine. *Engineering, 5*(1), 32–39.

Wang, H., Wei, S., He, Y., Wang, X., Li, Y., Wei, D., Wang, Z., Guo, L., Shaaban, M., & Hou, X. (2023). Characterization of agronomic and seed oil features for different cultivars of tree peony. *Plants (Basel), 12*(17), 1252.

Whitcomb, S. J., Rakpenthai, A., Brückner, F., Fischer, A., Parmar, S., Erban, A., Kopka, J., Hawkesford, M. J., & Hoefgen, R. (2020). Cysteine and methionine biosynthetic enzymes have distinct effects on seed nutritional quality and on molecular phenotypes associated with accumulation of a methionine-rich seed storage protein in rice. *Frontier Plant Sciences, 11*, 1118.

Wu, D., Feng, Z., Gu, W., Wang, Y., Liu, Z., Wang, W., Zhang, Y., Zhang, W., & Chen, W. (2023). Could continuous rice cropping increase soil fertility and rice productivity by rice straw carbonized utilization in cold areas? A 6-year field-located trial. *Environmental Science in Pollution Research, 30*(51), 110674–110686.

Xi, H., Shen, J., Qu, Z., Yang, D., Liu, S., Nie, X., & Zhu, L. (2019). Effects of long-term cotton continuous cropping on soil microbiome. *Science Report, 9*(1), 18297.

Xin-hui, Z., Lang, D.-Y., Zhang, E.-H., & Wang, Z.-S. (2015). Effect of autotoxicity and soil microbes in continuous cropping soil on angelica sinensis seedling growth and rhizosphere soil microbial population. *Chinese Herbal Medicine, 7*(1), 88–93.

Yang, Q. Q., Zhang, C. Q., Chan, M. L., Zhao, D. S., Chen, J. Z., Wang, Q., Li, Q. F., Yu, H. X., Gu, M. H., Sun, S. S., & Liu, Q. Q. (2016). Biofortification of rice with the essential amino acid lysine: Molecular characterization, nutritional evaluation, and field performance. *Journal of Experimental Botany, 67*(14), 4285–4296.

Yang, Q. Q., Zhao, D. S., Zhang, C. Q., Wu, H. Y., Li, Q. F., Gu, M. H., Sun, S. S., & Liu, Q. Q. (2018). A connection between lysine and serotonin metabolism in rice endosperm. *Plant Physiology, 176*(3), 1965–1980.
Yang, Q., Zhao, D., & Liu, Q. (2020). Connections between amino acid metabolisms in plants: Lysine as an example. *Frontier Plant Sciences, 11*, 928. https://doi.org/10.3389/fpls.2020.00928
Yoon, D. K., Ishiyama, K., Suganami, M., Tazoe, Y., Watanabe, M., Imaruoka, S., Ogura, M., Ishida, H., Suzuki, Y., Obara, M., Mae, T., & Makino, A. (2020). Transgenic rice overproducing rubisco exhibits increased yields with improved nitrogen-use efficiency in an experimental paddy field. *Nature Food, 1*(2), 134–139.
Yuan, Y., Zuo, J., Zhang, H., Zu, M., & Liu, S. (2022). The chinese medicinal plants rhizosphere: Metabolites, microorganisms, and interaction. *Rhizosphere, 22*, 100540.
Zeeshan-Ul-Haq, M., Yu, J., Yao, G., Yang, H., Iqbal, H. A., Tahir, H., Cui, H., Liu, Y., & Wu, Y. (2023). A systematic review on the continuous cropping obstacles and control strategies in medicinal plants. *International Journal of Molecular Sciences, 24*(15), 28.
Zhang, B., Li, X., Wang, F., Li, M., Zhang, J., Gu, L., Zhang, L., Tu, W., & Zhang, Z. (2016). Assaying the potential autotoxins and microbial community associated with rehmannia glutinosa replant problems based on its 'autotoxic circle.' *Plant and Soil, 407*(1), 307–322.
Zhang, C., Yang, Y., Chen, Z., Chen, F., Pan, L., Lu, Y., Li, Q., Fan, X., Sun, Z., & Liu, Q. (2020a). Characteristics of grain physicochemical properties and the starch structure in rice carrying a mutated alk/ssiia gene. *Journal of Agricultural Food Chemicals, 68*(47), 13950–13959.
Zhang, G., Ahmad, M. Z., Chen, B., Manan, S., Zhang, Y., Jin, H., Wang, X., & Zhao, J. (2020b). Lipidomic and transcriptomic profiling of developing nodules reveals the essential roles of active glycolysis and fatty acid and membrane lipid biosynthesis in soybean nodulation. *Plant Journal, 103*(4), 1351–1371.
Zhang, L., Han, Z., & Granato, D. (2021). Polyphenols in foods: Classification, methods of identification, and nutritional aspects in human health. *Advances in Food and Nutrition Research, 98*, 1–33.
Zhang, J., Hansen, L. G., Gudich, O., Viehrig, K., Lassen, L. M. M., Schrübbers, L., Adhikari, K. B., Rubaszka, P., Carrasquer-Alvarez, E., Chen, L., D'Ambrosio, V., Lehka, B., Haidar, A. K., Nallapareddy, S., Giannakou, K., Laloux, M., Arsovska, D., Jørgensen, M. A. K., Chan, L. J. G., … Keasling, J. D. (2022). A microbial supply chain for production of the anti-cancer drug vinblastine. *Nature, 609*(7926), 341–347.
Zhang, Y., Long, H., Chen, S., Ma, L., & Gan, M. (2023). The development of multifunctional agriculture in farming regions of china: Convergence or divergence? *Land Use Policy, 127*, 106576.
Zhang, P., Xia, L., Sun, Y., & Gao, S. (2024). Soil nutrients and enzyme activities based on millet continuous cropping obstacles. *Science Report, 14*(1), 17329.
Zhou, C., Pan, W., Peng, Q., Chen, Y., Zhou, T., Wu, C., Hartley, W., Li, J., Xu, M., Liu, C., Li, P., Rao, L., & Wang, Q. (2021). Characteristics of metabolites by seed-specific inhibition of fad2 in *Brassica napus* L. *Journal of Agricultural Food Chemistry, 69*(19), 5452–5462.
Zhou, Q., Wang, Y., Yue, L., Ye, A., Xie, X., Zhang, M., Tian, Y., Liu, Y., Turatsinze, A. N., Constantine, U., Zhao, X., Zhang, Y., & Wang, R. (2024). Impacts of continuous cropping on the rhizospheric and endospheric microbial communities and root exudates of astragalus mongholicus. *BMC Plant Biology, 24*(1), 340.

Soil Microbes for Controlling Insect Pests of Crop Plants

Aneela Younas and Muhammad Shaaban

Abstract This chapter explores the role of soil microbes in controlling insect pests of crop plants, highlighting their potential as a sustainable alternative to chemical pesticides. It discusses the mechanisms through which soil microbes interact with pests, including direct pathogenicity and the enhancement of plant resistance. Practical applications for incorporating these beneficial organisms into pest management strategies are examined, along with criteria for selecting effective microbial strains. The chapter also addresses the benefits, such as environmental sustainability and reduced pesticide use, while acknowledging the challenges posed by consistency in effectiveness and environmental factors. By integrating soil microbes into pest control practices, the chapter aims to promote a more resilient and ecologically balanced approach to agriculture, ultimately contributing to sustainable crop production.

Keyword Soil microbes · Insect pests · Crop protection · Sustainable agriculture

1 Introduction

Pest control is a critical component of agricultural practices, directly influencing crop yield, quality, and overall farm profitability. Insects, being one of the most prevalent agricultural pests, can cause significant damage to crops, leading to substantial economic losses. Effective pest management strategies are essential for maintaining healthy crops and ensuring food security. With the increasing demand for sustainable farming practices, the importance of developing eco-friendly pest management solutions has become more pronounced (Younas et al., 2017, 2022). Traditional pest management methods have relied heavily on synthetic chemical pesticides. While these chemicals have been effective in controlling pest populations, they pose risks

A. Younas · M. Shaaban (✉)
College of Agriculture, Henan University of Science and Technology, Luoyang, China
e-mail: shabanbzu@hotmail.com

M. Shaaban (ed.), *Soils and Sustainable Agriculture*, Frontier Studies in Soil Science,
https://doi.org/10.1007/978-3-031-91114-9_23

to the environment, human health, and non-target organisms. Additionally, the emergence of pesticide resistance among pest populations has spurred the need for alternative approaches. As a result, agricultural scientists and practitioners are increasingly exploring biological control methods, which utilize natural organisms to manage pest populations (Deka et al., 2021; Dent & Binks, 2020). One promising area within biological control is the use of soil microbes. Soil harbors a diverse community of microorganisms, including bacteria, fungi, and nematodes, which play crucial roles in ecosystem functioning. These microbes can interact with insect pests and plants in various ways, making them valuable allies in pest management strategies (Deka et al., 2021). By harnessing the natural abilities of soil microbes, farmers can develop sustainable, effective pest control methods that reduce reliance on chemical inputs and minimize environmental impacts.

This chapter aims to explore the principles and mechanisms by which soil microbes can be employed to control insect pests affecting crop plants. We will examine how these microbes operate, the advantages they offer over conventional methods, and the practical applications of integrating soil microbes into pest management regimens. By highlighting the potential of soil microbes as biocontrol agents, this chapter seeks to provide insights into innovative strategies that can enhance agricultural sustainability and productivity.

2 Mechanisms of Action

Soil microbes employ various mechanisms to control insect pests, which can be categorized into direct and indirect actions. Direct mechanisms involve interactions that directly affect the pest, while indirect mechanisms focus on the benefits imparted to the host plants, enhancing their defense against pest attacks. One of the primary direct mechanisms of action is the pathogenicity of certain soil microbes towards insect pests (Islam et al., 2021). Specific species of bacteria and fungi can infect and kill insect larvae, disrupting their development and reducing pest populations. For instance, *Bacillus thuringiensis* is a well-known bacterium that produces crystal proteins toxic to specific insect larvae when ingested. This natural insecticide has been utilized in agricultural practices for decades, demonstrating how soil microbes can directly target and control pest populations (Gassmann & Reisig, 2023). Another important direct mechanism is the antifeedant properties exhibited by some soil microbes. Certain fungal metabolites or bacterial by-products can deter insects from feeding on treated plants. For example, some species of fungi produce secondary metabolites that elicit a strong aversion response in phytophagous insects, thereby reducing their feeding activity. This not only decreases pest damage but also can lead to a healthier, more resilient crop (Islam et al., 2021).

Indirect mechanisms also play a crucial role in pest management. One key aspect is the ability of soil microbes to induce plant resistance to pests. When plants are colonized by beneficial microbes, they can activate their defense pathways, enhancing their ability to fend off insect attacks. For instance, specific mycorrhizal fungi can stimulate systemic acquired resistance (SAR) in plants, leading to an overall increase in resistance to herbivorous insects (Alfonso et al., 2021). Enhancing plant health and stress tolerance is another vital indirect mechanism. Soil microbes contribute to improved nutrient uptake, soil structure, and overall plant vigor, which can make plants more resilient to pest pressures. A healthy plant is better able to withstand stressors, including pest infestations, thus reducing the potential for significant damage. For example, rhizobacteria can improve nutrient availability and microbial diversity, bolstering plant growth and resilience in the face of pest challenges (Ruiu, 2020).

3 Practical Applications

A. *Methods for Incorporating Soil Microbes in Pest Management*

Incorporating soil microbes into pest management strategies involves various methods designed to enhance the effectiveness of these beneficial organisms. Two prominent approaches are inoculation techniques and soil amendment strategies.

Inoculation Techniques

Inoculation techniques refer to the deliberate introduction of selected beneficial microbes into the soil or on plant surfaces. This can be accomplished through several methods. Beneficial microbes can be applied to seeds before planting, creating a protective and beneficial microbial environment that promotes plant growth and resilience against pests. A solution containing specific microbes can be applied directly to the soil around plant roots. This method ensures that the microorganisms establish themselves in the rhizosphere, where they can exert their beneficial effects. Beneficial microbes may also be applied as foliar sprays on the above-ground parts of the plant. This approach can be useful for targeting pests directly and inducing plant resistance concurrently (del Rosario Cappellari et al., 2020; Qessaoui et al., 2020).

Soil Amendment Strategies

Soil amendment strategies involve integrating microbial inoculants with organic matter, compost, or other soil health amendments. This can enhance the microbial population's establishment and offer additional benefits to plant health. Mixing beneficial microbes with compost or decomposed organic matter not only facilitates better microbial survival but also improves soil structure and nutrient availability. Incorporating biochar into the soil can create a habitat for beneficial microbes, enhancing their viability and activity in the soil. This combination can promote positive interactions between the microbes and plant roots. Applying organic mulch enriched with beneficial microbes enhances moisture retention in soil and provides nutrients, which can synergistically improve the overall ecosystem within the root zone.

B. ***Selection Criteria for Effective Soil Microbes***

Selecting the appropriate soil microbes is crucial to ensure optimal pest management outcomes. Several key criteria should be considered when identifying effective microbial candidates.

The chosen strains must demonstrate proven effectiveness in controlling the specific insect pests of concern. This assessment should be based on rigorous scientific studies indicating their pathogenicity or pest deterrent properties. Effective microbes should be capable of withstanding the environmental conditions of the target soil, including variations in pH, temperature, moisture levels, and nutrient availability. This resilience will ensure that they remain active in the field. The selected microbes must be compatible with the crop species being grown. Non-pathogenic strains that support plant growth and health will contribute positively to overall plant vigor and resilience against pests. Candidates should have user-friendly application methods that are feasible for farmers and compatible with existing agricultural practices. Microbial products must adhere to agricultural regulations for safety and environmental impact, ensuring they do not pose any risks to human health or non-target organisms.

C. ***Monitoring and Assessing Effectiveness***

To ensure that the incorporation of soil microbes into pest management is successful, ongoing monitoring and assessment are vital. Various approaches can be employed for this purpose.

Prior to implementing any microbial treatments, farmers should collect baseline data on pest population levels, plant health, and general soil conditions. This information serves as a reference point for evaluating the impact of microbial applications. After applying soil microbes, consistent monitoring of pest populations and plant health should be conducted. This might include visual assessments, trapping methods for pests, and measuring key plant parameters such as growth rate and yield. Periodic evaluations of soil health metrics, such as microbial biomass, nutrient availability, and soil structure, can provide insights into the effectiveness of microbial amendments in enhancing soil ecology. Data collected from monitoring efforts should be analyzed to determine the correlation between microbial application and pest population dynamics. This analysis can help fine-tune pest management strategies for greater efficacy. Based on monitoring results, farmers may need to adjust inoculation rates, microbial strains, or application methods. Continuous improvement will be essential to optimizing pest management strategies utilizing soil microbes.

4 Benefits and Challenges

The use of soil microbes for pest control offers several significant benefits that contribute to more sustainable agricultural practices. One of the primary advantages is environmental sustainability. Soil microbes, as natural agents of pest control, can reduce the reliance on synthetic chemical pesticides, which are often associated with environmental degradation, soil and water contamination, and harm to non-target organisms, including beneficial insects and wildlife. By utilizing soil microbes, farmers can promote a healthier ecosystem that encourages biodiversity while maintaining effective pest management (Pineda et al., 2020). Additionally, reducing chemical pesticide use is not only beneficial for the environment but also for agricultural health. Lower pesticide application rates can lead to healthier crops and reduce the risk of pesticide resistance developing in pest populations. This shift toward biological control methods can enhance the long-term viability of pest management strategies and decrease associated costs for farmers over time (O'Callaghan et al., 2022). However, incorporating soil microbes into pest control strategies does present potential limitations and challenges. One major concern is consistency in effectiveness. The success of microbial applications can vary widely depending on numerous factors, including the specific strains used, application methods, and environmental conditions. Thus, the performance of microbial agents can be unpredictable, which may lead to challenges in pest management consistency and confidence for farmers (O'Callaghan et al., 2022). Environmental factors affecting microbial activity also pose challenges. Conditions such as soil temperature, moisture, pH levels, and the presence of competing microorganisms can significantly impact the survival, establishment, and efficacy of beneficial microbes. For instance, extreme weather events or drought conditions can inhibit microbial function, reducing their capacity to suppress pest populations. Therefore, understanding and managing these environmental variables is crucial for maximizing the effectiveness of soil microbes in pest control (O'Callaghan et al., 2022).

5 Conclusion

In conclusion, the integration of soil microbes into the management of insect pests represents a promising advancement in sustainable agriculture. As natural allies in the fight against pest populations, these microorganisms offer a range of benefits that align with the essential goals of reducing chemical pesticide reliance and promoting environmental health. The ability of soil microbes to directly combat pests, enhance plant resilience, and improve soil health positions them as valuable tools for farmers navigating the complex challenges of modern agriculture. The diverse mechanisms through which soil microbes operate, combined with effective application methods and robust selection criteria, provide a framework for successful pest control strategies. However, it is crucial to remain aware of the potential limitations, including variability in efficacy and the influence of environmental factors on microbial activity. By addressing these challenges through research and innovation, agricultural practices can be optimized, ultimately enhancing productivity while safeguarding ecosystems. The continued exploration of soil microbial communities and their interactions with pests and plants will play a significant role in shaping the future of pest management. As new strains are discovered and technologies improve, the potential for soil microbes to contribute to sustainable agricultural systems will expand. Therefore, fostering a deeper understanding of these microorganisms not only underscores their significance in pest control but also promotes a holistic approach to agriculture that values biodiversity and ecological balance. By embracing the use of soil microbes, farmers can actively participate in creating resilient agricultural systems that benefit both their livelihoods and the environment, paving the way for a more sustainable future in crop production.

Acknowledgements The work was supported by the "Fundamental Research Funds for Henan University of Science and Technology, 4024-13510002."

References

Alfonso, E., Stahl, E., Glauser, G., Bellani, E., Raaymakers, T. M., Van den Ackerveken, G., Zeier, J., & Reymond, P. (2021). Insect eggs trigger systemic acquired resistance against a fungal and an oomycete pathogen. *New Phytologist, 232*, 2491–2505.

Deka, B., Baruah, C., & Babu, A. (2021). Entomopathogenic microorganisms: Their role in insect pest management. *Egyptian Journal of Biological Pest Control, 31*, 1–8.

del Rosario Cappellari, L., Chiappero, J., Palermo, T. B., Giordano, W., & Banchio, E. (2020). Impact of soil rhizobacteria inoculation and leaf-chewing insect herbivory on Mentha piperita leaf secondary metabolites. *Journal of Chemical Ecology, 46*, 619–630.

Dent, D., & Binks, R. H. (2020). Insect pest management. Cabi.

Gassmann, A. J., & Reisig, D. D. (2023). Management of insect pests with Bt crops in the United States. *Annual Review of Entomology, 68*, 31–49.

Islam, W., Adnan, M., Shabbir, A., Naveed, H., Abubakar, Y. S., Qasim, M., Tayyab, M., Noman, A., Nisar, M. S., & Khan, K. A. (2021). Insect-fungal-interactions: A detailed review on entomopathogenic fungi pathogenicity to combat insect pests. *Microbial Pathogenesis, 159*, 105122.

O'Callaghan, M., Ballard, R. A., & Wright, D. (2022). Soil microbial inoculants for sustainable agriculture: Limitations and opportunities. *Soil Use and Management, 38*, 1340–1369.

Pineda, A., Kaplan, I., Hannula, S. E., Ghanem, W., & Bezemer, T. M. (2020). Conditioning the soil microbiome through plant–soil feedbacks suppresses an aboveground insect pest. *New Phytologist, 226*, 595–608.

Qessaoui, R., Amarraque, A., Lahmyed, H., Ajerrar, A., Mayad, E. H., Chebli, B., Walters, A. S., & Bouharroud, R. (2020). Inoculation of tomato plants with rhizobacteria suppresses development of whitefly *Bemisia tabaci* (GENNADIUS) (HEMIPTERA: ALEYRODIDAE): Agro-ecological application. *PLoS ONE, 15*, e0231496.

Ruiu, L. (2020). Plant-growth-promoting bacteria (PGPB) against insects and other agricultural pests. *Agronomy, 10*, 861.

Younas, A., Wakil, W., Khan, Z., Shaaban, M., & Prager, S. M. (2017). The efficacy of Beauveria bassiana, jasmonic acid and chlorantraniliprole on larval populations of Helicoverpa armigera in chickpea crop ecosystems. *Pest Management Science, 73*, 418–424.

Younas, A., Waris, M. I., Shaaban, M., Tahir-ul-Qamar, M., & Wang, M. Q. (2022). Appraisal of MsepCSP14 for chemosensory functions in *Mythimna separata. Insect Science, 29*, 162–176.

GPSR Compliance

The European Union's (EU) General Product Safety Regulation (GPSR) is a set of rules that requires consumer products to be safe and our obligations to ensure this.

If you have any concerns about our products, you can contact us on ProductSafety@springernature.com

In case Publisher is established outside the EU, the EU authorized representative is:

Springer Nature Customer Service Center GmbH
Europaplatz 3
69115 Heidelberg, Germany

Batch number: 10607196

Printed by Printforce, the Netherlands

Frontier Studies in Soil Science

Series Editor

Avelino Núñez-Delgado, Lugo, Spain

This series aims to illuminate the latest research, insights, and advancements in the ever-evolving world of soil science. This series is dedicated to fostering a comprehensive understanding of soils, from their micro to macro characteristics, their role in ecosystems, to their significance in global challenges. Our commitment is to present groundbreaking studies, carried out by esteemed experts and emerging scholars, ensuring that readers are consistently provided with the most up-to-date and relevant information in the realm of soil science.

Muhammad Shaaban
Editor

Soils and Sustainable Agriculture

Interplay of Soil, Plant, Water and Environmental Systems for Sustainable Agriculture

Volume 2

Springer

Editor
Muhammad Shaaban
College of Agriculture
Henan University of Science
and Technology
Luoyang, China

ISSN 3004-9865 ISSN 3004-9873 (electronic)
Frontier Studies in Soil Science
ISBN 978-3-031-91113-2 ISBN 978-3-031-91114-9 (eBook)
https://doi.org/10.1007/978-3-031-91114-9

This Springer imprint is published by the registered company Springer Nature Switzerland AG
The registered company address is: Gewerbestrasse 11, 6330 Cham, Switzerland

Preface

In an era marked by rapid environmental changes and growing concerns about food security, the relationship between soil health and sustainable agriculture becomes increasingly crucial. *Soils and Sustainable Agriculture: Interplay of Soil, Plant, Water, and Environmental Systems for Sustainable Agriculture* aims to elucidate this intricate relationship and offer insights into sustainable agricultural practices that honor and enhance the natural systems.

Soil is the foundation of terrestrial life, serving as a dynamic medium in which plants grow and as a vital component of our ecosystems. It plays a pivotal role in nutrient cycling, water retention, and biological diversity, all of which are essential for robust agricultural systems. However, the pressures of modern agriculture, such as soil degradation, pollution, and climate change, have put immense strain on this precious resource, highlighting the need for a new paradigm in agricultural practices.

This book is the result of collaborative efforts from researchers, academics, and practitioners in the fields of soil science, agronomy, and environmental studies. Each chapter addresses various aspects of the interplay between soil, plants, water, and environmental systems, providing a comprehensive overview that integrates theory with practical applications. Readers will find discussions on soil management practices, the role of organic matter in soil health, water conservation techniques, and the impact of climate change on agricultural productivity.

Our goal with this publication is to equip practitioners, educators, and students with the knowledge and tools necessary to foster sustainable agricultural systems. By embracing an interdisciplinary approach, we highlight the importance of holistic management practices that not only enhance soil fertility and crop yields but also promote ecological balance and resilience.

We hope that this book inspires readers to recognize the indispensable role of soil in sustainable agriculture and motivates them to adopt practices that preserve and restore it for future generations. In our quest for food security and environmental health, let us remember that sustainable agriculture begins with a commitment to nurturing the very foundation of life, our soil.

Thank you for joining us on this journey towards a more sustainable future.

Luoyang, China
December 2024

Prof. Dr. Muhammad Shaaban

Contents

The Role of Insects in Soil Health and Fertility

Aneela Younas and Muhammad Shaaban

Abstract This chapter explores the integral role of insects in promoting soil health and fertility, highlighting their contributions to various ecological processes essential for sustainable agriculture. Insects facilitate decomposition and nutrient cycling, breaking down organic matter and converting it into accessible nutrients for plants. Their interactions with soil microorganisms enhance microbial activity and diversity, promoting a rich soil ecosystem that supports plant growth. The chapter emphasizes the significance of insect biodiversity in maintaining ecosystem resilience, showcasing how diverse insect populations contribute to pest control, soil structure improvement, and moisture retention. By recognizing the vital roles insects play, the chapter advocates for the conservation and enhancement of insect diversity as a cornerstone of sustainable agricultural practices, ultimately fostering long-term soil health and ecological balance.

Keywords Insects · Soil health · Fertility · Decomposition · Nutrient cycling · Biodiversity

1 Introduction

Soil health and fertility are foundational concepts in agriculture that refer to the capacity of soil to function effectively within ecosystems, supporting the growth of healthy plants, maintaining water quality, and fostering biodiversity. Soil health encompasses several physical, chemical, and biological properties, including texture, structure, nutrient availability, moisture retention, and the presence of a diverse microbiome (Shaaban, 2024; Shaaban & Nunez-Delgado, 2024). It is a dynamic condition influenced by land management practices, climatic factors, and ecological interactions. Fertility, on the other hand, specifically relates to the soil's ability to supply essential nutrients to plants in suitable forms and quantities, ultimately determining agricultural productivity. Healthy soils are not only critical for crop yield

A. Younas · M. Shaaban (✉)
College of Agriculture, Henan University of Science and Technology, Luoyang, China
e-mail: shabanbzu@hotmail.com

M. Shaaban (ed.), *Soils and Sustainable Agriculture*, Frontier Studies in Soil Science,
https://doi.org/10.1007/978-3-031-91114-9_24

but also contribute to larger environmental objectives, such as carbon sequestration, water filtration, and habitat provision for various organisms (Shaaban, 2024; Weil and Brady, 2017).

Insects play a crucial role in maintaining soil health and fertility, acting as both agents of decomposition and nutrient cycling while also influencing soil structure and biodiversity. They are essential components of agricultural ecosystems, engaging in processes that promote the breakdown of organic matter and the recycling of nutrients back into the soil (Menta & Remelli, 2020). Various insect groups, including beetles, ants, earthworms, and other decomposers, contribute significantly to the transformation of organic residues into fertile soil (Menta & Remelli, 2020; Na et al., 2022). Their activities improve soil aeration, enhance water infiltration, and create microhabitats that foster diverse microbial communities. Furthermore, insects serve as natural pest regulators, helping to balance the populations of harmful organisms and reducing the need for chemical interventions, thereby supporting sustainable agricultural practices (Menta & Remelli, 2020; Wu et al., 2023).

The objectives of this chapter are to explore the multifaceted roles that insects play in soil health and fertility, highlighting their contributions to essential ecological functions and the implications for agricultural systems. We aim to provide a comprehensive overview of how insects interact with soil environments and the wider agricultural landscape, emphasizing the linkage between insect diversity and ecosystem resilience. Furthermore, this chapter will discuss the challenges facing soil insect populations due to modern agricultural practices and climate change, while also presenting strategies for enhancing their roles through conservation and sustainable land management. By understanding the integral relationship between insects and soil health, we can better appreciate the need for their conservation in promoting sustainable agricultural practices that ensure long-term food security and environmental sustainability.

2 Soil Composition and Structure

Soil composition is a critical factor that directly influences plant growth and overall ecosystem health. Composed of minerals, organic matter, water, and air, the specific proportions of these components determine soil fertility, structure, and the availability of nutrients essential for plant development. Minerals provide the base material for soil and supply vital nutrients such as nitrogen, phosphorus, and potassium. Organic matter, derived from decomposed plant and animal residues, is essential for enhancing soil fertility, improving moisture retention, and supporting diverse microbial communities. The balance within this composition influences factors like pH, cation exchange capacity, and biological activity, all of which are crucial for robust plant growth and agricultural productivity.

Insects play a significant role in enhancing soil structure, which is vital for maintaining soil health (Menta & Remelli, 2020). Soil structure refers to the arrangement of soil particles into aggregates, which defines porosity and influences how easily air and water can move through the soil. Many insects, such as ants and beetles, contribute to the formation of soil aggregates through their feeding and nesting behaviors (Barragán-Fonseca et al., 2022). The activities of insects help bind soil particles together, enhancing aggregate stability, which in turn improves water retention and reduces erosion. Well-structured soil fosters a diverse array of soil organisms, creating a thriving ecosystem that supports plant life. In addition to enhancing soil structure, insects significantly contribute to soil aeration and drainage. Soil aeration is essential for the respiration of roots and soil organisms, particularly beneficial bacteria and fungi that promote nutrient cycling and organic matter decomposition. Insects create burrows and tunnels as they move through the soil in search of food and habitat. These activities increase pore space in the soil, allowing air to penetrate deeper layers, which can be particularly important in compacted soils where air movement is restricted. Improved aeration enhances root development, resulting in better nutrient uptake and overall plant health (Neher & Barbercheck, 2019). Moreover, the enhanced drainage provided by insect activity mitigates the risks of waterlogging, which can harm root systems and reduce crop yields. By increasing the porosity of the soil, insects help facilitate effective water movement, allowing excess moisture to drain away and reducing the likelihood of anaerobic conditions. This contributes to a healthier root environment, promoting robust plant growth even during periods of heavy rainfall (Torgerson et al., 2021; Verma et al., 2023).

3 Decomposition and Nutrient Cycling

Decomposition is a fundamental process that occurs within the soil ecosystem, playing a critical role in recycling nutrients, maintaining soil health, and supporting plant growth. It involves the breakdown of organic matter, such as fallen leaves, decaying plant material, and animal remains, transforming them into simpler forms that can be assimilated by soil organisms and plants (Wu et al., 2018). This process is driven by a variety of microorganisms, fungi, and detritivores, all of which work together to decompose organic matter efficiently. Decomposition not only replenishes the nutrient pool in the soil but also enhances soil structure and promotes a thriving ecosystem (Shaaban & Nunez-Delgado, 2024; Wu et al., 2017).

Insects are integral to the decomposition process, acting as key players in the breakdown of organic matter. Various insect groups, including beetles, ants, and termites, contribute significantly to this process. For instance, certain beetles, known as detritivores, feed on decomposing plant and animal materials, breaking them down physically through their feeding activities. As these insects consume organic matter, they fragment it into smaller pieces, increasing the surface area available for microbial action. This physical breakdown facilitates the work of bacteria and fungi, which further decompose the organic materials into simpler compounds. In this way, insects enhance the efficiency of decomposition, enabling a more rapid recycling of nutrients within the soil (Verma et al., 2023). In addition to their role in breaking down organic matter, insects contribute significantly to nutrient cycling and availability. Nutrient cycling refers to the movement and transformation of essential elements, such as nitrogen, phosphorus, and carbon, between the soil, plants, and living organisms. Detritivorous insects are vital contributors to this cycle, as their feeding activities stimulate microbial growth by introducing fresh organic substrates into the soil. The excretion of insect waste products, such as frass (insect feces), enriches the soil with nutrients and organic matter, promoting the fertility of the soil ecosystem (Barragán-Fonseca et al., 2022). For example, ants play a crucial role in nutrient cycling as they gather organic materials from their surroundings, including dead insects and plant debris. As they transport these materials back to their nests, they inadvertently mix them into the soil, enhancing the decomposition process (Fig. 1). Furthermore, the microorganisms thriving within ant nests benefit from the nutrients provided by the decomposing matter, leading to an increase in nutrient availability not only for the ants but also for surrounding plants. This symbiotic relationship illustrates how insects can create microenvironments that enhance nutrient cycling (Neher & Barbercheck, 2019; Torgerson et al., 2021).

The impact of insects on nutrient release and soil fertility is profound. Their activities facilitate the mineralization of nutrients, a process through which complex organic compounds are broken down into inorganic forms that can be readily uptake by plants. For instance, nitrogen is released from decomposed organic matter in the form of ammonium and nitrates, essential for plant growth. Similarly, phosphorus is made available in forms that plants can absorb directly from the soil. Capturing and enhancing these nutrient cycling processes through natural insect activity contributes to overall soil fertility and crop productivity.

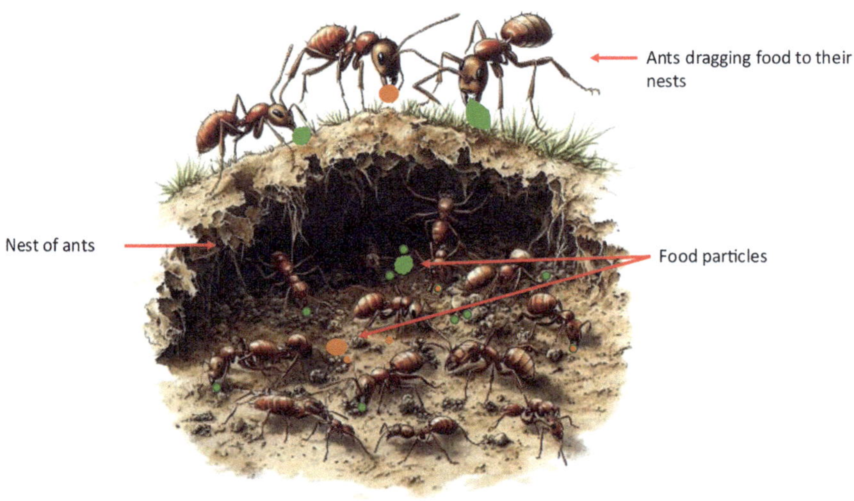

Fig. 1 Ants mixing and disintegrating food particles into soil

4 Interactions with Soil Microorganisms

Interactions between insects and soil microorganisms are crucial for understanding the dynamics of soil health and fertility. Soil is a complex habitat teeming with a diverse community of microorganisms, including bacteria, fungi, protozoa, and nematodes, all of which play significant roles in nutrient cycling, organic matter decomposition, and overall soil structure. Insects, as integral components of the soil ecosystem, engage in various interactions with these microorganisms, creating synergies that enhance soil processes and promote greater biological diversity (Verma et al., 2023). Insects facilitate microbial activity and diversity in several ways. Their activities can create physical changes in the soil that benefit microbial communities. For instance, when insects tunnel through the soil, they modify its structure, leading to increased porosity and improved aeration. This alteration allows for better gas exchange and water retention, creating favorable conditions for microbial growth and activity. Additionally, as insects feed on organic matter, they introduce new substrates into the soil environment, which provide essential nutrients for soil microbes. The breakdown of organic matter by insects not only increases the availability of nutrients but also supplies a continuous source of organic materials that sustain microbial populations. Moreover, insects play a key role in transporting microorganisms within the soil. Certain insect species, such as dung beetles, are known to collect and bury animal feces, which are rich in microbial life. By burying feces, these beetles facilitate the distribution of beneficial microbes throughout the soil profile, enhancing microbial diversity and promoting healthy soil ecosystems. Similarly, ants are known to harvest and transport spores of fungi, integrating these organisms into their nests and the surrounding soil. By doing so, they promote fungal diversity and enhance soil health.

One of the most notable examples of mutualistic relationships between insects and soil microorganisms is the dynamic between ants and fungi. Leafcutter ants, for example, are famous for cutting leaves and transporting them back to their nests (Fig. 2). Instead of consuming the leaves directly, these ants use the cut foliage to cultivate specific species of fungi, which serve as their primary food source. The ants provide the fungi with a continuous supply of organic material, while the fungi break down the leaves and convert them into a nutrient-rich form that the ants can effectively digest. In return, these fungi benefit from the protection and care provided by the ant colonies. This mutualistic interaction has profound implications for soil ecosystems, as the activities of leafcutter ants can significantly influence the composition of soil microbial communities and enhance soil fertility.

Another example of insect-microbe interactions is found in the relationships between termites and the microorganisms residing in their guts. Termites feed on cellulose from plant material, a complex carbohydrate that is challenging to digest. However, termites have a symbiotic relationship with various protozoa and bacteria in their gastrointestinal tracts that assist in breaking down cellulose into simpler sugars. This interaction not only allows termites to obtain nutrients from otherwise

Fig. 2 Leafcutter ants cutting leaves and transporting them to their nests

indigestible materials but also contributes to the microbial diversity within the soil as fecal matter from termites is deposited back into the soil, enriching it with microbes.

5 Soil Insect Diversity and Ecosystem Resilience

Soil insect diversity encompasses a wide range of species, each playing specialized roles that contribute to the health and fertility of the soil ecosystem. Among the key beneficial soil insect species are beetles, ants, and various decomposers that significantly enhance soil structure, nutrient cycling, and overall ecosystem functioning. Beetles, including dung beetles and predatory species, contribute to the decomposition of organic materials and the control of pest populations, while ants actively aerate soil and facilitate nutrient cycling. The diversity of insects within the soil ecosystem supports various biological functions that are essential for ecosystem resilience (Barragán-Fonseca et al., 2022; Menta & Remelli, 2020). A diverse insect community can effectively perform overlapping roles, ensuring that essential processes continue even when certain species are compromised due to environmental changes, such as drought or temperature fluctuations. For instance, if one species of predatory beetle declines, other insect predators can help fill the gap, maintaining the balance between herbivorous pests and their natural enemies. This functional redundancy is vital for ecosystem stability, as it allows the system to withstand shocks and recover from disturbances. Maintaining insect diversity is crucial for upholding ecosystem resilience, particularly in the context of agricultural systems that face increasing pressures from climate change, land-use practices, and pest infestations. A diverse soil insect population can enhance soil structure and fertility, leading to improved crop productivity. Various insects contribute to nutrient cycling by breaking down organic matter into forms that are more accessible to plants. For example, detritivorous insects such as certain beetles and ants feed on decaying plant residues, fragmenting them and accelerating their decomposition, which enhances nutrient availability (Barragán-Fonseca et al., 2022; Torgerson et al., 2021; Verma et al., 2023). Moreover, insects play a significant role in improving soil health through their interactions with soil microorganisms. Diverse insect populations can introduce a variety of organic substrates and promote microbial activity, enhancing nutrient cycling further. This interaction results in the release of essential nutrients such as nitrogen, phosphorus, and potassium in more available forms for plants. Thus, insect diversity is directly linked to soil fertility and, by extension, to crop yields.

In addition to enhancing nutrient cycling, soil insect diversity helps improve soil structure and moisture retention. Species such as ants and beetles create intricate burrow systems, which facilitate water infiltration and air movement within the soil. These structures aid in moisture retention, reducing the need for irrigation and making plants more resilient to drought conditions. Healthy, well-structured soils not only support crop growth but also mitigate erosion and promote the overall sustainability of agricultural practices. The decline of insect diversity poses a significant threat to soil health and agricultural productivity. Monoculture practices, habitat destruction, and the widespread use of pesticides can diminish insect populations, resulting in a reduced capacity for ecosystems to perform essential functions. With fewer beneficial insects in the soil, agricultural systems may rely more heavily on chemical inputs for pest control and nutrient supplementation, leading to a cycle of diminishing returns for soil health and crop yields.

6 Conclusion

In conclusion, insects play an indispensable role in maintaining soil health and fertility, contributing to various ecological functions that are critical for sustainable agricultural systems. Their involvement in decomposition and nutrient cycling serves as a foundation for soil fertility, facilitating the transformation of organic matter into vital nutrients that plants require for growth. By aiding in the breakdown of organic materials, insects help maintain the balance of nutrients in the soil, promoting a successful and productive agricultural environment. Moreover, the interactions between insects and soil microorganisms enhance the overall biodiversity of the soil ecosystem, enabling a dynamic interplay that supports plant health. Insects not only serve as agents of decomposition but also act as vectors for microbial life, ensuring the continuous replenishment of soil nutrients and fostering a thriving community of beneficial organisms. The mutualistic relationships observed, such as those between ants and fungi or termites and gut microorganisms, highlight the intricate connections that exist within the soil ecosystem, emphasizing the importance of these interactions in promoting soil resilience. Soil insect diversity is critical for the robustness and adaptability of agroecosystems. A diverse community of beneficial insects ensures that essential ecological functions can continue despite pressures from pests or environmental changes. This diversity provides a buffer against disturbances, allowing ecosystems to maintain their integrity and productivity. The presence of multiple insect species can enhance pest control services through predation, parasitism, and competition, reducing the reliance on chemical pesticides and promoting sustainable practices. Furthermore, healthy insect populations contribute to improved soil structure and moisture retention, which are vital for crop survival and productivity. Their burrowing activities enhance aeration and water infiltration, while their feeding habits promote organic matter decomposition. These processes collectively lead to a more resilient agricultural system that can thrive in the face of climate variability and increasing food demands. Recognizing the significant contributions of insects to soil health and fertility is essential for promoting practices that conserve and enhance insect biodiversity. Strategies such as integrated pest management, habitat preservation, and sustainable agricultural practices can play critical roles in supporting insect populations and ensuring their continued functionality within ecosystems. As the challenges of pest management and soil degradation intensify, prioritizing insect conservation will be paramount in fostering food security and environmental sustainability.

Acknowledgements The work was supported by the "Fundamental Research Funds for Henan University of
Science and Technology, 4024-13510002."

References

Barragán-Fonseca, K. Y., Nurfikari, A., Van De Zande, E. M., Wantulla, M., Van Loon, J. J., De Boer, W., & Dicke, M. (2022). Insect frass and exuviae to promote plant growth and health. *Trends in Plant Science, 27*, 646–654.

Menta, C., & Remelli, S. (2020). Soil health and arthropods: From complex system to worthwhile investigation. *Insects, 11*, 54.

Na, L., Hu, C., Jiang, Y., Hu, R., Shaaban, M., Younas, A., & Wu, Y. (2022). Earthworms promote the transfer of ^{15}N-urea to lettuce while limit appreciably increase ^{15}N losing to environment. *Environmental Research, 212*, Article 113423.
Neher, D. A., & Barbercheck, M. E. (2019). Soil microarthropods and soil health: Intersection of decomposition and pest suppression in agroecosystems. *Insects, 10*, 414.
Shaaban, M. (2024). Microbial pathways of nitrous oxide emissions and mitigation approaches in drylands. *Journal of Environmental Management, 354*, Article 120393.
Shaaban, M., & Nunez-Delgado, A. (2024). Soil adsorption potential: Harnessing Earth's living skin for mitigating climate change and greenhouse gas dynamics. *Environmental Research, 251*, Article 118738.
Torgerson, K., Meijering, J., Sok, J., Dicke, M., & Lansink, A. O. (2021). Towards circular agriculture–exploring insect waste streams as a crop and soil health promoter. *Journal of Insects as Food and Feed, 7*, 357–368.
Verma, R.C., Waseem, M.A., Sharma, N., Bharathi, K., Singh, S., Anto Rashwin, A., Pandey, S.K., & Singh, B.V., (2023). The role of insects in ecosystems, an in-depth review of entomological research. *International Journal of Environment and Climate Change 13*, 4340-4348
Weil, R.R., & Brady, N.C., 2017. The nature and properties of soils, 15 ed. Pearson Education Limited Edinburgh Gate Harlow Essex CM20 2JE England
Wu, L., Tang, S., He, D., Wu, X., Shaaban, M., Wang, M., Zhao, J., Khan, I., Zheng, X., & Hu, R. (2017). Conversion from rice to vegetable production increases N_2O emission via increased soil organic matter mineralization. *Science of the Total Environment, 583*, 190–201.
Wu, L., Wu, X., Lin, S., Wu, Y., Tang, S., Zhou, M., Shaaban, M., Zhao, J., Hu, R., & Kuzyakov, Y. (2018). Carbon budget and greenhouse gas balance during the initial years after rice paddy conversion to vegetable cultivation. *Science of the Total Environment, 627*, 46–56.
Wu, Y., Li, Q., Zheng, Y., Xiong, X., Chen, Y., Shaaban, M., & Hu, R. (2023). Optimizing biochar addition for vermicomposting: A comprehensive evaluation of earthworms' activity, N_2O emissions and compost quality. *Biochar, 5*, 1–4.

Impact of Climate Change on Cereal Crop Production: Challenges and Adaptive Strategies

Muhammad Naveed Riaz, Sumaira Maqsood, Muhammad Arif, Naila Batool, Muhammad Akram Riaz, Hina Safdar, Sumbal Nazir, Humaira Yasmin, and Sadia Khatoon

Abstract This chapter explores the effects of climate change on major cereal crops, including wheat, maize, and rice. Fluctuations in temperature, increased frequency of flooding, and rising global temperatures are shown to negatively impact crop growth, productivity, and resilience, leading to reduced yields and heightened susceptibility to pests and diseases. A comprehensive review of recent literature highlights the substantial influence of climate change on cereal crop systems. In response to these challenges, the chapter discusses a range of adaptive strategies such as optimized irrigation methods, improved nutrient management, advanced breeding approaches, and sustainable agronomic practices. These interventions are critical for sustaining cereal crop production under evolving climatic conditions.

M. N. Riaz
Department of Psychology, University of Sargodha, Sargodha, Pakistan

S. Maqsood
Department of Environmental Sciences, Kohsar University Murree, Murree, Pakistan

M. Arif (✉) · S. Khatoon
College of Agriculture, Guizhou University, Guiyang, China
e-mail: arif@gzu.edu.cn

N. Batool
Department of Psychology, Kohsar University Murree, Murree, Pakistan

M. A. Riaz
Department of Psychology, University of Home Economics, Lahore, Pakistan

H. Safdar
Department of Plant Pathology, College of Agriculture, University of Sargodha, Sargodha, Pakistan

S. Nazir
School of Zoology, Minhaj University Lahore, Lahore, Pakistan

H. Yasmin
Department of Psychology, University of Lahore, Sargodha Campus, Pakistan

M. Shaaban (ed.), *Soils and Sustainable Agriculture*, Frontier Studies in Soil Science,
https://doi.org/10.1007/978-3-031-91114-9_25

1 Introduction

Cereal refers to grass cultivated for its edible grain and is considered a staple food worldwide, including varieties like rice, oats, rye, wheat, millet, barley, and maize. Most cereals are annual plants, producing one crop per planting, although rice can sometimes be grown continuously. Winter cereals are sown in autumn, remain dormant through winter, and are harvested in spring or early summer, while spring cereals are planted in spring and harvested in late summer. The term "cereal" is derived from Ceres, the Roman goddess of grain and fertility (FAO. 2017).

Cereal crops belong to the Gramineae family and are typically grown for their digestible components and botanical structure composed of the bran, endosperm, and germ. Wheat, barley, oats, rye, maize, rice, millet, and sorghum are some of the major grain species within the cereal family, with wheat and rice being the most prevalent. Despite originating from various regions, each cereal crop is now extensively cultivated in diverse climatic conditions. Cereals are recognized as the most widely grown and consumed commodity worldwide (Adeboyejo et al., 2023).

Cereal crops can be described as plants grown for their seeds as human food, usually encompassing crops closely related to grass. These crops typically require physical processing like cutting, threshing, or combining to separate individual grains from the spike or ear where they grow. When harvested, barley, millet, oats, rice, rye, triticale, and wheat have grains resembling grass. Maize and grain sorghum seeds come from more robust plants less grass-like and more suited to sunny climates. Barley and most oat varieties retain their husks after threshing, while some naked oat varieties have husks removed during harvesting. Other cereals need husk removal through threshing before storage. Barley, oats, rye, triticale, and wheat grains have a crease due to their development process, leading to a complex milling process for white flour extraction from wheat. The more uniform shape of grains from other plants simplifies their processing (Adviento-Borbe, 2020).

The major component of human diets worldwide has been cereals and cereal-based products since ancient times. Cereal crops are rich in energy, providing about 10–20 times more energy than most juicy fruits and vegetables. Wheat, rice, corn, and barley are among the major cereal crops. Corn (or maize) is the most produced cereal crop (31%), but wheat and rice are more important nutritionally as they account for 55% of total cereal production. Wheat and rice are vital sources of dietary protein, carbohydrates, B group vitamins, vitamin E, iron, trace minerals, and fibers. Globally, cereals contribute around 45% of the required protein and energy for human diets and only about 7% of total fat (Rosell, 2011).

1.1 History

1.1.1 Origins

Wheat, barley, rye, and oats have a long history of being consumed by humans. In the early Neolithic period, people in the Fertile Crescent region gathered and ate these cereal grains. The discovery of 19,000-year-old cereal grains at the Ohalo II site in Israel, containing charred remains of wild wheat and barley, provides valuable insights into the ancient diets and agricultural practices of early human civilizations. It's incredible to see how these staple crops have been a crucial part of human nutrition for thousands of years (Renfrew et al., 2012).

In China, farmers started cultivating rice and millet around the same time. Their cultivation method includes handmade flooding and fires. Innovative agricultural practices demonstrate how early human creativity and resourcefulness led to the development of methods for cultivating staple crops like rice and millet. The methods of handmade floods and fires show how skilled these Chinese farmers are (Hancock, 2012). Eastern Asia, the Nile Valley, and Mesopotamia are the regions that started these practices independently. For successful crop cultivation in diverse measures of the biosphere, the method of soil enhancement highlights the importance and maintenance of fertile soil. In these regions, for the development and production of agriculture, these soiling methods and conditions play a very important role (Mayshar et al., 2022).

The people of the Fertile Crescent cultivated cereals like wheat and barley around 8,000 years ago (Purugganan & Fuller, 2009). Rice and millets were tame in East Asia, while sorghum and other millets were tame in sub-Saharan West Africa, mainly for livestock food. For the progress of human civilization and the development of agriculture in these regions, the process of crop taming plays a very important role. It is very interesting to know that different regions in the world are taming independently to fulfill their needs and for the development of agriculture (Henry & Kettlewell, 2012).

It's fascinating to see how religion and agriculture intertwined in these ancient civilizations. Many agricultural societies shaped their religious beliefs based on the divinity associated with grain and harvests (Standage, 2009). In Mesopotamia, the grain goddess Ashnan marked the beginning of an era of civilization in their creation myth (Grant & Hazel, 2004). Similarly, the Roman goddess Ceres was a significant figure presiding over fertility, agriculture, motherhood and grain crops. The word "cereal" itself is derived from the Latin word "cerealis," meaning "of grain," originally linked to the goddess Ceres ("JHOM–Bread–Hebrew," 2021).

1.1.2 Modern

The latter half of the twentieth century witnessed a remarkable surge in the manufacture of high-yield cereal crops globally, notably wheat and rice, thanks to the Green

Revolution. This transformative period was characterized by a technological revolution supported by development organizations ("16. After the world food summit," 2019). Some advanced agricultural techniques, such as the use of irrigation methods, high-yield crop diversities, and the application of fertilizers and insecticides developed by The Green Revolution, led to the development and production of crops that yield agriculture. Due to this advancement in agriculture, the problems regarding food shortages will be solved, and it will also increase food security at the global level (Sands et al., 2009). The Green Revolution also developed some strategies like monoculture, nitrogen fertilizers, mechanized tilling, and the growth of new seed strains, which play a vital role in enhancing agriculture. These strategies were mainly introduced to cope with hunger issues and to increase crop yields, which will collectively increase overall cereal grain production. Though nutritional quality is one of the main factors in yield enhancement, in modern times cereal production is high, but it has deficiencies in terms of amino acids, proteins, and other vital nutrients like vitamins, fatty acids, and minerals. This imbalance causes many problems and health issues. However, ancient traditional crops are unique in flavor and have nutritional benefits. One of the most important aspects of sustainable agriculture techniques is striking a balance between the necessity for adequate food production and the development of diversified, nutrient-rich crops (FAO. 2017).

1.2 Different Types of Cereals

Cereals indeed come in a wide array of varieties, each offering unique nutritional profiles and culinary uses. Some examples of different types of cereals include; Wheat, Rice, Oats, Barley, Quinoa, Millet, Rye, and Sorghum. These cereals play crucial roles in various cuisines worldwide and provide essential nutrients like carbohydrates, proteins, vitamins, and minerals. Incorporating a diverse range of cereals in your diet can contribute to a well-rounded and balanced nutritional intake (Rejman et al., 2020).

1.2.1 Wheat

Wheat is indeed a crucial crop globally, serving as a significant source of protein and dietary fiber. In India, wheat holds immense importance as it is commonly used for preparing bread, chapatis, and a variety of baked goods. The process of harvesting wheat involves threshing to separate the grain from the chaff (Javid Iqbal et al., 2022). The nutritional value of Wheat is the central constituent of foods (Wieser et al., 2020).

1.2.2 Rice

Globally, rice is a very important part of food. Because of its consumption, it is at the top of the list of cereals. Rice is a grain free of gluten, and it is known that after eating rice, we have a feeling of fullness for a long period of time. There are many varieties of rice, and it also satisfies hunger; therefore, it is a popular choice for eating around the world (Rathna Priya et al., 2019).

1.2.3 Oats

Oats, commonly consumed as breakfast cereal, are a multipurpose and healthy option in the cereal category. Instant, steel-cut, and rolled oats are different types of oats used for cooking. By choosing gluten-free oats, you can further improve the nutritional value of your muesli and make sure you're getting all the nutrients this hearty cereal has to offer (Holopainen-Mantila et al., 2024).

1.2.4 Barley

Globally, barley is considered as well-known cereal. It can be cultivated in similar temperature conditions as wheat, making it a versatile crop. Barley holds significant importance in Indian cuisine, being as popular as wheat and rice. Its nutritional value and adaptability in various dishes contribute to its widespread consumption and popularity in the country (Geng et al., 2022).

1.2.5 Quinoa

Quinoa is a fantastic choice if you're aiming to incorporate protein-rich cereals into your diet. Just one full bowl of quinoa can provide a significant amount of protein, helping you meet your recommended daily intake. Its impressive protein content makes quinoa a valuable addition to a balanced and nutritious diet (Bastian & Palmer-Keenan, 2022).

1.2.6 Millet

Millets are indeed packed with various essential nutrients like protein, minerals, vitamins, and fiber. One of the remarkable aspects of millets is that they require minimal water and fertile soil to grow, making them a sustainable and resource-efficient crop. Opting for millet muesli can provide you with substantial levels of protein and fiber, offering a nutritious and wholesome breakfast option. Incorporating millets into your diet can contribute to a well-rounded and healthy eating plan (Tripathi et al., 2023).

1.2.7 Rye

Rye flour is denser than wheat flour due to its lower gluten content. Traditional rye bread is typically made using the sourdough process, resulting in its characteristic mildly sour flavor. Rye stands out among other grains because both its endosperm and bran contain significant fiber content. This high fiber content contributes to rye products having a lower glycemic index (GI) compared to wheat and other grains, making them a favorable choice for managing blood sugar levels (Dziki, 2022).

1.2.8 Sorghum

Jowar, or sorghum, is often hailed as the "new quinoa" due to its whole grain benefits and gluten-free nature. As people shift away from refined flour products towards more nutritious options, this ancient grain from the millet family is making resurgence. For health benefits and wellbeing, jowar is a highly nutritional food. If we consume such beneficial cereals on a daily basis, it will promote our health, both physically and mentally (Rodriguez et al., 2020).

1.3 Uses

1.3.1 Direct Consumption

Rice is one of the cereals that are known as convenient cereals because it can be easily cooked and requires minimum preparation before eating. For example, to make plain rice, we just need to soak the rice in water to remove excess starch, and then it can be boiled. This easiest way of making rice makes it versatile and convenient for different cultures around the world (Solaimalai et al., 2021). Commercial breakfast cereals like granola can be highly processed and may contain added sugars, oils, and other ingredients to enhance flavor and texture, on the other hand, are highly processed and may contain added sugars, oils, and other ingredients to enhance flavor and texture. It's essential to be mindful of the ingredients in commercial cereals and opt for whole grain options, when possible, for a healthier breakfast choice (Robert, 2010).

1.3.2 Flour-Based Foods

Cereals can grind to produce flour, which is a key element in various food products. Wheat flour makes up a primary component of bread and pasta. Maize flour serves as a significant staple in Mesoamerican cuisine for centuries, finding use in dishes like Mexican tortillas and tamales. Rye flour is a common ingredient in bread in central and northern Europe, while rice flour is widely used in Asian cooking. A cereal grain consists of the germ, starchy endosperm, and bran. Whole meal flour contains all parts of the grain, offering more nutrients and fiber, while white flour may lack some or all of the germ or bran, resulting in a less nutritious product. Understanding the composition of different flours can help make informed choices when selecting ingredients for cooking and baking (Farmer, 2023).

1.3.3 Alcohol

Cereals with their high starch content are frequently utilized in the production of industrial alcohol and alcoholic beverages through the process of fermentation. On behalf of example, beer is crafted by making and agitating starch, primarily sourced from cereal grains, with malted barley being a popular choice. In Asia, rice wines like Japanese sake are brewed, showcasing the versatility of cereals in the production of alcoholic drinks. It's fascinating to see the long history of cereal-based alcoholic beverages, with evidence of honey wine and fermented rice being made in China as far back as 9,000 years ago (Solaimalai et al., 2021).

1.3.4 Animal Feed

Cereals and their byproducts like hay are commonly used as feed for farm animals. Maize, barley, wheat, and oats are popular cereal choices for animal nutrition. These grains can be processed in various ways for feeding livestock, such as chemical treatment, ensiling to make silage, mechanical flattening or crimping, and dry storage with low moisture content. Additionally, in commercial settings, grains are frequently mixed with other ingredients and compressed into food pellets for convenient and balanced animal nutrition (National Academies of Sciences Engineering and Medicine et al., 2021).

1.4 Nutrition

1.4.1 Whole-Grain and Processed

Whole grain cereals are rich sources of carbohydrates, polyunsaturated fats, vitamins, protein and minerals. When companies process them by removing the bran and germ, they leave behind the starchy endosperm. In many developing nations, people consume cereals as a significant portion of their daily sustenance. In developed countries, people consume a more diverse and moderate amount of cereal, often in the form of refined and processed grains. Understanding the nutritional composition of cereals in both whole and processed forms can help make informed dietary choices for optimal health and well-being (Allai et al., 2021).

1.4.2 Amino Acid Balance

Some cereals lack the essential amino acid lysine, so vegetarian cultures often combine cereal grains with legumes to ensure a balanced diet. While many legumes may be deficient in the essential amino acid methionine, which grains contain, combining legumes with grains creates a complementary protein profile, providing vegetarians with a well-balanced and complete source of essential amino acids for their dietary needs. This combination is a great way for vegetarians to ensure they are getting all the necessary nutrients for their health (Aksoylu Özbek et al., 2023). Such combinations include dal (lentils) with rice by South Indians and Bengalis, beans with maize tortillas, tofu with rice, and peanut butter with wholegrain wheat bread (as sandwiches) in several other cultures, including the Americas. For feeding animals, the amount of crude protein measured in grains is expressed as grain crude protein concentration (Greenhalgh et al., 2022).

2 Natural Disasters

The 2018 State of Food Security and Nutrition in the World report highlights the devastating impact of natural disasters, which are viewed as unavoidable events globally, causing significant property damage and loss of life (Organization, 2018), extreme weather conditions like natural disasters have adverse effects on agricultural output, food production, and current farming methods, consequently exacerbating the food supply deficit over an extended duration (Lesk et al., 2016). Climate elements such as temperature, precipitation (rain, snow, hail, and ice pellets, etc.), and the frequency of unpredictable events, like the rise in CO2 levels in the atmosphere and sea level, have a direct impact on livestock and agricultural yields (Agovino et al. 2019). Climate change can indeed have various impacts on agriculture, which can be either detrimental or beneficial, depending on the specific geographical location and the types of crops cultivated in that area (Kaye and Quemada 2017). Anticipating the severity of these natural disasters in advance enables communities to implement measures to maintain food production to the best of their ability (Wang et al., 2020) (Fig. 1).

2.1 Temperature

Tropical and sub-tropical regions are indeed more susceptible to elevated temperatures, which can harm crops and increase the water needs. This situation can lead to floods, famines, and ultimately contribute to socio-economic challenges within a country (Ali et al., 2017). Additionally, the higher temperatures can impact the maturity period of crops, as discussed (Hatfield & Prueger, 2015). Indeed, soil fertility can deteriorate due to erosion, pesticide use, and alterations in cropping patterns,

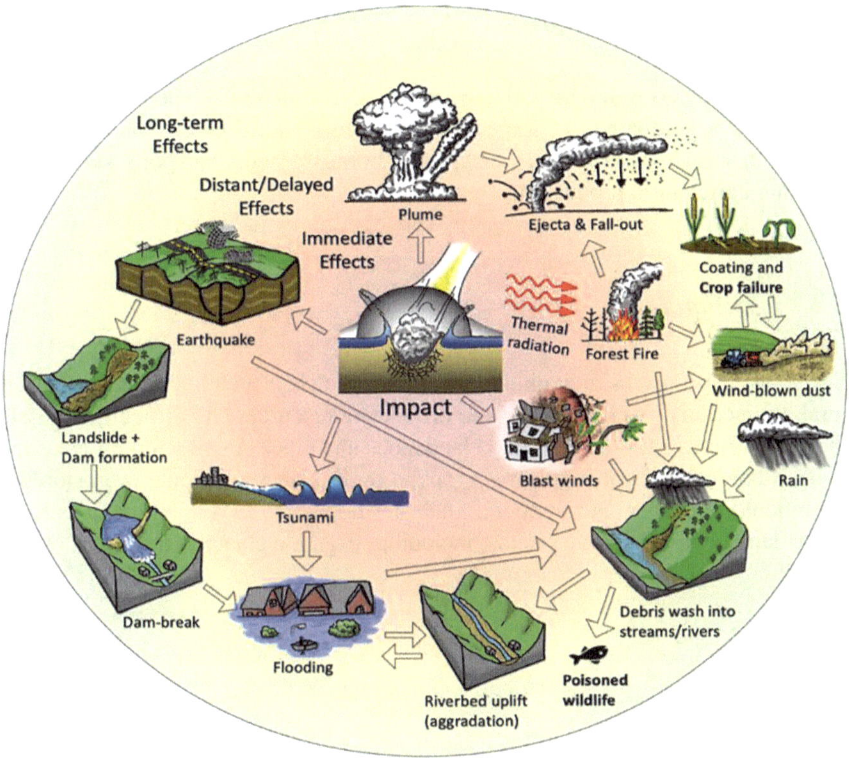

Fig. 1 Natural Disasters which effect cereal crops

harvest timing, and water availability. Furthermore, climatic variations and extreme events such as floods, droughts, and windstorms can impact both crop and livestock productivity (Quandt and Kimathi 2017).

2.2 *Frequency and Intensity of Rainfall*

The frequency and intensity of rainfall can indeed change in response to climate change, affecting the availability of water for crops, causing drought stress on crops, impacting animal production conditions, affecting forage supply for animals, and influencing irrigation facilities (Shankar and Shikha 2017).

2.3 CO_2 Concentration

The impact of CO_2 is anticipated to be more significant on C3 type like rice, wheat, and soybeans, compared to C4 species such as corn and sorghum. Extreme climate events can harm trees, crops, livestock, water-borne transport, and ports, significantly affecting agricultural productivity.

2.4 Earthquake

Due to the earthquakes' magnitude, there may be delays and disruptions in agricultural production as well as damage to productive resources and difficulty accessing them. Moreover, earthquake-related landslides and cracks in agricultural areas may make it harder for farmers to grow crops in those places and could even stop their operations altogether. Furthermore, such developments in agricultural areas may cause damage to irrigation systems, including irrigation channels and pipes (Yavuz, 2023).

2.5 Flooding

Most crops are indeed sensitive to waterlogging (root submergence) and submergence of aerial parts. Just a few days of flooding can severely harm plants, leading to significant agricultural losses. The main impact of flooding is the hindered gas diffusion between the plant and its surroundings. The rate of gas diffusion in water is about 10,000 times slower than in air, which restricts the supply of O_2 and CO_2. This limitation impairs cellular respiration and photosynthesis, causing an energy crisis that can ultimately result in the death of the plant (Voesenek and Bailey-Serres, 2015).

3 Impact of Climate Change on Cereal Crops

3.1 Rice (Paddy)

Numerous research studies have indicated that factors like irrigation, temperature, and fertilizer supply significantly impact the final harvest of paddy rice. One study specifically highlighted that the yield of paddy rice (variety: Xiushui 134) decreased notably when an alternate wetting and drying irrigation method was employed compared to continuous flooding irrigation (Ye et al., 2013). The study also discussed nitrogen fertilizer management under the other wetting and drying (AWD) irrigation system. This method reduced the quantity of irrigation water by 28% and 42% in the two years of the experiment, compared to flooding. The AWD technique increases the yields of rice and other plant biomass like root, shoot, and panicle growth. Through the two years of the study, grain yield increased by 5.7%–6.6% with AWD. Moreover, another study found that, as associated to measured irrigation treatments (6532–7868 kg ha − 1), the AWD irrigation method resulted in higher paddy yields (8516–8346 kg ha − 1) (Shao et al., 2014).

According to several studies, it has been identified that global warming has an adverse effect on paddy yields worldwide. There was an average rice temperature increase of 0.5 to 0.6 °C in the last century, due to which production in agriculture became low, especially rice production (Hansen et al., 2010). An increase in temperature increases the rate of plant respiration, leading to higher carbon metabolism and resulting in reduced yield of paddy crops. The correlation between the production of agriculture and climate challenges is highlighted by the increase in temperature especially when it is related to the production of paddy (rice) yields (Zhao & Fitzgerald, 2013). There is a high disruption in the process of reproduction and flower sterility due to an increase in temperature. IFPRI (International Food Policy Research Institute) predicts that because of these climate changes, yields of rice decreased by 10 to 15%. Because of this reduction in rice yields, market prices will increase to 32 and 37%. According to study data of 1999 and 2007, in Malaysia 2°C increased temperature results in 0.36 tons per hectare decrease in rice crops (Grote, 2018). So, it is essential to know the effect of climate change on the production of agriculture and economic growth (Vaghefi et al., 2010).

Another climate challenge is the increase in CO_2 concentration in the atmosphere. When CO_2 concentration increases, it impacts the productivity of rice yields in many parts of Asia. Rice crops decrease by 4% when the CO_2 level increases (Matthews et al., 2020). In some parts of Asia, when temperatures increase, production becomes high, but it is not enough to balance the loss of production, especially in those areas that are meant to be high in rice production around the world. So, these types of climate challenges, which include CO_2 concentration, increased temperature, and other growth changes that affect the overall rice production, draw attention to significant techniques. There should be some significant agricultural techniques and farming methods due to which we can prevent the loss of agriculture and ensure food security in the event of climate challenges. Many researchers say that due to the increase in temperature, agriculture production will become very low at the end of the century (Challinor et al., 2014). Moreover, studies also suggest that besides a decrease in production, there is also an increase in agriculture production by 2.9–34% due to global warming. It happens when warming takes place at night in those areas that are considered to be colder, but it is predicted that these areas will become warmer in the 2050s and 2080s due to climate change (Lobell et al., 2011a, 2011b) (Fig. 2).

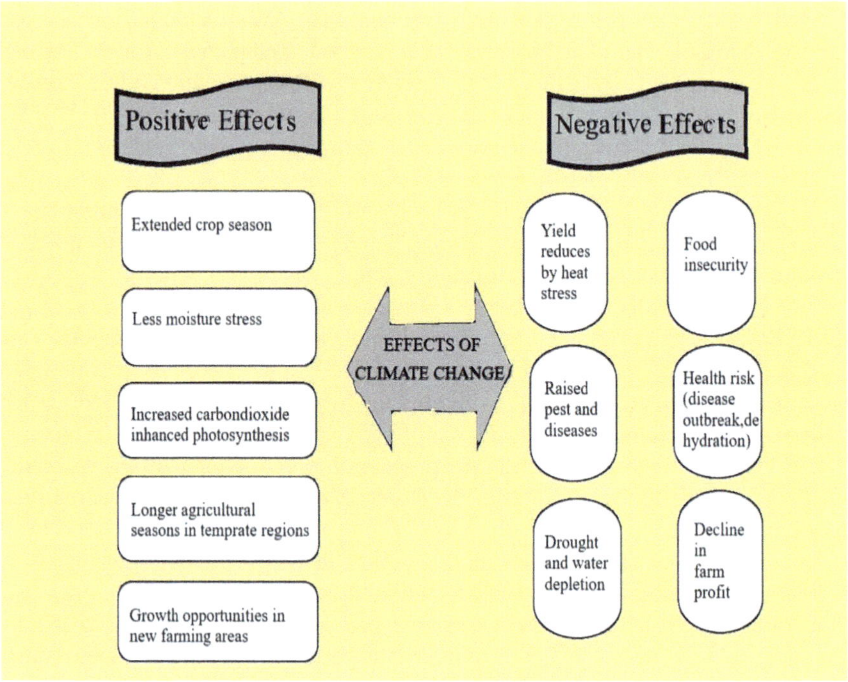

Fig. 2 Positive and negative effects of climate change and global warming on crops and humans (Raza et al., 2019)

3.2 *Millet*

The scientific name for pearl millet is Pennisetum glaucum. It is the type of crop that has a deep-routed system. This is a very unique characteristic of millet, due to which it can even survive climate challenges. Due to this system, it needs a very small amount of fertilizer. Even its plant weight, stems, and length are affected by nitrogen fertilizers if used in different amounts. Millet production is 50% global (Singh et al., 2017a, 2017b).

Because of the extensive root system of millet, it is also resistant to drought and flooding, and its production increases with an increase in temperature (Hadebe et al., 2016). The crop of millet is so resistant to different climate challenges that its production increases by 6% in drought challenges and 8% in heat challenges (Singh et al., 2017a, 2017b). So, together, millet production is increased by up to 15% under different climatic changes, which shows the flexibility of this plant. Millet has such diversity and flexibility that it can even be cultivated in hilly and mountainous areas where it is very difficult to cultivate the crops (Hussain et al., 2018). The quality of the millet crop is that it can be cultivated with minimal levels of nitrogen fertilizer in hilly and mountainous areas and is very flexible and resistant to drought and high temperatures, making it more important and sustainable s (Adhikari et al., 2017).

3.3 Maize

Maize is actually grown in two different conditions: one is a bed irrigation system, and the other is a drip irrigation system. These two systems are compared in a study that shows that bed irrigation systems will increase grains by 1%, plant height by 5%, and biological yield by 21% as compared to other systems (Amin et al., 2015). According to a study, maize production decreases by 10% with a 2 °C increase in temperature and a 20% decrease in precipitation (Lobell & Burke, 2010). Another study studies the effect of moisture on maize yields (Lobell et al., 2013). They discovered that in rain-fed and drought conditions, respectively, there was a 1% and 1.7% reduction in the crop's ultimate maize production for each day it was above 30 °C. Moisture plays an important role in enabling maize production with high temperatures and global warming, which show significant impacts (Lobell et al., 2011a, 2011b). The study conducted in 2014 investigated, by exploring yield sensitivity to temperature and atmospheric CO2 levels, how production of maize changes. In four different countries, like France, the USA, Brazil, and Tanzania, they tested a total of 23 models. The results showed that there is a negative correlation between temperature and yields. By every 1 °C rise in temperature, the harvests decreased by approximately 0.5 t ha − 1.

The study found that although atmospheric CO_2 levels doubled from 360 to 720 parts per million, there was a modest increase in yields. However, the possible effects of rising temperatures on maize yields are alarming. The study also concluded that if certain management strategies for rising temperatures are not applied, then the production of maize will decrease globally. The slight increase in production yields with a high CO_2 level is not sufficient to balance potential losses (Bassu et al., 2014).

In the cultivation of maize and in increasing grain yield, plant height, and kernel superiority, nitrogen directly plays a very important role (Woldesenbet & Haileyesus, 2016). The results show that there is a positive impact of nitrogen fertilization on various growth parameters and yield. By increasing nitrogen levels, traits such as the number of leaves per plant, the number of rows per ear, plant height, kernel quality, ear length, and grain yield also improve. Compared to the control and other nitrogen treatments, the number of kernels per ear and ear length reached their peak at 92 kg N ha − 1. In a similar study with the maize cultivar BH-660, researchers found that different nitrogen and phosphorus levels affect plant growth and yield. This study was directed to assess how the growth parameters and productivity of the BH-660 maize cultivar is influenced by different levels of phosphorus (0, 23, 46, 69, and 92 kg ha1) and nitrogen (0, 46, 92, 138, and 184 kg ha^{-1}) (Tesfaye et al., 2015).

The research findings show that there is a positive correlation between levels of nitrogen and phosphorus and yield production because the highest levels of nitrogen and phosphorus (184/92 kg ha^{-1}) resulted in the highest grain yield of 5497.5 kg ha^{-1} and biomass yield of 16,521 kg ha^{-1} for the maize crop.

3.4 Wheat

Climate change, driven by rising greenhouse gas emissions and global warming, poses a serious threat to global food security. The main concern is its impact on decreasing crop yield and wheat production (Fig. 3). It is predicted that crop yields will decline by 3.5 to 12.9% between 2037 and 2065 due to climate change. By the end of this century, winter wheat production is expected to decrease by 14.6 to 17.2% (Ortiz-Bobea & Tack, 2018). Rising temperatures, in particular, have a strong negative correlation with wheat yield every 1 °C increase leads to a 3 to 10% reduction. It is estimated that in the past two decades, wheat production has already declined by 4.5% due to temperature increases (You et al., 2009).

For the production, development, and growth of wheat, water plays a very important role. In a study on two winter wheat cultivars (Baviaans and 14SAWYT306), the group with irrigation was compared to the control group with no irrigation. Watering the crop up to stem extension, from stem extension to physiological maturity, and during the crop's growth stage were all part of the irrigation programs (Sagar & Naresh, 2001). Irrigation treatments had a significant positive impact on grain yield for both Baviaans and 14SAWYT306 wheat cultivars associated to no irrigation.

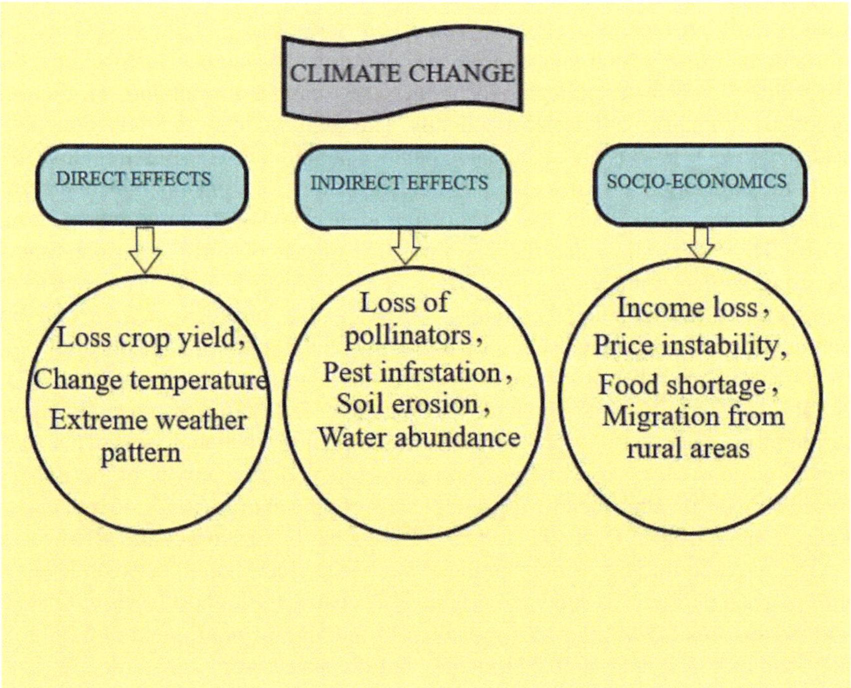

Fig. 3 Direct, indirect and socio-economic effects of climate change on agricultural production (Raza et al., 2019)

In particular, irrigation significantly increased field number, grain per spike, grain yield, harvest index, and grain protein content by 20.58%, 26.07%, 42.72%, 16.71%, respectively, during the entire growth period.

4 Possible Strategies to Increase Cereal Crops Production

Climate alteration can directly affect cereal yield through water stresses and heat. Moreover, it can indirectly impact pests, fertilizer supply, and diseases, further influencing crop productivity (Lobell, 2014). Researchers concur that global warming, stemming from the emission of carbon dioxide, is an irreversible trend that can span hundreds of years and have enduring consequences for the planet (Solomon et al., 2009). Refining the crops of different cereal crops and decreasing worldwide food damages are crucial to keep up with the growing people and to lessen the impact of climate change.

4.1 Breeding

The usual temperatures worldwide are projected to rise by 2.5–4.5 °C by the end of this century. (Péan & Romero, 2023). Maintaining the right temperature throughout the development stages is crucial for cereal crops as they are highly sensitive to environmental changes. High temperatures, for instance, can impact pollen viability, leading to reduced kernel numbers and ultimately affecting the crop through the flowering and heading stages (Stratonovitch & Semenov, 2015).

Traditionally, breeding has focused on enhancing crop yields rather than their resilience to high temperatures. However, there is a growing necessity to develop new crop varieties that can endure extreme weather conditions and rising temperatures while maintaining stable production levels (Fita et al., 2015). Studies have indeed focused on utilizing and expanding the available genetic diversity to create new crop genotypes that are both heat-tolerant and high-yielding. For example, under high temperatures to increase the production of yields, new varieties of cowpeas have been used. This is a major improvement over older cowpea types, whose output significantly decreased when nighttime temperatures rose beyond 20 °C (Ehlers & Hall, 2019). To improve scrap satisfying and crop larger scraps in BC1F6 wheat plants, a persistent selection method involving the ancestor T. tauschii as a genetic factor giver has been fruitfully employed (Eagles et al., 2002). A heat-tolerant gene found in vegetables was identified through a three-cycle regular selection breeding process, due to which there is and rise of up to 37.8% in yield (Figueiredo et al., 2015). According to certain research, transgenic tobacco plants with increased glycerol-3-phosphate acyltransferase overexpression showed increased thylakoid membrane lipid saturation. When exposed to heat stress, these transgenic plants recovered more quickly than their wild counterparts (Yan et al., 2008). The study shows that in improving the heat tolerance of plants, glycerol-3-phosphate acyltransferase plays an actual significant part. Molecular breeding techniques can achieve high expression of glycerol-3-phosphate acyltransferase-related enzymes in cereal crops such as wheat, maize and paddy. This approach holds the potential to enhance the heat tolerance of these crops significantly.

4.2 Irrigation and Fertilizer Efficiency

The presentation of water and fertilizer is crucial in influencing the crops of different yields at various phases of growth. The crop of produces tends to increase individual within an exact amount of water and fertilizer presentation. Unfortunately, few rising countries like Algeria, China, Egypt, India, Mexico, Iran, and Pakistan are encountering challenges with water scarcity due to factors such as weather alteration, rapid individual's growth, extensive farming practices, and industrial water consumption (Olmstead, 2014). Since 2003, China has experienced a 9% reduction in its wetlands, resulting in a loss of approximately 340,000 square kilometers (Junk et al., 2012). Improving irrigation efficiency is crucial as a key component of sustainable agriculture practices to enhance crop production. The International Food Policy Research Institute (IFPRI) highlighted the potential for a significant 21 percent increase in crop production by increasing irrigation efficiency (Gregory, 2014). Christian-Smith et al. (2012) reported that different farming techniques in California ranged from 60 to 85% water use efficiency for surface irrigation, 70 to 90% for sprinkler irrigation, and 88 to 90% for drip irrigation. Therefore, a drip irrigation system can effectively manage water use, improving yields and reducing water use by more than 50 percent (Ali, 2010).

The use of fertilizers is indeed a critical factor that has a direct influence on crop production. It has been estimated that ended the historical few periods, approximately 30–60% of the rise in crop manufacturing can be attributed to the usage of fertilizers (Ali, 2010). According to the Food and Agriculture Organization (FAO), fertilizer use is increasing in both developed and developing countries. In 2014, farmers used 186,900,000 tons of fertilizer, a 2.0 percent increase from the previous year. The worldwide request on behalf of fertilizers is projected to grow at a rate of 1.8% annually from 2014 to 2018. Also, it is noted that excessive use of fertilizers for a long period of time will decrease the quality of crops and their production. Studies concluded that the use of excessive fertilizers will increase the acidity of soil, resulting in less yield production (Fageria et al., 2010).

There will be some environmental changes in the regions where urea and nitrogen fertilizers are used due to the leakage of nitrogen and ammonia. Moreover, in Shouguang and Shandong provinces in China, a study highlighted that the pH of the soil reduced to 4.3 after 13 years of production. Due to this decline in pH, the quality and production of crops also decreased (Ju et al., 2007). So, it is very important that fertilizers be used at accurate levels to prevent environmental pollution and conserve resources. By using fertilizers at an appropriate level, we can increase crop production and protect our environment from negative effects.

4.3 Increasing the Cultivation Area of Tolerant Cereal Crops

Sorghum, particularly pearl millet, is known to be more drought-tolerant than other main cereal crops such as maize, wheat, and rice. Past research has shown that with warmer weather, rice cultivation may move to areas traditionally considered temperate regions, particularly in countries such as India and China. To cope with the probable loss of rice yields in these areas, farmers can consider sorghum cultivation in drylands as a practical alternative (Hadebe et al., 2016). This pliability is attributed to their capability to grow well in low-humidity environments. C4 crops have significantly greater water usage productively, ranging from 150 to 400% related to C3 crops, as studies have highlighted (Vogan & Sage, 2011). Moreover, under elevated atmospheric CO2 and temperature conditions, C4 crops exhibit lower levels of photorespiration. These factors are expected consequences of global warming (Lychuk et al., 2017).

Some modelling readings suggest that implementing better and water-efficient irrigation equipment in the forthcoming could increase C4 crop yields by up to 38% under projected climate circumstances in certain areas (Habiyaremye et al., 2017). In contrast, researchers do not expect significant changes in yield for C3 crops under comparable conditions. It emphasizes the potential of using enhanced irrigation approaches to increase the output of C4 crops such as maize and millet in response to changing climatic conditions (Lychuk et al., 2017). Sorghum, among the various C4 crops, stands out as the most ecologically friendly because it produces fewer conservatory gases than other cereal crops. Climate change needs a fewer significant influence on millet production. Sorghum, another resilient crop, is known for its ability to withstand drought conditions. Its flexibility stems from its solid and profuse root method, capability to keep great levels of stomatal conductance, capacity to normalize inner tissue water potential through osmotic adjustment, and phonological adaptation (Hadebe et al., 2017).

Instruction of local farmers, operating policy alterations, and the introduction of equipment to reduce the load on farmers can possibly increase sorghum manufacturing in these areas. Focusing on these zones is dangerous because a large number of the biosphere's population survives in the tropics, where weather alteration can directly affect crop production, yield levels, water balance, soil health, and temperature (Roudier et al., 2018). Increasing the cultivation of resilient cereal crops such as sorghum and millet is an important strategy for modifying the effects of weather variation, addressing water scarcity concerns, and improving food security ("Millet-seed sand grain," 2014).

5 Approaches to Combat Climate Changes

The environmental climate has a worldwide influence on food security and farming. It's not new for us to say that severe climate situations are a threat to food safety and security. In the previous era, problems were not considered severe. Therefore, combating these climate differences is a serious universal demand. New methods are needed to effectively adapt crops to changing environmental pressures.

5.1 Cultural Methodologies

Under environmental pressure, all these methods are really beneficial for crop adaptation. Farmers are experimenting with different strategies to cope with climate variability for plant adaptation. These policies include changing planting and gathering periods, calming produce with shorter life cycles, applying the yield revolution, adopting efficient irrigation methods, and increasing production. These practices play an important role in adapting crops to changing environmental conditions and increasing overall agricultural flexibility (Joyal et al., 2021). Changing the timing of sowing, using drought-resistant cultivars, and presenting innovative crops are necessary approaches to moderate the risks related to climate change. These measures improve the adaptability of crop plants, confirming good food production and security in altered ecological circumstances (Teixeira et al., 2020).

5.2 Conventional Techniques

Plant breeding actually plays an important role in crop growth and improvement under several ecological pressures. It suggests energetic methods to increase crop flexibility and efficiency, eventually ensuring food safety and security in extreme climate changes.

Rising stress-resistant cultivars through plant breeding allow plants to well endure altered pressures in serious growth stages, protecting an extra-maintainable and safe food supply (Henderson et al., 2018). Analysis of genetic divergence is really important to evaluate inbreeding, polymorphism, recombination, and assortment to attain plant development. These techniques play a very important role in successful breeding and increasing yield. Wheat landraces, which have deep inherited variances, are the most important hereditary sources. These landraces provide a hereditary pool that allows the crops to grow in different climatic conditions (Ali et al., 2020).

5.3 Genome Editing Strategies

Genome editing (GE) is a very useful method for plant development because it allows variations in the genetic material of plants by using sequence-specific nucleases. By applying the method of genome editing, we can combat food uncertainty and increase the production of agriculture globally (Chen et al., 2023). Traditional methods of plant breeding have identified genes associated with essential traits through change and predictable breeding methods. This approach is extensively recognized as a main technique for the generation of superior and high-yielding germplasm in plant species (Singh et al., 2017a, 2017b). Widespread cultivation of key crops has resulted in a decline in genetic diversity among elite varieties. A decreased genetic difference has increased vulnerability to both abiotic and biotic stresses under several circumstances (Hasanuzzaman et al., 2020). Plant breeding methods have indeed been significantly influenced by genome editing apparatus, leading to the search for novel plans for quick and accurate editing of crop genomes. The development goals are to increase crop flexibility against various pressures and to progress crop plants. Scientists have to make novel changes in the gene pool of a lot of plant germplasm under abiotic and biotic stress to breed superior crop variability, accomplished by manufacturing high-yielding plants. This method includes the use of genetic development to strengthen crop production and efficiency in inspiring environmental circumstances (Tripodi, 2022).

5.4 CRISPR/Cas9 System for Crop Advancement

CRISPR/Cas9, a novel genome editing policy, is founded on a prokaryotic defence mechanism initiated by the Type II RNA Institute that protects prokaryotes against invading worms. The CRISPR/Cas9 scheme developed genome excision by providing a multipurpose and exact instrument for embattled excision of DNA orders. Scientists adapted it for use in a wide range of organisms, including plants, for purposes of genetic manipulation of the bacterial immune system (Mousavi-Derazmahalleh et al., 2018). This system has genome editing, through which scientists create genetic mutants for detailed editing. This technology makes progress in genetic influence and research (Jiang et al., 2014). Moreover, CRISPR/Cas9 is the most important method in crop breeding, due to which agricultural production increases and crops can survive climatic changes. This system is also very important in ensuring food security and ruling out environmental tasks (Ma et al., 2023).

6 Conclusion

This chapter summarizes the impact of climate change on cereal crops such as wheat, maize, rice, and millet. The study reveals various climate challenges, including rising temperatures, increased CO_2 levels, and flooding, which can lead to fluctuating crop yields. While some conditions may temporarily boost production, they generally fail to offset overall losses. These climatic changes adversely affect not only crop yields but also plant growth and pest prevalence. To address these issues, it is crucial to adopt diverse strategies and farming methods to enhance agricultural output and ensure food security. Effective adaptation measures include breeding, improved irrigation, better nutrient management, expanding farmland, and implementing both cultural and conventional practices, which farmers globally recognize as vital for increasing total agricultural production.

References

Adeboyejo, F. O., Olatunde, S. J., Rustria, G. A., Azotea, A. N., Ostonal, J. M., Reyes, M. J., & Oyeyinka, S. A. (2023). Asian fermented cereal-based products. *Indigenous Fermented Foods for the Tropics*, 37-56. https://doi.org/10.1016/b978-0-323-98341-9.00002-5

Adhikari, P., Araya, H., Aruna, G., Balamatti, A., Banerjee, S., Baskaran, P., Barah, B. C., Behera, D., Berhe, T., Boruah, P., Dhar, S., Edwards, S., Fulford, M., Gujja, B., Ibrahim, H., Kabir, H., Kassam, A., Khadka, R. B., Koma, Y. S., & Verma, A. (2017). System of crop intensification for more productive, resource-conserving, climate-resilient, and sustainable agriculture: Experience with diverse crops in varying agroecologies. *International Journal of Agricultural Sustainability, 16*(1), 1–28. https://doi.org/10.1080/14735903.2017.1402504

Adviento-Borbe, M. A. (2020). An agronomic overview of US cereal cropping systems. *Breakfast Cereals and How They Are Made*, 39-71. https://doi.org/10.1016/b978-0-12-812043-9.00003-5

After the world food summit. (2019). *The Doubly Green Revolution*, 305–318. https://doi.org/10.7591/9781501722660-020

Agovino, M., Casaccia, M., Ciommi, M., Ferrara, M., & Marchesano, K. (2019). Agriculture, climate change and sustainability: The case of EU-28. *Ecological Indicators, 105*, 525–543. https://doi.org/10.1016/j.ecolind.2018.04.064

Aksoylu Özbek, Z., Taşkın, B., & Sözeri Atik, D. (2023). Fortification of plant-based food analogs. *Plant-Based Foods: Ingredients, Technology and Health Aspects*, 35-72. https://doi.org/10.1007/978-3-031-27443-5_3

Ali, M. (2010). Water application methods. *Practices of Irrigation & on-Farm Water Management:, 2*, 35–63. https://doi.org/10.1007/978-1-4419-7637-6_2

Ali, S., Liu, Y., Ishaq, M., Shah, T., Ilyas, A., & Din, I. U. (2017). Climate change and its impact on the yield of major food crops: Evidence from Pakistan. *Foods, 6*(6), 39. https://doi.org/10.3390/foods6060039

Ali, N., Rahman, I. U., Badakshi, F., Tariq, M. J., & Mujeeb-Kazi, A. (2020). Ensuring sustainable food security: Exploiting alien genetic diversity in wheat breeding for adaptation to emerging stresses. *Climate Change and Food Security with Emphasis on Wheat*, 31-42. https://doi.org/10.1016/b978-0-12-819527-7.00002-9

Allai, F. M., Azad, Z., Gul, K., & Dar, B. N. (2021). Wholegrains: A review on the amino acid profile, mineral content, physicochemical, bioactive composition and health benefits. *International Journal of Food Science & Technology, 57*(4), 1849–1865. https://doi.org/10.1111/ijfs.15071

Amin, M., Anjum, L., Alazba, A., & Rizwan, M. (2015). Effect of the irrigation frequency and quality on yield, growth and water productivity of maize crops. *Quality Assurance and Safety of Crops & Foods, 7*(5), 721–730. https://doi.org/10.3920/qas2014.0519

Bassu, S., Brisson, N., Durand, J., Boote, K., Lizaso, J., Jones, J. W., Rosenzweig, C., Ruane, A. C., Adam, M., Baron, C., Basso, B., Biernath, C., Boogaard, H., Conijn, S., Corbeels, M., Deryng, D., De Sanctis, G., Gayler, S., Grassini, P., & Waha, K. (2014). How do various maize crop models vary in their responses to climate change factors? *Global Change Biology, 20*(7), 2301–2320. https://doi.org/10.1111/gcb.12520

Bastian, G. E., & Palmer-Keenan, D. M. (2022). Simply plant based: Fabulous food for a healthy life. *Journal of Nutrition Education and Behavior, 54*(7), 702–703. https://doi.org/10.1016/j.jneb.2022.02.004

Bhardwaj, A., Misra, V., Mishra, A., Wootten, A., Boyles, R., Bowden, J. H., & Terando, A. J. (2018). Downscaling future climate change projections over Puerto Rico using a non-hydrostatic atmospheric model. *Clim Chang, 147*(1–2), 133–147. https://doi.org/10.1007/s10584-017-2130-x

Challinor, A. J., Watson, J., Lobell, D. B., Howden, S. M., Smith, D. R., & Chhetri, N. (2014). A meta-analysis of crop yield under climate change and adaptation. *Nature Climate Change, 4*(4), 287–291. https://doi.org/10.1038/nclimate2153

Chen, Z., Arango, J., Rao, I. M., & Liang, C. (2023). Editorial: Physiological, molecular and genetic mechanisms of abiotic stress tolerance in tropical crops. *Frontiers in Plant Science, 14*. https://doi.org/10.3389/fpls.2023.1150343

Christian-Smith, J., Cooley, H., & Gleick, P. H. (2012). Potential water savings associated with agricultural water efficiency improvements: A case study of California, USA. *Water Policy, 14*(2), 194–213.

Did quinoa get too popular for its own good? (2018, November 5). HowStuffWorks. https://recipes.howstuffworks.com/did-quinoa-get-too-popular-for-its-own-good.htm

Dziki, D. (2022). Rye flour and rye bran: New perspectives for use. *Processes, 10*(2), 293. https://doi.org/10.3390/pr10020293

Eagles, H. A., Hollamby, G. J., Gororo, N. N., & Eastwood, R. F. (2002). Estimation and utilisation of glutenin gene effects from the analysis of unbalanced data from wheat breeding programs. *Australian Journal of Agricultural Research, 53*(4), 367. https://doi.org/10.1071/ar01074

Ehlers, J., & Hall, A. (2019). Heat tolerance of contrasting cowpea lines in short and long days. *Field Crops Research, 55*(1–2), 11–21. https://doi.org/10.1016/s0378-4290(97)00055-5

Fageria, N. K., Dos Santos, A. B., & Moraes, M. F. (2010). Influence of urea and ammonium sulfate on soil acidity indices in lowland rice production. *Communications in Soil Science and Plant Analysis, 41*(13), 1565–1575. https://doi.org/10.1080/00103624.2010.485237

FAO. (2017). "Lessons from the green revolution: towards a new green revolution". Rome. https://web.archive.org/web/20170518074944, http://www.fao.org/docrep/003/w2612e/w2612e06a.htm

FAO-ESS. (2021). Crops Statistics - Concepts, Definitions and Classifications. FAO Statistics Division (ESS), Rome. https://www.fao.org/economic/the-statistics-division-ess/methodology/methodology-systems/crops-statistics-concepts-definitions-and-classifications/en/

Farmer, C. (2023). *The gestating and lactating sow* (3rd ed.). BRILL.

Figueiredo, I. C., Pinto, C. A., Ribeiro, G. H., Lino, L. D., Lyra, D. H., & Moreira, C. M. (2015). Efficiency of selection in early generations of potato families with a view toward heat tolerance. *Crop Breeding and Applied Biotechnology, 15*(4), 210–217. https://doi.org/10.1590/1984-70332015v15n4a37

Fita, A., Rodríguez-Burruezo, A., Boscaiu, M., Prohens, J., & Vicente, O. (2015). Breeding and domesticating crops adapted to drought and salinity: A new paradigm for increasing food production. *Frontiers in Plant Science, 6*. https://doi.org/10.3389/fpls.2015.00978

Food and Agricultural Organization (2007). http://www.fao.org/. Accessed December 2009.

Geng, L., Li, M., Zhang, G., & Ye, L. (2022). Barley: A potential cereal for producing healthy and functional foods. *Food Quality and Safety, 6*. https://doi.org/10.1093/fqsafe/fyac012

Grant, M., & Hazel, J. (2004). *Who's who in classical mythology*. Routledge.
Greenhalgh, S., Lemme, A., Dorigam, J. C., Chrystal, P. V., Macelline, S. P., Liu, S. Y., & Selle, P. H. (2022). Dietary crude protein concentrations, feed grains, and whey protein interactively influence apparent digestibility coefficients of amino acids, protein, starch, and performance of broiler chickens. *Poultry Science, 101*(11), Article 102131. https://doi.org/10.1016/j.psj.2022.102131
Gregory, P. J. (2014). Mark W. Rosegrant, I. Jawoo Koo, Nicola Cenacchi, Claudia Ringler, Richard Robertson, Myles Fisher, Condy cox, Karen Garrett, Nicostrato D. Perez and Pascale Sabbagh: Food security in a world of natural resource scarcity - the role of agricultural technologies. *Food Security, 6*(4), 603–604. https://doi.org/10.1007/s12571-014-0359-z
Grote, U. (2018). IFPRI: Global food policy report 2018. *Food Security, 10*(6), 1665–1667. https://doi.org/10.1007/s12571-018-0844-x
Guan, Y., Zheng, F., Zhang, P., & Qin, C. (2015). Spatial and temporal changes of meteorological disasters in China during 1950–2013. *Natural Hazards, 75*, 2607–2623. https://doi.org/10.1007/s11069-014-1446-3
Habiyaremye, C., Matanguihan, J. B., D'Alpoim Guedes, J., Ganjyal, G. M., Whiteman, M. R., Kidwell, K. K., & Murphy, K. M. (2017). Proso millet (Panicum miliaceum L.) and its potential for cultivation in the Pacific Northwest, U.S.: A review. *Frontiers in Plant Science, 7*. https://doi.org/10.3389/fpls.2016.01961
Hadebe, S. T., Modi, A. T., & Mabhaudhi, T. (2016). Drought tolerance and water use of cereal crops: A focus on sorghum as a food security crop in sub-saharan Africa. *Journal of Agronomy and Crop Science, 203*(3), 177–191. https://doi.org/10.1111/jac.12191
Hadebe, S. T., Mabhaudhi, T., & Modi, A. T. (2017). Water use of sorghum (<i>Sorghum bicolor</ i> L. Moench) in response to varying planting dates evaluated under rainfed conditions. *Water SA, 43*(1), 91. https://doi.org/10.4314/wsa.v43i1.12
Hancock, J. F. (2012). *Plant evolution and the origin of crop species*. Cabi.
Hansen, J., Ruedy, R., Sato, M., & Lo, K. (2010). Global surface temperature change. *Reviews of Geophysics, 48*(4). https://doi.org/10.1029/2010rg000345
Hasanuzzaman, M., Fujita, M., Filho, M. C., Nogueira, T. A., & Galindo, F. S. (2020). *Sustainable crop production*. BoD – Books on Demand.
Hatfield, J. L., & Prueger, J. H. (2015). Temperature extremes: Effect on plant growth and development. *Weather and Climate Extremes, 10*, 4–10. https://doi.org/10.1016/j.wace.2015.08.001
Henderson, B., Cacho, O., Thornton, P., Van Wijk, M., & Herrero, M. (2018). The economic potential of residue management and fertilizer use to address climate change impacts on mixed smallholder farmers in Burkina Faso. *Agricultural Systems, 167*, 195–205. https://doi.org/10.1016/j.agsy.2018.09.012
Henry, R., & Kettlewell, P. (2012). *Cereal grain quality*. Springer Science & Business Media.
Holopainen-Mantila, U., Vanhatalo, S., Lehtinen, P., & Sozer, N. (2024). Oats as a source of nutritious alternative protein. *Journal of Cereal Science, 116*, Article 103862. https://doi.org/10.1016/j.jcs.2024.103862
Hussain, A., Mahapatra, B., & Rasul, G. (2018). Adaptation in mountain agriculture: Food security in the Hindu-Kush himalayan (HKH) region. *Springer Climate*, 211-236. https://doi.org/10.1007/978-3-319-99347-8_10
Javid Iqbal, M., Shams, N., & Fatima, K. (2022). Nutritional quality of wheat. *Wheat*. https://doi.org/10.5772/intechopen.104659
JHOM - Bread - Hebrew. (2021). Jewish Heritage Online Magazine. https://www.jhom.com/topics/bread/hebrew.html
Jiang, D., Ma, G., Yang, R., Li, K., & Fang, M. (2014). Bayesian model selection for multiple QTLs mapping combining linkage disequilibrium and linkage. *Genetics Research, 96*. https://doi.org/10.1017/s0016672314000135

Joyal, T., Fremier, A., & Boll, J. (2021). Modeled forest conversion influences humid tropical watershed hydrology more than projected climate change. https://doi.org/10.22541/au.163873995.51016521/v1

Ju, X., Kou, C., Christie, P., Dou, Z., & Zhang, F. (2007). Changes in the soil environment from excessive application of fertilizers and manures to two contrasting intensive cropping systems on the north China plain. *Environmental Pollution, 145*(2), 497–506. https://doi.org/10.1016/j.envpol.2006.04.017

Junk, W. J., Piedade, M. T., Schöngart, J., & Wittmann, F. (2012). A classification of major natural habitats of Amazonian white-water river floodplains (várzeas). *Wetlands Ecology and Management, 20*(6), 461–475. https://doi.org/10.1007/s11273-012-9268-0

Kaye, J. P., & Quemada, M. (2017). Using cover crops to mitigate and adapt to climate change. *A Review. Agron Sustain Dev, 37*(1), 4. https://doi.org/10.1007/s13593-016-0410-x

Lesk, C., Rowhani, P., & Ramankutty, N. (2016). Influence of extreme weather disasters on global crop production. *Nature, 529*, 84–87. https://doi.org/10.1038/nature16467

Lobell, D. B. (2014). Climate change adaptation in crop production: Beware of illusions. *Global Food Security, 3*(2), 72–76. https://doi.org/10.1016/j.gfs.2014.05.002

Lobell, D. B., & Burke, M. B. (2010). On the use of statistical models to predict crop yield responses to climate change. *Agricultural and Forest Meteorology, 150*(11), 1443–1452. https://doi.org/10.1016/j.agrformet.2010.07.008

Lobell, D. B., Bänziger, M., Magorokosho, C., & Vivek, B. (2011a). Nonlinear heat effects on African maize as evidenced by historical yield trials. *Nature Climate Change, 1*(1), 42–45. https://doi.org/10.1038/nclimate1043

Lobell, D. B., Schlenker, W., & Costa-Roberts, J. (2011b). Climate trends and global crop production since 1980. *Science, 333*(6042), 616–620. https://doi.org/10.1126/science.1204531

Lobell, D. B., Hammer, G. L., McLean, G., Messina, C., Roberts, M. J., & Schlenker, W. (2013). The critical role of extreme heat for maize production in the United States. *Nature Climate Change, 3*(5), 497–501.

Lychuk, T. E., Hill, R. L., Izaurralde, R. C., Momen, B., & Thomson, A. M. (2017). Evaluation of climate change impacts and effectiveness of adaptation options on crop yield in the southeastern United States. *Field Crops Research, 214*, 228–238. https://doi.org/10.1016/j.fcr.2017.09.020

Ma, M., Zhong, W., Zhang, Q., Deng, L., Wen, J., Yi, B., Tu, J., Fu, T., Zhao, L., & Shen, J. (2023). Corrigendum: Genome-wide analysis of transcriptome and histone modifications in brassica napus hybrid. *Frontiers in Plant Science, 14*. https://doi.org/10.3389/fpls.2023.1232986

Matthews, R., Kropff, M., Horie, T., & Bachelet, D. (2020). Simulating the impact of climate change on rice production in Asia and evaluating options for adaptation. *Agricultural Systems, 54*(3), 399–425. https://doi.org/10.1016/s0308-521x(95)00060-i

Mayshar, J., Moav, O., & Pascali, L. (2022). The origin of the state: Land productivity or Appropriability? *Journal of Political Economy, 130*(4), 1091–1144. https://doi.org/10.1086/718372

Millet-seed sand grain. (2014). *Dictionary Geotechnical Engineering/Wörterbuch GeoTechnik*, 869–869. https://doi.org/10.1007/978-3-642-41714-6_131746

Mousavi-Derazmahalleh, M., Bayer, P. E., Hane, J. K., Valliyodan, B., Nguyen, H. T., Nelson, M. N., Erskine, W., Varshney, R. K., Papa, R., & Edwards, D. (2018). Adapting legume crops to climate change using genomic approaches. *Plant, Cell & Environment, 42*(1), 6–19. https://doi.org/10.1111/pce.13203

Munch, D. (2023). *New estimates reveal major 2022 weather disasters caused over $21 billion in crop losses*. American Farm Bureau Federation. https://www.fb.org/market-intel/new-estimates-reveal-major-2022-weather-disasters-caused-over-21-billion-in-crop-losses

National Academies of Sciences Engineering and Medicine, Division on Earth and Life Studies, Board on Agriculture and Natural Resources, & Committee on Nutrient Requirements of Dairy Cattle. (2021). *Nutrient requirements of dairy cattle: Eighth revised edition*.

OECD-FAO Agricultural Outlook. Paris: OECD Agriculture Statistics

Olmstead, S. M. (2014). Climate change adaptation and water resource management: A review of the literature. *Energy Economics, 46*, 500–509. https://doi.org/10.1016/j.eneco.2013.09.005

Organization, W. H. (2018). *The State of Food Security and Nutrition in the World 2018: Building Climate Resilience for Food Security and Nutrition.* Food & Agriculture Org.

Ortiz-Bobea, A., & Tack, J. (2018). Is another genetic revolution needed to offset climate change impacts for US maize yields? *Environmental Research Letters, 13*(12), Article 124009. https://doi.org/10.1088/1748-9326/aae9b8

Péan, C., & Romero, J. (2023). IPCC, 2023: Annex II: Acronyms, chemical symbols and scientific units [Fischlin, A., Y. Jung, N. leprince-ringuet, C. Ludden, C. Péan, J. Romero (EDS.)]. In: Climate change 2023: Synthesis report. Contribution of working groups I, II and III to the sixth assessment report of the intergovernmental panel on climate change [Core writing team, H. Lee and J. Romero (EDS.)]. IPCC, Geneva, Switzerland. *IPCC, 2023: Climate Change 2023: Synthesis Report. Contribution of Working Groups I, II and III to the Sixth Assessment Report of the Intergovernmental Panel on Climate Change [Core Writing Team, H. Lee and J. Romero (eds.)]. IPCC, Geneva, Switzerland*, 131–134. https://doi.org/10.59327/ipcc/ar6-9789291691647.003

Polonskiy, V., Loskutov, I., & Sumina, A. (2020). Biological role and health benefits of antioxidant compounds in cereals. *Biological Communications, 65*(1). https://doi.org/10.21638/spbu03.2020.105

Purugganan, M. D., & Fuller, D. Q. (2009). The nature of selection during plant domestication. *Nature, 457*(7231), 843–848. https://doi.org/10.1038/nature07895

Quandt, A., & Kimathi, Y. A. (2017). Perceptions of the effects of floods and droughts on livelihoods: Lessons from arid Kenya. *International Journal of Climate Change Strategies and Management, 9*, 337–351. https://doi.org/10.1108/IJCCSM-11-2014-0132

Rathna Priya, T. S., Eliazer Nelson, A. R., Ravichandran, K., & Antony, U. (2019). Nutritional and functional properties of coloured rice varieties of South India: A review. *Journal of Ethnic Foods, 6*(1). https://doi.org/10.1186/s42779-019-0017-3

Raza, A., Razzaq, A., Mehmood, S. S., Zou, X., Zhang, X., Lv, Y., & Xu, J. (2019). Impact of climate change on crops adaptation and strategies to tackle its outcome: A review. *Plants, 8*(2), 34. https://doi.org/10.3390/plants8020034

Rejman, K., Górska-Warsewicz, H., Czeczotko, M., & Laskowski, W. (2020). Nonalcoholic beverages as sources of nutrients in the average Polish diet. *Nutrients, 12*(5), 1262. https://doi.org/10.3390/nu12051262

Renfrew, Colin; Bahn, Paul (2012). *Archaeology: Theories, Methods, and Practice* (6th ed.). Thames & Hudson. p. 277. ISBN 978–0–500–28976–1.

Robert, C. P. (2010). James E. Gentle: Computational statistics (Statistics and computing series). *Statistics and Computing, 21*(2), 289–291. https://doi.org/10.1007/s11222-010-9189-9

Rodriguez, J. P., Jacobsen, S., Andreasen, C., & Sørensen, M. (2020). Cañahua (Chenopodium pallidicaule): A promising new crop for arid areas. *Environment & Policy*, 221-243. https://doi.org/10.1007/978-3-319-90472-6_9

Rosell, C. M. (2011). The science of doughs and bread quality. *Flour and Breads and their Fortification in Health and Disease Prevention*, 3-14. https://doi.org/10.1016/b978-0-12-380886-8.10001-7

Roudier, P., Sultan, B., Quirion, P., & Berg, A. (2018). The impact of future climate change on west African crop yields: What does the recent literature say? https://doi.org/10.31219/osf.io/y8p7k

Vipin Kumar Sagar, & R.K. Naresh. (2001). Effect of crop establishment methods and irrigation schedules on growth and yield of wheat (Triticum aestivum). *Indian Journal of Agronomy, 64*(2), 210–217. https://doi.org/10.59797/ija.v64i2.5256

Sands, D. C., Morris, C. E., Dratz, E. A., & Pilgeram, A. L. (2009). Elevating optimal human nutrition to a central goal of plant breeding and production of plant-based foods. *Plant Science, 177*(5), 377–389. https://doi.org/10.1016/j.plantsci.2009.07.011

Shankar S, Shikha (2017). Impacts of climate change on agriculture and food security. In Lakahan, R; Mondal, S (Ed.), Biotechnology for Sustainable Agriculture: Emerging Approaches and Strategies. https://doi.org/10.1016/B978-0-12-812160-3.00007-6

Shao, G., Deng, S., Liu, N., Yu, S., Wang, M., & She, D. (2014). Effects of controlled irrigation and drainage on growth, grain yield and water use in Paddy rice. *European Journal of Agronomy, 53*, 1–9. https://doi.org/10.1016/j.eja.2013.10.005

Singh, P., Boote, K., Kadiyala, M., Nedumaran, S., Gupta, S., Srinivas, K., & Bantilan, M. (2017a). An assessment of yield gains under climate change due to genetic modification of pearl millet. *Science of the Total Environment, 601–602*, 1226–1237. https://doi.org/10.1016/j.scitotenv.2017.06.002

Singh, R. K., Sahu, P. P., Muthamilarasan, M., Dhaka, A., & Prasad, M. (2017). Genomics-assisted breeding for improving stress tolerance of graminaceous crops to biotic and abiotic stresses: Progress and prospects. *Plant Tolerance to Individual and Concurrent Stresses*, 59-81. https://doi.org/10.1007/978-81-322-3706-8_5

Solaimalai, A., Anantharaju, P., Irulandi, S., & Theradimani, M. (2021). *Maize crop: Improvement, production, protection and post-harvest technology*. CRC Press.

Solomon, S., Plattner, G., Knutti, R., & Friedlingstein, P. (2009). Irreversible climate change due to carbon dioxide emissions. *Proceedings of the National Academy of Sciences, 106*(6), 1704–1709. https://doi.org/10.1073/pnas.0812721106

Standage, T. (2009). *An edible history of humanity*. Atlantic Books.

Stratonovitch, P., & Semenov, M. A. (2015). Heat tolerance around flowering in wheat identified as a key trait for increased yield potential in Europe under climate change. *Journal of Experimental Botany, 66*(12), 3599–3609. https://doi.org/10.1093/jxb/erv070

Teixeira, E., Ausseil, A., Burgueño, E., Brown, H., Cichota, R., Davy, M., Ewert, F., Guo, J., Holmes, A., Holzworth, D., Hu, W., De Ruiter, J., Hume, E., Jesson, L., Johnstone, P., Powell, J., Kersebaum, K. C., Kong, H., Liu, J., … Van der Weerden, T. (2020). A spatial analysis framework to assess responses of agricultural landscapes to climates and soils at regional scale. *Innovations in Landscape Research*, 495-508. https://doi.org/10.1007/978-3-030-37421-1_25

Tesfaye, K., Gbegbelegbe, S., Cairns, J. E., Shiferaw, B., Prasanna, B. M., Sonder, K., Boote, K., Makumbi, D., & Robertson, R. (2015). Maize systems under climate change in sub-Saharan Africa. *International Journal of Climate Change Strategies and Management, 7*(3), 247–271. https://doi.org/10.1108/ijccsm-01-2014-0005

Tripathi, G., Jitendrakumar, P. H., Borah, A., Nath, D., Das, H., Bansal, S., Singh, N., & Singh, B. V. (2023). A review on nutritional and health benefits of millets. *International Journal of Plant & Soil Science, 35*(19), 1736–1743. https://doi.org/10.9734/ijpss/2023/v35i193722

Tripodi, P. (2022). Development, preparation, and Curation of high-throughput phenotypic data for genome-wide association studies: A sample pipeline in R. *Methods in Molecular Biology*, 105-125. https://doi.org/10.1007/978-1-0716-2237-7_7

Vaghefi, N., Shamsudin, M. N., Makmom, A., & Bagheri, M. (2010). The economic impacts of climate change on the rice production in Malaysia. *International Journal of Agricultural Research, 6*(1), 67–74. https://doi.org/10.3923/ijar.2011.67.74

Voesenek, L. A., & Bailey-Serres, J. (2015). Flood adaptive traits and processes: An overview. *New Phytologist, 206*(1), 57–73.

VOGAN, P. J., & SAGE, R. F. (2011). Water-use efficiency and nitrogen-use efficiency of C₃-c₄ intermediate species of *Flaveria* Juss. (Asteraceae). *Plant, Cell & Environment, 34*(9), 1415–1430. https://doi.org/10.1111/j.1365-3040.2011.02340.x

Wang, Q., Zhang, Q.-P., Liu, Y.-Y., Tong, L.-J., Zhang, Y.-Z., Li, X.-Y., et al. (2020). Characterizing the spatial distribution of typical natural disaster vulnerability in China from 2010 to 2017. *Natural Hazards, 100*, 3–15. https://doi.org/10.1007/s11069-019-03656-7

Wieser, H., Koehler, P., & Scherf, K. A. (2020). The two faces of wheat. *Frontiers in Nutrition, 7*. https://doi.org/10.3389/fnut.2020.517313

Woldesenbet, M., & Haileyesus, A. (2016). Effect of nitrogen fertilizer on growth, yield and yield components of maize (Zea mays L.) in decha district, southwestern Ethiopia. *International Journal of Research -GRANTHAALAYAH*, *4*(2), 95–100. https://doi.org/10.29121/granthaalayah.v4.i2.2016.2817

Yan, K., Chen, N., Qu, Y., Dong, X., Meng, Q., & Zhao, S. (2008). Overexpression of sweet pepper *glycerol-3-Phosphate Acyltransferase* gene enhanced Thermotolerance of photosynthetic apparatus in transgenic tobacco. *Journal of Integrative Plant Biology, 50*(5), 613–621. https://doi.org/10.1111/j.1744-7909.2008.00652.x

Yavuz, F. (2023, March 20). The earthquakes' impact on agriculture. SETA. https://www.setav.org/en/opinion/the-earthquakes-impact-on-agriculture

Ye, Y., Liang, X., Chen, Y., Liu, J., Gu, J., Guo, R., & Li, L. (2013). Alternate wetting and drying irrigation and controlled-release nitrogen fertilizer in late-season rice. Effects on dry matter accumulation, yield, water and nitrogen use. *Field Crops Research, 144*, 212–224. https://doi.org/10.1016/j.fcr.2012.12.003

You, L., Rosegrant, M. W., Wood, S., & Sun, D. (2009). Impact of growing season temperature on wheat productivity in China. *Agricultural and Forest Meteorology, 149*(6–7), 1009–1014. https://doi.org/10.1016/j.agrformet.2008.12.004

Zhao, X., & Fitzgerald, M. (2013). Climate change: Implications for the yield of edible rice. *PLoS ONE, 8*(6), Article e66218. https://doi.org/10.1371/journal.pone.0066218

Cutting Edge Breeding Techniques for Soil Adaptability

Rabia Kalsoom, Saira Jabeen, Muhammad Arif, Azher Mustafa, Javeria Naz, Hina Abbas, Maryam Tanveer Akhtar, Rahmat Ali Khan, and Parvaiz Ahmad

Abstract In the past decade, genomics-assisted breeding (GAB) has emerged as a pivotal approach for leveraging modern genomic resources and enhancing allelic variation for germplasm improvement and varietal development. As global population growth intensifies the demand for agricultural output, there is an urgent need to enhance food security and feed availability. Future breeding tools will focus on developing climate-resilient genotypes to address global food security challenges. Recent advancements in GAB have enabled the creation of well-annotated crop pan-genomes, providing a comprehensive overview of genetic diversity and revitalizing gene catalogs for various species. The proposed GAB 2.0 aims to accelerate the manipulation of allelic variation, facilitating the rapid introduction of novel traits into crop improvement programs. This approach prioritizes the optimization of crop genomes by harnessing beneficial alleles and eliminating detrimental ones,

R. Kalsoom · A. Mustafa
Plant Virology Section, Ayub Agriculture Research Institute, Faisalabad, Pakistan

S. Jabeen
Department of Plant Pathology, Faculty of Agricultural Sciences, University of the Punjab, Lahore, Pakistan

M. Arif (✉)
College of Agriculture, Guizhou University, Guiyang, Guizhou, China
e-mail: arif@gzu.edu.cn

J. Naz
Department of Plant Sciences, Quaid-I-Azam University, Islamabad, Pakistan

H. Abbas
National Institute for Genomics and Advanced Biotechnology (NIGAB), National Agricultural Research Centre (NARC), Islamabad, Pakistan

M. T. Akhtar
Department of Environmental Science, International Islamic University, Islamabad, Pakistan

R. A. Khan
Department of Biotechnology, University of Science and Technology, Bannu, Khyber Pakhtunkhwa, Pakistan

P. Ahmad
Department of Botany, GDC-Pulwama, Pulwama, Jammu and Kashmir, India

M. Shaaban (ed.), *Soils and Sustainable Agriculture*, Frontier Studies in Soil Science,
https://doi.org/10.1007/978-3-031-91114-9_26

ultimately leading to the development of climate-smart cultivars with enhanced nutritional value. Furthermore, the application of CRISPR/Cas systems in crop research has shown success in enhancing resistance to abiotic and biotic stresses. Coupled with high-throughput phenotyping and advanced data analytics tools like AI and machine learning, agriculture is advancing towards greater mechanization and digitalization, setting the stage for a sustainable future in crop production.

Keywords Genomics-assisted breeding (GAB) · Allelic variation · CRISPR/Cas systems · Climate-resilient genotypes · Sustainable agriculture

1 Introduction

Forecasts indicate that global crop yields must double to sustain a projected population of 10 billion, yet climate change is creating increasingly unfavorable conditions for plant production in many agricultural regions (Razzaq et al., 2021). The challenges of addressing climate change, alongside diminishing nutritional needs and the pursuit of resilient energy sources, remain critical today. To tackle these issues, it is essential to enhance crop traits for the development of resilient varieties capable of withstanding abiotic and biotic stresses. For years, selective breeding has been a primary method for improving plant varieties. However, as urban sprawl encroaches on arable land and climate change along with soil degradation complicate traditional farming practices, researchers are exploring innovative strategies to help crops better adapt to their environments (Benitez-Alfonsoet al., 2023).

Before delving into methods, it's essential to understand soil adaptability, which refers to crop's ability to thrive and produce effectively in diverse soil types, including sandy, saline, and varying pH levels. As climate change intensifies, it introduces new threats that necessitate responsive soil health assessments (Ghimire et al., 2023). Progress in crop resilience may be undermined by more resilient pests, pathogens, and weeds, all exacerbated by changing conditions. Key soil health indicators for assessing climate change impacts include water penetration, stability, bulk density, soil organic matter (SOM) content, carbon and nitrogen cycling, and microbial diversity and activity (Ahmad & Wang, 2023). While this chapter does not provide empirical examples of these concepts in past or future experiments, it will outline how to select suites of indicators based on their sensitivity to manage changes and their integration with crucial soil functions. Additionally, soil health indicators must be responsive to global change drivers, such as rising temperatures, increased atmospheric carbon and nitrogen levels, and heightened variability in precipitation patterns.

Traditional crops frequently underperform in suboptimal soil conditions, resulting in lower yields and increased costs. By enhancing soil adaptability in crops, these issues can be mitigated by fostering more sustainable and resilient agricultural systems (Patil & Lamnganbi, 2018; Nigussie, 2025). This chapter explores advanced breeding practices aimed at improving soil adaptability, providing insights into the future of agriculture and its potential transformation.

2 Genomic Selection and Marker-Assisted Breeding

By focusing on key plant traits that affect soil stability and erosion resistance, genetic methods present innovative solutions for soil preservation. Altering root architecture is a particularly impactful strategy to enhance soil organic carbon formation (SOCF) and reduce soil erosion. Modified plants with larger, deeper root systems can better stabilize soil components, effectively preventing erosion and promoting healthier soil conditions (Beroueg et al., 2021).

Genetic manipulations that enhance plant resistance to ecological stressors such as soil erosion, drought, storm damage, and ultraviolet-B (UV-B) radiation can significantly aid in soil conservation. By maintaining plant cover and minimizing ground surface disturbance during extreme weather events, these genetic strategies can help reduce erosion risk and promote stability and sustainability in agroecosystems. Additionally, a genetic approach can enhance soil richness and stability by fostering beneficial plant–microbe interactions that mitigate erosion risks and complement other sustainable soil management practices (Mmbando & Ngongolo, 2024).

Genetic modifications (GMs) can enhance mutualistic relationships between plants and beneficial soil microorganisms, such as rhizobia and mycorrhizal fungi. These associations positively impact plant nutrition by promoting nitrogen fixation and mineral solubilization. GMs can improve the efficiency of these symbiotic interactions, thereby enhancing nutrient acquisition for plants. According to the Society for Experimental Biology, such strategies could lead to larger and more efficient root vascular networks, improving nutrient distribution throughout the plant. Overall, developing GMs that strengthen root systems offers viable strategies for increasing nutrient transport efficiency, which can enhance crop yield and quality. By using genetic approaches to improve root traits, increase soil aggregation, and optimize plant–microbe interactions, soil structure and conservation can be enhanced which ultimately fostering the long-term sustainability of agricultural ecosystems (Mmbando & Ngongolo, 2024).

2.1 Genomic Selection (GS)

Genomic selection (GS) is an innovative technique that employs molecular genetic markers to develop new breeding schemes and genome-wide association studies (GWAS) models for designing effective breeding programs. Some crop varieties exhibit a certain level of tolerance to abscisic acid (ABA), and breeders have harnessed this trait to develop varieties with enhanced stress adaptability. Furthermore, the overexpression of various molecularly characterized genes in plant transformation experiments has revealed numerous genes responsible for abiotic stress tolerance.

GS presents new opportunities for enhancing the genetic potential of complex traits in plant breeding, optimizing both time and cost. GS effectively balances cost and benefit, demonstrating its viability in crop plants. The availability of high-quality, genome-wide, and unbiased markers in a high-throughput and cost-effective manner is crucial for its application across large populations, whether for a model or non-model plant species, with or without reference genome sequences (Gidamo, 2022).

2.2 *Marker-Assisted Breeding (MAB)*

Agronomic traits, such as crop yield, are typically controlled by multiple polygenes with small effects and complex epistatic relationships. GS serves as a technological platform that utilizes genome-wide markers to estimate genomic breeding values (GEBV), which are essential for selecting individuals. Stress tolerance phenotypes are similarly complex traits influenced by many polygenes. The effective use of modern genomic tools, including molecular markers and site-directed mutagenesis techniques like CRISPR-Cas9, can expedite genetic introgression (Villalobos-López et al., 2022).

GS involves three major steps: training the prediction model, estimating GEBV, and selecting the best individuals based on their expected GEBV. The accuracy of GEBV predictions is evaluated by comparing them to observed traits (associations with phenotypes) and through cross-validation (Vasanthrao et al., 2023).

3 Genetic Engineering and CRISPR Technology

3.1 *Genetic Engineering*

Biotechnology has opened up significant opportunities for enhancing crop improvement, quality management, crop protection, and various horticultural traits. It offers a substantial potential for increasing fruit production of introducing novel genotypes for breeding, developing vigorous and disease-free planting materials, quality enhancement, shelf-life extension, and creating bio-pesticides and bio-fertilizers. Specific desired traits can be introduced into horticultural crops using genetic engineering, which involves isolating and ligating a gene of interest within an appropriate vector to construct a recombinant DNA (rDNA) molecule, then reintroducing this new genetic function into the plant genome.

Scientists in agricultural biotechnology often employ transgenic technology, which involves introducing desired genes from various sources such as plants, animals, or microorganisms into plants using non-conventional methods. Unlike conventional breeding, where tens of thousands of genes from both susceptible and resistant plants are mixed randomly, recombinant DNA technology allows for the selective transfer of specific genes to susceptible plants while retaining advantageous economic traits. Moreover, the genetic basis for resistance can extend beyond closely related plant species, further enhancing the potential for crop improvement.

The foundation of sustainable agriculture lies in addressing various types of biotic and abiotic stresses. While traditional and marker-assisted breeding methods are making strides in developing more resilient cultivars, these approaches may not be suitable for biennial or perennial horticultural crops due to their prolonged sexual generation periods. One of the key advantages of transgenic technology is the ability to isolate genes controlling essential traits from a wide range of organisms, including plants and microorganisms, facilitating the introduction of novel traits into target plants. However, for successful gene transfer to elite backgrounds, it is crucial to develop and refine procedures that increase the frequency of effective plant regeneration and genetic transformation. Numerous studies have established regeneration and genetic transformation protocols for various horticultural crops, including apple, pomegranate, chilly, cucumber, lily, sweet orange, broccoli, date palm, and chrysanthemum. These advancements are vital for enhancing the resilience and sustainability of horticultural production systems.

Genetic engineering offers vast applications for crop improvement, including enhancing biotic and abiotic stress tolerance and improving yield quality. Numerous transgenic crops have already been developed, with more expected in the next two decades. Resistance to insects and nematodes has been achieved by introducing genes such as natural and synthetic Cry toxins, protease inhibitors, and cystatin genes. Many horticultural crops have also been engineered for defense against fungal and bacterial diseases through the transfer of genes encoding chitinase, osmotin, glucanase, defensin, and other pathogenesis-related proteins. Additionally, the RNA interference (RNAi) technique has proven effective for virus resistance in horticultural crops. Abiotic stresses like drought, heat, and salinity significantly impact crop productivity, but engineered genes involved in the biosynthesis of stress-protecting compounds, such as mannitol and glycine betaine, along with heat shock proteins, can enhance tolerance. Antisense gene and RNAi technologies, which operate through post-transcriptional silencing mechanisms, have accelerated the development of horticultural crops, particularly ornamentals, by improving color variation, shelf life, and reducing post-harvest damage. Recently, genome editing tools, especially CRISPR/Cas9, have been successfully applied in tomatoes, petunia, citrus, grapes, potatoes, and apples for gene transmutation, suppression, stimulation, and epigenome editing (Parmar et al., 2017; Kumar, 2024; Imtiaz et al., 2024).

While GM techniques for soil preservation and nutrient uptake are gaining popularity and advancing rapidly, comprehensive literature reviews on these aspects remain scarce. Much of the available research tends to focus on specific topics, such as nutrient uptake or soil conservation, without providing a holistic overview of how GM techniques can address multiple challenges simultaneously. Additionally, further research is needed to explore the potential trade-offs and synergies associated with different genetic interventions in soil–plant systems. This gap highlights the need for future studies to better understand the role of GM in sustainable agriculture (Lebedev, 2021).

Conventional farming practices often fall short in combating erosion and improving nutrient uptake, which are crucial for achieving optimal crop yields and minimizing environmental impacts (Mirzabaev et al., 2023). Addressing these issues is essential for promoting ecologically sustainable food production and conservation. GM techniques have the potential to foster the development of innovative approaches for topsoil conservation, optimizing nutrient uptake, and enhancing nutrient availability for more efficient use. This knowledge can inform sustainable agricultural practices that ensure food security while reducing ecological footprints (Mmbando & Ngongolo, 2024).

3.2 *Difficulties and Aspects When Applying GM Techniques to Enhance Soil Conservation, Nutrient Uptake and Utilization*

Many ethical and environmental issues emerge in trying to implement GM techniques that mitigate and alleviate soil maintenance and nutrient uptake (Dima, 2023). The introduction of GMOs to the environment then poses unintended effects on the environment. Gene flow via cross-pollination between GMO crops and wild or non-GMO crops within natural ecosystems may lead to unintended genetic changes that disrupt the ecosystem. Third, GMO seeds and crop control and ownership are considered an ethical issue tagged "biopiracy" (Ayanoğlu, 2020). The patenting of GMOs by companies within the biotech industry raises concern over corporate monopolies in agriculture and access to agricultural resources. One issue also pertains to the enduring effects of genetic modification on the surroundings unintended effects on non-target species, soil microbiomes, and environmental biodiversity. All in all, though GM technology holds a promising future in solving some of the identified agricultural problems, it has to be approached with caution for its applications and implemented along with strong solutions to its ethical and environmental implications so that the result will be fair and sustainable (Ndudzo et al., 2024).

Furthermore, investigations concerning sustainability and the long-lasting effects of GM techniques in enhancing nutrient uptake and soil conservation are persistent. For that matter, although GM technology has the potential of combating the challenges in agriculture, its effects on health conditions of ecosystems and environmental sustainability, are long-term implications that must be well considered, in order not to let this technique inadvertently release genes to unintended species. Third, traits such as tolerance to maintain productivity under changing environmental conditions, impact on soil health and biodiversity, and resilience against new pests and diseases determine the sustainability of GMO crops (Mekonnen et al., 2022). Beyond these are the more extensive agricultural implements and socio-economic determinants like dynamics of the marketplace, policy or regulatory agendas, and availability of all forms of resources that essentially influence the sustainability of GM techniques. Sustainability demands a holistic approach considering social, economic, and environmental impacts of GMOs besides their technical problems (Ndudzo et al., 2024).

3.3 *CRISPR/Cas9 Technology*

CRISPR-mediated genome editing technology is a highly applied technique which permanently alters an organism's genome in a heritable way. The usage of CRISPR technology has become African small-holder farmers deemed the paramount approach for breeding climate-resilient crops for developing economies of the continent. CRISPR technology is more accurate than established methods of plant improvement, both faster, simpler, and less expensive. It improves crop quality through nutrient intake, resilient toward pests/diseases, and tolerance to water scarcity or drought conditions. CRISPR has been utilize to edit the genomes of numerous agricultural crops comprising grains, vegetables, and fruits by merely creating modifications at particular genetic loci. One of the advantages of CRISPR-Cas9 is that it generates more favorable nutrient profiles as amount of starch, protein, vitamin, and oleic acid in cereal crops is improved. Examples of these include rice, maize, and wheat cereal crops which have improved sweetness, shelf life, and scent with the usage of CRISPR-Cas9 technology (Kumar et al., 2023).

3.3.1 Overview of CRISPR

Several CRISPR-Cas proteins include fCas9, Cas12, Cas13a, and Cas13b which have recently opened new avenues for scientists involved in crop upgrading. CRISPR Cas systems have been engineered to edit eukaryotic DNA by designing a guide RNA, gRNA complementary to the marks sequence. This gRNA harbors a protospacer motif of about 20 base pairs in length and is flanked by the respective homologous sequences particular to interest site. In this CRISPR-Cas9 system, the Cas9 protein binds to this motif in the gRNA, and thereafter, the protein-DNA complex that results from binding to this motif moves forward to interact with the target site (Mekonnen et al., 2022). Furthermore, Cas9 contacts a proximal element in the host DNA, commonly referred to as PAM for protospacer adjacent motif, which leads to a DSB in DNA, three nucleotides upstream of the PAM. The DNA healing apparatus in the host cell will then heal the cut by using either HDR or NHEJ. The NHEJ mending may involve gene function knockout and knockdown via small deletions, whereas HDR repair can result in precise alterations of DNA sequence as gene knock-in. CRISPR-Cas systems can also induce epigenetic alterations of DNA sequences either by the modulation of chromatin structure or base editing thus bypassing the requirement to form a cut in the ds DNA. In return, it has made CRISPR-Cas technology the single most popularly used tool in genome engineering efforts (Ndudzo et al., 2024).

3.3.2 CRISPR-Cas9 in Abiotic Stress Management

Crop harvests drop due to climate change demanding the more forest land for farming to meet the food demand of rising population. The most important abiotic stress of soil is salinity that inhibits plant growth (Yadav et al., 2023). The development of salt-affected arable lands has been pinpointed noteworthy threats to world food security. In total, soil salinization damages approximately 6% of arable land globally and continues to increase by yet another 1 to 2% each year. Currently, crop production, particularly for maize, rice, and wheat, has been badly affected. For the most part, smallholder farmers can do nothing but irrigate because of drought-induced water shortages, which in turn increase soil salinity. While conventional breeding approaches have been used in the past to enhance crop production, the latter is increasingly in need of more efficient and streamlined crop development approaches. For example, lines of rice were developed using CRISPR technology, showing increases in yield of 11–68% (El-Sabagh et al., 2021) (Table 1).

Table 1 Utilization of CRISPR/Cas in prominent agricultural species

Crops	Target Gene/ Pathway	Objectives	Outcome	Reference
Wheat	TaHDT1	Increase in yield of grain	Yield enhancement in saline soils	(Mishra et al., 2024)
Rice	OsSWEET13	Resistance to bacterial blight disease	Reduction in % age disease incidence	(Rai et al., 2024)
Tomato	SlMYB12	Improve lycopene content	Increased nutritional value in terms of lycopene content	(Bhanu, 2023)
Maize	ZmCCT9	Decrease photoperiod sensitivity	Extended growing season adaptability	(Ahmad et al., 2024)
Potato	StGBSS	Modified starch biosynthesis	Enhanced food processing quality	(Rai et al., 2024)
Wheat	GBSSI	Enhanced resistant starch	Increased resistant starch	(Irshad et al., 2022)
Wheat	alfAFP	Disease resistance	Enhanced resistance against wheat stripe rust	(Zhang et al., 2013)

4 Root Phenotyping and Soil-Microbe Interactions

4.1 Root Phenotyping

Most breeding efforts concerning crops focus on aboveground expression and aim to increase yields and disease resistance and reduce the height of grains, enhance nutritional quality. Efforts toward these goals by roots have been largely ignored. Succeeding advancement in plant breeding programs must include roots because of the high demand for producing more food, feed, fiber, and fuel using less land and resources. Root water uptake is a significant hydrological process controlling photosynthesis and carbon cycling and is used for the estimation of gross primary productivity of ecosystems where experimental measurements have proven infeasible.

Root characters represent an almost unused source of phenotypic difference that will be crucial for the breeders working on the proliferation of yields. The structure and the configuration of different metabolites introduced by the roots in the rhizosphere are dynamic and change with plant development and in reaction to environmental, biotic, and edaphic factors. Assessing the physical properties of root system structural design will enable breeding for desirable root settlement according to the soil profile, such as deeper roots in drought-stricken non-till production systems or shallower root systems for better access to nutrients. Integration of physical attributes with chemical characteristics like enzymes and organic acid production will give knowledge about biogeochemical processes by which the roots obtain resources. Finally, structural and elemental root composition information can enhance predictions of root disintegration, their contribution organic carbon pool in soil, and the subsequent services reaped by the subsequent crop. Breeding can no longer target aboveground traits in narrow terms, and breeding for below-ground traits cannot be narrowed to only root system design. The integration of root bio-geochemical traits into breeding will make it possible for the germplasm with the needed traits to fulfil the demand of production on a wide variation of soil types and expected climate situations (Huang et al., 2023; McGrail et al., 2020).

4.2 Soil-Microbe Interactions

In the last decades, much work has focused on the interaction between plants and microbes in the rhizosphere which is the small soil zone contact with plant roots. It is taken as a hotspot for these interactions and is a major target area for improving the plant's efficacy for nutrient uptake. In this respect, it may be the major challenge to determine microbial activities for enhancing nutrient use efficiency in a sustainable agriculture system although usage of different methods like metabolomics and transcriptomic. It has been possible to characterize many mechanisms of interaction; still, much more remains largely unknown (Liao et al., 2023).

The significance of each partner in such types of interactions (soil-microbe-plant) is crucial for the system to function optimally, whereas, in stress, further actions are required to sustain maximum working. The growth promoter's microbes have different tools for alleviating the stress with either abiotic or biotic origin in agriculture. This may offer a conducive environment for attaining higher and safer plant production through the integration of proper plant variety, favorable soil conditions, and appropriate microbial populations (Nadarajah, 2021).

Microorganisms that thrive in a given stress circumstance may be helpful to plants since microbes are demonstrated to enhance soil productivity and fertility. These helpful microorganisms can be introduced into the soil or plant to enhance production and productivity and reduce the adverse effects of agrochemicals (Nadarajah, 2021). It is a practical substitute for chemical treatment and facilitates plant development, controls pests and diseases, and stabilizes soil structure being applied in the form of biocontrol agents, biopesticides, bioherbicides, and biofertilizers. In the past, there have been tremendous developments in the manufacturing, advertising, and use of these inoculants. These days, inoculants have gained widespread acceptance because of the excellent multifunctional strains that increase yield and are cheaper than artificial fertilizers. Rhizobia are the most commonly used microbes as inoculants and their symbiosis with legumes determines the mechanism of BNF that fulfils the N requirement of the plant. Plant growth-promoting bacteria (PGPB) would help the plant alone or in tandem with other factors for proper growth in various areas (Mohamed et al., 2021; Patel et al., 2024).

These are synthesized phytohormones and siderophores, phosphorus solubilization, and activation of an internal defense system in a plant to combat biotic and abiotic stressors.

Recent advances in new genome-based technologies enable further characterization of the genome, diversity and proteome of microorganisms living in association with plants or in soil. DNA/RNA analysis, proteome, transcriptome, metagenome, and other omics-based technologies allow us to investigate beneficial/ non-beneficial plant–microbe relations in deep and speed. These technologies will facilitate understanding the dynamics of systems and apply that information to enhance growth, yield, and disease reduction. If we can interpret what creates the microbial communities in the rhizosphere, it will enable us to design good, stable, and productive ecosystems for long periods (Benitez et al., 2021).

5 Precision Breeding and Phenomics Approaches

5.1 Precision Breeding

There is a critical necessity to transform agricultural production supporting long-term food security. The transformation needs to be crop-by-crop, for crops that either tolerate or are productive in low-input and low-fertility surroundings (Kalele et al.,

2021). Needed are gains across several crops. These new crops modified to reduced input systems should be sustainable and worldwide but preferably curative to the surroundings termed an "evergreen revolution" (Sarangthem et al., 2023). The secret of breeding these crops lies in the exploitation of yield genetic variation identified in precise target environments and an appreciation of how that yield variation intermingles with management, the environment, and a variety of yield-reducing stresses. The least research has been done on the traits accountable for yield, concerning the diversity in traits of roots, despite the insight that roots comprise the highest unexploited potential to convey forthcoming sustainable crops (Batungwanayo et al., 2023). Analysis of beneficial genotypes in vivo is critical in translating a genetic understanding of traits into sustainable production. This research can now benefit fully from the enormous wealth of genetic resources and information amassed over the past few years in model plants and crops (Kumar et al., 2023). In addition to growing stockpiles of genetic sequence information for the most important cereals such as wheat, rice, maize, and barley there is growing knowledge of SNPs. Further progress in high-throughput sequencing and methods for quantifying sequence polymorphisms along with falling costs for assays imply that high-resolution genotypic data can quickly and inexpensively be produced for most populations of targeted species and can also easily and rationally be extended to "orphan crops" or species for which very little has been done on the genetics. (Zafar et al., 2020). However, swift development has been made in genetics, and advancement has been made in relating these polymorphisms to plant traits.

Genetic engineering has significantly streamlined the developmental procedure of new and upgraded varieties with improved agronomic traits, such as disease resistance, tolerance to abiotic stress, enhanced shelf life, and value-added crop yield (Nerkar et al., 2022). The latest development of new breeding technologies such as genome editing has unlocked new doors for the accurate alteration of plant genomes deprived of introducing foreign DNA (Fiaz et al., 2022). Therefore, utilization of these technologies is urgently needed for the development of novel and better-quality crop lines/varieties to overwhelmed the troubles challenged in traditional breeding programs (Vu et al., 2023).

5.2 *Phenomics Approaches*

Plants will resist drought stresses through some morphological variations, like extremely established roots in some plants, and physiological alteration as osmotic regulation. Drought stress may affect the plant morphologically, including stomatal shutting and leaf drooping, while it can be physiological like difficulty in water absorption by the root, the decline of photosynthetic rate, and the change in the rate of respiration. The reaction of plants towards stress is quite diverse: some are sensitive to stress, while others are resistant. A general classification may be made in between plant resistance mechanisms based on the two forms: either by avoiding the stress or by withstanding it; adaptation or domestication of the plant could improve resistance. Durable evolution has allowed crops to formulate a set of physiological and biochemical tools that could aid them resist adversity from the surrounding hostile external environment including osmotic regulation, cell structure, protection from dehydration, antioxidant defense system, and metabolic pathways that are essential tools of plants to fight against drought (Seleiman et al., 2021). The higher plants have intricate retorts and adaptableness to abiotic stresses like drought to sustain their ideal growth under stresses. For those complicated physiological responses of the plants, a kind of diverse regulation mechanism at cellular and molecular level is needed by preventing water loss by transpiration. (Shinozaki, 2020; Nadarajah, 2021).

6 Integrating Traditional and Modern Techniques

6.1 Traditional Breeding

The growth of organic agriculture has increased the requirement for crop germplasm that are particularly well adjusted to organic farming systems. Most of the crops at present cultured in organic agriculture were developed from conventional breeding programs. Nevertheless, the future crop production was not satisfactory to the wants, thus there is a need for a new agriculture model, which incorporates integrated structures of up-to-date molecular breeding, multiple agronomic methods, and studying plant–microbe interactions. Thus, climate-smart agriculture is gaining relevance in the development of climate-resilient crops by using next-generation breeding approaches that can confront different types of stresses and make these crops an important element regarding food and nutritional safety (Naqvi et al., 2022).

Improved plant performance in drought conditions and direct selection over multiple sites have conventionally been used for yield. Small heritability and highly influenced G $\times$ E contact result from the dissimilarities owing to the heterogeneity of soil and other environmental factors. Indirect selection through secondary qualities has been successful in limited cases due to difficulties related to repeatability and lacking extraordinary phenotyping strategies. In recent years, significant efforts to find the physiological traits related to yield and drought resistance have been noticed. These integrated approaches based on molecular breeding and genetic engineering now suggest prospects for yield improvement under drought stress with many crops, considering the accessibility of whole genome sequences, genetics, physical maps, and useful genomics implements. Documentation of the genetic bases of main physiological characteristics and cloning of the genes maintaining yield in drought environments should shift the selection to a “breeding by design” tactic that will gather a growing number of valuable traits into exclusive genotypes, which will eventually cause a decrease in the gap concerning the potential of yield and actual yield (Varshney et al., 2021).

6.2 Hybrid Approaches

Traditional plant breeding practices make up the core of agriculture through superior crop varieties with desirable features. Genetic engineering or biotechnology relies on a natural genetic diversity that exists within a species for the selection and crossing of desirable traits. Main principles of traditional plant breeding include selection, hybridization, inbreeding, and backcrossing. These can be employed by plant breeders in identifying and propagating plants of better quality, cross combinations of complementary traits to produce hybrids, inbreeding desirable traits to fix them, and introducing specified traits through backcrossing. Traditional breeding relies heavily on phenotypic selection, where physically observable traits are assessed to produce wise selections. Among community-based methods are mass and pure line selections, through which the breeding process can form lines that are genetically uniform or multiply seeds from many individuals possessing desired characteristics. Pedigree selection, in which decisions are based on detailed recording of genetic ancestry, helps breeding programs to make rational decisions. Field testing and evaluation are also required to establish how well the plants can respond in a range of environmental situations as well as in the presence of potential pests and diseases. This approach, capitalizing on nature's diversity, provides sustainable ground for agricultural practices and combines with contemporary biotechnology techniques for generating hardy and productive crops. They will help make a sustainable and safe future for agriculture through proper knowledge and application of the methods (Varshney et al., 2021).

7 Conclusion

Advanced crop breeding technologies are transforming crops' ability to adapt to diverse soil conditions. Tools like genomics, genetic engineering, and advanced phenotyping empower breeders to tackle challenges posed by varying and changing soil environments. Ongoing improvements in these techniques are expected to strengthen crops, increase productivity, and significantly contribute to global food security and environmental sustainability.

References

Ahmad, K. W., & Wang, G. (2023). Evaluating the Crucial Relationships between Soil Health and Climate Change. *Journal of Environmental Impact and Management Policy (JEIMP) ISSN: 2799–113X*, *4*(01), 8–21.

Ahmad, A., Abd-Elsalam, K. A., Ali, M., Majeed, H. N., Arif, A., Munir, A., & Munawar, N. (2024). Regulatory triggers of CRISPR-edited crops. In *Global Regulatory Outlook for CRISPRized Plants* (pp. 91–112). Academic Press.

Ayanoğlu, F. B., Elçin, A. E., & Elçin, Y. M. (2020). Bioethical issues in genome editing by CRISPR-Cas9 technology. *Turkish Journal of Biology, 44*(2), 110–120.

Batungwanayo, P., Habarugira, V., Vanclooster, M., Ndimubandi, J., F. Koropitan, A., & Nkurunziza, J. D. D. (2023). Confronting climate change and livelihood: smallholder farmers' perceptions and adaptation strategies in northeastern Burundi. *Regional Environmental Change, 23*(1), 47

Benitez, M. S., Ewing, P. M., Osborne, S. L., & Lehman, R. M. (2021). Rhizosphere microbial communities explain positive effects of diverse crop rotations on maize and soybean performance. *Soil Biology and Biochemistry, 159*, Article 108309.

Benitez-Alfonso, Y., Soanes, B. K., Zimba, S., Sinanaj, B., German, L., Sharma, V., & Foyer, C. H. (2023). Enhancing climate change resilience in agricultural crops. *Current Biology, 33*(23), R1246–R1261.

Beroueg, A., Lecompte, F., Mollier, A., & Pagès, L. (2021). Genetic variation in root architectural traits in Lactuca and their roles in increasing phosphorus-use-efficiency in response to low phosphorus availability. *Frontiers in Plant Science, 12*, Article 658321.

Dima, O., Custers, R., De Veirman, L., & Inzé, D. (2023). EU legal proposal for genome-edited crops hints at a science-based approach. *Trends in Plant Science, 28*(12), 1350–1353.

Fiaz, N., Zahoor, I., Saima, S., & Basheer, A. (2022). Genomic landscape of alpha-variant of SARS-CoV-2 circulated in Pakistan. *PLoS ONE, 17*(12), Article e0276171.

Ghimire, R., Thapa, V. R., Acosta-Martinez, V., Schipanski, M., Slaughter, L. C., Fonte, S. J., & Noble Strohm, T. (2023). Soil health assessment and management framework for water-limited environments: Examples from the Great Plains of the USA. *Soil Systems, 7*(1), 22.

Gidamo, G. H. (2022). Genomic selection: A faster strategy for plant breeding. In *Case studies of breeding strategies in major plant species*. IntechOpen.

Huang, Z., Zhang, X., Ashton, R. W., Hawkesford, M. J., & Whalley, W. R. (2023). Root phenotyping and root water uptake calculation using soil water contents measured in a winter wheat field. *Agricultural Water Management, 290*, Article 108607.

Imtiaz, S., Shahid, S., Ishfaq, T., Ilyas, M., Nawaz, A. F., Shamshad, J., ... & Arif, M. (2024). Impact of Climate change on agriculture. In *Environment, Climate, Plant and Vegetation Growth* (pp. 285–305). Cham: Springer Nature Switzerland

Irshad, A., Guo, H., Ur Rehman, S., Gu, J., Wang, C., Xiong, H., & Liu, L. (2022). Screening of induced mutants led to the identification of starch biosynthetic genes associated with improved resistant starch in wheat. *International Journal of Molecular Sciences, 23*(18), 10741.

Kalele, D. N., Ogara, W. O., Oludhe, C., & Onono, J. O. (2021). Climate change impacts and relevance of smallholder farmers' response in arid and semi-arid lands in Kenya. *Scientific African, 12*, Article e00814.

Kumar, M., Prusty, M. R., Pandey, M. K., Singh, P. K., Bohra, A., Guo, B., & Varshney, R. K. (2023). Application of CRISPR/Cas9-mediated gene editing for abiotic stress management in crop plants. *Frontiers in Plant Science, 14*, 1157678.

Kumar, T., Wang, J. G., Xu, C. H., Lu, X., Mao, J., Lin, X. Q., & Liu, H. B. (2024). Genetic engineering for enhancing sugarcane tolerance to biotic and abiotic stresses. *Plants, 13*(13), 1739.

Lebedev, V. G., Popova, A. A., & Shestibratov, K. A. (2021). Genetic engineering and genome editing for improving nitrogen use efficiency in plants. *Cells, 10*(12), 3303.

Liao, Z., Fan, J., Lai, Z., Bai, Z., Wang, H., Cheng, M., & Li, Z. (2023). Response network and regulatory measures of plant-soil-rhizosphere environment to drought stress. *Advances in Agronomy, 180*, 93–196.

Mamatha Bhanu, L. S. (2023, November). Clustered Regularly Interspaced Short Palindromic Repeats-Cas: A Potential Genome Editing Tool in Crop Improvement. In *Biology and Life Sciences Forum* (Vol. 27, No. 1, p. 48). MDPI

McGrail, R. K., Van Sanford, D. A., & McNear, D. H., Jr. (2020). Trait-based root phenotyping as a necessary tool for crop selection and improvement. *Agronomy, 10*(9), 1328.

Mekonnen, T. W., Gerrano, A. S., Mbuma, N. W., & Labuschagne, M. T. (2022). Breeding of vegetable cowpea for nutrition and climate resilience in Sub-Saharan Africa: Progress, opportunities, and challenges. *Plants, 11*(12), 1583.

Mirzabaev, A., Strokov, A., & Krasilnikov, P. (2023). The impact of land degradation on agricultural profits and implications for poverty reduction in Central Asia. *Land Use Policy, 126*, Article 106530.

Mishra, S., Nayak, S., Tuteja, N., Poosapati, S., Swain, D. M., & Sahoo, R. K. (2024). CRISPR/Cas-Mediated Genome Engineering in Plants: Application and Prospectives. *Plants, 13*(14), 1884.

Mmbando, S. G., & Ngongolo, K. (2024). The recent genetic modification techniques for improve soil conservation, nutrient uptake and utilization. *GM Crops & Food, 15*(1), 233–247.

Mohamed, H. I., Sofy, M. R., Almoneafy, A. A., Abdelhamid, M. T., Basit, A., Sofy, A. R., ... & Abou-El-Enain, M. M. (2021). Role of microorganisms in managing soil fertility and plant nutrition in sustainable agriculture. *Plant growth-promoting microbes for sustainable biotic and abiotic stress management*, 93–114

Nadarajah, K., & Abdul Rahman, N. S. N. (2021). Plant–microbe interaction: Aboveground to belowground, from the good to the bad. *International Journal of Molecular Sciences, 22*(19), 10388.

Naqvi, R. Z., Siddiqui, H. A., Mahmood, M. A., Najeebullah, S., Ehsan, A., Azhar, M., & Asif, M. (2022). Smart breeding approaches in post-genomics era for developing climate-resilient food crops. *Frontiers in Plant Science, 13*, Article 972164.

Ndudzo, A., Makuvise, A. S., Moyo, S., & Bobo, E. D. (2024). CRISPR-Cas9 genome editing in crop breeding for climate change resilience: Implications for smallholder farmers in Africa. *Journal of Agriculture and Food Research*, 101132

Nerkar, G., Devarumath, S., Purankar, M., Kumar, A., Valarmathi, R., Devarumath, R., & Appunu, C. (2022). Advances in crop breeding through precision genome editing. *Frontiers in Genetics, 13*, 880195.

Nigussie, A. B., Ayeled, G. T., Endalew, A., Miheretu, B. A., Amognehegn, A. E., Adamu, A. Y., & Karuppannan, S. (2025). Integrating google earth engine and GIS for RUSLE-based soil erosion and sediment yield assessment in Borkena Watershed, Ethiopia. *Journal of Sedimentary Environments*, 1–23.

Parmar, N., Singh, K. H., Sharma, D., Singh, L., Kumar, P., Nanjundan, J., & Thakur, A. K. (2017). Genetic engineering strategies for biotic and abiotic stress tolerance and quality enhancement in horticultural crops: a comprehensive review. *3 Biotech, 7*, 1–35

Patel, M., Islam, S., Glick, B. R., Vimal, S. R., Bhor, S. A., Bernardi, M., & de los Santos Villalobos, S. (2024). Elaborating the Multifarious Role of PGPB for Sustainable Food Security under Changing Climate Conditions. *Microbiological Research*, 127895

Patil, A., & Lamnganbi, M. (2018). Impact of climate change on soil health: A review. *Int. J. Chem. Stud, 6*(3), 2399–2404.

Rai, N., Singh, N. A., Goyal, J., Upadhayay, P., Jain, V. (2024). CRISPR-Cas and Its Applications in Food Production. In *Gene Editing in Plants: CRISPR-Cas and Its Applications* (pp. 349–391). Singapore: Springer Nature Singapore

Razzaq, A., Kaur, P., Akhter, N., Wani, S. H., & Saleem, F. (2021). Next-generation breeding strategies for climate-ready crops. *Frontiers in Plant Science, 12*, Article 620420.

El Sabagh, A., Islam, M. S., Skalicky, M., Ali Raza, M., Singh, K., Anwar Hossain, M., ... & Arshad, A. (2021). Salinity stress in wheat (Triticum aestivum L.) in the changing climate: Adaptation and management strategies. *Frontiers in Agronomy, 3*, 661932

Sarangthem, I., Haldhar, S. M., Mishra, L. K., & Thakuria, D. (2023). The book of abstract: international conference on natural farming for revitalizing environment and resilient agriculture (NF-RERA, 2023). *Pub: College of Agriculture, CAU, Imphal, 374*

Seleiman, M. F., Al-Suhaibani, N., Ali, N., Akmal, M., Alotaibi, M., Refay, Y., & Battaglia, M. L. (2021). Drought stress impacts on plants and different approaches to alleviate its adverse effects. *Plants, 10*(2), 259.

Shinozaki, Y., Beauvoit, B. P., Takahara, M., Hao, S., Ezura, K., Andrieu, M. H., & Ariizumi, T. (2020). Fruit setting rewires central metabolism via gibberellin cascades. *Proceedings of the National Academy of Sciences, 117*(38), 23970–23981.

Varshney, R. K., Bohra, A., Yu, J., Graner, A., Zhang, Q., & Sorrells, M. E. (2021). Designing future crops: Genomics-assisted breeding comes of age. *Trends in Plant Science, 26*(6), 631–649.

Vasanthrao, T., Bandela, C., Yathish, K., & Vinay, I. (2023). Marker-assisted selection. *Genes Traits Mol Plant Breed, 31.*

Villalobos-López, M. A., Arroyo-Becerra, A., Quintero-Jiménez, A., & Iturriaga, G. (2022). Biotechnological advances to improve abiotic stress tolerance in crops. *International Journal of Molecular Sciences, 23*(19), 12053.

Vu, B. N., Vu, T. V., Yoo, J. Y., Nguyen, N. T., Ko, K. S., Kim, J. Y., & Lee, K. O. (2023). CRISPR-Cas-mediated unfolded protein response control for enhancing plant stress resistance. *Frontiers in Plant Science, 14*, 1271368.

Yadav, R. K., Tripathi, M. K., Tiwari, S., Tripathi, N., Asati, R., Chauhan, S., ... & Payasi, D. K. (2023). Genome editing and improvement of abiotic stress tolerance in crop plants. *Life, 13*(7), 1456.

Zafar, K., Sedeek, K. E., Rao, G. S., Khan, M. Z., Amin, I., Kamel, R., & Mahfouz, M. M. (2020). Genome editing technologies for rice improvement: Progress, prospects, and safety concerns. *Frontiers in Genome Editing, 2*, 5.

Zhang, L., Wang, J. F., Wang, X. J., Kang, Z. S., & Han, D. J. (2013). Wheat transformation of resistant genes to diseases using pollen-tube pathway. *Journal of Triticeae Crops, 33*, 29–33.

Crop Rotation; Cover Cropping & Soil Management-Sustainable Agricultural Practices

Kousar Parveen, Saima Nasreen, Saira Mansab, Ishrat Jamil, Rabail Urooj, and Afsar Khan

Abstract Nutrient-rich soil is essential for better agricultural productivity. With the rapid growth of the human population, the food demand has doubled. To fulfill the demand for food, intense farming is practiced worldwide. This Intense farming activity leads to nutrient depletion, land degradation, loss of soil biodiversity, application of agrochemicals, soil erosion, soil compaction, and waterlogging problems. In the current scenario, soil protection and conservation are essential to increase crop production and meet global food demand. Crop rotation, crop cover, and soil management are the foremost practices of sustainable agriculture. Sustainable agriculture can be defined as a method of agriculture that aims at fulfilling the present food requirements and at the same time conserving resources for the needs of future generations. It attempts to bring harmony among ecological sustainability, economic viability, and social equity. This chapter gives insight into the historical perspective, different crop species, benefits, and ecological roles. Thus, it shows how crop rotation, crop cover, and soil management are important for maintaining soil fertility.

Keywords Soil · Crop rotation · Crop cover · Soil management · Sustainable agriculture

K. Parveen (✉) · S. Nasreen · S. Mansab
Department of Environmental Sciences, The Women University Multan, Multan, Pakistan
e-mail: kousar.parveen@wum.edu.pk

I. Jamil
Department of Chemistry, University of Baltistan Skardu, Gilgit Baltistan, Pakistan

R. Urooj
Department of Environmental Sciences, Sardar Bahadur Khan Women's University, Quetta, Pakistan

A. Khan
School of Chemical and Environmental Engineering, Anhui Polytechnic University, Wuhu, China

M. Shaaban (ed.), *Soils and Sustainable Agriculture*, Frontier Studies in Soil Science,
https://doi.org/10.1007/978-3-031-91114-9_27

1 Introduction

Global food demand has increased due to the rapid growth in human population. To fulfill the demand for food, intense farming practices worldwide. Intense farming resulted in nutrient depletion, soil degradation, disturbed soil biodiversity, soil erosion, soil compaction, soil toxicity, etc. Soil protection and conservation are of utmost importance to maintain soil health and better agriculture productivity that help to meet global food security. Different conventional agricultural methods such as excessive use of synthetic agrochemicals, high-yielding cultivars, monoculture, and formation of transgenic crops, etc. have been practiced for a long time to increase agricultural productivity. However, these conventional practices have more consequences than its benefits. That's why there is a need to shift from conventional practices to sustainable practices which have economic, social, and environmental benefits. These sustainable practices are grouped under the umbrella of the word "Sustainable agriculture". Sustainable agriculture integrates biological, chemical, physical, ecological, economic, and social sciences comprehensively to develop new farming practices that are safe and sound for our environment. In 1988, the Food and Agriculture Organization defined sustainable agriculture as the management and conservation of the natural resource base, and the orientation of technological and institutional change in such a manner as to ensure the attainment and continued satisfaction of human needs for present and future generations (Ignaciuk et al., 2021). Crop rotation, crop cover, and soil management are the practices of sustainable agriculture for the conservation and protection of soil.

1.1 Principles of Sustainable Agriculture

Sustainable agriculture works on the following principles (Trigo et al., 2021):

- **Soil Conservation:** Soil is an important component of our ecosystem. It plays a different significant ecological role such as a habitat for living organisms, providing support for plant anchorage, and an important medium for plant nutrients, etc. Application of natural fertilizers (compost, animal, and green manure) helps to maintain soil fertility instead of using chemical fertilizers.
- **Water conservation:** Agriculture requires the same level of clean water as is required in real life. Local or native plant species are preferred to grow because they are compatible with the local environment and require less water in comparison to exotic plant species that require special soil conditioning. Furthermore, fertile soil (rich in soil organic matter) retains the soil moisture and reduces water loss. The adoption of different techniques such as drip irrigation, sprinkling water, terracing, rainwater harvesting, and contour farming helps in the conservation of water resources.
- **Disease Prevention:** Plants grown in natural fertilizer build resistance to different diseases. It encourages the beneficial insects and birds that prey on pests or insects,

instead of using pesticides and insecticide formulations. The practice of crop covering and crop rotation breaks the lifecycle of pests/insects and reduces the spread of plant diseases.

- **Diversification:** Growing a variety of crops in the same field promotes an ecological balance, enhances resilience, and ensures long-term agricultural productivity. Different types of crops require different nutrients from soil. The rotation of crops balances the depletion and replenishment of soil nutrients. It also focuses on producing a variety of crops that fulfill the food requirements of local communities, reducing dependence on single global commodities. Diversification also provides essential ecological service i.e., pollination that increases and preserves genetic diversity.
- **Social Equity:** it provides equitable food resources, fair income, and labour to all the local communities. It helps to engage all main stakeholders in one platform such as producers, and consumers.

Hence, adopting these key principles of sustainable agriculture enhanced crop productivity, soil health, and social well-being.

1.2 Crop Rotation

Crop rotation is one of the sustainable agriculture practices in which different crops are planted in a sequence in a field (Huang et al., 2001; Sumner, 2018). It increases the soil fertility and enhances crop productivity (Boincean et al., 2019). The evidence of crop rotation is evident in ancient civilizations such as the Han Dynasty of China, Roman civilization, etc. At that time, they practiced two field crop rotation systems where leguminous crops (such as beans, peas, and lentils) and grains crops (such as wheat, barley, and rye) grew in alternate ways. With time, the three-field system became widespread in medieval Europe, after Roman departure. In a three-field system, the fields were divided into three sections, each planted with a different crop with winter wheat or rye, spring oats or barley, and a fallow field that was left unplanted. The crops were rotated each year so that the field that had been left fallow in the previous year would be planted with a crop in the following year. During the fallow period, farmers would often allow their sheep or cattle to graze on the field and their manure would fertilize the soil simultaneously, providing nutrients that would help to improve the yield of the following year's crop. Then Norfolk four-crop rotation was introduced by the British Agricultural Revolution (1750 -1880). This four-crop rotation replaced two-field and three–field crop rotation systems. In this system, turnip, wheat, clover, and barley were grown in four parts in the field. Later on, peas or beans, and potatoes started to be used in place of clover and turnip (Ralls, 2013; Tariq et al., 2019).

Fig. 1 Ecological services of crop rotation (Yang et al., 2024)

1.2.1 Ecological Role of Crop Rotation

It maintains soil fertility, improves soil structure, enhances soil organic matter, reduces topsoil erosion, improves soil biodiversity, reduces the spread of pests, and controls weed germination as shown in Fig. 1.

1.2.2 Design of Crop Rotation

The crop rotation system is designed (Malik, 2010) as:

- Crops used in rotation should belong to different families
- Crops used in rotation should have well-defined root systems (deep and shallow)
- Crops used in rotation should have nutrient fixative, and uptake.

- Crops used in rotation should have the ability to disturb the pest and insect life cycle.

1.2.3 Types of Crop Rotation

Four different types of cop rotations are practiced nowadays based on soil nature, require nutrients, etc.

One-Year Crop Rotation

When crops are grown alternately in a year such as planting maize for a half year and then mustard for the rest of the year known as one-year crop rotation.

Two-Year Crop Rotation

When two different crops are grown alternately for two years such as planting maize followed by soybeans and then soybean followed the wheat.

Three-Year Crop Rotation

When three different crops are grown alternately for three years such as planting corn followed by soybeans and then wheat. The duration of rotation is completed in two years.

Four-Year Crop Rotation

When four different crops are grown alternately for four years such as planting corn followed by soybeans and then wheat followed by the fallow.

1.2.4 Crop Species in Crop Rotation System

Four different crops (legumes, leafy, fruiting, and root) are used in a crop rotation system because each crop shows different properties (see Fig. 2). Firstly, leguminous crops help in nitrogen fixation and carbon sequestration with the help of soil bacteria and fungi such as peas, beans, alfalfa, and clover, etc. Secondly, leafy crops uptake nitrogen as nitrates from the soil such as spinach, onion, and lettuce, etc. Thirdly, Fruiting crops require phosphorus from the soil for their growth and productivity such as cucumbers, tomatoes, and peppers, etc. Fourthly, root crops require potassium for better root production such as raddish, carrot, and beets, etc.

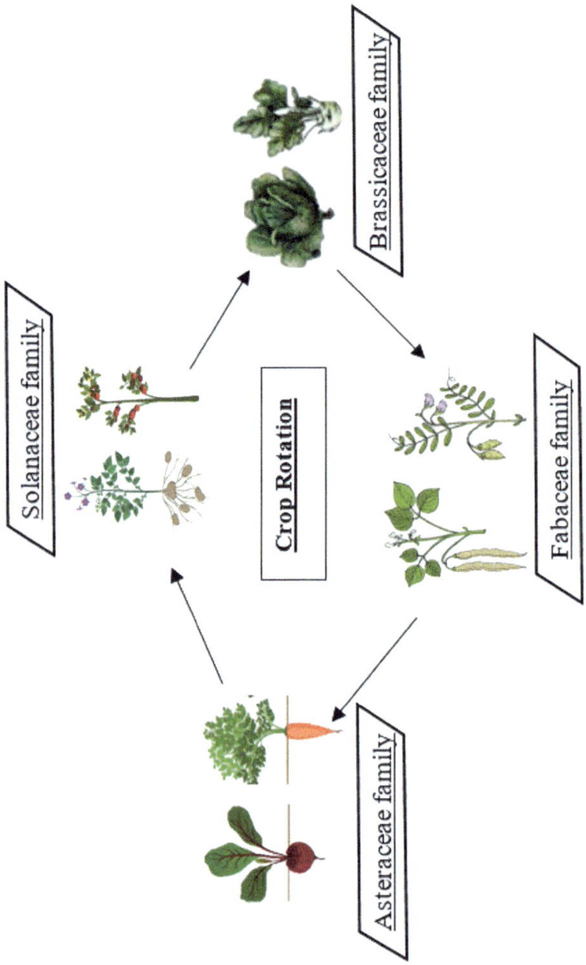

Fig. 2 Different crop families used in crop rotation system (Liz, 2020)

1.3 Cover Crops

Cover crops are those plant species that are grown purposely in on and off crop seasons to reduce soil erosion, soil compaction, and nutrient depletion while enhancing nitrogen fixation, and soil organic carbon content in the soil. Furthermore, it helps to decrease the soil temperature and increase the biological diversity with the control of weeds and pests. Cover crops have two categories.

- When the cover crop grows with cash crop for all or a portion of the growing season such as *Mucuna spp. and Crotalaria juncea L. Sorghum spp.* Known as on-season crop rotation.
- When the cover crop grows before planting the cash crop and is removed and left on the surface as residue or incorporated in the soil such as *Vicia villosa Roth, Secale cereale L.* known as off-season crop rotation.

These cover crops either release allelopathic chemicals that are toxic to weed germination (Scavo et al., 2019) or cut off the supply of temperature, light, and moisture for weed germination (Gerhards & Schapper., 2020). Other ecological role of cover crops is illustrated in Fig. 3.

1.3.1 Types of Cover Crops

There are three types of crop species (leguminous species, brassicaceous species, and gramineous species) that are commonly used as cover crops. They are very

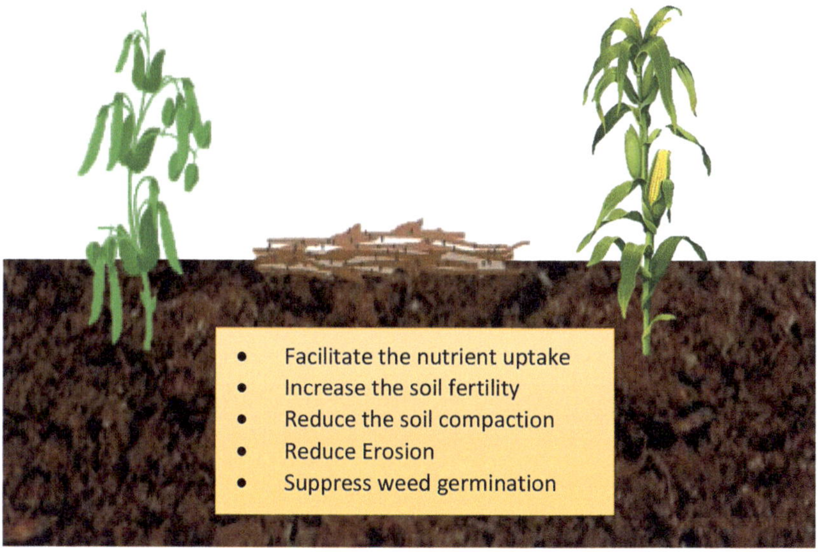

Fig. 3 Ecological services of cover crops (Quintarelli et al., 2022)

helpful in nitrogen fixation, reducing soil erosion, enhancing water retention, and weed suppression. Different plants fix nitrogen in different amounts in soil. The list of different types of plant species is listed in Table 1.

Gramineous Species

Grass has a fine, fibrous root system that grows quickly and is also removed easily by mechanical, and chemical methods (Ma et al., 2014). It also decays slowly and provides substantial soil cover that promotes the growth of earthworms, suppression of cotton aphids, Columbia lance, and root-knot, etc. Other ecological roles played by cover crops (buckwheat, flax, phacelia, radish, rapeseed, safflower, sunflower, and turnip) are illustrated in Fig. 4.

Leguminous Crops

Legume plants help in the nitrogen fixation from the atmosphere via a symbiotic relationship with bacteria in the root nodules (Koudahe et al., 2022). It includes clovers, cowpeas, faba bean, medics, peas, soybean, sunn hemp, and vetch. Different leguminous species fix different amounts of nitrogen in the soil.

Non-Leguminous Crop

Non-leguminous crop species have robust deep root structures that help reduce soil compaction, improve water infiltration, and improve soil aeration (Koudahe et al., 2022). It is easily decomposed in soil and releases nutrients in the soil. During decomposition, plants release glucosinolate compounds that suppress soil-borne diseases charcoal rot in soybeans, and act as natural pesticides. The list of cover crops with their common scientific names is listed in Table 2.

Table 1 Nitrogen fixation capability of different plant (Koudahe et al., 2022)

Plant Specie	Fixation of Nitrates and total Nitrogen (%)	Reference
Hairy vetch	80	Boselli et al., 2020
Pea	40	Zabtolowicz et al., 2011
Crimson Clover	67	Hubbard et al., 2013
Faba beans	14	Andersen et al.,

Table 2 List of cover crops with their common scientific names (Wofuru-Nyenke, 2023)

Types	Common name	Scientific name
Leguminous Species	Clovers Cowpea Faba bean Hairy vetch Lens culinaris	*Trifolium spp.* *Vigna unguiculata* *Vicia faba* *Vicia villosa* *Lens culinaris*
Non-leguminous	Buckwheat Flax Phacelia Raphanus sativus Rapeseed	*Fagopyrum esculentum* *Linum usitatissimum* *Phacelia tenacetifolia* *Raphanus sativus* *Brassica napus*
Grasses	Barley Forage sorghum Millet Oats Rye	*Hordeum vulgare* *Sorghum biocolor* *Pennisetum glaucum* *Avena sativa* *Secale cereal*

1.4 Soil Management

Soil forms the foundation of sustainable agriculture, serving as the cornerstone for food and energy production. It acts as a reservoir for water and nutrients essential for plant growth. The intense farming practices, the addition of fertilizers, and the use of heavy machinery disturb the soil health and its functioning. It caused problems such as compaction, pollution, soil erosion, nutrient depletion, and biodiversity loss. Thus, soil management is the utmost need of sustainable agriculture. It encompasses the practices employed to protect and conserve the soil for the proper function of the ecosystem. The key aspects of soil management (Cataldo et al., 2021) are given below:

1. **Natural fertilizers:** To maintain the soil nutrients and organic matter in the soil, the application of natural fertilizers such as compost, vermicomposting, green manure, and animal manure, etc. is essential for better crop productivity. The organic matter in soil not only improves soil fertility, but also helps in water retention, and improves biological activity.
2. **Soil Conservation Practices:** To reduce soil erosion and land degradation, different practices such as crop covering, crop rotation, terracing, no-till farming, contour farming, etc. need to be adopted.
3. **Mulching:** To reduce compaction, and maintain the soil structure, farmers should avoid the use of heavy machinery in the field and grow deep-rooted plants. On the other hand, the application of mulching helps to maintain soil moisture and prevent soil compaction.
4. **Pest Control:** To manage soil-borne pests and diseases, crop rotation, solarization, and promote the growth of biological organisms such as earthworms, fungi, and bacteria that attack harmful pests and enhance soil health.

5. **Monitoring of Soil Health:** To maintain soil quality, policies and regulations by government agencies should be implemented to keep check and balance on nutrient levels and help in making informed decisions about fertilizer application, irrigation, and crop planning.

Effective soil management practices ensure long-term agricultural productivity, reduce environmental impacts, and help maintain the ecological balance necessary for sustainable agriculture.

2 Conclusions

Crop rotation, crop cover, and soil management are sustainable agriculture practices that are essential to conserve soil fertility for the fulfillment of global food demand. Different plant species such as Legumes, leafy, fruiting, and grass are commonly used for crop rotation and cover crops. These plant species are helpful in nitrogen fixation, reducing soil erosion, enhancing water retention, and weed suppression and pest control. Thus enhancing the crop yield.

References

Boincean, B., Dent, D., Boincean, B., & Dent, D. (2019). Crop rotation. Farming the Black Earth: Sustainable and Climate-Smart Management of Chernozem Soils, 89–124.

Cataldo, E., Fucile, M., & Mattii, G. B. (2021). A review: Soil management, sustainable strategies and approaches to improve the quality of modern viticulture. *Agronomy, 11*(11), 2359.

Gerhards, R., & Schappert, A. (2020). Advancing cover cropping in temperate integrated weed management. *Pest Management Science, 76*, 42–46.

Ignaciuk, A., Ilicic, J., Asprooth, L., Sitko, N. J., Bernard, A., Maggio, G., & Mueller, M. (2021). Progress towards sustainable agriculture–Drivers of change.

Koudahe, K., Allen, S. C., & Djaman, K. (2022). Critical review of the impact of cover crops on soil properties. *International Soil and Water Conservation Research, 10*(3), 343–354.

Ma, J. F., Chen, Z. C., & Shen, R. F. (2014). Molecular mechanisms of Al tolerance in gramineous plants. *Plant and Soil, 381*, 1–12.

Malik, R. (2010, August). Soil quality benefits of break crops and/or crop rotations-a review. In 19th world congress of soil science, soil solutions for a changing world. Brisbane, Australia (pp. 95–98).

Quintarelli, V., Radicetti, E., Allevato, E., Stazi, S. R., Haider, G., Abideen, Z., & Mancinelli, R. (2022). Cover crops for sustainable cropping systems: A review. *Agriculture, 12*(12), 2076.

Ralls, K. M. (2013). Crop rotations have been around since Roman times. Central Oregon Agriculture Research and Extension Center.

Scavo, A., Abbate, C., Mauromicale, G. (2019) Plant allelochemicals: Agronomic, nutritional and ecological relevance in the soil system. *Plant Soil 442*, 23–48.

Sumner, D. R. (2018). Crop rotation and plant productivity. In: Handbook of agricultural productivity (pp. 273–314). CRC Press.

Tariq, M., Ali, H., Hussain, N., Nasim, W., Mubeen, M., Ahmad, S., & Hasanuzzaman, M. (2019). Fundamentals of crop rotation in agronomic management. Agronomic Crops: Volume 1: Production Technologies, 545–559.

Trigo, A., Marta-Costa, A., & Fragoso, R. (2021). Principles of sustainable agriculture: Defining standardized reference points. *Sustainability, 13*(8), 4086.

Wofuru-Nyenke, O. K. (2023). Mechanized cover crop farming: Modern methods, equipment and technologies. *Circular Agricultural Systems, 3*(1).

Yang, X., Xiong, J., Du, T., Ju, X., Gan, Y., Li, S., & Butterbach-Bahl, K. (2024). Diversifying crop rotation increases food production, reduces net greenhouse gas emissions and improves soil health. *Nature Communications, 15*(1), 198.

Liz Yugov, 2020. Crop rotation: Agriculture. https://psci.princeton.edu/tips/2020/3/21/agriculture-101. Retrieved on September 28th, 2024.

Role of Drought Tolerant Bacteria in Growth Promotion of Wheat Plant

Atia Iqbal, Shumaila Batool, Nimra Mushtaq, and Sidra Noureen

Abstract Drought is one of the significant environmental challenges, which impacts on global agriculture, particularly in staple crops like wheat (*Triticum aestivum*). Recent breakthroughs in plant–microbe interactions have underscored the potential of drought tolerant bacteria to augment wheat growth and resilience in water-limited environments. These advantageous bacteria, commonly called PGPRs, support plant survival by multiple processes, including osmotic adjustment, the synthesis of phytohormone (i.e. Auxin, ACC deaminase, Siderophore etc.), and the augmentation of nutrient absorption. Furthermore, drought tolerant bacteria can elicit systemic tolerance in plants by regulating stress-responsive pathways, encompassing antioxidant enzyme activities and the modulation of stress-related gene expression. This chapter explores the molecular and physiological mechanisms by which drought-tolerant bacteria facilitate growth promotion and drought resistance in wheat. This chapter incorporates recent research findings to examine the potential of drought tolerant bacteria as sustainable bio fertilizers, presenting a promising approach to enhance wheat yield in drought-prone areas. Ultimately, comprehending the symbiotic relationship between drought tolerant bacteria and wheat can facilitate novel agricultural strategies to mitigate the adverse impacts of climate change.

A. Iqbal (✉) · S. Batool · N. Mushtaq · S. Noureen
Department of Microbiology and Molecular Genetics, The Women University Multan, Multan, Pakistan
e-mail: atia.iqbal@wum.edu.pk

S. Batool
e-mail: shumaila.shakeel124@gmail.com

N. Mushtaq
e-mail: Nimramushtaq88@gmail.com

S. Noureen
e-mail: sidranoureen911@gmail.com

M. Shaaban (ed.), *Soils and Sustainable Agriculture*, Frontier Studies in Soil Science,
https://doi.org/10.1007/978-3-031-91114-9_28

1 Introduction

As global water scarcity tightens its grip on agriculture, the microscopic world beneath our feet may hold the key to unlocking the resilience of wheat in the face of drought. Global climate change trends indicate a prolonged fluctuation in temperature and precipitation patterns, resulting in increased warmth, environmental catastrophes including abiotic and biotic pressures, rising sea levels, and the melting of glacial caps (Nyaupane et al., 2024). Abiotic stressors are the primary constraints on yield in areas with limited water supply, particularly in dry and semi-arid climates. Drought stress constitutes a significant abiotic challenge confronting farmers globally due to climate change (Mansour et al., 2020). During drought, water resources diminish, and sources such as capillary water, groundwater, and irrigation water become limited, directly affecting agricultural productivity. The metabolic rate of plants typically decreases, impacting their development and growth (Fig. 1) (Kumar et al., 2024). Plants possess defense mechanisms that allow them to withstand both biotic and abiotic stressors. These mechanisms involve morphological alterations arising from complex physiological processes, rendering them difficult to comprehend (Goswami & Suresh, 2020). Droughts can range from mild and brief occurrences to severe, catastrophic, and extended episodes. This problem can severely impede agricultural food production on a broad scale. Estimates indicate that by 2050, 50% of the arable land necessary for agriculture will be negatively impacted by drought (Kumar et al., 2024). Drought stress represents a significant challenge for wheat production globally, impacting both agricultural yields and broader economic factors, including world GDP. Studies demonstrate that dryness can reduce wheat yields by 50–60%, especially in arid and semi-arid areas with restricted water availability (Sallam et al., 2019). This is crucial as wheat serves as a staple crop for around one-fifth of the global population, highlighting its significance for food security (Elnajar et al., 2024). Drought is thought to have decreased cereal production in the world by 9–10%, making it one of the main global food production constraints. According to study, wheat is farmed across an area of 220 million hectares globally. Wheat's yearly production reaches 778 million metric tons, contributing a 10% value addition to agriculture (Khan et al., 2020).The average annual productivity is 3.5 tons per hectare. It contains 1.8% fiber, 9.4% protein, 69% carbs, (Ahmad et al., 2022). It also significantly contributes to the animal feed and livestock industry (Nyaupane et al., 2024). The wheat crop is becoming increasingly important as the demand for food and nutrition rises in response to population growth (Solangi et al., 2021) for fulfill the world's present and future food requirements, production must expand in a sustainable manner. Before hunger and malnutrition could be eradicated, yield has to be increased by 50% to satisfy the world's population's food demands (Poudel et al., 2020).

It has resulted in decreased crop productivity, increased susceptibility to diseases and pests, and increased water scarcity. This can result in stunted growth, decreased leaf area, and premature senescence. Additionally, drought stress often exacerbates other environmental stresses like heat stress and salinity, further compromising wheat

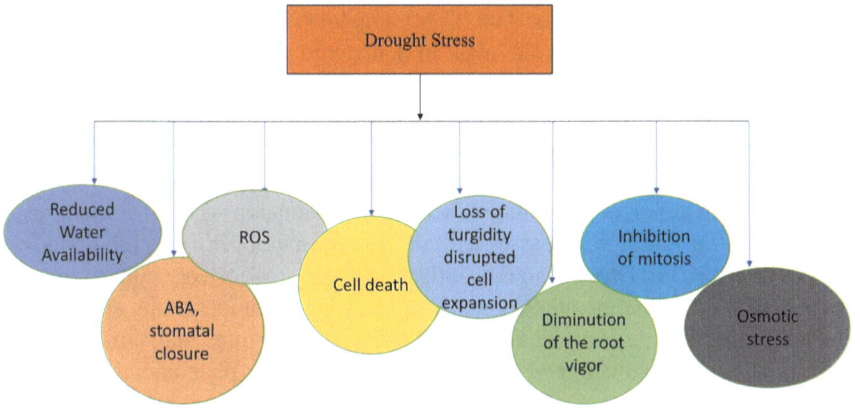

Fig. 1 Effects of drought stress on wheat plant

productivity (Nyaupane et al., 2024). Various mechanisms have evolved in wheat to cope with drought stress, including physiological adaptations such as stomatal regulation, osmotic mechanism and defense systems. However, the effectiveness of these mechanisms can be limited under severe and prolonged drought conditions (Nurhidayati et al., 2024). This has caused farmers to face economic losses and food security challenges, particularly in semiarid regions where rain-fed agriculture is predominant. To address the issue of drought stress in wheat and agriculture, researchers have been exploring various strategies to enhance drought tolerance in crops. It involves a combination of agronomic practices, breeding efforts, and technological innovations. These may include the use of drought-tolerant wheat varieties, improved irrigation techniques such as drip irrigation and rainwater harvesting, conservation tillage practices, the application of soil amendments to enhance water retention and application of drought tolerant plant growth promoting bacteria (Gupta & Pandey, 2023).

Numerous techniques utilized growth hormones and advantageous microbial strains have been formulated to mitigate the detrimental impacts of diverse stress conditions on plant health. PGPB are suggested to alleviate environmental challenges through two main mechanisms: plant water conservation and protection-recovery processes. As a result, there is heightened interest in transforming scientific information about drought-resistant bacteria into a sustainable solution for agroecosystems. At now, microbial inoculants are the most viable biological technology to meet plant development needs while providing protection against biotic and abiotic challenges.

2 Drought and Causes of Drought Stress

Alterations in rainfall patterns due to climate change, elevated atmospheric CO_2 levels, increasing temperatures, and hot, arid winds were the primary drivers of drought stress (Dai, 2013; Hossain et al., 2016; Nyaupane et al., 2024). The earth's temperature is rising at a rate of 0.06 °C, accompanied by an annual decrease precipitation (Paudel et al., 2020). The plant may undergo water stress despite adequate soil moisture, because to numerous edaphic conditions. Factors such as salt and soil temperatures impede the plant's roots. This state is termed faux drought or physiological drought (Arbona et al., 2013).

Drought is a naturally occurring complex phenomenon characterized by dry weather and insufficient precipitation, leading to a situation where the rate of evaporation and transpiration surpasses the precipitation rate over a specific period and area (Bianco, 2024). Drought significantly affects the economy, agriculture, environment, and society and its intensity and frequency has been increasing due to global warming and climate change, marked by rising temperatures and shifting precipitation patterns (Rahman et al., 2024). Extreme weather phenomena, aside from heatwaves and floods, are prevalent, intense, and frequent, varying by region and year. Drought stress arises from climate change-induced alterations in precipitation patterns, elevated atmospheric CO_2 concentrations, increasing temperatures, and hot, arid winds. The Earth's temperature is rising at a rate of 0.06 °C, accompanied by a yearly decrease in precipitation of 16.09 mm (Paul et al., 2022). The plant may undergo water stress despite adequate soil moisture, because to numerous edaphic conditions. These elements encompass salt, low soil temperatures, and flooding, which impede the absorption of water by the plant's roots. This state is termed faux drought or physiological drought. It triggers a series of physiological and biochemical adjustments in plants, including as diminished water absorption, stomatal closure, and modifications in hormonal equilibrium, all of which lead to compromised growth and development (Ngumbi & Kloepper, 2016).

3 Effect of Drought Stress on Wheat

Drought is a significant abiotic stress that threatens agricultural productivity globally (Nyaupane et al., 2024). Inadequate water availability for absorption leads to drought, which limits further nutrient uptake and adversely affects crop growth. Crops necessitate optimal environmental conditions to flourish and yield consistently. Crop growth and yield are generally diminished during droughts due to decreased nutrient absorption, impaired photosynthesis, and insufficient water availability. Drought stress in plants, influenced by timing, stage, and severity, induces diverse morphological, physiological, biochemical, and molecular abnormalities, resulting in diminished growth and development of crops (Chachar et al., 2016). AT theses stages i.e. tillering, jointing, booting, anthesis, and grain filling can result in yield reductions of up to 69% (Khan et al., 2023).

4 Effect of Drought Stress on Wheat Morphology

The morphology of wheat (*Triticum aestivum*) is greatly affected by drought stress, which causes a number of structural and developmental changes that have an effect on crop viability and output. Significant effects, including decreased germination %, seedling vigor, dry weight, root-shoot ratio, early maturity, leaf area, early leaf senescence, and seedling vigor, are brought about by drought (Anjum et al., 2017).

5 Effect of Drought on Seed Germination

Seed germination is crucial for the growth, reproduction, and yield of crops (Nyaupane et al., 2024). A deficiency in soil moisture postpones seed germination, leading to irregular seedling dispersion, inadequate plant population establishment, and reduced wheat grain yield (Hossain et al., 2012; Nyaupane et al., 2024). Under drought stress, the seed germination rate decreases by 9.4% over 1.91–2.05 days (Abdoli & Saeidi, 2012). The primary factor contributing to diminished seed germination under drought stress is the reduced water potential and seedling vigor index relative to normal conditions (Ahmad et al., 2022). Wheat is especially susceptible to drought stress throughout the reproductive and grain filling stages, leading to a significant decline in yield.

6 Effect of Drought on Leaf

Drought conditions result in decreased leaf area and early leaf senescence. This reduction in leaf size limits the plant's ability to photosynthesize effectively, further compounding stress effects (Ahmed et al., 2022). Additionally, drought can lead to increased leaf waxiness and trichome density as adaptive responses to minimize water loss (Ahmed et al., 2022). Drought results in diminished cell expansion of leaves, inhibited growth, and heightened abscission, culminating in a decrease in leaf area index (LAI) (McCree & Davis, 1974; Oladir et al., 1999). Insufficient water reduces nitrogen absorption by plants (Hajibarat & Saidi, 2022). A substantial reduction in photosynthesis, diminished cellular development and expansion, decreased nutrient absorption, heightened senescence, and a swift rise in stress hormones like as ABA during drought conditions lead to a decline in dry matter accumulation.

Wheat's flag leaf is a crucial source of assimilates during the critical period when the grain is forming. It provides between 30 and 50% of the plants total assimilates (Inoue et al., 2004). However, the rate of leaf senescence is accelerated by drought stress, which is mostly responsible for the decrease in wheat grain production. Early senescence and maybe shorter grain filling times are the results of drought stress during this phase. Due to a decrease in filling rate and length of time, the potential grain size is reduced when water is scarce early in grain growth. The most vulnerable stages effected by drought stress are flowering and grain filling. Plants' metabolisms can shift as a result of stress, resulting in death of cell. Stresses, on the other side, might have diverse effects on responses of plant at the level of whole plant and at cellular level. Drought stress has lowered grain yields by as much as 10% in the previous 40 years, and it is predicted that by the end of year 2050 the yield of more than half of irrigation area will be disrupted (Jochum et al., 2019). A plant's height will decrease as a result of drought stress. Drought was shown to have reduced spring wheat's biomass, which in turn affected output (Dhakal, 2021).

Wheat's chloroplasts rely heavily on chlorophyll, a key component of the photosynthesis process that is essential to the plant's survival. Wheat under drought stress has a lower chlorophyll content in its leaves (Duvnjak et al., 2024). When plants are under stress from drought, their chlorophyll levels drop because of changes in a number of metabolic processes. Leaf chlorophyll content was reduced by 49% when heat and drought stress were present together, and by just 9% when drought stress was present alone. Chlorophyll deficiency is often regarded as a primary cause of photosynthesis's inability to proceed. Wheat's photosynthetic response is governed by the nature, severity, and velocity of the stressor.

7 Effect of Drought Stress on Photosynthesis

Photosynthesis is a crucial process that plays a key part in CO_2 fixation and light absorption in leaves, essential for the optimal growth and development of plants (Ahmad et al., 2022) The effectiveness of photosynthesis depends on genetic capability for light energy absorption and its conversion into carbohydrates, but it is also directly influenced by dry conditions (Andrianasolo et al., 2016). Drought stress diminishes CO_2 diffusion via stomatal and non-stomatal mechanisms, hence directly reducing the rate of photosynthesis. In drought conditions, CO_2 diffusion leads to the excessive decrease of the photosynthetic electron transport chain, which is a significant source of reactive oxygen species (ROS) generation (Ahmad et al., 2022). In this way, the rates of photosynthesis, leaf senescence, and leaf expansion diminish, resulting in a decline in food production (Wahid & Close, 2007). Calvin cycle activity under drought circumstances ultimately reduces agricultural output. A significant effect of drought on photosynthesis is the disruption between the antioxidant defense system and reactive oxygen species (ROS) formation (Reddy et al., 2004), which eventually increases ROS generation and stress on cellular organelles, membrane lipids, and proteins.

8 Effect of Drought Stress on Length of Plant's Various Parts

Drought stress significantly impacted morphological and phenological measures, including plant height, tillers per plant, number of spikes per square meter, number of spikelets per spike, and spike length. The mean height of plants subjected to drought stress was seen to be decreased (Poudel et al., 2020). The substantial reduction in plant height may result from protoplasm dehydration, which ultimately diminishes cell division, cell expansion, and cell turgidity (Ozturk et al., 2022).

9 Effect of Drought Stress on Wheat Physiology

Drought stress significantly impacts grain output by disrupting numerous physiological processes, including diminished nutrition uptake, mobilization, stem reserve buildup, gametogenesis, fertilization, embryogenesis, endosperm development, and seed growth (Cakir, 2004). In plants, the gametogenesis phase is particularly susceptible to abiotic stressors, including drought and thermal stress. The reproductive organs subjected to these pressures exhibited meiotic abnormalities resulting from morphological, structural, and metabolic changes that cause premature gametes and reproductive sterility (Storme & Geelen, 2014). Drought diminishes the moisture needs of reproductive structures, including the style and stigma, and decreases pollen viability, resulting in inadequate fertilization. Drought stress diminishes the plant's capacity to absorb water from the soil, resulting in decreased turgor pressure. Reduced turgor pressure diminishes plant growth by lowering cell extensibility and impeding embryo development and embryogenesis (Sharma et al., 2022). The closure of stomata results in a diminished supply of CO2 for photosynthesis and a reduction in the photosynthetic rate (Ahmad et al., 2018).The drought directly impacted the availability of Ribulose 1,5-bisphosphate carboxylase/oxygenase (Rubisco) and reduced ATP generation (Farooq et al., 2009). In wheat, the stages of meiosis and microsporogenesis during pollen formation are particularly susceptible to dryness, resulting in pollen sterility, which adversely affects fertilization and reduces grain yield (Zhang et al., 2021a, 2021b). Drought stress induces the synthesis of plant stress hormones, primarily abscisic acid (ABA), which significantly influence stomatal behavior in response to water scarcity. This results in both beneficial and detrimental effects: stomatal closure conserves water during drought, but simultaneously hinders crop growth as an adverse consequence (Nyaupane et al., 2024).

10 Effect of Drought Stress on Wheat Biochemistry

Drought stress diminishes photochemical efficiency, elevates the generation of reactive oxygen species (ROS), exacerbates oxidative damage, reduces starch accumulation, enhances abscisic acid synthesis, promotes proline production, increases antioxidant enzyme activity, and decreases cytokinin levels (Sharma et al., 2022). The rhizosphere of wheat exhibits an elevated rate of respiratory carbon loss during drought conditions, resulting in diminished ATP synthesis (Akter & Rafiqul Islam, 2017). Drought stress also triggers the production of superoxide and hydrogen peroxide (H_2O_2) (Reddy et al., 2004). In plants, adenosine triphosphate (ATP) is synthesized in mitochondria by oxidative phosphorylation, which entails the translocation of protons through the electron transport chain. The oxygen molecule serves as the ultimate acceptor of electrons during the series of oxidation–reduction processes in the electron transport chain. During drought stress, merely 0.1–2% of electrons traverse the electron transport chain, resulting in the incomplete reduction of oxygen molecules to generate superoxide radicals ($\bullet O_2$), whereas under normal conditions, oxygen is fully reduced to water (Li et al., 2013). During drought stress, plant roots initially detect water-deficit signals, which are conveyed to the leaves via xylem sap. Abscisic acid is the primary stress signal generated in plants during drought conditions (Zhang & Davies, 1987).

11 Direct Effects on Plant Physiology

Drought resistance is the ability of plants to continue growing and surviving during times of severe stress. In order to survive drought stress, plants have evolved a number of defense mechanisms (Chaves et al., 2003). These defense mechanisms include morphological adaptations, osmotic adjustment, water resource optimization, antioxidant systems that lessen the negative effects of reactive oxygen species (ROS) associated with drought, and the induction of numerous stress-responsive genes and proteins (Camaille et al., 2021). Direct mechanism consist of two primary ways are either the manufacture of plant hormones by these bacteria or the enhancement of nutrient uptake via nitrogen fixation, phosphate or potassium solubilization, or both (Camaille et al., 2021). Plant hormones known as plant phytohormones such as abscisic acid (ABA), cytokinin (CK), ethylene, auxin (IAA), salicylic acid (SA), brassinosteroids (BR) and jasmonic acid (JA) (Peleg & Blumwald, 2011), are essential for a plant's ability to adjust to its environment by controlling growth, development, the distribution of nutrients, and the transition from source to sink. Wheat plants experience significant hormonal shifts and adjustments to their hormonal balance during shortage of water.

The quick, temporary build-up of abscisic acid is a crucial step in osmotic stress responses; it promotes stomatal closure and the activation of ABA-responsive genes, which shield plants from more water loss and harm (Kaushal & Wani, 2016). One of the first hormonal reactions to drought stress is the development of ABA. Usually, ABA is produced in the dry, dehydrated roots of plants that are exposed to dry soil. It is then transferred to the leaves via the xylem, enabling the plant to communicate its water deficit over great distances. The amount of ABA in wheat appears to be a factor in pollen sterility and is frequently adversely linked with photosynthetic efficiency and flag leaf chlorophyll concentration. In actuality, high amounts of ABA may cause pollen sterility in wheat, and the management of endogenous ABA level is crucial for pollen development (Dong et al., 2017). Numerous genes linked to dehydration stress, including those that minimize transpiration or are connected to oxidative stress, as well as certain genes involved in primary carbohydrate metabolism, have been shown to be induced by ABA in several transcriptome investigations (Camaille et al., 2021). Furthermore, the enzymes or proteins involved in drought stress tolerance are encoded by the ABA-responsive genes.

Since they promote stomatal opening and lessen stomata's sensitivity to ABA by delaying senescence, cytokinin's are the antagonists of ABA (Gujjar & Supaibulwatana, 2019). In wheat, the concentration of chlorophyll in the flag leaf and the degree of photosynthesis are frequently positively connected with the concentration of CKs (Li et al., 2016). The plant senescence brought on by environmental stressors like drought was postponed in transgenic plants that overexpressed cytokinin. When other plants wilt completely, transformed plants respond to dryness by staying green and healthy. In addition to increased phloem unloading and sugar import to endospermic cells via the cell wall-associated enzyme invertase, the increased number of CKs improves sink strength by over-expressing genes involved in cell division through sugar signaling (Peleg & Blumwald, 2011). Through an improvement in N metabolism, CK treatment leads to better biomass and yield in drought-stressed plants; nevertheless, during a drought stress, CK concentration frequently declines (Glick, 2014). A growing body of research indicates that CKs control plant adaptation to drought through a multistep phosphorus mechanism (Glick, 2014). However, a number of studies demonstrated the multifaceted character of cytokinins, since they can influence drought tolerance in both favorable and unfavorable ways (Glick, 2014).

All higher plants contain ethylene, which is involved in many different plant activities such as the growth of roots, leaves, flowers, and fruits as well as the rhizobia nodulation of legumes and the interactions between plants and mycorrhizal fungus (Ngumbi & Kloepper, 2016). When a plant is under drought stress, it produces a lot more ethylene until it hits a threshold of roughly 25 g. When this threshold is crossed, the ethylene known as "stress ethylene" causes senescence, chlorosis, and leaf abscission in addition to being detrimental to root and shoot growth.

Upon reaching this threshold, the ethylene known as "stress ethylene" becomes detrimental to the growth of roots and shoots, resulting in senescence, chlorosis, and leaf abscission (Peleg & Blumwald, 2011). Under extreme drought conditions, the concentration of ethylene (and its precursor ACC) doubled compared to the control group that was not subjected to stress. Additionally, the rate at which grains were filled and the final weight of the grains both saw considerable reductions. Ethylene can also stimulate the activation of drought-specific genes by interacting with ethylene responsive element binding proteins (ERF), which are transcription factors responsible for the response to water stress (refer to the "Specific proteins" section (Farooq et al., 2014).

Auxins are mostly recognized for their involvement in rhizo-genesis, which includes the start of lateral and adventitious roots as well as the encouragement of cell division (Dong et al., 2017). The interaction between auxins and ethylene plays a crucial role in regulating root development and architecture, making it an important factor in drought tolerance (Sadok & Schoppach, 2019). Genes that regulate root system architecture (RSA) have been the focus of molecular breeding efforts aimed at enhancing plant drought resistance (Barnawal et al., 2017). The study demonstrated a correlation between an enhanced root system and increased drought resistance in wheat. Hence, auxins play an essential albeit indirect role in enabling plants to withstand drought-induced stress. Auxins can also regulate root hydraulic properties by facilitating the expression of water-conserving characteristics, which are linked to increased crop yields during periods of drought (Chakraborty et al., 2013). Nevertheless, drought stress appears to diminish the production and communication channel of auxin in wheat. The plant auxin concentration reduced by over 30%, while there was an increase of more than twofold in the expression of the AUX/IAA1 gene. This gene encodes a transcriptional repressor of auxin-responsive genes (Vacheron et al., 2013).

Brassino-steroids (BR) have been found to stimulate the expression of genes that respond to stress. This leads to the preservation of photosynthetic activity, activation of antioxidative enzymes such as SOD, peroxidase, and catalase, accumulation of osmolytes like proline and soluble sugars, and the initiation of other hormonal responses. These effects can assist plants in enduring drought stress (Vacheron et al., 2013).

12 Strategies to Overcome Drought Stress

The challenge for 21st-century agriculture is to understand the effect of drought on crop output and, more crucially, to formulate clever ways to resist the drought while adhering to sustainable agricultural laws. Indeed, crops' resistance to water scarcity might be improved by a variety of approaches, including plant breeding, increased vegetation cover, genetic engineering, larger agricultural areas, and farm mechanization.

Although conventional breeding, management of water, and genetic engineering are regarded to be valuable strategies for alleviation of stress of drought but adopting and implementing these approaches requires a high level of technical expertise. But the majority of the remedies are expensive and time-consuming. They also require high cost for implementation.

One of the plant-growing technology is film farming which includes use of films of hydrogels. This hydrogel is a 3D polymer which entraps a lot of water in intramolecular spaces. It is a fine layer of film of width of 60 cm and a thickness of 0.06 mm

(Batista et al., 2019). Film farming takes very little amount of water, soil and fertilizers. This approach is applied for cultivation of a lot of crops. Still, this approach is not widely employed because it takes a lot of time. It is also not economical and necessitate to grow plants in greenhouse (Zhang, Rod, & Hosseinian, 2021a, 2021b). Transgenic plants have improved output and they can tolerate abiotic stresses (Zahra et al., 2021). However, there are many drawbacks to use these plants for alleviation of drought trauma. This method consume time and require intensive laboure, and it may cause the host's gene pool to lose other desirable features. Biochar's usefulness as a plant stress reliever has grown dramatically in recent years due to its potential to promote fertility of soil, absorption of nutrients and development in stressed plants (Haider et al., 2020).

Although application of biochar to alleviate stress of drought appears to be beneficial, the risks of technology are yet mostly unclear. Unavailability of feedstock, nutrient binding, increased soil pH, inefficient energy utilization during biochar production, heavy metal release in soil and demolition of the natural micro flora of soil are just a few of the known major disadvantages (Oni et al., 2019).

13 Use of PGPBs to Cope with Drought Stress

Despite the various strategies wheat plants have developed to manage the drought stress, considerable harms are usually recorded when wheat crop is subjected to severe drought stress. Soil microorganisms that have the potential to reduce abiotic stress as well as promote plant development might be particularly useful instruments for ensuring a sustainable agriculture (Camaille et al., 2021). As a result, in recent decades, significant emphasis is made to use plant growth-promoting rhizo-bacteria in cultivation as a possible sustainable alternative to upsurge yield and tolerance to drought. Plant characteristics are selected by microbial communities below the ground surface, which helps to lessen the influences of abiotic stress on plant populations. One of the most prevalent ways for increasing agricultural productivity is to improve soil fertility. Rhizo-bacteria are found in the rhizo-spheric soil around the plant, and plant growth-promoting rhizo-bacteria are those that have good influences on plant growth and production through diverse processes (Danish & Zafar-ul-Hye, 2019). Various rhizo-bacteria genera, including *Azospirillum, Klebsiella, Psuedomonas, Azotobacter, Alcaligenes, Burkholdaria, Enterobacter, Bacillus* and *Arthrobacter* have been studied to boost growth of plant and production under severe shortage of water (Danish et al., 2020).

Plant growth-promoting bacteria are considered a cost-effective and environmentally sustainable approach to mitigate drought stress and enhance crop output through many mechanisms. The utilization of growth-promoting rhizobacteria has proven to be an attractive and sustainable alternative for improving crop output under drought conditions by substituting pesticides and artificial fertilizers. Rhizospheric *Pseudomonas* have garnered significant attention due to their root colonization capability, catabolic versatility, and ability to produce many compounds and enzymes that aid plant survival in arid environments (Camaille et al., 2021).

14 Mechanisms of Action of Drought Tolerant Bacteria

Drought is a major environmental stressor that severely impacts agricultural productivity and ecosystem stability. Drought is thought to have decreased cereal production in the world by 9–10%, making it one of the main global food production constraints (Nyaupane et al., 2024). By 2050, drought is predicted to seriously impair agricultural growth on more than half of the planet's arable area. With the increasing frequency and severity of drought events due to climate change, there is an urgent need to develop strategies for enhancing drought tolerance in crops and soils. One promising approach is the utilization of drought tolerant bacteria, which can improve plant resilience to water scarcity through various mechanisms (Fig. 2).

Furthermore, beneficial rhizobacteria having potential to produce ACC deaminase, split ethylene precursor which is limiting factor for plant growth. Similarly, other rhizobacteria, *Psuedomonas* and *Actinobacte*r increase biomass of leaf and shoot and photosynthesis rate in drought effected plants. The importance of microbes in development of plant, availability of nutrients and role as biocontrol is widely known. Phyto-hormone production, production of ACC deaminase, induction of tolerance to drought by compounds released by bacteria and exopolysaccharide

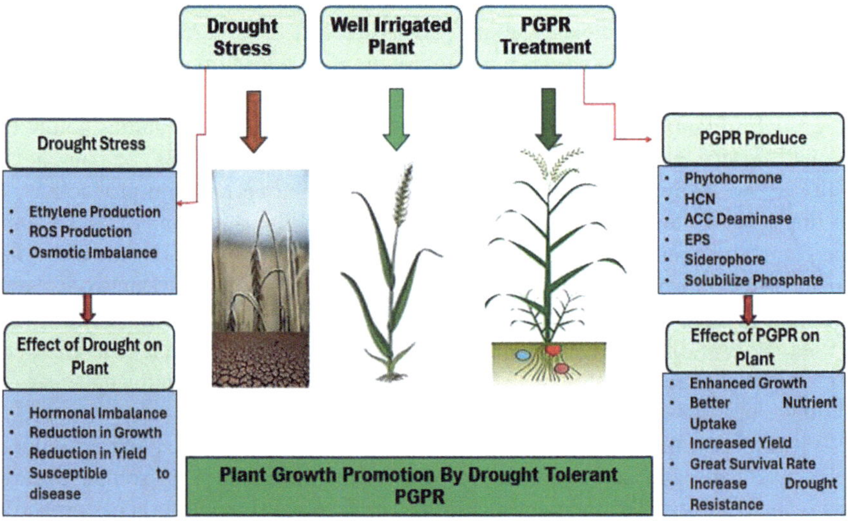

Fig. 2 Effects of drought stress on wheat mitigating by mechanism of action of PGPR on wheat plant growth

produced by bacteria are all possible explanations for plant drought tolerance mechanism induced when beneficial rhizobacteria are applied to plants (Saleem et al., 2021). Rhizobacteria can endure abiotic stresses by accumulating compatible solutes, synthesizing exopolysaccharides, producing antibiotics, and forming spores, among other physiological processes. Atmospheric nitrogen is fixed by rhizobacteria and it is made accessible to plants (Hakim et al., 2021). Phosphate solubilization is carried out by plant growth-promoting rhizobacteria and phosphate is made available for use of plant (Kabiraj et al., 2020). Siderophore production is also done by rhizobacteria which increase the availability of micronutrients like iron (Singh, 2020).

15 Production of Phytohormones

By producing and secreting phytohormones, rhizobacteria can help to standardize levels of hormones in plant, allowing for a rise in the total surface area and root and shoot length, and overall growth of root system (Gu et al., 2024). Phyto-hormone auxin (IAA) help plant in tissue elongation and growth of plant cell, which results in enhanced root's surface area to absorb water and food, as well as coordinating a cellular defense against the stress of drought. According to research by Frankenberger, indole acetic acid acts as a growth regulator in plants. In plants, indole acetic acid delays leaf senescence and fruit ripening as well as increases cell elongation, cell division, phloem and xylem differentiation, lateral root growth, stem initiation, and fruit set (El-Sabagh et al., 2021).

In reaction to drought, abscisic acid (ABA) is crucial. ABA is known as stress hormone because it accumulates in stress especially in drought stress conditions. ABA regulates water loss by controlling closing of stomata in leaves and pathways of stress signal transduction. Studies have demonstrated that treating plants with exogenous ABA increases their resistance to drought. Exogenous ABA treatment, for instance, promotes plant growth and improves drought resistance of a variety of crops including maize, wheat, sweet potato, pearl millet, and oats by regulating the processes of energy metabolism, amino acid metabolism, and lipid metabolism; promoting the accumulation of flavonoids, betaine, and other substances; improving enzymatic and non-enzymatic antioxidant regulation systems; and enhancing the photosynthetic performance and RWC of plants (Jiang & Asami, 2018).

Plants respond to abiotic stress by producing ethylene, a hormone that inhibits root development. In a series of events performed by ACC synthase, methionine is converted into the 1-aminocyclopropane-carboxylic acid (ACC), a precursor of ethylene, a phytohormone.

Root exudates in the rhizosphere may include ACC that rhizobacteria can then use. Bacteria that produce ACC deaminase rely only on ACC for their Nitrogen needs. Ethylene levels drop because ACC is broken down by ACC deaminase into ammonium and beta-ketobutyrate. Root morphology, especially in times of stress, has been shown to be significantly altered by the production of ACC deaminase by many different PGPR strains. Rhizobacterial inoculation producing ACC deaminase can alleviate plant growth inhibition caused by elevated ethylene levels during drought stress. Inducing growth promotion and stress tolerance in a variety of plant species is a well-documented effect of bacteria that produce ACC deaminase (Glick & Nascimento, 2021).

Since the 1950s, microbial inoculants/bio fertilizer derived from soil microorganisms have been utilized as an alternative to chemical fertilizers. They boost crop health and yield through chemical and biological processes without negatively impacting the surrounding environment. Rhizobia that break down phosphate into a form that may be taken up by plant roots are called phosphate solubilizing rhizobacteria. Phosphorus is primarily taken up during vegetative growth, and then it is stored again in the plant's reproductive structures like seeds and fruits. It has been widely documented that using phosphate-solubilizing bacteria inoculum improves the development and harvest of numerous crops, including wheat, maize, and sweet potato. As the third most widely grown cereal crop worldwide, *Triticum aestivum* is a staple meal in Pakistan. Wheat grows better when there is plenty of phosphorus available, but when there isn't, the crop fails.

16 Fixation of Nitrogen

As the most important nutrient for plant development and yield, nitrogen is a necessity for all forms of life. Even while nitrogen makes up around 78 percent of the air, it can't be used by plants. Regrettably, no singular plant species exists that can transform atmospheric dinitrogen into useable ammonia. Biological nitrogen fixation (BNF) is the process via which atmospheric nitrogen is converted into ammonia by nitrogen-fixing bacteria employing a complex enzyme system known as nitrogenase. Plant-growth-promoting rhizobacteria can fix atmospheric nitrogen and render it accessible to plants through two mechanisms. The mutualistic relationship between bacteria and a plant is fundamental to symbiotic nitrogen fixation. Nitrogen fixation occurs in nodules, which are created when the microbe initially infiltrates the root. Nonetheless, free-living diazotrophs can perform non-symbiotic nitrogen fixation, hence facilitating the growth of non-leguminous plants such as radish and rice. Non-symbiotic bacteria, including *Azoarcus, Azotobacter, Acetobacter, Azospirillum* etc. inhabit the rhizosphere and facilitate nitrogen fixation (Gupta & Pandey, 2023).

17 Exopolysaccharide Production

Exopolysaccharides are released by bacteria which are hydrated substances containing 97% water. EPS provides resistance against desiccation as they are rich in water. EPS assist them to withstand in harsh situations such as water scarcity by retaining huge quantities of water. Surface attachment, biofilm development, food delivery and bearing of abiotic stressors are all functions of exopolysaccharide produced by rhizobacteria (Pathania et al., 2020). By enhancing aggregate stability, EPS promotes microaggregates, which boost plant development under drought stress (Gontia et al., 2016; Awasthi et al., 2017).

18 Antioxidant Defense Enzymes

Reactive oxygen species (ROS), such as superoxide radicals, hydrogen peroxide, and hydroxyl radicals, are frequently generated in minimal quantities within various plant organelles under ideal development conditions. Insufficient water impairs photosynthesis and leads to increased production of reactive oxygen species (ROS) due to alterations in cellular homeostasis (Verma et al., 2019). The antioxidant defense systems in plants consist of enzymatic and non-enzymatic components that collaboratively scavenge reactive oxygen species (ROS) to mitigate oxidative damage caused by dehydration. Enzymes that combat free radical damage comprise catalase, superoxide dismutase, ascorbate peroxidase, peroxidase, and glutathione reductase. Ascorbic acid, glutathione, and cysteine constitute an additional non-enzymatic defense mechanism in conditions of water stress. Reactive oxygen species (ROS) signify the initiation of stress responses and defensive mechanisms. Consequently, it is essential to mitigate oxidative damage while simultaneously modulating signaling events by regulating reactive oxygen species (ROS) levels through the orchestration of ROS formation and scavenging processes. Plants exhibiting elevated antioxidant enzyme activity has an enhanced capacity to endure oxidative stress. A robust correlation exists between drought stress and the function of antioxidant enzymes; however, inoculation with PGPR diminishes the likelihood of drought stress adversely affecting antioxidant enzyme activity (Hasanuzzaman et al., 2021).

19 Production of Osmolytes

Bacteria can adjust to water scarcity by producing osmo-protectants like K, glutamate, trehalose, proline, and glycine betaine. Bacteria were able to adapt to drought stress and affect root colonization in part by producing secondary metabolites and volatile chemicals. Upregulation of genes involved in protein stabilization (heat shock proteins and chaperones) is just one example of the multistep process involved in the bacterial stress response. To cope with the oxidative stress generated by desiccation, bacteria upped their levels of hydroperoxide resistance proteins (Men et al., 2018).

Plants with plant growth-promoting rhizobacteria undergo comparable chemical, physical and biological changes which help them to survive and thrive in stressful situations like drought. Furthermore, its implementation is both cost-effective and time-consuming. As a result, plant growth-promoting rhizobacteria can provide additional benefits such as drought resistance and increased plant performance (Gupta et al., 2015).

20 Induced Systemic Resistance

A plant's natural defenses are strengthened after experiencing induced resistance, which may be thought of as a physiological condition triggered in response to certain environmental stressors (Kumar et al., 2024). Bio-priming plants with certain plant growth boosting rhizo-bacteria can confer resistance against a wide variety of plant diseases. Bio-priming is also known as inoculation. After applying rhizo-bacteria that encourage plant development, there is a decreased risk of plant diseases that are caused by fungi, bacteria, and viruses, and in certain cases there is also a decreased risk of damage caused by insects and nematodes. Furthermore, jasmonate and ethylene signaling inside the plant are involved in induced systemic resistance, and both hormones enhance the host plant's defense responses against a wide range of plant diseases (Gu et al., 2024).

21 Improve Interactions with Plant Root Systems

The root systems of crops actively interact with the soil to absorb water and nutrients, and are frequently the main victims of many soil-related pressures. The crucial function of the root system architecture in investigating soil zones for the collection of soil resources has been extensively recorded. The root system architecture is an intricate three-dimensional arrangement, displaying a distinct spatial and temporal configuration of different types of roots. The plasticity and dynamism of plants enable them to adapt and respond to their surroundings, thereby optimizing their ability to obtain nutrients and water (Hadier et al., 2020).

Plant roots release significant quantities of primary and secondary metabolites into the rhizosphere, which make up 5–21% of the plant's photosynthate. Root exudates have various functions such as changing the physical and chemical properties of the soil, aiding in beneficial symbiotic relationships, controlling the root microbiome and soil microbial communities, and regulating interactions with other organisms living in the soil. Especially fascinating are secondary metabolites that are released by roots and function as chemical signals in interactions between plants below the earth. Root signals initiate a sequence of adaptive responses including root detection and identification, chemical protection, and alterations in root behavior. Crucially, interactions between plants' underground signaling systems affect their aboveground performance, especially in terms of flowering and reproduction, ultimately impacting their overall fitness (Kumar et al., 2024).

22 Studies Showed Utilization of Plant Growth Promoting Bacteria in Drought Tolerance of Wheat and other Crops

A plethora of studies has been conducted to check the effects of PGPB on plant in different conditions. PGPBs are acknowledged for their effectiveness on several plants, including legumes, grains, and vegetables, in adverse environments. Multiple studies have shown the advantageous role of rhizobacteria in alleviating the negative impacts of diverse stresses on crop development (Ansari et al., 2023; Paul et al., 2022; Yadav et al., 2022).

The improvement of growth during drought stress may also be attributed to specific plant PGPBs that synthesis ACC enzyme that enhance nutrients uptake such as nitrogen, phosphorus, and potassium, thereby promoting plant growth in challenging environmental conditions. The application of the plant growth-promoting bacteria *Herbaspirillum seropedicae and Azospirillum brasilense* enhanced drought tolerance in maize, likely due to improved water use efficiency and increased antioxidant enzyme activity under drought conditions. Furthermore, PGPB augment growth attributes by producing plant growth hormones including gibberellins (GA), indole-3-acetic acid (IAA), and cytokinins, resulting in enhanced nitrogen fixation and superior nutrient uptake. Moreover, PGPB are crucial in enhancing drought stress tolerance by promoting the accumulation of ABA.

During drought stress, plant growth-promoting bacteria (PGPBs) can directly or indirectly enhance ACC deaminase activity, hence facilitating plant development. This approach depends on the use of ACC by PGPBs before its oxidation by ACC oxidases produced in plants. Thus, PGPBs may function as an enhanced source of growth promoters and stress tolerance owing to their capacity to reduce ethylene levels. The use of plant growth-promoting bacteria (PGPBs) in wheat enhanced plant development by increasing ACC deaminase activity, hence regulating ethylene levels. The study by Benmerid et al., (2024) demonstrated that Arthrobacter species, which possess drought tolerance and phosphorus solubilization capabilities, enhance seed germination rates, root and shoot lengths, and elevate phosphorus levels in plants. Researchers have also shown that several Bacillus species significantly influence seed germination, plant hormones, plant physiology, and yield rates in wheat plants (Paul et al., 2022; Yadav et al., 2022; Ansari et al., 2023; Noureen et al., 2024).

Different PGPB have been identified to improve the growth and yield of plants under drought stress. Recent studies which showed the effects of PGPBs on wheat plant under drought stress and explore their effects are listed below in (Tables 1 and 2).

Table 1 Drought tolerant plant growth promoting bacteria used for wheat growth and development

Bacterial name	Wheat Genotype	Treated Drought level	Treatment strategy	Effects on Wheat Health	Reference
Arthrobacter oryzae, Arthrobacter pascens	*Triticum durum*	Moderate	Drought tolerant and P solubilizing rhizobacteria	• Increase in seed germination, root length, volume and surface area, Increase in plant height, dry weight of shoot • P content 68.91%- and 2-folds increase in chlorophyll content	Benmerid et al., 2024
Bacillus subtilis and Bacillus tequilensis	*Triticum aestivum*	40% PEG	Drought tolerant rhizobacetria and biochar	• Increasing the root length from 17.0% to 70.0% and shoot length from 30.0% to 82.0% • Increase in chlorophyll and carotenoids content 477% and 423% respectively	Noureen et al., 2024

(continued)

Table 1 (continued)

Bacterial name	Wheat Genotype	Treated Drought level	Treatment strategy	Effects on Wheat Health	Reference
Bacillus subtilis-FAB1 and Pseudomonas azotoformans-FAP3	*Triticum aestivum*	40% PEG	Biofilm producing drought tolerant rhizobacteria	• Inhibitory effect on 2.3% to 3.6% of seeds germinating in both single and combined treatments; -improvement in shoot fresh weight 44.4%, shoot dry weight 23.3%, shoot number 30.8%, shoot length 5.8%, root length 7.4%, and root dry weight 21.8%, RWC 1.7% and RAS/RT 33.3% • Decreases in MDA 28.7%, proline content 76.1%, GR activity 85.3%, SOD activity 82.7%, and catalase 65.1%	Ansari et al., 2023
Enterobacter bugandensis WRS7	*Triticum aestivum L*	10% PEG	Inoculum of PGPR	• Reduced lipid peroxidation -enhance elevated osmolyte concentration (proline and total soluble sugar), relative water content, catalase, and superoxide dismutase activity	Aror et al., 2023
Actinobaciota, Variovorax, Streptomyces, and Conocybe	*Triticum aetivum spring*	Normal stress (25–28%); Moderate stress(15–18%); Severe stress (9–12%)	Consortium of PGPR	• Enhanced leaf area, photosynthetic rate, transpiration rate, stomatal conductance, internal CO_2, shoot and root length, and biomass	Fang et al., 2023

(continued)

Table 1 (continued)

Bacterial name	Wheat Genotype	Treated Drought level	Treatment strategy	Effects on Wheat Health	Reference
Piriformospora indica and arbuscular mycorrhiza fungi (AMF)	*Triticum aestivum (var. HD2967)*	30% and 75% PEG	Co- inoculum of AMF and PGPB	• Increased stomatal conductance, internal CO_2, photosynthetic rate, transpiration rate, leaf area, shoot and root biomass, and glycine betaine content; • Decreased lipid peroxidation, relative membrane permeability, and lipoxygenase enzyme activity; • Accumulated higher content of carotenoids, proteins, putrescine, spermidine, and proline;	Singh et al., 2023a, 2023b

(continued)

Table 1 (continued)

Bacterial name	Wheat Genotype	Treated Drought level	Treatment strategy	Effects on Wheat Health	Reference
Bacillus subtilis 104 and 26D	*Triticum aetivum L. (var. E70 and SY)*	4%, 8%, 12% PEG	Drought tolerant and herbicidal bacteria	• Increasing length of roots and shoots, biomass, photosynthetic pigments (chlorophyll a and b), and leaf area; • Reducing stress-caused LPO (i.e., malondialdehyde), and regulating proline biosynthesis, as well as contributing to a faster recovery of growth, photosynthetic pigments, and redox-status of plants in post-stress period	Lastochkina et al., 2023
Klebsiella aerogenes SH-8	*Triticum aetivum*	5%, 10%, 15%, 20%, 25% PEG stress	Drought tolerant bacteria	• Enhancing the drought tolerance of bioprimed seeds • 20–60% increase in seed germination, seed vigor index and drought tolerance	Shaffique et al., 2023
Azotobacter,Azospirillum,Microbacterium, Pseudomonas,	*Triticum aestivum GY*	Low level, moderate level 12%, High level 24%	Drought tolerant PGPB	• PGPB significantly increased root and shoot length, dry weight and yield	Rostamian et al., 2023

(continued)

Table 1 (continued)

Bacterial name	Wheat Genotype	Treated Drought level	Treatment strategy	Effects on Wheat Health	Reference
Bacillus sp. BT3 and Klebsiella sp. HA9	*Triticum aestivum (var. HD2967)*	30% FC	Drought tolerant PGPB	• Positively influenced the plant growth, chlorophyll and proline status of seedlings • Significantly improved plant height, fresh weight and dry weight of root and shoot, relative water content, chlorophyll content and root morphological parameters	Singh et al., 2022
Bacillus pseudomycoides	*Triticum aetivum L. (BARI gom-33)*	50%	Biofilm forming and drought tolerant bacteria	• Increase growth, germination rate, root-shoot length, RWC, and chlorophyll content • Improve the levels of soluble sugars, proteins, H_2O_2, NO, and antioxidant enzymes (CAT, SOD, POD, and APX)	Paul et al., 2022
Bacillus paramycoides DT 85 and Bacillus paranthracis DT 97	*Triticum aetivum L*	10–30% PEG	Drought tolerant bacteria	• Tolerance to high osmotic stress (10–30% polyethylene glycol 6000), salinity (5–15%), and temperature (45 °C), • Enhanced production of drought combating molecules like superoxide dismutase, peroxidase, catalase and proline to increase plant growth and yield	Yadav et al., 2022

(continued)

Table 1 (continued)

Bacterial name	Wheat Genotype	Treated Drought level	Treatment strategy	Effects on Wheat Health	Reference
Pseudomonas azotoformans FAP5	*Triticum aetivum L. var. 343*	0.05 MPa,–0.3 MPa PEG	Biofilm forming PGPR	• Improvement in growth attributes, photosynthetic pigment efciency and other physiological attributes • Bioflm development by FAP5 enhanced signifcantly and root colonization under drought stress	Ansari et al., 2021
Curtobacterium faccumfaciens (Cf D3-25) and Arthrobacter sp. (Ar sp. D4-14)	*Triticum aetivum (cv Bob White Red Haplotype)*	Mild and severe drought stress	Plant beneficial microbes	• Efficient in terms of increasing shoot length, root length, fresh weight and dry weight of wheat	Hone et al., 2021
Azospirillum brasilense NO40 and Stenotrophomonas maltophilia B11	*Triticum aestivum L. (cv.Gemiza9)*	70% FC	Plant growth promoting bacteria	• Diminish the effects of drought stress on the relative water content of roots, shoots and leaves; area of leaves; contents of pigments (chlorophyll a and b) and ascorbic acid; and on the protein patterns of roots • Reduced the drought • Induced damage indicated by lower leakage of electrolytes and less accumulation of Malondialdehyde and hydrogen peroxide,	Kasim et al., 2021

(continued)

Table 1 (continued)

Bacterial name	Wheat Genotype	Treated Drought level	Treatment strategy	Effects on Wheat Health	Reference
Pseudomonas baetica B21 (B1), and Pseudomonas helmanticensis B30	*Triticum aestivum L*	Moderate drought	Phosphate solubilizing drought tolerant rhizobacterai	• Efficient in terms of increasing shoot length, root length, fresh weight and dry weight of wheat • Reduce electrolyte leakage	Shirmohammadi et al., 2020
Pseudomonas libanensis EU-LWNA-33	*Triticum aetivum L*	50%,75%, and 100% FC	Phosphate solubilizing drought tolerant bacteria	• Efficient in terms of increasing shoot length, root length, fresh weight and dry weight of wheat • Enhanced proline, as well as the glycine betaine content	Kour et al., 2020

Table 2 Drought tolerant PGPB as a metabolic modulator in Wheat

Drought Tolerant PGPB Strains	Metabolic Modifications	Mode of action	References
Pseudomonas azotoformans FAP5	Increase proline content, soluble sugars and antioxidant enzymes	Induces osmotic balance and oxidative stress tolerance	Ansari et al., 2021
Bacillus megaterium	Improved antioxidant enzyme activities; upregulated drought tolerant genes	Modulate the stress tolerance gene expression	Rashid et al., 2022
Bacillus sp.	Improved root architecture; enhanced nutrient uptake	Promotes root development and nutrient assimilation	Li et al., 2023
Bacillus spp. MNS4; Burkholderia phytofirmans PsJN	Increased nutrient uptake and growth parameters of plant	Enhanced root development and nutrient accumulation	Jabran et al., 2024
Bacillus sp.	Improved antioxidant capacity; improved glyoxalate system and osmotic regulation	Enhances stress responsive pathways	Talaat & Adel-salam, 2024
Bacillus rugosus MMH01	Enhanced Chlorophyll and proline; decreased malondialdehyde (MDA)	Increased antioxidant activity and osmolyte accumulation	Salem et al., 2024
Providencia vermicola ME1, Pantoea agglomerans Pa, Pseudomonas knackmussi MR6, Bacillus sp.D13	Elevated chlorophyll; decreased Proline and malondialdehde	Enhanced antioxidant enzyme activities and plant root colonization	Bouremani et al., 2024
Bacillus halotolerans, Pseudomonas sihuiensis, Bacillus atrophaeus	Improved enzyme activity; enhanced water content and biomass	Consortium with folic acid synergistically improved drought response	Tanveer et al., 2024
Bacillus sp.	Enhanced protein and glycine betaine accumulation, Improved antioxidant enzyme activities	Improved osmotic adjustment and stress protection mechanism	Mutanda et al., 2024
Pseudomonas azotoformans JRBHU5, Enterobacter hormaechei JRBHU9	Increase Osmoprotectants and ROS enzymes; regulation of ABA signaling pathways	Molecular reprogramming of drought responsive pathways	Prasad et al., 2025

23 Drought Tolerant PGPBs Reprogram the Metabolic Structure of Plant

Numerous microorganisms have the ability to reprogrammed plants during drought stress by controlling their metabolism and molecular pathways. Under drought conditions, wheat exhibits significant reductions in several metabolites, including pyruvic acid (PA), succinic acid, thiamine pyrophosphate etc. However, treatment with *Bacillus velezensis* increases the availability of these metabolites to counteract drought stress. In situations of drought and salinity stress, *Azotobacter brasilense*, *A. chroococcum*, and *Bacillus sp.* enhance the accumulation of sugar and other compounds like Abscisic acid, and oxygenated monoterpenes in pennyroyal. A consortium of bacteria, comprising B. megaterium, B. subtilis, and B. thuringiensis, significantly elevated the concentrations of total sugar, 3-hydroxy-3-methylglutarate, glycerol, L-asparagine, nicotinamide, and riboflavin in drought-stressed chickpea leaves. Consequently, to alleviate the effects of drought, plant-associated bacteria might modulate the metabolic changes induced by the circumstance (Abdelaal et al., 2021).

24 Drought Tolerant PGPBs Induces the Biochemical Changes

Plants utilize adaptive methods to mitigate salt stress and drought, encompassing metabolite release, new protein synthesis, and genetic reprogramming. Specific Plant Growth-Promoting Bacteria (PGPBs) enhance the synthesis of essential phytohormones, including cytokinin's, gibberellins, and indole-3-acetic acid (Gupta & Pandey, 2023). These phytohormones are essential in modulating plant development and stress responses. This is achieved by activating numerous signaling pathways in the plants. Consequently, to enhance plant resilience to salinity and drought, several phytohormone-related methodologies must be investigated and refined. Cellular tolerance to dehydration is facilitated by LEA proteins. LCOs are synthesized by several bacteria and establish symbiotic partnerships with plants. Rhizobacteria release Node factors (NFs) in reaction to the flavonoids found in root exudates, which initiate nodule formation. High-affinity K^+ transporters (HKT) located on the plasma membranes of plants facilitate Na^+ transit; by excluding Na^+ ions from the shoots during photosynthesis, they inhibit the accumulation of Na^+ ions during this process.

Furthermore, they influence pH, fertility, soil structure, oxygen availability, and nutrient accessibility. These bacteria augment plants' ability to endure salinity and drought via several rhizosphere and phyllo sphere mechanisms, hence promoting plant growth (Dubey et al., 2021).

25 Drought Tolerant PGPBs Modulate the Molecular Structure of Plant

PGPR induces molecular changes in plants, boosting drought resilience. By altering gene expression and phytohormone production, PGPR helps plants adapt to stress. Research highlights PGPR's role in upregulating key genes, such as nitrogenase, and modifying IAA-related transcription patterns. Numerous studies have shown their ability to generate molecular alterations; for example, rhizobacteria that produce IAA have unique nifH transcriptional patterns, leading to the overexpression of the nitrogenase gene in Oryza sativa plants (Sonbrasi et al., 2020). Drought increases the expression of the 9-cis-epoxycarotenoid dioxygenase (NCEDs) gene, essential for abscisic acid (ABA) synthesis in plants (Ali et al. 2020). Alginate elicits comparable drought stress tolerance in wheat pre-treated with B. thuringiensis. Sen et al. (2018) noted that a recent study emphasized the importance of *Pseudomonas putida* GAP-P45 in modulating the expression of critical genes related to polyamine biosynthesis (AIH, ADC, CPA, SPDS, SPMS, and SAMDC) and increasing putrescine and spermidine levels in plant cells under drought conditions. A recent real-time PCR investigation demonstrated that the inoculation of *Bacillus pumilus* on Moringa oleifera increased the concentrations of carotenoids, tocopherols, and folate in the leaves. This was achieved by upregulating the genes for dihydrofolate reductase-thymidylate synthase (DHFR-TS), phytoene synthase (PSY), phytoene desaturase (PDS), lycopene β-cyclase (LBC), and phytoene synthase (PSY) (Sonbrasi et al., 2020).

26 Conclusion

The extensive dependence on chemical fertilizers and pesticides have catastrophic effects on ecosystems, beneficial creatures, water sources, human health, and soil integrity. Conversely, Plant Growth-Promoting Bacteria (PGPBs) provide a sustainable remedy by alleviating the detrimental impacts of drought, heat, and salinity. The favorable impacts of plant growth-promoting bacteria (PGPBs) enhance several physiological processes in plants, such as phosphorus solubilization, nitrogen fixation, nutrient availability, and the synthesis of vitamins, phytohormones, and growth regulators. PGPBs enhance plant growth and yield in reaction to rising population expansion, facilitating plant adaptation to drought stress and fulfilling human requirements. Utilizing the advantageous characteristics of PGPBs, farmers can improve crop resilience, productivity, and nutritional quality.

References

Abdelaal, K., AlKahtani, M., Attia, K., Hafez, Y., Király, L., & Künstler, A. (2021). The role of plant growth-promoting bacteria in alleviating the adverse effects of drought on plants. *Biology, 10*(6), 520.

Abdoli, M., & Saeidi, M. (2012). Effects of water deficiency stress during seed growth on yield and its components, germination and seedling growth parameters of some wheat cultivars. *International Journal of Agriculture and Crop Sciences, 4*(15), 1110–1118.

Ahmad, A., Aslam, Z., Javed, T., Hussain, S., Raza, A., Shabbir, R., & Ali, M. M. (2022). Screening of wheat (*Triticum aestivum L.*) genotypes for drought tolerance through agronomic and physiological response. *Agronomy, 12*(2), 287.

Ahmad, Z., Waraich, E. A., Akhtar, S., Anjum, S., Ahmad, T., Mahboob, W., & Rizwan, M. (2018). Physiological responses of wheat to drought stress and its mitigation approaches. *Acta Physiologiae Plantarum, 40*, 1–13.

Ahmed, H. G. M.-D., Zeng, Y., Shah, A. N., Yar, M. M., Ullah, A., & Ali, M. (2022). Conferring of drought tolerance in wheat (*Triticum aestivum L.*) genotypes using seedling indices. *Frontiers in Plant Science, 13*, 961049.

Akter, N., & Rafiqul Islam, M. (2017). Heat stress effects and management in wheat. *A Review. Agronomy for Sustainable Development, 37*, 1–17.

Ali, M., Gul, A., Hasan, H., Gul, S., Fareed, A., Nadeem, M., ... & Jamil, M. (2020). Cellular mechanisms of drought tolerance in wheat. In *Climate change and food security with emphasis on wheat* (pp. 155–167).

El-Sabagh, A. E., Islam, M. S., Hossain, A., Iqbal, M. A., Mubeen, M., Waleed, M., Reginato, M., Battaglia, M., Ahmed, S., & Rehman, A. (2021). Phytohormones as growth regulators during abiotic stress tolerance in plants. *Front. Agron., 254*, 706.

Andrianasolo, F. N., Debaeke, P., Champolivier, L., & Maury, P. (2016). Analysis and modelling of the factors controlling seed oil concentration in sunflower: A review. *OCL, 23*(2), D206.

Anjum, S. A., Ashraf, U., Zohaib, A., Tanveer, M., Naeem, M., Ali, I., & Nazir, U. (2017). Growth and development responses of crop plants under drought stress: a review.

Ansari, F. A., Ahmad, I., & Pichtel, J. (2023). Synergistic effects of biofilm-producing PGPR strains on wheat plant colonization, growth and soil resilience under drought stress. *Saudi Journal of Biological Sciences, 30*(6), Article 103664.

Ansari, F. A., Jabeen, M., & Ahmad, I. (2021). Pseudomonas azotoformans FAP5, a novel biofilm-forming PGPR strain, alleviates drought stress in wheat plant. *International Journal of Environmental Science and Technology, 18*(12), 3855–3870.

Arbona, V., Manzi, M., de Ollas, C., & Gómez-Cadenas, A. (2013). Metabolomics as a tool to investigate abiotic stress tolerance in plants. *International Journal of Molecular Sciences, 14*(3), 4885–4911.

Arora, S., Babele, P. K., & Jha, P. N. (2023). Biochemical and metabolic signatures are fundamental to drought adaptation in PGPR Enterobacter bugandensis WRS7. *Molecular Omics, 19*(8), 640–652.

Awasthi, S., Srivastava, P., & Mishra, P. K. (2017). Application of EPS in agriculture: An Important natural resource for crop improvement. *Agriculture Resource Technology Open Access Journal, 8*, 8–10.

Barnawal, D., Bharti, N., Pandey, S. S., Pandey, A., Chanotiya, C. S., & Kalra, A. (2017). Plant growth-promoting rhizobacteria enhance wheat salt and drought stress tolerance by altering endogenous phytohormone levels and TaCTR1/TaDREB2 expression. *Physiologia Plantarum, 161*(4), 502–514.

Batista, R. A., Otoni, C. G., & Espitia, P. J. (2019). Fundamentals of chitosan-based hydrogels: Elaboration and characterization techniques. *Materials for Biomedical Engineering, 55*(2), 61–81.

Benmrid, B., Ghoulam, C., Ammar, I., Nkir, D., Saidi, R., Staropoli, A., & Bargaz, A. (2024). Drought-tolerant rhizobacteria with predicted functional traits enhanced wheat growth and P uptake under moderate drought and low P-availability. *Microbiological Research*, 127795.

Bianco, C. (2024). Plant-Growth-Promoting Bacteria. *Plants, 13*(10), 1323. Retrieved from https://www.mdpi.com/2223-7747/13/10/1323

Bouremani, N., Cherif-Silini, H., Silini, A., Rabhi, N. E. H., Bouket, A. C., & Belbahri, L. (2024). Osmotolerant plant growth promoting bacteria mitigate adverse effects of drought stress on wheat growth. *AIMS microbiology, 10*(3), 507.

Cakir, R. (2004). Effect of water stress at different development stages on vegetative and reproductive growth of corn. *Field Crops Research, 89*(1), 1–16.

Camaille, M., Fabre, N., Clément, C., & Ait Barka, E. (2021). Advances in wheat physiology in response to drought and the role of plant growth promoting rhizobacteria to trigger drought tolerance. *Microorganisms, 9*(4), 687.

Chachar, M. H., Chachar, N. A., Chachar, Q., Mujtaba, S. M., Chachar, S., & Chachar, Z. (2016). Physiological characterization of six wheat genotypes for drought tolerance. *International Journal of Research-Granthaalayah, 4*(2), 184–196.

Chakraborty, U., Chakraborty, B., Chakraborty, A., & Dey, P. (2013). Water stress amelioration and plant growth promotion in wheat plants by osmotic stress tolerant bacteria. *World Journal of Microbiology and Biotechnology, 29*, 789–803.

Chaves, M. M., Maroco, J. P., & Pereira, J. S. (2003). Understanding plant responses to drought—From genes to the whole plant. *Functional Plant Biology, 30*(3), 239–264.

Dai, A. (2013). Increasing drought under global warming in observations and models. *Nature Climate Change, 3*(1), 52–58.

Danish, S., & Zafar-ul-Hye, M. (2019). Co-application of ACC-deaminase producing PGPR and timber-waste biochar improves pigments formation, growth and yield of wheat under drought stress. *Scientific Reports, 9*(1), 1–13.

Danish, S., Zafar-ul-Hye, M., Mohsin, F., & Hussain, M. (2020). ACC-deaminase producing plant growth promoting rhizobacteria and biochar mitigate adverse effects of drought stress on maize growth. *PLoS ONE, 15*(4), Article 230615.

De Storme, N., & Geelen, D. (2014). The impact of environmental stress on male reproductive development in plants: Biological processes and molecular mechanisms. *Plant, Cell & Environment, 37*(1), 1–18.

Dhakal, A. (2021). Effect of drought stress and management in wheat—A review. *Food and Agribusiness Management, 2*(2), 62–66.

Dong, B., Zheng, X., Liu, H., Able, J. A., Yang, H., Zhao, H., & Liu, M. (2017). Effects of drought stress on pollen sterility, grain yield, abscisic acid and protective enzymes in two winter wheat cultivars. *Frontiers in Plant Science, 8*, 1008.

Dubey, A.; Saiyam, D.; Kumar, A.; Hashem, A.; Abd_Allah, E.F.; Khan, M.L. (2021). Bacterial root endophytes: characterization of their competence and plant growth promotion in soybean (*Glycine max* (L.) Merr.) under Drought Stress. *Int. J. Environ. Res. Public Health, 18*, 931.

Duvnjak, J., Sarcevic, H., Vukovic, R., & Spanic, V. (2024). Effects of drought at anthesis on flag leaf physiology and gene expression in diverse wheat (*Triticum aestivum L.*) Genotypes. *Agronomy, 14*(7), 1522.

Elnajar, M., Aldesuquy, H., Abdelmoteleb, M., & Eltanahy, E. (2024). Mitigating drought stress in wheat plants (*Triticum Aestivum L.*) through grain priming in aqueous extract of spirulina platensis. *BMC Plant Biology, 24*(1), 233.

Fang, J., Wei, S., Gao, Y., Zhang, X., Cheng, Y., Wang, J., & Lu, Z. (2023). Character variation of root space microbial community composition in the response of drought-tolerant spring wheat to drought stress. *Frontiers in Microbiology, 14*, 1235708.

Farooq, M., Hussain, M., & Siddique, K. H. (2014). Drought stress in wheat during flowering and grain-filling periods. *Critical Reviews in Plant Sciences, 33*(4), 331–349.

Farooq, M., Wahid, A., Kobayashi, N., Fujita, D., & Basra, S. M. (2009). Plant drought stress: effects, mechanisms and management. *Sustainable agriculture*, 153–188.

Glick, B. R. (2014). Bacteria with ACC deaminase can promote plant growth and help to feed the world. *Microbiological Research, 169*(1), 30–39.

Glick, B. R., & Nascimento, F. X. (2021). Pseudomonas 1-Aminocyclopropane-1-carboxylate (ACC) Deaminase and Its Role in Beneficial Plant-Microbe Interactions. *Microorganisms, 9*, 2467.

Gontia-Mishra, I., Sapre, S., Sharma, A., & Tiwari, S. (2016). Amelioration of drought tolerance in wheat by the interaction of plant growth-promoting rhizobacteria. *Plant Biology, 18*, 992–1000.

Goswami, M., & Suresh, D. (2020). Plant growth-promoting rhizobacteria—alleviators of abiotic stresses in soil: A review. *Pedosphere, 30*(1), 40–61.

Gu, Z., Hu, C., Gan, Y., Zhou, J., Tian, G., & Gao, L. (2024). Role of microbes in alleviating crop drought stress: A review. *Plants, 13*, 384.

Gujjar, R. S., & Supaibulwatana, K. (2019). The mode of cytokinin functions assisting plant adaptations to osmotic stresses. *Plants, 8*(12), 542.

Gupta, G., Parihar, S. S., Ahirwar, N. K., Snehi, S. K., & Singh, V. (2015). Plant growth promoting rhizobacteria (PGPR): Current and future prospects for development of sustainable agriculture. *Journal Microbiology Biochemical Technology, 7*(2), 096–102.

Gupta, S., & Pandey, S. (2023). Plant growth promoting rhizobacteria to mitigate biotic and abiotic stress in plants. *Sustainable Agriculture Reviews 60: Microbial Processes in Agriculture*, 47–68.

Haider, I., Raza, M. A. S., Iqbal, R., Aslam, M. U., Habib-ur-Rahman, M., Raja, S., & Ahmad, S. (2020). Potential effects of biochar application on mitigating the drought stress implications on wheat (Triticum aestivum L.) under various growth stages. *Journal of Saudi Chemical Society, 24*(12), 974–981.

Hajibarat, Z., & Saidi, A. (2022). Senescence-associated proteins and nitrogen remobilization in grain filling under drought stress condition. *Journal of Genetic Engineering and Biotechnology, 20*(1), 101.

Hasanuzzaman, M., Parvin, K., Bardhan, K., Nahar, K., Anee, T. I., Masud, A. A. C., Fotopoulos, V. (2021). Biostimulants for the regulation of reactive oxygen species metabolism in plants under abiotic stress. *Cells, 10*, 2537.

Hakim, S., Naqqash, T., Nawaz, M. S., Laraib, I., Siddique, M. J., Zia, R., & Imran, A. (2021). Rhizosphere engineering with plant growth-promoting microorganisms for agriculture and ecological sustainability. *Frontiers in Sustainable Food Systems, 5*(2571), 1–16.

Hone, H., Mann, R., Yang, G., Kaur, J., Tannenbaum, I., Li, T., & Sawbridge, T. (2021). Profiling, isolation and characterisation of beneficial microbes from the seed microbiomes of drought tolerant wheat. *Scientific Reports, 11*(1), 11916.

Hossain, M. A., Wani, S. H., Bhattacharjee, S., Burritt, D. J., & Tran, L.-S. P. (2016). Drought stress tolerance in plants, Volume. *Switzerland: Springer, 1*, 105–143.

Inoue, T., Inanaga, S., Sugimoto, Y., An, P., & Eneji, A. (2004). Effect of drought on ear and flag leaf photosynthesis of two wheat cultivars differing in drought resistance. *Photosynthetica, 42*, 559–565.

Jabran, M., Ali, M. A., Acet, T., Zahoor, A., Abbas, A., Arshad, U., ... & Gao, L. (2024). Growth regulation in bread wheat via novel bioinoculant formulation. *BMC Plant Biology, 24*(1), 1039.

Jiang, K., & Asami, T. (2018). Chemical regulators of plant hormones and their applications in basic research and agriculture. *Bioscience, Biotechnology, and Biochemistry, 82*, 1265–1300.

Jochum, M. D., McWilliams, K. L., Borrego, E. J., Kolomiets, M. V., Niu, G., Pierson, E. A., & Jo, Y.-K. (2019). Bioprospecting plant growth-promoting rhizobacteria that mitigate drought stress in grasses. *Frontiers in Microbiology, 10*, 2106.

Kabiraj, A., Majhi, K., Halder, U., Let, M., & Bandopadhyay, R. (2020). Role of plant growth-promoting rhizobacteria (PGPR) for crop stress management. In *Sustainable Agriculture in the Era of Climate Change* (pp. 367–389): Springer.

Kasim, W. A., Osman, M. E., Omar, M. N., & Salama, S. (2021). Enhancement of drought tolerance in Triticum aestivum L. seedlings using Azospirillum brasilense NO40 and Stenotrophomonas maltophilia B11. *Bulletin of the National Research Centre, 45*(1), 95.

Kour, D., Rana, K. L., Sheikh, I., Kumar, V., Yadav, A. N., Dhaliwal, H. S., & Saxena, A. K. (2020). Alleviation of drought stress and plant growth promotion by Pseudomonas libanensis EU-LWNA-33, a drought-adaptive phosphorus-solubilizing bacterium. *Proceedings of the National Academy of Sciences, India Section b: Biological Sciences, 90*, 785–795.

Kaushal, M., & Wani, S. P. (2016). Plant-growth-promoting rhizobacteria: Drought stress alleviators to ameliorate crop production in drylands. *Annals of Microbiology, 66*, 35–42.

Khan, M. A., Waseem Akram, M., Iqbal, M., Ghulam Muhu-Din Ahmed, H., Rehman, A., Arslan Iqbal, H. S. M., & Alam, B. (2023). Multivariate and association analyses of quantitative attributes reveal drought tolerance potential of wheat (Triticum aestivum L.) genotypes. *Communications in Soil Science and Plant Analysis, 54*(2), 178–195.

Khan, M. A. U., Mohammad, F., Khan, F. U., Ahmad, S., Raza, M. A., & Kamal, T. (2020). Comparison among different stability models for yield in bread wheat. *Sarhad J. Agric, 36*(1), 282–290.

Kumar, A., Naroju, S. P., Kumari, N., Arsey, S., Kumar, D., Gubre, D. F., & Saini, P. (2024). The role of drought response genes and plant growth promoting bacteria on plant growth promotion under sustainable agriculture: A review. *Microbiological Research*, 127827.

Lastochkina, O., Yakupova, A., Avtushenko, I., Lastochkin, A., & Yuldashev, R. (2023). Effect of seed priming with endophytic Bacillus subtilis on some physio-biochemical parameters of two wheat varieties exposed to drought after selective herbicide application. *Plants, 12*(8), 1724.

Li, J., Wang, J., Liu, H., Macdonald, C. A., & Singh, B. K. (2023). Microbial inoculants with higher capacity to colonize soils improved wheat drought tolerance. *Microbial Biotechnology, 16*(11), 2131–2144.

Li, W., Herrera-Estrella, L., & Tran, L.-S.P. (2016). The Yin-Yang of cytokinin homeostasis and drought acclimation/adaptation. *Trends in Plant Science, 21*(7), 548–550.

Li, X., Fang, P., Mai, J., Choi, E. T., Wang, H., & Yang, X.-F. (2013). Targeting mitochondrial reactive oxygen species as novel therapy for inflammatory diseases and cancers. *Journal of Hematology & Oncology, 6*, 1–19.

Mansour, H. A., El Sayed Mohamed, S., & Lightfoot, D. A. (2020). Molecular studies for drought tolerance in some Egyptian wheat genotypes under different irrigation systems. *Open Agriculture, 5*(1), 280–290.

McCree, K., & Davis, S. (1974). Effect of water stress and temperature on leaf size and on size and number of epidermal cells in Grain Sorghum 1. *Crop Science, 14*(5), 751–755.

Men, Y., Wang, D., Li, B., Su, Y., & Chen, G. (2018). Effects of drought stress on the antioxidant system, osmolytes and secondary metabolites of *Saposhnikovia divaricata* seedlings. *Acta Physiologiae Plantarum, 40*, 191.

Mutanda, M., Figlan, S., Chaplot, V., Madala, N. E., & Shimelis, H. (2024). Selection of wheat (Triticum aestivum L.) genotypes using yield components, water use efficiency and major metabolites under drought stress. *Journal of Agronomy and Crop Science, 210*(5), e12766.

Ngumbi, E., & Kloepper, J. (2016). Bacterial-mediated drought tolerance: current and future prospects. *Applied Soil Ecology, 105*, 109–125.

Noureen, S., Iqbal, A., & Muqeet, H. A. (2024). Potential of drought tolerant rhizobacteria amended with biochar on growth promotion in wheat. *Plants, 13*(9), 1183.

Nurhidayati, T., Sari, L., Anggraeni, A., Luqman, A., Shovitri, M., Kuswytasari, N., & Rizki, H. (2024). The effect of mycorrhiza and plant growth-promoting rhizobacteria supplementation on zea mays saccharata sturt. growth and productivity grown on low nutrients soil. *Nature Environment and Pollution Technology, 23*(3), 1449–1459.

Nyaupane, S., Poudel, M. R., Panthi, B., Dhakal, A., Paudel, H., & Bhandari, R. (2024). Drought stress effect, tolerance, and management in wheat—A review. *Cogent Food and Agriculture, 10*(1), 2296094.

Oladir, G., Saeed, M., & Cheerna, M. A. (1999). Effect of water stress on growth and yield performance of four wheat cultivars. *Pakistan Journal of Biological Sciences, 5*(8), 56–98.

Oni, B. A., Oziegbe, O., & Olawole, O. O. (2019). Significance of biochar application to the environment and economy. *Annals of Agricultural Sciences, 64*(2), 222–236.

Öztürk, M. A., Altay, V., & Efe, R. (2022). *Biodiversity, Conservation and Sustainability in Asia*: Springer.

Pathania, P., Rajta, A., Singh, P. C., & Bhatia, R. (2020). Role of plant growth-promoting bacteria in sustainable agriculture. *Biocatalysis and Agricultural Biotechnology, 30*(1), Article 101842.

Paudel, B., Zhang, Y., Yan, J., Rai, R., Li, L., Wu, X., & Khanal, N. R. (2020). Farmers' understanding of climate change in Nepal Himalayas: Important determinants and implications for developing adaptation strategies. *Climatic Change, 158*, 485–502.

Paul, G. K., Mahmud, S., Dutta, A. K., Sarkar, S., Laboni, A. A., Hossain, M. S., & Saleh, M. A. (2022). Volatile compounds of bacillus pseudomycoides induce growth and drought tolerance in wheat (*Triticum aestivum L.*). *Scientific Reports, 12*(1), 19137.

Peleg, Z., & Blumwald, E. (2011). Hormone balance and abiotic stress tolerance in crop plants. *Current Opinion in Plant Biology, 14*(3), 290–295.

Poudel, M. R., Ghimire, S., Pandey, M. P., Dhakal, K. H., Thapa, D. B., & Poudel, H. K. (2020). Evaluation of wheat genotypes under irrigated, heat stress and drought conditions. *J. Biol. Today's World, 9*(1), 212.

Prasad, J. K., Dey, R., Pandey, P., & Raghuwanshi, R. (2025). Exploring drought tolerance in PGPR-protected wheat plants through physiological and molecular expression. *Journal of Plant Growth Regulation*, 1–16.

Rahman, K. U., Ejaz, N., Shang, S., Balkhair, K. S., Alghamdi, K. M., Zaman, K., & Hussain, A. (2024). A robust integrated agricultural drought index under climate and land use variations at the local scale in Pakistan. *Agricultural Water Management, 295*, Article 108748.

Rashid, U., Yasmin, H., Hassan, M. N., Naz, R., Nosheen, A., Sajjad, M., ... & Ahmad, P. (2022). Drought-tolerant Bacillus megaterium isolated from semi-arid conditions induces systemic tolerance of wheat under drought conditions. *Plant Cell Reports*, 1–21.

Reddy, A. R., Chaitanya, K. V., & Vivekanandan, M. (2004). Drought-induced responses of photosynthesis and antioxidant metabolism in higher plants. *Journal of Plant Physiology, 161*(11), 1189–1202.

Rostamian, A., Moaveni, P., Mozafari, H., & Rajabzadeh, F. (2023). Effective drought mitigation by rhizobacteria consortium in wheat field trials. Rhizosphere, 25, 100653.

Salem, M. A., Ismail, M. A., Radwan, K. H., & Abd-Elhalim, H. M. (2024). Unlocking the potential of plant growth-promoting rhizobacteria to enhance drought tolerance in Egyptian wheat (Triticum aestivum). *Sustainability, 16*(11), 4605.

Saleem, M., Nawaz, F., Hussain, M. B., & Ikram, R. M. (2021). Comparative effects of individual and consortia plant growth promoting bacteria on physiological and enzymatic mechanisms to confer drought tolerance in maize (*Zea mays L.*). *Journal of Soil Science and Plant Nutrition, 21*(4), 3461–3476.

Sallam, A., Alqudah, A. M., Dawood, M. F., Baenziger, P. S., & Börner, A. (2019). Drought stress tolerance in wheat and barley: Advances in physiology, breeding and genetics research. *International Journal of Molecular Sciences, 20*(13), 3137.

Sen, S., Ghosh, D., & Mohapatra, S. (2018). Modulation of polyamine biosynthesis in Arabidopsis thaliana by a drought mitigating Pseudomonas putida strain. *Plant Physiology and Biochemistry, 129*, 180–188.

Sharma, B., Yadav, L., Shrestha, A., Shrestha, S., Subedi, M., Subedi, S., & Shrestha, J. (2022). Drought stress and its management in wheat (*Triticum aestivum L.*): a review. *Agricultural Science and Technology (1313–8820), 14*(1).

Shaffique, S., Imran, M., Kang, S. M., Khan, M. A., Asaf, S., Kim, W. C., & Lee, I. J. (2023). Seed bio-priming of wheat with a novel bacterial strain to modulate drought stress in Daegu, South Korea. *Frontiers in Plant Science, 14*, 1118941.

Shirmohammadi, E., Alikhani, H. A., Pourbabaei, A. A., & Etesami, H. (2020). Improved phosphorus (P) uptake and yield of rainfed wheat fed with P fertilizer by drought-tolerant phosphate-solubilizing fluorescent pseudomonads strains: A field study in drylands. *Journal of Soil Science and Plant Nutrition, 20*(4), 2195–2211.

Singh, S. K. (2020). Microbes for iron chlorosis remediation in Peach. *Biostimulants in Plant Science, 49*(1), 777–780.

Singh, D., Thapa, S., Singh, J. P., Mahawar, H., Saxena, A. K., Singh, S. K., ... & Chakdar, H. (2024). Prospecting the Potential of Plant Growth-Promoting microorganisms for mitigating drought stress in crop plants. Current Microbiology, 81(3), 84.

Singh, D., Thapa, S., Yadav, J., Singh, D., Chakdar, H., Kumar, M., & Saxena, A. K. (2023a). Deciphering the mechanisms of microbe mediated drought stress alleviation in wheat. *Acta Physiologiae Plantarum, 45*(6), 81.

Singh, M., Sharma, J. G., & Giri, B. (2023b). Microbial inoculants alter resilience towards drought stress in wheat plants. *Plant Growth Regulation, 101*(3), 823–843.

Solangi, A. H., Solongi, N., Jatoi, W., Solangi, M. K., Solangi, S. K., Memom, S., & Soomro, S. (2021). Drought tolerance indices of wheat (*Triticum aestivum L.*) genotypes under water deficit conditions. *Plant Cell Biotechnol Mol Biol, 22*, 1–19.

Sonbarse, P. P., Kiran, K., Sharma, P., & Parvatam, G. (2020). Biochemical and molecular insights of PGPR application for the augmentation of carotenoids, tocopherols, and folate in the foliage of Moringa oleifera. *Phytochemistry, 179*, Article 112506.

Talaat, N. B., & Abdel-Salam, S. A. (2024). An innovative, sustainable, and environmentally friendly approach for wheat drought tolerance using vermicompost and effective microorganisms: Upregulating the antioxidant defense machinery, glyoxalase system, and osmotic regulatory substances. *BMC Plant Biology, 24*(1), 866.

Tanveer, Y., Yasmin, H., Nosheen, A., Farah, M. A., & Altaf, M. A. (2024). Synergizing Bacillus halotolerans, Pseudomonas sihuiensis and Bacillus atrophaeus with folic acid for enhanced drought resistance in wheat by metabolites and antioxidants. *BMC Plant Biology, 24*(1), 1003.

Vacheron, J., Desbrosses, G., Bouffaud, M.-L., Touraine, B., Moënne-Loccoz, Y., Muller, D., & Prigent-Combaret, C. (2013). Plant growth-promoting rhizobacteria and root system functioning. *Frontiers in Plant Science, 4*, 356.

Verma, G., Srivastava, D., Tiwari, P., Chakrabarty, D. (2019). ROS modulation in crop plants under drought stress. *React Oxyg Nitrogen Sulfur Species Plants Prod Metab Signal Def Mech* 2019, *2019*, 311–336

Wahid, A., & Close, T. (2007). Expression of dehydrins under heat stress and their relationship with water relations of sugarcane leaves. *Biologia Plantarum, 51*, 104–109.

Yadav, V. K., Yadav, R. C., Choudhary, P., Sharma, S. K., & Bhagat, N. (2022). Mitigation of drought stress in wheat (*Triticum aestivum L.*) by inoculation of drought tolerant Bacillus paramycoides DT-85 and Bacillus paranthracis DT-97. *J. Appl. Biol. Biotechnol, 10*, 59–69.

Zahra, N., Wahid, A., Hafeez, M. B., Ullah, A., Siddique, K. H., & Farooq, M. (2021). Grain development in wheat under combined heat and drought stress: Plant responses and management. *Environmental and Experimental Botany, 188*(2), 104517–104537.

Zhang, J., & Davies, W. (1987). Increased synthesis of ABA in partially dehydrated root tips and ABA transport from roots to leaves. *Journal of Experimental Botany, 38*(12), 2015–2023.

Zhang, Z., Hu, M., Xu, W., Wang, Y., Huang, K., Zhang, C., & Wen, J. (2021a). Understanding the molecular mechanism of anther development under abiotic stresses. *Plant Molecular Biology, 105*, 1–10.

Zhang, Z., Rod, M., & Hosseinian, F. (2021b). A comprehensive review on sustainable industrial vertical farming using film farming technology. *Sustainable Agriculture Research, 10*(1), 46–53.

Efficacy of Carrier-Mediated Salt-Tolerant Bacteria in Plant Growth Promotion

Atia Iqbal and Raina Rashid

Abstract The climate around us has been changing continuously that made stability in the agricultural environment way more difficult resulting in the generation of no. of environmental stresses including salinity. The following chapter deals with the significance of plant defenses against salinity stress including both hormonal defense as well as enzymatic defense. The defense systems in plants are not strong enough to cope alone with stress so there is a need for some kind of assistance for plants that help them to mitigate salt stress. This chapter further highlights the significance of rhizobacterial application along with carrier material to enhance the wheat growth parameters under salt stress conditions. Rhizobacteria are proven to be a promising strategy because of their ability to cope with salinity stress of a diverse range from 100 to 800 mM alone and along with a wide range of organic carriers that add up the fertility of the soil and provide a medium for these bacteria to be applied feasibly in fields as well.

Keywords Salinity · PGPR · Carrier · Agriculture · Rhizobacteria

1 Introduction

Wheat, the prime crop worldwide that supplies 40% of energy as well as 20% of protein to the world's population. Wheat is the second most cultivated crop after maize because of its high nutritional value and a staple crop for accomplishing the feeding of about 35% of the human population (Hussain et al., 2021).

There is a drastic increase in the human population and according to the report by the UN human population will increase to 9.3 billion by the year 2050. This increase will risk our natural resources including water, crops, vegetables, and much more. To overcome the risks there is a lot of need for food security and this could be possible by the collaborative efforts of scientists in the fields of genetics, plant

A. Iqbal (✉) · R. Rashid
Department of Microbiology and Molecular Genetics, The Women University Multan, Multan, Pakistan
e-mail: atia.iqbal@wum.edu.pk

M. Shaaban (ed.), *Soils and Sustainable Agriculture*, Frontier Studies in Soil Science,
https://doi.org/10.1007/978-3-031-91114-9_29

microbiology, plant breeding, and biotechnology (İbrahimova et al., 2021). Due to continuous climate change, the plant will face several stresses during its growth most of them are salinity, drought, changes in temperature and pH, etc. Stress reduces yield and limits plant growth and development (Zhang et al., 2021). About three-quarters of the cultivable area is affected by salinization. In the future, more agricultural land will face the same due to the production of industrial wastes, use of chemicals, and poor irrigation practices (Zulfiqar andAshraf, 2021) (Fig. 1).

The field application of pesticides and agrochemicals has an adverse effect that hurts a major contributor toward environmental pollution disturbing the agro-ecosystem. To discourage such practices an alternative, bio-fertilizers should be introduced to farmers for field application. rhizobacterial strains and carrier material can be used in the fields for different crops.

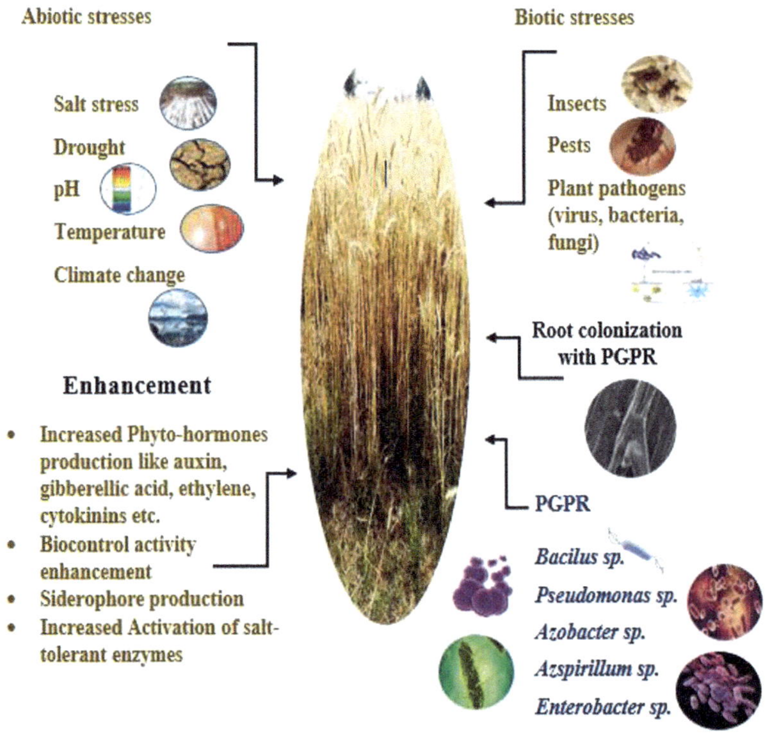

Fig. 1 The symbiotic relationship between plants and microorganisms in the natural Environment (Selosse et al., 2004)

2 Effect of Salinity on Plants

Increased salt concentration in soil greatly threatens the world's agriculture and economy. It will also affect the balance in the ecosystem. Different reports have suggested an immersive increase in population day by day. There is a need to improve crop production worldwide to feed such a huge number of living organisms around the globe. According to the USDA (United States Department of Agriculture), when the electric conductivity of soil reaches 4 dS m^{-1} or 40 mM of NaCl it will be considered as saline soil (Munns andTester, 2008). Several plants are unable to withstand salinity conditions and many show lesser growth and yield under salt stress. So, salinity is a threat to agriculture on one hand and acts as a driving force for evolution in plants on the other hand. Under such stress many plants adapt themselves to cope with stress by developing different mechanisms, symbiosis is also a great example of such adaptations (Kosová et al., 2013).

Salt stress negatively influences plant growth and yield in many ways including osmotic imbalance as well as nutritional instability in soil. Moreover, due to salt stress, reactive oxygen species such as hydrogen peroxide H_2O_2 production will be more which further affects the plant growth adversely and causes oxidative stress-induced damage to DNA (Kumar et al., 2021). Salinity also affects many crucial processes including photosynthesis, metabolic pathways, ionic imbalance, and activities of many enzymes. Certain plants such as maize show moderate tolerance against salt stress but that's not enough. There is a need to enhance the tolerance to maintain the internal physiology of the plants under such stressed conditions (Ahmad et al., 2021) (Fig. 2).

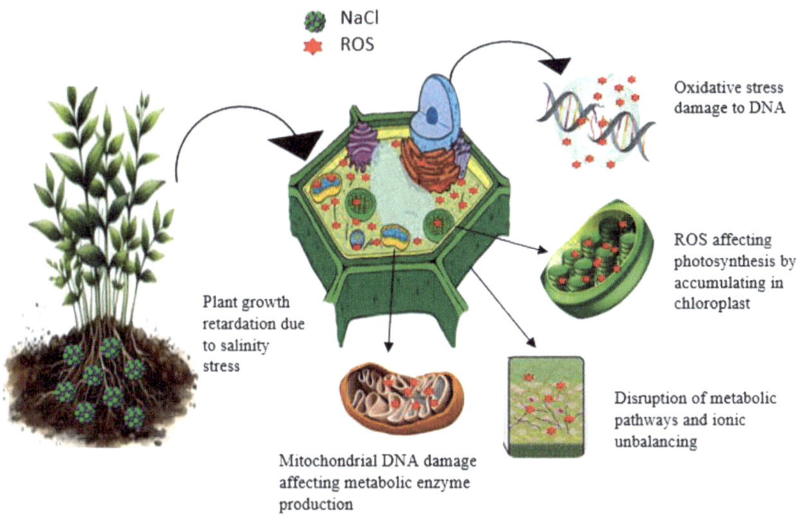

Fig. 2 Effect of salinity on plant physiology (Khan et al., 2019)

Under a saline environment, plant cells produce many ROS molecules, hydrogen peroxide (H_2O_2), and singlet oxygen O_2 negatively affect photosynthesis which will disturb the electron flow ultimately causing the oxidative burst, damaging plant tissues, and cell death. Moreover, such damages will also disturb the intracellular ionic balance that will affect the nutrient uptake from the soil causing nutrient deficiency in the plants (Sharma et al., 2021).

3 Plant's Defenses Against Salinity

Certain mechanisms in plants against ROS include enzymatic and non-enzymatic defenses. The non-enzymatic antioxidants include tocopherols, flavonoids, ascorbic acid, glutathione, phenolic, etc. (Rudenko et al., 2023). Along with these many enzymatic antioxidants work against ROS including ascorbate peroxidase (APX), superoxide dismutase (SOD), de-hydro ascorbate reductase (Azeem et al.), catalase (CAT), peroxidase (POX) and mono de-hydro ascorbate reductase (MDHAR) (Rajput et al., 2021).

4 Mechanism

Superoxide dismutase (SOD) converts superoxide (O_{2-}) into hydrogen peroxide (H2O2) and oxygen through a catalytic reaction, while catalase (CAT) enzyme decomposes H_2O_2 into water and oxygen. Peroxidase (POD) utilizes electron donors like ascorbate or glutathione for reducing H_2O_2 to water. Glutathione reductase (GR) maintains glutathione pools, and ascorbate peroxidase (APX) utilizes ascorbate to reduce H_2O_2 (Fig. 3).

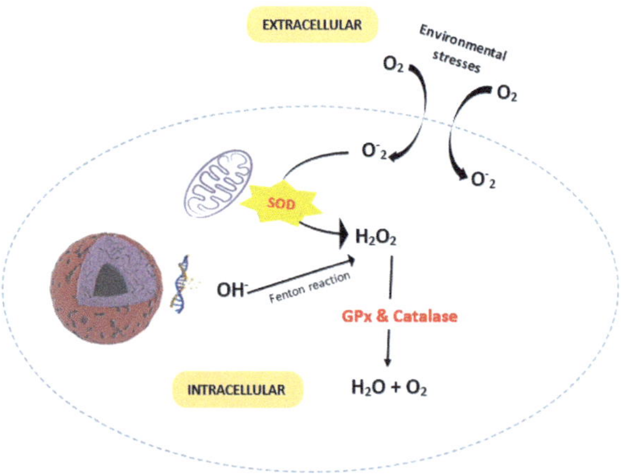

Fig. 3 A schematic representation of plant enzymatic defense system against ROS production as a result of environmental stresses (de Souza-Vieira et al., 2024)

These interconnected reactions scavenge reactive oxygen species (ROS), guard cells from oxidative injury, maintain redox balance, and regulate stress responses. Localized in various cellular compartments, these enzymes are regulated by hormones, transcription factors, and post-translational modifications (Mir & Khah, 2024). Their coordinated action ensures efficient ROS scavenging, protecting plants from salinity stress, and understanding these enzymatic reactions informs strategies to enhance plant tolerance to oxidative stress (César et al., 2024).

5 Hormonal Defense

Many plant hormones such as gibberellic acid, auxin (Indole Acetic Acid), cytokinins, and salicylic acid promote plant seed germination. Both plants, as well as rhizobacteria or soil bacteria, produce such hormones (Waris et al., 2024). Auxins, for instance, enhance cell elongation, cell differentiation, and division, ultimately increasing root length, shoot length, and apical dominance (Wu et al., 2024). Indole-3-acetic acid (Omara et al.) and Indole-3-butyric acid (IBA) are examples of auxins that regulate cell enlargement, root initiation, leaf abscission, and tropisms (Li et al., 2022).

Gibberellins, another important hormone for plants, regulate the germination of seeds, stem elongation, and expansion of leaves (Tian et al., 2024). Gibberellic acid (GA3) is a familiar gibberellin that enhances seed germination, fruit production, flowering, and breaking dormancy (Sakouhi et al., 2024). Ethylene is a volatile hormone often known for the regulation of fruit stress responses, ripening, and senescence (Liu et al., 2024). It promotes fruit ripening, senescence, and abscission while responding to environmental stresses like salinity, flooding, and drought (Mahto et al., 2024).

Cytokinins, such as Kinetin and benzylaminopurine (BAP) stimulate cell division and cell differentiation and delay senescence, which promotes root length and mobilization of nutrients (Nagar et al., 2024). Cytokinin regulatory mechanisms involve complex interactions with other plant hormones like auxins, abscisic acid (ABA), and ethylene (Jan et al., 2024). Cytokinins and auxin interaction helps in the regulation of root and shoot development by enhancing auxin-induced cell elongation, while auxins inhibit Cytokinin-induced cell division (Cruz et al., 2024). Similarly, Cytokinin induced cell division is inhibited by Ethylene, and ethylene production is reduced by cytokinins, delaying senescence (Magar et al., 2024). Many plant stresses including drought and dormancy can be regulated through ABA and Cytokinin interactions as ABA inhibits Cytokinin-induced cell growth, and cytokinins reduce ABA-induced stomatal closure (Abd-Elsalam andMohamed, 2024) (Fig. 4).

Understanding plant hormones is essential for agricultural applications and plant breeding. Hormone interactions and balance regulate plant growth and defense, making them chemical messengers that coordinate responses to environmental stimuli (Xie et al., 2024). PGPR (rhizobacterial strains) possess characteristics such as auxin production, solubilization of phosphate, nitrogen fixation, exopolysaccharides, hormones, and many important metabolic enzymes (Baha, 2021).

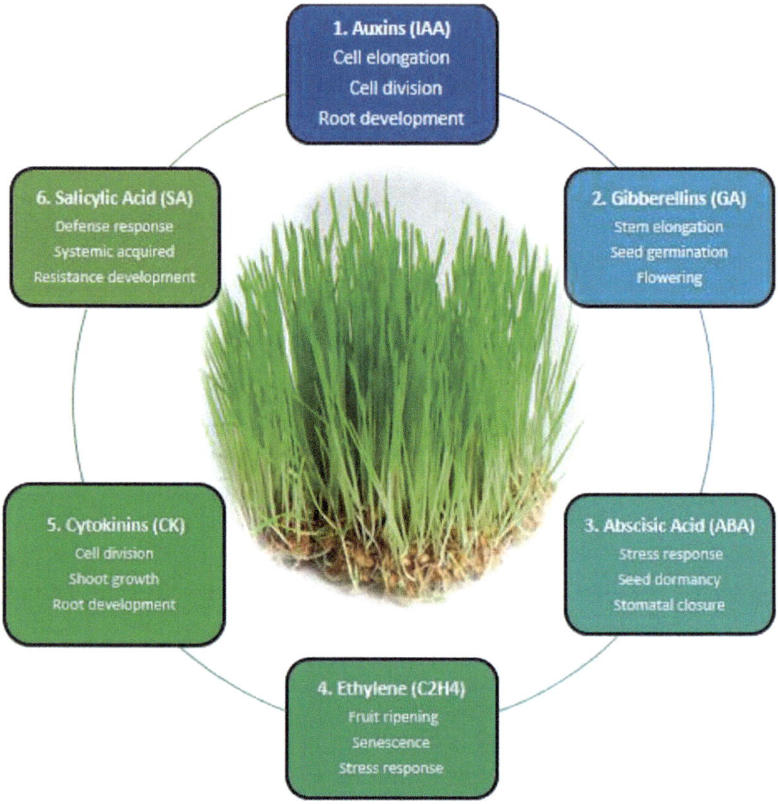

Fig. 4 Phytohormones and their roles in mitigating and regulating the plant growth and defense mechanisms

6 Use of Chemical Fertilizer and Their Drastic Effects on Soil

Fertilizers are the compounds or materials that are used to improve the growth of plants and fertility of soil. Chemical fertilizers include a number of chemicals that increase plant yields but also have adverse effects on the environment as well as the quality of crops. With the increasing environmental pollution, the use of chemical fertilizers is risky and there is a need to replace them with a safer approach is the use of rhizobacteria as bio-fertilizers. Bio-fertilizers are organic inoculants along with suitable microorganisms to enhance soil fertility and plant growth (Fasusi et al., 2021).

7 Rhizobacteria an Eco-Friendlier and Cost-Effective Approach

Rhizobacteria can be applied to seed and added to the soil to colonize the rhizosphere of plants by biofilm formation. They improve the crop yield and enhance the soil texture and nutrients. The biofilm-producing rhizobacteria proved to be more effective than non-biofilm producing bacteria (Ansari et al., 2024). Moreover, Rhizobacteria increase the growth, and survival of seedlings, root system, and chlorophyll content, improve defense mechanism by degrading many harmful compounds and enhancing the production of enzymes, produce several plants' growth-promoting hormones, fix environmental nitrogen, reduce flowering time, and speed up seed germination (Nosheen et al., 2021).

Several studies showed that rhizobacterial inoculation will increase the production and activity of antioxidant enzymes by decreasing the proline production as well as balancing of Na^+/ K^+ ratio. Rhizobacterial strains also accelerate the production of protein as well as the biosynthesis of many metabolic and defense enzymes (Zhao et al., 2021).

Rhizobacteria provide plants with essential molecules including N_2, phosphate, Potassium, Iron, and Zinc along with the direct production of many enzymes including ACC deaminase and phytohormones like auxin. The indirect way involves inducing defense mechanisms against many pathogens and promoting plant growth and development through the production of hydrogen cyanide (HCN), siderophore, and metabolites showing antimicrobial activity (Sharma et al., 2021) (Table 1).

8 Use of Carrier to Formulate a Bio-Fertilizer

The field application of bio-fertilizers requires long-term stability of rhizobacterial strains that can be provided through carrier materials. Liquid bio-fertilizers usually have a short shelf life as compared to solid ones. The solid carrier materials can be organic (peat, compost, biogas slurry, charcoal, biochar, lignite, husk of different plants, corn cob, etc.), inorganic (perlite, vermiculite, zeolite, etc.), and synthetic (nanoparticles, alginate beads, polyacrylamide gel, etc.) (Sohaib et al., 2020).

Organic fertilizers show the best results as compared to inorganic and synthetic carriers due to their high C and N content. Peat is the most effective form among organic carriers, but its availability is deficient in Pakistan. To combat the need many studies suggest that some other organic materials including coal, biogas slurry, and compost have shown satisfactory results as a carrier (Tabassam et al., 2015). The previous study showed the positive effect of *Pseudomonas* species along with sugarcane husk improved the salt-tolerant capacity of wheat as well as increased up to 25% of wheat yield as compared to control. The use of sugarcane husk has improved the stability and survival of rhizobacteria (Hassan and Bano, 2015).

Table 1 Effect of PGPR on different varieties of plants facing salinity stress

Sr. no.	Plant specie	PGPR	Salt tolerance by PGPR	Function	References
1	Rice	*Bacillus tequilensis* *Bacillus aryabhattai*	2 M of NaCl	Maintaining Na + /K + ionic balance by secreting exopolysaccharide (EPS)	(Shultana et al., 2020)
2	Maize	*Alcaligense sp. E1*, *Alacligense sp. E2*	150 mM	IAA production by both strains Antioxidant scavenging activities	(El-Amriti et al., 2024)
3	Wheat	*Bacillus thuringiensis*, *Bacillus tropicus*,	3% or 30 mM	Phosphate solubilization,	(Malik et al., 2024)
4	Tomato	*Bacillus subtilis*, *Bacillus mojavensis*	150 mM	Production of exopolysaccharides, ACC deaminase and polyamines	(Abdelkefi et al., 2024)
5	Quinoa	*Bacillus inaquosorum*, *Bacillus thuringiensis*, *Bacillus proteolyticus*	300 mM	Proline production	(Slatni et al., 2024)
6	Rice	*Pseudomonas atacamensis*	200 mM	IAA production, siderophore production, and extracellular enzyme production	(Arora et al., 2024)
7	Wheat	*Pantoea agglomerans*	500 mM	Regulation of photosynthesis, antioxidant system, and sulfur metabolism	(Ansari et al., 2024)
8	Wheat	*Bacillus velezensis*	10 ds/m or 100 mM	Balancing of Na^+ and K^+ levels, production of osmoprotectants like glycine betaine and proline, increase production of chlorophyll as well as carotenoids	(Mehrabi et al., 2024)
9	Tomato	*Pseudarthrobacter oxydans*, *Staphylococcus pasteuri*	150 mM	Increasing antioxidant enzyme production and total protein content	(Latif et al., 2024)
10	Soyabean	*Streptomyces lasalocidi*	N/A	Regulation of the expression genes (GmVSP, GmWRKY54, and GmPHD2,) involve in stress-response, and the secretion of indole-3-carboxaldehyde (ICA1d) to overcome stress and improve root architecture	(Lu et al., 2024)

(continued)

Table 1 (continued)

Sr. no.	Plant specie	PGPR	Salt tolerance by PGPR	Function	References
11	Arabidopsis thaliana	Beijerinckia *fluminensis*,	200 mM	ABA production, regulating genes that control Na^+ compartmentalization, and Na^+ recycling in plant tissues	(Giannelli et al., 2024)
12	Rice	*Enterobacter sp.*	200 mM	Improve chlorophyll content, antioxidant activity, photosynthetic efficiency, protease, and urease activity	(Ji et al., 2024)
13	Wheat	*Streptomyces sp., Pseudomonas sp.*	200 mM to 800 mM	Exopolysaccharides (EPS) production and biofilm formation	(Thakur & Yadav, 2024)
14	Maize	*Bacillus aryabhattai*	1.5 M	EPS production, antioxidant production, and secondary metabolites production	(Azeem et al., 2024)
15	*Phaseolus vulgaris (L.)*	*Rhizobium azibense*	300 mM	Upregulating the antioxidant defense mechanism and EPS production	(Shahid et al., 2024)

Carrier materials are used to stabilize the bacterial inoculum and also increase the shelf life of bio-fertilizer. For this purpose, inorganic materials like perlite, zeolite, etc., and synthetic materials like agarose gel, and alginate beads can also be used. These inorganic materials provide nutrients like K, P, and N to soil and also protect bacteria from environmental conditions. The combination of rhizobacterial strains and inorganic material should be compared with an organic one. The optimization of carrier material is necessary to construct the most suitable combination for the field application (Al-Jubouri andKhafaji, 2021).

The synthetic carriers including nanoparticles, and alginate beads also proved beneficial but are costly. The handling of these carriers is complicated as compared to organic and inorganic carriers but with less risk of contamination (Sparrow Jr andHam, 1983).

A good carrier material should be inexpensive, readily available, provide stability to the rhizobacterial strains, and also protect the inoculant from diverse climate change. It should be autoclave able to avoid contamination, be non-polluting, degradable, and have sufficient water holding and pH buffering capacity (Raj et al., 2021). However, previous studies showed that many organic fertilizers like compost, cow dung, and peat help in increasing soil moisture as well as fertility but very low effect on increasing plant tolerance towards abiotic stresses (Table 2). A combination of

Table 2 Combination of different organic carriers with plant growth-promoting microorganisms applied on a variety of plants facing salinity stress

Sr. no	Carrier material	Plant growth-promoting microbes	Plant	Stress type	Growth attributes	References
1	Peat moss	*Firmicutes sp., Proteobacteria sp.*	*Arabidopsis thaliana*	Salt stress and drought stress	Enhancement and modulation of root system architecture (RSA)	(Nourashrafeddin et al., 2024)
2	Coco-coir, Rice hull ash, and Composted sugarcane	*Bacillus, Enterobacter* and *Trichoderma species*	Ornamental plants Inca marigold and zinnia	No stress	Increase germination percentage (93.6%), root length, and leaf total chlorophyll content	(Rasool et al., 2024)
3	Biogas slurry	*Bacillus subtilis*	*Pisum sativum*	Salt stress	Increase stem length by 62.7%, root length by 5.93%, as well as chlorophyll content by 29.26% under stress	(Muslim et al., 2023)
4	Compost tea	*Azospirillum brasilense, Pseudomonas koreensis,* and *Bacillus circulans*	Wheat	Salt stress	Increased grain yield by 80% also improves photosynthesis and antioxidant activity	(Elbagory, 2023)
5	Biochar	*Pseudomonas putida, Azotobacter chroococcum*	Rapeseed *(Brassica napus L.)*	Salt stress	Improves root and shoot length. Enhances seed yield from 83 to 109% and seed oil from 69 to 115%	(Ghassemi-Golezani andAbdoli, 2023)

(continued)

Table 2 (continued)

Sr. no	Carrier material	Plant growth-promoting microbes	Plant	Stress type	Growth attributes	References
6	Organic manure	*Pseudomonas atacamensis*	Rice	Salt stress	Improve nutrient content by 35% and biochemical parameters by 32%, significantly reducing EC from 7.43 to 4.3 dS/m	(Arora et al., 2024)
7	Compost	*Arbuscular mycorrhizal fungi (AMF),*	Tomato	Salt stress	Increases plant photosynthetic pigment concentration and protein content	(Mekkaoui et al., 2024)
8	Biochar	*Bacillus thuringiensis, Bacillus tropicus,*	Wheat	Salt stress	Improves soil Na^+/K^+ ratio by 44%, number of tillers by 16–28.57%, grain weight by 34–37.9%, and roots by 5–21.5%	(Malik et al., 2024)
9	Biochar	N/A	Alfalfa, Amaranth and Maize	Salt stress	Improves plant height and root morphological traits, enhances chlorophyll content and antioxidant enzyme activities	(Jabborova et al., 2023)

(continued)

Table 2 (continued)

Sr. no	Carrier material	Plant growth-promoting microbes	Plant	Stress type	Growth attributes	References
10	Salicylic acid	*Pseudomonas aeruginosa*	Maize *(Zea mays L.)*	Salt stress	Enhances shoot and root dry weights by 41% and 56%, relative water contents by 32%. Improves chlorophyll contents by 27% and grain yield by 41%	(Ali et al., 2023)
11	Biochar	*Pseudomonas sp.*	Maize	Salt stress	Improves soil moisture cntet, proline content an peroxidase (POD) enzyme activity	(Fazal andBano, 2016)
12	Compost	*Azospirillum lipoferum, Bacillus coagulans, Bacillus circulance,* and *Bacillus subtilis*	Wheat	Salt stress	Plant growth and yield parameters showed significant enhancement, improved chlorophyll content and antioxidant activity of enzymes	(Omara et al., 2022)

bio-fertilizer with organic carrier material will surely enhance plant fertility and increase tolerance in wheat against abiotic stresses (Fatawi et al., 2021).

9 Conclusion

Salt tolerant bacteria plays a significant role in growth promotion of wheat. Amended with carrier material, the efficacy of these salt tolerant bacteria may be enhanced significantly that would be helpful in improving crop yield and productivity especially in saline conditions. This chapter elucidated the role of salt tolerant bacteria along with carrier material to enhance plant health that would be helpful in future agriculture practices and application of bio-fertilizers.

References

Abd-Elsalam, K., & Mohamed, H. I. (2024). *Plant growth regulators to manage biotic and abiotic stress in agroecosystems*. CRC Press.

Abdelkefi, N., Louati, I., Mechichi, H.-Z., Sayahi, N., El-Sayed, W. S., El Nayal, A., Ismail, W., Hanin, M., & Mechichi, T. (2024). Enhanced salt stress tolerance in tomato plants following inoculation with newly isolated plant growth-promoting rhizobacteria. *Scientia Horticulturae, 328*, 112921.

Ahmad, S., Cui, W., Kamran, M., Ahmad, I., Meng, X., Wu, X., Su, W., Javed, T., El-Serehy, H. A., & Jia, Z. (2021). Exogenous application of melatonin induces tolerance to salt stress by improving the photosynthetic efficiency and antioxidant defense system of maize seedling. *Journal of Plant Growth Regulation, 40*(3), 1270–1283.

Al-Jubouri, A. K. and Khafaji, H. A. (2021). The effect of type and method of immobilizing bacillus megaterium bacteria inoculation in increasing potassium available and growth of wheat plant *Triticum Aestivum L.* IOP conference series: earth and environmental science

Ali, Q., Ahmad, M., Kamran, M., Ashraf, S., Shabaan, M., Babar, B. H., Zulfiqar, U., Haider, F. U., Ali, M. A., & Elshikh, M. S. (2023). Synergistic effects of Rhizobacteria and salicylic acid on maize salt-stress tolerance. *Plants, 12*(13), 2519.

Ansari, F. A., Ahmad, I., Pichtel, J., & Husain, F. M. (2024). Pantoea agglomerans FAP10: A novel biofilm-producing PGPR strain improves wheat growth and soil resilience under salinity stress. *Environmental and Experimental Botany, 222*, 105759.

Arora, N. K., Mishra, J., Singh, P., & Fatima, T. (2024). Salt-tolerant plant growth-promoting Pseudomonas atacamensis KSS-6 in combination with organic manure enhances rice yield, improves nutrient content and soil properties under salinity stress. *Journal of Basic Microbiology, 64*(6), 2300767.

Azeem, M. A., Khan, S., Ali, F., Ahmad, S., Rahim, G., Iftikhar, M., Iqbal, M., Aljaloud, R. S., Alarjani, K. M. & Chaudhary, H. J. (2024). Halotolerant bacillus aryabhattai strain PM34 mitigates salinity stress and enhances the physiology and growth of maize. *Journal of Plant Growth Regulation*, 1–19.

Baha, N. (2021). Impact of bio-fertilizers on germination and early seedling growth of Alfalfa (*Medicago sativa L.*) under salt stress. *Acta Physiologiae Plantarum, 43*(5), 1–11.

César, A. P., Lopes, F. E., Azevedo, F. F., Pinto, Y. O., Andrade, C. R., Mesquita, F. P., Silva, G. O., Freitas, C. D., & Souza, P. F. (2024). Antioxidant peptides from plants: A review. *Phytochemistry Reviews, 23*(1), 95–104.

Cruz, K. Z. C. M., da Silva Alencar, A. A., Santana, J. G. S., & Alves, L. E. O. (2024). The auxin and cytokinin promotes the regeneration of shoots in Phalaenopsis Golden Peoker 'BL'HCC/ AOS (Orchidaceae). *Brazilian Journal of Development, 10*(4), e68953–e68953.
de Souza-Vieira, Y., Felix-Mendes, E., Galhego, V., Bastos, G. A., Felix-Cordeiro, T., Ding, X., Zhang, Y., Corrêa, R. L., Wang, X., & Sachetto-Martins, G. (2024). Euphorbiaceae superoxide dismutase, catalase, and glutathione peroxidase as clues to better comprehend high drought tolerance in castor bean. *Industrial Crops and Products, 222*, 119510.
El-Amriti, F. A., Ouf, S. A., Abu-Elghait, M., Desouky, S. E. & Mohamed, M. S. (2024). Alleviation of salt stress on Zea mays L. plant by PGPR isolates as an effective sustainable strategy. *Biocatalysis and Agricultural Biotechnology*, 103346.
Elbagory, M. (2023). Reducing the adverse effects of salt stress by utilizing compost tea and effective microorganisms to enhance the growth and yield of wheat (*Triticum aestivum L.*) plants. *Agronomy, 13*(3), 823.
Fasusi, O. A., Cruz, C., & Babalola, O. O. (2021). Agricultural sustainability: Microbial biofertilizers in rhizosphere management. *Agriculture, 11*(2), 163.
Fatawi, A., Pujiasmanto, B., Zaki, M. & Noda, K. (2021). Application of organic amendments and PGPR on Salibu Rice yield for drought adaptation. IOP conference series: earth and environmental science
Fazal, A., & Bano, A. (2016). Role of plant growth-promoting rhizobacteria (PGPR), biochar, and chemical fertilizer under salinity stress. *Communications in Soil Science and Plant Analysis, 47*(17), 1985–1993.
Ghassemi-Golezani, K. & Abdoli, S. (2023). Alleviation of salt stress in rapeseed (*Brassica napus L.*) plants by biochar-based rhizobacteria: new insights into the mechanisms regulating nutrient uptake, antioxidant activity, root growth and productivity. *Archives of Agronomy and Soil Science, 69*(9), 1548–1565.
Giannelli, G., Mattarozzi, M., Gentili, S., Fragni, R., Maccari, C., Andreoli, R., & Visioli, G. (2024). A novel PGPR strain homologous to Beijerinckia fluminensis induces biochemical and molecular changes involved in Arabidopsis thaliana salt tolerance. *Plant Physiology and Biochemistry, 206*, 108187.
Hassan, T. U., & Bano, A. (2015). Role of carrier-based biofertilizer in reclamation of saline soil and wheat growth. *Archives of Agronomy and Soil Science, 61*(12), 1719–1731.
Hussain, D., Asrar, M., Khalid, B., Hafeez, F., Saleem, M., Akhter, M., Ahmed, M., Ali, I., & Hanif, K. (2021). Insect pests of economic importance attacking wheat crop (*Triticum aestivum L.*) in Punjab, Pakistan. *International Journal of Tropical Insect Science, 42*, 1–12. https://doi.org/10.1007/s42690-021-00574-9
İbrahimova, U., Kumari, P., Yadav, S., Rastogi, A., Antala, M., Suleymanova, Z., Zivcak, M., Tahjib-Ul-Arif, M., Hussain, S., & Abdelhamid, M. (2021). Progress in understanding salt stress response in plants using biotechnological tools. *Journal of Biotechnology, 329*, 180–191. https://doi.org/10.1016/j.jbiotec.2021.02.007
Jabborova, D., Abdrakhmanov, T., Jabbarov, Z., Abdullaev, S., Azimov, A., Mohamed, I., AlHarbi, M., Abu-Elsaoud, A., & Elkelish, A. (2023). Biochar improves the growth and physiological traits of alfalfa, amaranth and maize grown under salt stress. *PeerJ, 11*, e15684.
Jan, M., Muhammad, S., Jin, W., Zhong, W., Zhang, S., Lin, Y., Zhou, Y., Liu, J., Liu, H., & Munir, R. (2024). Modulating root system architecture: cross-talk between auxin and phytohormones. *Frontiers in Plant Science, 15*, 1343928.
Ji, J., Zhang, J., Wang, X., Song, W., Ma, B., Wang, R., Li, T., Wang, G., Guan, C. & Gao, X. (2024). The alleviation of salt stress on rice through increasing photosynthetic capacity, maintaining redox homeostasis and regulating soil enzyme activities by Enterobacter sp. JIV1 assisted with putrescine. *Microbiological Research, 280*, 127590.
Khan, A., Khan, A. L., Muneer, S., Kim, Y.-H., Al-Rawahi, A., & Al-Harrasi, A. (2019). Silicon and salinity: crosstalk in crop-mediated stress tolerance mechanisms. *Frontiers in Plant Science, 10*, 1429.

Kosová, K., Vítámvás, P., Urban, M. O., & Prášil, I. T. (2013). Plant proteome responses to salinity stress—comparison of glycophytes and halophytes. *Functional Plant Biology, 40*(9), 775–786. https://doi.org/10.1071/FP12375

Kumar, S., Li, G., Yang, J., Huang, X., Ji, Q., Liu, Z., Ke, W. & Hou, H. (2021). Effect of salt stress on growth, physiological parameters, and ionic concentration of water dropwort (*Oenanthe javanica*) cultivars. *Frontiers in plant science, 12.*

Latif, A., Ahmad, R., Ahmed, J., Mueen, H., Khan, S. A., Bibi, G., Mahmood, T., & Hassan, A. (2024). Novel halotolerant PGPR strains alleviate salt stress by enhancing antioxidant activities and expression of selected genes leading to improved growth of Solanum lycopersicum. *Scientia Horticulturae, 338*, 113625.

Li, C., Lei, C., Wang, K., Tan, M., Xu, F., Wang, J., & Zheng, Y. (2022). MADS2 regulates priming defence in postharvest peach through combined salicylic acid and abscisic acid signaling. *Journal of Experimental Botany, 73*(11), 3787–3806.

Liu, M., Wang, C., Ji, H., Sun, M., Liu, T., Wang, J., Cao, H., & Zhu, Q. (2024). Ethylene biosynthesis and signal transduction during ripening and softening in non-climacteric fruits: an overview. *Frontiers in Plant Science, 15*, 1368692.

Lu, L., Liu, N., Fan, Z., Liu, M., Zhang, X., Tian, J., Yu, Y., Lin, H., Huang, Y., & Kong, Z. (2024). A novel PGPR strain, Streptomyces lasalocidi JCM 3373T, alleviates salt stress and shapes root architecture in soybean by secreting indole-3-carboxaldehyde. *Plant, Cell & Environment, 47*(6), 1941–1956.

Magar, M. M., Liu, H. & Yan, G. (2024). Wheat TaAP2/ERF genes regulate heat tolerance through ethylene signaling at grain-filling stage. *Journal of Plant Growth Regulation*, 1–18.

Mahto, R. K., Tripathi, S., Singh, D. P., Lal, K., Kumar, D., Kumar, R., Ramlal, A., Singh, S. K. & Sharma, S. S. (2024). Ethylene in abiotic stress. In *Phytohormones in Abiotic Stress* (pp. 73–78). CRC Press.

Malik, L., Sanaullah, M., Mahmood, F., Hussain, S., & Shahzad, T. (2024). Co-application of biochar and salt tolerant PGPR to improve soil quality and wheat production in a naturally saline soil. *Rhizosphere, 29*, 100849.

Mehrabi, S. S., Sabokdast, M., Bihamta, M. R., & Dedičová, B. (2024). The coupling effects of PGPR inoculation and foliar spraying of strigolactone in mitigating the negative effect of salt stress in wheat plants: insights from phytochemical, growth, and yield attributes. *Agriculture, 14*(5), 732.

Mekkaoui, F., Ait-El-Mokhtar, M., Zaari Jabri, N., Amghar, I., Essadssi, S., & Hmyene, A. (2024). The use of compost and arbuscular mycorrhizal fungi and their combination to improve tomato tolerance to salt stress. *Plants, 13*(16), 2225.

Mir, R. A. & Khah, M. A. (2024). Recent progress in enzymatic antioxidant defense system in plants against different environmental stresses. *Improving Stress Resilience in Plants*, 203–224.

Munns, R., & Tester, M. (2008). Mechanisms of salinity tolerance. *Annual Review of Plant Biology, 59*(1), 651–681. https://doi.org/10.1146/annurev.arplant.59.032607.092911

Muslim, N., Akhtar, N., Saddiq, S., Imran, M., Yousaf, K., & andSharif, N. (2023). Effect of PGPR and biogas slurry on growth and yield parameters of pea under salt affected conditions. *Pakistan Journal of Biotechnology, 20*(02), 154–162.

Nagar, S., Meena, S., Karwa, S., Dhakar, R., Raju, D., Kumar, S. andUmesh, D. K. (2024). Regulatory role of cytokinin in abiotic stress tolerance of plants. In *Phytohormones in Abiotic Stress* (pp. 64–72). CRC Press.

Nosheen, S., Ajmal, I., & Song, Y. (2021). Microbes as biofertilizers, a potential approach for sustainable crop production. *Sustainability, 13*(4), 1868.

Nourashrafeddin, S. M., Ramandi, A., & Seifi, A. (2024). Rhizobacteria isolated from xerophyte Haloxylon ammodendron manipulate root system architecture and enhance drought and salt tolerance in Arabidopsis thaliana. *International Microbiology, 27*(2), 337–347.

Omara, A.E.-D., Hafez, E. M., Osman, H. S., Rashwan, E., El-Said, M. A., Alharbi, K., Abd El-Moneim, D., & Gowayed, S. M. (2022). Collaborative impact of compost and beneficial rhizobacteria on soil properties, physiological attributes, and productivity of wheat subjected to deficit irrigation in salt affected soil. *Plants, 11*(7), 877.

Raj, R., Preethy, H. A., & Rex, K. G. R. (2021). Development of banana peel powder as organic carrier based bioformulation and determination of its plant growth promoting efficacy in rice Cr100g. *Journal of Pure and Applied Microbiology, 15*(3), 1279–1291.

Rajput, V. D., Harish, Singh, R. K., Verma, K. K., Sharma, L., Quiroz-Figueroa, F. R., Meena, M., Gour, V. S., Minkina, T. & Sushkova, S. (2021). Recent developments in enzymatic antioxidant defence mechanism in plants with special reference to abiotic stress. *Biology, 10*(4), 267

Rasool, S., Ahmad, I., Ziaf, K., & Naveed, M. (2024). Plant growth promoting rhizobacteria incorporated soilless substrates–A potential arena for high quality nursery production of ornamentals. *Scientia Horticulturae, 326*, 112745.

Rudenko, N. N., Vetoshkina, D. V., Marenkova, T. V., & Borisova-Mubarakshina, M. M. (2023). Antioxidants of non-enzymatic nature: Their function in higher plant cells and the ways of boosting their biosynthesis. *Antioxidants, 12*(11), 2014.

Sakouhi, L., Hussaan, M., Murata, Y., & Chaoui, A. (2024). Role of calcium signaling in cadmium stress mitigation by indol-3-acetic acid and gibberellin in chickpea seedlings. *Environmental Science and Pollution Research, 31*(11), 16972–16985.

Selosse, M.-A., Baudoin, E., & Vandenkoornhuyse, P. (2004). Symbiotic microorganisms, a key for ecological success and protection of plants. *Comptes Rendus Biologies, 327*(7), 639–648.

Shahid, M., Altaf, M. & Danish, M. (2024). Halotolerant exopolysaccharide-producing Rhizobium azibense increases the salt tolerance mechanism in Phaseolus vulgaris (L.) by improving the growth, ion homeostasis, and antioxidant defensive enzymes. *Chemosphere*, 142431.

Sharma, A., Singh, R. K., Singh, P., Vaishnav, A., Guo, D.-J., Verma, K. K., Li, D.-P., Song, X.-P., Malviya, M. K., & Khan, N. (2021). Insights into the bacterial and nitric oxide-induced salt tolerance in sugarcane and their growth-promoting abilities. *Microorganisms, 9*(11), 2203.

Shultana, R., Kee Zuan, A. T., Yusop, M. R., & Saud, H. M. (2020). Characterization of salt-tolerant plant growth-promoting rhizobacteria and the effect on growth and yield of saline-affected rice. *PLoS ONE, 15*(9), e0238537.

Slatni, T., Ben Slimene, I., Harzalli, Z., Taamalli, W., Smaoui, A., Abdelly, C., & Elkahoui, S. (2024). Enhancing quinoa (Chenopodium quinoa) growth in saline environments through salt-tolerant rhizobacteria from halophyte biotope. *Physiologia Plantarum, 176*(4), e14466.

Sohaib, M., Zahir, Z. A., Khan, M. Y., Ans, M., Asghar, H. N., Yasin, S., & Al-Barakah, F. N. (2020). Comparative evaluation of different carrier-based multi-strain bacterial formulations to mitigate the salt stress in wheat. *Saudi Journal of Biological Sciences, 27*(3), 777–787.

Sparrow, S., Jr., & Ham, G. (1983). Survival of rhizobium phaseoli in six carrier materials 1. *Agronomy Journal, 75*(2), 181–184.

Tabassam, T., Sultan, T., Akhtar, M. E., Hassan, M. M.-U. & Ali, A. (2015). Suitability of different formulated carriers for sustaining microbial shelf life. *Pakistan Journal of Agricultural Research, 28*(2).

Thakur, R. & Yadav, S. (2024). Biofilm forming, exopolysaccharide producing and halotolerant, bacterial consortium mitigates salinity stress in Triticum aestivum. *International Journal of Biological Macromolecules, 262*, 130049

Tian, Z., Chen, B., Li, H., Pei, X., Sun, Y., Sun, G., Pan, Z., Dai, P., Gao, X. and Geng, X. (2024). The strigolactone-gibberellin crosstalk mediated by a distant silencer fine-tunes plant height in upland cotton. *Molecular Plant*.

Waris, I., Saleem, M., Sher, N., Shahid, M., Abbas, M. J., Aziz, S. U., Nabi, M. M. U., & Noor, H. (2024). The regulation of stress responses in fruit crops is influenced by plant hormones: A review. *Asian Journal of Research in Agriculture and Forestry, 10*(1), 72–78.

Wu, M., Zhou, J., Li, Q., Quan, D., Wang, Q. & Gao, Y. (2024). Auxin–brassinosteroid crosstalk: Regulating rice plant architecture and grain shape. *The Crop Journal*.

Xie, P., Liu, F. andXie, Q. (2024). Manipulating hormones to mitigate trade-offs in crops. *Plant, Cell and Environment*.

Zhang, H., Zhu, J., Gong, Z. & Zhu, J.-K. (2021). Abiotic stress responses in plants. *Nature Reviews Genetics*, 1–16.

Zhao, Y., Zhang, F., Mickan, B., Wang, D. & Wang, W. (2021). Physiological, proteomic, and metabolomic analysis provide insights into Bacillus sp.-mediated salt tolerance in wheat. *Plant Cell Reports, 41*(1), 1–24. https://doi.org/10.1007/s00299-021-02788-0

Zulfiqar, F. & Ashraf, M. (2021). Nanoparticles potentially mediate salt stress tolerance in plants. *Plant Physiology and Biochemistry, 160*, 257–268. https://doi.org/10.1016/j.plaphy.2021.01.028

Biochar for Sustainable Development

Manita Aryal, Bhim Kumar Dahal, and Baidya Purnima

Abstract Biochar (BC) was used in Amazon Basin about 2000 year ago. It was extensively used in the soil as an ancient indigenous culture to increase the soil nutrients and soil fertility. Despite leaching caused by heavy tropical rain, this region continues to be extremely productive, balancing pH, protecting and preserving the natural environment. BC are produced by the process of pyrolysis. It has gained significant attention recently due to its potential benefits in improving soil fertility by recycling nutrients, mitigating greenhouse gas emissions by sequestering carbon, promoting sustainable agricultural practices by increasing water-holding capacity and reducing erosion. In addition to it's agricultural benefits, it has also been used in the remediation of environmental pollutants, showing resistance to biotic and abiotic stress to increase the plant growth and yield. It helps to meet the needs of the present without compromising the ability of future generations to meet their own needs. It encompasses three main pillars which includes economic growth, social inclusion and environmental protection. Implementing BC based strategies requires careful consideration of production methods, application rates and long-term effects, paving the way for a more sustainable, and resilient agricultural future. This chapter discusses the uses of BC for sustainable development.

Keywords Biochar · Greenhouse gas · Nutrient recycling · Sustainable development · Agriculture

---

M. Aryal · B. Purnima (✉)
Central Department of Microbiology, Tribhuvan University, Kirtipur, Nepal
e-mail: purnima.baidya@cdmi.tu.edu.np

B. K. Dahal
Civil Department of Engineering, Pulchowk Campus, Tribhuvan University, Kathmandu, Nepal

M. Shaaban (ed.), *Soils and Sustainable Agriculture*, Frontier Studies in Soil Science,
https://doi.org/10.1007/978-3-031-91114-9_30

1 History of Biochar

Biochar has deep historical roots, with its use dating back thousands of years. Notably, ancient civilizations in the Amazon Basin utilized it to create "Terra Preta" (Portuguese for "black earth"). (Schmidt et al., 2021). Terra Preta soils are characterized by high fertility and dark coloration, resulting from the incorporation of charcoal, organic matter, and pottery shards. These soils were created between 450 BCE and 950 CE by indigenous people and remain highly fertile even after centuries, providing early evidence of BC's long-term benefits to soil health (Shareef & Zhao, 2017).

Various traditional agricultural practices across the world have incorporated forms of BC, though not always under that name. In Japan, for instance, "Kuntan" (rice husk charcoal) has been used for centuries to enhance soil quality and crop yields (Koyama et al., 2015). Similarly, in parts of Africa and Asia, farmers have used charcoal and ash from cooking fires to improve soil fertility (Hamidi et al., 2023; Neina, Faust, & Joergensen, 2020).

The modern scientific interest in BC began in earnest in the late twentieth century, spurred by environmental concerns and the search for sustainable agricultural practices. Researchers began to study "Terra Preta" soils more closely in the 1960s and 1970s, leading to a better understanding of how BC contributes to soil fertility (H. P. Schmidt et al., 2021). In the early 2000s, the concept of BC gained significant attention as part of discussions on climate change mitigation (Bo et al., 2023). BC was recognized not only for its ability to enhance soil quality but also for its potential to sequester carbon and reduce greenhouse gas emissions (Lefebvre et al., 2020). This renewed interest was driven by the work of scientists such as Johannes Lehmann, who helped to elucidate the mechanisms through which BC interacts with soil and affect the microbiota (Lehmann et al., 2011).

2 Recent Developments

In the twenty-first century, BC has become a focus of extensive research and commercial interest. The formation of the International BC Initiative (IBI) in 2006 marked a significant step in promoting BC research and development globally (H. P. Schmidt et al., 2021). IBI has worked to standardize BC production and application methods, and to disseminate knowledge about its benefits and potential uses (Meyerkord, 2023). Besides this, BC production technologies have also advanced, with various methods developed to optimize the process for different feedstocks and intended uses (Murtaza et al., 2024). Pyrolysis, the thermal decomposition of organic material in the absence of oxygen, is the most common method for producing BC (Lehmann et al., 2011; Schmidt et al., 2021). Advances in pyrolysis technology have made it possible to produce BC more efficiently and with greater control over its properties and different companies have also been established throughout the world to produce BC from different kinds of substrates (Murtaza et al., 2024).

3 Application of Biochar

Today, BC has been used in a variety of applications beyond traditional agriculture. It has been utilized as silage agent, feed additive/supplement, litter additive, slurry treatment, manure composting, water treatment in fish farming, carbon fertilizer, compost additive, substitute for peat in potting soil, plant protection, compensatory fertilizer for trace elements, insulation, air decontamination, decontamination of earth foundations, humidity regulation, protection against electromagnetic radiation, biomass additive, biogas slurry treatment, active carbon filter, pre-rinsing additive, soil substrate for organic plant beds, composting toilets, micro-filters in developing countries, exhaust filters in controlling emission and room air filters, industrial materials for carbon fibers and plastics, semiconductors and batteries, in reduction of metals, in cosmetics such as soap, skin cream and therapeutic bath additives, etc. It is also used in paints and coloring agents as food colorants and industrial paints, in energy production, such as pellets, and as a substitute for lignite. It is used for detoxification of toxic compounds. It is also used as the carrier for active pharmaceutical ingredients. It can be used as fabric additive for functional underwear. It is used for thermal insulation for functional clothing. It can be used as the deodorant for shoe soles. Similarly, it can be used as the cataplasm for insect bites, abscesses and eczema. It is also used to fill mattresses and pillows. It can also be used in soil remediation, waste management, water filtration, and even as a component in building materials (Schmidt, 2013). Research continues to explore new uses and to optimize BC production processes for maximum environmental and economic benefit (Ahmad Bhat et al., 2022; Schmidt, 2013; Singh Yadav et al., 2023). The properties of BC have been illustrated in the Fig. 1.

The potential use of BC to contribute to climate change mitigation remains a significant driving force behind ongoing research and development (Woolf et al., 2010). By sequestering carbon in a stable form and improving soil health, BC offers a promising tool for addressing some of the pressing environmental challenges of our time (Singh Yadav et al., 2023). As interest in sustainable and regenerative agriculture grows, the historical legacy of BC and its modern rediscovery are likely to play an increasingly important role in shaping the future of farming and environmental stewardship (Bo et al., 2023; Murtaza et al., 2024). The characteristics and applications of BC have been illustrated in the Fig. 2.

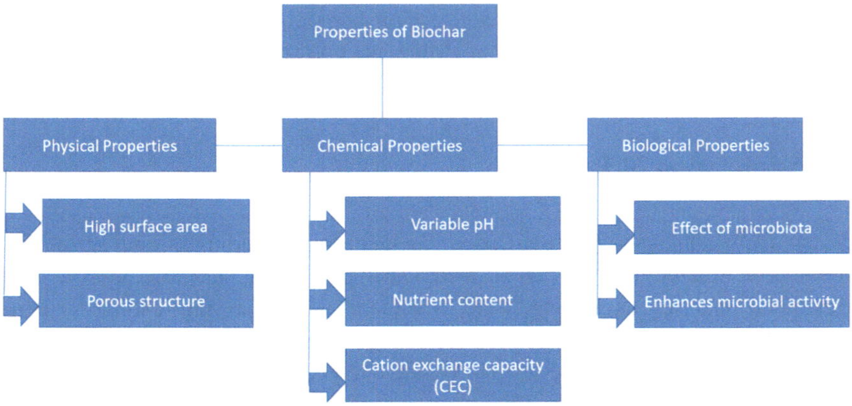

Fig. 1 Properties of biochar

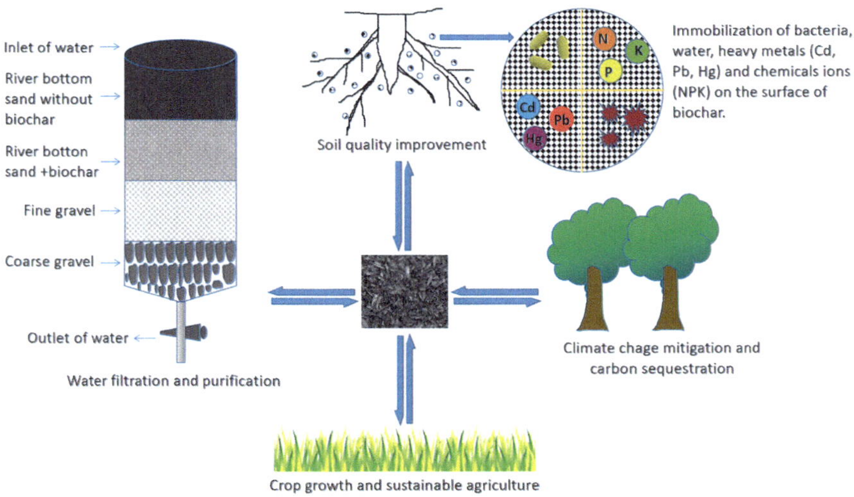

Fig. 2 Sustainable development using biochar

4 Biochar for Sustainable Development

The United Nations adopted 2030 Agenda for Sustainable Development in 2015, which includes 17 Sustainable Development Goals (SDGs) providing a universal framework to address global challenges and promote sustainable development. BC can be directly linked with the SDG 6, SDG 7, SDG 12 and SDG 13 (Ahmad Bhat et al., 2022) and also with SDG 1, SDG 2, SDG 3, SDG 6, SDG 7, SDG 8, SDG 9, SDG 10, SDG 11, SDG 12, SDG 13, SDG 14 and SDG 15 (Abhishek Kumar & Bhattacharya, 2021). Some of the most common effects of BC include:

4.1 *Environmental Benefits*

4.1.1 Soil Health Improvement

Biochar can significantly enhance soil fertility and productivity, leading to higher crop yields. It can improve soil structure, increases water retention, and enhances nutrient availability leading to better crop growth and resistant to drought and poor soil condition (Wang et al., 2022). BC has a highly porous structure and a large surface area, which allows it to hold water within its pores. This porosity enhances the soil's ability to retain water, making it available to plants over a longer period (Obour et al., 2019). By improving soil aggregation, BC enhances soil structure. Better-structured soils have larger pore spaces that can hold water more effectively (He et al., 2020). This leads to increased infiltration and reduced runoff. BC can contribute to the increase of organic matter content in the soil. Organic matter has high water-holding capacity and can absorb and retain significant amounts of water (Batista et al., 2018). BC can increase the cation exchange capacity (CEC) of the soil (Shaaban et al., 2018). Higher CEC improves the soil's ability to retain water and nutrients, making them more available to plants (Yuan et al., 2023). BC can reduce soil compaction by improving soil structure and porosity. Less compacted soils have better water infiltration and retention capacities (Blanco-Canqui, 2021). BC promotes the growth of beneficial soil microorganisms, which helps to improve soil structure and increase the soil's ability to retain water through their activity and the production of organic compounds (Singh Yadav et al., 2023). BC can influence the hydraulic conductivity of the soil, allowing it toretain water during dry periods better and improve drainage during wet periods (Libutti et al., 2019). The surface chemistry of BC can attract and hold water molecules. This hygroscopic nature helps in retaining water in the soil matrix. BC helps to create the microclimates that are favorable for crop growth (Shaaban et al., 2018).

4.1.2 Carbon Sequestration and Climate Change Mitigation

Biochar sequesters carbon in the soil, helping to mitigate climate change and improve soil health over the long period. BC consists of carbon that is in a stable form (Vaccari et al., 2011). When organic materials are pyrolyzed (thermally decomposed in the absence of oxygen), the carbon content is converted into a form that resists decomposition. This stability allows BC to persist in soils for hundreds to thousands of years, effectively sequestering carbon that would otherwise be released back into the atmosphere as CO_2 (Mulabagal et al., 2020). The application of BC to soils has been shown to reduce emissions of methane (CH_4) and nitrous oxide (N_2O), two potent greenhouse gases (Efosa et al., 2023). This reduction occurs because BC can enhance soil aeration, which reduces anaerobic conditions that favor methane production. Additionally, BC can interact with nitrogen compounds in the soil, reducing nitrous oxide emissions (Efosa et al., 2023). By improving soil structure, water retention, and nutrient availability, BC can enhance plant growth. Healthier plants can absorb more CO_2 through photosynthesis, thus increasing the amount of carbon sequestered in plant biomass and subsequently in the soil when plant residues decompose (Brar et al., 2024). Converting waste biomass into BC through pyrolysis prevents the release of CO_2 that would occur if the biomass were left to decompose or were burned. This process not only sequesters carbon but also provides a valuable soil amendment (Mulabagal et al., 2020). BC improves soil structure and reduces erosion, preventing the loss of soil organic carbon and maintaining the soil's carbon storage capacity (Khan et al., 2024). BC can be combined with other sustainable agricultural practices, such as no-till farming, cover cropping, and organic farming, to enhance overall carbon sequestration and soil health (Nogués et al., 2023). The pyrolysis process used to produce BC also generates bioenergy in the form of syngas or bio-oil, which can be used as renewable energy sources (Aboelela et al., 2023). This offsets the use of fossil fuels, further reducing greenhouse gas emissions.

4.1.3 Pollution Remediation

Biochar's large surface area and porous structure enable it to adsorb heavy metals from soil and water (Shaaban et al., 2018). Metals like lead (Pb), cadmium (Cd), and arsenic (As) can bind to BC surfaces, reducing their mobility and bioavailability (Zang et al., 2023). Functional groups on BC surfaces (e.g., carboxyl, hydroxyl) can form complexes with heavy metals, enhancing their immobilization (Zeng et al., 2021). BC can adsorb various organic pollutants, including pesticides, herbicides, and polycyclic aromatic hydrocarbons (PAHs). Its porous structure traps these compounds, reducing their presence in soil and water (Qiu et al., 2022). BC can support microbial communities that degrade organic pollutants. The enhanced microbial activity around BC particles helps break down contaminants more effectively (Mukherjee et al., 2022). BC can adsorb nutrients like nitrogen (N) and phosphorus (P), preventing their leaching into waterways (Wu et al., 2024) reducing the risk of eutrophication (Brar et al., 2024). By immobilizing contaminants, BC improves soil health, making it safer for plant growth and reducing the risk of contaminants entering the food chain (Zang et al., 2023). BC can be used in water filtration systems to remove contaminants from drinking water or wastewater. Its adsorptive properties help capture and retain pollutants, providing cleaner water (Qiu et al., 2022). BC can adsorb volatile organic compounds from the air and soil, reducing their release into the environment and potential harm to human health (Kumar et al., 2020). BC can absorb oil and other hydrocarbons from soil and water, aiding in the cleanup of oil spills and reducing environmental damage. BC can be used in conjunction with plants that uptake contaminants through phytoremediation (Zahed et al., 2021). The combination enhances the overall efficiency of contaminant removal from the environment. Bioremediation techniques can be integrated with BC, where microorganisms are used to degrade pollutants (Manikandan et al., 2023). BC offers a habitat and boosts the activity of these microorganisms, improving the remediation process.

4.2 Agricultural Applications

4.2.1 Soil Amendment

Biochar helps in the formation of soil aggregates, providing better soil structure, which enhances aeration, water infiltration, and root penetration (Manikandan et al., 2023). Adding BC to soil reduces bulk density, decreasing soil compaction and making it easier for roots to grow and for water and air to move through the soil (Khan et al., 2024). BC has a high surface area and porous structure, which allows it to hold water within its pores (Wang et al., 2022) increasing its ability to retain moisture, making water available to plants during dry periods (He et al., 2020). Improved soil porosity from BC application enhances the soil's hydraulic conductivity, balancing water retention and drainage (Wang et al., 2022) with high cation exchange capacity (CEC) allowing adsorption and retention of essential nutrients such as potassium, calcium, magnesium, and ammonium, making them available to plants over time (Yuan et al., 2023). This minimizes nutrient leaching, ensuring more nutrients remain in the root zone for plant uptake (Alkharabsheh et al., 2021). BC can neutralize acidic soils due to its alkaline nature, leading to a more favorable pH range for many crops (Shi et al., 2019). This pH buffering capacity helps in creating optimal conditions for nutrient availability and microbial activity. The porous structure of BC provides a habitat for beneficial soil microorganisms, including bacteria and fungi (Hossain et al., 2020). These microbes play crucial roles in nutrient cycling and organic matter decomposition. BC supports diverse microbial communities, which can lead to healthier soil ecosystems and improved plant growth (Bolan et al., 2023). BC helps bind soil particles together, reducing the risk of soil erosion by wind and water. Stable soils are less likely to lose valuable topsoil and nutrients (Sharma, 2024). BC contributes to the organic matter content of soil. Unlike other organic amendments that decompose rapidly, BC is stable and persists in the soil for long periods, continually benefiting soil health (Nogués et al., 2023). BC can adsorb and immobilize soil contaminants such as heavy metals and pesticides, reducing their bioavailability and toxicity to plants and soil organisms (Qiu et al., 2022). Enhanced soil structure and nutrient availability from BC application support better root development, leading to healthier and more vigorous plants (Khan et al., 2024). The cumulative benefits of BC on soil health often result in increased crop yields and better quality.

4.2.2 Reduces Dependency on Chemical Fertilizers

Biochar creates a stable agricultural system that relies less on external chemical inputs by building up soil health and fertility. BC has a high CEC, which allows it to retain and exchange essential nutrients like potassium, calcium, and magnesium (Yuan et al., 2023), preventing leaching. BC can adsorb nutrients and release them slowly, providing a steady nutrient supply to plants (Wang et al., 2022) which gradually reduces the need for frequent fertilizer applications. BC provides a habitat for beneficial soil microorganisms such as nitrogen-fixing bacteria and mycorrhizal fungi, which enhance nutrient cycling and availability, reducing the need for external nutrient inputs. However a synergistic effect of chemical and BC was observed to increase plant growth (Lei et al., 2023). Enhanced microbial activity promotes the decomposition of organic matter, releasing nutrients that plants can use, thus reducing the need for chemical fertilizers (Hu et al., 2024). Besides that, BC provides a stable carbon source, improving soil organic matter content and better soil structure, improving root growth and nutrient uptake efficiency, reducing the need for external fertilizers (Li et al., 2023). By retaining nutrients within its structure, BC minimizes nutrient leaching, increasing the efficiency of any applied fertilizers (Oni et al., 2019). BC can neutralize acidic soils, reducing the need for pH-adjusting fertilizers and amendments (Shi et al., 2019). Reduced dependency on chemical fertilizers leads to lower nutrient runoff into water bodies, mitigating the environmental impact, such as eutrophication (Bare et al., 2023).

4.3 Water Management

4.3.1 Increases Water Retention in Soil

Biochar has a highly porous structure with a significant number of micro and macropores which can trap and hold water, making it available to the plants over a longer period of time (Hossain et al., 2020). The large surface area of BC increases the soil's ability to adsorb and retain water. Water adheres to the BC particles, enhancing overall soil moisture content (Khan et al., 2024). However, improved soil aggregation, increased organic matter content, and enhanced cation exchange capacity also play a significant role in increasing the water-holding capacity (Wang et al., 2022). These properties collectively improve the soil's ability to retain and manage water, leading to better plant growth, reduced irrigation needs, and increased resistance to drought.

4.3.2 Pest and Disease Control

Biochar creates a conducive environment for beneficial soil microorganisms, including bacteria and fungi. These microbes compete with pathogens for resources and space, reducing the prevalence of harmful organisms (Bolan et al., 2023). BC supports beneficial microbes such as mycorrhizal fungi and rhizobacteria, which can enhance plant health and adoption (Anbuganesan et al., 2024; Bolan et al., 2023; Jabborova et al., 2021). Some of these microbes produce antibiotics or other compounds that inhibit pathogen growth (Cycoń et al., 2019). BC can induce systemic acquired resistance (SAR) in plants, a defense mechanism where the plant's immune system is activated throughout its tissues (Jaiswal et al., 2020). Similar to SAR, induced systemic resistance (ISR) is triggered by beneficial microbes supported by BC (Choudhary et al., 2007). ISR helps plants resist attacks from various pests and pathogens. This makes the plant more resistant to a wide range of pathogens (Bolan et al., 2023). The presence of BC can suppress soil-borne pathogens through physical and chemical interactions. For example, BC can adsorb allelopathic compounds released by pathogens, reducing their harmful effects. BC can adsorb and immobilize toxins and harmful compounds in the soil, reducing stress on plants and decreasing the likelihood of disease (Iacomino et al., 2022). BC can adsorb and release phenolic compounds that have allelopathic effects, inhibiting the growth of certain pests and pathogens. BC can be combined with other pest and disease management practices, such as crop rotation, cover cropping, and the use of organic amendments, to enhance overall effectiveness (Chen et al., 2022).

4.4 *Economic Aspects*

4.4.1 Cost–Benefit Analysis

A cost–benefit analysis (CBA) of using BC involves evaluating the financial, environmental, and social costs and benefits associated with its production and application in various contexts, particularly in agriculture (Wrobel-Tobiszewska et al., 2015). The cost that is liable in the BC production and its utilization in the field includes; the cost of biomass feedstock (e.g., agricultural residues, forestry waste) used to produce BC, investment in the equipment, technology, and energy required during pyrolysis. Also, labor involved in collecting biomass, operating pyrolysis units, and applying BC to fields, transportation, storing, as well as application in the field (Dickinson et al., 2014).

4.4.2 Market Potential

The market potential of BC is significant, driven by its diverse applications and benefits in different fields. It commonly includes agriculture sustainability (soil amendment, sustainable farming practices), environmental management (carbon sequestration, pollution remediation), industry (waste management, construction and materials), market trend and growth (growing awareness, government policies and incentives), research and development (ongoing research and development), market segmentation (geographical segmentation, application segmentation), challenges and opportunities, market size and projections (current market size, growth rate) (Garcia et al., 2022).

4.4.3 Economic Incentives

Economic incentives for BC introduction and utilization include government policies and subsidies provided to users, and carbon offset programs (Pourhashem et al., 2018). Advancing the BC production process from biomass residues, such as agricultural waste or forestry by-products, which might otherwise be discarded or burned, contributing to a circular economy, can improve efficiency and reduce carbon emissions compared to existing traditional methods, leading to more sustainable industrial processes. Investments in BC production facilities can contribute to infrastructure development, particularly in rural or agricultural areas. For the establishment of BC production, infrastructure such as processing units, transportation networks, and storage facilities, which can stimulate economic development and create job opportunities, is required. Investments in research and development (R&D) of BC production and application technologies can enhance technological capacity, particularly in regions where the adoption of BC is still emerging. Building local R&D capacity contributes to technological innovation and fosters sustainable industrial development.

4.5 Biochar in Waste Management

4.5.1 Conversion of Waste Biomass

Biochar are stable carbon-rich material (Bo et al., 2023). Different kinds of biomass can be used to produce BC. It has a range of applications that transform waste biomass into useful products, contributing to sustainability and waste management (Murtaza et al., 2024). Especially, cities generate substantial amounts of organic waste, which can be transformed into BC, reducing landfill use and associated greenhouse gas emissions, promoting a circular economy (Möslinger et al., 2023). This can reduce the volume of waste that needs to be managed and lower the environmental footprint of urban areas.

4.5.2 Synergies with Composting and Anaerobic Digestion

Biochar is different from composting and anaerobic digestion. However, the integration of BC with composting and anaerobic digestion processes offers multiple benefits, including enhanced microbial activity, improved nutrient retention, odor reduction, and increased biogas production (Möslinger et al., 2023). These synergies result in higher-quality end products, more efficient waste management, and reduced environmental impact. The properties of BC add more advantages for composting and anaerobic digestion processes, which can become more sustainable and effective (Chiappero et al., 2020).

4.6 Technological Innovations

4.6.1 Advances in Production Technologies

Biochar are produced by pyrolysis. Precise control over pyrolysis temperatures can optimize BC properties for specific applications, as lower temperatures produce BC with high bioavailability of nutrients, while high temperatures yield BC with better carbon stability (Ghorbani et al., 2022). Similarly, the duration that biomass remains in the pyrolysis reactor can influence the quality and yield of BC, as shorter times generally produce higher yields, whereas longer times can enhance stability and surface area (Alperay et al., 2024). Developing reactors that allow continuous feeding of biomass can increase production efficiency and throughput. This method utilizes microwave energy to heat the biomass internally, resulting in faster and more uniform pyrolysis (Allende et al., 2023). These reactors enhance heat transfer and can handle a diverse range of feedstocks, thereby improving the efficiency and scalability of BC production. Exploring various types of biomass waste, including agricultural residues, forestry waste, and organic municipal waste, can make BC production more sustainable and economically viable (Murtaza et al., 2024). Mixing different types of biomass can optimize the properties of the resulting BC for specific applications. Secondly, the heat generated during pyrolysis can reduce energy consumption and operational costs and can also act as a source of renewable energy (Hasan et al., 2021). The technological advancement of sensor and automation over pyrolysis can lead to consistent BC quality. Understanding the pyrolysis parameters based on real-time data analysis and machine learning can enhance production efficiency (Hossain et al., 2011). Developing a modular pyrolysis unit that can be easily scaled up and down based on demand can make BC production more flexible and accessible (Javaid et al., 2024). Small-scale as well as large-scale pyrolysis units can promote local waste management as well as soil enhancement (Ghorbani et al., 2022). Identification of specific catalyst to be added can enhance the properties of BC during pyrolysis (Renda, Di Stasi, Manyà, & Palma, 2021). Techniques such as activation can increase the surface area and porosity of BC, making it more effective for a specific applications (Leng et al., 2021). Encouraging funding, collaboration between academia, industry, and government can drive innovation and practical advancements in BC production technologies (Ibitoye et al., 2024).

4.6.2 Development of Biochar Based Products

The beneficial effect of BC has been explored throughout the world, and research throughout the world has been ongoing to enhance its effect on agriculture, environment, water purification as well as carbon sequestration (Bo et al., 2023). BC based products are available throughout the world and some of these products are available for commercial purpose and available in pure form, blended form as well as pelltized form (Ibitoye et al., 2024).

4.7 Social and Policy Dimensions

4.7.1 Community Engagement and Education

The social and policy dimensions of BC refer to the various societal impacts and regulatory frameworks associated with the production, use, and promotion of BC as a technology for carbon sequestration and soil enhancement (Rittl et al., 2015). Engaging local communities in BC projects is crucial for their success. Acceptance can be influenced by awareness, perceived benefits, and potential risks associated with BC use. The economic viability of BC projects can affect local economies, potentially creating jobs and stimulating local businesses. Ensuring that the benefits of BC technologies are accessible to all, especially marginalized and low-income communities, is important. This includes addressing disparities in access to resources and knowledge. Understanding and mitigating potential health risks associated with BC production and application are necessary to protecting workers and communities. Providing education and training on BC technology and its benefits can empower communities and stakeholders to effectively implement and manage BC projects (Pourhashem et al., 2018).

4.7.2 Policy and Regulatory Frameworks

Developing regulations that ensure the safe production and application of BC is essential (Barbhuiya et al., 2024). This includes setting standards for BC quality, environmental safety, and emission controls. Policies that provide financial incentives, such as subsidies or tax credits, can encourage the adoption of BC technologies (Pourhashem et al., 2018). These can help offset initial costs and make BC more economically viable. Recognizing BC as a tool for climate change mitigation in national and international climate policies can promote its use (Lehmann, Amonette, & Roberts, 2010). This includes incorporating BC into carbon credit schemes and emissions reduction targets. Government funding and support for research and development can advance BC technology, making it more efficient and cost-effective. Encouraging collaboration between governments, the private sector, academia, and non-governmental organizations can foster innovation and share best practices in BC implementation (Pourhashem et al., 2018). Establishing systems for monitoring and evaluating the impacts of BC projects can help assess their effectiveness and make necessary adjustments to policies and practices.

4.7.3 Ethical Considerations and Social Justice

Ethical considerations and social justice issues related to BC encompass the fair and responsible implementation, distribution of benefits, and the potential impacts on various communities (Tisserant & Cherubini, 2019). Ensuring that BC production and application practices are environmentally sustainable is critical. This includes using feedstocks that do not compete with food production, maintaining biodiversity, and minimizing pollution. Stakeholders and communities should be fully informed about the benefits and potential risks associated with BC projects (Feng et al., 2021). Transparent communication and obtaining informed consent are essential to maintaining trust and accountability (Pourhashem et al., 2018). Conducting rigorous and unbiased research on BC is important for generating reliable data. Researchers should avoid conflicts of interest and ensure that their findings are shared openly and used responsibly. Given the relatively new nature of BC technology, applying the precautionary principle, erring on the side of caution in the face of scientific uncertainty, is crucial to avoid unintended negative consequences (Jatav et al., 2021).

Ensuring that the benefits of BC, such as improved soil fertility and carbon sequestration, are distributed equitably among different communities is important (Bo et al., 2023). This includes addressing disparities in access to resources and technology. Engaging marginalized and vulnerable communities in decision-making processes related to BC projects helps ensure that their needs and perspectives are taken into account. This promotes inclusivity and empowerment. BC projects should be designed to support low-income and rural communities, who might benefit the most from soil improvement and economic opportunities but often have the least access to technology and investment (Scholz et al., 2014). Ensuring that BC projects respect local land rights and do not displace communities or disrupt traditional land use practices is crucial. Projects should be developed in a way that supports local autonomy and livelihoods. Protecting the health and safety of workers and communities involved in BC production and application is essential (Ayaz et al., 2021). This includes addressing potential exposure to harmful substances and ensuring safe working conditions. Recognizing the role of BC in climate mitigation, it is important to ensure that its implementation does not exacerbate existing inequalities (Amonette et al., 2021). Climate justice involves making sure that those who are least responsible for climate change, but most affected by it, benefit from climate solutions like BC.

5 Case Studies and Success Stories

There are several successful stories demonstrating the benefits of using BC in many parts of the world, mainly in agricultural and environmental contexts. "Terra Preta" in the Amazon Basin is one of the success stories of using BC. This soil remains highly fertile for centuries, supporting robust agricultural activities (Schmidt et al., 2021). A study conducted in New South Wales, Australia, found that wheat yields increased by up to 25% and improved soil water retention and nutrient availability (Blackwell et al., 2010). Similarly, the Kenya Agricultural and Livestock Research Organization (KALRO) documented cases where farmers have reported an increase in maize yields and reduced dependency on chemical fertilizers (Kätterer et al., 2022). In Greece and China, BC has been used to remediate soils contaminated with heavy metals (Golia et al., 2022; Zhang et al., 2013). They have reported that BC could effectively immobilize heavy metals in the soil, reducing their bioavailability and mitigating environmental pollution. In China, BC has been used to improve rice yields, reduce greenhouse gas emissions from paddy fields, and enhance overall soil health, showing positive results on sustainable agriculture (Bo et al., 2023; Liu et al., 2019). BC has been applied in both forestry and agriculture in the United States (Krounbi et al., 2021; Lehmann et al., 2011). BC has been used to restore forest soil after wildfires, improving soil structure and aiding vegetation recovery. Increased crop yields and better soil moisture retention have been reported with BC. A study conducted in Finland reported BC as carbon negative product, reporting that willow BC produced in fallow land can be used to compensate agricultural greenhouse gas emissions (Leppäkoski et al., 2021). Practices implementing BC have highlighted the multifaceted benefits, emphasizing its importance in context-specific application, economic considerations, and community involvement for its successful use.

6 Future Research and Development

The long-term impact of BC on soil health, crop productivity, and carbon sequestration has not been clearly known. So, longitudinal studies are necessary to assess these effects over decades (Tan & Kuebbing, 2023). BC influencing soil microbial communities, nutrient cycles, and plant growth are not clearly understood. So, research directing specific mechanisms is necessary to elucidate these processes (Ling et al., 2022). Research is required to develop guidelines for optimal application in diverse environmental conditions, soil types, climates, and agricultural systems (Ntawuruhunga et al., 2023). BC are produced from different feedstocks (wood, crop residues, manure) with varying properties (Novotný et al., 2023). However, further research is crucial to understand the properties that affect soil and plant responses for selecting the most appropriate BC types for specific uses. The economic viability and social acceptance of BC technologies are not well documented (Campion et al., 2023). Further research on cost–benefit analysis, market development, and farmer adoption barriers is essential. Standardized methods should be developed for the production, characterization, and application of BC to ensure quality and facilitate comparison across studies (Yaashikaa et al., 2020). Synergistic effects of BC should be investigated in combination with soil amendments, compost, fertilizers, and organic matter on optimum soil management practices. A study on production of BC influencing pyrolysis conditions (temperature, time, and atmosphere) will aid in producing BC tailored for specific agricultural and environmental applications (Lataf et al., 2022). Research on the impact of BC on soil microbial communities and their functions should be further explored to enhance soil health and promote plant growth. Comprehensive environmental impact assessments of BC production and application, including life cycle analysis will help determine its overall sustainability (Amalina et al., 2022). Evaluating the potential of BC to mitigate climate change through carbon sequestration, reduction of greenhouse gas emissions, and resistance to extreme weather events will be a critical research area. Policies and incentives should be developed to promote BC production and use (Pourhashem et al., 2018). Innovative applications of BC beyond agriculture should be explored, such as water purification, waste management, and construction materials for its utility and benefits.

7 Conclusion

Biochar serves as a multifaceted soil amendment, enhancing the physical, chemical, and biological properties of soil. Its use can lead to healthier soils, more resilient agricultural systems, improved plant growth, and productivity. By integrating BC into soil management practices, farmers and land managers can achieve sustainable agricultural development and better environmental stewardship.

References

Aboelela, D., Saleh, H., Attia, A. M., Elhenawy, Y., Majozi, T., & Bassyouni, M. (2023). Recent advances in biomass pyrolysis processes for bioenergy production: Optimization of operating conditions. *Sustainability, 15*(14), 11238.

Ahmad Bhat, S., Kuriqi, A., Dar, M. U. D., Bhat, O., Sammen, S. S., Towfiqul Islam, A. R. M., & Heddam, S. (2022). Application of biochar for improving physical, chemical, and hydrological soil properties: A systematic review. *Sustainability, 14*(17), 11104.

Alkharabsheh, H. M., Seleiman, M. F., Battaglia, M. L., Shami, A., Jalal, R. S., Alhammad, B. A., & Al-Saif, A. M. (2021). Biochar and its broad impacts in soil quality and fertility, nutrient leaching and crop productivity: A review. *Agronomy, 11*(5), 993.

Allende, S., Brodie, G., & Jacob, M. V. (2023). Breakdown of biomass for energy applications using microwave pyrolysis: A technological review. *Environmental Research, 226*, Article 115619. https://doi.org/10.1016/j.envres.2023.115619

Alperay, A., Hakkı, A. M., Aysun, A., Emin, B. M., & Sefa, A. (2024). A comprehensive study of biochar yield and quality concerning pyrolysis conditions: A multifaceted approach. *Sustainability, 16*(2), 937.

Amalina, F., Razak, A. S. A., Krishnan, S., Zularisam, A. W., & Nasrullah, M. (2022). A comprehensive assessment of the method for producing biochar, its characterization, stability, and potential applications in regenerative economic sustainability—A review. *Cleaner Materials, 3*, Article 100045. https://doi.org/10.1016/j.clema.2022.100045

Amonette, J. E., Blanco-Canqui, H., Hassebrook, C., Laird, D. A., Lal, R., Lehmann, J., & Page-Dumroese, D. (2021). Integrated biochar research: A roadmap. *Journal of Soil and Water Conservation, 76*(1), 24A-29A. https://doi.org/10.2489/jswc.2021.1115A

Anbuganesan, V., Vishnupradeep, R., Bruno, L. B., Sharmila, K., Freitas, H., & Rajkumar, M. (2024). Combined application of biochar and plant growth-promoting rhizobacteria improves heavy metal and drought stress tolerance in zea mays. *Plants, 13*(8), 1143.

Ayaz, M., Feizienė, D., Tilvikienė, V., Akhtar, K., Stulpinaitė, U., & Iqbal, R. (2021). Biochar role in the sustainability of agriculture and environment. *Sustainability, 13*(3), 1330.

Barbhuiya, S., Bhusan Das, B., & Kanavaris, F. (2024). Biochar-concrete: A comprehensive review of properties, production and sustainability. *Case Studies in Construction Materials, 20*, Article e02859. https://doi.org/10.1016/j.cscm.2024.e02859

Bare, W. F. R., Struhs, E., Mirkouei, A., Overturf, K., Chacón-Patiño, M. L., McKenna, A. M., & Raja, K. S. (2023). Controlling eutrophication of aquaculture production water using biochar: Correlation of molecular composition with adsorption characteristics as revealed by FT-ICR mass spectrometry. *Processes, 11*(10), 2883.

Batista, E. M. C. C., Shultz, J., Matos, T. T. S., Fornari, M. R., Ferreira, T. M., Szpoganicz, B., & Mangrich, A. S. (2018). Effect of surface and porosity of biochar on water holding capacity aiming indirectly at preservation of the Amazon biome. *Scientific Reports, 8*(1), 10677. https://doi.org/10.1038/s41598-018-28794-z

Blackwell, P., Krull, E., Butler, G., Herbert, A., & Solaiman, Z. (2010). Effect of banded biochar on dryland wheat production and fertiliser use in South-Western Australia: An agronomic and economic perspective. *Soil Research, 48*(7), 531. https://doi.org/10.1071/sr10014

Blanco-Canqui, H. (2021). Does biochar application alleviate soil compaction? *Review and Data Synthesis. Geoderma, 404*, Article 115317. https://doi.org/10.1016/j.geoderma.2021.115317

Bo, X., Zhang, Z., Wang, J., Guo, S., Li, Z., Lin, H., Zou, J. (2023). Benefits and limitations of biochar for climate-smart agriculture: A review and case study from China. *Biochar, 5*(1). https://doi.org/10.1007/s42773-023-00279-x

Bolan, S., Hou, D., Wang, L., Hale, L., Egamberdieva, D., Tammeorg, P., & Bolan, N. (2023). The potential of biochar as a microbial carrier for agricultural and environmental applications. *Science of the Total Environment, 886*, Article 163968. https://doi.org/10.1016/j.scitotenv.2023.163968

Brar, B., Saharan, B. S., Seth, C. S., Kamboj, A., Surekha, Bala, K., Duhan, J. S. (2024). Nanobiochar: Soil and plant interactions and their implications for sustainable agriculture. *Biocatalysis and Agricultural Biotechnology, 57*, 103077. https://doi.org/10.1016/j.bcab.2024.103077

Campion, L., Bekchanova, M., Malina, R., & Kuppens, T. (2023). The costs and benefits of biochar production and use: A systematic review. *Journal of Cleaner Production, 408*, Article 137138. https://doi.org/10.1016/j.jclepro.2023.137138

Chen, Q., Lan, P., Wu, M., Lu, M., Pan, B., & Xing, B. (2022). Biochar mitigates allelopathy through regulating allelochemical generation from plants and accumulation in soil. *Carbon Research, 1*(1). https://doi.org/10.1007/s44246-022-00003-7

Chiappero, M., Norouzi, O., Hu, M., Demichelis, F., Berruti, F., Di Maria, F., & Fiore, S. (2020). Review of biochar role as additive in anaerobic digestion processes. *Renewable and Sustainable Energy Reviews, 131*, Article 110037. https://doi.org/10.1016/j.rser.2020.110037

Choudhary, D. K., Prakash, A., & Johri, B. N. (2007). Induced systemic resistance (ISR) in plants: Mechanism of action. *Indian J Microbiol, 47*(4), 289–297. https://doi.org/10.1007/s12088-007-0054-2

Cycoń, M., Mrozik, A., & Piotrowska-Seget, Z. (2019). Antibiotics in the soil environment—Degradation and their impact on microbial activity and diversity. *Frontiers in Microbiology, 10*. https://doi.org/10.3389/fmicb.2019.00338

Dickinson, D., Balduccio, L., Buysse, J., Ronsse, F., van Huylenbroeck, G., & Prins, W. (2014). Cost-benefit analysis of using biochar to improve cereals agriculture. *Global Change Biology Bioenergy, 7*(4), 850–864. https://doi.org/10.1111/gcbb.12180

Efosa, N., Krause, H.-M., Hüppi, R., Krauss, M., Vaucher, N., Zourek, F., & Bünemann, E. K. (2023). Emissions of nitrous oxide and methane after field application of liquid organic fertilizers and biochar. *Agriculture, Ecosystems and Environment, 356*, Article 108642. https://doi.org/10.1016/j.agee.2023.108642

Feng, Q., Wang, B., Chen, M., Wu, P., Lee, X., & Xing, Y. (2021). Invasive plants as potential sustainable feedstocks for biochar production and multiple applications: A review. *Resources, Conservation and Recycling, 164*, Article 105204. https://doi.org/10.1016/j.resconrec.2020.105204

Garcia, B., Alves, O., Rijo, B., Lourinho, G., & Nobre, C. (2022). Biochar: Production, applications, and market prospects in Portugal. *Environments, 9*(8), 95.

Ghorbani, M., Amirahmadi, E., Neugschwandtner, R. W., Konvalina, P., Kopecký, M., Moudrý, J., & Murindangabo, Y. T. (2022). The impact of pyrolysis temperature on biochar properties and its effects on soil hydrological properties. *Sustainability, 14*(22), 14722. https://doi.org/10.3390/su142214722

Golia, E. E., Aslanidis, P.-S. C., Papadimou, S. G., Kantzou, O.-D., Chartodiplomenou, M.-A., Lakiotis, K., Tsiropoulos, N. G. (2022). Assessment of remediation of soils, moderately contaminated by potentially toxic metals, using different forms of carbon (charcoal, biochar, activated carbon). Impacts on contamination, metals availability and soil indices. *Sustainable Chemistry and Pharmacy, 28*, 100724. https://doi.org/10.1016/j.scp.2022.100724

Hamidi, N. H., Ahmed, O. H., Omar, L., Ch'ng, H. Y., Johan, P. D., Paramisparam, P., Jalloh, M. B. (2023). Co-Application of inorganic fertilizers with charcoal and sago bark ash to improve soil nitrogen availability, uptake, use efficiency, and dry matter production of sorghum cultivated on acid soils. *Sustainability, 15*(1), 827. https://doi.org/10.3390/su15010827

Hasan, M. M., Rasul, M. G., Khan, M. M. K., Ashwath, N., & Jahirul, M. I. (2021). Energy recovery from municipal solid waste using pyrolysis technology: A review on current status and developments. *Renewable and Sustainable Energy Reviews, 145*, Article 111073. https://doi.org/10.1016/j.rser.2021.111073

He, L., Zhao, J., Yang, S., Zhou, H., Wang, S., Zhao, X., & Xing, G. (2020). Successive biochar amendment improves soil productivity and aggregate microstructure of a red soil in a five-year wheat-millet rotation pot trial. *Geoderma, 376*, Article 114570. https://doi.org/10.1016/j.geoderma.2020.114570

Hossain, M. K., Strezov, V., Chan, K. Y., Ziolkowski, A., & Nelson, P. F. (2011). Influence of pyrolysis temperature on production and nutrient properties of wastewater sludge biochar. *Journal of Environmental Management, 92*(1), 223–228. https://doi.org/10.1016/j.jenvman.2010.09.008

Hossain, M. Z., Bahar, M. M., Sarkar, B., Donne, S. W., Ok, Y. S., Palansooriya, K. N., & Bolan, N. (2020). Biochar and its importance on nutrient dynamics in soil and plant. *Biochar, 2*(4), 379–420. https://doi.org/10.1007/s42773-020-00065-z

Hu, W., Zhang, Y., Rong, X., Zhou, X., Fei, J., Peng, J., & Luo, G. (2024). Biochar and organic fertilizer applications enhance soil functional microbial abundance and agroecosystem multifunctionality. *Biochar, 6*(1), 3. https://doi.org/10.1007/s42773-023-00296-w

Iacomino, G., Idbella, M., Laudonia, S., Vinale, F., & Bonanomi, G. (2022). The suppressive effects of biochar on above and belowground plant pathogens and pests: A review. *Plants (Basel), 11*(22). https://doi.org/10.3390/plants11223144

Ibitoye, S. E., Loha, C., Mahamood, R. M., Jen, T. C., Alam, M., Sarkar, I., & Akinlabi, E. T. (2024). An overview of biochar production techniques and application in iron and steel industries. *Bioresources and Bioprocessing, 11*(1), 65. https://doi.org/10.1186/s40643-024-00779-z

Jabborova, D., Annapurna, K., Paul, S., Kumar, S., Saad, H. A., Desouky, S., & Elkelish, A. (2021). Beneficial features of biochar and arbuscular mycorrhiza for improving spinach plant growth, root morphological traits, physiological properties, and soil enzymatic activities. *Journal of fungi (Basel, Switzerland), 7*(7). https://doi.org/10.3390/jof7070571

Jaiswal, A. K., Alkan, N., Elad, Y., Sela, N., Philosoph, A. M., Graber, E. R., & Frenkel, O. (2020). Molecular insights into biochar-mediated plant growth promotion and systemic resistance in tomato against Fusarium crown and root rot disease. *Scientific Reports, 10*(1), 13934. https://doi.org/10.1038/s41598-020-70882-6

Jatav, H. S., Rajput, V. D., Minkina, T., Singh, S. K., Chejara, S., Gorovtsov, A., & Kalinitchenko, V. P. (2021). Sustainable approach and safe use of biochar and its possible consequences. *Sustainability, 13*(18), 10362.

Javaid, S. F., Dai, M., Wu, Y., Luo, H., Amjed, M. A., Ali, I., & Naz, I. (2024). Production of biochar by slow and solar-biomass pyrolysis: Focus on the output configuration assessment, adaptability, and barriers to market penetration. *Arabian Journal for Science and Engineering, 49*(6), 7731–7750. https://doi.org/10.1007/s13369-023-08549-3

Kätterer, T., Roobroeck, D., Kimutai, G., Karltun, E., Nyberg, G., Sundberg, C., & de Nowina, K. R. (2022). Maize grain yield responses to realistic biochar application rates on smallholder farms in Kenya. *Agronomy for Sustainable Development, 42*(4). https://doi.org/10.1007/s13593-022-00793-5

Khan, S., Irshad, S., Mehmood, K., Hasnain, Z., Nawaz, M., Rais, A., & Ibrar, D. (2024). Biochar production and characteristics, its impacts on soil health, crop production, and yield enhancement: A review. *Plants, 13*(2), 166.

Koyama, S., Inazaki, F., Minamikawa, K., Kato, M., & Hayashi, H. (2015). Increase in soil carbon sequestration using rice husk charcoal without stimulating CH_4 and N_2O emissions in an Andosol paddy field in Japan. *Soil Science and Plant Nutrition, 61*(5), 873–884. https://doi.org/10.1080/00380768.2015.1065511

Krounbi, L., Enders, A., Gaunt, J., Ball, M., & Lehmann, J. (2021). Plant uptake of nitrogen adsorbed to biochars made from dairy manure. *Scientific Reports, 11*(1), 15001. https://doi.org/10.1038/s41598-021-94337-8

Kumar, A., & Bhattacharya, T. (2021). Biochar: A sustainable solution. *Environment, Development and Sustainability, 23*(5), 6642–6680. https://doi.org/10.1007/s10668-020-00970-0

Kumar, A., Singh, E., Khapre, A., Bordoloi, N., & Kumar, S. (2020). Sorption of volatile organic compounds on non-activated biochar. *Bioresource Technology, 297*, Article 122469. https://doi.org/10.1016/j.biortech.2019.122469

Lataf, A., Jozefczak, M., Vandecasteele, B., Viaene, J., Schreurs, S., Carleer, R., & Vandamme, D. (2022). The effect of pyrolysis temperature and feedstock on biochar agronomic properties. *Journal of Analytical and Applied Pyrolysis, 168*, Article 105728. https://doi.org/10.1016/j.jaap.2022.105728

Lefebvre, D., Williams, A., Meersmans, J., Kirk, G. J. D., Sohi, S., Goglio, P., & Smith, P. (2020). Modelling the potential for soil carbon sequestration using biochar from sugarcane residues in Brazil. *Scientific Reports, 10*(1), 19479. https://doi.org/10.1038/s41598-020-76470-y

Lehmann, J., Amonette, J. E., & Roberts, K. (2010). Role of biochar in mitigation of climate change *Handbook of climate change and agroecosystems, 1*, 343–363. Cornell University, Ithaca NY, USA.

Lehmann, J., Rillig, M. C., Thies, J., Masiello, C. A., Hockaday, W. C., & Crowley, D. (2011). Biochar effects on soil biota—A review. *Soil Biology and Biochemistry, 43*(9), 1812–1836. https://doi.org/10.1016/j.soilbio.2011.04.022

Lei, W., Leghari, S. J., Jiajun, W., Na, W., Min, P., & Liang, J. (2023). Interactive effects of biochar and chemical fertilizer on water and nitrogen dynamics, soil properties and maize yield under different irrigation methods. *Frontiers in Plant Science, 14*. https://doi.org/10.3389/fpls.2023.1230023

Leng, L., Xiong, Q., Yang, L., Li, H., Zhou, Y., Zhang, W., & Huang, H. (2021). An overview on engineering the surface area and porosity of biochar. *Science of the Total Environment, 763*, Article 144204. https://doi.org/10.1016/j.scitotenv.2020.144204

Leppäkoski, L., Marttila, M., Uusitalo, V., Levänen, J., Halonen, V., & Mikkilä, M. (2021). Assessing the carbon footprint of biochar from willow grown on marginal lands in Finland. *Sustainability, 13*, 10097. https://doi.org/10.3390/su131810097

Li, H., Yang, L., Mao, Q., Zhou, H., Guo, P., Agathokleous, E., & Wang, S. (2023). Modified biochar enhances soil fertility and nutrient uptake and yield of rice in mercury-contaminated soil. *Environmental Technology and Innovation, 32*, Article 103435. https://doi.org/10.1016/j.eti.2023.103435

Libutti, A., Cammerino, A. R. B., Francavilla, M., & Monteleone, M. (2019). Soil amendment with biochar affects water drainage and nutrient losses by leaching: Experimental vidence under field grown conditions. *Agronomy, 9*(11), 758.

Ling, L., Luo, Y., Jiang, B., Lv, J., Meng, C., Liao, Y., & Xu, J. (2022). Biochar induces mineralization of soil recalcitrant components by activation of biochar responsive bacteria groups. *Soil Biology and Biochemistry, 172*, Article 108778. https://doi.org/10.1016/j.soilbio.2022.108778

Liu, X., Zhou, J., Chi, Z., Zheng, J., Li, L., Zhang, X., & Pan, G. (2019). Biochar provided limited benefits for rice yield and greenhouse gas mitigation six years following an amendment in a fertile rice paddy. *CATENA, 179*, 20–28. https://doi.org/10.1016/j.catena.2019.03.033

Manikandan, S. K., Pallavi, P., Shetty, K., Bhattacharjee, D., Giannakoudakis, D. A., Katsoyiannis, I. A., & Nair, V. (2023). Effective usage of biochar and microorganisms for the removal of heavy metal ions and pesticides. *Molecules, 28*(2). https://doi.org/10.3390/molecules28020719

Meyerkord, A. (2023). *International biochar initiative*. Retrieved from Philadelphia, The USA

Möslinger, M., Ulpiani, G., & Vetters, N. (2023). Circular economy and waste management to empower a climate-neutral urban future. *Journal of Cleaner Production, 421*, Article 138454. https://doi.org/10.1016/j.jclepro.2023.138454

Mukherjee, S., Sarkar, B., Aralappanavar, V. K., Mukhopadhyay, R., Basak, B. B., Srivastava, P., & Bolan, N. (2022). Biochar-microorganism interactions for organic pollutant remediation: Challenges and perspectives. *Environmental Pollution, 308*, Article 119609. https://doi.org/10.1016/j.envpol.2022.119609

Mulabagal, V., Baah, D. A., Egiebor, N. O., Sajjadi, B., Chen, W.-Y., Viticoski, R. L., & Hayworth, J. S. (2020). Biochar from biomass: A strategy for carbondioxide sequestration, soil amendment, power generation, CO2 utilization, and removal of perfluoroalkyl and polyfluoroalkyl substances (PFAS) in the environment. In M. Lackner, B. Sajjadi, & W.-Y. Chen (Eds.), *Handbook of Climate Change Mitigation and Adaptation* (pp. 1–64). New York, NY: Springer New York.

Murtaza, G., Ahmed, Z., Valipour, M., Ali, I., Usman, M., Iqbal, R., & Tariq, A. (2024). Recent trends and economic significance of modified/functionalized biochars for remediation of environmental pollutants. *Science and Reports, 14*(1), 217. https://doi.org/10.1038/s41598-023-50623-1

Neina, D., Faust, S., & Joergensen, R. G. (2020). Characterization of charcoal and firewood ash for use in African peri-urban agriculture. *Chemical and Biological Technologies in Agriculture, 7*(1). https://doi.org/10.1186/s40538-019-0171-2

Nogués, I., Mazzurco Miritana, V., Passatore, L., Zacchini, M., Peruzzi, E., Carloni, S., & Marinari, S. (2023). Biochar soil amendment as carbon farming practice in a Mediterranean environment. *Geoderma Regional, 33*, Article e00634. https://doi.org/10.1016/j.geodrs.2023.e00634

Novotný, M., Marković, M., Raček, J., Šipka, M., Chorazy, T., Tošić, I., & Hlavínek, P. (2023). The use of biochar made from biomass and biosolids as a substrate for green infrastructure: A review. *Sustainable Chemistry and Pharmacy, 32*, Article 100999. https://doi.org/10.1016/j.scp.2023.100999

Ntawuruhunga, D., Ngowi, E. E., Mangi, H. O., Salanga, R. J., & Shikuku, K. M. (2023). Climate-smart agroforestry systems and practices: A systematic review of what works, what doesn't work, and why. *Forest Policy and Economics, 150*, Article 102937. https://doi.org/10.1016/j.forpol.2023.102937

Obour, P. B., Danso, E. O., Yakubu, A., Abenney-Mickson, S., Sabi, E. B., Darrah, Y. K., & Arthur, E. (2019). Water retention, air exchange and pore structure characteristics after three years of rice straw biochar application to an acrisol. *Soil Science Society of America Journal, 83*(6), 1664–1671. https://doi.org/10.2136/sssaj2019.07.0230

Oni, B. A., Oziegbe, O., & Olawole, O. O. (2019). Significance of biochar application to the environment and economy. *Annals of Agricultural Sciences, 64*(2), 222–236. https://doi.org/10.1016/j.aoas.2019.12.006

Pourhashem, G., Hung, S. Y., Medlock, K. B., & Masiello, C. A. (2018). Policy support for biochar: Review and recommendations. *GCB Bioenergy, 11*(2), 364–380. https://doi.org/10.1111/gcbb.12582

Qiu, M., Liu, L., Ling, Q., Cai, Y., Yu, S., Wang, S., & Wang, X. (2022). Biochar for the removal of contaminants from soil and water: A review. *Biochar, 4*(1), 19. https://doi.org/10.1007/s42773-022-00146-1

Renda, S., Di Stasi, C., Manyà, J. J., & Palma, V. (2021). Biochar as support in catalytic CO_2 methanation: Enhancing effect of CeO_2 addition. *Journal of CO_2 Utilization, 53*, 101740. https://doi.org/10.1016/j.jcou.2021.101740

Rittl, T. F., Arts, B., & Kuyper, T. W. (2015). Biochar: An emerging policy arrangement in Brazil? *Environmental Science and Policy, 51*, 45–55. https://doi.org/10.1016/j.envsci.2015.03.010

Schmidt, H.-P. (2013). 55 Uses of biochar *ithakajournal*, 286–289.

Schmidt, H. P., Kammann, C., Hagemann, N., Leifeld, J., Bucheli, T. D., Sánchez Monedero, M. A., & Cayuela, M. L. (2021). Biochar in agriculture – A systematic review of 26 global meta-analyses. *GCB Bioenergy, 13*(11), 1708–1730. https://doi.org/10.1111/gcbb.12889

Scholz, S. M., Sembres, T., Roberts, K., Whitman, T., Wilson, K., & Lehmann, J. (2014). Biochar systems for smallholders in developing countries leveraging current knowledge and exploring future potential for climate smart agriculture. https://doi.org/10.1596/978-0-8213-9525-7. From The World Bank

Shaaban, M., Van Zwieten, L., Bashir, S., Younas, A., Núñez-Delgado, A., Chhajro, M. A., & Hu, R. (2018). A concise review of biochar application to agricultural soils to improve soil conditions and fight pollution. *Journal of Environmental Management, 228*, 429–440. https://doi.org/10.1016/j.jenvman.2018.09.006

Shareef, T. M. E., & Zhao, B. (2017). Review paper: The fundamentals of biochar as a soil amendment tool and management in agriculture scope: An overview for farmers and gardeners. *Journal of Agricultural Chemistry and Environment, 06*(01), 38–61. https://doi.org/10.4236/jacen.2017.61003

Sharma, P. (2024). Biochar application for sustainable soil erosion control: a review of current research and future perspectives. *Frontiers in Environmental Science, 12*. https://doi.org/10.3389/fenvs.2024.1373287

Shi, R.-Y., Li, J.-Y., Ni, N., & Xu, R.-K. (2019). Understanding the biochar's role in ameliorating soil acidity. *Journal of Integrative Agriculture, 18*(7), 1508–1517. https://doi.org/10.1016/s2095-3119(18)62148-3

Singh Yadav, S. P., Bhandari, S., Bhatta, D., Poudel, A., Bhattarai, S., Yadav, P., & Oli, B. (2023). Biochar application: A sustainable approach to improve soil health. *Journal of Agriculture and Food Research, 11*, Article 100498. https://doi.org/10.1016/j.jafr.2023.100498

Tan, S. S. X., & Kuebbing, S. E. (2023). A synthesis of the effect of regenerative agriculture on soil carbon sequestration in Southeast Asian croplands. *Agriculture, Ecosystems & Environment, 349*, Article 108450. https://doi.org/10.1016/j.agee.2023.108450

Tisserant, A., & Cherubini, F. (2019). Potentials, limitations, co-benefits, and trade-offs of biochar applications to soils for climate change mitigation. *Land, 8*(12), 179.

Vaccari, F. P., Baronti, S., Lugato, E., Genesio, L., Castaldi, S., Fornasier, F., & Miglietta, F. (2011). Biochar as a strategy to sequester carbon and increase yield in durum wheat. *European Journal of Agronomy, 34*(4), 231–238. https://doi.org/10.1016/j.eja.2011.01.006

Wang, C., Luo, D., Zhang, X., Huang, R., Cao, Y., Liu, G., & Wang, H. (2022). Biochar-based slow-release of fertilizers for sustainable agriculture: A mini review. *Environmental Science and Ecotechnology, 10*, Article 100167. https://doi.org/10.1016/j.ese.2022.100167

Woolf, D., Amonette, J. E., Street-Perrott, F. A., Lehmann, J., & Joseph, S. (2010). Sustainable biochar to mitigate global climate change. *Nature Communications, 1*, 56. https://doi.org/10.1038/ncomms1053

Wrobel-Tobiszewska, A., Boersma, M., Sargison, J., Adams, P., & Jarick, S. (2015). An economic analysis of biochar production using residues from Eucalypt plantations. *Biomass and Bioenergy, 81*, 177–182. https://doi.org/10.1016/j.biombioe.2015.06.015

Wu, X., Quan, W., Chen, Q., Gong, W., & Wang, A. (2024). Efficient adsorption of nitrogen and phosphorus in wastewater by biochar. *Molecules, 29*(5), 1005.

Yaashikaa, P. R., Kumar, P. S., Varjani, S., & Saravanan, A. (2020). A critical review on the biochar production techniques, characterization, stability and applications for circular bioeconomy. *Biotechnology Reports, 28*, Article e00570. https://doi.org/10.1016/j.btre.2020.e00570

Yuan, Y., Liu, Q., Zheng, H., Li, M., Liu, Y., Wang, X., & Xing, B. (2023). Biochar as a sustainable tool for improving the health of salt-affected soils. *Soil and Environmental Health, 1*(3), Article 100033. https://doi.org/10.1016/j.seh.2023.100033

Zahed, M. A., Salehi, S., Madadi, R., & Hejabi, F. (2021). Biochar as a sustainable product for remediation of petroleum contaminated soil. *Current Research in Green and Sustainable Chemistry, 4*, Article 100055. https://doi.org/10.1016/j.crgsc.2021.100055

Zang, Y., Wang, M., Shohag, M. J. I., Lu, L., He, T., Liao, C., & Tian, S. (2023). Biochar performance for preventing cadmium and arsenic accumulation, and the health risks associated with mustard (Brassica Juncea) grown in co-contaminated soils. *Ecotoxicology and Environmental Safety, 263*, Article 115216. https://doi.org/10.1016/j.ecoenv.2023.115216

Zeng, B., Xu, W., Khan, S. B., Wang, Y., Zhang, J., Yang, J., & Lin, Z. (2021). Preparation of sludge biochar rich in carboxyl/hydroxyl groups by quenching process and its excellent adsorption performance for Cr(VI). *Chemosphere, 285*, Article 131439. https://doi.org/10.1016/j.chemosphere.2021.131439

Zhang, X., Wang, H., He, L., Lu, K., Sarmah, A., Li, J., & Huang, H. (2013). Using biochar for remediation of soils contaminated with heavy metals and organic pollutants. *Environmental Science and Pollution Research, 20*(12), 8472–8483. https://doi.org/10.1007/s11356-013-1659-0

Integrated Insect-Pest Management and Sustainable Agriculture

Aneela Younas and Muhammad Shaaban

Abstract This chapter explores the multifaceted approach of Integrated Pest Management (IPM) in sustainable agriculture, emphasizing the importance of combining various pest control strategies. It highlights the role of biodiversity in enhancing natural pest controllers, cultural practices such as crop rotation and soil health management, and biological control methods involving natural predators and biopesticides. Furthermore, it addresses the responsible use of chemical pesticides and the need for pesticide resistance management. The integration of chemical and non-chemical methods forms the foundation of an effective IPM strategy, contributing to agricultural productivity while minimizing environmental impact. The chapter underscores the necessity of adopting IPM practices for a sustainable agricultural future.

Keywords Integrated pest management · Sustainability · Biodiversity · Biological control · Cultural practices · Chemical control · Agricultural productivity

1 Introduction

Integrated Pest Management (IPM) is a comprehensive approach to managing agricultural pests that focuses on long-term prevention and control strategies, rather than relying solely on chemical interventions. IPM integrates various management practices based on the understanding of pest life cycles, ecological interactions, and environmental considerations. This method emphasizes the use of multiple strategies such as biological control, habitat manipulation, and the careful application of pesticides only when necessary, creating a balanced ecosystem conducive to agricultural production (Younas et al., 2017). The importance of sustainable agriculture has grown significantly as global populations rise and environmental concerns become more pressing. Sustainable agriculture seeks to meet current food demands while preserving resources for future generations. This approach emphasizes practices that

A. Younas · M. Shaaban (✉)
College of Agriculture, Henan University of Science and Technology, Luoyang, China
e-mail: shabanbzu@hotmail.com

M. Shaaban (ed.), *Soils and Sustainable Agriculture*, Frontier Studies in Soil Science,
https://doi.org/10.1007/978-3-031-91114-9_31

protect ecological health, promote biodiversity, and improve the resilience of agricultural systems. Sustainable agriculture not only aims for the productive capacity of the land but also considers economic viability and social equity, thus ensuring holistic benefits for communities and ecosystems alike (Shaaban et al., 2018).

The relationship between Integrated Pest Management and sustainable agricultural practices is pivotal. IPM contributes to sustainability by reducing reliance on chemical pesticides, which can lead to environmental degradation and health hazards (Younas et al., 2017). By implementing IPM strategies, farmers can manage pest populations effectively while minimizing the impact on non-target organisms, such as beneficial insects and pollinators. Furthermore, the integration of IPM within sustainable agriculture enhances biodiversity, promotes soil health, and supports the overall ecosystem balance. Together, IPM and sustainable agricultural practices foster a resilient agricultural framework capable of addressing the multifaceted challenges posed by pests and environmental change (Younas et al., 2017).

2 Principles of Integrated Pest Management

The principles of Integrated Pest Management (IPM) are grounded in a thorough understanding of pest biology and ecology. By studying the life cycles, behavior, and ecological interactions of pests, practitioners can better predict and manage pest populations effectively. Knowledge of the natural enemies of pests, such as predators and parasitoids, is also critical, as these organisms can play a vital role in controlling pest populations in a sustainable manner.

Monitoring and identification of pests are crucial components of IPM. Regular field observations and scouting enable farmers to detect pest presence early and accurately identify the species involved (Younas et al., 2022). This information informs decision-making processes and helps distinguish between harmful pests and beneficial organisms that contribute to the ecosystem. Effective monitoring tools, such as traps and field surveys, aid in collecting data on pest populations and assessing their potential impacts on crops (Younas et al., 2017). Another key principle of IPM is the concept of economic threshold levels for pest management. This entails determining the pest population density at which the cost of pest damage exceeds the cost of management efforts. By establishing these thresholds, farmers can make informed decisions about when to intervene, thereby avoiding unnecessary treatments and reducing unnecessary costs. This approach promotes economic sustainability while minimizing the environmental impact associated with pest control (Waris et al., 2022).

Combining multiple control tactics is a fundamental aspect of IPM that enhances its effectiveness (Fig. 1). A diverse array of methods can be utilized, including biological, cultural, mechanical, and chemical strategies. Biological control involves the use of natural predators or parasites to manage pests, while cultural practices focus on modifying farming techniques, such as crop rotation and habitat management, to

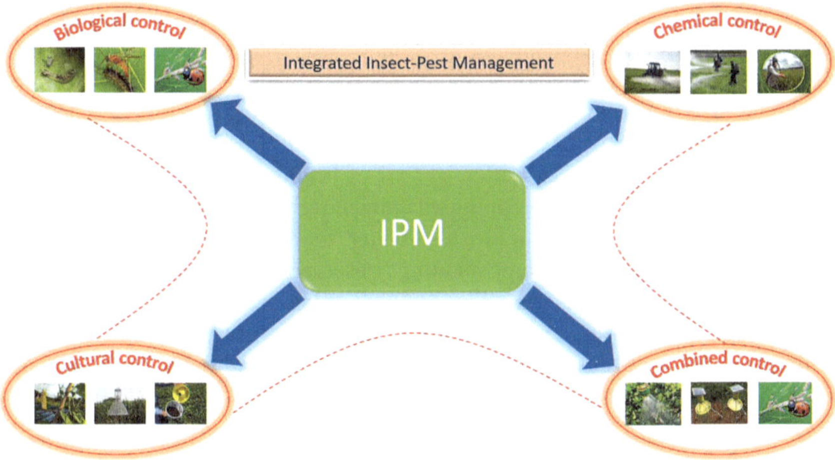

Fig. 1 Various approaches for integrated insect-pest management

reduce pest establishment. Mechanical methods, such as traps or barriers, can physically remove pests or prevent their access to crops. Chemical methods, although used judiciously within IPM, can serve as a last resort when pest populations exceed economic thresholds. The integration of these various tactics creates a synergistic approach that enhances pest control effectiveness while promoting sustainability in agricultural systems (Waris et al., 2020).

3 Role of Biodiversity in IPM

Biodiversity plays a crucial role in Integrated Pest Management (IPM) by contributing to the resilience and stability of agroecosystems. Maintaining a diverse array of plant and animal species can lead to more sustainable agricultural practices, as biodiversity supports ecosystem functions, enhances soil health, and provides vital services that help mitigate the impacts of pests. Diverse ecosystems are more capable of withstanding and recovering from disturbances, thereby promoting long-term agricultural productivity. In essence, a rich biodiversity can act as a buffer against pest pressures, reducing the reliance on chemical pesticides and promoting healthier farming systems (Waris et al., 2022; Younas et al., 2017).

Beneficial insects are key players in pest control within IPM frameworks (Younas et al., 2017). These organisms, which include predatory insects such as ladybugs, lacewings, and parasitic wasps, help keep pest populations in check through natural predation and parasitism. For instance, ladybugs can consume vast numbers of aphids, which are common agricultural pests. Additionally, pollinators, like bees and butterflies, contribute to the productivity of crops by facilitating pollination, which is essential for fruit and seed development. Recognizing and conserving these beneficial insects is vital for maintaining the balance within agroecosystems, ultimately leading to reduced pest populations and improved crop yields (Deguine et al., 2021).

Habitat management is a strategic approach to enhancing natural pest controllers in agricultural landscapes. This involves creating and maintaining environments that promote the presence and effectiveness of beneficial insects (Deguine et al., 2021). Practices such as intercropping, planting cover crops, and preserving natural vegetation around farmland can provide shelter, food, and breeding sites for these valuable allies. Additionally, implementing conservation practices, such as creating hedgerows or maintaining buffer strips, can support the diversity of beneficial species while simultaneously reducing pest pressures. By fostering appropriate habitats, farmers can enhance the natural pest control services provided by biodiversity, contributing to a more sustainable and resilient agricultural system (Dent & Binks, 2020).

4 Cultural Practices for Pest Management

Cultural practices play a significant role in Integrated Pest Management (IPM) by promoting healthy crop growth and minimizing pest pressures through strategic agricultural techniques (Dent & Binks, 2020). One of the most effective methods within this realm is crop rotation and diversity. By alternating the types of crops grown in a given area, farmers can disrupt the life cycles of pests that are specific to certain plants. This practice not only helps prevent the buildup of pest populations but also enhances soil health and fertility. A diverse crop system supports a variety of biological processes that enrich the soil and contribute to a balanced ecosystem, making crops more resilient to pest attacks (Dent & Binks, 2020). Maintaining soil health and fertility management is another critical component of cultural practices in pest management. Healthy soils support robust plant growth and improve the overall vigor of crops, which can help mitigate the impacts of pests. Practices such as the application of organic matter, cover cropping, and reduced tillage contribute to soil structure and nutrient availability, fostering a thriving agricultural environment. The healthier the crops are, the less alluring they become to potential pests, thereby reducing the likelihood of infestations (Menta & Remelli, 2020; Verma et al., 2023). The use of resistant crop varieties is also an essential cultural practice within IPM. Breeding and selecting plants that are naturally resistant or tolerant to specific pests and diseases can be an effective way to reduce pest pressure. These resistant varieties often require fewer chemical inputs, aligning well with sustainable agricultural practices. By incorporating these varieties into their crop management strategies, farmers can significantly decrease the efficacy of pests and pathogens while simultaneously promoting biodiversity within their fields (Neher & Barbercheck, 2019; Younas et al., 2022). Timing of planting and harvesting is another cultural practice that can help reduce pest infestations. By strategically scheduling planting and harvesting, farmers can avoid peak pest populations and environmental conditions that favor pest establishment. For example, early or late planting can sometimes help crops escape pest pressure associated with certain life cycles. Additionally, synchronizing harvest times can minimize the opportunities pests have to invade and damage the crop. By implementing these timing strategies, farmers can enhance crop resilience and reduce reliance on pest control measures, contributing to a more sustainable agricultural system.

5 Biological Control Methods

Biological control methods are integral to Integrated Pest Management (IPM), leveraging natural processes to manage pest populations sustainably. One of the primary strategies involves the introduction of natural predators and parasitoids. By carefully selecting and releasing these beneficial organisms into the agricultural environment, farmers can effectively reduce pest populations without the need for chemical pesticides. For example, the release of predatory insects such as ladybugs or lacewings can help control aphid populations, while parasitoids like wasps can target caterpillar pests. These approaches not only aid in pest control but also promote a balanced ecosystem that supports biodiversity (Dent & Binks, 2020).

The use of biopesticides and microbial agents is another important biological control method in IPM (Malinga & Laing, 2022). These environmentally friendly alternatives to synthetic pesticides are derived from natural organisms, including bacteria, fungi, and plant extracts. For instance, Bacillus thuringiensis (Bt), a soil-dwelling bacterium, produces toxins that target specific insect larvae without harming beneficial species. Biopesticides offer effective pest control options while reducing the environmental impact associated with conventional chemical treatments. Additionally, microbial agents can enhance plant health and resilience, creating a more robust agricultural system that can withstand pest pressures (Dent & Binks, 2020; Malinga & Laing, 2022).

Conservation of beneficial species is a cornerstone of biological control methods. Rather than solely relying on the introduction of new predators or microbial agents, promoting the natural populations of existing beneficial organisms within the agroecosystem is equally vital. This can be achieved through habitat management practices, such as planting cover crops, creating hedgerows, or maintaining natural habitats in and around agricultural fields. By providing food, shelter, and breeding locations for beneficial species, farmers can enhance their effectiveness in controlling pests. Supporting these natural allies not only contributes to pest management but also fosters ecosystem services such as pollination, nutrient cycling, and soil health, which are essential for sustainable agriculture. Together, these biological control methods form a robust framework for reducing reliance on chemical interventions and promoting ecological balance within farming systems.

6 Chemical Control and Its Role in IPM

Chemical control, while often viewed with caution in sustainable agriculture, plays a significant role within Integrated Pest Management (IPM) when applied responsibly. The use of chemical pesticides involves several key principles aimed at minimizing negative impacts on the environment, human health, and beneficial organisms. This includes selecting the appropriate pesticide for the specific pest in question, applying it at the right time and dosage, and ensuring targeted application methods that limit exposure to non-target species. Farmers are encouraged to use pesticides as a supplementary measure rather than a primary solution, employing them only when monitoring indicates that pest populations have reached economic threshold levels (Dent & Binks, 2020). Pesticide resistance management is another critical aspect of chemical control in IPM. Over-reliance on specific chemical classes can lead to the development of resistance among pest populations, rendering those pesticides ineffective over time. To combat this issue, farmers can implement strategies such as rotating different classes of pesticides, mixing compounds with different modes of action, and integrating non-chemical methods into their pest management practices. These strategies help to preserve the effectiveness of chemical controls and prolong their usability, thus maintaining a viable toolkit for pest management (Dent & Binks, 2020; Torgerson et al., 2021; Waris et al., 2020). Integration of chemical and non-chemical methods is fundamental to the effectiveness of IPM. By combining chemical controls with cultural, biological, and mechanical tactics, farmers can create a multifaceted approach that enhances overall pest management. For instance, a farmer might utilize pest-resistant crop varieties and biological controls while reserving chemical pesticides for situations where these measures fall short. This holistic strategy not only addresses pest problems more effectively but also reduces the potential ecological footprint of agricultural practices. Overall, when used judiciously and in conjunction with other pest management techniques, chemical controls can contribute to a balanced and sustainable approach to agriculture that safeguards productivity while protecting the environment (Deguine et al., 2021).

7 Conclusion

In conclusion, Integrated Pest Management (IPM) represents a holistic and sustainable approach to pest control that integrates multiple strategies for effective agricultural production. By understanding the principles of pest biology and ecology, employing essential cultural practices, leveraging the benefits of biodiversity, and responsibly utilizing chemical controls, farmers can manage pests in a way that promotes environmental health and sustains productivity. The emphasis on biological control methods further enhances the resilience of agricultural ecosystems, allowing natural processes to play a significant role in pest management. The convergence of these diverse approaches within IPM not only helps to minimize the reliance on chemical pesticides but also fosters biodiversity, soil health, and the overall well-being of agroecosystems. By adopting IPM practices, farmers can enhance their capacity to respond to pest challenges while simultaneously supporting the sustainability of their farming operations. As global agricultural demands continue to rise, the implementation of IPM becomes increasingly crucial. It offers a pathway to meet the needs of current and future generations while preserving the natural resources and ecosystems upon which agriculture depends. Through education, research, and collaboration, the continuous evolution and adaptation of IPM strategies will contribute to a more sustainable and resilient agricultural future.

Acknowledgements The work was supported by the "Fundamental Research Funds for Henan University of Science and Technology, 4024-13510002."

References

Deguine, J.-P., Aubertot, J.-N., Flor, R. J., Lescourret, F., Wyckhuys, K. A., & Ratnadass, A. (2021). Integrated pest management: Good intentions, hard realities. A review. *Agronomy for Sustainable Development, 41*, 38.

Dent, D., & Binks, R. H. (2020). *Insect pest management*. Cabi.

Malinga, L. N., & Laing, M. D. (2022). Role of microbial biopesticides as an alternative to insecticides in integrated pest management of cotton pests. In *Insecticides-impact and benefits of its use for humanity*. IntechOpen.

Menta, C., & Remelli, S. (2020). Soil health and arthropods: From complex system to worthwhile investigation. *Insects, 11*, 54.

Neher, D. A., & Barbercheck, M. E. (2019). Soil microarthropods and soil health: Intersection of decomposition and pest suppression in agroecosystems. *Insects, 10*, 414.

Shaaban, M., Van Zwieten, L., Bashir, S., Younas, A., Núñez-Delgado, A., Chhajro, M. A., Kubar, K. A., Ali, U., Rana, M. S., & Mehmood, M. A. (2018). A concise review of biochar application to agricultural soils to improve soil conditions and fight pollution. *Journal of Environmental Management, 228*, 429–440.

Torgerson, K., Meijering, J., Sok, J., Dicke, M., & Lansink, A. O. (2021). Towards circular agriculture–exploring insect waste streams as a crop and soil health promoter. *Journal of Insects as Food and Feed, 7*, 357–368.

Verma, R. C., Waseem, M. A., Sharma, N., Bharathi, K., Singh, S., Anto Rashwin, A., Pandey, S. K., & Singh, B. V. (2023). The role of insects in ecosystems, an in-depth review of entomological research. *International Journal of Environment and Climate Change, 13*, 4340–4348.

Waris, M. I., Younas, A., Adeel, M. M., Duan, S. G., Quershi, S. R., Kaleem Ullah, R. M., & Wang, M. Q. (2020). The role of chemosensory protein 10 in the detection of behaviorally active compounds in brown planthopper, Nilaparvata lugens. *Insect Science, 27*, 531–544.

Waris, M. I., Younas, A., Ullah, R. M. K., Rasool, F., Adeel, M. M., & Wang, M.-Q. (2022). Molecular and in vitro biochemical assessment of chemosensory protein 10 from brown planthopper Nilaparvata lugens at acidic pH. *Journal of Integrative Agriculture, 21*, 781–796.

Younas, A., Wakil, W., Khan, Z., Shaaban, M., & Prager, S. M. (2017). The efficacy of Beauveria bassiana, jasmonic acid and chlorantraniliprole on larval populations of Helicoverpa armigera in chickpea crop ecosystems. *Pest Management Science, 73*, 418–424.

Younas, A., Waris, M. I., Shaaban, M., Tahir ul Qamar, M., & Wang, M. Q. (2022). Appraisal of MsepCSP14 for chemosensory functions in Mythimna separata. *Insect Science, 29*, 162–176.

Fundamentals of Wastewater Treatment and Sludge Processing

Rana Muhammad Asif Kanwar, Zahid Mahmood Khan, Muhammad Shoaib, and Hafiz Umar Farid

Abstract Wastewater (WW) reuse in agriculture is a sustainable technique for developing agrarian countries like Pakistan. Despite the various WW treatment technologies are available but their selection and application for agricultural purposes involve some complications. So, because of their WW treatment status, the present paper provides a fundamental review on different aspects of WW treatment technologies regarding their description, abilities, merits and, demerits to assess their application potential. It recommends complete WW treatment system by a comprehensive review of different aspects of selected treatment technologies from primary to advanced level along with sludge processing. It mainly focuses on their treatment principles, factors affecting treatment, application potential, designing and performance calculations. The recommended WW treatment technologies such as sedimentation, trickling filter, adsorption filter, anaerobic digestion have critical fundamental status under the domain of WW treatment. Thus, this paper will provide a foundation approach towards the development of WW treatment systems for developing countries.

Keywords Agriculture · WW treatment system · Treatment principles · Design and performance calculations

1 Introduction

Most of the developing, as well as agrarian countries like Pakistan are facing tremendous challenges in the field of water management. Due to agricultural applications, their water demands are very high (Hamid et al., 2013). The composite task of sustaining growing demand for water while escaping degradation of ecologies needs feasible solutions following communal, economic and ecological criteria according to their local settings (Macleod & Haygarth, 2010; Molinos-Senante et al., 2012).

Rana Muhammad Asif Kanwar and Zahid Mahmood Khan are contributed equally to this work.

R. M. A. Kanwar (✉) · Z. M. Khan · M. Shoaib · H. U. Farid
Department of Agricultural Engineering, Bahauddin Zakariya University, Multan, Pakistan
e-mail: muhammadasifkanwar@bzu.edu.pk

M. Shaaban (ed.), *Soils and Sustainable Agriculture*, Frontier Studies in Soil Science,
https://doi.org/10.1007/978-3-031-91114-9_32

Worldwide more than 70% of water withdrawn is consumed for agricultural irrigation (Al-Jassim et al., 2015; UNESCO, 2003). So, search for other non-conventional and additional sources of water that can be used effectively and economically for agriculture is becoming necessary for planers. As, wastewater (WW) has unique importance towards agricultural applications being cheap and easily available round the year and due to the high concentration of nutrients (Maqbool et al., 2014; Dalkmann et al., 2014). Notably, WW irrigation is justified on economic and agronomic grounds because WW reuse is shared with nutrient reclaim especially N and P but treatment is essential to avoid environmental impacts (Gatto et al., 2015; Meda & Cornel, 2010; Odigie, 2014).

WW treatment refers to the removal of contaminants from WW for generating environmentally friendly treated effluent being discharged in receiving water bodies or for effective utilization in agricultural activities (Khan et al., 2015). In Pakistan there is a dire requirement of WW treatment because 26% of total national vegetables produced are irrigated by WW, approximately 2000 million gallons of sewage per day is used directly to a drain or open agricultural land and about 35–40% deaths were also occurred due to water-borne infections (Naz et al., 2016; Norton-Brandão et al., 2013). The most common constraints for developing countries in WW treatment are lack of financial resources, the existence of combined sewers for both domestic and industrial WW, lack of well-defined policies, and the shortage of professionals in the field of WW treatment (Sato et al., 2013). The treatment status of Pakistan is poor being of about 1–2% (Hamid et al., 2013; Murtaza et al., 2010; Shah & Hashmi, 2012) and biological treatment is confined to only two cities (Karachi and Islamabad) whose working is also not satisfactory.

It is essential to implement the efficient, economic and conceptual selection of WW treatment systems to cope with increasing energy and water crisis. In most cases, developing countries leads to the selection of unsuitable technologies according to local metrological and physical conditions, economic and human resource proficiencies, and communal or ethnic acceptability due to deficiency in activities of research and development. The effluent permissible limits, local conditions, environmental concerns, investment cost, and operational cost, etc. increases the complexity of the selection of treatment technology. So, this book chapter presents a critical review of basic treatment technologies from primary to advance along with sludge treatment. It will provide a fundamental approach towards the development of WW treatment systems. The discussed WW treatment technologies such as sedimentation, trickling filter, adsorption filter, anaerobic digestion have critical fundamental status under the domain of WW treatment.

2 Wastewater Treatment Methods

The treated effluent quality depends on the quality of influent water supply, the characteristic of waste being added and employed unit of treatments (Pedrero et al., 2010). The comparison mostly focuses on the application of treatment technologies for municipal WW for agrarian use and safe disposal in receiving water bodies. The main groups of concern pollutants in WW are oxygen demanding materials, pathogens, nutrients, heavy metals and other hazardous materials (Seow et al., 2016; Gatto et al., 2015; Topare et al., 2011). The review of different WW treatment approaches was considered that varies from conventional WW treatment systems to onsite limited or site-specific treatment systems (Table 1). The conventional systems were found feasible for treatment of large WW volume for large communities. These were usually located far from generation point and having high capital investment for large pipes and excavations. On the other hand, onsite limited regionalized treatment systems were found to collect, treat and dispose WW near generation point and favorable regarding cost, for areas of adverse topography and densely populated at small scale (Zhang et al., 2015; Jones et al., 2001). The onsite limited regionalized treatment systems were found more applicable to be modified according to local settings and influent loading to promote the output of treated WW within the community and evades the implementation of costly conventional systems (Massoud et al., 2009). The aspects of treatment technologies that should be considered for their development and adoption are given in (Table 2). The methodology used for development of fundamental approach towards WW treatment technologies adoption and application was based on the treatment technologies selection, their treatment principles with schematic representation, factors affecting on treatment, designing and performance calculations. The major objective of this study was to establish comprehensive understanding of WW treatment technologies to facilitate the researchers for adoption, development and performance assessment of selected treatment technologies. The selected treatment technologies were observed to have foundation status among existing WW treatment technologies. Most modern WWT technologies principle was also based on examined treatment technologies.

Table 1 Comparison of WW treatment technologies regarding their removal efficiencies, merits and demerits (Seow et al., 2016; Buha et al., 2015; Molinos-Senante et al., 2012; Massoud et al., 2009; Norton Brandão et al., 2013)

Technology	Description	Removal efficiencies	Merits	Demerits
Constructed wetlands	Land-based WW treatment by filtration and settling followed by microbial decomposition under anoxic condition	Bacteria 1 to 6 log units, N 30–70%; P 20–60%; COD 55–80%; SS 60–98%, NH_4 70%, Cr 55%, Ni 30%; Cu 9%, Zn 25–87%, and Co 75%	Low energy usage, less operational and maintenance cost, no formation of by-products, Tendency to remove emerging contaminants such as estrogens, PCPs and, pesticides. Easily handle variable WW loadings	Meteorologically dependent, Entrapment of solids and sediments, growth of biofilm and chemical precipitation, less formation of by-products. Destruction can occur due to overloads of ammonia and solids level

(continued)

Table 1 (continued)

Technology	Description	Removal efficiencies	Merits	Demerits
Pond system	Treatment of WW is done in man-made structures naturally under the influence of wind, microorganisms, solar light and, algae	N 20–40%, P 60–70%, COD 60–96%, SS 50–90%	Preferred WW treatment process in developing countries. The effluent of high nutrient and low pathogen content. Cost-effective in areas where land is cheap. Low energy requirement than most other WW treatment systems. Easy and simple to operate and maintain	Less effective in removing heavy metals. Odor, mosquitoes, and insects may be a problem if vegetation is not controlled. Require a large land area. Effluent does not meet state standards consistently thus requires additional treatments or disinfection. Problematic in cold areas due to longer retention time
Trickling filter	A reactor filled with support media over which WW flows downward developing a biological layer (biofilm) that metabolized the organic matter	N 35–50%, P 35–55%, COD: 75–90%, SS & BOD_5: 50–90, fecal coliform 0.6–0.9. PO_4-P 4–10 mg/L	The higher concentration of biomass, microbial activity, and toxicity resistance, Require small land area. May be developed with indigenous filter media such as maize cobs, tire derived rubber, etc. Low investment and operating cost	Electricity requirements, Nitrification can take place only at low loading rates in warm climates. Assessment of physical stability, life and efficiency of filter media is not an easy task. Low fecal coliform removal
Intermittent sand filter	Treatment of WW is done under well-built aerobic biological community linked with the surface of filter media	COD 75–90%, SS 85–95%, N 65–95%, P 75–99%	Effluent of high quality particularly for BOD and TSS. Under aerobic condition, it can transform nitrogen completely to Nitrate. Minimum and easy to operate and maintain. It requires no chemical addition	If media is not indigenously available, cost may increase. Its clogging potential is high. Electricity is needed. The limiting factor is the requirement of land area

(continued)

Table 1 (continued)

Technology	Description	Removal efficiencies	Merits	Demerits
Rotating biological reactor	Fixed film reactors with rotating radial and concentric passages that used to develop a thin layer of biomass by alternate exposure to oxygen/ sewage	COD 70–93%, SS 75–98%, N 20–80%, P 10–30%	Better capturing of suspended solids. It requires less complexity than extended aeration system. It requires minimal operation	Less commercially fixed film systems are available for onsite application. It is mostly applicable to cluster systems. Nitrification can take place only at low loading rates in warm climates. Less efficient than bio filters
Extended aeration	The activated sludge process is modified for small loads having mixed liquor where low operating efficiency is compensated by mechanical simplicity	COD 70–90%, SS 85–99%, N 50–90%, P 15–70%	Produces effluent of high degree nitrification due to high hydraulic and solid retention time Easily available in market	To some extent odor and noise issue. Less capturing of suspended solids. Require electricity as well as regular operation and maintenance. High initial capital cost relatively than other systems
Sequencing batch reactor	Here aeration tank serves all operations: fill, react, settle and draw in single batch reactor i.e. fill and draw activated sludge system	BOD > 95%, COD 70–90%, SS 85–99%, N 55–90%, P 25–70%	Economy of construction and stable operations. Due to large surface area of aeration tank, it can easily handle peak flows and shock loads. Suitable for site where enhanced treatment is needed including nitrogen removal for local surface/ground water	Insurance of operational control and periodic routine maintenance is essential for proper functioning of SBR
Moving bed biofilm reactor	Built up on the principle of aerobic biofilm principle having large surface and optimal conditions for development of better bacterial culture due to polyethylene carriers	COD 20–40%, SS 60–80%, N 10–20%, P 30–40%	Total solids complete removal with no problem of sludge bulking, not require backwashing and filter channeling, settling characteristics improved	Having fine screening at upstream level due to retention of media screen assemblies, limited process control due to less common involved processes

(continued)

Table 1 (continued)

Technology	Description	Removal efficiencies	Merits	Demerits
Membrane bioreactors	It combined the conventional activated sludge with membrane filtration	Nutrient setting according to pore size, bacterial removal > 5 log, COD 70–90%, SS 85–99%,, N 50–90%, P 20–70%	Ability to remove heavy metals, dye stuff, PCPs, EDCs, pharmaceuticals. Disinfection and removal of EC is done simultaneously	High investment and operating cost. Issue of membrane fouling and disposal

Table 2 Factors affecting development and adoption of WW treatment technologies

Engineering aspects	Design period, topography, available hydraulic head, groundwater depth that affect sewer infiltration, bearing capacity of soil and disposal facilities
Environmental aspects	Assimilative capacity of receiving water bodies, surface, ground and coastal water quality, mosquito and odor nuisance, effluent standards, permissible limits, nutrients control and bioaccumulation in food chain for public health concerns
Treatment aspects	WW flow and contaminants of interest, level of treatments needed, process optimization, performance and reliability characteristics, land, equipment's and skilled staff availability and power requirements
Economically aspects	Capital cost: land, construction, equipment's etc. Operating cost: repair and maintenance, staff, fuels, chemicals and electricity etc. Population density, residuals management and technology efficiency
Socially aspects	Protection of public health, government rules and regulations, planning and human settlements

3 Wastewater Characterization

WW characterization is important for WWTP design and development such as TSS is important for sedimentation tank designing, BOD loading is important for biological treatment, heavy metals assessment for adsorption system requirements etc. WW characterization refers to analytical procedures that are used to assess and measure various WW characteristics. These characteristics are used to indicate contamination strength, mitigation plan and treatment requirements of WW. Determination of WW flow and characteristics is prerequisite for influent loadings determination and technology selection due to widely variation in WW composition (Hamid et al., 2013). So, WW samples should be collected from disposal site/WW channels according to standard sampling procedure and analyzed immediately or stored at 4^0C in container with preservative to avoid biological and chemical changes depending upon the preservation techniques and holding time of parameters analyzed (APHA, 2012) for test results validity and samples integrity.

4 Primary Treatment (Screening and Sedimentation)

Primary treatment of WW usually based on physical forces mainly gravitational force governing settling, skimming, screening, flotation, flocculation, filtration, mixing, and gas transfer operations for coarse solids removal with associated fractions (organic matter, nitrogen, phosphorus, etc.) for about limited extent by mainly septic tank and imhoff tank technologies (Topare et al., 2011). The principal objectives of the primary sedimentation tank are coarsely dispersed phase removal, flocculated and coagulated impurities removal, and precipitated impurities removal after chemical treatment to reduce the organic and inorganic suspended load based on of specific gravity and viscosity of particles with water (Fig. 1). Normally primary treatment can be expected to remove settleable solids 90–95%, total suspended solids 40–60%, and particulate biological oxygen demand 25–35% (Table 3). Poor primary treatment (solids removal) leads to organic overloading of secondary treatment (biological treatment).

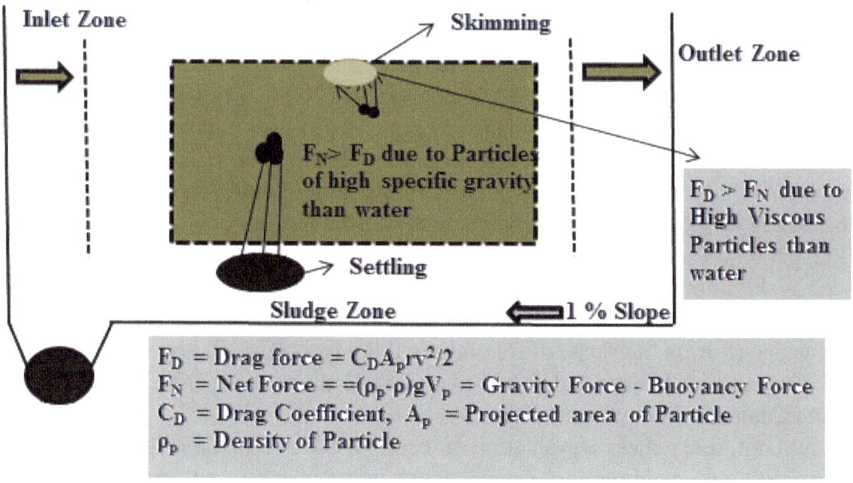

Fig. 1 Sedimentation theory of primary clarifier (Metcalf & Eddy, 2003)

Table 3 Design and performance calculations of primary clarifier

Designing steps and calculations	Performance calculations
Estimation of design flows = Q	Weir loading rate = Q/weir length
Determination of influent suspended solids = TSS_{inf}	% biosolids removal = (Settled biosolid volume/Total sample volume) × 100
Estimation of overflow rate = Vo = Q/L × W	$SS_{dry\ mass}$ = Q × ▼ SS
Determination of area = A = L × W or πd2/4 or Q/Vo	SS wet mass = SS dry mass/% solids by weight
Estimation of retention time = t_r = V/Q Determination of depth = D = t_r × Q/A	Q_{sludge} = SS wet mass/specific gravity Solids pumped = Pump rate × pump time × % solids Volatile solids = Pump rate × pump time × % solids × % volatile matter

5 Secondary Limited Treatment (Cascade Aeration)

Different research studies and reports on water quality monitoring point out that depleting DO level is a major problem for water bodies in developing countries. Cascade aeration is crucial technology for increasing DO level of water bodies, pre-aeration before secondary treatment to increase aerobic bacterial decomposition in WW treatment systems, reduce CO_2 concentration to decrease its corrosiveness, reduce odor caused by dissolved gases such as H_2S and CH_4, oxidation of iron and manganese from soluble to insoluble states and leads to remove certain volatile organic compounds (Kumar et al., 2013; Bagatur, 2009). The principle of cascade aeration based on stair-like mechanism (steps series) over which thin film of WW flows from one step level to lower one generating turbulence that leads to raise oxygen transfer/diffusion from atmosphere to WW due to air bubble formation and hence increasing DO level (Fig. 2) (Moulick et al., 2010; Hanbay et al., 2009; Metcalf & Eddy, 2003). The water flow over cascade steps usually involves three flow phases such as skimming (skims of water over edges at narrow steps or higher flows), transition (somewhat turbulent due to rigorous splash and significant free-surface aeration) and nappe (flow in the shape of free-falling thin layers governed by wider steps and low flow rates) (Chanson and Toombes (2004); Hanbay et al., 2009; Toombes et al., 2008; Chanson, 2004; Hager, 2002).

Different researches on evaluation of the influence of cascade heights and inward bound BOD on aeration performance (such as provision of sill, prismatic rectangular stepped chute, nappe flow condition, skimming flow with different seven-step geometrics, etc.) give satisfactory results related to aeration capability of cascades (El Monayeri et al., 2006; Emiroglu & Baylar, 2003, 2006; Moulick et al., 2010). Cascade aeration structures can be used for various applications like degassing/ VOC removal (methane, chlorine, dissolved iron and manganese, carbon dioxide, hydrogen sulfide taste and color due to volatile oil presence) re-oxygenation, and denitrification (Moulick et al. (2010), Hanbay et al., 2009). Basic calculations for the design and performance of the cascade aeration system are presented (Table 4).

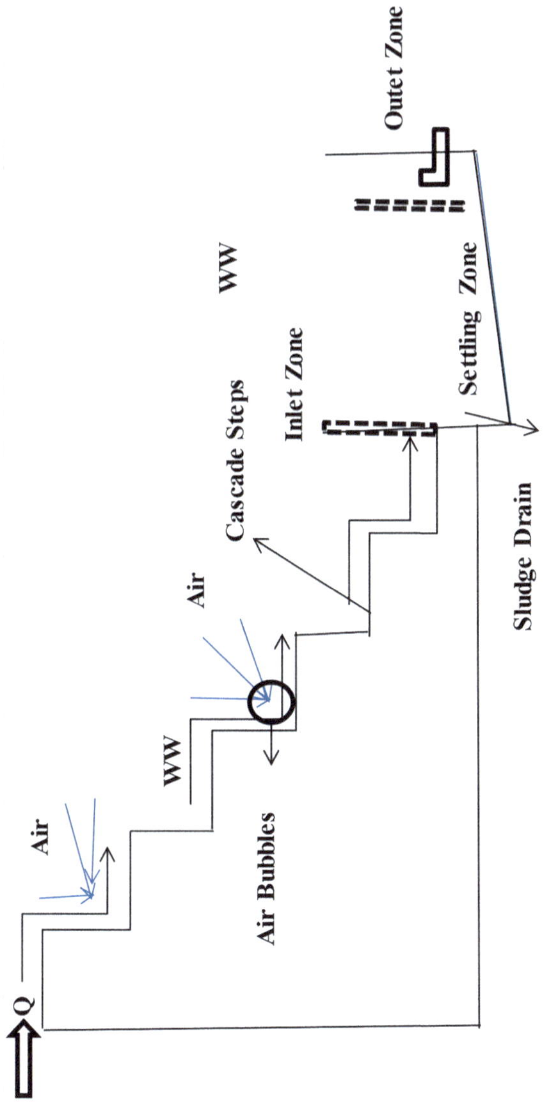

Fig. 2 Schematic representation of cascade aeration system (Hanbay et al., 2009; Metcalf & Eddy, 2003)

The Provision of settling tank for cascade system is essential for solid–liquid separation through sedimentation. The settled solids are collected at the bottom of the tank and supernatant water from outlets (Bajcar et al., 2011).

The cascade aeration system is a low-cost method of aeration that will be used to aerate the raw domestic or primary treated effluents to further reduce the biodegradable organics or BOD (in terms of COD) as required. The system efficiency should be evaluated in terms of DO, COD, BOD, TSS and, TN before and after the treatment process as treatment scheme for dilute WW (Fig. 3).

After sedimentation and cascade aeration, the physical and chemical characteristics of WW should be determined. If the characteristics of WW are within the permissible limit then it will be recommended for irrigation otherwise further processing of WW will be carried out by passing through the trickling filter and adsorption filter (Fig. 4).

Table 4 Mathematical calculations for design and performance of cascade aeration

Designing steps and equations	Performance calculations
Determine the DO of influent = Co	Aeration Efficiency = $E_T = (C_f - C_i)/(C_s - C_i) = 1 - 1/r$ ((Bagatur, 2009)
Determine the DO at temperature T = C_s	$\beta = C_{sw}/C_s$
Estimate the required final DO after post aeration = C	Percent Removal = $[(C_i - C_f)/C_i] \times 100$
Determine deficit ratio = $R = C_s - Co/ C_s - C$	$E_{20} = 1\text{-} (1 - E_T)\ 1/(1 + 0.02103(T - 20) + 8.261 \times 10^{-5})\ (T\text{-}20)^2)$
Determine the cascade height = H = R-1/0.11 a b (1 + 0.046 T)	
According to cascade height determine the step length, height and inclination angle from table 5–34 (Metcalf & Eddy, 2003)	

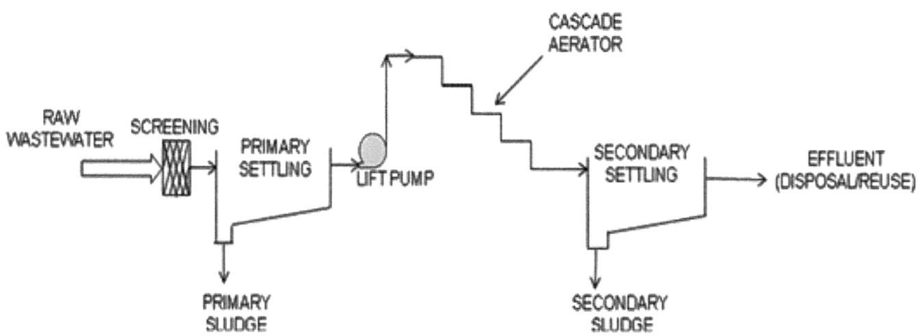

Fig. 3 Proposed treatment system for dilute WW (Kumar et al., 2013)

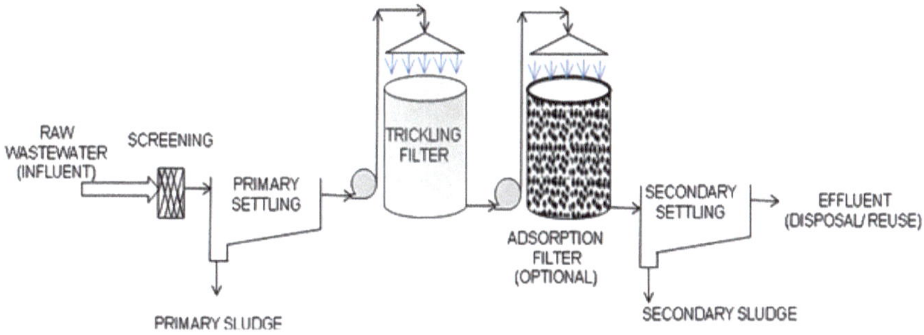

Fig. 4 Treatment scheme for concentrated WW (Metcalf & Eddy, 2003; Zhang et al., 2015)

6 Biological Secondary Treatment (Trickling Filter System)

Trickling filter is an attractive attached growth WW treatment technology for developing countries due to its less mechanical complexity, better process stability and well sludge thickening characteristics than other biological treatment systems. It is resistant to shock loads and power failures. It has simple considerations of design and having less ecological footprint, installation cost, land, energy, and maintenance requirements (Eding et al., 2006; Gikas, 2016; Mahmoud et al., 2010; Naz et al., 2016). Here the organic matter stabilization is accomplished by biologically means using various microorganisms (aerobic and anoxic bacteria, fungi, protozoans and, algae) (Eding et al., 2006; Metcalf & Eddy, 2003). At the same time, natural aeration transfers oxygen to the open spaces to keep the outer layer of slime aerobic. The dissolved and colloidal organic matter is converted by microorganisms aerobically into protoplasm and various gases. As protoplasm specific gravity is slightly more than water, so it can be removed by gravity settling from treated effluent (Fig. 5). So, a secondary clarifier is complementary for protoplasm settling that is itself organic in nature and otherwise will be considered as BOD of effluent. Basic calculations for the design and performance of the Trickling filter system are presented (Table 5).

Performance of developed biofilm is affected by various factors such as media selection (size, porosity, weight, and specific surface area), organic substrate, temperature, aeration, pH, oxygen concentration, etc. The most critical parameter in trickling filter research is media selection for biofilm support because it provides contact surface between biomass growths (microbial attachment) with contaminants to be treated (Godoy-Olmos et al., 2016; Kamstra et al., 1998). Several trickling filter medias have been utilized to evaluate and increase the efficiency of trickling filter such as calcitic gravel (Eding et al., 2006), rocks/plastic (Metcalf & Eddy, 2003), nylon pan scrubber (Sanjib et al., 2011), geotextile (Alimahmoodi et al., 2012), commercial and pall rings, stone (Khan et al., 2015), coal (Zhao et al., 2013), tire drubber (Naz et al., 2014), plastic sheet (corrugated) (Kim et al., 2014), gravel (Zhang et al., 2015), ceramsite and zeolite (Zhang et al., 2016), oyster shell, plastic (Li et al.,

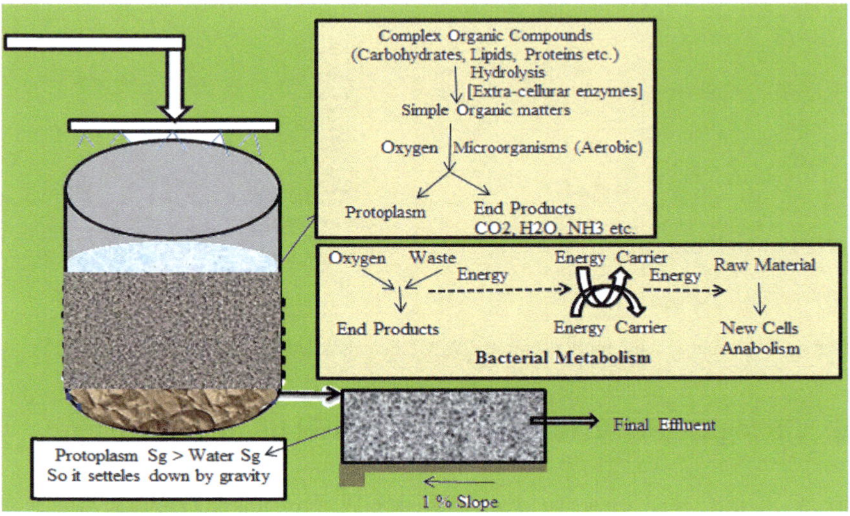

Fig. 5 Biochemistry of trickling filter WWT system (Ali et al., 2017; Eding et al., 2006)

Table 5 Steps of design and performance evaluation trickling filter WWT system

Designing steps and equations	Performance equations
Determine actual BOD loading = Q × BOD_{inf}	Hydraulic loading rate = Q/A
Estimate design BOD loading	Organic loading rate = weight of BOD or COD/day/m^3 of media
Volume of TF = Actual BOD loading/Design BOD loading	Recirculation ratio = recirculated flow/ primary effluent flow
Determine the dimensions of TF reactor by assuming depth of filter media	$E = 100/1 + 0.4432\sqrt{W/VF}$
Vol = Area × depth	E = removal efficiency. W = BOD loading
Dia = $\pi d^2/4$	F = recirculation factor, V = volume of TF
	η (%) = (Cin − Cout/Cin)100

2016), chips of bio char (Li et al., 2016), ceramic particles (Wu et al., 2016), cylindrical luffa (Vianna et al., 2012), sponge (Wang et al., 2015), etc. The author also works on an indigenously developed maize cobs trickling filter (MCTF) that has BOD removal efficiency of about 79% on an average basis (Ali et al., 2016).

7 Adsorption Filter

Along with biodegradable pollutants, it is also essential to remove non-biodegradable constituent's especially heavy metals due to their toxicity, persistence, bioaccumulation, and biomagnification by food chain that leads to potential threat to human health and ecological system (Lin et al., 2013; Salem et al., 2014; Yu et al., 2012). Adsorption is a mass transfer process of a substance from liquid phase to solid surface bound by intermolecular attraction forces (physiosorption) and/or chemical interaction between adsorbant and adsorbate (chemisorption) (Fig. 6) (Abas et al., 2013; Ghaedi et al., 2015; Kushwaha et al., 2013; Yadla et al., 2012). Adsorption is gaining a long term promising approach for heavy metals removal due to its effectiveness and economy of design, installation and maintenance than other conventional methods such as ion exchange, chemical precipitation, electrodialysis, membrane filtration, solvent extraction, oxidation and evaporation due to more initial capital and maintenance cost, large sludge production and flow rate restriction (Lakherwal et al., 2014; Keng et al., 2014).

In current years, research for low-cost adsorbents (which are present in huge amounts and easily available such as agricultural wastes, natural materials or by-products of industries) is rapidly increasing (Bhatnagar et al., 2015; Ramaraju et al., 2014; Sadaf et al., 2015). The materials which are found abundantly in nature, or considered as waste (mostly agricultural) or by product of industries and require economical processing, modifications and preparation are considered as low-cost adsorbents. Previous adsorption researches and studies have been investigated plant wastes, industrial by-products, clay and, siliceous materials as an alternative treatment to substitute commercial activated carbon that is listed (Table 6) (Ali et al., 2012). Agricultural wastes are the potential low-cost adsorbents due to their different adsorption mechanisms such as surface adsorption, diffusion, precipitation, transport,

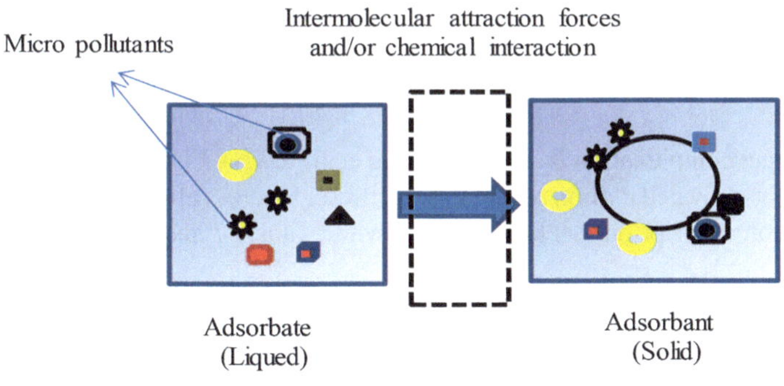

Fig. 6 Schematic of adsorption principle (Ghaedi et al., 2015; Kushwaha et al., 2013; Roque-Malherbe, 2018)

Table 6 List of potentially and indigenous low-cost adsorbents (Mohammed et al., 2014; Ali et al., 2012; Kushwaha et al., 2013)

Sawdust	Papaya wood	Coffee waste	Bagasse of sugar cane
Rice husk	Fly ash	Nuts	Wheat waste
Walnut modified sawdust	Orange peels	Peanuts	Cotton waste
Tea	Peels of grape fruit	Olive wastes	Maize and corn
Almonds	Apricot and date stones	Cherries	Jute stick

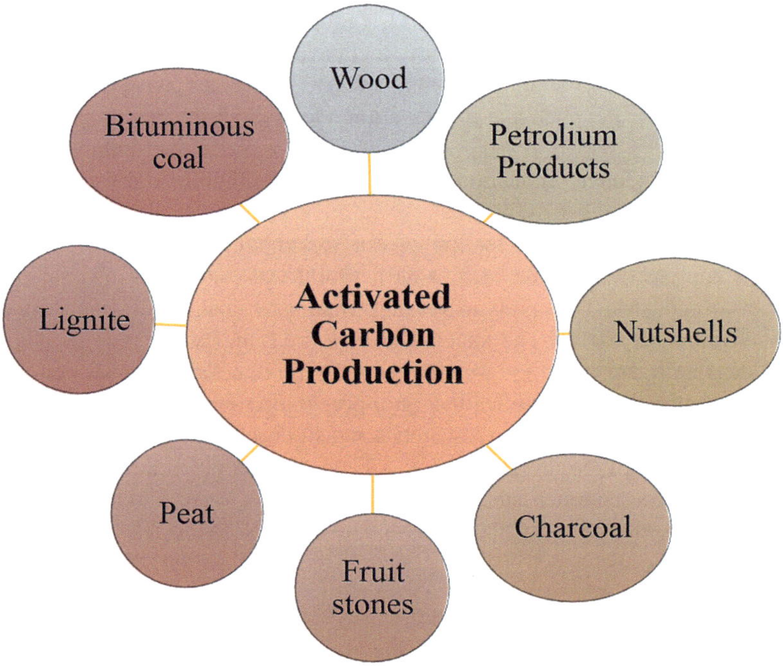

Fig. 7 Byproducts used for activated carbon production (Kushwaha et al., 2013)

ion exchange, complexation & chelation (Sud et al., 2008). Figure 7 shows the materials used for activated carbon preparation. Different factors affecting adsorption and its adsorption description mechanism is shown in Tables 7 and 8.

8 Sludge Digestion

Anaerobic conversion/digestion of organic compounds and pollutants is a technologically established process with less energy requirement for environmental protection. This process is used to govern both WW treatment and biogas production (CH_4 +

Table 7 Factors affecting the adsorption process (Ramaraju et al., 2014, Kushwaha et al., 2013; Abas et al., 2013)

pH	Metal adsorption is influenced by pH because it determines the adsorbate surface charge, ionization degree, adsorbate species, biomass cell wall functional group, metal chemistry and system equilibrium
Temperature	Temperature affect the adsorption equilibrium process being exothermic or endothermic nature as entropy, enthalpy and Gibbs free energy that determines the adsorption system spontaneity are the functions of temperature
Contact time	To complete the adsorption process, there should be equilibrium interaction between the functional group of solute solution and adsorbent surface
Adsorbent dose	Optimum adsorbent dosage is essential to determine adsorbent capacity by given quantity of adsorbate
Ionic strength	Initial ionic concentration alters the surface potential and so plays the role of driving force of metal for mass transfer resistance between liquid and solid phases
Aerobic conditions	Aerobic conditions in aqueous phase leads to promote extraction capability of activated carbon and chemical transformation such as organic compounds oligomerization than anaerobic isotherm
Natural organic matter	With natural organic matter in water source leads to decrease adsorption efficiency of target chemical due to decrease in porosity (pores blockage and constriction) and impact on surface polarity

Table 8 Adsorption isotherm and kinetic

Mechanism	Description	Equation	References
Langmuir isotherms	Adsorption of molecules on well-defined fixed number of active sites being homogeneously distributed over adsorbent that leads to same attraction mono molecular layer adsorption	$q_e = q_{max}bCa/1 + bCa$	Langmuir (1918)
Freundlich isotherms	It interprets heterogeneous surfaces adsorption with interaction between adsorbed molecules and not confined to monolayer development	$q_e = K_fC_e^{1/n}$	Freundlich (1906)
Pseudo-first order	The solute uptake rate of change with time is directly related with saturation concentration difference and solid uptake amount with time	$d_{qt}/d_t = k(q_e - q_t)$	Lagergren (1898)
Pseudo second order	It is rate limiting step mechanism significantly Chemisorption relates to valency forces sharing between adsorbate and adsorbent	$d_{qt}/d_t = k(q_e - q_t)^2$	Singh et al. (1988)

CO_2) that is a renewable useful energy source (Mao et al., 2015, Jang et al., 2014). Sludge generated from WWTPs is primary and secondary sludge. Primary sludge produced from the primary clarifier is easily biodegradable and generally consists of proteins, carbohydrates and long-chain hydrocarbons with methane production potential 315–400 Nm^3/t DOM (dry organic matter) can be predicted under optimum conditions. Secondary sludge comes from secondary clarifier due to microorganism's overproduction or protoplasm produced due to secondary treatment (trickling filter, RBC, activated sludge) which is more difficult to digest than primary one and methane produce of 190–240 Nm^3/t DOM (dry organic matter) can be estimated. The environmental benefits of anaerobic sludge digestion are production of green energy, safe disposal of organic waste, environmental protection, agro-system linkages with biogas and substitute for conventional energy sources (Abubaker et al., 2012; Gori et al., 2011; Tambone et al., 2010; Qi et al., 2005).

Biogas produced from human excreta generally contains CH_4 (65–66%), CO_2 (32–34%) and other trace gases while municipal sludge biogas composition is CH_4 (68–72%), CO_2 (18–20%) and other trace gases. The other basic objectives of anaerobic digestion are organic matter stabilization, volume reduction and pathogenic organism's eliminations (Caporgno et al., 2015). The types of equipment used for anaerobic digestion are either fixed cover or floating type, sludge pumps for addition and withdrawal as well as heating and mixing equipments for recirculation. Other ancillaries consist of cleaning equipment and gas storage. Anaerobic digestion involves generally a series of biological as well as chemical steps (hydrolysis, acidogenesis, acetogenesis and, methanogenesis) in which microbes degrade organic matter under anaerobic conditions to waste management or/and release energy/biogas for industrialized or domestic purposes by processes of mass reduction, gas production and improved manuring and dewatering characteristics of treated sludge (Fig. 8) (Bharathiraja et al., 2014; Demirbas et al., 2016).

The widely used parameters to predict the biogas potential are COD and BOD because they indicate the anaerobic biodegradability and hydrolysis constant. Other factors affecting the gas yield are organic matter digestibility, retention time (Time required for complete degradation of organic matter), microbial activity digestion kinetics and temperature, pH (ideal range: 6.8–7.4), C/N ratio (digestion substrate nutrient level, ideal range: 20–35), and organic loading rate (volatile acids amount fed into digester daily under continuous operation) (Table 9) (Zhen et al., 2017, Caporgno et al., 2015; Mao et al., 2015). Different additives are used to enhance the biogas potential of anaerobic digestion process such as greenery biomass (crop residues, weeds, plants and, their extracts), biological (fungi, microbial consortium and, enzymes), inorganic (alkali, acid and oxidative reagents, inorganic salts, trace elements and macronutrients).

The raw sludge from treatment sources should be fed into the digester well up to 2/3rd of its depth. A floating plastic drum (closed at the top and open at the bottom) with an outer diameter about 2 inches less than the inner diameter of the digester well, should be used for the collection and storage of the biogas generated in the digester. The proposed digester should be operated in a continuous or batch mode, depending upon the nature of the raw sludge received at the digester and the ambient

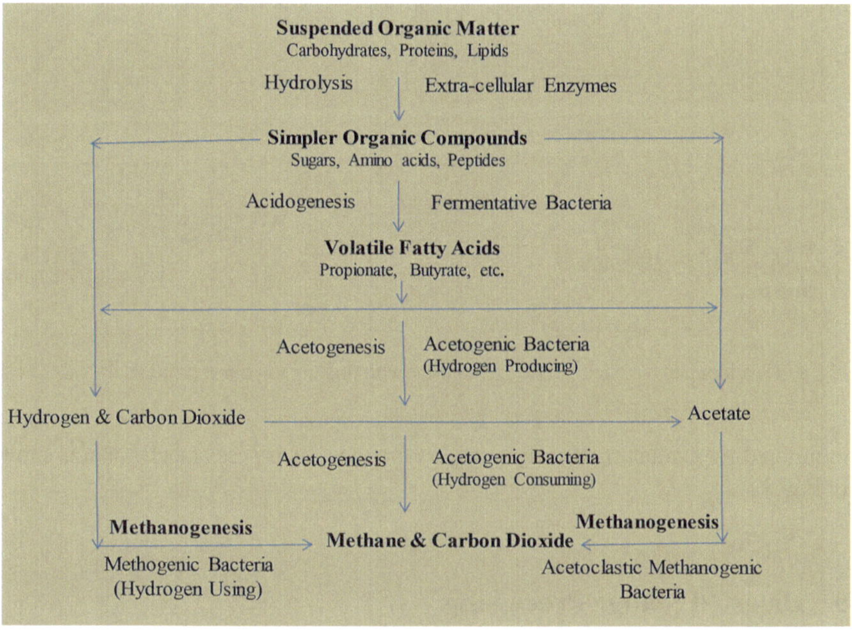

Fig. 8 Biochemistry of anaerobic sludge digestion

Table 9 Controlled calculations process control of sludge digestions (anaerobic)

Estimating Daily Biosolids Production = (TSS_{inf} − TSS_{eff}) × flow/%solids in sludge **Surface Loading** = Biosolids applied to thickener/thickener area **Thickening effectiveness** = % thickened biosolids/% influent Biosolids **Volatile Solids Loading** = Volatile Solids added/Digester Volume **Digestion time** = Digester Volume/Biosolids added **Chemical required (pH adjustment)** = Chemical used in Lab test × Digester Volume/Sample Volume **Seed volume** = digester volume × % seed To indicate the potential change in digester operating conditions, volatile acids/alkalinity ratio is used: **Ratio** = volatile acids concentration/alkalinity concentration **Biosolids Retention Time** = Digester volume/Biosolids volume added per day **Gas production** = volatile matter × % volatile reduction × production rate **Volatile Matter Reduction (%)** = (% volatile $matter_{in}$ − % volatile $matter_{out}$) 100/% volatile $matter_{in}$ − (% volatile $matter_{in}$ × % volatile $matter_{out}$) **Percent Moisture Reduction In Digested Biosolids** = (% $moisture_{in}$ − % $moisture_{out}$) 100/% $moisture_{in}$ − ((% $moisture_{in}$ × % $moisture_{out}$)

temperature. An inlet chute leading into the digester well should be provided for raw sludge feeding at appropriate intervals. Periodic mixing should be provided through a mechanical stirrer operated manually from the top of the digester. Similarly, a bucket type conveyer should be arranged to pull out the digested sludge from the digester bottom. The digestion period/solids retention time (SRT), solids reduction, digester loading, rate of biogas production, alkalinity consumption and pH should be

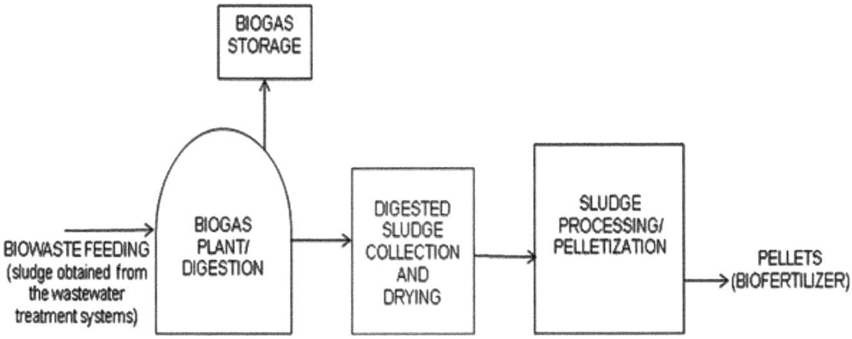

Fig. 9 Development of anaerobic sludge digester with floating top bio-gas collector

monitored for evaluating the digester performance. Its process mechanism is shown in Fig. 9.

9 Digested Sludge Processing

The final step of the WW and the associated sludge/biosolids treatment will include sludge processing and pelletizing. The continuously increasing generation, mismanagement and improper disposal of solid waste are serious problems faced by Pakistan today. After the unit processes of dewatering and drying, sludge is pelletized for preparation of class A-bio solids of acceptable quality for handling and land application (Sial et al., 2007). It is the process of densifying the organic solid waste by converting it into small pellets in order to economize the handling, transportation, storage and land application. It is based on the principle of crushing and compressing using different sieve sizes of the pelletizing machine. The pellets quality is evaluated on the basis of compost characteristics, raw material, molding machine power, sieve size and management of moisture during pelletization as moisture content is directly related to friction coefficient, bulk density and angle of repose (Table 10). Pelletizer plant provides the basis of technology for converting sludge into acceptable bio-solids for land application (Fig. 10).

Table 10 Parameters of performance evaluation of Pelletizer system

Moisture Content = (Initial Weight-Oven Dry Weight)/(Oven Dry Weigh) × 100%
Individual Pellet Volume = Vu = $\pi/4$ D^2 L
Individual Pellet density = Pu = Mu/Vu [Zhijia et al.]
Bulk density = Pb = Mb/Vb
Pellets durability = D (%) = [(Wi − Wf)/Wi] × 100 W_i = pellets weight before sieving
W_f = pellets weight after sieving
Fine contents of Pellets = P_f (%) = (mi − mf)/mi × 100 i.e. i & f for initial and final mass

Fig. 10 Pelletizer plant for palletization of digested sludge

The Pelletizer machine operates in batches. For initiating the process, raw sludge is manually shoveled into the sump at the footing of the elevator, which conducts material into pan of vibrating sieve. The oversized material including plastic, metallic, glass pieces, debris, etc. are separated and disposed. However, the sieved product enters into hammer mill where it is crushed into almost a powder form. This is cut off point for a batch. The material received above is again manually shoveled into same sump at the footing of elevator. A chain-controlled lever is pulled to actual which now conducts the material into hopper of the mixer. The bio fertilizers will be available for raising various vegetables, floriculture and ornamental plants.

10 Conclusions

This book chapter provides a comprehensive and fundamental approach for WW treatment plant design, development and, operation by critically reviewing the different aspects of treatment levels along with sludge processing targeting their indigenous application. It also presents the important design and performance calculations that will help the WWT system researchers and operators. Proper designing and operation of primary treatment are essential to avoid organic overloading of biological treatment. The selection of overflow rate and retention time is critical in its design and operation. Cascade aeration system is a simple, low-cost and suitable approach for raising of DO level and removing of suspended solids for safe disposal and/or possible agricultural reuse of domestic WW. It involves simple designing and functioning approach as described in present work. This work also proposed a treatment scheme for dilute WW. After evaluating its performance and reliability, further treatment levels will be decided. The Provision of biological treatment in terms of trickling filter with indigenous filter media leads to enhance WW treatment level of developing countries. Biofilm media selection and estimation of design BOD loading is critical in its designing and operation. Further polishing of trickling filter effluents may be done by low cost agricultural/industrial waste by slow filtration. For high industrial load effluents, the provision of adsorption filters with low-cost adsorbents is essential for the economic point of view and treatment enhancement. This work also proposed a treatment scheme for concentrated WW. Sludge digestion is an environmentally friendly technique for organic waste safe disposal, production of renewable energy and bio-fertilizer. However, it involves some complications that can be resolved by its effective designing and following its precautions. Pelletizer is a promising technology for sludge/organic waste reuse in agriculture. But, it involves complex mechanical linkages that limit its application and adoptability. However, its provision on district-level leads to raising bio-fertilizer production for soil fertility and crop productivity enhancement. This work has generated valuable WW data and related WW processing and treatment information for WW management departments and personals for designing and developing WW Treatment Systems/facilities according to the regional conditions and existing WW streams contamination levels. Thus, the present study can potentially play an important role in managing not on the WW pollution in developing countries but also help a relatively safe re-use of the WW for peri urban agriculture and protect our receiving environment. This is particularly significant in the current water resource shortage scenario in the country and the increasing use of WW for peri-urban food agriculture in Pakistan and other developing countries.

References

Abas, S. N. A., Ismail, M. H. S., Kamal, M. L., & Izhar, S. (2013). Adsorption process of heavy metals by low-cost adsorbent: A Review. *World Applied Sciences Journal, 28*(11), 1518–1530.

Abubaker, J., Risberg, K., & Pell, M. (2012). Biogas residues as fertilisers–Effects on wheat growth and soil microbial activities. *Applied Energy, 99*, 126–134.

Ali, I., Asim, M., & Khan, T. A. (2012). Low cost adsorbents for the removal of organic pollutants from wastewater. *Journal of Environmental Management, 113*, 170–183.

Ali, I., Khan, Z. M., Peng, C., Naz, I., Sultan, M., Ali, M., Mahmood, M. H., & Niaz, Y. (2017). Identification and Elucidation of the designing and operational issues of trickling filter systems for wastewater treatment. *Polish Journal of Environmental Studies* 26(6).

Ali, I., Khan, Z. M., Sultan, M., Mahmood, M. H., Farid, H. U., Ali, M., & Nasir, A. (2016). Experimental study on maize cob trickling filter-based wastewater treatment system: design, development, and performance evaluation. *Polish Journal of Environmental Studies, 25*(6), 2265–2273.

Alimahmoodi, M., Yerushalmi, L., & Mulligan, C. N. (2012). Development of biofilm on geotextile in a new multi-zone wastewater treatment system for simultaneous removal of COD, nitrogen and phosphorus. *Bioresource Technology, 107*, 78–86.

Al-Jassim, N., Ansari, M. I., Harb, M., & Hong, P. Y. (2015). Removal of bacterial contaminants and antibiotic resistance genes by conventional wastewater treatment processes in Saudi Arabia: Is the treated wastewater safe to reuse for agricultural irrigation? *Water Research, 73*, 277–290.

APHA. (2012). *Standard methods for the examination of water and wastewater* (22nd ed.).

Bagatur, T. (2009). Evaluation of preaeration with V-notch weir and cascade structures in clarifiers. *Journal of Environmental Engineering, 135*(3), 176–184.

Bajcar, T., Steinman, F., Širok, B., & Prešeren, T. (2011). Sedimentation efficiency of two continuously operating circular settling tanks with different inlet-and outlet arrangements. *Chemical Engineering Journal, 178*, 217–224.

Bharathiraja, B., Yogendran, D. R., Kumar, R., Chakravarthy, M., & Palani, S. (2014). Biofuels from sewage sludge-A review. *International Journal of ChemTech Research, 6*(9), 4417–4427.

Bhatnagar, A., Sillanpää, M., & Witek-Krowiak, A. (2015). Agricultural waste peels as versatile biomass for water purification–A review. *Chemical Engineering Journal, 270*, 244–271.

Buha, D. M., Atalia, K. R., Baagwala, W. Y., & Shah, N. K. (2015). Review on wastewater treatment technologies for nitrogen removal. *Global Journal of Multidisciplinary Studies, 4*(6), 299–318.

Caporgno, M. P., Taleb, A., Olkiewicz, M., Font, J., Pruvost, J., Legrand, J., & Bengoa, C. (2015). Microalgae cultivation in urban wastewater: Nutrient removal and biomass production for biodiesel and methane. *Algal Research, 10*, 232–239.

Chanson, H. (2004). *Hydraulics of open channel flow*. Butterworth-Heinemann.

Chanson, H., & Toombes, L. (2004). Hydraulics of stepped chutes: The transition flow. *Journal of Hydraulic Research, 42*(1), 43–54.

Dalkmann, P., Siebe, C., Amelung, W., Schloter, M., & Siemens, J. (2014). Does long-term irrigation with untreated wastewater accelerate the dissipation of pharmaceuticals in soil? *Environmental Science & Technology, 48*(9), 4963–4970.

Demirbas, A., Taylan, O., & Kaya, D. (2016). Biogas production from municipal sewage sludge (MSS). *Energy Sources, Part a: Recovery, Utilization, and Environmental Effects, 38*(20), 3027–3033.

Eding, E. H., Kamstra, A., Verreth, J. A. J., Huisman, E. A., & Klapwijk, A. (2006). Design and operation of nitrifying trickling filters in recirculating aquaculture: A review. *Aquacultural Engineering, 34*(3), 234–260.

El Monayeri, O., Bayoumi, S., & Khalifa, A. (2006). Enhancement of self-purification of streams using stepped aeration. In *Tenth international water technology conference IWTC10*, pp. 307–316.

Emiroglu, M. E., & Baylar, A. (2006). Self-aeration in smooth and stepped chutes. *International Journal of Science and Technology, 1*(2), 105–113.

Emiroglu, M. E., & Baylar, A. H. M. E. T. (2003). An investigation of effect of stepped chutes with end sill on aeration performance. *Water Quality Research Journal of Canada, 38*(3), 527–539.

Freundlich, H. M. F. (1906). Over the adsorption in solution. *Journal of Physics and Chemistry, 57*(385), 470.

Gatto D'Andrea, M. L., Salas Barboza, A. G. J., Garcés, V., Rodriguez-Alvarez, M. S., Iribarnegaray, M. A., et al. (2015). The use of (treated) domestic wastewater for irrigation: Current situation and future challenges. *International Journal of Water and Wastewater Treatment, 1*(1).

Ghaedi, A. M., Ghaedi, M., Vafaei, A., Iravani, N., Keshavarz, M., Rad, M., Tyagi, I., Agarwal, S., & Gupta, V. K. (2015). Adsorption of copper (II) using modified activated carbon prepared from Pomegranate wood: Optimization by bee algorithm and response surface methodology. *Journal of Molecular Liquids, 206*, 195–206.

Gikas, P. (2016). Towards energy positive wastewater treatment plants. *Journal of Environmental Management*.

Godoy-Olmos, S., Martínez-Llorens, S., Tomás-Vidal, A., & Jover-Cerdá, M. (2016). Influence of filter medium type, temperature and ammonia production on nitrifying trickling filters performance. *Journal of Environmental Chemical Engineering, 4*(1), 328–340.

Gori, R., Jiang, L. M., Sobhani, R., & Rosso, D. (2011). Effects of soluble and particulate substrate on the carbon and energy footprint of wastewater treatment processes. *Water Research, 45*(18), 5858–5872.

Hager, W. H. (2002). The hydraulics of stepped chutes and spillways By Hubert Chanson. *Canadian Journal of Civil Engineering, 29*(4), 634–634.

Hamid, A., Zeb, M., Mehmood, A., Akhtar, S., & Saif, S. (2013). Assessment of wastewater quality of drains for irrigation. *Journal of Environmental Protection, 4*(9), 937.

Hanbay, D., Baylar, A., & Batan, M. (2009). Prediction of aeration efficiency on stepped cascades by using least square support vector machines. *Expert Systems with Applications, 36*(3), 4248–4252.

Jang, H. M., Cho, H. U., Park, S. K., Ha, J. H., & Park, J. M. (2014). Influence of thermophilic aerobic digestion as a sludge pre-treatment and solids retention time of mesophilic anaerobic digestion on the methane production, sludge digestion and microbial communities in a sequential digestion process. *Water Research, 48*, 1–14.

Jones, K. B., Neale, A. C., Wade, T. G., Wickham, J. D., Cross, C. L., Edmonds, C. M., Loveland, T. R., Nash, M. S., Riitters, K. H., & Smith, E. R. (2001). The consequences of landscape change on ecological resources: An assessment of the United States Mid-Atlantic Region, 1973–1993. *Ecosystem Health, 7*(4), 229–242.

Kamstra, A. V., der Heul, J. W., & Nijhof, M. (1998). Performance and optimisation of trickling filters on eel farms. *Aquacultural Engineering, 17*(3), 175–192.

Keng, P. S., Lee, S. L., Ha, S. T., Hung, Y. T., & Ong, S. T. (2014). Removal of hazardous heavy metals from aqueous environment by low-cost adsorption materials. *Environmental Chemistry Letters, 12*(1), 15–25.

Khan, Z. U., Naz, I., Rehman, A., Rafiq, M., Ali, N., & Ahmed, S. (2015). Performance efficiency of an integrated stone media fixed biofilm reactor and sand filter for sewage treatment. *Desalination and Water Treatment, 54*(10), 2638–2647.

Kim, B., Gautier, M., Prost-Boucle, S., Molle, P., Michel, P., & Gourdon, R. (2014). Performance evaluation of partially saturated vertical-flow constructed wetland with trickling filter and chemical precipitation for domestic and winery wastewaters treatment. *Ecological Engineering, 71*, 41–47

Kumar, A., Moulick, S., Singh, B. K., & Mal, B. C. (2013). Design characteristics of pooled circular stepped cascade aeration system. *Aquacultural Engineering, 56*, 51–58.

Kushwaha, S., Soni, H., Ageetha, V., & Padmaja, P. (2013). An insight into the production, characterization, and mechanisms of action of low-cost adsorbents for removal of organics from aqueous solution. *Critical Reviews in Environmental Science and Technology, 43*(5), 443–549.

Lagergren, S. (1898). *About the theory of so-called adsorption of soluble substances.*

Lakherwal, D. (2014). Adsorption of heavy metals: A review. *International Journal of Environmental Research and Development, 4*(1), 41–48.

Langmuir, I. (1918). The adsorption of gases on plane surfaces of glass, mica and platinum. *Journal of the American Chemical Society, 40*(9), 1361–1403.

Li, W., Loyola-Licea, C., Crowley, D. E., & Ahmad, Z. (2016). Performance of a two-phase biotrickling filter packed with biochar chips for treatment of wastewater containing high nitrogen and phosphorus concentrations. *Process Safety and Environmental Protection, 102*, 150–158.

Lin, Y. C., Chang-Chien, G. P., Chiang, P. C., Chen, W. H., & Lin, Y. C. (2013). Multivariate analysis of heavy metal contaminations in seawater and sediments from a heavily industrialized harbor in Southern Taiwan. *Marine Pollution Bulletin, 76*(1), 266–275.

Macleod, C. J. A., & Haygarth, P. M. (2010). Integrating water and agricultural management under climate change.

Mahmoud, M., Tawfik, A., & El-Gohary, F. (2010). Simultaneous organic and nutrient removal in a naturally ventilated biotower treating presettled municipal wastewater. *Journal of Environmental Engineering, 136*(3), 301–307.

Mao, C., Feng, Y., Wang, X., & Ren, G. (2015). Review on research achievements of biogas from anaerobic digestion. *Renewable and Sustainable Energy Reviews, 45*, 540–555.

Maqbool, T., Khan, S. J., & Lee, C. H. (2014). Effects of filtration modes on membrane fouling behavior and treatment in submerged membrane bioreactor. *Bioresource Technology, 172*, 391–395.

Massoud, M. A., Tarhini, A., & Nasr, J. A. (2009). Decentralized approaches to wastewater treatment and management: Applicability in developing countries. *Journal of Environmental Management, 90*(1), 652–659.

Meda, A., & Cornel, P. (2010). Aerated biofilter with seasonally varied operation modes for the production of irrigation water. *Water Science and Technology, 61*(5), 1173–1181.

Metcalf and Eddy H. (2003). *Wastewater engineering: Treatment, disposal, and reuse* (4th ed., pp 887–981). The McGraw-Hill Companies.

Mohammed, M. A., Shitu, A., & Ibrahim, A. (2014). Removal of methylene blue using low cost adsorbent: A review. *Research Journal of Chemical Sciences*, p. 606X, ISSN, 2231.

Molinos-Senante, M., Garrido-Baserba, M., Reif, R., Hernández-Sancho, F., & Poch, M. (2012). Assessment of wastewater treatment plant design for small communities: Environmental and economic aspects. *Science of the Total Environment, 427*, 11–18.

Moulick, S., Tambada, N. V., Singh, B. K., & Mal, B. C. (2010). Aeration characteristics of a rectangular stepped cascade system. *Water Science and Technology, 61*(2), 415–420.

Murtaza, G., Ghafoor, A., Qadir, M., Owens, G., Aziz, M. A., & Zia, M. H. (2010). Disposal and use of sewage on agricultural lands in Pakistan: A review. *Pedosphere, 20*(1), 23–34.

Naz, I., Khatoon, N., Ali, M. I., Saroj, D. P., Batool, S. A. U., Ali, N., & Ahmed, S. (2014). Appraisal of the tire derived rubber (TDR) medium for wastewater treatment under aerobic and anaerobic conditions. *Journal of Chemical Technology and Biotechnology, 89*(4), 587–596.

Naz, I., Ullah, W., Sehar, S., Rehman, A., Khan, Z. U., Ali, N., & Ahmed, S. (2016). Performance evaluation of stone-media pro-type pilot-scale trickling biofilter system for municipal wastewater treatment. *Desalination and Water Treatment, 57*(34), 15792–15805.

Norton-Brandão, D., Scherrenberg, S. M., & van Lier, J. B. (2013). Reclamation of used urban waters for irrigation purposes—A review of treatment technologies. *Journal of Environmental Management, 122*, 85–98.

Odigie, J. O. (2014). Harmful effects of wastewater disposal into water bodies: A case review of the Ikpoba River, Benin city, Nigeria. *Tropical Freshwater Biology, 23*, 87

Pedrero, F., Kalavrouziotis, I., Alarcón, J. J., Koukoulakis, P., & Asano, T. (2010). Use of treated municipal wastewater in irrigated agriculture—Review of some practices in Spain and Greece. *Agricultural Water Management, 97*(9), 1233–1241.

Qi, B. C., Aldrich, C., Lorenzen, L., & Wolfaardt, G. W. (2005). Acidogenic fermentation of lignocellulosic substrate with activated sludge. *Chemical Engineering Communications, 192*(9), 1221–1242.

Ramaraju, B. M., Kumar Reddy, P., & Subrahmanyam, C. (2014). Low cost adsorbents from agricultural waste for removal of dyes. *Environmental Progress & Sustainable Energy, 33*(1), 38–46.

Roque-Malherbe, R. M. (2018). *Adsorption and diffusion in nanoporous materials*. CRC Press.

Sadaf, S., Bhatti, H. N., Arif, M., Amin, M., & Nazar, F. (2015). Adsorptive removal of direct dyes by PEI-treated peanut husk biomass: Box-Behnken experimental design. *Chemistry and Ecology, 31*(3), 252–264.

Salem, D. M. A., Khaled, A. E., Nemr, A., & El-Sikaily, A. (2014). Comprehensive risk assessment of heavy metals in surface sediments along the Egyptian Red Sea coast. *The Egyptian Journal of Aquatic Research, 40*(4), 349–362.

Sanjib, M., Mohd, T., & Mukherjee, C. K. (2011). Evaluation of nitrification performance of a trickling filter with nylon pot scrubber as media. *IJSN, 2*(3), 515.

Sato, T., Qadir, M., Yamamoto, S., Endo, T., & Zahoor, A. (2013). Global, regional, and country level need for data on wastewater generation, treatment, and use. *Agricultural Water Management, 130*, 1–13.

Seow, T. W., Lim, C. K., Nor, M. H. M., Mubarak, M. F. M., Lam, C. Y., Yahya, A., & Ibrahim, Z. (2016). Review on wastewater treatment technologies. *International Journal of Applied Environmental Sciences, 11*(1), 111–126.
Shah, M., & Hashmi, H. N. (2012). Macrophyte waste stabilization ponds: An option for municipal wastewater treatment. *International Journal of Physical Sciences, 7*(30), 5162–5166.
Sial, J. K., Randhawa, I. A., Shafi, A., & Hussain, K. A. (2007). Development and testing of a sludge pelletizer. *Pakistan Journal of Agriculture Research, 44*, 4.
Singh, D. B., Prasad, G., Rupainwar, D. C., & Singh, V. N. (1988). As (III) removal from aqueous solution by adsorption. *Water, Air, & Soil Pollution, 42*(3), 373–386.
Sud, D., Mahajan, G., & Kaur, M. P. (2008). Agricultural waste material as potential adsorbent for sequestering heavy metal ions from aqueous solutions—A review. *Bioresource Technology, 99*(14), 6017–6027.
Tambone, F., Scaglia, B., D'Imporzano, G., Schievano, A., Orzi, V., Salati, S., & Adani, F. (2010). Assessing amendment and fertilizing properties of digestates from anaerobic digestion through a comparative study with digested sludge and compost. *Chemosphere, 81*(5), 577–583.
Toombes, L., Wagner, C., & Chanson, H. (2008). Flow patterns in nappe flow regime down low gradient stepped chutes. *Journal of Hydraulic Research, 46*(1), 4–14.
Topare, N. S., Attar, S. J., & Manfe, M. M. (2011). Sewage/wastewater treatment technologies: A review. *Science Revives Chemeistry Communication, 1*(1), 18–24.
Unesco. (2003). *Water for people-water for life-The United Nations World Water Development Report.*
Vianna, M. R., de Melo, G. C., & Neto, M. R. V. (2012). Wastewater treatment in trickling filters using luffa cyllindrica as biofilm supporting medium. *Journal of Urban and Environmental, 6*(2).
Wang, H., Ji, G., Bai, X., & He, C. (2015). Assessing nitrogen transformation processes in a trickling filter under hydraulic loading rate constraints using nitrogen functional gene abundances. *Bioresource Technology, 177*, 217–223.
Wu, H., Yin, Z., Quan, Y., Fang, Y., & Yin, C. (2016). Removal of methyl acrylate by ceramic-packed biotrickling filter and their response to bacterial community. *Bioresource Technology, 209*, 237–245.
Yadla, S. V. Sridevi, V., & Lakshmi, M. C. (2012). A review on adsorption of heavy metals from aqueous solution. *Journal of Chemical, Biological and Physical Sciences (JCBPS). 2*(3), 1585.
Yu, S., Zhu, Y. G., & Li, X. D. (2012). Trace metal contamination in urban soils of China. *Science of the Total Environment, 421*, 17–30.
Zhang, X., Li, J., Yu, Y., Xu, R., & Wu, Z. (2016). Biofilm characteristics in natural ventilation trickling filters (NVTFs) for municipal wastewater treatment: Comparison of three kinds of biofilm carriers. *Biochemical Engineering Journal, 106*, 87–96.
Zhang, Y., Cheng, Y., Yang, C., Luo, W., Zeng, G., & Lu, L. (2015). Performance of system consisting of vertical flow trickling filter and horizontal flow multi-soil-layering reactor for treatment of rural wastewater. *Bioresource Technology, 193*, 424–432.
Zhao, Q., Zhong, H., Wang, K., Wei, L., Liu, J., & Liu, Y. (2013). Removal and transformation of organic matters in domestic wastewater during lab-scale chemically enhanced primary treatment and a trickling filter treatment. *Journal of Environmental Sciences, 25*(1), 59–68.
Zhen, G., Lu, X., Kato, H., Zhao, Y., & Li, Y. Y. (2017). Overview of pretreatment strategies for enhancing sewage sludge disintegration and subsequent anaerobic digestion: Current advances, full-scale application and future perspectives. *Renewable and Sustainable Energy Reviews, 69*, 559–577.

Wastewater Treatment Systems for Peri-Urban Agriculture

Rana Muhammad Asif Kanwar, Zahid Mahmood Khan, and Muhammad Shoaib

Abstract The largest source of water supply and agricultural product contamination is wastewater from homes, businesses, and municipalities. As a result, this book chapter focused on wastewater treatment methods for safe agriculture, notably the Trickling Filter (TF) technology. Next, the kinetics and physical mechanisms of mass change in biofilm reactors were examined. Biological processes that contribute to pollution biodegradation include the removal of organic waste, nitrification or partial nitrification, denitrification, and anammox. Wastewater treatment also requires an analysis of model data. The activated sludge and biofilm diffusion models are completed, the Retention Time Distribution analysis for hydrodynamic modeling is examined, and some of the most recent hydraulic models are presented. Furthermore, to build and design new TF systems and retrofit current TFs, the answers to the TF system's operational and performance issues are extensively analyzed and resolved.

Keywords Agriculture · Trickling filter · Activated sludge · Design and hydrodynamic modelling · Operation and performance

1 Introduction

The most significant and priceless natural resource on the planet is water. Even though most of the Earth is covered in water, the unequal distribution of clean water makes it more and more scarce. Reusing wastewater effluents from industrial, agricultural, and municipal sources is becoming essential to reducing the amount and quality of declining water supplies (Dalkmann et al., 2014; Maqbool et al., 2014). As a result, planners must treat wastewater prior to reuse or disposal (Gatto et al., 2015; Odigie, 2014). Furthermore, the primary motivation for wastewater treatment is to prevent groundwater and soil contamination, safeguard aquatic life, reuse for agriculture, groundwater recharge, and industries, and address social issues caused

R. M. A. Kanwar (✉) · Z. M. Khan · M. Shoaib
Department of Agricultural Engineering, Bahauddin Zakariya University, Multan, Pakistan
e-mail: muhammadasifkanwar@bzu.edu.pk

M. Shaaban (ed.), *Soils and Sustainable Agriculture*, Frontier Studies in Soil Science,
https://doi.org/10.1007/978-3-031-91114-9_33

by sewage accumulation (Khan et al., 2019; Sato et al., 2013). Wastewater treatment consists essentially of pre-primary, secondary, and advanced treatment. During the preliminary phase, coarse materials and grit can be removed using screening and gravity setting, whilst suspended solids (TSS) and particle biological oxygen demand (BOD) are removed by primary treatment. According to Shelley et al. and Rivas et al., primary treatments eliminate around 65% of TSS, 40% of BOD, 30% of COD, 20% N, 60% of oil and grease, and 10% of P. Mechanical or physical unit operations are commonly used to remove contaminants from the primary treatment terrain.

The biological microbiological processes that remove suspended and dissolved organic pollutants under regulated conditions account for most of the secondary treatment. Secondary clarifiers are used to settle protoplasm and damaged biological flocks. This phase typically removes 90% of TS and BOD, as well as the majority of P, N, and metals. The wastewater treated can be recycled for non-potable purposes such as gardening, toilet flushing, and agriculture, or it can be safely discharged into receiving water bodies. The ultimate purpose of tertiary/advanced treatment is to further purify wastewater, notably by removing pollutants such as metals, nutrients, and dangerous chemicals that remain in secondary treated wastewater. In addition to the, disinfection can be utilized to improve water quality at the end of each treatment stage (Quach-Cu et al., 2018; Mittal, 2011).

Wastewater treatment systems are classified as biological or physiochemical based on the unit processes used. For urban/industrial wastewater, biological treatment provides a simple, promising, and cost-effective alternative to a variety of physiochemical procedures. Numerous researchers have made significant contributions to wastewater bioremediation (Costley & Wallis, 2001; Abou-Elela et al., 2010; Mononen et al., 2010; Mosa et al., 2016; De Beer et al., 2017; Ghimire & Wang, 2018). Microorganisms in biological therapy require organic matter and nutrients for metabolic processes to alter structural biomass organization and thrive. Biological wastewater treatment systems are classed as attached or suspended growth systems based on the structural arrangement of their biomass.

Attached growth systems provide several advantages over suspended growth systems, including improved reaction kinetics, shorter retention durations, increased decomposition capacity, less sludge production, and higher microbial densities. Trickling filter (TF) systems have been shown to be more effective than all other attached growth treatment systems due to their low energy and maintenance costs, simplicity, reduced biomass washout, ease of operation, and environmental compatibility (Shah & Hashmi, 2012). According to Kumar and Kundu, it uses a variety of contaminant removal strategies, including biodegradation, biosorption, bioaccumulation and biomineralization. However, when carbonaceous and nitrogenous elements, as well as pathogens, must be removed to reach rigorous acceptable limits, the treated effluent quality must improve. Furthermore, it was discovered that these systems are incapable of providing adequate treatment efficiency under either high or low temperature circumstances (Ali et al., 2017). Correlating microbial dynamical structure to organic digestion, nitrification, denitrification, and phosphate removal necessitates substantial research into TF models.

1.1 Wastewater Treatment Plants

According to Metcalf and Eddy Inc., wastewater treatment facilities are critical in removing or reducing the pollution risks related to the disposal of raw wastewater. Wastewater treatment facilities protect public health by cleaning and treating waste, whereas untreated wastewater disposal offers a major risk due to the high concentration of bacteria, germs, and viruses it contains. By removing sediments and organic waste and increasing oxygen levels, they minimize pollution and health concerns to acceptable levels. The primary goal of wastewater treatment is to reduce hazardous substances in wastewater to an acceptable level that will not affect the environment or public health after disposal. The following describes the steps of wastewater treatment.

1.2 Preliminary Treatment

According to Metcalf & Eddy Inc., it is the process of removing wastewater components such as grit, muck, and grease that might clog treatment methods, ancillary systems, and operations.

1.3 Primary Treatment

Wastewater is directed to a quiescent basin in this type of treatment in order to remove organic trash and settle heavy solid particles. The residual wastewater can be sent for secondary treatment once the settled and floating particles have been removed.

1.4 Advanced Primary Treatment

This method eliminates organic compounds and suspended particles from wastewater by using chemicals or filtering.

1.5 Secondary Treatment

It removes both suspended and dissolved organic and inorganic contaminants from sewage. Microorganisms perform the biological process of secondary therapy. A suitable medium is used to manage bacteria in a habitat, and when sewage passes through this medium, the microbes start to break down the contaminants in wastewater.

1.6 Secondary Treatment with Removal of Nutrient

It is the elimination of suspended particulates, nutrients like phosphorus and nitrogen or phosphorus and nitrogen together, and biodegradable organics.

1.7 Tertiary Treatment

Before the effluent is released to the receiving body, it undergoes this treatment, which uses a variety of chemicals to eliminate any remaining suspended material components.

1.8 Advanced Tertiary Treatment

Here lift-over dissolved or suspended pollutants are eliminated after biological treatment.

1.9 Biological Treatment of Wastewater

Organic micropollutants and biological elements that are a significant issue to receiving bodies can be efficiently addressed by biological wastewater treatment (Pedrero et al., 2010; Solon et al., 2019). By creating the right circumstances for wastewater treatment, microorganisms are produced in biological treatment. The contaminants are broken down by microorganisms that stick to this medium when wastewater passes through it. Reducing the levels of chemical and biological oxygen demand (BOD and COD) is the main objective of this treatment. Two phases of immobilization are distinguished in biological treatment.

- Self microbial attachment to filter medium/support material (Attached)
- Artificial organisms' immobilization to filter medium/support material (Suspended).

1.10 Biological Treatment Objectives

The objectives of this kind of wastewater treatment are to eliminate nitrogen and phosphorus, oxidize particulate and dissolved biodegradable materials, incorporate suspended and non-settleable solids into biofilm, and eliminate trace organic materials.

2 Wastewater Treatment Methods

Wastewater treatment systems are frequently built to execute a variety of treatment operations and processes, starting with the removal of papers, wood branches, dead animal carcasses, plastics and heavy settleable materials during the preliminary treatment stage. The preliminary effluent is additionally treated in the primary and secondary phases. Wastewater treatment methods can be classified as physical, chemical, or biological based on their nature, with the goal of improving the water for safe disposal or reuse (Sen, 2015; Solon et al., 2019) (Table 1).

2.1 *Physicochemical Wastewater Treatment Methods*

Many traditional physical unit operations, such as screening, grit removal, comminution, sedimentation, flotation, filtration, adsorption, and membranes, use naturally occurring driving forces such as gravity, van der Waal forces, energy gradients, and electrical attraction to separate contaminants during wastewater treatment (Hamid et al., 2013; Kesari et al., 2011). Physical unit actions often remove pollutants while preserving the chemical properties of the target contaminants. Silica gel and activated carbon have been used extensively as adsorbents for pesticides, detergents, phenols, and dangerous compounds (Grégorio, 2005). Slaughterhouse effluent has been treated using a variety of membrane techniques, including diffusion-dialysis, ultrafiltration, reverseosmosis, and nanofiltration, which can reduce contaminants by up to 90% (Kabuk et al., 2015).

Chemical treatment techniques, unlike physical unit activities, take advantage of pollutants' reactivity with different chemicals. For example, reduction or oxidation reactions can produce an insoluble precipitate, non-biodegradable constituents can be converted into biodegradable ones, and surface charge reduction can cause flocculation of colloidal suspension, facilitating the removal of target pollutants (Lin & Chen, 1997). These chemical reactions include ozonation, coagulation, ion exchange, photocatalytic oxidation and hydrodynamic cavitation (Chou et al., 2011; Musmarra et al., 2016; Oppong et al., 2019; Yin et al., 2019). Advanced oxidation is an innovative and promising treatment method for pollutants such organic dyes, detergents, hazardous chemicals, salts, and recalcitrant contaminants. Zhu et al. (2018) demonstrated the efficiency of ultrasound-enhanced electrochemical oxidation of textile effluent in completely removing dyes and demineralizing them.

Furthermore, physicochemical processes and operations have been validated for wastewater treatment and decontamination (Dimoglo et al., 2004). For example, Mittal (2011) reported employing coagulation and Fenton's reagent accelerated oxidation. Physicochemical procedures for removing fine heterocyclic compounds, suspensions, tar materials, and inorganic pollutants such as sulfates, cyanides, sulfides, thiosulphate, and heavy metals from industrial effluent. These approaches each have a number of advantages and disadvantages (Table 2). For instance,

Table 1 Different methods of wastewater treatment followed by preliminary treatment

Physical	Chemical	Physicochemical	Biological
Reverse osmosis_ (Jiang et al., 2018; Kesari et al., 2011) **Adsorption_** (Grégorio, 2005; Sarai Atab et al., 2018) **Coagulation/ Flocculation_** (Kesari et al., 2011; Piekutin, 2018) **Membrane separation_** (Kabuk et al., 2015) **Diffusion_** (Quach-Cuet al., 2018)	**Oxidation_** (Boczkaj & Fernandes, 2017) **Ozonation_** (Chou et al., 2011; Wei et al., 2017) **ChemicalPrecipitation_** (Oncel et al., 2013; Ye et al., 2016) **Ammonia stripping_** (Kinidi et al., 2018)	**Ion exchange_** (Amini et al., 2015) **Electrochemical_** (Dheeravath et al., 2018; Dimoglo et al., 2004) **Electroflotation_** (de Oliveira da Mota et al., 2015) **Electrodialysis**(Mahmoudi, 2016) **Electrocoagulation** (Islam, 2019; Oncel et al., 2013)_	**Suspended growth systems_** (Wollmann et al. 2019; Yuan et al., 2012a, 2012b) **Attached/Biofilm growth_ systems** (Miranda et al., 2017; Vergine et al., 2018; Zhao et al., 2018)

Table 2 Benefits and drawbacks of physicochemical methods of treatment

Methods	Merits	Demerits
NaOCl	Bactericidal action, operating costs	Operability, by products formation, investment costs
Ozone	Bactericidal action, low by-products	Operability, investment costs
UV	High bactericidal action, No formation of by-products, low operating costs	High operability, moderate investment costs
TiO_2	High inactivation of coliforms, likely use of renewable energy in the case of solar photocatalysis, no formation of by-products, use of inexpensive catalysts and facilities	Lack of residual bactericidal action and slow kinetic behaviour
Medium filtration	Low investment and operating costs, good effluent quality	Low fecal coliform removal
Membrane filtration	Simultaneous EC removal and disinfection	High investment and operating
Electrolysis	Effecient in killing a wide ofmicroorganisms spectrum	Formation of significant perchlorates

biofouling membrane modules shortens membrane service life, and the enormous burden of disposing of chemical sludge has negative effects on the environment and the economy. Other issues include power fluctuations, high turbidity, and the difficulty of recovering common catalysts, all of which lead to inefficiencies in system performance (Chun et al., 2017). It is dynamic to define that chemical adsorbents, chemical additives and catalyst such as $FeCl_3$ $Al_2(SO_4)_3$ and $FeSO_4$ are typically more expensive (Bayramoğlu et al., 2004).

2.2 *Biological Wastewater Treatment Methods*

According to Pedrero et al. (2010), biological wastewater treatment systems not only overcome the disadvantages of physicochemical technologies, but also have lower operating and capital costs, produce less sludge, and emit less dangerous chemicals into the environment. Examples of such technologies include anaerobic lagoons, oxidation ponds, aerated lagoons, rotating biological contactors, activated sludge systems, and trickling filters (Table 3). The kind of sewage effluent and the technological and financial feasibility of the proposed treatment facility all have a substantial impact on treatment technology selection (Delgadillo et al., 2016; Mittal, 2011; Sehar & Naz, 2016). Heavy metals, fertilizers, pathogens, oxygen-demanding chemicals, and other potentially hazardous substances are among the most common contaminants found in wastewater. Biological activities are classified into two types based on their structural organization or mode of microbial development: suspended growth and established attached growth (Fig. 1). Without substrate support, microbial cells develop planktonic in bulk liquid media in suspended growth systems.

3 Suspended Growth Systems

Numerous research suggest that biological treatment could be used for any type of wastewater. Lefebvre et al. (2005) treated hypersaline effluents from the tannery industry using an aerobic sequencing batch reactor (SBR) loaded with halophilic bacteria. They assessed the treatment's effectiveness at different organic loadings and salt concentrations. Despite the variations in the effluent properties, the SBR system achieved optimal removal with a hydraulic retention length of five days. Similarly, Abou-Elela et al. (2010) used an activated sludge system inoculated with Staphylococcus xylosus to treat industrial effluent, including pickled vegetables, and they estimated that the COD removal rate was close to 90%. Recently, an effective method for treating wastewater from a range of industrial sectors, including the textile, food processing, pulp and paper, and pharmaceutical industries, has been proposed: a membrane bioreactor with a UF or NF membrane design. For example, a submerged membrane reactor successfully eliminated toxic contaminants such hydrocarbons, phenol, ester derivatives, and benzene from petrochemical wastewater, allowing effluent water to be reused for various industrial processes (Llop et al., 2009). A similar membrane configuration was used to extract antibiotics, analgesics, anti-inflammatory, and lipid regulator drugs from wastewater, yielding high-quality biosolids with an estimated removal rate of more than 80%. According to Radjenovic et al. (2007) and Medina et al. (2015), nutrient-enriched sludge or biosolids were used to nourish about 70% of the soil. Recently, the discharge limits have placed strict restrictions on the pulp and paper sectors' effluent. An integrated membrane bioreactor and continuous membrane filtration were used to treat wastewater discharged from the effluent in order to meet the threshold values of treated water for the paper industry process (Zhang et al., 2009).

More effective and successful treatment techniques have been made possible by the recovery of resources or nutrients from wastewater and biosolids during wastewater treatment. Wastewater treatment may be able to recover P and N, which are bioessential components used as agriculture fertilizers. Currently, synthetic fertilizers are made from nonrenewable resources like phosphate rock and natural gas using conventional energy-demanding methods. Much attention has been focused on transforming the abundant P- and N-based nutrients present in untreated wastewater into more suitable forms in order to control nutrient demand in sustainable ways (Kodera et al., 2013, Mayer et al. 2016; Yuan et al., 2012a, 2012b). Additionally, recovering materials from wastewater could reduce energy consumption and increase the overall efficiency of wastewater treatment plants. The "fit-for-purpose" strategy, which allows integrated nutrient harvesting designs to fulfill energy needs indicated by a developing civilization, is supported by increases in resource recovery.

Table 3 Wastewater treatment technologies' description, removal efficiencies, benefits, and drawbacks (Massoud et al., 2009; Seow et al., 2016, Norton-Brandão et al., 2013)

Technique	Removal efficiencies	Merits	Demerits
Constructed wetlands	N 30–70%, P 20–60%, COD 55–80%; SS 60–98%, NH_4 70%	Easy to manage shock loadings, low energy consumption, low cost, and no byproducts	Meteorologically dependent, solids Entrapment, Imperfactions can occur due to ammonia and solids overloads
Pond system	COD 60–96%,N 20–40%, P 60–70%, SS 50–90%	Preferred medical technology in underdeveloped nations. low pathogen effluent, economical, requires less energy, and is easy to operate and maintain	Less effective in eliminating heavy metals, odor, and issues with insects and mosquitoes, need a huge land area, Because of its prolonged retention duration, it is not suitable for cold climates
Rotating biological reactor	COD 70–93%, N 20–80%, P 10–30%, SS 75–98%	Improved suspended solids removal, minimal operation and less complexity than an extended aeration system	Fixed film systems are less readily available, though they might work with cluster systems. limitations in nitrification, Not as good as biofilters
Extended aeration	COD 70–90%, N 50–90%, P 15–70%, SS 85–99%	Good effluent quality, good nitrification, commercially available with ease	Issue with noise and stink, need electricity, routine upkeep of operations, high cost of capital
Sequencing batch reactor	COD 70–90%, BOD > 95%, N 55–90%, P 25–70%, SS 85–99%	Stable operations, low building costs, and shock load handling ease	Periodic maintenance and operational control are necessary for proper operation
Trickling filter	COD 75–90%, BOD_5: 50–90, N 35–50%, P 35–55%	High levels of microbial activity, toxicity resistance, and biomass require a modest amount of land, support media for indigenous biofilms, inexpensive	Low nitrification, evaluation of biofilm life and physical stability, and difficulty little elimination of fecal coliform
Moving bed biofilm reactor	COD 20–40%, N 10–20%, P 30–40%, SS 60–80%	Total solids removal, no thickening of the sludge, backwashing and channeling of the filter, enhanced settling properties	Limited treatment proces control
Membrane bioreactors	COD 70–90%, N 50–90%, P 20–70%, SS 85–99%	Elimination of PCPs, EDCs, dyes, heavy metals, and concurrent EC and disinfection	Membrane fouling, high operational and capital costs

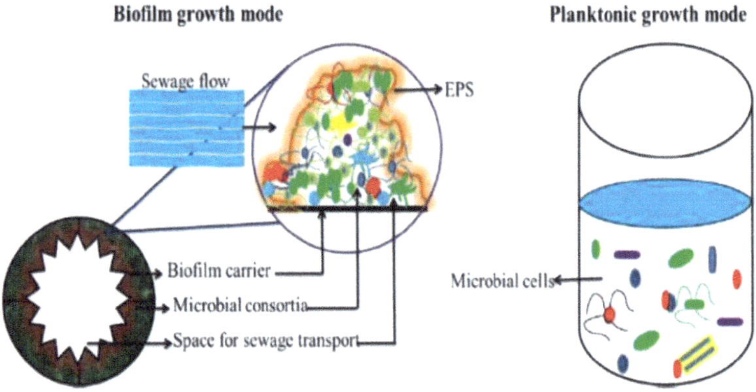

Fig. 1 Biomass structural arrangement in connected and suspended growth modes

The presence of phosphorus in phosphate-accumulating bacteria such as Candidatus Accumulibacter and Candidatus Competibacter was a significant indicator of improved biological phosphorus removal from wastewater. Following dissolution, P-enriched biosolids can be applied straight to land as manure or recovered as minerals (Yuan et al., 2012a, 2012b). Numerous kinds of microalgae have recently been identified as environmentally beneficial microorganisms for wastewater treatment in municipalities, aquaculture, and the dairy sector due to their capacity to remove C, CO_2, N, P, and harmful metals. Wang et al. (2009) revealed that the potential removal effectiveness of Chlorella sp. microalgae in municipal wastewater was 90% for TP as phosphate and 83% for TN as ammonium. The microbial consortium with microalgae was able to reduce eutrophication by absorbing C, N, and P from wastewater and converting flourishing biomass into helpful fertilizer (Delgadillo et al., 2016). Growing Chlorella vulgaris in wastewater resulted in bioremediation and subsequent protein production, according to another study (Blanco-Carvajal et al., 2018).

The discharge of untreated wastewater from mining, metallurgical, cement, petrochemical, and electroplating operations is one of the primary causes of metals in water sources. Biological treatment techniques were studied to recover valuable metals such as silver, gold, iridium, gallium, cadmium, and palladium; however, these studies used bioremediation to extract metal from mining sources instead of reclamation from wastewater sources (Johnson, 2014; Mosa et al., 2016). Microbial biomass can replenish heavy metals in untreated water. For example, Chlorella vulgaris was observed to lower 60% and 65% of the lead and cadmium, respectively, with 60 min of HRT (Kikuchi & Tanaka, 2012). According to Ju et al. (2016), the algae Galdieria sulphuraria has been able to biosorb 90% of the palladium and gold from wastewater based on aqua regia. In wastewater, the filamentous fungus Aspergillus Niger showed a 91% copper and 70% zinc removal efficiency (Price et al., 2001). Similarly, Goncalves et al. evaluated the removal efficiency of heavy metals from metallurgical industry effluent using metal sulfate reducing bacteria (SRB) in an upflow anaerobic sludge blanket reactor and achieved more than 99% percentage removal (>) of Zn and Cd. The anaerobic biological treatment of acid mine effluent using the SRB system has shown a successful removal of approximately 99.2% of zinc, copper, and nickel (Sierra-Alvarez et al., 2006).

The use of suspended growth biological processes is a cost-effective and ecologically friendly approach of resource recovery and wastewater bioremediation. However, dispersed or submerged growth approaches create several major concerns. For example, bacteria may be "washed out" of the biological reactor due to their short hydraulic retention period for proliferation. Furthermore, there is a low concentration of floating biomass near the effluent section, which may lead cells to migrate at the same rate. Finally, this occasionally leads in the loss of active biomass, which influences how efficiently suspended biomass treatments the material. Therefore, the development of a multifunctional attachment growth system called biofilm is crucial for the concurrent removal of many pollutants and wastewater resource reclamation (Kesaano & Sims, 2014; Sehar & Naz, 2016). Multiple contaminants would be removed from sludge simultaneously by mixed microbial syndicates. According to Miranda et al. (2017), biofilm systems have drawn a lot of interest and are widely employed for the degradation of contaminants in sludge and wastewater.

4 Attached Growth Systems

The majority of bacteria in humid conditions live on biotic or abiotic substrates, forming complicated, resilient, three-dimensional colonies known as biofilms. A self-produced glue-like extracellular polymeric substance (EPS) matrix surrounds the microclusters created by the various biofilm species, such as bacteria, algae, yeast, and fungi. As the home of "biofilm cells," EPS fulfills a number of functions, including improving the chemical and mechanical stability of biofilms, acting as a natural metal sequester and absorber, and protecting against a variety of physiochemical stresses. The mechanism of biofilm formation can benefit from expressing these unique characteristics in single multi-microbial communities. Metabolic support, nutrient gradients, organic and inorganic materials, water-filled cavities, cell–cell interactions, enlarged genes with more effective DNA sharing, and a variety of other interactions provide a competitive advantage (Dang & Lovell, 2016; Machineni et al., 2018). For example, unlike pure culture biofilms, robust biofilm growth was observed when a bacterial combination—specifically, Acinetobacter calcoaceticus, Pseudomonas aeruginosa, and Comamonas denitrificans—with varying properties were developed and refined together, providing synergistic assistance (Land et al., 2008). Microorganisms adhere to carriers such as stone, plastic, pebbles, peat, fabric, plant materials, or polyethylene to create biofilms (Limoli et al., 2015). The two most popular biofilm technologies for wastewater treatment are trickling filters and rotating biological contactors (RBCs) (Fig. 2).

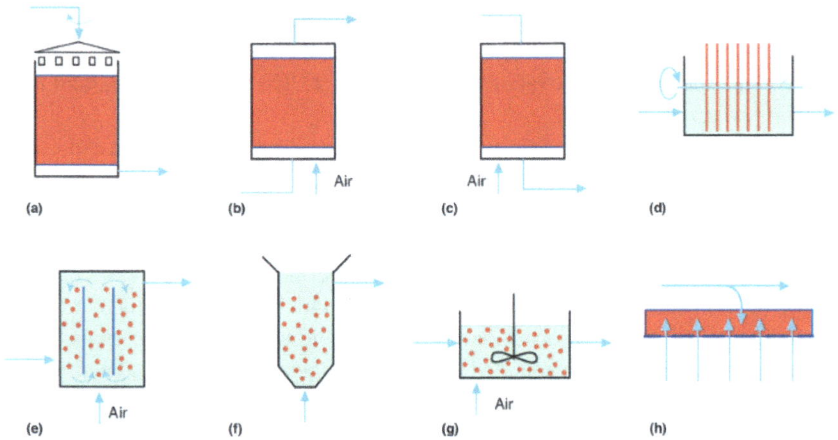

Fig. 2 Biofilm reactors configurations

Several research have used biofilms to treat wastewater, recover nutrients, remove heavy metals, and filter out other contaminants (Costley & Wallis, 2001; Andersson et al., 2008; Chavan & Mukherji, 2008; Schneider & Topalova, 2013; Sehar & Naz, 2016). Costley and Wallis (2001) evaluated the efficiency of RBC biofilm in extracting sorbed metals from concentrated industrial effluents and then reusing them. Extracellular sorption and re-exposure to microalgal biofilms successfully removed metals such as Zn, Cu, and Cd from the RBC biofilm. Biofilm systems can recover both phosphorus and nitrogen from wastewater (Boelee et al., 2011). In an anaerobic sequencing batch biofilm reactor, planktonic biomass and biofilm were used to study the effective removal of organic matter and nitrates from dairy effluent. Nitrate and COD levels were found to be two and four times lower than in free planktonic biomass (Schneider & Topalova, 2013).

It is well understood that biofilms containing EPS offer their colonizers with extremely durable habitats that can endure biological and physiochemical shocks from the outside world. The spinning disk biofilm reactor proved 600 times more resistant to heavy metal toxicity than suspended Pseudomonas aeruginosa. The biofilm was able to remove metals from wastewater, including lead, copper, and zinc, by adhering to metal ions in the EPS matrix (Teitzel & Parsek, 2003). Microbial biofilm-based solutions have proven to be a cost-effective and efficient replacement, particularly in peri-urban regions (Mononen et al., 2010). Stability, low maintenance, and reusability (Ghimire & Wang, 2018). Biofilm reactors, such as aerobic submerged fixed bed reactors and moving bed biofilm reactors, have recently been used in the treatment of municipal and agricultural wastewater due to their many advantages, which include higher reaction rates, stability, low maintenance, and reusability (Ghimire & Wang, 2018).

Trickling filter (TF) systems are considered the most successful biofilm reactors because to their low energy and maintenance costs, simplicity, minimal biomass washout, ease of operation, and environmental friendliness. Phosphorus was recovered from phosphate-containing wastewater using microorganisms that accumulate polyphosphate in a trickling filter biofilm (Kodera et al., 2013). Zhao et al. (2018) investigated the performance of pilot-scale rotating algal biofilm reactors for removing metals and nutrients from urban wastewater. Ironically, a rotating algal biofilm reactor was able to remove metals while simultaneously increasing the capacity to remove nutrients by decreasing the hydraulic retention period (Fig. 3). Increasing the diversity of metabolism in biological systems enhanced the efficacy of biofilm wastewater treatment, according to a new metagenomics technique. Localized nutrient gradients caused growth rates to vary across biofilm depth, according to fluorescent hybridization with microscopy (De Beer et al., 2017).

Unlike previous procedures, biosorption and bioremediation using biofilm technology do not incur high operating and maintenance expenses; a wide range of potential biofilm carriers are simply and reasonably available. The primary benefits of biofilm systems include improved HRT tolerance, minimal space requirements, rapid environmental change adaptation, active heterogeneous biomass concentration with improved capacity to break down high-strength contaminants, reduced sludge production, and easy removal and subsequent reclaim of attached biomass. Following impurity removal, treated wastewater is either recycled for agricultural use or safely discharged back into the environment (Sehar & Naz, 2016). In order to recover resources and bioenergy from wastewater and optimize the treatment process, it is critical to understand the growth microbiological dynamics of heterogeneous microbe populations in biofilm.

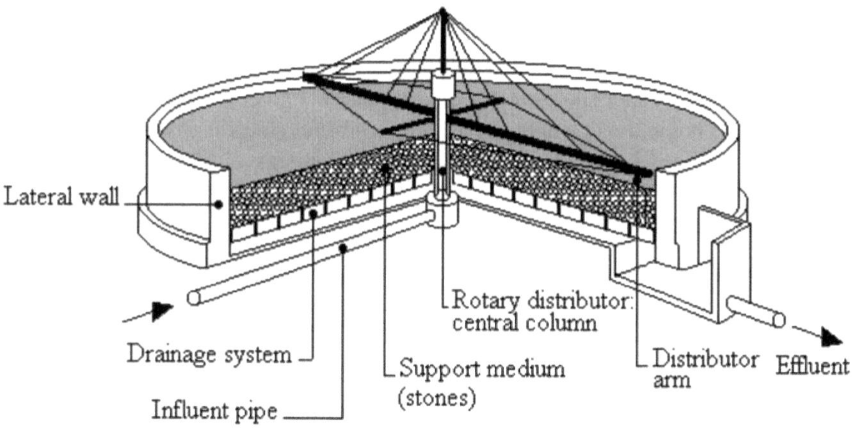

Fig. 3 Typical configuration and main parts of a typical circular TF

5 Merits of Attached Growth Process Over Suspended Growth Process

5.1 Higher Biomass Concentration

Microorganisms grow more in submerged media than in free water in both natural and man-made aquatic habitats. As a result, the suspended growth process contains fewer microorganisms than the connected growth process. According to reports, biomass concentrations in connected growth systems range from 2000 to 100,000 mg (MLVSS) per liter, whereas suspended growth processes retain concentrations of 700 to 2500 mg (MLVSS) per liter. The main advantage of attached growth processes over suspended growth processes is higher biomass output, which leads to successful wastewater treatment.

5.2 Higher Metabolic Activity

In most treatment settings, the attached growth process outperforms the suspended growth process in terms of metabolism. Furthermore, it has been demonstrated that, unlike suspended growth processes, connected growth processes may efficiently treat larger loading waste streams in shorter retention times. This suggests that connected activities are more successful than suspended processes at decomposing waste.

Waste decomposes more quickly when there is a greater concentration of active microorganisms in the attached growth process. The elevated quantity of nutrients in the accompanying biofilm could explain the microorganisms' heightened activity. This higher nutrient concentration may increase the activity of microbial biofilms and their ability to break down waste. According to several studies, physiological differences between attached and suspended microorganisms improve biodegradation functions (Senthilnathan & Ganczarczyk, 1990; Lazarova & Manem, 1995).

5.3 Higher Toxicity Resistance

Microorganisms in connected growth processes have been shown to be more resistant to harmful influents than bacteria in suspended growth processes. In a hospital context, connected growth bacteria can tolerate up to 20,000 mg/liter of chlorhexidine for 27 months, whereas suspended growth germs can only withstand 500 mg/liter. Because there is a physiological difference between connected growth microorganisms and suspended growth bacteria, several investigations have investigated this phenomenon. Another explanation for this phenomenon is that the concentration of nutrients surrounding the biofilm increased, allowing microorganisms to survive at higher levels of toxicity.

5.4 Good Sludge Properties

Sludge produced by a floating growth approach and sludge produced by an attached growth method have slightly different qualities. For every unit of biomass, the suspended growing process produces a substantial amount of sludge. Filamentous microorganisms, the most frequent in wastewater treatment, perform bulking and foaming roles in treatment facilities. Bulking, foaming, and settling issues are considerably decreased by the dense, biofilm-based sludge formed during the attachment growth phase. In general, connected growth processes provide better fluid treatment than suspended growth processes.

5.5 Other Advantages of Attached Growth Process

- Low installation and operation cost
- Simple design
- Small installation land area
- Less energy
- Less complexity in mechanical
- Operate simple
- No mix-liquor
- No sludge bulking
- Improved sludge thickening
- Fewer maintenance required
- Even better effluent against high load
- 80–90% BOD removal efficiency
- Ability to treat odour and VOCs
- Less pressure drop.

6 Biofilm Development, Maturation and Systems

Biofilm is a colony of microorganisms that grows on the surface of porous material. The biofilm is made up of water (97% mass ratio), a variety of microorganisms (bacteria, algae, protozoa, and fungi), and extracellular polymeric substances (EPS) (carbohydrates and mineral compounds). These EPS are usually made up of polysaccharides, proteins, and extracellular DNA. Planktonic cells are solitary cells that can float or swim in an aquatic medium and differ physiologically from the proliferation of microbial cells in a biofilm (Fig. 4). Biofilm types include autotrophic and heterotrophic, entrapped and self-attached, aerobic, anaerobic, and anoxic.

The biofilm life cycle consists of medium adherence, surface development, and environmental dispersal. The adhesion, adsorption, and adherence mechanisms of biofilm attachment are heavily impacted by bacterial characteristics as well as the

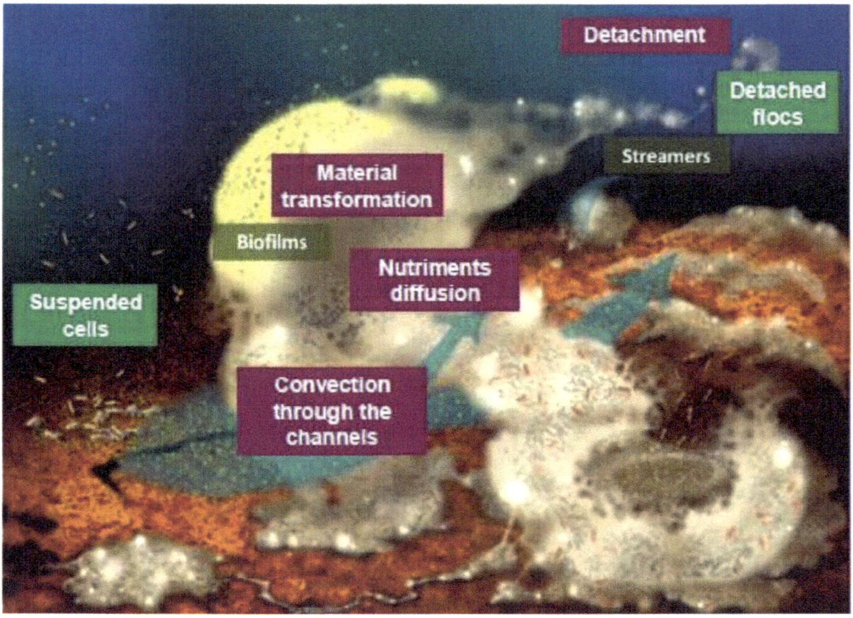

Fig. 4 Biofilm processes and different morphs of bacterial aggregates

surface nature of the medium. After several weeks, a mature biofilm is produced, consisting of complex matrix exopolysaccharides. Biofilm features can be measured using light microscopy (microbiology type and biofilm thickness), confocal scanning laser-microscopy (biofilm-matrix), transmission electron-microscopy, or scanning electron-microscopy. A biofilm's biological activity can be determined by its mass, porosity, density, total organic carbon, oxygen uptake rate, concentration of exopolysaccharides, and adenosine triphosphate levels. Because of their slower microbial growth rate and higher microbial concentrations/active biomass for contaminants removal, biofilm treatment systems offer better control over reaction kinetics, population dynamics, shorter retention times, increased operational flexibility, a better ability to break down resistant compounds, and less sludge production. Additionally, the extended sludge retention time (SRT) of biofilm encourages the development of nitrifying bacteria.

6.1 Conventional Biofilm Reactors

Traditional WWT biofilm reactors include a constructed wetland, sand filter, trickling filter, and rotating biological contactor (RBC). Trickling filter systems are believed to be the most successful biofilm treatment systems because to their low energy costs, little maintenance requirements, simplicity, minimized biomass

Table 4 Performance parameters of biofilm reactors

Parameters		TF[1]	RBC[2]	MBBR[3]	FBBR[4]	BAF[5]
HLR	$m^3\ m^{-2}\ day^{-1}$	3.5–28.8	0.04–0.15	0.2–7	0.6–0.8	1–24
HRT	hour	1.5–15	1.5–14	0.5–24	1–15	4.5–23
BOD loading	$kg\ m^{-3}\ day^{-1}$	0.2–1.0	0.1–0.8	1–1.4	6.4–20	0.5–15
NH^{4+}-N leading	$Kg\ m^{-3}\ day^{-1}$	0.6–1.3	0.6–1	4–8	0.2–2.6	0.1–1.5
BOD removal efficiency	%	40–90	85–95	70–80	90–95	33–93
NH^{4+}-N removal efficiency	%	84–90	70–80	41–86	60–90	46–95
Backwash		Yes	No	No	No	Yes

[1,2]Metcalf and Eddy (2003), Henze et al. (2001)
[3,4,5]Hajipour et al., Hassimi and Abdullah, Mendoza-Espinosa and Stephenson

washout, and environmental compatibility. It employs a variety of pollutant removal methods, including biosorption, biodegradation, biomineralization, and bioaccumulation. These technologies can handle WW's nitrogenous and carbonaceous materials. The trickling filter concept, which incorporates plants, acts as the foundation for the constructed wetland treatment process. It encompasses all processes involved in pollutant removal, such as volatilization, plant absorption, phytodegradation, and microbial degradation. Table 4 summarizes the biofilter performance metrics. Rotating disks with a shaft are used in rotating biological contactors that are partially submerged in wastewater and support biofilms. Thus, by changing working conditions such as disk format, rotation speed, and loading rate, aerobic, anaerobic, and anoxic biological processes can be easily established and regulated. According to Bounds et al., MBBR consumed 70% more electricity than TF.

6.2 *Advanced Biofilm Reactors*

Moving bed bioreactors (MBBR), biological aerated reactors (BAF), fluidized bed bioreactors (FBBR), membrane aerated biofilm reactors, and granule sludge reactors are some examples of recent modern/advanced biofilm reactors that require more energy than traditional reactors. The biological aerated filter (BAF) is a popular and simple technology for simultaneously eliminating suspended particles, COD, and total ammonia nitrogen. To reach the submerged porous media, air is fed into the reactor's bottom. Although downflow BAF provides a high rate of oxygen mass transfer to water, it also confines the anaerobic gas within the reactor. The water upflow mode of BAF is generally used under the brand name Biostyr. However, based on the headloss, this filter requires backwash to remove blocking biomass and sediments. This backwash is accomplished by pumping both water and air from the bottom at the same time. The Moving Bed Bioreactor (MBBR) simultaneously removes pollutants with suspended sludge and fixed films on plastic media. Based on the various treatment needs, Odegaard established that MBBR was viable for nitrogen and BOD removal. Using biofilm on fixed/suspended solid media (synthetic mesh), the integrated fixed film activated (mesh/plastic). Additionally, the fluidized bed bioreactor is a kind of three-stage circulation bioreactor with two faller and riser portions that is driven by air velocity. According to Tavares et al., air velocity has a major impact on the development of filamentous bacteria and biofilm. It also caused that filamentous bacterium to be developed at a low air velocity (0 to 3 m^3 m^{-2} h^{-1}). The protein and carbohydrate masses that stuck to the medium were described by this type of bacterium. Granular sludge was found in aerobic UASB (upflow anaerobic sludge blanket) and aerobic SBR (sequencing batch reactor) (Morgenroth et al., 1997; Schmidt & Ahring, 1996). Because of the advantageous treatment processes in biological WWT, it is becoming more and more popular. Granular sludge is frequently thought of as an alternative illustration of unsupported, self-immobilized biofilm cells. The membrane aerated biofilm reactor (MABR) was created to overcome oxygen diffusion constraints in biofilm thickness and boost the specific surface area of biofilm support media. The counter-diffusion biofilm made this approach more effective than another typical reactor for biological nitrogen removal. In a counter-diffusion biofilm, autotrophs and heterotrophs can coexist and be segregated in their respective zones. However, the majority of MABR research has thus far been undertaken at the laboratory scale, and the technique's field application is still not well recognized. Table 5 illustrates the range of biofilm thickness and concentration, as well as the benefits and drawbacks.

7 Trickling Filter

Compared to other biological treatment systems, trickling filters have better process stability, reduced mechanical complexity, and well-sludge thickening properties, making them an appealing attached growth wastewater treatment solution for developing nations. It can withstand power outages and stress loads. Its design is straightforward, and it requires less land, energy, maintenance, installation costs, and ecological footprint (Eding et al., 2006; Gikas, 2016; Mahmoud et al., 2010; Naz et al., 2016). Here, a variety of microorganisms, including anoxic and aerobic bacteria, fungi, protozoans, and algae, are used to stabilize the organic matter biologically (Eding et al., 2006, Metcalf & Eddy, 2003). Natural aeration gives oxygen to the open sections while maintaining the aerobic nature of the slime's outer layer. Microorganisms aerobically convert dissolved and colloidal organic molecules into protoplasm and other gases. Because protoplasm has a slightly higher specific gravity than water, it can be separated from treated wastewater by gravity settling. A secondary clarifier is recommended for protoplasm settling, which is organic in nature; otherwise, it will be considered the effluent's BOD.

Numerous parameters, including organic substrate, temperature, aeration, pH, oxygen concentration, and medium selection (size, porosity, weight, and specific

Table 5 Biofilm thickness and concentration of biofilm reactors; Hajipour et al., Syron and Casey, Verma et al., Ni et al., Dangcong et al. and Matsumoto et al.

Reactors	Merits	Demerits	Biofilm thickness (μm)	Biofilm concentration (g_VAS.L^{-1})
TF	Increased biomass concentration, resistance to toxicity, only a minimal amount of land. may be created using locally available filter material, including tire-derived rubber, corn cobs, etc. economical Easy, steady, and dependable functioning without the need for artificial aeration	Uneven wastewater distribution, odor, flies and mosquitoes, bioclogging, and challenges evaluating physical stability, longevity, and efficiency medium for biofilms	0.2–100	50–70
RBC	Reduced building footprint, many biological processes, minimal operation, and improved suspended solids capture	Shaft failure brought on by an overabundance of biofilm There are fewer commercially available fixed film systems for onsite application, and they are less aesthetically pleasing, difficult to maintain, and have a low nitrification rate	10–20	500–4000
MBBR or IFAS	Sludge bulking, clogging, backwashing, and filter channeling issues are eliminated, and the settling properties are improved	High operating expenses from aeration and pumping, and limited process control because of less often used processes	0.2–3	100–240
FBBR	Long sludge retention period for xenobiotic and hazardous compound breakdown, high-quality effluent, and toxic compound removal	High operating costs due to pumping and aeration	10–50	Less than 100

(continued)

Table 5 (continued)

Reactors	Merits	Demerits	Biofilm thickness (μm)	Biofilm concentration (g_VAS.L^{-1})
Granule sludge	High compactness and structural strength, good separation and settling, and no need for support media	Operational issues that are hard to regulate for complex granulation processes	15–30	300–1200 for aerobic 2000–5000 for anoxic
MABR	High efficiency, specific surface area, and oxidation capacity of the membrane. optimal dispersion of several microbes	Maintaining the ideal biofilm thickness is challenging, and scaling up presents challenges	–	60–1200

surface area), influence the performance of the formed biofilm. Because it offers a surface for contact between biomass growths (microbial attachment) and the contaminants to be treated, medium selection for biofilm support is the most important factor in trickling filter research (Godoy-Olmos et al., 2016; Kamstra et al., 1998). To assess and improve trickling filter efficiency, a variety of trickling filter media have been used such as calcitic gravel (Eding et al., 2006), rocks/plastic (Metcalf & Eddy, 2003), nylon pan scrubber (Sanjib et al., 2011), geotextile (Alimahmoodi et al., 2012), commercial and pall rings, stone (Khan et al., 2015), coal (Zhao et al., 2013), tire drubber (Naz et al., 2014), plastic sheet (corrugated) (Kim et al., 2014), gravel (Zhang et al., 2015), ceramsite and zeolite (Zhang et al., 2016), oyster shell, plastic, chips of bio char (LI et al., 2016), ceramic particles (Wu et al., 2016), cylindrical luffa (Vianna et al., 2012), sponge (Wang et al., 2015) etc.

7.1 *Selection of Trickling Filter Media*

The packing or biofilm support medium is a critical component of any TF system. Metcalf and Eddy classify biofilm support media into two types: structured and random. The medium provides a surface for the formation of biofilms, which are composed of bacteria and other microorganisms. The support media must be durable, insoluble, and substance resistant. It is chosen based on size, density, porosity, erosion resistance, and chemical resistance. Ideal biofilm media have high porosity, specific surface area, durability, affordability, and the ability to offer ventilation without clogging. Matos et al. investigated the effect of support media surface area and geometry on reactor hydrodynamic conditions, biofilm growth, and wastewater treatment.

In connected growth reactors, a variety of synthetic and natural materials have been studied to support biofilm. Early modular plastic media experiments were started in the 1950s and 1960s by Dow Chemical Co. (Midland, Michigan) (Bryan, 1955;

Bryan & Moeller, 1960). According to Daigger and Harrison and Parker and Bradtby, plastic media can be found in a variety of shapes, including cross flow, vertical flow, honeycomb appearance, rectangular redwood, horizontal, and flat modules. It has been demonstrated that both vertical and crossflow medium may efficiently remove biological oxygen demand (BOD) and total suspended solids (TSS) (Aryan & Johnson, 1987; Harrison & Daigger, 1987). Medium-density crossflow media were suggested by Parker et al. for nitrification in trickling filter systems. Higher nitrification rates were also discovered by Gujer and Boller and Boller and Gujer when wastewater was treated at pilot size using low density media of modular plastic. The formation of dry areas beneath the flow disruption locations may be the cause of the reduced nitrification rates of high-density support medium. Using plastic balls as biofilm-supporting media, Mofokeng et al. created a large-scale trickling filter for the nitrification of secondary treated wastewater. The highest ammonia removal rate of 71% was achieved at high hydraulic loading rates, and the COD decreased by 40–50 mg/L as a result.

In order to provide sufficient ventilation from the voidage, Boltz et al. recommended the use of inexpensive, locally available media containing uniform particles. Using wool fiber or plastic pall rings as support media, Hu et al. developed a trickling filter for wastewater treatment on a lab scale. They discovered that wool fiber outperformed plastic pall support medium in the removal of suspended particles and total organic carbon from wastewater. However, both medium types provided complete nitrification with a load of 0.4 kg COD/m^3/day. In 2012, Neto et al. examined the efficacy of stone-media and Luffa cyllindrica as media in laboratory-scale TF plants for the treatment of domestic wastewater. They concluded that because Luffa cylindrica media is organic, it outperformed stone media in the TF system at removing organic materials as well as settleable and suspended particles. Spychała et al. tested two types of geotextiles for treating residential wastewater using TF. TC/PP 300 and TS 50 showed similar carbonaceous contaminant removal efficiencies (BOD5: 80%, COD: 64%). However, TS 50 (40%) eliminated more ammonium nitrogen than TC/PP 300 (35%), possibly due to their differing structural makeup. Another difficulty in emerging countries is the handling of solid waste. Waste materials such as date palm fiber (DPF), maize cob (MC), and various polymers such as polyvinyl chloride (PVC), polypropylene (PP), low density polyethylene (LDPE), and polystyrene (PS) can be recycled as support media in TF systems for two reasons: wastewater treatment and solid waste management. To boost treatment efficacy, the TFs were designed with diverse biofilm support materials (Table 6). The growth and health of the biofilm on the support material are the major factors influencing TF performance.

7.2 Trickling Filter Problems and Solutions

Predators and filter flies impede TF systems' operational efficiency (Learner, 2000). In addition to the receiving environment and the surrounding community, they pose a threat to the operational personnel. In TFs, species such as Parischnogaster alternata, Astraeus hygrometricus, and Pseudocolasp is severini were widespread. It was discovered that filter flies can be efficiently controlled with Bacillus thuringiensis israelensis insecticide without affecting system functionality (Coombs et al., 1991). Filter flies can currently be eradicated from TF media by routine flodding and chemical and biological treatments (Learner, 2000). Snails have a negative impact on both nitrifying organism populations and TF function. Snail populations are regulated in a variety of ways, including pH dosing, saline chlorinated water dosing, copper sulphate dosage (0.4 g/L), slow distributor speed, high flushing rate, and periodic flooding. The foul odor issue in TF can be addressed by managing loading concentrations with oxygen transfer rate, chemical scrubbers, sodium hypochlorite dosage, and ideal moisture in TF media (Henze et al., 2001; Patria et al., 2001; Raquel et al., 2012; Shareefdeen & Singh, 2005; Tsang et al., 2008). For WW that is substantially loaded, the oxygen transfer rate should be high, and vice versa. For bacterial metabolic processes and to avoid chenneling and short-circuiting, the TF media bed's ideal moisture content is essential (Schroeder & Tchobanglous, 1976). TF performance is affected by climate variables such water and air temperatures (Zielinski et al., 2013). The formation of biofilm layers and the rate at which pollutants decompose can both be slowed by cold temperatures. The TF system may have bonding effects from ice blockage at very low temperatures. Different trickling filter problems, their causes and proposed solutions are given in Table 7.

Table 6 Trickling filter system for WWT application

Wastewater (WW)	Support media	Pollutant removed	References
Municipal	Gravel	BOD and TN	Simsek et al.
Textile	Polyurethane foam	Color, TOC and Dyes	Novotný et al.
Domestic	Coal cinder	COD, TP, NH_4^+ and SS	Zhao et al. (2013)
Mine water	Plastic	Iron	Jarvis and Younger
Municipal	Corrugated plastic sheet	TSS, COD, BOD5, TKN, TP and NO3–N,	Kim et al. (2014)
Domestic	Stones	pH, odor, turbidity, alkalinity, COD, BOD_5, TDS, TSS, EC, PO4, $SO_2{}^{-4}$, NO^{-2}, NO^{-3} and bacterial count	Khan et al. (2015)
Municipal	Oyster shell and plastic balls	COD, NH3-N, and TP	Liu et al. (2010)
Combined	Oyster shell	COD, BOD, NH3-N, TP and TSS	Luo et al.
Domestic	Luffa cyllindrica	BOD5,20, COD, SS, and settleable solids	Vianna et al. (2012)
Synthetic	Geotextile	Organic nitrogen and phosphorus COD	Alimahmoodi et al. (2012)

(continued)

Table 6 (continued)

Wastewater (WW)	Support media	Pollutant removed	References
Domestic	Plastic balls	COD, BOD5, TSS, turbidity, NO_3, NO_2, SO_4, PO_4 and pathogenic indicator microbes	Rehman et al.
Domestic	Tire derived rubber (TDR)	BOD, COD, Pathogen indicators, pH, NO^{2-}, NO^{3-}, $PO4^{3}$- and $SO3^{2-}$	Naz et al. (2014)
Municipal	Stones	BOD5, COD, NH_4-N, and pathogen	Naz et al. (2016)
Municipal	Rubber, polystyrene, plastic and stone	COD and BOD, faecal coliforms	Naz et al.
Municipal	Stones	Odor, alkalinity, pH, turbidity, BOD_5, COD, TDS, TSS, EC, $PO4^{3-}$, $SO3^{2}$, NO^{2-}, NO^{3-}, and pathogens	Morton and Auvermann
Municipal	Stones	Di-(2-ethylhexyl) phthalate and diethyl phthalate	Oliver et al.
Synthetic	Sponge	COD and TN	Ahammad et al.
Industria	Ceramic foams	Phenol	Zapico et al.
Textile	TM foam	Organic dyes	Skybová et al.
Industrial	Ceramic material	COD and ammonia nitrogen ($NH_4{}^+$-N)	Wu et al.
Synthetic gold mill	Plastic rings of polypropylene	COD, copper, thiocyanate, free cyanide, iron and zinc	Evangelho et al.
Landfill leachate	Polypropylene plastic media	Ammonium and organic carbon (TOC)	Matthews et al.
Domestic	Sponge, zeolite and ceramsite	COD and ammonia	Zhang et al. (2016)
Municipal	Corrugated plastic sheet	TSS, BOD5, COD, TKN, $NO_3{}^-$–N and TP	Kim et al. (2014)
Municipal	Plastic media	Phosphate	Kim et al.

The sloughing of biomass from biofilm, high hydraulic loading rate, and particle displacement from primary clarifier to trickling filter may all contribute to the poor quality of TF effluent in terms of TSS concentrations (Ali et al., 2017). Because of the blockage, high solids concentrations, and limited ventilation, biomass sloughing destabilizes the treatment process. Because anaerobic conditions are common in WWT as pH levels rise, pH can be viewed as the key indicator of TF performance. The type of influent WW source impacts the pH value change. pH changes may have an impact on microbial activity. According to Naz et al., biofilm is a population of microorganisms that produces metabolic activity and degrades pollutants into harmless byproducts. The type of pollutants, the support media used, the temperature of the wastewater and surrounding air, the moisture content, the humidity, the configuration, and the system design are all factors that influence the thickness and structure of the biofilm (Maureen & Ronald, 2014).

Table 7 Operational problems, reasons and solutions for trickling filter systems

Problem	Reason	Solution	References
Support media	• Clogging • Uneven biofilm growth • Decrease efficiency	• Use proper media • Media uniformity • Increase the filter media installation width and decrease the depth • Apply active aeration to increase the efficiency of the operation	Metcalf and Eddy
Bio-film slough off	• Clogging the filter media • Changes in waste load • Insufficient nutrients • Uneven distribution of influent • Low moisture content and High hydraulic loading rate	• Optimize organic and hydraulic loading or to use parallel or series TF system to handle shock variations	Metcalf and Eddy, Chaudhary et al., Das and Naga
Filter clogging and ponding	• Improper media selection during targeted pollutants removal • Slough off biomass accumulation into the void pores • Loss of open area in the filter • Excessive organic loading • Lack of good primary clarification • Excessive growth of insect larvae or snails • Shock load and Lower transport of air	• Use proper flashing with low doses of chlorine to remove deposited solids and kill excess biomass • Enhance the recirculation rate • Optimize the organic loading and apply low organic loading by enhancing the performance of primary settling tank • Replace the TF media if needed	Iliuta, Metcalf and Eddy

(continued)

Table 7 (continued)

Problem	Reason	Solution	References
Odors	• Predominant anaerobic conditions in the system • Excessive biomass accumulation in the media surface • Improper oxygen transfer rate in the filter	• Pre-aerating, pre-chlorinating the TF influent • Increase recirculation rate to provide more oxygen to the filter bed and increase sloughing • Clean the nozzles on weekly basis for proper influent distribution • Use of chemical scrubbers and sodium hypochlorite dosage • Minimize the incoming organic loading by reducing the BOD concentration • Remove slough off biomass by increasing hydraulic loading	Raquel et al. (2012), Henze et al. (2001), Patria et al. (2001), Shareefdeen and Singh (2005), Tsang et al. (2008)
Shock loads	• Strom events will increase the hydraulic loading, which slough off the biofilm • Industrial effluents upset the biological processes and increase the organic loading or toxic loading	• Neutralized the toxic shock load by using TF system in series, this technique will save the biofilm growth • Dilute the toxic shock load by increasing the recirculation rate	Metcalf and Eddy, Zhang et al. (2015), Parker and Newman
Distributor arm	• Uneven oxygen utilization • Uneven growth of slime layer • Low BOD & COD removal	• Adjust the proper hydraulic flow rate according to system design • Use innovative nozzles in distributor arms to increase the system efficiency	Metcalf and Eddy, Maulik, Wu et al. (2016), Kumar et al. (2013)

References

Abou-Elela, S., Kamel, M. M., & Fawzy, M. E. (2010). Biological treatment of saline wastewater using a salt-tolerant microorganism. *Desalination, 250*(1), 1–5.

Ali, I., Khan, Z. M., Peng, C., Naz, I., Sultan, M., Ali, M., Mahmood, M. H., & Niaz, Y. (2017). Identification and elucidation of the designing and operational issues of trickling filter systems for wastewater treatment. *Polish Journal of Environmental Studies, 26*(6).

Alimahmoodi, M., Yerushalmi, L., & Mulligan, C. N. (2012). Development of biofilm on geotextile in a new multi-zone wastewater treatment system for simultaneous removal of COD, nitrogen and phosphorus. *Bioresource Technology, 107*, 78–86.

Amini, A., Kim, Y., Zhang, J., Boyer, T. & Zhang, Q. (2015). Environmental and economic sustainability of ion exchange drinking water treatment for organics removal. *Journal of Cleaner Production, 104*.

Andersson, S., Rajarao, G., Johan Land, C., & Dalhammar, G. (2008). Biofilm formation and interactions of bacterial strains found in wastewater treatment systems. *FEMS Microbiology Letters, 283*(1), 83–90.

Bayramoğlu, M., Kobya, M., Can, O., & Sozbir, M. (2004). Operating cost analysis of electrocoagulation of textile dye wastewater. *Separation and Purification Technology, 37*(2), 117–125.

Blanco-Carvajal, E., Sanchez-Galvis, E., González-Delgado, A., Garcia Martinez, J., & Barajas, A. (2018). Cultivation of Chlorella vulgaris in aquaculture wastewater for protein production. *Contemporary Engineering Sciences, 11*(2), 93–100.

Boczkaj, G., & Fernandes, A. (2017). Wastewater treatment by means of advanced oxidation processes at basic pH conditions: A review. *Chemical Engineering Journal, 320*, 608–633.

Boelee, N. C., Temmink, H., Janssen, M., Buisman, C. J., & Wijffels, R. H. (2011). Nitrogen and phosphorus removal from municipal wastewater effluent using microalgal biofilms. *Water Research, 45*(18), 5925–5933.

Chavan, A., & Mukherji, S. (2008). Treatment of hydrocarbon-rich wastewater using oil degrading bacteria and phototrophic microorganisms in rotating biological contactor: Effect of N:P ratio. *Journal of Hazardous Materials, 154*(1–3), 63–72.

Chou, C., Yu, Y., Chang, C., Lin, C., & Shang, N. (2011). The combination of ozonation and ion exchange processes for treatment of a municipal wastewater plant effluent. *International Conference on Electric Technology and Civil Engineering (ICETCE)*. https://doi.org/10.1109/ICETCE.2011.5775819

Chun, Y., Mulcahy, D., Zou, L., & Kim, I. S. (2017). A short review of membrane fouling in forward osmosis processes. *Membranes, 7*(2), 30.

Coombs R. M., Dancer B. N., Davies D. H., Houston J., & Learner, M. A. (1991). The use of Bacillus thuringiensis var. israelensis to control the nuisance fly Sylvicola fenestralis (Anisopodidae) in sewage filter beds. *Water Research, 25*(5), 605.

Costley, S., & Wallis, F. (2001). Bioremediation of heavy metals in a synthetic wastewater using a rotating biological contactor. *Water Research, 35*(15), 3715–3723.

Dalkmann, P., Siebe, C., Amelung, W., Schloter, M., & Siemens, J. (2014). Does long-term irrigation with untreated wastewater accelerate the dissipation of pharmaceuticals in soil? *Environmental Science & Technology, 48*(9), 4963–4970.

Dang, H., & Lovell, C. R. (2016). Microbial surface colonization and biofilm development in marine environments. *Microbiology and Molecular Biology Reviews, 80*(1), 91–138.

De Beer, D., Botes, M., & Cloete, T. (2017). The microbial community of a biofilm contact reactor for the treatment of winery wastewater. *Journal of Applied Microbiology, 124*(2), 598–610.

de Oliveira da Mota, I., de Castro, J. A., de Góes Casqueira, R., & de Oliveira Junior, A. G. (2015). Study of electroflotation method for treatment of wastewater from washing soil contaminated by heavy metals. *Journal of Materials Research and Technology, 4*(2), 109–113.

Delgadillo, L., Lopes, F., Taidi, B., & Pareau, D. (2016). Nitrogen and phosphate removal from wastewater with a mixed microalgae and bacteria culture. *Biotechnology Reports (Amst), 11*, 18–26. https://doi.org/10.1016/j.btre.2016.04.003

Dheeravath, D., Chandan, V., Srilatha, K., Shankar, G., Rani, Y., & Vurimindi, H. (2018). Industrial wastewater treatment using electrochemical process. *IOP Conference Series: Earth and Environmental Science, 191*, Article 012022.

Dimoglo, A., Akbulut, H. Y., Cihan, F., & Karpuzcu, M. (2004). Petrochemical wastewater treatment by means of clean electrochemical technologies. *Clean Technologies and Environmental Policy, 6*, 288–295.

Eding, E. H., Kamstra, A., Verreth, J. A. J., Huisman, E. A., & Klapwijk, A. (2006). Design and operation of nitrifying trickling filters in recirculating aquaculture: A review. *Aquacultural Engineering, 34*(3), 234–260.

Gatto D.'Andrea, M. L., Salas Barboza, A. G. J., Garcés, V., Rodriguez-Alvarez, M. S., Iribarnegaray, M. A., et al. (2015). The use of (treated) domestic wastewater for irrigation: Current situation and future challenges. *International Journal of Water and Wastewater Treatment, 1*(1).

Ghimire, N., & Wang, S. (2018). Biological treatment of petrochemical wastewater. In *Petroleum chemicals*. IntechOpen. https://doi.org/10.5772/intechopen.79655

Gikas, P. (2016). Towards energy positive wastewater treatment plants. *Journal of Environmental Management*.

Godoy-Olmos, S., Martínez-Llorens, S., Tomás-Vidal, A., & Jover-Cerdá, M. (2016). Influence of filter medium type, temperature and ammonia production on nitrifying trickling filters performance. *Journal of Environmental Chemical Engineering, 4*(1), 328–340.

Grégorio, C. (2005). Recent developments in polysaccharide based materials used as adsorbents in waste water treatment. *Progress in Polymer Science, 30*(1), 38–70.

Hamid, A., Zeb, M., Mehmood, A., Akhtar, S., & Saif, S. (2013). Assessment of wastewater quality of drains for irrigation. *Journal of Environmental Protection, 4*(9), 937.

Henze, M., Harremoes, P., La Cour Jansen, J., & Arvin, E. (2001). *Wastewater treatment: Biological and chemical processes* (pp. 367–478). Springer Science & Business Media.

Islam, S.M.D.-U. (2019). Electrocoagulation (EC) technology for wastewater treatment and pollutants removal. *Sustainable Water Resources Management, 5*(1), 359–380.

Jiang, L., Tu, Y., Li, X., & Li, H. (2018). Application of reverse osmosis in purifying drinking water. *E3SWeb of Conferences 38*, 01037.

Johnson, D. B. (2014). Biomining-biotechnologies for extracting and recovering metals from ores and waste materials. *Current Opinion in Biotechnology, 30*, 24–31.

Ju, X., Igarashi, K., Miyashita, S., Mitsuhashi, H., Inagaki, K., Fujii, S., Sawada, H., Kuwabara, T., & Minoda, A. (2016). Effective and selective recovery of gold and palladium ions from metal wastewater using a sulfothermophilic red alga, Galdieria sulphuraria. *Bioresource Technology, 211*, 759–764.

Kabuk, H. A., Coskun, T., Debik, E., Manav Demir, N., Baştürk, I., Yildirim, B., Temizel, D., & Kucuk, S. (2015). Treatment of poultry slaughterhouse wastewater using a membrane process, water reuse, and economic analysis. *Desalination and Water Treatment, 57*(11), 4944–4951.

Kamstra, A., Van der Heul, J. W., & Nijhof, M. (1998). Performance and optimisation of trickling filters on eel farms. *Aquacultural Engineering, 17*(3), 175–192.

Kesaano, M., & Sims, R. C. (2014). Algal biofilm based technology for wastewater treatment. *Algal Research, 5*, 231–240.

Kesari, K., Verma, H., & Behari, J. (2011). Physical methods in wastewater treatment. *International Journal of Environmental Technology and Management, 14*(1/2/3/4), 43–66.

Khan, Z. M., Asif Kanwar, R. M., Farid, H. U., Sultan, M., Arsalan, M., Ahmad, M., Shakoor, A., & Ahson Aslam, M. M. (2019). Wastewater evaluation for Multan, Pakistan: Characterization and agricultural reuse. *Polish Journal of Environmental Studies, 28*(4).

Khan, Z. U., Naz, I., Rehman, A., Rafiq, M., Ali, N., & Ahmed, S. (2015). Performance efficiency of an integrated stone media fixed biofilm reactor and sand filter for sewage treatment. *Desalination and Water Treatment, 54*(10), 2638–2647.

Kikuchi, T., & Tanaka, S. (2012). Biological removal and recovery of toxic heavy metals in water environment. *Critical Reviews in Environmental Science and Technology, 42*(10), 1007–1057.

Kim, B., Gautier, M., Prost-Boucle, S., Molle, P., Michel, P., & Gourdon, R. (2014). Performance evaluation of partially saturated vertical-flow constructed wetland with trickling filter and chemical precipitation for domestic and winery wastewaters treatment. *Ecological Engineering, 71*, 41–47.

Kinidi, L., Tan, I. A. W., Abdul Wahab, N., Tamrin, K., Nolasco-Hipolito, C., & Faridah Salleh, S. (2018). Recent development in ammonia stripping process for industrial wastewater treatment. *International Journal of Chemical Engineering, 12*, 1–14.

Kodera, H., Hatamoto, M., Abe, K., Kindaichi, T., Ozaki, N., & Ohashi, A. (2013). Phosphate recovery as concentrated solution from treated wastewater by a PAO-enriched biofilm reactor. *Water Research, 47*(6), 2025–2032.

Kumar, A., Moulick, S., Singh, B. K., & Mal, B. C. (2013). Design characteristics of pooled circular stepped cascade aeration system. *Aquacultural Engineering, 56*, 51–58.

Land, C. J., Dalhammar, G., Andersson, S., & Kuttuva Rajarao, G. (2008). Biofilm formation and interactions of bacterial strains found in wastewater treatment systems. *FEMS Microbiology Letters, 283*(1), 83–90.

Learner M. A. (2000). Egression of flies from sewage filter-beds. *Water Research, 34*(3), 877, 615.

Lefebvre, O., Vasudevan, N., Torrijos, M., Thanasekaran, K., & Moletta, R. (2005). Halophilic biological treatment of tannerysoak liquor in a sequencing batch reactor. *Water Research, 39*(8), 1471–1480.

Li, W., Loyola-Licea, C., Crowley, D. E., & Ahmad, Z. (2016). Performance of a two-phase biotrickling filter packed with biochar chips for treatment of wastewater containing high nitrogen and phosphorus concentrations. *Process Safety and Environmental Protection, 102*, 150–158.

Limoli, D. H., Jones, C. J., & Wozniak, D. J. (2015). Bacterial extracellular polysaccharides in biofilm formation and function. *Microbiology Spectrum, 3*(3). https://doi.org/10.1128/microbiolspec.MB-0011-2014

Lin, S. H., & Chen, M. L. (1997). Treatment of textile wastewater by chemical methods for reuse. *Water Research, 31*(4), 868–876.

Liu, Y. X., Yang, T. O., Yuan, D. X., & Wu, X. Y. (2010). Study of municipal wastewater treatment with oyster shell as biological aerated filter medium. *Desalination, 254*(1–3), 149–153.

Llop, A., Pocurull, E., & Borrull, F. (2009). Evaluation of the removal of pollutants from petrochemical wastewater using a membrane bioreactor treatment plant. *Water, Air, and Soil Pollution, 197*, 349–359.

Machineni, L., Reddy, C. T., Nandamuri, V., & Pawar, P. D. (2018). A 3D individual-based model to investigate the spatially heterogeneous response of bacterial biofilms to antimicrobial agents. *Mathematical Methods in the Applied Sciences, 41*(18), 8571–8588.

Mahmoud, M., Tawfik, A., & El-Gohary, F. (2010). Simultaneous organic and nutrient removal in a naturally ventilated biotower treating presettled municipal wastewater. *Journal of Environmental Engineering, 136*(3), 301–307.

Mahmoudi, H. (2016). In E. Drioli & L. Giorno (Eds.), *Encyclopedia of membranes* (pp. 2015–2015). Springer.

Maqbool, T., Khan, S. J., & Lee, C. H. (2014). Effects of filtration modes on membrane fouling behavior and treatment in submerged membrane bioreactor. *Bioresource Technology, 172*, 391–395.

Massoud, M. A., Tarhini, A., & Nasr, J. A. (2009). Decentralized approaches to wastewater treatment and management: Applicability in developing countries. *Journal of Environmental Management, 90*(1), 652–659.

Maureen, K., & Ronald, C. S. (2014). Algal biofilm based technology for wastewater treatment. *Algal Research, 5*, 231.

Mayer, B. K., Baker, L., Boyer, T. H., Drechsel, P., Gifford, M., Hanjra, M. A., Parameswaran, P., Stoltzfus, J., Westerhoff, P., & Rittmann, B. (2016). Total value of phosphorus recovery. *Environmental Science and Technology, 50*(13). https://doi.org/10.1021/acs.est.6b01239

Medina, V., Scholze, R. J., Waisner, S., & Griggs, C. S. (2015). Energy and resource recovery from wastewater treatment: State of the art and potential application for the army and the DoD. Available from: https://usace.contentdm.oclc.org/digital/collection/p266001coll1/id/3840/

Metcalf and Eddy, H. (2003). *Wastewater engineering: Treatment, disposal, and reuse* (4th ed., pp. 887–981). The McGraw-Hill Companies.

Miranda, A. F., Ramkumar, N., Andriotis, C., Höltkemeier, T., Yasmin, A., Rochfort, S., Wlodkowic, D., Morrison, P., Roddick, F., Spangenberg, G., Lal, B., Subudhi, S., & Mouradov, A. (2017). Applications of microalgal biofilms for wastewater treatment and bioenergy production. *Biotechnology for Biofuels, 10*(1), 120.

Mittal, A. (2011). Biological wastewater treatment. *Water Today*, 32–44. Available from: http://www.watertoday.org/Article%20Archieve/Aquatech%2012.pdf

Mononen, T., Coloma, S., Kapanen, A., Vikman, M., Itävaara, M., & Romantschuk, M. (2010). Biodegration of BPA and HHCB from waste water by multi-stage biofilm processes. Linnaeus ECO-TECH' 10 Kalmar, Sweden, November 22–24. https://doi.org/10.15626/Eco-Tech.2010.016

Mosa, K. A., Saadoun, I., Kumar, K., Helmy, M., & Dhankher, O. P. (2016). Potential biotechnological strategies for the cleanup of heavy metals and metalloids. *Frontiers in Plant Science, 7*, 303–303.

Musmarra, D., Prisciandaro, M., Capocelli, M., Karatza, D., Iovino, P., Canzano, S., & Lancia, A. (2016). Degradation of ibuprofen by hydrodynamic cavitation: Reaction pathways and effect of operational parameters. *Ultrasonics Sonochemistry, 29*, 76–83.

Naz, I., Khatoon, N., Ali, M. I., Saroj, D. P., Batool, S. A. U., Ali, N., & Ahmed, S. (2014). Appraisal of the tire derived rubber (TDR) medium for wastewater treatment under aerobic and anaerobic conditions. *Journal of Chemical Technology and Biotechnology, 89*(4), 587–596.

Naz, I., Ullah, W., Sehar, S., Rehman, A., Khan, Z. U., Ali, N., & Ahmed, S. (2016). Performance evaluation of stone-media pro-type pilot-scale trickling biofilter system for municipal wastewater treatment. *Desalination and Water Treatment, 57*(34), 15792–15805.

Norton-Brandão, D., Scherrenberg, S. M., & van Lier, J. B. (2013). Reclamation of used urban waters for irrigation purposes—A review of treatment technologies. *Journal of Environmental Management, 122*, 85–98.

Odigie, J. O. (2014). Harmful effects of wastewater disposal into water bodies: A case review of the Ikpoba river, Benin City, Nigeria. *Tropical Freshwater Biology, 23*, 87.

Oncel, M. S., Muhcu, A., Demirbas, E., & Kobya, M. (2013). A comparative study of chemical precipitation and electrocoagulation for treatment of coal acid drainage wastewater. *Journal of Environmental Chemical Engineering, 1*(4), 989–995.

Oppong, S.O.-B., Opoku, F., & Govender, P. P. (2019). Tuning the electronic and structural properties of Gd-TiO_2-GO nanocomposites for enhancing photodegradation of IC dye: The role of Gd^{3+} ion. *Applied Catalysis B: Environmental, 243*, 106–120.

Patria, L., Cathelain, M., Laurens, P., & Barbere, J. P. (2001). Odour removal with a trickling filter at a small WWTP strongly influenced by the tourism season. *Water Science and Technology, 44*(2–3), 243.

Pedrero, F., Kalavrouziotis, I., Alarcón, J. J., Koukoulakis, P., & Asano, T. (2010). Use of treated municipal wastewater in irrigated agriculture—Review of some practices in Spain and Greece. *Agricultural Water Management, 97*(9), 1233–1241.

Piekutin, J. (2018). The use of coagulation and reverse osmosis for petroleum hydrocarbons removal from water. *Desalination and Water Treatment, 128*, 437–441.

Price, M. S., Classen, J. J., & Payne, G. A. (2001). Aspergillus niger absorbs copper and zinc from swine wastewater. *Bioresource Technology, 77*(1), 41–49.

Quach-Cu, J., Herrera-Lynch, B., Marciniak, C., Adams, S., Simmerman, A., & Reinke, R. (2018). The effect of primary, secondary, and tertiary wastewater treatment processes on antibiotic resistance gene (ARG) concentrations in solid and dissolved wastewater fractions. *Water, 10*(1). https://doi.org/10.3390/w10010037

Radjenovic, J., Petrovic, M., & Barcelo, D. (2007). Analysis of pharmaceuticals in wastewater and removal using a membrane bioreactor. *Analytical and Bioanalytical Chemistry, 387*(4), 1365–1377. https://doi.org/10.1007/s00216-006-0883-6

Raquel, L., Elisa, R., Jose, M. E., García-Encina, P. A., & Muñoz, R. (2012). Odor abatement in biotrickling filters: Effect of the EBRT on methyl mercaptan and hydrophobic VOCs removal. *Bioresource Technology, 109*(38), 2012.

Sanjib, M., Mohd, T., & Mukherjee, C. K. (2011). Evaluation of nitrification performance of a trickling filter with nylon pot scrubber as media. *IJSN, 2*(3), 515.

Sarai Atab, M., Smallbone, A. J., & Roskilly, A. P. (2018). A hybrid reverse osmosis/adsorption desalination plant for irrigation and drinking water. *Desalination, 444*, 44–52.

Sato, T., Qadir, M., Yamamoto, S., Endo, T., & Zahoor, A. (2013). Global, regional, and country level need for data on wastewater generation, treatment, and use. *Agricultural Water Management, 130*, 1–13.

Schneider, I., & Topalova, Y. (2013). Microbial structure and functions of biofilm during wastewater treatment in the dairy industry. *Biotechnology & Biotechnological Equipment, 27*(3), 3782–3786.

Schroeder, E. D., & Tchobanglous, G. (1976). Mass transfer limitations on trickling filter design. *Journal of Water Pollution Control Federation, 48*, 771.

Sehar, S., & Naz, I. (2016). Role of the biofilms in wastewater treatment. In *Microbial biofilms—Importance and applications*. https://doi.org/10.5772/63499

Sen, T. K. (2015) Physical, chemical and biological treatment processes for water and wastewater. In *Chemical engineering methods and technology*, pp. 29–56.

Seow, T. W., Lim, C. K., Nor, M. H. M., Mubarak, M. F. M., Lam, C. Y., Yahya, A., & Ibrahim, Z. (2016). Review on wastewater treatment technologies. *International Journal of Applied Environmental Sciences, 11*(1), 111–126.

Shah, M., & Hashmi, H. N. (2012). Macrophyte waste stabilization ponds: An option for municipal wastewater treatment. *International Journal of Physical Sciences, 7*(30), 5162–5166.

Shareefdeen Z., & Singh, A. (2005). *Biotechnology for odor and air pollution control*. Biotech (pp. 378–1056. 634). Springer.

Sierra-Alvarez, R., Karri, S., Freeman, S., & Field, J. A. (2006). Biological treatment of heavy metals in acid mine drainage using sulfate reducing bioreactors. *Water Science and Technology, 54*(2), 179–185.

Solon, K., Volcke, E., Spérandio, M., & van Loosdrecht, M. (2019). Resource recovery and wastewater treatment modelling. *Environmental Science: Water Research & Technology, 5*, 631–642.

Teitzel, G. M., & Parsek, M. R. (2003). Heavy metal resistance of biofilm and planktonic Pseudomonas aeruginosa. *Applied and Environmental Microbiology, 69*(4), 2313–2320.

Tsang, Y. F., Chua, H., Sin, S. N., & Chan, S. Y. (2008). Treatment of odorous volatile fatty 635 acids using a biotrickling filter. *Bioresource Technology, 99*(3).

Vergine, P., Salerno, C., Berardi, G., & Pollice, A. (2018). Sludge cake and biofilm formation as valuable tools in wastewater treatment by coupling integrated fixed-film activated sludge (IFAS) with self forming dynamic membrane bioreactors (SFD-MBR). *Bioresource Technology, 268*, 121–127.

Vianna, M. R., de Melo, G. C., & Neto, M. R. V. (2012). Wastewater treatment in trickling filters using luffa cyllindrica as biofilm supporting medium. *Journal of Urban and Environmental, 6*(2).

Wang, H., Ji, G., Bai, X., & He, C. (2015). Assessing nitrogen transformation processes in a trickling filter under hydraulic loading rate constraints using nitrogen functional gene abundances. *Bioresource Technology, 177*, 217–223.

Wang, L., Min, M., Li, Y., Chen, P., Chen, Y.-F., Liu, Y., Wang, Y., & Ruan, R. (2009). Cultivation of green algae Chlorella sp. in different wastewaters from municipal wastewater treatment plant. *Applied Biochemistry and Biotechnology, 162*(4), 1174–1186.

Wei, C., Zhang, F., Hu, Y., Feng, C., & Wu, H. (2017). *Ozonation in water treatment: The generation, basic properties of ozone and its practical application* (p. 49).
Wollmann, F., Dietze, S., Ackermann, J. U., Bley, T., Walther, T., Steingroewer, J., & Krujatz, F. (2019). Microalgae wastewater treatment: Biological and technological approaches. *Engineering in Life Sciences*. https://doi.org/10.1002/elsc.201900071
Wu, H., Yin, Z., Quan, Y., Fang, Y., & Yin, C. (2016). Removal of methyl acrylate by ceramic-packed biotrickling filter and their response to bacterial community. *Bioresource Technology, 209*, 237–245.
Ye, L., Chai, L., Li, Q., Yan, X., Wang, Q., & Liu, H. (2016). Chemical precipitation granular sludge (CPGS) formation for copper removal from wastewater. *RSC Advances, 6*(115), 114405–114411.
Yin, H., Qiu, P., Qian, Y., Kong, Z., Zheng, X., Tang, Z., & Guo, H. (2019). Textile wastewater treatment for water reuse: A case study. *Processes, 7*(34). https://doi.org/10.3390/pr7010034
Yuan, Z., Pratt, S., & Batstone, D. J. (2012a). Phosphorus recovery from wastewater through microbial processes. *Current Opinion in Biotechnology, 23*(6), 878–883.
Yuan, Z., Pratt, S., & Batstone, D. J. (2012b). Phosphorus recovery from wastewater through microbial processes. *Current Opinion in Biotechnology, 23*(6), 878–883.
Zhang, X., Li, J., Yu, Y., Xu, R., & Wu, Z. (2016). Biofilm characteristics in natural ventilation trickling filters (NVTFs) for municipal wastewater treatment: Comparison of three kinds of biofilm carriers. *Biochemical Engineering Journal, 106*, 87–96.
Zhang, Y., Cheng, Y., Yang, C., Luo, W., Zeng, G., & Lu, L. (2015). Performance of system consisting of vertical flow trickling filter and horizontal flow multi-soil-layering reactor for treatment of rural wastewater. *Bioresource Technology, 193*, 424–432.
Zhang, Y., Ma, C., Ye, F., Kong, Y., & Li, H. (2009). The treatment of wastewater of paper mill with integrated membrane process. *Desalination, 236*(1), 349–356.
Zhao, Q., Zhong, H., Wang, K., Wei, L., Liu, J., & Liu, Y. (2013). Removal and transformation of organic matters in domestic wastewater during lab-scale chemically enhanced primary treatment and a trickling filter treatment. *Journal of Environmental Sciences, 25*(1), 59–68.
Zhao, X., Kumar, K., Gross, M. A., Kunetz, T. E., & Wen, Z. (2018). Evaluation of revolving algae biofilm reactors for nutrients and metals removal from sludge thickening supernatant in a municipal wastewater treatment facility. *Water Research, 143*, 467–478.
Zhu, C., Jiang, C., Chen, S., Mei, R., Wang, X., Cao, J., Ma, L., Zhou, B., Wei, Q., Ouyang, G., Yu, Z., & Zhou, K. (2018). Ultrasound enhanced electrochemical oxidation of Alizarin Red S on boron doped diamond (BDD) anode: Effect of degradation process parameters. *Chemosphere, 209*, 685–695.
Zielinski, M., Zielinska, M., & Dębowski, M. (2013). Application of microwave radiation to biofilm heating during wastewater treatment in trickling filters. *Bioresource Technology, 127*, 223.

Agricultural Renewable Residues for a Sustainable Agro-Industrial Production

María Carla Groff, Luis Fernando Flores, Khayali Vaidya, and Patra Priyadarshini Priyanka

Abstract Considering a projected world population of 10 billion by 2050, global agriculture and the millions of tons of annually generated renewable residues are expected to increase. Currently, these untreated and underutilized resources are disposed of by dumping, burning, or landfilling, which has detrimental effects on soil, air, and water resources. It also contributes to greenhouse gas emissions and affects both ecosystems and human health. These low-cost, renewable by-products include crop residues, fruit and vegetable wastes, and livestock manure, which have an abundant availability and unique chemical composition that makes them an ideal feedstock for the development of value-added bioproducts for sustainable agricultural production, from small-scale to large-scale agro-industrial operations. The latent benefits of these wastes have been harnessed through their conversion into biofertilizers, biostimulants, biopesticides, bioenergy, and bioadsorbents and coagulants for agricultural wastewater treatment. Such products are potentially greener, safer for plants, the environment, and, most importantly, for human consumers. Their production requires harvesting, transport, storage, and bioconversion. Pre-treatment methods, upstream and downstream processing, process optimization and scale-up

M. C. Groff
Facultad de Ingeniería, Instituto de Biotecnología, Universidad Nacional de San Juan (IBT-FI-UNSJ), San Juan, Argentina

Consejo Nacional de Investigaciones Científicas y Técnicas (CONICET), Buenos Aires, Argentina

Facultad de Ingeniería, Instituto de Ingeniería Química, Universidad Nacional de San Juan (IIQ-FI-UNSJ), San Juan, Argentina

L. F. Flores (✉)
Club de Biotecnología Agrícola y Ambiental, Villa Regina, Río Negro, Argentina
e-mail: lfflores.lab@gmail.com

K. Vaidya
Department of Environment and Life Sciences, Krantiguru Shyam Ji Krishna Verma Kachchh University, Kachchh, Gujarat, India

P. P. Priyanka
Department of Animal Biology, School of Life Sciences, University of Hyderabad, Hyderabad, India

M. Shaaban (ed.), *Soils and Sustainable Agriculture*, Frontier Studies in Soil Science,
https://doi.org/10.1007/978-3-031-91114-9_34

are involved during the formulation of these bio-based products. In addition, assessments of plant design and location, availability of renewable resources, logistics, and the most environmentally friendly conversion methods are essential. Bringing innovation to the market also implies partnerships between R&D and industry, government incentives, and support through bioeconomic policies. In this context, the future integration of bio-products derived from agricultural residues into agro-industrial systems could mitigate waste generation, improve resource efficiency and sustainability throughout the value chain, and foster circular economies, minimizing environmental impact and improving agricultural productivity. In conclusion, this chapter reviews the latest developments to design a closed loop whereby agricultural waste materials are reused as a resource stream in the sector, leading to a sustainable agricultural scenario.

Graphical Abstract

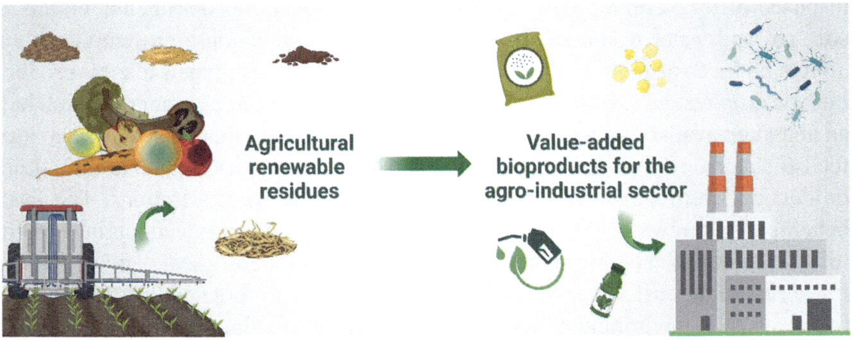

Keywords Renewable residues · Sustainable agriculture · Value added-products · Bioproducts

1 Introduction

Agriculture has long been an integral part of human society. Around 12,000 years ago, amid the Neolithic period, agriculture transformed patterns of existence and survival, allowing for the construction of permanent settlements and sedentary lifestyles, which were critical for the formation of sophisticated societies. Over time, agricultural techniques have evolved significantly, from using simple hand tools and domesticating plants and animals to applying techniques such as plowing, crop rotation, and irrigation systems, which allowed for increased productivity. However, a significant increase in production occurred with the Industrial Revolution, when mechanization was introduced, with tractors and harvesters, transforming agricultural efficiency (Larsen, 2023). Nevertheless, this progress has also led to increased soil contamination and degradation, which poses a major risk to environmental sustainability and global food security, making soil management strategies essential. Thus, this issue represents a worldwide challenge due to various forms of degradation, such as erosion by water or wind, pollution affecting soil or water, a reduction in soil organic matter, imbalances in nutrients, the formation of crusts, salinization, acidification, and declines in soil biodiversity (Matisic et al., 2024).

With the fast rise of the world's population and rising need for food, fiber, and fuel, the need for sustainable farming techniques has never been greater. On the one hand, obtaining high agricultural yields makes it possible to obtain products with marketable value for feeding an ever-increasing population. However, it also generates a significant amount of waste, which is often discarded or utilized with minimal added value (Sadh et al., 2023). Thus, the agricultural industry is also facing challenges in resource utilization, waste management, and environmental sustainability. Therefore, efficiently utilizing agricultural renewable residues is one viable way to overcome these difficulties (Divyabharathi et al., 2024).

Crop stalks, husks, leaves, and other agricultural byproducts have historically been underused or considered trash. However, technological advancements and an increasing emphasis on sustainability have shown their potential as significant resources in agro-industrial production. By transforming these wastes into other value-added products, the agricultural industry can minimize waste, lower its carbon footprint, and create a sustainable agro-industrial production system within a circular economy framework. Therefore, this chapter describes the chemical and physical qualities of agricultural renewable residues as feedstock for the production of new value-added products with application in agro-industrial processing. It also investigates the technical advances and procedures for efficiently converting agricultural renewable leftovers into a wide range of valuable products for the agro-industrial sector, including biofuels, biofertilizers, bioenergy, biomaterials, bioadsorbents, and microbial enzymes. Consequently, the environmental and economic pros are highlighted, such as lower greenhouse gas emissions, improved soil health, and the establishment of new revenue streams for farmers. Drawbacks such as obstacles of residue collection, transportation, and processing are also presented, as well as ideas for overcoming them. By tapping into the potential of agricultural renewable residues, the agro-industrial sector may transition to more sustainable production techniques that not only fulfill society's expanding demands but also protect the environment for future generations.

2 Agricultural Renewable Residues

Agricultural waste feedstocks, as shown in Fig. 1, encompass crop residues, fruit and vegetable wastes, and livestock manure, and present both opportunities and challenges in sustainable bioproduct development. Most of these wastes can be considered renewable; although, determining the renewability of each waste depends on the processes to which it was subjected within the industry from which it comes, and their specific characteristics (Flores et al., 2024). However, these materials often contribute

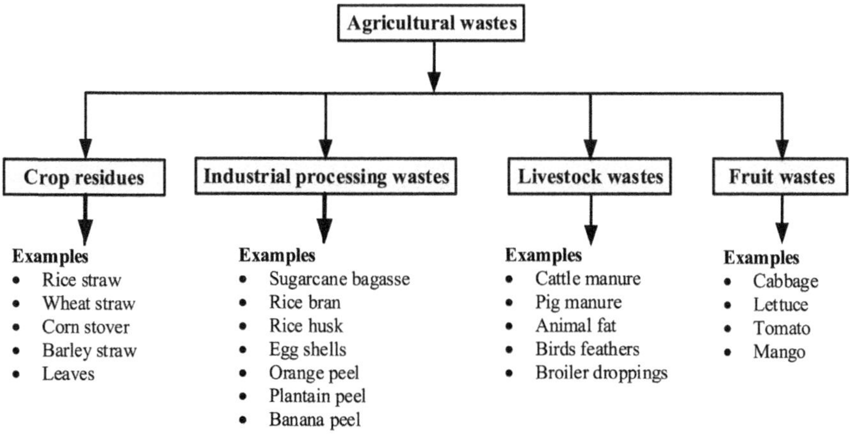

Fig. 1 Types and key examples of agricultural renewable residues. Adopted from (Awogbemi and Kallon 2022)

to pollution problems due to their decomposition and improper management (Ezeorba et al., 2024). When these wastes are discarded or inadequately processed, they can lead to environmental issues such as:

- **Soil and water contamination**: Improper disposal of agricultural waste can introduce harmful substances into the soil and water systems. Pesticides, herbicides, and heavy metals present in the waste can leach into the groundwater, affecting both human health and ecosystems (Setyaningsih & Sanjaya, 2022). Additionally, the decomposition of organic waste in landfills or open fields can produce leachate, a liquid that carries pollutants into surrounding soils and water bodies (Abdel-Shafy et al., 2024).
- **Greenhouse gas emissions**: Decomposing agricultural waste emits considerable volumes of greenhouse gases, notably methane (CH_4) and nitrous oxide (N_2O), which are far more powerful than carbon dioxide (CO_2) in terms of their impact on global warming (Lee et al., 2022). Manure and other organic waste materials, when left to decompose anaerobically (without oxygen), produce methane, contributing to climate change (He et al., 2023).
- **Nutrient runoff**: Excess nutrients, especially nitrogen and phosphorus from livestock dung and crop wastes, can run off into bodies of water during heavy rains. This fertiliser runoff contributes to eutrophication, which promotes hazardous algal blooms, depletes oxygen in the water, and creates dead zones where aquatic life cannot survive. Eutrophication disrupts aquatic ecosystems and can harm fisheries and recreational water use (Ladapo & Aminu, 2017).

The latent benefits of the agricultural renewable residues can be harnessed through their conversion into a wide range of value-added bioproducts:

Crop residues: These materials are rich in cellulose, hemicellulose, and lignin, which are essential for various bioconversion processes. For instance, they can be converted into biofuels such as bioethanol and biodiesel, which provide renewable energy sources that reduce reliance on fossil fuels. Additionally, the lignocellulosic content in crop residues can be processed to produce bioplastics, which are biodegradable and offer an environmentally friendly alternative to conventional plastics (Arora et al., 2023).

Fruit and vegetable wastes: This classification includes peels, seeds, and pulp discarded during processing, contain high levels of sugars, vitamins, and minerals, and are excellent feedstocks for producing biogas through anaerobic digestion. This process not only generates renewable energy but also produces digestate, a nutrient-rich by-product that can be used as a fertilizer to enhance soil fertility. Furthermore, the high sugar content in these wastes can be utilized to produce organic acids, enzymes, and other biochemicals that have applications beyond the agricultural sector, such the food, pharmaceutical, and cosmetic industries (Kaur et al., 2023).

Livestock manure: Manure is rich in organic matter and nutrients such as nitrogen, phosphorus, and potassium. The abundance and diverse chemical composition of these residues make them highly suitable for conversion into bioproducts. Its rich nutrient profile can be composted to create high-quality organic fertilizers. These fertilizers improve soil structure, increase microbial activity, and enhance the overall health of the soil. Additionally, livestock manure can be processed through anaerobic digestion to produce biogas, contributing to the energy needs of agricultural operations (He et al., 2023).

3 Agricultural Renewable Residues-Derived Bioproducts for the Agro-Industrial Sector

The circular bioeconomy is increasingly recognized as a viable approach to reducing reliance on petroleum-derived substances in chemical synthesis. It focuses on utilizing and revalorizing available agricultural biomass, with the primary goals of promoting sustainable development and advancing the circular economy (Kumar Sarangi et al., 2023). Thus, renewable by-products attracted attention as they are abundantly available, relatively cheaper, and possess a unique chemical composition, making them ideal feedstocks for developing value-added bioproducts for sustainable agro-industrial production (Khunnonkwao et al., 2024). Consequently, the conversion of agricultural renewable residues into value-added bioproducts plays an essential role in promoting sustainable agricultural production.

Renewable agri-waste can be transformed into valuable bioproducts through various processes. On small-scale operations, farmers can convert these by-products into compost or biofertilizers, enhancing soil health and reducing the need for chemical inputs. On a larger scale, advanced technologies can process these materials into bioenergy, bioplastics, and other biochemicals, contributing to a circular economy (Kaur et al., 2023). In this context, biorefineries play an important role in this new paradigm by converting agricultural waste into a variety of useful products (Annavelink et al., 2022). This approach might be enhanced by advances in Industry 4.0, which integrates advanced technologies such as artificial intelligence and automation into processes, optimizing efficiency and sustainability of these resource recovery practices (Chávez et al., 2024).

For a successful conversion into new bioproducts, agricultural renewable residues may require the application of pretreatments to improve the characteristics and nature of the biomass in order to break down complex structures and remove impurities, improving the efficiency of subsequent stages. The various pretreatment methods can be grouped into conventional methods, which include physical (milling, chopping, thermochemical, hammering and grinding), chemical (alkaline, acid, ionic liquid and organic solvent), and biological (fungal, bacterial, algal and enzymatic) treatments, and unconventional methods, which involve the use of ionic solvents and pulsed electric field methodologies (Kumar et al., 2023). In this way, the structure of agro-industrial wastes is readily available for the application of various technological bioprocesses.

Bioprocessing and fermentation are key stages where the pretreated feedstock is converted into valuable products, in which microorganisms are used to transform organic materials into biofuels, biochemicals, or other bioproducts (El Sheikha & Ray, 2023). This process is sensitive to factors like substrate composition (carbohydrate, lipid and protein content), microbial strain (bacterial, yeast or fungal), and fermentation conditions (temperature, pH, among others), which can significantly impact yield and efficiency (Mujtaba et al., 2023). The viability of the bioprocess is examined by different techno-economic evaluations at scale, taking into consideration the ecological principles of circular economy and biorefinery (Blasi et al., 2023).

Once the bioprocess to be applied to the particular agricultural residue has been chosen, it must be taken into account that it is essential to apply recovery and purification stages of the bioproducts obtained from fermentation, known as downstream process. This stage includes separation, concentration, and purification processes such as filtration, centrifugation, and distillation. Effective downstream processing is critical for achieving high-quality and commercially viable products (El Sheikha & Ray, 2023). Formulation encompasses the development of the final product from the purified bioproducts. This may involve blending with other ingredients or adjusting the product's physical and chemical properties to meet specific application needs (Cuadrado-Osorio et al., 2022), such as in biostimulants, fertilizers, biofuels, or enzymatic cocktails production, as shown in the next subsections.

3.1 Biostimulants

The widespread use of synthetic fertilizers in agricultural areas has caused significant ecological harm. Serious human health issues, reduced crop fertility, climate fluctuations, and the high costs of soil maintenance are the primary negative impacts caused by inorganic fertilizers, requiring urgent attention. Recent studies highlight the increasing market for biostimulants derived from organic waste. These biostimulants encompass a wide array of biological extracts, synthetic and natural organic compounds, inorganic molecules, and microorganisms, all contributing positively to crops (Zuzunaga-Rosas et al., 2024). Various studies have explored the development of biostimulants from different organic waste sources. Importantly, distinct types of organic waste present different compositions and mechanisms as biostimulants, requiring interdisciplinary analysis from fields such as chemistry and biology (Zhang et al., 2024). For instance, advancements in the use of agri-waste biomass for organic biostimulants have enhanced agricultural cultivation by regulating phytohormones. This organic waste is rich in nutrients, boosts phytohormone production, and generates beneficial chemicals (Devi et al., 2024).

3.2 Biofertilizers

Biofertilizers have demonstrated significant potential to improve crop productivity by enhancing nutrient availability and promoting beneficial microbial interactions in the soil. Studies show that biofertilizers can increase crop yields by enhancing nutrient uptake, improving soil health, and reducing the need for chemical fertilizers. However, their efficiency can be limited by environmental factors such as soil pH, temperature, and moisture, which can affect microbial activity and survival. Therefore, large-scale application of biofertilizers requires careful consideration of these factors to ensure consistent results across different agricultural systems (Du et al., 2020; Puglia et al., 2021).

Using agricultural residues as substrates for biofertilizer production can significantly enhance nutrient cycling. They can be used to cultivate nitrogen-fixing bacteria like Rhizobium and Azotobacter, phosphorus-solubilizing bacteria, and mycorrhizal fungi that improve nutrient availability to plants. Thus, the selection of appropriate renewable agri-waste feedstock, microbial strains, and fermentation conditions is essential for effectively producing effective biofertilizers from renewable resources. Moreover during separation and concentration of microbial cultures to generate a concentrated biofertilizer product, viability and activity should be ensured and secured during storage and transportation, and is effective upon application to soils or plants (Bagga et al., 2024; Puglia et al., 2021; Thadiyan et al., 2024).

3.3 Biopesticides

Large-scale production of biopesticides presents challenges such as the consistent supply of quality feedstock, scalability of the fermentation process, and cost-effective downstream processing. Thus, using agricultural waste as feedstock significantly improves the environmental sustainability of biopesticides by reducing waste, repurposing organic materials, and reducing reliance on non-renewable resources. Moreover, the sustainability of biopesticide production is enhanced when feedstocks are sourced locally, reducing transportation emissions and costs. *Bacillus thuringiensis* is commonly used in the fermentation process to produce bioinsecticides, which are effective against various insect pests. The efficiency of biopesticides is generally high, with specific advantages such as target specificity, reduced environmental impact, and lower resistance development in pests compared to chemical pesticides. However, there are limitations, including the relatively narrow spectrum of activity, susceptibility to environmental conditions, and the need for precise application timing (Priya et al. 2023; Mattedi et al., 2023; Kumar et al., 2024).

3.4 Bioenergy

Biomass is considered one of the most adaptable and crucial energy sources on the planet. Agricultural renewable residues hold significant energy potential due to their biochemical composition, notably macromolecules such as lignin and cellulose, which makes them ideal for sustainable and clean energy production through biomass-based bioenergy. Most agricultural wastes are the result of solar energy collected by plants during photosynthesis, and the byproducts can be reconverted into various kinds of energy, such as biofuels (Ufitikirezi et al., 2024). Biofuels refer to fuels derived from biomass, which can be utilized in transportation within the agro-industrial sector and beyond (Guo et al., 2024). Agricultural residues can be transformed into solid (biochar), liquid (bioethanol, biobutanol, biodiesel, and bio-oil), or gaseous biofuels (biohydrogen, biomethane, and biogas) (Zielińska & Bułkowska, 2024). Various biomass conversion pathways, such as thermochemical and biochemical processes, are employed to produce biofuels. For instance, biochemical processes involve enzymes and microorganisms breaking down renewable agricultural residues into precursor molecules such as sugars, fatty acids, and amino acids, which are then converted into biofuels through aerobic or anaerobic pathways. These biofuels provide energy from biomass to various sectors, including agriculture, industry, and electricity generation, and are considered sustainable alternatives to conventional fossil fuels (Zielińska & Bułkowska, 2024). Bioethanol is primarily utilized as a fuel for internal combustion engines, often mixed with gasoline. Using biohydrogen as a fuel in internal combustion engines produces no CO_2 emissions, avoids releasing gaseous pollutants, and prevents the formation of greenhouse gases. In recent years, biomethane has gained traction as a renewable fuel, replacing natural gas and biogas in powering internal combustion engines to help reduce CO_2 emissions (Awogbemi & Kallon, 2023). By recovering energy from agricultural renewable residues, environmental pollution and health issues related to poor waste management can be reduced, minimizing greenhouse gas (GHG) emissions and promoting a circular economy (Roudneshin & Sosa, 2024; Zielińska & Bułkowska, 2024).

3.5 *Biochar*

Agricultural residues generated on farms as byproducts of crop production are commonly used to make biochar (Shaaban & Abid, 2021). Biochar is a carbon-based solid material, often created by pyrolysis of organic wastes such as coconut husk and wood chips, among other agricultural wastes, under complete or partial oxygen exclusion (Enaime & Lübken, 2021). Biochar production collects CO_2 and CH_4 that would otherwise be released into the environment as a result of biomass degradation in waste sites, which helps to mitigate climate change. This strategy may lessen the added load of green waste and associated pollution factors such as greenhouse gas emissions, disposal, and landfill issues (Sirohi et al., 2023).

Biochar is made up of a mixture of carbon (C), hydrogen (H), oxygen (O), and nitrogen (N), as well as S, P, K, Ca, Mg, Na, and Si in the ash element. (Chen et al., 2019a). The physicochemical characteristics of biochar, which govern its application diversity, rely heavily on the type of thermal treatment, processing parameters (heating temperature, retention time, heating rate, pressure), and feedstock (Enaime & Lübken, 2021). In addition, physico-chemical components of biochar primarily depend upon the selection of raw material in the preparation of biochar. Biochars are often alkaline in nature because ashes are formed during the pyrolysis process, making them particularly effective for bioremediation of acidic soils (Chen et al., 2019a).

Biochar are highly effective in improving micro-ecological structures of soil by amending bulk density and water holding capacity (Shaaban et al., 2018). This property enables soils to retain water for extended periods of time, eliminating the need for regular watering and creating a more stable environment for plant roots (Jiang et al., 2017). Furthermore, biochar improves soil structure by reducing compaction and encouraging the establishment of beneficial soil microbes, which are essential for nutrient cycling and plant health. Another key advantage of biochar is its ability to retain nutrients (Chen et al., 2019a). The negatively charged surface of biochars attracts the cations such as sodium, potassium, calcium, which improves the cation exchange capacity of soil and enhances bioavailability of the nutrients to the roots of the plants (Chen et al., 2019a). This leads to reduction in nutrient leaching and plant roots get the excess of nutrients over an extended period. Subsequently, the requirement of the fertilizer gets reduced which leads to lower agriculture runoff and lower contamination in nearby waterbodies. Biochar also contributes significantly to climate change mitigation efforts. When biomass decomposes spontaneously, carbon dioxide is released into the atmosphere. However, when biomass is turned into biochar, the carbon inside it is stabilized and trapped in a solid state, perhaps for millennia. This makes biochar a viable method for carbon sequestration since it stores carbon that would otherwise contribute to greenhouse gas emissions. Furthermore, by improving soil fertility, biochar can indirectly contribute to increased carbon uptake by plants, which aids in climate change mitigation. In this approach, biochar serves a dual purpose: it boosts agricultural output while simultaneously promoting environmental sustainability (Enaime & Lübken, 2021).

In livestock production, biochar can be used as a feed supplement to promote animal health and lower methane emissions (Man et al., 2021). Overall, Biochar is a versatile agriculture solution for fighting against issues of climate change and soil degradation and to promote environmental sustainability.

3.6 Veterinary Antibiotic/Pharmaceuticals Removal

In recent decades, the presence of antibiotics, non-steroidal anti-inflammatory drugs, and other pharmaceuticals in the environment has garnered significant attention due to their indiscriminate use in treating infectious diseases in both animals and humans, as well as their introduction into livestock feed additives (Juela, 2022; Nordin et al., 2023). The incomplete metabolism of these compounds in animals and humans leads to the release of unmetabolized substances into the environment. Consequently, these emerging contaminants have been detected in various environmental matrices such as sediments, soil, potable water, and wastewater from both livestock and domestic sources (Chen et al., 2019b; Li et al., 2018).

The incomplete removal of these pharmaceuticals by conventional treatment plants further exacerbates their accumulation in the environment, posing a significant risk due to their high persistence (Chen et al., 2019a, 2019b; Juela, 2022; Nordin et al., 2023). As a result, animals and humans are constantly exposed to these contaminants through the consumption of food crops irrigated with polluted water or through the consumption of contaminated water itself. This poses a serious threat not only to food safety but also to human and animal health, as antibiotics can travel through the food chain, potentially leading to antibiotic resistance (Chen et al., 2019b; Juela, 2022; Nordin et al., 2023).

The conventional biological treatment of wastewater contaminated with antibiotics has proven to be inadequate due to the toxic and bacteriostatic effects of these compounds on microorganisms, as well as the migration of antibiotics and their prolonged half-lives. Consequently, new methods for the removal of these emerging contaminants have become necessary. Recent approaches have employed the adsorption of antibiotics, anti-inflammatory drugs, and other pharmaceuticals using adsorbents derived from agricultural renewable residues. This technique is considered safe, cost-effective, and simple (Chen et al., 2019b; Li et al., 2018).

The principal structural constituents of agricultural residues (cellulose, hemicellulose, and lignin) are particularly suitable for creating biochar, activated carbon, and other adsorbents. These adsorbents are often porous, with cavities capable of trapping compounds such as antibiotics. Additionally, the surface of these adsorbents can be modified due to the presence of significant amounts of proteins, lipids, starch, and other functional groups (Bankole et al., 2023; Nordin et al., 2023). For instance, the adsorption of pharmaceutical compounds on renewable agricultural residue-based adsorbents is driven by donor/acceptor interactions and the formation of hydrogen bonds. Several factors influence the adsorption process, including the nature of the emerging contaminant, the pore structure of the adsorbent, and the conditions of adsorption, such as pH, temperature, and the presence of competing ions. In the agricultural sector, optimizing adsorption could be a viable technology for cleaning environments contaminated by pharmaceutical compounds (Ajala et al., 2023; Bankole et al., 2023).

As mentioned, biochar is gaining interest due to its vast range of uses in agriculture. Because of its porous nature, biochar not only sequesters carbon but also assists in the remediation of toxic contaminants from soil (Sirohi et al., 2023) and wastewater (Awogbemi & Kallon, 2023), which would otherwise be adversely impacting to soil and crop growth. Biochar's capacity to restore contaminated soils is another significant agricultural use. It can adsorb heavy metals, herbicides, and other contaminants, decreasing their availability to plants and keeping them out of the food chain. This feature is especially worthwhile in places with a history of industrial contamination or intense agricultural practices that have caused soil deterioration. Biochar preserves plant health by immobilizing harmful pollutants while also improving soil quality (Jiang et al., 2017).

3.7 Others Bioproducts

A wide range of bioproducts with applications in the agro-industrial sector are being developed from renewable agricultural residues, as demonstrated in Table 1.

4 Selecting the Best Agricultural Renewable Residue Feedstock

To contribute to sustainable agro-industrial production, the environmental impact of using a particular agricultural renewable waste should be the primary factor in the selection process. This ensures both minimal environmental harm and economic gains while maximizing ecological benefits (Flores et al., 2024).

Table 1 Renewable residues as feedstock for agriculture-targeted microbial bioproducts

	Fermentation process	Microorganism	Waste	Efficiency/yield	Key companies	Price	References
Organic acids	Acetic acid (C2)	*Aspergillus niger, Acetobacter, Gluconobacter, Gluconoacetobacter*	Agro – / fruit-waste, corn steep liquor, hydrolyzed soy flour, ultrafiltered ethanol silage and whey lactose	69.38%	Zea2; Godavari Biorefineries Ltd.; Sucroal SA; Novozymes AS; Wacker Chemie AG	4 USD/kg	MordorIntelligence (2024a) Morales-Vera et al. (2020) Vashisht et al. (2019)
	Lactic acid (C3)	*Lactobacillus Paracasei, Lactococcus lactis, Sporolactobacillus Inulinus, Trichoderma reesei + Lactobacillus pentosus, Clostridium thermocellum* *Rhizopus oryzae*	Cottonseed meal and corncob residues Grape stalk	62.4–89% 200%	NatureWorks LLC, Corbion-Purac, Galactic S.S	1.3–4 USD/kg	Groff et al. (2024) Groff et al. (2022) Mazzoli (2021)
	Propionic acid (C3)	*Propionibacterium*	Whey permeates, hydrolyzed corn meal, corn steep liquor, glycerol, sugarcane bagasse hydrolysate, apple pomace	40–86%	BASF SE, Eastman Chemical Company, Dow, Perstorp, Daicel Corporation	2–3 U$D/kg	Piwowarek et al. (2021) MordorIntelligence (2024b)

(continued)

Table 1 (continued)

	Fermentation process	Microorganism	Waste	Efficiency/yield	Key companies	Price	References
	Butyric acid (C4)	*Clostridium tyrobutyricum*, *Bacillus* sp., *C. thermocellum and C. thermobutyricum*	Wheat straw, wheat bran, rice polishings, whey, cane molasses, corn cob, corn husk, corn fiber, rice straw, oilseed rape straw, soybean hull, sugarcane bagasse, switchgrass, wheat straw, waste paper, and paper mill sludge	88–99.5%	Eastman Chemical Company, Perstorp Holding AB, Merck KGaA, Tokyo Chemical Industry Co., Ltd, OQ Chemicals GmbH	3.50–4.78 U$D/kg	MordorIntelligence (2024c) Lu et al. (2020) Baroi et al. (2017) Akhtar et al. (2020) Jiang et al. (2018)
	Fumaric acid (C4)	*Trichoderma Reesei, Rhizopus delemar, Rhizopus oryzae, Rhizopus Formosa, Rhizopus arrhizus. Aspergillus niger*	Cassava bagasse, corn straw and pulp and paper solid waste	37–91%	Bartek Ingredients Inc., XST Biological Co. Ltd., Changmao Biochemical Engineering Company Limited, Fuso Chemical Co. Ltd, Polynt	1.4 USD/kg	Mazzoli (2021) Lu et al. (2020) Guo et al. (2020) MordorIntelligence (2024d)

(continued)

Table 1 (continued)

	Fermentation process	Microorganism	Waste	Efficiency/yield	Key companies	Price	References
	Succinic acid (C4)	*Actinobacillus succinogenes, Anaerobiospirillum succiniciproducens, Mannheimia succiniciproducens* and recombinant *Escherichia coli*	Food waste, vegetables, fruits, meat, rice, corncob and noodles	10–97%	BASF SE, Roquette Frères, Mitsubishi Chemical Corporation, DSM, Mitsui Chemicals, Inc	6.4 USD/kg	Mazzoli (2021) Merrylin et al. (2020) MordorIntelligence (2024e) Lu et al. (2021)
	Citric acid (C6)	*Aspergillus niger, Mucor pyriformis, Penicillium lactum, Tricoderma viride, Saccharomyces, Candida, and Yarrowia*	Pineapple, waste grape pomace, apple pomace, corn cobs, date pulp, kiwifruit peel, cassava bagasse, banana peel, and wheat bran and soya bean meal	8.2–93%	Jungbunzlauer Suisse AG, Cofco Biochemical, Cargill, Incorporated, Gadot Biochemical Industries Ltd, Shandong Ensign Industry Co. Ltd.	0.5–1.2 USD/kg	Merrylin et al. (2020) Amato et al. (2020) MordorIntelligence (2024f) Wang et al. (2020)

(continued)

Table 1 (continued)

	Fermentation process	Microorganism	Waste	Efficiency/yield	Key companies	Price	References
Industrial enzymes	Alpha amylases	*Aspergillus niger and Rhizopus Bacillus Licheniformis Bacillus subtilis, Lactobacillus fermentum*	Rice husk, apple peel, banana peel, rice bran, wheat bran, green gram husk, pineapple peel, lentil husk, cotton stalk, coarse meal of corn, coarse meal of millet	14–34 U/g	Novozymes, DuPont, AB Enzymes, BASF SE, DSM	145 (100 U)–991 (1000 U) U$D/kg	Merrylin et al. (2020) (MordorIntelligence) Deljouc et al. (2018) Lee Biosolutions (2024)
	Lactases (β-galactosidases)	*Polyporus versicolor Lactobacillus* sp., *Kluyveromyces lactis, Kluyveromyces fragilis*	Food wastes	111.61–159.62 U/mL		0.37 €/capsule LactroZym®	(MordorIntelligence) Shen et al. (2023) Amin et al. (2023) Biogena (2024)
	Laccases	*Phanerochaete chrysosporium, Agaricus bisporus, Cerrena unicolor, Trametes versicolor Azospirillum lipoferum, Streptomyces lavendula, Streptomyces cyaneus, Bacillus subtilis, Pseudomonas aeruginosa, Streptomycetes cyaneus*	Oil cakes, wheat bran, bagasse, rice bran, Soybean meal, rice husk, groundnut oil cake paddy straw	1.2–400 U/mL		0.04 €/U	Mahuri et al. (2023) Gutierrez-Rangel et al. (2024) (MordorIntelligence)

(continued)

Table 1 (continued)

	Fermentation process	Microorganism	Waste	Efficiency/yield	Key companies	Price	References
	Lipases	*Aspergillus* sp., *Thermomyces lanuginosus* *Candida antarctica*	Banana peel, potato peel, and cassava peel through			15–174 €/kg	Mata et al. (2017) (MordorIntelligence)
Biopackaging	Polyhydroxyalkanoates (PHAs)	*Cupriavidus necator, Pseudomonas oleovorans, Bacillus megaterium, Escherichia coli, Alcaligenes latus, Enterobacter aerogenes*	Wheat, corn, sugar beet, potatoes and vegetable oils	80–95%	Mondi Group, Tetra Pak International SA, Sealed Air Corporation, International Paper Company, Amcor Group GmbH	2.4–5.5 USD/kg	Faizan Muneer et al. (2021) Malagón et al. (2017) MordorIntelligence (2024g) Crutchik et al. (2020)
	Polyhydroxybutyrate (PHB)	*Ralstonia eutropha, Alcaligenes spp., Azotobacter spp., Bacillus spp., Nocardia spp., Pseudomonas spp., Rhizobium spp., Ralstonia eutropha*	Dairy waste, date seeds, grain crops, starch and damaged food grains	0.33–0.50 g/L			McAdam et al. (2020) Crutchik et al. (2020)
Natural gums	Xanthan gum	*Xanthomonas campestris*	Whey, citrus wastes, corn steep liquor, chicken feathers, molasses and glucose syrup and olive oil wastewaters	1–53 g/L	Archer Daniels Midland Company, Fufeng Group, MeiHua Holdings Group Co Ltd, J. M. Huber Corporation, Cargill Incorporated	7.6 USD/kg	Murad et al. (2019) MordorIntelligence (2024h) Dey & Chatterji (2023) Malone et al. (2023)

(continued)

Table 1 (continued)

	Fermentation process	Microorganism	Waste	Efficiency/yield	Key companies	Price	References
Microbial proteins	Microalgae	*Arthospira platensis (Spirulina maxima) Chlorella pyrenoidosa Chlorella sorokiana Euglena gracilis*	Orange peel, sugarcane bagasse, rice husk, wheat straw, cassava residues, sawdust, corn cobs, coconut residues, mandarin residues, beet pulp	Protein content: 45–70%	Bega Cheese ltd., Blue Green Foods, Cyanotech, BioProcess Algae LLC, Calysta, Algaeon Inc., Nucelis Inc., Unibio A/S, Euglena Co. ltd., Biomin Holding, Evonik, BlueBioTech, Nutreco NV, Lallemand Inc, Marlow Foods ltd, Vagan Pharma ltd, Angel Yeast Co. Ltd, LeSaffre	5.16–9 €/kg	Cardoso Alves et al. (2023) Voutilainen et al. (2021)
	Fungi and yeast	*Aspergillus niger Candida tropicalis Debaryomyces hansenii Kluyveromyces marxianus Yarrowia lipolytica*		Protein content: 32–9%			
	Bacteria	*Bacillus cereus Escherichia coli Haloarcula sp. Rhodopseudomonas palustris*		Protein content: 55–76%			
Phenolic extracts		*Lactobacillus lactic, Lactobacillus plantarum, Aspergillus awamori, Aspergillus niger, Aspergillus oryzae, Rhizopus oligosporus, Rhizopus oryzae, Aspergillus fumigatus*	Rice bran, peanut press cake, hass avocado seeds Mango seed waste, peanut press cake, apricot pomace, olive mill waste, orange peel waste	–	ADM, Aquanova AG., Ajinomoto Co., Inc., Chr. Hansen Holding A/S, CPC Ingredients Ltd	4100 U$D/kg (hydroxytyrosol in capsules form)	Astudillo et al. (2023) MordorIntelligence (2024i) Zhao et al. (2022)

Availability and accessibility are key factors when choosing an agricultural residue for bioconversion. The ideal residue should be abundant and locally available, minimizing transportation costs and reducing the overall carbon footprint associated with collection. Consequently, residues that are scarce or located in remote areas, which incur higher logistical costs, are less attractive for their conversion into new bioproducts. Availability is also influenced by seasonality, as some residues are only available during specific harvesting periods, affecting their continuous supply for bioconversion processes (Flores et al., 2024). Thus, a consistent and reliable supply of the chosen residue is essential to maintain stable bioproduct-based agroindustrial production throughout the year. Alternatively, multiple residues can be integrated or stored to bridge the gap between harvest seasons. Other factors impacting the supply chain of renewable residues include their inherent characteristics and variability, geographical distribution, climate conditions, volume of raw material, delivery networks, supplier contracts, and local or international regulations (Roudneshin & Sosa, 2024; Ufitikirezi et al., 2024).

The biochemical composition of agricultural renewable residues is also a critical factor for successful bioconversion into agro-industrial value-added bioproducts. Residues rich in cellulose, hemicellulose, and lignin are well-suited to be broken down into fermentable sugars and other intermediates, which can then be converted into microbial biomass, biofuels, chemicals, or other valuable products. Nutrient content is another factor that influences bioconversion potential, which can be optimized through supplementation with microelements, pH adjustment, or the removal of inhibitory compounds (Cuadrado-Osorio et al., 2022). For instance, nitrogen-rich residues are ideal for producing biofertilizers or biostimulants, which enhance soil fertility and crop productivity. On the other hand, the presence of secondary metabolites or inhibitory compounds, can negatively impact microbial fermentation processes (Bhavana et al., 2024; Bishop et al., 2024). In such cases, additional steps to remove or neutralize these inhibitors may be needed, although this adds time and cost to the conversion process. Moreover, compatibility with the available infrastructure is another key factor. In terms of scalability, the chosen residue should support large-scale production without requiring significant modifications to existing equipment or processes and should be adaptable to different bioconversion pathways to offer flexibility in response to changing market demands or technological advancements (Ufitikirezi et al., 2024; Zielińska & Bułkowska, 2024).

The economic viability of using a particular residue is also a key consideration. Agricultural renewable residues should be assessed in terms of the costs associated with collection, transportation, and processing, as well as the potential market value of the resulting products in the agroindustry. Therefore, residues that are readily available and require minimal pretreatment are more economically attractive. However, since market demand can fluctuate, conducting a thorough market analysis is important to assess the potential return on investment for valorizing a particular residue. Subsidies or incentives for sustainable practices should also be considered, as there is growing governmental support for financially backing the valorization of certain residues (Cuadrado-Osorio et al., 2022; Mehariya et al., 2020; Ufitikirezi et al., 2024).

5 Benefits

By revalorizing agricultural renewable residues, negative environmental impacts are reduced, new economic opportunities are generated and sustainable social development is promoted, in addition to creating a more resilient and environmentally friendly agro-industrial system (Roy et al., 2023). Biofertilizers, for example, increase nutrient availability while also stimulating beneficial microbial activity in soil (Mahmud et al., 2021). This minimizes the demand for synthetic fertilizers, which are frequently linked to environmental concerns including soil quality deterioration, water pollution and greenhouse gas emissions. Similarly, biostimulants promote plant development and resistance to stresses such as drought and pests, possibly improving agricultural yields and lowering reliance on artificial growth enhancers (Franzoni et al., 2022). Biopesticides are an environmentally benign alternative to traditional chemical pesticides, frequently targeting specific pests while not affecting beneficial insects, people, or the larger ecosystem (Liu et al., 2021). These targeted biopesticides contributes to biodiversity conservation and the promotion of sustainable pest control strategies. Another by-product, biochar, improves soil structure, nutrient retention, and agricultural production while also sequestering carbon, helping to mitigate climate change (Chen et al., 2019a). Bioenergy obtained from agricultural wastes is a renewable energy source that can minimize dependency on fossil fuels and carbon emissions (Jiang et al., 2017). Furthermore, bioenergy production may offer economic advantages by creating jobs in rural regions and giving farmers with an extra revenue stream from biomass sales. Bioadsorbents, which are used in water purification, may successfully remove heavy metals and organic contaminants (Aguilar-Rosero et al., 2022), improving water quality and providing a sustainable alternative to traditional treatment techniques.

6 Limitations

The use of agricultural renewable residue-derived bioproducts has limits that must be addressed to fully fulfill their potential. The first major difficulty to be considered is the environmental and economic challenges of agricultural residue management techniques, particularly the absence of effective and timely prediction tools capable of providing unambiguous information to policymakers and end users through Life Cycle Assessment (LCA), which is a widely used approach for calculating the environmental impact of products and services (Gontard et al., 2018). Moreover, the primary barriers to turning agricultural leftovers into bio-products and materials include high energy consumption, degrading processes, waste chemical complexity and fluctuation, contamination, and societal perception (Uçkun Kıran et al., 2015). Another key challenge is low economic value and inadequate strategies for the storage and handling of agri-based bio products (Ahring et al., 2015). Valuing waste and by-products across the supply chain is another complex issue due to resource heterogeneity, volume fluctuations, and diverse conversion and end-use industries. Despite several well-researched benefits of biofertilizers, higher cost, requirement of large amount and inadequate nutrient load are significant challenges of usage of biofertilizers. Overcoming these problems will require a mix of technological innovation, supporting regulations, and education activities to encourage sustainable behaviors and expand agri-waste-based bioproduct utilization.

7 Future Sustainable Agro-Industrial Production

Sustainable agro-industrial production is a comprehensive approach that aims to balance agricultural productivity with environmental stewardship and social responsibility. This model aims to minimise waste, optimise resource utilisation, and reduce the environmental impact of agricultural processes while preserving its financial sustainability and ensuring social equity (Bose et al., 2023).

Bringing innovation to the market in the context of sustainable agro-industrial production requires robust partnerships between research and development (R&D) entities and the industry. These collaborations are essential for translating scientific discoveries into practical applications that can be scaled up for industrial use. R&D partnerships enable the development of innovative technologies and processes that improve sustainability, efficiency, and production in the agro-industrial sector (Dürr et al., 2024).

Government incentives and bioeconomic policies play a critical role in supporting these innovations. By offering financial incentives such as grants, subsidies, and tax breaks, governments can reduce the risk associated with adopting new technologies. Additionally, bioeconomic policies that promote the sustainable use of biological resources ensure that economic growth is aligned with environmental preservation. These regulations encourage firms to adopt activities that contribute to the bioeconomy, such as producing bioplastics, biofuels, and other bio-based goods, thus creating a circular economy (Dietz et al., 2024).

Green technologies are at the core of sustainable agro-industrial production. These technologies include renewable energy sources, such as solar and wind power, which can be utilised to power agricultural operations and minimise reliance on fossil fuels (Hamed & Alshare, 2022). Additionally, advancements in precision agriculture allow for the efficient use of water, fertilizers, and pesticides, reducing waste and lowering the environmental impact of farming activities (Shafi et al., 2019). Green technologies also encompass innovations in waste management, such as the conversion of agricultural residues into bioenergy or bioproducts, thus closing the loop in production processes and contributing to a circular economy (Ramos et al., 2017). These technologies not only improve productivity but also reduce resource consumption and emissions.

8 Conclusion

The utilization of agricultural residues for sustainable agro-industrial production represents a significant opportunity to enhance environmental stewardship and resource efficiency within the agricultural sector. The increasing demand for bio-based, sustainable products creates substantial market opportunities for developing value-added products derived from agricultural residues. The effective use of these renewable feedstocks addresses waste management challenges by minimizing environmental pollution associated with agricultural waste and providing renewable, eco-friendly alternatives to conventional products. Continued investment in research and innovation in this area is needed for achieving a more sustainable agro-industrial future that benefits both the environment and society. By integrating agricultural waste into the circular economy, it is possible to reduce environmental impacts, decrease reliance on non-renewable resources, and promote sustainable agricultural practices. Therefore, this chapter presents a comprehensive overview of current knowledge in this field and advocates for increased research and innovation to fully unlock the potential of agricultural renewable residues.

References

Abdel-Shafy, H. I., Ibrahim, A. M., Al-Sulaiman, A. M., & Okasha, R. A. (2024). Landfill leachate: Sources, nature, organic composition, and treatment: An environmental overview. *Ain Shams Engineering Journal, 15*(1), Article 102293. https://doi.org/10.1016/j.asej.2023.102293

Aguilar-Rosero, J., Urbina-López, M. E., Rodríguez-González, B. E., León-Villegas, S. X., Luna-Cruz, I. E., & Cárdenas-Chávez, D. L. (2022). Development and characterization of bioadsorbents derived from different agricultural wastes for water reclamation: A review. *Applied Sciences, 12*(5), 2740. https://doi.org/10.3390/app12052740

Ahring, B. K., Biswas, R., Ahamed, A., Teller, P. J., & Uellendahl, H. (2015). Making lignin accessible for anaerobic digestion by wet-explosion pretreatment. *Bioresource Technology, 175*, 182–188. https://doi.org/10.1016/j.biortech.2014.10.082

Ajala, O. A., Akinnawo, S. O., Bamisaye, A., Adedipe, D. T., Adesina, M. O., Okon-Akan, O. A., Adebusuyi, T. A., Ojedokun, A. T., Adegoke, K. A., & Bello, O. S. (2023). Adsorptive removal of antibiotic pollutants from wastewater using biomass/biochar-based adsorbents. *RSC Advances, 13*(7), 4678–4712. https://doi.org/10.1039/D2RA06436G

Akhtar, T., Hashmi, A. S., Tayyab, M., Anjum, A. A., Saeed, S., & Ali, S. (2020). Bioconversion of agricultural waste to butyric acid through solid state fermentation by *Clostridium tyrobutyricum*. *Waste Biomass Valorization, 11*(5), 2067–2073. https://doi.org/10.1007/s12649-018-0475-7

Amato, A., Becci, A., & Beolchini, F. (2020). Citric acid bioproduction: The technological innovation change. *Critical Reviews in Biotechnology, 40*(2), 199–212. https://doi.org/10.1080/07388551.2019.1709799

Amin, A. A., Olama, Z. A., & Ali, S. M. (2023). Characterization of an isolated lactase enzyme produced by *Bacillus licheniformis* ALSZ2 as a potential pharmaceutical supplement for lactose intolerance. *Frontiers in Microbiology, 14*(September), 1–11. https://doi.org/10.3389/fmicb.2023.1180463

Annavelink, B., Chaves, L. G., van Ree, R., Gursel, I. V. (2022). IEA bioenergy task 42 biorefining in a circular economy: global biorefinery status report 2022

Arora, J., Ramawat, K. G., Mérillon, J.-M. (2023). Disposal of agricultural waste and its effects on the environment, production of useful metabolites and energy: Potential and challenges, 3–20. https://doi.org/10.1007/978-981-19-8774-8_1

Astudillo, Á., Rubilar, O., Briceño, G., Diez, M. C., & Schalchli, H. (2023). Advances in agroindustrial waste as a substrate for obtaining eco-friendly microbial products. *Sustainability, 15*(4), 1–15. https://doi.org/10.3390/su15043467

Awogbemi, O., Kallon, D. V. V. (2022). Valorization of agricultural wastes for biofuel applications. *Heliyon, 8*(10). https://doi.org/10.1016/j.heliyon.2022.e11117

Awogbemi, O., & Kallon, D. V. V. (2023). Progress in agricultural waste derived biochar as adsorbents for wastewater treatment. *Applied Surface Science Advances, 18*, Article 100518. https://doi.org/10.1016/j.apsadv.2023.100518

Bagga, D., Chauhan, S., Bhavanam, A., GN, N., Meena, S. S., Mohanty, A. (2024). Recent advancements in fermentation strategies for mass production and formulation of biofertilizers: Towards waste valorization. *Journal of Soil Science and Plant Nutrition*. https://doi.org/10.1007/s42729-024-01947-y

Bankole, D. T., Oluyori, A. P., & AdejumokeA, I. (2023). The removal of pharmaceutical pollutants from aqueous solution by Agro-waste. *Arabian Journal of Chemistry, 16*(5), Article 104699. https://doi.org/10.1016/j.arabjc.2023.104699

Baroi, G. N., Gavala, H. N., Westermann, P., Skiadas, I. V. (2017). Fermentative production of butyric acid from wheat straw: Economic evaluation. *Industrial Crops and Products, 104*(October 2016), 68–80. https://doi.org/10.1016/j.indcrop.2017.04.008

Bhavana, B. K., Mudliar, S. N., Bokade, V. V., Debnath, S. (2024). Effect of furfural, acetic acid and 5-hydroxymethylfurfural on yeast growth and xylitol fermentation using Pichia stipitis NCIM 3497. *Biomass Conversion and Biorefinery, 14*(4), 4909–4923. https://doi.org/10.1007/s13399-022-02758-w

Biogena (2024) Lactrozym. https://biogena.com/en-gb/products/lactrozym_p_51902. Accessed Auguest 9 2024

Bishop, B. D. C., Sanusi, I. A., & Kana, G. E. B. (2024). Enhanced substrate suitability of autoclave-assisted acid pre-treated waste sugarcane molasses: Pre-treatment optimization, sequential nano-based detoxification strategies, and bioproduct production. *Biomass Convers Biorefinery*. https://doi.org/10.1007/s13399-024-06127-7

Blasi, A., Verardi, A., Lopresto, C. G., Siciliano, S., & Sangiorgio, P. (2023). Lignocellulosic agricultural waste valorization to obtain valuable products: an overview. *Recycling, 8*(4), 1–46. https://doi.org/10.3390/recycling8040061

Bose, S., Biswas, S., Pakhira, R., Nandi, S. (2023). Strategies for agricultural sustainability. Chapter 5. In Sinha, G., Arya, D., Pal, S., Indira, G., & Yadav, A. (Eds.), *Strategy for agricultural sustainability* (1st ed.). AkiNik Publications. https://doi.org/10.22271/ed.book.2419

Cardoso Alves, S., Díaz-Ruiz, E., Lisboa, B., Sharma, M., Mussatto, S. I., Thakur, V. K., Kalaskar, D. M., Gupta, V. K., Chandel, A. K. (2023). Microbial meat: A sustainable vegan protein source produced from agri-waste to feed the world. *Food Research International, 166*(November 2022), 112596. https://doi.org/10.1016/j.foodres.2023.112596

Chávez, A. C. F., Meraz, K. Y. C., García, P. C., Escareño, M. P. L., Balagurusamy, N. (2024). Economic, social, and organizational challenges in biorefineries. In Chandel, A. K. (Ed.), Biorefinery and industry 4.0: Empowering sustainability (pp. 293–313). Springer Nature Switzerland.

Chen, Y., Shi, J., Du, Q., Zhang, H., Cui, Y. (2019b). Antibiotic removal by agricultural waste biochars with different forms of iron oxide. *RSC Advances, 9*(25), 14143–14153. https://doi.org/10.1039/C9RA01271K

Chen, W., Meng, J., Han, X., Lan, Y., Zhang, W. (2019a). Past, present, and future of biochar. *Biochar, 1*(1), 75–87. https://doi.org/10.1007/s42773-019-00008-3

Crutchik, D., Franchi, O., Caminos, L., Jeison, D., Belmonte, M., Pedrouso, A., Val del Rio, A., Mosquera-Corral, A., & Campos, J. L. (2020). Polyhydroxyalkanoates (PHAs) production: A feasible economic option for the treatment of sewage sludge in municipalwastewater treatment plants? *Water, 12*(4), 1–12. https://doi.org/10.3390/W12041118

Cuadrado-Osorio, P. D., Ramírez-Mejía, J. M., Mejía-Avellaneda, L. F., Mesa, L., & Bautista, E. J. (2022). Agro-industrial residues for microbial bioproducts: A key booster for bioeconomy. *Bioresource Technology Reports, 20*, Article 101232. https://doi.org/10.1016/j.biteb.2022.101232

Deljouc, A., Arezi, I., Khanahmadi, M. (2018). Scale-up thermostable α-Amylase production in lab-scale fermenter using rice husk as an elicitor by *Bacillus licheniformis*-AZ2 isolated from qinarje hot spring (Ardebil prov. of Iran). *Periodicum biologorum, 120*(1), 11–21. https://doi.org/10.18054/pb.v120i1.6775

Devi, K., Sharma, N., Bakshi, P., Dhiman, S., Kour, J., Singh, A.D., Bhardwaj, T., Ohri, P., Singh, A. P., Bhardwaj, R. (2024). Utilization of agri-waste biomass and its mechanism in enhancing the functioning of biostimulants in plants. In Pal, D. B., Rai, A. K., Siddiqui, S. (Eds.), *Sustainable clean energy production using waste biomass: sustainable energy production and utilization* (pp. 1–21). Springer Nature

Dey, R., Chatterji, B. P. (2023) Sources and methods of manufacturing xanthan by fermentation of various carbon sources. *Biotechnology Progress, 39*(6). https://doi.org/10.1002/btpr.3379

Dietz, T., Bogdanski, A., Boldt, C., Börner, J., von Braun, J., Ní Choncubhair, Ó., Durham, B., Ecuru, J., Lang, C., Li, Y., Lund, M., MacRae, E., Maxon, M., Chavarría Miranda, H., Mizunashi, W., Mungeyi, P., Pittaluga Fonseca, L., Popov, V., Regúnaga, M., Rodríguez, A., Teitelbaum, L., Barcelos Vargas, D. (2024). Bioeconomy globalization: Recent trends and drivers of national programs and policies a report by the International Advisory Council on Global Bioeconomy (IACGB) April 2024

Divyabharathi, R., Kalidasan, B., JS, S. S, Chinnasamy, S. (2024). Recent advances in sustainable agro residue utilisation, barriers and remediation for environmental management: Present insights and future challenges. *Industrial Crops and Products, 216*, 118790. https://doi.org/10.1016/j.indcrop.2024.118790

Du, C., Munir, S., Abad, R., & Lu, D. (2020). Chapter 7—Valorization of organic waste into biofertilizer and its field application. In T. Bhaskar, A. Pandey, E. R. Rene, & D. C. W. Tsang (Eds.), *Waste biorefinery* (pp. 179–198). Elsevier.

Dürr, J., Sili, M., Mac Clay, P., & Sellare, J. (2024). Bioeconomic innovations breeding more sustainable innovations: A value chain perspective from Argentina. *Business Strategy and the Environment*. https://doi.org/10.1002/bse.3845

El Sheikha, A. F., & Ray, R. C. (2023). Bioprocessing of horticultural wastes by solid-state fermentation into value-added/innovative bioproducts: a review. *Food Review International, 39*(6), 3009–3065. https://doi.org/10.1080/87559129.2021.2004161

Enaime, G., & Lübken, M. (2021). Agricultural waste-based biochar for agronomic applications. *Applied Sciences, 11*(19), 8914. https://doi.org/10.3390/app11198914

Ezeorba, T. P. C., Okeke, E. S., Mayel, M. H., Nwuche, C. O., Ezike, T. C. (2024). Recent advances in biotechnological valorization of agro-food wastes (AFW): Optimizing integrated approaches for sustainable biorefinery and circular bioeconomy. *Bioresource Technology Reports, 26*(October 2023), 101823. https://doi.org/10.1016/j.biteb.2024.101823

Flores, L. F., Osorio Gonzalez, C. S., Saini, R., Brar, S. K. (2024). Renewable residues as feedstock for drop-in biofuel production. In Brar, S. K., Osorio Gonzalez, C. S., Saini, R., Soccol, C. R. (Eds.), *The microbiology of the drop-in biofuel production*. Springer.

Franzoni, G., Cocetta, G., Prinsi, B., Ferrante, A., & Espen, L. (2022). Biostimulants on crops: Their impact under abiotic stress conditions. *Horticulturae, 8*(3), 189. https://doi.org/10.3390/horticulturae8030189

Gontard, N., Sonesson, U., Birkved, M., Majone, M., Bolzonella, D., Celli, A., Angellier-Coussy, H., Jang, G.-W., Verniquet, A., Broeze, J., Schaer, B., Batista, A. P., & Sebok, A. (2018). A research challenge vision regarding management of agricultural waste in a circular bio-based economy. *Critical Reviews in Environment Science and Technology, 48*(6), 614–654. https://doi.org/10.1080/10643389.2018.1471957

Groff, M. C., Scaglia, G., Gaido, M., Kassuha, D., Ortiz, O. A., Noriega, S. E. (2022). Kinetic modeling of fungal biomass growth and lactic acid production in *Rhizopus oryzae* fermentation by using grape stalk as a solid substrate. Biocatal Agric Biotechnol, *39*. https://doi.org/10.1016/j.bcab.2021.102255

Groff, M. C., Noriega, S. E., Gil, R. M., Pantano, N., & Scaglia, G. (2024). Dynamic optimization of lactic acid production from grape stalk solid-state fermentation with *Rhizopus oryzae* applying a variable temperature profile. *Fermentation, 10*(2), 101. https://doi.org/10.3390/fermentation10020101

Guo, F., Wu, M., Dai, Z., Zhang, S., Zhang, W., Dong, W., Zhou, J., Jiang, M., Xin, F. (2020) Current advances on biological production of fumaric acid. *Biochemical Engineering Journal, 153*(June 2019), 107397. https://doi.org/10.1016/j.bej.2019.107397

Guo, J., Zhang, Y., Fang, J., Ma, Z., Li, C., Yan, M., Qiao, N., Liu, Y., & Bian, M. (2024). Reduction and reuse of forestry and agricultural bio-waste through innovative green utilization approaches: A review. *Forests, 15*(8), 1372. https://doi.org/10.3390/f15081372

Gutierrez-Rangel, P. C., Mayolo-Deloisa, K., Torres-Acosta, M. A. (2024) Decisional tool development and application for techno-economic analysis of fungal laccase production. *Bioresource Technology*, *402*. https://doi.org/10.1016/j.biortech.2024.130781

Hamed, T. A., Alshare, A. (2022). Environmental impact of solar and wind energy-a review. *Journal of Sustainable Development of Energy, Water and Environment Systems*, *10*(2), 1–23. https://doi.org/10.13044/j.sdewes.d9.0387

He, X., Wang, Y., Zhang, Y., Wang, C., Yu, J., Ohtake, H., & Zhang, T. (2023). The potential for livestock manure valorization and phosphorus recovery by hydrothermal technology—A critical review. *Materials Science for Energy Technologies, 6*, 94–104. https://doi.org/10.1016/j.mset.2022.11.008

Jiang, L., Fu, H., Yang, H. K., Xu, W., Wang, J., & Yang, S. T. (2018). Butyric acid: Applications and recent advances in its bioproduction. *Biotechnology Advances, 36*(8), 2101–2117. https://doi.org/10.1016/j.biotechadv.2018.09.005

Jiang, L., Han, G., Lan, Y., Liu, S., Gao, J., Yang, X., Meng, J., & Chen, W. (2017). Corn cob biochar increases soil culturable bacterial abundance without enhancing their capacities in utilizing carbon sources in biolog eco-plates. *Journal of Integrative Agriculture, 16*(3), 713–724. https://doi.org/10.1016/S2095-3119(16)61338-2

Juela, D. M. (2022). Promising adsorptive materials derived from agricultural and industrial wastes for antibiotic removal: A comprehensive review. *Separation and Purification Technology, 284*, Article 120286. https://doi.org/10.1016/j.seppur.2021.120286

Kaur, M., Singh, A. K., & Singh, A. (2023). Bioconversion of food industry waste to value added products: Current technological trends and prospects. *Food Bioscience, 55*(June), Article 102935. https://doi.org/10.1016/j.fbio.2023.102935

Khunnonkwao, P., Thitiprasert, S., Jaiaue, P., Khumrangsee, K., Cheirsilp, B., & Thongchul, N. (2024). The outlooks and key challenges in renewable biomass feedstock utilization for value-added platform chemical via bioprocesses. *Heliyon, 10*(10), Article e30830. https://doi.org/10.1016/j.heliyon.2024.e30830

Kumar, J. A., Sathish, S., Prabu, D., Renita, A. A., Saravanan, A., Deivayanai, V.C., Anish, M., Jayaprabakar, J., Baigenzhenov, O., Hosseini-Bandegharaei, A. (2023). Agricultural waste biomass for sustainable bioenergy production: Feedstock, characterization and pre-treatment methodologies. *Chemosphere, 331*(November 2022), 138680. https://doi.org/10.1016/j.chemosphere.2023.138680

Kumar, R., Singh, A., Srivastava, A. (2024). Production of biopesticides from agricultural waste as an alternative to chemical pesticides. In Saha, S. P., Mazumdar, D., Roy, S., Mathur, P. (Eds.), *Agro-waste to microbe assisted value added product: Challenges and future prospects: Recent developments in agro-waste valorization research* (pp. 365–379). Springer Nature Switzerland.

Kumar Sarangi, P., Subudhi, S., Bhatia, L., Saha, K., Mudgil, D., Prasad Shadangi, K., Srivastava, R. K., Pattnaik, B., & Arya, R. K. (2023). Utilization of agricultural waste biomass and recycling toward circular bioeconomy. *Environmental Science and Pollution Research, 30*(4), 8526–8539. https://doi.org/10.1007/s11356-022-20669-1

Ladapo, H. L., & Aminu, F. O. (2017). Agriculture and eutrophication of freshwaters: A review of control measures. *Journal of Research in Forestry, Wildlife and Environment, 9*(1), 67–74.

Larsen, C. S. (2023). The past 12,000 years of behavior, adaptation, population, and evolution shaped who we are today. *Proceedings of the National Academy of Sciences, 120*(4), 1–4. https://doi.org/10.1073/pnas.2209613120

Lee Biosolutions, I. (2024). Alpha-amylase. https://www.leebio.com/product/112/a-amylase-human-pancreas-120-15. Accessed August 7, 2024.

Lee, K. T., Chen, W. H., Sarles, P., Park, Y. K., & Ok, Y. S. (2022). Recover energy and materials from agricultural waste via thermalchemical conversion. *One Earth, 5*(11), 1200–1204. https://doi.org/10.1016/j.oneear.2022.10.010

Li, R., Zhang, Y., Chu, W., Chen, Z., & Wang, J. (2018). Adsorptive removal of antibiotics from water using peanut shells from agricultural waste. *RSC Advances, 8*(24), 13546–13555. https://doi.org/10.1039/C7RA11796E

Liu, X., Cao, A., Yan, D., Ouyang, C., Wang, Q., & Li, Y. (2021). Overview of mechanisms and uses of biopesticides. *International Journal of Pest Management, 67*(1), 65–72. https://doi.org/10.1080/09670874.2019.1664789

Lu, J., Li, J., Gao, H., Zhou, D., Xu, H., Cong, Y., Zhang, W., Xin, F., & Jiang, M. (2021). Recent progress on bio-succinic acid production from lignocellulosic biomass. *World Journal of Microbiology and Biotechnology, 37*(1), 1–8. https://doi.org/10.1007/s11274-020-02979-z

Lu, J., Lv, Y., Qian, X., Jiang, Y., Wu, M., Zhang, W., Zhou, J., Dong, W., Xin, F., & Jiang, M. (2020). Current advances in organic acid production from organic wastes by using microbial co-cultivation systems. *Biofuels, Bioproducts and Biorefining, 14*(2), 481–492. https://doi.org/10.1002/bbb.2075

Mahmud, A. A., Upadhyay, S. K., Srivastava, A. K., & Bhojiya, A. A. (2021). Biofertilizers: A nexus between soil fertility and crop productivity under abiotic stress. *Current Research in Environmental Sustainability, 3*, Article 100063. https://doi.org/10.1016/j.crsust.2021.100063

Mahuri, M., Paul, M., & Thatoi, H. (2023). A review of microbial laccase production and activity toward different biotechnological applications. *Systems Microbiology and Biomanufacturing, 3*(4), 533–551. https://doi.org/10.1007/s43393-023-00163-6

Malagón, M., López, S., & Martínez, Á. (2017). Síntesis de bioplásticos a partir de microorganismos. *América Semilleros Form Investig, 3*(1), 135.

Malone, A. B., Greer, J., Thomas, L., Moore, L. (2023) Xanthan gum isolation alternative via ultrafiltration. Chancellor's Honors Program Proj, 1–31.

Man, K. Y., Chow, K. L., Man, Y. B., Mo, W. Y., & Wong, M. H. (2021). Use of biochar as feed supplements for animal farming. *Critical Reviews in Environment Science and Technology, 51*(2), 187–217. https://doi.org/10.1080/10643389.2020.1721980

Mata, T. M., Correia, D., Pinto, A., Andrade, S., Trovisco, I., Matos, E., Martins, A. A., & Caetano, N. S. (2017). Fish oil acidity reduction by enzymatic esterification. *Energy Procedia, 136*(October), 474–480. https://doi.org/10.1016/j.egypro.2017.10.306

Matisic, M., Dugan, I., & Bogunovic, I. (2024). Challenges in sustainable agriculture—the role of organic amendments. *Agriculture, 14*(4), 643. https://doi.org/10.3390/agriculture14040643

Mattedi, A., Sabbi, E., Farda, B., Djebaili, R., Mitra, D., Ercole, C., Cacchio, P., Del Gallo, M., & Pellegrini, M. (2023). Solid-State fermentation: Applications and future perspectives for biostimulant and biopesticides production. *Microorganisms, 11*(6), 1408. https://doi.org/10.3390/microorganisms11061408

Mazzoli, R. (2021). Current progress in production of building-block organic acids by consolidated bioprocessing of lignocellulose. *Fermentation, 7*(4). https://doi.org/10.3390/fermentation7040248

McAdam, B., Fournet, M. B., McDonald, P., Mojicevic, M. (2020). Production of polyhydroxybutyrate (PHB) and factors impacting its chemical and mechanical characteristics. *Polymers, 12*(2908). https://doi.org/10.3390/polym12122908

Mehariya, S., Iovine, A., Casella, P., Musmarra, D., Chianese, S., Marino, T., Figoli, A., Sharma, N., Molino, A. (2020). Bio-based and agriculture resources for production of bioproducts. In Current trends and future developments on (Bio-) Membranes (pp. 263–282). Elsevier.

Merrylin, J., Kannah, R. Y., Banu, J. R., Yeom, I. T. (2020). Production of organic acids and enzymes/biocatalysts from food waste. INC

Morales-Vera, R., Crawford, J., Dou, C., Bura, R., & Gustafson, R. (2020). Techno-economic analysis of producing glacial acetic acid from poplar biomass via bioconversion. *Molecules, 25*(18), 1–16. https://doi.org/10.3390/molecules25184328

MordorIntelligence (2024e) Bio based succinic acid. source: https://www.mordorintelligence.com/industry-reports/global-bio-based-succinic-acid-market-industry. Accessed August 6, 2024

MordorIntelligence (2024d) Fumaric acid market. source: https://www.mordorintelligence.com/industry-reports/fumaric-acid-market. Accessed August 6, 2024

MordorIntelligence (2024g) Biodegradable packaging market. source: https://www.mordorintelligence.com/industry-reports/gobal-biodegradable-packaging-solutions-market-industry. Accessed August 8, 2024

MordorIntelligence (2024i) Polyphenol market. source: https://www.mordorintelligence.com/industry-reports/polyphenol-market. Accessed August 8. 2024

MordorIntelligence (2024h) Xanthan gum market. source: https://www.mordorintelligence.com/industry-reports/xanthan-gum-market. Accessed August 8, 2024

MordorIntelligence (2024a) Bio acetic acid companies. https://www.mordorintelligence.com/industry-reports/bio-acetic-acid-market/companies. Accessed August 5, 2024

MordorIntelligence (2024f) Citric acid market. https://www.mordorintelligence.com/industry-reports/citric-acid-market. Accessed August 6, 2024

MordorIntelligence (2024b) Propionic acid market. source: https://www.mordorintelligence.com/industry-reports/propionic-acid-market. Accessed August 6, 2024

MordorIntelligence (2024c) Butyric acid market. source: https://www.mordorintelligence.com/industry-reports/butyric-acid-market. Accessed August 6, 2024

MordorIntelligence Industrial enzymes market. https://www.mordorintelligence.com/industry-reports/industrial-enzymes-market

Mujtaba, M., Fernandes Fraceto, L., Fazeli, M., Mukherjee, S., Savassa, S. M., Araujo de Medeiros, G., do Espírito Santo Pereira, A., Mancini, S. D., Lipponen, J., Vilaplana, F. (2023). Lignocellulosic biomass from agricultural waste to the circular economy: a review with focus on biofuels, biocomposites and bioplastics. *Journal of Cleaner Production, 402*(March), 136815. https://doi.org/10.1016/j.jclepro.2023.136815

Muneer, F., Nadeem, H., Arif, A., & Zaheer, W. (2021). Bioplastics from biopolymers: An eco-friendly and sustainable solution of plastic pollution. *Polymer Science, Series C, 63*(1), 47–63. https://doi.org/10.1134/S1811238221010057

Murad, H. A., Abo-Elkhair, A. G., & Azzaz, H. H. (2019). Production of xanthan gum from nontraditional substrates with perspective of the unique properties and wide industrial applications. *JSMC Microbiol, 1*(6), 1–6.

Nordin, A. H., Norfarhana, A. S., Noor, S. F. M., Paiman, S. H., Nordin, M. L., Husna, S. M. N., Ilyas, R. A., Ngadi, N., Bakar, A. A., Ahmad, Z., Azami, M. S., Nawawi, W. I., & Nabgan, W. (2023). Recent advances in using adsorbent derived from agricultural waste for antibiotics and non-steroidal anti-inflammatory wastewater treatment: A review. *Separations, 10*(5), 300. https://doi.org/10.3390/separations10050300

Piwowarek, K., Lipińska, E., Hać-Szymańczuk, E., Pobiega, K. (2021). Propionic acid production from apple pomace in bioreactor using Propionibacterium freudenreichii: an economic analysis of the process. *3 Biotech, 11*(2), 1–15. https://doi.org/10.1007/s13205-020-02582-x

Priya, A. K., Alagumalai, A., Balaji, D., Song, H. (2023). Bio-based agricultural products: a sustainable alternative to agrochemicals for promoting a circular economy. RSC Sustainablity, *1*(4), 746–762. https://doi.org/10.1039/D3SU00075C

Puglia, D., Pezzolla, D., Gigliotti, G., Torre, L., Bartucca, M. L., & Del Buono, D. (2021). The opportunity of valorizing agricultural waste, through its conversion into biostimulants, biofertilizers, and biopolymers. *Sustainability, 13*(5), 2710. https://doi.org/10.3390/su13052710

Ramos, J. L., García-Lorente, F., Valdivia, M., & Duque, E. (2017). Green biofuels and bioproducts: Bases for sustainability analysis. *Microbial Biotechnology, 10*(5), 1111–1113. https://doi.org/10.1111/1751-7915.12768

Roudneshin, M., & Sosa, A. (2024). Optimising agricultural waste supply chains for sustainable bioenergy production: A comprehensive literature review. *Energies, 17*(11), 2542. https://doi.org/10.3390/en17112542

Roy, P., Mohanty, A. K., Dick, P., & Misra, M. (2023). A review on the challenges and choices for food waste valorization: Environmental and economic impacts. *ACS Environmental Au, 3*(2), 58–75. https://doi.org/10.1021/acsenvironau.2c00050

Sadh, P. K., Chawla, P., Kumar, S., Das, A., Kumar, R., Bains, A., Sridhar, K., Duhan, J. S., Sharma, M. (2023). Recovery of agricultural waste biomass: A path for circular bioeconomy. *Science of the Total Environment, 870*(December 2022). https://doi.org/10.1016/j.scitotenv.2023.161904

Setyaningsih, W., Sanjaya, R. S. (2022). The impact of agricultural waste on river water quality of kreo watershed in Semarang city. *IOP Conference Series: Earth and Environmental Science, 1041*(1). https://doi.org/10.1088/1755-1315/1041/1/012083

Shaaban, M., & Abid, M. (2021). Chapter 9—Biochar as a sorbent for organic and inorganic pollutants. In A. Núñez-Delgado (Ed.), *Sorbents materials for controlling environmental pollution* (pp. 189–208). Elsevier.

Shaaban, M., Van Zwieten, L., Bashir, S., Younas, A., Núñez-Delgado, A., Chhajro, M. A., Kubar, K. A., Ali, U., Rana, M. S., Mehmood, M. A., & Hu, R. (2018). A concise review of biochar application to agricultural soils to improve soil conditions and fight pollution. *Journal of Environmental Management, 228*, 429–440. https://doi.org/10.1016/j.jenvman.2018.09.006

Shafi, U., Mumtaz, R., García-Nieto, J., Hassan, S. A., Zaidi, S. A. R., & Iqbal, N. (2019). Precision agriculture techniques and practices: From considerations to applications. *Sensors, 19*(17), 1–25. https://doi.org/10.3390/s19173796

Shen, X., Liao, L., Zhang, G., Zhou, J., Li, J., & Du, G. (2023). Characterization of putative mannoprotein in Kluyveromyces lactis for lactase production. *Synthetic and Systems Biotechnology, 8*(1), 168–175. https://doi.org/10.1016/j.synbio.2023.01.001

Sirohi, R., Vivekanand, V., Pandey, A. K., Tarafdar, A., Awasthi, M. K., Shakya, A., Kim, S. H., Sim, S. J., Tuan, H. A., & Pandey, A. (2023). Emerging trends in role and significance of biochar in gaseous biofuels production. *Environmental Technology and Innovation, 30*, Article 103100. https://doi.org/10.1016/j.eti.2023.103100

Thadiyan, V., Kaur, M., Gupta, R. (2024). Agricultural waste for biofertilizers and soil amendments. In Arora, J., Joshi, A., Ray, R. C. (Eds.) *Transforming agriculture residues for sustainable development: from waste to wealth* (pp. 183–204). Springer Nature Switzerland.

Uçkun Kıran, E., Trzcinski, A. P., & Liu, Y. (2015). Platform chemical production from food wastes using a biorefinery concept: Platform chemical production from food waste. *Journal of Chemical Technology and Biotechnology, 90*(8), 1364–1379. https://doi.org/10.1002/jctb.4551

Ufitikirezi, J. D. D. M., Filip, M., Ghorbani, M., Zoubek, T., Olšan, P., Bumbálek, R., Strob, M., Bartoš, P., Umurungi, S. N., Murindangabo, Y. T., Heřmánek, A., Tupý, O., Havelka, Z., Stehlík, R., Černý, P., & Smutný, L. (2024). Agricultural waste valorization: exploring environmentally friendly approaches to bioenergy conversion. *Sustainability, 16*(9), 3617. https://doi.org/10.3390/su16093617

Vashisht, A., Thakur, K., Kauldhar, B. S., Kumar, V., & Yadav, S. K. (2019). Waste valorization: identification of an ethanol tolerant bacterium *Acetobacter pasteurianus* SKYAA25 for acetic acid production from apple pomace. *Science of the Total Environment, 690*, 956–964. https://doi.org/10.1016/j.scitotenv.2019.07.070

Voutilainen, E., Pihlajaniemi, V., Parviainen, T. (2021). Economic comparison of food protein production with single-cell organisms from lignocellulose side-streams. *Bioresource Technology Reports, 14*(December 2020). https://doi.org/10.1016/j.biteb.2021.100683

Wang, J., Cui, Z., Li, Y., Cao, L., & Lu, Z. (2020). Techno-economic analysis and environmental impact assessment of citric acid production through different recovery methods. *Journal of Cleaner Production, 249*, Article 119315. https://doi.org/10.1016/j.jclepro.2019.119315

Zhang, X., Yin, J., Ma, Y., Peng, Y., Fenton, O., Wang, W., Zhang, W., & Chen, Q. (2024). Unlocking the potential of biostimulants derived from organic waste and by-product sources: Improving plant growth and tolerance to abiotic stresses in agriculture. *Environmental Technology and Innovation, 34*, Article 103571. https://doi.org/10.1016/j.eti.2024.103571

Zhao, H., Avena-Bustillos, R. J., Wang, S. C. (2022). Extraction, Purification and in vitro antioxidant activity evaluation of phenolic compounds in California olive pomace. *Foods, 11*(2). https://doi.org/10.3390/foods11020174

Zielińska, M., & Bułkowska, K. (2024). Agricultural wastes and their by-products for the energy market. *Energies, 17*(9), 2099. https://doi.org/10.3390/en17092099

Zuzunaga-Rosas, J., Boscaiu, M., & Vicente, O. (2024). Agroindustrial by-products as a source of biostimulants enhancing responses to abiotic stress of horticultural crops. *International Journal of Molecular Sciences, 25*(6), 3525. https://doi.org/10.3390/ijms25063525

Alternative Sources of Fodder: An Indian Perspective

Aakash Kumar, Taniya Bohra, and Sanjay Singh

Abstract The growing interest in alternative sources of fodder reflects an important need to address the challenges of sustainable agriculture. This chapter explores the potential of unconventional fodder resources, including crop residues, seaweed, Azolla, and non-traditional fodder crops. Emphasis is placed on the innovative approaches such as agro-industrial waste utilization and hydroponics, which not only contribute to fodder production but also align with the principles of renewable energy and sustainability. Despite the promise of these alternatives, their adoption is constrained by factors such as supply–demand imbalances, high initial costs, and limited access to knowledge and technology. Also, the chapter highlights Indian government programs like the Accelerated Fodder Development Programme (AFDP), which aims to equip farmers with the necessary tools and knowledge to integrate these alternative sources into their farming practices. By overcoming these challenges, the fodder industry has the potential to enhance environmental sustainability and boost agricultural income.

Keywords Alternative fodder resources · Sustainable agriculture · Agro-industrial waste utilization · Hydroponics · Accelerated fodder development programme (AFDP)

A. Kumar
Forest Research Institute (Deemed to be) University, Dehradun, Uttarakhand, India

T. Bohra
School of Agriculture, Uttaranchal University, Dehradun, Uttarakhand, India

S. Singh (✉)
Centre of Excellence on Sustainable Land Management (CoE-SLM), Indian Council of Forestry Research and Education (ICFRE), Dehradun, Uttarakhand, India
e-mail: sanjaysingh83@gmail.com

M. Shaaban (ed.), *Soils and Sustainable Agriculture*, Frontier Studies in Soil Science,
https://doi.org/10.1007/978-3-031-91114-9_35

1 Introduction

The food used for feeding livestock, including cattle, sheep and other domesticated animals is known as animals feed or fodder. It includes a wide range of plant materials such as hay, silage, straw, and various grains, which are specifically cultivated and harvested in order to fulfill the dietary requirements of these animals. The lone objective of the fodder is to ensure that animals receive the essential nutrients which help in growth, reproduction and also helps in production of important outputs like milk, wool, meat or other animal based products. This animal feed can be occupied from numerous sources which includes cultivated crops, pasture grasses, and crop residues (Mudzengi et al., 2020), each of which is having its significant role in the feeding system.

According to the heterogeneity in the animal species and their unlike feeding pattern, it is necessary to narrow down their feeds in order to fulfill nutritional requirements (da Silva et al., 2015), consequently enhancing their health and productivity (Baris, 2023; Bonneau & Laarveld, 1999). For example, hay and silage which are rich in fibre are essential for grass eating animals (ruminants) like cattle and sheep, which helps in proper digestion and nutrients absorption. Similarly, grains provide energy and often used as supplement in animal diet (Loy & Lundy, 2019), requiring higher calorie intake for growth and lactation. Other than providing proteins, carbohydrates, vitamins, fats and minerals, fodder plays an important role in maintaining the well-being of the livestock. High quality fodder improves the animals' immune systems and also reduces the incidence of diseases (Tona, 2018), and enhances their resilience against various environmental stressors. Right balance of fodder in diets of animals not only improves the physical health but also helps livestock farmers to achieve economic sustainability. Proper management of fodder, reduces feeding costs, improves the feed conversion ratios and also increases profits for farmers (Devendra & Leng, 2011; Makkar, 2016).

The selection and management of fodder have significant implications for environmental sustainability, which protect the ability of pasturelands to sequester carbon dioxide, ensure vegetative cover and prevent soil erosion (Tulu et al., 2023). Type of fodder grown, methods of cultivation and efficiency of its use can probably have its significant impact on soil health, usage of water and also greenhouse gas emissions associated with livestock farming. Therefore, the choice of right fodder is not only a matter of nutrition of animals but also an inseparable entity of sustainable agriculture practices.

2 Fodder Consumption in India

With the livestock sector being integral to India's rural economy and food security, the demand for fodder significantly impacts how land is utilized and managed (Datta et al., 2024). This relationship between fodder consumption and land use has far-reaching implications for sustainable agriculture and environmental conservation in India.

India is home to one of the largest livestock populations in the world (Singh et al., 2012), including cattle, buffalo, sheep, and goats. This substantial livestock population requires a continuous and large supply of fodder. In India, cultivated fodders take up merely 4% of the total cultivable land (Singh et al., 2022). Traditional fodder sources in India include cultivated fodder crops like sorghum, maize, and alfalfa, as well as crop residues from cereals and pulses. Also, grazing on natural pastures and common lands accounts for a significant portion of fodder consumption, particularly in rural areas. The demand for fodder has profound effects on land use and land change practices in India (Semwal et al., 2004; Singh et al., 2004). One of the primary impacts is the conversion of agricultural land to fodder cultivation. Farmers often devote significant portions of their arable land to grow fodder crops to meet the dietary needs of their livestock. This practice may lead to the displacement of food crops, altering the overall agricultural output and potentially affecting food security.

In the state of Punjab, a major agricultural hub in India, the intensive dairy farming industry has led to a substantial portion of agricultural land being dedicated to fodder crops like berseem (*Trifolium alexandrinum*) and maize (*Zea mays*). This shift is induced by the need to maintain high milk production levels, which requires a consistent supply of fodder. While this practice helps the dairy industry, it also emphasises the trade-offs between fodder and food crop production.

Fodder consumption in India has its significant impact on common lands and pastures. In many regions, especially in arid and semi-arid areas, livestock grazing on communal lands is a traditional practice. However, the increasing livestock population has led to overgrazing, resulting in land degradation and desertification. For example, in Rajasthan, overgrazing of common lands has aggravated soil erosion and reduced the productivity of these lands, challenging the sustainability of traditional grazing practices. The pressure on land resources due to fodder demand also drives land change practices such as deforestation and conversion of forests to grazing lands (Mekuria et al., 2018; Weindl et al., 2017). In the north-eastern states of India, shifting cultivation and encroachment into forest areas for grazing have contributed to deforestation and habitat loss. This threatens the biodiversity and disrupts the ecological balance of these regions.

Several initiatives and sustainable practices are being promoted to mitigate the challenges (Dinesha et al., 2023; Tomar et al., 2021). Agroforestry, which integrates trees and shrubs into agricultural systems, is one of the practices that can provide multiple benefits. For example, in Karnataka, the adoption of agroforestry systems combining fodder crops with trees like Leucaena and Gliricidia has improved fodder availability (Shahapurmath et al., 2020), soil fertility and reduced soil erosion. The government of India has also implemented schemes to support fodder development and management. Accelerated Fodder Development Programme (AFDP) focuses on increasing the availability of quality fodder through the promotion of high-yielding fodder crop varieties and efficient fodder management practices. These initiatives are essential to balancing the needs of the livestock sector with sustainable land use.

3 Potential for Use of Alternate Source of Fodder in India

The potential alternative sources of fodder may play a significant role in transforming agricultural practices, reducing land use changes, and mitigating land degradation. As India face dual challenges of sustaining its vast livestock population and managing its finite land resources, exploration and adoption (Paroda, 2018) of alternative fodder sources becomes important. These alternatives have nutritional benefits for livestock as well as contribute to sustainable land management practices, preserving the environment and enhancing agricultural productivity.

One of the significant alternative fodder source is crop residues, which include the stalks, leaves, and husks left after harvesting crops such as rice, wheat, and maize. These residues are often burned or left to decompose in the fields which contributes to air pollution and loss of valuable organic matter. By salvaging crop residues as fodder, farmers can reduce the need for dedicated fodder crops which further help in freeing up arable land for food production. This practice can possibly enhance soil health through the return of organic matter, reducing the need for chemical fertilizers and mitigating soil erosion.

Another significant alternative is the use of agro-industrial by-products, such as brewers' spent grain (BSG), distillers' dried grains with solubles (DDGS), and oilseed cakes. These by-products are rich in nutrients and are currently underutilized in India. Incorporating them into livestock diets can significantly reduce the demand for conventional fodder crops. For example, BSG, a by-product of beer production, is high in protein and fibre, making it an excellent feed supplement for cattle. Utilisation these by-products supports waste reduction and promotes a circular economy, where industrial waste is efficiently repurposed for agricultural use.

Seaweed and algae also hold great potential as alternative fodder sources. These marine and aquatic plants are rich in essential nutrients, including proteins, vitamins, and minerals. They can be cultivated in coastal and freshwater environments, reducing the pressure on terrestrial land resources. The integration of seaweed in livestock diets has been shown to improve animal health and reduce methane emissions, contributing to more sustainable livestock farming. The cultivation of algae and seaweed can also provide additional income opportunities for coastal communities.

By reducing the need for dedicated fodder crops, farmers can maintain more land under food production, thereby enhancing food security. This shift can also prevent the conversion of forests and grasslands into agricultural fields, protecting biodiversity and preserving ecosystems. Also, using crop residues and agro-industrial by-products helps retain organic matter in the soil, improving soil structure and fertility, and reducing the risk of erosion and degradation.

The adoption of alternative fodder sources aligns with sustainable land management practices. For instance, integrating crop residue management into existing farming systems promotes nutrient cycling and reduces the reliance on external inputs. This practice can be complemented by agroforestry, where trees and shrubs are incorporated into agricultural landscapes. Trees such as Leucaena and Gliricidia can provide high-quality fodder while enhancing soil fertility and offering additional benefits such as carbon sequestration and habitat for biodiversity.

4 Potential Role of Alternate Fodder to Fill the Gap Between Demand and Supply

The potential role of alternate fodder in bridging the gap between demand and supply in the livestock sector is substantial. As India's livestock population continues to grow, so does the demand for nutritious and sustainable feed. Traditional fodder sources, such as green fodder and crop residues, are often insufficient to meet this demand (Akram & Fırıncıoğlu, 2019), leading to a critical imbalance that impacts both livestock productivity and agricultural sustainability. Alternate fodder sources, including crop residues, agro-industrial by-products, seaweed, algae, and unconventional plants, can play a crucial role in addressing this gap.

Crop residues, such as straw from rice, wheat, and maize, are abundant and underutilized resources in India (Kumar et al., 2018, 2023). These residues, often discarded or burned, can be effectively repurposed as fodder. For instance, rice straw, which is typically left to decompose or set ablaze, can be treated and used as a high-fiber feed for ruminants. By utilizing these residues, farmers can significantly supplement the traditional fodder supply, reducing dependency on cultivated fodder crops and improving feed availability during lean periods.

Agro-industrial by-products, such as Brewers' Spent Grain (BSG), Distillers' Dried Grains with Solubles (DDGS), and oilseed cakes, are rich in nutrients and readily available from various industries. BSG, a by-product of beer production, is high in protein and fiber, making it an excellent feed for cattle. DDGS, a by-product of ethanol production, offers a balanced mix of protein, fat, and fiber, suitable for various livestock species. Incorporating these by-products into livestock diets can effectively bridge the fodder gap, ensuring a steady supply of nutritious feed throughout the year. This practice also supports waste reduction and promotes a circular economy, where industrial by-products are repurposed for agricultural use.

Seaweed and algae represent innovative and sustainable fodder alternatives. These marine and aquatic plants are rich in essential nutrients, including proteins, vitamins, and minerals. Seaweed, for instance, can be cultivated in coastal areas without competing for arable land. Studies have shown that incorporating seaweed into livestock diets can improve animal health and reduce methane emissions, a significant contributor to climate change. Algae, on the other hand, can be grown in freshwater and wastewater environments, providing a high-protein feed source that does not rely on traditional agricultural inputs. These alternatives can significantly supplement the conventional fodder supply, particularly in regions with limited land and water resources.

Unconventional plants, such as moringa, Leucaena, and Gliricidia, also hold promise as alternate fodder sources. Moringa, known for its high nutritional value, can be cultivated in arid and semi-arid regions, providing a reliable feed source during dry seasons. Leucaena and Gliricidia are fast-growing trees that can be integrated into agroforestry systems, offering high-protein fodder while enhancing soil fertility and providing additional ecological benefits. These plants can help diversify the fodder supply, making it more resilient to climatic variations and reducing the pressure on traditional fodder resources.

Incorporating alternate fodder sources into livestock diets can effectively bridge the gap between demand and supply, ensuring a consistent and nutritious feed supply throughout the year. This approach also promotes sustainable land management by reducing the need for dedicated fodder crops, freeing up arable land for food production, and mitigating land degradation. Additionally, utilizing crop residues and agro-industrial by-products helps retain organic matter in the soil, improving soil health and reducing erosion risks.

Government initiatives and policies are crucial in promoting the adoption of alternate fodder sources. Programs that support research and development, provide incentives for sustainable practices, and facilitate knowledge transfer to farmers can drive the widespread use of these alternatives. For example, expanding the Accelerated Fodder Development Programme (AFDP) to include training and support for using crop residues and agro-industrial by-products as fodder can enhance feed availability and support sustainable agricultural practices.

The potential role of alternate fodder in filling the gap between demand and supply is significant. By repurposing crop residues, utilizing agro-industrial by-products, and incorporating seaweed, algae, and unconventional plants into livestock diets, India can ensure a steady supply of nutritious feed, promote sustainable land management, and support the resilience of its agricultural systems.

5 Current Achievement in the Field of Alternate Fodder

Various innovative approaches and technologies are being developed to address the growing demand for sustainable livestock feed. These advancements not only enhance the nutritional profile of animal diets but also contribute to environmental sustainability and economic efficiency. One of the significant achievements is the effective utilization of crop residues as fodder. Techniques such as ammoniation and urea treatment have been developed to enhance the digestibility and nutritional value of residues like rice straw and wheat straw. These methods break down the lignocellulosic structure of the residues, making them more palatable and nutrient-rich for livestock. In regions like Punjab and Haryana, these treatments have been widely adopted, helping to reduce fodder shortages and improve livestock productivity.

The use of agro-industrial by-products as animal feed has also seen considerable advancements. Brewers' spent grain (BSG), a by-product of the brewing industry, is now being processed into high-protein animal feed. Innovations in drying and pelletizing technologies have made it easier to store and transport BSG, increasing its use as a fodder supplement. Similarly, distillers' dried grains with solubles (DDGS), a by-product of the ethanol industry, have been optimized for use in livestock diets. These by-products are now an integral part of feed formulations, providing essential nutrients and reducing feed costs.

Significant progress has been made in the cultivation and use of seaweed and algae as alternative fodder. Research has shown that certain species of seaweed, such as Asparagopsis taxiformis, can reduce methane emissions in ruminants (Tulu et al., 2023) by up to 80%. This has led to pilot projects and commercial ventures exploring large-scale seaweed farming for livestock feed. Algae, particularly microalgae like spirulina and chlorella, are being cultivated for their high protein content and rich nutrient profile. Innovations in photobioreactor technology have improved the efficiency of algae production, making it a viable feed alternative. The cultivation of unconventional fodder crops like moringa, Leucaena, and Gliricidia has also gained momentum. These crops are drought-resistant and can be grown in marginal lands, providing a reliable source of fodder in regions prone to water scarcity. Moringa, for example, is rich in protein, vitamins, and minerals, making it an excellent supplement for livestock. Research institutions and agricultural universities have developed high-yielding varieties and agronomic practices to maximize the productivity and nutritional value of these crops.

Agroforestry and silvopastoral systems, which integrate trees and shrubs into agricultural landscapes, have been successfully implemented in various regions. These systems provide multiple benefits, including fodder, fuelwood, and improved soil health. For instance, the integration of Leucaena in pasturelands has enhanced fodder availability while improving soil fertility through nitrogen fixation. These practices have been promoted by various government and non-governmental organizations, contributing to sustainable land management and increased fodder production.

Government initiatives and institutional support have played a crucial role in advancing the field of alternate fodder. Programs like the Accelerated Fodder Development Programme (AFDP) in India have been instrumental in promoting research, development, and dissemination of technologies related to alternative fodder. Collaborations between research institutions, universities, and the private sector have led to the development of innovative feed solutions and their widespread adoption.

The field of alternate fodder has seen remarkable achievements, driven by advancements in technology, research, and sustainable practices. The utilization of crop residues, agro-industrial by-products, seaweed, algae, and unconventional fodder crops has significantly contributed to closing the gap between demand and supply of livestock feed. These innovations not only enhance animal nutrition and productivity but also promote environmental sustainability and economic efficiency. Continued support from government and research institutions will be essential in scaling these solutions and ensuring the long-term sustainability of the livestock sector.

6 Alternative Sources of Fodder

1. **Crop Waste**:

The utilization of crop waste as fodder is a sustainable and economically viable alternative to traditional feed sources. Crop residues, often considered agricultural waste, can be used to feed livestock. Crop waste encompasses a variety of residues left after the harvest of crops. Key types include straw and stover, hulls and husks, and pulp and peels. Residues from cereals such as rice, wheat, barley, and maize are commonly referred to as straw and stover. These materials are rich in cellulose and hemicellulose which makes them suitable for ruminant digestion (Zhang et al., 2018). Hulls and husks, by-products of processing crops like rice, oats, and sunflower seeds, are often rich in fiber. Pulp and peels, residues from fruits and vegetables such as citrus pulp, potato peels, and sugar beet pulp, can be valuable fodder due to their nutritional content. The nutritional value of crop residues varies, but they generally provide essential fibers, carbohydrates, and sometimes proteins. Wheat straw contains about 3–4% protein, 30–40% fiber, and is low in digestible energy. Maize Stover offers higher digestibility and energy as compared to wheat straw, making it a better choice for feeding ruminants. Rice straw is low in protein and high in silica, which can affect palatability and digestibility but can be treated to improve its nutritional value. To enhance the nutritional value of crop waste, several treatments can be applied. Physical treatments like chopping, grinding, and pelleting can improve palatability and intake. Chemical treatments, including ammoniation and urea treatment, can break down lignocellulosic bonds, increasing digestibility and protein content. Biological treatments, like fermentation and the use of fungi, can enhance the nutritional profile by breaking down complex carbohydrates and increasing protein availability. Utilizing crop waste as fodder offers significant environmental and economic benefits. Repurposing agricultural residues reduces the volume of waste that would possibly be burned or left to decompose, mitigating air pollution and greenhouse gas emissions. Farmers can lower feed costs by using readily available crop residues, enhancing their overall profitability. By integrating crop and livestock systems, farmers can create a more sustainable agricultural cycle, optimizing the use of land and water resources.

The use of crop waste as fodder faces several challenges. Many crop residues are low in essential nutrients, but solutions include supplementing with concentrates or treating the residues to enhance their nutritional value. Proper storage to prevent spoilage and contamination is important, and techniques such as silage making and proper drying can help in preserving the quality of crop waste. There is also a need for greater awareness and acceptance among farmers regarding the benefits and methods of utilizing crop residues. In this regard, extension services and training programs can play a vital role.

2. **Sea Weed**:

Seaweed, an abundant marine resource, is increasingly recognized as a viable alternative source of fodder for livestock. As the agricultural sector struggles with the challenges of sustainable feed production amidst diminishing natural resources, seaweed offers a favourable solution. Its rich nutritional profile, ease of cultivation, and environmental benefits position seaweed as an alternative to traditional fodder.

Seaweed comprises a diverse group of marine algae, broadly categorized into brown, red, and green seaweeds (Baweja et al., 2016). Each type has its unique properties and nutritional values. Brown seaweeds, such as kelp, are rich in minerals like iodine and calcium. Red seaweeds, including dulse and nori, are high in protein and vitamins, while green seaweeds, such as sea lettuce, offer a balanced mix of proteins, carbohydrates, and essential fatty acids. This diverse nutrient composition makes the seaweed a highly beneficial supplement for the livestock diets. The nutritional benefits of seaweed are substantial. It provides a rich source of vitamins, including A, C, E, and various B vitamins (Kumar et al., 2021). Seaweed is also rich in minerals such as magnesium, potassium, and trace elements like selenium and zinc. These nutrients are essential for animal health which contributes to improved growth rates, enhanced immune function, and overall well-being. Seaweed also contains bioactive compounds like antioxidants and prebiotics, which can improve gut health and reduce the occurrence of diseases in the livestock.

Beyond its nutritional advantages, seaweed cultivation and utilization as fodder offer significant environmental benefits. Seaweed farming does not require arable land, freshwater, or synthetic fertilizers which makes it a highly sustainable practice. It can be grown in marine environments, utilizing space that would otherwise be unproductive. Seaweed also acts as a bio filter, absorbing excess nutrients and carbon dioxide from the water, thereby mitigating the effects of ocean eutrophication and contributing to carbon sequestration. This process enhances the overall health of marine ecosystems and supports biodiversity. Including seaweed into livestock diets can reduce the environmental impact of animal farming. Studies have shown that certain seaweeds, particularly red varieties like Asparagopsis, can significantly reduce methane emissions from ruminants (Black et al., 2021; Camer-Pesci et al., 2023). Methane is a potent greenhouse gas, and its reduction is crucial for mitigating climate change. By incorporating seaweed into feed, farmers can lower the carbon footprint of livestock production, contributing to more sustainable and environmentally friendly agricultural practices.

Rather than these advantages, there are challenges to the widespread adoption of seaweed as fodder. One primary concern is the variability in nutritional content among different seaweed species and even within the same species due to environmental factors. Ensuring a consistent and balanced nutrient supply requires careful selection and blending of seaweed types. While seaweed is generally safe, certain species can accumulate heavy metals and other contaminants, necessitating active monitoring and quality control.

Economic considerations should be kept in mind while adoption of seaweed fodder. While seaweed farming can be cost-effective in the long term, initial setup and harvesting can be labour-intensive and require significant investment. However, advancements in aquaculture technology and economies of scale are expected to reduce these costs over time (García-Poza et al., 2020), making seaweed a more accessible option for farmers.

3. **Algae**:

Algae, a diverse group of photosynthetic organisms, are emerging as a promising alternative source of fodder for livestock (Rajauria et al., 2015). As the global agricultural sector faces the challenge of producing sustainable and nutritious feed while minimizing environmental impacts, algae offer a viable solution. Their high nutritional content, rapid growth rates, and ability to thrive in a variety of environments (Diaz et al., 2023) make algae an attractive option for supplementing traditional livestock diets. Algae can be broadly classified into microalgae and macroalgae. Microalgae, such as spirulina and chlorella, are single-celled organisms known for their rich protein content and essential fatty acids. Macro algae, commonly referred to as seaweeds, include species like marine algae wakame (*Undaria pinnatifida*), kelp and nori (*Porphyra purpurea*), which are abundant in minerals and vitamins (Taboada et al., 2012). These algae offer significant nutritional benefits that can enhance livestock health and productivity. Algae are rich in proteins, often containing all essential amino acids, making them a complete protein source. For example, spirulina contains up to 70% protein by dry weight, making it comparable to traditional protein-rich feeds like soybean meal. Additionally, algae are abundant in essential fatty acids, including omega-3 and omega-6, which are crucial for maintaining animal health. Algae also provide a wealth of vitamins, such as A, B, C, and E, and minerals like calcium, magnesium, and iron. These nutrients support various physiological functions in livestock, including growth, reproduction, and immune response.

One of the most advantages of using algae as fodder is their environmental sustainability. Algae can be cultivated on non-arable land (Diaz et al., 2023), making them an efficient use of space. They can grow in freshwater, marine, and even wastewater environments, reducing the demand on freshwater resources. Algae cultivation also has a low carbon footprint; they absorb carbon dioxide during photosynthesis, helping to mitigate greenhouse gas emissions. Certain algae species can be used to bio-remediate polluted water, further enhancing their environmental benefits. Including algae into livestock diets can also address some of the ecological challenges associated with traditional animal farming. For example, supplementing ruminant diets with algae has been shown to reduce methane emissions, a potent greenhouse gas. Research has demonstrated that specific compounds in algae, such as bromoform found in red algae, can inhibit methane production in the digestive systems of ruminants (Abbott et al., 2020). This reduction in methane emissions contributes to more sustainable livestock production and aligns with global efforts to combat climate change. Despite these benefits, the adoption of algae as fodder is not without challenges. One significant hurdle is the cost of production. Algae cultivation and processing can be expensive due to the need for controlled environments and specialized equipment. However, advancements in cultivation techniques and economies of scale are expected to lower these costs over time. Another challenge is the variability in nutrient content among different algae species and batches, necessitating thorough analysis and quality control to ensure consistent feed quality. Additionally, there are regulatory and safety considerations, particularly regarding the potential for algae to accumulate toxins or heavy metals from their growing environments.

4. **Industry Waste**:

Industries generating vast quantities of by-products that often go to waste, repurposing these materials as animal feed can contribute to a circular economy, reducing environmental impact and enhancing resource efficiency. Industrial waste suitable for fodder includes a wide array of by-products from the food, beverage, and biofuel industries. Common examples include brewers' spent grain (BSG), distillers' dried grains with solubles (DDGS), fruit and vegetable pulp, and oilseed cakes. These by-products are often rich in nutrients that can support livestock health and productivity.

Brewers' spent grain, a by-product of the beer brewing process, is high in protein and fiber. It typically contains around 20% protein and is a valuable source of energy due to its high digestible fiber content. Similarly, distillers' dried grains with solubles, a by-product of ethanol production, are rich in protein, fat, and fiber, making them an excellent feed supplement for cattle, swine, and poultry. These by-products not only enhance the nutritional profile of livestock diets but also reduce feeding costs for farmers. Fruit and vegetable pulp, derived from juice and puree production, is another valuable source of fodder. This pulp is rich in carbohydrates, fiber, and various vitamins and minerals. For instance, citrus pulp, a by-product of orange juice production, provides a good source of digestible carbohydrates and fiber, supporting energy intake and digestive health in livestock. Oilseed cakes, produced during the extraction of oils from seeds like soybeans and sunflowers, are high in protein and essential fatty acids. These cakes are commonly used as a protein supplement in animal feeds, promoting growth and development.

The environmental benefits of utilizing industrial waste as fodder are significant. By diverting these by-products from landfills or incineration, the agricultural sector can reduce greenhouse gas emissions and the environmental footprint of both industries and agriculture. Additionally, this practice promotes the efficient use of resources, aligning with the principles of a circular economy. Industrial by-products that are repurposed as animal feed contribute to waste reduction and resource conservation, creating a more sustainable agricultural system. Incorporating industrial waste into animal diets also offers economic advantages. By utilizing readily available by-products, farmers can reduce feed costs, improving their overall profitability. Industries, in turn, can reduce disposal costs and potentially create additional revenue streams by selling these by-products to the agricultural sector. This symbiotic relationship benefits both parties and supports the development of sustainable business practices. Despite these benefits, several challenges must be addressed to ensure the safe and effective use of industrial waste as fodder. One primary concern is the variability in the nutritional composition of by-products, which can fluctuate based on production processes and raw materials. Regular testing and quality control are essential to ensure that the by-products meet the nutritional requirements of livestock. Additionally, there are safety considerations, particularly regarding the potential for contamination with harmful substances such as mycotoxins, heavy metals, or chemical residues. Rigorous monitoring and adherence to safety regulations are crucial to mitigate these risks.

6.1 Alternative Source of Fodder in India

Feed shortage in India is a major concern in livestock and agriculture especially in the dry season when agriculture resources are limited. The productivity and production have to be increased to meet the needs of the increasing livestock population. The alternative for fodder is also import because of the decreasing land area of fodder in comparison to cereal and pulses crops. Here are some major and minor alternatives of fodder that India can adopt.

6.2 Major Alternative

1. Hydroponics

Hydroponics is the science of growing crop fodder or vegetables with the help of nutrient-rich solutions without soil this method is efficient both way economically and environmentally. The fodder crop which are adopted to grown by this method are maize, cowpea, ragi, bajra, jowar, and foxtail millet. It give complete utilization of plant part (root, stem, grain, roots) in green fodder (Reddy & Harani, 2023). Forage green shoots grow up to 20–30 cm long in 7–10 days, but sometimes it also depends on the type of crop chosen and factors like pH, nutrient supplies, grain variety and soaking time etc. Hydroponics gained much attention in fodder production due to the following reason.

- Precise use of space
- Conservation of water
- Reduce growth period
- Limited use of pesticides and herbicides
- Reduce carbon footprints
- Maintain the flow of fodder (Shit, 2019)

Sprouting breaks down starch into sugar that is soluble throughout the plant's biochemical processing. This means that the amount of starch, dry matter, and organic matter content declines. But as the vegetation produces the ether extract the rise in hydroponic fodder because to the rise in structural fats and green leafy algae. There is moreover a rise in focus accompanies the emergence. The evolution of growth in structural carbohydrates is the amount of crude fiber (CF), fiber from neutral detergent (NDF), fibers from acid detergents (ADF), and linoleic acid, although nitrogen-free extract (NFE) is reduced. The emergence procedure raises the amount of ash overall connected to the decline of organic materials not in the roots only boosts the absorption of minerals but also raises the mineral content of the fourth-day sprout and more in terms of the nutritional.

2. Silage

Silage is the end result of many procedures that include fermenting high-moisture-content chopped fodder to create a steady supply that in anaerobic storage prevents further decomposition. The goal is to produce a silage that animals will consume while preserving or increasing the nutrients found in the original feed. This is typically accomplished by anaerobic fermentation that is dominated by lactic acid bacteria. High-quality silage made from tropical forages has the following characteristics: pH less than 5.0, ammonia (NH_3 N: N) concentration of less than 15%, lactic acid (50 percent or more of total organic acids), and butyric acid (not more than 0.5% of total dry matter). When forage is placed in a closed space, such as a plastic-covered pit, a drum, or a plastic bag, the container is called a silo. To guarantee the quick establishment of anaerobic conditions and a rapid drop in pH, a silo must be sealed entirely against air and its feed material cut and crushed inside. Under these circumstances, lactic acid bacteria—which turn part of the plant's carbohydrates into delicious lactic acid—predominate over other bacteria that aren't so good for animals, such as clostridia, which make butyric acid—which isn't very tasty—and moulds, which make the silage rot (Chaudhary et al., 2012).

3. Crop Residue

The term "crop residues" describes the part of the harvested crop that is left behind after the grain, or marketable component, has been taken. Stems, leaves, and chaff are the main parts of the agricultural waste that is separated in this manner. The biomass of above-ground plants remaining in the field following harvest makes up the other 12 dimensions of agricultural residue, sometimes referred to as the crop aftermath. In addition to grain, major field crops yield considerable amounts of agricultural residues, such as stoves, haulms, and straws. Crop residue kinds that are accessible in the nation differ depending on the agroecology or farming practices used in various regions. The leftovers from pulse crops vary depending on the agro-ecology. The main crop residues utilized in the highlands for animal Tef, wheat, barley, oats, faba bean, field pea, and chickpea straws are used for feeding (Tesfaye et al., 2011). Some of the most important factors that alter the quality of crop residue are leaching, mode of harvest, and plant density (Table 1).

4. Agro Industrial by Product

Natural crop and pasture leftovers are inherently of low quality and do not provide the nutrients that animals need. Thus, animals with high productivity, like dairy.

Table 1 Crop residue of the different crops (adapted from Bhandari & Bahadur, 2019)

Crop residue for feed	Crop
Straw and bran	Rice and wheat
Straw	Barley, maize, sorghum
Stover, cob and bran	Maize plant
Bagasse, tops, molasses	Sugarcane
Hay and meal	Pulses
Cake, straw, pruning	Sunflower, olive, linseed, mustard
Roots and tubers	Waste

High energy and/or protein concentrates should be added to cattle and fattening animals, the most popular concentrate feeds come from a variety of agro-industrial by-products, such as: Grinding leftovers, such as wheat bran Cake flavors include nougat, cottonseed, peanut, linseed, sesame, and sunflower. Molasses derived from sugar refineries complete cotton seeds Recycled beer. Occasionally, there may be excess grain or damaged grain due to processing. Agro-industrial by-products are nutritionally better, but because of their high cost, unstable supply, and limited accessibility for smallholder farmers, they have been employed on a very small scale. For the effective use of agro-industrial by-products, market-oriented animal businesses like dairy and fattening surrounding the processing facilities may be crucial. Additionally, smallholder dairy farmers can form cooperatives, making obtaining various inputs and by-products from the agro-industrial sector easier (Tesfaye et al., 2011).

5. Fodder Trees and Shrubs

A few more feed sources that may be used in addition to natural forages are farmed fodder trees and bushes, both leguminous and non-leguminous. The selection of a substitute feed source need to be non-discriminatory and compatible with the current agricultural systems. For many years, India's farming systems have relied on shrubs and trees for fodder, as they are versatile resources that provide vital nutrition to cattle. The primary purpose of fodder trees is to feed dairy cows. In times of harsh weather, such as droughts, they can satisfy output shortfalls. Growing fodder trees is simple, requires little space, labor, or cash, produces a variety of by-products, and may frequently provide feed within a year after planting. Some examples of tree and shrubs used as fodder Acacia tortilis, Acacia nilotica, Acacia senegal, Prosopis cineraria, Prosopis juliflora, Azadirachta indica, Ziziphus nummularia, Colophospermum mopane, Dichrostachys nutans, Colligonum polygonoides.

6.3 Minor Alternative

1. Azolla

Azolla, *Azolla pinnata,* an aquatic fern from the Azollacea plant family, is largely used as cattle fodder. It has a significant amount of protein, vital amino acids, vitamins, minerals, and growth promoter intermediates, making it a promising feed for animals. In terms of dry weight azolla is composed of 25–35% protein, 10–15% mineral content, and 7–10% amino acid, bioactive ingredient, and biopolymer mixture. Through the holding of demonstrations in 10 communities, this Azolla farming method was spread. The fat content and milk output of milch cows increased when Azolla was fed as livestock feed. Azolla's bio composition makes it one of the most cost-effective and efficient feed alternatives for animals. Furthermore, because Azolla has a low lignin concentration and a high protein content, it is easily digested by animals (Kumar et al., 2020)

2. Non-conventional Fodder Crop

Moringa oleifera

Moringa is a small, slender, fast-growing, deciduous shrub or tree 9–15 m in height, with an open canopy-shaped crown. Use of M. oleifera complete plants as unconventional feed for ruminants, as well as the use of leaves or leaf extract as a source of protein for laying hens and broilers. It is unclear how M. oleifera is used in animal feed to produce functional food (meat, eggs, and milk) with the right amounts of nutrients that are good for human health, like organic selenium and omega Moreover, adding M. oleifera to animal feed may lengthen the shelf life of animal products while they are being processed and stored. M. oleifera contains a variety of pharmacologically active chemical component groupings. Alkaloids, glycosides, flavonoids, phenols, phytosterols, saponins, tannins, coumarins, and terpenoids were some of the chemical substances that were present (Nouman et al., 2014)

6.4 Achievement of Fodder Industries

The alternative fodder industry has made big progress lately, tackling the growing need for sustainable livestock feed solutions. Here are some cool achievements and trends:

Using Opuntia spp. (Prickly Pear Cactus): Opuntia getting props for thriving in dry areas where water is scarce. This cactus is packed with nutrients, giving animals a top-notch forage option that can boost performance and improve animal products' quality. Growing Opuntia needs little investment and gives way more biomass than typical forage crops, making it a green choice for dry areas.

Making Compound Feeds: In places like Ethiopia, compound feed production has almost doubled in recent years. This growth comes from high demand for commercial livestock farming, even with challenges like pricey feed ingredients and limited feed processing plants nearby. People are thinking about using local maize and soybean in feed production to cut back on imports and up feed availability.

Using Agro-Industrial By-Products: By-products from sugar mills (like molasses and bagasse) and breweries are more common in animal feed now. These materials are an economical and eco-friendly feed option, helping reduce waste impact while giving animals key nutrients.

Creating Feed Supplements: More focus is on making feed supplements like premixes, minerals, and vitamins locally instead of always importing them. This change aims to cut costs and boost the nutritional value of animal feed. Also, effective microbial inoculants is showing promise in improving feed efficiency and animal health.

Better Forage Varieties: Advances in forage research, improved varieties that are tougher and more productive are being developed. Efforts to bring these varieties to market are ongoing, ensuring they meet different livestock needs and regional conditions with a friendly approach.

7 Strengths and Weaknesses of Alternative Fodder Use

Strengths

Sustainability:

Growing hydroponic fodder needs less water & land than traditional farming. It can grow year round, no matter the weather. Using agro-industrial by-products helps reduce waste & promotes a circular economy by using leftovers from other industries like brewery waste & fruit pulp. This is great for the environment.

Nutritional Value:

Legumin fodder is packed with protein & essential amino acids, which are good for livestock health and productivity. Algae & seaweed are rich in minerals, vitamins, and omega-3 fatty acids, making a nutrient-packed supplement for animals.

Cost-Effectiveness:

Crop residues like straw, husks, and leaves are often cheaper as they come from crop production by-products. Preserving surplus forage as silage during peak seasons can save money by reducing the need to buy expensive commercial feeds later.

Adaptability:

Plants like cactus and sorghum can survive in dry regions, ensuring a stable food supply even during droughts.

Environmental Benefits:

Cover crops boost soil health & reduce erosion, making farming more sustainable in the long run. Perennial grasses help store carbon & improve soil structure, aiding in fighting climate change.

Weaknesses

Initial Investment:

Setting up hydroponic fodder systems can cost a lot initially due to needing equipment and technology. Scaling algae cultivation might be complicated & expensive in terms of technology requirements.

Consistency and Quality:

Quality and availability of agro-industrial by-products can fluctuate, which might affect animal performance negatively. Crop residues may lack protein & digestibility, so additional supplements or treatments may be needed to improve nutritional value.

Technical Knowledge and Skills:

Managing hydroponic and aquaponic systems requires specialized knowledge for efficient operation and fixing issues that may arise. Producing silage needs careful handling to prevent spoilage & ensure the feed's quality.

Acceptance and Adoption:

Farmers might be unsure about new fodder sources due to a lack of info or skepticism about their benefits. Some regions have deep-rooted traditions regarding traditional fodder sources that might hinder adopting alternative options.

Potential Health Risks:

Agro-industrial by-products could contain contaminants or residues from primary production processes that could harm livestock health. Certain legumes may have anti-nutritional factors if not processed correctly or fed excessively.

8 Conclusion

The exploration of alternative fodder sources in India underscores a promising path toward sustainable livestock production which addresses both feed shortages and environmental challenges. With a diverse array of unconventional resources such as crop residues, seaweed, algae, and agro-industrial by-products, the potential to transform agricultural practices is substantial. These alternatives not only supplement traditional feed but also improve soil health, reduce greenhouse gas emissions, and contribute to a circular economy. Technologies like hydroponics and silage production further enhance fodder availability, optimizing resource use and supporting year-round supply. The integration of fodder trees and minor alternatives like Azolla reflects innovative approaches to bolstering livestock nutrition, particularly in regions prone to resource scarcity. Despite of the advantages some of the challenges includes high initial costs, quality variability, and limited farmer awareness. Addressing these barriers through government initiatives, capacity building, and technological advancements is crucial. Programs like the Accelerated Fodder Development Programme exemplify effective strategies for scaling these solutions. By adopting alternative fodder sources, India can mitigate land degradation, improve livestock productivity, and enhance rural livelihoods, ensuring a balanced approach to food security and environmental conservation. Continued research, policy support, and stakeholder collaboration will be essential in unlocking the full potential of these sustainable practices.

References

Abbott, D. W., Aasen, I. M., Beauchemin, K. A., Grondahl, F., Gruninger, R., Hayes, M., Huws, S., Kenny, D. A., Krizsan, S. J., Kirwan, S. F., Lind, V., Meyer, U., Ramin, M., Theodoridou, K., von Soosten, D., Walsh, P. J., Waters, S., & Xing, X. (2020). Seaweed and seaweed bioactives for mitigation of enteric methane: Challenges and opportunities. In *Animals* (Vol. 10(12), pp. 2432). MDPI AG. https://doi.org/10.3390/ani10122432

Akram, M. Z., & Fırıncıoğlu, S. Y. (2019). The use of agricultural crop residues as alternatives to conventional feedstuffs for ruminants: A review. *Eurasian Journal of Agricultural Research, 3*(2), 58–66.

Baris, A. (2023). Impact of feed quality on livestock productivity. *Journal of Livestock Policy, 2*(1), 1–8.

Baweja, P., Kumar, S., Sahoo, D., & Levine, I. (2016). Biology of seaweeds. In *Seaweed in health and disease prevention* (pp. 41–106). Academic Press.

Bhandari, B., & Bahadur. (2019). Crop residue as animal feed. https://doi.org/10.13140/RG.2.2.20372.04486

Black, J. L., Davison, T. M., & Box, I. (2021). Methane emissions from ruminants in Australia: Mitigation potential and applicability of mitigation strategies. In *Animals* (Vol. 11(4), pp. 951). MDPI AG. https://doi.org/10.3390/ani11040951

Bonneau, M., & Laarveld, B. (1999). Biotechnology in animal nutrition, physiology and health. In *Livestock production science* (Vol. 59(2–3), pp. 223–241). Elsevier BV. https://doi.org/10.1016/s0301-6226(99)00029-9

Camer-Pesci, B., Laird, D. W., van Keulen, M., Vadiveloo, A., Chalmers, M., & Moheimani, N. R. (2023). Opportunities of Asparagopsis sp. cultivation to reduce methanogenesis in ruminants: A critical review. In *Algal Research* (Vol. 76, p. 103308). Elsevier BV. https://doi.org/10.1016/j.algal.2023.103308

Chaudhary, D., Kumar, A., Langyan, S., Mandhania, S., Srivasta, P., & Kumar, R. S. (2012). Maize as fodder? An alternative approach.

da Silva, R. F., Kitagawa, A., & Sánchez Vázquez, F. J. (2015). Dietary self-selection in fish: A new approach to studying fish nutrition and feeding behavior. In *Reviews in fish biology and fisheries* (Vol. 26(1), pp. 39–51). Springer Science and Business Media LLC. https://doi.org/10.1007/s11160-015-9410-1

Datta, P., Behera, B., & Rahut, D. B. (2024). Assessing the role of agriculture-forestry-livestock nexus in improving farmers' food security in South Asia: A systematic literature review. In *Agricultural Systems* (Vol. 213, pp. 103807). Elsevier BV. https://doi.org/10.1016/j.agsy.2023.103807

Devendra, C., & Leng, R. A. (2011). Feed resources for animals in Asia: Issues, strategies for use, intensification and integration for increased productivity. *Asian-Australasian Journal of Animal Sciences*. Asian Australasian Association of Animal Production Societies. https://doi.org/10.5713/ajas.2011.r.05

Diaz, C. J., Douglas, K. J., Kang, K., Kolarik, A. L., Malinovski, R., Torres-Tiji, Y., Molino, J. V., Badary, A., & Mayfield, S. P. (2023). Developing algae as a sustainable food source. In *Frontiers in Nutrition* (Vol. 9). Frontiers Media SA. https://doi.org/10.3389/fnut.2022.1029841

Dinesha, S., Hosur, S. R., Toushif, P. K., Bodiga, D., Deepthi Dechamma, N. L., Ashwath, M. N., & Pradhan, D. (2023). Sustaino-Resilient agroforestry for climate resilience, food security and land degradation neutrality. In *Land and environmental management through forestry* (pp. 217–245). Wiley. https://doi.org/10.1002/9781119910527.ch9

García-Poza, S., Leandro, A., Cotas, C., Cotas, J., Marques, J. C., Pereira, L., & Gonçalves, A. M. M. (2020). The evolution road of seaweed aquaculture: Cultivation technologies and the industry 4.0. In *International journal of environmental research and public health* (Vol. 17(18), pp. 6528). MDPI AG. https://doi.org/10.3390/ijerph17186528

Kumar, N., Chaudhary, A., Ahlawat, O. P., Naorem, A., Upadhyay, G., Chhokar, R. S., & Singh, G. P. (2023). Crop residue management challenges, opportunities and way forward for sustainable food-energy security in India: A review. *Soil and Tillage Research, 228*, Article 105641.

Kumar, S., Paritosh, K., Pareek, N., Chawade, A., & Vivekanand, V. (2018). De-construction of major Indian cereal crop residues through chemical pretreatment for improved biogas production: An overview. *Renewable and Sustainable Energy Reviews, 90*, 160–170.

Kumar, A. S., Murugesan, S., & Balamurugan, P. (2020). Feeding of Azolla as a green fodder feed supplement on productive performance and milk composition of crossbred dairy cows in Theni District of Tamil Nadu, India. *International Journal of Current Microbiology and Applied Sciences, 9*(6), 1388–1382. https://doi.org/10.20546/ijcmas.2020.906.171

Kumar, Y., Tarafdar, A., & Badgujar, P. C. (2021). Seaweed as a source of natural antioxidants: Therapeutic activity and food applications. *Journal of Food Quality, 2021*(1), 5753391.

Loy, D. D., & Lundy, E. L. (2019). Nutritional properties and feeding value of corn and its coproducts. In *Corn* (pp. 633–659). Elsevier. https://doi.org/10.1016/b978-0-12-811971-6.00023-1

Makkar, H. P. S. (2016). Smart livestock feeding strategies for harvesting triple gain—the desired outcomes in planet, people and profit dimensions: a developing country perspective. In *Animal Production Science* (Vol. 56(3), pp. 519). CSIRO Publishing. https://doi.org/10.1071/an15557

Mekuria, W., Mekonnen, K., Thorne, P., Bezabih, M., Tamene, L., & Abera, W. (2018). Competition for land resources: driving forces and consequences in crop-livestock production systems of the Ethiopian highlands. In *Ecological processes* (Vol. 7(1)). Springer Science and Business Media LLC. https://doi.org/10.1186/s13717-018-0143-7

Nouman, W., Basra, S. M. A., Siddiqui, M. T., Yasmeen, A., Gull, T., & Alcayde, M. A. C. (2014). Potential of *Moringa oleifera* L. as livestock fodder crop: a review. In *Turkish journal of agriculture and forestry* (Vol. 38, pp. 1–14). The Scientific and Technological Research Council of Turkey (TUBITAK-ULAKBIM)—Digital Commons Journals. https://doi.org/10.3906/tar-1211-66

Paroda, R. S. (Ed.). (2018). Reorienting Indian agriculture: Challenges and opportunities. *CAB International*. https://doi.org/10.1079/9781786395177.0000

Princess Mudzengi, C., Dahwa, E., & Simbarashe Kapembeza, C. (2020). Livestock feeds and feeding in semi-arid areas of Southern Africa. In *Livestock health and farming*. IntechOpen. https://doi.org/10.5772/intechopen.90109

Rajauria, G., Cornish, L., Ometto, F., Msuya, F. E., & Villa, R. (2015). Identification and selection of algae for food, feed, and fuel applications. In *Seaweed sustainability* (pp. 315–345). Elsevier. https://doi.org/10.1016/b978-0-12-418697-2.00012-x

Reddy, P. & Harani, M. (2023). Hydroponics: A sustainable way of green fodder production. *Indian Farming, 73*, 02–05.

Semwal, R. L., Nautiyal, S., Sen, K. K., Rana, U., Maikhuri, R. K., Rao, K. S., & Saxena, K. G. (2004). Patterns and ecological implications of agricultural land-use changes: a case study from central Himalaya, India. In Agriculture, ecosystems &; environment (Vol. 102(1), pp. 81–92). Elsevier BV. https://doi.org/10.1016/s0167-8809(03)00228-7

Shahapurmath, G., Inamati, S. S., & Mutanal, S. M. (2020). Growth and yield performance of fodder tree species with intercrops under agro forestry systems in Dharwad Karnataka, India. In *Journal of Tree Sciences* (Vol. 39(1), pp. 26). Diva Enterprises Private Limited. https://doi.org/10.5958/2455-7129.2020.00004.7

Shit, N. (2019). Hydroponic fodder production: An alternative technology for sustainable livestock production in India. *Exploratory Animal and Medical Research, 9*, 108–119.

Singh, H. P., Sharma, K. D., Subba Reddy, G., & Sharma, K. L. (2004). Dryland agriculture in India. In *Challenges and strategies of dryland agriculture* (pp. 67–92). Crop Science Society of America and American Society of Agronomy. https://doi.org/10.2135/cssaspecpub32.c6

Singh, K. M., Meena, M., & Singh, R. (2012). Livestock value chains: Prospects, challenges and policy implications for Eastern India. *Challenges and Policy Implications for Eastern India (March 13, 2012).*

Singh, D. N., Bohra, J. S., Tyagi, V., Singh, T., Banjara, T. R., & Gupta, G. (2022). A review of India's fodder production status and opportunities. In *Grass and forage science* (Vol. 77(1), pp. 1–10). Wiley. https://doi.org/10.1111/gfs.12561

Taboada, M. C., Millán, R., & Miguez, M. I. (2012). Nutritional value of the marine algae wakame (Undaria pinnatifida) and nori (Porphyra purpurea) as food supplements. In *Journal of Applied Phycology* (Vol. 25(5), pp. 1271–1276). Springer Science and Business Media LLC. https://doi.org/10.1007/s10811-012-9951-9

Tesfaye, A., Tsigie, A., & Feyissa, F. (2011). Producing and Using Alternative Feeds to Crop Residues. *User Manual, EIAR.*

Tomar, J. M. S., Ahmed, A., Bhat, J. A., Kaushal, R., Shukla, G., & Kumar, R. (2021). Potential and opportunities of agroforestry practices in combating land degradation. *Agroforestry-small landholder's tool for climate change resiliency and mitigation*, 1–20.

Tona, G. O. (2018). Current and future improvements in livestock nutrition and feed resources. *Animal Husbandry and Nutrition, 18*, 73088.

Tulu, D., Gadissa, S., Hundessa, F., & Kebede, E. (2023). contribution of climate-smart forage and fodder production for sustainable livestock production and environment: Lessons and challenges from ethiopia. In Z. Wang (Ed.), *Advances in agriculture* (Vol. 2023, pp. 1–11). Hindawi Limited. https://doi.org/10.1155/2023/8067776

Weindl, I., Popp, A., Bodirsky, B. L., Rolinski, S., Lotze-Campen, H., Biewald, A., Humpenöder, F., Dietrich, J. P., & Stevanović, M. (2017). Livestock and human use of land: Productivity trends and dietary choices as drivers of future land and carbon dynamics. In *Global and planetary change* (Vol. 159, pp. 1–10). Elsevier BV. https://doi.org/10.1016/j.gloplacha.2017.10.002

Zhang, H., Zhang, P., Ye, J., Wu, Y., Liu, J., Fang, W., & Zeng, G. (2018). Comparison of various pretreatments for ethanol production enhancement from solid residue after rumen fluid digestion of rice straw. *Bioresource Technology, 247*, 147–156.

Plant Pathology and Its Implications on Soil Health

Muhammad Arif, Rabia Kalsoom, Faisal Khalid, Parvaiz Ahmad, Wenqi Shi, W. M. W. W. Kandegama, Hina Abbas, Haris Khurshid, Jamal Nasar, and Muhammad Amir Zia

Abstract This chapter provides an overview of the complex relationships between plant pathogens and soil health, emphasizing the significance of soil-borne pathogens in agricultural systems. It discusses how these pathogens affect soil structure, nutrient dynamics, and various ecosystem functions. Pathogens often alter the soil microflora, leading to the destruction of beneficial microorganisms and biocontrol agents, which contributes to soil degradation and reduced plant productivity. The chapter concludes that integrated management strategies are essential for maintaining soil health and mitigating the negative impacts of pathogens. These approaches aim to foster a balanced and resilient soil microbial community, thereby sustaining agricultural productivity while reducing reliance on chemical controls. The research advocates for a holistic management approach that considers the interactions among plants, soil, and pathogens to promote sustainable agricultural practices.

M. Arif (✉) · W. Shi · J. Nasar
College of Agriculture, Guizhou University, Guiyang, Guizhou, China
e-mail: arif@gzu.edu.cn

R. Kalsoom
Plant Virology Section, Ayub Agriculture Research Institute, Faisalabad, Pakistan

F. Khalid
Department of Forestry, Shaheed Benazir Bhutto University, Sheringal, Khyber Pakhtunkhwa, Pakistan

P. Ahmad
Department of Botany, GDC-Pulwama, Jammu and Kashmir, India

W. M. W. W. Kandegama
State Key Laboratory of Green Pesticides, Key Laboratory of Green Pesticide and Agricultural Bioengineering, Ministry of Education, Center for Research and Development of Fine Chemicals, Guizhou University, Guiyang, China

H. Abbas · M. A. Zia
National Institute for Genomics and Advanced Biotechnology (NIGAB), National Agricultural Research Centre (NARC), Islamabad, Pakistan

H. Khurshid
Oilseeds Research Program, Crop Sciences Institute, NARC, Islamabad, Pakistan

M. Shaaban (ed.), *Soils and Sustainable Agriculture*, Frontier Studies in Soil Science,
https://doi.org/10.1007/978-3-031-91114-9_36

Keywords Soil health · Pathogens · Agricultural productivity · Sustainable agriculture

1 Introduction

The term "soil health" is defined as the capability of the soil to endure plant and animal productivity, to maintain or enhance water and air quality, and to support plant and animal health (Chiranjeeb et al., 2024) It is an integral part of sustainable agriculture affecting crop productivity, environmental quality, and ecosystem resilience. It is not only described by the physical and chemical nature of the soil but also by its biological components, principally microbial communities playing a key role in nutrient cycling, organic matter decomposition, and overall plant growth and health (Srivastava et al., 2023).

Plant pathology deals with the diseases of plants, the causal agents of these diseases, and their management using environment-friendly approaches. Plant pathology has a profound effect on soil health because soil is not merely a growth medium or substrate for plants, but also has a complex living system composed of microorganisms, organic matter, and nutrients that interact intimately with the roots of plants. It can be stated that plant development and soil health are intertwined. The presence of plant diseases usually leads to changes in soil microbial communities, nutrient cycling, and sometimes even soil structure (Bruna et al., 2023). Understanding the relationships between plant diseases and soil is required to develop sustainable agricultural practices and conserve environmental health (Guzmán, 2021).

This chapter will discuss the interactions of a diverse group of plant pathogens and their impacts on soil ecosystems and the latest research towards strategies for the management of plant diseases, in a way that ensures soil health is retained or enhanced. It specifically emphasizes the role that pathogens play in microbial communities, the disturbance of nutrient cycling, the decomposition of soil organic matter, and the overall structural integrity of soil (Maria et al., 2014). Plant pathogens may enhance nutrient cycling, but in general, their effects are more negative, associated with decreased soil fertility and generally lower ecosystem stability (de Vries et al., 2020). Understanding these interactions will be required to design sustainable agricultural ecosystems and for practical implementation in the strategies of soil management (Spadaro & Gullino, 2019). This chapter also will consider the impact plant pathology has on the health of soils, posing an integrative treatment of the good and bad outcomes from these associations.

1.1 Plant Pathogens

In plant pathology, the disease triangle comprises three factors, host, pathogen and environmental conditions which favor the disease development or suppression. After analyzing the organisms and environmental factors that induce diseases and the mechanisms of underlying disease progression, plant pathology enables us to develop better strategies for their management (Lucas, 2011). Plant pathology has extensive scope, ranging from the analysis of minute entities, such as molecular biology, to the consideration of extensive systems like ecosystems or ecology, with profound ramifications for global food security. Plant diseases may arise from a diverse array of pathogens, which include fungi, bacteria, viruses, nematodes, and parasitic plants. Each category of pathogens possesses unique characteristics and distinct modes of interaction with plants and the surrounding soil environment (Table 1).

Table 1 Types of plant pathogens and their effects on soil health

Pathogen type	Examples	Key characteristics	Impact on soil health	References
Fungal pathogens	*Phytophthora spp., Fusarium spp.*	Persistent in soil, produces mycotoxins	Disrupt root function, reduce microbial diversity, impair nutrient cycling	(Wang et al., 2020; Yan et al., 2022)
Bacterial pathogens	*Ralstonia solanacearum*	Soilborne, survives in soil without a host	Alter microbial communities, increase disease incidence in subsequent crops	(Ahmed et al., 2022)
Viral pathogens	*Tobacco mosaic virus (TMV), Tomato yellow leaf curl virus (TYLCV)*	Obligate parasites, transmitted by vectors	Alter soil microbial composition, and affect soil health through plant stress	(Hussain et al., 2022)
Nematodes	*Meloidogyne spp. (root-knot nematodes)*	Microscopic, damage plant roots	Create entry points for secondary pathogens, reduce nutrient uptake	(Gowda et al., 2019; Sonam & Rakesh, 2019)
Parasitic plants	*Dodder (Cuscuta spp.), Broomrape (Orobanche spp.)*	Extract water and nutrients from host plants	Compromise host plant health affects soil nutrient dynamics	(Jat & Singh, 2018; Negewo et al., 2022)

1.1.1 Fungal Pathogens

Fungi are the most common and diverse group of plant pathogens. They are accountable for many plant diseases, like rusts, smuts, blights, wilts, etc. The fungal pathogen *Phytophthora* spp. causes root rot and remains in the soil for years, acting relentlessly on soil health through the disruption of root functioning and microbial equilibrium (Benkeblia, 2022) Mycotoxins produced by fungi are both phytotoxic and toxic to microorganisms in the soil. Nearly 50% of post-harvest losses are reported due to mycotoxin contamination worldwide (Atungulu & Mohammadi-Shad, 2019).

1.1.2 Bacterial Pathogens

They are single-celled organisms that can lead to bacterial wilt, blight, and soft rots. For example, *Ralstonia solanacearum* is a kind of soil-borne bacterium that can lead to bacterial wilt in a wide range of plants. This pathogen can endure in the soil for years without the availability of any host and can play a vital role in taking major parts during the modification of microbial communities that exist in the soil (Hayes et al., 2022). Many bacterial diseases in plants are incited or enhanced by high soil moisture combined with warm temperatures, a combination that offers ideal conditions for high bacterial multiplication.

1.1.3 Viral Pathogens

They are smaller than bacteria and fungi; they consist of a nucleic acid enclosed in a protein coat. They are obligate parasites, able to reproduce only inside a living host. Viral diseases, such as *TMV* (*Tobacco Mosaic Virus*) or *TYLCV* (*Tomato Yellow Leaf Curl Virus*), cause very serious losses by reducing both crop yield and quality. These viruses are always in the form of insect-borne transmission; however, few viruses could remain on the plant debris in the soil and therefore impact the health of the soil through a shift in the community composition of microbes (Liang et al., 2024).

1.1.4 Nematodes

These are very small, worm-like organisms which cause severe damage to plant roots from their residence within the soil. *Meloidogyne* spp., more commonly known as root-knot nematodes, are among the most damaging plant-parasitic nematodes. These can form galls on the root system and reduce nutrient uptake of the plant, thus causing stunted plant growth. This could further alter the ecological balance concerning other soil microorganisms since the infestation of the plants with these nematodes can inflict wounds through which entry of secondary pathogens, like fungi and bacteria, may occur and worsen the deterioration of soil health (Khan et al., 2023).

1.1.5 Parasitic Plants

Parasitic plants, such as dodder (*Cuscuta* spp.) and broomrape (*Orobanche* spp.), attach themselves to host plants and extract water and nutrients, weaken their hosts and prone them to other pathogens (Yuan & Li, 2022). These parasitic interactions can also change soil nutrient dynamics, as the host plants' ability to contribute to soil organic matter and nutrient cycling is compromised.

2 The Impact of Plant Pathogens on Soil Health

Soil health is mainly dependent on microbial communities' diversity and activity, including bacteria, fungi, archaea, and protozoa. These microorganisms play key roles in nutrient cycling, organic matter decomposition, and the formation of soil structure. Now the effect of pathogens on soil health will be discussed in detail.

2.1 Soil Microbial Communities

The presence of plant pathogens can significantly alter these microbial communities, often in ways that are detrimental to soil health.

2.1.1 Pathogen-Mediated Shifts

When a plant becomes infected, it often releases specific exudates into the rhizosphere (the soil region closes to plant roots). These exudates can modify the composition and activity of soil microbial communities. For example, plants infected by root pathogens like *Phytophthora* spp. and *Fusarium* spp. may exude compounds that suppress beneficial soil bacteria and fungi, such as mycorrhizal fungi, which are critical for nutrient uptake. These shifts can decrease microbial diversity, making the soil more susceptible to further pathogen assaults and less resilient to environmental stressors (Batista et al., 2024).

2.1.2 Antagonistic Interactions

Beneficial microorganisms in the soil, in most instances, are invaded by plant pathogens due to the production of inhibitory compounds against the pathogens. Coming into play here is the competition by utilization of resources between the beneficial microorganism and the pathogen as well as the elicitation of resistance mechanisms in the plant. The dominance of a pathogen reduces the effectiveness of these valuable organisms, causing an effect on soil health (Liang et al., 2024). This antagonistic association shows that to be opposed to soil health, it's important to maintain a balance between pathogens and useful microbes in the soil.

Sometimes, appearance of the plant pathogens may lead to an increase in the population of the presence of microbial as the soil ecosystem responds towards stress. For instance, some pathogens stimulate the production of root exudates that attract a much wider diversity of soil microbes and thus enhance the overall biodiversity in the soil. Some soils contain soil microbiota that can elicit a response against some diseases of the plants. For example, the pathogen attack in some cases may lead to the recruitment of antagonistic microbes—for example, members of the genera *Trichoderma* or *Bacillus*—with the ability to suppress such pathogens, thus promoting soil health by enhancing nutrient cycling and plant growth (Tong et al., 2024).

2.2 *Nutrient Cycling and Soil Fertility*

Plant pathogens can have profound effects on nutrient cycling in soil, often leading to imbalances that can negatively impact both plant and soil health.

2.2.1 Disruption of Root Function

The extent of nutrient and water uptake impairment by pathogenic roots, such as nematodes and root-rotting fungi, is, in most cases, severe. Such weakening not only debilitates the plant but also changes nutrient dynamics in the soil. For example, a reduction in nitrogen uptake has been manifested for nematode-infected roots, resulting in nitrogen accumulation in the soil. This could alter microbial processes involved in the nitrogen cycle, such as nitrification and denitrification, and may lead to increased gaseous losses of greenhouse gaseous, for example, nitrous oxide (Meister et al., 2023).

2.2.2 Altered Organic Matter Decomposition Rates

The presence of microbial pathogens could also impact the rate as well as nature of organic matter decomposition in soil. Degraded plants tend to have a different amount and type of litter production than healthy plants, which would influence the decomposers' activity (Jakubcsiková et al., 2023). The development of plant pathogens reduces the decomposition rate of litter, changes the chemical composition of leaf litter, and increases soil organic matter in the form of more prevalent, undecomposed organic matter. This can develop into a decreasing fertility status of soil as nutrients available for the progression growth of subsequent plants are exhausted (Wiesenbauer et al., 2024).

On the flip side, sometimes, pathogens might enhance nutrient mobilization indirectly develop to nutrient mobilization by breaking down the plant tissues thus releasing nutrients to the soil. For example, certain fungal pathogens decompose organic matter rapidly, potentially increasing the short-term availability of nutrients like nitrogen and phosphorus. Plant diseases can sometimes induce systemic resistance in plants, leading to changes in root exudation patterns. These changes can enhance nutrient cycling by promoting the activity of specific soil microbes that assist in the breakdown of organic matter and nutrient mineralization.

2.3 *Effects on Soil Structure*

Soil structure, including the formation and stability of soil aggregates, is crucial for maintaining soil health, water retention, and aeration. In this regard, healthy plants are major contributors as they stabilize the soil aggregates and facilitate the soil pores by their root systems. On the other hand, when these plants are afflicted by disease, their root systems frequently exhibit signs of deterioration, which results in diminished root biomass eventually deterioration in soil structure is observed.

2.3.1 Impact on Soil Aggregation and Compaction

The pathogens that cause root rot such as *Rhizoctonia solani* are capable of inducing a decrease in root biomass and consequently precipitating the disintegration of soil aggregates Soil aggregates play an essential role in maintaining soil porosity and water retention along with the safeguarding organic matter from degradation. The deterioration of soil aggregation can result in compaction, impaired water infiltration, and susceptibility to erosion (Even & Cotrufo, 2024; Yang et al., 2024).

2.3.2 Increased Erosion Risk

Diseased plants with diminished root systems make the soil more vulnerable to erosion. This is particularly challenging in regions with high rainfall or wind exposure (Zichun et al., 2022). Erosion not only leads to the loss of topsoil, which is rich in organic matter and nutrients but also contributes to the degradation of downstream ecosystems by increasing sedimentation in water bodies.

In contrast, certain plant pathogens stimulate the decomposition of organic matter, which can lead to the formation of more steady soil aggregates. This can improve soil aeration and water infiltration, contributing positively to soil structure, especially in soils with high clay content (Du et al., 2024). Some plant pathogens can stimulate compensatory root growth in plants as a survival mechanism. This increased root biomass can boost soil structure by promoting aggregate formation and improving soil porosity.

2.4 *Long-Term Soil Degradation*

Repeated or severe plant disease outbreaks can lead to long-term soil degradation. The persistent presence of pathogens can result in a cycle of deteriorating plant health and soil quality, making it increasingly difficult to sustain productive agriculture.

2.4.1 Persistent Pathogens

Some soil-borne pathogens can survive in the soil for years, continuously affecting new plantings. This can lead to a buildup of pathogens in the soil, creating a hostile environment for healthy plant growth and further degrading soil health (Bakker & Berendsen, 2022).

2.4.2 Reduced Soil Fertility

Over time, the combined effects of reduced microbial diversity, disrupted nutrient cycling, and weakened soil structure can lead to a decline in soil fertility. This not only affects current crop yields but also has long-term implications for the sustainability of agricultural systems. For example, soils that are repeatedly exposed to *Fusarium* spp. may become less fertile and more prone to erosion, leading to a decline in overall soil health (van der Sloot et al., 2024).

In some cases, over time, exposed soils may get natural suppressiveness due to the evolution of a soil microbiome hostile to the pathogen. This can enhance soil health in the meaning of long-term soil health if resistance to specific diseases in the soil builds up. This process of constant exposure at the site could, in the longer run, result in more robustly healthy soil ecosystems from the evolution of disease-resistant plants and resilient microbial communities (Fujita et al., 2023).

3 Management Strategies to Mitigate Negative Effects of Plant Pathology on Soil Health

3.1 *Integrated Disease Management (IDM)*

Integrated Disease Management (IDM) is a holistic line of attack that combines biological, cultural, chemical, and physical approaches to manage plant diseases while curtailing the impact on soil health. IDM promotes the use of sustainable practices that guard and improve soil health while reducing plant pathogens (Table 2).

Table 2 Integrated disease management practices and their impact on soil health

Management practice	Target pathogens	Impact on soil microbial diversity	Mechanism of disease control	Effects on soil health	Challenges	References
Crop rotation	*Fusarium, Pythium, Rhizoctonia*	Increases microbial diversity by disrupting pathogen cycles	Rotating crops with non-host species reduces pathogen load in the soil	Improves soil structure, enhances nutrient cycling	Requires changes in farming practices, may be complex to implement; potential for crop rotation to be less effective in monocultures	(Sanzovo et al., 2023; Shah et al., 2021)
Biological control (*Trichoderma* Spp.)	*Rhizoctonia, Verticillium, Phytophthora*	Promotes beneficial microbes, suppresses pathogens	Antagonistic fungi outcompete pathogens for resources and space	Enhances soil fertility, reduces disease incidence	Requires careful application, and potential resistance development; may require frequent reapplication and monitoring	(Álvarez & Biosca, 2017; Peralta et al., 2018; Thakur et al., 2022)
Cover crops	General soilborne pathogens	Increases in organic matter support diverse soil biota	Cover crops provide organic residues that feed beneficial microbes	Improves soil stability, reduces erosion	Competition with main crops for nutrients and water; termination of cover crops can be labour-intensive	(Bailey & Lazarovits, 2017; Dawadi et al., 2019)
Solarization	*Fusarium, Verticillium*	Temporarily reduces microbial diversity but suppresses pathogens	Uses solar heat to reduce pathogen load, allowing beneficial microbes to recolonize	Enhances soil health by reducing pathogen pressure	High labour costs, and limited effectiveness in cooler climates; may temporarily harm beneficial soil organisms	(Caihua et al., 2022; Elshahawy et al., 2023; Funahashi et al., 2022)

(continued)

Table 2 (continued)

Management practice	Target pathogens	Impact on soil microbial diversity	Mechanism of disease control	Effects on soil health	Challenges	References
Organic amendments	*Rhizoctonia, Pythium, Phytophthora*	Increases microbial activity, promotes beneficial fungi	Adds organic matter that supports a healthy soil food web	Improves soil fertility, enhances organic matter content	The cost and availability of amendments can be limiting; uneven distribution in the soil may reduce effectiveness	(Elshahawy et al., 2023; Gilardi et al., 2014)
Resistant varieties	Specific pathogens (e.g., *Fusarium* spp.)	Reduces pathogen pressure, allows beneficial microbes to thrive	Breeding plants that resist pathogen attacks, indirectly support microbial health	Maintains soil structure, and promotes sustainable cropping systems	Limited availability of resistant varieties; potential for pathogen evolution and resistance breakdown	(Jacott et al., 2024; Li et al., 2021; Précigout et al., 2023)

3.1.1 Biological Control

The use of beneficial organisms, such as mycorrhiza fungi, *Trichoderma* spp., and *Bacillus* spp., can help suppress plant pathogens and promote soil health. Biological control is considered a viable and sustainable strategy because it enhances soil health along with outcompeting pathogens for vital resources, synthesize antimicrobial substances, or stimulate resistance mechanisms within plants. *Trichoderma* spp. are the most famous biocontrol agents that not only demonstrated efficacy in promoting plant growth and augmenting nutrient absorption but also enhanced soil structure by fostering the aggregation of soil particles. Many researchers have highlighted the importance of integrated strategies including biological control along with cultural practices to attain sustainable soil health over extended periods. The potential of beneficial microbes (*Trichoderma* spp. and *Bacillus* spp.,) in suppressing soil-borne pathogens while simultaneously improving soil health is studied by (Asghar et al., 2024).

3.1.2 Soil Amendments

Introducing exogenous organic matter in the form of compost, cover crops, or biochar may improve soil structure, enhance the diversity of microbes, and suppress soil-borne pathogens. The added organic amendments are used as a food source by such beneficial microorganisms, possibly outcompeting the pathogens. In particular, for biochar, there are probable benefits for improved fertility, increased microbial activity, and specific suppression of certain soil-borne diseases.

3.1.3 Chemical Control

While this application of chemical pesticides in the treatment of plant diseases might be quite useful, one ought to be very cautious not to harm the soil. Over-reliance on chemical control is likely to breed pesticide-resistant pathogens and consequently interfere with other helpful soil organisms. Integrated approaches that join chemical control with biological and cultural practices are often more effective and sustainable in the long run.

3.1.4 Cultural Practices and Disease Suppression

Cultural practices, such as crop rotation and cover cropping, have been shown to improve soil health by reducing the prevalence of soil-borne diseases. Crop rotation with the resistance variety in the same area is the key strategy in a sustainable agriculture system as it significantly reduces the frequency of root pathogens along with improved soil microbial diversity and nutrient cycling. This approach can effectively disrupt the life cycles of specific pathogens, particularly those exhibiting host specificity like the alternation of a susceptible crop with a non-host crop significantly reducing the population of soil-borne pathogens such as nematodes and certain fungi in many cases. Furthermore, crop rotation contributes to the preservation of soil fertility and structure by diversifying the types of organic matter reintroduced into the soil ecosystem.

3.2 Soil Health Monitoring

The systematic and continual assessment of soil health is vital for the prompt detection of potential challenges and for appraising the effectiveness of disease management strategies. Metrics of soil health, such as microbial diversity, nutrient availability, and soil structure, can provide profound insights into the ramifications of plant diseases on the soil ecosystem.

3.2.1 Microbial Diversity

Monitoring of diversity and abundance of microbes present in soil can aid in the detection of changes within microbial communities that may indicate the arrival of phytopathogens or the decline of beneficial organisms. For the assessment of diversity in microorganisms, different technologies like deoxyribonucleic acid sequencing and metagenomic analysis are being used (Edet et al., 2017).

3.2.2 Nutrient Levels

An analysis of the nutrient composition of soil can clarify deficiencies that may arise due to the presence of phytopathogens. For example, an overabundance of nitrogen was detected in soil due to its impaired absorption by diseased plants. Subsequently, soil assessment regularly is needed to plan the application of fertilisers and other soil amendments to rectify these nutritional imbalances (Hughes et al., 2023).

3.2.3 Soil Structure

Assessment of soil structure, comprising porosity and aggregate stability of soil can give an idea of the damage caused by plant pathogens to soil health. Poor soil structure can signal the presence of root pathogens or the breakdown of organic matter, which may result in enduring consequences for soil vitality (Muramoto et al., 2022).

Nevertheless, the intricacy and cost involved in implementing integrated disease management strategies are significant barriers. The requirement for constant monitoring, the deployment and application of new biological control agents, and the adoption of specialized cultural practices may necessitate substantial resources and expertise. At times, pathogens may evolve resistance to biological control agents or adapt to mitigate the effects of crop rotation and other cultural practices. This phenomenon can complicate disease management and diminish the efficacy of these strategies in safeguarding soil integrity.

4 Future Prospects of Plant Pathology and Soil Health

The role of plant pathology in maintaining soil health will, therefore, continue to rise as the agricultural sector struggles with the challenges of climate change, population expansion, and degradation of the environment. The combination of plant pathology with soil science, ecology, and agronomy will be paramount in the formulation of sustainable disease management practices that encourage soil resilience, foster biodiversity, and sustain robust ecosystems. The role of plant pathology in maintaining soil health remains a dynamic area of study, with significant implications for sustainable agriculture or ecosystem and food security. As our comprehension of these interactions' advances, numerous prospects and research trajectories are anticipated to arise.

4.1 Advances in Soil Microbiome Research

4.1.1 Microbiome Engineering

Future work will likely be geared toward manipulating the soil microbiome to enhance disease suppression and soil health. Scientists can bioengineer soils that are more resistant to plant pathogens and more conducive to healthy plant growth by detecting and stimulating beneficial microbes.

4.1.2 Precision Agriculture

The incorporation of modern technologies in the agriculture sector like metagenomics, machine learning, and remote sensing, will facilitate in accurate monitoring, soil health maintenance and pathogen dynamics. Such advancements could pave the way for the making of targeted interventions aimed at reducing the adverse effects of plant diseases and their management on soil health. The advancements in high-throughput sequencing and metagenomics analysis facilitate precise monitoring of soil microbial populations and the prompt identification of pathogens. Besides these, these technologies can contribute significantly to the proactive management of soil health and the prevention of diseases.

4.2 Sustainable Disease Management Practices

4.2.1 Sustainable Agriculture Practices

The future approach of plant pathology will always be oriented toward sustainable farming practices which describe the need to maintain soil ecosystem health and preserve the environment. This includes the development of disease resistance varieties of crops, the use of cover crops and crop rotation, in addition to minimizing the application of chemicals by applying integrated systems of managing pests and diseases. Sustainable practices assist in managing plant diseases while at the same time promoting soil health, water quality improvement, and biodiversity.

4.2.2 Biological Control Innovations

New biological control agents like microbial consortia, phage therapy precise predator species, and their implementation strategies for the future are also expected to be discovered. These innovative biological agents are supposed to be far more effective, sustainable and well-targeted predator species for soil-borne pathogens without damaging beneficial or environment-friendly organisms.

4.2.3 Enhanced Soil Amendments

Further refinement and usage of different organic and inorganic soil amendments for the improvement of soil health and diminishing the levels of pathogenic organisms, will continue. Further research is needed for the optimization of such soil amendments to enhance their performances while lessening any possible adverse impacts on the ecosystem or environment.

4.3 Climate Change and Plant Pathology

4.3.1 Climate Change Influence

Climate changes have a deep relationship with plant pathology as any climatic alterations in temperature and precipitation regimes are anticipated to modify the distribution and dynamics of soil-borne pathogens. Future scholarly inquiries must investigate the implications of these transformations on the interplay between plant pathology and soil health, as well as the methods by which agricultural practices can develop to alleviate these consequences. A comprehensive review conducted by Garrett et al. (2006) elucidated how shifting temperature and precipitation patterns could potentially affect the distribution and pathogenicity of soil-borne pathogens, thereby intensifying soil health dilemmas. Future investigations ought to concentrate on revealing these interactions and formulating adaptive approaches for soil and disease management within the context of climate change.

4.3.2 Adaptation Strategies

The primary focus will be on the development of disease-resistant germplasm of crops and soil health management practices that enhance resilience against both climate change and pathogen pressures. This may necessitate the breeding of plant varieties that exhibit not only resistance to diseases but also have the potential to perform better in changing soil conditions.

4.4 Integration of Plant Pathology into Holistic Soil Health Models

4.4.1 Systems Approach

The trajectory of future research in plant pathology and soil health is likely to embrace a more integrated, systems-oriented methodology that takes into account the intricate interactions among plants, pathogens, soil microbes, and environmental variables. Such a broad paradigm will be essential to furthering sustainable agricultural systems.

4.4.2 Policy and Education

The assimilation of information derived from plant pathology into the governing frameworks of soil health policies and educational curricula is profoundly essential. Improving awareness regarding the essential function of soil health in disease management might lead to more extensive adoption of sustainable agricultural approaches.

4.5 Genomic and Biotechnological Advances

4.5.1 CRISPR and Gene Editing

Improvements in gene editing technologies like CRISPR may also be used to develop crops with improved resistance to soil-borne pathogens. In addition, gene editing technologies can be used to modify the soil microbes to increase their disease-suppressing capability.

4.5.2 Synthetic Biology

The field of synthetic biology may provide novel avenues for engineering soil microbes or plants to interact better with their surroundings, thereby enhancing both soil health and plant disease resistance.

4.6 Economic and Environmental Considerations

4.6.1 Cost-Effective Solutions

Subsequent investigations must prioritize the formulation of economically viable disease management methodologies that are readily available to agricultural practitioners globally, especially in contexts characterized by limited resources.

4.6.2 Environmental Impact

Ensuring that new strategies for managing plant diseases and improving soil health are environmentally sustainable will be critical. This includes minimizing the use of chemical inputs and reducing the environmental footprint of agricultural practices.

5 Conclusion

This chapter offers a comprehensive examination of the mechanisms by which soil-borne pathogens upset soil architecture, microbial diversity, and nutrient cycling while emphasizing the formidable challenges facing modern agriculture in maintaining soil health in the presence of pathogenic pressure. The findings accentuate the absolute need for integrated soil management frameworks that harmonize disease mitigation with enhancing robust soil ecosystems. Implementation of sustainable agricultural practices such as crop rotation, cover cropping, use of organic amendments, and development of disease-resistant cultivars are crucial to alleviate detrimental impacts caused by soil-borne pathogens. By adopting these strategies soil vitality or health can be enhanced and long-lasting agricultural sustainability can be achieved while reducing reliance on chemical inputs. Therefore, with the increasing demand for food in light of growing global populations, there is an emerging importance of soil conservation amidst these challenges regarding pathogens. The formulation and implementation of comprehensive soil management practices that focus on agricultural productivity and the resilience of ecosystems are imperative in future research undertakings and policy frameworks. These approaches would help them to protect soil health and promote sustainable agriculture that can ensure food security for generations to come (Table 3).

Table 3 Future research directions in plant pathology and soil health

Research area	Focus	Potential impact	References
Microbiome engineering	Manipulation of soil microbiomes for disease suppression	Enhanced soil resilience, improved plant health	(Gomez-Exposito, 2017; Luo et al., 2022)
Precision agriculture	Integration of technologies for monitoring soil and pathogens	More targeted interventions, improved disease management	(Hillnhütter et al., 2010; Jayalakshmi et al., 2023; Shireshi & Raman, 2024)
Sustainable practices	Development of practices that balance disease management and soil health	Increased sustainability, reduced environmental impact	(Atwood et al., 2022; Maharjan et al., 2024; Moulik et al., 2023)
Climate change adaptation	Understanding the effects of climate change on pathogens and soil health	Better preparedness for changing disease dynamics and soil conditions	(Egon & Oloyede, 2023; Lahlali et al., 2024)
Genomic and biotechnological advances	Use of CRISPR and synthetic biology for disease resistance and soil health	Enhanced plant and soil resilience	(Egon & Oloyede, 2023; Erdoğan et al., 2023)

References

Ahmed, W., Yang, J., Tan, Y., Munir, S., et al. (2022). Ralstonia solanacearum, a deadly pathogen: Revisiting the bacterial wilt biocontrol practices in tobacco and other Solanaceae. *Rhizosphere, 21*, Article 100479.

Álvarez, B., & Biosca, E. G. (2017). Bacteriophage-based bacterial wilt biocontrol for an environmentally sustainable agriculture. *Frontiers in Plant Science, 8*, 1218.

Asghar, W., Craven, K. D., Kataoka, R., Mahmood, A., et al. (2024). The application of Trichoderma spp., an old but new useful fungus, in sustainable soil health intensification: A comprehensive strategy for addressing challenges. *Plant Stress*, 100455.

Atungulu, G., & Mohammadi-Shad, Z. (2019). Reference on mycotoxins occurrence, prevalence, and risk assessment in food systems. In Novel technologies and systems for food preservation (pp. 294–343). *IGI Global.*

Atwood, L. W., Racette, K. A., Diggelmann, M., Masala, C. A., et al. (2022). Soil health: New opportunities to innovate in crop protection research and development. *Frontiers in Environmental Science, 10*, Article 821742.

Bailey, K. L., & Lazarovits, G. (2017). Suppressing soil-borne diseases with organic amendments. *Applied Soil Ecology, 15*, 165–175.

Bakker, P. A., & Berendsen, R. L. (2022). The soil-borne ultimatum, microbial biotechnology and sustainable agriculture. *Microbial Biotechnology, 15*(1), 84.

Batista, B. D., Wang, J., Liu, H., Kaur, S., et al. (2024). Biotic and abiotic responses to soilborne pathogens and environmental predictors of soil health. *Soil Biology and Biochemistry, 189*, Article 109246.

Benkeblia, N. (2022). Fungal diseases in agriculture: significance, management, and control In *Mycoagroecology* (pp. 161–174). *CRC Press.*

Bruna, D., Batista., Jun-Tao, W., Hongwei, L., et al. (2023). Biotic and abiotic responses to soil-borne pathogens and environmental predictors of soil health. *Soil Biology & Biochemistry, 189*, 109246.

Caihua, S. H. I., Linlin, S. H. I., Qingjun, W. U., Shaoli, W. A. N. G., et al. (2022). Effect of solarization to kill Bradysia cellarum on chinese chive growth and soil microbial diversity. *Frontiers of Agricultural Science & Engineering, 9*(1).

Chiranjeeb, K., Kumar, S., & Rana, R. S. (2024). Soil: A potential source for mitigating food, water and bioenergy crisis. *Environmental Nexus for Resource Management* (pp. 287–301). CRC Press.

Dawadi, S., Baysal-Gurel, F., Addesso, K., Oliver, J., et al. (2019). Impact of cover crop usage on soilborne diseases in field nursery production. *Agronomy, 9*(11), 753.

de Vries, F. T., Griffiths, R. I., Knight, C. G., Nicolitch, O., et al. (2020). Harnessing rhizosphere microbiomes for drought-resilient crop production. *Science, 368*(6488), 270–274.

Du, M., Minasny, B., & Rabbi, S. M. (2024). Carbon to nitrogen stoichiometry of organic amendments influences the improvement of aggregate stability of a cropping vertisol. *Soil Use and Management, 40*(2), Article e13087.

Edet, U., Antai, S., Brooks, A., Asitok, A., et al. (2017). An overview of cultural, molecular and metagenomic techniques in description of microbial diversity. *Journal of Advances in Microbiology, 7*(2), 1–19.

Egon, K., & Oloyede, J. O. (2023). The impact of climate change on plant pathogens. https://doi.org/10.31219/osf.io/e4xkz

Elshahawy, I., Saied, N., Abd-El-Kareem, F., & Abd-Elgawad, M. (2023). Enhanced activity of Trichoderma asperellum introduced in solarized soil and its implications on the integrated control of strawberry-black root rot. *Heliyon, 10*(17).

Erdoğan, İ, Cevher-Keskin, B., Bilir, Ö., Hong, Y., et al. (2023). Recent developments in CRISPR/Cas9 genome-editing technology related to plant disease resistance and abiotic stress tolerance. *Biology, 12*(7), 1037.

Even, R. J., & Cotrufo, M. F. (2024). The ability of soils to aggregate, more than the state of aggregation, promotes protected soil organic matter formation. *Geoderma, 442*, Article 116760.

Fujita, H., Yoshida, S., Suzuki, K., & Toju, H. (2023). Alternative stable states in soil microbiomes of agroecosystems. *bioRxiv*. https://doi.org/10.1101/2022.08.23.505048

Funahashi, F., Myrold, D. D., & Parke, J. L. (2022). The effects of soil solarization and application of a Trichoderma biocontrol agent on soil fungal and prokaryotic communities. *Soil Science Society of America Journal, 86*(2), 369–383.

Garrett, K. A., Dendy, S. P., Frank, E. E., Rouse, M. N., et al. (2006). Climate change effects on plant disease: Genomes to ecosystems. *Annual Review of Phytopathology, 44*(1), 489–509.

Gilardi, G., Demarchi, S., Gullino, M. L., & Garibaldi, A. (2014). Effect of simulated soil Solarization and organic amendments on f usarium wilt of rocket and basil under controlled conditions. *Journal of Phytopathology, 162*(9), 557–566.

Gomez-Exposito, R. (2017). *Microbiome dynamics of disease suppressive soils*. Wageningen University.

Gowda, M. T., Rai, A. B., & Singh, B. (2019). Root knot nematodes menace in vegetable crops and their management in India: A review. *Vegetable Science, 46*(1and 2), 1–16.

Guzmán, M. D. P. R. (2021). Soil biodiversity and root pathogens in agroecosystems. *In Biodiversity of Ecosystems*. IntechOpen.

Hayes, M. M., Dewberry, R. J., Babujee, L., Moritz, R., et al. (2022). Validating methods to eradicate plant-pathogenic Ralstonia strains reveals that growth in planta increases bacterial stress tolerance. *Microbiology Spectrum, 10*(6), e02270-e12222.

Hillnhütter, C., Schweizer, A., Kühnhold, V., & Sikora, R. A. (2010). Remote sensing for the detection of soil-borne plant parasitic nematodes and fungal pathogens. *Precision crop protection-the challenge and use of heterogeneity*, 151–165.

Hughes, H. M., Koolen, S., Kuhnert, M., Baggs, E. M., et al. (2023). Towards a farmer-feasible soil health assessment that is globally applicable. *Journal of Environmental Management, 345*, Article 118582.

Hussain, I., Farooq, T., Khan, S. A., Ali, N., et al. (2022). Variability in indigenous Pakistani tomato lines and worldwide reference collection for Tomato Mosaic Virus (ToMV) and Tomato Yellow Leaf Curl Virus (TYLCV) infection. *Brazilian Journal of Biology, 84*, Article e253605.

Jacott, C. N., Schoonbeek, H. J., Sidhu, G. S., Steuernagel, B., et al. (2024). Pathogen lifestyle determines host genetic signature of quantitative disease resistance loci in oilseed rape (Brassica napus). *Theoretical and Applied Genetics, 137*(3), 65.

Jakubcsiková, M., Demková, L., Renčo, M., & Čerevková, A. (2023). Evaluation of the effect of organic matter from invasive plants on soil nematode communities. *Plants, 12*(19), 3459.

Jat, R. S., & Singh, D. (2018). Broomrape intimidation to rapeseedmustard production in semi-arid regions. *Advances in Plants and Agriculture Research, 8*, 560–562.

Jayalakshmi, K., Manzar, N., Namriboi, B.K., Kashyap, A.S., et al. (2023). Detection and diagnosis of important soil-borne pathogens. In *Soil-borne Phytopathogens* (101–126). *Springer Nature Singapore.*

Khan, A., Ansari, S. A., Haris, M., Hussain, T., et al. (2023). Meloidogyne species: threat to vegetable produce. In *Root-Galling disease of vegetable plants* (pp. 61–83). Springer Nature Singapore.

Lahlali, R., Taoussi, M., Laasli, S. E., Gachara, G., et al. (2024). Effects of climate change on plant pathogens and host-pathogen interactions. *Crop and Environment, 3*(3), 159–170.

Li, Q., Wang, B., Yu, J., & Dou, D. (2021). Pathogen-informed breeding for crop disease resistance. *Journal of Integrative Plant Biology, 63*(2), 305–311.

Liang, J., Wei, C., Song, X., Wang, R., et al. (2024). Bacterial wilt affects the structure and assembly of microbial communities along the soil-root continuum. *Environmental Microbiome, 19*(2), 6.

Lucas, J. A. (2011). Advances in plant disease and pest management. *The Journal of Agricultural Science, 149*(S1), 91–114.

Luo, L., Zhang, J., Ye, C., Li, S., et al. (2022). Foliar pathogen infection manipulates soil health through root exudate-modified rhizosphere microbiome. *Microbiology Spectrum, 10*(6), e02418-02422.

Maharjan, B., Das, S., Thapa, V. R., & Sharma Acharya, B. (2024). Soil health cycle. *Agrosystems, Geosciences & Environment, 7*(2), Article e20504.

Maria, R., Finckh, & Christian, B. (2014). Soil health management and biodiversity: the central pillars of plant disease management in organic agriculture.

Meister, A., Bohm, K., Gutiérrez-Ginés, M. J., Gaw, S., et al. (2023). Effects of native plants on nitrogen cycling microorganisms in soil. *Applied Soil Ecology, 191*, Article 105031.

Moulik, A., Bhowmik, A., & Bhattacharya, P. (2023). Soil-health: Concepts, implications and management. *International Journal of Advanced Biochemistry Research*. https://doi.org/10.33545/26174693.2024.v8.i4sc.965

Muramoto, J., Parr, D. M., Perez, J., & Wong, D. G. (2022). Integrated soil health management for plant health and one health: Lessons from histories of soil-borne disease management in California strawberries and arthropod pest management. *Frontiers in Sustainable Food Systems, 6*, Article 839648.

Negewo, T., Ahmed, S., Tessema, T., & Tana, T. (2022). Biological characteristics, impacts, and management of crenate broomrape (Orobanche crenata) in faba bean (Vicia faba): A review. *Frontiers in Agronomy, 4*, Article 708187.

Peralta, A. L., Sun, Y., McDaniel, M. D., & Lennon, J. T. (2018). Crop rotational diversity increases disease suppressive capacity of soil microbiomes. *Ecosphere, 9*(5), Article e02235.

Précigout, P. A., Renard, D., Sanner, J., Claessen, D., et al. (2023). Crop mixtures outperform rotations and landscape mosaics in regulation of two fungal wheat pathogens: A simulation study. *Landscape Ecology, 38*(1), 77–97.

Sanzovo, A. W. S., Silvestre, D. A., Goes, K. C. G. P., Volsi, B., et al. (2023). Crop rotation and inoculation increase soil bradyrhizobia population, soybean grain yields, and profitability. *Brazilian Journal of Microbiology, 54*(4), 3187–3200.

Shah, K. K., Modi, B., Pandey, H. P., Subedi, A., et al. (2021). Diversified crop rotation: An approach for sustainable agriculture production. *Advances in Agriculture, 2021*(1), 8924087.

Shireshi, S. S., & Raman, R. (2024). Soil-borne pathogens management in cloud-connected agriculture: An IoT and decision tree algorithmic approach. *International Conference on Emerging Smart Computing and Informatics (ESCI), 2024*, 1–5.

Sonam, A., & Rakesh, K. (2019). Biocontrol of root-knot nematodes through PGPR. *International Journal of Current Microbiology and Applied Sciences*. https://doi.org/10.20546/IJCMAS.2019.811.062

Spadaro, D., & Gullino, M. L. (2019). Sustainable management of plant diseases. *Innovations in sustainable agriculture*, 337–359.

Srivastava, P., Sachan, K., Baskar, P., Saikanth, D. R. K., et al. (2023). Soil microbes expertly balancing nutrient demands and environmental preservation and ensuring the delicate stability of our ecosystems-a review. *International Journal of Plant & Soil Science, 35*(18), 989–1000.

Thakur, R., Verma, S., Gupta, S., Negi, G., et al. (2022). Role of soil health in plant disease management: A review. *Agricultural Reviews, 43*(1), 70–76.

Tong, Y., Zheng, X., Hu, Y., Wu, J., et al. (2024). Root exudate-mediated plant–microbiome interactions determine plant health during disease infection. *Agriculture, Ecosystems and Environment, 370*, Article 109056.

van der Sloot, M., Maerowitz-Mcmahan, S., Postma, J., Limpens, J., et al. (2024). Soil-borne disease suppressiveness after short and long term application of fermented, composted or fresh organic amendment treatments in arable soils. *Applied Soil Ecology, 195*, Article 105268.

Wang, J., Shen, D., Ge, C., Du, Y., et al. (2020). Filamentous Phytophthora pathogens deploy effectors to interfere with bacterial growth and motility. *Frontiers in Microbiology, 11*, Article 581511.

Wiesenbauer, J., König, A., Gorka, S., Marchand, L., et al. (2024). A pulse of simulated root exudation alters the composition and temporal dynamics of microbial metabolites in its immediate vicinity. *Soil Biology and Biochemistry, 189*, Article 109259.

Yan, D., Wang, Q., Li, Y., Guo, M., et al. (2022). Efficacy and economics evaluation of seed rhizome treatment combined with preplant soil fumigation on ginger soilborne disease, plant growth, and yield promotion. *Journal of the Science of Food and Agriculture, 102*(5), 1894–1902.

Yang, Q., Peng, J., Ni, S., Zhang, C., et al. (2024). Soil erosion-induced decline in aggregate stability and soil organic carbon reduces aggregate-associated microbial diversity and multifunctionality of agricultural slope in the Mollisol region. *Land Degradation and Development, 35*(11), 3714–3726.

Yuan, Y., & Li, J. (2022). Dodder parasitism limited the effect of arbuscular mycorrhizal fungi on litter decomposition. *Soil Biology and Biochemistry, 174*, Article 108837.

Zichun, G. U. O., Wei, L. I., ISLAM, M. U., Yuekai, W. A. N. G., et al. (2022). Nitrogen fertilization degrades soil aggregation by increasing ammonium ions and decreasing biological binding agents on a Vertisol after 12 years. *Pedosphere, 32*(4), 629–636.

Trichoderma a Holistic Fungus with Biocontrol Potential for Agricultural Diseases

Laith Khalil Tawfeeq Al-Ani, Ivana Mitrović, Edson Luiz Furtado, Pranab Nanda Bhattacharyya, Lobna Hajji-Hedfi, Dalia Sukmawati, Liliana Aguilar-Marcelino, Rungtiwa Phookamsak, Francesca Garganese, Rafaela Araújo Guimarães, Monica Esther Maria Roca, Franca Carrasco, Carolina Gracia Poitevin, Mitzuko Dautt-Castro, Mavis Agyeiwaa Acheampong, and Surono Surono

Abstract *Trichoderma* is a genus, belonging to the division Ascomycota (Sordariomycetes, Hypocreales, Hypocreaceae) that have been popularly used as a biocontrol agent in agriculture because of their ability to interact and antagonize phytopathogenic organisms such as bacteria, fungi, nematodes, and oomycetes.

The original version of the chapter has been revised. A correction to this chapter can be found at https://doi.org/10.1007/978-3-031-91114-9_48

L. K. T. Al-Ani (✉)
School of Biology Science, Universiti Sains Malaysia, 11800 Minden, Pulau Pinang, Malaysia
e-mail: cmv_virus2002@yahoo.com

I. Mitrović
Department of Biotechnology and Pharmaceutical Engineering, Faculty of Technology Novi Sad, University of Novi Sad, Novi Sad, Serbia

E. L. Furtado
Departamento de Proteção Vegetal, Faculdade de Ciências Agronômicas de Boturcatu, Universidade Estadual Paulista, São Paulo, Brazil

P. N. Bhattacharyya
Institutional Biotech Hub (IBT-HUB), Nanda Nath Saikia College, Titabar, Jorhat 785630, Assam, India

L. Hajji-Hedfi
Regional Centre of Agricultural Research of SidiBouzid, CRRA, SidiBouzid, Tunisia

D. Sukmawati
Department of Biology, Faculty of Mathematics and Natural Sciences, Universitas Negeri Jakarta, Rawamangun, Jakarta Timur, Indonesia

Universitas Negeri Jakarta Culture Collection (UNJCC), Faculty of Mathematics and Natural Sciences, Universitas Negeri Jakarta, Rawamangun, Indonesia

L. Aguilar-Marcelino
CENID-Salud Animal e Inocuidad, INIFAP. Carretera Federal Cuernavaca-Cuautla, 62550 Jiutepec, Mexico

M. Shaaban (ed.), *Soils and Sustainable Agriculture*, Frontier Studies in Soil Science, https://doi.org/10.1007/978-3-031-91114-9_37

Trichoderma can easily be isolated from natural soil, dead and decaying plant organic matter, and wood and are classified as imperfect fungi. Production of antibiotics, competition for nutrients, and parasitism are considered as direct antagonistic mechanisms of *Trichoderma* during microbe-pathogen interactions. The production of secondary metabolites as well as the secretion of hydrolytic cell wall degrading enzymes also plays inevitable roles during microbial antagonism. The indirect mechanisms involve the interaction between *Trichoderma* with host plants. Working as a plant growth promoter, inducing systemic resistance against biotic and abiotic stressors, and enhancing the uptake of nutrients are other beneficial roles exhibited by *Trichoderma* during plant–microbe interactions. The current chapter is intended to discuss in detail on the mechanisms of *Trichoderma*-mediated management of plant pathogenesis. The knowledge and understanding on the tripartite symbiotic interaction between *Trichoderma* with other beneficial plant microorganisms and host plants would explore new avenues for improving sustainable agriculture.

Keywords Bioremediation · Endophytic fungi · Hyperparasitism · Mycoparasitism · Mycotoxin · Siderophores

---

R. Phookamsak
Key Laboratory for Plant Diversity and Biogeography of East Asia, Kunming Institute of Botany, CAS, Kunming 650201, Yunnan, China

F. Garganese
Dipartimento Di Scienze del Suolo, Della Pianta E Degli Alimenti, University of Bari Aldo Moro, Bari, Italy

R. A. Guimarães
Laboratório de Controle Biológico, Departamento de Fitopatologia, Universidade Federal de Lavras, Lavras, MG, Brazil

M. E. M. Roca
University National of La Rioja, La Rioja, Argentina

F. Carrasco
Estación Experimental Agropecuaria Catamarca, Instituto Nacional de Tecnología Agropecuaria, Valle Viejo, Catamarca, Argentina

C. G. Poitevin
Laboratory of Microbiology and Molecular Biology (LabMicro), Department of Basic Pathology, Federal University of Paraná, Curitiba, Paraná, Brazil

M. Dautt-Castro
Laboratorio de Genética y Biología Molecular de Plantas, Centro de Investigación en Alimentación y Desarrollo, A.C. (CIAD), Hermosillo, Mexico

M. A. Acheampong
Department of Crop Science,, University of Ghana, Legon, Accra, Ghana

S. Surono
Balai Penelitian Tanah, Jawa Barat, Indonesia

1 Introduction

Trichoderma was reported as the biocontrol agent in the early 1930s (Tyskiewicz et al., 2022) . This genus is classified as imperfect fungi previously, then found belongs to the order Hypocreales of Ascomycota after detected the sexual stage (Howell, 2003; Küçük & Kivanç, 2003). *Trichoderma* is known for its abilities to parasite other pathogenic fungal microflora especially associated with root rots and wilt diseases (Santoro et al., 2014; Verma et al., 2007). *Trichoderma* species are reported as ubiquitous and cosmopolitan soil-borne Ascomycetes that is found in almost all types of soil habitats and other environments such as manure, decaying plant tissues, and wood (Singh et al., 2014) (Fig. 1). According to Chaverri et al. (2011) *Trichoderma* spp. can be considered as endophytic fungi that generally found to be reported in all types of soils such as agricultural, orchard, and forest as opportunistic as plant symbionts, and is considered as a successful competitor of plant pathogens (Woo et al., 2006; Kim et al., 2012; Hajji-Hedfi et al. 2024). *Trichoderma* is one of the largest and hyperdiverse genera and contains approximately 524 accepted species based on the morpho-molecular taxonomic approach (387 species as *Trichoderma* and 137 species as *Hypocrea*) (Hyde et al., 2020)

The importance of *Trichoderma* as an antagonist is due to its ability to survive in adverse environmental conditions and its working efficiency in the utilization of nutrients. High reproductive potential, the capability to engineer the rhizosphere, intense aggressiveness against phytopathogenic fungi, effectiveness in plant growth promotion, and induction of systemic resistance (ISR) (Al-Ani, 2018a; Hermosa et al., 2012; Mukherjee et al., 2012; Verma et al., 2007), among others are associated with beneficial roles played by *Trichoderma* in soil and agriculture. Based on its potential in plant protection, different commercial *Trichoderma* formulations has recently been developed for their use in agriculture (Sanjeev et al., 2014).

Definitely for inhibiting the colony of fungi could by b antifungal compounds as natural compounds (Polymers) and secondary compounds producing by microbes (Al-Ani et al., 2012; Mohammed et al., 2012, 2013; Jatoi et al., 2020, 2022; Sukmawati et al., 2024a, b), as well as, physical method as gamma radiation (De Souza and Al-Ani, 2023). Members of *Trichoderma* produce active secondary metabolites and bioactive compounds that have immense potential in agricultural managements and biotechnological applications (Schuster & Schmoll, 2010; Mukherjee et al., 2013; Singh et al., 2017; Sharma et al., 2019). More than 300 secondary metabolites and mycotoxins such as gliovirin, harzianic, peptaibols, polyketides, polyketides, pyrones, siderophores, steroids, terpenes, and trichodermin, are obtained from *Trichoderma* species with significant antagonistic properties against other organisms (Schuster & Schmoll, 2010; Malmierca et al., 2013; Garnica-Vergara et al., 2016; Sharma et al., 2019).

As eco-friendly approaches of disease management (Villar-Luna et al., 2022), *Trichoderma* species are widely being used as fungal biocontrol agent against a series of soil-borne pathogens and plant-parasitic nematodes. The fungus is effective against phytopathogens as it can exert egg parasitism, as well as can be used

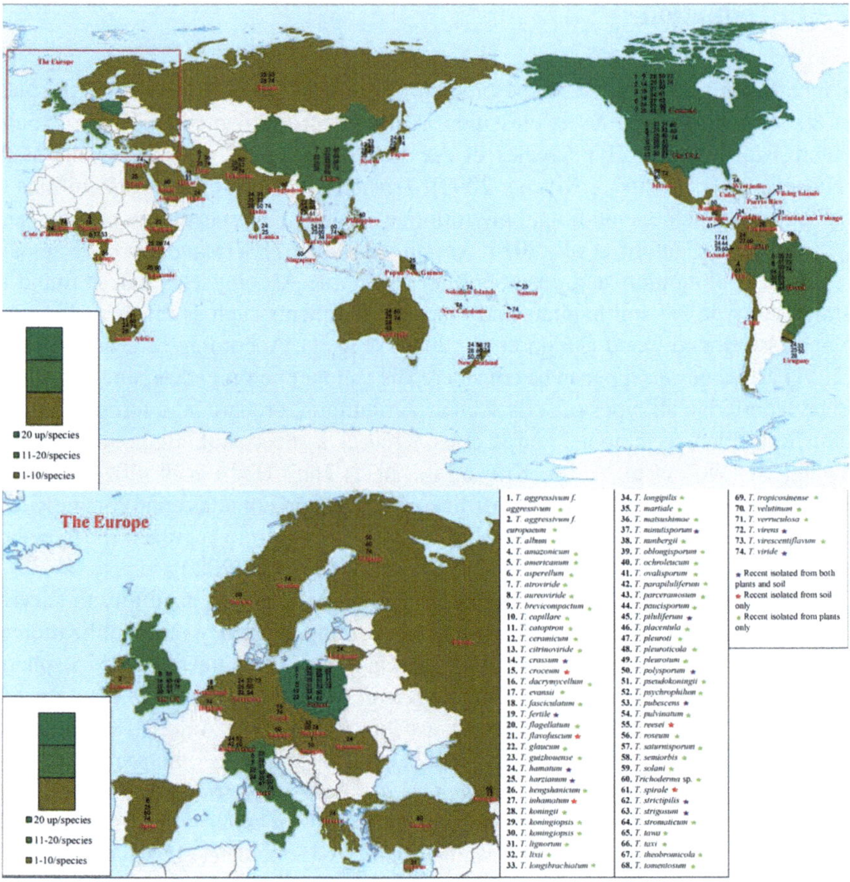

Fig. 1 World distribution of *Trichoderma* species based on U.S. National fungus collections fungus-host database (https://nt.ars-grin.gov/fungaldatabases/)

as commercial bioformulations against devastating pests and pathogens (Samolski et al., 2012; Szabó, 2014; Singh et al., 2017; Mahato et al., 2018; Sharma et al., 2019; Poveda et al., 2020). *Trichoderma* are also known to secrete hydrolytic cell wall degrading enzymes (CWDE) such as cellulases, chitinases, glucanases, and proteases, which is essential in biomass degradation. These hydrolytic enzymes can inhibit the growth of phytopathogens and enhance plant disease resistance through invading and destroying the fungal prey's cell walls using mycoparasitism and other antagonistic activities (Vinale et al., 2008; Bischof et al., 2016; Guzmán- Guzmán et al., 2019; Sharma et al., 2019).

However, there are some biological factors and natural products in supporting plants to confront different abiotic stress with enhancing the plant growth (Lalarukh et al., 2022a, b). As plant growth promoters, *Trichoderma* is an excellent biological tool to enhance plant growth and increasing the plant tolerance in ecologically harsh conditions such as biotic and abiotic stresses including the drought, extreme temperature, salinity, light, nutrient and water stress conditions (Stewart & Hill, 2014; Balestrini et al.,2017; Al-Ani, 2018b; Sharma et al., 2019; Rebolledo-Prudencio et al., 2020). *Trichoderma* species are known to assist the plants to improve the root architecture for enhanced nutrient mineralization, and uptake and enhanced leaf growth and photosynthetic efficiency (Schuster & Schmoll, 2010; Guzmán-Guzmán et al., 2019; Sharma et al., 2019). The role of *Trichoderma* in enhancing the plant growth as a biofertilizer is possible due to its abilities to form a biofilm (Aguilar-Marcelino et al., 2021; Triveni et al., 2013). *Trichoderma*-enriched biofertilizers are recently being used to increase the bioavailability of root phosphorous, increasing crop yield and lowering the essentially for conventional agrochemicals and pesticides that can decrease pesticides residues (Aguilar-Marcelino et al., 2023; Al-Ani et al. 2021a).

Trichoderma comprises different economically essential fungal species including the *T. asperellum*, *T. atroviride*, *T. harzianum*, *T. reesei*. *T. virens*, and *T. viride*. Recent advancements in molecular techniques have included *Trichoderma* into the following four groups: (1) *T. asperellum*, (2) *T. atroviride –T. koningii* complex, (3) *T. harzianum—T. inhamatum* complex, and (4) *T. longibrachiatum*, (Tyskiewicz et al., 2022). As the suppression mechanisms of phytopathogenic fungi by *Trichoderma* may vary, it is of practical importance for agriculture to study the synergistic effects of *Trichoderma* culture in microbial consortium.

This chapter focuses on the possibility of using *Trichoderma* strains in biocontrol of significant plant diseases caused by plant pathogenic microbes including the bacteria, fungi, nematodes, and other devastating plant pests. Special emphasis has, therefore, been given to highlight the putative mechanisms of action of *Trichoderma* biocide against various phytopathogens including their abilities to produce secondary metabolites and exhibition of tripartite interactions between *Trichoderma*-host–pathogen. In addition to microbial antagonism, the importance of *Trichoderma* as plant growth promoter has also been discussed in the current review.

2 The Role of *Trichoderma* in Plant Disease Control

Many scientists around the world have used the potential of beneficial microorganisms in enhancement of agricultural sustainability (Singh et al., 2021) and control of many plant pests and pathogens (Al-Ani et al., 2013a; Al-Ani, 2017a, 2019a; Sharma et al., 2020). Instances of *Trichoderma* applications in agriculture can be best explained due to their immense potential in reducing the symptoms of cucumber mosaic virus (Al-Ani, 2006; Al-Ani & Al-Ani, 2011), reductions in symptoms of *Fusarium* diseases (Al-Ani & Salleh, 2010; Al-Ani, 2017b; Mitrović et al., 2021a), control of eggs and *Meloidogyne* spp. adults (Al-Ani et al., 2020a; Sarker et al., 2020), induction of plant defense mechanisms (Al-Ani et al., 2020b; Rhouma et al., 2024), and utilization of secondary metabolites (Aguilar-Marcelino et al., 2022; Al-Ani et al., 2024; Castañeda-Ramírez et al., 2020; Comans-Pérez et al., 2021).

Species of *Trichoderma* exhibit antagonistic behavior against several phytopathogenic organisms, including the bacteria, fungi and nematodes. According to Hajieghrari et al., (2008) *Trichoderma* can act as microbial parasite against a wide range of phytopathogens, especially the soil-borne fungi. This genus shows antagonistic activity by inhibiting their growth either by direct interactions or an indirect mechanism. The direct mechanism reduces the pathogen population through microbial antagonism such as the competition for nutrients and space, antibiotics, and parasitism (Zhang et al., 2017). The direct antagonism results from the physical contact between the biocontrol agent and pathogen, including the mycoparasitism or hyperparasitism, which is a complex process of interaction (Harman et al., 2004b) of two or more fungi, in which one parasite the other (Druzhinina et al., 2011).

The indirect mechanism is achieved by the reaction between *Trichoderma* colonized-plant *versus* infecting fungus. *Trichoderma* can act as sustainable tools in agriculture due to their inevitable roles in enhancing the plant growth and vigor, stress tolerance and active uptake of nutrients, bioremediation of contaminated rhizosphere, as well as providing plants with several secondary metabolites, enzymes, and pathogenesis-related proteins (Kumar, 2013). This interaction is indicated as cross-protection or induced resistance and is based on creating the host's own defense system (Marois, 1990).

Trichoderma has been reported as endophytes due to their ability to colonize internal plant tissues. For instance, *Trichoderma* species have been shown to persist within the plant roots and in above-ground tissues through endophytic fungi association. The addition of different species of *Trichoderma* in a plant rhizosphere is known to improve the plant defense mechanism against several phytopathogens such as viruses, bacteria, and fungi. *Trichoderma* is effective against these microbial pathogens as because of its abilities in stimulating the initiation of different resistance mechanisms mainly encompassing for induced systemic resistance, hypersensitive response, and systemic acquired resistance (Harman et al., 2004a).

Furthermore, roots serve as a resort of plant pathogenic fungi and nematodes, the target for *Trichoderma* as mycoparasite and nematophagous. Results indicate that bacterium-degrading ability is common in *Trichoderma* biocides, however, the value highly among different *Trichoderma* strains. According to Pandey et al. (2016), Kumar et al. (2023), and de Freitas Soares et al. (2023) *Trichoderma*-mediated mycoparasitism is associated with the production of proteases, Extracellular N-acetylglucosaminidase (NAGases) and muramidases that seemed to have great potential in the degradation of pathogenic bacterial cells (Rau et al., 2001).

2.1 **Trichoderma***—Plant—Pathogen Interactions*

Microbes that are currently being marketed and studied worldwide for promoting trend of eco-friendly, and sustainable agriculture against the toxic use of agrochemicals and pesticides include *Agrobacterium, Ampelomyces, Bacillus, Candida, Coniothyrium, Pseudomonas* (Haas & Défago, 2005; Rahman et al., 2023), *Streptomyces*

(Mitrović et al., 2021b), Frankia species. (Bhattacharyya et al., 2023), dark septate endophytic (Dalimunthe et al., 2023), and *Trichoderma* (Al-Ani & Mohammed, 2020). These biological control agents (BCA) can exhibit beneficial plant–microbe interactions, as a result of which a suppress in pathogen growth appears through direct and indirect mechanisms including hyperparasitism, competition, induction of host resistance, production of antibiotics, lytic enzymes (chitinases, proteases, and glucanases), and non-regulatory waste products (hydrogen cyanide, ammonia, carbon dioxide). Globally, *Trichoderma* covers about 60% of all fungal-based BCAs, and it may be noted that *T. harzianum* is the most effective microbial biocontrol agent used as biopesticides (Kumar & Khurana, 2021; Lorito et al., 2010). The inherent qualities of *Trichoderma*-based BCAs are cumulative as driving factors for constant success of *Trichoderma*-mediated pest control (Verma et al., 2007). However, according to Harman (2011) the role of *Trichoderma* in crop management is not restricted to biocontrol alone. It extends beyond plant disease management to abiotic stress management and improved overall vigor of the plants even in the absence of phytopathogens.

The knowledge about *Trichoderma*-fungi-plants evolution will certainly improve and expand our knowledge and understandings on their efficacious applications in microbial biocontrol of pests and diseases in agriculture. Topics related to plant disease resistance through induced systemic resistance (ISR) to systemic acquired resistance (SAR) induced by the *Trichoderma*-root symbiosis (Harman et al., 2004a; Hermosa et al., 2012) are seemed to be very interesting. According to the researchers, antagonistic fungi like *Trichoderma* can exert numerous advantageous influences on plants, the plant development and production, the increase in nutrient absorption, rhizosphere modification, and tolerance improvement to biotic/abiotic stresses (Qasim et al., 2018; Samuels, 2006; Verma et al., 2007).

Disease control facilitated by biocontrol mediators including *Trichoderma* biocide is an outcome of the interactions among the plant symbiotic and pathogenic microorganisms. *Trichoderma* has been extensively analyzed and exploited commercially as a biocontrol agent, soil improve, and biofertilizer for the capability to defend plants and control pathogen populations, placing *Trichoderma* among the most explored fungal BCAs (Al-Ani, 2019b, c, d; Hermosa et al., 2000; Verma et al., 2007). Several species of this genus are rhizosphere competent and can also decompose polysaccharides, hydrocarbons, chlorophenols compounds and the xenobiotic pesticides employed in cultivation (Li et al., 2009). The key biocontrol strategies that *Trichoderma* develops in direct conflict with fungal pathogens are mycoparasitism (Howell, 2003; Karuppiah et al., 2019), competition (Ahluwalia et al., 2015) and antibiosis (Druzhinina et al., 2011; Juliatti et al., 2019).

Mycoparasitism is the fungal ability to attack other species of fungi and utilizing their nutrients (Atanasova et al., 2013). This process involves several stages, constituting of host fungus recognition followed by an attack, penetration, and killing (Rabea et al., 2003). The process is carried out via the production of various enzymes that have their abilities to degrade the pathogen fungus cells wall, such as cellulases and chitinases (Horbach et al., 2011; Rabea et al., 2003). The *Trichoderma*—Adenosine triphosphate (ATP)—binding cassette (ABC) transporter proteins are involved in the process of nutrient uptake and mycoparasitism (Locher, 2004).

Trichoderma atroviride is a biocontrol agent for a wide range of economically important aerial and soil-borne pathogens. The mycoparasitic activity of this microorganism is attributed to a synergism of nutrient competition (Chet, 1987), the production of cell wall-degrading enzymes (Woo et al., 1999), and antibiosis (Grayston et al., 1996; Schirmbock et al., 1994). However, the full-scale application of *Trichoderma* for the biological control of pathogens is not been fully investigated. At the molecular level, attempts to increase the biocontrol ability of *Trichoderma* are focused on increasing chitinase or proteinase activity.

Recently, studies on the mechanisms of biocontrol of plant diseases via *Trichoderma* has evolved to a new stage with the proposal of plant basic immunology theory with pathogen-/microbe-/damage-associated molecular pattern (PAMPs/MAMPs/DAMPs) as the core content and the application of omics technology. The three models are evidenced to support maize systemically induced resistance. Systemic resistance generated in maize through seed or soil treatment with *Trichoderma* agents is significant for multiple plant diseases along the development stage. To establish a dialogue with plants, *Trichoderma* releases DAMPs (Damage-associated molecular patterns) and expresses a collection of MAMPs (Microbial-associated molecular patterns) and elicitors that, after recognition by NBS-LRR (Nucleotide-binding and leucine-rich repeat) proteins are up-regulated in plants challenged with *Trichoderma* and can induce different types of beneficial responses (Hermosa et al., 2013).

The versatility of *Trichoderma* strains in suppressing pathogen-induced diseases and their broad environmental opportunism is evident (Daoubi et al., 2009). Several strains are included as active bio-fungicide matter in registered biological control, although many *Trichoderma* formulations against plant pathogens are not registered and they are marketed plant inoculants (Vey et al., 2001). The *Trichoderma* beneficial actions are not limited to fighting pathogens, but are opportunistic plant symbionts, enhancing the systemic resistance of plants (Shoresh et al., 2010; Yedidia et al., 1999).

Studies of the early invading fungi, *Trichoderma* showed that root colonization stimulated plant defense responses such as induction of chitinases, peroxidases, β-1, 3 glucanase, phenylalanine, and hydroperoxidase lyase, causing accumulation of low-molecular-weight phytoalexins (Yedidia et al., 2003). Moreover, the interaction appears to be a symbiotic relationship in which *Trichoderma* inhabits in the nutritional niche provided by the plant, and the plant is protected against diseases.

Trichoderma has the ability to colonize plant roots both externally and internally. This interaction is the result of chemical signals between two partners, *Trichoderma* and the plant, yet studied less in comparison to that normally followed, attachment, penetration, and internal colonization of plant roots. *Trichoderma* spp. can produce hormonal signals for facilitating the root colonization. This microorganism secretes auxins that accelerate root growth that may facilitate more colonization due to the increased surface area (Contreras-Cornejo et al., 2009). Plant root anchoring is facilitated by cysteine-rich proteins known as hydrophobin (Samolski et al.,

2012; Viterbo & Chet, 2006). *Trichoderma harzianum*-derived chitinase (Chit42), expressed in potato and tobacco plants, led to the development of tolerant or resistant transgenic lineages towards soil-borne pathogen like *Rhizoctonia solani* as well as pathogens such as *Alternaria solani* and *Botrytis cinerea* (Howell, 2003). Yedidia et al. (1999) confirmed that cucumber roots inoculated with *T. harzianum* are characterized by a higher expression of peroxidase and chitinase activities, which improved plant resistance to pathogenic attacks.

Plant defense is established by different *Trichoderma* spp. through microbe-associated molecular patterns (MAMPs) (Maaløe, 2002; Mathys et al., 2012). *Trichoderma* spp. have a massive repertoire of genes putatively implicated in the biosynthesis of secondary metabolites such as non-ribosomal peptides, polyketides, terpenoids, and pyrones, but at the moment, few of them have been characterized as MAMPs (Garganese et al., 2016; Wenderoth et al., 2019).

It is known that the ability of numerous *Trichoderma* species to promote plant growth is largely due to their stimulation of the plant root system. For example, Yedidia et al., (2001) studied an inoculation of cucumber plants with the T-203 strain of *T. harzianum* in a hydroponic system produced a 95% increase in the plant's root area and a 75% increase in root length relative to non-inoculated plants. According to Yedidia et al., (2001), this was possibly due to the fungal production of plant hormones (Yedidia et al., 2001). Root colonization by *Trichoderma* damages the cell wall and produces damage-associated molecular patterns (DAMP) molecules, which are responsible for the ISR and conferred resistance to foliar and soil borne pathogens.

Trichoderma is known to produce a diversity of secondary metabolites (Aguilar-Marcelino et al., 2020a, b). For example, trichothecenes are sesquiterpenes produced by *Trichoderma*, inducing systemic resistance in plants. Trichodermin produced by *T. brevicompactum* and *T. harzianum* produced by *T. arundinaceum* possess antifungal activity against some *Fusarium* isolates, *Rhizoctonia solani* and *Botrytis cinerea* (Tijerino et al., 2011; Malmierca et al., 2013).

The *Trichoderma* hyphae are able to colonize the root cortex, but callose-enriched wall appositions efficiently restricted fungal growth. Thus, a very active and symbiotic interaction evolves between *Trichoderma* and the host plant, whereby *Trichoderma* occupies a nutritional habitat, and the host plant is protected from phytopathogens. Plants respond to fungal invasion by producing and accumulating antimicrobial compounds including the phytoalexins, flavonoids, terpenoids, phenolic derivatives, and aglycones. The colonization of plant roots by *Trichoderma* depends strongly on the competence of each strain to tolerate these compounds. This resistance is interconnected with the presence of ATP-binding cassette transport systems in the fungi, which are required for plant colonization (Ruocco et al., 2009; Sun et al., 2019).

A group of elicitors produced by the *Trichoderma* hyphae induces distinctive sorts of defensive signals inside the plants through production of salicylic acid (SA), jasmonic acid (JA), brassinolide (BR), reactive oxygen species (ROS), and activation of defense enzymes. Hyd1 and cellulase are the elicitors obtained from *T. harzianum* that is found to recognize root target proteins. The activation of these genes induces the plant to produce compounds with the reduction of pathogens invasion and improve the biochemical activities for the plant growth and development (Chen et al., 2020).

Also, *Trichoderma* can trigger plant resistance towards pathogen attack by inducing plant immune response. *Trichoderma asperellum, T. aureoviride, T. harzianum, T. virens*, and *T. viride* are used as microbial inducers of plant immunity. An immunity-inducing protein (Sm1/Ep11) of the cerato-platanin family and elicitors produced by *Trichoderma* increase the expression of genes involved in defense, which in turn induces immunity (Gaderer et al., 2015).

2.2 Plant Fungal Diseases

Plant diseases represent a severe economic threat to crops in the world. Several different strategies are being employed to control diverse fungal diseases. Among these, the chemical fungicides are considered as the popular strategy employed by small scale farmers throughout the globe (Islam, 2017), but now the regulations on products pose attention to remove the hazardous chemicals.

The beginnings of studying the concept of biocontrol, as protection against plant pathogens, date back to the 1920s (Choudhary and Johri 2009). Since then, scientists are investigating the biological control strategy related to plant protection by utilizing useful microbes (Al-Ani, 2019e, f; Idrees et al., 2017; Islam, 2017; Qasim et al., 2018; Sukmawati et al., 2024b). Employing biological control *versus* fungal diseases means the use of antagonistic fungi (AF) such as *Trichoderma* spp, a highly diverse genus presents in all types of climates (Raza et al., 2023; Samuels, 2006; Wijayawardene et al., 2020), but also present in different plant materials (Lee et al., 2012). The most important feature of *Trichoderma* are the varying mechanisms for phytopathogens management and plant diseases, in which pathogen antagonized by biocontrol agent results from different types of interaction between organisms (Pal & Gardener, 2006).

Trichoderma species control and antagonize a broad range of phytopathogenic fungi as well as bacteria and viruses (Yedidia et al., 2003). Chemical activator and some microbes can control several plant pathogens as control of *Fusarium oxysporum* f. sp. *cubense* tropical race 4 (*Foc*TR4) (Al-Ani & Furtado, 2020; Mohammed et al., 2011, 2014; Soumare et al., 2021). But the abilities of isolates of *Trichoderma* sp. represented high efficiency in inhibition of *Foc*TR4 (Al-Ani & Albaayit, 2018a, b) from others. This genus can be used as biological control agent and have the ability to compete with other parasitic and saprophytic pathogens due to high metabolic activities and competitive nature that make them decomposer of herbaceous and woody material (Elad, 2000) in which the sexual perfect stage (Teleomorph genus *Hypocrea*) has found usually.

The success of *Trichoderma* colonization is owing to their rapid growth, intensive production of spores in different environmental conditions, efficiency in the utilization of nutrients, and capacity to communicate with plants through chemical signals (Badri et al., 2009; López-Bucio et al., 2015). Strains of *T. harzianum, T. atroviride,*

T. asperellum, *T. longibrachiatum*, *T. virens*, and *T. viride* are the most widely used against fungal diseases of plants (Guzmán-Guzmán et al., 2023).

A mechanism of secondary metabolites to control phytopathogenic fungi is their role in the competition for nutrients. Siderophores are a complex compound utilizing in Fe-chelating by many species of fungi to compete for nutrients (Al-Ani et al., 2021b). The fast-growing ability of *Trichoderma* spp. makes them potential competitors for nutrients and space. *Trichoderma* spp. makes iron unavailable for the competing microorganisms by releasing siderophores, which scavenge iron from the environment. The iron competition has a role in the antagonistic activity of *T. asperellum* against *F. oxysporum* (Segarra et al., 2010). In additional, *Phytophthora infestans* cause very epidemic disease for potato called potato late blight and rot for potato tuber (Raza et al., 2022). *T. atroviride* caused degradation of cytoplasmic organelles and deformation for cells by secreting different volatile compounds including 3-methyl-1-butanol, acetoin and 6-pentyl-2-pyrone, 2-methyl-1-propanol (Elsherbiny et al., 2020).

Trichoderma brevicompactum produce trichothecenes which are sesquiterpenes (C15 terpene compounds). This fungus also produces trichodermin, which is an antifungal compound, and is used as a biocontrol agent (Tijerino et al., 2011; Hajji-Hedfi et al., 2025a). Harzianum A, produced by *T. arundinaceum*, is also a trichothecene with antifungal activity and has been shown to be active against *B. cinerea* and *R. solani* in tomato plants probably by eliciting systemic priming responses and defense (Malmierca et al., 2012).

Trichoderma is a potential alternative to control charcoal rot in soybean (Khalili et al., 2016). Also, maize roots colonized by *T. harzianum* or *T. virens* inhibited the proliferation of soil-borne disease caused by *Colletotrichum graminicola* (Djonović et al., 2007; Harman et al., 2004b). *Arabidopsis* seedlings inoculated with either *T. virens* or *T. atroviride* and then challenged with *B. cinerea* developed fewer leaf necrotic lesions compared to *Trichoderma*-untreated plants (Contreras-Cornejo et al., 2016). Like-wise, *Trichoderma* is effective in controlling significant pathogenesis in tea cultivation under integrated disease management (IDM) programme (Bhattacharyya and Sarmah 2018), and this microbe is more popularly known as miracles of tea industry. *Trichoderma* is used as effective BCA against significant tea diseases including the Blister blight (causal agent: *Exobasidium vexans*) (Adhikari & Rai, 2022), black rot (causal agent: *Corticium theae*) (Islam et al., 2018), Fusarium die-back (causal agent: *Fusarium solani*) (Bhattacharyya et al., 2016), Charcoal rot (causal agent: *Macrophomina phaseolina*) (Martínez-Salgado et al., 2021), white root rot (causal agent: *Rosellinia necatrix*) (Sawant et al., 2024) under field conditions. As conventional chemical management practices seemed to be harmful for both the human and environment, the current approach of *Trichoderma* exploitation as plant growth promoter (PGP) and BCA is beneficial to improve the soil health and quality tea bush for maintaining an eco-friendly and sustainable tea cultivation and practices. Also, many pathogenic fungi produce different types of mycotoxins (Attitalla et al., 2010a, b; Mohammed and Al-Ani, 2021). *Trichoderma atroviride* suppressed more than 70% of mycotoxins including DON (Deoxynivalenol) and ZEN (Zearalenone) (Błaszczyk et al., 2017). The detoxification of the DON through

Trichoderma strains by means of a biological control-guided procedure reached the incredible consideration. As potential antagonists, *Trichoderma* strains inhibited the mycelial development and mycotoxin synthesis of *Fusarium. graminearum* and *Aspergillus* (Mitrovic et al., 2021a; Mitrović et al., 2023). The conversion of DON to deoxynivalenol-3-glucoside (D3G) was considered as the final product of detoxification.

Trichoderma success of the antifungal activities *vs.* phytopathogenic fungi could be attributed to the synergy action of secondary metabolites and hydrolytic enzymes (Druzhinina et al., 2011). The inhibition of *Botrytis cinerea* spore germination is due to gliotoxin and endochitinase enzymes. *Trichoderma* is known to release elicitor-like substances that induce a systemic or localized resistance response in plants (Khan et al., 2020). Also, the seeds of tomato plants treated with *T. harzianum* decreased the intensity of leaf blight caused by *B. cinerea* (Moran-Diez et al., 2009).

The peptaibols trichokonins VI-VII-VIII obtained from *Trichoderma koningii*, showed a broad-spectrum antimicrobial activity against essential pathogens, such as *Rhizoctonia solani, Fusrium oxysporum, Verticillium dahliae* and *Botrytis cinerea.* The peptaibols trichorzianine A1 and B1 from *T. harzianum* could inhibit the spore germination, as well as hyphal elongation of plant fungi (Lee et al., 1999), and there is a synergistic interaction between hydrolytic enzymes and peptaibols (Schirmbock et al., 1994). The 6-Pentyl pyrones is a volatile metabolite of *Trichoderma*, that acts as an auxin-like compound, and inhibits the growth and proliferation of *F. oxysporum* through the induction of PR1 genes (Rubio et al., 2014) that have beneficial responses in plants (Contreras-Cornejo et al., 2009; Viterbo et al., 2010).

Although, conventional agrochemicals possess significant benefits to plant protection and crop yield, however, injudicious and unlawful application of these chemicals have significant toxicity to soil, plant and environment. Now-a-days, exploiting the potential of sustainable disease management strategies becoming more popular and an emergent need to reduce the load of toxic chemicals and pesticides. *Trichoderma* species reduce the pathogenic fungi efficiency at the site of inoculation. This type of resistance is called induced localized acquired resistance. These methods reduce the diseases caused by pathogens. The application of chemical fungicides and consumer acceptance of resistant cultivars are seemed to be very complex, which makes biological control of fungi a valid alternative against conventional chemical practices in agriculture (Al-Ani et al., 2020a).

2.3 Plant Nematode Diseases

Plant parasitic nematodes are versatile in their mode of occurrence and distribution. They are more commonly found in in soil, and representing one of the most important pests worldwide in various agricultural crops, mainly in tropical and subtropical countries (Back et al., 2002; Mandal et al., 2021; Al-Ani et al., 2025 a, b). These organisms are phytonematodes, causing various symptoms that will depend on the genus and species of nematodes, and can affect different parts of the plant. Likewise, the degree of pathogenicity will depend on the aggressiveness of the strain, as well as the anatomical and physiological adaptations of each plant to parasitism (Dutta et al., 2011).

However, plant parasitic nematodes are small transparent worms, with bilateral symmetry, pseudocoelomate, multicellular, free-living, or parasitic, and can be predatory, aquatic, terrestrial, endopathogenic, ectoparasite, endoparasite or semi-endoparasite (Mandal et al., 2021). Jones et al. (2013) have reported ten nematodes according to their pathogenicity: root-knot (*Meloidogyne* spp.), cyst (*Heterodera* and *Globodera* spp.), root lesion (*Pratylenchus* spp.), burrowing (*Radopholus similis*), the stem and bulb nematode (*Ditylenchus dipsaci*), pine wilt (*Bursaphelenchus xylophilus*), reniform (*Rotylenchulus reniformis*), virus vector (*Xiphinema index*), *Nacobbus aberrans* and *Aphelenchoides besseyi*.

Indeed, symptoms induced by phytopathogenic nematodes vary according to gender, and some in aerial parts and others in subterranean parts (Wesemael et al., 2011). However, current studies describe the mechanism of infection of parasitic nematodes in plants through the activation of different immune response systems. Plants recognize PAMPs (Pathogen-associated molecular patterns) derived from plant parasitic nematodes by cell surface-localized pattern recognition receptors (PRRs), leading to pattern-triggered immunity (PTI). Plants can also recognize tissue and cellular damage caused by the invasion or migration of plant parasitic nematodes through PRR-based recognition of DAMPs (Sato et al., 2019).

Plant immune responses against plant parasitic nematodes include the secretion of anti-nematode enzymes, the production of anti-nematode compounds, cell wall reinforcement, the production of reactive oxygen species and nitric oxide, and hypersensitive response-mediated cell death (Sato et al., 2019).

In addition to the pronounced antagonistic activity shown on phytopathogenic fungi, *Trichoderma* species also exert their antagonistic activity on nematodes (Al-Ani et al., 2022). Interestingly, the group of gall-forming nematodes are those of the genus *Meloidogyne* spp. These root-knot nematodes are considered worldwide as one of the most important pests in various crops, causing severe economic losses estimated at US$ 118 billion annually (Khan et al., 2020; Safdar & McKenry, 2012).

The most important species of *Meloidogyne* genus, *M. javanica* and *M. javanica* are the most studied, globally. According to de Medeiros et al., (2017), *T. atroviride* reduces the number of galls, eggs and the amount of adult nematodes *M. javanica* observed on tomatoes under greenhouse conditions. On the other hand, activity of *T. harzianum* against *M. javanica* has been confirmed by many scientists (Al-Hazmi & TariqJaveed, 2016; Sahebani & Hadavi, 2008; Sharon et al., 2001). In addition to *T. harzianum* as the most frequently investigated, *T. viride* also shows significant bioefficacy against *M. javanica* (Al-Hazmi & TariqJaveed, 2016).

On the other side, antagonistic activity of *Trichoderma* spp. against *M. incognita* is also widely studied. Research by Fan et al. (2020) showed that the *T. citrinoviride* Snef1910 strain shows high virulence in *M. incognita* second stage juveniles while at the same time showing a high percentage of inhibition of hatched eggs. Also, in addition to the nematocidal effect, this *Trichoderma* strain showed a promoter effect in the growth of the tomato plants that were analysed in the paper (Fan et al.,

2020). Pocurull et al. (2020) have carried out similar research with other *Trichoderma* isolates. Namely, they investigated the possibility of *T. asperellum* (T34) and *T. harzianum* (T22) to induce systemic resistance in cucumber and tomato against *M. incognita*. In their work, they determined that both tested isolates induce resistance to the tested nematode on tomato, but not on cucumber (Pocurull et al., 2020). Potential in the control of *M. incognito* also showed *T. viride* and *T. harzainum* in the work of Mukhtar (2018).

However, some *Trichoderma* species, in addition to nematicidal activity, simultaneously show antibacterial activity. One of the examples is certainly *T. pseudoharzianum* (T113) and *T. viride*. Khan et al., (2020) proved that the secondary metabolites of these fungi show a high percentage of mortality of *M. incognita* nematode while, at the same time showing pronounced antibacterial activity on *Xanthomonas campestris* and *Ralstonia solanacearum*. Therefore, in addition to antifungal and nematocidal, *Trichoderma* species can produce secondary metabolites with pronounced antibacterial action, while some species of this genus exhibit cumulative antagonistic actions.

2.4 Plant Bacterial Diseases

Trichoderma species was more described based on biological control fungal disease (Howell, 2003). On the order hand, the broad-spectrum relation of mechanism action and ability to produce several secondary metabolites, make *Trichoderma* spp. a complete multiple BCA (Xiao-Yan et al., 2006; Leelavathi et al., 2014). Concerning control of plant bacterial diseases, *Trichoderma* is a huge opportunity, in the sense that, no other genus has available and efficient tools in the management of bacterial diseases (Sundin et al., 2016). Moreover, the management of bacterial soil-borne diseases is more arduous since bacteria are lasting and survive in plant debris association (Yuan et al., 2016; Zanon et al., 2011).

Regarding the mechanism of biocontrol by *Trichoderma* in bacterial diseases, the inside response based by ISR is more assertive (Sundin et al., 2016). The priming effect is key to ensure that plant defence is signalled by plant hormones (Pieterse et al., 2014), and consequently reduces the severity of bacterial soil-borne such as *Ralstonia solanacearum* (Konappa et al., 2018), *Pseudomonas syringae* (Morán-Diez et al., 2020) and *Xanthomonas oryzae* pv. *oryzae* (Shobha et al., 2020). Besides bacterial diseases biocontrol, strains of *Trichoderma* spp., can promote plant growth (Zachow et al., 2016), improving production yield and quality (Molla et al., 2012). Additionally, may reduce production costs in systems that are difficult and/or expensive to manage, as bacterial diseases (Gupta et al., 2016).

When looking at core *Trichoderma* species, *T. asperellum* reduces severity of *Ralstonia solanacearum* when early treated (central point for reducing bacterial severity), in the time of seedlings or seed tuber. Thus, when transplanting the young plants these bring a defence memory for protection against bacterial wilt.

The response is mediated by enzymes (peroxidase, polyphenol oxidase, phenylalanine ammonia-lyase and β-1,3-glucanase), total phenol activities, salicylic acid, siderophore, indole-3-acetic acid and hydrogen cyanide content that show the ability for long-term protection across ISR in tomato and potato (Konappa et al., 2018; Mohamed et al., 2020; Yendyo et al., 2017). These enzymes, siderophore, indole-3-acetic acid and hydrogen cyanide are considered as suppression against plant plant pathogens (Hajji-Hedfi et al., 2025b)

Leelavathi et al. (2014) investigated the effect of *T. harzianum* against bacteria. They determined that *T. harzianum* acts against three human bacterial isolates, *Staphylococcus aureus*, *Escherichia coli* and *Klebsiella* sp., while against *Proteus* sp. did not exhibit antibacterial activity.

Another *Trichoderma* isolate, *T. hamatum* FB10, showed antibacterial activity against *Acidovorax avenae*, *Erutimacarafavora*, and *Xanthomonas campestris* (Baazeem et al., 2021). The antibacterial activity of *Trichoderma* in the treatment of tomato bacterial wilt caused by *Ralstonia solanacearum* was investigated by Konappa et al. (2018), Kariuki et al. (2020) and Yan and Khan (2021). The use of *T. asperellum* (T4 and T8) isolates in the control of *R. solanacearum* on tomatoes in the field showed that in addition to reducing the development of the disease, this beneficial microorganism has the key role as a successful growth promoter (Konappa et al., 2018). Similar results were obtained by Yan and Khan (2021) in using *T. harzianum* under in vitro and greenhouse conditions.

2.5 *A Tripartite Symbiotic Relationship Among* **Trichoderma**, *Other Beneficial Microbes, and Plants*

The symbiotic relationship is not only limited to two mutually beneficial individuals, for example between *Rhizobium* and vegetables. There are reports on the symbiosis involving more than two mutually beneficial individuals, such as *Diaporthe arecae (= Phomopsis liquidambari)*, *Rhizobium* and peanuts, with increasing peanut nodulation and growth (Zhang et al., 2017). A tripartite symbiotic relationship also occurs between *Piriformospora indica* and *Paenibacillus lentimorbus,* which stimulates root nodulation and plant growth of *Cicer arietinum* L. (chickpea) (Nautiyal et al., 2010). Associations between plants, fungi, and N_2-fixing bacteria in endomycorrhizae and ectomycorrhizae have also been widely reported (Ruiz-Herrera et al., 2015; Yelikbayev et al., 2024). The tripartite symbiosis between fungi and bacterial endosymbionts that have a positive effect on their host plants reveals their microbial ecology, especially for mitigating plant diseases. Likewise, *Trichoderma* also has the potential to be involved in a tripartite symbiotic relationship with fungi, bacteria, or other microbes that are beneficial to the host plants.

Trichoderma is capable of tripartite symbiosis with dark septate endophytic fungi (DSE). The combination of *T. viride* with DSE has a positive impact on licorice (*Glycyrrhiza uralensis*) plants, especially in increasing root biomass, length, and surface area, so nutrient absorption by the host plant is also increased (He et al., 2020). DSE is known as a root endophytic fungus that has symbiosis with various plants, especially under biotic and abiotic stress conditions, so that the combination with *Trichoderma* has the potential to further improve host plant performance under environmental stress conditions, although studies and applications have not been widely reported.

In addition to tripartite symbiosis with DSE, *Trichoderma* is also known for establishing symbiotic interactions with mycorrhizae to jointly increase the growth of host plants and protect them from diverse pathogen attack. In sunflower, the combination of *T. viride* with *Pseudomonas fluorescens*, and arbuscular mycorrhizal fungi (AMF) increases AMF colonization, thereby increasing phosphorus uptake and plant growth hormones, and changing the physiology of the host plants (Yadav et al., 2015; Mitra et al., 2023). AMF synergized with *T. harzianum* in a symbiotic relationship with melon reduced the incidence of *Fusarium* melon wilt, and the combination of *G. intraradices* with *T. harzianum* gave the best results (Martínez-Medina et al., 2009). This indicates that the potential of *Trichoderma* species can be enhanced when combined with mycorrhizae and the microbial consortium can be used in controlling plant diseases in an environmentally friendly and multifunctional way because the two microbes do not only have one function in supporting the fitness and productivity of their host plants under biotic stress.

In addition to tripartite symbiosis with other beneficial fungi and plants, *Trichoderma* also interacts synergistically with *Bacillus subtilis* by increasing the plant growth (Santoyo et al, 2024). Bravo et al. (2013) verified the spread of bacteria by fungi, especially *Trichoderma*, stating that there was migration of *Stenotrophomonas*, *Achromobacter*, *Lysobacter*, *Pseudomonas*, *Agrobacterium*, *Cohnella*, and *Variovorax* through *Trichoderma* hyphae on oxalate-rich media. The synergy in the consortium relationship between *Pseudomonas aeruginosa* and *T. harzianum* was able to increase growth and induce biochemical changes in banana plants under the challenge of *Fusarium* wilt (Wong et al., 2021).

Tripartite, even multipartite symbiosis, such as various plants associated with fungi, including *Trichoderma*, mycorrhizae, DSE, and other beneficial fungi with rhizobia is expected to be more effective in environmental conditions with limited nutrient availability such as in most natural ecosystems (van Der Heijden et al., 2015). This symbiotic relationship needs to be revealed to better understand the synergy between each of the functional groups of individuals in a tripartite symbiosis, especially in the face of biotic stress. However, more in-depth study in relation to understanding the role of *Trichoderma* in a tripartite or multipartite symbiotic relationship is important because as we know, *Trichoderma* can live and survive in various biotic stress conditions as well as to increase the fitness of the host plants in synergy with other microbes in adapting in stressful environments.

2.6 Other Plant Diseases

Trichoderma plays an irreplaceable role in the control of various important plant diseases such as fungi, bacteria, and nematode, as possible useful to control other plant diseases. In this section, the mechanisms useful against other plant diseases (plant viral and viroids diseases, parasitic higher plant, and phytoplasma) are possible into two ways including direct and indirect (Al-Ani, 2018a, b). The effect of *Trichoderma* against other plant diseases is confronting directly by inducing the plant immunity, and production of secondary compounds. Indirect effect of *Trichoderma* on other plant diseases occurs by enhancing the plant growth and tolerance, which decrease the occurrence of diseases (Al-Ani, 2018b).

Another skill of *Trichoderma* spp., and their secondary metabolites regarding the management of plant disease is their potential against plant virus (Luo et al., 2010). The response of plant defense by long signal transduction pathways mediated by *Trichoderma* spp. has the potential to prevent virus infection. The specific action is related to the coat protein, for example, during *Tobacco Mosaic Virus* (TMV) pathogenesis in cucumber (Vitti et al., 2015).

Furthermore, *Trichoderma* species also produce antimicrobial peptides that interact with viruses, thereby directly disturbing or destabilizing them or competing for space in plant tissues when colonized endophytically (Salas-Marina et al., 2015). These barriers help the host to avoid the virus infection, apart from reducing the virus multiplication and consequently its survival and dissemination, since it had no contact with a live host (Silva et al., 2019). The trichodermin compound produced by *T. albolutescens* showed active against *Pepper Mottle Virus* (PepMoV) by inactivation systemic of virus-based and are related as an antiviral agent (Ryu et al., 2017).

Interestingly, *Trichoderma pseudokoningii* SMF2 produces trichokonins (As antimicrobial peptaibols) inducing the resistance and defenses in *Nicotiana tabacum* var. Samsun NN (Tobacco) (Luo et al., 2010). Trichokonins showed high significant effect against spreading TMV on tobacco by inhibiting more than 50% of lesion and reduction in area of lesion (Luo et al., 2010). The treatment of *Arabidopsis thaliana* with *T. asperellum* could induce the systemic resistance by signaling salicylic acid against CMV (*Cucumber mosaic virus*) (Elsharkawy et al., 2013).

3 The Role of *Trichoderma* in Industry of Biopesticides

The production of biopesticides is dependent on utilizing the natural biological agents in the preparation of biopesticides (Adetunji et al., 2019; Bhattacharyya et al., 2024). Due to the capability to defend plants and control pathogen populations, *Trichoderma* spp. have been extensively analyzed and exploited commercially as BCAs, placing this genus amongst the most explored fungi (Harman, 2000; Harman et al., 2010; Lorito et al., 2010; Al-Ani et al., 2018). Their agricultural importance is due to good antagonistic abilities against plant pathogenic fungi, thanks to different mechanisms of antagonism: the production of antifungal metabolites (antibiosis), competition for space and nutrients, induction of defense responses in plant and mycoparasitism.

Since the challenges in coping with issues in the agriculture industry have grown tremendously, sustainable strategies by using biological control approach are necessary. Thus, the use of *Trichoderma* spp., as a biological agent seems to be an excellent approach (Sood et al., 2020; Hajji-Hedfi et al., 2025c). According to PAN Pesticide Database, there are 33 *Trichoderma* strains certified, globally covering 11 species: *T. afroharzianum*, *T. asperellum*, *T. atroviride*, *T. fertile*, *T. gamsii*, *T. hamatum*, *T. harzianum*, *T. ovalisporum*, *T. polysporum*, *T. virens* and *T. viride* determined as as biofungicides. There are more than 250 products commercially available, for use in conventional and organic agricultural production, representing for about 60% of the international biofungicide market (Topolovec-Pintaric 2019). Their use is distributed worldwide, with prominent distribution in India, South and Central America. Most of the products consist of a single species, whereas the remaining products are developed as microbial consortium, which can be a mix of *Trichoderma* strains or a combination with antagonistic bacteria (Woo et al., 2014). *Trichoderma harzianum* and *T. viride* are mostly widely used as microbial antagonists for soil treatment in around 87 various crops against 70 soil-borne and 18 foliar-borne pathogens, mostly fungi (Sharma et al., 2014; Topolovec-Pintaric 2019). The efficiency of products containing various species or strains may act differently within the similar field and climate conditions (Zin & Badaluddin, 2020).

Trichoderma-based pesticides are available worldwide as biofungicide, bioinoculant, plant growth promoter, and increasing crop yield. Final products are used in the greenhouse, nursery, field, and orchard as foliar spray, seed treatment, and soil drench. The active ingredient can be incorporated in different formulations as pure conidia, in suspension or liquid culture filtrates, to facilitate storage and use. As any BCA, production of viable *Trichoderma* must be rapid and low cost, and the products with relatively long shelf-life, low toxicity and adapted to adverse environmental conditions (Aguilar-Marcelino et al., 2020a, b; Martinez et al., 2023; Teixidó et al., 2022).

Data about industrial production of *Trichoderma* preparations are modest, mainly because it is a trade secret or is patented. What is known from the scientific literature is that *Trichoderma* can be cultivated using two types of production: solid state cultivation (SSC) or submerged cultivation (SmC). SSC produces aerial conidia and uses a solid substrate in the near absence of free water. It is relatively inexpensive since low cost agro-industrial residues can be used as nutrient source. Most of the products available consist primarily of aerial conidia from SSC. Sala et al. (2020) used rice husk to grow *T. harzianum* in agitated batch reactor, and Naeimi et al. (2020) obtained significant results using rice husk, rice straw, broom sorghum grain, cow dung, and sugar beet pulp for *T. harzianum* growth. However, SSC have some constraints, as long fermentation times, need of drying, and difficult scale-up, which favors the study of SmC.

On the other hand, in SmC, the microorganism is completely immersed in the cultivation medium, which prevents its contact with the gas phase. This problem was solved by applying aeration and mixing of cultivation liquid that allow constant distribution of nutrients, heat and dissolved oxygen in all parts of the vessel equally (Grahovac et al., 2020). In the past, this has been a problem in production however, the development of automation in the industry has allowed the SmC type of production to stand out (Verma et al., 2007).

Lately, scientists are increasingly engaged in SmC production since this type of cultivation allows better control of bioprocesses, easier and more successful scale-up and higher productivity. SSC is considered a cheaper type of production from the aspect of labor force however, this type of production does not provide high yields like SmC (Verma et al., 2007). Also, the disadvantages of SSC are the difficulty of controlling the bioprocess parameters (temperature, pH), and the difficulty of extracting biomass. SmC gave good production results for *T. harzianum* microsclerotia and *T. asperelloides* submerged conidia (de Rezende et al., 2020; Kobori et al., 2015).

About 35–40% of the production cost goes to the raw material i.e. the production medium (Verma et al., 2007). Standard media generally contain the most important nutrients important for the growth and production of each microorganism. These are primary sources of carbon and nitrogen and their mutual combination in the medium are also significant (Mitrović et al., 2021a). For the cultivation of *Trichoderma* spp. the most important sources of carbon are glucose, starch, glycerol, lactose, fructose, and others whereas, the most important sources of nitrogen included both organic and inorganic sources such as yeast extract, soybean flour, NH_4NO_3, $(NH_4)_2SO_4$ and others (Mitrović et al., 2021a).

In order to produce *Trichoderma* biopesticide with a commercial application, it is, however, necessary to reduce the production cost in order to be more competitive in the market. One of the ways to reduce the cost of production is to replace standard synthetic media with media of low market value (Mitrović et al., 2021b). The fact that many synthetic or semi-synthetic media for the production of *Trichoderma* BCAs or various enzymes can be replaced by different wastes has great significance in terms of sustainability. For instances, different wastewater can be used as medium in SmC (Coradi et al., 2013; Emerson & Mikunthan, 2015; Ezekiel et al., 2010) while different solid wastes can be used as raw material in SSC (Dorta-Vásquez et al., 2019; Molla et al., 2004; Pandey et al., 2015). Coradi et al. (2013) examined lipase production using *Trichoderma harzianum* on different media of low market value (agro-industrial residues) using SmC and SSC. Research has shown that the highest lipase activity (1.4 U/mL) was obtained in SmC. As a medium for lipase production using SmC, the medium with olive oil proved to be the best in relation to the medium containing soybean, corn, and sunflower oil. On the other hand, the usage of olive oil in castor/cane medium by SSC was the most economical bioprocess to produce lipase, under the conditions compared.

In order to protect themselves, but also the environment, it is necessary for people to introduce changes that will be sustainable. This is primarily reflected in the possibility of using microorganisms for the purpose of profit-motivated production, which is safe for the environment, improves the quality of life, prolongs human life, and in addition is economical.

4 Conclusion

Now-a-days, it's becoming an emergent need to explore novel, eco-friendly and non-chemical agricultural strategies to reduce the use of toxic chemicals and pesticides that possess deleterious effects to soil, plant and the environment including the human beings. The current approach significantly defines the roles played by *Trichoderma* as potential BCA against major pathogenesis in agriculture. The abilities of *Trichoderma* to antagonize plant pathogens as well as to provide benefits to plants in promoting their growth and enhancing their immune responses makes this fungal genus as a unique biological tool under field situations. Besides, the relationships that *Trichoderma* maintains with other symbiotic microorganisms in supporting the growth, proliferation and plant protection seemed to be vital while working with such miracles of greener approach in agriculture. Further research pertaining to the improved efficiency of *Trichoderma* bioformulation needs to be properly carried out under the field situations to assist the marginal farmers and to reduce the inputs of toxic chemicals and products for developing climate-smart agriculture.

References

Adetunji, C. O., Egbuna, C., Tijjani, H., Adom, D., Al-Ani, L. K. T., & Patrick-Iwuanyanwu, K. C. (2019). Homemade preparations of natural biopesticides and applications. In C. Egbuna & B. Sawicka (Eds.), *Natural remedies for pest, disease and weed control* (pp. 179–186). Academic Press.

Adhikari, K., & Rai, A. (2022). In vitro efficacy of Trichoderma isolates and commercial fungicides against Exobasidium vexans, causal agent of blister blight in tea. *International Journal of Applied Sciences and Biotechnology, 10*(3), 182–189. https://doi.org/10.3126/ijasbt.v10i3.47524

Aguilar-Marcelino, L., Al-Ani, L. K. T., Castañeda-Ramirez, G. S., Garcia-Rubio, V., & Ojeda-Carrasco, J. J. (2020a). Microbial technologies to enhance crop production for future needs. In A. A. Rastegari, A. N. Yadav, A. K. Awasthi, N. Yadav (Eds.), *New and future developments in microbial biotechnology and bioengineering* (pp. 29–47). Elsevier.

Aguilar-Marcelino, L., Mendoza-de-Gives, P., Al-Ani, L. K. T., López-Arellano, M. E., Gómez-Rodríguez, O., Villar-Luna, E., & Reyes-Guerrero, D. E. (2020b). Using molecular techniques applied to beneficial microorganisms as biotechnological tools for controlling agricultural plant pathogens and pest. In V. Sharma, R. Salwan, L. K. T. Al-Ani (Eds.), *Molecular aspects of plant beneficial microbes in agriculture* (pp. 333–349). Elsevier Science.

Aguilar-Marcelino, L., Al-Ani, L. K. T., Soares, F. E. F., Moreira, A. L. E., Téllez-Téllez, M., Castañeda-Ramírez, G. S., Acosta-Urdapilleta, M. L., Díaz-Godínez, G., & Pineda-Alegría, J. A. (2021). Formation, resistance, and pathogenicity of fungal biofilms: Current trends and future challenges. In A. N. Yadav (Ed.), *Recent trends in mycological research* (Vol. 1, pp. 411–438). Agricultural and medical perspective. Springer Nature.

Aguilar-Marcelino, L., Al-Ani, L. K. T., Soares, F. E. D. F., Braga, F. R., Padilla, A. V. V., & Sharma, A. (2022). Wild mushrooms: Characteristics, nutrition, and processing. In: S. B. Dhull, A. Bains, P. Chawla, & P. K. Sadh (Eds.), *Wild mushrooms: characteristics, nutrition, and processing* (pp. 471–492). Taylor & Francis, CRC Press.

Aguilar-Marcelino, L., Al-Ani, L. K. T., Wong-Villarreal, A., & César Sotelo-Leyva, C. (2023). Persistence of pesticides residues with chemical food preservatives in fruits and vegetables.

In: J. Singh, A. Pandey, S. Singh, V. K. Garg, & P. Ramamurthy (Eds.), *Current developments in biotechnology and bioengineering pesticides: human health, environmental impacts and management* (1st ed., pp. 99–118). Elsevier, India.

Ahluwalia, V., Kumar, J., Rana, V. S., Sati, O. P., & Walia, S. (2015). Comparative evaluation of two *Trichoderma harzianum* strains for major secondary metabolite production and antifungal activity. *Natural Product Research, 29*, 914–920. https://doi.org/10.1080/14786419.2014.958739

Al-Ani, L. K. T. (2006). *Induce resistance against Cucumber mosaic virus by Pseudomonas fluorescens Migula* (p. 90). Master thesis, Department of Plant Protection, College of Agriculture, University of Baghdad.

Al-Ani, L. K. T. (2017a). PGPR: a good step to control several of plant pathogens. In Singh, H. B., Sarma, B. K., Keswani, C. (Eds.), Advances in PGPR Research (pp. 398–410). CABI, Oxfordshire.

Al-Ani, L. K. T. (2017b). *Potential of utilizing biological and chemical agents in the control of Fusarium wilt of banana* (pp. 259). PhD thesis, School of Biology Science, Universiti Sains Malaysia.

Al-Ani, L. K. T. (2018a). Trichoderma: beneficial role in sustainable agriculture by plant disease management. In Egamberdieva, D., Ahmad, P. (Eds.), *Plant Microbiome: Stress Response. Microorganisms for Sustainability* (vol. 5, pp. 105–126) Springer.

Al-Ani, L. K. T. (2018b). Trichoderma from extreme environments: physiology, diversity, and antagonistic activity. In Egamberdieva, D., Birkeland, N.-K., Panosyan, H., & Li, W.-J. (Eds.), Extremophiles in Eurasian ecosystems: ecology, diversity, and applications. Microorganisms for sustainability, (Vol. 8, pp. 388–403). Springer.

Al-Ani, L. K. T. (2019a). Secondary metabolites of nonpathogenic Fusarium spp.; scope in agriculture. In Singh, H. B., Keswani, C., Reddy, M. S., Royano, E. S., García-Estrada, C. (Eds.), *Secondary metabolites of plant growth promoting rhizomicroorganisms—discovery and applications* (pp. 59–76). Springer.

Al-Ani, L. K. T. (2019b). Entomopathogenic fungi in IP landscape. In Singh, H. B., Keswani, C., Singh, S. P. (Eds.), Intellectual property issues in microbiology (pp. 223–238). Springer.

Al-Ani, L. K. T. (2019c). Recent patents on endophytic fungi and their international market. In Singh, H. B., Keswani, C., Singh, S. P. (Eds.), Intellectual property issues in microbiology (pp. 271–284). Springer.

Al-Ani, L. K. T. (2019d). The importance of endophytic fungi from the medicinal plant: Diversity, natural bioactive compounds, and control of plant pathogens. In Egamberdieva, D., Tiezzi, A. (Eds.), Medically important plant biomes source of secondary metabolites (pp. 189–238). Springer.

Al-Ani, L. K. T. (2019e). Bioactive secondary metabolites of Trichoderma spp. for efficient management of phytopathogens. In Singh, H. B., Keswani, C., Reddy, M. S., Royano, E. S., García-Estrada, C. (Eds.), *Secondary metabolites of plant growth promoting rhizomicroorganisms—discovery and applications* (pp. 125–143). Springer.

Al-Ani, L. K. T. (2019f). A patent survey on Trichoderma spp. (from 2007–2017). In Singh, H. B., Keswani, C., Singh, S. P. (Eds.), *Intellectual property issues in microbiology* (pp. 163–192). Springer.

Al-Ani, L. K. T., & Salleh, B. (2010). *Control of Fusarium wilt of banana by non pathogenic Fusarium oxysporum* (p. 10). PPSKH colloquium, Pust Pengajian Sains Kajihayat/School of Biological Sciences, USM, June 2–4.

Al-Ani, R. A., & Al-Ani, L. K. T. (2011). Induced of systemic resistance in cucumber plants against Cucumber mosaic virus (CMV) by *Pseudomonas fluorescens* Migula. *Arab Journal of Plant Protection, 29*, 36–42.

Al-Ani, L. K. T., & Albaayit, S. F. A. (2018a). Antagonistic of some Trichoderma against Fusarium oxysporum sp. f. cubense tropical race 4 (FocTR4). International conference on Research in Education & Science, ICRES April 28—May 1, Marmaris, Turkey, pp. 271 (Abstract).

Al-Ani, L. K. T., & Albaayit, S. F. A. (2018b). Antagonistic of some Trichoderma against Fusarium oxysporum sp. f. cubense tropical race 4 (FocTR4). The Eurasia Proceedings of Science, Technology, Engineering & Mathematics (EPSTEM), 2, 35–38.

Al-Ani, L. K. T., Negim, E.-S., Mohammed, A. M., Salleh, B., & Saleh, M. I. (2012). Antifungal activity of novel binary grafting polymers. 1st USM—KAZNU International Conference on Challenges of Teaching and Chemistry Research in Institutions of Higher Learning (p. 44), 11–13 July.

Al-Ani, L. K. T., Salleh, B., Mohammed, A. M., Ghazali, A. H. A., Al-Shahwany, A. W., & Azuddin, N. F. (2013a). Biocontrol of *Fusarium* wilt of banana by non-pathogenic *Fusarium* spp. *International symposium on tropical fungi* (pp. 50–51) September 2023. ISTF, IPB International Convention Center.

Al-Ani, L. K. T., Salleh, B., & Ghazali, A. H. A. (2013b). Biocontrol of Fusarium wilt of banana by Trichoderma spp. 8th PPSKH colloquium, Pust Pengajian Sains Kajihayat/School of Biological Sciences (P. 13), USM, 5–6 June.

Al-Ani, L. K. T., Yonus, M. I., Mahdii, B. A., Omer, M. A., Taher, J. K., Albaayit, S. F. A., & Al-Khoja, S. B. (2018). First record of use *Fusarium proliferatum* fungi in direct treatment to control the adult of wheat flour *Tribolium confusum*, as well as use the entomopathogenic fungi *Beauveria bassiana*. *Ecology, Environment and Conservation, 24*(3), 29–34.

Al-Ani, L. K. T., & Furtado, E. L. (2020a). The effect of incompatible plant pathogen on the host plant. In V. Sharma, R. Salwan, & L. K. T. Al-Ani (Eds.), *Molecular aspects of plant beneficial microbes in agriculture* (pp. 47–57). Elsevier Science.

Al-Ani, L. K. T., & Mohammed, A. M. (2020b). Versatility of *Trichoderma* in plant disease management. In V. Sharma, R. Salwan, & L. K. T. Al-Ani (Eds.), *Molecular aspects of plant beneficial microbes in agriculture* (pp. 159–168). Elsevier Science.

Al-Ani, L. K. T., Aguilar-Marcelino, L., Fiorotti, J., Sharma, V., Sarker, M. S., Raza, W., Furtado, E. L., Wijayawardene, N. N., & Herrera-Estrella, A. (2020a). Biological control agents and their importance for the plant health. In J. S. Singh & S. R. Vimal (Eds.), *Microbial services in ecological restoration* (pp. 13–36). Elsevier.

Al-Ani, L. K. T., Franzino, T., Aguilar-Marcelino, L., Haichar, F. Z., Furtado, E. L., Raza, W., Jatoi, G. H., & Raza, M. (2020b). The role of microbial signals in plant growth and development: Current status and future prospects. In A. A. Rastegari, A. N. Yadav, A. K. Awasthi, & N. Yadav (Eds.), *New and future developments in microbial biotechnology and bioengineering* (pp. 225–242). Elsevier.

Al-Ani, L. K. T., Surono Aguilar-Marcelino L., Salazar-Vidal, V. E., Becerra, A. G., & Raza, W. (2021a). Role of useful fungi in agriculture sustainability. In A. N. Yadav (Ed.), *Recent trends in mycological research, fungal biology* (pp. 1–44), Springer Nature Switzerland.

Al-Ani, L. K. T., Aguilar-Marcelino, L., Becerra, A. G., & Salazar-Vidal, V. E. (2021b). Fe-chelating compounds producing fungal communities and their applications. In A. N. Yadav (Ed.), *Recent trends in mycological research, Volume 1: Agricultural and medical perspective* (pp. 135–157). Springer Nature.

Al-Ani, L. K. T., Soares, F. E. D. F., Sharma, A., Villalobos, S. D. l. S., Valdivia-Padilla, A. V., & Aguilar-Marcelino, L. (2022). Strategy of nematophagous fungi in determining the activity of plant parasitic nematodes and their prospective role in sustainable agriculture. *Frontiers in Fungal Biology, 3*, 863198. https://doi.org/10.3389/ffunb.2022.863198

Al-Ani, L. K. T., Mitrović, I., Zimowska, B., Sarker, M. S., Dautt-Castro, M., Amjad, S. F., Jatoi, G. H., Surono, S., Hajji-Hedfi, l., & Sukmawati, D. (2024). Metabolites of fungi and applications in the agricultural. In S. Siddiqui, V. Meshram, R. S. Patil, & M. Gupta (Eds.), *Present status and prospects of plant associated fungi* (pp. 351–397). Cambridge Scholars Publishing.

Al-Ani, L. K. T., Aguilar-Marcelino, L., Santos-Villalobos, S. D. L., Montoya-Martínez, A. C., Pineda-Alegría, J. A., & Mitrovic, I. (2025a). Identification and characterization of plant parasitic nematodes using molecular techniques. In M. S. Akhtar & K. R. Hakeem (Eds.), Plant-nematode interactions and their management strategies. CRC Press.

Al-Ani, L. K. T., Aguilar-Marcelino, L., Cruz-Arévalo, J., Grifaldo-Alcántara, P. F., & Palomares-Pérez, M. (2025b). Progress and development of plant nematology and future consequences. In M. S. Akhtar & K. R. Hakeem (Eds.), Plant-nematode interactions and their management strategies. CRC Press.

Al-Hazmi, A. S., & TariqJaveed, M. (2016). Effects of different inoculum densities of *Trichoderma harzianum* and *Trichoderma viride* against *Meloidogyne javanica* on tomato. *Saudi Journal of Biological Sciences, 23*, 288–292.

Atanasova, L., Le Crom, S., Gruber, S., Coulpier, F., Seidl-Seiboth, V., Kubicek, C. P., & Druzhinina, I. S. (2013). Comparative transcriptomics reveals different strategies of *Trichoderma mycoparasitism. BMC Genomics, 14*, 1471–2164. https://doi.org/10.1186/1471-2164-14-121

Attitalla, I. H., Mansour, S. E., Mohamed, W. S., Al-Ani, L. K. T., Mohammed, A. M., Faturi, M. Y., Balal, I. A. A., & El-Maraghy, S. S. M. (2010a). Influence of *Aspergillus flavus* and *Aspergillus terreus* on the protein value of the two varieties of peanut grains. International Conference, International Mycotoxin Conference (p. 177), MycoRed, Penang, Malaysia.

Attitalla, I. H., Al-Ani, L. K. T., Nasib, M. A., Balal, I. A. A., Zakaria, M., El-Maraghy, S. S. M., & Karim, S. M. R. (2010b). Screening of fungi associated with commercial grains and animal feeds in Al-Bayda Governorate, Libya. *World Applied Sciences Journal, 9*(7), 746–756.

Baazeem, A., Almanea, A., Manikandan, P., Alorabi, M., Vijayaraghavan, P., & Abdel-Hadi, A. (2021). In vitro antibacterial, antifungal, nematocidal and growth promoting activities of *Trichoderma hamatum* FB10 and its secondary metabolites. *Journal of Fungi, 7*, 331. https://doi.org/10.3390/jof7050331

Back, M. A., Haydock, P. P. J., & Jenkinson, P. (2002). Disease complexes involving plant parasitic nematodes and soil borne pathogens. *Plant Pathology, 51*, 683–697. https://doi.org/10.1046/j.1365-3059.2002.00785.x

Badri, D. V., Weir, T. L., Van der Lelie, D., & Vivanco, J. M. (2009). Rhizosphere chemical dialogues: Plant-microbe interactions. *Current Opinion in Biotechnology, 20*, 642–650. https://doi.org/10.1016/j.copbio.2009.09.014

Balestrini, R., Chitarra, W., Fotopoulos, V., & Ruocco, M. (2017). Potential role of beneficial soil microorganisms in plant tolerance to abiotic stress factors. In R. Balestrini, W. Chitarra, V. Fotopoulos, & M. Ruocco (Eds.), *Soil biological communities and ecosystem resilience* (pp. 191–207). Springer International Publishing.

Bhattacharyya, P. N., & Sarmah, S. R. (2018). The role of microbes in tea cultivation. In *Global tea science* (pp.155–188). Burleigh Dodds Science Publishing.

Bhattacharyya, P. N., Dutta, P., Sarmah, S. R., Tanti, A., Madhab, M., & Begum, R. (2016). An in vitro assessment on fusarium decomposition in tea. *International Journal of Current Microbiology and Applied Sciences, 5*(11), 121–125. https://doi.org/10.20546/ijcmas.2016.511.013

Bhattacharyya, P. N., Islam, N. F., Sarma, B., Nath, B. C., Al-Ani, L. K. T., & Lesueur, D. (2023). Frankia-actinorhizal symbiosis: A non-chemical biological assemblage for enhanced plant growth, nodulation and reclamation of degraded soils. *Symbiosis*. https://doi.org/10.1007/s13199-023-00956-2

Bhattacharyya, P. N., Nath, B. C., Sarma, B., Al-Ani, L. K.T., Borgohain, D. J., Garganese, F., Ahmed, S., Batsya, S., Mudoi, A., & Kumari, R. (2024). Interaction of Metarhizium anisopliae against emergent insect pest problems in the North-Eastern tea industry. In S. K. Deshmukh & K. R. Sridhar (Eds.), *Entomopathogenic fungi: prospects and challenges* (pp. 319–357). Springer Nature Singapore. https://doi.org/10.1007/978-981-97-5991-0_13

Bischof, R. H., Ramoni, J., & Seiboth, B. (2016). Cellulases and beyond: The first 70 years of the enzyme producer *Trichoderma reesei. Microbial Cell Factories, 15*, 1–13.

Błaszczyk, L., Basińska-Barczak, A., Ćwiek-Kupczyńska, H., Gromadzka, K., Popiel, D., & Stępień, Ł. (2017). Suppressive effect of *Trichoderma* spp. on toxigenic *Fusarium* species. *Journal of Applied Microbiology, 66*, 85–100. https://doi.org/10.5604/17331331.1234996

Bravo, D., Cailleau, G., Bindschedler, S., Simon, A., Job, D., Verrecchia, E., & Junier, P. (2013). Isolation of oxalotrophic bacteria able to disperse on fungal mycelium. *FEMS Microbiology Letters, 348*(2), 157–166.

Castañeda-Ramírez, G. S., Torres-Acosta, J. F. D. J., Sánchez, J. E., Mendoza-de-Gives, P., González-Cortázar, M., Zamilpa, A., Al-Ani, L. K. T., Sandoval-Castro, C., Soares, F. E. D. F., Aguilar-Marcelino, L. (2020). The possible biotechnological use of edible mushroom bioproducts for controlling plant and animal parasitic nematodes. *BioMed Research International, 2020*, 1–12. https://doi.org/10.1155/2020/6078917

Chaverri, P., Gazis, R. O., & Samuels, G. J. (2011). *Trichoderma amazonicum*, a new endophytic species on *Hevea brasiliensis* and *H. guianensis* from the Amazon basin. *Mycologia, 103*, 139–151. https://doi.org/10.3852/10-078

Chen, J., Vallikkannu, M., & Karuppiah, V. (2020). Systemically induced resistance against maize diseases by Trichoderma spp. In A. Sharma & P. Sharma (Eds.) *Trichoderma. Rhizosphere biology*. Springer. https://doi.org/10.1007/978-981-15-3321-1_6

Chet, I. (1987). *Trichoderma-application, mode of action, and potential as a biocontrol agent of soilborne plant pathogenic fungi*. Wiley.

Choudhary, D. K., & Johri, B. N. (2009). Interactions of *Bacillus* spp. and plants-with special reference to induced systemic resistance (ISR). *Microbiological Research, 164*, 493–513. https://doi.org/10.1016/j.micres.2008.08.007

Comans-Pérez, R. J., Sánchez, J. E., Al-Ani, L. K. T., González-Cortázar, M., Castañeda-Ramírez, G. S., Gives, P. M.-d., Sánchez-García, A. D., Orozco, J. M., & Aguilar-Marcelino, L. (2021). Biological control of sheep nematode *Haemonchus contortus* using edible mushrooms. *Biological Control, 152*, 104420. https://doi.org/10.1016/j.biocontrol.2020.104420

Contreras-Cornejo, H., Ortiz-Castro, R., & López-Bucio, J. (2009). Promotion of plant growth and the induction of systemic defense by *Trichoderma*: Physiology, genetics, and gene expression. In P. K. Mukherjee, B. A. Horwitz, U. S. Singh, M. Mukherjee, & M. Schmoll (Eds.) *Trichoderma: Biology and applications* (pp. 175–196). CABI International.

Contreras-Cornejo, H. A., Macías-Rodríguez, L., Del-Val, E. K., & Larsen, J. (2016). Ecological functions of *Trichoderma* spp. and their secondary metabolites in the rhizosphere: Interactions with plants. *FEMS Microbiology Ecology, 92*(4), fiw036.

Coradi, G. V., da Visitação, V. L., de Lima, E. A., Saito, L. Y. T., Palmieri, D. A., Takita, M. A., de Oliva Neto, P., & Gomes de Lima, V. M. (2013). Comparing submerged and solid-state fermentation of agro-industrial residues for the production and characterization of lipase by *Trichoderma harzianum*. *Annals of Microbiology, 63*, 533–540.

Dalimunthe, C. I., Soekarno, B. P. W., Al-Ani, L. K. T., Munif, A., Sriherwanto, C., & Nurdebyandaru, N. (2023). First report of inhibitory abilities of dark septate endophytic fungi against white root rot disease on Hevea brasiliensis seedlings in nursery conditions. *Egyptian Journal of Biological Pest Control, 33*(1), 1–12.

Daoubi, M., Pinedo-Rivilla, C., Rubio, M. B., Hermosa, R., Monte, E., Aleu, J., & Collado, I. G. (2009). Hemisynthesis and absolute configuration of novel 6-pentyl-2H-pyran-2-one derivatives from *Trichoderma* spp. *Tetrahedron, 65*, 4834–4840. https://doi.org/10.1016/j.tet.2009.04.051

de Freitas Soares, F. E., Ferreira, J. M., Genier, H. L. A., Al-Ani, L. K. T., & Aguilar-Marcelino, L. (2023). Biological control 2.0: Use of nematophagous fungi enzymes for nematode control. *Journal of Natural Pesticide Research, 4*, 100025.

de Medeiros, H. A., de Araújo Filho, J. V., de Freitas, L. G., Castillo, P., Rubio, M. B., Hermosa, R., & Monte, E. (2017). Tomato progeny inherit resistance to the nematode *Meloidogyne javanica* linked to plant growth induced by the biocontrol fungus *Trichoderma atroviride*. *Scientific Reports, 7*, 40216. https://doi.org/10.1038/srep40216

de Rezende, L. C., Carvalho, A. L. A., Costa, L. B., Halfeld-Vieira, B. A., Silva, L. G., Pinto, Z. V., Morandi, M. A. B., Medeiros, F. H. V., Mascarin, G. M., & Bettiol, W. (2020). Optimizing mass production of *Trichoderma asperelloides* by submerged liquid fermentation and its antagonism against *Sclerotinia sclerotiorum*. *World Journal of Microbiology and Biotechnology, 36*(8), 1–14. https://doi.org/10.1007/s11274-020-02882-7

de Souza, G. F., & Al-Ani, L. K. T. (2023). Effects of gamma radiation and electron beam on samples of the peanut candies contaminated with Aspergillus flavus. Conference: VIII. International Scientific and Vocational Studies Congress—(BILMES 2023)At: Turkey.

Djonović, S., Vargas, W. A., Kolomiets, M. V., Horndeski, M., Wiest, A., & Kenerley, C. M. (2007). A proteinaceous elicitor Sm1 from the beneficial fungus *Trichoderma virens* is required for induced systemic resistance in maize. *Plant Physiology, 145*, 875–889. https://doi.org/10.1104/pp.107.103689

Dorta-Vásquez, R., Valbuena, O., & Pavone-Maniscalco, D. (2019). Solid-state fermentation of paper sludge to obtain spores of the fungus *Trichoderma asperellum. The EuroBiotech Journal, 3*(2), 71–77.

Druzhinina, I. S., Seidl-Seiboth, V., Herrera-Estrella, A., Horwitz, B. A., Kenerley, C. M., Monte, E., Mukher¬jee, P. K., Zeilinger, S., Grigoriev, I. V., & Kubicek, C. P. (2011). *Trichoderma*: the genomics of opportunistic success. *Nature Reviews Microbiology, 9*(10), 749–759. https://doi.org/10.1038/nrmicro2637

Dutta, T. K., Stephen, J. P., Kerry, B. R., Gaur, H. S., & Curtis, R. H. C. (2011). Comparison of host recognition, invasion, development and reproduction of *Meloidogyne graminicola* and *M. incognita* on rice and tomato. *Nematology, 13*(5), 509–520. https://doi.org/10.1163/138855410X528262

Elad, Y. (2000). Biological control of foliar pathogens by means of *Trichoderma harzianum* and potential modes of action. *Crop Protection, 19*(8–10), 709–714. https://doi.org/10.1016/S0261-2194(00)00094-6

Elsharkawy, M. M., Shimizu, M., Takahashi, H., Ozaki, K., & Hyakumachi, M. (2013). Induction of systemic resistance against cucumber mosaic virus in *Arabidopsis thaliana* by *Trichoderma asperellum* SKT-1. *Plant Pathology Journal, 29*(2), 193–200. https://doi.org/10.5423/ppj.si.07.2012.0117

Elsherbiny, E. A., Amin, B. H., Aleem, B., Kingsley, K. L., & Bennett, J. W. (2020). Trichoderma volatile organic compounds as a biofumigation tool against late blight pathogen Phytophthora infestans in postharvest potato tubers. *Journal of Agriculture and Food Chemistry, 68*, 8163–8171.

Emerson, F. L., & Mikunthan, G. (2015). Small scale production of *Trichoderma viride* on locally available liquid waste and other substrates. *American-Eurasian Journal of Agricultural & Environmental Sciences, 15*(8), 1666–1671.

Ezekiel, O. O., Aworh, O. C., Blaschek, H. P., & Ezeji, T. C. (2010). Protein enrichment of cassava peel by submerged fermentation with *Trichoderma viride* (ATCC 36316). *African Journal of Biotechnology, 9*(2), 187–194.

Fan, H., Yao, M., Wang, H., Zhao, D., Zhu, X., Wang, Y., Liu, X., Duan, Y., & Chen, L. (2020). Isolation and effect of *Trichoderma citrinoviride* Snef 1910 for the biological control of root-knot nematode *Meloidogyne Incognita. BMC Microbiology, 20*, 299. https://doi.org/10.1186/s12866-020-01984-4

Gaderer, R., Lamdan, N. L., Frischmann, A., Sulyok, M., Krska, R., Horwitz, B. A., & Seidl-Seiboth, V. (2015). Sm2, a paralog of the Trichoderma cerato-platanin elicitor Sm1, is also highly important for plant protection conferred by the fungal-root interaction of Trichoderma with maize. *BMC Microbiology 16*, 15(1), 2. https://doi.org/10.1186/s12866-014-0333-0. PMID: 25591782; PMCID: PMC4320488.

Garganese, F., Schena, L., Siciliano, I., Prigigallo, M. I., Spadaro, D., De Grassi, A., Ippolito, A., & Sanzani, S. M. (2016). Characterization of citrus-associated *Alternaria* species in Mediterranean areas. *PLoS ONE, 11*(9), Article e0163255. https://doi.org/10.1371/journal.pone.0163255

Garnica-Vergara, A., Barrera-Ortiz, S., Muñoz-Parra, E., Raya-González, J., Méndez-Bravo, A., Macías-Rodríguez, L., Ruiz-Herrera, L. F., & López-Bucio, J. (2016). The volatile 6-pentyl-2H-pyran-2-one from Trichoderma atroviride regulates Arabidopsis thaliana root morphogenesis via auxin signaling and ETHYLENE INSENSITIVE 2 functioning. *New Phytologist, 209*(4), 1496–1512.

Grahovac, J., Mitrović, I., Dodić, J., Grahovac, M., Rončević, Z., Dodić, S., & Jokić, A. (2020). Biocontrol agent for apple *Fusarium* rot: Optimization of production by *Streptomyces hygroscopicus. Zemdirbyste-Agriculture, 107*(3), 263–270.

Grayston, S. J., Vaugham, D., & Jones, D. (1996). Rhizosphere carbon flow in trees, in comparison with annual plants: The importance of root exudation and its impact on microbial activity and nutrient availability. *Applied Soil Ecology, 5*, 29–56. https://doi.org/10.1016/S0929-1393(96)00126-6

Gupta, V. K., Steindorff, A. S., de Paula, R. G., Silva-Rocha, R., Mach-Aigner, A. R., Mach, R. L., & Silva, R. N. (2016). The post-genomic era of *Trichoderma reesei*: What's next? *Trends in Biotechnology, 34*(12), 970–982. https://doi.org/10.1016/j.tibtech.2016.06.003

Guzmán-Guzmán, P., Porras-Troncoso, M. D., Olmedo-Monfil, V., & Herrera-Estrella, A. (2019). Trichoderma species: Versatile plant symbionts. *Phytopathology, 109*(1), 6–16.

Guzmán-Guzmán, P., Kumar, A., de Los Santos-Villalobos, S., Parra-Cota, F. I., Orozco-Mosqueda, M. D. C, Fadiji, A. E., Hyder, S., Babalola, O.O., & Santoyo, G. (2023). Trichoderma species: Our best fungal allies in the biocontrol of plant diseases-a review. *Plants (Basel) 17, 12*(3), 432. https://doi.org/10.3390/plants12030432. PMID: 36771517; PMCID: PMC9921048.

Haas, D., & Défago, G. (2005). Biological control of soil-borne pathogens by fluorescent pseudomonads. *Nature Reviews Microbiology, 3*, 307–319. https://doi.org/10.1038/nrmicro1129

Hajieghrari, B., Torabi-Giglou, M., Mohammadi, M. R., & Davari, M. (2008). Biological potential of some Iranian *Trichoderma* isolates in the control of soil borne plant pathogenic fungi. *African Journal of Biotechnology, 7*, 967–972.

Hajji-Hedfi, L., Rhouma, A., Al-Ani, L. K. T., Bargougui, O., Tlahig, S., Jaouadi, R., Zaouali, Y., Degola, F., & Abdel-Azeem, A. M. (2025a). The second life of citrus: Phytochemical characterization and antifungal activity bioprospection of C. limon and C. sinensis peel extracts against potato rot disease. *Waste Biomass Valor*. https://doi.org/10.1007/s12649-025-03095-5

Hajji-Hedfi, L., Wannassi, T., Tawfeeq Al-Ani, L. K., Balbool, B. A., Tissaoui, S., Mougou-Hamdane, A., Hamdi, W., Azeem, A. M. A., & Rebouh, N. Y. (2025b). Investigating the potential role of beneficial rhizobacteria for protecting grapevine health and promoting growth. *Frontiers in Sustainable Food Systems, 9*, 1619801. https://doi.org/10.3389/fsufs.2025.1619801

Hajji-Hedfi, L., Dali, S., Rhouma, A., Al-Ani, L. K. T., Chihani-Hammas, N., Khlif, A., & Bargougui O. (2025c). Microbial inoculants for plant resilience performance: roles, prospects and challenges. *Journal of Plant Diseases and Protection, 132*(2), 88. https://doi.org/10.1007/s41348-025-01080-7

Hajji-Hedfi, L., Rhouma, A., Al-Ani, L. K. T., El-Abeid, S. E., Ibrahim, D. S. S., Acheampong, M. A., Bousselma, A., Ortega-Villaizán, A. G., & Pollmann, S. (2024). Exploring endophytic fungal diversity: Review of traditional and molecular techniques. In A. M. Abdel Azeem, A. N. Yadav, & N. Yadav (Eds.), *Developments in applied microbiology and biotechnology, endophytic fungi* (pp. 1–24). Academic Press. https://doi.org/10.1016/B978-0-323-99314-2.00005-X

Harman, G. E. (2000). Myths and dogmas of biocontrol: Changes in perceptions derived from research on Trichoderma harzianum T-22. *Plant Disease, 84*, 377–393.

Harman, G. E. (2011). *Trichoderma*-not just for biocontrol anymore. *Phytoparasitica, 39*(2), 103–108. https://doi.org/10.1007/s12600-011-0151-y

Harman, G. E., Howell, C. R., Viterbo, A., Chet, I., & Lorito, M. (2004a). *Trichoderma* species-opportunistic, avirulent plant symbionts. *Nature Reviews Microbiology* 2(1), 43–56. https://doi.org/10.1038/nrmicro797

Harman, G. E., Petzoldt, R., Comis, A., & Chen, J. (2004b). Interactions between *Trichoderma harzianum* strain T22 and maize inbred line Mo17 and effects of these interactions on diseases caused by *Pythium ultimum* and *Colletotrichum graminicola. Phytopathology 94*, 147–153. https://doi.org/10.1094/PHYTO.2004.94.2.147

Harman, G. E., Ma, O., Samuels, G. J., & Lorto, M. (2010). Changing models for commercialization and implementation of biocontrol in the developing and the developed world. *Plant Disease, 94*(8), 928–939. https://doi.org/10.1094/PDIS-94-8-0928

He, C., Wang, W., & Hou, J. (2020). Plant performance of enhancing licorice with dual inoculating dark septate endophytes and *Trichoderma viride* mediated via effects on root development. *BMC Plant Biology, 20*, 1–14.

Hermosa, M. R., Grondona, I., Iturriaga, E. T., Diaz-Minguez, J. M., Castro, C., Monte, E., & Garcia-Acha, I. (2000). Molecular characterization and identification of biocontrol isolates of *Trichoderma* spp. *Applied and Environment Microbiology, 66*(5), 1890–1898. https://doi.org/10.1128/AEM.66.5.1890-1898.2000

Hermosa, R., Viterbo, A., Chet, I., & Monte, E. (2012). Plant-beneficial effects of *Trichoderma* and of its genes. *Microbiology, 158*, 17–25. https://doi.org/10.1099/mic.0.052274-0

Hermosa, R., Rubio, M. B., Cardoza, R. E., Nicolás, C., Monte, E., & Gutiérrez, S. (2013). The contribution of Trichoderma to balancing the costs of plant growth and defense. *International Microbiology, 16*(2), 69–80.

Horbach, R., Navarro-Quesada, A. R., Knogge, W., & Deising, H. B. (2011). When and how to kill a plant cell: Infection strategies of plant pathogenic fungi. *Journal of Plant Physiology, 168*, 51–62. https://doi.org/10.1016/j.jplph.2010.06.014

Howell, C. R. (2003). Mechanisms employed by *Trichoderma* species in the biological control of plant diseases: The history and evolution of current concepts. *Plant Disease, 87*(1), 4–10. https://doi.org/10.1094/PDIS.2003.87.1.4

Hyde, K. D. (2020). Refined families of Sordariomycetes. *Mycosphere, 11*, 305–1059

Idrees, A., Zhang, H., Luo, M., Thu, M., Cai, P., Islam, W., Hussain, M., Chen, J., & Ji, Q. (2017). Protein baits, volatile compounds and irradiation influence the expression profiles of odorant-binding protein genes in *Bactrocera dorsalis* (Diptera: Tephritidae). *Applied Ecology and Environmental Research 15*, 1883–1899. https://doi.org/10.15666/aeer/1504_18831899

Islam, W. (2017). Management of plant virus diseases; farmer's knowledge and our suggestions. *Hosts Viruses, 4*, 28–33.

Islam, M. S., Ahmad, I., & Ali, M. (2018). Biocontrol studies on rizpspheric microorganisms against black rot disease of tea caused by Corticium theae Bernard. *Bangladesh Journal of Botany, 47*(4), 985–991. https://doi.org/10.3329/bjb.v47i4.47399

Jatoi, G. H., Muhammad, S., Metlo, W. A., Al-Ani, L. K. T., Haseenullah Abro, M. A., Gadhi, M. A., Awan, N. W., & Reki, M. A. (2020). Efficacy of different essential oils, fungicides and biocontrol agents against Aspergillus niger the causal agent of fruit rot in Pomegranate. *International Journal of Biosciences, 16*(3), 51–65.

Jatoi, G., Abro, M., Ahmed, S., Al-Ani, L. K. T., Ali, U., Jatoi, M., Figari, I., Qambrani, J., Ahmed, I., & Soomro, A. (2022). Preliminary selection and evaluation of fungicides and natural compounds to control grey mold disease of rose caused by botrytis cinerea. *International Journal of Phytopathology, 11*(1), 49–58. https://esciencepress.net/journals/index.php/phytopath/article/view/4181

Jones, J. T., Haegeman, A., Danchin, E. G. J., Gaur, H. S., Jones, M. G. K., Kikuchi, T., Manzanilla-López, R., Palomares-Rius, J. E., Wesemael, W. M. L., & Perry, R. N. (2013). Top 10 plant-parasitic nematodes in molecular plant pathology. *Molecular Plant Pathology, 14*(9), 946–961. https://doi.org/10.1111/mpp.12057

Juliatti, F. C., Rezende, A. A., Juliatti, B. C. M., & Morais, T. P. (2019). Trichoderma as a biocontrol agent against Sclerotinia stem rot or white mold on soybeans in Brazil: Usage and technology. In M. M. Shah, U. Sharif, & T. R. Buhari (Eds.) Trichoderma—The most widely used fungicide. IntechOpen.

Kariuki, C. K., Mutitu, E. W., & Muiru, W. M. (2020). Effect of Bacillus and *Trichoderma* species in the management of the bacterial wilt of tomato (*Lycopersicum esculentum*) in the field. *Egyptian Journal of Biological Pest Control, 30*, 109. https://doi.org/10.1186/s41938-020-00310-4

Karuppiah, V., Sun, J., Li, T., Vallikkannu, M., & Chen, J. (2019). Co-cultivation of *Trichoderma asperellum* GDFS1009 and *Bacillus amyloliquefaciens* 1841 causes differential gene expression and improvement in the wheat growth and biocontrol activity. *Frontiers in Microbiology, 10*, 1068.

Khalili, E., Javed, M. A., Huyop, F., Rayatpanah, S., Jamshidi, S., & Wahab, R. A. (2016). Evaluation of *Trichoderma* isolates as potential biological control agent against soybean charcoal rot disease caused by *Macrophomina phaseolina. Biotechnology & Biotechnological Equipment, 30*(3), 479–488. https://doi.org/10.1080/13102818.2016.1147334

Khan, R. A. A., Najeeb, S., Mao, Z., Ling, J., Yang, Y., Li, Y., & Xie, B. (2020). Bioactive secondary metabolites from *Trichoderma* spp. against phytopathogenic bacteria and root-knot nematode. *Microorganisms, 8*, 401. https://doi.org/10.3390/microorganisms8030401

Kim, C. S., Park, M. S., Kim, S. C., Maekawa, N., & Yu, S. H. (2012). Identification of Trichoderma, a competitor of shiitake mushroom (Lentinula edodes), and competition between *Lentinula edodes* and *Trichoderma* species in Korea. *Plant Pathology Journal, 28*, 137–148. https://doi.org/10.5423/PPJ.2012.28.2.137

Kobori, N. N., Mascarin, G. M., Jackson, M. A., & Schisler, D. A. (2015). Liquid culture production of microsclerotia and submerged conidia by *Trichoderma harzianum* active against damping-off disease caused by *Rhizoctonia solani. Fungal Biology, 119*(4), 179–190. https://doi.org/10.1016/j.funbio.2014.12.005

Konappa, N., Krishnamurthy, S., Siddaiah, C. N., Ramachandrappa, N. S., & Chowdappa, S. (2018). Evaluation of biological efficacy of *Trichoderma asperellum* against tomato bacterial wilt caused by *Ralstonia solanacearum. Egyptian Journal of Biological Pest Control, 28*(1), 1–11. https://doi.org/10.1186/s41938-018-0069-5

Küçük, Ç., & Kivanç, M. (2003). Isolation of *Trichoderma* spp. and determination of their antifungal, biochemical and physiological features. *Turkish Journal of Biology, 27*, 247–253.

Kumar, S. (2013). *Trichoderma*: A biological weapon for managing plant diseases and promoting sustainability. *International Journal of Agriculture Science and Medical Veterinary, 1*, 106–121.

Kumar, N., & Khurana, S. P. (2021). Trichoderma-plant-pathogen interactions for benefit of agriculture and environment. *Biocontrol agents and secondary metabolites* (pp. 41–63). Woodhead Publishing.

Kumar, R., Samanta, P., Vijay Raj, S., Bera, P., & Naimuddin, M. (2023). Potential and prospects of Trichoderma in plant protection. *Advances in Agriculture, 2023*(1), 5573662.

Lalarukh, I., Al-Dhumri, S. A., Al-Ani, L. K. T., Hussain, R., Al Mutairi, K.A., Mansoora, N., Amjad, S.F., Abbas, M. H. H., Abdelhafez, A. A., Poczai, P., Meena, K. R., & Galal, T. M. (2022a). A combined use of rhizobacteria and moringa leaf extract mitigates the adverse effects of drought stress in wheat (*Triticum aestivum* L.). *Frontiers in Microbiology, 13*, 813415. https://doi.org/10.3389/fmicb.2022.813415

Lalarukh, I., Al Dhumri, S. A., Al-Ani, L. K. T., Shahbaz, M., Amjad, S. F., Mansoora, N., Al Shammari, W. B., Alhusayni, F. S., & Almutari, M. M. (2022b). Combined application of poultry litter biochar with NPK fertilizer mitigates the effects of stress in wheat (Triticum aestivum L.) grown on chromium contaminated soil. *Plant and Soil, 480*, 201–212. https://doi.org/10.1007/s11104-022-05574-3

Lee, S. J., Yeo, W. H., Yun, B. S., & Yoo, I. D. (1999). Isolation and sequence analysis of new peptaibol, boletusin, from *Boletus* spp. *Journal of Peptide Science, 5*(8), 374–378. https://doi.org/10.1002/(SICI)1099-1387(199908)5:8%3c374::AID-PSC211%3e3.0.CO;2-X

Lee, J., Huh, N., Hong, J. H., Kim, B. S., Kim, G. H., & Kim, J. J. (2012). The antagonistic properties of *Trichoderma* spp. inhabiting woods for potential biological control of wood-damaging fungi. *Holzforschung, 66*, 883–887. https://doi.org/10.1515/hf.2011.187

Leelavathi, M. S., Vani, L., & Reena, P. (2014). Antimicrobial activity of Trichoderma harzianum against bacteria and fungi. *International Journal of Current Microbiology and Applied Sciences, 3*(1), 96–103.

Li, G.-H., Zheng, L. J., Liu, F. F., Dang, L. Z., Li, L., Huang, R., & Zhang, K. Q. (2009). New cyclopentenones from strain *Trichoderma* sp. YLF-3. *Natural Product Research, 23*, 1431–1435.

Locher, K. P. (2004). Structure and mechanism of ABC transporters. *Current Opinion in Structural Biology, 14*, 426–431. https://doi.org/10.1016/j.sbi.2004.06.005

López-Bucio, J., Pelagio-Flores, R., & Herrera-Estrella, A. (2015). *Trichoderma* as biostimulant: Exploiting the multilevel properties of a plant beneficial fungus. *Scientia Horticulturae, 196*, 109–123. https://doi.org/10.1016/j.scienta.2015.08.043

Lorito, M., Woo, S. L., Harman, G. E., & Monte, E. (2010). Translational research on *Trichoderma*: From omics to the field. *Annual Review of Phytopathology, 48*, 395–417. https://doi.org/10.1146/annurev-phyto-073009-114314

Luo, Y., Zhang, D. D., Dong, X. W., Zhao, P. B., Chen, L. L., Song, X. Y., Wang, X. J., Chen, X. L., Shi, M., & Zhang, Y. Z. (2010). Antimicrobial peptaibols induce defense responses and systemic resistance in tobacco against tobacco mosaic virus. *FEMS Microbiology Letters, 313*(2), 120–126.

Maaløe, S. (2002). Physical behavior of the plume source during intermittent eruptions of Hawaiian basalts. *Contributions to Mineralogy and Petrology, 142*, 653–665. https://doi.org/10.1007/s00410-001-0323-8

Mahato, S., Bhuju, S., & Shrestha, J. (2018). Effect of Trichoderma viride as biofertilizer on growth and yield of wheat. *Malaysian Journal of Sustainable Agriculture (MJSA), 2*(2), 1–5.

Malmierca, M. G., Cardoza, R. E., Alexander, N. J., McCormick, S. P., Hermosa, R., Monte, E., & Gutierrez, S. (2012). Involvement of *Trichoderma trichothecenes* in the biocontrol activity and induction of plant defense-related genes. *Applied and Environment Microbiology, 78*, 4856–4868. https://doi.org/10.1128/AEM.00385-12

Malmierca, M. G., Cardoza, R. E., Alexander, N. J., McCormick, S. P., Collado, I. G., Hermosa, R., Monte, E., & Gutiérrez, S. (2013). Relevance of trichothecenes in fungal physiology: Disruption of tri5 in Trichoderma arundinaceum. *Fungal Genetics and Biology, 53*, 22–33.

Mandal, R. H., Katel, S., Subedi, S., & Shrestha, J. (2021). Plant parasitic nematodes and their management in crop production: A review. *Journal of Agriculture and Natural Resources, 4*(2), 327–338. https://doi.org/10.3126/janr.v4i2.33950

Marois, J. J. (1990). Biological control of diseases caused by *Fusarium oxysporum*. In R. C. Ploetz (Ed.), *Fusarium wilt of banana* (pp. 77–81). APS Press.

Martínez-Salgado, S. J., Andrade-Hoyos, P., Parraguirre Lezama, C., Rivera-Tapia, A., Luna-Cruz, A., & Romero-Arenas, O. (2021). Biological control of charcoal rot in peanut crop through strains of *Trichoderma* spp., in Puebla, Mexico. *Plants, 10*(12), 2630. https://doi.org/10.3390/plants10122630

Martinez, Y., Ribera, J., Schwarze, F. W., & De France, K. (2023). Biotechnological development of Trichoderma-based formulations for biological control. *Applied Microbiology and Biotechnology, 107*(18), 5595–5612.

Mathys, J., De Cremer, K., Timmermans, P., Van Kerckhove, S., Lievens, B., Vanhaecke, M., Cammue, B. P. A., & De Coninck, B. (2012). Genome-wide characterization of ISR induced in Arabidopsis thaliana by *Trichoderma hamatum* T382 against *Botrytis cinerea* infection. *Frontiers in Plant Science, 3*, 108. https://doi.org/10.3389/fpls.2012.00108

Mitra, D., Nayeri, F. D., Sansinenea, E., Ortiz, A., Bhatta, B. B., Adeyemi, N. O., Janeeshma, E., Tawfeeq Al-Ani, L. K., Sharma, S. B., Boutaj, H., et al. (2023). Unraveling arbuscular mycorrhizal fungi interaction in rice for plant growth development and enhancing phosphorus use efficiency through recent development of regulatory genes. *Journal of Plant Nutrition, 46*, 1–37.

Mitrović, I., Tančić Živanov, S., Purar, B., Trivunović, Z., & Mitrović, B. (2021a). Effect of different carbon and nitrogen sources combination in medium for production of biocontrol agent *Trichoderma harzianum*. *Ratarstvo i povrtarstvo/Field and Vegetable Crops Research, 58*(1), 1–6.

Mitrović, I., Grahovac, J., Hrustić, J., Jokić, A., Dodić, J., Mihajlović, M., & Grahovac, M. (2021b) Utilization of waste glycerol for the production of biocontrol agents nigericin and niphimycin by *Streptomyces hygroscopicus*: bioprocess development. *Environmental Technology*. https://doi.org/10.1080/09593330.2021.1913241

Mitrović, I., Vučurović, D., Al-Ani, L. K. T., Mitrović, B., Bajić, B., Dodić, S., & Živanov, S. T. (2023). Production of Trichoderma harzianum K179 bioagent for maize diseases control:

complete laboratory stage bioprocess development. *Journal of Applied Microbiology, 134*(6), lxad115.

Mohammed, A. M., Al-Ani, L. K. T., Bekbayeva, L., & Salleh, B. (2011). Biological control of fusarium oxysporum f. sp. cubense by pseudomonas fluorescens and BABA in vitro. *World Applied Sciences Journal 15*(2), 189–191.

Mohammed, A. M., Negim, E.-S., Al-Ani, L. K. T., Salleh, B., & Saleh, M. I. (2012). Utilization of amino-azines polymers as antifungal activity for banana. 1st USM—KAZNU International Conference on: Challenges of Teaching and Chemistry Research in Institutions of Higher Learning (p. 29). July11–13.

Mohammed, A. M., Al-Ani, L. K. T., & Salleh, B. (2013). Potential management of Fusarium oxysporum f. sp. cubense, the banana wilt pathogen by using pseudomonas and beta-amino-butyric acid (BABA). *International Symposium on Tropical Fungi* (p. 37), ISTF, IPB International Convention Center, Bogor, Indonesia, September 2013.

Mohammed, A. M., Al-Ani, L. K. T., Salleh, B., & Ghazali, A. M. A. (2014). Determining plant growth promoting and biocontrol factor of bacterial culture media. *The 3rd conference on pests management* (p.103). Crop Protection Research Centre, Sudan, February 3–4.

Mohamed, B. F., Sallam, N. M., Alamri, S. A., Abo-Elyousr, K. A., Mostafa, Y. S., & Hashem, M. (2020). Approving the biocontrol method of potato wilt caused by *Ralstonia solanacearum* (Smith) using *Enterobacter cloacae* PS14 and *Trichoderma asperellum* T34. *Egyptian Journal of Biological Pest Control, 30*, 1–13. https://doi.org/10.1186/s41938-020-00262-9

Mohammed, A. M., & Al-Ani, L. K. T. (2021). Identification and production of beauvericin by *Fusarium subglutinans* and *F. sacchari* from sugarcane. *Brazilian Archives of Biology and Technology 64*, e21200088. https://doi.org/10.1590/1678-4324-2021200088

Molla, A. H., Fakhru'l-Razi, A., Hana, M. M., & Md Alam, Z. (2004). Optimization of process factors for solid-state bioconversion of domestic wastewater sludge. *International Biodeterioration & Biodegradation, 53*, 49–55.

Molla, A. H., Haque, M. M., Haque, M. A., & Ilias, G. N. M. (2012). *Trichoderma*-enriched biofertilizer enhances production and nutritional quality of tomato (*Lycopersicon esculentum* Mill.) and minimizes NPK fertilizer use. *Agricultural Research 1*(3), 265–272. https://doi.org/10.1007/s40003-012-0025-7

Moran-Diez, E., Hermosa, R., Ambrosino, P., Cardoza, R. E., Gutierrez, S., Lorito, M., & Monte, E. (2009). The ThPG1 endopolygalacturonase is required for the *Trichoderma harzianum*-plant beneficial interaction. *Molecular Plant-Microbe Interactions, 22*, 1021–1031. https://doi.org/10.1094/MPMI-22-8-1021

Morán-Diez, M. E., Tranque, E., Bettiol, W., Monte, E., & Hermosa, R. (2020). Differential response of tomato plants to the application of three T*richoderma* species when evaluating the control of *Pseudomonas syringae* populations. *Plants, 9*(5), 626. https://doi.org/10.3390/plants9050626

Mukherjee, M., Mukherjee, P. K., Horwitz, B. A., Zachow, C., Berg, G., & Zeilinger, S. (2012). *Trichoderm*a-plant-pathogen interactions: Advances in genetics of biological control. *Indian Journal of Microbiology, 52*, 522–529. https://doi.org/10.1007/s12088-012-0308-5

Mukherjee, P. K., Horwitz, B. A., Herrera-Estrella, A., Schmoll, M., & Kenerley, C. M. (2013). *Trichoderma* research in the genome era. *Annual Review of Phytopathology, 51*(1), 105–129.

Mukhtar, T. (2018). Management of root-knot nematode, *Meloidogyne incognita*, in tomato with two *trichoderma* species. *Pakistan Journal of Zoology, 50*(4), 1589–1592. https://doi.org/10.17582/journal.pjz/2018.50.4.sc15

Naeimi, S., Khosravi, V., Varga, A., Vágvölgyi, C., & Kredics, L. (2020). Screening of organic substrates for solid-state fermentation, viability and bioefficacy of *Trichoderma harzianum* AS12-2, a biocontrol strain against rice sheath blight disease. *Agronomy, 10*(9), 1258. https://doi.org/10.3390/agronomy10091258

Nautiyal, C. S., Chauhan, P. S., DasGupta, S. M., Seem, K., Varma, A., & Staddon, W. J. (2010). Tripartite interactions among *Paenibacillus lentimorbus* NRRL B-30488, *Piriformospora indica* DSM 11827, and *Cicer arietinum* L. *World Journal of Microbiology and Biotechnology, 26*, 1393–1399.

Pal, K. K., & Gardener, B. M. (2006). Biological control of plant pathogens. *The Plant Health Instructor, 2*, 1117–1142. https://doi.org/10.1094/PHI-A-2006-1117-02

Pandey, S., Srivastava, M., Shahid, M., Kumar, V., Singh, A., Trivedi, S., & Srivastava, Y. K. (2015). *Trichoderma* species cellulases produced by solid state fermentation. *Journal of Data Mining in Genomics & Proteomics, 6*, 170. https://doi.org/10.4172/2153-0602.1000170

Pandey, V., Shukla, A., & Kumar, J. (2016). Physiological and molecular signalling involved in disease management through *Trichoderma*: An effective biocontrol paradigm. In P. Kumar, V. K. Gupta, A. K. Tiwari, & M. Kamle (Eds.), Current trends in plant disease diagnostics and management practices (pp. 317–346). Springer International Publishing.

Pieterse, C. M., Zamioudis, C., Berendsen, R. L., Weller, D. M., Van Wees, S. C., & Bakker, P. A. (2014). Induced systemic resistance by beneficial microbes. *Annual Review of Phytopathology, 52*, 347–375. https://doi.org/10.1146/annurev-phyto-082712-102340

Pocurull, M., Fullana, A. M., Ferro, M., Valero, P., Escudero, N., Saus, E., Gabaldón, T., & Sorribas, F.J. (2020). Resistance to *Meloidogyne incognita* in tomato and the effect is additive to that of the Mi-1.2 resistance gene. *Frontiers in Microbiology, 10*, 3042. https://doi.org/10.3389/fmicb.2019.03042

Poveda, J., Abril-Urias, P., & Escobar, C. (2020). Biological control of plant-parasitic nematodes by filamentous fungi inducers of resistance: *Trichoderma*, mycorrhizal and endophytic fungi. *Frontiers in Microbiology, 11*, 992.

Qasim, M., Husain, D., Islam, S. U., Ali, H., Islam, W., Hussain, M., Wang, F., & Wang, L. (2018). Effectiveness of *Trichogramma chilonis Ishii* against spiny bollworm in Okra and susceptibility to insecticides. *Journal of Entomology and Zoology Studies, 6*, 1576–1581.

Rabea, E. I., Badawy, M. E. T., Stevens, C. V., Smagghe, G., & Steurbaut, W. (2003). Chitosan as antimicrobial agent: Applications and mode of action. *Biomacromolecules, 4*, 1457–1465. https://doi.org/10.1021/bm034130m

Rahman, V., Meena, K. R., Al-Ani, L. K. T., Singh, A., & Kumar, A. (2023). Bacillus velezensis strain improvement to control Helminthosporium maydis causing southern corn leaf blight disease in maize. *European Journal of Plant Pathology*, 1–13.

Rau, A., Hogg, T., Marquardt, R., & Hilgenfeld, R. (2001). A new lysozyme fold. Crystal structure of the muramidase from Streptomyces coelicolor at 1.65 Å resolution. *Journal of Biological Chemistry, 276*, 31994–31999. https://doi.org/10.1074/jbc.M102591200

Raza, W., Ghazanfar, M. U., Asif, M., Ikram-ul-Haq Zakria, M., & Al-Ani, L. K. T. (2022). Morphological characterization of phytophthora infestans and its growth on different growth media. *Sarhad Journal of Agriculture, 38*(4),1189–1202.

Raza, M., Razaghi, P., Al-Ani, L. K. T., & Aguilar-Marcelino, L. (2023). Underpinning the phylogeny and taxonomy of phytofungi through computational biology. In: I. U. Haq, S. Ijaz, & I. A. Khan, (Eds 1.), *Phytomycology and molecular biology of plant pathogen interactions* (pp. 149–169). Taylor & Francis Group, CRC Press.

Rebolledo-Prudencio, O. G., Dautt-Castro, M., Estrada-Rivera, M., del Carmen González-López, M., Jijón-Moreno, S., & Casas-Flores, S. (2020). Trichoderma in the rhizosphere: An approach toward a long and successful symbiosis with plants. In: Gupta, V. K., Zeilinger, S., Singh, H. B., & Druzhinina, I. (Eds.), *New and future developments in microbial biotechnology and bioengineering* (pp. 3–38). Elsevier. ISBN 9780128194539.

Rhouma, A., Hajji-Hedfi, L., Bouqellah, N. A., Khaire, P. B., Dali, S., Bargougui, O., Khlif, A., & Al-Ani, L. K. (2024). Uniting the role of Entomopathogenic fungi against *Rhizoctonia solani* JG Kühn, the causal agent of cucumber damping-off and root rot diseases. *Phyton, 93*(11), 2857–2881.

Rubio, M. B., Quijada, N. M., Pérez, E., Domínguez, S., Monte, E., & Hermosa, R. (2014). Identifying beneficial qualities of *Trichoderma parareesei* for plants. *Applied and Environment Microbiology, 80*(6), 1864–1873. https://doi.org/10.1128/AEM.03375-13

Ruocco, M., Lanzuise, S., Vinale, F., Marra, R., Turrà, D., Woo, S. L., & Lorito, M. (2009). Identification of a new biocontrol gene in *Trichoderma atroviride*: The role of an ABC transporter membrane pump in the interaction with different plant-pathogenic fungi. *Molecular Plant-Microbe Interactions, 22*, 291–301. https://doi.org/10.1094/MPMI-22-3-0291

Ryu, S. M., Lee, H. M., Song, E. G., Seo, Y. H., Lee, J., Guo, Y., Kim, B. S., Kim, J. J., Hong, J. S., Ryu, K. H., & Lee, D. (2017). Antiviral activities of trichothecenes isolated from *Trichoderma albolutescens* against Pepper Mottle Virus. *Journal of Agriculture and Food Chemistry, 65*(21), 4273–4279. https://doi.org/10.1021/acs.jafc.7b01028

Safdar, A. A., & McKenry, M. V. (2012). Incidence and population density of plant-parasite nematodes infecting vegetable crops and associated yield losses in Punjab Pakistan. *Pakistan Journal of Zoology, 44*(2), 327–333.

Sahebani, N., & Hadavi, N. (2008). Biological control of the root-knot nematode *Meloidogyne javanica* by *Trichoderma harzianum. Soil Biology and Biochemistry, 40*, 2016–2020.

Sala, A., Artola, A., Sánchez, A., & Barrena, R. (2020). Rice husk as a source for fungal biopesticide production by solid-state fermentation using *B. bassiana* and *T. harzianum. Bioresource Technology 296*, 122322. https://doi.org/10.1016/j.biortech.2019.122322

Salas-Marina, M. A., Isordia-Jasso, M. I., Islas-Osuna, M. A., Delgado-Sánchez, P., Jiménez-Bremont, J. F., Rodríguez-Kessler, M., Rosales-Saavedra, M. T., Herrera-Estrella, A., & Casas-Flores, S. (2015). The Epl1 and Sm1 proteins from *Trichoderma atroviride* and *Trichoderma virens* differentially modulate systemic disease resistance against different life style pathogens in *Solanum lycopersicum. Frontiers in Plant Science, 6*, 77.

Samolski, I., Rincón, A. M., Pinzón, L. M., Viterbo, A., & Monte, E. (2012). The qid74 gene from *Trichoderma harzianum* has a role in root architecture and plant biofertilization. *Microbiology, 158*(1), 129–138. https://doi.org/10.1099/mic.0.053140-0

Samuels, G. J. (2006). *Trichoderma*: Systematics, the sexual state, and ecology. *Phytopathology, 96*, 195–206. https://doi.org/10.1094/PHYTO-96-0195

Sanjeev, K., Manibhushan, T., & Archana, R. (2014). *Trichoderma*: Mass production, formulation, quality control, delivery and its scope in commercialization in India for the management of plant diseases. *African Journal of Agricultural Research, 9*, 3838–3852.

Santoro, P. H., Cavaguchi, S. A., Alexandre, T. M., Zorzetti, J., & Neves, P. M. O. J. (2014). In vitro sensitivity of antagonistic *Trichoderma atroviride* to herbicides. *Brazilian Archives of Biology and Technology, 57*, 238–243. https://doi.org/10.1590/S1516-89132014000200012

Santoyo, G., Orozco-Mosqueda, M. D. C., Afridi, M. S., Mitra, D., Valencia-Cantero, E., & Macías-Rodríguez, L. (2024). *Trichoderma* and *Bacillus* multifunctional allies for plant growth and health in saline soils: Recent advances and future challenges. *Frontiers in Microbiology, 15*, 1423980.

Sarker, M. S., Mohiuddin, K. M., Al-Ani, L. K. T., Hassan, M. N., Akter, R., Hossain, M. S., & Khand, M. N. M. (2020). Effect of bio-nematicide and bau-biofungicide against root-knot (*Meloidogyne* spp.) of soybean. *Malaysian Journal of Sustainable Agriculture 4*(2), 44–48. https://doi.org/10.26480/mjsa.02.2020.44.48

Sato, K., Kadota, Y., & Shirasu, K. (2019). Plant immune responses to parasitic nematodes. *Frontiers in Plant Science, 1165*. https://doi.org/10.3389/fpls.2019.01165

Sawant, S. S., Bhapkar, S. R., Choi, E., Lee, B., Song, J., Cho, Y. S., Park, Y., & Seo, H. J. (2024). Potential of *Trichoderma* species to control *Rosellinia necatrix*, the etiological agent of white root rot. *Biological Control, 199*, 105664. https://doi.org/10.1016/j.biocontrol.2024.10566

Schirmbock, M., Lorito, M., Wang, Y. L., Hayes, C. K., Arisan-Atac, I., Scala, F., Harman, G. E., & Kubicek, C. P. (1994). Parallel formation and synergism of hydrolytic enzymes and peptaibol antibiotics, molecular mechanisms involved in the antagonistic action of *Trichoderma harzianum* against phytopathogenic fungi. *Applied and Environment Microbiology, 60*, 4364–4370. https://doi.org/10.1128/aem.60.12.4364-4370.1994

Schuster, A., & Schmoll, M. (2010). Biology and biotechnology of *Trichoderma. Applied Microbiology and Biotechnology, 87*, 787–799.

Segarra, G., Casanova, E., Aviles, M., & Trillas, I. (2010). *Trichoderma asperellum* strain T34 controls *Fusarium* wilt disease in tomato plants in soilless culture through competition for iron. *Microbial Ecology, 59*, 141–149. https://doi.org/10.1007/s00248-009-9545-5

Sharma, P., Sharma, M., Raja, M., & Shanmugam, V. (2014). Status of *Trichoderma* research in India: A review. *Indian Phytopathology, 67*, 1–119.

Sharma, S., Kour, D., Rana, K. L., Dhiman, A., Thakur, S., Thakur, P., Thakur, S., Thakur, N., Sudheer, S., Yadav, N., & Yadav, A. N. (2019). *Trichoderma*: Biodiversity, ecological significances, and industrial applications. In A. N. Yadav, S. Mishra, S. Singh, & A. Gupta (Eds.), *Recent advancement in white biotechnology through fungi: Vol. 1: Diversity and enzymes perspectives* (pp. 85–120). Springer International Publishing.

Sharma, V., Salwan, R., & Al-Ani, L. K. T. (2020). *Molecular aspects of plant beneficial microbes in agriculture*. Elsevier Science.

Sharon, E., Bar-Eyal, M., Chet, I., Herrera-Estrella, A., Kleifeld, O., & Spiegel, Y. (2001). Biological control of the root-knot nematode *Meloidogyne javanica* by *Trichoderma harzianum*. *Biological Control, 91*(7), 687–693.

Shobha, B., Lakshmeesha, T. R., Ansari, M. A., Almatroudi, A., Alzohairy, M. A., Basavaraju, S., Alurappa, R., Niranjana, S. R., & Chowdappa, S. (2020). Mycosynthesis of ZnO nanoparticles using *Trichoderma* spp. isolated from rhizosphere soils and its synergistic antibacterial effect against *Xanthomonas oryzae pv. oryzae*. *Journal of Fungi 6*(3), 181. https://doi.org/10.3390/jof6030181

Shoresh, M., Harman, G. E., & Mastouri, F. (2010). Induced systemic resistance and plant responses to fungal biocontrol agents. *Annual Review of Phytopathology, 48*, 21–43. https://doi.org/10.1146/annurev-phyto-073009-114450

Silva, R. N., Monteiro, V. N., Steindorff, A. S., Gomes, E. V., Noronha, E. F., & Ulhoa, C. J. (2019). *Trichoderma*/pathogen/plant interaction in pre-harvest food security. *Fungal Biology, 123*(8), 565–583. https://doi.org/10.1016/j.funbio.2019.06.010

Singh, A., Shahid, M., Srivastava, M., Pandey, S., Sharma, A., & Kumar, V. (2014). Optimal physical parameters for growth of *Trichoderma* species at varying pH, temperature and agitation. *Virology and Mycology, 3*, 1–7. https://doi.org/10.4172/2161-0517.1000127

Singh, U. B., Singh, S., Malviya, D., Chaurasia, R., Imran, M. O. H. D., & Rai, J. (2017). Harnessing biocontrol potential of *Trichoderma harzianum* for control of *Meloidogyne incognita* in tomato. *Indian Phytopathol, 70*(3), 331–335.

Singh, S., Kumar, V., Dhanjal, D. S., Sonali Dhaka, V., Thotapalli, S., Singh, J., Al-Ani, L. K. T., & Aguilar-Marcelino, L. (2021). Rhizosphere biology: A key to agricultural sustainability. In: A. N. Yadav, J. Singh, C. Singh, & N. Yadav (Eds.), *Current trends in microbial biotechnology for sustainable agriculture* (pp. 161–182). Springer Nature. https://doi.org/10.1007/978-981-15-6949-4_7

Sood, M., Kapoo, D., Kumar, V., Sheteiwy, M. S., Ramakrishnan, M., Landi, M., Araniti, F., & Sharma, A. (2020). *Trichoderma*: The "secrets" of a multitalented biocontrol agent. *Plants, 9*(6), 762. https://doi.org/10.3390/plants9060762

Soumare, A., Diédhiou, A. G., Arora, N. K., Al-Ani, L. K. T., Ngom, M., Fall, S., Hafidi, M., Ouhdouch, Y., Kouisni, L., & Sy, M. O. (2021). Potential role and utilization of plant growth promoting microbes in plant tissue culture. *Frontiers in Microbiology, 12*, Article 649878. https://doi.org/10.3389/fmicb.2021.649878

Stewart, A., & Hill, R. (2014). Applications of Trichoderma in plant growth promotion. In V. G. Gupta, M. Schmoll, A. Herrera-Estrella, R. S. Upadhyay, I. Druzhinina, & M. Tuohy (Eds.), *Biotechnology and biology of Trichoderma* (pp. 415–428). Elsevier.

Sukmawati, D., Adisyahputra, A., Al-Ani, L. K. T., Al Husna, S. N., Afifah, Z. N., Sriherwanto, C., Surono, S., Setiarto, R. H. B., Nurjayadi, M., & Rahman, R. A. (2024a). Pichia kudriavzevii UNJCC Y-137 and Candida tropicalis UNJCC Y-140 isolated from Durio kutejensis as potential probiotic agents. *Food Science and Biotechnology*. https://doi.org/10.1007/s10068-024-01609-w

Sukmawati, D., Nurmayanti, F., Supiyani, A., Kurniati, T. H., Adisyahputra, A., Tawfeeq Al-Ani, L. K., Widowati, R., Nida Al Husna, S., Hidayat, I., Bimo Setiarto, R. H., Sulistiani, S., & Abdul Rahman, R. (2024b). Molecular Identification and Antifungal Activity of Candida chrysomelidarum and Rhodotorula toruloides Isolated from Indonesian Indigenous Fruit Flacourtia inermis Roxb. (Lobi-lobi) against Pathogenic Mold Aspergillus flavus. *Trends in Sciences, 20*(10), 8321.

Sun, J., Zhang, T., Li, Y., Wang, X., & Chen, J. (2019). Functional characterization of the ABC transporter TaPdr2 in the tolerance of biocontrol the fungus *Trichoderma atroviride* T23 to dichlorvos stress. *Biological Control, 129*, 102–108. https://doi.org/10.1016/j.biocontrol.2018.10.004

Sundin, G. W., Castiblanco, L. F., Yuan, X., Zeng, Q., & Yang, C. H. (2016). Bacterial disease management: Challenges, experience, innovation and future prospects: Challenges in bacterial molecular plant pathology. *Molecular Plant Pathology, 17*(9), 1506–1518. https://doi.org/10.1111/mpp.12436

Szabó, M. (2014). Potential of *Trichoderma* species and nematodetrapping fungi to control plant–parasitic nematodes: In vitro confrontation and gene expression assays using *Caenorhabditis elegans* model system. Ph. D. thesis. Szent Istv án University, Gödöllo, Hungary.

Teixidó, N., Usall, J., & Torres, R. (2022). Insight into a successful development of biocontrol agents: Production, formulation, packaging, and shelf life as key aspects. *Horticulturae, 8*(4), 305. https://doi.org/10.3390/horticulturae8040305

Tijerino, A., Cardoza, R. E., Moraga, J., Malmierca, M. G., Vicente, F., Aleu, J., Collado, I. G., Gutiérrez, S., Monte, E., & Hermosa, R. (2011). Overexpression of the trichodiene synthase gene tri5 increases trichodermin production and antimicrobial activity in *Trichoderma brevicompactum. Fungal Genetics and Biology, 48*, 285–296. https://doi.org/10.1016/j.fgb.2010.11.012

Topolovec-Pintarić, S. (2019). Trichoderma: invisible partner for visible impact on agriculture. In Shah, M. M., Sharif, U., Buhari, T. R. (Eds.) *Trichoderma-the most widely used fungicide*, IntechOpen.

Triveni, S., Prasanna, R., Shukla, L., & Saxena, A. K. (2013). Evaluating the biochemical traits of novel Trichoderma-based biofilms for use as plant growth-promoting inoculants. *Annals of Microbiology, 63*, 1147–1156. https://doi.org/10.1007/s13213-012-0573-x

Tyskiewicz, R., Nowak, A., Ozimek, E., & Jaroszuk-Sciseł, J. (2022). *Trichoderma*: The current status of its application in agriculture for the biocontrol of fungal phytopathogens and stimulation of plant growth. *International Journal of Molecular Sciences, 2022*(23), 2329. https://doi.org/10.3390/ijms23042329

van Der Heijden, M. G., Martin, F. M., Selosse, M. A., & Sanders, I. R. (2015). Mycorrhizal ecology and evolution: The past, the present, and the future. *New Phytologist, 205*(4), 1406–1423.

Verma, M., Brar, S. K., Tyagi, R. D., Surampalli, R. Y., & Valéro, J. R. (2007). Antagonistic fungi, *Trichoderma* spp.: Panoply of biological control. *Biochemical Engineering Journal, 37*, 1–20. https://doi.org/10.1016/j.bej.2007.05.012

Vey, A., Hoagland, R., & Butt, T. (2001). Toxic metabolites of fungal biocontrol agents. In T. M. Butt, C. Jackson, & N. Magan (Eds.), *Fungi as biocontrol agents progress, problems and potential* (pp. 311–346). CABI.

Villar-Luna, E., Gómez-Rodríguez, O., Villar-Luna, H., Aguilar-Marcelino, L., Al-Ani, L. K. T., & Fernández-Herrera, E. (2022). Eco-friendly management of false root-knot nematode Nacobbus Aberrans: An overview. In K. K. Chaudhary & M. K. Meghvansi (Eds.), *Sustainable management of nematodes in agriculture, Vol.1: Organic management* (pp. 327–342). Springer Nature Switzerland.

Vinale, F., Sivasithamparam, K., Ghisalberti, E. L., Marra, R., Woo, S. L., & Lorito, M. (2008). *Trichoderma*–plant–pathogen interactions. *Soil Biology and Biochemistry, 40*(1), 1–10.

Viterbo, A. D. A., & Chet, I. (2006). TasHyd1, a new hydrophobin gene from the biocontrol agent *Trichoderma asperellum,* is involved in plant root colonization. *Molecular Plant Pathology 7*, 249–258.https://doi.org/10.1111/j.1364-3703.2006.00335.x

Viterbo, A., Landau, U., Kim, S., Chernin, L., & Chet, I. (2010). Characterization of ACC deaminase from the biocontrol and plant growth-promoting agent *Trichoderma asperellum* T203. *FEMS Microbiology Letters, 305*(1), 42–48.

Vitti, A., La Monaca, E., Sofo, A., Scopa, A., Cuypers, A., & Nuzzaci, M. (2015). Beneficial effects of *Trichoderma harzianum* T-22 in tomato seedlings infected by Cucumber mosaic virus (CMV). *BioControl, 60*(1), 135–147. https://doi.org/10.1007/s10526-014-9626-3

Wenderoth, M., Garganese, F., Schmidt-Heydt, M., Soukup, S. T., Ippolito, A., Sanzani, S. M., & Fischer, R. (2019). Alternariol as virulence and colonization factor of *Alternaria alternata* during plant infection. *Molecular Microbiology, 112*(1), 131–146. https://doi.org/10.1111/mmi.14258

Wesemael, W. M. L., Viaene, N., & Moens, M. (2011). Root-knot nematodes (*Meloidogyne* spp.) in Europe. *Nematology, 13*, 3–16. https://doi.org/10.1163/138855410X526831

Wijayawardene, N. N., Hyde, K. D., Al-Ani, L. K. T., Tedersoo, L., Tedersoo, L., Haelewaters, D., Rajeshkumar, K. C., et al. (2020). Outline of fungi and fungi-like taxa. *Mycosphere, 11*(1), 1060–1456. https://doi.org/10.5943/mycosphere/11/1/8

Woo, S. L., Donzelli, B., Scala, F., Mach, R. L., Harman, G. E., Kubicek, C. P., Del Sorbo, G., & Lorito, M. (1999). Disruption of the ech42 (endochitinase encoding) gene affects biocontrol activity in *Trichoderma harzianum* P1. *Molecular Reproduction and Development, 12*, 419–429. https://doi.org/10.1094/MPMI.1999.12.5.419

Woo, S. L., Scala, F., Ruocco, M., & Lorito, M. (2006). The molecular biology of the interactions between *Trichoderma* spp., phytopathogenic fungi, and plants. *Phytopathology, 96*, 181–185. https://doi.org/10.1094/PHYTO-96-0181

Woo, S. L., Ruocco, M., Vinale, F., Nigro, M., Marra, M., Lombardi, N., Pascale, A., Lanzuise, S., Manganiello, G., & Lorito, M. (2014). *Trichoderma*-based products and their widespread use in agriculture. *The Open Mycology Journal, 8*, 71–126.

Xiao-Yan, S., Qing-Tao, S., Shu-Tao, X., Xiu-Lan, C., Cai-Yun, S., & Yu-Zhong, Z. (2006). Broad-spectrum antimicrobial activity and high stability of Trichokonins from *Trichoderma koningii* SMF2 against plant pathogens. *FEMS Microbiology Letters, 260*(1), 119–125. https://doi.org/10.1111/j.1574-6968.2006.00316.x

Yan, L., & Khan, R. A. A. (2021). Biological control of bacterial wilt in tomato through the metabolites produced by the biocontrol fungus, *Trichoderma harzianum*. *Egyptian Journal of Biological Pest Control, 31*, 5. https://doi.org/10.1186/s41938-020-00351-9

Yedidia, I., Benhamou, N., & Chet, I. (1999). Induction of defense responses in cucumber plants (*Cucumis sativus* L.) by the biocontrol agent *Trichoderma harzianum*. *Applied and Environmental Microbiology, 65*(3), 1061–1070. https://doi.org/10.1128/AEM.65.3.1061-1070.1999

Yedidia, I., Srivastva, A. K., Kapulnik, Y., & Chet, I. (2001). Effect of *Trichoderma harzianum* on microelement concentrations and increased growth of cucumber plants. *Plant and Soil, 235*(2), 235–242. https://doi.org/10.1023/A:1011990013955

Yedidia, I., Shoresh, M., Kerem, Z., Benhamou, N., Kapulnik, Y., & Chet, I. (2003). Concomitant induction of systemic resistance to *Pseudomonas syringae* pv. *lachrymans* in cucumber by *Trichoderma asperellum* (T-203) and accumulation of phytoalexins. *Applied and Environment Microbiology, 69*, 7343–7353. https://doi.org/10.1128/AEM.69.12.7343-7353.2003

Yelikbayev, B., Correa, E. J. A., Al-Ani, L. K., Duarte, N. F., & Pagano, M. C. (2024). Revitalizing degraded rural landscapes: Approaches and strategies for ecological restoration. *Environmental Reports*. https://doi.org/10.51470/ER.2024.6.2.23

Yendyo, S., Ramesh, G. C., & Pandey, B. R. (2017). Evaluation of *Trichoderma* spp., *Pseudomonas fluorescens* and *Bacillus subtilis* for biological control of *Ralstonia* wilt of tomato. F1000Res, 6. https://doi.org/10.12688/f1000research.12448.3

Yuan, S., Li, M., Fang, Z., Liu, Y., Shi, W., Pan, B., Wu, K., Shi, J., Shen, B., & Shen, Q. (2016). Biological control of tobacco bacterial wilt using *Trichoderma harzianum* amended bioorganic fertilizer and the arbuscular mycorrhizal fungi *Glomus mosseae*. *Biological Control, 92*, 164–171. https://doi.org/10.1016/j.biocontrol.2015.10.013

Zachow, C., Berg, C., Müller, H., Monk, J., & Berg, G. (2016). Endemic plants harbour specific *Trichoderma* communities with an exceptional potential for biocontrol of phytopathogens. *Journal of Biotechnology, 235*, 162–170. https://doi.org/10.1016/j.jbiotec.2016.03.049

Zanon, M. J., Font, M. I., & Jordá, C. (2011). Use of tomato crop residues into soil for control of bacterial wilt caused by *Ralstonia solanacearum*. *Crop Protection, 30*(9), 1138–1143. https://doi.org/10.1016/j.cropro.2011.03.025

Zhang, J., Chen, G. Y., Li, X. Z., Hu, M., Wang, B. Y., Ruan, B. H., Zhou, H., Zhao, L. X., Zhou, J., Ding, Z. T., & Yang, Y. B. (2017). Phytotoxic, antibacterial, and antioxidant activities of mycotoxins and other metabolites from *Trichoderma* sp. *Natural Product Research, 31*, 2745–2752. https://doi.org/10.1080/14786419.2017.1295235

Zin, N. A., & Badaluddin, N. A. (2020). Biological functions of *Trichoderma* spp. for agriculture applications. *Annals of Agricultural Science, 65*, 168–178.

Investigating the Potential of Climate-Smart Pest Management in Sustainable Agriculture

Dilawar Abbas, Umer Liaqat, Muhammad Adeel Ghafar, Kamran Haider, Naveed Akhtar, Mubashir Mulazim, Najeeb Ullah Khan, Yasmin Kausar, and Muhammad Sufian

Abstract Climate change poses significant challenges to sustainable agriculture and innovative pest management approaches. Climate Smart Pest Management (CSPM) integrates climate adaptation and mitigation practices with pest management, aiming to reduce pest-related issues with minimal environmental impact. This chapter includes different CSPM techniques, including climate-resilient crop varieties, biological control agents, habitat manipulation, and predictive modeling for pest outbreaks. It also emphasizes the role of advanced technologies, such as remote sensing and precision agriculture, in improving the efficiency and effectiveness of CSPM. The implementation requirements of CSPM include a multi-stakeholder approach, researchers, policymakers, and extension services. It also highlights the

D. Abbas (✉)
Key Laboratory for Biology of Plant Diseases and Insect Pests, Institute of Plant Protection, Chinese Academy of Agricultural Sciences, Beijing, China
e-mail: dilawar72@yahoo.com

U. Liaqat · Y. Kausar
Department of Zoology, Wildlife and Fisheries, University of Agriculture, Faisalabad, Pakistan
e-mail: umer.liaqat@uaf.edu.pk

M. A. Ghafar
State Key Laboratory of Ecological Pest Control for Fujian and Taiwan Crops, Key Laboratory of Bio Pesticide and Chemical Biology, MOE, College of Plant Protection, Fujian Agriculture and Forestry University, Fuzhou, China

K. Haider
Institute of Urban and Horticultural Entomology, College of Plant Science and Technology, Huazhong Agricultural University, Wuhan, China

N. Akhtar
Department of Zoology, GC University Lahore, Punjab, Pakistan

M. Mulazim · M. Sufian (✉)
Department of Entomology, University of Agriculture Faisalabad, Faisalabad, Pakistan
e-mail: muhammad.sufyan@uaf.edu.pk

N. U. Khan
Department of Forestry and Wildlife Management, University of Haripur, Khyber Pakhtunkhwa, Pakistan

M. Shaaban (ed.), *Soils and Sustainable Agriculture*, Frontier Studies in Soil Science,
https://doi.org/10.1007/978-3-031-91114-9_38

importance of tailored CSPM strategies considering pest dynamics, local climate conditions, and socio-economic factors. Additionally, it identifies critical barriers to adoption, such as limited awareness, financial constraints, and inadequate policies. Recommendations for overcoming these challenges include increasing investment in research and development, training, fostering collaborative networks, and promoting farmer education. Overall, this chapter also underscores the potential of CSPM as a vital component of sustainable agriculture and environmental conservation, contributing to food security and climate change resilience.

Keywords Climate change · Sustainable agriculture · Insect pests · Biological control · Pest management

1 Introduction

Extreme weather events, global shifts, and fluctuations in precipitation patterns or temperature all contribute to an increase in irregular and severe weather patterns, which considerably impact global agriculture and agricultural yields (Lobell & Field, 2007) or the future it will continue to do (Beddington et al., 2012; Challinor et al., 2014). Climate change substantially influences the severity and spread of insect pests, directly or indirectly. For example, high temperatures and humidity increase the whitefly population in cotton crops, reducing crop yield (Lamichhane et al., 2015; Macfadyen et al., 2018; Pathania et al., 2020). Climate change also facilitates invasive species to invade various new areas of the world, like *Spodoptera furgiperda*, which originates from America and migrates to Asia due to a favorable climate, causing severe damage to the maize crop. Thus, invasive species can become significant pests in these new locations if effective monitoring and control measures are absent (Hellmann et al., 2008).

Even though climate change is harmful, some changes have benefitted agro-crops; for example, small insects like jassids, mites, aphids, whiteflies, and many other minute insect pests were washed away due to heavy rain (Pathak et al., 2012). Furthermore, evidence indicates pest issues are expected to become more unpredictable and more significant in amplitude (Gregory et al., 2009). Predicting how climate change affects insect populations is complicated because of complex interactions between rising atmospheric CO_2 concentration, extreme weather events' high frequency and severity, and shifting climate regimes (Bebber et al., 2013). According to the last report of the Intergovernmental Panel on Climate Change (IPCC), greenhouse gasses (GHG) emissions have already increased temperature by 1.1 °C as compared to the 1850–1900 era, and studies exhibited this enhancement expected to reach 1.5 °C in the following decades and will severely affect Earth (Change, 2023), and negatively affect the crop yield; for example, 10.8 and 16.7% reductions in yield were recorded in maize and soybean crops due to climate change, respectively (Porter et al., 2019). Thus, food security faces a significant threat due to climate change in quantity and quality worldwide.

Due to a rapid increase in the global population by 2050, food demands need to be enhanced by 70% of the world population's needs (Sheets, 2018). As the last few decades have witnessed a steady advancement in agricultural practices to increase productivity, modern farming systems that rely heavily on new technology, automation, and extensive use of chemicals have severe drawbacks. The primary environmental issues imposed by current farming techniques include, for instance, soil degradation, dangers to human health, contamination of soil, ground, and surface waters, and depletion of biodiversity (Devlet, 2021). Moreover, these farming practices promoted the deterioration of working places and the disintegration of rural communities.

Therefore, the global agricultural sector must use innovative farming and landscape management practices to counter the risks of growing insect pests and intensified climate change. The actions should include multiple levels rather than just the agricultural level (Newell et al., 2019), such as national and international food security, regional scales, environmental economics, and social sustainability. In agriculture, sustainability is the ability of agricultural systems to withstand shocks and endure for extended periods without endangering or exhausting the resources necessary to satisfy the demands of the present and coming generations. Resource conservation is essential to agricultural output. Sustainable farming practices improve soil health, conserve water, and encourage reduced air, groundwater, and greenhouse gas pollution (Brodt et al., 2011). The Food and Agriculture Organization of the United Nations (FAO) introduced a novel term for pest management known as Climate-Smart Agriculture (CSA) to designate a new approach to farming. The ultimate goal is to ensure food security by establishing actions and guiding entire agricultural systems for sustainable development, resilient activities, and adoptable strategies under climate change (Lipper & Zilberman, 2018).

As discussed earlier, Climate Smart Pest Management (CSPM) is important in the CSA approach. CSPM has several advantages in making agricultural systems more sustainable by reducing chemical inputs. Thus, while CSPM is pretty advantageous, it has noticeable challenges. In this chapter, the concept of CSPM is introduced, together with the challenges of climate change, to identify research gaps and where more effective solutions can be developed.

2 Impact of Climate Change on Crop Pests

Small increments in temperature can increase the disease severity and pest population (Barzman et al., 2015), which leads to more application of insecticides and fungicides, e.g., in the USA, where high temperatures led to the development of pest populations in the southern areas, the former pest's used more insecticides as compared to other regions. A specific example is where pesticides sprayed for control of lepidopteran insect pests in sweet corn are higher in Florida (15–32 applications) compared to New York (0–5 applications). This excessive pesticide use can negatively impact human health and the environment (Hatfield et al., 2011; Larsen et al.,

2017). Furthermore, high temperatures can also diminish the effectiveness of insecticides, e.g., the toxic effect of bifenthrin and lambda-cyhalothrin (two pyrethroids) and Spinosad (a spinosyn) were found to reduce against *Ostrinia nubilalis* as the increase of temperature to a specific limit (Musser & Shelton, 2005). Research has also demonstrated that slight variations in thermal conditions have also been observed to affect the ability of parasitoids to suppress pest species (Thomson et al., 2010) and alter the genetic expression of defense attributes of insect pests, utilizing them against larval parasitoids (Iltis et al., 2018). Similarly, de Sassi and Tylianakis (Sassi & Tylianakis, 2012) synthesize biomass production, where high temperatures were recorded to have had an influence. Parasitoids displayed slight fluctuation but did not significantly develop in reaction to the temperature change, and thus, the total and relative biomass of herbivores sharply increased. The study also pointed out that the overall changes in temperature at the multidimensionality of the trophic level remain challenging; however, the reduction in the top-down option could be closely linked to an increase in herbivores, which would affect the other processes in the ecosystem.

2.1 Precipitation

(Insufficient or excessive) is another factor that significantly affects the insect pest population. For example, humid and warm climatic conditions favor various species of insect pests and pathogens; however, induced water stress can make the crop more vulnerable to damage caused by pests (Rosenzweig et al., 2001). Changes in rainfall patterns also intensify the outbreaks of desert locusts. Studies exhibited that invasion of locusts was recorded higher in 10 countries of northern and western Africa than in normal rainfall countries, resulting in crop and yield losses. It was predicted that increased intensity and frequency of rainfall would lead to more future locust outbreaks (Masters & Norgrove, 2010).

2.2 Increasing CO_2 levels

Are positively related to maximum crop yield; any yield gain may be nullified entirely or partly due to losses caused by plant pathogens, weeds, and phytophagous insects (Coakley et al., 1999). For example, research studies exhibited that the cabbage *looper Trichoplusia* ni (Hubner) (Lepidoptera: Noctuidae) is found to ingest a large number of cabbage leaves under high CO_2 levels in North America; this may be due to the increased level of CO_2 reduced the nitrogen level in cabbage leaves growing under certain conditions. Extreme weather events significantly affect the relationships among diseases, pests, and crops impulsively, possibly reducing crop protection strategies' efficacy and yield loss (Chakraborty & Newton, 2011). Slovakia in Europe experienced sweltering and dry climatic conditions in May 1993, and this hot and dry weather reduced the population of *Trichogramma evanescens* (egg parasitoid) (Cagan et al., 1998). Moreover, due to extreme weather, disease-causing agents like insects or fungus spores can be carried by strong air currents during storms from their overwintering locations to new locations where they might create more issues. Furthermore, disrupted ecosystems brought about by catastrophic weather events are inherently more prone to invasions by both native and alien species.

3 Climate Smart Agriculture (CSA)

The abbreviation CSA refers to the complex Climate Smart Agriculture, which dates back to 2009 and defines it as the means and process of developing more climate-resilient agriculture systems intended primarily for climate change and food security linkages. This paper defined Climate Smart Agriculture and proposed it as the means for addressing agriculture in the context of climate change through and covering policies, practices, and financing for food security, adaptation, and mitigation in 2010 by the FAO. It has been in operation for several decades now, and many professionals have participated in creating and spreading CSA methods (FAO, 2010).CSA promotes the international development of agricultural management concepts to attain food security in climate change. CSA depends on the three pillars of strategic management: (i) reducing greenhouse gas emissions, (ii) viable improvements in agriculture and household incomes, and (iii) adopting and structuring household resilience to resolve climate change (Sekabira et al., 2022). It is still necessary to highlight plant conservation in light of climate change and sustainability.

Nevertheless, despite endorsing IPM as a sensible tactic, more discussion of pest control should be given to the subject. At the end of November 2017, at the 23rd Climate Change Conference of the Parties (COP23) under the UNFCCC, all participating countries restated their commitment to fight climate change and increase adaptation. Agriculture, land use, and pest management were often pinpointed as priority sectors in adaptation and emission reduction in countries' NDCs. Thus, at the right time, there is a notion of Climate-Smart Pest Management (CSPM) that can help turn these commitments into national action plans (Allouche et al., 2019) (Fig. 1).

4 Concept of Climate Smart Pest Management (CSPM)

CSPM is a relatively new approach that involves a range of interrelated techniques and strategies that are critical for the growth of agriculture, given climate change. Climate change and land use characteristics affect the distribution of crops and pests to their natural enemies. Hence, the question of how to make primary production sustainable calls for using sound strategies associated with functions that include extension

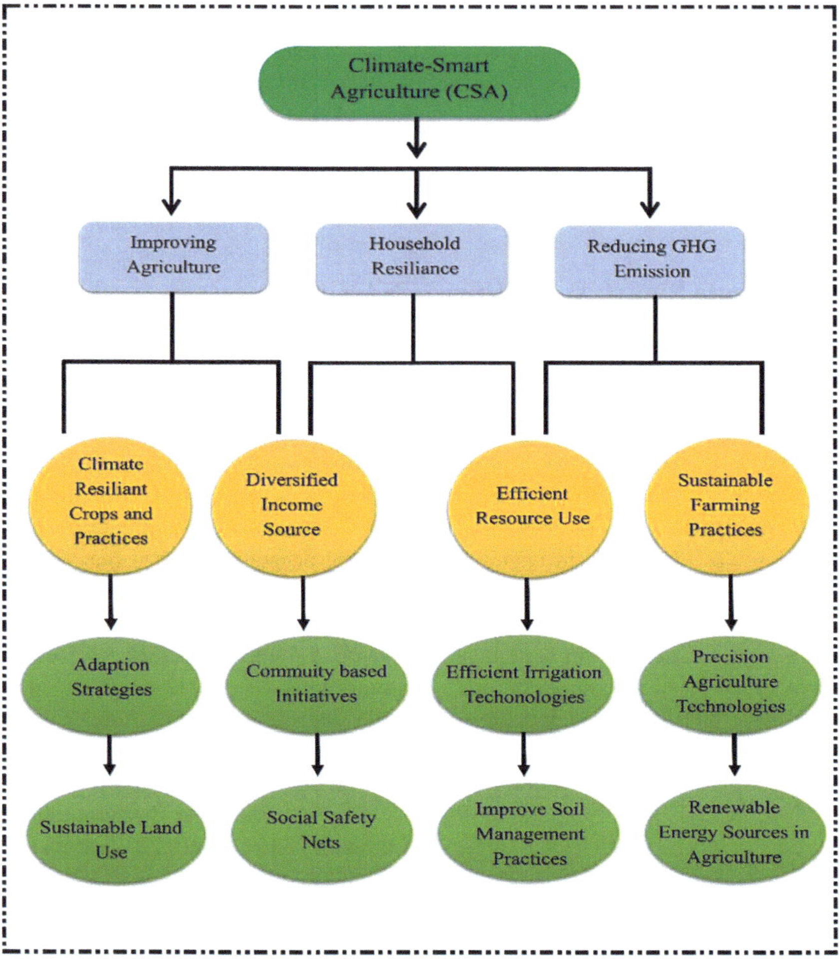

Fig. 1 A flowchart explaining the three pillars of CSA (reducing GHG emissions, improving agriculture, and household resilience)

services and research and an appropriate enabling environment that includes policies, infrastructure, and social human capital (Egan et al., 2021).

Most IPM strategies conventionally focus on the logical and adequate application of pesticides to control pests and minimize the effects of such on climate and people's well-being. However, while CSPM builds on the IPM concept, it goes a notch higher regarding climate-smart practices. CSPM's definition was recently given by (Heeb et al., 2019). CS-IPM must be incorporated into climate-smart agriculture. CS-IPM is a comprehensive strategy utilizing environmentally friendly and compatible techniques for controlling plant pathogens, weeds, and insect pests without chemicals. An essential benefit of CS-IPM practices is that they may successively decrease damage to agrobiodiversity, ecosystems, and human health. The CSPM also used biological and natural control methods for pests and produced suitable crop species with promising alternatives to mother crops (Heeb et al., 2019).

When multiple terminologies refer to pest control using a climate-smart integrated pest management strategy, it may take time to mobilize information and novel changes. Thus, the CS-IPM changed to a simple "CSPM," a novel term used to introduce CSPM, which is defined as a set of interlinked strategies and approaches needed for primary production to adapt to changing climatic conditions. Therefore, to reduce the risk to people and the environment, information about the biology of pests and methods for controlling them must be coordinated with practical and affordable solutions (Hobbs et al., 2008).

5 The Benefits of Climate-Smart Pest Agriculture

5.1 *Food Security*

A novel, cost-effective approach to lowering crop damage, food security, and raising farmer's earrings is CSPM. CSPM effectively manages the existing and newly emerged crop pests. By reducing detrimental effects on the larger ecosystem, adaptation CSPM strengthens farmed systems' ability to withstand climate change. Regarding pest forecasting, monitoring, detection, and control—all essential services for boosting resilience—CSPM revitalizes the significant roles of extension, research, and the public and commercial sectors nationally and internationally (Hobbs et al., 2008).

5.2 *Mitigation*

CSPM reduces the intensity of crop production's greenhouse gas emissions, helps increase agricultural productivity, and encourages the sensible use of inputs.

5.3 *Double/Triple-Wins*

CSPM combines integrated pest management as a natural barrier against insect pests, providing more effective mitigation and pest prevention strategies. These strategies improve soil and biomass organic carbon sequestration and heighten plant resistance to specific pests (Bate et al., 2016; FAO, 2010; Hobbs et al., 2008). This is important because if adaption plans are designed independently of mitigation strategies, there is a possibility that emissions would deteriorate (for instance, as a result of increased usage of agrochemicals). When these actions are coupled with the decline in crop losses caused by pests, a triple-win scenario of lowering greenhouse gas emissions, improving food security, and adapting to shifting pest threats may be minimized.

6 Implementation of Climate-smart Pest Management

Thus, weather changes impact people in societies across the globe. Given the dynamics of climate change that affect disease conditions, pest control methodologies must address such changes. The climate-smart pest management (CSPM) approach, depicted in a pyramid structure (Fig. 2), comprises three primary methods: Hence, the role of OSCE includes (i) forecasting or risk assessment, (ii) early diagnosis, and (iii) efficiency of interventions. However, using these methods is accompanied by preventive measures to enhance the success rate of pest control mechanisms. Protection measures are indispensable to keep pests and plant diseases below the ability level of society to control them or to avoid sudden and drastic outbreaks. This includes using certified seeds and champ that are highly tolerant or disease resistant; the others are natural enemies. One of the protective measures in preventing plant diseases is establishing crop production using pathogen-free seeds and plant breeding.

6.1 Forecasting and Risk Assessment

Epidemiology is crucial to understanding climate change's impact on pest biology and forecasting and analyzing pest situations or population dynamics. The CSPM program provides an effective way of monitoring and predicting the time of appearance of an outbreak and pest development. Forecasting and risk management comprise an early warning system that gives information about the physiology of plants and pathogens' susceptibility. In the same regard, CSPM provides a time frame for when plants are most susceptible to pathogens, as mentioned by Hein. There is always a completion of mathematical models, especially for predicting the emergence of diseases (Bate et al., 2016). Thus, studying climate factors is essential for predicting changes in pest populations, determining the location with the best environment, creating maps that show the possible spread of disease, and supporting response efforts to outbreaks (Velasco-Amo et al., 2022).

Consequently, defensive measures, such as crop alternating and change in the date of plant sowing, can be implemented to reduce the severity of the infection through early treatments or preventative interventions. Therefore, early warning techniques and computer technology have led to the development of forecasting and software (Gleason et al., 2008), which provide in-time management of situations. Several weather-based disease warning systems have been created and proven effective for dozens of crops in recent decades. However, considering the complexity of climatic

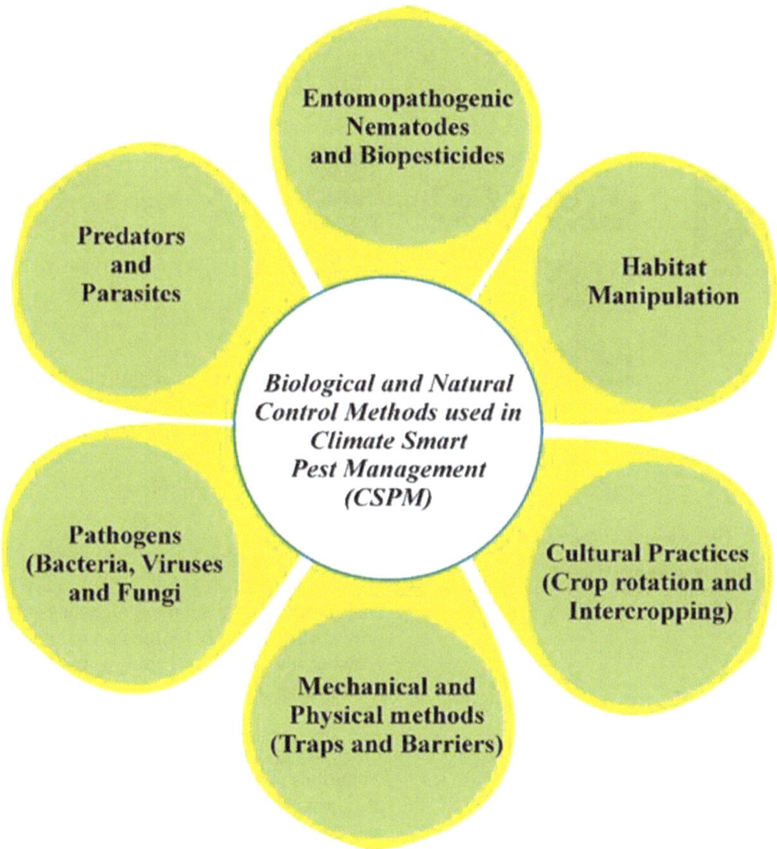

Fig. 2 Chart of biological and natural control methods used in CSPM

and geographical elements, many more models are still required to increase accuracy, i.e., in Alicante, Spain, the spread of *Xylella fastidiosa* was recorded using nonstationary and stationary models, which provide the impact of meteorological and geographical circumstances on the outbreak of this pest (Cendoya et al., 2020). Still, more extensive work is necessary to develop them into realistic scenarios, as existing analysis is sometimes based on the hypothetical concept that barriers are nonporous to the disease. This assumption is invalid for *X. fastidiosa* and most other plant disease pathogens (Fig. 3; Table 1).

However, to successfully establish such new programs based on public observations and reports, more work has to be done to develop well-thought-out analytical techniques that are practical instruments for handling potential uncertainties and evaluating these unstructured data.

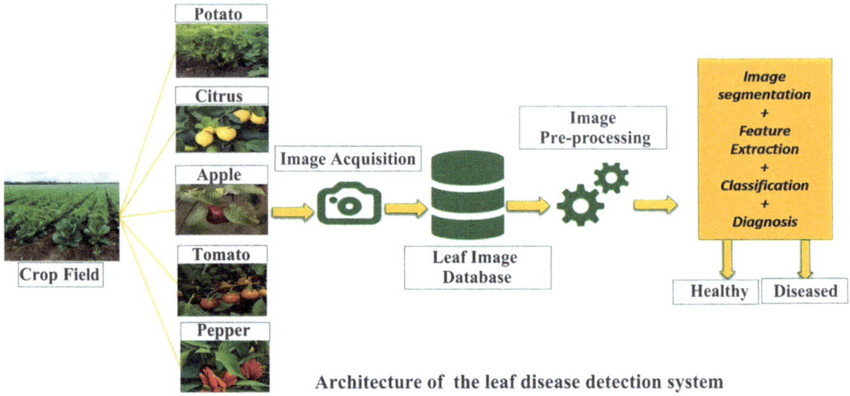

Fig. 3 Example of a weather-based disease warning system model, possibly including screenshots or flowcharts of software used

Table 1 Historical examples of disease warning systems

Crop disease warning system	Aim
TOM-CAST	Distributing information to processing-tomato growers once weekly for the first month of the growing season and ascending to three times weekly after applying fungicides (Pitblado, 1992)
SkyBit Inc.	Considering current site-specific weather conditions and/or assessment of pest risks, it also provides site-specific weather forecasts up to 3 days in advance for a spatial grain of 1 sq km (Russo, 2000)
A network of more than 2000 automated weather stations of private farm management companies	Offering color-coded risk maps for the western part of the United States to possibly attack targeted agricultural risks such as 12 diseases and six insect pests (Thomas et al., 2002)

6.2 *Early Diagnosis*

Diagnosing the early arrival of insect pests, invasion, and pest outbreaks is essential to support farmers in these areas. For example, an international group of stakeholders predicted armyworm and locust outbreaks in Africa. Similarly, the national agriculture research group predicted and monitored early diagnosis of aphid invasion in Europe. The annual northward spread of major migratory pests in China (Cock et al., 2013, 2016) is monitored and forecast by the National Agricultural Technology Extension and Service Center (NATESC), a non-profit institution under the Ministry of Agriculture. Five tiers of public extension agencies receive their projections and recommendations: national, provincial, prefectural, county, and township levels. The extension service offers a complete agricultural service center at the township level. The county-level and township-level extension and plant protection stations would receive new data from remote sensing and other monitoring and forecasting techniques. These stations would then cascade the information to village heads and farmers via various channels (such as meetings, training, phone calls, SMS, leaflets, village broadcast systems, local TV, and mass media) with the slightest delay. Though transmission may be less quickly than ideal, these routes may notify 90% of farmers about which pests require which treatment methods. Experiences from the more than 30 countries involved in the Plantwise approach to plant health system development show that National Plant Protection Organisations (NPPOs; see *Countries List:—International Plant Protection Convention*) are critical stakeholders for obtaining high-level buy-in and ensuring country-wide coordination. However, as the development of the plant health system goes well beyond the purview of the NPPO, strong ownership by and cooperation with other governmental and non-governmental organizations, such as extension or research divisions, are equally crucial. This will also apply to CSPM, the success of which is heavily reliant on political leadership, stakeholder coordination, public backing, and role and responsibility sharing. It is essential to bear in mind that some developing nations, such as Pakistan, Kenya, and Nepal, have either completed or are in the process of devolving authority to their provinces. When implementing CSPM, this delegation of powers must be considered since, via this process, numerous sub-national agencies may become possible sources of financing and will require proof that investments in strategies like CSPM will pay off (Fig. 4).

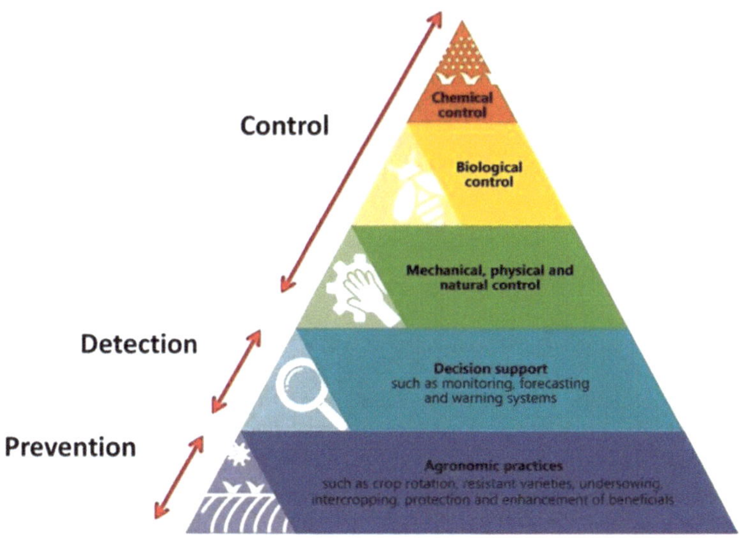

Fig. 4 The CSPM pyramid illustrates the three primary methods: forecasting/risk assessment, early diagnosis, and efficient interventions.

6.3 Efficient Interventions

Appropriate decision-making for any interventions in CSPM must be ecologically and economically justified (38). Innovations seeking to decrease the chemicals used in pest management programs are considered environmentally friendly, improving food security. CSPM approach adheres to proper prevention measures and other controls that will lead to little negative impact. These include pheromone tracing, parasites and pathogens, and other microorganisms, which are environmentally friendly methods (Kadoić Balaško et al., 2020). Other climate-sensitive biological control organisms include the novel strains of climate-resilient bacteria and highly competitive entering into CSPM strategies. One method is the CSPM, which usually comprises strains specific to some geographical regions and crops (Strand, 2000).

Changing crop planting or other sowing dates to meet new disease conditions potentially related to climate change is also a part of CSPM. Some of the existing effects of changes in temperature and rainfall include the pesticide's degradation rate, the pesticide's ability in terms of wettability, percentage retention of contact, or capacities for air-borne dispersal. Supervising pesticide application during the prevailing weather conditions improves the mode of management and measures. In the last decade alone, the frequency of pest treatment has been altered due to climatic changes in many countries. Thermal and photo-optical sensors are intelligent agricultural technologies forming specific treatments' foundations. CSPM technologies can provide sensors that take measurements of early changes in plants due to biotic stress. Such changes may consist of transitions in the size and color of the leaves, transpiration rates, plant population, and light repellence (West et al., 2010). This accurate data on imaging results in the proper identification of diseased plant areas that can be treated with pesticides, as well as determining the required pesticide amounts (Nabaei et al., 2018). Adequate use of chemicals has a positive impact on the abolishment of pesticide resistance and minimization of the effects of chemicals on the environment and human health. Similarly, pesticide application through drones decreases chemicals' raw impact and dispersion (Faiçal et al., 2014).

6.4 Successful CSPM Story from the Rice-Potato Farming System in Asia

The primary obstacle to sustainable intensification in Asia is rice-based systems. Numerous studies have addressed issues with soil degradation in these Asian agricultural systems, excessive pesticide usage, carbon emissions from burning rice waste, and overuse of water for irrigation (Bhatt et al., 2016; Faiçal et al., 2014). One viable strategy for mitigating the effects of climate change has been suggested: introducing sustainable intensification methods into rice fields. The measures taken in various countries of Asia have ensured quite a success in the diversification of potato production under a no-tillage system coupled with organic mulching with rice straw. This approach has been appreciated on various levels, including salty grounds (Bouri et al., 2023). A successful example of such intervention through CSPM is the pilot intercrop project by FAO that pushed for sustainable rice-potato farming in Thai Binh province in Vietnam between 2009 and 2011.

The FAO describes the mentioned technique as 'zero-tillage potato IPM' because the potato seed tubers were directly sown on beds formed by the drainage furrows. The rice straw crop residue was mulched, and there was no tilling. As opposed to the traditional method of tilling the soil and placing the potatoes within the soil mulch, this zero tillage technique offers satisfying climate change results at all levels. This can be illustrated by measures aimed at the efficient use of rice straw: with its help, irrigation requirements decreased from 5000 m^3/ha to 900 m^3/ha, and the use of straw for mulching of the potato seed tubers instead of burning it, which is done traditionally (Lindqvist-Kreuze et al., 2024).

However, this method has proved that straw mulch can increase the biological diversity necessary for insect control. It can be established that the temperature, moisture level, and organic matter supply of a straw mulch simulate the proper environment for the growth of favorable microbes in the soil (Du et al., 2022). This has been prompted by the use of pesticides by the farmers, mainly due to a pest invasion such as the fall armyworm threatening crop production in Pakistan and other disease species affecting crops in Asia, which have informed the use of the CSPM approach. This strategy deals with the effects of climate change and protects the environment and health of farmers (Gatto, 2022).

7 Challenges to the Adoption of Climate-smart Pest Management

Organizational obstacles at the expansion, research, and regulation levels are present. If discussed, they might significantly speed up the implementation of more climate-smart and sustainable pest management approaches on multiple levels, national and even global. Farm-level issues often impede the immediate widespread use of more sustainable strategies.

7.1 Crop Production (Farm and Landscape)

Several challenges affect the application of CSPM methods: inadequate structures and infrastructure, small plots of land, low understanding, and possibly associated costs. Another problem is little knowledge about climate-smart practices (Abid et al., 2016). Among these, factors like the level of preparedness and capacity to invest in available technologies, the degree of risk involved in technology adoption, the cost of implementation, demographics of households, the size of farms, farmer's literacy

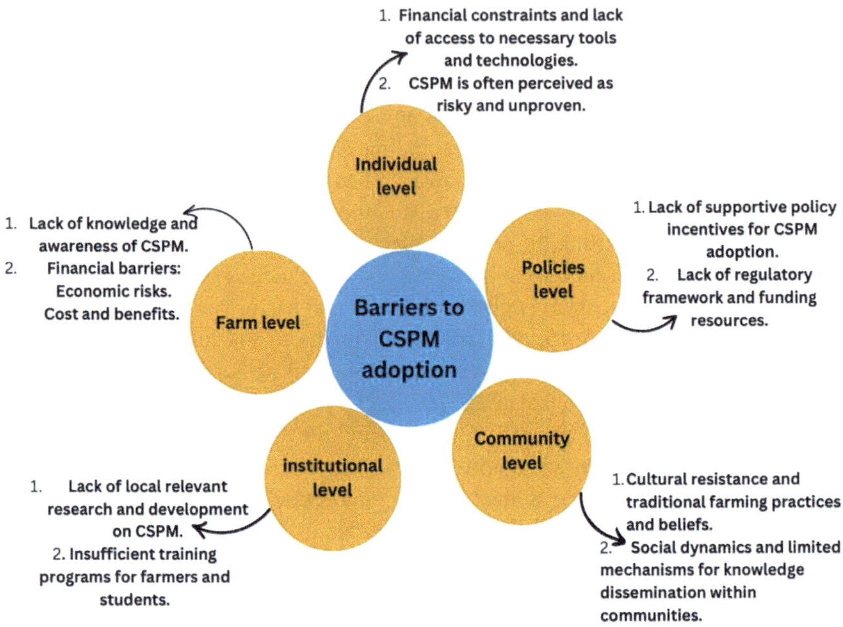

Fig. 5 A graph showing the barriers to CSPM adoption at different levels (farm, research, public sector)

levels, especially on climate change (in this study the household head); the size of the household, income; gender of the head of the households; market accessibility, access to credits, and current and expected weather patterns in their locations are Some of these challenges can be managed through a participatory, bottom-up process involving the farming community, extension and other scientists to come up with CSPM strategies for the country adapted to farm and landscape. This approach will help in increasing the knowledge base of the local agricultural population on the financial, social, and physical capital available to them and also on consciousness of climate change impacts. Additionally, it will foster a sense of ownership and motivation for implementing jointly developed mitigation and adaptation measures. To strengthen farmers' ability while promoting the adoption of innovative practices, it should also be a top priority to encourage the formation of public or private finance channels (such as access to microcredits) (Fig. 5).

7.2 Research

The assistance given by research institutions through need-based research is an essential precondition for developing and implementing innovative and adaptable CSPM methods in a functional and practical plant health system. However, in many nations, some obstacles keep this essential service from operating as needed. A few instances of these obstacles are the following: insufficient funding and a reduction in pertinent specialized knowledge, insufficient crop protection education and training to produce competent crop protection specialists, a lack of research focus on particular sustainable management techniques; and inadequate regional research collaboration and coordination (Machekano et al., 2017). High-quality, cooperative, and need-based research will be more readily accessible to all customers if funds and resources are allocated to national and international research institutions for capacity building and knowledge sharing. Additionally, these institutions should be connected to diagnostic centers, extension, and climate information services.

7.3 Public Sector

Government policies and regulatory tools must give all stakeholders clear direction and advice for proper planning for climate change adaptation and efficient resource and funding targeting. Making this a reality, however, will be difficult in many nations due to the need for more research knowledge on local requirements and limits informing current policy (Ampaire et al., 2017). In addition, there is a dearth of finance and a paucity of individuals with the necessary scientific background and quantity of experience in crop protection, all of which are necessary to create suitable plans or strategies for adaptation (Liu et al., 2016). Government information transfer and awareness-building regarding the effects of climate change and response strategies will improve the ability to adopt educated, coherent policies and regulatory tools that address adaptation and mitigation. Giving these policies some natural substance would involve setting up national earmarked funds to aid in the creation of strategies for mitigation and adaptation. This would encourage the financial industry, corporations, civil society, and global organizations to contribute money to implement these ideas at various levels. Although they belong to distinct customer groups, the numerous implementation-related problems of CSPM should only sometimes be considered and handled separately. Multidisciplinary activities should be organized for optimum efficiency in removing obstacles and creating opportunities for implementing CSPM ideas and technology because many difficulties are interconnected and interdependent. In most situations, the constraints may be overcome by careful targeting of information transfer, awareness-raising, increased funding, and policy reform.

7.4 *Costs, Benefits, and Funding for Climate-Smart Pest Management*

CSPM implementation costs would need to be assessed against the short and long-term monetary and non-monetary advantages of crop production (field and whole production system), extension, research, and public and private sectors. Thus, CSPM can affect the farm-level agriculture yield and cost of inputs. Unfortunately, while detailed cost/benefit studies of CSPM have yet to be made, published research does exist on Integrated Pest Management (or IPM), a very similar strategy in the sphere of agricultural production itself. This guidance suggests that the adoption of IPM increases crop yields by as much as 41%, entails the average use of an opposite number of pesticides by 31%, and more frequently leads to better gross returns (Norton & Mullen, However, the evaluation of the financial return on IPM is not easy because, in most cases, IPM projects entail other practices not restricted to pest control. Moreover, CSPM and IPM help gain additional advantages in mitigating negative external costs. For example, presently, in the global farming sector, 1–3 farming employees among 100 are exposed to acute pesticide poisoning, resulting in thousands of deaths. While the developing nations apply only a quarter of the total global pesticides, they constitute 99% of the pesticide deaths. Looking at the cost effects of pesticide utilization, the damages caused by the contamination of groundwater by pesticides alone are two billion dollars annually in the USA.

The costs and advantages of CSPM may be substantial on a national scale. Encouraging the public and private sectors to invest heavily in infrastructural and human resources is necessary to establish a national extension system that can perform the dual function previously outlined. Regretfully, despite data showing that returns on investments may be significant, extensions have seen a decline in investment in recent decades. For instance, between 2014 and 2018, Plantwise Kenya conducted a multi-year RCT. *Randomized controlled trials* are the most rigorously scientific approach to evaluating hypotheses and are regarded as the gold standard for assessing effect. Plantwise Kenya's initiative primarily targeted developing national plant health systems by increasing organizational ability to address problems, including early identification and response and high-quality pest management guidance. The Plantwise intervention had significant overall expenditures (the United States$669,000 in 2017 alone). However, the Randomized Control Trial showed that programmatic advantages (realized through increased revenue and productivity) were even more significant, coming in at the United States$1,914,000. The American Institution of Research (2018) reports that this results in a benefit–cost ratio of 2.9:1 and an inner rate of return of 54%, demonstrating that while significant expenses are associated with improving the adaptability of extending systems, these are outweighed by the realized advantages.

Most pest problems call for a multi-sectoral intervention and international support to finance CSPM since it needs the support of public and private entities and regional and local companies. The recently developed country policies analysis shows that some countries, including Myanmar, Rwanda, Sri Lanka, Bolivia, and Cambodia, have considered pest management important enough to be included in national adaptation and mitigation strategies. As a start towards implementing CSPM, there is a need to search for definite ways of financing in the existing frameworks and financial instruments like the Green Climate Fund and Nationally Appropriate Mitigation Actions. Large-scale aid contributions from international donors will be required in the future years and decades. A guideline on "the process to develop additional objectives on finance from 2025 upwards to follow on from the present goal of utilizing 100 billion United States dollars per year from 2020 to help developing countries" forming the part of the "Katowice Climate Package" adopted within the framework of COP 24 (November 2018) represents a positive indication for more ambitious goals from the developed countries for financing climate action in the developing world. This can be regarded as a significant development that may prove helpful in applying approaches such as CSPM.

The other approach to financing the implementation of CSPM is through adopting regulatory tools, which are part of the public policy that encourages adoption through incentive-based systems. It is possible to use the research results to set up these systems, as they facilitate more accurate identification of efficient management practices adapted to the local environment and contribute to the overall benefit. The well-known European Union program is the Common Agricultural Policy (CAP), a partnership between agribusiness and the community and between Europe and its farmers. It is focused on supporting environmentally friendly and sustainable farming methods. That is why the CAP admits that the quest for cost-focused motives may shape the farmer's outlook on higher inputs and inefficient production. Therefore, by rewarding landowners for delivering societal benefits (such as air, biodiversity, and clean water) not directly covered by the market, the CAP encourages farmers to engage in sustainable farming through income support programs. Despite claims to the contrary from detractors, the CAP includes several valuable tools that support the long-term utilization of plant protection products (such as cross-compliance regulations, agri-environmental measures, direct payments, and "green" direct payments) and align with the CSPM approach mentioned here. A valid criticism of the CAP is that it lacks obligatory tools to promote the change to low-input farming, even though the Overview Report on Appropriate Use of Chemicals, Such as Pesticides, lists several beneficial achievements. The CAP must still incorporate the European Union Regulation on Appropriate Application of Pesticides (Drozdzynski, 2020). It is also important to remember that the CAP cost €58.82 billion in 2018; income assistance alone accounts for €41.74 billion. Nevertheless, it is crucial to compare these expenses to other expenses associated with farming, which have been estimated to be as high as the United States$257 million and the United States$166 million, respectively, for the United Kingdom and Germany, just about pesticides and their effects on human health and the environment (Pretty & Waibel, 2012).

As mentioned earlier, governmental financing for initiatives encouraging the widespread use of CSPM still needs to be reduced in many developing countries. This necessitates using alternate financing strategies, such as those involving the business sector or market recognition (i.e., food standards). By including farm-based CSPM techniques into voluntary standards based on products (like Fairtrade and Rainforest

Alliance), consumers can encourage farmers to use sustainable practices. A different approach would be to use supply chain certification standards (like The Responsible Palm Oil Organization), which would be especially appealing for agricultural products and encourage more funding from private companies to adopt CSPM. Retailers also have considerable control over the kind and caliber of food items they want to offer for sale, which significantly impacts farmers and other upward value chain players. The adoption of CSPM approaches may be favored by transactions resulting from this choice, which is motivated by profit and demand and corporate social responsibility. It is necessary for customers to be prepared to pay a premium price and for the public sector to be willing and able to compensate for positive or negative externalities for both of the well-established IPM systems in Europe, public policies, and food market recognition (Lefebvre et al., 2015).

7.5 Interaction with Other Climate-Smart Agriculture Practices

CSPM is not a 'package deal' but links with current farm- and nationwide CSA approaches and concepts. The adoption of CSPM concepts by farmers thereby becomes a challenge embedded with weaknesses, including the following: to implement CSPM concepts, farmers encounter several challenges. They must identify all-inclusive, no-prohibitive adaptation techniques and harmonize them with mitigation measures. This indicates that CSPM at the farm level includes crop husbandry that includes but is not limited to soil management and abiotic factors, including crop nutrition and biotic factors. Therefore, there are many points where CSPM and CSA converge, as pointed out below;

CSPM generally depends on CSA practices such as Integrated Soil Fertility Management, Conservation Agriculture, climate-resilient crops, Site-specific nutrient management, and Crop Diversification methods. However, integrating CSPM with CSA approaches to improve the gender responsiveness of the interventions is essential. Even though women produce more than half of the world's food, this yields significantly less than men, mainly producing 20–30% less due to constrained access to production inputs and information on options available (Organization & Development, 2008). Hence, gender-sensitive planning is needed for the efficiency of CSPM. Evaluating a low-cost strategy to reduce maize stem borer (*Chilo* spp., Lepidoptera: Crambidae) and witchweed (*Striga* spp., Orobanchaceae) while improving soil fertility in East Africa yielded favorable findings. According to the research done by (Murage et al., 2015), women showed a higher inclination towards adopting and further using this particular technology, adding to the fact that it consists of all the characteristics that women appreciate (Fig. 6).

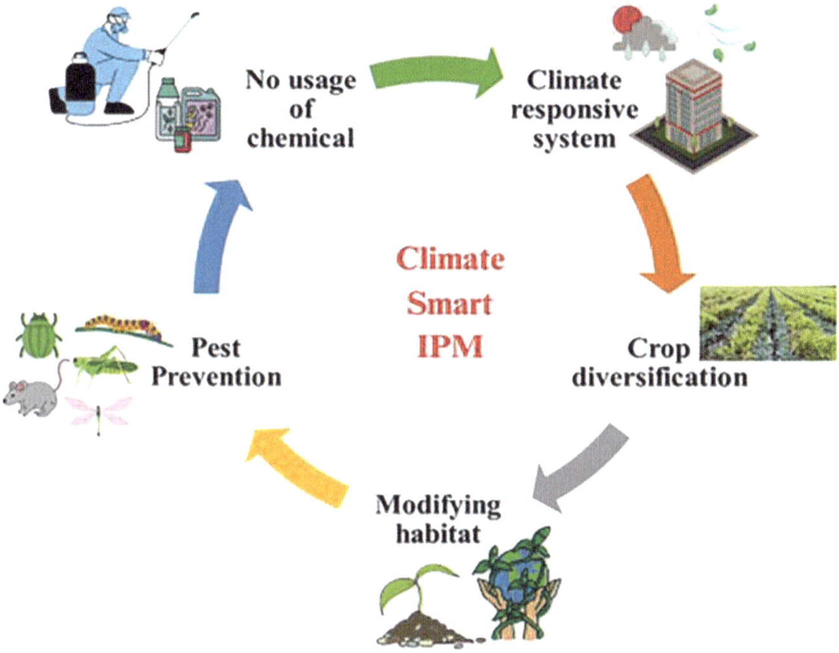

Fig. 6 Diagrams showing the integration of CSPM with other CSA practices, highlighting synergies and complementarities

8 Conclusions

Climate-smart pest management represents a promising approach to advancing sustainable agriculture amidst the challenges posed by climate change. Climate-smart pest management strategies aim to optimize crop protection while minimizing environmental impacts. The potential of CSPM is to synchronize the integration of climate data, pest dynamics, and agronomic practices for effective management of pest problems. Studies reveal that adopting climate-smart pest management practices can enhance resilience in agroecosystems. Strategies such as diversified cropping systems, biological control, and precision agriculture techniques contribute to reducing reliance on chemical pesticides. CSPM mitigates the risk of pest resistance and minimizes adverse effects on beneficial organisms and overall ecosystem health. Moreover, incorporating climate information into pest forecasting enables farmers to anticipate pest outbreaks and adjust management strategies accordingly, thus enhancing efficiency and reducing economic losses. However, challenges such as variability in climate patterns and socio-economic constraints remain; the evidence suggests climate-smart pest management holds significant promise for sustainable agriculture. Its holistic approach addresses current pest challenges and aligns with border goals of environmental stewardship and food security. Furthermore, continued research, education, and policy support are vital to realizing the full potential of innovative climate pest management in safeguarding agriculture's productivity and resilience in a changing climate. Adopting these tactics may make agricultural systems more productive, resilient, and sustainable for future generations.

References

Abid, M., Schilling, J., Scheffran, J., & Zulfiqar, F. (2016). Climate change vulnerability, adaptation and risk perceptions at farm level in Punjab Pakistan. *Science of the Total Environment, 547*, 447–460.

Allouche, J., Middleton, C., Gyawali, D. 2019. *The water–food–energy nexus: Power, politics, and justice*. Routledge.

Ampaire, E. L., Jassogne, L., Providence, H., Acosta, M., Twyman, J., Winowiecki, L., & Van Asten, P. (2017). Institutional challenges to climate change adaptation: A case study on policy action gaps in Uganda. *Environmental Science and Policy, 75*, 81–90.

Barzman, M., Bàrberi, P., Birch, A. N. E., Boonekamp, P., Dachbrodt-Saaydeh, S., Graf, B., Hommel, B., Jensen, J. E., Kiss, J., & Kudsk, P. (2015). Eight principles of integrated pest management. *Agronomy for Sustainable Development, 35*, 1199–1215.

Bate, A. M., Jones, G., Kleczkowski, A., MacLeod, A., Naylor, R., Timmis, J., Touza, J., & White, P. C. (2016). Modelling the impact and control of an infectious disease in a plant nursery with infected plant material inputs. *Ecological Modelling, 334*, 27–43.

Bebber, D. P., Ramotowski, M. A., & Gurr, S. J. (2013). Crop pests and pathogens move polewards in a warming world. *Nature Climate Change, 3*, 985–988.

Beddington, J. R., Asaduzzaman, M., Clark, M. E., Bremauntz, A. F., Guillou, M. D., Jahn, M. M., Lin, E., Mamo, T., Negra, C., & Nobre, C. A. (2012). The role for scientists in tackling food insecurity and climate change. *Agriculture and Food Security, 1*, 1–9.

Bhatt, R., Kukal, S. S., Busari, M. A., Arora, S., & Yadav, M. (2016). Sustainability issues on rice–wheat cropping system. *International Soil and Water Conservation Research, 4*, 64–74.

Bouri, M., Arslan, K. S., & Şahin, F. (2023). Climate-smart pest management in sustainable agriculture: Promises and challenges. *Sustainability, 15*, 4592.

Brodt, S., Six, J.; Feenstra, G., Ingels, C., Campbell, D. 2011. Sustainable agriculture. *National Education Knowledge, 3*.

Cagan, L., Tancik, J., Hassan, S. 1998. Natural parasitism of the European corn borer eggs Ostrinia nubilalis Hbn.(Lep., Pyralidae) by Trichogramma in Slovakia—need for field releases of the natural enemy. *Journal of Applied Entomology, 122*, 315–318.

Cendoya, M., Martínez-Minaya, J., Dalmau, V., Ferrer, A., Saponari, M., Conesa, D., López-Quílez, A., & Vicent, A. (2020). Spatial Bayesian modeling applied to the surveys of Xylella fastidiosa in Alicante (Spain) and Apulia (Italy). *Frontiers in Plant Science, 11*, 1204.

Chakraborty, S., & Newton, A. C. (2011). Climate change, plant diseases and food security: An overview. *Plant Pathology, 60*, 2–14.

Challinor, A. J., Watson, J., Lobell, D. B., Howden, S. M., Smith, D., & Chhetri, N. (2014). A meta-analysis of crop yield under climate change and adaptation. *Nature Climate Change, 4*, 287–291.

Change, C. 2023. Synthesis Report. URL: https://www.ipcc.ch/report/ar6/syr/downloads/report. IPCC_AR6_SYR_LongerReport.pdf

Coakley, S. M., Scherm, H., & Chakraborty, S. (1999). Climate change and plant disease management. *Annual Review of Phytopathology, 37*, 399–426.

Cock, M. J., Biesmeijer, J. C., Cannon, R. J., Gerard, P. J., Gillespie, D., Jimenez, J. J., Lavelle, P. M., Raina, S. K. 2013. The implications of climate change for positive contributions of invertebrates to world agriculture. *CABI Reviews*, 1–48.

Cock, M., Tang Rui, T. R., Liu Zhi, L. Z., Wan HuanHuan, W. H., McGillivray, L., Thomas, S., Cameron, K., Zhang Feng, Z. F. 2016. The main agricultural insect and disease pests of China and implications for the use of remote sensing for their management. *CABI Reviews*, 1–23.

de Sassi, C., & Tylianakis, J. M. (2012). Climate change disproportionately increases herbivore over plant or parasitoid biomass. *PLoS ONE, 7*, Article e40557.

Devlet, A. (2021). Modern agriculture and challenges. *Frontiers in Life Sciences and Related Technologies, 2*, 21–29.

Drozdzynski, F. 2022. The common agricultural policy post 2020: An analysis of the beliefs of selected key stakeholders. University of Twente.

Du, C., Li, L., & Effah, Z. (2022). Effects of straw mulching and reduced tillage on crop production and environment: A review. *Water, 14*, 2471.

Egan, P. A., Chikoye, D., Green, K. K., Tamò, M., Feit, B., Kumar, P., Bandyopadhyay, R., Tepa-Yotto, G.T., Ortega-Beltran, A., Sæthre, M.-G. 2021. Harnessing nature-based solutions for smallholder plant health in a changing climate.

Faiçal, B. S., Pessin, G., Geraldo Filho, P., Carvalho, A. C., Furquim, G., Ueyama, J. 2014. Fine-tuning of UAV control rules for spraying pesticides on crop fields. In *Proceedings of the 2014 IEEE 26th international conference on tools with artificial intelligence*, pp. 527–533.

FAO, F. 2010. Climate smart agriculture: policies, practices and financing for food security, adaptation and mitigation. In *Proceedings of the hague conference on agriculture, food security and climate change.*

Gatto, M. 2022. Potato production through zero-tillage with straw mulch (PZTM).

Gleason, M. L., Duttweiler, K. B., Batzer, J. C., Taylor, S. E., Sentelhas, P. C., Monteiro, J. E. B. A., & Gillespie, T. J. (2008). Obtaining weather data for input to crop disease-warning systems: Leaf wetness duration as a case study. *Scientia Agricola, 65*, 76–87.

Gregory, P. J., Johnson, S. N., Newton, A. C., & Ingram, J. S. (2009). Integrating pests and pathogens into the climate change/food security debate. *Journal of Experimental Botany, 60*, 2827–2838.

Hatfield, J. L., Boote, K. J., Kimball, B. A., Ziska, L., Izaurralde, R. C., Ort, D., Thomson, A. M., & Wolfe, D. (2011). Climate impacts on agriculture: Implications for crop production. *Agronomy Journal, 103*, 351–370.

Heeb, L., Jenner, E., & Cock, M. J. (2019). Climate-smart pest management: Building resilience of farms and landscapes to changing pest threats. *Journal of Pest Science, 92*, 951–969.

Hellmann, J. J., Byers, J. E., Bierwagen, B. G., & Dukes, J. S. (2008). Five potential consequences of climate change for invasive species. *Conservation Biology, 22*, 534–543.

Hobbs, P. R., Sayre, K., & Gupta, R. (2008). The role of conservation agriculture in sustainable agriculture. *Philosophical Transactions of the Royal Society B: Biological Sciences, 363*, 543–555.

Iltis, C., Martel, G., Thiéry, D., Moreau, J., & Louâpre, P. (2018). When warmer means weaker: High temperatures reduce behavioural and immune defences of the larvae of a major grapevine pest. *Journal of Pest Science, 91*, 1315–1326.

Kadoić Balaško, M., Bažok, R., Mikac, K.M., Lemic, D., Pajač Živković, I. 2020. Pest management challenges and control practices in codling moth: A review. *Insects, 11*, 38.

Lamichhane, J. R., Barzman, M., Booij, K., Boonekamp, P., Desneux, N., Huber, L., Kudsk, P., Langrell, S. R., Ratnadass, A., & Ricci, P. (2015). Robust cropping systems to tackle pests under climate change A Review. . *Agronomy for Sustainable Development, 35*, 443–459.

Larsen, A. E., Gaines, S. D., & Deschênes, O. (2017). Agricultural pesticide use and adverse birth outcomes in the San Joaquin Valley of California. *Nature Communications, 8*, 302.

Lefebvre, M., Langrell, S. R., & Gomez-y-Paloma, S. (2015). Incentives and policies for integrated pest management in Europe: A review. *Agronomy for Sustainable Development, 35*, 27–45.

Lindqvist-Kreuze, H., Bonierbale, M., Grüneberg, W. J., Mendes, T., De Boeck, B., & Campos, H. (2024). Potato and sweetpotato breeding at the international potato center: Approaches, outcomes and the way forward. *Theoretical and Applied Genetics, 137*, 12.

Lipper, L., Zilberman, D. 2018. A short history of the evolution of the climate smart agriculture approach and its links to climate change and sustainable agriculture debates. *Climate Smart Agriculture: Building Resilience to Climate Change*, 13–30.

Liu, T., Ma, Z., Huffman, T., Ma, L., Jiang, H., & Xie, H. (2016). Gaps in provincial decision-maker's perception and knowledge of climate change adaptation in China. *Environmental Science and Policy, 58*, 41–51.

Lobell, D. B., & Field, C. B. (2007). Global scale climate–crop yield relationships and the impacts of recent warming. *Environmental Research Letters, 2*, Article 014002.

Macfadyen, S., McDonald, G., & Hill, M. P. (2018). From species distributions to climate change adaptation: Knowledge gaps in managing invertebrate pests in broad-acre grain crops. *Agriculture, Ecosystems and Environment, 253*, 208–219.

Machekano, H., Mvumi, B. M., Nyamukondiwa, C. 2017. Diamondback moth, *Plutella xylostella* (L.) in Southern Africa: Research trends, challenges and insights on sustainable management options. *Sustainability, 9*, 91.

Masters, G., Norgrove, L. 2010. Climate change and invasive alien species.

Murage, A., Pittchar, J., Midega, C., Onyango, C., & Khan, Z. (2015). Gender specific perceptions and adoption of the climate-smart push–pull technology in eastern Africa. *Crop Protection, 76*, 83–91.

Musser, F. R., & Shelton, A. M. (2005). The influence of post-exposure temperature on the toxicity of insecticides to Ostrinia nubilalis (Lepidoptera: Crambidae). *Pest Management Science: Formerly Pesticide Science, 61*, 508–510.

Nabaei, A., Hamian, M., Parsaei, M. R., Safdari, R., Samad-Soltani, T., Zarrabi, H., & Ghassemi, A. (2018). Topologies and performance of intelligent algorithms: A comprehensive review. *Artificial Intelligence Review, 49*, 79–103.

Newell, P., Taylor, O., Naess, L. O., Thompson, J., Mahmoud, H., Ndaki, P., Rurangwa, R., & Teshome, A. (2019). Climate smart agriculture? Governing the sustainable development goals in Sub-Saharan Africa. *Frontiers in Sustainable Food Systems, 3*, 55.

Organization, A., Development, I. F. f. A. 2008. *Gender in agriculture sourcebook*. World Bank Publications.

Pathak, H., Aggarwal, P. K., & Singh, S. (2012). Climate change impact, adaptation and mitigation in agriculture: Methodology for assessment and applications. *Indian Agricultural Research Institute, New Delhi, 302*, 19.

Pathania, M., Verma, A., Singh, M., Arora, P. K., & Kaur, N. (2020). Influence of abiotic factors on the infestation dynamics of whitefly, Bemisia tabaci (Gennadius 1889) in cotton and its management strategies in North-Western India. *International Journal of Tropical Insect Science, 40*, 969–981.

Pitblado, R. 1992. *TOM-CAST: A weather-timed fungicide spray program for field tomatoes*. Ridgetown College.

Porter, J. R., Challinor, A. J., Henriksen, C. B., Howden, S. M., Martre, P., & Smith, P. (2019). Invited review: Intergovernmental panel on climate change, agriculture, and food—a case of shifting cultivation and history. *Global Change Biology, 25*, 2518–2529.

Pretty, J., & Waibel, H. (2012). *Paying the price: The full cost of pesticides* (pp. 39–54). Routledge.

Rosenzweig, C., Iglesius, A., Yang, X.-B., Epstein, P. R., Chivian, E. 2001. Climate change and extreme weather events-Implications for food production, plant diseases, and pests.

Russo, J. M. 2000. Weather forecasting for IPM.

Sekabira, H., Tepa-Yotto, G. T., Djouaka, R., Clottey, V., Gaitu, C., Tamò, M., Kaweesa, Y., & Ddungu, S. P. (2022). Determinants for deployment of climate-smart integrated pest management practices: A meta-analysis approach. *Agriculture, 12*, 1052.

Sheets, F. F. B. 2018. Available online: http://www.fao.org/faostat/en/#data. *FBS (accessed November 15, 2018)*.

Strand, J. F. (2000). Some agrometeorological aspects of pest and disease management for the 21st century. *Agricultural and Forest Meteorology, 103*, 73–82.

Thomas, C., Skinner, P., Fox, A., Greer, C., & Gubler, W. (2002). Utilization of GIS/GPS-based information technology in commercial crop decision making in California, Washington, Oregon, Idaho, and Arizona. *Journal of Nematology, 34*, 200.

Thomson, L. J., Macfadyen, S., & Hoffmann, A. A. (2010). Predicting the effects of climate change on natural enemies of agricultural pests. *Biological Control, 52*, 296–306.

Velasco-Amo, M. P., Vicent, A., Zarco-Tejada, P. J., Navas-Cortés, J. A., & Landa, B. B. (2022). Recent research accomplishments on early detection of Xylella fastidiosa outbreaks in the Mediterranean Basin. *Phytopathologia Mediterranea, 61*, 549–561.

West, J. S., Bravo, C., Oberti, R., Moshou, D., Ramon, H., McCartney, H.A. 2010. Detection of fungal diseases optically and pathogen inoculum by air sampling. *Precision Crop Protection-the Challenge and Use of Heterogeneity*, 135–149.

Ecoenzymes: Nature's Keyholders to Global Food Security

Khayali Vaidya

Abstract This chapter presents a comprehensive overview of the role of ecoenzymes also known as bioenzymes or garbage enzymes in advancing sustainable practices across various sectors. Ecoenzymes, which are biocatalysts derived from natural sources such as kitchen and agriculture wastes, exhibit remarkable catalytic efficiency and substrate specificity under mild reaction conditions. Their inherent ability to facilitate biochemical transformations with minimal energy input and without the need for harsh chemicals positions, define them as a critical component in the transition toward greener and more sustainable growth. The chapter begins by elucidating the fundamental biochemical properties of ecoenzymes, including their structure–function relationship, catalytic mechanisms, and environmental adaptability. Emphasis is placed on the advantages of ecoenzymes over traditional chemical catalysts, particularly in terms of their lower activation energy requirements, higher reaction rates, and reduced generation of hazardous byproducts. The use of ecoenzymes as sustainable alternatives in industrial applications is explored, with specific examples drawn from sectors such as agriculture, waste management, and environmental remediation in the agricultural sector, ecoenzymes are increasingly employed to enhance soil health, improve nutrient availability, and promote sustainable crop production. The chapter details the mechanisms by which ecoenzymes contribute to the degradation of organic matter, the solubilization of nutrients, and the suppression of phytopathogens, thereby reducing the reliance on synthetic agrochemicals. Additionally, the potential of ecoenzymes in organic farming systems is discussed, highlighting their role in fostering environmentally sustainable food production practices. The chapter also delves into the utilization of ecoenzymes in wastewater treatment and environmental clean-up. Ecoenzymes are presented as efficient agents for the biodegradation of organic pollutants, including recalcitrant compounds found in industrial effluents and municipal wastewater. The capacity of ecoenzymes to catalyze the breakdown of complex pollutants into less harmful intermediates is explored, alongside their application in the bioremediation of contaminated soils and aquatic ecosystems. Moreover, the chapter examines the role of ecoenzymes in addressing the environmental challenge posed by plastic waste, through enzymatic depolymerization and

K. Vaidya (✉)
Department of Environment and Life Sciences, KSKV Kachchh University, Bhuj, Gujarat, India
e-mail: Vaidyakhayali20@gmail.com

M. Shaaban (ed.), *Soils and Sustainable Agriculture*, Frontier Studies in Soil Science,
https://doi.org/10.1007/978-3-031-91114-9_39

subsequent biodegradation. Economic considerations associated with the adoption of ecoenzymes in industry are also addressed. The chapter discusses the potential cost savings resulting from the reduced need for energy-intensive processes, lower raw material consumption, and decreased waste disposal costs. Finally, the chapter outlines the current challenges facing the widespread implementation of bioenzyme technology, such as enzyme stability, scalability, and integration into existing industrial systems, while also identifying future research directions aimed at enhancing the efficiency and applicability of ecoenzymes in sustainable practices. This chapter positions ecoenzymes as pivotal tools in the pursuit of sustainability, offering a scientifically sound and economically viable approach to reducing the environmental impact of industrial activities.

Keywords Ecoenzymes · Sustainability · Agriculture · Bioremediation · Organic matter

1 Introduction

The escalating environmental challenges posed by industrial activities have necessitated a paradigm shift toward sustainable practices. Central to this shift is the development and implementation of technologies that minimize environmental impact while maintaining or enhancing economic viability. Among these emerging technologies, ecoenzymes stand out as powerful tools that offer an environmentally friendly alternative to traditional chemical processes. Ecoenzymes, or biocatalysts derived from natural sources such kitchen and agriculture wastes have garnered significant attention due to their ability to catalyze specific biochemical reactions under mild conditions with high efficiency. In recent years, ecoenzymes have been increasingly recognized for their potential to revolutionize various industries by offering sustainable solutions that align with the principles of green chemistry. Unlike conventional chemical catalysts, ecoenzymes are formulated after controlled fermentation process and require less energy input. These properties not only contribute to a reduction in the carbon footprint of industrial processes but also offer substantial economic benefits through resource efficiency and waste minimization.

The integration of ecoenzymes into industrial practices has shown promise across multiple sectors, including agriculture, waste management, and environmental remediation. In agriculture, ecoenzymes have been employed to enhance soil fertility, promote plant growth, and manage pests in an eco-friendly manner. By facilitating the breakdown of organic matter and improving nutrient availability, ecoenzymes contribute to sustainable farming practices that reduce reliance on synthetic fertilizers and pesticides. In waste management, ecoenzymes have demonstrated effectiveness in the treatment of industrial effluents and domestic wastewater, offering a biological approach to pollutant degradation that minimizes the environmental burden of waste disposal. Moreover, the application of ecoenzymes in the biodegradation of persistent pollutants, such as plastics, presents a promising avenue for addressing one of the most pressing environmental issues of our time.

This chapter aims to provide a detailed exploration of the role of ecoenzymes in promoting sustainable practices in multiple areas including household products. It will examine the biochemical properties and mechanisms of ecoenzymes, their current applications across various industries, and the economic and environmental benefits they offer. Additionally, the chapter will address the challenges associated with the large-scale implementation of ecoenzymes and discuss potential future developments in this field. By highlighting the significance of ecoenzymes in the context of sustainability, this chapter seeks to underscore their importance as a cornerstone of green technology and their potential to drive the transition toward more sustainable ecosystems.

2 Understanding Ecoenzymes

Ecoenzymes are generally dark brown color solution prepared from kitchen and agricultural garbage. They are fascinating solution for management of organic waste and convert them into a powerful substance with multiple benefits. It requires jaggery, molasses or brown sugar as a carbon source for the microbes. It is prepared under controlled fermentation process, generally in room temperature in water as a solution. The ratio of the components is 1(carbon source): 3(garbage): 10(water) (Pasalari et al., 2024). The process of preparation of an ecoenzyme is shown in Fig. 1. The fermentation process is generally within 3 months, however modifications like introduction of yeast reduces the fermentation time (Vaidya et al. 2024). Ecoenzymes are typically a mixture of beneficial enzymes like protease, amylase, lipase etc. with the pH typically ranging from 1.8 to 2 which makes them highly acidic in nature. Ecoenzymes were created by Dr. Rosukon Poompanvong, a farmer and the founder of the Organic Agriculture Association in Thailand. The components used to make ecoenzymes interact with one another during fermentation to form a stable, complex enzymatic environment that effectively suppresses the growth of pathogenic organisms.

Owing to their numerous benefits, ecoenzymes have caught the attention of scientists and researchers who intend to conduct in-depth research on them. Table 1 lists the general traits of the ecoenzymes that have been researched and observed in different studies.

Similar to conventional enzyme activity, ecoenzymes function on a lock and key basis, with the active site of the enzyme serving as the "lock" and particular organic molecules or pollutants serving as the "key.“ Natural enzymes like protease, lipase, and amylase are found in ecoenzymes, which are created by fermenting organic waste and target particular molecules. Every enzyme possesses an active site that is specific to a certain substrate, enabling highly efficient breakdown of proteins, lipids, or carbohydrates. Because of their specificity, ecoenzymes are able to break down

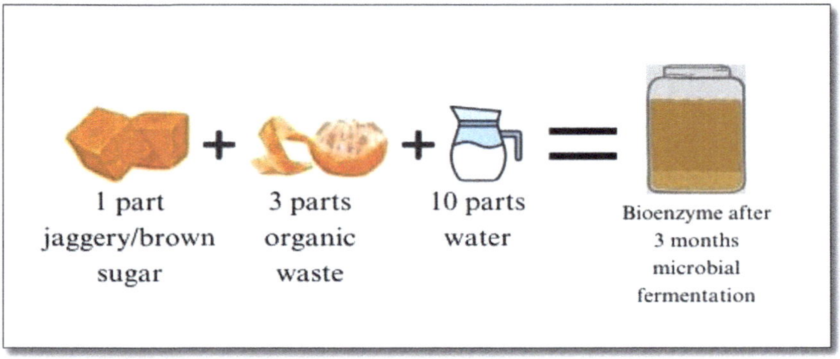

Fig. 1 Complete process of making ecoenzymes

Table 1 General characteristics of ecoenzymes

S. No.	Parameter	Observation	References
1	pH	2	Vaidya et al. (2024)
2	Colour	Light brown	
3	Odour	Fruity	
4	TDS	1040 mg/l	Arun and Sivashanmugam (2017)
5	BOD	79 mg/l	
6	COD	158 mg/l	

specific molecules without causing harm to the environment, effectively cleaning and decomposing organic waste while also improving soil health. Figure 2 shows the general mechanism of any ecoenzyme functioning.

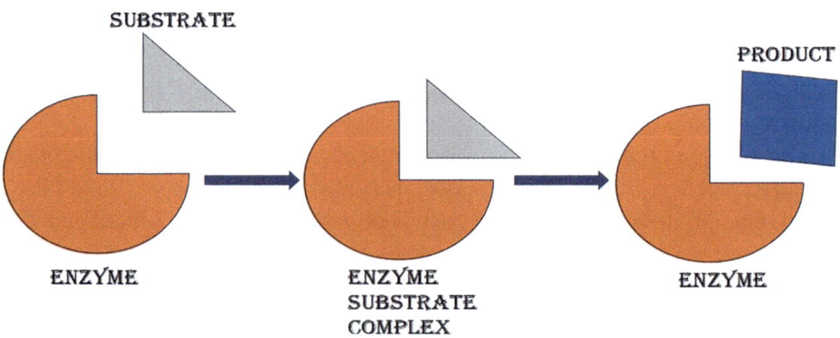

Fig. 2 General "lock and key" model for ecoenzyme functioning

3 Applications of Ecoenzymes

Due to their ability of decomposing organic materials, ecoenzymes have a wide range of uses in diverse industries. They are utilized in agriculture as organic insecticides and fertilizers, improving soil quality and lowering the demand for chemical inputs in the process. In addition to being efficient household cleaners, enzymes are a greener option to chemical cleansers. They also aid in the decomposition of contaminants in wastewater treatment, improving the quality of the water. By accelerating decomposition, their usage in composting improves soil health and reduces waste.

3.1 3.1. Ecoenzymes in Agriculture

Since ecoenzymes naturally break down organic material and increase the bioavailability of key nutrients for plants, they are important for the promotion of sustainable agriculture. By stimulating microbial activity, ecoenzymes when added to soil improve soil health and promote plant uptake of nutrients. Better agricultural yields and increased plant resistance to disease result from this. Ecoenzymes can also be employed as bio-pesticides, which lessens the need for artificial chemicals that over time may harm beneficial insects and deteriorate soil quality. Due to their acidic nature ecoenzymes have beneficial impacts on rose production (Naik et al. 2009). Ecoenzymes are helpful in composting because they can accelerate the decomposition process and create nutrient-rich compost by breaking down organic waste. A study (Bulai et al. 2021) also showed the ability of the garbage enzymes to remediated the soil contaminated with motor oil through biocatalytic decomposition. The soil can then be improved using this and further promoting plant growth.

3.2 3.2. Applications of Ecoenzymes in Everyday Utilities

Bioenzymes, which are produced by fermenting organic waste like fruit peels and vegetable scraps, are naturally occurring enzymes that efficiently break down proteins, lipids, and carbohydrates. These enzymes include protease, lipase, and amylase (Arun and Sivashanmugam 2017). Because of this, they are perfect for doing surface cleaning, stain removal, and odor removal without the use of harsh chemicals (IRJET). They help to break down oil (Bulai et al. 2021), grime, and food residues, leaving surfaces spotless and safe for both people and the environment. They are frequently used to clean kitchen countertops, floors, bathrooms, and drains. Bioenzymes are a safe alternative to traditional chemical cleansers for households with children, pets, and sensitive individuals because they are non-toxic, biodegradable, and emit no hazardous smells.

Furthermore, these powerful natural liquids have proven their efficacy on rejuvenation of water bodies by creating a significant change in value of DO, BOD and COD (Penmatsa et al. 2019).

4 Challenges and Future Directions

While ecoenzymes offer promising solutions for global food security, several challenges remain, including scalability, cost-effectiveness, and regulatory hurdles. The quality and effectiveness of the finished product can vary because ecoenzymes are frequently made by small-scale or home fermentation processes utilizing organic waste. Ecoenzymes with varying enzymatic activity and efficacy might result from changes in the fermentation process, raw materials, temperature, and microbial activity. In large-scale applications like industrial cleaning or agricultural, this irregularity may lessen their dependability. Furthermore, Due to lack of standardization it is difficult to measure the efficacy of these bioenzymes.

Another key challenge is lack of public awareness. Public education initiatives that highlight the advantages of ecoenzymes for both human use and environmental protection are necessary to increase their uptake. Ecoenzymes can be promoted by businesses, governments, and environmental organizations via social media campaigns, online platforms, and seminars.

The extensive commercialization of ecoenzymes is impeded by the absence of well-defined regulatory frameworks. Therefore, performing extensive investigation on these natural substances is vital to improving their efficiency and optimizing their potential for wider application.

5 Conclusion

A promising solution to some of the most urgent environmental problems brought on by industrialization and contemporary agriculture is found in ecoenzymes. These natural biocatalysts, which are derived from the fermentation of organic waste, have several advantages in a variety of industries, such as waste management, agriculture, and everyday utilities. They are an environmentally beneficial substitute for traditional chemical processes because of their capacity to accelerate biochemical reactions in mild circumstances with little energy input. Ecoenzymes help with waste management, crop production, and soil health, which in turn helps with pollution control and sustainable agricultural methods.

Despite its potential, ecoenzymes have several problems, including uneven quality, a lack of standardization, and low public awareness. These barriers impede their widespread application and commercialization, highlighting the need for additional research, regulatory frameworks, and public education. To realize their full potential, rigorous scientific research is required to optimize production, increase efficiency, and ensure efficacy across industries. Furthermore, awareness campaigns and education can highlight the benefits of ecoenzymes, hastening their uptake.

As the worldwide focus changes to sustainable practices and green technologies, ecoenzymes are well-positioned to help drive this transition. Their integration into industrial and home practices has the potential to significantly decrease environmental impact while providing cost-effective, natural alternatives to potentially hazardous chemical procedures. Moreover, ecoenzymes show enormous promise as a cornerstone of future sustainability initiatives, necessitating collaborative research and policy support to fulfill their extensive applications and potential.

References

Arun, C., & Sivashanmugam, P. (2017). Study on optimization of process parameters for enhancing the multi-hydrolytic enzyme activity in garbage enzyme produced from preconsumer organic waste. *Bioresource Technology, 226*, 200–210.

Bulai, I. S., Haruna, H., Umar, Y. A., & Sabo, A. (2021). Optimization of fruit garbage enzymes requirements for biocatalytic remediation of used motor oil-contaminated soil. *Journal of Korean Society of Environmental Engineers, 43*, 241–256.

IRJET_Bringing_Lab_to_Our_Home_Bio_Enzym.pdf

Naik, P. G., Jature, S. D. ,& Phad, G. N. 2009. Effect of bioenzymes on growth and yield of rose (*Rosa indica* L.) cv. *GLADIATOR*, 289–291.

Pasalari, H., Moosavi, A., Kermani, M., Sharifi, R., & Farzadkia, M. (2024). A systematic review on garbage enzymes and their applications in environmental processes. *Ecotoxicology and Environmental Safety, 277*, Article 116369.

Penmatsa, B., Sekhar, D. C., Diwakar, B. S., & Nagalakshmi, T. 2019. Effect of bio-enzyme in the treatment of fresh water bodies. *International Journal of Recent Technology and Engineering*, *8*.

Vaidya, K., Sharma, S. B., Sharma, A., & Solanki, M. 2024. Exploring the efficacy of organic citrus formulations for salinity amelioration in contrasting agro climatic regions. *Biosciences Biotechnology Research Asia, 21*, 705–711.

The Environment Security with Nontraditional Security Management Approaches: Case Study of Hanoi

To The Nguyen, Nguyen Thi Thanh Huyen, Le Thuy Trang, Luu The Anh, and Tuan-Dung Hoang

Abstract Environment Security Issues have been emphasized and adopted as national policies in many countries. However, the concept of management of environment security (MES) has not yet been clarified either in research papers or in policies so that it could help stakeholders, players, and managers increase their awareness and boost their actions in protecting environment security in their respective countries and the world. This paper reviews concepts of environmental security, introduces basic concepts of MES as well, and provides some current practices of MES for problems of environmental security in Hanoi, Vietnam from 2010–2020. Based on the approach of Management of Nontraditional Security (MNS), the basic equation of MES is introduced and applied to analyze some environmental insecurity cases

The original version of the chapter has been revised. A correction to this chapter can be found at https://doi.org/10.1007/978-3-031-91114-9_48

T. T. Nguyen
University of Economics and Business, Vietnam National University, Hanoi, Cau Giay, Hanoi, Vietnam

N. T. T. Huyen
Faculty of International Economics, Foreign Trade University, Hanoi, Vietnam

L. T. Trang
Faculty of Environmental and Natural Sciences, Duy Tan University, Da Nang, Vietnam

School of Engineering and Technology, Duy Tan University, Da Nang, Vietnam

L. T. Anh
Central Institute for Natural Resources and Environmental Studies, Vietnam National University, 19 Le Thanh Tong, Hoan Kiem, Hanoi, Vietnam

T.-D. Hoang (✉)
School of Interdisciplinary Sciences and Arts, Vietnam National University, Hanoi, Cau Giay, Hanoi, Vietnam
e-mail: tuandunghoang@gmail.com

School of Chemistry and Life Sciences, Hanoi University of Science and Technology, Hanoi, Vietnam

M. Shaaban (ed.), *Soils and Sustainable Agriculture*, Frontier Studies in Soil Science, https://doi.org/10.1007/978-3-031-91114-9_40

in Hanoi, Vietnam. This article also provides a new approach or pathway that plays a key role in enhancing the MES for national sustainable development goals.

Keywords Environment security (ES) · Management of environment security (MES) · Policy implications · National sustainable development (NSD) · Climate change, circular economy

1 Introduction

The concept of Environment Security (ES) is ambiguous and open to appropriation (Jon Barnett, 2001). The debate over the meanings and connotations of this concept stems from the combination of two vast but equally abstract concepts—environment and security. A variety of disciplines in this field and diverse approaches to thought contribute to the richness of this concept. The way a person a research group or an organization understands this concept differently reflects the viewpoints, properties, and characteristics of that individual or organization.

There are some approaches to explaining and defining the term environment security. Jon Barnett (2019) defines environmental security as broadly about the threats to nations or state scopes rather than to the environment per se, or to individuals. According to Zurlini & Müller (2008), the environmental security definition emerged from the early 1960s stemming from an important concept in security studies. This is followed by an increase in the level of environmental awareness in developed countries, typically environmental events and environmental movements that occurred around the 1960s. Rachel Carson's famous book "Silent Spring" (Rachel Carson, 1962) is one of the special publications of the time and has brought a higher level of environmental awareness to everyone. The book warned of the dangers to all natural systems including animals, plants, and food chains caused by the overuse of Dichlorodiphenyltrichloroethane, known as DDT, a toxic chemical pesticide welcomed in developing countries but not in developed countries. Carson's book contributed to the public debate at that time. She was said to be one of the people who helped urge the public to be more aware of environmental damage issues. Subsequently, a large number of environmental NGOs were formed and developed which brought more attention to environmental issues such as the World Wildlife Fund WWF (1961), Friends of the Earth (1969), and Greenpeace (1971). These activities gradually formed the concept of "environment security".

In 1987, the book "Our Common Future" published by the United Nations (UN) through Oxford University Press summarizing the success of the Stockholm Human—Environment Conference 1972 officially used this term. This marked the first time the "environmental security" concept' was officially used by the United Nations (UN).

Later on, when climate change became a global issue and gained more attention globally, climate security was used and this is understood as an extended term of the environmental security definition. In the context of global change where global problems such as energy security, and global warming appear acute, environmental security has become more important than it ever could be. Smil (1997), even affirms that environment security is now equivalent to and replaces the threat of nuclear security. Both environmental security and nuclear security are of global scale and their destruction to Earth if those issues are mismanaged and happen will be enormously serious. Therefore, it is essential to review the research and key debates on this term and topic.

2 Materials and Methods

2.1 Literature Selection

To review the Environment Security (ES) and Management of ES (MES), we searched the articles using the Web of Science Core Collection with the combinations of the following terms in the subject: environment security or management of environment security, and "environment security management" or "nontraditional security management approach". Then the articles generated from the initial search were checked by reading through the title, keywords, and abstract. In addition, important gray literature such as research and technical reports, government documents, and environmental strategic plans were also included in this work. Finally, we got 55 publications on which the systematic literature review is based.

2.2 Concepts of Management of Environment Security (MES) and the Equation of Management of Nontraditional Security (MNS)

There is not a consistent definition for ES, and terminologies for Environment (E) and Security (S) depict two main definition approaches of ES under different perspectives. Dower (1995) defines security as concerned with safety and certainty from contingency. Allenby (2000), Rwabizambuga (2007), and Swanström (2010) prove that security had conventionally been associated with military and state power to ensure sovereignty, stability, and peace as well as to pursue armed conflict. As noted in the introduction, the United Nations Development Programme (UNDP, 1994) has been instrumental in advancing the human dimensions of security. In its broadened conceptualization, environmental security is viewed as ecological processes (anthropogenically influenced or not) as well as natural resources (water, land, forests, fisheries, air, etc.) as direct or contributing factors that cause or induce conflicts and insecurity.

Detraz (2009) discusses three nexuses between environment and security concepts: (1) environmental degradation directly or collaterally causing conflicts,

(2) environmental degradation inducing negative impacts on human security, and (3) ecological security posing negative impacts on the environment.

Trombetta (2008) and Graeger (1996) illustrate different perspectives on environmental security terms by discussing seminal work on environmental conflicts and environmental cooperation, a human-security approach to solving the issues of environmental security, and climate security.

Arnold Wolfers (1952) stated that environmental security qualifies the absence of threats to obtained values. Galgano and Rose (2021) define environmental security as typically a scope of security issues intensified by environmental factors in which environmental stress has the potential to trigger violent conflicts within vulnerable populations. Peter Hough (2014), provides an attested and critical analysis of the causes and effects of environmental degradation, both natural and human-induced degradation.

Historically, according to various sources, several other environment security concepts and definitions are documented in Table 1.

Environment security is a key element of regional and national security and is of transboundary and even global scales in terms of its importance. Environment security deals with the minimization and prevention of threats to energy, water resources, natural resources, energy, and climate (Bandh, et al., 2024; Hoang and Nghiem, 2021; Malla, et al., 2023; Nguyen, et al., 2021).

2.3 Environment Security in National Policies of Some Countries

Environmental risks and stress related to the environment can cause instability or political, and economic conflicts in a country or a region (Rademarker et al., 2016). Environment security addresses energy concerns and is directly related to a country's foreign policy. The threats to environmental security may not directly damage the territory of one country, but they can still threaten the security of another country. Environmental security concerns include those related to nuclear pollution, used fuels, and waste; threats of energy sources; pollution, degradation, or depletion of essential environmental resources; or environmental problems caused by damaged infrastructure that can threaten the security of a country or damage the stability of the region or the world. Climate change (Hoang, 2023), water pollutions by heavy metals (Tran TK, 2024a, b), nanoparticles (Tran TK, 2024a, b), micro plastics (Hoang, 2024; Hoang et al., 2024, 2025), all are of environmental security issues.

In the US, President G. Bush in 1991 had identified environmental security concerns in his National Security Strategy address. The 2000 US National Security Strategy Report (page 13, section environment & health) clearly stated, "Decisions today regarding the environment and possible natural resources affecting our security (USA) for generations. Environment threats don't just stop at national borders;

Table 1 Some definitions of environmental security

Environment security concept definitions	Defined by organizations or authors
“”is a relative general safety, free from environmental hazards caused by natural or human processes from ignorance, accident, poor management or design, and having origin within or across national borders”	AC/UNU Millennium Project (1998)
“”is the state of human–environment dynamics that includes restoration of the environment damaged by military actions, and amelioration of resource scarcities, environmental degradation, and biological threats that could lead to social disorder and conflict”	AC/UNU Millennium Project (1998)
“”is the proactive minimization of anthropogenic threats to the functional integrity of the biosphere and thus to its interdependent human component”	Barnett J (1997) “Environmental Security: Now What?”, seminar, Department of International Relations, Keele University, 4 Dec. 1997
“”By ensuring environmental security we mean guarding against environmental degradation to preserve or protect human, material, and natural resources at scales ranging from global to local”	Belluck et al. (2006)
“”is one in which social systems interact with ecological systems in sustainable ways, all individuals have fair and reasonable access to environmental goods, and mechanisms exist to address environmental crises and conflicts”	By AC/UNU Millennium Project (1998)
“”is the protection of the natural environment and vital interests of citizens, society, the state from internal and external impacts, adverse processes and trends in development that threaten human health, biodiversity and sustainable functioning of ecosystems, and survival of humankind. Environmental security is an integral part of Russia’s national security”	‘’Environmental Security of Russia”, issue 2, The Security Council of the Russian Federation, Moscow, 1996, p. 55
“”consists of restoration, compliance, conservation, pollution prevention, environmental security technology, and international activities”	US Department of Defense (1996)
“”is the state of protection of vital interests of the individual, society, and the natural environment from threats resulting from anthropogenic and natural impacts on the environment”	“On Principles of Environmental Security in the Commonwealth States”, Dec. 4, 1997

(continued)

Table 1 (continued)

Environment security concept definitions	Defined by organizations or authors
“”Scientific problems related to environmental security including the reclamation of contaminated military sites, regional environmental problems, and natural and man-made disasters; affordable cleanup technologies are of particular interest”	NATO Science Program (1997)
“”When people do not have enough food, water, shelter or natural resources needed to live, it creates unsettled situations. Environmental security has always been implicitly recognized. Over the past 25 years, this concept has been understood more clearly, both in defense strategy and in the environmentalist community. There is a convergence of traditional national security thinking with environmental protection and policymaking. They combined to form a new field that is said to ‘’environmental security “”	American security and safety research and analysis organization CNA: https://www.cna.org/ (2012)
“”is a process that creates cooperation between countries and regions to address environmental problems and promote the goals of political stability, economic development, and peace”	US Environmental Protection Agency—USEPA (1999)
“”is the process of establishing security for important environmental factors (e.g. water, soil, air, vegetation, biodiversity, climate, and other factors) that are the essential component of a country’s environmental background and socio-economic activities, and thus its political stability”	United Nations Environment Program (UNEP) (2017)
“”refers to a range of security issues enhanced by environmental factors such as resource competition, natural disasters, climate change, and other adverse environmental conditions”	U.S. State Department (1996)

environmental threats abroad (also) are dangerous for the security and well-being of the United States”.

The report noted that responding positively to environmental threats remains part of the mainstream US foreign policy and concern about environmental threats has been part of the US National Security Strategy over the past decade.

Svensson, (1988) pointed out that the environmental security concept was presented at the International Conference on Environmental Stress and Security in Sweden, by the Royal Swedish Academy of Sciences, Stockholm, Sweden, in December 1988.

According to Matthew (1995), the Australia-Canada-U.S. trilateral commission is an attempt to address environmental problems cooperatively when these countries realize the environment as a national security. Bilateral agreements for cooperation on environmental security issues were negotiated with Germany, Norway, Sweden, Russia, Poland, the Czech Republic, and Hungary in 1990 (Mathew, 1995).

In Vietnam, the concept of environmental security is new but this concept has been included in the “Laws on Environmental Protection” that was issued and ratified by the Congress in 2014. The Document of the XII Congress of the Communist

Party of Vietnam (CPV) also emphasizes the importance of ensuring environmental security. In the CPV's documents, it is stated that "Preventing and gradually overcoming the degradation of the natural environment caused by humans, development and economic projects; to limit and proceed to fundamentally overcome the destruction and depletion of natural resources and environmental pollution of production facilities, industrial parks and urban areas; to complete the legal system, enact strong enough sanctions; to protect the environment, prevent and handle according to law to end environmental pollution; to strengthen prevention and control of sources of pollution and environmental contamination." The party document also affirmed "ready to respond to traditional and non-traditional security threats.

In "Management of Nontraditional Security: A New Approach", the authors classified and emphasized the Management of environmental security (MES) as an important pillar of the Management of Nontraditional Security (MNS) field that reflects the cognitive process of the actors and players in making and implementing policies, strategies and action plans in preventing the risks and dealing with the pollutions to prevent the environment insecurity or to protect the environment security for the National Sustainable Development (NSD) of each country.

From the literature review and the basic concept and MNS equation, the authors have tested a new framework of MES or MES Analysis Framework of 6 Dimensions as below:

$$\text{MES} = (S1 + S2 + S3) - (C1 + C2 + C3)$$

The Usage of 6 Dimensions of MES Analysis for Some Time:

S1, S2, and S3 are 3 main factors or dimensions that reflect the result of MES of the actor/actors during a period (1–2-3-n years). C1, C2, and C3 are 3 main factors or dimensions or activities of MES of the actor/actors during a period (1–2-3-n years).

Using the Likert 5 scale, the total score for S1 + S2 + S3 = 15, and the total score for C1 + C2 + C3 = 15. The MES final score = (15)–(15) = 0.

If the MES final score is a plus figure, it is a good sign of MES. It is a bad practice of MES if the MES final score is a minus figure. The MES final score reflects the level of success or failure in MES of the management subject of management actor during a period and thus helps the actors and stakeholders to review for lessons learned as well as for new policy and strategy/plans making for the next period to pursue a better MES.

S1 represents the safety of the main environmental factors (quality of soil, water, air, noise, biodiversity...). Soft data can be taken from interviews and surveys of main players and stakeholders by using Likert 5-scale measurement (1: very low—5: very high). Hard data can be taken from reliable reports to support the soft data. The degree or level of S1 is evaluated by 2 main groups of correspondents: (1) the referent objects (households or residents/individuals); and (2) environment experts (independent, not working for government offices).

S2 represents the stability of main environmental factors (increase or decrease of quality of a subjective study). Soft data can be taken from interviews and surveys of main players and stakeholders by using Likert 5-scale measurement (ranging from 1: very low to 5: very high). Hard data can be taken from reliable reports to support the soft data. The degree or level of S2 is evaluated by 2 main groups of correspondents: (1) the reference objects who are households or residents/individuals living in the studied areas; (2) environment experts who are independent and are neither state officials nor staff of the government.

S3 represents sustainability (state & corporate investments in environmental protection). Soft data are taken from interviews and surveys of main players and stakeholders by using Likert 5-scale measurement (1: very low—5: very high). Hard data are taken from reliable reports to support the soft data. The degree or level of S3 is evaluated by 2 main groups of correspondents: (1) the reference objects (households or residents/individuals); and (2) environment experts (independent, not state officials or staff).

C1 represents the cost for all risk management activities. As the main aim of most policies, strategies, and action plans in MES is for the protection of the environment by maximizing risk management and minimizing any environmental crisis that may happen from a potential risk, C1 is the most important factor in the equation, and for analysis of MES. C1 is the starting point of any process of MNS, including MES. If C1 is effectively and efficiently managed, there would be no crisis or at least no serious crisis, and therefore there would not be any environmental insecurity in the country.

The cost here is a relative figure of time and money that are spent for all risk management activities in comparison with the GDP year by year or state budget/ local budget allocated year by year for MES by using experts' methods. The level of C1 ranges from very low (1) to very high (5). Soft data can be taken from interviews and surveys of main players and stakeholders. Hard data can be taken from reliable reports of the main actors and organizations.

C2 represents the cost of all crisis management activities. When C1 is not managed properly, C2 may happen any time that is named a crisis at a level that might be from 1 to 5 on a Likert scale (ranges of very small, small, medium, big, and very big). The environmental crises (C2) can be determined by experts' methods based on the size, the loss, and the impact…, and according to various practices. For example: a weak MES has caused To Lich River in Hanoi to be so heavily polluted for the last 20 years that no fish or plant can survive in this river. Then this should be called "a crisis".

But the main indicators of any environmental crisis are the economic loss, human loss, the political loss, and the natural loss. Let's take examples of the environmental crisis of craft villages in the country. According to the ''National Report of Environmental Status, from 2011 to 2015", the whole country has 5096 craft villages, of which, traditional craft villages 1748; employ about 10 million workers and is contributing significantly to the economy. However, the downside of rural craft village development is its environmental pollution: wastewater and waste generated by craft villages in rural areas pollute water, air, and land. The percentage of craft villages using wastewater and hazardous waste treatment equipment is only 4.1%.

Most craft villages have not built wastewater treatment systems. Untreated gases and wastewater are discharged directly into the environment. In particular, wastewater in metal recycling craft villages, and processing agricultural and aquatic products is a serious issue that seriously affects the environment. Solid waste in most craft villages has not been collected and treated thoroughly, causing adverse impacts on the landscape, and polluting the air, water, and soil environment.

Air pollution arising from industrial production facilities and craft villages is causing diseases and health problems for local people and livestock living in craft villages, mostly rural areas, especially children. In metal recycling craft villages, the emission of toxic gases, high heat sources, and metal dust from processes such as foundries, smelters, etc. during the production process has caused common diseases such as respiratory, pulmonary hypertension, and neurological diseases. By estimate, lung diseases, digestive diseases, eye diseases, and gynecological diseases are some of the most common diseases and the percentage of people with pulmonary tuberculosis is 0.4–0.6%, of lung cancer is 0.35–1%.

When C2 happens, it is very serious and dangerous and often leads to consequences for the MES actors according to the criminal laws of many countries. For example, in Vietnam, the criminal court will punish both individuals and enterprises that caused heavy pollution by prison terms. But so far, law enforcement is still weak and only a few among thousands of cases have been charged.

For C2 it is hard to get soft data as the main players who caused pollution often try to cover the offence and not to confess. Instead, the soft data can be taken by interviews with the reference objects or stakeholders such as households or those individuals who have directly suffered from that pollution or environmental insecurity.

Hard data can be taken from reliable reports of the main actors and organizations.

C3 represents costs for all after-crisis recovery activities. In any management process, C2 is always followed by C3 as it is often costly for the management subject/actor/actors to rehabilitate and recover the environmental security as it was before the crisis. The level of C3 ranges from very low (1) to very high (5). Soft data can be taken from interviews and surveys of main players and stakeholders. Hard data can be taken from reliable reports of the main actors and organizations.

3 Using the MES Analysis Framework to Assess Some MES Practices in Vietnam During 2010–2020

3.1 MES Analysis Framework

MES is a very new concept in Vietnam. Currently, most state and non-state actors and players are starting to be concerned about environmental security, but, for them, the concept of MES is yet politicized for many reasons.

In the past 10 years due to economic development priorities, Vietnam has not paid enough attention to environmental security. Many examples of environmental insecurity occurred (such as fish death incidents in coastal areas in the central region, the problem of household waste, and air pollution in big cities). It is necessary to have comprehensive solutions and the participation of many actors who are responsible for that (Table 2).

According to national reports on the status of the environment in Vietnam in 2019, total domestic and municipal solid waste in the country is 64,658 tons/day (of which the amount of solid waste in urban areas is 35,624 tons/day and that in rural areas is 28,394 tons/day), an increase of 46% compared to 2010. Around 70% of domestic and municipal solid wastes are not treated thoroughly with modern technologies such as recycling, and waste to energy /electricity generation technologies but simply being disposed of in landfills (Table 3).

About 80% of residential apartments, buildings, and offices in Vietnam do not have appropriate wastewater treatment systems meeting type B national standards. About 80% of big cities cannot control emissions. Air Quality Indexes (AQIs) of big cities in Vietnam are usually ranked ''poor''. The World Bank in its report entitled ''Toward a Safe, Clean, and Resilient Water System'' (WB, 2019), reported that urban wastewater is the largest contributor to water pollution in Vietnam, and only

Table 2 Waste and pollution in Vietnam in recent years

	Vietnam in total	Hanoi and HCM City
Solid waste	64,658 tons/day	–
Urban areas	35,624 tons/day	–
Rural areas	28,394 tons/day	12,000 tons of domestic solid waste a day (2019)
Domestic waste	–	
Plastic waste and nylon bags	25 million tons	80 tons/day
Air pollution rankings	Poor	Poor
The main form of solid waste treatment	Landfill	Landfill

Table 3: Solid waste generation (million tonnes per year) in Vietnam per year

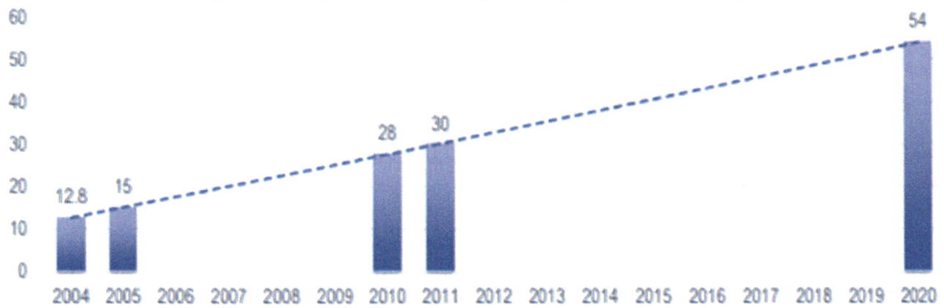

13% of municipal wastewater in Vietnam was collected and treated properly before discharge into water bodies.

In Vietnam, only 60 percent of households are connected to public water systems, and only 10 percent of collected wastewater is treated. The national environmental status reports published annually by the Ministry of Natural Resources and Environment have pointed out that domestic waste in total in the country is increasing, especially in big cities. Each year the country is generating about 25 million tons of domestic waste, doubling that of the past 15 years. Among these, solid waste is always problematic and a "headache" for managers.

About 7% of the total 25 million tons of household waste each year is plastic waste. On average, each household uses about 1 kg of plastic bags/month. Hanoi and Ho Chi Minh City alone discharge about 80 tons of plastic and nylon bags into the environment every day, of which the rate of garbage collected and recycled is very low. Landfilling is still the main form of disposal (71% of the volume of collection) but only about 20% of landfills meet sanitary requirements. This leads to the problem of environmental pollution in landfills.

In 2020 alone, the inner city of Hanoi experienced two floods of garbage. Garbage from the house piled atop the street, piled up on small alleys, and in yards in vacant lots in the city.

As per the latest Vietnam Provincial Governance and Public Administration Performance Index (PAPI, 2020), the Environmental Governance of two major cities in Vietnam, namely Hanoi and Ho Chi Minh cities, ranked lowest among 63 provinces and cities surveyed. This result on the one hand is quite surprising and unexpected due to the dense concentration of governance offices in these two cities, on the other hand sets new requirements for better governance of environmental security in those main cities. In sum, the World Bank also estimated that water pollution is a main problem in Vietnam, causing an estimated 3.5% loss of GDP each year.

3.2 *Assessing Some Typical MES Practices in Hanoi Capital City Such as West Lake Water Pollution, to Lich River, and Nam Son Landfill.*

Calculation of Environment Insecurity of to Lich River

Tolich is a small river that flows through districts in Hanoi city. In the past decades, this river pollution has become an issue and insecurity in Hanoi. During 2019–2020, more than 300 direct interviews with households, living on the banks of the Tolich River, have been made. Assessing was made by combined use of both soft data and hard data and by using experts' methods. Participants were asked about their perceptions of Tolich water river quality on a 5-point Likert scale (1-Totally disagree, 2- Disagree, 3- Disagree, 4-Agree, and 5-Totally disagree). Table 4. Summary of statistically significant survey responses about the Tolich River water quality.

Table 4 Survey responses about the Tolich river water quality

Question:	Disagree (%)	Disagree (%)	Neither agree nor disagree (%)	Agree (%)	Agree (%)
Tolich River water does not have bad smells	72	75	40	2	0
Tolich River water is safe and of good quality for fish and aquatic species to live in	34	55	76	21	3
Readiness to pay for the improvement of the river quality	4	9	41	91	44

S1: stands for the safety or the level of safety of the reference object, reflecting the first aim of any MNS of any subject/actor. In this study survey, S1 reflects the safety indicator of the Tolich River.

S2: represents the level of stability of the health of the reference object, namely stability or health concern of people living alongside the river.

S3: reflects the sustainability of the reference object (Tolich River) in the context of a non-security environment.

C1: effectiveness and efficiency of risk management in terms of activities, time, cost, and results.

C2: cost of crisis management and the loss or impact on human lives and/or assets and properties when the risk or threat has occurred and caused a crisis.

C3: cost to be paid for recovery or rehabilitation activities paid to revitalize the river after a pollution crisis or disaster.

Authors have found that most households live around the banks of To Lich River in 4 inner districts. All respondents have a certain interest in river water quality, are aware of the pollution level of the river, and expressed the need for wastewater treatment. The majority of respondents are willing to contribute to the improvement of river quality so that future generations also have the opportunity to use it. In this case study of To Lich River pollution, the survey results reveal MES final score is – 1.25.

In addition, a previous study under the Asian Water Environment Research Collaboration in 2015 showed that heavy metals (μg/l) were present in the analyzed water samples of To Lich River (Table 5).

Heavy metals in the analyzed samples varied considerably among collected water samples. Concentration of Nickel (Ni), chromium (Cr), Lead (Pb), Copper (Cu), Zinc (Zn), Cadmium (Cd), Cobalt (Co), Molybdenum (Mo), Tin (Sn), Vanadium (V) ranged from 1.24–7.74, 1.51–6.45, 0.28–1.97, 1.66–2.31, 25.19–77.94, 0.06–0.14, 0.07–0.22, 0.46–0.79, 0.07–3.13, 0.12–0.25 (μg/l), respectively. These heavy

Table 5 The concentration of heavy metals in Tolich water samples

	Nickel (Ni)	Chromium (Cr)	Lead (Pb)	Copper (Cu)	Zinc (Zn)	Cadmium (Cd),	Cobalt (Co)	Molybdenum (Mo)	Tin (Sn)	Vanadium (V)
Water samples	1.24–7.74	1.51–6.45	0.28–1.97,	0.28–1.97	1.66–2.31	25.19–77.94	0.07–0.22	0.46–0.79	0.07–3.13	0.12–0.25
Sediment samples	34–66	-	32–45	40–50	98–237	0.16–0.36	–	–	–	–

Table 6 Distribution of Cr and Ni in Nhue and To Lich rivers

Sampling sites	Cr (ng/ml)	Ni (ng/ml)
S1	1.51	1.24
S2	1.04	1.49
S3	3.29	3.27
S4	0.45	7.74
S5	3.25	5.76
S6	3.31	4.92

metals came from both industrial and municipal waste sources discharged directly or indirectly into the river (Table 6).

Analyzing sediment samples of To Lich River, it was found that the heavy metal concentrations of Cu, Pb, Zn, Ni, and Cd ranged from 0.16–0.36 (Cd), 34–66 (Ni), 32–45 (Pb), 98–237 (Zn), 40–50 (Cu). These parameters were all higher than the concentration of heavy metals in normal agricultural soil samples. From the analyzed results, it is evident that both the water and sediments of the To Lich River are highly polluted with high concentrations of heavy metals. This pollution source is harmful and toxic to plants and animals in areas nearby and in vast areas where this river flows through.

In addition to that, the To Lich River pollution also causes unpleasant odors for residents living along this river and deteriorates the capital's landscape image and its beauty. This is a typical case of environmental insecurity in Hanoi city.

Environment Insecurity at Nam Son Waste Storage and Treatment Zone

Looking at some news sources in the Hanoi newspaper, we can observe the blocking and obstructing activities organized by the local people of Nam Son landfill areas to the city's urban waste disposal trucks. This insecure incident occurred from time to time few years ago, with a frequency of a few times per year.

On October 23rd, 2020, the Nam Son landfill was blocked by people in the two communes of Nam Son and Hong Ky of Soc Son district of Hanoi capital city. The local people opposed and stopped trucks carrying garbage into the municipal solid waste treatment zone due to the area's serious pollution. Foul smells generated from the treatment area, especially in the area of village 2 (Hong Ky commune) affected the local people and triggered the local people to do this. The site clearance policies in the area of a 500 m radius and several sound policies that have not yet been resolved over the years in the area of the complex such as health compensation and clean water supply delays have also triggered the local people in two communes to stop garbage trucks to go inside the Nam Son garbage disposal area.

Environmental Problems of the West Lake

West Lake is located northwest of old Hanoi. The lake is of oxbow shape, has a long history, and is said to have been linked to the Red River. This is the largest natural lake in the inner city of Hanoi, with an area of more than 500 hectares, and a circumference of about 14.8 km. The lake acts as a scenic spot, and a spiritual place, and functions as an ecological lung conditioning the air of Hanoi's capital city. The West Lake is expected to become the centerpiece of the new Hanoi capital in the coming years.

In October 2016, Hanoi people witnessed the mass death of fish in the lake. Floating on the surface of the West Lake, dead fish washed ashore causing an unpleasant stench. Fish die-off on the West Lake is not the first time, however, the frequency and scale of fish deaths as much as in October 2016 had never happened previously.

According to a report by the chairman of the People's Committee of Hanoi, the number of dead fish collected in West Lake is up to 200 tons. The number of dead fishes is enormous, fishes die on a large scale both in surface water and in mezzanine water depth, about 2 m. The deadest fish species are cyprinid fishes, tilapia, and oily fishes. Dead fishes of the West Lake in the morning of October 3, 2016, are mainly of large sizes of 3–4 kg; carp with the size of 5–6 kg had also died, floating and bobbing on the current and surface of the lake. According to the People's Committee of Hanoi City, the quick test results show that all surface water in West Lake has no oxygen.

There are many reasons why the West Lake is polluted such as approximately 30 sluices are still discharging directly into the lake every day, and for many years the lake has not been dredged. In previous years, toxic and organic wastes were poured into the West Lake causing organic waste to be abundant in the lake to absorb oxygen to decompose, leading to the mass death of the West Lake fish.

In summary, some most urgent tasks in solving environmental problems in Vietnam are the proper collection and treatment of urban wastewater, municipal solid wastes, and the reduction and appropriate control of emissions. These are the most critical and challenging activities in managing environmental security in Vietnam. Implementation of AI technologies (Hoang TD, 2022), biofuels, public transportation, green technologies and circular economy approaches might significantly solve these challenges.

4 Conclusions and Policy Implications for MES Practices

As analyzed and alluded to above, the final result of MES at any level reflects the awareness, actions, and management skills of the subject/actor. Ensuring environmental security will be very challenging for any management team, city, or even a country to take on. To ensure environmental security, human security, and sustainable development, a multidisciplinary approach and integrating collective efforts of multiple agencies, international institutions, private and other sectors, and further social and technological research and pilot projects for determining a package of solutions for MES practices are needed.

The security and insecurity scope is by no means limited to violent conflict or its absence but includes the roots of sustainable livelihoods, health, and well-being.

Environmental security is also related to national security, is of paramount importance, and is essential to maintaining the earth's life-supporting ecosystems generating water, food, and clean air. The relationship between human security and environmental problems is correlated. Mitigation of environmental security risks depends fundamentally on improving resource governance and social resilience to natural resource shocks and stresses. The environment is better protected and secured when there are no domestic and cross-border conflicts and when stable, effective management is included. In this sense, conflict prevention through environmental cooperation such as transboundary water government (Le Anh & Hoang, 2024; Hoang, 2024), should be contributed by all countries in the world.

In the case of Vietnam, to ensure environmental security in the next 10 years, MES and integrated management solutions should be implemented. On a national scale, strengthening environmental education, and environmental awareness programs ranging from primary to higher levels in the education system such as high schools and universities nationwide could help; Pursuing a knowledge-based economy, a circular economy, or less resource-based economy and cooperation among nations can maintain and ensure better environmental security. All areas of hydrology, biology, chemistry, interdisciplinary technical engineering expertise (such as renewable energy, and clean energy), and emerging fields in behavioral sciences, as well as analytics, and traditional national security sciences, need to be mobilized to solve the problems of environmental insecurity. There is also no doubt that this field will require both social and technological innovations.

At a local scale, applying non-traditional security governance approaches and using environmental performance indexes could be also a sound solution approach. Public perceptions and determinants need to be investigated for future implementation of MES practices. In addition, public education should be taken as soon as possible to raise public concerns about environmental awareness and behaviors. In the short term, the government could improve the collection, sharing, analysis, and visualization of environmental security data to identify priorities for more systemic risk monitoring. Over the long term, building capacity to monitor resource trends, to increase transparency and access to environmental information, together with measures to enable proactive stakeholder engagement in assessing risks and developing shared action plans, can build forms of cooperation that may prove critical when environmental insecurities emerge.

4.1 Declaration of Competing Interest

The authors declare that they have no known competing financial interests or personal relationships that could have appeared to influence the work reported in this paper.

Acknowledgements The authors would like to thank all participants involved in surveys for this study. Authors also wishes to thank anonymous reviewers for their thoughtful comments on this work.

References

Environmental Security—ResearchGate. https://www.researchgate.net/publication/235698902_Environmental_Security

China's Environment and Security. Simple Myths and Complex.

About PAPI. http://papi.org.vn/eng/gioi-thieu/

Allenby, B. R. (2000). Environmental security: Concept and implementation. *International Political Science Review, 21*(1), 5–21.

AMCEN/16/EGM/5: Environmental Security | UNEP - UN https://www.unep.org/resources/report/amcen16egm5-environmental-security

Le Anh, T., & Hoang, T. D. 2024. Trans-boundary water governance in the Mekong countries. In *The Mekong river basin* (pp. 237–272). Elsevier.

Arnold, W. (Dec 1952). *Political science quarterly*. (Vol. 67(4), pp. 481–502). Stable URL: http://links.jstor.org

Bandh, S. A., Malla, F. A., Hoang, T. D., Qayoom, I., Mohi-Ud-Din, H., Bashir, S., Betts, R., Le, T. T., Nguyen Le, D. T., Linh Le, N. V., & Le, H.C. 2024. Track to reach net-zero: Progress and pitfalls. *Energy and Environment*, pp. 0958305X241260793.

Barnett, J (2001). The meaning of environmental security: *Ecological politics and policy in the new security era*. Zed Books.

Barnett, J (2019). Global environmental change I: Climate resilient peace? *Progress in Human Geography* 43(5), 927–936.

Belluck, D. A., Hull, R.N., Benjamin, S.L., Alcorn, J., & Linkov, I. (2006). Environmental security, critical infrastructure and risk assessment: Definitions and current trends. *Environmental security and environmental management: The role of risk assessment*. 3–16.

Jon Barnett. 2003. *Security and climate change.*

Blocked landfill queues Hanoi garbage crisis—VnExpress International.https://e.vnexpress.net/photo/news/blocked-landfill-queues-hanoi-garbage-crisis-4131454.html

https://www.indoeng.com/Data/Sites/1/media/case-study/ICE_PPP_CaseStudy-VietnamSolidWasteSector-RevA.pdf

Chỉ số chất lượng không khí (AQI) và thông tin về ô nhiễm ... - IQAir. https://www.iqair.com/vi/vietnam

Climate and Environment. United States Department of State. https://www.state.gov/policy-issues/climate-and-environment/

CNA: Home. https://www.cna.org/

CECODES, VFF-CRT, RTA & UNDP (2021). The 2020 Viet Nam Governance and Public Administration Performance Index (PAPI 2020): Measuring Citizens' Experiences. A Joint Policy Research Paper by Centre for Community Support and Development Studies (CECODES), Centre for Research and Training of the Viet Nam Fatherland Front (VFF-CRT), Real-Time Analytics, and United Nations Development Programme (UNDP). Ha Noi, Viet Nam.

Comparison between traditional security and nontraditional .... https://www.researchgate.net/figure/Comparison-between-Traditional-Security-and-Nontraditional-Security_tbl1_334290018

Detraz, N. 2009. Environmental security and gender: Necessary shifts in an evolving debate. *Security Studies 18*(2), 345–369.

Dower, N. (1995). Peace and security: Some conceptual notes. In M. Salla, W. Tonetto, & E. Martinez (Eds.), *Essays on peace: Paradigms for global order* (pp. 18–23). Central Queensland University Press.

Environmental Security. Dec 2019. International encyclopedia of human geography (2nd ed.). Jon Barnett.

Environmental security, critical infrastructure.https://link.springer.com/chapter/10.1007/1-4020-3893-3_01

Environmental Security—an overview|ScienceDirect Topics. https://www.sciencedirect.com/topics/earth-and-planetary-sciences/environmental-security

Galgano, F. A., Edward, Rose, P. F. 2021. *Encyclopedia of geology* (2nd ed.).

GEF. 2015. *Environmental security: dimensions and priorities*
Graeger, N. 1996. Environmental security? *Journal of Peace Research*, 109–116.
Hà Nội báo cáo kết quả sơ bộ việc cá chết ở Hồ Tây—Baochinhphu.vn.http://baochinhphu.vn/Hoat-dong-dia-phuong/Ha-Noi-bao-cao-ket-qua-so-bo-viec-ca-chet-o-Ho-Tay/288066.vgp
Hoang, V. H., Nguyen, M. K., Hoang T. D., Ha, M. C., Huyen, N. T., Bui, V. K., Pham, M. T., Nguyen, C. M., Chang, S. W., & Nguyen, D. D. (2024). Sources, environmental fate, and impacts of microplastic contamination in agricultural soils: A comprehensive review. *Science of the Total Environment*. 3:175276.
Hoang, V. H., Nguyen, M. K., Hoang, T. D., Ha, M. C., Huyen, N. T. T., Bui, V. K. H., Pham, M. T., Nguyen, C. M., Chang, S. W., & Nguyen, D. D. (2024). Sources, environmental fate, and impacts of microplastic contamination in agricultural soils: A comprehensive review. *Science of the Total Environment*. 175276.
Hoang, Van-Hiep, et al. (2025) Microplastic characteristics, transport, risks, and remediation in groundwater: A review. *Environmental Chemistry Letters* 1–21.
Hoang, T. D., Ky, N. M., Thuong, N. T., Nhan, H. Q., & Ngan, N.V. (2022). Artificial intelligence in pollution control and management: Status and future prospects. *Artificial Intelligence and Environmental Sustainability: Challenges and Solutions in the Era of Industry 4.0*. 23–43.
Hoang, T. D, Ha, P. T., Halog, A. B., Malla, F. A., & Bandh, S. A. (2023). Biofuel projects and current environmental policies: Vietnam's case and neighboring Asian countries. In: Biofuels in Circular Economy (Vol. 1, pp. 73–88). Springer Nature Singapore.
Hoang, T. D., Van Binh, D., Umans, L., Kantoush, S. A., & Ngan, N. V. C. 2024. Regional collaboration for water development projects for strengthening the governance of transboundary rivers in the Mekong countries. In *The Mekong river basin* (pp. 211–235). Elsevier.
Hoang, T. D., Bandh, S. A., Malla, F. A., Qayoom, I., Bashir, S., Peer, S. B., & Halog, A. (2023). Carbon-based synthesized materials for CO_2 adsorption and conversion: Its potential for carbon recycling. *Recycling, 8*(4), 53.
Hoang, T. D., & Nghiem, N. (2021). Recent developments and current status of commercial production of fuel ethanol. *Fermentation, 7*(4), 314.
Peter Hough. 2014. *Environmental security: An introduction.*
JonBarnett.2020. *Environmental security, international encyclopedia of human geography* (2nd ed.).
Levy, M. A. 1995. Is the environment a national security issue? *International Security*, *20*(2) (Fall 1995).
LUẬT BẢO VỆ MÔI TRƯỜNG—Cong bao chinh phu—Chính phủ. http://congbao.chinhphu.vn/tai-ve-van-ban-so-55-2014-qh13-4894-7352?format=pdf
Malla, F. A., Dung, T., Bandh, S. A., Malla, A. A., & Wani, S. A. (2023). Circular economy to decarbonize electricity. *Renewable energy in circular economy* (pp. 71–87). Springer International Publishing.
(PDF) Management of Nontraditional Security: A New Approach. https://www.researchgate.net/publication/334290018_Management_of_Nontraditional_Security_A_New_Approach
Minh H. N. 2017. Hong, master's thesis. Tolich river management, HSB, VNU
https://muse.jhu.edu/article/30338/summary
National Security Strategy. 2000. National security strategy archive.https://nssarchive.us/national-security-strategy-2000-2/
NATO Review No. 6. Nov–Dec 1997 (pp. 30–33). https://www.nato.int/docu/review/1997/9706-09.htm
https://nepis.epa.gov/Exe/ZyPURL.cgi?Dockey=91018N4S.TXT
Nguyen, V. C. N., Dung, H. T., Nguyen, N. S. H., & Nguyen, T. H. (2021). Green technology for sustainable development: Practice and experience of renewable energy in Vietnam. *Research Nepal Journal of Development Studies (RNJDS), 4*(2), 79–92.
Nhue—To Lich river basin—Water Environment Partnership in Asia. http://www.wepa-db.net/policies/cases/nhue_to_lich/04.htm

Nhue—To lich river basin—Water environment partnership in Asia.http://www.wepa-db.net/policies/cases/nhue_to_lich/04.htm
Our common future (Book, 1987) [WorldCat.org]. https://www.worldcat.org/title/our-common-future/oclc/15489268
Quan điểm, chủ trương của Đảng về bảo vệ môi trường - Nghiên .... https://kinhtetrunguong.vn/web/guest/nghien-cuu-trao-doi?p_p_id=101&p_p_lifecycle=0&p_p_state=normal&p_p_mode=view&p_p_col_id=column-1&p_p_col_count=1&_101_struts_action=%2Fasset_publisher%2Fview_content&_101_assetEntryId=2394651&_101_type=content&_101_urlTitle=quan-%C4%91iem-chu-truong-cua-%C4%91ang-ve-bao-ve-moi-truong
Rachel Carson. Silent Spring. http://www.rachelcarson.org/SilentSpring.aspx
Rademaker. M., Jans, K., Rõõs, H., Borum, A., van Schaik, L., Slingerland, S., della Frattina, C. F. (2016). The economics of planetary security: Climate change as an economic conflict factor. The Hague Centre for Strategic Studies.
Rwabizambuga, A. (2007). Environmental security and development. *Conflict, Security and Development, 7*(1), 201–225.
Environmental Security, by Christopher A. Scott, Bhuwan Thapa. https://www.oxfordbibliographies.com/view/document/obo-9780199363445/obo-9780199363445-0012.xml
https://e.vnexpress.net/news/news/vietnam-s-solid-waste-up-46-pct-in-decade-report-4193656.html
Svensson, U. 1988. Environmental security: A concept. *Presented at the international conference on environmental stress and security, the royal swedish academy of sciences*, Stockholm, Sweden, December 1988.
Swanström, N. (2010). Traditional and non-traditional security threats in central Asia: Connecting the new and the old. *China and Eurasia Forum Quarterly, 8*(2), 35–51.
The Concept of Environmental Security—The solutions journal. 2016. https://thesolutionsjournal.com/2016/02/22/the-concept-of-environmental-security-2/
The Meaning of Environmental Security. Ecological politics and policy in the new security era. https://books.google.com/books/about/The_Meaning_of_Environmental_Security.html?id=jheOra9RUhIC
The Millennium Project. http://www.millennium-project.org/
The Millennium Project. http://www.millennium-project.org/. (7) "(PDF) Environmental Security—ResearchGate." https://www.researchgate.net/publication/235698902_Environmental_Security
Thuong, N.T., Yoneda, M., Shimada, Y., & Matsui, Y., (2015). Assessment of trace metal contamination and exchange between water and sediment systems in the To Lich River in inner Hanoi, Vietnam. *Environmental Earth Sciences,* 73, 3925–3936.
Toàn văn Báo cáo hiện trạng môi trường quốc gia giai đoạn ...—CEID. http://ceid.gov.vn/toan-van-bao-cao-hien-trang-moi-truong-quoc-gia-giai-doan-2011-2015/
Tran, T. K., Nguyen, M. K., Lin, C., Hoang, T. D., Nguyen, T. C., Lone, A. M., Khedulkar, A. P., Gaballah, M. S., Singh, J., & Chung, W. J. (2024a). Review on fate, transport, toxicity and health risk of nanoparticles in natural ecosystems: Emerging challenges in the modern age and solutions toward a sustainable environment. *Science of the Total Environment* 912,169331.
Tran, T. K., Huynh, L., Nguyen, H. L., Nguyen, M. K., Lin C., Hoang, T. D., Hung, N. T., Nguyen, X. H., Chang, S. W., & Nguyen, D. D. (2024b) Applications of engineered biochar in remediation of heavy metal (loid) s pollution from wastewater: Current perspectives toward sustainable development goals. Science of the Total Environment. 20:171859.
Trombetta, M. J. (2008). Environmental security and climate change: Analyzing the discourse. *Cambridge Review of International Affairs, 21*(4), 585–602.
UNEP—UN Environment Programme. https://www.unep.org/
United Nations Development Programme (UNDP). 1994. *Human development report 1994*. Oxford Univ. Press.
US EPA. https://www.epa.gov/. Strengthening National Security Through the Environment ... - epa nepis.

Vietnam: Toward a Safe, Clean, and Resilient Water System. https://documents.worldbank.org/curated/en/589341559130979599/Vietnam-Toward-a-Safe-Clean-and-Resilient-Water-System
Vietnam : Toward a Safe, Clean, and Resilient Water System.https://documents.worldbank.org/curated/en/589341559130979599/Vietnam-Toward-a-Safe-Clean-and-Resilient-Water-System
Vietnam: Urban wastewater review—World Bank Group. https://www.worldbank.org/en/country/vietnam/publication/vietnam-urban-wastewater-review
World Bank Group, (2019). Vietnam: Toward a safe, clean, and resilient water system. World Bank.
Zurlini G, Müller, F (2008). Environmental security. *Encyclopedia of Ecology, 1*(2):1350–1356.

Microalgae and Cyanobacteria-Based Sustainable Solutions for Agriculture

Denzel Christopher Makepa, Luis Fernando Flores, and Arun T. Ram

Abstract Climate change, the increasing environmental impacts of agricultural intensification, and the growing global demand for food highlight the need for sustainable alternatives to maintain and boost agricultural productivity. As a result, microalgae and cyanobacteria have emerged as renewable resources with attractive potential for modern agriculture. On one hand, they can sequester carbon and contribute to sustainable agriculture by reducing greenhouse gas emissions. Moreover, they can be produced from various agricultural residues and wastewater. In terms of enhancing agricultural productivity, these microorganisms improve crop growth and soil health by fixing nitrogen, solubilizing potassium and phosphorus, releasing trace elements from the soil, and converting organic matter into usable nutrients. They also release bioactive substances, vitamins, and hormones that promote plant growth and control pests. Therefore, this chapter focuses on the effects of cyanobacteria and microalgae to improve soil fertility and soil quality while promoting plant growth. Moreover, their production is described, including formulation and global market aspects the technical and economic challenges of this environmentally friendly microbial alternative are also analyzed. In this context, the chapter summarizes the advances in microalgae and cyanobacteria technology as sustainable bioproduct platforms for the future of climate-ready agriculture.

D. C. Makepa
Department of Fuels and Energy Engineering, Chinhoyi University of Technology, Chinhoyi, Zimbabwe

L. F. Flores (✉)
Club de Biotecnología Agrícola y Ambiental, Villa Regina, Río Negro, Argentina
e-mail: lfflores.lab@gmail.com

A. T. Ram
PG & Research Department of Botany, St. Peter's College, Kolenchery, Ernakulam District, Kerala, India

M. Shaaban (ed.), *Soils and Sustainable Agriculture*, Frontier Studies in Soil Science,
https://doi.org/10.1007/978-3-031-91114-9_41

Graphical Abstract

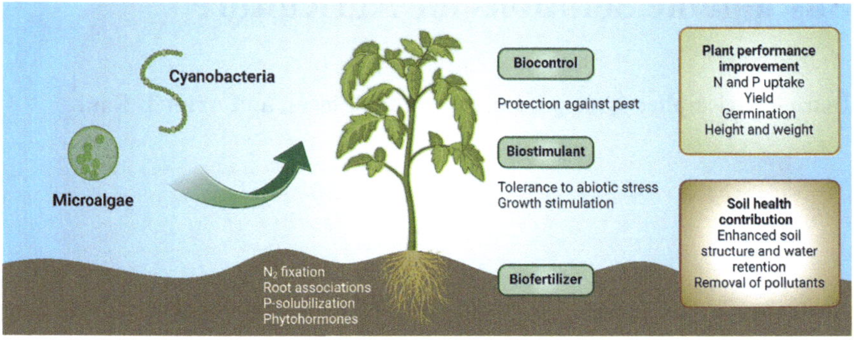

Keywords Microalgae · Cyanobacteria · Sustainable agriculture · Sustainable production

1 Introduction

The increase in CO_2 emissions has been impacted by changes in agriculture and land use, deforestation, wetland drainage, and inadequate soil management practices, all of which have disrupted the carbon sequestration capacity of agricultural soils. Moreover, the growing global population, both biotic and abiotic stressors, soil degradation, and climate change, are posing serious challenges to food security. Therefore, the agricultural sector is under enormous pressure to develop innovative solutions that can withstand the negative impacts of climate change (Kamyab et al., 2024; Shao, 2024).

Since reducing agricultural greenhouse gas (GHG) emissions is essential to achieving overall climate policy goals, several strategies have been developed to mitigate the effects of climate change, such as composting and nutrient recycling, conservation agriculture, and agricultural residue management. All of these practices promote a more sustainable and resilient agricultural system and can contribute to the reduction of greenhouse gas emissions (Kreft et al., 2024; Nazir et al., 2024).

Recently, alternative bio-based solutions to boost crop yields and reduce GHG emissions have gained attention in sustainable agriculture. In this regard, an emerging and promising path to solving the CO_2 emissions issue lies in the untapped potential of cyanobacteria (prokaryotic) and microalgae (eukaryotic), both photosynthetic microorganisms that produce a variety of bioactivities from atmospheric carbon dioxide utilization (Mutale-Joan et al., 2023). Moreover, certain species can transform atmospheric nitrogen into amino acids, which are then converted into proteins (Gul et al., 2024) reducing the need for synthetic fertilizers that contribute to additional greenhouse gas emissions (Bhuyan et al., 2023; Jose et al., 2024). By exploiting the unique properties of cyanobacteria and microalgae, researchers have developed innovative strategies to improve plant stress tolerance, nutrient management, and even carbon sequestration (Gangl et al., 2015; Shahinnia et al., 2021). Therefore, this chapter explores the recent advances in the sustainable production of microalgae and cyanobacteria-based products for more resilient agriculture systems.

2 Potential of Microalgae and Cyanobacteria for Climate-Ready Agriculture

Microalgae and cyanobacteria represent a diverse collection of eukaryotic and prokaryotic photosynthesizing organisms capable of thriving in numerous aquatic habitats, including ponds, rivers, lakes, oceans, and wet soils, encompassing a diversity of up to 30,000 species. Due to their adaptability to thrive in diverse and often harsh conditions, they can be prevalent across a wide range of ecosystems and environmental stressors, including fluctuations in temperature, water availability, and nutrient levels (Sarsekeyeva et al., 2024).

These microorganisms typically hold notable nitrogen (N) levels, ranging between 6–8%, and phosphorus (P) levels, between 1–3%, measured by dry weight. By gradually releasing nutrients, microalgae can efficiently reduce nutrient losses. In addition, beyond providing vital nutrients, microalgal compounds have shown the ability to enhance soil fertility, strengthen plant defenses against abiotic stress such as drought, salinity, and extreme temperatures, activate defense systems against pathogens, and improve nutrient absorption from the soil. Microalgae contain numerous advantageous biomolecules, including polysaccharides, phytohormones, soluble amino acids (AA), micronutrients, and additional signaling molecules for plant growth and soil health promotion. Moreover, microalgae and cyanobacteria can be used to produce high-quality animal feed to meet the challenges of livestock nutrition in a changing climate (Santini et al., 2021; Zhang et al., 2024).

Both green microalgae and cyanobacteria can enhance soil organic carbon by absorbing atmospheric CO_2 and using it during photosynthesis to produce nutrient-dense biomass. They also have the potential to stabilize soil organic carbon levels, contributing to the mitigation of farmland degradation. As a result, they have an important role in enhancing soil structure, boosting fertility, and improving crop yield, along with increasing tolerance to abiotic stress in agriculture (Mutale-Joan et al., 2023). Regarding CO_2 fixation, they play key roles in **mitigating greenhouse gas** emissions from agriculture (Morales et al., 2019). Furthermore, some studies have investigated the use of these microorganisms in biochar production to increase long-term carbon sequestration in soils (Hussain et al., 2018). Some species can be used to treat wastewater and reduce methane emissions from animal husbandry (Suresh Kumar et al., 2015; Wollmann et al., 2019). In addition, microalgal biomass can efficiently trap other greenhouse gases, such as nitrous oxide, contributing to the

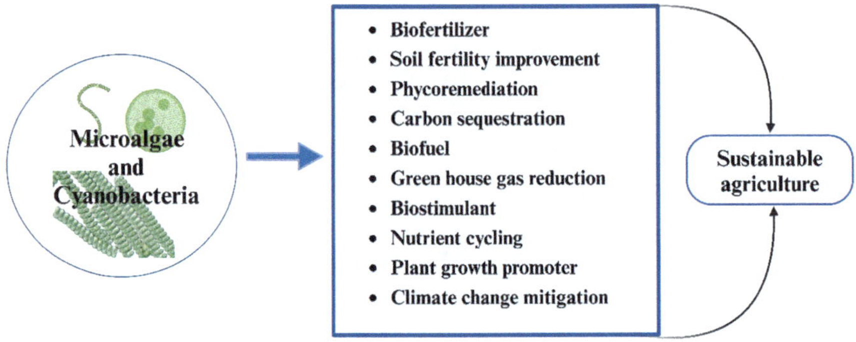

Fig. 1 Potential applications of microalgae and cyanobacteria in sustainable agriculture

mitigation of climate change effects (Zhang et al., 2024). For instance, the symbiotic relationship between the cyanobacterium *Anabaena azollae* and the water fern *Azolla* has been successfully utilized in paddy fields to increase the nitrogen content (Adhikari et al., 2020; Yadav et al., 2014). Moreover, microalgae and cyanobacterial biomass and extracts can be used as fertilizers, and represent a promising alternative for enhancing soil fertility and water retention to promote plant growth, boost crop yields, and help alleviate salt stress, in a cost-effective and environmentally friendly way (Gul et al., 2024). Consequently, with increasing environmental stresses caused by climate change, microalgae and cyanobacteria plays significant roles in improving plant resilience (Gonçalves, 2021; González-Pérez et al., 2022; Lutzu et al., 2024; Parmar et al., 2023; Sarsekeyeva et al., 2024).

Figure 1 illustrates the potential applications of cyanobacteria and microalgae in sustainable agriculture.

3 Emerging Microalgae and Cyanobacteria-Based Products for Agriculture

Microalgae and cyanobacteria hold significant potential to revolutionize agricultural practices through their diverse and sustainable applications. These microorganisms are rich in bioactive compounds and essential nutrients, making them valuable resources for various agricultural products. One of the primary products derived from microalgae and cyanobacteria is biofertilizers, which enhance soil fertility through nitrogen fixation and nutrient cycling (Ramakrishnan et al., 2023). Biostimulants derived from these microorganisms promote plant growth and stress tolerance by producing phytohormones and other growth-enhancing substances (Sible et al., 2021). Pesticides based on microalgae and cyanobacteria offer a natural alternative to chemical pesticides, leveraging bioactive compounds to protect crops from pests and diseases (Gonçalves et al., 2023). Additionally, these microorganisms are used as animal feed additives, providing a high-protein, nutrient-rich supplement that improves livestock growth and health (Andrade et al., 2018). However, the potential of cyanobacteria and microalgae applications extends beyond the agricultural sector. The cultivation of these organisms can produce biofuels, which can help create a more sustainable source of energy and reduce the agricultural sector's dependence on fossil fuels. Biofuels such as biogas, biohydrogen, and biodiesel derived from microalgae and cyanobacteria present renewable energy solutions that can power agricultural machinery and on-farm operations (Chisti, 2019; Ng et al., 2024). Moreover, microalgae biomass can be used to develop biobased materials and chemicals, fostering the growth of a circular economy in agriculture. Detailed descriptions of the most important cyanobacterial and microalgae-based bioproducts are presented below.

3.1 Biofertilizers

Microalgae and cyanobacteria have emerged as promising biofertilizers due to their ability to enhance soil fertility through various mechanisms. These microorganisms play a key role in nutrient cycling and nitrogen fixation and nutrient cycling, making essential nutrients more available to plants. Nitrogen fixation involves converting atmospheric nitrogen into ammonia, which plants can readily absorb (Ramakrishnan et al., 2023). Cyanobacteria, such as *Anabaena* and *Nostoc*, have specialized cells called heterocysts that facilitate this process. In addition to nitrogen fixation, these microorganisms help in decomposing organic matter, thereby releasing essential nutrients like phosphorus and potassium into the soil. As a result, nutrient availability and overall soil health are enhanced (Bhardwaj et al., 2024).

Biofertilizers derived from microalgae and cyanobacteria are available in various formulations, including liquid and solid forms. Liquid biofertilizers are often sprayed directly onto the soil or plants, providing immediate nutrient availability. Solid formulations, such as granules or powders, are mixed with the soil and offer a slow-release of nutrients over time. Both types of biofertilizers are designed to improve soil structure, increase water retention, and enhance microbial activity (Gonçalves et al., 2023).

Biofertilizers offer several advantages over conventional chemical fertilizers. Firstly, they are sustainable and environmentally friendly, reducing the reliance on synthetic chemicals that can cause soil degradation and water pollution. The use of biofertilizers also promotes the growth of beneficial soil microorganisms, improving soil health and plant resilience. Furthermore, biofertilizers contribute to the reduction of greenhouse gas emissions associated with the production and application of chemical fertilizers (Chakraborty & Akhtar, 2021).

Numerous field trials and studies have demonstrated the efficacy of microalgae and cyanobacteria-based biofertilizers. For instance, a study conducted in South China evaluated the use of cyanobacterial biofertilizers in rice fields. The results showed that the combined treatment of cyanobacteria and 50% topdressing fertilizer increased rice yields by 13.9–22% compared with the conventional fertilizer treatment (Bao et al., 2021). Similarly, research in China demonstrated that applying microalgal biofertilizers to wheat crops enhanced grain yield and nutrient content, highlighting their potential in sustainable agriculture (Zhou et al., 2023). In another study, the application of Spirulina-based biofertilizers in tomato cultivation resulted in increased plant growth, higher fruit yield, and improved nutrient uptake (Gonçalves et al., 2023). These field trials and research studies underscore the potential of microalgae and cyanobacteria-based biofertilizers to enhance crop productivity while promoting sustainable agricultural practices.

3.2 *Biostimulants*

Biostimulants are substances or microorganisms that, when applied to plants or the rhizosphere, stimulate natural processes to enhance nutrient uptake, nutrient efficiency, stress tolerance, and crop quality. Unlike traditional fertilizers, which directly provide nutrients to plants, biostimulants work by improving the plant's overall health and growth capacity, while supporting plant resilience against various stressors (Sible et al., 2021). Microalgae and cyanobacteria act as biostimulants through several mechanisms, including the production of phytohormones and enhancing stress tolerance, such as auxins, gibberellins, cytokinins, and abscisic acid, which regulate plant growth and development (Ramakrishnan et al., 2023). For instance, auxins promote root elongation, while cytokinins stimulate cell division and shoot formation. Additionally, microalgae and cyanobacteria can produce polysaccharides and other bioactive compounds that improve soil structure and water retention, thereby enhancing the plant's ability to withstand abiotic stresses such as drought and salinity (Gonçalves et al., 2023).

Traditional biostimulants are often derived from plant extracts, seaweed, or humic substances (Bell et al., 2022). While these biostimulants are effective, microalgae and cyanobacteria offer distinct advantages. They can be cultivated on a large scale using non-arable land and wastewater, making them highly sustainable. Their rapid growth rate and the ability to produce biomass year-round further enhance their potential as reliable sources of biostimulants (González-Pérez et al., 2022).

Several studies have demonstrated the effectiveness of microalgae and cyanobacteria-based biostimulants in various crops. For instance, a study on tomato plants treated with microalgal extracts showed significant improvements in growth, fruit yield, and nutrient content compared to untreated plants (Mutale-joan et al., 2020). Another study on wheat demonstrated that the application of cyanobacterial biostimulants enhanced grain yield and improved stress tolerance under drought conditions (Elnajar et al., 2024). In rice cultivation, the use of cyanobacterial biostimulants has been shown to increase biomass production and improve grain quality (Singh et al., 2022). Similarly, in maize, the application of microalgae-based biostimulants resulted in increased root and shoot growth, leading to higher overall crop productivity (Martini et al., 2021). The latest research shows that bio-stimulant effects derived from microalgae varies according the application approaches tested, being the foliar spray method the most efficient (Parmar et al., 2023). Thus, microalgae and cyanobacteria-based biostimulants are valuable tools for supporting the shift towards more eco-friendly and efficient farming practices in modern agriculture.

3.3 Biopesticides

Climate change can lead to altered pest and disease dynamics in agricultural systems. In this regard, microalgae and cyanobacteria offer potential solutions through the production of bioactive compounds with pesticidal properties (Costa et al., 2019; Gonçalves, 2021; Ramakrishnan et al., 2023). These compounds include secondary metabolites such as toxins, enzymes, and antimicrobial substances that can inhibit the growth of or kill pests. The mechanisms by which microalgae and cyanobacteria-based biopesticides act against pests are diverse. Insecticidal activity is achieved through compounds such as alkaloids, terpenoids, and polyketides, which disrupt the nervous system of insects, leading to paralysis and death. Antimicrobial activity is provided by antimicrobial peptides and enzymes produced by these organisms, which inhibit the growth of pathogenic bacteria and fungi, protecting plants from infections (Akmukhanova et al., 2023). Therefore, biopesticides derived from microalgae and cyanobacteria offer several advantages over conventional chemical pesticides. One of the primary benefits is biodegradability, which reduces the risk of environmental contamination and persistence in the ecosystem. Additionally, these biopesticides are generally less toxic to non-target organisms, including humans, beneficial insects, and wildlife, promoting a safer agricultural environment (Costa et al., 2019).

Biopesticides derived from microalgae and cyanobacteria are available in various forms, including algal extracts and cyanobacterial metabolites. Microalgae, such as *Chlorella*, *Scenedesmus*, and cyanobacteria like *Nostoc* and *Anabaena*, have been identified as sources of these potent biopesticides. The diversity and efficacy of these compounds offer a promising alternative to conventional chemical pesticides, providing a more sustainable and eco-friendly approach to pest management (Gonçalves et al., 2023). Algal extracts contain a mixture of bioactive compounds and can be used as foliar sprays or soil treatments to control pests. Cyanobacteria produce a range of metabolites, such as allelochemicals, that have pesticidal properties. These metabolites can be used directly or formulated into biopesticides (Asimakis et al., 2022; Gonçalves et al., 2023). For instance, *Nostoc muscorum* produces nostocyclopeptides, which have demonstrated antifungal and insecticidal activities (Yadav et al., 2022).

Numerous studies have demonstrated the efficacy of microalgae and cyanobacteria-based biopesticides in various crops. For instance, a study by (Bao et al., 2021) evaluated the potential applicability of *Anabaena variabilis SCAU30* as a biopesticide in rice fields of South China. Their results showed that the control efficacies of *A. variabilis* SCAU30 against rice sheath blight and bacterial blight were

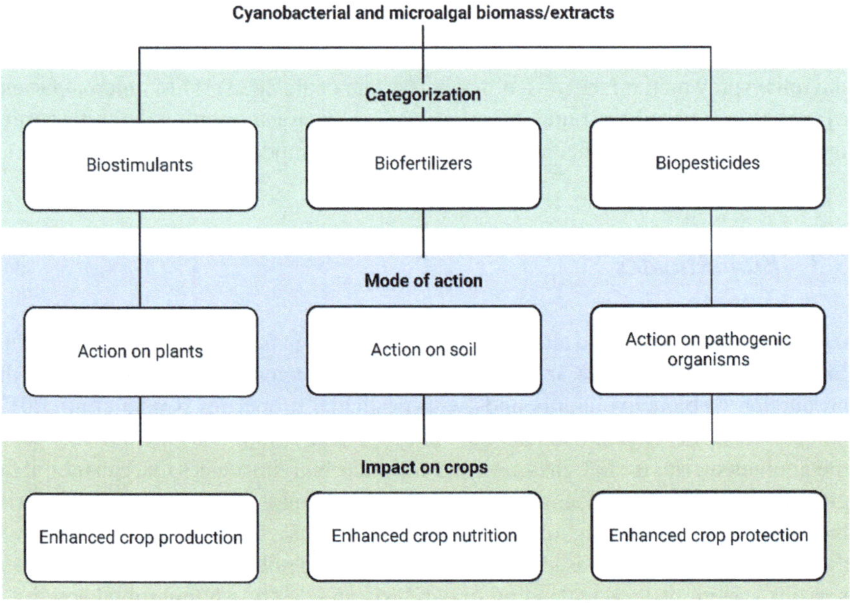

Fig. 2 Categorization of primary activities associated with algal and cyanobacterial biomass and extracts in crop production. Adapted from (Gonçalves, 2021)

62.3 ± 8.9% and 45.3 ± 5.6%, respectively. Another study by (Rong et al., 2020) investigated the antifungal properties of *Bacillus safensis* B21, a bacterial strain isolated from the fruit of sweet osmanthus. This strain produces metabolites such as antibiotics and biosurfactants, which confer antibacterial, antifungal, and cytotoxic properties. The culture filtrate of *B. safensis* B21 exhibited antifungal activity against *Magnaporthe oryzae*, the pathogen responsible for rice blast disease.

By incorporating these natural biopesticides into integrated pest management strategies, farmers can potentially reduce their impact on the environment while adapting to changing pest pressures, thereby contributing to the development of climate-resilient agricultural systems.

The impact of green and blue algae biomass and extracts on agricultural crops is illustrated in Fig. 2, which categorizes the bioproducts into biostimulants, biofertilizers, and biopesticides.

3.4 Animal Feed Additives

Microalgae are rich in proteins, essential amino acids, vitamins, and minerals, making them an excellent alternative to conventional livestock feed sources due to their rich nutritional profile (Bischoff, 2024). Microalgae such as *Chlorella*, are particularly notable for their high protein content, which can range up to 70% of their dry weight (Muys et al., 2019). Additionally, they are abundant sources of vitamins A, B, C, and E, as well as minerals like iron, magnesium, and calcium. Cyanobacteria, like *Arthrospira* (commonly known as *Spirulina*), also provide these vital nutrients, enhancing the overall dietary intake for animals (Andrade et al., 2018).

The benefits of incorporating microalgae and cyanobacteria into livestock feed are substantial. Studies have shown that these feed additives can significantly improve growth rates and overall health in animals. For example, the high protein content supports muscle development and weight gain, while the vitamins and minerals boost immune function and reduce the incidence of diseases. Additionally, the presence of bioactive compounds in microalgae and cyanobacteria can enhance gut health, leading to better nutrient absorption and improved feed efficiency (Bature et al., 2022).

When compared to conventional feed additives, microalgae and cyanobacteria offer several advantages. Traditional feed additives often rely on fishmeal and soy protein, which can be expensive and environmentally taxing due to overfishing and deforestation. In contrast, microalgae and cyanobacteria can be sustainably cultivated using non-arable land and wastewater, reducing the environmental footprint. Moreover, these microorganisms have a faster growth rate and higher yield potential, making them a more economical and reliable source of feed additives (Mutale-Joan et al., 2023).

Several case studies illustrate the successful application of these feed additives in livestock. For instance, a study on broiler chickens supplemented with *Spirulina* showed improved growth performance, enhanced immune response, and better meat quality compared to those fed conventional diets (El-Shall et al., 2023). Similarly, dairy cows fed with microalgal supplements exhibited increased milk yield and improved milk composition, highlighting the benefits for dairy production (Orzuna-Orzuna et al., 2024). In aquaculture, fish fed with *Chlorella* supplements demonstrated higher growth rates and improved health status, proving the efficacy of microalgae in enhancing fish production (Alagawany et al., 2021). As research and development continue, the use of microalgae and cyanobacteria-based feed additives might revolutionize animal nutrition, while supporting more sustainable and efficient livestock farming practices.

3.5 *Biofuels for the Agricultural Sector*

Biofuels play a key role in advancing sustainable energy solutions within the agricultural sector. These renewable energy sources are derived from biological materials, offering a cleaner and more sustainable alternative to fossil fuels. The agricultural sector, which is energy-intensive, can benefit significantly from biofuels by reducing its carbon footprint and enhancing energy security. Among the various sources of biofuels, microalgae and cyanobacteria have emerged as promising candidates due to their high biomass productivity and versatile applications (Singh et al., 2023).

Microalgae and cyanobacteria can be used to produce several types of biofuels, including biogas for electricity generation, biohydrogen, and biodiesel. Biogas, primarily composed of methane and carbon dioxide, is produced through the anaerobic digestion of algal biomass. This biogas can be used to generate electricity, providing a renewable energy source for on-farm use and reducing reliance on grid electricity. Biohydrogen, produced through photobiological processes, is another promising fuel derived from microalgae and cyanobacteria. Hydrogen gas is a clean fuel with water as its only by-product, making it an environmentally friendly energy source for agricultural machinery and equipment. Biodiesel, produced through the transesterification of algal lipids, can be used to power tractors, harvesters, and other farm machinery, providing a sustainable alternative to conventional diesel (Jabłońska-Trypuć et al., 2023; Sathya et al., 2023).

The production processes and technologies for biofuels from microalgae and cyanobacteria are diverse and evolving. For biogas production, algal biomass is subjected to anaerobic digestion in bioreactors, where microorganisms break down the organic matter to produce methane-rich biogas (De Jesús Martínez-Roldán et al. 2024). This biogas can be used to generate electricity for farm operations, including lighting, heating, and irrigation systems (De Jesús Martínez-Roldán et al. 2024). Biohydrogen production involves cultivating microalgae or cyanobacteria under specific conditions that promote hydrogenase activity, leading to the generation of hydrogen gas (Li et al., 2022). Biodiesel production typically involves the extraction of lipids from algal cells, followed by a chemical process called transesterification, where the lipids are converted into biodiesel and glycerol (Vasistha et al., 2021). Resulting biodiesel can be used as fuel for tractors, harvesters, and other agricultural machinery, reducing dependence on fossil fuels and lowering greenhouse gas emissions (Li et al., 2022).

Several case studies highlight the successful implementation of biofuels derived from microalgae and cyanobacteria in agriculture. For example, a study conducted on a dairy farm in Northern Ireland demonstrated the feasibility of using biogas produced from biomass to generate electricity for farm operations, resulting in significant energy savings and a reduction in carbon emissions (Purdy et al., 2018). Another study in India explored the use of algal biodiesel to power an agricultural tractor, showing promising results in terms of fuel efficiency and emission reductions (Karthikeyan et al., 2020). These examples underscore the potential of biofuels from microalgae and cyanobacteria to transform the agricultural sector by providing sustainable and eco-friendly energy solutions and supports the transition to more sustainable and resilient agricultural energy systems.

4 Commercial Development of Microalgae and Cyanobacteria-Based Agricultural Products

The growth prospects for bioproducts based on cyanobacteria and microalgae to remain optimistic. This positive outlook is reinforced by the taxonomy diversity of these microorganisms, their biochemical compositions, and associated bioactivities, which present significant potential applications in various markets. The majority of the microalgal biomass produced globally is currently utilized in food and feed sectors. However, emerging markets are beginning to develop, such as cosmetics, agriculture, pharmaceuticals, and biomedicine. For instance, the leading five producing countries of microalgae in the fertilizers sector include France, Germany, China, Japan, and the United States. Notably, companies like Allmicroalgae, Algenol, Cellana LLC, Algatec, and Nikken Sohonsha Corp are located in three of these nations, as illustrated in Table 1. Emerging strains of microalgae and cyanobacteria that serve as the basis for biofertilizers include species like *Spirulina platensis*, *Chlorella vulgaris*, *Euglena*, *Nannochloropsis*, and *Nostoc* (Cunha et al., 2024; Lutzu et al., 2024; Parmar et al., 2023).

5 Benefits

The sustainability and eco-friendly benefits of microalgae and cyanobacteria-based products are substantial. Their cultivation requires minimal land and can utilize renewable wastewater resources, making it an environmentally sustainable option. Additionally, the cultivation of microalgae in marginal land areas, although not directly offers the opportunity to diversify agricultural income without competing with food crops (Salam et al., 2016). Integrating microalgae cultivation into modern agricultural systems, such as aquaponics or cover crops, can increase carbon sequestration while providing additional benefits such as biofertilization and soil improvement (Ravindran et al., 2016; Renganathan et al., 2024). The ability of biofertilizers from microalgae and cyanobacteria to fix nitrogen, cycle nutrients, and improve soil health makes them a valuable alternative to conventional chemical fertilizers. Consequently, through the incorporation of microalgae and cyanobacteria into agricultural practices, farmers can potentially reduce their carbon footprint and contribute to efforts to mitigate climate change. Moreover, the versatility of microalgae and cyanobacteria extends beyond their applications in crop improvement. These organisms are also being studied for their potential to produce a variety of valuable compounds, from biofuels to high-value nutraceuticals (Krohn et al., 2022). Biofuels derived from microalgae and cyanobacteria offer several benefits over fossil fuels. These algae-based fuels are renewable and can be produced sustainably using non-arable land and wastewater, thereby avoiding competition with food crops and reducing environmental impact. They also have a lower carbon footprint compared to fossil fuels, as the carbon dioxide released during their combustion is

Table 1 Microalgae and cyanobacteria-based bioproduct-producing companies. Adapted from (Lutzu et al., 2024)

Country	Company	References
China	Panpan Industry	Panpan Industry, 2024
	New-nutri	New-nutri, 2024
France	Terra Aquatica T.A	Terra Aquatica T.A. 2024
Germany	Growland	Growland, 2024
Guatemala	Enlasa	Enlasa, 2024
Israel	Haifa Group	Haifa Group, 2024
Mexico	Algamar	Algamar, 2024
Norway	Algea	Algea, 2024
Poland	Growtent	Growtent, 2024
Portugal	Algatec	Algatec, 2024
	Allmicroalgae	Allmicroalgae, 2024
Spain	AlgaEnergy	AlgaEnergy, 2024
	Hispanagar S A	Hispanagar S A 2024
	Aqualgae	Aqualgae, 2024
	CARTIF	CARTIF, 2024
	Biorizon Biotech	Biorizon Biotech, 2024
United Kingdom	Envii	Envii, 2024
United States	Algenol	Algenol, 2024
	Cellana LCC	Cellana LCC 2024
	TrueAlgae	TrueAlgae, 2024
	Heliae	Heliae, 2024
	BOC sciences	BOC Sciences, 2024
	Enlightened soil corp	EnlightenedSoil, 2024

offset by the carbon dioxide absorbed during the growth of the algae or cyanobacteria. This results in a closed carbon cycle that helps mitigate climate change (Bhardwaj et al., 2020). Moreover, certain species of microalgae and cyanobacteria can be used for phytoremediation of contaminated soils and water, thereby contributing to the recovery of degraded agricultural land. This application is especially relevant in areas where increased pollution or salinization occurs due to changing climatic conditions (Hooda et al., 2023).

6 Challenges

To fully unlock the agricultural potential of microalgae and cyanobacteria, several significant challenges must be addressed. Chief among these is the high cost of production, which remains a major barrier to the widespread application of microalgae. Utilizing recycled nutrient solutions from wastewater for microalgae cultivation can help reduce production costs and minimize environmental pollution (Zhang et al., 2024). Additionally, advancements are needed in optimizing cultivation systems for large-scale production, improving downstream processing technologies to ensure the economic feasibility of microalgal-based products, and tailoring strains for specific agricultural uses (Vo et al., 2019). Expensive and energy-intensive unit operations also contribute significantly to overall production costs. For this reason, higher-value applications, such as producing bioactive compounds for nutraceutical, pharmaceutical, and cosmetic industries, are often prioritized over agricultural uses (Gonçalves, 2021). Moreover, the transition from laboratory settings to field applications presents further obstacles, as strains or consortia that perform well in vitro often exhibit poor results in real-world agricultural environments. Key challenges include ensuring the survival, colonization, and interaction of microalgae and cyanobacteria with host plants. Field trials are complicated by various edaphic factors, such as adverse biotic interactions with resident soil microbiomes that can lead to competition or antagonism, as well as fluctuating soil conditions, including pH, moisture, temperature, and salinity (Parmar et al., 2023). Thus, the benefits of using microalgae and cyanobacteria in agriculture depends on optimal concentrations for promoting plant growth or providing biocontrol. Incorrect dosing can negatively impact plants, potentially leading to toxicity, delayed growth, or death. Finally, further investigation is needed to evaluate the long-term effects of these organisms on ecosystem dynamics, soil health, and potential risks associated with their use in agriculture. The vast diversity of described photosynthetic organisms, along with the wide array of metabolites they produce, complicates a thorough understanding of these interactions, thereby limiting their agricultural applications (Gonçalves, 2021). Biosafety considerations, particularly for human health, are critical, as some strains produce toxins, such as microcystins, nodularins, and anatoxins, which can adversely affect human neurological functions and pose significant risks. As such, the toxicological properties of new algae-based bioproducts must be rigorously evaluated before they can be widely adopted (Lutzu et al., 2024; Parmar et al., 2023).

7 Future Prospects

The future prospects for the use of microalgae and cyanobacteria in sustainable agriculture are promising, but there are still several technical and economic challenges that need to be addressed. One key area is the diversification and improvement of the product portfolio. While biofertilizers, biostimulants, and animal feed additives are well-established applications, there is significant potential to develop new value-added products, such as soil amendment materials, and high-value compounds for the pharmaceutical and nutraceutical industries. To achieve this, these resilient microorganisms can be engineered to increase the range of valuable compounds for sustainable industries beyond the agricultural sector (Shahinnia et al., 2021). Thus, targeted genetic engineering and metabolic engineering of microalgae and cyanobacteria could enhance the production of specific compounds of interest and expand the range of applications. Another key factor for future development is the optimization of production processes to improve the cost-effectiveness and scalability of microalgae and cyanobacteria-based products. This includes advances in cultivation systems, harvesting, and processing technologies, and the valorization of byproducts to maximize resource efficiency. Moreover, integrating microalgae and cyanobacteria-based solutions into precision agriculture and smart farming practices will also be crucial for the future. Through leveraging digital technologies, sensors, and data analytics, these microbial platforms can be tailored to specific crop needs, soil conditions, and environmental factors, optimizing their performance and reducing the environmental footprint of agriculture.

8 Conclusion

As climate change increasingly threatens global food security, the integration of microalgae and cyanobacteria into agricultural systems represents a promising avenue for enhancing sustainability, efficiency, and resilience in food production. Their capacity for carbon sequestration directly addresses the critical need to mitigate greenhouse gas emissions. Moreover, the inherent stress tolerance mechanisms of these microorganisms can be harnessed to bolster crop resilience against extreme weather events, soil salinization, and water scarcity challenges that are intensifying due to climate change. Beyond their agricultural benefits, microalgae and cyanobacteria play vital roles in phytoremediation, biofuel production, and as nutritional supplements, underscoring their broader significance. However, widespread adoption will require careful attention to regulatory frameworks and public acceptance. By capitalizing on the unique properties of these microorganisms, we can advance an agricultural paradigm that not only adapts to the realities of climate change but also contributes to its mitigation, thereby ensuring food security for future generations.

References

Adhikari, K., Bhandari, S., Acharya, S. (2020). An overview of azolla in rice production: A review. *Reviews in Food and Agriculture (RFNA)*, *2*(1), 04–08. https://doi.org/10.26480/rfna.01.2021.04.08

Akmukhanova, N. R., Leong, Y. K., Seiilbek, S. N., Konysbay, A., Zayadan, B. K., Sadvakasova, A. K., Sarsekeyeva, F. K., Bauenova, M. O., Bolatkhan, K., Alharby, H. F., Chang, J.-S., & Allakhverdiev, S. I. (2023). Eco-friendly biopesticides derived from CO_2-Fixing cyanobacteria. *Environmental Research, 239*, Article 117419. https://doi.org/10.1016/j.envres.2023.117419

Alagawany, M., Taha, A. E., Noreldin, A., El-Tarabily, K. A., & Abd El-Hack, M. E. (2021). Nutritional applications of species of Spirulina and Chlorella in farmed fish: A review. *Aquaculture, 542*, Article 736841. https://doi.org/10.1016/j.aquaculture.2021.736841

AlgaEnergy. (2024). https://ag.algaenergy.com/

Algamar. (2024). https://grupoalgamar.com/division/agriculture/

Algatec. (2024). Algatec. https://a4f.pt/en/projects/algatec

Algea. (2024). https://www.algea.com/

Algenol. (2024). https://www.algenol.com/

Allmicroalgae. (2024). https://www.allmicroalgae.com/en/

Andrade, L. M., Andrade, C. J., Dias, M., Nascimento, C. A., Mendes, M. A. (2018). Chlorella and spirulina microalgae as sources of functional foods, nutraceuticals, and food supplements; An overview. *MOJ Food Process Technol, 6*(1). https://doi.org/10.15406/mojfpt.2018.06.00144

Aqualgae. (2024). https://aqualgae.com/es/empresa/

Asimakis, E., Shehata, A. A., Eisenreich, W., Acheuk, F., Lasram, S., Basiouni, S., Emekci, M., Ntougias, S., Taner, G., May-Simera, H., Yilmaz, M., & Tsiamis, G. (2022). Algae and their metabolites as potential bio-pesticides. *Microorganisms, 10*(2), 307. https://doi.org/10.3390/microorganisms10020307

Bao, J., Zhuo, C., Zhang, D., Li, Y., Hu, F., Li, H., Su, Z., Liang, Y., & He, H. (2021). Potential applicability of a cyanobacterium as a biofertilizer and biopesticide in rice fields. *Plant and Soil, 463*(1–2), 97–112. https://doi.org/10.1007/s11104-021-04899-9

Bature, A., Melville, L., Rahman, K. M., & Aulak, P. (2022). Microalgae as feed ingredients and a potential source of competitive advantage in livestock production: A review. *Livestock Science, 259*, Article 104907. https://doi.org/10.1016/j.livsci.2022.104907

Bell, J. C., Bound, S. A., & Buntain, M. (2022). Biostimulants in agricultural and horticultural production. In I. Warrington (Ed.), *Horticultural reviews* (1st ed., pp. 35–95). Wiley.

Bhardwaj, A., Singh, P., Gupta, N., Bhattacharjee, S., Srivastava, A., Parida, A., Mishra, A. K. (2024). Cyanobacteria: a key player in nutrient cycling. In *Cyanobacteria* (pp. 579–596). Elsevier.

Bhardwaj, N., Agrawal, K., Verma, P. (2020). Algal biofuels: an economic and effective alternative of fossil fuels. In Srivastava, N., Srivastava, M., Mishra, P. K., Gupta, V. K. (Eds.), *Microbial strategies for techno-economic biofuel production* (pp. 207–227). Springer Singapore.

Bhuyan, P. P., Nayak, R., Jena, M., & Pradhan, B. (2023). Convoluted role of cyanobacteria as biofertilizer: An insight of sustainable agriculture. *Vegetos, 36*(2), 309–321. https://doi.org/10.1007/s42535-022-00415-1

Biorizon Biotech. (2024). https://www.biorizon.es/en/

Bischoff, L.-S. (2024). Algae-based food technologies. In *Algae as a natural solution for challenges in water-food-energy nexus: Toward carbon neutrality* (pp. 639–649).

BOC Sciences. (2024). https://bio-fermen.bocsci.com/

CARTIF. (2024). https://www.cartif.es/en/about-us/

Cellana, L. C. C. (2024). https://cellana.com/

Chakraborty, T., Akhtar, N. (2021). Biofertilizers: prospects and challenges for future. In Inamuddin Ahamed, M. I., Boddula, R., Rezakazemi, M. (Eds.), *Biofertilizers* (1st edn, pp. 575–590). Wiley.

Chisti, Y. (2019). Introduction to algal fuels. In *Biofuels from algae: biomass, biofuels, biochemicals* (2nd edn, pp. 1–31).

Costa, J. A. V., Freitas, B. C. B., Cruz, C. G., Silveira, J., & Morais, M. G. (2019). Potential of microalgae as biopesticides to contribute to sustainable agriculture and environmental development. *Journal of Environmental Science and Health, Part B: Pesticides, Food Contaminants, and Agricultural Wastes, 54*(5), 366–375. https://doi.org/10.1080/03601234.2019.1571366

Cunha, P., Costa, M., Silva, J. L., Cotas, J., Gonçalves, A. M. M., Pereira, L. (2024). A road to the microalgae market: Allmicroalgae case study. In *Algal farming systems*. Apple Academic Press.
De Jesús, M.-R., Rodríguez Rosales, M. D. J., & Valle Cervantes, S. (2024). Biogas obtained from the anaerobic digestion of microalgal biomass. In A. D. J. Martínez-Roldán (Ed.), *Biotechnological processes for green energy, and high value bioproducts by microalgae, and cyanobacteria cultures* (pp. 181–193). Springer International Publishing.
Elnajar, M., Aldesuquy, H., Abdelmoteleb, M., Eltanahy, E. (2024). Mitigating drought stress in wheat plants (*Triticum Aestivum* L.) through grain priming in aqueous extract of spirulina platensis. *BMC Plant Biology, 24*(1), 233. https://doi.org/10.1186/s12870-024-04905-z
El-Shall, N. A., Jiang, S., Farag, M. R., Azzam, M., Al-Abdullatif, A. A., Alhotan, R., Dhama, K., Hassan, F., & Alagawany, M. (2023). Potential of Spirulina platensis as a feed supplement for poultry to enhance growth performance and immune modulation. *Frontiers in Immunology, 14*, 1072787. https://doi.org/10.3389/fimmu.2023.1072787
Enlasa. (2024). https://grupoenlasa.com/algatec/
EnlightenedSoil. (2024). https://www.enlightenedsoil.com/
Envii. (2024). https://www.envii.co.uk/
Gangl, D., Zedler, J. A. Z., Rajakumar, P. D., Martinez, E. M. R., Riseley, A., Włodarczyk, A., Purton, S., Sakuragi, Y., Howe, C. J., Jensen, P. E., & Robinson, C. (2015). Biotechnological exploitation of microalgae. *Journal of Experimental Botany, 66*(22), 6975–6990. https://doi.org/10.1093/jxb/erv426
Gonçalves, A. L. (2021). The use of microalgae and cyanobacteria in the improvement of agricultural practices: A review on their biofertilising Biostimulating and Biopesticide Roles. *Applied Sciences, 11*(2), 871. https://doi.org/10.3390/app11020871
Gonçalves, J., Freitas, J., Fernandes, I., & Silva, P. (2023). Microalgae as biofertilizers: A sustainable way to improve soil fertility and plant growth. *Sustainability, 15*(16), 12413. https://doi.org/10.3390/su151612413
González-Pérez, B. K., Rivas-Castillo, A. M., Valdez-Calderón, A., & Gayosso-Morales, M. A. (2022). Microalgae as biostimulants: A new approach in agriculture. *World Journal of Microbiology and Biotechnology, 38*(1), 4. https://doi.org/10.1007/s11274-021-03192-2
Growland. (2024). https://www.growland.net/
Growtent. (2024). https://www.growtent.eu/eng_m_LINKS_Algae-based-fertilizers-2808.html
Gul, S., Shahnaz, L., Raiz, S., Nawaz, M. F. (2024). Microalgae: Production, consumption and challenges. In Kurniawan, T. A., Anouzla, A. (Eds.), *Algae as a natural solution for challenges in water-food-energy nexus: Toward carbon neutrality*. Springer Nature (pp. 31–59), Singapore.
Haifa Group. (2024). https://www.haifa-group.com/about-us-0
Heliae. (2024). https://phycoterra.com/future-of-soil-health/
Hispanagar, S. A. (2024). https://www.hispanagar.com/en/company
Hooda, S., Malik, G., Saini, P., Grewall, A., & Pandey, V. C. (2023). Cyanobacteria as a potential bioasset for restoring degraded land. *Land Degradation and Development, 34*(12), 3435–3450. https://doi.org/10.1002/ldr.4711
Hussain, F., Hussain, I., Khan, A. H. A., Muhammad, Y. S., Iqbal, M., Soja, G., Reichenauer, T. G., & Zeshan, Y. S. (2018). Combined application of biochar, compost, and bacterial consortia with Italian ryegrass enhanced phytoremediation of petroleum hydrocarbon contaminated soil. *Environmental and Experimental Botany, 153*, 80–88. https://doi.org/10.1016/j.envexpbot.2018.05.012
Jabłońska-Trypuć, A., Wołejko, E., Ernazarovna, M. D., Głowacka, A., Sokołowska, G., & Wydro, U. (2023). Using algae for biofuel production: A review. *Energies, 16*(4), 1758. https://doi.org/10.3390/en16041758
Jose, S., Malla, M. A., Renuka, N., Bux, F., & Kumari, S. (2024). Cyanobacteria-green microalgae consortia enhance soil fertility and plant growth by shaping the native soil microbiome of *Capsicum annuum*. *Rhizosphere, 30*, Article 100892. https://doi.org/10.1016/j.rhisph.2024.100892

Kamyab, H., SaberiKamarposhti, M., Hashim, H., & Yusuf, M. (2024). Carbon dynamics in agricultural greenhouse gas emissions and removals: A comprehensive review. *Carbon Letters, 34*(1), 265–289. https://doi.org/10.1007/s42823-023-00647-4

Karthikeyan, S., Periyasamy, M., Prathima, A., & Yuvaraj, M. (2020). Agricultural tractor engine performance analysis using Stoechospermum marginatum microalgae biodiesel. *Materials Today: Proceedings, 33*, 3438–3442. https://doi.org/10.1016/j.matpr.2020.04.908

Kreft, C., Finger, R., & Huber, R. (2024). Action-versus results-based policy designs for agricultural climate change mitigation. *Applied Economic Perspectives and Policy, 46*(3), 1010–1037. https://doi.org/10.1002/aepp.13376

Krohn, I., Menanteau-Ledouble, S., Hageskal, G., Astafyeva, Y., Jouannais, P., Nielsen, J. L., Pizzol, M., Wentzel, A., & Streit, W. R. (2022). Health benefits of microalgae and their microbiomes. *Microbial Biotechnology, 15*(7), 1966–1983. https://doi.org/10.1111/1751-7915.14082

Li, S., Li, F., Zhu, X., Liao, Q., Chang, J.-S., & Ho, S.-H. (2022). Biohydrogen production from microalgae for environmental sustainability. *Chemosphere, 291*, Article 132717. https://doi.org/10.1016/j.chemosphere.2021.132717

Lutzu, G. A., Parsaeimehr, A., Ozbay, G., Ciurli, A., Bacci, L., Rao, A. R., Ravishankar, G. A., Concas, A. (2024). Microalgae and cyanobacteria role in sustainable agriculture: From wastewater treatment to biofertilizer production. In *Algae mediated bioremediation* (pp. 565–618). Wiley Ltd.

Martini, F., Beghini, G., Zanin, L., Varanini, Z., Zamboni, A., & Ballottari, M. (2021). The potential use of *Chlamydomonas reinhardtii* and *Chlorella sorokiniana* as biostimulants on maize plants. *Algal Research, 60*, Article 102515. https://doi.org/10.1016/j.algal.2021.102515

Morales, M., Hélias, A., & Bernard, O. (2019). Optimal integration of microalgae production with photovoltaic panels: Environmental impacts and energy balance. *Biotechnology for Biofuels, 12*(1), 239. https://doi.org/10.1186/s13068-019-1579-4

Mutale-joan, C., Redouane, B., Najib, E., Yassine, K., Lyamlouli, K., Laila, S., Zeroual, Y., & Hicham, E. A. (2020). Screening of microalgae liquid extracts for their bio stimulant properties on plant growth, nutrient uptake and metabolite profile of *Solanum lycopersicum* L. *Science and Reports, 10*(1), 2820. https://doi.org/10.1038/s41598-020-59840-4

Mutale-Joan, C., Sbabou, L., & Hicham, E. A. (2023). Microalgae and cyanobacteria: how exploiting these microbial resources can address the underlying challenges related to food sources and sustainable agriculture: A review. *Journal of Plant Growth Regulation, 42*(1), 1–20. https://doi.org/10.1007/s00344-021-10534-9

Muys, M., Sui, Y., Schwaiger, B., Lesueur, C., Vandenheuvel, D., Vermeir, P., & Vlaeminck, S. E. (2019). High variability in nutritional value and safety of commercially available Chlorella and Spirulina biomass indicates the need for smart production strategies. *Bioresource Technology, 275*, 247–257. https://doi.org/10.1016/j.biortech.2018.12.059

Nazir, M. J., Li, G., Nazir, M. M., Zulfiqar, F., Siddique, K. H. M., Iqbal, B., & Du, D. (2024). Harnessing soil carbon sequestration to address climate change challenges in agriculture. *Soil and Tillage Research, 237*, Article 105959. https://doi.org/10.1016/j.still.2023.105959

New-nutri. (2024). New-nutri. https://www.new-nutri.com/

Ng, Z. Y., Ajeng, A. A., Cheah, W. Y., Ng, E.-P., Abdullah, R., & Ling, T. C. (2024). Towards circular economy: Potential of microalgae—bacterial-based biofertilizer on plants. *Journal of Environmental Management, 349*, Article 119445. https://doi.org/10.1016/j.jenvman.2023.119445

Orzuna-Orzuna, J. F., Godina-Rodríguez, J. E., Garay-Martínez, J. R., Reséndiz-González, G., Joaquín-Cancino, S., Lara-Bueno, A. (2024). Milk yield, composition, and fatty acid profile in milk of dairy cows supplemented with *Microalgae Schizochytrium* sp.: A meta-analysis. *Agriculture, 14*(7), 1119. https://doi.org/10.3390/agriculture14071119

Panpan Industry. (2024). https://panpanindustry.goldsupplier.com/

Parmar, P., Kumar, R., Neha, Y., Srivatsan, V. (2023). Microalgae as next generation plant growth additives: Functions, applications, challenges and circular bioeconomy based solutions. *Frontiers in Plant Science*, 14. https://doi.org/10.3389/fpls.2023.1073546

Purdy, A., Pathare, P. B., Wang, Y., Roskilly, A. P., & Huang, Y. (2018). Towards sustainable farming: Feasibility study into energy recovery from bio-waste on a small-scale dairy farm. *Journal of Cleaner Production, 174*, 899–904. https://doi.org/10.1016/j.jclepro.2017.11.018

Ramakrishnan, B., Maddela, N. R., Venkateswarlu, K., & Megharaj, M. (2023). Potential of microalgae and cyanobacteria to improve soil health and agricultural productivity: A critical view. *Environmental Science: Advances, 2*(4), 586–611. https://doi.org/10.1039/D2VA00158F

Ravindran, B., Gupta, S., Cho, W.-M., Kim, J., Lee, S., Jeong, K.-H., Lee, D., & Choi, H.-C. (2016). Microalgae potential and multiple roles—Current progress and future prospects—An overview. *Sustainability, 8*(12), 1215. https://doi.org/10.3390/su8121215

Renganathan, P., Puente, E. O. R., Sukhanova, N. V., & Gaysina, L. A. (2024). Hydroponics with microalgae and cyanobacteria: emerging trends and opportunities in modern agriculture. *Biotech, 13*(3), 27. https://doi.org/10.3390/biotech13030027

Rong, S., Xu, H., Li, L., Chen, R., Gao, X., & Xu, Z. (2020). Antifungal activity of endophytic Bacillus safensis B21 and its potential application as a biopesticide to control rice blast. *Pesticide Biochemistry and Physiology, 162*, 69–77. https://doi.org/10.1016/j.pestbp.2019.09.003

Salam, K. A., Velasquez-Orta, S. B., & Harvey, A. P. (2016). A sustainable integrated in situ transesterification of microalgae for biodiesel production and associated co-product-a review. *Renewable and Sustainable Energy Reviews, 65*, 1179–1198. https://doi.org/10.1016/j.rser.2016.07.068

Santini, G., Biondi, N., Rodolfi, L., & Tredici, M. R. (2021). Plant biostimulants from cyanobacteria: An emerging strategy to improve yields and sustainability in agriculture. *Plants, 10*(4), 643. https://doi.org/10.3390/plants10040643

Sarsekeyeva, F. K., Sadvakasova, A. K., Sandybayeva, S. K., Kossalbayev, B. D., Huang, Z., Zayadan, B. K., Akmukhanova, N. R., Leong, Y. K., Chang, J.-S., & Allakhverdiev, S. I. (2024). Microalgae- and cyanobacteria-derived phytostimulants for mitigation of salt stress and improved agriculture. *Algal Research, 82*, Article 103686. https://doi.org/10.1016/j.algal.2024.103686

Sathya, A. B., Thirunavukkarasu, A., Nithya, R., Nandan, A., Sakthishobana, K., Kola, A. K., Sivashankar, R., Tuan, H. A., & Deepanraj, B. (2023). Microalgal biofuel production: Potential challenges and prospective research. *Fuel, 332*, Article 126199. https://doi.org/10.1016/j.fuel.2022.126199

Shahinnia, F., Carrillo, N., & Hajirezaei, M.-R. (2021). Engineering climate-change-resilient crops: new tools and approaches. *International Journal of Molecular Sciences, 22*(15), 7877. https://doi.org/10.3390/ijms22157877

Shao, H. (2024). Agricultural greenhouse gas emissions, fertilizer consumption, and technological innovation: A comprehensive quantile analysis. *Science of the Total Environment, 926*, Article 171979. https://doi.org/10.1016/j.scitotenv.2024.171979

Sible, C. N., Seebauer, J. R., & Below, F. E. (2021). Plant biostimulants: A categorical review, their implications for row crop production, and relation to soil health indicators. *Agronomy, 11*(7), 1297. https://doi.org/10.3390/agronomy11071297

Singh, K. B., Kaushalendra, V. S., Lalnunpuii, R., & Rajan, J. P. (2023). Current issues and developments in cyanobacteria-derived biofuel as a potential source of energy for sustainable future. *Sustainability, 15*(13), 10439. https://doi.org/10.3390/su151310439

Singh, P. K., Kumar, A., Yadav, M. K., Kaushalendra Singh, S. (2022). Cyanobacteria as biostimulants in the paddy fields. In Cyanobacterial lifestyle and its applications in biotechnology (pp. 281–306). Elsevier

Suresh Kumar, K., Dahms, H.-U., Won, E.-J., Lee, J.-S., & Shin, K.-H. (2015). Microalgae—A promising tool for heavy metal remediation. *Ecotoxicology and Environmental Safety, 113*, 329–352. https://doi.org/10.1016/j.ecoenv.2014.12.019

Terra Aquatica, T. A. (2024). https://www.terraaquatica.com/fr/

TrueAlgae (2024). https://truealgae.com/

Vasistha, S., Khanra, A., Clifford, M., & Rai, M. P. (2021). Current advances in microalgae harvesting and lipid extraction processes for improved biodiesel production: A review. *Renewable and Sustainable Energy Reviews, 137*, Article 110498. https://doi.org/10.1016/j.rser.2020.110498

Vo, H. N. P., Ngo, H. H., Guo, W., Nguyen, T. M. H., Liu, Y., Liu, Y., Nguyen, D. D., & Chang, S. W. (2019). A critical review on designs and applications of microalgae-based photobioreactors for pollutants treatment. *Science of the Total Environment, 651*, 1549–1568. https://doi.org/10.1016/j.scitotenv.2018.09.282

Wollmann, F., Dietze, S., Ackermann, J., Bley, T., Walther, T., Steingroewer, J., & Krujatz, F. (2019). Microalgae wastewater treatment: Biological and technological approaches. *Engineering in Life Sciences, 19*(12), 860–871. https://doi.org/10.1002/elsc.201900071

Yadav, P., Singh, R. P., Patel, A. K., Pandey, K. D., & Gupta, R. K. (2022). Cyanobacteria as a biocontrol agent. In A. Kumar (Ed.), *Microbial biocontrol: Food security and post harvest management* (pp. 167–185). Springer International Publishing.

Yadav, R. K., Abraham, G., Singh, Y. V., Singh, P. K. (2014). Advancements in the utilization of azolla-anabaena system in relation to sustainable agricultural practices. *Proceedings of the Indian National Science Academy, 80*(2), 301. https://doi.org/10.16943/ptinsa/2014/v80i2/55108

Zhang, Z., Xu, M., Fan, Y., Zhang, L., & Wang, H. (2024). Using microalgae to reduce the use of conventional fertilizers in hydroponics and soil-based cultivation. *Science of the Total Environment, 912*, Article 169424. https://doi.org/10.1016/j.scitotenv.2023.169424

Zhou, L., Liu, W., Duan, H., Dong, H., Li, J., Zhang, S., Zhang, J., Ding, S., Xu, T., & Guo, B. (2023). Improved effects of combined application of nitrogen-fixing bacteria *Azotobacter beijerinckii* and microalgae *Chlorella pyrenoidosa* on wheat growth and saline-alkali soil quality. *Chemosphere, 313*, Article 137409. https://doi.org/10.1016/j.chemosphere.2022.137409

Advances in Agronomy for Soil Conservation and Plant Health

Ishrat Perveen, Ahsan-ul-Haq, Adnan Ahmad, Ayesha Ayaz, Zoya Ahmad, Aizah Amin, Naaz Abbas, Yasar Saleem, Sania Mazhar, Shaista Nawaz, Quratulain Syed, and Syed Hussain Imam Abidi

Abstract Sustainable agriculture is significant in meeting the growing demand for food, fibre, and fuel while maintaining environmental integrity. This chapter entails the advancements in agronomic practices to conserve soil and promote plant health to ensure long-term agricultural productivity. This study's primary goals are to appraise soil management methods, including cropping, cover cropping, crop rotation, mulching and conservation tillage to reduce soil erosion, facilitate soil structure maintenance and increase soil organic matter. In addition, integrated nutrient management and biological control agents for pest management and water-conserving strategies such as drip irrigation and rainwater harvesting have also been discussed. Sustainable soil health depends on the role of microbial communities, and introducing disease-resistant plant varieties is essential to reduce chemical inputs. This suggests that using these advanced and sustainable techniques will aid in producing healthy soil ecosystems and resilient crop growth and curtail environmental impacts. The strategies to be further explored in future research include the global application of these techniques across various ecosystems and improving farmer education and outreach programs to promote the adoption of these strategies across varied ecosystems. These integrative approaches are essential to achieving sustainable agriculture that also satisfies ecological and economic objectives.

I. Perveen (✉) · Ahsan-ul-Haq · A. Ahmad · A. Ayaz · N. Abbas · Y. Saleem · S. Mazhar · S. Nawaz · Q. Syed
GenEd and Molecular Biology Labs, Food and Biotechnology Research Centre, Pakistan Council of Scientific and Industrial Research Centre, Lahore, Pakistan
e-mail: drishratperveen51214@outlook.com

Z. Ahmad · A. Amin
Dr Ikram ul Haq Institute of Industrial Biotechnology, Government College University, Lahore, Pakistan

S. H. I. Abidi
Pakistan Council of Scientific and Industrial Research Centre, Islamabad, Pakistan

M. Shaaban (ed.), *Soils and Sustainable Agriculture*, Frontier Studies in Soil Science,
https://doi.org/10.1007/978-3-031-91114-9_42

1 Introduction

Practices supporting sustainable agriculture are based on balancing productivity with environmental protection and conservation techniques. There is a stringent need for agricultural systems globally to produce more food, fibres, and fuel while also preserving soil health and the natural resources of the ecosystem, which have never been observed to grow in proportion to the expanding global population. This balancing act is highly dependent on agronomy, that is, the practice of developing, utilizing and propagating methods to protect the soil from erosion, improve nutrient availability and maintain plant health without overreliance on chemical inputs. This chapter explores the key developments in agronomy for soil conservation and plant health, focusing on sustainable approaches which support agroecological and economic objectives.

Since the end of World War II, underexplored technological advancements in agriculture have been witnessed. The Green Revolution of the 1960s brought high-yielding varieties of wheat and rice; and huge quantities of nitrogenous fertilizers and irrigation (Ajl & Sharma, 2022). For agricultural researchers and practitioners, the long-term implications of lack of management of pests, diseases, and weeds and excessive application of fertilizer are often ignored. In developed countries, crop yields were improved by low-input agricultural systems such as crop rotation, machinery, and organic matter incorporation (Friedmann, 2021).

A higher crop yield in developed nations is due to the existence of new crop varieties that respond to elevated nitrogen levels. One should first manipulate or understand how organic farming can use fewer chemicals and produce high-component (Mohanty et al., 2024). Regenerative agriculture and sustainable agriculture developed in the 1980s, and the word sustainable first appeared in 1987. Always, change in agriculture is based on an image of what should exist which some people label the referent the desirable or the belief system (Gojon et al., 2023). All that separates this goal from what we do today in agriculture is that there is a development factor. Early 1900s US agriculture is a prime example of how technology has affected agriculture (Moerman, 2021).

Land conversion and increased crop production contribute to climate change altering the Earth's atmosphere, and in turn, dramatically impact global food security (Muluneh, 2021). Greenhouse gases like CO_2 have increased the world average temperature since 1850 by over 1 °C. Uncontrolled climate deterioration can become irreversible and include more forest fires and melting ice sheets in the Arctic (Laamrani et al., 2020). In addition, global warming leads to unbalanced distribution of rain and worsening flood and drought-prone areas. It also can affect crop growth patterns, poor growth, seed germination, and reproduction, increasing diseases and pests on plants. All require proper control to prevent irreversible climate deterioration (Tian et al., 2021). The carbon cycle is critical in understanding how climate change may affect larger land conversion and crop production. Various anthropogenic activities like intensive agriculture, deforestation, and fossil fuel combustion resulted in an increase in CO_2 emissions. CO_2 absorption from the atmosphere in photosynthesis is indicated, as shown in the figure, but land conversion reduces the number of plants that can sequester carbon and crop production, and livestock adds more respiration and waste. With this imbalance, higher levels of atmospheric CO_2 result, greatly aggravating the greenhouse effect responsible for global warming. This, along with rising diseases, pest infestations, and changing weather and rainfall patterns, further highlights the perseverance of sustainable agricultural practices, which may ensure no more disruptions in the carbon cycle and its undeniable repercussions on global food security. Figure 1 represents the interconnection between carbon cycling, climate change, and agricultural practices.

The regulation of soil's genetic, biological and ecosystem components down to the ecosystem level is defined as soil biodiversity. The Convention on Biological Diversity (CBD) defines soil biodiversity as regulating those components at the gene to landscape level. Higher human population, climate change, degraded soils, and reduced agricultural land quality have stressed natural resources and the processes that keep the world of sustainability in place (Guerra et al., 2021). Bioindicators of specific functions in the soil ecosystem are soil microorganisms, linking roots to soil and recycling nutrients with the breakdown of organic material (Sakin et al., 2024).

Soil microbial community functions and interaction with soil and plants can generate a more harmonious soil ecological structure for crop production, growth, development and sustainable yield (Shah et al., 2021). The knowledge of microbial communities' functions, behavioural, and communication processes is vital to avoid

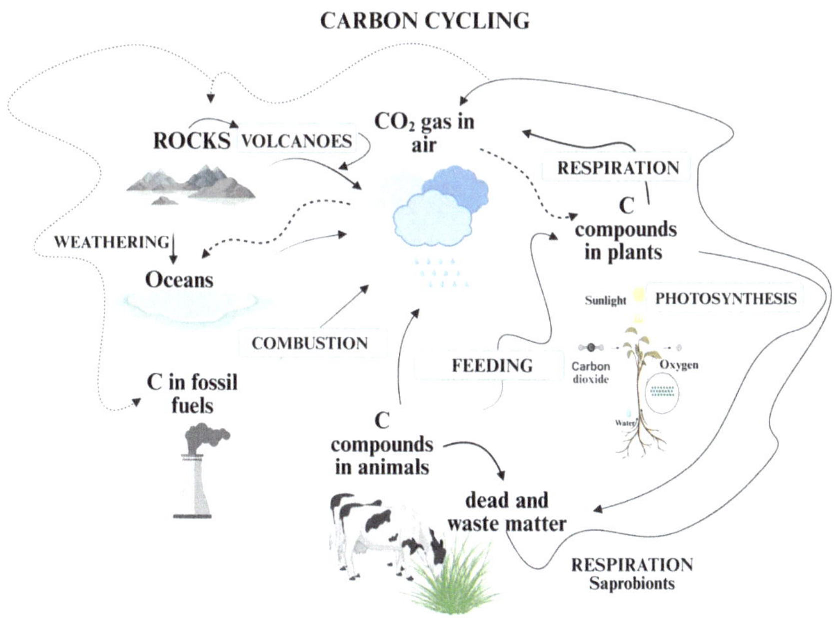

Fig. 1 Carbon cycle—its impact on climate changes and sustainable agricultural practices

undesired management practices and evaluate the sustainable state of soils and the production of crops (Chernov & Semenov, 2021; Das et al., 2022).

Global soil biota is one of the largest components of total biological diversity known to mankind. It is essential that we understand the world's soil resources and the distribution of soil biodiversity and functions to drive forward the global sustainability agenda. Ecosystem stability depends on the rate of change in plant species richness, aboveground net primary productivity and species synchrony, all of which are influenced by soil biota (Tahat et al., 2020).

Global agriculture, after the green revolution, came to depend to a great extent on the agrochemical partnerships embodied in liberal partnerships between agribusinesses and chemical corporations like those that depended on synthetic fertilizers and pesticides, and in the transformation of soil health permanently. This paradox has been addressed by efforts to develop strategies to increase agri-food production whilst preserving soil quality as a future resource. Agricultural production, which involves farming crops and managing soil and farmland, is closely related to the utilization and management of farmland and crops and is considered a threat to soil fertility. Although strategies such as converting atmospheric nitrogen to plant-available nitrogen, using organic and inorganic mulch, diverse crop rotations, no-till farming with straw retention, and managing the interaction between plants, soil, microorganisms, and anthropogenic activity can increase soil health (Wang et al., 2024).

This book chapter aims to assess the progress of agronomic practices that improve soil conservation and plant health and proposes future research and implementation. In the future of sustainable agriculture, however, it is essential to integrate technology into mainstream farming practices, increasing the basic understanding of the soil microbiome and improving pest and disease control methods. These practices offer the potential to create more resilient and more productive worldwide farming systems that simultaneously provide for both immediate agricultural needs as well as long-term ecological health.

2 Soil Conservation

Some of the modern agricultural practices have been known to have adverse impacts on the environment and human health, for example, climate change, deforestation and zoonotic disease emergence. Intervention plans to affect those health effects include encouraging plant-based diets, nutrition and food security and safety, biosecurity, surveillance, and antibiotic management. Various issues like AMR, zoonoses, food availability and sustainability, and global warming are all tied to soil health (Pitt & Gunn, 2024). Land use change and intensive agriculture over resources have negative effects on health. Soil health relates to the soil's capacity to promote plant, animal, and human life, ecosystem goods and services, and food production. Nutrient management remains imperative for improving both soil health and crop productivity in crop production (Mrabet, 2023; Pisante et al., 2023).

Sustainable agriculture seeks to sustain or enhance natural ecosystem processes and to sustain or enhance social and economic well-being; be economically viable with reliance on local means for addressing social security needs; use technologies that are compatible with natural processes; stress food security and maintain or improve the quality of the physical and biological environment. These involve practices to accumulate organic matter, rebuild soil fertility and conserve biocultural diversity (Montgomery et al., 2024).

A basic understanding of the soil layers is essential for plant health and conservation. The layers of soil have been given in Fig. 2. The O-layer consists of decomposing leaves and organic matter. The A-layer, or topsoil, mixes minerals and organics necessary for plant growth. The B-layer (subsoil) is sand, silt, or clay, containing most water retention and nutrient storage functions. The partially disintegrated parent rock of the C-layer and the un-weathered bedrock of the R-layer form the basis for all the other upper layers. These layers further increase the complexity required for sustainable agronomic practices.

Sustainable agriculture is built upon two fundamental pillars: soil protection and reinforcement of plants against erosion and the promotion of plant health (Banerjee & Van Der Heijden, 2023). A nutrient-rich, well-draining, well-draining soil is needed for the long-term productivity of the aquatic agri-food industry. The strategies involved in the conservation of soil include cover cropping, crop rotation, and organic matter (compost or manure). These practices increase soil organic matter, enhance soil structure, improve soil water retention, and increase nutrient availability. It also

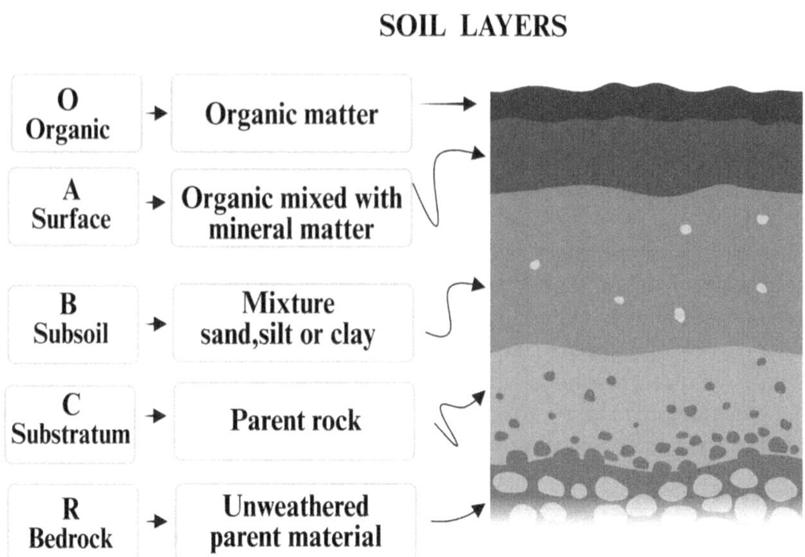

Fig. 2 Basic distribution of soil layers—and their composition for agronomic insight

involves soil conservation from erosion or soil degradation. Agronomists also recommend contour farming, terracing and planting trees as wind brakes or shelter belts to minimize soil degradation. These conservation measures keep the topsoil layer, which is more fertility-productive than deeper layers.

The Agronomists are responsible for sustainable agriculture by applying fertilizers, both artificial and natural, for the correct application of soil nutrients. Nutrient management is a focal point for achieving proper nutrients for crop production without contaminating the environment due to leaching. Conversely, plant health is important for sustainable agriculture, and agronomists use several methods for supporting crop growth and development through Integrated Pest Management (IPM) and promoting crops that are resistant to and free of infection by disease (Vialatte et al., 2021). Diversification of cropping systems is also encouraged by agronomists as it helps steer a more favourable path for other species interfering with pests and diseases on plants. By focusing on these aspects, agronomists can protect nature and the planet and lead to sustainable and safe production, taking into account modern agriculture principles (Sharma et al., 2021). Agronomists develop a comprehensive plan to guide the sustainable management of agricultural systems. The objectives of an agronomy plan are multifaceted, addressing the complex interplay between environmental, economic, and social factors.

3 Soil Conservation Practices

3.1 Soil Erosion Prevention Techniques

Soil erosion is one of several environmental issues caused by the loss of topsoil, decreased crop yields on agricultural fields, and soil slipping off into flood peaks, avalanches, or landslides. Some of the procedures to fix it include cover crops, crop terracing, and contour farming.

Grasses, shrubs, and trees are called cover crops and can help hold the soil in place so that it is not washed or blown away. Additionally, they play a role in soil consolidation through root systems that form channels and pores in the soil, allowing water to penetrate the soil and reduce runoff and soil erosion. This can be done by growing cover crops, grasses, shrubs and planting of trees. Some of the conservation techniques include growing legumes or grasses during periods that are not suitable for the growth of main crops. Crops such as grasses, shrubs and trees can be planted as windbreakers or in the direction of slopes to slow the rate of water and wind currents (Aneseyee et al., 2020; Weeraratna, 2022).

Another method to reduce or inhibit soil erosion is **terracing**. This is a process of forming constructive forms on the slopes, which lead to the formation of flat or slight features that slalom water and prevent the removal of fertile layers of ground. Terracing allows the slope's disruption to slow the movement of water, as the movement of water in a slope can be high velocity and very nascent. Terracing produces a slower water flow, and more water enters the ground than if there were no terracing at all (Deng et al., 2021).

Contour farming is a farming practice in which crops, mainly food crops, are sown at the natural twofold slope angles instead of up and down the slope. So, this method greatly reduces the rate at which water washes the surface and carries away the rocks and gold along with it, which means a great reduction in soil erosion. This technique directs the formation of small mounds and troughs parallel to the slope of the ground (acting as small terraces) in order to slow the flow of water and encourage it to work its way through the sub-surface instead of acting as rills and running off. It reduces water speed and thus the amount of silt, or soil, that the erosion process transports. This technique is best suited for moderate to steep slopes because it controls soil erosion to a high degree. Also, it can be combined with other soil conservation practices such as plant cover crops. But whereas contour farming involves more labours involvement and implementations than conventional farming techniques, it does have the potential to contribute favourably towards soil conservation and the continuation of the productivity of the farmlands for future generations. Contour farming can play a positive role in soil conservation and the maintenance of farmlands' productivity for future generations (Hills, 2024).

The modern techniques of soil conservation shown are crop rotation, no-till farming, terracing, contour farming and mulching. Figure 3 indicates that modern soil conservation techniques ensure sustainable agronomy and advanced plant health. The objectives of these practices are to decrease the emissions of greenhouse gases, CO_2, N_2O, and NH_3, along with increasing soil organic nitrogen (SON), soil organic carbon (SOC), and phosphorus availability (P-available). Surplus and balanced supply of nutrients like nitrogen (N) and phosphorus (P) balance that, and which balances crop yield and plant health, will depend on best management practices. The judicious use of fertilizers prevents soil depletion and nutrient overload, preserving soil fertility and food security. Sustainable practices can help control plant diseases and lower greenhouse gases contributing to global warming.

4 Mulching

Mulching is a method of simply covering the soil using materials like wood chips, leaves, straw or compost to the soil around the plants. For gardeners or in the compound, this practice is good as it helps the garden become weed-free; this further helps the environment since it controls the growth and spread of the various weeds. Mulching builds a physical barrier that will keep the seeds from seeing light, thus denying the weed the ability to germinate or grow. This ensures that plants don't require to be manually weeded by gardeners, thereby saving time and money (Prem et al., 2020).

Mulching also helps lower water evaporation from the soil surface, leading to reduced water usage frequency, especially in hot and dry areas. It also allows the

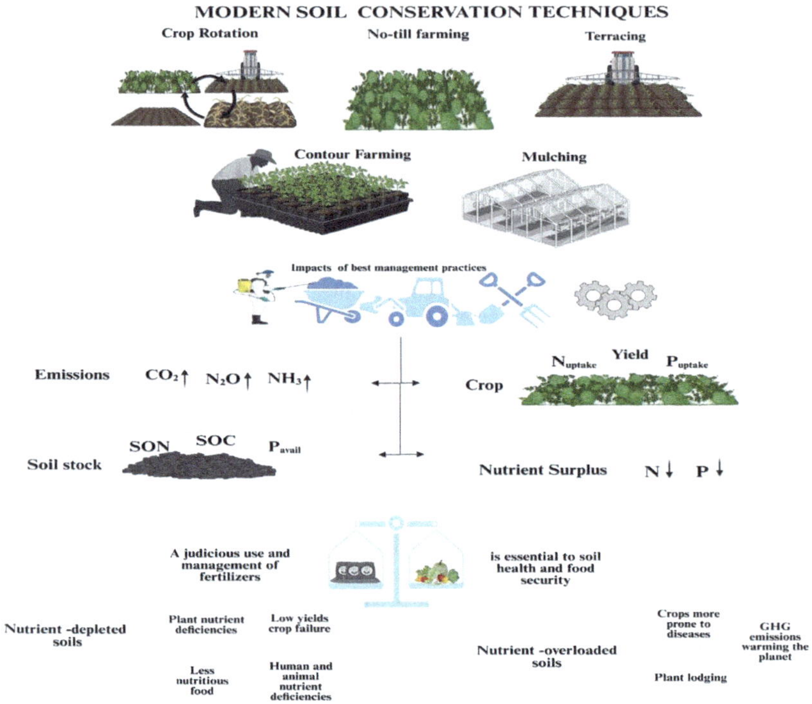

Fig. 3 Sustainable agronomy and advanced plant health enhanced by modern soil conservation techniques

soil to gain structure and nutrient value. The degradation of mulch affects soil structure and fertility, which in turn minimizes the effect on plant roots at high or low temperatures (Naorem et al., 2023; Walia et al., 2024).

Conservation tillage is used to mitigate the challenges of soil disturbance and enhance crop residue on the soil surface, which helps improve soil structure and reduce erosion. It has many environmental and economic benefits. Leaving a significant portion of crop residue on the soil surface reduces soil erosion and protects the soil from wind and water forces. It helps to hold the structure of soil by reducing erosion (Shaikh et al., 2023). It reduces water evaporation from the soil and increases water infiltration. By keeping the crop residue in the soil, a man hurts soil erosion and decreases the risk of a dust storm. This practice is critical, particularly during times of little or no rainfall in areas with scarce rainfall. Conservational tillage also takes less time and less fuel, saving farmers time and fuel and increasing profitability and efficiency within an agricultural system (Chopra & Koul, 2020; El-Beltagi et al., 2022).

5 Soil Organic Matter Management Technique

Soil organic matter (SOM) is vital to healthy and productive soils. It cleans the air by improving soil structure, water-holding capacity, nutrient cycling, and microbial activity. Maintaining and enhancing SOM levels is required for sustainable agriculture and land use. The key techniques for managing SOM include composting, crop rotation, green manure, agroforestry, and perennial crops.

Composting is the biological breakdown of organics by bacteria and fungi to break down organics, resulting in the release of nutrients and a stable material called compost. The compost is high in organic material structures, nitrogen, phosphate, and potash for plants. The compost thus includes plant remains, animal droppings and food wastes, which are turned into nutrient-enriched soil. The addition of organic residues into soil improves the SOM so that it increases the volumetric weight of soil, increases the availability of water and nutrients for the plants, improves soil structure and water holding capacity and serves as a source of nutrition for soil organisms and hence, promotes the soil micro-ecosystem. In addition, the soil structure and water-holding capacity are improved (Cooper et al., 2020; Jain & Kalamdhad, 2020) (Fig. 4).

Crop rotation is a succession of different crops that keep pace with building and maintaining soil organic matter. Second, this approach starts with different rootstock, nutrient consumption patterns, and biomass levels, which change the soil's organic matter. Growing crops sequentially aids in nitrogen fixation, increasing soil nitrogen (N_2) levels and decreasing their dependency on anthropogenic or synthetic nitrogen.

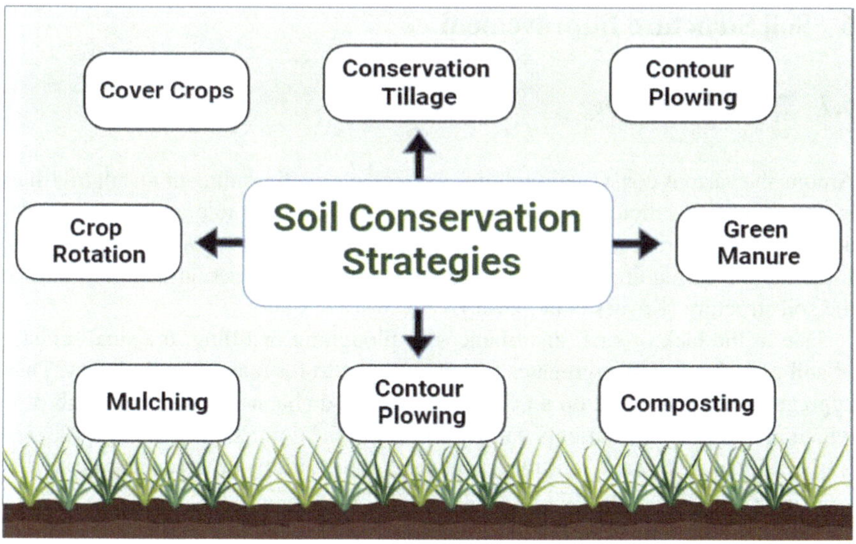

Fig. 4 Soil conservation strategies essential for sustainable agriculture

Leguminous crop plants like soybeans and alfalfa tap aerial nitrogen into their soil nitrogen content, increasing a future crop's nitrogen supply. Cover crops and annual crops, aerate the soil and pull in nutrients from lower layers, increase soil organic matter and add fertility to it (Farmaha et al., 2022).

Green manuring is one of the best practices that should be done to manage the soil and increase organic matter. Green manure is growing plants, weeds or cover crops that are later planted green to increase soil fertility. Replenishing nutrients, improving soil structure, and using environment-friendly farming methods are all reasons that they can help. Nitrogen fixation by green manure and soil organic matter (SOM) management are sufficiently green. These plants will decompose in the soil and provide nitrogen and organic matter for the subsequent crops. Legumes, such as clover or alfalfa, or grasses, such as rye oats or sorghum, are commonly used as green manure plants. They can be grown as an intercrop or part of crop rotation. They decompose into soil microorganisms, which break organic compounds, such as carbohydrates, protein and lignin, further enriching the soil and thus promoting humus formation (Özbolat et al., 2023).

Another method for improving SOM quality in the agroforestry system is agroforestry and perennial crops. Agroforestry is the concurrent use of trees and shrubs associated with crops and/or livestock, with and without yields. Incorporating deep-rooted, long-lived woody plants can serve as a competent means for increasing SOM accumulation and stabilisation. It provides a food source for microorganisms living in the soil, food nutrient cycling, and the formation of stable humus substances (Angst et al., 2021).

6 Soil Structure Improvement

6.1 No-Till Farming

Among the various conservation tillage strategies, no-till farming or no-till till tillage is considered an efficient technique for inducing the desirable soil structure. The aggregation of parent materials through multiple processes results in soil structure formation. This requires no-till practices to ensure the reduction in disturbance of the soil structure (Kovács et al., 2023).

Due to the lack of such disturbances as ploughing or tilling, the small clusters of soil particles called aggregates are encouraged to form and remain intact. These aggregates join and build up a network of pore and channel systems through or by which air, water, and nutrients can pass through the soil (Kovács et al., 2023)This leads to increased water infiltration into the soils, aiding in water retention. This will further allow the inward and outward movement of gases to ensure the nutrient enrichment activities within the plant root systems and soil microorganisms (Hussain et al., 2021). The accumulation of crop residues on the soil surface also helps hold the soil particles together, making the soil structure stronger. This keeps off the instances of soil erosion and increases the general health and productivity of the soil (Rahman et al., 2020).

6.2 *Soil Amendments*

Adding certain materials, such as lime and gypsum, can improve the soil structure. A compound called calcium sulfate (or gypsum) can flocculate the clay particles and improve the plants' soil structure, porosity, aeration and water intake. Soil structure is influenced by pH control measures taken by the chemical addition of lime, which lowers acidity and promotes the aggregation and presence of beneficial organisms such as earthworms. Due to human activities, these are most effective in soil with high clay content or compacted soil. Using soil amendments can improve aggregates, water, nutrients, and overall fertility of soils and represent a beneficial management practice of sustainable agriculture. Human activities such as soil compaction can add to the benefits that accrue from these amendments, particularly in high clay-content soils (Abdolvand & Sadeghiamirshahidi, 2024). Table 1 enumerates various strategies for soil conservation and plant Health: Techniques and Purposes.

7 Plant Health Management

7.1 *Disease Control Measures*

Controlling diseases is crucial in keeping plants healthy and productive. Mitigation measures may consist of the use of resistant plant varieties, sanitation practices, fungicides, or other suitable treatments (Panth et al., 2020). Crop rotation, elimination of plant debris that has disease, and vernalization are equally crucial in halting the transmission of diseases. It is also essential to control various environmental factors such as moisture of soil and nutrient content of the soil to make the plant healthy and stern so that diseases cannot quickly attack it (Dolatabadian et al., 2022) Such anticipatory measures, when used in combination, guarantee that the impact of plant diseases will be well dealt with and that the crop will remain healthy and productive (Ek, 2023; Rawat et al., 2021).

7.1.1 Crop Rotation

The prevention of plant diseases is made easier by crop rotation, a classical crop management technique. To reduce the attack of pathogens in soil, it is pertinent to grow two different kinds of crops in that soil, which will thus break the cycle of pathogens in that soil (Rawat et al., 2021)For example, planting a cereal crop followed

Table 1 Strategies for soil conservation, plant health, techniques, and purposes

Strategy	Purpose	Techniques	Citations
Soil conservation practices	Soil erosion prevention techniques	Cover crops	Koudahe et al. (2022)
		Conservation tillage	Ogieriakhi and Woodward (2022)
		Sustainable grazing	Hancock et al. (2020)
		Soil amendments	
		Afforestation	
		Terraces	Fernández-Cabezón and Nikel (2020)
		Contour plowing	Fantappiè et al. (2020)
	Soil organic matter management	Composting	Manea et al. (2024)
		Crop rotation	Shah et al. (2021)
		Agroforestry	Fahad et al. (2022)
		Green manure/cover crop	Carr et al. (2020)
		Perennial forage crops	Kim et al. (2022)
	Soil structure improvement	No-till farming	Somasundaram et al. (2020)
		Vermi-compost	Rehman et al. (2023)
		Use crop residue	Mirzaei et al. (2021)
		Soil amendments (e.g., gypsum, lime)	Garbowski et al. (2023)
		Mulching	El-Beltagi et al. (2022)
Plant health management	Disease control measures	Crop rotation	Jalli et al. (2021)
		Eradication	Maloy (2005)
		Exclusion	Kahn (1991)
		Resistant varieties	Panth et al. (2020)
		Biological control agents	Ons et al. (2020)
	Pest Management Strategies	Integrated pest management (IPM)	Dent and Binks (2020)
		Trap crops	Zhang et al. (2020)
		Natural predators	Beaumelle et al. (2021)
	Nutrient management	Soil testing and analysis	Havlin (2020)
		Fertiliser application techniques	Pawase et al. (2023)
		Balanced fertilization	Kishore et al. (2021)

(continued)

Table 1 (continued)

Strategy	Purpose	Techniques	Citations
Sustainable practices integration	Agroforestry	Alley cropping	Hombegowda et al. (2022)
		Forest farming	Trozzo et al. (2021)
		Riparian buffers	Udawatta and Gantzer (2022)
	Precision agriculture technologies	GPS guided machinery	Gawande et al. (2023)
		Remote sensing	Radočaj et al. (2022)
		Variable rate technology	Sishodia et al. (2020)
	Water management	Irrigation efficiency	Gupta et al. (2023)
		Rainwater harvesting	Huang et al. (2020)
		Drip irrigation	Fan et al. (2020)
		Smart irrigation system	Obaideen et al. (2022)
	Crop diversification	Intercropping/ agroforestry	Alcon et al. (2020)
		Mixed farming system	Sharma et al. (2021)
		Crop rotation	Shah et al. (2021)
Monitoring and evaluation	Soil health assessment	Soil nutrition testing	Lehmann et al. (2020)
		Soil microbial activity	Fierer et al. (2021)
		Soil pH and salinity monitoring	Singh (2022)
	Plant disease and pest monitoring	Pest population tracking	
		Disease spread modelling	Dong et al. (2020)
		Remote sensing for early detection	Terentev et al. (2022)
	Yield analysis	Crop harvest data collection	Lobell et al. (2020)
		Post-harvest less assessment	Florkowski et al. (2021)
	Environmental impact assessment	Carbon footprint calculation	Lannelongue et al. (2021)
		Water usage impact	de Mello et al. 2020)
		Biodiversity monitoring	Stephenson (2020)
Education and outreach	Farmer training programs	On-farm demonstration	Adamsone-Fiskovica and Grivins (2022)

(continued)

Table 1 (continued)

Strategy	Purpose	Techniques	Citations
		Technology adoptive workshops	Wang et al. (2020)
		Sustainable farming practices training	Zakaria et al. (2020)
	Knowledge sharing platforms	Online learning modules	Adamashvili et al. (2020)
		Mobile applications for farmers	Tiwari (2022)
		Farmer-to-farmer knowledge exchange	Adamashvili et al. (2020)

by a legume is an effective way of breaking the life cycle of many pathogens of fungal and bacterial origin (Yu et al., 2022).

Rotating crops can help improve soil for plants, assisting them in resisting diseases. Crop rotations may include rather diverse compositions, improving the microorganisms present in the soil, which may compete with or produce toxic substances for plant pathogens. When selecting for crop rotation, consider what is susceptible, where the crops get their pathogens, and how the soil acts. Crop rotation ensures an innovative approach to not using chemicals but providing proper crop nutrition and health solutions, reducing chemical use (Abbas et al., 2022).

7.1.2 Resistant Varieties

Disease-resistant varieties are one method of combating plant diseases in a more sustainable manner. Resistant varieties are developed with the hope of containing genes that impart natural abilities to withstand particular pathogens, which may include fungi, bacteria, viruses, or nematodes (Tariq et al., 2020; Van Esse et al., 2020).

If the farmers or cultivators initiate growing disease-resistant varieties, disease load can be reduced in their crops. This approach hinders the excessive utilization of chemical pesticides, which adversely affect the environment and people's health (Van Esse et al., 2020). The resistant varieties slow down the pathogen resistance, leading to more effective control over time. New resistant varieties can breed quickly for specific diseases because of genomics and molecular genetics advances. Stacking offers long life against rapidly evolving pathogens by utilizing the enlarged number of resistant genes. It is essential to first introduce the developed resistant varieties and other sustainable insecticide control methods (ICM) in agricultural systems. This will decrease chemical dependency, directly reduce losses, and improve plant health (Duhaithem, 2024).

7.1.3 Biological Control Agents

Biological control is a technique used to get plant diseases and pests under control using living organisms, such as microbes and natural enemies. *Bacillus*, *Pseudomonas* and *Trichoderma* antagonistic microorganisms antagonize or produce antagonists to plant pathogens. These agents can be used for seed treatment, soil inoculants or foliar sprays. The use of hyperparasites or mycoparasites, such as mycoparasites, as bioagents for control of other fungal plant pathogens can be seen. Predatory/parasitic organisms (flags) attack or parasitize plant pathogens and insect pests (e.g. spider mites) (Wenninger et al., 2020).

The presence of recurrent or self-permitted antagonists in an agroecosystem affords durable and environment-friendly disease control through biological control. It is often used with resistant varieties, crop rotation, and the appropriate use of chemicals. Optimal results require proper choice and preparation of agents and appropriate use. There is a stringent need to involve additional research to ensure the further addition of biocontrol possibilities (Tanda, 2024).

7.2 Pest Management Strategies

7.2.1 Integrated Pest Management (IPM)

Integrated Pest Management (IPM) is an economical management method involving the individual and/or population control of various organisms. The method may include cultural, mechanical, biological, and chemical methods that are selected depending on the type of pest and crop. An important element of IPM is to manage the pest population below the levels that cause significant crop damage.

Trap cropping uses plants grown primarily to attract pests away from any main crop and reduce pest pressure on it. The target pest prefers these trap crops compared to the principal crop; therefore, these trap crops divert the pests away from the principal crop (Yadav et al., 2023).

Nutrient management is another method of IPM. Nutrient levels should be tested to ensure the deficiencies or excessive levels at least once a season. This may be followed by adding organic manure (i.e., compost) to enhance soil fertility. The addition of synthetic fertilizers can thus be curtailed or applied whenever they are needed most by the plants or soils, which would prevent nutrient leaching and/or accumulation. Some additional strategies to optimise nutrient status include the introduction of region-specific crop varieties coupled with plant tissue analysis, which may reduce the requirement for pesticide application (Walia & Walia, 2021).

Nutrient management is highly dependent on **soil testing**. Regular and multiple-site collection of soil samples, followed by weighing crop history, soil types, and practices applied, would collectively help identify nutrient imbalances. Furthermore, soil analysis can offer information about nutrient levels, such as the amount of nitrogen, phosphorus, potassium, calcium, magnesium, and micronutrients, and help develop a nutrient management plan (Bungau et al., 2021).

Fertilizer application profoundly affects the utilization of nutrients, crop yield, and ecosystem. Fertilizers are applied uniformly throughout the field. Some scientists found the application of fertilizer in narrow strips around crop roots to be an effective method. Fertigation is the application of soluble fertilizers through an irrigation system at fixed timings. Foliar application of fertilisers, which is the application through the plant leaves, has also been an effective technique. The appropriate selection of the fertiliser's application technique is suggested by various factors, which include crop type, soil conditions, availability of equipment and environmental considerations (Chen et al., 2021).

The **balanced fertilization** is thus an essential way of improving crop yield. Fertilization of crops at the correct rates of crucial nutrients—nitrogen, phosphorus and potassium—is made possible by balanced fertilization based on soil conditions. The use of moderate fertilizers can enhance nutrient efficiency, reduce nutrient loss, mitigate environmental impacts, and optimise plant growth and yield (Hopkins et al., 2020).

8 Sustainable Practices Integration

Agricultural operations with sustainable practices include using renewable sources of energy, managing water resources and using seeds that favor biodiversity. A reduction in adverse environmental effects and improvement in the sustainability of farm production systems result from these practices. **Agroforestry** is about growing trees, shrubs or conserving natural stands of trees and other woody species in the same area as annual crops or pastures, improving fertility, controlling moisture, growing plant life diversely and multiple income streams for growers (Korir et al., 2022).

Precision agriculture technologies are systematic and multi-disciplinary processes which apply efficient multiple-technology management to agricultural production systems. This allows GPS-controlled tractors and drones to 'apply inputs' (fertilizers, seeds etc.) only on specific regions, reducing inputs and their environmental impact. Real time analysis of soil moisture content, nutrient value, and crop condition are reviewed by **satellite data and soil monitoring devices**. However, in the context of sustainable agriculture, these technologies are useful since they help the farmers to utilize practically all the necessary money, machines and devices up to the optimum level without compromising on the quality and quantity of their crop yields (Raj et al., 2022).

Conservation agriculture includes water management. The crop and farm yield calculations depend on the quality of water. With new irrigation devices like drips and smart sprinklers, farmers can now deliver water to their crops when they should, at the appropriate level, without wasting water. Real-time water and water use information is relayed by hydrological information systems and satellite-supported integrated information systems to ensure timely irrigation and equitable water distribution (Metta, 2023).

Irrigation efficiency is a commonly utilized criterion for judging irrigation system water used in crop production. Such advanced measures require the use of standard languages in the communication of information and the determination of the best time or period for irrigation. Dripping and sprinkling can make micro-irrigation methods more efficient in vegetable gardens to reduce or curtail the water evaporation from the soil. Even thorough evaluation of irrigation requirements and subsequent application of sensible solutions, however, is obstructed by various factors like inadequate expertise of farmers, unequipped cultivators, shortage of facilities, along with the financial and political hindrances (Khorsandi et al., 2022).

Rainwater harvesting is a sustainable method for harvesting rainwater for irrigation, borehole recharge and household use. It is considered a particularly appropriate method in areas with limited water resources or during changing weather patterns (i.e., arid or seasonal regions). Rainwater harvesting systems can be simple local systems or more complex community systems. It can be used for many purposes to take pressure off of groundwater and surface water. It also helps to manage drought and flood disasters and reduces the price burden on farmers (Loh, 2022).

Drip irrigation is a water-efficient technology whereby plants are irrigated at the roots by pipes fitted with emitters, especially in arid and semi-arid areas, to meet the water shortage reasons. Drip irrigation systems reduce the water wastage through evaporation. Fertigation is the process of combining fertilizers with water and tillage. Integration of fertigation may also ensure rational utilization of water resources in agriculture (Bhavsar et al., 2023).

Crop diversification is an excellent agricultural management strategy in which farmers grow various crops rather than eliminating thinning crops. The advantages of this practice are high soil status, effective pest and disease control and return on investment to farmers. Through crop diversification in different climatic conditions, farmers can fulfill the demands of different markets, with resultant better risk distribution and better production. Adopting diversified cropping systems will go a long way in providing alternative options in food production. A wide variety of crops, thus, can be produced which will expand the food production system. These sustainable practices must be applied to combat various challenges of climate, land utilization, and soil structures, to realize the benefits of such practices fully (Bhavsar et al., 2023).

Marketing agricultural sustainability involves monitoring and evaluating soil health, disease and pest surveillance, yield analyses and environmental impact assessment. The maintenance of proper health conditions of production sites requires soil health assessment so that farmers may become capable of deciding when and how to use organic inputs, cover crops, or change tillage practices. This information is useful to preserve and increase the capacity of the soil to allow plant growth, hold water and prevent soil erosion (Sivaraman et al., 2023).

Supervision of plants by the use of biomarkers for diseases and pests that may potentially affect plant health is included in planned pest management (i.e., IPM) in crop production. Early detection of these diseases is important for farmers to control them at the right time (Liu & Wang, 2021).

Today's **agriculture is a functioning model** in which output is analyzed and monitored, in terms of data accumulation and crop production, productivity, quality and performance. It also enables to locate the impact of activities, for example, the release of new plant varieties, precise farming approaches, or sustainable agricultural practices, on yields. This enables farmers to analyze and improve farming processes and boost productivity, profitability, and conservation of the environment (Lee & Shin, 2020).

Environmental Impact Assessment (EIA) is a fundamental process designed to describe the environmental impacts associated with a planned project or activity. The typical contents of EIA include scoping, impact identification, impact prediction and impact evaluation. Authorities and interested parties remain informed of the collection and assessment of the information, and plans and measures to reduce or avoid the adverse impacts on the environment are decided. An EIA offers an informed decision on whether a project is viable and has sustainable impacts. It is used to identify potential environmental impacts, evaluate alternatives, and determine how to manage actions that will limit adverse effects. The use of the EIA framework is deemed an important tool for developments to be based on sustainable development, in which there is a balance between rapid economic development and environmental conservation. It provides a broader view of the potential long-term cumulative effect of a project enabling more professional decisions that the weight of the project (Enríquez-de-Salamanca, 2021b).

9 Education and Outreach

Farmer training programs are important to educate the farmers to improve farming practices, yielding, and sustainable methods. These programs are knowledge-based on advanced technologies, crop formation techniques and resource preservation. Resources and location-specific information on soil health, pest and disease management, efficient water use and postharvest management are imparted through field days, group discussions and exposure visits (Antwi-Agyei & Stringer, 2021).

Information exchange opportunities like workshops and conferences provide unique platforms for the global research and scientific communities to share innovative ideas to meet emerging global challenges. However, these events provide the farmers, the researchers and other stakeholders with an opportunity to learn from each other and develop strategies to cope with problems faced by various farming communities (Raj et al., 2022).

Knowledge-sharing platforms, which include decision support systems, apply information and communication technologies to share agricultural raw materials, innovation and experiences. They are used to create networks of people and entities that can give (or receive) information about many things in the field of agriculture. There are support groups on social networking sites, newsgroups, mailing lists, and bulletin boards. Research publications, technical manuals, and extension literature are compiled by agricultural universities/publishers and put in the database and digital libraries to update the knowledge base and extension messages (Ashraf et al., 2020).

Community engagement initiatives, including school, local community and policy maker engagement, aimed at increasing regional food production. Such support for the long-term sustainability of organic agriculture is seen in the active participation of the school community in health and local agricultural matters through program implementation. Support includes free and open access to urban agriculture and the development of relevant policies by community organizations and local governments. Whereas the involvement of youth in the agricultural value chain is critical, yet must also deal with many constraints (Borrero & Mariscal, 2022). These programs are meant to promote sustainable horticultural efforts in the community. The individual experiences of individuals working on agriculture projects at different scales can help illustrate how local civil society operates across organizational silos, using different 'technologies of government.'

10 Conclusion

Modern agronomy addresses the problem of soil degradation and plant health decline by providing techniques such as crop rotation, mulching, and conservation tillage, together with precision agriculture technologies. In the future, the main ways to optimize resource use will be artificial intelligence (AI) tools, satellite-based monitoring and precision agriculture technologies. Agriculture data can be fed into vast quantities of data by AI algorithms to predict crop performance, optimize pest management and guide fertilizer application. Real-time monitoring of soil moisture, crop health, and nutrient levels is provided by satellites and drones, so we can intervene and reduce waste and increase yield with timely increases. In addition, remote sensing technologies adapted to machine learning will transform how farmers measure soil conditions and health across large areas. Future research is needed to refine these modern techniques and expand their use to farmers worldwide to reach truly sustainable agriculture. It is important to put ample emphasis on educational IT initiatives and outreach programs to extend the use of these tools in regions where traditional methods still dominate. Changing climate and increasing demand for food of increasing quality will require a close collaboration among scientists, policymakers and farmers for progress in global food security and protection of natural resources to benefit future generations.

References

Abbas, M., Saleem, M., Hussain, D., Ramzan, M., Jawad Saleem, M., Abbas, S., Hussain, N., Irshad, M., Hussain, K., & Ghouse, G. (2022). Review on integrated disease and pest management of field crops. *International Journal of Tropical Insect Science, 42*(5), 3235–3243.

Abdolvand, Y., & Sadeghiamirshahidi, M. (2024). Soil stabilization with gypsum: A review. *Journal of Rock Mechanics and Geotechnical Engineering.*

Adamashvili, N., Colantuono, F., Conto, F., & Fiore, M. (2020). Investigating the role of community of practice for sharing knowledge in agriculture sector. *Journal for Global Business Advancement, 13*(2), 162–184.

Ajl, M., & Sharma, D. (2022). The green revolution and transversal countermovements: Recovering alternative agronomic imaginaries in Tunisia and India. *Canadian Journal of Development Studies/revue canadienne d'études du développement, 43*(3), 418–438.

Alcon, F., Marín-Miñano, C., Zabala, J. A., de-Miguel, M.-D., & Martínez-Paz, J. M. (2020). Valuing diversification benefits through intercropping in Mediterranean agroecosystems: A choice experiment approach. *Ecological Economics, 171*, 106593.

Aneseyee, A. B., Elias, E., Soromessa, T., & Feyisa, G. L. (2020). Land use/land cover change effect on soil erosion and sediment delivery in the Winike watershed, Omo Gibe Basin, Ethiopia. *Science of the Total Environment, 728*, 138776.

Angst, G., Mueller, K. E., Nierop, K. G., & Simpson, M. J. (2021). Plant-or microbial-derived? A review on the molecular composition of stabilized soil organic matter. *Soil Biology and Biochemistry, 156*, 108189.

Antwi-Agyei, P., & Stringer, L. C. (2021). Improving the effectiveness of agricultural extension services in supporting farmers to adapt to climate change: Insights from northeastern Ghana. *Climate Risk Management, 32*, 100304.

Ashraf, E., Sarwar, A., Junaid, M., Baig, M. B., Shurjeel, H. K., & Barrick, R. K. (2020). An assessment of in-service training needs for agricultural extension field staff in the scenario of climate change using borich needs assessment model. *Sarhad Journal of Agriculture, 36*(2).

Banerjee, S., & Van Der Heijden, M. G. (2023). Soil microbiomes and one health. *Nature Reviews Microbiology, 21*(1), 6–20.

Beaumelle, L., Auriol, A., Grasset, M., Pavy, A., Thiéry, D., & Rusch, A. (2021). Benefits of increased cover crop diversity for predators and biological pest control depend on the landscape context. *Ecological Solutions and Evidence, 2*(3), e12086.

Bhavsar, D., Limbasia, B., Mori, Y., Aglodiya, M. I., & Shah, M. (2023). A comprehensive and systematic study in smart drip and sprinkler irrigation systems. *Smart Agricultural Technology, 5*, 100303.

Blair, J., & Mataraarachchi, S. (2021). A review of landfills, waste and the nearly forgotten nexus with climate change. *Environments, 8*(8), 73.

Borrero, J. D., & Mariscal, J. (2022). A case study of a digital data platform for the agricultural sector: A valuable decision support system for small farmers. *Agriculture, 12*(6), 767.

Bungau, S., Behl, T., Aleya, L., Bourgeade, P., Aloui-Sossé, B., Purza, A. L., Abid, A., & Samuel, A. D. (2021). Expatiating the impact of anthropogenic aspects and climatic factors on long-term soil monitoring and management. *Environmental Science and Pollution Research, 28*(24), 30528–30550.

Carr, P. M., Cavigelli, M. A., Darby, H., Delate, K., Eberly, J. O., Fryer, H. K., Gramig, G. G., Heckman, J. R., Mallory, E. B., & Reeve, J. R. (2020). Green and animal manure use in organic field crop systems. *Agronomy Journal, 112*(2), 648–674.

Chen, Y., Hu, S., Guo, Z., Cui, T., Zhang, L., Lu, C., Yu, Y., Luo, Z., Fu, H., & Jin, Y. (2021). Effect of balanced nutrient fertilizer: A case study in Pinggu District, Beijing, China. *Science of the Total Environment, 754*, 142069.

Chernov, T., & Semenov, M. (2021). Management of soil microbial communities: Opportunities and prospects (a review). *Eurasian Soil Science, 54*, 1888–1902.

Chopra, M., & Koul, B. (2020). Comparative assessment of different types of mulching in various crops: A review. *Plant Arch, 20*, 1620–1626.

Cooper, J., Greenberg, I., Ludwig, B., Hippich, L., Fischer, D., Glaser, B., & Kaiser, M. (2020). Effect of biochar and compost on soil properties and organic matter in aggregate size fractions under field conditions. *Agriculture, Ecosystems & Environment, 295*, 106882.

Das, P. P., Singh, K. R., Nagpure, G., Mansoori, A., Singh, R. P., Ghazi, I. A., Kumar, A., & Singh, J. (2022). Plant-soil-microbes: A tripartite interaction for nutrient acquisition and better plant growth for sustainable agricultural practices. *Environmental Research, 214*, 113821.

de Mello, K., Taniwaki, R. H., de Paula, F. R., Valente, R. A., Randhir, T. O., Macedo, D. R., Leal, C. G., Rodrigues, C. B., & Hughes, R. M. (2020). Multiscale land use impacts on water quality: Assessment, planning, and future perspectives in Brazil. *Journal of Environmental Management, 270*, 110879.

Deng, C., Zhang, G., Liu, Y., Nie, X., Li, Z., Liu, J., & Zhu, D. (2021). Advantages and disadvantages of terracing: A comprehensive review. *International Soil and Water Conservation Research, 9*(3), 344–359.

Dolatabadian, A., Cornelsen, J., Huang, S., Zou, Z., & Fernando, W. D. (2022). Sustainability on the farm: Breeding for resistance and management of major canola diseases in Canada contributing towards an IPM approach. *Canadian Journal of Plant Pathology, 44*(2), 157–190.

Dong, E., Du, H., & Gardner, L. (2020). An interactive web-based dashboard to track COVID-19 in real time. *The Lancet Infectious Diseases, 20*(5), 533–534.

Duhaithem, S. (2024). *Engineering and functional assessment of native and pruned versions of the Sr26 stem rust resistance gene in wheat.*

Ek, L. C. (2023). *Replenishment of essential natural resources: An assessment of the ability of freshwater algae to enhance crop quality and soil health.*

El-Beltagi, H. S., Basit, A., Mohamed, H. I., Ali, I., Ullah, S., Kamel, E. A., Shalaby, T. A., Ramadan, K. M., Alkhateeb, A. A., & Ghazzawy, H. S. (2022). Mulching as a sustainable water and soil saving practice in agriculture: A review. *Agronomy, 12*(8), 1881.

Enríquez-de-Salamanca, Á. (2021b). Simplified environmental impact assessment processes: Review and implementation proposals. *Environmental Impact Assessment Review, 90*, Article 106640.

Fahad, S., Chavan, S. B., Chichaghare, A. R., Uthappa, A. R., Kumar, M., Kakade, V., Pradhan, A., Jinger, D., Rawale, G., & Yadav, D. K. (2022). Agroforestry systems for soil health improvement and maintenance. *Sustainability, 14*(22), 14877.

Fantappiè, M., Lorenzetti, R., De Meo, I., & Costantini, E. A. C. (2020). How to improve the adoption of soil conservation practices? Suggestions from farmers' perception in western Sicily. *Journal of Rural Studies, 73*, 186–202.

Farmaha, B. S., Sekaran, U., & Franzluebbers, A. J. (2022). Cover cropping and conservation tillage improve soil health in the southeastern United States. *Agronomy Journal, 114*(1), 296–316.

Fernández-Cabezón, L., & Nikel, P. I. (2020). Advanced metabolic engineering strategies for the development of sustainable microbial processes. In *New and future developments in microbial biotechnology and bioengineering* (pp. 225–246). Elsevier. https://doi.org/10.1016/B978-0-444-64301-8.00011-1

Fierer, N., Wood, S. A., & de Mesquita, C. P. B. (2021). How microbes can, and cannot, be used to assess soil health. *Soil Biology and Biochemistry, 153*, 108111.

Florkowski, W. J., Banks, N. H., Shewfelt, R. L., & Prussia, S. E. (2021). *Postharvest handling: A systems approach.* Academic press.

Friedmann, H. (2021). Changes in the international division of labor: agri-food complexes and export agriculture. In *Towards a new political economy of agriculture* (pp. 65–93). Routledge.

Garbowski, T., Bar-Michalczyk, D., Charazińska, S., Grabowska-Polanowska, B., Kowalczyk, A., & Lochyński, P. (2023). An overview of natural soil amendments in agriculture. *Soil and Tillage Research, 225*, 105462.

Gawande, V., Saikanth, D. R. K., Sumithra, B. S., Aravind, S. A., Swamy, G. N., Chowdhury, M., & Singh, B. V. (2023). Potential of precision farming technologies for eco-friendly agriculture. *International Journal of Plant & Soil Science, 35*(19), 101–112.
Gojon, A., Nussaume, L., Luu, D. T., Murchie, E. H., Baekelandt, A., Rodrigues Saltenis, V. L., Cohan, J. p., Desnos, T., Inzé, D., & Ferguson, J. N. (2023). Approaches and determinants to sustainably improve crop production. *Food and Energy Security, 12*(1), e369.
Guerra, C. A., Bardgett, R. D., Caon, L., Crowther, T. W., Delgado-Baquerizo, M., Montanarella, L., Navarro, L. M., Orgiazzi, A., Singh, B. K., & Tedersoo, L. (2021). Tracking, targeting, and conserving soil biodiversity. *Science, 371*(6526), 239–241.
Gupta, P., Gupta, N., Dash, S., & Singh, M. (2023). Role of algae in biodegradation of plastics. In *Next-generation algae* (pp. 125–145). Wiley. https://doi.org/10.1002/9781119857839.ch5
Hancock, G. R., Ovenden, M., Sharma, K., Rowlands, W., Gibson, A., & Wells, T. (2020). Soil erosion – The impact of grazing and regrowth trees. *Geoderma, 361*, 114102. https://doi.org/10.1016/j.geoderma.2019.114102
Havlin, J. L. (2020). Soil: Fertility and nutrient management. In *Landscape and land capacity* (pp. 251–265). CRC Press.
Hills, E. (2024). Arid lands and human problems. *Arid Lands: A Geographical Appriasal, 13.*
Hombegowda, H. C., Adhikary, P. P., Jakhar, P., & Madhu, M. (2022). Alley cropping agroforestry system for improvement of soil health. In *Soil health and environmental sustainability: Application of geospatial technology* (pp. 529–549). Springer.
Hopkins, B. G., Stark, J. C., & Kelling, K. A. (2020). Nutrient management. In *Potato production systems*, 155–202.
Huang, X., Liang, P., Jiang, B., Zhang, P., Yu, W., Duan, M., Guo, L., Cui, X., Huang, M., & Huang, X. (2020). Hyperbaric oxygen potentiates diabetic wound healing by promoting fibroblast cell proliferation and endothelial cell angiogenesis. *Life Sciences, 259*, 118246. https://doi.org/10.1016/j.lfs.2020.118246
Hussain, S., Hussain, S., Guo, R., Sarwar, M., Ren, X., Krstic, D., Aslam, Z., Zulifqar, U., Rauf, A., & Hano, C. (2021). Carbon sequestration to avoid soil degradation: A review on the role of conservation tillage. *Plants, 10*(10), 2001.
Jain, M. S., & Kalamdhad, A. S. (2020). Soil revitalization via waste utilization: Compost effects on soil organic properties, nutritional, sorption and physical properties. *Environmental Technology & Innovation, 18*, 100668.
Jalli, M., Huusela, E., Jalli, H., Kauppi, K., Niemi, M., Himanen, S., & Jauhiainen, L. (2021). Effects of crop rotation on spring wheat yield and pest occurrence in different tillage systems: A multi-year experiment in Finnish growing conditions. *Frontiers in Sustainable Food Systems, 5*, 647335.
Kahn, R. P. (1991). Exclusion as a plant disease control strategy. *Annual Review of Phytopathology, 29*(1), 219–246. https://doi.org/10.1146/annurev.py.29.090191.001251
Khorsandi, M., Homayouni, S., & Van Oel, P. (2022). The edge of the petri dish for a nation: Water resources carrying capacity assessment for Iran. *Science of the Total Environment, 817*, 153038.
Kim, Y. E., Cho, K. H., Bang, I., Kim, C. H., Ryu, Y. S., Kim, Y., Choi, E. M., Nong, L. K., Kim, D., & Lee, S. K. (2022). Characterization of an Entner–Doudoroff pathway-activated Escherichia coli. *Biotechnology for Biofuels and Bioproducts, 15*(1), 120. https://doi.org/10.1186/s13068-022-02219-6
Korir, K. E., Sirmah, P. K., Matonyei, T. K., & Nampushi, J. S. O. (2022). Classification and socio-economic benefits of agroforestry systems in soin ward, Kericho County, Kenya.
Koudahe, K., Allen, S. C., & Djaman, K. (2022). Critical review of the impact of cover crops on soil properties. *International Soil and Water Conservation Research, 10*(3), 343–354.
Kovács, G. P., Simon, B., Balla, I., Bozóki, B., Dekemati, I., Gyuricza, C., Percze, A., & Birkás, M. (2023). Conservation tillage improves soil quality and crop yield in Hungary. *Agronomy, 13*(3), 894.

Laamrani, A., Valeria, O., Chehbouni, A., & Bergeron, Y. (2020). Analysis of the effect of climate warming on paludification processes: Will soil conditions limit the adaptation of northern boreal forests to climate change? A synthesis. *Forests, 11*(11), 1176.

Lannelongue, L., Grealey, J., & Inouye, M. (2021). Green algorithms: Quantifying the carbon footprint of computation. *Advanced Science, 8*(12), 2100707.

Lee, Y.-J., & Shin, B.-S. (2020). Development of potato yield monitoring system using machine vision. *Journal of Biosystems Engineering, 45*, 282–290.

Lehmann, J., Bossio, D. A., Kögel-Knabner, I., & Rillig, M. C. (2020). The concept and future prospects of soil health. *Nature Reviews Earth & Environment, 1*(10), 544–553.

Liu, J., & Wang, X. (2021). Plant diseases and pests detection based on deep learning: A review. *Plant Methods, 17*, 1–18.

Lobell, D. B., Azzari, G., Burke, M., Gourlay, S., Jin, Z., Kilic, T., & Murray, S. (2020). Eyes in the sky, boots on the ground: Assessing satellite-and ground-based approaches to crop yield measurement and analysis. *American Journal of Agricultural Economics, 102*(1), 202–219.

Loh, Y. W. (2022). *Design of an automated rainwater harvesting system for urban agriculture.* UTAR.

Luna, J. M., & House, G. J. (2020). Pest management in sustainable agricultural systems. In *Sustainable agricultural systems* (pp. 157–173). CRC Press.

Lykogianni, M., Bempelou, E., Karamaouna, F., & Aliferis, K. A. (2021). Do pesticides promote or hinder sustainability in agriculture? The challenge of sustainable use of pesticides in modern agriculture. *Science of the Total Environment, 795*, 148625.

Maloy, O. C. (2005). Plant disease management. *The Plant Health Instructor.* https://doi.org/10.1094/PHI-I-2005-0202-01

Manea, E. E., Bumbac, C., Dinu, L. R., Bumbac, M., & Nicolescu, C. M. (2024). Composting as a sustainable solution for organic solid waste management: Current practices and potential improvements. *Sustainability, 16*(15), 6329.

Metta, M. (2023). *Digitalisation and on-farm diversification. Challenges and opportunities for multifunctional and diversified agriculture.* Ghent University.

Mirzaei, M., Gorji Anari, M., Razavy-Toosi, E., Asadi, H., Moghiseh, E., Saronjic, N., & Rodrigo-Comino, J. (2021). Preliminary effects of crop residue management on soil quality and crop production under different soil management regimes in corn-wheat rotation systems. *Agronomy, 11*(2), 302. https://doi.org/10.3390/agronomy11020302

Moerman, M. (2021). *Agricultural change and peasant choice in a Thai village.* University of California Press.

Mohanty, L. K., Singh, N., Raj, P., Prakash, A., Tiwari, A. K., Singh, V., & Sachan, P. (2024). Nurturing crops, enhancing soil health, and sustaining agricultural prosperity worldwide through agronomy. *Journal of Experimental Agriculture International, 46*(2), 46–67.

Montgomery, D. R., Rabinowitz, P., Sipos, Y., & Wheat, E. E. (2024). Soil health: A common focus for one health and planetary health interventions. *One Health, 18*, 100673.

Mrabet, R. (2023). Sustainable agriculture for food and nutritional security. In *Sustainable agriculture and the environment* (pp. 25–90). Elsevier.

Muluneh, M. G. (2021). Impact of climate change on biodiversity and food security: A global perspective—A review article. *Agriculture & Food Security, 10*(1), 1–25.

Naorem, A., Jayaraman, S., Dang, Y. P., Dalal, R. C., Sinha, N. K., Rao, C. S., & Patra, A. K. (2023). Soil constraints in an arid environment—Challenges, prospects, and implications. *Agronomy, 13*(1), 220.

Obaideen, K., Yousef, B. A. A., AlMallahi, M. N., Tan, Y. C., Mahmoud, M., Jaber, H., & Ramadan, M. (2022). An overview of smart irrigation systems using IoT. *Energy Nexus, 7*, 100124. https://doi.org/10.1016/j.nexus.2022.100124

Ogieriakhi, M. O., & Woodward, R. T. (2022). Understanding why farmers adopt soil conservation tillage: A systematic review. *Soil Security, 9*, 100077.

Ons, L., Bylemans, D., Thevissen, K., & Cammue, B. P. A. (2020). Combining biocontrol agents with chemical fungicides for integrated plant fungal disease control. *Microorganisms, 8*(12), 1930.

Özbolat, O., Sánchez-Navarro, V., Zornoza, R., Egea-Cortines, M., Cuartero, J., Ros, M., Pascual, J. A., Boix-Fayos, C., Almagro, M., & de Vente, J. (2023). Long-term adoption of reduced tillage and green manure improves soil physicochemical properties and increases the abundance of beneficial bacteria in a Mediterranean rainfed almond orchard. *Geoderma, 429*, 116218.

Panth, M., Hassler, S. C., & Baysal-Gurel, F. (2020). Methods for management of soilborne diseases in crop production. *Agriculture, 10*(1), 16.

Pawase, P. P., Nalawade, S. M., Bhanage, G. B., Walunj, A. A., Kadam, P. B., Durgude, A. G., & Patil, M. R. (2023). Variable rate fertilizer application technology for nutrient management: A review. *International Journal of Agricultural and Biological Engineering, 16*(4), 11–19.

Pisante, M., Gogoi, N., & Farooq, M. (2023). Sustainability, sustainable agriculture, and the environment. In *Sustainable agriculture and the environment* (pp. 3–24). Elsevier.

Pitt, S. J., & Gunn, A. (2024). The one health concept. *British Journal of Biomedical Science, 81*, 12366.

Prem, M., Ranjan, P., Seth, N., & Patle, G. T. (2020). Mulching techniques to conserve the soil water and advance the crop production—A Review. *Current World Environment: An International Research Journal of Environmental Sciences, 15*, 10–30.

Radočaj, D., Jurišić, M., & Gašparović, M. (2022). The role of remote sensing data and methods in a modern approach to fertilization in precision agriculture. *Remote Sensing, 14*(3), 778.

Rahman, M. M., Alam, M. S., Kamal, M. Z. U., & Rahman, G. M. (2020). Organic sources and tillage practices for soil management. *Resources Use Efficiency in Agriculture*, 283–328.

Raj, E. F. I., Appadurai, M., & Athiappan, K. (2022). Precision farming in modern agriculture. In *Smart agriculture automation using advanced technologies: Data analytics and machine learning, cloud architecture, automation and IoT* (pp. 61–87). Springer.

Rawat, L., Bisht, T., & Naithani, D. C. (2021). Plant disease management in organic farming system: strategies and challenges. In *Emerging trends in plant pathology* (pp. 611–642).

Rehman, S. U., De Castro, F., Aprile, A., Benedetti, M., & Fanizzi, F. P. (2023). Vermicompost: Enhancing plant growth and combating abiotic and biotic stress. *Agronomy, 13*(4), 1134. https://doi.org/10.3390/agronomy13041134

Sakin, E., Yanardağ, H. İ., Firat, Z., Çelik, A., Beyyavaş, V., & Suat, C. (2024). Some indicators for the assessment of soil health: A mini review. *MAS Journal of Applied Sciences, 9*(2), 297–310.

Shah, K., Tripathi, S., Tiwari, I., Shrestha, J., Modi, B., Paudel, N., & Das, B. (2021). Role of soil microbes in sustainable crop production and soil health: A review. *Agricultural Science and Technology, 13*(2), 109–118.

Shaikh, T., Patel, J. R., & Patel, B. (2023). Mulching effect on water management in agro-ecological systems: A review. *AQUA—Water Infrastructure, Ecosystems and Society, 72*(9), 1664–1680.

Sharma, G., Shrestha, S., Kunwar, S., & Tseng, T.-M. (2021). Crop diversification for improved weed management: A review. *Agriculture, 11*(5), 461.

Sharma, I. P., Kanta, C., Dwivedi, T., & Rani, R. (2020). Indigenous agricultural practices: A supreme key to maintaining biodiversity. *Microbiological Advancements for higher Altitude Agro-ecosystems & Sustainability*, 91–112.

Singh, A. (2022). Soil salinity: A global threat to sustainable development. *Soil Use and Management, 38*(1), 39–67.

Singh, V. K., Malhi, G. S., Kaur, M., Singh, G., & Jatav, H. S. (2022). Use of organic soil amendments for improving soil ecosystem health and crop productivity. *Ecosystem Services.*

Sishodia, R. P., Ray, R. L., & Singh, S. K. (2020). Applications of remote sensing in precision agriculture: A review. *Remote Sensing, 12*(19), 3136.

Sivaraman, K., Thankamani, C., & Srinivasan, V. (2023). Crop diversification: Cropping/system approach for enhancing farmers' income. In *Handbook of spices in India: 75 years of research and development* (pp. 3847–3926). Springer.

Somasundaram, J., Sinha, N., Dalal, R. C., Lal, R., Mohanty, M., Naorem, A., Hati, K., Chaudhary, R., Biswas, A., & Patra, A. (2020). No-till farming and conservation agriculture in South Asia–issues, challenges, prospects and benefits. *Critical Reviews in Plant Sciences, 39*(3), 236–279.

Stephenson, P. J. (2020). Technological advances in biodiversity monitoring: Applicability, opportunities and challenges. *Current Opinion in Environmental Sustainability, 45*, 36–41.

Tahat, M. M., Alananbeh, K. M., Othman, Y. A., & Leskovar, D. I. (2020). Soil health and sustainable agriculture. *Sustainability, 12*(12), 4859.

Tanda, A. S. (2024). *Advances in biological control pest management technology*. IK International Pvt Ltd.

Tariq, M., Khan, A., Asif, M., Khan, F., Ansari, T., Shariq, M., & Siddiqui, M. A. (2020). Biological control: a sustainable and practical approach for plant disease management. *Acta Agriculturae Scandinavica, Section B—Soil & Plant Science, 70*(6), 507–524.

Terentev, A., Dolzhenko, V., Fedotov, A., & Eremenko, D. (2022). Current state of hyperspectral remote sensing for early plant disease detection: A review. *Sensors, 22*(3), 757.

Tian, Z., Wang, J. W., Li, J., & Han, B. (2021). Designing future crops: Challenges and strategies for sustainable agriculture. *The Plant Journal, 105*(5), 1165–1178.

Tiwari, S. P. (2022). *Information and communication technology initiatives for knowledge sharing in agriculture*. arXiv preprint arXiv:2202.08649.

Trozzo, K. E., Munsell, J. F., Chamberlain, J. L., Gold, M. A., & Niewolny, K. L. (2021). Forest farming: Who wants in? *Journal of Forestry, 119*(5), 478–492.

Udawatta, R. P., & Gantzer, C. J. (2022). Soil and water ecosystem services of agroforestry. *Journal of Soil and Water Conservation, 77*(1), 5A–11A.

Van Esse, H. P., Reuber, T. L., & van der Does, D. (2020). Genetic modification to improve disease resistance in crops. *New Phytologist, 225*(1), 70–86.

Vialatte, A., Tibi, A., Alignier, A., Angeon, V., Bedoussac, L., Bohan, D. A., Bougherara, D., Carpentier, A., Castagneyrol, B., & Cordeau, S. (2021). Promoting crop pest control by plant diversification in agricultural landscapes: A conceptual framework for analysing feedback loops between agro-ecological and socio-economic effects. In *Advances in ecological research* (Vol. 65, pp. 133–165). Elsevier.

Walia, S., & Walia, U. (2021). *Crop management*, 2nd ed. Scientific Publishers.

Walia, S. S., Kaur, K., & Kaur, T. (2024). *Rainfed agriculture and watershed management*. Springer.

Wang, L., Lu, P., Feng, S., Hamel, C., Sun, D., Siddique, K. H., & Gan, G. Y. (2024). Strategies to improve soil health by optimizing the plant–soil–microbe–anthropogenic activity nexus. *Agriculture, Ecosystems & Environment, 359*, 108750.

Weeraratna, S. (2022). Control of land degradation. In *Understanding land degradation: An overview* (pp. 39–51). Springer.

Wenninger, E. J., Rashed, A., Rondon, S. I., Alyokhin, A., & Alvarez, J. M. (2020). Insect pests and their management. *Potato Production Systems*, 283–345.

Yadav, A., Singh, D., & Yadav, A. (2023). Insect pest management through integrated pest tactics. *Advanced Trends in Agricultural Entomology, 1*, 89–110.

Yu, T., Mahe, L., Li, Y., Wei, X., Deng, X., & Zhang, D. (2022). Benefits of crop rotation on climate resilience and its prospects in China. *Agronomy, 12*(2), 436.

Zhang, X., Davidson, E., Zou, T., Lassaletta, L., Quan, Z., Li, T., & Zhang, W. (2020). Quantifying nutrient budgets for sustainable nutrient management. *Global Biogeochemical Cycles, 34*(3), e2018GB006060.

The Impact of Waste on Psychological Health in Urban Environments

Qui Thanh Hoai Ta, Van Nam Huynh, Anh Tuan Nguyen, Ninh Hai Nguyen, Quoc-Minh Truong, Minh Hai Tran, Le Thuy Trang, Nguyen Thi Thanh Huyen, Van-Giang Le, and Tuan-Dung Hoang

Abstract In urban settings worldwide, the increase of waste presents a multifaceted challenge with profound implications for public health, particularly in the realm of psychological well-being. This chapter offers a synthesis of existing research examining the intricate relationship between waste and psychological health in urban environments. The chapter discusses diverse forms of waste, including solid waste,

Q. T. H. Ta
Institute of Advanced Technology, Vietnam Academy of Science and Technology, Thanh Loc Ward, District 12, Ho Chi Minh City, Vietnam

V. N. Huynh
Faculty of Natural Sciences, Quy Nhon University, Quy Nhon, Vietnam

A. T. Nguyen · N. H. Nguyen
Department of Environment and Natural Resources, Nong Lam University of Ho Chi Minh City, Ho Chi Minh City, Vietnam

Q.-M. Truong
Faculty of Management Science, Thu Dau Mot University, Thu Dau Mot, Binh Duong, Vietnam

M. H. Tran
Materials and Chemical Engineering Department, Concordia University, Montreal, Canada

L. T. Trang
Faculty of Environmental and Natural Sciences, Duy Tan University, Da Nang, Vietnam

School of Engineering and Technology, Duy Tan University, Da Nang, Vietnam

N. T. T. Huyen
Faculty of International Economics, Foreign Trade University, Hanoi, Vietnam

V.-G. Le
Central Institute for Natural Resources and Environmental Studies, Vietnam National University, Hanoi, Vietnam

T.-D. Hoang (✉)
School of Chemistry and Life Science, Hanoi University of Science and Technology, Hanoi, Vietnam
e-mail: tuandunghoang@gmail.com

Vietnam National University, Hanoi-School of Interdisciplinary Sciences and Arts, Cau Giay District, Hanoi 100000, Vietnam

M. Shaaban (ed.), *Soils and Sustainable Agriculture*, Frontier Studies in Soil Science,
https://doi.org/10.1007/978-3-031-91114-9_43

air pollution, pointing out their collective impacts on mental health. Exploring the psychological effects of waste to human health reveals a complex interplay of stress, anxiety, and diminished quality of life among urban residents. Factors such as exposure to noxious odors, visual blight, and concerns about environmental degradation contribute to heightened levels of psychological distress. Additionally, waste burdens across socio-economic aspects affected mental health outcomes, underscoring the need for equitable waste management policies. This chapter examines the mediating role of green spaces and community engagement in mitigating the adverse psychological effects of waste in urban environments. Research suggests that access to greenery and opportunities for social cohesion can buffer the negative impact of waste, promoting resilience and fostering a sense of well-being among urban dwellers. Furthermore, this paper reviews concepts of environment security. This chapter also provides a new approach or pathway that plays a key role in enhancing sustainable development goals. Ultimately, this chapter underscores the urgent imperative for multiple approaches to waste management that prioritize both environmental sustainability and psychological health. By addressing the psychological dimensions of waste in urban planning and policy formulation, stakeholders can cultivate healthier, more resilient communities and enhance the overall well-being of urban populations.

Keywords Psychological dimensions · Urban planning · Solid waste · Well-being; urban environments · Mental health

1 Introduction

Waste in an urban environment can be solid waste, wastewater, and air pollution (Khajuria et al., 2010). Some of the primary contributors to environmental pollution are wastewater, solid waste, and waste gas, which constitute pressing social and environmental concerns. Specifically, these pollutants directly impact human psychological health. According to existing literature, these types of waste are closely interrelated, leading to a cascade of effects and repercussions.

Solid waste encompasses various forms of refuse, including garbage, sludge from water treatment plants, and materials from air pollution control facilities, among others (Muthuraman & Ramaswamy, 2019). This category includes solid, liquid, and semisolid waste originating from urban environments, commercial activities, and mining operations. Meanwhile, gaseous wastes predominantly stem from vehicular activities and chemical industries as manufacturing byproducts. The first category of common gaseous waste comprises carbon dioxide (CO_2), carbon monoxide (CO), methane (CH_4), ethane (C_2H_6), and others (González et al., 2024). The second category includes nitrogen-containing gaseous waste such as nitrogen dioxide (NO_2), nitric oxide (NO), nitrous oxides (N_2O), and others. Wastewater, on the other hand, refers to water generated after its use in various processes or applications, including drinking water, freshwater, and raw water (Habibi-Yangjeh & Pournemati, 2024).

Exposure to even moderate concentrations of toxic gases, such as nitrogen dioxide (NO_2), over time, can potentially induce changes in the respiratory system and lung function. For instance, concentrations ranging from 0.01 to 0.02% of NO_2 have been associated with mild irritation of the respiratory system and nasal passages, while levels between 0.025 and 0.05% may lead to edema, consequently increasing the risk of bronchitis or pneumonia (Ta, 2024). In more severe cases, where the concentration of NO_2 in the air exceeds 0.1%, fatalities have been reported due to asphyxiation and pulmonary edema (Piersanti et al., 2024).

Despite the gravity of these health risks, symptoms of NO_2 exposure are often not immediately evident, with individuals experiencing only transient coughing, fatigue, or nausea at the time of exposure. However, after approximately 60 min, inflammation in the lungs may become apparent due to edema. Additionally, skin and ocular tissues are susceptible to adverse effects upon exposure, manifesting as hemoglobin alteration, airway inflammation, heightened asthma risk, and various respiratory symptoms (Abbah et al., 2023).

The infiltration of contaminants from urban soils into the human food and water supply represents a significant issue. Trace elements, nano particles, microplastic originating from soil, including heavy metals such as chromium, cadmium, lead, arsenic, and mercury, can permeate the human food chain via soil-to-water and soil-to-crop pathways (Tziourrou & Golia, 2024, Tran et al., 2024b). Once soil becomes contaminated due to waste residues, microplastic or heavy metal deposits, it poses substantial risks to human health (Hoang V.-H et al., 2024). These risks include heightened levels of stress, anxiety, depression, mood disorders, and cognitive impairment. Increased human exposure to these contaminants occurs through inhalation, dermal contact, ingestion of soil particles, as well as through dietary intake or consumption of contaminated drinking water. The pervasive impact of waste on psychological health underscores the urgent need for comprehensive measures to mitigate environmental contamination and save communities.

The occurrence of viral pathogens in drinking water sources such as raw surface water or poorly treated wastewater systems was important to guarantee promising transmission risks (Dong et al., 2024). This phenomenon can be explained by the wastewater from the urban environment being discharged into the natural environment without any adequate treatment, even basic sanitation facilities (Fig. 1). As a result, exposure to low doses of viral pathogens can pose a potential risk of illness, infection, and mortality to susceptible citizens. There are several kinds of research have reported that the health risks can infected with the probable enteric and non-enteric viruses by facing different aquatic environments, such as recreational water, wastewater, and drinking water (Kumthip et al., 2019).

Moreover, chemical poisoning resulting from inhalation of uncollected water chemicals can impede stormwater runoff, leading to flooding, low birth weight, cancer, congenital malformations, and neurological diseases (Singh, 2023; Le Thuy et al., 2025). Various symptoms manifest in the human body when exposed to hazardous solid waste. For instance, individuals may experience nausea and vomiting, alongside an increase in hospitalization among diabetic residents residing near

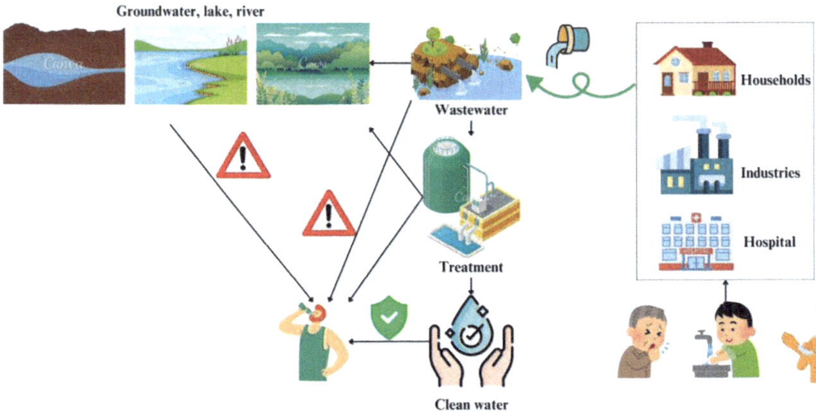

Fig. 1 The possible way untreated wastewater comes into water bodies and human health

hazardous waste sites. Furthermore, mercury toxicity may arise from the consumption of fish containing elevated mercury levels, which can be attributed to the disposal of mercury in rivers. Plastic pollution in oceans poses risks to birds through ingestion, consequently resulting in a proliferation of algal populations in rivers and seas, thereby degrading water and soil quality.

Besides, noise pollution represents a serious and escalating concern in urban settings. With the advancement of science and technology, urban residents are increasingly exposed to a plethora of high-tech devices, which serve as significant sources of noise (Jhanwar, 2016). For example, the sounds emanating from industrial activities and vehicular traffic exert a tangible impact on both the psychological and physical well-being of individuals. Exposure to sound levels exceeding 60 dB can lead to adverse effects on mental health, potentially resulting in hearing impairment due to increased pressure on the ears (Hahad, 2024). Over the long term, such exposure can manifest in behavioral disturbances and contribute to sleep disturbances, along with exacerbating blood pressure issues.

While noise pollution may not directly cause mental illness, it is widely recognized for its capacity to exacerbate and amplify latent mental disorders. Furthermore, it is implicated in a host of deleterious effects, including but not limited to headaches, anxiety, nausea, stress, mood swings, neurosis, nervousness, sexual aggression and impotence, hysteria, psychosis, emotional instability, and heightened social conflict (Fig. 2) (Singh et al., 2022). The multifaceted impact of noise pollution on psychological health underscores the need for comprehensive interventions aimed at mitigating its deleterious effects on urban populations (Nguyen et al., 2021a). Furthermore, research efforts should be directed towards elucidating the intricate mechanisms through which noise pollution influences mental well-being, thereby informing the development of targeted interventions and policy initiatives aimed at promoting healthier urban environments.

Fig. 2 Impacts of waste on psychological health

2 Vulnerable Populations

The recent data reveals the rapid expansion of urban areas approximately twice in the last decade, which has introduced a critical problem relative to substantial environmental and well-being of citizens, especially significant impact the vulnerable groups of society such as children, seniors, and low-income dwellers (Vardoulakis et al., 2016). The risk of environmental pollution in residential areas comes from inappropriate waste disposal practices from residents in the area causing water, soil, and air pollution, posing a serious threat to people and the surrounding ecological balance. In addition, mishandling hazardous waste endangers waste management workers and neighboring communities (Abubakar et al., 2022). To thoroughly understand the impact of residential waste on population health, we must first understand the extent of its impact on vulnerable population groups. Understanding these aspects will enable us to develop appropriate measures and reasonable plans to solve the residential waste problem and make sustainable development decisions to promote the urban environment. healthier and fairer for everyone, in line with the SGD 11 goal of building a resilient city (Bang, 2023; Siri, 2016).

1. ***Children***

Due to the immaturity of physiological conditions, children are the group most easily affected by living environmental conditions (Carroquino et al., 2013). Their higher metabolic uptake means exposure to waste and polluted environments carries more health risks than to adults (Weinstangel et al., 2016). According to the World Health

Organization (WHO), each year about 3,000,000 children can not live longer than 5 because of environmental diseases (UNICEF, 2023). Specifically, environmentally caused diseases cost more than 55,000,000,000 USD per year in the United States alone (Weinstangel et al., 2016). As reported by the United Nations Children's Fund (UNICEF), an estimated 26% of deaths in children under preschool age could be prevented by solvable environmental threats (UNICEF, 2023). This indicates it is essential to secure environmental quality for children's health in our society.

- *Developmental impacts of exposure to environmental toxins*

One of the common reasons used to consider the impact of the surrounding environment on children's health is that infants and children, due to their small size put them at higher risk of absorbing environmental toxins through their skin. This is because their surface area to total body volume ratio is about two to three times greater than that of adults, making exposure and absorption of toxins through the skin high (Carroquino et al., 2013). In addition, due to the close distance from the head to the ground, the tendency to explore the surrounding environment with the mouth and the higher water consumption per body weight increases the likelihood of exposure to airborne pollutants from air, water, and food (Weinstangel et al., 2016). At the early stage of life, infants and toddlers experience rapid growth and development, and any disruption that occurs is potentially harmful. During this stage of development, crucial organs are developing with irreversible progress such as brain development, so any intervention in this process will have long-term consequences (Carroquino et al., 2013). Because their brains are undergoing rapid growth and development, they are especially vulnerable to exposure to toxic substances. According to Hannal's research (Weinstangel et al., 2016), during the early stages of fetal development, As well as infancy and toddlerhood, children are especially vulnerable to the absorption of environmental toxins through the placenta and skin. At this stage of early body formation, their cells are differentiating and growing vigorously, and their organs are still developing, making them unable to effectively remove toxins from toxic environments. The level of exposure to toxic environments during this period poses a significant risk of developing cancer in children (Bang, 2023; Weinstangel et al., 2016). For school-age children, exposure to harmful environments lies in the high risk of absorbing harmful environmental toxins and chemicals through daily diet and consumption of fresh foods such as green vegetables, and fruits.

Metals, especially heavy metals such as lead, mercury, cadmium, and chromium are widely used in manufacturing and exist in mostly products in human activities. However, those ingredients are significantly dangerous to human well-being due to their wide environmental dispersion and their tendency to accumulation on living organisms. Most of the heavy metals are categorized as neurotoxic, nephrotoxic, fetotoxic, and teratogenic types. Exposure to these toxins can lead to cognitive and physical impairments (Hauptman, 2017). For children, heavy metals are especially dangerous for their nervous system and kidneys which can lead to irreversible complications (Carroquino et al., 2013). The metals spread into human habitats by particulate particles of waste incinerators and penetrate the body via water, air, or contact with metal-contaminated soil. Among them, lead and mercury are considered as significant risks as environmental toxins because of their prevalence in residential waste and their particularly severe neurotoxic properties (Carroquino et al., 2013). The exposure risk to these toxins can happen from early pregnancy, with the ability of transmittance from mother to baby through the placenta and breast milk. These heavy metals disrupt all the critical stages of the body's development including nervous system development, hematological function, cardiovascular health, and kidney function. Although the adult metabolism system can tolerate certain levels of heavy metals, there appears to be no safe "threshold" of exposure during childhood (Carroquino et al., 2013). Accordingly, the metal toxins in need to be strictly controlled in residental areas and neighbouring industrial zones to protect the long-term effect on youngsters.

One of the current major problems from the urban environment that has risen during the last decades is plastics because plastic products are widely used in human activities. Especially, there is a large quantity of plastics that are used frequently for single-use purposes. This increases the burden of urban waste. A rise in social awareness regarding the potential harm caused by plastic waste to the environment due to some common compounds in plastics such as bisphenol A, phthalates, and phytoestrogens (Carroquino et al., 2013). Those are known to interfere with hormone production in children. Similarly, the release of plasticizer compounds containing polychlorinated biphenyls (PCBs) and the creation of dioxins when burning plastic are both endocrine disruptors that can lead to systemic effects in children. Some petroleum-based waste such as polybrominated diphenyl ethers (PBDEs) usually exists in flame retardants, which also links to endocrine disruption and developmental effects in children (Weinstangel et al., 2016). Formaldehyde, another common household chemical, can trigger severe asthma attacks in sensitive people such as children, seniors, and people with a weak immune system (Hauptman, 2017). In addition, plastics with chemically complex chain structures often release nano- and micro-particles and chemicals during production, use, and decomposition. Although the impact of microplastics is still being researched (Sripada et al., 2022; Hoang et al., 2025a, b), there are several types of research reveal the evidence that microplastics have been found in human placentae, meconium, and infant stool, highlighting potential risks associated with microplastic waste during early childhood (Sripada et al., 2022). Early exposure to hazardous chemicals, even in small quantities, can have lasting impacts on human health.

Pesticides are crucial for farming, securing food for society, and raising human nutrition. However, they come with health risks due to their toxic effects. Young children are frequently exposed to pesticides through a variety of sources, including food, water, and milk. Also, babies and toddlers with prenatal and childhood exposures are especially worried because their developing organ systems are vulnerable (Carroquino et al., 2013). Pesticides such as organophosphates, carbamates, and organochlorines, which are commonly used in pesticide applications, can affect nervous system function and disrupt neuronal membrane transport. These neurotoxic pesticides, in conjunction with pyrethrins and pyrethroids utilized in residential settings, have the potential to exacerbate allergies and asthma (Hauptman, 2017).

The build-up of pesticides in municipal garbage poses considerable health risks, particularly for youngsters (Carroquino et al., 2013).

Electrical waste also known as e-waste indicates the discard of electronic devices like TVs, computers, smartphones, and other personal gadgets, posing a new concern for the environment. Recent data from WHO reveals about 54,000,000 tons of e-waste were generated in 2019 globally, with the projection of a significant increase up to 75,000,000 tons by 2030 due to technological advancements and increased market demand (Jutta, 2017; World Health, 2021). E-waste has been proven may contain hazardous materials such as heavy metals (like lead, mercury, and cadmium) and flame retardants. Improper disposal or recycling of e-waste can lead to the leaching of toxic elements into soil, water, and air, posing serious health risks, particularly for children, from exposure to heavy metals (World Health, 2021) and other hazardous substances such as polychlorinated biphenyls (PCBs). These substances can cause a range of health issues, including neurological damage, reproductive disorders, and cancer (Chen et al., 2011). Table 1 is a summary of common environmental health hazards impacting children.

- *Increased susceptibility to respiratory issues and cognitive impairment*

According to the United Nations Children's Fund (UNICEF), approximately 17 million babies worldwide are currently breathing toxic air, highlighting the widespread impact of air pollution on children's health (UNICEF, 2018). Children, who are particularly vulnerable to the health effects of pollution despite being the least responsible for its causes, face significant risks due to exposure to harmful pollutants. In the UK alone, Unicef UK estimates that one in three children, totaling 4.5 million,

Table 1 Summary of common environmental toxins in urban wastes and health impact on children

Urban waste sources	Common toxins	Health impacts
Waste from households	Plastics, lead, phthalates, bisphenol A	Endocrine disruption, developmental issues, neurological damage, developmental delays
Demolition and Building wastes	Asbestos, lead, mercury	Lung diseases (e.g., mesothelioma), neurological damage, developmental delays
Biomedical waste	Anatomical wastes, pathological and infectious wastes	Infectious diseases (if not properly handled and disposed of)
Gardening and slaughterhouse waste	Pesticides, pathogens	Neurological damage, developmental delays (from pesticides), infectious diseases (from pathogens)
Industrialized waste	Chemical wastes, heavy metals, industrial solvents	Neurological damage, respiratory problems, developmental delays, various health issues

reside in urban areas with unsafe levels of particulate pollution (UNICEF, 2018). Children are highly vulnerable to respiratory issues from urban waste, with pollution having more severe effects on them compared to adults (Kenagy et al., 2016). Factors like their developmental stage and shorter stature increase their risks. Exposure to air pollution during childhood can hinder lung growth, impair brain development, and raise the likelihood of conditions like asthma. Children inhale more pollutants relative to their body weight due to their proximity to vehicle exhausts and faster breathing rate. Spending more time outdoors, especially while walking to school and playing, exposes them to higher pollution levels (Kenagy et al., 2016; UNICEF, 2018). Studies indicate that children walking on busy roads may face up to a third more air pollution than adults, emphasizing significant health risks, especially for those who walk to school daily (Kenagy et al., 2016).

Mental health and well-being are consistently impacted, either directly or indirectly, by various environmental exposures. Recent research has aimed to demonstrate how certain biophysical environmental exposures interact with the central structures and functions of the brain, thereby influencing the neurobiology of depression (Bosch & Meyer-Lindenberg, 2019). In a cross-sectional study conducted in a Brazilian city (Dumcke et al., 2020), researchers investigated the emotional and behavioral well-being of schoolchildren aged 6–13 from disadvantaged communities. They categorized the children into two groups based on their proximity to a waste recycling site: those within 100 m and those more than 150 m away. Using the Strengths and Difficulties Questionnaire, they assessed emotional and behavioral difficulties and employed logistic regression to analyze contributing factors. Results revealed a higher prevalence of emotional and behavioral problems among children living near the recycling sites. Additionally, families in the exposed group had lower income and educational levels, indicating their vulnerability to the effects of urban waste. The study highlights the need for further research into potential neurodevelopmental delays associated with proximity to waste sites, as limited studies have explored this connection (Dumcke et al., 2020).

2. **Elderly**

The elderly are highly vulnerable to urban waste and pollution due to age-related physiological changes, such as weakened immune systems and decreased lung capacity, making them more susceptible to health risks. Existing health conditions, like cardiovascular or respiratory disorders, can worsen with pollution exposure. Prolonged exposure to pollutants over their lifetime increases the risk of chronic diseases and cognitive decline. Additionally, spending more time indoors exposes them to household pollutants, further heightening their vulnerability (Madani Hosseini et al., 2024).

- *Compromised health due to cumulative exposure to pollution*

Environmental factors contribute to accelerated aging processes, heightened biological age, and the onset of various age-associated diseases. Specifically, pollutants like ambient particulate matter, pesticides, heavy metals, and organic solvents significantly impact cardiovascular and brain aging. These toxicants induce macro- and microvascular damage, cross the blood–brain barrier, and trigger neurotoxic effects, inflammation, and neuronal dysfunction. Alzheimer's and Parkinson's diseases, the most prevalent age-related neurodegenerative conditions, are affected by environmental factors. These diseases involve complex interactions between genetics, environment, and lifestyle, with hallmarks such as microvascular pathologies and abnormal protein aggregates. Moreover, longitudinal studies link exposure to fine particulate matter with increased hospital admissions for Parkinson's and Alzheimer's diseases, underscoring the role of environmental pollutants in exacerbating neurodegenerative pathogenesis (Pandics et al., 2023).

Environmental quality factors are one of the significant influences impacting the aging process and relating to age-related non-communicable diseases. Those problems can be cardiovascular problems (heart failure, stroke), cancer, and chronic respiratory diseases. Surrounding habitat toxicants speed vascular aging, resulting in the development of micro and macrovascular pathologies, such as atherosclerosis and cerebral small vessel disease. This also relates to the expansion of neurodegenerative problems like Alzheimer's and Parkinson's diseases. The environment quality relates to the containing toxicants ambient particulate matter, pesticides, heavy metals, and organic solvents. Those toxicants can disrupt the blood–brain barrier, evoking neurotoxic effects, neuroinflammation, and neuronal dysfunction, thereby exacerbating cognitive decline and promoting the genesis of neurodegenerative diseases (Pandics et al., 2023). To demonstrate the influence of environmental factors on aging in the elderly, a conceptual model depicts how these drivers contribute to unhealthy aging in Fig. 3.

- *Limited mobility leads to higher exposure to waste-related stressors*

As older adults often face challenges in mobility due to age-related conditions or disabilities, they may spend more time close to waste disposal sites or areas with high pollution levels. Meanwhile, their vulnerable immune systems associated with aging make their well-being condition more fragile against vector-borne diseases from the environment. Consequently, this sustained exposure to environmental toxicants around urban wastes puts them at high risk of health problems (Madani Hosseini et al., 2024). For example, the data from the Public Health Agency of Canada reveals that in 2018 most of the severe and fatal cases of West Nile Virus patients happened in older Canadian adults. Similarly, the treatment results for Lyme disease were not as expected for the seniors (Maresova et al., 2023).

Moreover, reduced mobility can hinder the elderly from accessing proper waste management services or facilities, leading to improper waste disposal practices in their immediate surroundings. This lack of access may result in the accumulation of waste near their homes, increasing the risk of contamination and exposure to harmful substances. The disparities in infrastructure, such as outdated water treatment systems combined with their dependence on others, increase the risk of waterborne and foodborne illnesses among older people. Those also face a heightened risk of air pollution, leading to respiratory problems (Maresova et al., 2023).

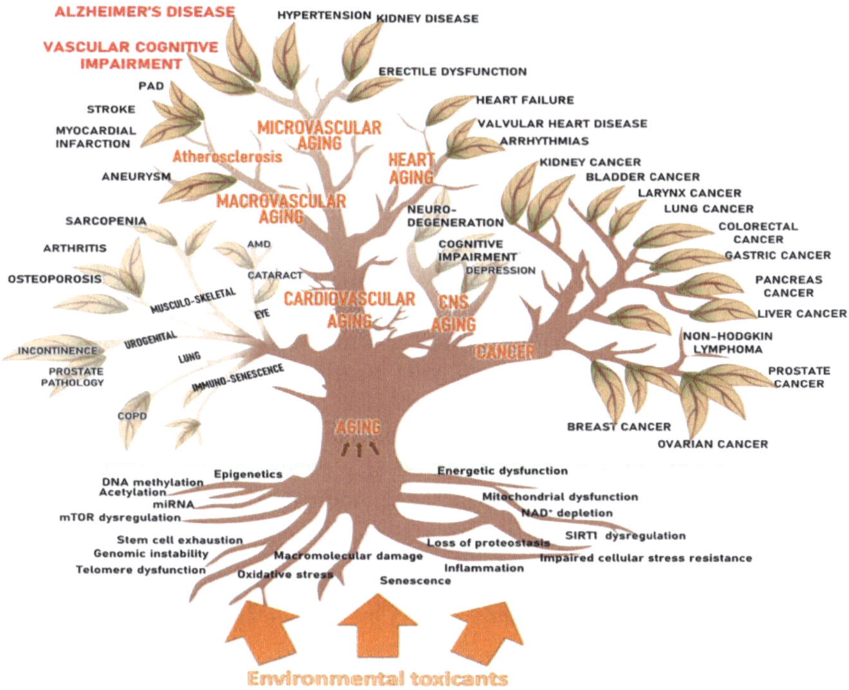

Fig. 3 Illustration of the impact of environmental factors on aging processes, accelerating them and increasing biological age, thus leading to the development of age-related diseases. Environmental toxicants, depicted in orange, worsen fundamental aging mechanisms at the molecular and cellular levels, thereby contributing to the onset of specific age-related ailments. Copyright at Springer (Pandics et al., 2023)

3. **Low-Income Communities**

Poor people are considered one of the fragile groups under the stress tensors of urban waste. World Health Organization reported that about 6,500,000 premature deaths in 2016 resulted from household and environmental pollution. Besides, approximately 90% of the environmental death cases occurred in low- and middle-income countries around Southeast Asia and Western Pacific regions (WHO Releases Country Estimates on Air Pollution).

- *Disproportionate burden of waste-related health impacts*

According to Brown (Exposure & Health Impact, 2016), "the overwhelming bulk of evidence supports the "environmental justice" belief that environmental hazards are inequitably distributed by class, and especially race" which was confirmed by recent studies (Brown, 1995). Low-income communities often bear a disproportionate burden of the negative health impacts associated with urban waste. Firstly, these communities frequently lack adequate waste management infrastructure, leading to the accumulation of waste in their neighborhoods. This can result in increased exposure to harmful substances present in the waste, such as toxins and pathogens, which can contaminate the air, soil, and water. As a result, residents may experience higher rates of respiratory problems, skin infections, and other health issues.

Specifically, household air pollution poses a significant public health challenge in developing countries, particularly affecting pregnant women who spend the majority of their time at home. Globally, household air pollution is responsible for nearly 4 million premature deaths, including over 400,000 deaths of children under 5 years, and results in 85.5 million disability-adjusted life years annually. More than 90% of these health impacts occur in low and middle-income countries, with Sub-Saharan African nations experiencing the highest mortality rates (Devakumar et al., 2018). In Ethiopia alone, household air pollution contributes to over 50,000 deaths and accounts for nearly 5% of the national disease burden. This pollution primarily stems from biomass fuel use and vehicle emissions, leading to respiratory and eye problems among pregnant women and adverse effects on fetal health, including asphyxia and low birth weight. Furthermore, household air pollution is associated with psychosocial effects such as feelings of inferiority and depression (Brown, 1995).

- *Limited access to healthcare exacerbates psychological effects*

Low-income communities often face challenges accessing healthcare services, which exacerbates the psychological impact of urban waste. The lack of adequate medical care leaves residents struggling to cope with mental health issues worsened by exposure to waste-related pollution. Additionally, socioeconomic hardships such as substandard housing and food insecurity further compound the psychological strain on these communities in the face of environmental hazards. De Siqueira's review (Shine et al., 2023) emphasizes the disparities in healthcare access and economic burdens among urban residents in low- and middle-income countries, particularly those living in slum areas24. Furthermore, poor waste management is closely linked to increased health risks among low-income populations, leading to heightened exposure to harmful pollutants (Siqueira Filha et al., 2022). The heightened exposure to environmental waste and constrained healthcare access significantly elevate the risk of various adverse health outcomes, such as liver and bladder cancers, as well as low birth weight and pre-term birth (Parin, 2023) among vulnerable populations. These circumstances intensify the psychological impacts experienced by low-income individuals (Fazzo et al., 2017).

- *Community Engagement and Empowerment*

Poor waste and wastewater management could result in the pollution of water, soil, and the atmosphere, and this can constitute a risk to human health and the environment. In addition, poor garbage collection, improper waste disposal, and a lack of infrastructure can lead to environmental and public health effects at both regional and local levels (Giebel et al., 2024). Emissions from open burning, dumping and leachate from landfills can not only pollute the environment, but also significantly

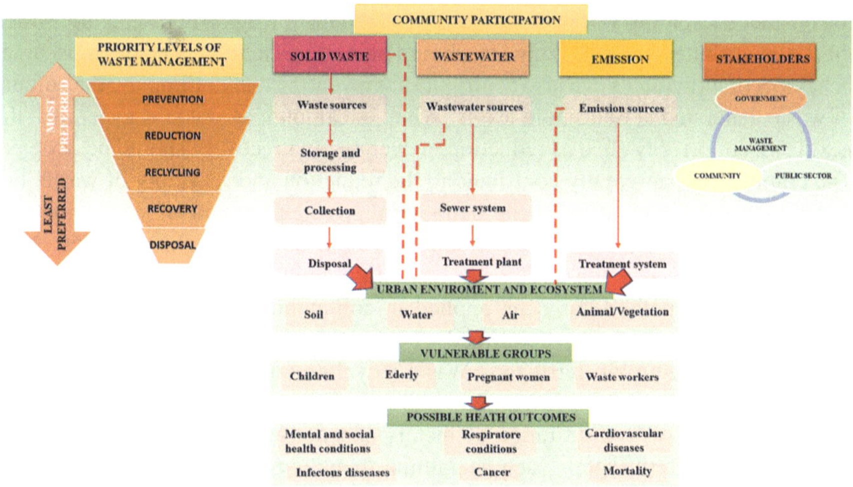

Fig. 4 Community participation and possible effects of waste sources in urban waste management system

affect public health. Various pollutants found in wastewater, including nitrogen, bacteria, heavy metals, and emerging contaminants such as endocrine-disrupting compounds cause significant health concerns in communities. Preventive intervention by building healthier environments should thus be a major component of most disease control efforts. Figure 4 shows community participation and possible effects of waste sources in urban waste management system.

Community engagement in the urban environmental management system encompasses a range of actions, including planning, enforcement, and monitoring, in cooperation with the government, private sector, and non-governmental entities. For instance, the community should be engaged in the consultation process of policy-making and implementation related to urban environmental aspects. Its importance has been emphasized in previous studies. The process of consulting communities and stakeholders on policy proposals and urban environment planning helps establish consensus in society and implement those plans effectively. In addition, education is a key to success in the change of personal behaviors and awareness, which underpins individuals' contribution to environmental protection. Individuals can contribute to improving the air quality through personal behaviors such as saving energy at the household level and daily commuting because vehicle emissions are a significant contributor to urban air pollution. Also, saving fresh water or reusing wastewater helps reduce wastewater released into the surrounding environment, contributing to preventing water pollution.

Community participation in the waste management process from source to disposal is important for sustainable municipal solid waste management. The level of community engagement should be prioritized for prevention and reduction because of the minimal amount of waste going to be disposed of. Manufacturers also need to consider preventing waste from the steps of product design and packaging, which greatly contributes to minimizing the amount of waste generated after use. Consumer behaviors of individuals can also contribute to preventing and minimizing the amount of waste generated at household levels. When the amount of waste is generated, the next levels of priority such as reduction, recycling, and recovery should be adopted. The community can greatly contribute to the reduction and recycling of waste. For instance, waste segregation onsite helps minimize the amount of waste going to disposal, and some can be reused or input for other processing or purposes. After sorting, organic waste can undergo composting, which can enhance soil fertility, and this simple method has been applied in communities around the world. The Vermicompost method was confirmed to have the potential for treating organic waste and producing bio-fertilizers. Waste recycling and recovery are intended to extract useful substances or materials and then create new products. For detail, e-waste consists of a wide range of metals and non-metals, and it is useful for recovering rare earth metals such as gallium, tantalum, and platinum groups, and noble metals (e.g. silver, gold, and palladium). Informal recycling, which is mainly happening in developing countries such as India, Ghana, and China, significantly contributes to mitigating the environmental impact of e-waste. However, the byproducts and hazardous substances in e-waste such as lead, cadmium, mercury, and nickel have a high potential to be harmful for ecosystems and human health. In addition, municipal waste open burning at household levels has a negative impact on the air environment because some byproducts, such as polycyclic aromatic hydrocarbons, polychlorinated dibenzo-p-dioxins, polychlorinated dibenzofurans, polychlorinated biphenyls, and other harmful compounds can be created during this process, significantly contributing to air pollution in the surroundings. These hazardous subtances can cause respiratory and cardiovascular diseases (Chinh et al., 2021).

- *Community clean-up initiatives to improve local environments*

Active involvement of the community is crucial for the successful implementation of any activities aimed at managing urban waste. The urban population, particularly poor communities, faces exacerbated waste management issues due to a lack of awareness and low engagement of communities. Citizens can help create both short-term and long-term solutions for their neighborhoods by taking part in cleanups. Cleanups can encourage communities to adopt good habits like recycling and reuse, which have a significant impact on waste management in a community, and they can also act as catalysts for long-lasting changes in behavior and attitude. The efficiency of urban waste management systems can be improved through the promotion of energy and water conservation at both the household level and community levels. Community engagement can facilitate the coordination of actions from many sectors of the community, such as non-governmental organizations (NGOs), community-based organizations (CBOs), private entities, and local government, to allocate resources effectively and achieve favorable outcomes. The good examples of effective urban waste management are expanding waste collection through microenterprises in Thailand, community-based approaches through channeling waste for recycling and composting in Philippine communities, public–private partnerships in Japan and Singapore, and encouraging stakeholders and education in Nepal.

3 Green Space and Urban Planning to Reduce Pollution Effects (Solid Waste, Air Pollution, Noise Pollution, Water Contamination), Stress Relief, and Increase Social Interaction

Reducing the risk of urban environmental waste is a crucial strategy for cities to mitigate the increasingly serious impacts on community mental health, environmental security, and vulnerable groups. Identifying suitable coping mechanisms and interventions is crucial, particularly amid the increasing impacts of climate change and continuous urban population growth. The advancement of urban resilience through the enhancement of green spaces has highlighted its effectiveness as an important tool to address these challenges. Therefore, urban planning emerges as a comprehensive alternative solution and a significant catalyst in cities' adaptation planning towards sustainable development. Urban green spaces (UGS) planning harnesses nature's potential by restoring degraded ecosystems. This approach aids in effective mitigation and adaptation to natural disasters and pandemics, enhances climate resilience, and addresses social challenges, thereby enhancing cities' resilience. These solutions will also be linked and considered from various perspectives on the United Nations' Sustainable development Goals (SDGs), including Goal 11.6, which is "decrease the individual environmental footprint of urban parts, particularly by prioritizing air quality, municipal waste, and other waste management"; Goal 11.7, which is "ensure that everyone, including women, children, the elderly, and individuals with disabilities, has equal access to safe, inclusive, and easily accessible green public spaces"; and Goal 11.8, which is "strengthen national and regional development planning to foster positive economic, social, and environmental connections between urban, peri-urban, and rural areas".

Urban green spaces or green infrastructure (GI) have different definitions and classifications based on their recreational function and public nature. They can be wetlands, urban forests, street trees, bioswales, green walls, roofs, etc.. It may also include parks, squares, children's playgrounds, sports facilities, vacant lots (where children can play), and adventure playgrounds. They play a crucial role in improving ecological conditions through enhancing air quality, such as by reducing local air temperatures and filtering air pollutants. To illustrate, in Zimbabwe, research indicates that the transformation of green spaces into low-density residential zones may lead to a potential rise in ground temperature by approximately 0.2 °C. Additionally, it also creates thermal advantages in reducing the effect of the Urban Heat Island (UHI) through extensive vegetation cover and sustainable carbon sequestration, providing water retention capabilities, and various ecosystem services to sustain biodiversity. In addition, it aids in reducing disease transmission and enhancing both physical and mental health, fostering community well-being. As an effective tool contributing to downtown resilience improvement, it promotes biodiversity and enhances the welfare of residents.

Incorporating GI into master urban planning as a widespread tool for ecological sustainability involves adjusting ecosystem services and climatic conditions, safeguarding biodiversity, and delivering diverse socio-ecological profits for city residents. This strategy is being adopted. The application of various design strategies and adjustments of urban designers and officials on various factors includes the extent of metropolitan green coverage, the urban development phases, the city's location, and prevailing climatic situations. It is vital that planning for sustainable cities is informed by a clear understanding of the scale and pace of ecological and carbon footprint reduction needed to prevent widespread climate change and address other urgent ecological challenges. The results yielded from forecasting that integrates GI is most likely to be effective when the plans are provided by those with the most comprehensive understanding, those with the deepest expertise and experience in urban planning, as well as contributions from a wide range of residents and stakeholders. In this way, policies will be adjusted to suit the people they affect, and the benefits will therefore be more effective.

However, it can be observed that the integration of master planning with the ecological network towards sustainability, where the basic functions of the city (activities of living, economics, culture, and entertainment) with ecological functions supporting humans (climate conditioning, pollutant removal, water balance, land reclamation) can interact with each other, is currently a challenge. The establishment of a city-wide green space network with connecting corridors needs to be considered and valued as an important land use type, along with other important land use sectors. Moreover, integrated planning also needs to consider other factors, for example, cost–benefit, legal aspects, implementation time, and the project life cycle. Thus, to achieve these goals, integrated GI planning requires ensuring public benefits, democratizing territories, and designing new institutions.

1. **Integrating green infrastructure into master urban planning to mitigate the impact of environmental pollution**

Municipal environmental pollution is becoming increasingly severe due to the unstoppable process of urbanization in most countries worldwide. The rapid inhabitant growth and urban sprawl have reduced the green coverage of cities, thereby increasing surface temperature and causing irreversible loss of biodiversity. To address these challenges, multi-scale approaches toward sustainable development is exceedingly noteworthy, and mitigation strategies are necessary. So, urban space design must be careful, incorporating further flora, boosting construction materials and colors, and encouraging human-nature interaction are urgent necessities. Hence, a scientific methodology is imperative to recognize and implement the most effective mitigation processes.

Firstly, the economic boom, rapid urbanization, and shift in consumption patterns have led to an increase in solid waste generation. Currently, the world generates 2.01 billion tons of waste per year, of which 33% is improperly managed. By 2050, it is projected that the annual global waste generation rate could soar to 20 billion tons in case of the present trend persists.

If the current trend continues, the global waste generation rate could reach 20 billion tons per year by 2050. Thus, solid waste management is extremely necessary and of major importance for municipal areas to prevent the spread of diseases, reduce contamination, promote resource recovery, and create job chances with the support of appropriate urban development platforms. However, the harsh reality is that waste management is still an aspect neglected by policymakers and planners.

The optimal solution from both an ecological and socio-economic perspective is to avoid waste from the outset by reducing emissions to minimize the demand for energy and resources. Consequently, cities have a fundamental importance in preventing and managing waste by being able to determine how to handle waste with preventive measures or appropriate classification systems, with the close interaction of local stakeholders. Also, it is possible to change societal behavior to reduce, classify, and recycle waste at its source through urban planning. An example is the planning of zero-waste cities, which also encourages residents to participate in implementing measures, such as using more biodegradable plastics which are environmentally-friendly (Khedulkar et al., 2025). In this manner, the city can exploit the potential to reduce waste and completely avoid this waste, such as plastics, in the long term.

Secondly, the risk of being affected by air pollution is present in urban zones worldwide, for instance, in some cities in Asia. It can harm human health, the environment, and property, creating an Urban Pollution Island (UPI) effect. Air pollution may harm the brain and may be associated with memory loss, according to emerging evidence, cognitive deterioration heart disease, respiratory diseases, and lung cancer. Air pollution has been linked to early-life effects as well. For this reason, designing green space areas in cities is key to ensuring aids against adverse controls and maximizing the probable of the profits that GI brings, contributing to reducing air pollution when dust particles and smoke, especially from vehicle emissions, are retained by trees and vegetation, helping to protect against heavy rains. To cite an example, research in the southern United States shows that both cities have similar green coverage, with the number of particles removed annually by trees being 4.7 and 3.2 tons per square mile, respectively. In reality, inner-city plants in the United States alone have removed 711,000 tons of pollutants each year, estimated to be worth $3.8 billion. In other words, research in Beijing, China demonstrations that greenery impassive 1261.4 tons of contaminants from the air in 2002.

Because the influence of GI on air quality is surely complex, therefore, careful planning must be undertaken to have a progressive overall effect on air quality. With appropriate design methods, UGS can reduce the level of urban air pollution, bringing remarkable health benefits. On one hand, good traffic network planning is needed to encourage the community to use public transport instead of personal vehicles while also reducing the reflectance of the urban surface, reducing traffic congestion, transitioning to clean energy sources, and advocating for eco-friendly public transportation. On the other hand, planning more green spaces (green belts, urban forests) as an indispensable part of the city landscape hold a noteworthy position in reducing air pollution and other pollutants. Furthermore, planning concentrated industrial zones and having policies to relocate factories out of residential areas. For this reason, it will help create greener, healthier, and more sustainable and smart cities.

Thirdly, clean water is indispensable for the survival of every living organism. However, each year, up to 3.5 million people fall ill due to exposure to polluted wastewater. Good planning of GI will contribute to improving water quality and environmental sanitation (SDG6), particularly focusing on mitigating heavy metal emissions into the air to prevent water tank contamination, preventing and reducing runoff, helping to provide rainwater, reducing the risk factors of public, ecological, and technical, dropping pollutants entering the water area, and thus helping to decrease the influences of economic losses. Furthermore, GI holds a significant position in the hydrological cycle. Moreover, protecting GI to support the hydrological cycle strengthens community health at the most basic level, possibly through solutions of green roofs and green walls. The application of green buildings in urban regions is not only a technical strategy but also a flexible solution, as it can transform some existing spaces into UGS that have social value and improve urban health (Fig. 5).

Ultimately, noise pollution from road traffic, aviation, railway operations, and commercial activities poses significant challenges in urban areas. At elevated levels, this noise can directly impact physical health, mental well-being, and overall happiness among residents (Ramadan et al., 2022), affecting the quality of residents lives (Ky et al., 2021). Noise stands as the second most significant environmental factor affecting health, trailing only behind air quality. The WHO estimates that approximately 40% of Europe's population is subjected to road traffic noise levels around 55 A-weighted decibels (dB(A)), with 20% experiencing levels surpassing 65 dB(A) during the day, and more than 30% enduring levels above 55 dB(A) at night. Noise is also a common factor associated with a high risk of cardiovascular disease, hypertension, myocardial infarction and stroke. The link between noise exposure and adverse effects on the cardiovascular system has been established not only in adults but also in children and newborns.

To control the level of environmental noise, we need to rely on comprehensive resolutions that can scientifically discourse city issues. Especially, the solution of integrating master urban design with UGS should also be considered through the application of Nature-based Solutions (NBS) such as green walls to decrease the temperature, absorb sound, and reduce noise, biohar for heavy metal removal (Nguyen, 2014), activated carbon derived from biomass for carbon dioxide removal (Hoang T.-D et al., 2024), constructed wetlands for microplastic removal (Hoang et al., 2025a, b). Similarly, design strategies could be taken into account, for example, designing green barriers to combat outdoor noise with a density of at least 30.48 m in width the sound of a fountain or rustling leaves can be designed to distract and mask unpleasant noises.

In short, UGS needs to be comprehensively planned and maintained. Designing towards sustainable development will help people have a healthy life in their spirits of daily habits. Therefore, the approach to GI in design planning will provide urban

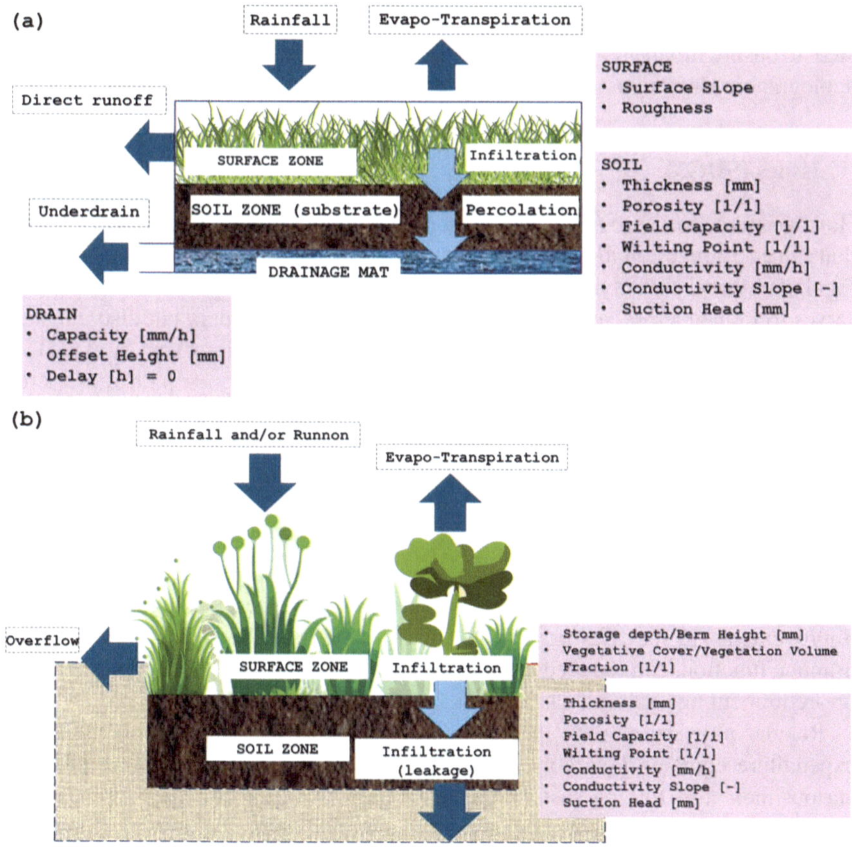

Fig. 5 Schematic view of a green roof system (**a**) and rain garden system (**b**)

planners with a broader scope to consider the interaction between humans, landscapes, and resources in a specific urban area Also, there needs to be a dedicated land area for GI, and official planning programs must have suitable capital funding as well as community participation to ensure success. To achieve this, of course, from the start of the design process, planners must identify the preferences, abilities, and favorite activities of residents groups, while also creating opportunities to connect UGS with sidewalks, bicycle networks, and other locations. This will promote a better understanding of ecological resources and reduce competition between the development of gray and green land areas.

Even though, integrating GI into master urban planning cannot substitute for reducing contamination at its source. And, of course, the crucial plan to diminish urban environmental infection releases is to control its source. Nevertheless, the benefits of GI not only contribute to health but also to the environment. In consequence, urban planners must rely on increasingly expanded databases and more comprehensive information on results from research surveys, considering the interventions of

GI as well as its characteristics and the context in which the contaminants are generated, to ensure urban areas are not just livable but also resilient and sustainable amid demographic shifts and meteorological pressures.

2. **Enhancing access to green infrastructure to improve health and happiness issues (SDG3)**

Human interaction with nature is recognized as a potential supportive environment that can encourage countless valuable benefits for community mental health through regular physical activity. Enhancing access to parks and natural areas not only serves to reduce stress, anxiety, depression, and mood disorders but also enhances social interaction and cohesion. Green spaces and smart cities (Tran et al., 2024a, Malla et al., 2023) can deliver health benefits through various pathways, across diverse populations. These include the ability to reduce stress by walking around parks, creating a positive emotional state and improving cognitive function providing local communities and tourists with opportunities for physical relaxation and social bonding.

Several other studies link GI to health benefits, potentially including reductions in cardiovascular disease, stroke, diabetes, and overall mortality rates, as well as decreases in circulatory diseases, obesity the incidence of respiratory diseases, improvements in the lifespan of the elderly, cognitive development in children, immune function, and reduced risk of stress, anxiety, and clinical depression among pre-retirement age individuals.

Regular physical activity in green spaces has the potential to shift the caloric expenditure equation favorably. Moreover, numerous other studies have proposed various indicators linking sustainable development with health and happiness. The spread of COVID-19, which has exacerbated air quality issues worldwide, has underscored the positive role of UGS for physical and mental health during extended periods of social distancing. In densely populated cities, these restrictions have emphasized the need to enhance UGS and their accessibility.

To sum up, it is evident that the relationship between humans and nature holds significant and profound importance for daily joy, work productivity, and overall mental health. The value of green spaces for urban communities is undeniable. What's crucial is taking action, and UGS can be utilized as a catalyst towards realizing sustainable development goals because they have the prospective to promote conservational, communal, and financial stability. The interaction between humans, infrastructure systems, and the natural environment forms a complex adaptive social-ecological scheme.

4 Mental Health Support Services

1. **Provision of counseling and support for individuals affected by waste-related stress**

In the following section, the models of service provision for individuals with waste-related stress will be depicted. In European countries, affected cases of individuals suffering from stress due to living and working near polluted areas (e.g. suicides) attracting great media attention may increase concern for waste stress. The main features of the scheme are described in Table 2.

2. **Education on coping strategies and resilience-building techniques**

Education on coping strategies

Most education about health psychology due to waste pollution will take place by attending seminars and conferences. While specialized education is at the university level, teaching of environmental health and environmental pollution psychology will be included. As part of their extensive certification training by the Board of Healthcare Professionals, providers are trained in awareness and detection of clinical problems (Malla et al., 2023). All health therapists are trained in a Mental Health First Aid course. New initiatives with health associations provide enhanced education programs on awareness of health issues and psychological stress related to waste. National health authorities issue guidance documents and accurately handle screening results and treatments in paraclinical cases. In-depth knowledge of stressors and stressful situations and the use of psychological diagnostic tools will be provided at universities specializing in medical training. Other universities offering bachelor's degrees in psychology provide advanced education in environmental pollution and waste management. To be registered as psychologists on waste-related stress, professionals are required to complete specialized courses offered by universities granted by the Ministry of Health after obtaining a Master's or Doctoral degree in environmental health psychology.

Resilience-building techniques

A synthesis of the literature on mental health disorders and an analysis of individual service delivery systems produced practical technical recommendations for coping with waste-related stress. These techniques will hopefully inspire stakeholders to enhance mental health support systems for individuals with waste-related stress, helping them live safely and healthily.

- Educate individuals living and working near waste-contaminated areas about the symptoms of mental health conditions in order to prevent and promptly detect psychological problems. An essential aid in seeking help is the training of psychologists in the psychology of waste-related stress.

Table 2 Models of service provision for individuals with waste-related mental health disorders in some European countries (Malla et al., 2023)

Professional consulting and support providers	Assist program	Diagnostic evaluation	Clinical therapy	Treatment monitoring	Timely detection of mental health issues	Education program
Psychological doctor	Regularly interview people living and working in waste-contaminated areas to assess and screen	If there are problems with stress after the interviews, the interviewee can request a consultation with one of the psychologists	Depending on the mental disorder, different treatment methods are offered	Monitor symptoms and disorders of the treatment process and make decisions	All individuals are required to have an annual interview	Psychologists with experience in waste-related mental health will be on the priority list
Mental health first aid providers	People who often experience psychological stress due to noise have health insurance. Support includes inpatient and outpatient assistance	Medical doctors will provide mental health support. Anyone who suspects they are suffering from stress caused by noise pollution should talk to a medical doctor	Treatment will vary depending on the diagnosis	Maintaining cooperation with a medical doctor will help people suffering from stress regain mental stability and integrate into life	This will vary between people living near and far from the noise source. Individuals will undergo screening performed by a medical doctor	Medical doctors receive training focusing on awareness of clinical psychological issues and mental health first aid

(continued)

Table 2 (continued)

Professional consulting and support providers	Assist program	Diagnostic evaluation	Clinical therapy	Treatment monitoring	Timely detection of mental health issues	Education program
Clinical psychotherapists collaborate with professional service providers	People experiencing stress can contact a clinical therapist to receive additional support from a speciallized network	Highly dependent on the supervising SP/SPE or therapist	The clinical approach depends on the expertise of the supervisor and may involve medication interventions prescribed by a psychiatrist	Stressed people can be referred to the clinic if necessary	The screening system is not implemented at the national level, but is implemented at the regional level, in which short screening tools are used to detect psychological problems early in people under stress	Advanced education for psychotherapists, focusing on knowledge of waste pollution, understanding and dealing with waste pollution problems
Clinically trained psychologist	Established a psychology treatment organization that provides experts to provide professional support to all athletes in the country	Research experts develop a plan for a standardized assessment system to improve psychological aspects	Depends on the clinical approach of the psychotherapist and the expertise of the supervisor	Depends on the problem of the person suffering from stress to join the group of monitoring experts	Establish a standard screening system for people at risk of waste-related stress	Psychology education related to mental health due to waste pollution is trained at universities

(continued)

Table 2 (continued)

Professional consulting and support providers	Assist program	Diagnostic evaluation	Clinical therapy	Treatment monitoring	Timely detection of mental health issues	Education program
Psychotherapists and clinical psychologists collaborate with professional service providers	People with mental health problems go to a psychologist on their own to seek professional support. The first people to contact are usually the family doctor	Depends on the service provider or supervising therapist. Psychophysiological assessment using biofeedback equipment	Depends on the expertise of the professional who is supposed to be working with them. Treatment regimens require integrated techniques. Pharmacological intervention can be prescribed by a psychiatrist	Based on the disorder and severity of symptoms. If mild, you can follow the service provider's instructions. If severe, prioritize treatment of clinical problems	Health monitoring is guaranteed by law, including physiological assessment and interpretation, with priority given to physical and cardiovascular assessment	Advanced education for psychologists in environmental health is offered by private educational institutions. Psychologists must follow a required course and several programs of study in mental health interventions
Clinical psychologist	Each country establishes medical organizations specializing in mental health. Each organization employs clinical therapists to provide psychological services to residents	The diagnostic screening process is performed by a licensed psychologist	Symptom-based treatment for the specific disorder identified. Can be treated with drugs combined with Cognitive Behaviour Therapy	Efficacy is integral to the treatment evaluation process. Cases of patients with complex symptoms will be discussed in the multidisciplinary treatment team	Research screening issues on the nationwide system	A Cognitive Behaviour Therapy-based training program for service providers with extensive clinical training

- Organize conferences to investigate and identify the problem between diagnosable clinical conditions and subclinical conditions as well as the relationship to the health and well-being of the stressed individual (Moesch et al., 2018).
- Acknowledging a mental health disorder is an important first step for individuals experiencing stress. Providing this information and seeking early help requires individual and professional education. Try to intervene at an early stage to prevent a more serious condition leading to a pathological condition (Gulliver et al., 2012). Creating a healthier and more sustainable living environment will help individuals feel comfortable and easily find support from others.
- Specialized investigations and studies are needed to determine specific diagnostic criteria for waste-related mental health disorders. Thoroughly examine existing diagnostic tools for general stress psychopathology (Margot, 2016). From there, develop tools specifically for psychological stress caused by waste pollution.
- Strengthen treatment regimens for individuals facing serious waste-related problems such as living too close to noise-polluted areas, working next to landfills, and having frequent contact with waste. It is necessary to monitor the effectiveness of the treatment regimen.
- Relatives working with individuals with stressful mental health disorders should monitor them carefully. If their psychology is harmed, notify service providers immediately. Ensure that service providers understand the nature of environmental pollution and waste to discuss paraclinical treatment.
- A simple understanding of mental health and psychotherapy is not enough when treating people with varying degrees of stress. Therefore, strengthen multidisciplinary collaboration between psychiatrists, psychotherapists, and service providers while monitoring and caring for people under stress (American Psychiatric Association & Association, 2013).
- Establish an accessible waste-related mental health referral program for individuals experiencing stress and increase education for medical support teams to improve the quality of services in the country. Implement many educational packages such as Mental Health Diagnosis, seminars on psychological health promotion on stress caused by environmental pollution for service providers, and use the media to introduce the service packages.

5 Conclusion

In conclusion, the article focuses on spatializing the multifunctionality of UGS as well as creating sustainable living spaces and environments by establishing a relationship between the city and the landscape through integrated space planning solutions. The comprehensive approaches, including biofuels (Busner & Targum, 2007), renewable energies and AI technologies (Busner & Targum, 2007; Hoang et al., 2023a, Bandh et al., 2024, Hoang et al., 2022), smart city, carbon recycling approaches (Nguyen et al., 2021b), circular economy or waste utilization (Hoang et al., 2023b), energy application from bio sources (Abi Bianasari et al., 2024), transboundary governance

(Sezali et al., 2024), multifunctional conceptual framework of the urban planning process integration with the green space provides a unique platform for planners to minimize the impact of environmental waste on the mental health of municipal residents. With important support from metropolitan planners, policymakers, and researchers in guiding future research efforts, those approaches can be applied to increase UGS areas. In this way, GI will make significant contributions in leveraging nature in the city and using the ecosystem services that provide advantages for the town on the path towards sustainable development. As a result, GI can be used as a strategy to connect green and municipal systems towards resilience.

References

Abbah, A.P., Xu, S., & Johannessen, A. (2024). Long-term health effects of outdoor air pollution on asthma and respiratory symptoms among adults in low-and middle-income countries (LMICs): a systematic review and metaanalysis. *Frontiers in Environmental Health, 3*, 1352786.

Abi Bianasari, A., et al. (2024) Influence of combined catalysts on the catalytic pyrolysis process of biomass: A systematic literature review. *309*, 118437.

Abubakar, I. R., et al. (2022). Environmental sustainability impacts of solid waste management practices in the Global South. *International Journal of Environmental Research and Public Health, 19*. https://doi.org/10.3390/ijerph191912717.

American Psychiatric Association, D., & Association, A. P. (2013) *Diagnostic and statistical manual of mental disorders: DSM-5*. Vol. 5. American Psychiatric Association Washington, DC.

Bandh, S. A., et al. (2024). Track to reach net-zero: Progress and pitfalls. *Energy and Environment*, 0958305X241260793.

Bang, K.-S. (2023). The environment and children's health. *Child Health Nursing Research, 29*(1), 1–4.

van den Bosch, M., & Meyer-Lindenberg, A. (2019). Environmental exposures and depression: Biological mechanisms and epidemiological evidence. *Annual Review of Public Health, 40*, 239–259.

Brown, P. (1995). Race, class, and environmental health: A review and systematization of the literature. *Environmental Research, 69*(1), 15–30.

Busner, J., & Targum, S. D. (2007). The clinical global impressions scale: Applying a research tool in clinical practice. *Psychiatry (Edgmont), 4*(7), 28.

Carroquino, M. J., Posada, M., & Landrigan, P. J. (2013). Environmental toxicology: Children at risk. In E. A. Laws (Ed.), *Environmental toxicology: Selected entries from the encyclopedia of sustainability science and technology* (pp. 239–291). New York: Springer New York

Chen, A., et al. (2011). Developmental Neurotoxicants in e-waste: An emerging health concern. *Environmental Health Perspectives, 119*(4), 431–438.

Chinh, P. C., et al. (2021). Willingness to Pay for improving household solid waste management in Vietnam. *Applied Environmental Research, 43*(2), 1–14.

de Siqueira Filha, N. T., et al. (2022). The economics of healthcare access: A scoping review on the economic impact of healthcare access for vulnerable urban populations in low- and middle-income countries. *International Journal for Equity in Health, 21*(1), 191.

Devakumar, D., et al. (2018). Women's ideas about the health effects of household air pollution, developed through focus group discussions and artwork in Southern Nepal. *International Journal of Environmental Research and Public Health, 15*. https://doi.org/10.3390/ijerph15020248.

Dong, Y., et al. (2024). Pathogen contamination of groundwater systems and health risks. *Critical Reviews in Environmental Science and Technology, 54*(4), 267–289.

Dumcke, T. S., et al. (2020). Association between exposure to urban waste and emotional and behavioral difficulties in schoolchildren. *Jornal de Pediatria, 96*(3), 364–370.

Exposure and Health Impact. (2016).

Fazzo, L., et al. (2017). Hazardous waste and health impact: A systematic review of the scientific literature. *Environmental Health, 16*(1), 107.

Giebel, C., et al. (2024). Community-based mental health interventions in low- and middle-income countries: A qualitative study with international experts. *International Journal for Equity in Health, 23*(1), 19.

González, D., et al. (2024). Addressing the gaseous and odour emissions gap in decentralised biowaste community composting. *Waste Management, 178*, 231–238.

Gulliver, A., Griffiths, K. M., & Christensen, H. (2012). Barriers and facilitators to mental health help-seeking for young elite athletes: A qualitative study. *BMC Psychiatry, 12*(1), 157.

Habibi-Yangjeh, A., & Pournemati, K. (2024). A review on emerging homojunction photocatalysts with impressive performances for wastewater detoxification. *Critical Reviews in Environmental Science and Technology, 54*(4), 290–320.

Hahad, O., et al. (2024). Noise and mental health: evidence, mechanisms, and consequences. *Journal of Exposure Science & Environmental Epidemiology*, 1–8.

Hauptman, M., & Woolf Alan, D. (2017). Childhood ingestions of environmental toxins: what are the risks? *Pediatric Annals, 46*(12), e466–e471.

Hoang, T.-D., et al. (2022). Artificial intelligence in pollution control and management: status and future prospects. In: *Artificial intelligence and environmental sustainability: Challenges and solutions in the era of industry 4.0* (pp. 23–43).

Hoang, T.-D., et al. (2023a). Biofuel projects and current environmental policies: Vietnam's case and Neighboring Asian countries. In *Biofuels in circular economy* (pp. 73–88), Springer.

Hoang, T.-D., et al. (2023b). Carbon-based synthesized materials for CO_2 adsorption and conversion: Its potential for carbon recycling. *8*(4), 53.

Hoang, V.-H., et al. (2024). Sources, environmental fate, and impacts of microplastic contamination in agricultural soils: A comprehensive review. *Science of the Total Environment,* 175276.

Hoang, T.-D., Liu, Y., & Le, M. T. (2024). Performance of CO_2 adsorption on modified activated carbons derived from okara powder waste: Impacts of ammonia impregnation. *Processes, 12*(9), 2024.

Hoang, V.-H., Nguyen, M. K., Hoang, T. D., Rangel-Buitrago, N., Lin, C., Pham, M. T., Ha, M. C., Nguyen, T. P., Shaaban, M., Chang, S. W., & Nguyen, D. D. (2025a). Microplastic characteristics, transport, risks, and remediation in groundwater: A review. *Environmental Chemistry Letters*, 1–21.

Hoang, V.-H., Nguyen, M. K., Hoang, T. D., Nguyen, H. L., Lin, C., Yousaf, B., Ha, M. C., Bui, V. K. H., Pham, M. T., Chang, S. W., & Nguyen, D. D. (2025b). Global occurrence and environmental fate of microplastics in stormwater runoff: Unlock the in-depth knowledge on nature-based removal strategies. *Reviews of Environmental Contamination and Toxicology, 263*(1), 1–15.

Jhanwar, D. (2016). Noise pollution: A review. *Journal of Environment Pollution and Human Health, 4*(3), 72–77.

Jutta, G. (2017). Waste in the City: Challenges and opportunities for urban agglomerations. In E. Mustafa (Ed.), *Urban Agglomeration* (p. Ch. 9). Rijeka: IntechOpen.

Kenagy, H. S., et al. (2016). Greater nitrogen dioxide concentrations at child versus adult breathing heights close to urban main road kerbside. *Air Quality, Atmosphere & Health, 9*(6), 589–595.

Khajuria, A., Yamamoto, Y., & Morioka, T. (2010). Estimation of municipal solid waste generation and landfill area in Asian developing countries. *Journal of Environmental Biology, 31*(5), 649–654.

Khedulkar, A. P., et al. (2025). Bio-based nanomaterials as effective, friendly solutions and their applications for protecting water, soil, and air. *Materials Today Chemistry, 46*, 102688.

Kumthip, K., et al. (2019). Enteric and non-enteric adenoviruses associated with acute gastroenteritis in pediatric patients in Thailand, 2011 to 2017. *PLoS ONE, 14*(8), e0220263.

Ky, N. M., et al. (2021). Investigation and assessment of road traffic noise: A case study in Ho Chi Minh City, Vietnam. *Water, Air, & Soil Pollution, 232*(7), 259.
Le Anh, T., & Hoang, T.-D. (2024). Trans-boundary water governance in the Mekong countries. In *The Mekong River Basin* (pp. 237–272). Elsevier.
Le Thuy, T., et al. (2025). A review on flame retardants in soils: Occurrence, environmental impact, health risks, remediation strategies, and future perspectives. *Toxics, 13*(3), 228.
Madani Hosseini, M., Zargoush, M., & Ghazalbash, S. (2024). Climate crisis risks to elderly health: Strategies for effective promotion and response. *Health Promotion International, 39*(2).
Malla, F. A., et al. (2023). Circular economy to decarbonize electricity. In *Renewable energy in circular economy* (pp. 71–87). Springer.
Maresova, P., et al. (2023). Challenges and opportunity in mobility among older adults—Key determinant identification. *BMC Geriatrics, 23*(1), 447.
Margot, P. (2016). The psychological response to injury in student athletes: A narrative review with a focus on mental health. *British Journal of Sports Medicine, 50*(3), 145.
Moesch, K., et al. (2018). FEPSAC position statement: Mental health disorders in elite athletes and models of service provision. *Psychology of Sport and Exercise, 38*, 61–71.
Muthuraman, L., & Ramaswamy, S. (2019). *Solid waste management*. MJP Publisher.
Nguyen, V. C. N., et al. (2021). Green technology for sustainable development: Practice and experience of renewable energy in Vietnam. *4*(2), 79–92.
Nguyen, M. K. (2014). Community response to road traffic noise in Hue City, Vietnam. *Environment and Natural Resources J, 12*(2), 24–28.
Nguyen, Q. T., et al. (2021a). Psychological stress and associated factors among municipal solid waste collectors in Hanoi, Vietnam: A cross-sectional study. *PLoS ONE, 16*(7), e0254355.
Pandics, T., et al. (2023). Exposome and unhealthy aging: Environmental drivers from air pollution to occupational exposures. *GeroScience, 45*(6), 3381–3408.
Parin, S., Health impacts of poor solid waste management in the 21st century. In Pengzhong, L. (Ed.), *Solid waste management* (p. Ch. 5), IntechOpen: Rijeka.
Piersanti, V., et al. (2024). Sudden death due to butane abuse—An overview. *Journal of Forensic and Legal Medicine*, 102662.
Ramadan, B. S., et al. (2022). A comprehensive review of domestic-open waste burning: recent trends, methodology comparison, and factors assessment. *24*(5), 1633–1647.
Sezali, N. A. A., et al. (2024). *Bio-based nanomaterials for energy application: A review*.
Shine, S., et al. (2023). Pregnant women's perception on the health effects of household air pollution in Rural Butajira, Ethiopia: A phenomenological qualitative study. *BMC Public Health, 23*(1), 1636.
Singh, O. (2023). *Environmental science*. OrangeBooks Publication.
Singh, V., Kumar, A., & Gupta, S. (2022). Mental health prevention and promotion—A narrative review. *Frontiers in Psychiatry, 13*, 898009.
Siri, J. G. (2016). Sustainable, healthy cities: Making the most of the urban transition. *Public Health Reviews, 37*(1), 22.
Sripada, K., et al. (2022). A Children's health perspective on nano- and microplastics. *Environmental Health Perspectives, 130*(1), 015001.
Ta, Q. T. H., et al. (2024) *In situ growth of* TiO_2 *on* $Ti_3C_2T_x$ *MXene for improved gas-sensing performances*. Ceramics International,
Tran, T.-K., et al. (2024a). *Applications of engineered biochar in remediation of heavy metal (loid) s pollution from wastewater: Current perspectives toward sustainable development goals* (p. 171859).
Tran, T.-K., et al. (2024b). Review on fate, transport, toxicity and health risk of nanoparticles in natural ecosystems: Emerging challenges in the modern age and solutions toward a sustainable environment. *Science of The Total Environment, 912*, 169331.
Tziourrou, P., & Golia, E. E. (2024). Plastics in agricultural and urban soils: interactions with plants, micro-organisms, inorganic and organic pollutants: An overview of polyethylene (PE) litter. *Soil Systems, 8*(1), 23.

UNICEF, (2018). *A Breath of Toxic Air*, UK.
UNICEF. (2023). *Healthy environments for healthy children*. UNICEF.
Vardoulakis, S., Dear, K., & Wilkinson, P. (2016). Challenges and opportunities for urban environmental health and sustainability: The HEALTHY-POLIS initiative. *Environmental Health, 15*(1), S30.
Weinstangel, H., Buka, I., & Campbell, S. (2016). Environmental pediatrics: an introduction and evaluation of online resources. *31*(4), 435–446.
WHO. *WHO Releases Country Estimates on Air Pollution.*
World Health, O. (2021). *Children and digital dumpsites E-waste exposure and child health: WEB ANNEX Literature review on the health effects of exposure to E-waste*. World Health Organization.

Water Resources and Bioenergy

Van Toi Pham, Hiep Hoang Van, Oanh Nguyen Thi, Nguyen Vo Chau Ngan, Dang Thi Thom, Le Van Giang, Luu The Anh, Nguyen Thi Thanh Huyen, Le Thuy Trang, and Tuan-Dung Hoang

Abstract The intersection of bioenergy production and water resources presents a complex and multifaceted challenge that requires careful consideration and sustainable management. The impact assessment at the bioenergy-water nexus highlights the crucial role water plays in bioenergy development and the need to address local

---

The original version of the chapter has been revised. A correction to this chapter can be found at https://doi.org/10.1007/978-3-031-91114-9_48

V. T. Pham
Department of Energy and Environment, Faculty of Environmental Engineering, Hanoi University of Civil Engineering, 55 Giai Phong, Hanoi, Vietnam

H. H. Van · O. N. Thi · T.-D. Hoang (✉)
VNU School of Interdisciplinary Studies, Vietnam National University, Hanoi, 144 Xuan Thuy, Cau Giay, Hanoi, Vietnam
e-mail: tuandunghoang@gmail.com

N. V. C. Ngan
College of Environment and Natural Resources, Can Tho University, Can Tho, Vietnam

D. T. Thom
Graduate University of Science and Technology, Vietnam Academy of Science and Technology, Hanoi, Vietnam

D. T. Thom · L. V. Giang · L. T. Anh
Central Institute for Natural Resources and Environmental Studies, Vietnam National University, 19 Le Thanh Tong, Hoan Kiem, Hanoi, Vietnam

N. T. T. Huyen
Faculty of International Economics, Foreign Trade University, Hanoi, Vietnam

L. T. Trang
Faculty of Environmental and Natural Sciences, Duy Tan University, Da Nang, Vietnam

School of Engineering and Technology, Duy Tan University, Da Nang, Vietnam

L. T. Trang
e-mail: lethuytrang@dtu.edu.vn

T.-D. Hoang
School of Chemistry and Life Science, Hanoi University of Science and Technology, 1 Dai Co Viet, Hanoi, Vietnam

M. Shaaban (ed.), *Soils and Sustainable Agriculture*, Frontier Studies in Soil Science, https://doi.org/10.1007/978-3-031-91114-9_44

and regional water implications. Alternative energy sources and carbon capture and sequestration processes are essential to cope with the immense global challenges of resource scarcity and climate change. Biomass is a potential alternative carbon-based fuel that can reduce the environmental impacts of various applications. Many countries have increased the production of biomass crops for use as bioenergy sources. Biomass can include wood, food crops, energy crops, waste from forests, yards, or farms, and even municipal solid waste. Bioenergy can be converted into fuels, heat, electricity, and products, which can be used in various applications, such as heating, electricity generation, transportation, and bio-products. Applying bioenergy which comes from biomass crops, is a risk to food insecurity, and causes water scarcity for crop irrigation. This study aims to review recent documents on the relationship between the cultivation of biomass crops and the amount of water exploited, thereby gaining a more scientific perspective on this link.

Keywords Bioenergy · Energy crops · Hydro cycle · Water demand · Water security

1 Bioenergy and Water Footprint

To evaluate how water resources are used across a product's life cycle, which includes manufacture, transportation, marketing, consumption, and reuse, the term "water footprint" (WF) was used (Hoekstra & Hung, 2005). The Water Footprint Network defines the direct WF of a consumer or producer as the pollution and freshwater consumption directly related to the consumer's or producer's usage of water (Hoekstra et al., 2011). To supplement the life cycle evaluation, the International Organization for Standardization (ISO) also suggested an international WF standard in parallel (International Organization for Standardization, 2014). All things considered, WF offers a crucial scientific foundation for quantifying environmental studies and water resource management.

WF combines the two characteristics of water and monitors the agricultural water footprint in terms of crop production's blue, green, and grey waters. The blue water is representative of water that may be extracted for human use, such as irrigation, from rivers, lakes, wetlands, and aquifers. The unsaturated zone of soil, created by precipitation and accessible to plants, is indicated by the green water. Water that is contaminated during a production process is referred to as grey water. According to several studies, the WF of bioenergy is significantly higher than that of other primary energy carriers since the production of biomass necessitates a lot of water for processing, fertilizing, and irrigation.

With its increasing industrialization and urbanization, China has emerged as one of the world's top users of water. Due to China's scarcity of water and other resources as a result of its rapid economic expansion and expanding population, WF flows are particularly important. Zhu et al., (2019) conducted a bibliometric analysis of WFs research from 2003 to 2018 by categorizing the agricultural sector among the studied areas through a review of the literature. The climate, farming practices, and available water resources in the area all affect the agricultural WF. While the average water footprints in the South are 900 L/kg, 1210 L/kg, 1660 L/kg, and 2510 L/kg, respectively, the average water footprints in the Northeast region of China are 670 L/kg for maize, 830 L/kg for rice, 860 L/kg for wheat, and 1520 L/kg for soybeans.

In Penglai City, Shandong Province, China, Li et al. (2021) used a life cycle assessment (LCA) model to analyze the WFs for three categories of nonfood crops (hybrid Pennisetum, sorghum, and maize) on marginal arable land. These three unique energy crops are useful for producing biogas. When compared to maize and sorghum, hybrid Pennisetum appears to have the greatest impact on water resources, according to the assessment results obtained using the Cropwat software.

Yu and Qingshan (2014) investigated the relationship between crop production's environmental impact and water usage in China's Heilongjiang land reclamation area. This region is a globally recognized blackland area that produces the most rice, maize, soybeans, and wheat among other green grains. The climate data, which includes average monthly rainfall, evapotranspiration, temperature, sunshine, wind speed, humidity, and radiation, crop production data, which includes planting and harvest dates, crop coefficients, rooting depths, and the lengths of each crop's developmental stages, and soil parameters, which include maximum infiltration rate, values of available soil water content, maximum rooting depth, and initial soil moisture depletion, are all used by the authors to account for their water footprint using Cropwat software. The findings demonstrate that in the study area, agricultural water consumption levels did not rise in direct proportion to crop output. However, the expansive coupling state serves as a warning sign for the issue of over-exploitation of groundwater. Furthermore, of all the crops grown in this region, rice cultivation accounted for 56% of the gray footprint and 50% of the blue-green water footprint.

Using Hoekstra's method, Xie et al. (2017) calculated the water consumption and effects of three different biofuel crops: cassava, sweet sorghum, and Jatropha curcas. The study determined the WF of each crop in various Chinese locations using the LCA approach. The life cycle WF of the seeds of Jatropha curcas, sweet sorghum, and cassava was estimated to be 64.7–182.3 L/MJ of biofuel, 115.9–210.4 L/MJ, and 73.9–222.2 L/MJ, respectively. An additional Gerbens-Leenes et al. (2009a) studied from a worldwide perspective, the water footprints (WFs) required to produce cassava ethanol range from 783 to 2926 L per L of ethanol, whereas the WFs for sweet sorghum ethanol were found to have a water footprint between 4394 and 13,541 L per L of ethanol.

Research by Gerbens-Leenes (2018) on the various forms of contemporary bioenergy in the global energy system toward 2040 states that, with an average of 162 m^3/GJ, the WF of bioenergy from various feedstocks and conversion technologies ranges from 5 to 1400 m^3/GJ. Because energy crops like wheat, rapeseed, and soybean require extensive irrigation and fertilization, their WF is significantly greater. However, since they don't need further water inputs, crop leftovers like bagasse or straw have the lowest WF of bioenergy.

The most ecologically friendly and water-efficient alternatives for bioenergy supply can be found by comparing the WF of bioenergy with the WF of other energy sources and taking into account the availability and scarcity of water in various places. However, other aspects like land resources, carbon footprint, biodiversity, social and economic implications, greenhouse gas emissions, etc. should also be taken into account when evaluating the performance of bioenergy systems; the WF of bioenergy is not the only criterion for doing so.

The optimal willow cultivars for producing bioenergy were determined by Richard et al. (2019) by taking into account both yield and water use efficiency. The study assesses the trade-offs between water use and carbon harvest by simulating various soil, climate, and willow canopy types in the British using the process-based model LUCAS. According to the results, narrow-leaf cultivars are more productive and require less water than broad-leaf cultivars, making them more appropriate for marginal soils and drought-prone areas.

In a humid subtropical watershed in Entre Rios, Argentina, Heidari et al. (2021) examined the effects of Eucalyptus (*E. grandis*) plants for biofuel production on water resources and biomass productivity. The main issue impeding eucalyptus development was shallow soils resulting in seasonal water stress, which was simulated using the Soil and Water Assessment Tool (SWAT) model. Here, 0.8 m^3 of water is needed to produce 1 kg of eucalyptus biomass on average. The outcome demonstrates that, in comparison to other land cover types, eucalyptus plantations require more water and have lower streamflow. Additionally, biomass production varies geographically based on soil depth and water availability.

Mekonnen et al. (2015) assessed that the global WF of heat and electricity consumption from biomass ranges from 48,000 to 500,000 m^3/TJe. This range is greater than the wind energy's consumptive WF of 0.2–12 m^3/TJ for electricity and heat, 6.4–303 m^3/TJ for solar energy, 7.3–759 m^3/TJ for geothermal energy, and 300–850,000 m^3/TJ for hydropower. The ranges of 214–1190 m^3/TJ for conventional oil, 76–1240 m^3/TJ for natural gas, 81–1270 m^3/TJ for shale gas, 419–1340 m^3/TJ for oil sand, 93–1580 m^3/TJ for lignite, 316–1830 m^3/TJ for oil shale, 79–2100 m^3/TJ for coal, and 18–1450 m^3/TJ for nuclear energy are comparable for electricity and heat from fossil fuels and nuclear energy. Global production weighted WF per unit of heat and electricity produced was 4241 m^3/TJ on average. Because hydropower contributes more to energy in these regions, Africa has the largest WF among them, with a weighted average of 22,800 m^3/TJ, followed by Latin America with 11,200 m^3/TJ. Europe came next with 3240 m^3/TJ, followed by the USA, Canada, and Oceania with 2520 m^3/TJ and 1900 m^3/TJ. The Middle East region's WF ranges from 200 to 500 m^3/TJ because it is mostly dependent on natural gas and oil.

When considering the supply of fresh water globally and the capability of the ground to absorb it, the amount of water used in the production of bioenergy is not sustainable. Gheewala et al. (2011) proposed reducing non-beneficial consumptive water use throughout the supply chain and/or improving the management and planning of beneficial consumptive and non-consumptive water use, including increasing productivity across a range of agricultural management regimes, from rainfed to irrigated crops, to lessen the impacts of used water in the production of bioenergy.

Additionally, there are studies on the use of WF (wheat, barley, maize, and sunflower) in Iran for the generation of bioenergy. Ababaei and Ramezani Etedali

(2017) computed the WFs of Iran's primary cereal production, which includes wheat, barley, and maize, using the AGWAT model. Similar WFs are used for farming barley and wheat: 2640 L/kg (green WF ranged from 302 to 798 L/kg, and blue WF from 435 to 1305 L/kg) and 2, 57 L/kg (499–1023 L/kg for green WF, and 521–1402 L/kg for blue WF), respectively. The highest WFs ever recorded were 1958 L/kg on the maize crop, where the blue WF varied from 567 to 1172 L/kg and the green WF from 5 to 130 L/kg. These measured values were significantly greater than the WFs used in the western Iranian region of Kermanshah to produce sunflowers. Ababaei and Etedali (2014) assessed the WF components of wheat output from 236 plains throughout fifteen main wheat-producing provinces in Iran using AGWAT software as well. Across all chosen provinces and irrigated areas, the average total WF is roughly 3188 L/kg of wheat grain. To account for the irrigation losses, a new component (white WF) was proposed in this research.

The different crops cultivated in Thailand were compared in terms of water requirements by Gheewala et al. (2014). The two crops with the lowest WFs per crop weight are sugarcane and cassava, at roughly 150–174 L/kg and 394–413 L/kg, respectively. While soybeans have one of the greatest water requirements of any crop, requiring between 1676 and 2344 L/kg, maize farming has an average WF use of 837–1084 L/kg. The report made recommendations for future crop production in Thailand that would be sustainable and would use less agricultural water.

The most significant agricultural sector in Malaysia is the palm oil sector, although oil palm is associated with considerable water usage during the plantation stage (Muhammad-Muaz & Marlia, 2014). In Pahang, Malaysia, oil palm fields and nurseries were the site of an initial investigation. The overall WF for cultivating oil palm, according to the Cropwat model, is 243 L/kg; the corresponding WFs for green, blue, and grey palm are 229.0 L/kg, 6.6 L/kg, and 7.0 L/kg, respectively. Chalvantharan et al. (2023) recently implemented smart irrigation systems in oil palm plantations as a result of the crops' excessive water use. The WF for oil palm farming was roughly 680 m^3/ha/year, which reduced the agriculture sector's expenses and environmental impact.

2 Bioenergy and Water Quality

Generally speaking, many indicators can be used to evaluate the quality of water based on its type and intended purpose, such as household use, hygienic practices, agricultural output, or manufacturing. Regarding bioenergy farming, runoff from intensive agricultural production using fertilizers and various pesticides (herbicides, insecticides, fungicides), along with other agricultural malpractices like tillage of improper soils, pollutes the water supply directly.

Because agrochemicals and fertilizers are employed intensively downstream, bioenergy production may have an impact on the quality of the water, such as contaminating water with heavy metals (Tran et al., 2024a), nanoparticles (Tran et al., 2024b) or micro plastics (Hoang et al., 2024; Hoang et al., 2025a, b) The material contaminates the uppermost layer of the water, endangering the health of the local population and leading to excessive eutrophication, which can harm aquatic ecosystems, human health, and water quality. One key component of biofuels from energy crops' influence is that they contribute more to eutrophication than fossil fuels do.

Research has been conducted on the nitrogen loading that occurs in the Double Mountain Fork Brazos watershed in the Texas High Plains when switchgrass (*Panicum virgatum* L.) and Miscanthus × giganteus are planted instead of cotton, a non-food crop. According to the SWAT-CUP 2012 software, perennial grasses absorb between 30 and 40% more nitrogen from the soil than cotton does, so less nitrogen is available for loss through surface runoff (Chen et al., 2017). Perennial grasses also practically eliminate the need for pesticides, which is good for the surrounding water bodies' quality.

Guo et al. (2018) investigated how bioenergy crops affected streamflow quantity and quality in a heavily farmed area in the US Little Vermilion River watershed. The simulated outcomes from bioenergy crop scenarios were compared with the baseline using the scenario analysis with the SWAT2012 model. The findings demonstrate that removing maize stover with bioenergy crops can lower nutrient losses that are advantageous to water quality at the watershed scale, even on highly erodible areas and marginal land.

Kumar et al. (2020) used the SWAT model to model the long-term effects of gigantic miscanthus and switchgrass on water quality in the Cache River Watershed in Arkansas, United States. The conversion of all marginal cropland to switchgrass resulted in a reduction of 13.7%, 17.3%, and 15.7% in sediment, TP, and TN loadings, respectively; on the other hand, the regular-SWAT model showed a decrease of 13.7%, 17.2%, and 15.8% in sediment, TP, and TN loadings at the watershed outlet when the marginal land was converted to miscanthus. Similarly, at the location-SWAT model's watershed outlet, switching all of the marginal cropland to switchgrass resulted in reductions in sediment, TP, and TN loadings of 11.9%, 15%, and 13.6%, respectively; conversely, switching the same land to miscanthus resulted in reductions in sediment, TP, and TN loadings of 12.1%, 15.2%, and 13.3%, respectively. This study also showed that the location-SWAT model, or dynamic land-use representation approach, was a better tool for assessing the environmental effects of bioenergy crops grown in marginal areas.

Using the SWAT model, Ha and Wu (2015) investigated the impact of agricultural best management practices and land conversion on water quality. Sediments considerably decreased by 70.7%, 68.0% for organic N, 63.7% for organic P, 55.5% for total N, and 59.0% for total P at a 30 m vegetable buffer. Sediment reductions of 42.2%, 39.9% organic nitrogen, 38.6% organic phosphorus, 30% total nitrogen, and 35.2% total phosphorus may be accomplished even with a 5 m buffer width. According to the study, a vegetative barrier reduced sediment loadings in the watershed the greatest, followed by organic N and P. Nitrate control was shown to be less efficient. In addition to bioenergy crops, other factors that may have an impact on the reduction of sediment and nutrient loss include watershed geology, soil characteristics, hydrology, land use, etc.

In the Tennessee River Basin of North America, Wang et al. (2018) investigated the potential effects of specialized perennial bioenergy crops (miscanthus, willow, and switchgrass) on hydrology and water quality. Two future cultivation situations were simulated using an auto-calibration feature of SWATopt created by the authors. The findings demonstrated that while runoff, sediment, and phosphorus saw only minor changes, nitrate and total nitrogen loadings and concentrations were significantly lower than those of the current landscape. Because the amount of N fertilizer decreased by 46% when hay was switched to miscanthus and willow crops. Aside from site-specific factors, the crop, planted area, and management techniques all affected how much and how well biomass was produced.

In the Arkansas-White-Red River basin of the United States, Jager et al. (2015) investigated the effects of switching from conventional crops to bioenergy crops of switchgrass, poplar, and high-yield sorghum. The scientists used the SWAT model to simulate changes in sediment and nutrient concentrations for the year 2030, based on a USDA-projected annual production increase. Nutrient loadings and concentrations simulated for the years 2022 and 2030 in the subbasins differed from the 2009 landscape conditions. While phosphorus concentrations fell in 75% of subbasins in 2022, nitrate, total nitrogen, and total suspended sediment concentrations increased in 60%, 55%, and 60% of subbasins, respectively. By 2030, 90%, 95%, 95%, and 70% of subbasins had lower amounts of nitrate, total nitrogen, total phosphorus, and total suspended sediment, respectively. The 2030 scenario shows more widespread improvements in water quality than the 2022 scenario, indicating that more land conversion is needed to satisfy the sustainability targets related to bioenergy and better water quality.

Ha and Wu (2017) looked at how land management techniques and climate change can affect the sediment, nitrogen, and phosphorus levels in the South Fork Iowa River watershed in Iowa, which is predominantly covered in annual corn and soybean crops. To encourage sediment, nitrate, and phosphorus concentrations to pass through the riparian buffer and land conversion to the watershed, SWAT was used in this study. The highest reduction in sediment loss, 56.7%, was observed in land converted to switchgrass and winter cover crops, according to the results. For all climate scenarios, the sediment loads were lower than those in the baseline scenario (Scenario 1) by 24,030 tons in the scenario of riparian buffer, 41,340 tons in the scenario of cropland conversion with switchgrass, and 64,000 tons in the scenario of stover harvest, cover crop, and cropland conversion with switchgrass. After stimulating scenarios from 2 to 4, the land management techniques can lower the amount of nitrate loadings to the watershed from 62 to 566 tons. In parallel, for all historical climatic situations, phosphorus was reduced by 7.5 tons in scenario 2, 3.9 tons in scenario 3, and 13.8 tons in scenario 4.

The majority of earlier research has found that water quality is not significantly impacted by the growing of energy crops. The amount of fertilizer needed and soil leaching can be decreased by switching to the cultivation of bioenergy crops, according to the results of long-term scenarios evaluated using the SWAT model. Furthermore, the aforementioned water-related issues can be managed by choosing the right crop species and practicing optimal management, suggesting that the production of bioenergy with the preservation of water resources can coexist in harmony.

3 Land Footprint for Bioenergy Production

3.1 The Land Footprint Concept

The use of renewable energy is becoming more widely accepted as a key component of climate change mitigation strategies (Holmatov et al., 2019). Energy from biological, geological, or solar sources that naturally replenish more quickly or at a rate equivalent to its consumption is referred to as renewable energy (Holmatov et al., 2019). The availability, conversion efficiency, and rate of usage of necessary natural resources determine the viability and efficiency of various renewable energy sources. The main renewable energy sources—biomass, solar photovoltaics (PV), solar thermal, hydro, wind, ocean, and geothermal—have very different resource allocations and must be carefully considered for specific locations or objectives.

Globally, renewable energy sources like bioenergy currently account for a sizeable portion of final energy consumption (Holmatov et al., 2019), and many nations have established national goals to switch to renewable energy production shortly. Without taking into account minerals that are buried in geological formations and converted into fossil fuels, the term "bioenergy" suggests that energy is generated from materials with a biological origin (Holmatov et al., 2019). By 2015, there were 164 nations with national targets for renewable energy, up from 43 in 2005. Furthermore, in 2015, 59 nations established national targets for renewable transport energy (Holmatov et al., 2019). We must comprehend how energy decisions affect the sustainable use of natural resources and the appropriate use of those resources, given the rapid shift to renewable energy.

Evaluating the "footprint" of a specific product or human activity is a useful way to compare and quantify the impact of humans on the environment. To comprehend environmental changes and the effects that come from human strain on the environment, footprints are used as "indicators of human pressure on the environment" (Hoekstra & Wiedmann, 2014). The land footprint (LF), water footprint (WF), and carbon footprint (CF) are the three common footprints. Originally, the LF was a component of the "ecological footprint," which quantifies the amount of land needed to meet consumption as well as the requirements for absorbing the waste output of a specific population (Rees & Wackernagel, 2008). However, the term "LF" is typically used to refer to the amount of land needed to meet consumption alone (Hoekstra & Wiedmann, 2014; Steen-Olsen et al., 2012).

The land footprint is a technique for estimating the entire amount of land, both domestic and foreign, needed to meet a nation's or a region's final consumption of goods and services (De Schutter & Giljum, 2014). From a consumption standpoint, the land footprint shows the true land demand for all other parts of the world as well as for home territory. Thus, embodied land can be included in imports and exports to compute the total amount of land needed for final consumption and/or to show how dependent various countries or regions of the world are on foreign land.

3.2 *Energy from Biomass*

All material flows originating from the biosphere are collectively referred to as biomass. These include energy crops, food and feed crops, and organic wastes like crop leftovers and manure. This part solely considered crops for the assessment of the LF of biomass; wastes were not included in the report. Generally speaking, crops are grown in agriculture for their reproductive or storage organs, which are valuable when used to produce food, feed, or materials. Crop yield, or the harvested production per unit of harvested area for crop goods, is the word used to describe the harvested organs (Winnie Gerbens-Leenes et al., 2008). However, for these organs to grow, entire plants with stems and leaves must first grow. The harvest index (HI), which measures crop yield as a percentage of total biomass yield, varies greatly amongst crops (Goudriaan et al., 2001). Agriculture aims at the agricultural yield for food or feed. However, total biomass yield, as opposed to crop yield, can be used for energy and other purposes. For instance, biomasses have been used to synthesize nanocelluloses for energy storage devices (Sezali et al., 2024a), for medical application (Sezalia et al., 2024b) or for the removal of heavy metals and organic pollutants from contaminated water and wastewater (Khedulkar et al., 2025) or to produce biochars and activated carbons (Hoang et al., 2024).

The universal process of photosynthesis, which stores solar energy in chemical bonds, is the foundation for biomass-based energy production. Under conditions of appropriate water and nutrient availability, this process, though not always efficient, demonstrates a linear relationship between intercepted global radiation and the biomass of above-ground plants (Winnie Gerbens-Leenes et al., 2008). Glucose serves as the building block for all other organic molecules that comprise plant tissues and is the molecule that stores energy from photosynthesis in all plants (Goudriaan et al., 2001). Organic compounds can be divided into five primary categories: lignin, carbohydrates, proteins, lipids, and organic acids. This implies that the energy content of a given biomass type varies depending on its composition, leading to variations in combustion energy. The quantity of heat generated during combustion at 25 °C and 1 bar is the definition of a fuel's energy content according to energy analysis.

3.3 *Bioenergy Technologies and Feedstock*

Bioenergy is a term used to describe renewable energy derived from biological material through a variety of transformation processes, including gasification, burning, fermentation, and pyrolysis. For this purpose, biomass feedstock is derived from forests, farms, and waste streams. The trash comes from retail or organic home waste, agricultural residues, and waste products from industrial processes, such as those in the paper and pulp industries.

Several conversion techniques can provide a variety of energy products to power transportation as well as heat and electricity, depending on the biomass source. Conventional (i.e., already available) and advanced (i.e., not yet proved at scale and/or far from cost-effectiveness) conversion technologies are frequently divided into groups, or alternately, first and second (or even third-generation technologies).

The majority of solid biomass is found in woody resources, such as sawdust, logs, bark, branches, and leaves, which come from forests and the wood processing industry (Hoang et al., 2024a; Ngan et al., 2024). While a growing portion is burned to produce energy, the majority of woody resources are utilized for heating. For the sake of bioenergy, we have separated primary wood resources (mainly timber and harvesting residues) from secondary wood resources (such as sawmill leftovers) in this section. Wood chips, which come from leftovers or entire trees, and greater density pellets, which are frequently manufactured from sawdust, are goods that are traded frequently (De Schutter & Giljum, 2014) (Fig. 1).

First and foremost, common food and feed crops—mostly rapeseed, sugar beet, wheat, corn, and rapeseed in the EU—are processed into biofuels using standard conversion processes (De Schutter & Giljum, 2014; Hoang et al., 2021). These agricultural resources are then employed for bioenergy purposes. Apart from domestic feedstock, foreign crops like oil palm, sugar cane, or soya are imported as feedstock or final products, mostly because of their cost-competitiveness.

Perennial energy grasses like miscanthus, reed canary grass, and others, along with short rotation coppice from quickly growing woody species like poplar and

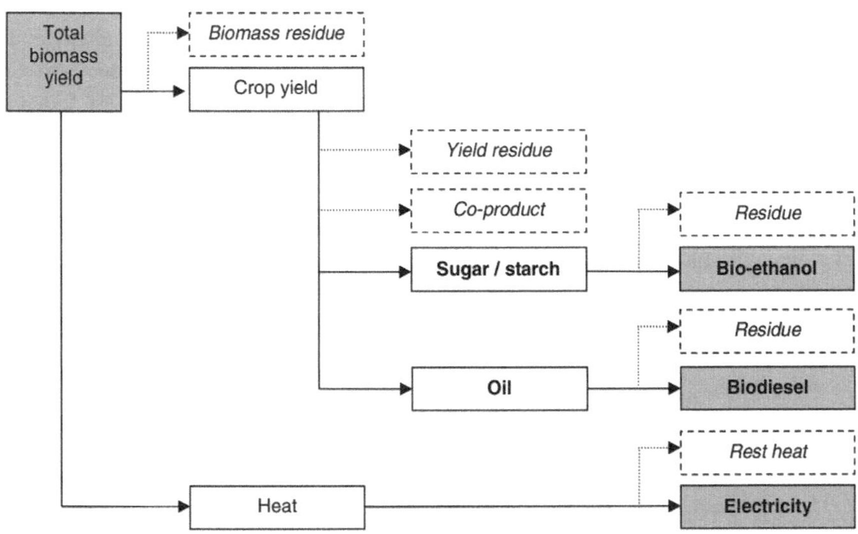

Fig. 1 From biomass to bioenergy (Gerbens-Leenes et al., 2009a)

willow, comprise a more innovative generation of dedicated, non-food energy crops. Growing some crops on "marginal," lower-quality lands has some potential, although there is typically a yield penalty involved (Searle & Malins, 2014). Nevertheless, they are frequently viewed as a sustainable way to enhance the supply of bioenergy with a small environmental footprint. The lack of widespread cultivation of dedicated energy crops is now caused by several obstacles and problems, such as longer-term investments, ambiguous technical incentives, and less farmer flexibility.

The fermentation of starch or sugar crops into bioethanol and the esterification of oil crops into biodiesel are the two main technologies employed with agricultural resources (Hoang & Nghiem, 2021). An increasingly popular technique for producing biogas for combined heat and power (CHP), heating, and electricity generation is anaerobic digestion of biogenic waste, which includes food and agricultural wastes. Fodder maize is a significant energy crop for the production of biogas in some nations, including Germany, with related land use implications. Larger facilities (like co-firing ones) can also burn agricultural crop leftovers. Similar to the aforementioned dedicated energy crops, they also can provide liquid fuel using biochemical or sophisticated conversion technologies (De Schutter & Giljum, 2014). Since waste and residue streams—which include secondary woody biomass—do not inherently indicate a land footprint, they are not taken into further consideration within the framework of this study. It is crucial to acknowledge that a variety of current applications, such as animal bedding, composting, and soil penetration, depend on these resources, potentially restricting the achievable rates of harvesting and gathering for energy conversion.

3.4 Methodologies for Calculating the Land Footprint

3.4.1 The Steps to Calculate Land Footprint

The process essentially uses a coefficient calculation approach to convert the final demand for bioenergy into the quantity of land needed for feedstock. The following steps are included in the calculations, which are based on the facts, projections, and assumptions mentioned in the preceding sections (De Schutter & Giljum, 2014):

Step 1: The bioenergy demand (in MJ) for biofuels, bioelectricity, and bioheating is the starting point. Energy consumed by machinery for six types of field activities (stubble cultivation, shallow ploughing, pre-sowing cultivation, conventional drilling, pesticide spraying, and harvesting) is assumed to be similar for all crops with a small adjustment for sugarcane (see Table 1). Six agricultural productions are maize, sugar beet, sugarcane, soybean, and rapeseed.

Step 2: The second step further specifies and breaks down the bioenergy demand (in Mtoe) by feedstock type for each technology. For instance, it displays the amount of fuel or electricity (in MJ) that is made from maize, sugar beet, sugarcane, soybean, and rapeseed.

Step 3: The final gross bioenergy is calculated in this step.

Table 1 Overview of energy requirements in agricultural production (Holmatov et al., 2019)

Input	Category	Energy equivalent	
Ag. machinery		MJ L^{-1}	MJ ha^{-1}
	Stubble cultivation	35.9	384
	Shallow ploughing		592
	Pre-sowing cultivation		165
	Conventional drilling		83
	Pesticide spraying		32
	Harvesting		833
	Total		2089
Fertilizer[b]		MJ kg^{-1}	–
	Nitrogen	78.2	
	Phosphate	17.5	
	Potash	13.8	
Pesticide		MJ kg^{-1}	–
	Insecticide	127.9	
	Herbicide	146.7	
Transport		MJ t^{-1} km^{-1}	–
	Crop/biomass processing	3	
	Fuel delivery		

Note: For sugarcane, the diesel requirements for each category is divided by six because it is sowed every six years except for the 'pesticide spraying' and 'harvesting' categories that are performed annually. [b] Original values that still include the non-energy use of fossil fuels in fertilizer production

The final gross bioenergy production of circular bioenergy production systems is calculated according to the scheme depicted in Fig. 2. The scheme is an adapted form of the internal loop (Holmatov et al., 2019) and is expressed as:

$$\begin{cases} X - \alpha_1 X - \beta_1 Y = NBF \\ X - \alpha_2 X - \beta_2 Y = NBE \end{cases}$$

where X and Y are the gross biofuel and bioelectricity production in GJ. α and β are the production efficiencies calculated as the ratio of input to output in biofuel and bioelectricity production. NBF and NEB refer to the net biofuel and bioelectricity production, respectively. Solving equations for X and Y yields the gross biofuel and bioelectricity production for a particular feedstock in a given scenario. The final gross biofuel and bioelectricity generated in the CHP and power-only routes are added and represent the total gross energy (Table 3).

Step 4: The fourth step calculates the energy yield of crop c ($E_{heat}(c)$) (GJ/ton) in the form of biofuel, heat, and bioelectricity (See part 3.4.2).

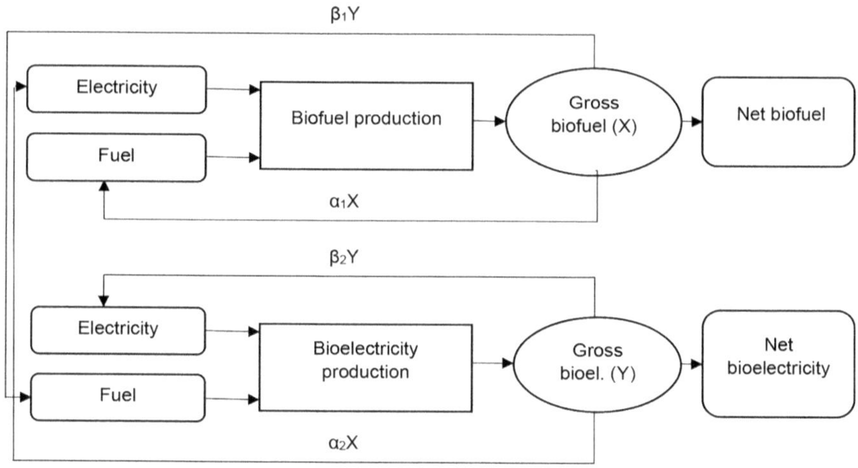

Fig. 2 Schematic representation of the gross energy calculation (Gerbens-Leenes et al., 2009a)

Step 5: The fifth step calculates the land requirement for gross biofuel production using a crop c.

The land requirement for gross biofuel production using crop c (ha) is calculated as the sum of the gross biofuel requirements across crop c (Table 2) divided by the gross biofuel output per hectare for crop c (Table 3).

Step 6: The land footprints (LF) calculation (ha) is calculated in this step.

The total LF is calculated by summing land requirement for gross biofuel and gross bioelectricity production for a given crop c.

3.4.2 Energy Calculations

a. Calculation of the energy of heat and electricity from biomass.

Combustion values are used to express the energy content of biomass. The quantity of heat generated during burning at 25 °C and 1 bar is what energy analysis uses to determine a substance's energy content (Gerbens-Leenes et al., 2009a). It makes a distinction between the higher heating values (HHV) and lower (LHV) (Blok, 2006). In the case of the HHV (Table 4), energy analysis analyzes the liquid form of the water's heat content, while for the LHV, it measures the gaseous form of the water's heat content. The method of Gerbens-Leenes et al. (2009b), which determines the energy yield of a crop [GJ/ton] by combining data on the heat of combustion of plant components with information on composition, harvest index (HI), and dry-mass fraction of a crop as shown in Tables 5, has been followed in this study for the computation of the LF of heat from biomass.

Table 2 Energy input for bioelectricity production for the two scenarios (GJ ha^{-1}) (Gerbens-Leenes et al., 2009a)

	Fertilizer		Insecticide		Herbicide		Ag. machinery	Tran	Total	
	100% scenario[a]									
	Fuel	Elec	Fuel	Elec	Fuel	Elec	Elec	Fuel	Fuel	Elec
Maize	1.8	3.7	–	0.0	–	0.4	2.1	1.9	3.7	6.1
Sugar beet	1.9	3.2	–	0.2	–	0.0	2.1	13.7	15.6	5.4
Sugarcane	1.2	1.7	–	0.0	–	0.3	1.1	17.1	18.3	3.1
Soybean	0.3	0.4	–	0.0	–	0.2	2.1	1.0	1.3	2.6
Rapeseed	0.9	2.0	–	0.0	–	0.1	2.1	1.0	1.8	4.2
	SDS-bio scenario[b]									
	Fuel	Elec	Fuel	Elec	Fuel	Elec	Fuel	Fuel	Fuel	Elec
Maize	2.3	3.2	0.0	0.0	0.2	0.2	2.1	1.9	6.4	3.4
Sugar beet	2.3	2.7	0.1	0.1	0.0	0.0	2.1	13.7	18.2	2.8
Sugarcane	1.5	1.4	0.0	0.0	0.1	0.2	1.1	17.1	19.8	1.6
Soybean	0.4	0.3	0.0	0.0	0.1	0.1	2.1	1.0	3.5	0.4
Rapeseed	1.1	1.8	0.0	0.0	0.0	0.0	2.1	1.0	4.2	1.9

Note: Tran. - refers to biomass transport from the field to the power plant. [a] The 100% scenario is a complete shift to bioenergy. [b] The SDS-bio scenario is adapted from the sustainable development scenario (SDS) of the International Energy Agency (IEA), with a 9.77% bioenergy share in the final consumed energy mix

$$E(c) = \left(HI(c) \times DM_y(c) \times \sum_{i=1}^{5} f_{y,i} \times HHV_i \right) + \left((1 - HI(c)) \times DM_r(c) \times \sum_{i=1}^{5} f_{r,i} \times HHV_i \right)$$

$E_{heat}(c)$ is the energy yield of crop c in the form of heat (GJ/ton), HI(c) the harvest index of crop c (g/g), $DMF_y(c)$ the dry-mass fraction of the crop yield (g/g), $DMF_r(c)$ the dry-mass fraction in the rest fraction (i.e., in the residue biomass), $f_{y,i}$ the fraction of component i in the dry mass of the crop yield (g/g), $f_{r,i}$ the fraction of component i in the dry mass of the rest fraction (g/g), and HHV_i the higher heating value of component i [kilojoule (kJ)/g].

Industry can use the heat released during the combustion of the entire biomass to generate power from biomass. The amount of energy from crop c that is converted into electricity (GJ/ton) is determined by how well biomass-heat energy can be converted into electricity:

$$E_{electr}(c) = \eta \times E_{heat}(c)$$

Table 3 Gross energy production by scenario and feedstock crops (Gerbens-Leenes et al., 2009a)

Feedstock	Output type	Energy output (GJ)				
		CHP		Electricity		Total
		Net	Gross	Net	Gross	Gross
		100% scenario				
Sugar beet	Biofuel	0	4.7E+08	4.7E+10	8.6E+10	8.6E+10
	Bioelectricity	4.0E+09	4.3E+09	3.1E+11	3.3E+11	3.3E+11
Sugarcane	Biofuel	0	5.2E+08	4.7E+10	9.1E+10	9.1E+10
	Bioelectricity	4.0E+09	4.1E+09	3.1E+11	3.2E+11	3.2E+11
Maize	Biofuel	0	3.3E+08	4.7E+10	7.4E+10	7.5E+10
	Bioelectricity	4.0E+09	4.8E+09	3.1E+11	3.6E+11	3.6E+11
Soybean	Biofuel	0	4.5E+08	4.7E+10	8.1E+10	8.2E+10
	Bioelectricity	4.0E+09	5.2E+09	3.1E+11	3.7E+11	3.8E+11
Rapeseed	Biofuel	0	6.5E+08	4.7E+10	8.7E+10	8.7E+10
	Bioelectricity	4.0E+09	8.0E+09	3.1E+11	4.5E+11	4.6E+11
		SDS-bio scenario				
Sugar beet	Biofuel	0	7.4E+08	1.6E+10	1.8E+10	1.8E+10
	Bioelectricity	5.6E+09	5.8E+09	5.5E+08	9.7E+08	6.8E+09
Sugarcane	Biofuel	0	7.8E+08	1.6E+10	1.8E+10	1.8E+10
	Bioelectricity	5.6E+09	5.6E+09	5.5E+08	6.6E+08	6.3E+09
Maize	Biofuel	0	8.1E+08	1.6E+10	1.8E+10	1.9E+10
	Bioelectricity	5.6E+09	6.4E+09	5.5E+08	5.1E+09	1.1E+10
Soybean	Biofuel	0	1.8E+09	1.6E+10	2.1E+10	2.3E+10
	Bioelectricity	5.6E+09	6.5E+09	5.5E+08	5.5E+09	1.2E+10
Rapeseed	Biofuel	0	1.7E+09	1.6E+10	2.1E+10	2.2E+10
	Bioelectricity	5.6E+09	8.3E+09	5.5E+08	8.7E+09	1.7E+10

Table 4 HHV for major plant components (Gerbens-Leenes et al., 2009a; Holmatov et al., 2019)

Component	HHV (GJ tonne^{-1})
Fats (plant)	37.7
Lignins (plant)	29.9
Proteins (plant)	22.7
Carbohydrates (plant)	17.3
Organic acids (plant)	13.9
Minerals (K, Ca, P, S) (plant)	0
Biodiesel	37.7
Ethanol	29.7

Table 5 Harvest index and dry mass for several plants (Holmatov et al., 2019)

Plant	Harvest index	Dry mass	Plant	Harvest index	Dry mass
Maize	0.45	0.85	Barley	0.42	0.85
Sugar beet	0.66	0.21	Maize	0.45	0.85
Sugar cane	0.6	0.27	Paddy rice	0.42	0.85
Rapeseed	0.32	0.74	Potato	0.7	0.25
Soybean	0.4	0.92	Rye	0.42	0.85
Cassava	0.7	0.38	Sorghum	0.42	0.85
Wheat	0.42	0.85			

where η is the converted biomass-heat energy efficiency.

b. Calculation of the energy of biofuels from biomass

At the moment, sugars from sugar cane or sugar beet, or starch hydrolyzed into sugars from cassava, wheat, or maize, are used to generate bioethanol (Hoekstra & Chapagain, 2008). Sugar naturally ferments into acids and alcohols (mostly ethanol) when anaerobic conditions are met. Humans have been using yeast to speed up fermentation for thousands of years. Glycolysis is the primary metabolic process in ethanol fermentation, resulting in the synthesis of two pyruvate molecules and the metabolism of one glucose molecule (FAO, 2003). Pyruvate is further converted to ethanol under anaerobic conditions, releasing CO_2. In total, $C_6H_{12}O_6 \rightarrow 2\ C_2H_5OH + 2CO_2$ occurs. Assuming a kilogram of glucose is digested, the highest yield of ethanol is 511 g of ethanol and 489 g of carbon dioxide (or 530 g of ethanol per kilogram of starch).

Numerous byproducts are frequently also generated, such as glycerol (FAO, 2003). Stresses like ethanol buildup that occurs during ethanol fermentation prevent yeast cells from growing and producing ethanol (Hoang et al., 2021). Approximately 10–12% ethanol is present at the end (Gerbens-Leenes et al., 2009a). To reduce product inhibition, the fermentation industry employs a tanks-in-series technology. As of right now, it can yield between 90 and 93 percent of the glucose to ethanol theoretical value (Berndes, 2002).

Oilseed crops, such as soybean, jatropha, and rapeseed, are used to make biodiesel or pure vegetable oil. The oil that has been taken from an oilseed crop and utilized straight for energy production is known as straight vegetable oil (Hoang et al., 2023; Hoekstra & Chapagain, 2008). Olive oil is one example of lighting. Vegetable oil by itself is frequently a challenging biofuel to employ in diesel engines due to its chemical characteristics, such as its high viscosity at low temperatures (Hoekstra & Chapagain, 2008). Because of the comparatively high temperatures in warm-climate countries, raw vegetable oil can be used as a fuel without thickening. To produce a biodiesel that is less susceptible to colder temperatures, the oil needs further processing in nations with moderate climates. Glycerine molecules are the main byproduct of the transesterification chemical reaction, which turns oil and alcohol into an alkyl ester of the fatty acid and produces biodiesel. The most common feedstock for biodiesel in Europe is rapeseed oil, however, some sunflower oil is also utilized. Soybean oil serves as the primary feedstock in the United States, whereas palm, coconut, and jatropha oils are utilized in tropical and subtropical regions (Hoang et al., 2023; Hoekstra & Chapagain, 2008).

It is important to consider a product's entire life cycle when estimating the consumption of natural resources. However, agriculture—the initial link in the industrial chain—uses water most of the time. For instance, producing one liter of ethanol requires about 21 L of water, although this water is frequently recycled (Gerbens-Leenes et al., 2009a). As a result, this study solely considered the amount of water needed in agriculture and neglected the amount of water used in the industrial links of the production chain. A crop's ethanol-energy yield (measured in GJ/ton) was computed using the formula below:

$$E_{ethanol}(c) = DMF_y(c) f_{carbohydr}(c) f_{ethanol}(c) \times HHV_{ethanol}$$

where $DMF_y(c)$ is the dry-mass fraction in the crop yield (g/g), $f_{carbohydr}(c)$ is the fraction of carbohydrates in the dry mass of the crop yield (g/g), $f_{ethanol}$ is the amount of ethanol obtained per unit of carbohydrate (g/g), and $HHV_{ethanol}$ the higher heating value of ethanol (kJ/g). For the amount of ethanol per unit of sugar, we assumed the theoretical maximum value of 0.51 g/g, and for starch, 0.53 g/g (Berndes, 2002).

The biodiesel-energy yield of a crop (in GJ/ton) was calculated as follows:

$$E_{diesel}(c) = DMF_y(c) f_{fat}(c) f_{diesel}(c) \times HHV_{diesel}$$

where $DMF_y(c)$ is the dry-mass fraction in the crop yield (g/g), $f_{fat}(c)$ is the fraction of fats in the dry mass of the crop yield (g/g), f_{diesel} is the amount of biodiesel obtained per unit of fat (g/g), and HHV_{diesel} the higher heating value of biodiesel (kJ/g). For the fraction of biodiesel per fat weight, we assumed the value of 1. The fractions of carbohydrates and fats in the dry mass of crop yields are given in Table 5. Table 6 gives the HHVs of ethanol and biodiesel.

4 Impact Assessment at the Bioenergy-water Nexus

Water plays a vital role in bioenergy production. However, conventional bioenergy potential studies overlook the local and regional water implications of shifting land use for biomass production. It is crucial to consider the impact of biomass production on freshwater resources in the future.

Water is a key issue in bioenergy development and land use. Indicators and assessment tools have been proposed to include the water perspective in analyses to assist in strategy development and land-use planning. If done correctly, integrating bioenergy into the agricultural or forest landscape can optimize water utilization and mitigate water quality impacts associated with current land use. However, if done poorly, biomass production may create new problems and worsen already unsustainable land and water use practices.

Water use in bioenergy production is mainly associated with cultivating biomass feedstock (Fingerman et al., 2010). The source of this water can be rainwater or surface and groundwater used for irrigation. The use of agrochemicals can indirectly cause water pollution. Feedstock production for bioenergy may increase competition with food and fiber for water resources. From a life cycle perspective, it is also important to consider the "embedded" water in agrochemicals and post-harvest processing of the feedstock.

Bioenergy strategies prioritize high land use efficiency, which means maximizing bioenergy output per unit of land. However, this may lead to high-yielding systems that require large amounts of resources and can cause pollution, leading to trade-offs between managing climate change mitigation and preserving resources. To mitigate climate change while ensuring resource longevity, we need to adopt sustainable bioenergy systems that use fewer resources (Bayart et al., 2010).

There are two methods to analyze the relationship between bioenergy and water. The first approach is product-oriented, which assesses the environmental impact of various ways of meeting consumption needs. The second approach focuses on the watershed and evaluates how land use affects catchment water yield and environmental flow requirements. It is important to note that the definition of "water use" varies depending on the scope and purpose of the study and whether a product-oriented or a watershed approach is used.

4.1 The Value of Bioenergy Sources in Water Treatment

Establishing large-scale bioenergy plantations in areas with sparse vegetation may lead to an increase in evapotranspiration, which can divert water from runoff to surface water and reduce groundwater recharge. However, if bioenergy plantations are strategically placed in the landscape, it can help to decrease run-off, resulting in lower soil erosion at the site. This can also help to lessen the risk of flooding and sedimentation in rivers, which is a positive effect. To reduce the non-productive part of evapotranspiration, techniques such as rainwater harvesting, and improved soil, and land management have been proposed.

Bioenergy production has the potential to increase plant productivity and decrease negative effects on freshwater sources, like rivers, lakes, and aquifers. However, to fully understand the impact of bioenergy production on water resources, it's important to analyze water usage and its consequences at different spatial and temporal scales. This will enable the development of strategies to minimize negative effects and maximize positive outcomes. When considering the interaction between water resources and bioenergy production, it is essential to bear in mind key points:

1. The relationship between bioenergy production and water resources is not constant and can vary depending on the location and time.

2. Water moves not only vertically, through evaporation and transpiration, but also horizontally, through hill slopes, soils, groundwater, and rivers. Any changes in water quantity or quality can have a significant impact downstream and beyond the basin as water virtually moves through space and time due to the trade of agricultural products.
3. When thresholds are reached, it can result in a complete ecosystem transition to new states, which may happen quickly, making it difficult to return to the previous ecosystem state. Therefore, it is crucial to understand the thresholds beyond which far-reaching consequences can occur.

Natural systems are composed of a complex network of subsystems that are interconnected in both linear and nonlinear ways. To maintain these systems, one must consider the complexity and the uncertainties involved. This poses significant challenges in developing guiding principles, criteria, and indicators for those engaged in activities that directly or indirectly impact natural resources such as soil, water, and ecosystem health. Many indicators based on volumetric aggregation of water use are found inadequate in measuring these impacts. In the context of bioenergy, different evaluation frameworks, criteria, and indicators have been proposed. One of these frameworks is the "water footprint", also known as the 'virtual water footprint' to avoid confusion with other approaches. This framework is based on the concepts of green water and blue water. Green water refers to rainwater that is available in the root zone and flows due to evaporation and plant transpiration. Bluewater, on the other hand, is the surface and groundwater used for irrigation or other purposes such as industrial or residential water supply. The framework also includes a third component called gray water, representing the water required to dilute pollutant emissions to reach defined water quality requirements. By combining these three components, a single index—the virtual water footprint—can be obtained (Hoekstra et al., 2011). The virtual water footprint representation has gained popularity as an effective communication tool. However, it has also been criticized for its oversimplification of complex water contexts. By reducing everything to a single value based on average spatial and temporal conditions, it fails to account for the strongly varying local water contexts, such as the temporal and spatial variation of water resource availability and competing uses. This leads to a risk that non-experts may not understand the differences between the different types of water use, which can result in a loss of conceptual clarity and scientific rigor. To address this, some have suggested modifying the concept by separating water use by source and using regionalized water stress indices. This will allow for a more meaningful presentation of water use that reflects the nuances of local water contexts.

4.2 Life Cycle Assessment Approach

Water engineering approaches have been developed to estimate water use by identifying and quantifying all water flows within a defined system. These approaches utilize water balances, which consider the inputs and outputs of water. However, if the system boundaries are too narrow, downstream consequences may be missed, and water quality aspects may be left out, leading to a failure to consider the reality that water can become degraded beyond the point of critical pollution. This limits further use to those who can use such degraded water or require additional water for dilution to meet quality requirements for other uses. Therefore, it is necessary to consider varying spatial and temporal scales while defining the studied system to avoid such issues.

There has been a recent push to better understand water and its impact on the environment. This has resulted in significant efforts in both inventory and impact assessment stages, particularly regarding water accounting in LCA. Initiatives from organizations like the World Business Council on Sustainable Development and the UNEP.SETAC Life Cycle Initiative has been driving the development of new frameworks for assessing the consequences of freshwater consumption (Bayart et al., 2010). The proposed framework provides a way to assess the potential impact of freshwater consumption on human health, ecosystems, and resources. It suggests that parameters should be taken into account when establishing characterization models for describing these impact pathways on the midpoint and endpoint levels. Furthermore, the assessment method should be regionalized in the hydrological context. Freshwater consumption can have significant impacts on human health, ecosystems, and resources by lowering groundwater levels and reducing water availability. By modeling the consequences of water consumption at the midpoint and endpoint levels, we can better understand the effects of resource depletion, human health, and reduction in biodiversity (Bayart et al., 2010). To address the inconsistencies around different water footprinting approaches, the International Organization for Standardization (ISO) is considering the development of a new standard. The ISO is currently in the Preliminary Work Item (PWI) stage of developing ISO 14046, Water Footprint—Requirements and Guidelines, which is intended to complement existing LCA standards and ongoing work on carbon footprint metrics.

LCAs, or life cycle assessments, are used to evaluate the various environmental impacts of a product or process. LCAs take into account multiple impact categories and sometimes include a weighting system to compare the trade-offs between different impacts, such as water use versus other environmental impacts. However, it should be noted that the weighting system is controversial, as value judgments come into play.

To better integrate local and regional conditions within the impact assessment scheme, assessments are being made at the watershed, national, or cross-level depending on the scope and conditions (Anh & Hoang, 2024; Hoang et al., 2024a). These assessments take into account factors such as water quality, water availability, socio-economy, and water allocation requirements. Furthermore, temporal variations, such as seasonality, are also important for assessing water stress and should be considered in future modeling efforts.

4.3 The Value of Bioenergy Sources in Water Treatment

The efficient operation of a wastewater treatment system involves not only meeting regulatory standards but also streamlining processes and minimizing costs in the long run. In developing countries, several factories remain idle due to frequent power outages. Apart from this, financial constraints and maintenance issues during operation pose complex challenges that can affect treatment efficiency and sustainability. Therefore, it is imperative to explore alternative energy sources and opt for solutions that reduce operational costs.

A lot of individuals consider that wastewater is a valuable resource with high economic potential. Biosludge is a useful carbon source that makes efficient use of energy sources during the anaerobic treatment stage, biological silos, and biogas energy generation systems. These systems can produce green energy to cater to the energy requirements of factories and other activities.

Wastewater is treated through a two-stage process. Firstly, solid particles are removed in the primary treatment stage. Next, biological treatment takes place in a conventional activated sludge system. This process produces primary and residual biological sludge which is used as a thickener and to maintain biomass density during digestion in an anaerobic sludge digester. The digester produces methane-rich biogas. The biological sludge decomposes well so it can be used as a fertilizer after undergoing the sterilization process.

Biogas is produced during wastewater treatment and collected in a gas chamber. The collected biogas is then used as fuel to power a biogas engine. Some wastewater treatment plants focus on sludge treatment and generate electricity from the biogas.

By adopting a resource recovery model, wastewater treatment plants can operate more efficiently and reduce operating costs by installing bioenergy recovery equipment. AI can assist this (Hoang, 2022). This model, when implemented effectively, can result in profitable returns for businesses within 3–4 years. It also eliminates the need to transport waste sludge to remote landfills, saving on associated costs.

New technologies can produce energy from waste more efficiently than direct combustion. Thermal and heatless technology, gasification, heat sterilization, pyrolysis (Abi Bianasari et al., 2024), plasma gasification, anaerobic digestion, fermentation, esterification, and biological treatment are some of these technologies (Hoang et al., 2024b). They create flammable gas, hydrogen, synthetic fuel, synthetic crude oil, tar gas, syngas, biogas, ethanol, lactic acid, hydrogen, biodiesel, and abundant fuel sources.

Producing bioenergy from organic waste and residue is a great way to reduce greenhouse gas emissions caused by human activities (Bandh et al., 2024; Hoang et al., 2023, 2024c, 2024d). There are various methods to generate methane gas, which is a clean and renewable source of energy. One of these methods involves using wastewater treatment plants, which produce a large amount of methane during the sludge fermentation process. Another approach is to clean industrial organic waste streams through anaerobic digestion, which not only purifies water sources but also produces abundant methane gas that can be used to generate electricity and heat. Additionally, thermophilic organic matter can be decomposed at high temperatures ranging from 35 to 55 °C, producing methane gas. By using bioenergy produced from organic waste and residue, we can reduce our dependence on fossil fuels and prevent the natural decomposition of raw materials, which would otherwise contribute to greenhouse gas emissions, proves circular economy concept (Malla et al., 2023). Bioenergy from crops requires sustainability assessment based on factors like resource efficiency, competition for resources, and environmental impact. The EU's Renewable Energy Directive (REDII) sets ambitious targets, including 32% renewable energy in total consumption and 14% renewable fuels in transport and railway by 2030. REDII strengthens criteria for bioenergy sustainability by introducing import and biomass use rules. Unsustainable farming methods, such as those causing Land Use Change (LUC), can also be indirectly affected by the rules. One of the most debated topics in the twenty-first century relates to the availability and production of energy (Johansson, 2013). The demand for increased energy, driven by population growth and global economic expansion, is conflicting with the need for reduced CO_2 emissions and a shift away from fossil fuels concerns regarding food security, water scarcity, and land consumption are growing more important by the day. These aspects are closely linked to energy production and use and are not independent of each other. The relationship between energy, food, and water is commonly known as the water-food-energy nexus. It is not feasible to confine the evaluation of energy generation to a single restricted sector, as measures aimed at enhancing local efficiency can ultimately lead to the deterioration of the overall system beyond its original state (Dubreuil et al., 2013).

In recent times, there has been a heated debate among scholars about the production of biofuels. Policymakers have recently introduced various incentives to promote the production of energy from renewable sources (Nguyen et al., 2021; Parraga et al., 2018), finding that biofuels offer a great opportunity to reduce CO_2 emissions. Technologies even have been researched to convert CO_2 into fuel (Hoang et al., 2023). Biofuels help to reduce carbon dioxide emissions during energy production by utilizing feedstocks that have a shorter life cycle. Many countries have the potential to simultaneously increase their domestic energy production. The practice of using edible crops as a source of energy production is in direct competition with food production. Agricultural demand for food and energy crops impacts water and land management. Designing energy production systems is challenging due to low water resource utilization efficiency despite the increased crop yield achieved by massive irrigation.

4.4 *The Water Required to Produce a Reference Quantity of Energy from Biofuels*

Water is an essential resource in both energy conversion and crop production. In this field, understanding the concept of virtual water is crucial. Virtual water is a measure of the total amount of water used to create goods and services, taking into account the water used in all stages of production. The term virtual water trade is used to describe the embedded water in traded goods.

Defining the precise amount of water required to produce bioenergy is a fundamental concept, and there are many studies devoted to this topic. While biofuel production facilities can calculate the amount of water required at each step, they cannot determine the renewability of the water used or its effective usage in production.

The sustainability of water sources has been ranked according to their value, including economically. Understanding the overall sustainability of a process from a water perspective requires the separation of blue and green water. This is particularly important when analyzing the impact of biofuel production on renewable resources, as the blue water footprint enables us to quantify this impact.

After distinguishing between the blue and green water footprint, the resulting water efficiency indicator does not provide any information on the availability or scarcity of water in the production location. To account for water availability, the World Data Bank Water Withdrawal Index is taken into consideration as an additional factor in determining the index.

Bioenergy plantations (BPs) can be categorized as completely rainfed or partially irrigated. The former type of plantation relies solely on "green" precipitation water stored in soils, while the latter type involves the use of "blue" water from lakes, rivers, reservoirs, and aquifers in addition to precipitation water (Hogeboom et al., 2020).

The debate on whether large-scale irrigation of bioenergy plantations is a viable option or not depends on the economic and environmental tradeoffs involved. To make an informed decision, it is necessary to accurately quantify the water use of bioenergy systems. Meeting ambitious climate targets requires high biomass productivity, which may necessitate irrigation to reduce land requirements and tradeoffs with food production. However, this would come at the cost of damaging freshwater ecosystems. Water stress is increasing in both natural environments and human societies, leading to a rise in overall water stress or the action may result in undesired changes to the way water cycles on land. Investing in irrigation systems may become economically feasible due to increased biomass value from carbon pricing. At least 15% (and potentially more, as most studies omit this parameter) of field experiments involving lignocellulosic bioenergy crops are irrigated (Li et al., 2018).

The quantity of water available for bioenergy modeling is influenced by the climate data used. Different models can lead to variations in the data considered. Anticipated changes in water availability due to climate change should also be considered when analyzing future water abstraction estimates in a region. Understanding bioenergy systems' water requirements is crucial. We need to identify ways to increase efficiency and decrease blue water abstractions. The current review will provide valuable information on green and blue water demands of large-scale bioenergy. This will aid in better decision-making and policy development.

5 Conclusion and Prospect

Biomass is commonly used for bioenergy through traditional methods like charcoal, wood, and manure. Modern bioenergy is gaining popularity in industry, power generation, heat, and transportation. It currently accounts for 20% of bioenergy and is growing rapidly. Biofuels for transportation use food/feed crops, but second-generation biofuels like alcohols, diesel substitutes, and methane are being researched from lignocellulosic material. In the next decade, lignocellulose is expected to become the primary feedstock for biofuel production, replacing conventional food crops. This shift will significantly affect water resources due to the use of specialized plants and agricultural residues for biofuel production.

Growers who have invested in first-generation crop production will likely start to consider the value of their residues. Crop residues may soon become a part of biofuel feedstock, just like residues and processing by-products in the food and forest sectors are currently used to produce solid biofuels. This is especially important for conventional (first-generation) biofuels, where the primary product only represents a fraction of the actual biomass produced. Using residues and processing by-products can reduce water usage for bioenergy. This is because the same water used for food and forest products can create these residues and by-products. It also increases water productivity.

Residues play a crucial role in hydrology. A favorable mulch cover increases soil moisture by facilitating infiltration and reducing evaporation. However, excessive biomass extraction leads to decreased soil moisture storage, which can adversely affect plant growth, increase soil evaporation, and result in intensive erosion and runoff. Residues are pivotal to soil health and carbon sequestration. When using residues for bioenergy, it's crucial to conduct studies to determine their value.

Bioenergy demand may lead to high-yielding plantations, reducing downstream water availability, especially in water-scarce areas. Increased evapotranspiration can address soil salinity problems in some areas. It is important to manage the trade-off between upstream benefits and downstream costs. When planning for bioenergy production at a larger scale, it's crucial to consider the impact of climate change on water resources. Climate change will significantly affect the hydrological cycle, leading to changes in precipitation that will reduce water availability in rivers and aquifers. This will affect water quality and increase seasonal water demand for crops. Therefore, it's important to discuss and address these issues to prevent negative consequences on our water resources. Bioenergy crops can help adapt to difficult water situations and promote sustainable water management. Some crops are drought-tolerant and water-efficient, making them suitable for areas unsuitable for conventional food and feed crops. Bioenergy crops can also be used to treat nutrient-bearing water and limit water erosion, reduce runoff, trap sediment, enhance infiltration, and reduce landslide risks. It is essential to study the hydrological effects of bioenergy systems to understand their impact on water resources.

References

Ababaei, B., & Etedali, H. R. (2014). Estimation of water footprint components of Iran's wheat production: Comparison of global and national scale estimates. *Environmental Processes, 1*(3), 193–205.

Ababaei, B., & Ramezani Etedali, H. (2017). Water footprint assessment of main cereals in Iran. *Agricultural Water Management, 179*, 401–411.

Abi Bianasari, A., et al. (2024). Influence of combined catalysts on the catalytic pyrolysis process of biomass: A systematic literature review. *309*, 118437.

Bandh, S. A., et al. (2024) *Track to reach net-zero: Progress and pitfalls* (p. 0958305X241260793).

Bayart, J.-B., Cécile, B., Louise, D., Manuele, M., Stephan, P., François, V., & Annette, K. (2010). Accounting for the water impacts of ethanol production. *The International Journal of Life Cycle Assessment, 15*(5), 439–453.

Berndes, G. (2002). Bioenergy and water the implications of large-scale bioenergy production for water use and supply. *Global Environmental Change, 12*, 253–271.

Blok, K. (2006). *Introduction to energy analysis*. Techne Press.

Chalvantharan, A., Lim, C. H., & Ng, D. K. S. (2023). Economic feasibility and water footprint analysis for smart irrigation systems in palm oil industry. *Sustainability, 15*. https://doi.org/10.3390/su15108069.

Chen, Y., et al. (2017). Modeling the effects of land use change from cotton (*Gossypium hirsutum L.*) to perennial bioenergy grasses on watershed hydrology and water quality under changing climate. *Agricultural Water Management, **192***, 198–208.

De Schutter, L., Giljum, S. (2014). *A calculation of the EU Bioenergy land footprint*.

Dubreuil, A., Assoumou, E., Bouckaert, S., Selosse, S., & Mai, N. (2013). Water modeling in an energy optimization framework—The water-scarce middle east context. *Applied Energy, 101*, 268–279.

FAO. (2003). *World agriculture towards 2015/2030. An FAO perspective*, J. Bruinsma, Ed. Earthscan, London, UK.

Fingerman, K. R., Torn, M. S., O'Hare, M. H., & Kammen, Accounting for the water impacts of ethanol production. *Environmental Research Letters, 5*.

Gerbens-Leenes, P. W. (2018). Green, blue and grey bioenergy water footprints, a comparison of feedstocks for bioenergy supply in 2040. *Environmental Processes, 5*(1), 167–180.

Gerbens-Leenes, P. W., Hoekstra, A. Y., & van der Meer, T. (2009b). The water footprint of energy from biomass: A quantitative assessment and consequences of an increasing share of bio-energy in energy supply. *Ecological Economics, 68*(4), 1052–1060.

Gerbens-Leenes, W., Hoekstra, A. Y., & van der Meer, T. H. (2009a). The water footprint of bioenergy. *Proceedings of the National Academy of Science U S A, 106*(25), 10219–10223.

Gheewala, S. H., et al. (2014). Water footprint and impact of water consumption for food, feed fuel crops production in Thailand. *Water, 6*, 1698–1718. https://doi.org/10.3390/w6061698

Gheewala, S. H., Berndes, G., & Jewitt, G. (2011). The bioenergy and water nexus. *Biofuels, Bioproducts and Biorefining, 5*(4), 353–360.

Goudriaan, J. G., Rob, J. J., & Uithol, Peter, W. J. Productivity of agro-ecosystems. In B. S. Jacques Roy & H. A. Mooney (Eds.), *Terrestrial global productivity. Terrestrial global productivity*, (pp. 301–313), San Diego: Academic Press.

Guo, T., et al. (2018). Evaluation of bioenergy crop growth and the impacts of bioenergy crops on streamflow, tile drain flow and nutrient losses in an extensively tile-drained watershed using SWAT. *Science of the Total Environment, 613–614*, 724–735.

Ha, M., & Wu, M. (2015). Simulating and evaluating best management practices for integrated landscape management scenarios in biofuel feedstock production. *Biofuels, Bioproducts and Biorefining, 9*(6), 709–721.

Ha, M., & Wu, M. (2017). Land management strategies for improving water quality in biomass production under changing climate. *Environmental Research Letters, 12*(3), 034015.

Hoang, T. D., & Nghiem, N. (2021). Recent developments and current status of commercial production of fuel ethanol. *Fermentation, 7*(4), 314.
Heidari, A., et al. (2021). Spatially variable hydrologic impact and biomass production trade-offs associated with Eucalyptus (*E. grandis*) cultivation for biofuel production in Entre Rios, Argentina. *GCB Bioenergy, 13*(5), 823–837.
Hoang, A. T., et al. (2021). Progress on the lignocellulosic biomass pyrolysis for biofuel production toward environmental sustainability. *223*, 106997.
Hoang, T.-D., et al. (2022). Artificial intelligence in pollution control and management: status and future prospects. 23–43.
Hoang, T.-D., et al. (2023). Carbon-based synthesized materials for CO_2 adsorption and conversion: Its potential for carbon recycling. *8*(4), 53.
Hoang, T.-D., et al. (2023). Biofuel projects and current environmental policies: Vietnam's case and neighboring Asian countries. *Biofuels in Circular Economy* (pp. 73–88). Springer.
Hoang, T.-D., et al. (2024a). Regional collaboration for water development projects for strengthening the governance of transboundary rivers in the Mekong countries. *The Mekong River Basin* (pp. 211–235). Elsevier.
Hoang, T. D., et al. (2024b). Valorization of agriculture residues into value-added products: a comprehensive review of recent studies. *24*(8), e202300333.
Hoang, T.-D., Liu, Y., & Le, M. T. J. S. (2024c). Synthesis and characterization of biochars and activated carbons derived from various biomasses. *16*(13), 5495.
Hoang, T.-D., Liu, Y., & Le, M. T. J. P. (2024d) Performance of CO_2 adsorption on modified activated carbons derived from Okara powder waste: impacts of Ammonia impregnation. *12*(9), 2024.
Hoang, T.-D., Liu, Y., & Le, M. T. (2024e). Synthesis and characterization of biochars and activated carbons derived from various biomasses. *Sustainability, 16*(13), 5495.
Hoang, V.-H., et al. (2024). Sources, environmental fate, and impacts of microplastic contamination in agricultural soils: A comprehensive review. 175276.
Hoang, V.-H., et al. (2025a). Microplastic characteristics, transport, risks, and remediation in groundwater: A review. *Environmental Chemistry Letters*, 1–21.
Hoang, V. H., Nguyen, M. K., Hoang, T. D., Nguyen, H. L., Lin, C., Yousaf, B., Ha, M. C., Bui, V. K. H., Pham, M. T., Chang, S. W., & Nguyen, D. D. (2025b). Global occurrence and environmental fate of microplastics in stormwater runoff: Unlock the in-depth knowledge on nature-based removal strategies. *Reviews of Environmental Contamination and Toxicology, 263*(1), 1–15.
Hoekstra, A. Y., & Chapagain, A. (2008). *lobalization of water: Sharing the planet's freshwater resources*, 1st edn. Wiley-Blackwell.
Hoekstra, A. Y., Chapagain A. K., Aldaya, M. M., Mekonnen, M. M. (2011). *The water footprint assessment manual: Setting the global standard* 1st ed. London, UK.
Hoekstra, A. Y., Chapagain A. K., Aldaya, M. M., & Mekonnen, M. M. (2011). *The water footprint assessment manual*. Earthscan.
Hoekstra, A. Y., & Hung, P. Q. (2005). Globalisation of water resources: International virtual water flows in relation to crop trade. *Global Environmental Change, 15*(1), 45–56.
Hoekstra, A. Y., & Wiedmann, T. O. (2014). Humanity's unsustainable environmental footprint. *Science, 344*(6188), 1114–1117.
Hogeboom, R. J., De Bruin, D., Schyns, J. F., Krol, M. S., & Hoekstra, A. Y. (2020). Capping human water footprints in the world's river basins. *Earth's Future, 8*, e2019EF001363.
Holmatov, B., Hoekstra, A. Y., & Krol, M. S. (2019). Land, water and carbon footprints of circular bioenergy production systems. *Renewable and Sustainable Energy Reviews, 111*, 224–235.
International Organization for Standardization. (2014). *Environmental management—Water footprint—Principles, requirements and guidelines*. Geneva, Switzerland.
Jager, H. I., et al. (2015). Forecasting changes in water quality in rivers associated with growing biofuels in the Arkansas-White-Red river drainage, USA. *GCB Bioenergy, 7*(4), 774–784.
Johansson, B. (2013). Security aspects of future renewable energy systems—A short overview. *Energy, 61*, 598–605.

Khedulkar, A. P. et al. (2025). Bio-based nanomaterials as effective, friendly solutions and their applications for protecting water, soil, and air. *Materials Today Chemistry, 46*, 102688.
Kumar, E., Saraswat, D., & Singh G. (2020). Comparative analysis of bioenergy crop impacts on water quality using static and dynamic land use change modeling approach. *Water, 12*. https://doi.org/10.3390/w12020410.
Le Anh, T., & Hoang, T.-D. (2024). Trans-boundary water governance in the Mekong countries. In *The Mekong river basin* (pp. 237–272). Elsevier.
Li, W, Ciais, P., Makowski, D., & Peng, S. (2018). A global yield dataset for major lignocellulosic bioenergy crops based on field measurements. *Scientific Data, 5*(180169).
Li, J., Xiong, F., & Chen, Z. (2021). An integrated life cycle and water footprint assessment of nonfood crops based bioenergy production. *Scientific Reports, 11*(1), 3912.
Malla, F. A., et al. (2023). Circular economy to decarbonize electricity. In *Renewable energy in circular economy* (pp. 71–87). Springer.
Mekonnen, M. M., Gerbens-Leenes, P. W., & Hoekstra, A. Y. (2015). The consumptive water footprint of electricity and heat: A global assessment. *Environmental Science: Water Research & Technology, 1*(3), 285–297.
Muhammad-Muaz, A., & Marlia, M. H. (2014). Water footprint assessment of oil palm in Malaysia: A preliminary study. *AIP Conference Proceedings, 1614*(1), 803–807.
Ngan, N. V. C., et al. (2024). Studying on forest biomass in Tra Su Melaleuca forest landscape protection Area, An Giang Province. *Mangroves in a Changing World: Adaptation and Resilience* (pp. 167–181). Springer.
Nguyen, V. C. N., et al. (2021). Green technology for sustainable development: Practice and experience of renewable energy in Vietnam. *4*(2), 79–92.
Parraga, J., Khalilpour, K. R., & Vassallo, A. (2018). Polyfeed and polyproduct integrated gasification systems. In *Polygeneration with polystorage chemical and energy hubs* (pp. 175–212).
Rees, W., & Wackernagel, M. (2008). Urban ecological footprints: why cities cannot be sustainable—And why they are a key to sustainability. In J. M. Marzluff, et al. (Eds.), *Urban ecology: An international perspective on the interaction between humans and nature* (pp. 537–555). Springer US: Boston, MA.
Richard, B., et al. (2019). Optimizing the bioenergy water footprint by selecting SRC willow canopy phenotypes: Regional scenario simulations. *Annals of Botany, 124*(4), 531–542.
Searle, S. Y., & Malins, C. J. (2014). Will energy crop yields meet expectations? *Biomass and Bioenergy, 65*, 3–12.
Sezali, N. A. A., et al. (2024a). Bio-based nanomaterials for energy application: A review. *Vietnam Journal of Chemistry 62*(1), 1–12.
Sezalia, N. A. A., Ong, H. L., & Hoang, T. D. (2024b). Bio-based nanomaterials and their biomedical applications: A short review. *Vietnam Journal of Science and Technology, 62*(4), 615–634.
Steen-Olsen, K., et al. (2012). Carbon, land, and water footprint accounts for the European Union: Consumption, production, and displacements through international trade. *Environmental Science and Technology, 46*(20), 10883–10891.
Tran, T.-K., et al. (2024a). Applications of engineered biochar in remediation of heavy metal (loid) s pollution from wastewater: Current perspectives toward sustainable development goals. 171859.
Tran, T.-K., et al. (2024b). Review on fate, transport, toxicity and health risk of nanoparticles in natural ecosystems: Emerging challenges in the modern age and solutions toward a sustainable environment. *Science of The Total Environment, 912*. 169331.
Wang, G., et al. (2018). Hydrologic and water quality responses to biomass production in the Tennessee river basin. *GCB Bioenergy, 10*(11), 877–893.
Winnie Gerbens-Leenes, A. Y. H., Theodorus, H., van der Meer (2008). *Water footprint of bioenergy and other primary energy carriers.*
Xie, X., et al. (2017). Regional water footprints of potential biofuel production in China. *Biotechnology for Biofuels, 10*(1), 95.

Yu, Z., & Qingshan, Y. (2014). Decoupling agricultural water consumption and environmental impact from crop production based on the water footprint method: A case study for the Heilongjiang land reclamation area, China. *Ecological Indicators, 43*, 29–35.
Zhu, Y., et al. (2019). A Bibliometrics review of water footprint research in China: 2003–2018. *Sustainability, 11*. https://doi.org/10.3390/su11185082.

Sustainable Approaches for Enhancing Crop Production

Muhammad Moaz Khursheed, Muhammad Abdullah, Tooba Mumtaz, Azmat Qadir Maria, Muhammad Farhan Qadir, Iftikhar Ali Ahmad, Ahmad Jawad, and Muhammad Tahir Murad

Abstract In the modern world, sustainable methods for enhancing agricultural output are crucial to ensuring food security, protecting natural resources, and reducing agriculture's negative environmental effects. These methods seek to increase yields while minimizing the use of artificial inputs, reducing greenhouse gas emissions, and enhancing ecosystem health. Adopting agro ecological practices is a crucial component of a sustainable strategy. This entails incorporating ecological ideas into agricultural systems so that we may cooperate with nature rather than against it. Crop rotation, cover crops, and agroforestry are three practices that farmers may use to increase soil fertility, manage pests and diseases, and boost biodiversity. Agro ecological practices lessen the need for chemical pesticides and fertilizers by constructing adaptable farming systems that can thrive climate change. Precision agriculture is a crucial sustainable strategy. For the purpose to best utilize the available resources, technology and data-driven strategies are used. Remote sensing, geographic information systems (GIS), and GPS are tools that farmers may use to monitor crop development, evaluate soil conditions, and accurately administer components including irrigation, fertilizer, and pesticides. Precision agriculture minimizes pollution and the hazardous effect on environment by only using resources where and when they are required. A key component of producing crops sustainably is effective water

M. M. Khursheed (✉) · M. Abdullah · T. Mumtaz
Institute of Soil and Environmental Sciences, University of Agriculture, Faisalabad, Pakistan
e-mail: moazkhursheed3@gmail.com

A. Q. Maria
University of Agriculture, Faisalabad, Sub-Campus Burewala, Vehari, Punjab, Pakistan

M. F. Qadir
College of Resources and Environment, Xinjiang Agricultural University, Urumqi, Xinjiang, China

I. A. Ahmad
College of Resources and Environment, Huazhong Agricultural University, Wuhan, China

A. Jawad
PARC Rice Programme, Kala Shah Kaku, Lahore, Pakistan

M. T. Murad
Department of Plant Pathology, University of Agriculture, Faisalabad, Pakistan

M. Shaaban (ed.), *Soils and Sustainable Agriculture*, Frontier Studies in Soil Science,
https://doi.org/10.1007/978-3-031-91114-9_45

management. Effective irrigation methods are crucial due to the growing shortage of water in many locations. For instance, drip irrigation sends water straight to the roots of the plant, decreasing water loss due to evaporation and increasing water usage effectiveness. To gather and store rainfall for irrigation during dry spells, rainwater harvesting and conservation techniques may also be used. Lastly, the utilization of organic and biological inputs may improve sustainable agricultural production. Compost and manure are examples of organic fertilizers that feed the soil with nutrients, strengthen its structure, and promote microbial activity. While using fewer synthetic pesticides, biological pest management techniques, such as the utilization of beneficial insects and microbiological agents, may efficiently manage pests. A path ahead for agriculture that strikes a balance between productivity and environmental stewardship is provided by sustainable methods for increasing crop yield. Farmers may grow healthier crops, save resources, and Assist development of a more robust and sustainable food system by employing agro ecological practices, precision agricultural methods, better water management, and using organic and biological inputs.

Keywords Precision agriculture · Integrated nutrient management · Soil health · Crop production

1 Introduction to Sustainable Approaches

A sustainable agricultural method aims to use resources from nature so effectively whereby they can restore their production potential while also minimizing negative effects on habitats exceeding the land's boundary. Sustainable agriculture also entails increasing soil production while limiting negative impacts on the environment, soil, atmosphere, hydrosphere, biosphere, and human health. The primary notion driving sustainable agricultural production is to achieve an equilibrium amongst the requirement for food and the protection of environmental resources from depletion and destruction. This sustainable strategy also has many other goals, such as using fewer inorganic fertilizers, protecting water resources, conserving biodiversity, reducing waste output etc. (Velten et al., 2015). Green fertilizers have been shown to enhance crop productivity and improve soil conditions (Rai et al., 2017; Singh et al., 2010). Empowering farmers financially, implementing novel methods, and offering better amenities to improve their standard of life are all part of the sustainable approach aims (Campbell et al., 2014). Figure 1 illustrates the various components of sustainable agriculture.

In the context of rising global issues that include global warming, economic scarcity, food insecurity and imperative need to evolve sustainable practices has become more urgent than ever. In this pursuit, the convergence of science and technology has produced a powerful array of sustainable practices that have the potential to shape industries and societies alike. Among these transformative solutions, nanotechnology, precision agriculture, geographic information systems (GIS),

Fig. 1 Different components of sustainable agriculture

global positioning systems (GPS), and remote sensing emerge as beacons of hope, offering unprecedented opportunities to address environmental, social, and economic challenges. This comprehensive exploration explores the multifaceted world of these innovative tools, unveiling their profound impact on sustainability and their promise to promote harmonious coexistence between humanity and the natural world. Through a thorough examination of their applications, benefits and potential drawbacks, this article aims to prepare for the approaches for more sustainable and flourishing future for coming generations.

At the forefront of these sustainable practices is the field of nanotechnologies, a field that explores the manipulation and use of matter at the nanoscale. By exploiting the unique properties exhibited by materials at such small dimensions, nanotechnologies offer a multitude of applications in diverse fields. One of the most important contributions of nanotechnology is in the realm of environmental protection. Nanomaterials have proven their value in water purification, air filtration, and waste treatment, where their high surface area and reactivity facilitate the removal of contaminants with remarkable efficiency. Additionally, in the pursuit of renewable energy, nanotechnology serves a significant impact in development of more efficient solar cells and energy storage devices, driving the transition to clean and green energy sources. However, the widespread adoption of nanotechnologies also raises concerns about their potential environmental and health impacts, warranting careful regulation and responsible development.

Another important pillar of sustainable landscaping is health agriculture, a data-driven approach to farming that seeks to optimize resource use while minimizing environmental impact (Kamble et al., 2020; Mohamed, 2023). Leveraging advances in sensors, robotics, and data analysis, precision agriculture empowers farmers to reach smart choices about irrigation, fertilization, and pest management, modifying interventions to fulfill the certain requirements of each crop and even individual plants. By minimizing inputs and reducing wastage, precision agriculture not merely improves agricultural production and farmers' incomes, however also minimizes the negative impacts of conventional farming on land, water and biodiversity (Ahmad & Dar, 2020). Globally, the population continues to expand, adopting healthy agriculture becomes increasingly important to Improve food security and encourage sustainable foods production systems that can withstand the challenges of a changing climate (Dwivedi et al., 2017; Umesha et al., 2018).

2 Different Sustainable Approaches

2.1 Agro Ecological Practices

For decades High crop yield and productivity was one of the main focuses of agriculture but now the main attention of the world has shifted to achieve similar goals with sustainable means. The main aim in doing so is to not only solve the issues like climate change but also ensure the food security for the world even meeting the potential population bloom that is expected. According to UN report by population division 2022 The population of the world will reach 9.6 billion in 2050. This creates a big question regarding food security and sustainable crop production. Therefore, practices are being proposed that can provide the same high production in most sustainable way possible. This will help shift conventional means of agriculture to high technology based, ecology based sustainable agriculture. Agro ecological practices are a way to achieve the future goals of sustainable agriculture (Dalgaard et al., 2003).

Agro ecology is a recent terminology developed to study the science behind the integrated effect of agriculture on social and biological aspects of ecosystem. It can be used to help maintain the ecosystem while enhancing the productivity of agriculture. Agro ecological studies puts additional emphasis on protection of biodiversity, resource conservation, food security, rural development, sustainable food production and many other aspects for the upcoming generations of the future. Agro ecology serves a vital part in strengthening the resilience of agronomic systems so that it can withstand the potential hazards that reduces agricultural productivity and are becoming more and more frequent like floods, un even rain fall patterns, heat waves, change in growing seasons, water availability, erosion, less availability of nutrients, etc. most of which are related to climate change (Anderson et al., 2021).

Agro ecological practices: are an integrated approach to using sustainable agricultural practices, keeping in view the ecological concepts and principles, to enhance food production, ensuring food security for the world. Practices such as crop rotation, high yielding varieties, relay cropping, organic farming, reduced tillage and cover crops are around for many decades now these practices are being studied under agro ecological practices based on their positive impacts on agricultural productivity and ecological preservance. Agro ecological practices also improves the soil characteristics like soil health fertility penetrability water holding capacity infiltration percolation and many other physical and chemical properties (Gliessman, 2018).

2.1.1 Crop Rotation

Crop rotation is a technique of utilizing same land to grow different combination of crops in a series. It is one of the oldest practices that some farmers invented as a solution to solve many issues like pest and disease control. This practice has many positive returns like breaking pest and disease cycles, in this the host crop on which a pest feed on is removed from an area where extensive out breaks are recorded by this depletes the amount of host hence reducing the amount of pest for the next growing season. It only works for obligate parasites or those pests that have either short over-winter period of a smaller number of secondary hosts (Brust & King., 1994). It can also reduce greenhouse emissions of Nitrous Oxides as they are required as an essential nutrient. Growing multiple exhaustive crops in series will lead to volatilization losses of Nitrogen fertilizers. Leguminous crops in crop rotation reduces the emissions by volatilization by 35% (Behnke et al., 2018). Selection of crops to be grown is done by keeping in view the locality, climate, soil and environmental features, a proper crop rotation cycle should be 3–4 years including leguminous crops to increase the fertility status of soils and to improve soil health by utilizing more biofertilizers (Dhaliwal et al., 2021).

2.1.2 High Yielding Varieties

As the name suggests it simply means growing those varieties that have the potential to perform best in the field. There are many factors to be considered apart from harvest or economical yield of the crop which included pest, disease, drought, salinity and many more stresses that can affect the yield of crop. The variety to be grown should have the capacity to withstand any stress that can reduce the yield of the crop. More efficient selection of variety will result in maximum yield and more efficient utilization of available resources for a particular field. The only problem that is faced by the growers is lack of new and adapted to stress varieties and their cost (Swarbreck et al., 2019).

2.1.3 Relay Cropping

Relay cropping is an agro ecological practice in which the next crop is sown in standing previous crop before its harvest. This practice can increase the productivity of the area under cultivation it reduces the time requirement for next crop sowing. This practice can help reduce weeds pests and diseases even in some cases it can provide better germination condition for seeds of next crop. Not all crops can be grown this way it depends on crop to be grown, time of the year, soil and environmental conditions (Gesch et al., 2023). There are some draw backs as it can cause competition between crops and hinders the availability of nutrients for the crop. It can create issues in harvesting as extra care is needed to protect the seedlings of 2nd crop. The standing crop will also interrupt light therefore it can lead to uneven growth of seedlings. This agro-ecological practice is highly specific to the field conditions so it is necessary to opt for this where it is suitable to avoid any yield loss (Tanveer et al., 2017).

2.1.4 Organic Farming

Organic farming means utilization of organic fertilizers as a mean of nutrition for crops. There are many benefits of utilization of organic fertilization for crop production like no residues of heavy metals that originate from fertilizers like Cd from rock phosphate in phosphatic fertilizers. This leads to reusing of nutrients in waste there for helps in sustainable nutrient source for crops. Organic fertilizers enhance the nutrient holding capacity of the in addition to the water holding capacity of the soil. They enhance the soil microbiota and enhances soil health. The main problem faced by organic farming is bulking, huge quantities of organic fertilizers like farm yard manure, poultry manure and many more are required to complete the recommended dose of NPK for the crop. Besides that, Organic farming is the best method to produce cleaner sustainable and better yield crops. It is the definite future direction of agriculture in coming years (Shepherd et al., 2002).

2.1.5 Reduced Tillage

Reduced tillage is an agro ecological practice in which tillage practice in minimized to avoid disturbing the soil. It is one the most beneficial agro ecological practices as it conserves the fuel and labor cost, it reduces soil erosion by wind or rain fall, it increases the soil biota and conserves organic matter of the soil (Shepherd et al., 2002). Less physical manipulation of soil can develop hard pan in soil if the soil is continuously being tilled before therefore in order to reduce tillage slowly reduce the amount of tillage while adding organic fertilizers or growing a green manuring crop for some seasons. Herbicide utilization may increase accordingly as the weed control by tilling them deep in soil is reduced.

2.1.6 Cover Crops

Cover crops are a type of poly cultural practice in which non-economical crops are grown next to main crop not for harvest purpose but to provide additional benefit to the main crop. Cover crops can control erosion, prevent lodging of crops, it can help reduce soil compaction, it can act as a bait crop to lure beneficial insects as biological control, it can improve soil health, enhance the microbial biodiversity in soil as two types of rhizospheres will co-exist enriching the soil with different microbes and enhance other properties of Soil. Cover cropping with leguminous crops can also serve as a nitrogen source to closely grown plants or to next sown crop (Scholberg et al., 2010).

3 Precision Agriculture

Precision agriculture (also known as precision farming or site-specific crop management) is an approach to farming which utilizes advanced technologies and data analysis techniques to enhance agricultural practices. It includes the application of information technology, remote sensing, and various tools to monitor, measure, and manage agricultural processes with high efficiency to increase the production and quality of the crops (Kumar & Ilango, 2018).

Precision agriculture idea originated in the United States in the beginning of the 1980s but it is difficult to point out the specific event or dates as the beginning of its adaptation worldwide. One of the early pioneers of precision agriculture was Dr. John Deere, an American engineer and inventor who developed the commercially successful self-scanning seed planter in the late 1970s. This innovation allowed farmers to plant crops more precisely and efficiently. In the 1990s, precision agriculture gained more attention and momentum as technology advanced. The use of GPS and GIS became more widespread, enabling farmers to collect and analyze data about their fields, like soil properties, nutrient levels, and crop growth patterns. This data-driven method allowed Farmers can make better educated judgments on products like fertilizer, pesticides and irrigation, optimizing their resource use and increasing yields. Since then, precision agriculture has continued to develop and incorporate modern technologies such as drones, satellite imagery, and advanced sensors. It has been adopted in numerous regions of the world, including North America, Europe, Australia and parts of Asia, where agriculture plays a significant impact in the economy (Fig. 2).

3.1 Importance of Precision Agriculture

Precision agriculture enables farmers to make optimal the use of natural resources like water and fertilizers, pesticides and energy. Using modern technologies such as GPS, remote sensing, and data analysis, Farmers can effectively focus use of these resources based on the particular requirements of multiple domains within their fields. It reduces waste material, its environmental impact, and increases general efficacy in resource use. Precision agriculture helps farmers maximize crop productivity and yield. By using advanced technology to monitor and analyze factors such as soil characteristics, variations in the weather as well as yields so farmers can make rational choices about irrigation, fertilization, and pest control. It leads to improved crop

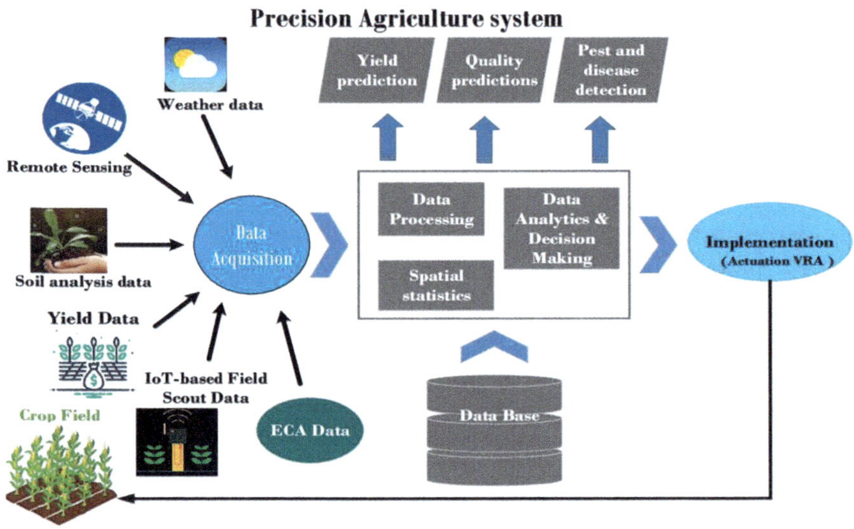

Fig. 2 Precision agriculture system

health, better conditions for growth and ultimately higher yields. Precision agriculture enables timely identification and targeted management of pests and diseases. Using sensors, drones and data analysis, farmers can recognize pest or disease hot-spots, assess the severity of infestations, and apply interventions where needed. This blanket reduces the need for pesticide application and reduces chemical usage, resulting in more efficient pest management and reduced environmental impact. Precision agriculture enables farmers to adopt site-specific approaches to farming. By analyzing field variables and crop management practices accordingly, farmers can account for differences in soil types, nutrient levels, moisture content, and other variables. This allows for customized treatments to improve crop performance in different areas of the field, including applying variable rates of inputs, seed selection, and irrigation scheduling.

Precision agriculture promotes environmental sustainability by reducing the negative impacts of farming on ecosystems. By optimizing resource use, reducing chemical inputs, and minimizing soil erosion, precision agriculture helps protect water quality, preserve biodiversity, and mitigate climate change. It enables farmers to adopt more sustainable farming practices, aligning with conservation goals and reducing the overall environmental impact of agriculture. Precision agriculture can lead to economic benefits for farmers. By optimizing inputs, reducing waste, and improving crop productivity, precision agriculture can increase profitability and financial stability. It helps farmers make sensible choices regarding investments, crop choice, and resource allocation, resulting in better returns on investment. Precision agriculture facilitates better traceability and the standard of quality all through the agricultural supply chain. By collecting and managing data on crop inputs, growth conditions, and production methods, precision agriculture enables farmers to provide accurate information about their products. It promotes transparency, builds consumer confidence, and meets the growing demand for traceable and high-quality agricultural products. Overall, precision agriculture offers a transformative approach to integrating farming, technology, data analysis, and agricultural expertise. Its importance lies in its ability to improve resource utilization, increase productivity, improve sustainability and contribute to the economic viability of farming operations (Thompson et al., 2019).

3.2 Precision Agriculture Techniques:

There are several precision farming techniques and technologies that are used to improve the farming practices by which farmer can get more benefit in the form of increased yield of crop than the use of conventional methods.

3.2.1 Global Positioning System (GPS)

GPS (Global Positioning System) receivers offer a way to locate oneself on the globe. Automated accurate, location tracing with GPS users can be used by farmers and agricultural service providers to monitor information automatically and relate changing rates of inputs to tiny locations in greater areas (Pfost et al., 1998). The U.S. Department of Defense is in charge of the Global Positioning System (GPS), a network of satellites used to assist ground-based equipment locate themselves in real time using latitude and longitude coordinates. In agricultural applications, GPS is utilized for field scouting, soil sampling, harvesting operations, machine guidance, controller (variable-rate-input applications, discussed below), and data collecting (Morgan & Ess, 1997).

There are three segments in the GPS. The first is made up of 24 satellites (21 operational, 3 spares), which are in a line-of-sight orbit 20,200 km above the Earth and offer round-the-clock service. The satellites go through one of six orbital planes and finish an orbit in just under 12 h. Each satellite transmits a pseudo random noise (PRN) code that identifies its specific location and time of signal transmission. The second part of the software is composed of ground-based control centers that calculate each satellite's orbit about a week in advance, forecast ionosphere situations during that time, and transfer the information to each satellite's computer. Each satellite conveys a pseudo random noise (PRN) code that identifies its specific location and time of signal transmission. For two-dimensional (latitude and longitude) and three-dimensional (latitude, longitude, and altitude) placement, the receiver needs a minimum of three satellites and four satellites, respectively. A GPS user requires a direct Range of recognition for the antenna. Either area hindrance (trees, buildings, steep slopes, etc.) that may block sunlight can likewise block a GPS signal (Lange, 1996). An observer who has a GPS receiver at their disposal can navigate or gather location data whether they are moving or still.

3.2.2 Geographic Information System (GIS)

Applications of GIS can significantly enhance the accuracy of crop production estimation, enabling the industry to better balance supply and demand. Additionally, it aids in detecting soil erosion, repairing damaged soil, and evaluating the soil. Farmers gain from all of these factors. GIS is useful for precision farming as well. It makes it possible to gather timely geo-spatial data that agriculturists can use to determine what the soil and the plants require in order to thrive. With that knowledge, they can also implement user-friendly, cost-effective site-specific treatments.

A Geographic Information System (GIS) is a set of computer hardware, software, and operating systems created to allow gathering, archiving, retrieval, analysis and presentation of geographically relevant data which may support planning and management choices (Aronoff, 1989). A GIS for crop production may comprise data from multiple origins such as aerial photography, satellite images, pest or pathogen scouting data, input operations data, GPS-based yield maps, soil surveys, and historical field information. The information is shown spatially (Geo-referenced) above which of a field base map, enabling the layering of information to offer precise analysis of crop health and maturity. The farm manager is able to gather and add data (such as weather, insect and weed issues, nematode concentrations, seed varieties, and planting populations) after the base map has been created in order Offer details regarding this particular crop, evaluate treatments, and even make predicted harvest maps.

Application of GIS in Agriculture

For GIS to be effective for agriculture, there are several types of mapping that RS provides:

Soil type mapping: It is crucial to understand various soil types, the crops they can sustain, and the places where each type of soil is present in order to plan agriculture on a large scale. If employed wisely, this knowledge has the potential to significantly raise the output and quality of agriculture. Humus, nutrition, and name are only a few characteristics of soils.

Crop coverage mapping: The production of different crops can be estimated fairly efficiently utilizing maps that show crop coverage. These aid in the analysis of current agricultural practices including crop rotation and terracing, among others. It's crucial to research the current trends and approaches in order to make any improvements. Crop characteristics include things like the right time of year or climate, the need for water or manure or fertilizer, the amount of time needed for planting, ripening, and harvesting, etc.

Rivers/distributaries mapping: The locations of the many river and distributary systems inside and surrounding an area are depicted on these maps. Finding the closest supply of water and creating the most cost-effective canal systems are made easier with this knowledge. And we can know about the quality of that water is fit or not for the irrigation purposes. Name, volume, depth, width, etc. are attributes.

Land use mapping: These maps reveal information about the current uses of the land. It assists in assessing how efficiently land resources are being used and in developing more cost-effective solutions for the same, in addition to maintaining up-to-date records required for various other calculations. Attributes include things like the type of use, amount of forest cover, amount of agriculture produced, and amount of predicted urbanization.

Contour mapping: The elevation information of the numerous geographical features on the surface of the earth is displayed on these maps. This knowledge aids us in making critical estimations, such as the most cost-effective canal laying method taking into account excavation and leveling expenses, etc.

Irrigation system mapping: This type of mapping focuses on the specifics of more established or older systems, such dams and canals. By analyzing this data, one can develop extension projects and maintenance plans based on how effectively the systems function. Details on efficiency, service areas, water volume carried, advantages, etc. are among the attributes.

Meteorological mapping: Under this type of mapping, areas such as rainfall patterns, climatic conditions, seasonal fluctuations, and forecasts are covered. This is crucial information that farmers and agriculturists need in order to plan the most efficient and optimal style of farming. It aids in determining how much water should be released into canal systems in accordance with forecasts of rainfall intensities.

Identification, area estimation and monitoring: Major components of plantation crops include identification, estimation of the growing stock, analysis of distribution, and periodic monitoring. In comparison to other agricultural crops, the spread of plantation crops is rather more localized due to specialized management techniques.

Crop condition assessment: The physiological alterations brought on by stress in a plant may alter its spectral reflectance/remittance characteristics, making it possible to detect stress via remote sensing techniques. It is crucial to monitor the crop at regular intervals throughout the crop growth cycle in order to take the necessary action and determine any potential output losses.

Crop yield modeling and production forecasting: The planning of the government's national food policy as well as the economy of the nation both heavily rely on information about crop productivity before harvest. One of the most crucial elements of estimating agricultural production is a reliable crop yield estimate.

Agricultural draught assessment: Draught assessment is yet another field in which operational use of remote sensing data has been made drought monitoring and assessment at the district level.

Reflectance modeling: Crop physical reflectance models are crucial for understanding the intricate relationship among sun radiation and plant canopy. Crop growth simulators need to mimic growth of crops so that yield may be estimated accurately. Crop growth models describe how environmental variables such as temperature, water, and nutrient availability interact with physiological procedures in plants.

3.2.3 Remote Sensing

According to its classification, remote sensing is the art and science of learning details regarding things or places over an extent lacking physically being there. The fundamental idea in remote sensing is that it utilizes the electromagnetic spectrum (visible, infrared, and microwaves) to evaluate physical features of the earth. These wavelength regions' targets typically respond in distinct ways, making them useful for differentiating between vegetation, bare soil, water, and other comparable phenomena. Two of the most common kinds of remote sensing are passive and active remote sensing. Passive remote sensing is used as apparatus for measuring energy from the environment. For instance, the Sun can only be taken when it is lighting the Earth. Active provides its own source of power for illumination. The radiation emitted by the sensor is directed on the target to be analyzed. The sensor detects and quantifies the radiation that returned from target. No matter any time of day or season, they take measurements. Radar and laser, for instance.

Resources from agriculture are valuable, dynamic, renewable natural resources. In India, the agriculture sector alone provides a living for almost 70% of the people and contributes for about 35% of the nation's gross domestic production. As there is minimal room for expanding the area under cultivation, enhancing agriculture production has been the primary goal. This necessitates the judicious and impactful organization of water and land resources. Since the past two decades, remote sensing methods have been used to investigate agriculture applications like crop yield tracking. This requires complete and consistent data on land use cover, forest area, soils, geological data, the magnitude of wastelands, agriculture crops, water resources both surface and underground, and hazards natural disasters like drought and flood. Seasonal data on crops, their land, strength, and yield facilitate the nation to execute appropriate support and procurement policies and adopt reasonable measures to address shortages, if any.

- Ground bases: Spectral radiometer, Pilot-Balloons, radars, and infrared thermometer
- Air Bases: Aircraft air based remote sensing tools.
- Satellite based: The digital image processing, using powerful computers, is the key tool for analyzing and interpretation of remotely sensed data.

The limited use and high expense of ground- and air-based platforms has made space-based satellite technology useful for a larger range of remote sensing applications.

In comparison to the conventional method, remote sensing technology offers various benefits for the survey of agricultural resources. The benefits include having a synoptic perspective, the potential for quick surveys, the capacity to detect changes through repetitive coverage, minimal cost involvement, improved accuracy, and the utilization of hyper-spectral data to expand knowledge.

3.2.4 Variable Rate Technology (VRT)

According to an assessment support mechanism and/or a management approach, variable rate technology (VRT) relates to the data employed to control the implementation frequency of fertilizer, lime, insecticides, and seed that has been applied over a field. With shifting contributions over a particular area based on the quantity of fields and yield characteristics, VRT resembles a return to the fundamentals of farming. The data required to support VRT may come from a variety of sources, including maps created using GIS, RS pictures, and GPS linked data. A specific to the site implementation design depends upon strong agronomic rules are created using all of the data. Automated application tools respond fast to modifications in the environment and make adaptations to field procedures (seeding rates, fertilizer and chemical application rates, etc.), current VRT equipment enables the user to watch machine operations. When used in conjunction with a GPS receiver, VRT offers adjusting process built upon the positioning of obscure lines chosen in advance by the farm manager or machine operator. With VRT, it is possible to control yield which depends upon a variety of variables, including soil type, soil texture, organic matter, nutrients, pH, weed and insect inhabitants, ailment, the distribution of nematodes, required production, and others.

3.2.5 Unmanned Aerial Vehicles (UAVs) or Drones

Unmanned aerial vehicles (UAVs) are a category of aircraft that can fly independently. UAVs appears in a diversity of varieties which are employed for a number of tasks. The military initially used the technology for aim practice, collecting information, and monitoring of specific antagonistic territories. The potential for using UAVs in agriculture is enormous. A type of use is the precise and factual prediction of agricultural yield utilizing spatial information gathered by the UAV. Agriculturalists can observe their fields from the air using UAVs. This aerial perspective can expose a variety of difficulties on the farm, some of which are common: troubles with irrigation, variations in the soil, pest and fungal invasions. UAVs can display additional data involving to water availability, shifting weather patterns, wind, soil quality, the occurrence of weeds and insects, varying growing seasons etc. (Banjo & Ajayi, 2019).

Soil and field analysis: In the early stages of a crop cycle UAV is used for ecological data. Gathered information is helpful for early soil investigation and for organizing seed planting patterns. This information can help the farmers to plan irrigation systems and calculate how much fertilizer is required for the soil or field after planting. A strategy based on information can help farmers enhance productivity in agriculture whilst reducing money on pesticides and fertilizers. All of this becomes feasible by analyzing aircraft drone-shot distant pictures. UAV photography has enormous promise for creating weed management strategies that are site-specific as well. Farmers may quickly and precisely identify weeds using the high-quality photographs nearly as soon as they appear and use a small amount of pesticide to contain them. On a sequence of UAV photos taken with six-band multi-spectral cameras on a corn field in Spain, Pena et al. (2013) established an Object-Based Image Analysis (OBIA). While the mechanical details and design of a UAV that might be utilized to gather remote photos for Early Season Site-Specific Weed Management (ESSWM) were provided by (Torres-Sánchez et al., 2013). An unmanned aerial vehicle (UAV) was launched fitted with multiple spectrum cameras and examined its configuration and technical parameters to produce images at various angles with the high magnification needed for the recognition and localization of weed seeds in a sunflower crop. The experiment's findings may be useful in choosing a suitable sensor and setting up the flight mission for ESSWM.

Planting: Crop planting is an expensive and time-consuming task that generally demands a lot of labor. UAVs have transformed crop cultivation simpler for farmers, as they can cover enormous expanses of land fast and correctly whilst retaining the greatest amount of accuracy. With today's advanced UAV farming technology, planting costs can be cut by up to 85% using UAV-powered approaches. The UAV's capacity to complete several jobs at once has led to a decrease in planting expenses. In recent years, UAVs have grown more and more common in agricultural research applications. In agriculture, it has been discovered that they are capable of attaining pictures with great spatial and temporal resolutions. Six sorghum hybrids' crop growth parameters were quantified using a UAV-based remote sensing coordination by Shafian et al. (2018) they took into account factors including yields, fractional vegetation, and the Leaf Area Index (LAI).

Crop and spot spraying: For farmers and agricultural production organizations, crop spraying is frequently a challenging and burdensome chore. To ensure good agronomic growth, it entails completely covering extraordinarily huge areas of land. Producers' spraying of crops has been considerably simpler by agricultural UAVs, which can swiftly cover vast expanses of land. Sensors allow UAVs to autonomously alter their altitude while operating over rough terrain. This improves spraying accuracy while saving energy. The use of UAVs for agricultural spraying has a number of profits for farmers, comprising reduced time and expense, effective spraying because Both of the vegetation and the soil beneath can be accessed, and protection Persistent exposition to possibly toxic compounds which have previously been linked with manual spraying. Modern topographical scanning techniques are used by agricultural UAVs to disperse just the right quantity of liquid for healthy crop growth. This guarantees uniform treatment with minimal waste. The feasibility of infrared thermal imaging in assessing particle accumulation in the atmosphere of airborne spray. In this experiment, the impact The UAV flying performance on drops of spray was examined, and a varied spray test using an aerial spray system in a controlled environment was carried out via a UAV simulation platform. The study's findings included the following: drop width (i.e., the distribution uniformity of particle size) declines with increased flight speed, leading to decreased homogeneity of sprayed droplets; deposition density falls with flight speed. Therefore, in order to increase plant production, the scientists provide theoretical help for adjusting the spraying UAV settings.

3.3 Nano Technologies

Nanotechnologies in agriculture refer to the use of nanoscale materials, techniques, and devices to addressing difficulties and improving many areas of agriculture practices. This includes the use of nanomaterials, Nano devices, and Nano sensors to increase crop productivity, protect crops from pests and diseases, improve soil quality, improve resource management, and address other agricultural concerns included. Examples of nanotechnologies in agriculture include the development of nanofertilizers for controlled nutrient release, nanopesticides for targeted pest management, nanosensors are used to detect soil quality environmental factors, and post-harvest monitoring of agricultural production. Contains nano-coatings for protection. Agrochemicals such as insecticides and fertilizers are essential to optimal output and productivity in modern agriculture. Nevertheless, each herbicide has potential side effects that could harm both humans and the environment, like pollution of water or traces on dietary supplements. Thus, proper administration and regulation of supplies might assist to limit such hazards (Kah, 2015). To modernize agricultural methods, decrease or eliminate the detrimental effects of contemporary agriculture on the atmosphere, and enhance the quantity as well as quality of crops, the development of a modern farming system employing developed intelligent Nano tools might serve as an excellent option (Sekhon, 2014).

3.3.1 Nano Fertilizers

Nano-fertilizers are products of nanotechnology that are used to alter and create substances that are utilized to enrich soil for increased output and improve yield and value of crop. They are crucial in increasing nutrient efficiency through regulated nutrient release based on plant requirement and lowering environmental safety costs through judicious use of fertilizers. With the aid of nanotechnology and a variety of physical, chemical, or biological processes, nano-fertilizers are specialized versions of traditional fertilizers that are either unrefined constituents or generated from vegetative or reproductive portions of plants. The small size of nanoparticles and higher ratio of surface and volume result in fundamentally distinct molecular characteristics from those of their bulk substance. A significant step toward sustainable agriculture could be to substitute nano-fertilizer for conventional fertilizer.

Various nanoparticles may be employed as soil conditioners such as nano-fertilizers based on nitrogen, phosphate, iron, zinc, titanium, copper, and silver. They contribute significantly to the modification of fertilizers, which increases crop yield and aids in nurturing the expanding people of world. There are various several forms of nano-fertilizers having exclusive potentials play a significant part in carrying revolt in the agricultural field.

Nano-fertilizer use increases crop production and lowers environment pollution. The use of nano-fertilizers is more important for civilization. Although nano-fertilizers have positive impacts, they can have the opposite effect on plants and animals if used in higher concentrations. Future studies in the field of nano-fertilizers are required, as described below, in order to improve the effectiveness of nano-fertilizers in agriculture sector.

(a) Create nano-composite to give critical nutrients in the right amounts by intelligent distribution systems, that may aid in supplying the crop with a balanced supply of nutrients.
(b) Research on the uptake of nano-nutrients by plant systems.
(c) Examine how nano-fertilizers affect soil and helpful soil microorganisms.
(d) Research the impacts of nano-fertilizers on plants and soil.
(e) It is necessary to adjust the concentration and dosage of nano-fertilizers for various crops.
(f) Using nano-fertilizers in precision agriculture with site-specific management.

3.3.2 Nano Pesticide

Nano-pesticides refers to the use of nanotechnology in the development and use of pesticides for agricultural purposes. These nano formulations aim to enhance the effectiveness, efficiency and targeted delivery of active ingredients used in pest management. Some important aspects of nano-pesticide technology in agriculture are:

Nano-pesticides enable targeted delivery of active ingredients to insects, reducing off-target effects and reducing the amount of pesticide required. Nanoparticles can encapsulate or carry active compounds, allowing them to be put in a precise way directly at the site of insect infestation. This targeted delivery increases pesticide application efficiency and reduces environmental pollution. Nanotechnology can improve the efficacy of pesticides by increasing their bio-availability and stability. Nano-formulations can protect active ingredients from degradation and provide controlled release ensuring sustained efficacy against pests. This improved stability and bio-availability can lead to better pest control results with lower pesticide doses. Nano-pesticides can help reduce environmental impacts associated with conventional pesticides. By precisely delivering active ingredients to targeted pests, nano-pesticides reduce exposure to non-target organisms and decrease the overall quantity of pesticides discharged openly to the atmosphere. It can help protect beneficial insects, soil organisms and wildlife while reducing water and soil pollution. Nanoparticles can be designed to release the active ingredients gradually over time, providing a prolonged and sustained effect. This controlled release ensures that the pesticides remain effective for an extended period, reducing the need for frequent applications.

It may enhance the effectiveness of pest management and decrease the employment and expenditures associated with repeated pesticide applications.

Nano-pesticides have the potential to reduce pesticide resistance in insects. By providing targeted and sustained release of active ingredients, they can help reduce selection pressures that lead to the development of resistance. Additionally, nanotechnology can be used to develop compound formulations, where multiple active ingredients are incorporated with different routes, further reducing the risk of resistance. Nanotechnology offers the potential to reduce the toxicity of pesticides. By encapsulating or modifying active ingredients at the nano scale, their toxicity to non-target organisms, including humans, can potentially be reduced. However, thorough research and risk assessment is necessary to ensure the safety of these nano formulations. Although nano-pesticide technologies offer significant potential benefits in agriculture, it is vital to investigate the safety, regulatory aspects and possible environmental impacts linked with their use. Adequate risk assessment, regulation and responsible implementation are essential to ensure effective and sustainable use of nano-pesticides in agricultural practices.

3.3.3 Nano-biosensor

Nano-biosensor technology serves as an essential part in the agricultural sector through live observing and detection of numerous factors and analyzes in agriculture organizations. Some of the important roles of nano-biosensor technology in agriculture are:

- Nano-biosensors can detect the presence of pathogens, like bacteria, viruses and fungi, in crops, soil and water. These biosensors can identify specific molecular markers or antigens associated with plant diseases, enabling early diagnosis and timely implementation of disease management strategies.
- Nano-biosensors can assess soil parameters like nutrient levels, pH, moisture content, and organic matter. By supplying real-time data on soil conditions, farmers can optimize fertilizer application, irrigation, and additional soil managing practices to improve crop health and harvest.
- Nano-biosensors can detect pollutants and contaminants in irrigation water and aquifers. They can identify the existence of heavy metals, pesticides and other detrimental constituents, ensure the usage of clean water for irrigation and prevent potential undesirable effects on crops and the atmosphere.
- Nano-biosensors can measure environmental factors such as temperature, humidity, light intensity and air quality. This information can be used to optimize growing conditions in greenhouses or controlled environments, ensuring optimal crop growth and reducing resource wastage.
- Nano-biosensors can detect pesticide residues in crops and soil or water samples. These biosensors can provide quick and precise analysis, letting farmers to ensure compliance with safety regulations and reduce potential health risks.

- Nano-biosensors can observe many variables of plant stress, like nutrient deficiency, drought, or heat stress. By detecting these stress factors early, farmers can take appropriate measures to minimize their impact, thereby improving plant health and productivity.

Nano-biosensor technology offers several advantages, including high sensitivity, specificity, and portability. It enables real-time and on-site monitoring, providing farmers with timely and actionable information for decision-making. By early disease detection, improved resource management, and improved crop health, nano-biosensors contribute to sustainable and efficient agricultural practices.

3.4 Nutrient Management Strategies

Nutrient management strategies refer to deliberate and systematic methods used in agricultural systems to improve nutrient uptake, utilization and cycling, with the aim of maximizing crop productivity while minimize negative environmental impacts. These strategies include the careful management of vital nutrients, such as nitrogen (N), phosphorus (P), potassium (K), and other micro-nutrients, to ensure their availability in adequate amounts and at right time for plant development (Tiwari, 2002). The main objectives of nutrient management strategies are:

- Promoting efficient and balanced nutrient uptake by crops, minimizing losses through leaching, runoff, or volatilization, and minimizing nutrient waste.
- Selecting appropriate nutrient sources, such as organic or inorganic fertilizers, based on their nutrient content, availability, cost-effectiveness, and environmental considerations.
- Applying nutrients at optimal times during crop growth to meet plant demand and minimize nutrient losses.
- Develop nutrient application rates and practices taking into account soil characteristics, nutrient levels, and crop changing requirements on the basis of field to field.
- Promoting nutrient recycling and conservation within agricultural systems like crop residue management, cover cropping, and the employment of organic amendments.
- Regularly monitoring soil fertility, nutrient levels, and crop performance to assess the efficacy of nutrient management strategies and to make necessary modifications.

The expanding food requirements of increasing numbers of people, as well as the requirement for an ecologically appropriate plan for sustainable agricultural development, need major consideration while tackling the problem of crop yield (Wu et al., 2015).

3.4.1 Integrated Nutrient Management

Integrated Nutrient Management (INM) is a comprehensive method to nutrient management in agriculture that associates the use of various organic and inorganic nutrient sources, with appropriate soil and crop management practices, to improve nutrient availability. To improve soil fertility, increase crop productivity, and ensure sustainable agricultural production (Wu et al., 2015). Integrated Nutrient Management recognizes that a single source of nutrients cannot meet all the nutrient requirements of crops and soils. Therefore, it emphasizes the integration of multiple nutrients including synthetic fertilizers, organic fertilizers, crop residues, biofertilizers, and microbial inoculants to provide a balanced and diverse nutrient supply to plants.

3.4.2 Coated Nitrogenous Fertilizers

Another nutrient management technique is the use of coated nitrogenous fertilizers because the agricultural industry will inevitably demand ever-increasing amounts of fertilizers to supplement the food supply because of the rapid expansion of the global population, which will raise the cost of food production. When applied to crops, urea is susceptible to leaching and volatilization losses. The efficiency with which plants utilize nitrogen (NUE) is also decreased by current methods, which limits crop yields and, in addition, contributes to environmental contamination through harmful gas emissions and water eutrophication. To minimize the losses of urea to the atmosphere in the form of nitrous oxide and reducing the leaching of nitrate to ground water, the coated fertilizers can be used as a slow-release nitrogenous fertilizers in the soil which reduces these environments related issues effectively. For the coating of commercially available urea different sources like Zinc coated urea, plant growth promoting bacteria (PGPR) coated urea, neem oil coated urea (Ali et al., 2020) and polymer coated urea will be used and these coated fertilizers also increase the nitrogen use efficiency. Because only one application is necessary for the growing season, controlled release fertilizers result in savings in fertilizer quantity and labor associated with application frequency. They also prevent the loss of nutrients, the toxicity of seeds, dangerous emissions, leaf burning, cutaneous irritation, and breathing issues. They also increase germination rates, handling capabilities, and soil quality. On the other side, they cost a lot of money and present marketing challenges. Additionally, some of the coating components used to make CRFs are hazardous to the soil and non-biodegradable. The release pattern is typically unpredictable in field applications as well. Some CRFs can significantly alter the pH of the soil, which is not desired. Modifications to storage facilities are also necessary to prevent premature nutrient release brought on by moisture absorption through fissures created by granule attrition (Davidson & Gu, 2012).

3.5 Improved Irrigation Methods

Water plays a key role in crop production; all crops require some amount of water for their growth and development. Some plants require more water than others some can sustain at a higher EC (electrical conductivity) level than the rest. Sustainable crop production is a burning issue which is impossible to achieve without the use of sustainable and improved irrigation methods. Fresh water is a limited resource agriculture is the major user of fresh water, about 70% of all freshwater withdrawals throughout the world goes to agricultural use (Alexandratos & Bruinsma, 2012). There are many sources of water being used nowadays for irrigation which include both natural sources like rain, snow, dew, or any other means of precipitation or artificial sources like canals, dams, water tanks, manmade glaciers etc. The application strategy of these water resources also holds great importance when it comes to sustainability. Throughout the world, several methods are being commonly used for irrigation purposes and hold sustainability status like drip irrigation, use of wastewater, use of salty water, wastewater utilization and rainwater harvesting (Nikolaou et al., 2020).

3.5.1 Drip Irrigation

The drip irrigation technique is the most sustainable methods of precise irrigation. It holds its supremacy over other irrigation methods as this method has the highest water use efficiency. It plays a key role as a sustainable irrigation method. Reducing the water losses, especially application losses, conserves soil, optimizes nutrient delivery and reduces the nutrient losses. It has very low water demand which can help farmers to cultivate lands and adopt precision agriculture. The amount of water required is very low in drip irrigation technique thus it can convert barren non arable land to use which increases the total area under cultivation and in return gives rise to more crop production (Kavianand et al., 2016). In this irrigation method, long plastic tubes are laid in fields with many small openings or emitters. From these small openings, water is pushed out through pumps in these tubes. Water slowly drips from each emitter point and hence irrigates the area around the plant which is mostly just the root zone. The main constraints keeping this method from its adaptation are high infrastructure cost and high maintenance cost of parts (Camp, 1998).

3.5.2 Use of Salty Water for Crop Production

Salty or brackish water refers to water which has more salt concentration or higher EC than fresh water but less than that of sea or marine water. Canals are the major source of fresh water for agriculture but these canals aren't close to being enough to sustain current agriculture. Therefore, groundwater utilization is becoming a popular means to supply water. The main issue with using groundwater is the addition of salts in soil and salinity development. This is due to the fact that most of the groundwater resources are brackish. Certain practices are required for utilizing this water for crop production (Wang et al., 2023).

The first step is identifying the extent of salts present in the water and selecting crops that can sustain under these set of conditions. The other important point to consider is the soil salinity level, there must be some heavy irrigations with fresh water in order to leach excess salts below the root zone. If the availability of fresh water is low, then low delta water crops should be preferred and rain will leach the excess salts when it happens. High-efficiency irrigation methods should be preferred so the least number of salts are introduced in the soil per irrigation (Cheng et al., 2021).

3.5.3 Waste Water Utilization

Wastewater means water that has been used for some purpose and contains impurities. Wastewater especially municipal wastewater is used in some parts of the world as irrigation water as it contains many nutrients organic matter and various other materials. This helps to boost crop production, help reduce fertilization requirements, and improve soil quality and soil ecology. The use of wastewater is controversial as it also contains some toxic agents which can accumulate and eventually cause bio magnification in humans. Keeping in view these problems we can easily use this type of water in forestry and agroforestry to produce timber and wood.

Wastewater is of great importance as most of the nutrients are being lost by disposing of this type of water in rivers streams and oceans. In order to gain sustainability, there must be some means to reduce this loss, to remove nutrients from wastewater and to make fertilizers or compost from it. This can be done by chelation, precipitation, active uptake or growth in these waters (Jaramillo & Restrepo, 2017).

3.5.4 Rainwater Harvest

Rainwater harvest as the name suggests is the assemblage and storing of rainwater and its utilization meant for various purposes including domestic and for agriculture. Rainwater harvesting prevents water from reaching drains sewers rivers or getting absorbed in the ground which in return saves water. In the current era where climate change is affecting rainfall patterns, it is required to conserve fresh water. It can help provide fresh water in places where fresh water is scarce and if it is available, it is either polluted or very hard to get deep wells etc. This can also provide water in places where agriculture is very low and depends mostly on water from rain as irrigation, they can tolerate dry spells by collecting rainwater. It can boost the productivity of lands where saline water was the only source by combining this with high-efficiency irrigation systems crop production could be enhanced (Oweis et al., 2012).

4 Issues Related to the Adaptation of Sustainable Approaches

The adaptation of precision farming, like any technological advancement, is not without its challenges and problems. Here are some important concerns related to adopting precision farming:

Cost: Cost is an important factor Which may impact uptake and performance of sustainable agriculture practices. The implementing of sustainable agriculture technology can be expensive because initial investments on equipment, sensors, software, and data management systems can be a significant obstacle for small, limited in resources farmers. Besides the initial investment, there are ongoing operational costs associated with sustainable farming. These costs include maintenance, repairs, software updates and data storage. Farmers need to budget for these costs to ensure the continued functionality and effectiveness of their precision farming systems.

Natural and environmental concerns: Precision agriculture relies on accurate weather data for decision making. However, weather patterns can be unpredictable, and extreme occasions such as droughts, floods, storms, or heat waves can disturb farming operations. These changes in weather can affect crop growth, soil conditions, and pest/disease prevalence, making it difficult to accurately predict and manage crop information. The spread of pests or the emergence of new diseases can challenge the effectiveness of sanitary agricultural systems. Rapid changes in pest populations or the appearance of new pathogens may require adjustments in monitoring strategies, application rates, or selection of resistant crop varieties. The rising temperatures, altered rainfall patterns, and changing climatic conditions can affect traditional farming practices and affect the effectiveness of precision agriculture. The rapid development of technology may result in shorter lifespans of precision agricultural equipment, leading to an increase in electronic waste (e-waste). Improper disposal of old or damaged equipment can pose environmental hazards. Additionally, the production and disposal of health-related agricultural technologies can have associated environmental impacts, including carbon emissions, water pollution, and resource depletion.

Lack of technical expertise: Lack of technical expertise can significantly affect the successful implementation and use of precision agriculture practices. Precision agriculture relies on the proper operation and use of various technologies, such as GPS, remote sensing, and data analysis tools. Without the essential technical expertise and information, farmers may struggle to function and use these technologies effectively. This can result in imprecise data collection, misinterpretation of results, and, poor decision-making regarding to the crop production. Farmers who lack technical skills may have limited knowledge and understanding of precision agricultural concepts, technologies and their potential benefits. This can lead to reluctance to adopt or invest in these practices, as they perceive them as complicated or uncertain.

Inadequate infrastructure: Rural infrastructure includes roads, facilities for storage, transportation, electricity, education, hygiene, and medical services, etc. Sustainable agricultural technologies want a firm and reliable power source. In regions which have inconstant or inadequate supply of power, the functionality of sensors, automation systems, and data processing equipment can be compromised. Power outages can lead to data loss, equipment malfunctions, and disruptions to automated farming processes. Precision agriculture generates a large quantity of information that is required to be stored, treated and examined in real time. In regions with inadequate data storage and processing facilities, farmers struggle to effectively manage and interpret large datasets. This can delay the implementation of data-driven agricultural practices and bound the potential benefits of precision agriculture.

Compatibility and Integration: Different precision farming techniques and tools may not always be compatible or seamlessly integrated with each other. This can create challenges when farmers want to take a holistic approach and integrate different technologies for optimal decision-making. Inconsistency and lack of integration can result in fragmented data analysis and decision-making processes. If data from different sources cannot be easily aggregated and analyzed, farmers may struggle to gain comprehensive insights and make informed decisions. This can reduce the efficient use of precise agriculture in optimizing resource utilization and enhancing overall farm productivity.

Improper crop rotation: All of the crops have different nutrient necessities and contribute in their own way to soil fertility. Proper crop rotation helps maintain nutrient balance in the soil by diversifying the nutrient requirements of different crops and by adding legumes that fix atmospheric nitrogen. When crop rotation is not implemented correctly, soil nutrient levels can become imbalanced, leading to nutrient deficiencies or excesses. Precision agriculture relies on accurate soil nutrient information for targeted fertilization, and incorrect crop rotations can complicate nutrient management strategies.

Consumption oriented: The farmers do not keep accurate records of their revenue and costs. They primarily invest on a variety of regional festivities and gatherings. A consumption-oriented approach may emphasize short-term profit maximization, potentially leading to a narrow focus on immediate returns rather than long-term sustainability. Precision agriculture requires upfront investments in technology, data management systems, and training, which may not be consistent with short-term profit-driven decision-making. This may hinder the adoption of sound agricultural practices among farmers who prioritize immediate financial gains.

Addressing these issues requires collaboration between farmers, technology providers, policy makers and researchers. This includes investing in research and development, providing access to training and education, improving infrastructure, and creating supportive policies and incentives to encourage the adoption of widespread and sustainable farming practices.

5 Conclusion

Sustainable agriculture approaches are critical for solving modern-day concerns such as food security, resource conservation, and preservation of the environment. Farmers can grow systems that balance production and environmental sustainability by implementing agroecological methods, precision agriculture, efficient irrigation, and organic materials fertilization. These strategies not only minimize the need for chemical fertilizers, but they also promote ecology, improve the health of soils, and lessen risks associated with climate change. Implementing these measures strengthens the farming industry, enhances efficacy of resources, and lays the groundwork for a sustainable and balanced worldwide food system. Agriculture's future depends on collaborating with nature, harnessing technology, and adopting methods that are beneficial to both the globe and its inhabitants.

References

Ahmad, S. F., & Dar, A. H. (2020). Precision farming for resource use efficiency. *Resources Use Efficiency in Agriculture*, 109–135.

Alexandratos, N., & Bruinsma, J. (2012). *World agriculture towards 2030/2050: the 2012 revision*. ESA Working paper No. 12-03. Rome, FAO

Ali, M., Maqsood, M. A., Azizl, T., & Awan, M. I. (2020). Neem (Azadirachta indica) oil coated urea improves nitrogen use efficiency and maize growth in an alkaline calcareous soil. *Pakistan Journal of Agricultural Sciences, 57*(3).

Anderson, C. R., Bruil, J., Chappell, M. J., Kiss, C., & Pimbert, M. P. (2021). *Agroecology now!: Transformations towards more just and sustainable food systems* (p. 199). Springer Nature.

Aronoff, S. (1989). *Geographic information stems: A management perspective*. WDL Publications.

Banjo, C., & Ajayi, O. (2019). Sky-Farmers: applications of unmanned aerial vehicles (UAV) in agriculture. *Unmanned Aerial Vehicles 2019.*

Behnke, G. D., Zuber, S. M., Pittelkow, C. M., Nafziger, E. D., & Villamil, M. B. (2018). Long-term crop rotation and tillage effects on soil greenhouse gas emissions and crop production in Illinois, USA. *Agriculture, Ecosystems & Environment, 261*, 62–70.

Brust, G. E., & King, L. R. (1994). Effects of crop rotation and reduced chemical inputs on pests and predators in maize agroecosystems. *Agriculture, Ecosystems & Environment, 48*(1), 77–89.

Camp, C. R. (1998). Subsurface drip irrigation: A review. *Transactions of the ASAE, 41*(5), 1353–1367.

Campbell, B. M., Thornton, P., Zougmoré, R., Van Asten, P., & Lipper, L. (2014). Sustainable intensification: What is its role in climate smart agriculture? *Current Opinion in Environmental Sustainability, 8*, 39–43.

Cheng, M., Wang, H., Fan, J., Wang, X., Sun, X., Yang, L., Xiang, Y., & Zhang, F. (2021). Crop yield and water productivity under salty water irrigation: A global meta-analysis. Agricultural Water Management, 256, 107105

Dalgaard, T., Hutchings, N. J., & Porter, J. R. (2003). Agroecology, scaling and interdisciplinarity. *Agriculture, Ecosystems & Environment, 100*(1), 39–51.

Davidson, D., & Gu, F. X. (2012). Materials for sustained and controlled release of nutrients and molecules to support plant growth. *Journal of Agricultural and Food Chemistry, 60*(4), 870–876.

Dhaliwal, S. S., Sharma, V., Mandal, A., Naresh, R. K., & Verma, G. (2021). Improving soil micronutrient availability under organic farming. In *Advances in organic farming* (pp. 93–114). Woodhead Publishing.

Dwivedi, S. L., Van Bueren, E. T. L., Ceccarelli, S., Grando, S., Upadhyaya, H. D., & Ortiz, R. (2017). Diversifying food systems in the pursuit of sustainable food production and healthy diets. *Trends in Plant Science, 22*(10), 842–856.

Gesch, R. W., Berti, M. T., Eberle, C. A., & Weyers, S. L. (2023). Relay cropping as an adaptive strategy to cope with climate change. *Agronomy Journal.*

Gliessman, S. (2018). Defining agroecology. *Agroecology and Sustainable Food Systems, 42*(6), 599–600. https://www.giz.de/en/downloads/giz2015-en-what-is-sustain-agric.pdf

Jaramillo, M. F., & Restrepo, I. (2017). Wastewater reuse in agriculture: A review about its limitations and benefits. *Sustainability, 9*(10), 1734.

Kah, M. (2015). Nanopesticides and nanofertilizers: Emerging contaminants or opportunities for risk mitigation? *Frontiers in Chemistry, 3*, 64.

Kamble, S. S., Gunasekaran, A., & Gawankar, S. A. (2020). Achieving sustainable performance in a data-driven agriculture supply chain: A review for research and applications. *International Journal of Production Economics, 219*, 179–194.

Kavianand, G., Nivas, V. M., Kiruthika, R., & Lalitha, S. (2016). Smart drip irrigation system for sustainable agriculture. In *2016 IEEE Technological Innovations in ICT for Agriculture and Rural Development (TIAR)* (pp. 19–22). IEEE.

Kumar, S. A., & Ilango, P. (2018). The impact of wireless sensor network in the field of precision agriculture: A review. *Wireless Personal Communications, 98*, 685–698.

Lange, A. F. (1996). Introduction to differential GPS for precision agriculture applications. In *Proceedings of the 1996 Information Agriculture Conference.* Atlanta, GA: Potash and Phosphate Institute.

Mohamed, M. (2023). Agricultural sustainability in the age of deep learning: Current trends, challenges, and future trajectories. *Sustainable Machine Intelligence Journal, 4*, 1–2.

Morgan, M. T., & Ess, D. R. (1997). *The precisionfarming guide for agriculturists.* John Deere Publishing.

Nikolaou, G., Neocleous, D., Christou, A., Kitta, E., & Katsoulas, N. (2020). Implementing sustainable irrigation in water-scarce regions under the impact of climate change. *Agronomy, 10*(8), 1120.

Oweis, T. Y., Prinz, D., & Hachum, A. Y. (2012). *Rainwater harvesting for agriculture in the dry areas.* CRC Press.

Peña, J. M., Torres-Sánchez, J., de Castro, A. I., Kelly, M., & López-Granados, F. (2013). Weed mapping in early-season maize fields using object-based analysis of unmanned aerial vehicle (UAV) images. *PLoS ONE, 8*(10), e77151.

Pfost, D. L., Casady, W. W., & Shannon, K. (1998). *Precision agriculture: Global positioning system (GPS)* (1998).

Rai, A., Kumar, S., Bauddh, K., Singh, N., & Singh, R. P. (2017). Improvement in growth and alkaloid content of Rauwolfia serpentina on application of organic matrix entrapped biofertilizers (Azotobacter chroococcum, Azospirillum brasilense and Pseudomonas putida). *Journal of Plant Nutrition, 40*(16), 2237–2247.

Scholberg, J. M., Dogliotti, S., Leoni, C., Cherr, C. M., Zotarelli, L., & Rossing, W. A. (2010). Cover crops for sustainable agrosystems in the Americas. *Genetic Engineering, Biofertilisation, Soil Quality and Organic Farming*, 23–58.

Sekhon, B. S. (2014). Nanotechnology in agri-food production: An overview. *Nanotechnology, Science and Applications, 7*, 31–53.

Shafian, S., Rajan, N., Schnell, R., Bagavathiannan, M., Valasek, J., Shi, Y., & Olsenholler, J. (2018). Unmanned aerial systems-based remote sensing for monitoring sorghum growth and development. *PLoS ONE, 13*(5), e0196605.

Shepherd, M. A., Harrison, R., & Webb, J. (2002). Managing soil organic matter–implications for soil structure on organic farms. *Soil Use and Management, 18*, 284–292.

Singh, R. P., Manish, S., Kuldeep, B., Sengar, R. S., & Jaiwal, P. K. (2010). Sustained nutrient supply, reduced nutrient loss and high plant productivity with slow release fertilizers. In *Stable food production and sustainable agriculture: A challenge ahead in 21st Century* (pp. 62–79).

Swarbreck, S. M., Wang, M., Wang, Y., Kindred, D., Sylvester-Bradley, R., Shi, W., Bentley, A. R., & Griffiths, H. (2019). A roadmap for lowering crop nitrogen requirement. *Trends in Plant Science, 24*(10), 892–904.

Tanveer, M., Anjum, S. A., Hussain, S., Cerdà, A., & Ashraf, U. (2017). Relay cropping as a sustainable approach: Problems and opportunities for sustainable crop production. *Environmental Science and Pollution Research, 24*, 6973–6988.

Thompson, N. M., Bir, C., Widmar, D. A., & Mintert, J. R. (2019). Farmer perceptions of precision agriculture technology benefits. *Journal of Agricultural and Applied Economics, 51*(1), 142–163.

Tiwari, K. N. (2002). Nutrient management for sustainable agriculture. *Journal of the Indian Society of Soil Science, 50*(4), 374–397.

Torres-Sánchez, J., López-Granados, F., De Castro, A. I., & Peña-Barragán, J. M. (2013). Configuration and specifications of an unmanned aerial vehicle (UAV) for early site specific weed management. *PLoS ONE, 8*(3), e58210.

Umesha, S., Manukumar, H. M., & Chandrasekhar, B. (2018). Sustainable agriculture and food security. In *Biotechnology for sustainable agriculture* (pp. 67–92). Woodhead Publishing.

Velten, S., et al. (2015). What is sustainable agriculture? A systematic review. *Sustainability, 7*(6), 7833–7865.

Wang, X., Tian, W., Zheng, W., Shah, S., Li, J., Wang, X., & Zhang, X. (2023). Quantitative relationships between salty water irrigation and tomato yield, quality, and irrigation water use efficiency: A meta-analysis. *Agricultural Water Management, 280*, 108213.

Wu, W., Ma, B., & Uphoff, N. (2015). *A review of the system of rice intensification in China.*

Biochemistry in Soil Science: Enzyme Activities, Nutrient Cycling

Ishrat Perveen, Khizra Ismail, Naaz Abbas, Yasar Saleem, Sana Riaz, Ayesha Ayaz, Tariq Javed, Taha Ikram, Amna Mumtaz, Sania Mazhar, Shaista Nawaz, Quratulain Syed, and Syed Hussain Imam Abidi

Abstract Biochemistry is the science of biology, chemistry, or both; it helps to understand the relationship between microbial community and organic matter content that affects nutrient cycling and soil structure. The biochemicals consist of β-glucosidase, urease and phosphatase that act as catalysts in the C, N, P, and S compulsory rings that are fundamental to soil health and bio equilibrium. Soil enzyme resources are sensitive to conditions; for instance, the pH, temperature, moisture level and human behaviour, such as deforestation or pollution, affect the processes involved in soil. Traditional biochemical methods such as spectrophotometry, fluorescence assays, and zymography can be adopted to measure enzyme kinetics to generate substantial information on the state of the soil. This chapter also analyses the involvement of enzymes in nutrient cycling, where Cd, N-transformation, and P-solubilization are central to plant growth. Climate and land use alter overall biospheric enzyme activity and nutrient dynamics, highlighting the importance of sustainability. This chapter aims to analyze the biochemical dynamics of the soil ecosystems with specific reference to enzyme actions and nutrient cycling as beneficial aspects for maintaining soil healthiness and fertility for agricultural purposes.

I. Perveen (✉) · K. Ismail · N. Abbas · Y. Saleem (✉) · S. Riaz · A. Ayaz · S. Mazhar · S. Nawaz · Q. Syed
GenEd and Molecular Biology Labs, Food and Biotechnology Research Centre, Pakistan Council of Scientific and Industrial Research Centre, Lahore, Pakistan
e-mail: drishratperveen51214@outlook.com

Y. Saleem
e-mail: ysaleem73@gmail.com

Y. Saleem · S. H. I. Abidi
Pakistan Council of Scientific and Industrial Research Centre, Islamabad, Pakistan

T. Javed
Department of Pharmaceutical Sciences GCU, Lahore, Pakistan

T. Ikram
Dr Ikram Ul Haq Institute of Industrial Biotechnology, Government College University, Lahore, Pakistan

A. Mumtaz
Department of Biochemistry, Government College University, Faisalabad, Pakistan

M. Shaaban (ed.), *Soils and Sustainable Agriculture*, Frontier Studies in Soil Science,
https://doi.org/10.1007/978-3-031-91114-9_46

The study is critical in enhancing the conservation of soil, the yield in agriculture, and the management of environmental effects in agricultural production. Subsequent research should seek to expand knowledge about enzyme effects on soil microorganisms using prospective strategies in soil evaluation, management and biodegradation.

Keywords Soil biochemistry · Nutrient cycling · Enzymes · Analytical techniques · Soil health · Environmental impact

1 Introduction

Biochemical processes are significant for soil quality and influence practically all soil properties, i.e., fertility, the ability to stimulate plant growth and maintenance of the stability of the bottom layer of the production. Hence, microorganisms and enzymes are of considerable importance following the decomposition of organic matter and transformations of nutrients that plants can assimilate. Practitioners consider, in particular, that knowing these processes is intellectually enriching and indispensable. It offers great insight into sustainable agricultural practices, effective soil management and measures of practicality to reduce undesirable environmental impacts. According to previous literature, soil maintenance is very crucial for the maintenance of soil quality and also to increase agricultural productivity (Jansson & Hofmockel, 2020). Thus, soil biochemistry will offer strategies for healthier soil, yield increase, and environmental conservation for future generations.

1.1 Definition of Soil Biochemistry

"Soil biochemistry" is the analysis of processes that take place in the soil together with organic matter, biochemical transformations of which occur under the influence of enzymes and microorganisms (Paul, 2007). Microbial cells contain enzymes either as part of their cellular structures, embedded within the plasma membrane or as extracellular entities present in the soil solution or associated with organic matter and clay minerals. These enzymes play a crucial role in catalyzing the breakdown of large organic molecules, decomposing them into smaller molecular structures that microbes can readily utilize (Fanin et al., 2022) (Fig. 1).

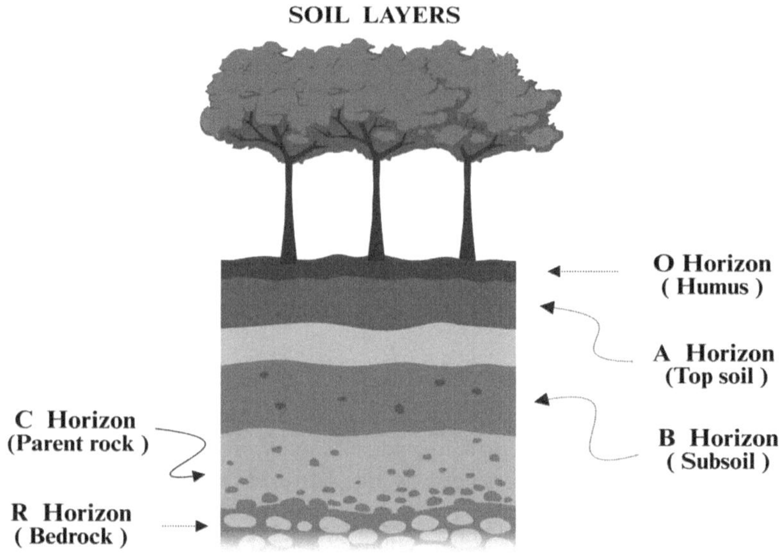

Fig. 1 Soil layers

1.2 Importance of Biochemistry in Understanding Soil Processes

Soil is a large, active component that serves as a link between agriculture and the environment. Some functions of soil enzymes are to carry out the biochemical activities within microbes, plants and animals. Decomposition of organic matter, stabilization of soil structure and nutrient cycling play a pivotal role. Carbon, nitrogen, phosphorus and sulfur are cycling in the ecosystems through biogeochemical processes involving microorganisms and enzymes. There are nutrient cycles that originate from biological and physiochemical processes acting in the soil. These include amidases, aryl-sulphatases, cellulases, dehydrogenases, glucosidases, laccases, phosphatases and ureases involved in these processes and act as bio-indicators. This implies shifts in the chemical, physical and biological properties of the soil (Neemisha & Sharma, 2022).

1.3 Overview of Enzyme Activities and Nutrient Cycling

The ability of soils to handle biogeochemical reactions and activities of enzymes in the soil are critical and serve as vital indicators of soil health. It is essential to monitor human activities like deforestation, agriculture, forestry, and soil system pollution. Enzyme assays are particularly important due to their simplicity, rapidness, high reproducibility, and relatively low cost (Nannipieri et al., 2018). Further, the biochemical factors regarding the phytochemical composition involved in carbon, nitrogen, phosphorus, and sulfur cycles are used to make an estimate of the quality of the soil. These include microbial biomass carbon, dehydrogenase activity, nitrogen mineralisation potential and specific enzyme activities like phosphatases, ureases, and β-glucosidases. These indicators can work independently or concurrently, along with the help of intricate mathematical-statistical equations, in order to offer an integrated assessment of soil health (Gil-Sotres et al., 2005).

2 Enzyme Activities in Soil

Two components form microsite enzyme activity, including the accumulated enzymes and the enzymes produced by the active microbial biomasses. Most of the enzymes in the soil originate from microbial biomass, but they can also be found in plant and animal residues. These activities include free enzymes like exoenzymes released by metabolically active cells, endoenzymes excreted by cellular debris, and structurally bound enzymes (Tabatabai & Fu, 2021) (Fig. 2).

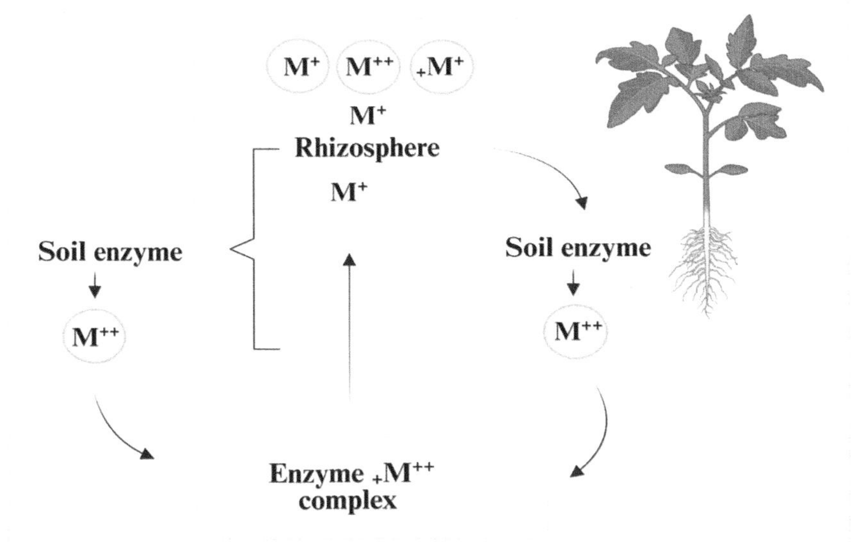

Fig. 2 Enzyme activity in soil

2.1 Types of Enzymes Found in Soil

Soil enzymes participate in numerous biochemical processes, enabling the transformation and cycling of vital nutrients for ecosystem health. Major types of soil enzymes include hydrolases, oxidoreductases, and transferases.

2.1.1 Hydrolases

Enzymes represent a diverse and essential group of hydrolases that catalyze the hydrolysis of peptide, ether, ester, or glycosidic bonds. The activity of soil microorganisms and their growth depends upon the breakdown of these complex organic compounds into smaller and metabolisable subunits. Microbial soil proteases are extracellular proteases functioning to degrade proteins to oligopeptides, which ensures amino acids availability as nutrients and sustains soil fertility (Greenfield et al., 2021). Amylases are enzymes that catalyse the hydrolysis of starch to glucose, maltose, malt triose, and dextrin. Cellulases are enzymes that combine to convert cellulose into glucose (de Souza Vandenberghe et al., 2020). Phosphatases catalyse the degradation of organic phosphorus into inorganic types, and urease shifts urea into carbon dioxide and ammonia (Karaca et al., 2011).

2.1.2 Oxidoreductases

Oxidoreductases execute oxidation–reduction reactions in soil, contributing to respiration, methane oxidisation, and nitrogen oxide reduction. For instance, phenol oxidases and peroxidases enhance the bio-degradation of electron-rich substrates such as phenols and lignin in the degradation of various aromatic compounds in soil (Khosrozadeh et al., 2022). Microbial oxidoreductase metabolism is connected with catalase, which is implicated in the metabolism of aerobic organisms. NAD-dependent dehydrogenases, widespread in all life forms, are involved in the reactions necessary for microbial energy transfer metabolism (Karaca et al., 2011).

2.1.3 Transferases

Transferases are enzymes that move functional groups of molecules from one molecule to the other. They help change the soil's sulfur, nitrogen, and carbon compounds. For instance, aminotransferases catalyse nitrogen translocation from amino acids to oxoacids. Hence, they play a role in protein catabolism and gluconeogenesis (Vroon & Israili, 2011).

2.2 *Factors Influencing Enzyme Activity*

Several environmental factors influence the activity of soil enzymes, including soil pH, temperature, and moisture.

2.2.1 Soil pH

Microbes such as bacteria and fungi are affected by changes in pH, which can alter enzyme structure and charge. Soil pH influences decide nutrient dissolution and nutrient availability. It also has a direct impact on how microbes contribute to nutrient cycling. When soil pH drops, it can disrupt microbial activity, make nutrients less available, and result in nutrient deficiencies for plants. Extremely acidic soil (pH < 5.5) can also harm plant growth because of the presence of toxic heavy metals, affecting overall soil health (Naz et al., 2022).

2.2.2 Temperature

Temperature regulates soil enzyme production. When a sufficient amount of moisture is present in the soil, then the warm, humid environment will enhance all enzyme activities, hence increasing nutrient mineralization. Soil invertase, made by microorganisms, indicates changes in carbon turnover. Ureases release ammonium nitrogen (NH_4^+_N) by breaking down amino compounds from plants (Song et al., 2021).

2.2.3 Moisture Content

Moisture content significantly influences microbes and enzyme activity in soils. Major effects in soil moisture, like precipitation events, have a substantial impact on soil microorganisms. Microbial growth and respiration can sharply increase within hours to days following such events. Climatic factors, such as precipitation or snowmelt, boost microbial activity by alleviating drought conditions and facilitating nutrient transfer in the soil profile (Baldrian, 2017).

2.3 *Methods for Measuring Enzyme Activities in Soil*

Information regarding enzyme activities in the soil is crucial to explaining the biochemical processes of the soil and the detailed functioning of microbes. Three principal methods are applied to measure these activities. These include functional protein assays (done by methods such as spectrophotometry), fluorescence assays, and zymography. They all provide valuable information on enzyme dynamics and their significance in soil chemistry.

2.4 *Spectrophotometry*

UV spectrophotometry, a process known as the optical method, is now widely used instead of chemical colourimetry to determine nitrates present in the soil. This method is based on the UV absorption of nitrates at almost a wavelength of 220 nm, where the data reflects the availability of nitrate models. Due to the interferences of soluble organic materials, the absorbance is also determined at 275 nm wavelength (Liu et al., 2023).

2.4.1 Fluorescence Assays

Fluorescence assays are methods that detect the intensity of light generated by certain materials when certain wavelengths illuminate them. These assays employ fluorescent dyes of labels on the target molecules, emitting light detectable with the help of special instruments. They are often employed in the analysis of biomolecules in biochemical and molecular biological and environmental sciences because they are very sensitive and accurate.

2.4.2 Zymography

The zymography method used for imaging the distribution of the enzyme activities in the soil and rhizosphere represents the spatial organisation connected with plant species. It is used to study the effect of temperature on enzyme activity in the rhizosphere. This method develops the mapping of enzyme activity at the surface of the soil, rhizosphere, and detritus here. However, its application in heavy metal soil contamination still has some limitations, especially for understanding enzyme and heavy metal interactions (Duan et al., 2018).

3 Role of Enzymes in Nutrient Cycling

Soil enzymes are crucial for the biogeochemical cycles of carbon (C), nitrogen (N), and phosphorus (P). They act as early, sensitive indicators of changing the soil nutrients due to climate change (Liu et al., 2021).

3.1 Carbon Cycling

Carbon cycling is the continuous variation of inorganic and organic carbon compounds by plants, micro- and macro-organisms between soil, plants, and the atmosphere. The enzymes β-glucosidase, invertase, and hydrolases are primarily responsible for carbon cycling in soils (Wang et al., 2023).

3.1.1 Decomposition of Organic Matter

Organic matter decomposition is a natural biological process influenced by soil organisms, the physical environment, and the quality of the organic material. Products like carbon dioxide (CO_2), energy, water, plant nutrients, and reformed organic carbon compounds are released during decomposition. Carbohydrates—cellulose, a frequent carbohydrate in plant residues, is leached quickly if coupled with pentosans such as xylans and mannans, whereas the decomposition is slow if the cellulose is combined with lignin. Cellulose decomposition occurs in two stages: hydrolysis (i.e., breaking the complex into glucose) and oxidation (that involves the conversion of glucose into CO_2 and water) (Khatoon et al., 2017).

3.1.2 Mineralization and Immobilization

Mineralisation is the biological activity that transforms the organic matter into simpler inorganic constituents. This comprises redox (oxidation–reduction) reactions where organisms release stored energy in organic compounds. Its conversion to more oxidised states liberates energy, which, in turn, is used by organisms for growth and cellular differentiation (Grzyb et al., 2020). Mineralisation is a critical process for balancing carbon in ecosystems and cycling nutrients present in soil (Zhang et al., 2021). Mineralisation and immobilisation in the soil cause nutrient cycling in the soil with the help of saprophytic and pathogenic fauna and microorganisms. They are organisms that feed on plant material that has become dead. The extent to which energy and nutrients are released from crop residues depends on a residue's biochemical characteristics (Grzyb et al., 2020).

3.2 Nitrogen Cycling

The nitrogen (N) cycle in the soil is an important geochemical transformative process that regulates agroecosystems' stability, productivity, and sustainability. The variety of microbial communities involved in the nitrogen cycle may vary based on the soil type (Ma et al., 2024). Urease is essential for the degradation of nitrogen from organic compounds, which releases mineral nitrogen back into the soil, which replenishes

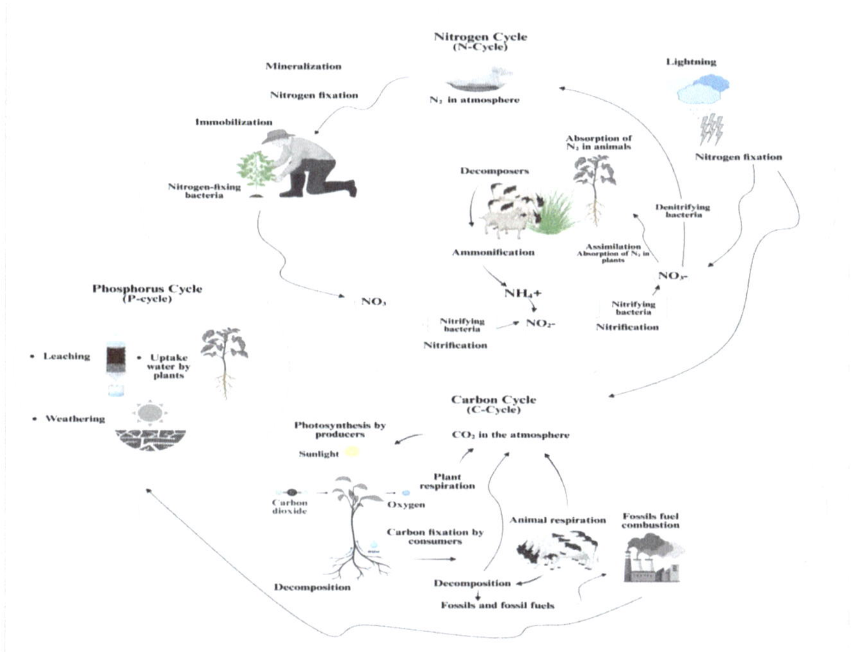

Fig. 3 Nutrient cycle

the soil with mineral nitrogen (Wang et al., 2023). The present forms of nitrogen removal include the nitrification and denitrification processes, which are efficient and cost-effective biological methods (Arumugham et al., 2024) (Fig. 3).

3.2.1 Nitrification

Nitrification is when ammonia (NH_3) in succession is oxidised by microbes to nitrite and then to nitrate. This process utilizes dissolved oxygen (DO) which significantly assists in determining the rate of N_2O production. The low levels of DO above the optimum nitrification hinder the nitrification process and cause the production of N_2O.

3.2.2 Denitrification

Nitrate is converted to nitrogen gas during denitrification and released into the atmosphere. However, nitrous oxide (N_2O), a greenhouse gas with a much more significant warming effect than CO_2, is produced as an intermediate. Reducing DO levels creates longer anoxic conditions that allow denitrifying microbes to utilize available carbon sources and complete the denitrification process, which helps to reduce the accumulation of N_2O before it is emitted (Arumugham et al., 2024).

3.3 Phosphorus Cycling

The phosphorus (P) cycle is different from the nitrogen and carbon cycle because it does not involve the formation of stable gaseous phosphorus compounds under the earth's normal temperatures and atmospheric pressures, unlike nitrogen and carbon, which have significant gaseous phases (e.g., nitrogen and carbon dioxide gases). Phosphorus remains primarily in solid or liquid forms, such as phosphate minerals or dissolved phosphate in water. This absence of gaseous forms means that phosphorus cycling occurs mainly through terrestrial and aquatic processes rather than atmospheric exchanges (Tian et al., 2021) (Fig. 4).

3.3.1 Phosphatase Activity

Microbes are essential for phosphorus (P) cycling in soils, as they help manage the availability of phosphorus in soil. They produce various enzymes that form free orthophosphate from stubborn organic phosphorus formulations. Important enzymes involved in this process include alkaline phosphatase, acid phosphate, phosphatase, C-P lyase, and phytase.

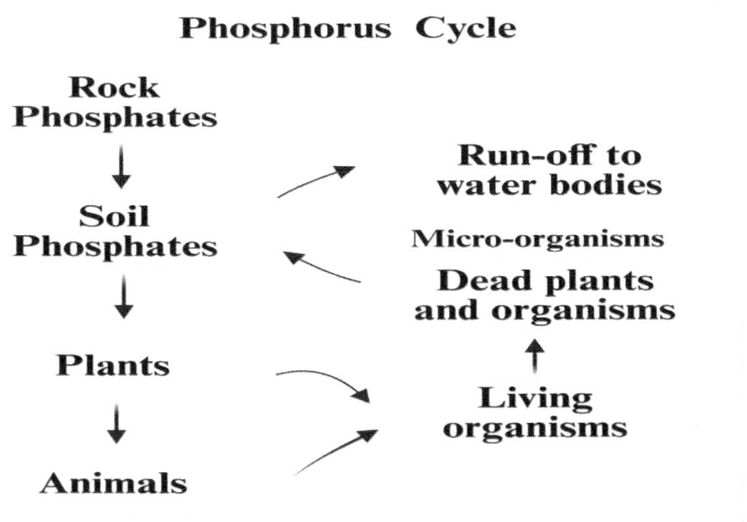

Fig. 4 Phosphorus cycle

3.3.2 Phosphate Solubilization

Microbes also facilitate the solubilisation of resistant inorganic phosphorus compounds by producing organic acids. These include formic acid, citric acid, gluconic acid, oxalic acid and malic acid, with gluconic acid being particularly significant. Glucose dehydrogenase (GCD) is a key enzyme in gluconic acid production (Liang et al., 2020).

3.4 Sulfur Cycling

The sulfur cycle includes sulfur oxidation, reduction, mineralisation, immobilisation, and assimilation. Elemental sulfur is oxidised to sulfate, which plants and microbes can use, or reduced to hydrogen sulfide (H_2S) and leached into groundwater (Chaudhary et al., 2023).

3.4.1 Sulfatase Activity

Sulfatase is responsible for the cleavage of sulfur from organic compounds and thus aids in the provision of a mineralized form of sulfur which is the only form of sulfur plants can utilize (Wang et al., 2023).

3.4.2 Sulfur Reduction

Sulfur reduction happens via two main processes: assimilatory sulfate reduction, where sulfate is incorporated into proteins, and dissimilatory sulfate reduction, where sulfate is converted to hydrogen sulfide (H_2S) by sulfate-reducing bacteria under anaerobic conditions. Key bacterias involved in the process include desulfovibrio, desulfobacter and desulfotomaculum. These bacteria can help to reduce sulfate, lower soil fertility, and impact water quality. Sulfur gases like hydrogen sulfide, released from both natural and human sources, play an important role in climate regulation as well as the nutrient cycle (Chaudhary et al., 2023).

4 Interactions Between Enzyme Activities and Soil Microorganisms

Enzymes and soil microorganisms interact in complex ways, significantly influencing soil health, nutrient cycling, and ecosystem productivity. These interactions can be grouped into mutualistic, competitive relations and, in general, the effects of the soil microorganisms' community composition on the enzyme activities.

4.1 Mutualistic Relationships

Mutualistic relation is support or collaboration between the microorganisms and enzymes in the soil formation. Extracellular enzymes disintegrate organic matter, and nutrients are, in turn, returned to the environment by microorganisms. For example, mycorrhizal fungi improve nutrient acquisition by decomposing organic forms of phosphorus with the help of symbiosis (Smith & Read, 2010). Even soil bacteria develop symbiotic partnerships where they decompose plant residues using several enzymes such as cellulases and hemicellulases to enable them to break down the residues. Such a process brings nutrients to the availability of bacteria and enhances plant growth and the activeness of microbes, too. This mutual benefit forms a virtuous circle and hence retains the soil fertility (Baldrian, 2017).

4.2 Competitive Interactions

In soil ecosystems, micro-organisms uptake the nutrients, which in turn affects the production of enzymes. For example, bacterial and fungal microbes can fight for cellulose resources; therefore, the equipped enzyme can differ due to capacity and competition (Trivedi et al., 2016). Some bacteria release chemicals that inhibit the activities of competitors, such as fungi. For instance, bacterial-producing antibiotics inhibit fungal growth, meaning that few fungal enzymes decompose organic matter (Garbeva et al., 2008). This alters microbial populations and so affects the health of soils.

4.3 Influence of Soil Microbial Community Composition on Enzyme Activities

The soil microbial communities influence the decomposition of organic matter and the nutrient cycling enzyme. Some bacterial phylotypes, like actinobacteria and acidobacteria, are often associated with elevated levels of enzymes active in carbon and phosphorus turnover (Pratibha et al., 2023). pH and moisture of the environment also affect microbial status and enzyme activities and, hence, soil health.

5 Environmental Impacts on Enzyme Activities and Nutrient Cycling

Environmental factors have great significance in determining enzyme activities and the nutrient cycle of the soil's ecosystem. They consist of climatic variation, operation techniques, land utilisation and pollution (Fig. 5).

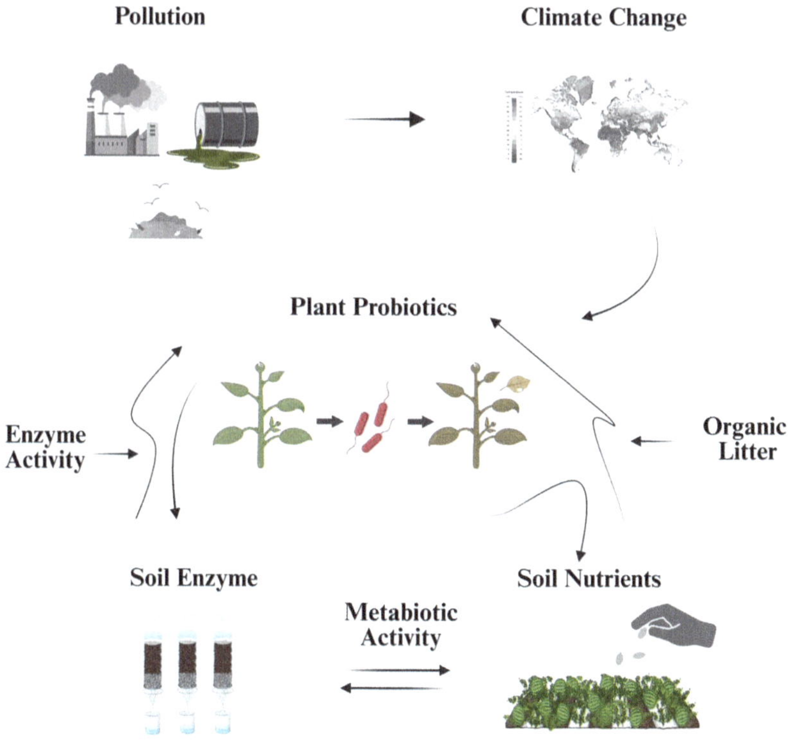

Fig. 5 Environmental impacts on enzyme activities and nutrient cycle

5.1 Climate Change Effects

Depending on whether the soil is organic or mineral, it stores three to four times more carbon than the atmosphere and, therefore, is the most significant source of natural CO_2 gas. Global warming perturbations in the temporal change of soil carbon stocks can profoundly influence these climate paths. The two enzymes, cellulases and ligninases, play an essential role in carbon decomposition into soil organic matter. Cellulase can degrade easily available carbon sources, and ligninase can degrade rigid carbon sources. This led to the ligninase activity that might facilitate microbial access to the litter and SOM, thereby enhancing the rate of soil carbon loss during long-term warming enhanced the microbial enzyme interactions in soil carbon dynamics (Chen et al., 2020).

5.2 Land Use and Management Practices

The management of land practices enables us to determine if the soil is acting as a carbon source or sink. Practices that reduce levels of tillage include conservation tillage, under which soil organic carbon stock is also likely to be increased. However, the conversion of native forests to agricultural land results in the loss of this carbon, accounting for the thinning or removal of forests for agriculture. A common practice in soil management called tillage is a process of soil inversion, and the disintegration of aggregates, compaction and inversion affects the soil's microbial and fauna activity (Ramesh et al., 2019).

5.3 Pollution and Contaminant Impacts

Heavy metal pollutants include cadmium, lead, arsenic, mercury, and chromium. Metalloids are considered very toxic and cause harm to crop and soil health in agricultural fields. The effectiveness of these inputs varies with the type of crop grown, the soil, and available metal ions. These metals can adversely affect plant cellular activities and metabolic rate, reducing productivity and growth (Rashid et al., 2023).

6 Applications and Future Directions

Subsequent developments plan to include complete characterisation of soil conditions, improvement in the utilisation of conservation tillage, and improvement in bioremediation strategies. Further research will improve these techniques and assess their extended consequences on soil composition.

6.1 *Soil Health Assessment*

Soil quality is the extent of the capability of the soil to facilitate plant growth, support the ecosystem and maintain the physical condition of the soil. 'Soil health' is a term that considers the soil to be a living entity, while the term 'soil quality' refers to several physical, chemical, and biological properties. Recent research based on the Cornell Comprehensive Assessment of Soil Health (CASH) and Haney Soil Health Test (HSHT) measures the health of the soil by specific parameters, including microbial properties, enzyme activity, and nutrient availability (Yang et al., 2020).

6.2 *Soil Management Strategies*

To mitigate both soil degradation and climate change challenges, it is essential to continue and enhance rainfed climate-smart technologies such as Conservation Agriculture (CA). Most CA techniques like crop rotation and no-till cropping practices afford efficient and sustainable practices and develop soil cover. These improve the chemical, physical and biological properties of the soil, decrease soil erosion, improve soil status, and increase food production (Pratibha et al., 2023).

6.3 *Bioremediation Approaches*

Bioremediation is the process of using microorganisms to degrade environmental pollutants to produce less toxic substances. Toxic compounds are broken up into simpler forms (i.e., CO_2, CH_4 and H_2O) by microbial enzymes like lipases and esterases. This method is economical and efficient for environmental remediation as it is safer (Sharma et al., 2019).

7 Conclusion

Soil biochemistry, which assesses the interactions between soils and microorganisms, is an important concept that ensures the conversion of organic matter decomposition, structure, and nutrient cycling. Enzymes participating in nitrogen, carbon, sulfur, and phosphorus cycling are helpful for soil health and crop productivity. Some biochemical assays include spectrophotometry, fluorescence assays, and zymography, which show the enzyme activities and how they may be affected by environmental changes. Further studies on soil biochemistry underscore the significance of improving knowledge in soil processes to deal with challenges like climate change, emission of pollution and soil erosion. Subsequent knowledge would enhance soil utilisation and nourishment strategies while expanding the efficiency of soil evaluation and restoration. The research progress in soil biochemistry and its potential impacts on the future of green agriculture and soil preservation are imminent. Through these advancements, the health of soil and fertility can be optimised, the farming practices can be made sustainable, the environmental impacts can be reduced, and overall yield stability and productivity of ecosystems can enhance.

References

Arumugham, T., Khudzari, J., Abdullah, N., Yuzir, A., Iwamoto, K., & Homma, K. J. (2024). Research trends and future directions on nitrification and denitrification processes in biological nitrogen removal. 111897.

Baldrian, P. J. (2017). Microbial activity and the dynamics of ecosystem processes in forest soils. *37*, 128–134.

Chaudhary, S., Sindhu, S. S., Dhanker, R., & Kumari, A. J. (2023). Microbes-mediated sulphur cycling in soil: Impact on soil fertility, crop production and environmental sustainability. *271*, 127340.

Chen, J., Elsgaard, L., van Groenigen, K. J., Olesen, J. E., Liang, Z., Jiang, Y., Lærke, P. E., Zhang, Y., Luo, Y., & Hungate, B. A. (2020). Soil carbon loss with warming: New evidence from carbon-degrading enzymes. *26*(4), 1944–1952.

de Souza Vandenberghe, L. P., Karp, S. G., Pagnoncelli, M. G. B., von Linsingen Tavares, M., Junior, N. L., Diestra, K. V., Viesser, J. A., & Soccol, C. R. (2020). Classification of enzymes and catalytic properties. In *Biomass, biofuels, biochemicals* (pp. 11–30). Elsevier.

Duan, C., Fang, L., Yang, C., Chen, W., Cui, Y., & Li, S. (2018). Reveal the response of enzyme activities to heavy metals through in situ zymography. *Ecotoxicology and Environmental Safety, 156*, 106–115.

Fanin, N., Mooshammer, M., Sauvadet, M., Meng, C., Alvarez, G., Bernard, L., Bertrand, I., Blagodatskaya, E., Bon, L., & Fontaine, S. (2022). Soil enzymes in response to climate warming: Mechanisms and feedbacks. *36*(6), 1378–1395.

Garbeva, P., Van Elsas, J., & Van Veen, J. J. P. (2008). Rhizosphere microbial community and its response to plant species and soil history. *302*, 19–32.

Gil-Sotres, F., Trasar-Cepeda, C., Leirós, M. C., & Seoane, S. (2005). Different approaches to evaluating soil quality using biochemical properties. *Soil Biology and Biochemistry., 37*(5), 877–887.

Greenfield, L. M., Puissant, J., & Jones, D. L. J. S. B. (2021). Synthesis of methods used to assess soil protease activity. *158*, 108277.

Grzyb, A., Wolna-Maruwka, A., & Niewiadomska, A. J. A. (2020). Environmental factors affecting the mineralisation of crop residues. **10**(12), 1951.
Jansson, J. K., & Hofmockel, K. S. J. N. R. M. (2020). Soil microbiomes and climate change. *18*(1), 35–46.
Karaca, A., Cetin, S. C., Turgay, O. C., & Kizilkaya, R. J. (2011). Soil enzymes as indication of soil quality. 119–148.
Khatoon, H., Solanki, P., Narayan, M., Tewari, L., Rai, J., & Hina Khatoon, C. J. (2017). Role of microbes in organic carbon decomposition and maintenance of soil ecosystem. *5*(6), 1648–1656.
Khosrozadeh, S., Guber, A., Kravchenko, A., Ghaderi, N., & Blagodatskaya, E. J. S B. (2022). Soil oxidoreductase zymography: Visualizing spatial distributions of peroxidase and phenol oxidase activities at the root-soil interface. *167*, 108610.
Liang, J.-L., Liu, J., Jia, P., Yang, T.-T., Zeng, Q.-W., Zhang, S.-C., Liao, B., Shu, W.-S., & Li, J.-T. (2020). Novel phosphate-solubilizing bacteria enhance soil phosphorus cycling following ecological restoration of land degraded by mining. *14*(6), 1600–1613.
Liu, C., Song, Y., Dong, X., Wang, X., Ma, X., Zhao, G., & Zang, S. J. (2021). Soil enzyme activities and their relationships with soil C, N, and P in peatlands from different types of permafrost regions. *Northeast China, 9*, 670769.
Liu, J., Cai, H., Chen, S., Pi, J., & Zhao, L. J. A. (2023). A review on soil nitrogen sensing technologies: challenges, progress and perspectives. *13*(4), 743.
Ma, T., Yang, K., Yang, L., Zhu, Y., Jiang, B., Xiao, Z., Shuai, K., Fang, M., Gong, J., & Gu, Z. (2024). Different rotation years change the structure and diversity of microorganisms in the nitrogen cycle, affecting crop yield. *193*, 105123.
Nannipieri, P., Trasar-Cepeda, C., & Dick, R. P. (2018). Soil enzyme activity: A brief history and biochemistry as a basis for appropriate interpretations and meta-analysis. *Biology and Fertility of Soils., 54*(1), 11–19.
Naz, M., Dai, Z., Hussain, S., Tariq, M., Danish, S., Khan, I. U., Qi, S., & Du, D. (2022). The soil pH and heavy metals revealed their impact on soil microbial community. *321*, 115770.
Neemisha, S. S. (2022) Soil enzymes and their role in nutrient cycling. In B. Giri, R. Kapoor, Q.-S. Wu & A. Varma (Eds.), *Structure and functions of pedosphere* (pp. 173–188). Singapore: Springer Nature Singapore.
Paul, E. A. (2007). 1—Soil microbiology, ecology, and biochemistry in perspective. In E. A. Paul (Ed.), *Soil microbiology, ecology and biochemistry* (3rd ed., pp. 3–24). San Diego: Academic Press.
Pratibha, G., Manjunath, M., Raju, B., Srinivas, I., Rao, K., Shanker, A. K., Prasad, J., Rao, M. S., Kundu, S., & Indoria, A. J. (2023). Soil bacterial community structure and functioning in a long-term conservation agriculture experiment under semi-arid rainfed production system. *14*, 1102682.
Ramesh, T., Bolan, N. S., Kirkham, M. B., Wijesekara, H., Kanchikerimath, M., Rao, C. S., Sandeep, S., Rinklebe, J., Ok, Y. S., & Choudhury, B. U. (2019). Soil organic carbon dynamics: Impact of land use changes and management practices: A review. *156*, 1–107.
Rashid, A., Schutte, B. J., Ulery, A., Deyholos, M. K., Sanogo, S., Lehnhoff, E. A., & Beck, L. J. A. (2023). Heavy metal contamination in agricultural soil: environmental pollutants affecting crop health. *13*(6), 1521.
Sharma, A., Sharma, T., Sharma, T., Sharma, S., Kanwar, S. S. J. M. (2019). Role of microbial hydrolases in bioremediation. 149–164.
Smith, S. E., & Read, D. J. (2010). *Mycorrhizal symbiosis*. Academic Press.
Song, Y., Liu, C., Song, C., Wang, X., Ma, X., Gao, J., Gao, S., & Wang, L. J. C. (2021). Linking soil organic carbon mineralization with soil microbial and substrate properties under warming in permafrost peatlands of Northeastern China. *203*, 105348.
Tabatabai, M. A., & Fu, M. (2021). Extraction of enzymes from soils. In *Soil biochemistry* (pp. 197–227). CRC Press.

Tian, J., Ge, F., Zhang, D., Deng, S., & Liu, X. J. B. (2021). Roles of phosphate solubilizing microorganisms from managing soil phosphorus deficiency to mediating biogeochemical P cycle. *10*(2), 158.

Trivedi, P., Delgado-Baquerizo, M., Anderson, I. C., Singh, B. K. (2016). Response of soil properties and microbial communities to agriculture: Implications for primary productivity and soil health indicators. *7*, 990.

Vroon, D. H., & Israili, Z. (2011). Aminotransferases.

Wang, L., Hamel, C., Lu, P., Wang, J., Sun, D., Wang, Y., Lee, S.-J., & Gan G. Y. (2023). Using enzyme activities as an indicator of soil fertility in grassland-an academic dilemma. *14*, 1175946.

Yang, T., Siddique, K. H. M., & Liu, K. (2020). Cropping systems in agriculture and their impact on soil health—A review. *Global Ecology and Conservation., 23*, e01118.

Zhang, S., Fang, Y., Luo, Y., Li, Y., Ge, T., Wang, Y., Wang, H., Yu, B., Song, X., Chen, J., & Zhou, J. (2021). Linking soil carbon availability, microbial community composition and enzyme activities to organic carbon mineralization of a bamboo forest soil amended with pyrogenic and fresh organic matter. *801*, 149717.

Conclusion

Muhammad Shaaban

Abstract In the conclusion of this book, it is essential to bring together the various threads that have been woven throughout the chapters, reinforcing the importance of the interplay between soil, plants, water, and environmental systems in achieving sustainable agriculture. As we navigated the complexities and nuances of these interrelationships, we witnessed how each element contributes uniquely to the agricultural landscape and how their interactions can either foster or hinder sustainability. The critical role of soil has been emphasized continuously, serving as the foundation for not only plant growth but also broader ecological health. Healthy soils harbor diverse microorganisms, store essential nutrients, and retain moisture, all of which are fundamental to supporting robust agricultural systems. Furthermore, the methods we employ to manage and amend soil directly impact its health and productivity. By adopting practices such as organic amendments, reduced tillage, and careful crop rotation, farmers can cultivate soils that are not only resilient but also capable of supporting increased agricultural output in a sustainable manner. Additionally, the interdependence of soil and water is crucial. Water, a finite resource, becomes both a challenge and an asset in agricultural practices. Efficient water management strategies have been highlighted as key to ensuring that crops receive adequate moisture without depleting local water resources or harming surrounding ecosystems. Implementing techniques such as rainwater harvesting, drip irrigation, and contour farming can significantly enhance water retention and minimize runoff, illustrating the need for an integrated approach that considers both soil characteristics and water availability. The influence of broader environmental systems on agriculture cannot be understated. Climate change, land degradation, and biodiversity loss present significant challenges to farming practices everywhere. Understanding these environmental dynamics allows us to tailor our agricultural approaches effectively, fostering resilience and adaptability in the face of shifting conditions. By aligning agricultural practices with ecological principles, we can help mitigate adverse environmental impacts and maintain the balance necessary for both agricultural productivity and

M. Shaaban (✉)
College of Agriculture, Henan University of Science and Technology, Luoyang, China
e-mail: shabanbzu@hotmail.com

M. Shaaban (ed.), *Soils and Sustainable Agriculture*, Frontier Studies in Soil Science,
https://doi.org/10.1007/978-3-031-91114-9_47

environmental preservation. Moreover, the importance of collaboration among stakeholders has emerged as a recurring theme. Engaging farmers, researchers, policymakers, and educators in a shared vision for sustainable agriculture is imperative for fostering innovation and disseminating best practices. Continuous education about soil health, water management, and environmental stewardship will equip all stakeholders with the knowledge and tools necessary to make informed decisions that benefit both the agricultural sector and the planet. This book serves as a call to action, urging all players in the agricultural field to embrace a holistic perspective that recognizes the interdependence of soil, plants, water, and environmental systems. It is only through this understanding and commitment to sustainable practices that we can envision a future where agriculture not only meets human needs but does so in a manner that is harmonious with the earth and its ecosystems. The journey toward sustainable agriculture requires unwavering dedication, innovation, and collaboration, paving the way for a resilient agricultural system that can support current and future generations.

1 Significance of Soil Health

The significance of soil health stands as a cornerstone of sustainable agriculture. Without a doubt, soil serves as the vital medium for plant growth, acting as both a foundation and a reservoir of resources that support agricultural ecosystems. Healthy soil is characterized by its rich organic matter, diverse microorganisms, and balanced nutrient profile, all of which work in concert to foster robust plant development and crop yields. The health of the soil is not merely an agricultural concern; it is the linchpin in the intricate web of interactions that define our environmental systems. As we have explored throughout this book, the diverse microbial communities present in healthy soils play a crucial role in breaking down organic matter, cycling nutrients, and enhancing soil structure. These microorganisms contribute to soil fertility by decomposing organic residues and facilitating nutrient exchange between soil and plants, which ultimately leads to improved crop performance. Additionally, healthy soils have enhanced water retention capabilities, allowing them to absorb and store moisture effectively. This characteristic is increasingly important in the face of climate variability and changing weather patterns, where droughts and heavy rainfall can become more frequent and severe.

Soil management practices have immense implications for maintaining and enhancing soil health. Strategies such as the application of compost, cover cropping, and reduced tillage can profoundly impact soil quality over time. Compost enriches the soil with organic matter, promoting microbial activity and improving nutrient availability. Cover crops, on the other hand, act as protective blankets that prevent erosion, suppress weeds, and enhance soil structure. Furthermore, reduced tillage minimizes soil disturbance, allowing for the natural development of soil structure and preserving the habitat for beneficial organisms. Recognizing the intricate relationships between soil health and agricultural productivity is essential for farmers.

Sustainable practices rooted in maintaining soil quality not only enhance crop yields but also lead to long-term benefits for the environment. When farmers prioritize soil health, they inadvertently contribute to the establishment of ecosystems that are capable of withstanding environmental shocks, reducing the need for chemical fertilizers and pesticides, and fostering biodiversity, which further enriches the soil. Moreover, the economic benefits of investing in soil health become increasingly evident. Healthier soils lead to increased crop resilience, reducing the likelihood of crop failure and the associated financial burdens. Farmers who utilize sustainable practices that prioritize soil health often see improved productivity and better marketability for their products, as consumers increasingly demand responsibly grown food. In summary, the significance of soil health cannot be overstated in the quest for sustainable agriculture. Healthy soils are fundamental to enhancing agricultural productivity, ensuring food security, and preserving environmental integrity. The interconnectedness of soil health, crop yields, and ecosystem resilience underscores the importance of adopting practices that foster soil vitality. As we look to the future, prioritizing soil health will not only benefit individual farmers but also contribute to the collective effort of creating sustainable agricultural systems that can thrive in harmony with the environment. This understanding paves the way for informed decision-making and responsible stewardship of one of our most precious natural resources.

2 Water Management in Agriculture

The relationship between soil and water is one of the most critical aspects of sustainable agriculture, as it fundamentally affects crop growth, soil health, and overall ecosystem functionality. Water is an essential resource for plants, serving not only for hydration but also as a medium for nutrient transport in the soil. However, what often goes unrecognized is how soil quality and structure influence water retention and drainage, which can ultimately determine the success or failure of agricultural systems. Effective water management is paramount in ensuring that crops receive adequate moisture while also preserving this vital resource. The management techniques employed by farmers can significantly enhance the interaction between soil and water. Practices such as contour farming, where plowing follows the land's contours, help reduce erosion and promote water retention in fields. By slowing down water runoff and increasing infiltration, this method not only protects soil quality but also ensures that more water is available for crops.

Another important technique in water management is the use of cover crops, which offer dual benefits. During the off-season, cover crops can prevent soil erosion and improve soil structure while also capturing moisture and enhancing its infiltration into the soil profile. These plants can provide a protective cover that minimizes evaporation losses during hot months when water is scarce. When cover crops are terminated, their biomass improves soil fertility, contributing to healthier soils that can better retain water. Drip irrigation is another innovative practice that optimizes water use, allowing water to be delivered directly to the root zone of plants in a controlled manner. This method minimizes waste and ensures efficient water usage while reducing waterlogged soils that can lead to root disease. When combined with healthy soil management practices, such as mulching and the addition of organic amendments, drip irrigation can significantly enhance both soil moisture retention and plant health. Moreover, the integration of agroforestry systems further exemplifies the interdependence of soil and water. By planting trees alongside crops, farmers can enhance moisture retention and reduce the impact of wind and water erosion. Trees contribute to improved microclimates, which help moderate temperature extremes and improve local hydrology. This interplay between trees, soil, and water creates a more sustainable agricultural environment that promotes resilience and biodiversity.

The challenges posed by climate change also heighten the necessity for efficient water management practices. As precipitation patterns become more erratic and instances of drought and flooding increase, it is vital for farmers to adopt sustainable techniques that can mitigate these effects. Enhanced water retention practices and cohesive soil management strategies can help buffer against climate variability, ensuring that crops remain viable even in extreme weather conditions. Furthermore, the importance of water quality cannot be overlooked. Practices that protect and enhance soil health naturally led to better water quality by minimizing runoff and reducing the leaching of harmful substances. Healthy soils act as natural filters, improving the overall integrity of water resources and protecting local ecosystems from contamination. In conclusion, the interplay between soil and water is foundational to sustainable agriculture. Effective water management practices, when coupled with sound soil health strategies, create an agro-system that is resilient, productive, and environmentally sound. As agriculture faces the daunting challenges of climate change and resource scarcity, emphasizing the relationship between soil and water will become even more critical. By implementing innovative techniques that enhance water retention and improve soil structure, farmers can ensure both the sustainability of their operations and the health of the ecosystems upon which they depend. The ongoing commitment to exploring and refining these interactions will ultimately pave the way for a more sustainable agricultural future.

3 Environmental Systems and Agriculture

The impact of environmental systems on agriculture is profound and multifaceted, highlighting the need for an integrated approach to sustainable farming. Climate change, land use changes, and biodiversity loss have significant repercussions for agricultural practices, influencing productivity, resilience, and ecological balance. As the climate continues to shift, understanding its effects becomes essential for farmers seeking to adapt and thrive in an ever-changing landscape. Climate change brings increased variability in weather patterns, leading to more frequent and intense droughts, floods, and temperature extremes. These changes necessitate the adoption of adaptive agricultural practices that promote resilience against climate impacts. For instance, selecting climate-resilient crop varieties, implementing water-saving irrigation technologies, and enhancing soil carbon sequestration through organic practices are all critical strategies. By focusing on building resilience, farmers can mitigate the risks associated with unpredictable climate conditions while maintaining productivity. Land use changes, driven by urbanization, deforestation, and industrial expansion, also challenge agricultural systems. The conversion of natural ecosystems to agricultural land can disrupt the balance of local ecosystems, leading to soil degradation, loss of biodiversity, and altered water cycles. Sustainable land management practices, such as agroecology and responsible land-use planning, can help preserve natural habitats and maintain the vital services these ecosystems provide.

Biodiversity is essential for healthy, functioning agricultural systems, contributing to pest control, pollination, and soil fertility. The loss of biodiversity negatively impacts ecosystem resilience, making crops more vulnerable to diseases and pests. Emphasizing the importance of crop diversity, including intercropping and maintaining genetic diversity, can enhance the stability of agricultural systems and foster a more sustainable approach to food production. Ultimately, recognizing the interplay between agriculture and environmental systems is crucial for achieving sustainable development goals. By adopting holistic management practices that account for ecological dynamics, farmers can create agricultural systems that support not only their livelihoods but also the health and well-being of the environment. This focus on integrated, sustainable approaches paves the way for a brighter future, safeguarding both agricultural productivity and the ecological integrity of our planet.

4 Collaboration and Knowledge Sharing

Collaboration among stakeholders is essential for the advancement of sustainable agriculture and the effective management of agricultural systems. Farmers, researchers, policymakers, and educators all play critical roles in fostering innovation, sharing knowledge, and implementing best practices. Engaging these diverse groups creates a multi-faceted approach to addressing the complex challenges facing agriculture today.

Farmers are at the forefront of sustainable agricultural practices and possess invaluable experiential knowledge. Their insights into local conditions, soil health, and water management are critical for developing effective strategies that are applicable in their specific contexts. By facilitating platforms for farmers to connect and share their experiences, we can encourage the exchange of successful techniques and foster a sense of community that prioritizes sustainability.

Researchers contribute to this collaborative effort by providing scientific knowledge and data that can inform agricultural practices. By working closely with farmers, researchers can ensure that their findings are applicable and relevant, ultimately leading to the development of innovative solutions tailored to real-world challenges. Participatory research approaches, where farmers actively engage in the research process, can strengthen the connection between science and practice, yielding more effective results.

Policymakers are crucial in creating an enabling environment for sustainable agriculture. By implementing supportive policies, providing incentives for sustainable practices, and investing in agricultural research and education, they can foster a landscape that encourages innovation and resilience. Collaborative partnerships between farmers and policymakers can ensure that regulations are both practical and beneficial, promoting sustainability without imposing undue burdens on agricultural producers.

Educators play a vital role in raising awareness about sustainable agricultural practices and disseminating knowledge to future generations. By incorporating sustainability principles into educational curricula and offering training programs for farmers on the latest techniques, educators can empower individuals to adopt practices that enhance soil health, water management, and environmental stewardship.

In short, the collaboration between farmers, researchers, policymakers, and educators is essential for promoting sustainable agriculture and addressing the pressing challenges of our time. By combining diverse perspectives and expertise, we can create integrated solutions that enhance agricultural resilience and productivity while safeguarding the environment. This collaborative approach is vital for cultivating a sustainable agricultural future that benefits both people and the planet.

5 Future Directions

As we look to the future of sustainable agriculture, it is clear that a comprehensive vision must guide our efforts. First and foremost, the focus should be on enhancing the interplay between soil, plants, water, and environmental systems. This holistic approach is essential for creating resilient agricultural systems that can withstand the challenges posed by climate change, resource depletion, and ecological degradation. To achieve this vision, it is crucial to set clear goals that prioritize sustainable practices. These goals should encompass improving soil health through innovative management techniques, optimizing water usage to promote efficiency, and maintaining biodiversity to safeguard ecosystems. Embracing regenerative agricultural practices will not only enhance productivity but also restore ecological balance and promote the health of our natural resources. Furthermore, fostering collaboration among all stakeholders will be critical in driving progress. Engaging farmers, researchers, policymakers, and educators in cooperative efforts will facilitate knowledge sharing and the co-creation of solutions that are practical, effective, and context-specific. Creating networks that encourage the exchange of insights and promote dialogue will pave the way for innovations that align with local conditions and needs.

Investing in research and development will also play a pivotal role in shaping the future of sustainable agriculture. Prioritizing studies that explore new technologies, crop varieties, and soil amendments will provide the necessary data to inform best practices and adapt to changing environmental conditions. Public and private sectors must work together to fund research initiatives that address the pressing challenges of agriculture today. Finally, a call to action for stewardship and sustainability is critical. Encouraging individuals to take responsibility for their agricultural practices and environmental impact will create a culture of sustainability. This involves not only adopting sustainable methods but also advocating for policies that support ecological health and promote equitable access to resources. In conclusion, the future of sustainable agriculture is bright if we remain committed to a vision that emphasizes integration, collaboration, innovation, and stewardship. By working collectively towards shared goals, we can cultivate agricultural systems that not only meet the needs of current generations but also safeguard the planet for future generations. Achieving sustainability is not merely an aspirational goal; it is a necessity for ensuring food security, environmental health, and overall well-being in a rapidly changing world.

Correction to: Soils and Sustainable Agriculture

Muhammad Shaaban

Correction to: M. Shaaban (ed.), *Soils and Sustainable Agriculture*, Frontier Studies in Soil Science, https://doi.org/10.1007/978-3-031-91114-9

In the original version of this book, the affiliation corrections for the author Luu The Anh in chapter 40 have been updated and removed the sentence "This research was funded by The Ministry of Education and Training of Vietnam, grant number: B2023-NTH-02" from the acknowledgement. The affiliation corrections for Le Van Giang and Luu The Anh have been updated in chapter 44. In addition, Pranaba Nanda Bhattacharyya's name has been updated to Pranab Nanda Bhattacharyya in chapter 37.

The updated versions of these chapters can be found at
https://doi.org/10.1007/978-3-031-91114-9_37
https://doi.org/10.1007/978-3-031-91114-9_40
https://doi.org/10.1007/978-3-031-91114-9_44

M. Shaaban (ed.), *Soils and Sustainable Agriculture*, Frontier Studies in Soil Science,
https://doi.org/10.1007/978-3-031-91114-9_48

GPSR Compliance

The European Union's (EU) General Product Safety Regulation (GPSR) is a set of rules that requires consumer products to be safe and our obligations to ensure this.

If you have any concerns about our products, you can contact us on ProductSafety@springernature.com

In case Publisher is established outside the EU, the EU authorized representative is:

Springer Nature Customer Service Center GmbH
Europaplatz 3
69115 Heidelberg, Germany

Batch number: 10607196

Printed by Printforce, the Netherlands